A TEXTBOOK OF

AUTOMOBILE ENGINEERING

VOLUME-III

PRITAM SINGH GILL

BY THE SAME AUTHOR

1. A Textbook of Machine Drawing
2. A Text Book of Engineering Drawing (Geometrical Drawing)
3. A Text Book of Engineering Graphics and Drafting
4. A Text Book of Graphic Statics
5. Work-Book in Engineering Drawing
6. A Textbook of Automobile Engineering – Vol. I (Automotive IC Engines)
7. A Textbook of Automobile Engineering – Vol. II (Automobile Power Transmission, Controls, Safety Comfort, Lighting & other Auxiliary Systems— including Body)
8. Tolerance Stack-ups Using G.D. & T.
9. Geometric Dimensioning and Tolerancing
10. A Textbook of Automobile Engineering – Vol. III (Auto-Engines Testing, Problem Diagnosis and Service)

A TEXTBOOK OF
AUTOMOBILE ENGINEERING

VOLUME-III

(AUTO-ENGINES TESTING, PROBLEM DIAGNOSIS AND SERVICE)

(FOR ENGINEERING STUDENTS, PRACTICING MECHANICS, TECHNICIANS AND TRAINERS)

By

PRITAM SINGH GILL

*M. Sc. Engg. (Mechanical) Pb., M.E. (Prod. Engg) Pb.,
Grad. I.E. (Mech. Engg.) India, Dip. Mech. Engg Distinction),
M.I.E. (India), M.C.I. (India Section), M.I.S.T.E., A.M.I.A.E.*

S.K. KATARIA & SONS®

Publisher of Engineering Books

4885/109, Prakash Mahal, Dr. Subhash Bhargav Lane,
Opposite Delhi Medical Association, Daryaganj, New Delhi-110002
Phone: +91-11-43551243; Mobile: +91-9871775858, 7678425209
e-mail: katariabook@gmail.com, katariabooks@yahoo.com
Website: www.skkatariaandsons.com

(e-books, subscriptions and bulk buying also available on our website)

Published by:

S.K. Kataria & Sons®

4885/109, Prakash Mahal, Dr. Subhash Bhargav Lane,
Opposite Delhi Medical Association, Daryaganj, New Delhi-110002
Phone: +91-11-43551243; Mobile: +91-9871775858, 7678425209
e-mail: katariabook@gmail.com, katariabooks@yahoo.com
Website: www.skkatariaandsons.com

First Edition: 2012, Reprint: 2026

Laser Type Setting by:

SARA ASSIGNMENTS, Patparganj, Delhi-110092.

Printed at :
A.S. OFFSET, Delhi

Preface...

In recent years, there have been significant changes in the design and construction of automobiles. Although these changes have increased dependability, improved fuel economy, reduced emissions, and enhanced safety, they have made it more difficult to troubleshoot and repair late-model vehicles. Therefore, a thorough knowledge of the fundamentals of vehicle construction, operation, and service is now more important than ever.

Automobile Manufacturers are now using advanced engine designs to improve power, efficiency, and dependability. Late model passenger cars and SUVs are equipped with computer-controlled, high performance, 4, 6, and 8-cylinder engines. Many use 4-valves per cylinder, Overhead Camshafts, Aluminium alloy Heads, Aluminium alloy Cylinder Blocks, Turbochargers, Plastic Parts, and many other technological advances.

"Automobile Engine Diagnosis and Service" applies the fundamentals of manufacturing, operation of various engine components and systems described in "Principles of Automotive Engines" Volume I. It explores the many sciences involved in automotive operation, e.g., hydraulics pneumatics, electricity and electronics, fuels, lubricants, internal combustion, power transmission, steering, braking, suspension geometry, computers, metallurgy, and strength of materials. This book covers troubleshooting, and service of modern automotive engines.

With the ever increasing competition as a result of open markets and electronic advances made in the automobile engines control systems, it becomes necessary to educate the engineering students about all the new changes that the late model vehicles are incorporating. With this point in mind this book is written to keep the students abreast with the latest technological developments in the field of automobile engineering, even though the developments will keep on coming.

The book establishes the foundation on which a thorough knowledge of automotive technology is based. Once these fundamentals are mastered, knowledge gained through experience will enable troubleshooting and service of late-model vehicles. This book tends to cover most recent changes in the automotive technology, and is meant to meet the educational requirements of basic automotive engine, maintenance, troubleshooting and repair of automotive engines. Troubleshooting charts have been added to help diagnose and repair engines to facilitate repairs and help the mechanics to lay the ground work to efficiently diagnose and economically repair the engines to restore their performance, emission controls, and economy.

Late model vehicles manufactured worldwide with latest technologies, e.g., Compact, reliable, durable and economical engines; Electronic control of air-fuel ratios; Computer Command Control (CCC) system; Multi-point Fuel Injection (MPFI) System; Electronic Fuel Injection (EFI) System; Group Gasoline Injection System; Sensors and Electronic Control Unit (ECU); Distributor Type Fuel Injection Pump, Electronic Ignition System; New development of emission control; and Catalytic Converters are included here.

The book is written to meet the requirements of undergraduate and diploma students in the field of "Automobile Engineering." It will meet the requirements of the syllabi of different Indian Universities and State Boards of Technical Education. Numerous clear, simple line illustrations have been provided, which are easy to reproduce by the students. The author and Publishers would welcome any suggestions for the improvement and for enhancing the value of this text.

Author is thankful to various researchers, training centres, and manufacturing organizations whose technologies have been used as examples in this text and to all those authors in the field of automobile engineering whose publications have been referred as basic source for some illustrations in this text. Diagnostics requires in understanding of the fundamentals of a systematic approach to determine the problem and its source. When conditions indicate that the problem may be in the engine itself, the engine must be tested and the test results systematically analyzed to pinpoint the problem source. Then the remedial measures are taken to restore the engine to its proper condition. The engine requires disassembly for internal service when adjustments and service of the engine accessories do not correct the engine problem.

When most people think about earning a living in the automotive industry, they probably think about repairing or servicing automobiles. This volume is an endeavor to put together useful information required to develop skills and proficiency in testing, problem diagnosis and servicing the automobile engines.

Chapter 1 deals with the "Tools of Automotive Trade", both hand tools and power tools to fit the many automotive jobs. A mechanic or a vehicle owner must know their names, safe procedure to use tools, and the selection of proper tool for the job at hand. Power tools make the mechanic's job easier and faster. You must know when and how to use specialized engine service tools to be a productive mechanic/technician. Basic hand tools are covered in this chapter. This chapter will enable you to: identify the common hand tools used in automotive repair jobs; learn the proper use of common automotive hand tools; learn about the tool care; and learn safe and proper use of hand tools.

Chapter 2 - Every repair job involves fasteners, and frequently need the use of keys, cotters, machine screws, studs, castle nuts, washers, retainers, hoses, and belts, so the mechanic must understand their selection, care and use. This chapter overviews the various types of hardware found on automotive engines. A skilled mechanic/technician must know how to properly handle these devices when servicing an engine. This chapter describes: the purpose of various fasteners used on automotive engines; classification of threads and other bolt specifications; how to torque fasteners to specs; how to identify various types of nuts and screw heads; repair fastener and thread damage; remove and install gaskets; use the chemical sealers; diagnose gasket and seal failures; replace metal lines and rubber hoses properly; service and replace engine drive belts.

Chapter 3 - "Service information, tools and shop safety". Service information is needed by service mechanics to determine specifications and service procedures and learn about any necessary special tools needed. This chapter introduces you to some diagnostic test equipment, theory behind them and their most efficient use. Mechanics and technicians should know how to take various measurements, and should know the correct use of the tools that take measurements. Shop safety rules and safety in the use of service of tools and equipment, and disposal of waste are required to be observed at all times.

To diagnose if a problem is an internal, mechanical problem or a problem with an engine system, requires the use of very specialized testing equipment, and specialized tools for engine rebuilding. This chapter overviews the selection and utilization of engine service tools, cleaning tools, measuring tools, electrical test equipment, and engine test equipment. This will give you a sound background when learning to diagnose, service, and repair modern "high tech" engines. The tools discussed here are meant for specialized engine repairs.

Modern automobile engines use a wide variety of electrical/electronic devices. Some monitor engine conditions and others control engine related systems. It is thus important to understand test equipment, used in engine service operations.

Hundreds of engine mechanics/technicians are injured or killed while on the job each year. A majority of these injuries are caused by broken shop safety rules. A mechanic/technician is exposed to many dangerous situations and must learn to control these dangerous situations and maintain a safe environment. This chapter also outlines the basic shop safety rules that need to be understood and adopted by every mechanic working in repair shops. An auto shop can be a safe and enjoyable place to work if safety rules are followed. You must prevent fires, explosions, chemical burns, electrocution, and other physical injuries. Hazardous wastes are often generated in the auto shop. These wastes are regulated by governmental laws and must be disposed off properly.

In this chapter you will learn to: locate correct service information for an automobile in an owner's manual, in an automobile repair/service manual, and manufacturer's shop manual; identify and describe the purpose of service bulletins, and online repair information. You will learn to: demonstrate your knowledge of safety practices in a shop; recognize common hazards in the use of hand and power tools; identify the hazards of improper use of air and hydraulic equipment; describe the hazards of automotive work connected with running engines, storage batteries, and cleaning equipment; explain how fires are prevented and extinguished. Also you will be able to learn in more detail. The use of specialized engine service tools and equipment; select the right service tool when doing actual repairs; use the most common engine cleaning tools; use engine measuring tools; use typical engine electrical testing devices; describe the function of commonly used engine test equipment; use engine service tools and equipment.

Chapter 4 "Cooling and Lubrication Systems Diagnosis and Service" discusses typical cooling and lubricating systems maintenance, problems diagnosis, parts repair and replacement as necessary. An engine mechanic/technician must be able to service and maintain an engine's cooling and lubrication systems. Both are extremely important to the service life and dependability of an engine. Engine overheating is a serious problem that can cause major damage. Lubrication system oil should be of the right type, grade and quality. Use the type of oil recommended by the engine manufacturer. Oil filter that removes impurities in oil should be changed with the engine oil at regular, prescribed intervals. Always check the service manual for detailed procedure to service oil pan and replace the pan gasket, oil pump, engine bearings, etc. Do not overtighten the oil drain plug.

Here you will learn to: troubleshoot cooling system problems; perform basic cooling system tests; replace or repair defective cooling system components; diagnose lubrication system problems; measure engine oil pressure; replace or repair faulty lubrication system parts.

Chapter 5 reviews basic testing, maintenance, diagnosis and service procedures for starting and charging system components. An engine mechanic/technician must be able to do fundamental service operations on batteries, starting motors, alternators and other engine related components. During engine removal, the mechanic/technician must remove the starter motor, and alternator and properly install them when the engine is reinstalled in the vehicle after necessary service. This chapter summarizes the service procedures most commonly done by an engine mechanic/technician. Battery, charging system and starting system are interacting. A malfunction in one can adversely affect the operation of the other.

In this chapter you will learn to be able to: Diagnose battery, starting system, and charging system; perform common battery tests, charging system tests, and starting system tests; inspect for battery, starting system, and charging system troubles; use a load tester; replace a battery, starting motor, and alternator.

Chapter 6 "Carbureted-Fuel-System Diagnosis and Service" reviews carbureted engine's fuel supply system maintenance, testing, problem diagnosis and service procedures. An engine relies on clean air and fuel. Always check fuel filters for clogging before testing fuel pump output. Measure fuel pump pressure and volume to check its performance. Normally replace a fuel pump if it fails performance test. Technician/mechanic should be able to visualize the operation of each of the seven-carburetor circuits to effectively troubleshoot any carburetor problems. When trouble occurs, always inspect the carburetor for problems: fuel leaks, vacuum leaks, missing wires, etc.

The study of this chapter will enable you to: service the air and fuel filters; troubleshoot and replace fuel pumps; perform typical carburetor service tasks; testing of computer controlled carburetor systems.

Chapter 7, "Fuel-Injection Systems Diagnosis and Service", reviews fuel injection of automotive engines. Most new vehicle engines come equipped with electronic fuel injection systems. Therefore, automotive mechanics must know how to do common tests and repairs done on fuel injection systems. EFI systems use engine sensors, a computer, and electrically operated injectors. The sensors monitor engine temperature, air-fuel ratio, airflow, throttle position, and other variables. The sensors produce different voltage or current values with changes in operating conditions. The computer uses these signals from the sensors to control engine functions. The computer control unit is the "brain" for the system. It contains IC's or computer chips that store information electronically. After analyzing data from the sensors, it uses power transistors to energize the fuel injectors. To find EFI problems, begin by inspecting the engine compartment. Follow service manual instructions when adjusting a fuel injection system. Typical adjustments are idle speed, throttle plate stop, air-fuel mixture, and throttle cable.

Information documented in this chapter will enable you to: Troubleshoot typical fuel injection system problems; activate and utilize the self-diagnosis mode found on most EFI systems; pressure test fuel injection system; adjust engine idle speed on a fuel injected engine; diagnose common diesel injection problems; perform basic tests on a diesel injection system; remove and replace major diesel engine injection components; adjust injection timing and diesel engine speed settings.

Chapter 8 "Ignition Systems Diagnosis and Service" reviews the normal adjustments, maintenance needed and service that could be needed on both point and electronic ignitions and describes the few tools needed. Even though ignition systems are fairly simple collection of electrical and mechanical parts, yet ignition failures are a major cause of no-start or engine performance problems.

This chapter enable you to: Troubleshoot basic ignition system problems; perform basic ignition system tests; gap and replace spark plugs; adjust ignition timing; adjust contact points and pickup coil gaps; replace major components of an ignition system.

Chapter 9 "Engine Condition Diagnosis and Tune-up" reviews the testing and service procedures and efficient use of service equipment. One of the most difficult aspects of being an engine mechanic/technician is troubleshooting before actual engine teardown. You must be able to quickly and accurately locate mechanical problems and correct their source to restore the engine normal condition. An engine tune-up is not only important to engine power and fuel economy, it is equally critical to air quality and conservation of our natural resources. A tune-up assures that the engine, ignition system, fuel system, and emission control systems are within factory specifications. The procedures for a tune-up vary from shop-to-shop.

Study of this chapter text will enable you to: list the visual checks to determine engine condition; describe engine noise and its relation to engine condition; perform dry and wet compression tests; perform a cylinder leakage test; measure the timing chain slack; describe how an oil sample analysis can be used to determine engine condition; list and perform fundamental steps for a tune-up; describe the hazards of doing a tune-up; describe the maintenance operations of an engine.

Chapter 10 "Engine Performance and Mechanical Problems" reviews testing, diagnosis and service procedures to restore the performance of engine to factory conditions. An engine performance problem is any problem that lowers the engine power, fuel economy, driveability, or dependability. Performance problems can be caused by problems in the fuel system, ignition system, emission control system, or engine itself. This makes troubleshooting very challenging. You must use your knowledge of system operation and basic testing techniques to quickly and accurately pinpoint and correct the source of performance problem. This chapter also summarizes the causes, symptoms, and results of mechanical part failures in an engine.

Study of this chapter will enable you to: describe the most common engine performance problems; explain the symptoms of common engine performance problems; explain how engine support systems might cause many engine performance problems; use a systematic approach when predicting engine performance problems; describe typical mechanical failures of in an engine; explain the causes of engine mechanical failures; summarize the symptoms for common engine mechanical breakdowns; diagnose engine mechanical problems.

Chapter 11 "Engine Removal, Disassembly and Cleaning" outlines the common steps for a vehicle's engine removal, teardown, and cleaning in order to inspect and repair or replace various components, diagnose their wear, distortion, etc. in order to service them to restore the engine condition. When an engine is disassembled, you can diagnose as to what might need to correct when you rebuild it. Study the wear patterns on parts and find out what is wrong with the engine. The different methods for engine parts need to be carefully studied to get the best results in individual parts cases. This chapter overviews the most important or typical steps for engine removal, teardown, and engine cleaning. An engine overhaul involves cleaning, inspecting, testing, measuring, and replacing parts as necessary. When part measurements are not within specs, the parts must be machined, repaired, or replaced. Usually, piston rings, bearings, seals, gaskets, core plugs, timing chain, belt or gears, oil pump, filters, and any other worn parts are replaced during an overhaul.

Study of this chapter will enable you to: list common engine repairs that do not require engine removal; explain why engine removal would be necessary; summarize the steps to prepare for engine removal; list the safety rules to follow during engine removal; describe how to keep parts organized during engine removal and teardown; explain how to attach and use engine lifting devices; describe the major steps for engine disassembly; describe various methods used to clean engine parts.

Chapter 12 describes "Engine Reconditioning – Short Block Service". Measuring and machining engine parts are the most important parts of a rebuild. Unless these steps are performed accurately, the rest of your rebuild efforts will be wasted. This chapter overviews overhaul of an engine short block, inspect, measure, and install the crankshaft, main bearings, seals, pistons, rings, connecting rods, piston pins, big-end bearings, and cam bearings into the cylinder block. Inspect cylinder block closely for any problems before assembly. Follow service manual procedures to service the short block components where available.

Study of this chapter will prepare you to: check for cylinder block damage, cracks, and wear; describe the machine shop methods to repair cracks or flaws in a cylinder block; check for main bore and deck warpage of cylinder block; measure cylinder bore wear and compare it to manufacturer's specs; describe how a machine shop bores and sleeves a worn or damaged cylinder; hone or deglaze a cylinder; adequately clean a cylinder after honing; check a crankshaft for wear, cracks, and runout; install camshaft and balancer shaft bearings; install a rear main oil seal; install main bearings and the crankshaft; Plastigauge rod and main bearings to check their oil clearance; check crankshaft end play; service pistons and piston rings; service connecting rods and piston pins; measure ring and piston clearances; properly assemble the short block components after necessary service.

Chapter 13 "Engine Reconditioning – Top End Service" outlines the most important steps for reconditioning a cylinder head and parts mounted on and in it. This chapter overviews methods to check a cylinder head for cracks, warpage, and other problems, service diesel engine combustion chambers, measure valve guide and valve stem wear, recondition and replace valve guides, recondition or replace valve seats, grind engine valves, test and correctly install valve springs, valve stem seals, measure camshaft wear, camshaft end play, and bearing clearance, service rocker arms and rocker arm shafts, service push rods and valve lifters.

Study of this chapter will prepare you to: Check a cylinder head for cracks, warpage, and other problems; service diesel engine combustion chambers; measure valve guide and valve stem wear; knurl, ream, or replace valve guides; replace and recondition valve seats; grind engine valves as necessary; test, shim, and install valve springs; correctly install both O-ring and umbrella type valve stem seals; reassemble a head using a valve spring compressor; measure camshaft wear, camshaft end play, and camshaft bearing clearance; service rocker arms and rocker arm shafts; service push rods and valve lifters.

Chapter 14 "Front End – Timing Mechanism Service" overviews the most critical information concerning the service and repair of timing chains, timing belts, and timing gears. Here you will learn to: check a timing chain for wear; install a timing chain assembly; install timing gears; check timing gear runout and backlash; install a timing sprocket and belt; adjust timing belt tension; replace front cover oil seal; install an engine front cover; correctly torque timing gear or sprocket fasteners; service timing chain tensioners; install a timing belt cover.

Chapter 15, – "Engine Reassembly - Installation - Break-in,"- overviews the basic methods for reassembling the major components of an engine after necessary service. Here you will learn to: use general steps to prepare for engine reassembly; properly install an oil pump and oil pan; install and torque a cylinder head assembly; install and torque exhaust and intake manifolds; correctly install engine gaskets; adjust engine valves; correctly install valve covers; install engine accessory units; check and install a clutch assembly; refit the engine into the engine compartment; prepare for engine start-up; use a good engine break-in procedure.

Chapter 16 "Engine Maintenance and Troubleshooting" deals with a set regular and scheduled maintenance that are discussed first and are followed by engine troubleshooting. Preventive maintenance is as important as servicing a vehicle after parts have failed. Here you will learn to: Use SI engine troubleshooting charts; list the basic engine troubles; describe the checks and corrections for SI engine troubles; to make quick checks of the fuel and ignition systems; identify the source of various engine noises; describe the causes of different types of exhaust smoke; diagnose the causes of problems in SI engines.

After studying this chapter, you will be able to: describe the process of engine troubleshooting; explain why modern engine analyzers should be used to supplement visual inspection and manual checks; describe function of "plug-in" diagnosis; understand "self-diagnosis"; list various engine problems and identify their possible causes; understand the more common causes of engine overheating, excessive oil consumption, and engine noises.

PRITAM SINGH GILL

Disclaimer...

Acknowledgements...

The author and publishers are deeply thankful and indebted to numerous manufacturing companies, organizations, agencies, and individuals who have provided useful materials from their training manuals, service manuals, design handbooks, and other non-confidential materials that have gone a long way in the completion of this book. Thanks are also due to many Engineers (students from "Guru Nanak Dev Engineering College" Ludhiana where the author had taught from 1961 to 1992) working for US automotive companies, especially General Motors, Ford and Chrysler, who have shared valuable knowledge and provided useful materials and illustrations, many of which have been used to support the text in this book. Material from a number of Automotive Parts and Systems Suppliers has been used to support the text, presented in two volumes. Volume-I primarily deals with the "Principles of Automotive Engines" and explores the relevant text to explain the construction, operating principle and function of each sub-system involved in engine operation and control. Volume-II deals with the function, construction and operating principles of each of the sub-system used in transmission of engine power to vehicle wheels, and control of the vehicle on the road. Acknowledgements for this book are difficult to make because many books and manuals, and individuals have contributed to the completion of this book.

The author wishes to express sincere thanks to the following organizations for their special contributions:

Acura
American Hammered Piston Rings.
American Honda Motors Co.
American Isuzu Motors Inc.
Ammco Tools, Inc.
American Honda Motor Corporation, Inc.
American Motors Corporation.
Applied Power, Inc.
Automatic Transmission Parts Inc.
Autolite
AC - Rochester
AC - Delco Ltd.
American Motor Company
ATW
Automotive Div of Applied Power Inc.
Audi of America Inc.
Automotive Engine Rebuilders Association
Battery Council International
Bendix Corporation
Belden
Black & Decker Inc.
Binks Tools
BMW of North America
Build-All Corp.
Burgess Ltd.
Caterpillar Inc.
Champion Spark Plug Company
Citroen UK
Chloride Batteries
Chicago Pneumatic Tool Company.
Chrysler Motors Corporation
Clayton Manufacturing Company.
Clevite Industries Inc.
Daimler Benz AG (Germany)
Dana Corp. Service Parts Group.
Deere & Co.
Earl Manufacturing Co.
Echlin Manufacturing Company, Inc.
Electro-mechanical Division, Exide Corporation.
Edelbrock and Offenhauser.
Fedral Mogul Corporation.
Fel-Pro Gaskets.
Fiat Motors
Florida Dept of Voc. Ed.
Ford Motor Company
Gates Rubber Co.
General Motors Corporation
    AC Delco
    AC Spark Plug Division
    Buick Motor Division
    Cadillac Motor Car Division
    Chevrolet Motor Division

Pontiac Motor Division
Delco Remy Division
'Saginaw Steering Div.'
General Motors Corp., Service & Parts Operations
GM Products Service Training, General Motors Corp.
Goodson Auto machine Shop Tools and Supplies
Greenfield Tap & Die.
Honda Motors Co.
Heyco and Klein Tools
Hyundai Motor America.
Jaguar
Jasper Engines and Transmissions
K-Line Tools.
K-D Tools
Kent-Moore Tool Group.
Lisle Tools.Corporation.
Lotus Engineering Ltd.
L.S. Starret Company
Lucas Automotive Ltd.
Mack Trucks. Inc.
Magnaflux Corp.
Mazda of North America, Inc.
Mercedes Benz
Midwest Crankshaft
Mitsubishi Heavy Industries, Japan.
Mitsubishi Motors Sales of America, Inc.
Monroe Auto Equipment Company,
Mopar
Morgan Motors Co. Limited (England)
Motorola
Motor Vehicle Manufacturers Association.
National Institute for Automotive Service Excellence.
Neway Manufacturing Inc.
Nissan Motor Corporation, USA.
Owatonna Tools
OTC Tools
Peerless Instruments
Parker Tools
Perfect Circle
Permatex
Plew Tools
Plymouth
Peterson Publishing Company.
Peugeot Motors of America, Inc.
Prestolite Company
Proto Tool Company
Repco Corporation Ltd.
Renault USA Limited
Ricardo & Co. U.K.
Robert Boshch UK Ltd.
Robert Bosch Sales Corporation.
Rolls Roice Motors Limited (England)
Rover Cars Ltd.
Rockwell International Corp.
Rubber Manufacturers Association
Saab-Scania of America, Inc.
Sealed Power Corporation.
Sioux Tools Inc.
Snap-on Tools Corporation.
Society o Automotive Engineers
Sonoco
Stanadyne Diesel Systems
Storm Vulcan
Subaru
Sun Electric Corporation
Sunnen Products Company
The Eastwood Company
Texas Instruments, Incorporated
Toyota Motor Sales, USA, Inc.
The L.S Starret Company
The Prestolite Company
Trojan Battery Company
TRW Inc.
Varta Batteries
Volkswagen of America, Inc.
Volexvagon AG (Germany)
Volvo

CONTENT

Tools of Automotive Trade

1

1.1 INTRODUCTION

Tools play an important part in any automotive service operation. This chapter briefly describes the most common conventional tools used by mechanics in automotive service shops. Both hand tools and power tools will be described. A wide variety of tools are used to fit varied jobs. A mechanic should know the name of each tool, the safe procedure for using each tool, and be trained to choose a proper tool for the job at hand. The units of measure used will be metric but in certain cases based upon the source of the illustrations and manufacturers, English system united States Customary measurements, as adopted by SAE (Society of Automotive Engineers), are sometimes used.

User must be aware of the safe use of his/her tools. Using the right tool for the job is the first step in using tools safely. This not only will help perform the work as quickly as possible, but also to complete the job with maximum accuracy and safety. Obviously, using the tool for the type of job it was designed to do requires the user to know more about a tool than just its name. Therefore, the user needs to know as to what the tool is used for and how to use it safely. A user should always keep the safety precautions in mind and observe some general safety rules when working with tools.

Keep all tools clean and dry. After using them wipe them off. Greasy tools can easily slip from hand grasp. Some common sense and a little forethought can save eye, hand finger or even life.

1.2 HAND TOOLS

Automobiles have become increasingly complex, and a larger variety of tools are needed to service them. Every mechanic's tool kit should contain a good selection of hand tools, probably in both metric and fps units. Without an extensive collection of good quality tools, an auto mechanic may not be able to turn out precision jobs quickly and safely. When a mechanic has a complete tool kit, it will include the right tool for virtually every job, *Fig. 1.1*.

1.3 WRENCHES

Many automotive components are fastened together with bolts and nuts. Wrenches are tools designed to tighten or loosen bolts and nuts. Since there are many different sizes of bolts and nuts, wrenches must also be made in different sizes.

Each mechanic's tool kit should contain a complete set of wrenches. The size of a wrench is determined by the size of the nut or bolt head it fits on. The wrench shown in *Fig. 1.2* has 10 mm stamped on it. This means that the opening of the wrench measures 10 mm across the flats and that it will fit on a bolt or nut that measures 10 mm across the flats.

Wrench sizes are given either in metric system units or English system units. Metric wrench sizes are given in millimeters – 10, 11, 12 mm etc. Sizes of English system wrenches are given in fractions of an inch – 5/16, 3/8, 7/16, 1/2, 9/16 etc.

Open-End Wrench

Open-end wrenches have an opening at the end, which is placed on the bolt or nut. The opening is usually at an angle of 15° to the handle. This makes turning in a tight space easier. Open-end wrenches are made in many different sizes and shapes, *Fig. 1.3* (*a*) to (*c*). Most open-end wrenches have two open ends, one

Fig. 1.1 *Automobile machanics have to usually buy an assortment of good quality tools to be able to perform quality jobs quickly (Snap-on Tools Corp.)*

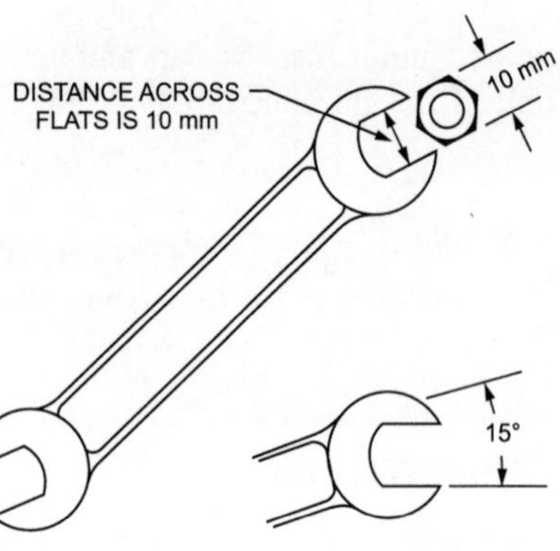

Fig. 1.2 *The size of a wrench.*

end one size different from the other. This allows the owner to have a full set of wrenches at nearly half the investment and weight. On wrenches shown in *Fig. 1.3* (*b*) the opening in one end is set at 90° to the handle rather than 15°. Wrench at (*c*) has one end specially fitted for fuel and brake tube fittings. It is called a *tubing wrench.* A *crowfoot wrench* can also be used on tubing fittings, *Fig. 1.3* (*d*).

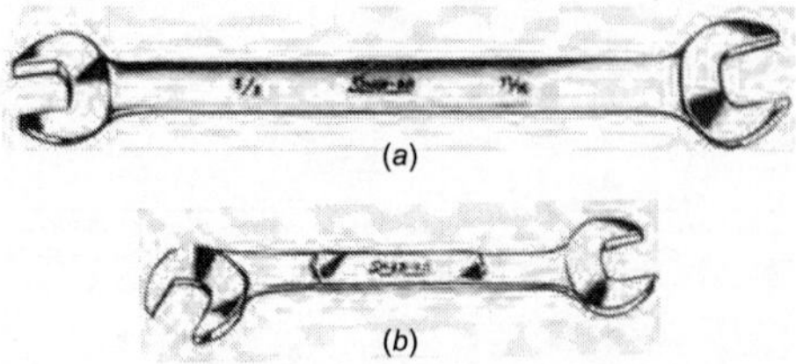

Fig. 1.3 *(a) and (b) Open-end wrenches. (Snap-on tools corp.)*

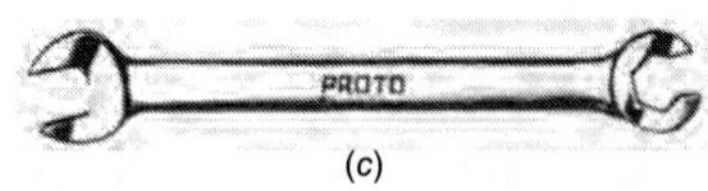

Fig. 1.3 *(c) Open-end wrenches (Proto).*

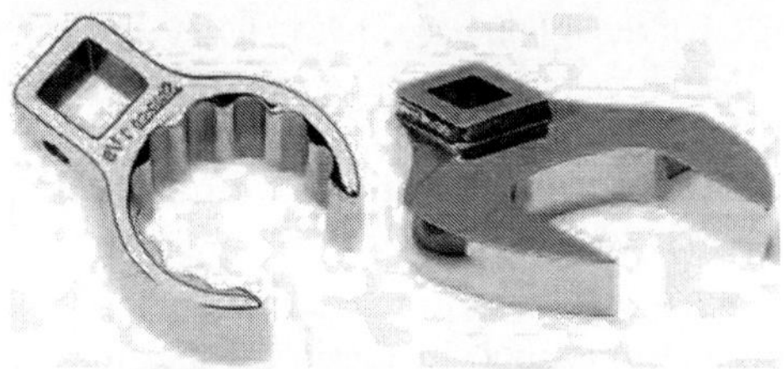

Fig. 1.3 *(d) The crowfoot wrench must be used with an extension. (Snap-on Tools Corp.)*

Box-End Wrenches

Box-end wrenches, *Fig. 1.4,* are designed to fit around a bolt of nut. They cannot be used in as tight a space as an open-end wrench. But they permit the mechanic to apply more force with less chance of the wrench slipping off the bolt head or nut. Both ends have either a 6-point or 12-point box with a different size at each end, *Fig. 1.4* (*a*), (*b*). A 12-point box wrench permits us to work in tighter areas than equivalent 6-point wrenches. Like other wrenches they come in sets of many sizes, *Fig. 1.4* (*c*). The handles of most are offset 15° in relation to the head to give turning room in a tight space, *Fig. 1.4* (*a*).

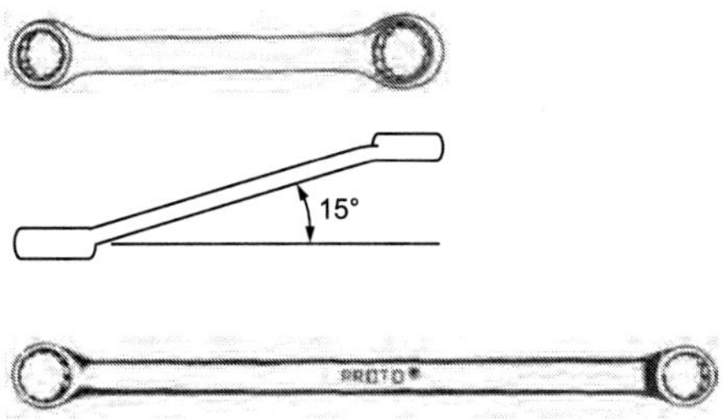

Fig. 1.4 *(a) Box-end wrench. (Proto)*

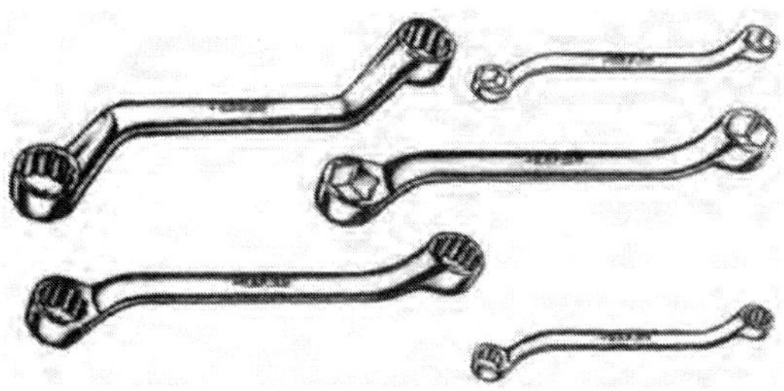

Fig. 1.4 *(b) Six-point and 12 point box end wrenches (Snap-on Tools Corp.)*

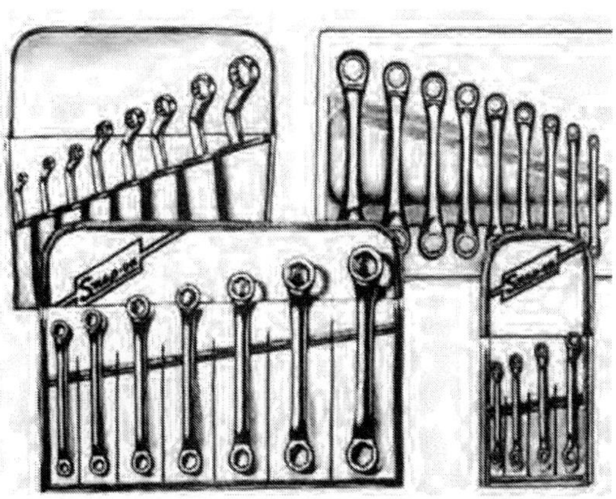

Fig. 1.4 *(c) Box-end wrench. (Snap-on Tools Corp.)*

Combination Wrenches

A *combination wrench* has an *open end* at one end and a *box end* at the other. *Fig. 1.5 (a)*. The box end will have 6-points or 12-points. These combination wrenches come in two lengths: standard – for working in tight areas, and long – for working where extra leverage is needed. Combination wrenches are usually the same size on both ends. Combination wrenches are general-purpose wrenches and are preferred by many mechanics.

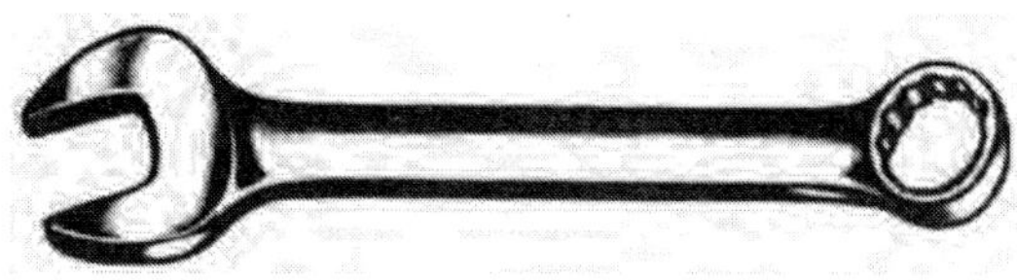

Fig. 1.5 *(a) Combination wrench. (Snap-on Tools Corp.)*

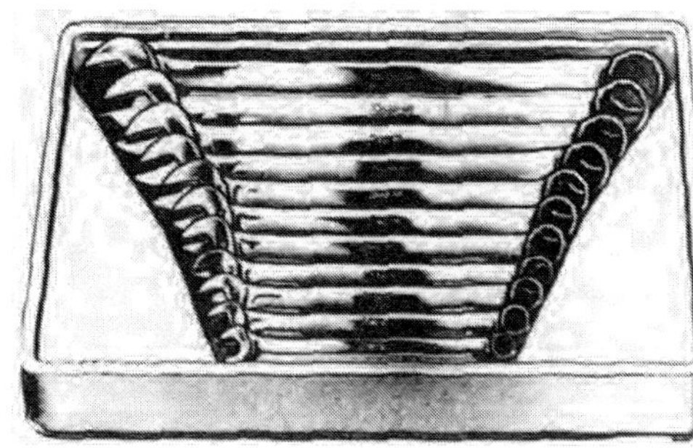

Fig. 1.5 *(b) A set of combination wrenches. (Snap-on Tools Corp.)*

The open-end side is used where space is limited. The box-end side is used for final tightening or to begin loosening. Its construction makes it less likely to slip off a nut or bolt head. The open end is usually best suited for running the nut down or holding a bolt head. As a safety precaution, always pull a wrench toward your body, to avoid getting hurt if the wrench slips off the nut.

Sockets and Ratchets

The most frequently used tool in a mechanic's tool kit is the *ratchet and socket, Fig. 1.6 (a), (b)*. Sockets are like box wrenches in that they go all the way around the bolt head or nut. But socket can be removed from the ratchet handle. Sockets are made in all the Metric and English sizes. Sockets of many different sizes can be used with one ratchet handle.

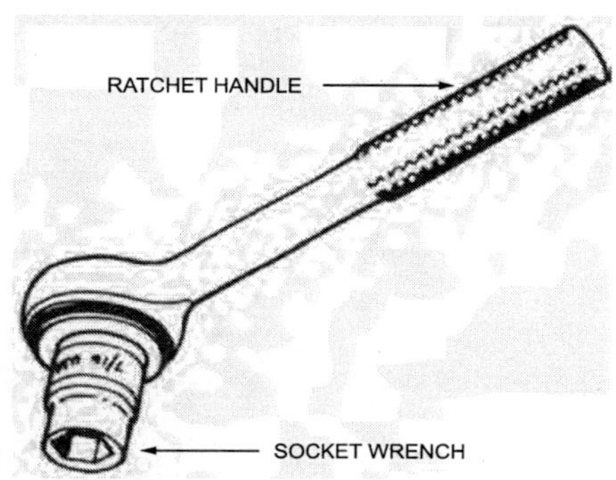

Fig. 1.6 *(a) Socket wrench and ratchet handle. (Proto)*

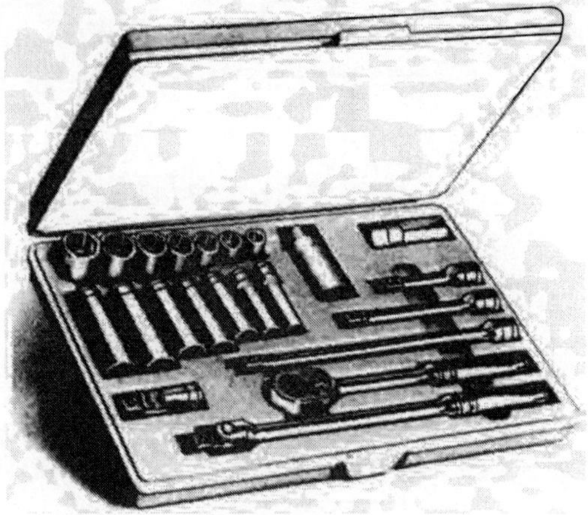

Fig. 1.6 *(b) Ratchet set. (Snap-on Tools Corp.)*

Sockets are attached to a handle by a square hole at one end. These drive holes are also made in different sizes. For small bolts and nuts, such as automobile trim parts, socket sets with a 1/4 inch square drive are useful. For general-purpose work, a 3/8-inch drive set is popular. Heavier work requires a 1/2 inch drive socket set. Even larger drives, 3/4 inch and 1 inch sizes are made for very large nuts and bolts.

Besides different opening sizes and drive sizes, sockets have different numbers of points and different lengths, Figs. 1.7, 1.8. The sockets, shown in 2D and 3D illustrations, have 6-points, 8-points, and 12-points. The fewer the points the more strength the socket has but the harder it is to slip over a bolt head or nut.

Figure 1.8 (*a*), (*b*) shows examples of long, or deep, sockets, with 8-points and 12-points. These are useful for driving nuts on long bolts and studs or working in a deep hole. A special type of the deep socket is shown in *Fig. 1.8* (*c*). This is a spark plug socket. The cutaway view shows that this socket has a rubber part inside to protect the spark plug when it is loosened or tightened.

Extensions are available in a variety of sizes. Extensions allow sockets to be used in restricted areas. Socket drivers are often used with sockets to facilitate fastener removal. Both extensions and sockets drivers are available with flexible shafts. *Extensions* are rods that extend the distance between the socket and ratchet, *Fig. 1.8* (*d*). They range from 25 mm to about 900 mm long.

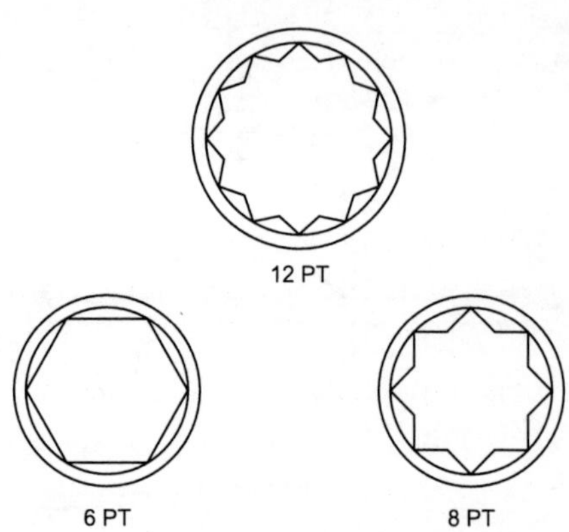

Fig. 1.7 *(a) Sockets with different numbers of points. (Proto)*

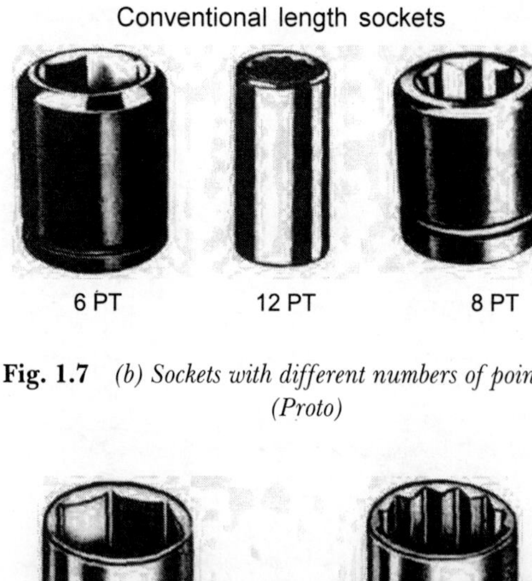

Fig. 1.7 *(b) Sockets with different numbers of points. (Proto)*

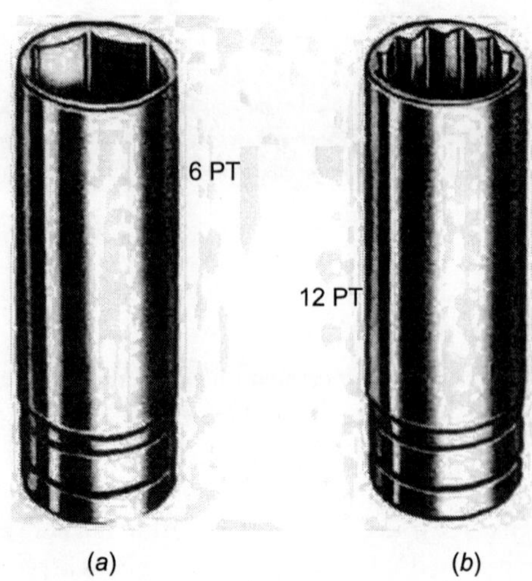

Fig. 1.8 *(a-c) Deep-well sockets are used for removing sparkplugs*

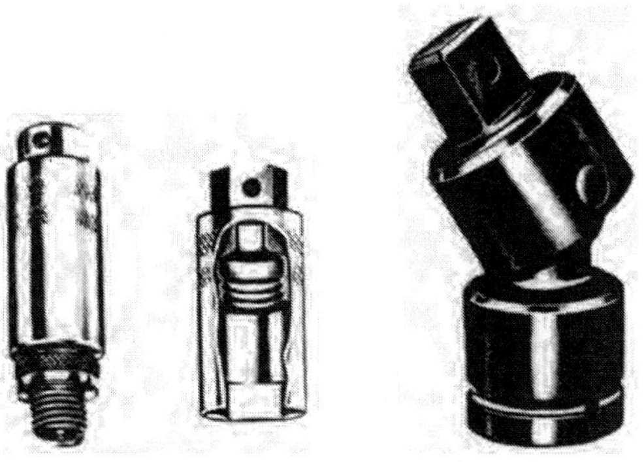

(c) (d)

Fig. 1.8 *(a-d) Different styles of sockets (Proto and Snap-on Tools Corp.)*

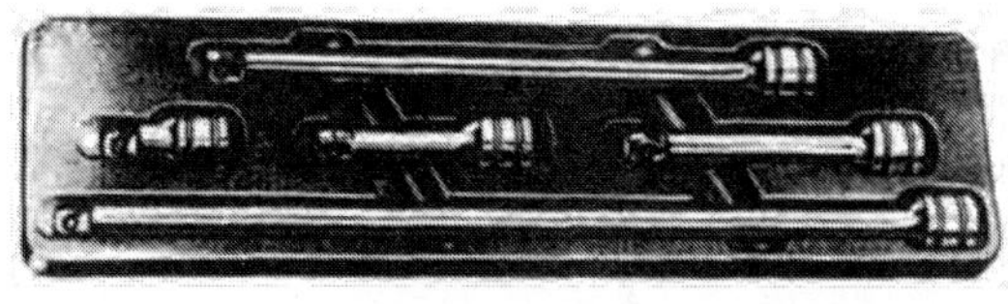

Fig. 1.8 *(d) Extensions. (Snap-on Tools Corp.)*

If it is possible to get a direct pull when removing a fastener, a *universal joint* is available for use between the ratchet handle and the socket. Sockets complete with universal joints are also available, *Fig. 1.9.* Note that these are 6-points sockets. Universal sockets also come in 8-points and 12-points sets. These are called universal sockets because of the hinge or universal joint in the middle that allows it to be used at an angle. This type of socket is very useful in a small space.

A large number of handles and attachments are available to drive socket wrenches. For all the tools shown in *Fig. 1.10* we have a ½ inch drive. They can be used only with sockets that have a ½ inch hole.

Fig. 1.9 *A set of swivel sockets is useful when removing bolts in hard-to-reach locations. (Snap-on Tools Corp.)*

Figure 1.10 (*a*), (*b*) shows two designs of ratchets in common use. A socket wrench is attached to the square drive on the ratchet handle. The socket is then placed over a bolt or nut. The bolt or nut is tightened or loosened by rotating the socket handle. A freewheeling or ratchet mechanism inside the ratchet allows it to drive the nut in one direction and to move freely in the other direction without driving the nut. This permits fast work in a small space because the socket does not have to be removed from the nut each time it is turned. A lever on the ratchet handle allows the mechanic to choose which direction the ratchet will drive and which direction it will turn free.

The socket drive shown in *Fig. 1.10* (*c*) is known as a *speeder* or *speed handle.* The combination of a swivel handle and crank allows very quick driving of a socket. It is preferred when a large number of bolts or nuts have to be replaced or removed. The driver in *Fig. 1.10* (*d*), a *sliding T-handle,* will slide to any position along the handle. *Fig. 1.10* (*e*) has a hinge handle or breaker bar. Its drive end has a hinge that will allow driving at different angles. Its long handle permits the mechanic to use large force to loosen a tight bolt or nut.

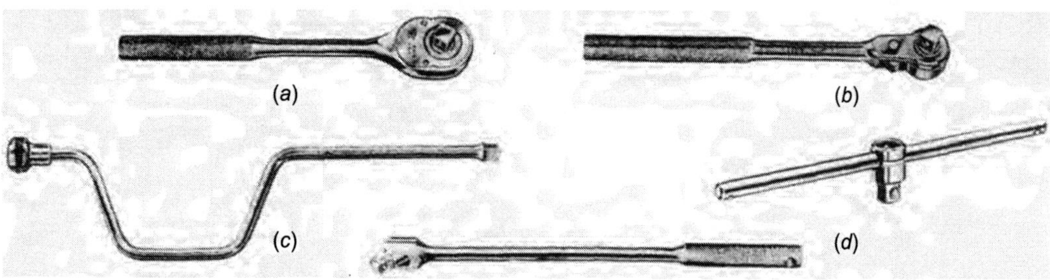

(e)

Fig. 1.10 *Socket handles. (Proto)*

Specialty Wrenches

Many *specialty wrenches* are designed to do one specific job. *Fig. 1.11* shows wrenches used when servicing automobile brake systems.

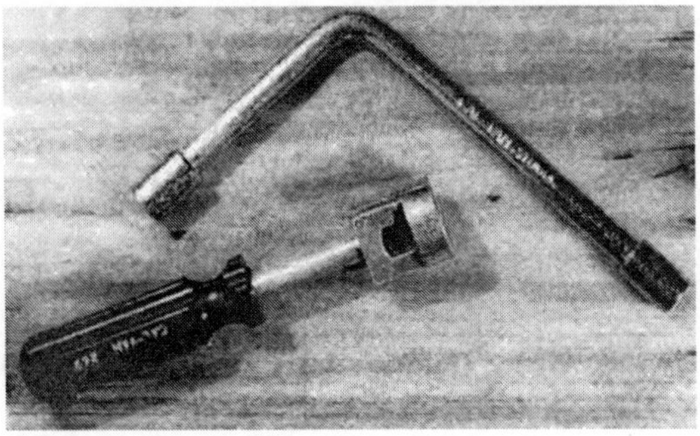

Fig. 1.11 *These specialty wrenches are designed for use when working on vehicle's brake system.*

Allen Wrenches

Some automotive components are fastened with hollow head Allen screws. These screws require special Allen wrenches. The *Allen wrench,* or hex wrench, *Fig. 1.12,* is a hex-shaped wrench and is available in a wide variety of sizes. The speed hex wrench, *Fig. 1.13* (*a*), has a ball at one end. The ball allows the wrench to swivel as the fastener is loosened or tightened, expediting removal or installation.

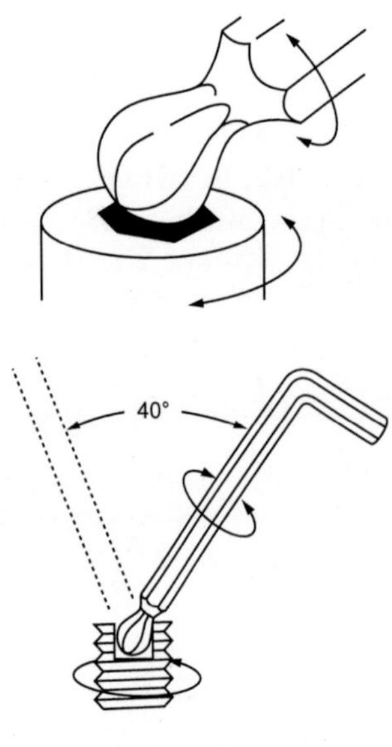

Fig. 1.12 *A speed hex wrench can swivel in tight spots. (Lisle Co,)*

Allen wrenches are available in sets, *Fig. 1.13* (*b*), according to the size of the Allen screw in which they fit. They are made in both metric system and English system sizes.

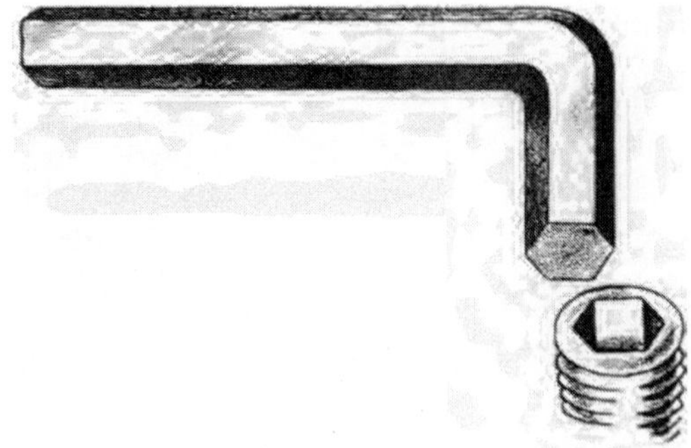

Fig. 1. 13 *(a) Allen wrench and Allen head screw (GM Corp)*

Fig. 1. 13 *(b) Allen wrench set. (Proto)*

Torque Wrench

Often, a fastener must be tightened to a specified torque. A special socket handle called a torque wrench is used for this purpose. A torque wrench measures the resistance to turning of a threaded fastener. This resistance is called torque.

If a fastener is overtightened, mating parts may be distorted and leakage of oil, coolant, compression, etc. may occur. Cylinder head bolts and engine bearing nuts and bolts must be 'torqued' correctly. When tightening these fasteners, a *torque wrench* must be used. *Fig. 1.14* (*a*) shows a *Click-type torque wrench* used to tighten fasteners to a specified torque. It has a ratchet drive head. Its adjustable handle and scale allows the mechanic to adjust the wrench to a certain torque setting. An audible 'click' is heard when the preset torque has been reached. This type has on one side of the wrench metric scale, while the other side has English units scale.

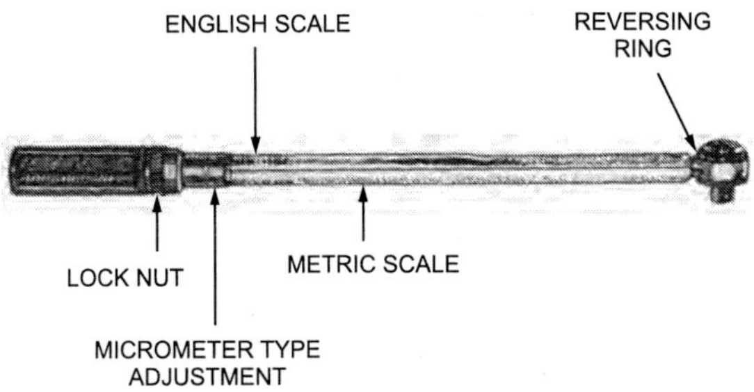

Fig. 1.14 *(a) A click-type torque wrench can be used to tighten fasteners to a specified torque. An audible "click" is heard when the preset torque has been reached. (Central Tools)*

Another popular type, *Fig. 1.14* (*b*) uses a beam and pointer assembly. During tightening, the beam on the wrench bends as the resistance to turning increases. The torque is displayed on a scale near the handle. The *dial type* has a scale that the mechanic can read directly. Digital-readout torque wrenches are also available.

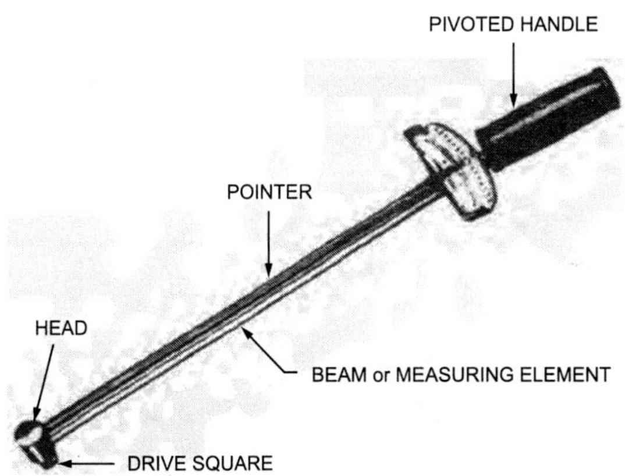

Fig. 1.14 *(b) Beam torque wrench. (P. A. Sturtevant Co.)*

Torque wrenches indicate torque in Newton meters, foot-pounds or inch-pounds, *Fig. 1.14* (*c*). The mechanic must use a torque wrench scaled to the measurement given or use a conversion chart.

When using a torque wrench, clean and oil the threads of the fastener so that no additional friction will be present.

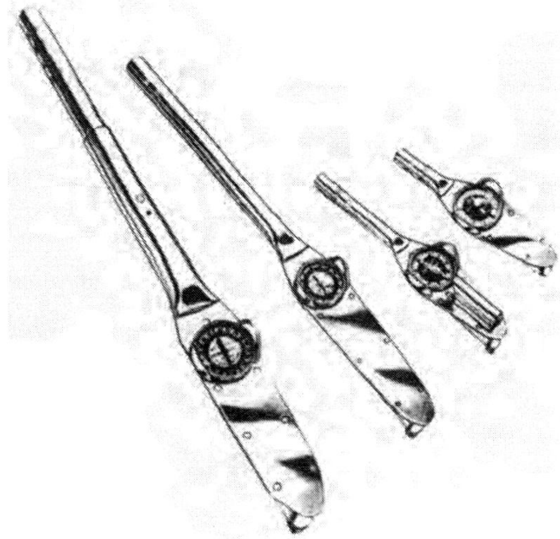

Fig. 1.14 *(c) Torque wrenches. (Snap-on Tools, Corp.)*

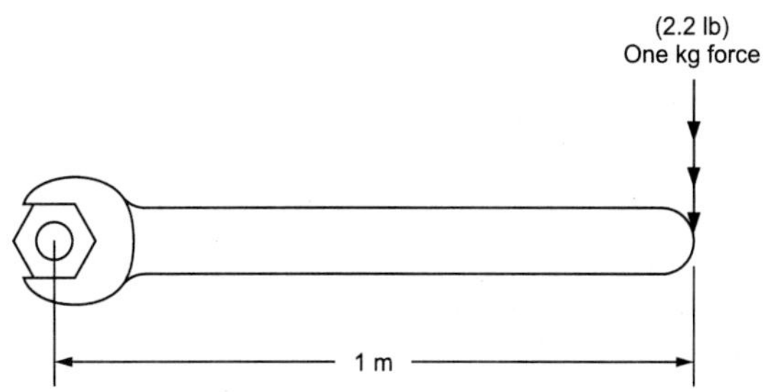

Fig. 1.14 *(d) Newton-meter*

Adjustable Wrench

An *adjustable wrench, Fig. 1.15* (*a*) adjusts to fit bolt heads and nuts of different sizes. There are adjustable wrenches available in different sizes. The longer the wrench the larger the opening it will adjust.

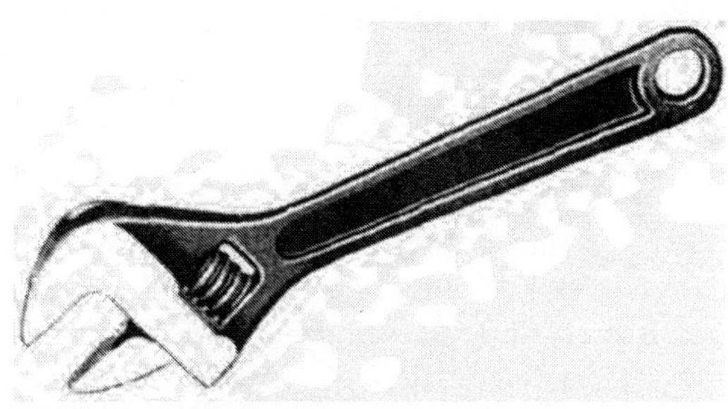

Fig. 1.15 *(a) Adjustable wrench. (Proto)*

The right and wrong ways to use an adjustable end wrench are shown in *Fig. 1.15* (*b*). For tightening, the wrench must be placed on the bolt head or nut so that stress falls on the stationary jaw. If the wrench is used wrong way, the adjustable jaw can be damaged.

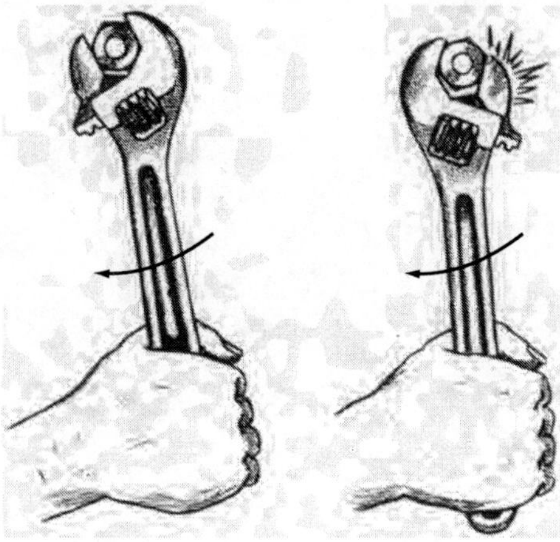

Fig. 1.15 *(b) Put the stress on the stationary jaw. (General Motors Corp.)*

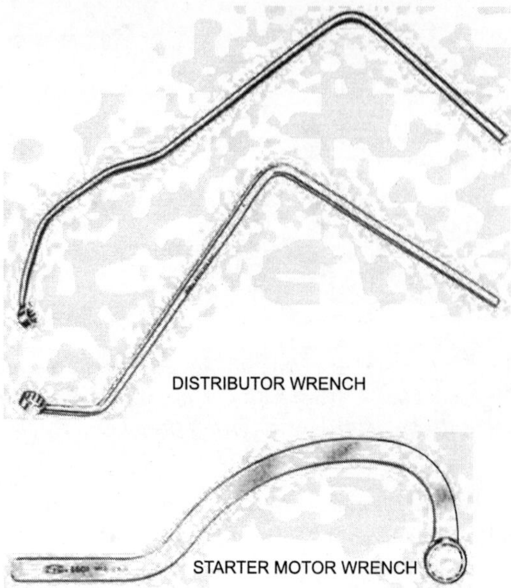

Fig. 1.16 *Distributor and starter wrenches. (Proto)*

Special Purpose Wrenches

Most automotive repairs can be made with regular open-end, box-end, combination, and socket wrenches. Some jobs, however, require special wrenches. Three types of special purpose wrenches are shown in *Fig. 1.16*. The box-end wrenches with special curved handles shown at (*a*) and (*b*) are called distributor wrenches. They are used to loosen and tighten distributor hold-down bolts. The one shown at (*c*) is called a starter wrench. Its special curved handle makes it useful for removing the bolts that hold on the starter motor.

In some cases more force is required to tighten or loosen a bolt or nut than can be applied by hand. The *striking wrench, Fig. 1.17*, is a special purpose wrench whose handle is designed to be hit with a hammer.

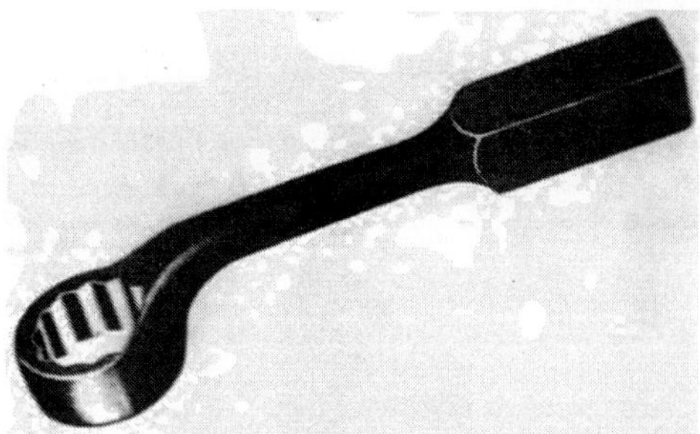

Fig. 1.17 *Striking wrench, (proto)*

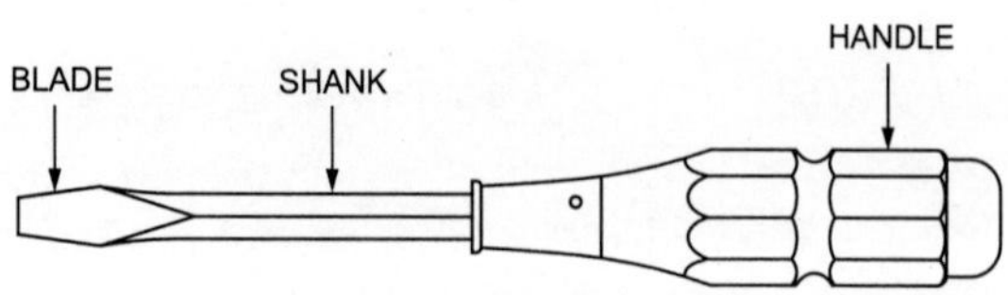

Fig. 1.18 *Parts of a screwdriver.*

1.4 SCREW DRIVERS

Many automotive parts are held together with screws. A *screwdriver, Fig. 1.18*, is a tool used to turn or drive a screw. Some screwdriver handles are still made of wood, but usually they are plastic. Shanks are made in different lengths and shapes, some round, others square. Screwdrivers have different types of blades to drive different types of screws. A screwdriver should never be used as a pry bar.

Common Screwdrivers

The most used are the *blade* type screwdrivers, *Fig. 1.19* (*a*). They are used to drive screws with a straight slot in the top. The stated length of these screwdrivers is sometimes the length of the shank alone and sometimes the screwdriver's overall length. The larger the screwdriver the larger is its blade. Use care to use a screwdriver whose blade fits the slot in the screw snugly, *Fig. 1.19* (*b*). If the fit is loose, the head of the screw may be damaged.

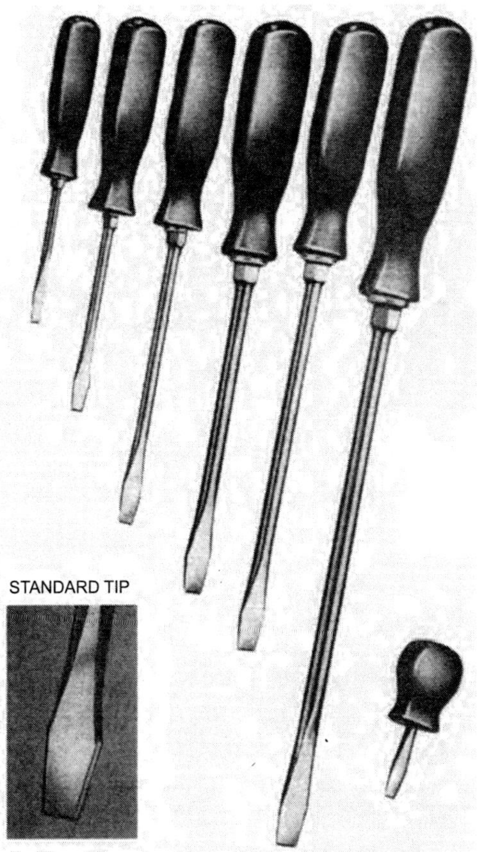

Fig. 1.19 *(a) Common screwdriver set. (Snap-on Tools Corp.)*

Fig. 1.19 *(b) A standard screwdriver must fit screw head slot.*

Phillips and Similar Screwdrivers

A *Phillips screwdriver* has a cross point, *Fig. 1.20*, to drive a corresponding Phillips screw. This makes the Phillips just a little more stable than the blade screwdriver. The blades are sized on a numbering system from 0 to 6, 0 being the smallest and 6 the largest. There are many variations on the Phillips head screw. A number of other screws with recessed heads are also shown in *Fig. 1.20*. Each of these screws needs a different type of screwdriver.

Nut-drivers

Nut-drivers are a combination of a screwdriver and a socket. The nut-driver shown in *Fig. 1.21* is part screwdriver and part wrench. Its handle and shank are the same as a screwdriver's, but in place of a blade it has a small wrench. Nut-drivers in small wrench sizes are useful for driving small nuts in tight places.

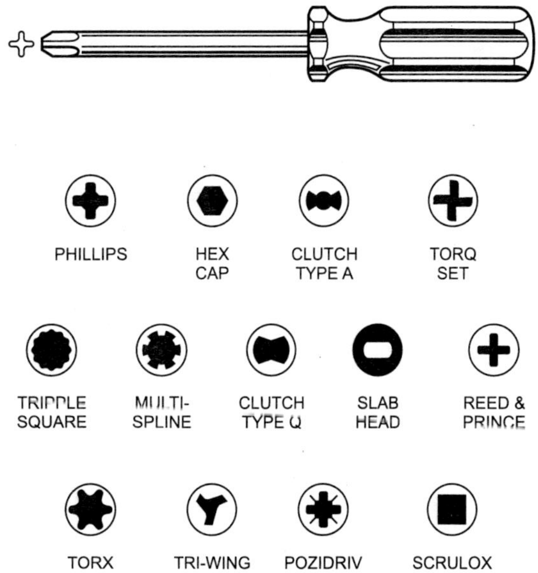

Fig. 1.20 *Phillips screwdriver and other screw types. (Snap-on Tools Corp.)*

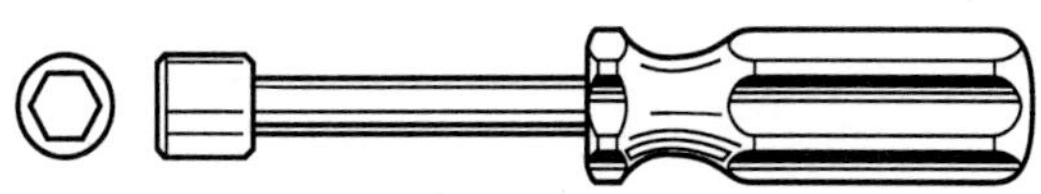

Fig. 1.21 *Nut driver*

1.5 PLIERS

Pliers are made for gripping fasteners that wrenches or screwdrivers do not fit. In addition, some pliers are made for cutting things like wire or cotter keys.

Slip-joint Pliers

The pliers shown in *Fig. 1.22* have a slip joint where the two jaws are attached. The slip joint can be set for either of two jaw openings, one for small objects and one for holding larger objects. These pliers are used for pulling out pins, bending wire, and removing cotter keys. They come in many different sizes and are grouped by their overall length.

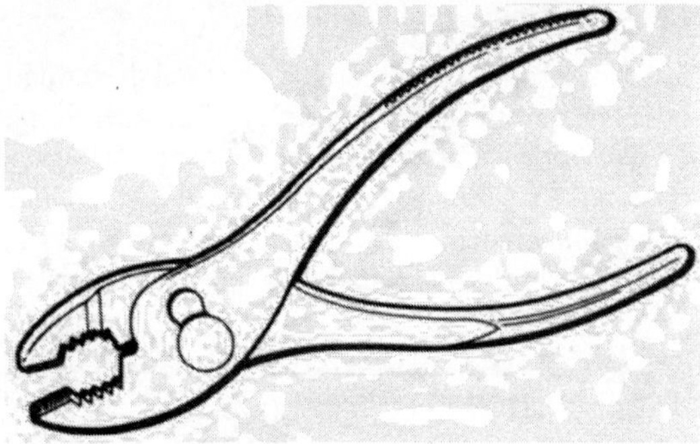

Fig. 1.22 *Slip joint pliers.*

Channel-lock Pliers

Channel-lock pliers, *Fig. 1.23*, are used to grip large objects. They get their name from the channels that allow the jaws to be set at many different openings. These pliers are made in a number of sizes and are grouped by their overall length.

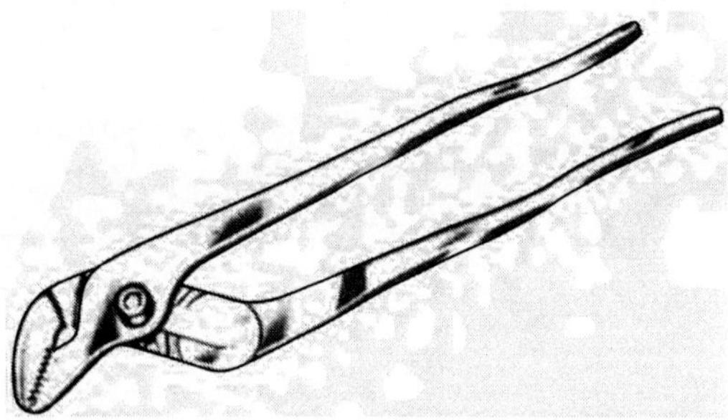

Fig. 1.23 *Channel-lock pliers. (Proto)*

Diagonal Cutting Pliers

Diagonal cutting pliers, Fig. 1.24, are used to cut electrical wire and cotter keys. Their jaws have hardened cutting edges. Like other pliers, they are made in many different sizes and are grouped by their overall length.

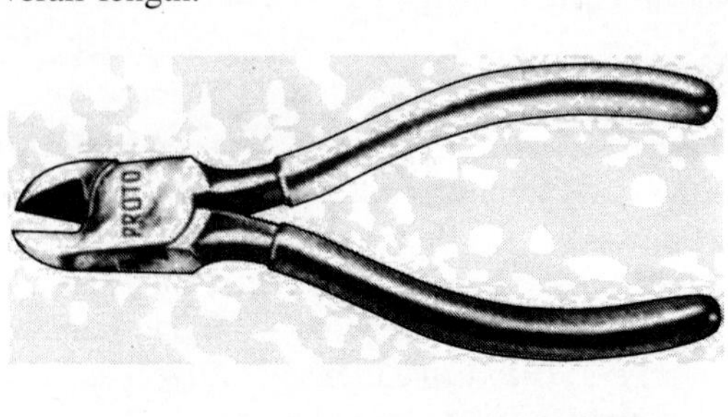

Fig. 1.24 *Diagonal cutting pliers, (Proto)*

Needle-nose Pliers

Needle-nose pliers, *Fig. 1.25*, have very long, thin jaws. They are used where there is not enough room for other types of pliers. Some needle-nose pliers have curved or bent jaws for reaching around objects.

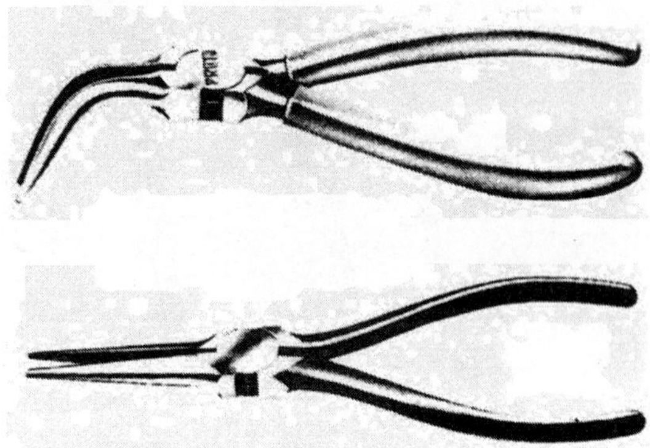

Fig. 1.25 *Needle nose pliers, (Proto)*

Special Purpose Pliers

Many types of special purpose pliers are designed for certain repair jobs, *Fig. 1.26*. The pliers at (*a*), (*b*) and (*c*) are used to replace brake shoe springs. Shown at (*d*) and (*e*) are battery pliers. The jaws on these pliers are made to grip terminal nuts on battery connections. The pliers, shown at (*f*) are used to remove hubcaps and grease caps. The round pad on one jaw is used as a hammer to replace grease caps.

Many automotive components, especially transmission assemblies, are held together with lock rings or snap rings. Shown at (*g*) and (*h*) are the snap-ring or lock-ring pliers that are designed to remove and replace snap-rings.

1.6 HAMMERS AND MALLETS

Automotive mechanics require *hammers* of various types and sizes in many repair jobs. The most important is the ball-peen hammer, *Fig. 1.27* (*a*). The flat portion of the head used for most hammering, is called the face. The other end is the peen. When the peen is ball-shaped, it is called a ball peen. A ball peen hammer is usually classified according to the weight of its head without the handle. A good hammer set for automotive work would include, 4 oz, 1 lb. and 3 lb. hammers. A small hammer is very handy for light work.

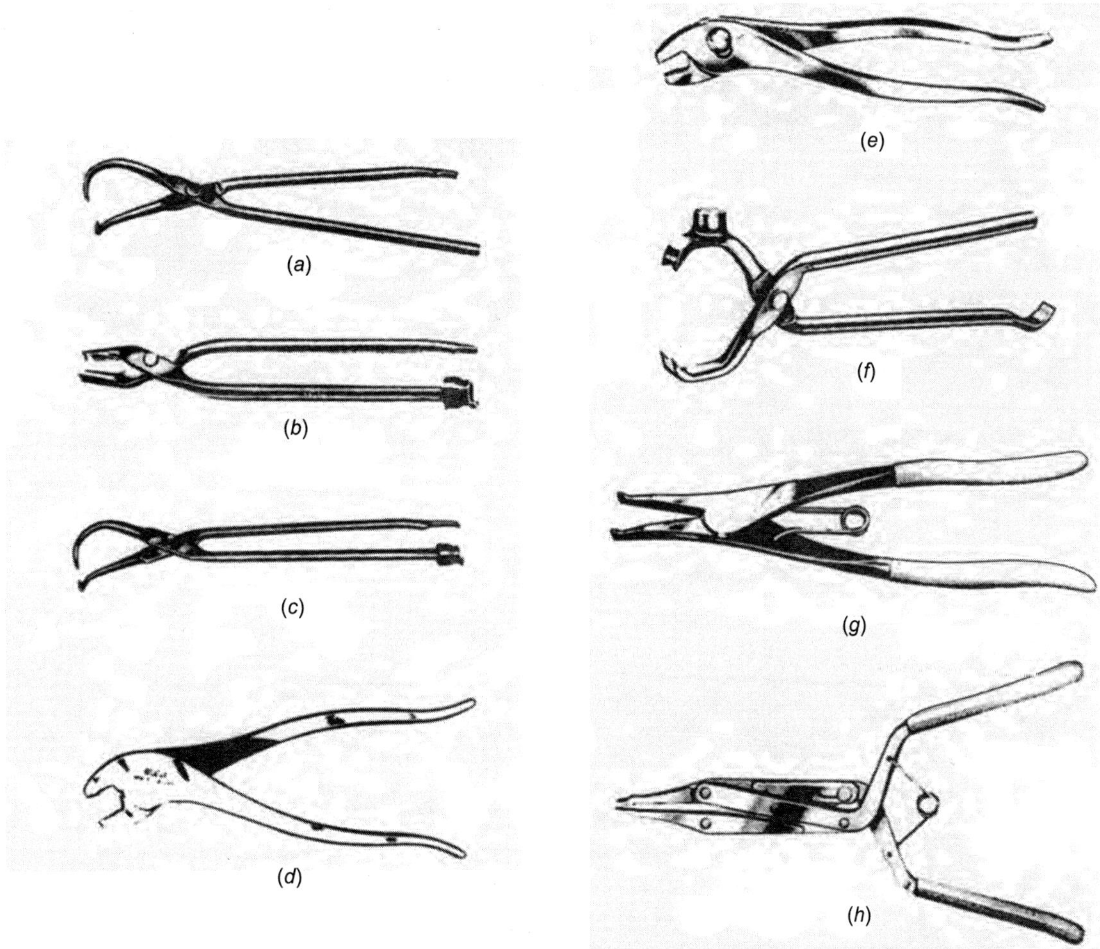

Fig. 1.26 *Special purpose pliers, (Proto)*

Fig. 1.27 *(a) Hammers. (Snap-on Tools Corp.)*

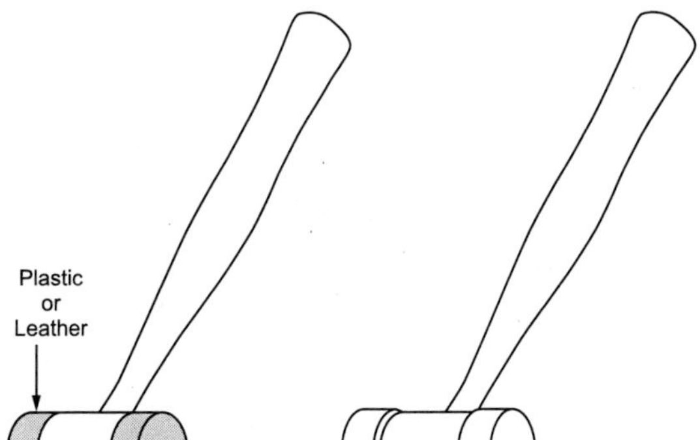

Fig. 1.27 *(b) Mallets (soft-face hammers)*

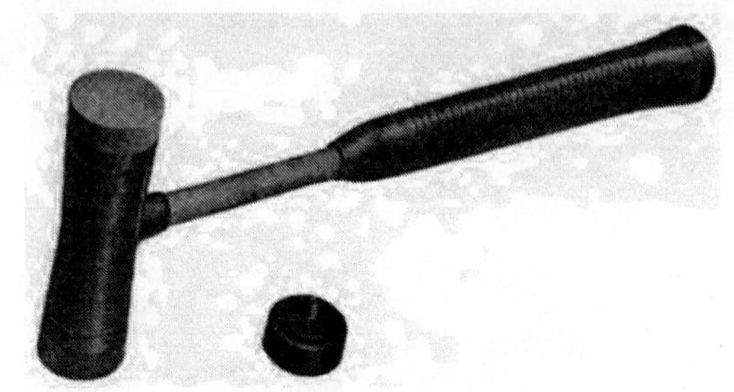

Fig. 1.27 *(c) Soft-faced hammer prevents marring the work surface*

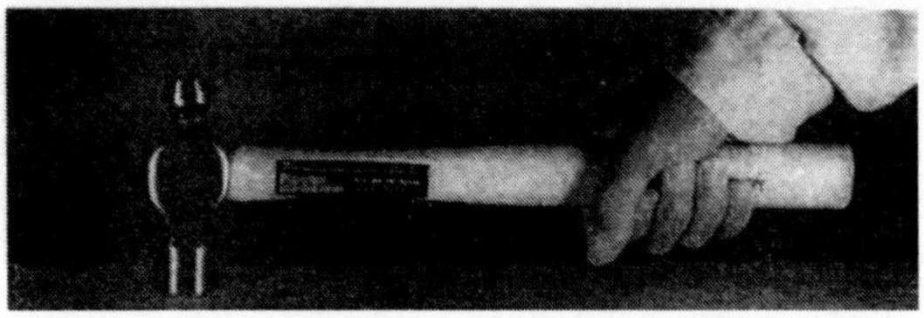

Fig. 1.27 *(d) The hammer should be gripped near the end of its handle.*

Hammers are used with punches and chisels. The ball peen hammer should never be used to hammer on an automotive part. Its hardened head could easily dent or otherwise damage the part.

Be careful not to damage an automotive part, which must be hammered. A number of hammers have heads softer than auto parts. Mallets or soft face hammers are used for tapping parts apart or for aligning parts together, *Fig. 1.27* (*b*). A soft-faced mallet will not harm the part it strikes, hammers will. Soft-faced hammers, *Fig. l.27* (*c*), have faces of rawhide, plastic, brass, or lead.

The hammer should be gripped close to the end of the handle, *Fig. 1.27* (*d*). In this way, a heavier blow can be struck with minimal effort. Be careful not to strike the work with the edge of the hammer face, *Fig. 1.27* (*e*). The full face of the hammer should contact the work-piece.

1.7 PUNCHES

Some automotive components are held together with pins. Pins are removed with a punch and hammer. The different kinds of punches may be required to remove a pin, *Fig. 1.28* (*a*). A *starting punch* is used to break the pin loose. Then a *pin punch* smaller than the hole is used to drive the pin out. A long tapered

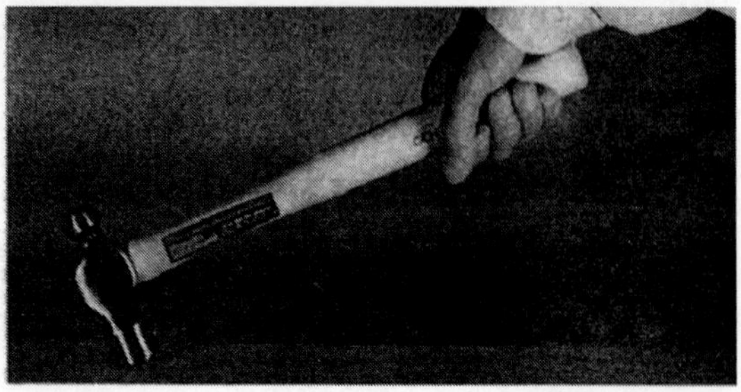

Fig. 1.27 *(e) Never strike work with the edge of the the hammer.*

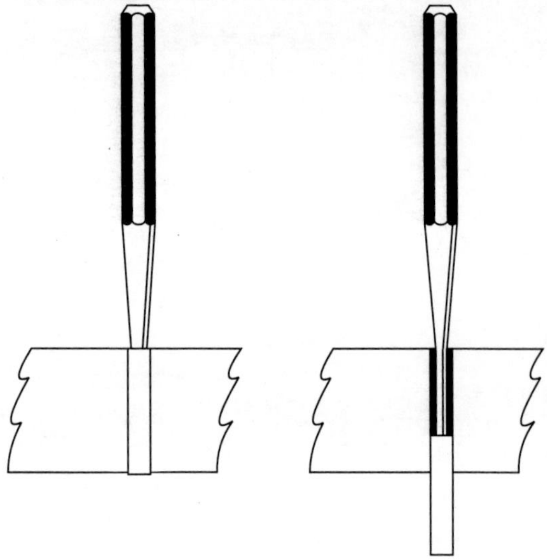

Fig. 1.28 *(a) Pin removed with a stating and then a pin punch*

point on the punch, *Fig. 1.28* (*b*), is used to align two parts with a hole in them. It is known as an *aligning punch.* Each type of punch is available in many different diameters and lengths to offer a choice for different size jobs.

A *centre punch* is ground to a fine point. It is used to mark the locations of the holes that are to be drilled. Without such a mark, the drill may wander over the surface and start the hole at the wrong position.

1.8 METAL-WORKING TOOLS

Many automotive repair jobs require the use of metal-working tools, to cut or shape metal. Metal shears, chisels, files, hacksaws, drills, reamers, taps, and dies are metal-working tools that need careful study and to learn their proper use to 'enhance productivity and quality of work'.

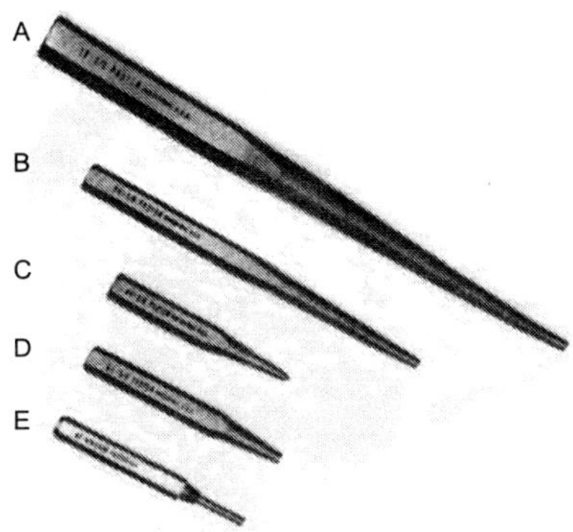

Fig. 1.28 *(b) A-Aligning punch. B-Short aligning punch. C-Centre punch. D-Starting punch. E-Pin punch.*

Chisels

A variety of *chisels, Fig. 1.29,* are used in an automotive repair shop. Cold chisels, shown at (*a*), are used for cutting metal, removing the heads from rivets, chipping metal, and splitting rusted nuts. In most cases, mechanics use the cold chisel, or *flat chisel.* The sizes most frequently used are 3/8", 1/2", and 3/4" or their metric counter parts.

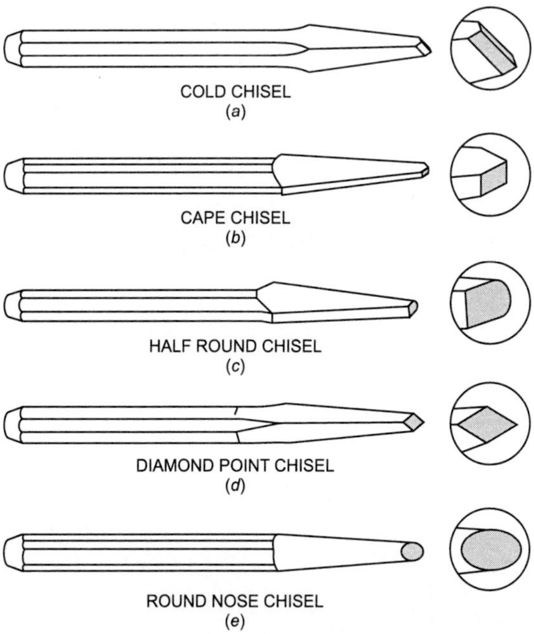

Fig. 1.29 *Cold chisel have a variety of cutting edges.*

The *cape chisel,* shown at (*b*), has a narrow cutting edge. It is used primarily for cutting keyways and narrow grooves.

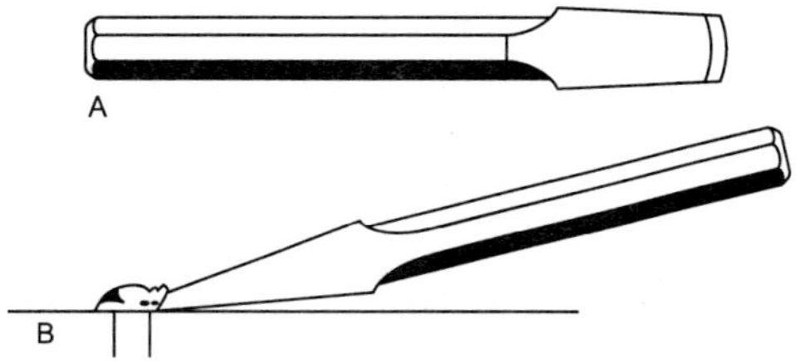

Fig. 1.30 *A rivet buster is used to shear the heads from rivets.*

A *rivet buster, Fig. 1.30,* is a special form of chisel designed particularly for cutting the heads from rivets. It differs from the conventional cold chisel in that only one side of the cutting edge is ground.

The chisel should be held rather loosely, with fingers curled around the chisel about 25 mm from the head of the chisel. A special tool can also be used to hold chisel during use.

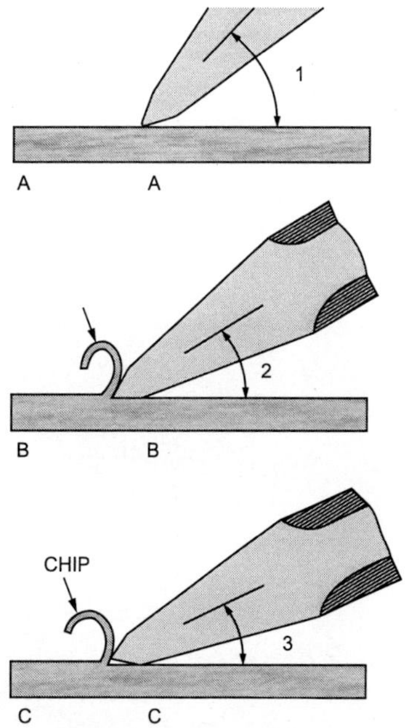

Fig. 1.31 *(a) The depth of a cut is controlled by angle of chisel.*

While chipping metal, the depth of cut is controlled by the angle of the chisel, *Fig. 1.31*. Deeper cuts are taken as the angle of the chisel approaches the vertical position. When chipping, the mechanic should watch the cutting edge, not the head of the chisel.

Wear protective goggles when chipping. Never use a chisel with mushroom end, *Fig. 1.31* (*b*). When struck with a hammer, portions of the mushroomed head may break and fly off and cause severe injuries to the mechanic or the bystanders. If needed, the mushroom end of the chisel should be ground smooth, *Fig. 1.31* (*c*) and (*d*).

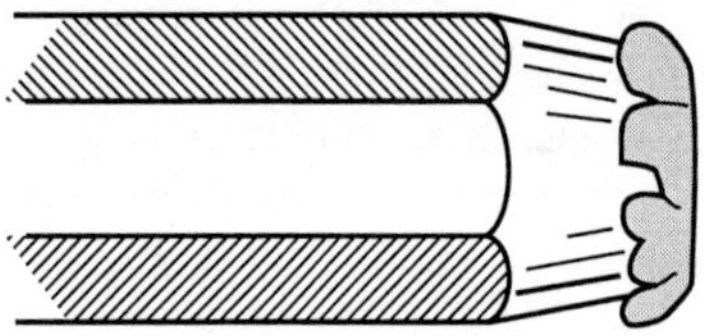

Fig. 1.31 *(b) After striking the end of a chisel repeatedly, the end becomes mushroomed.*

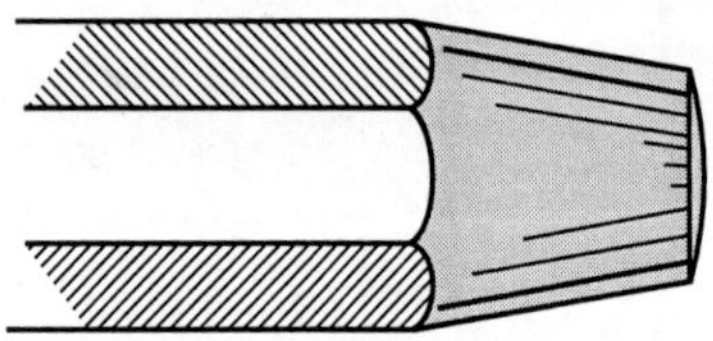

Fig. 1.31 *(c) The mushroom end of a chisel should be ground to prevent tiny pieces of metal from flying off when it is struck with a hammer.*

Fig. 1.31 *(d) The cutting edge of a chisel can be sharpened on an electric grinder.*

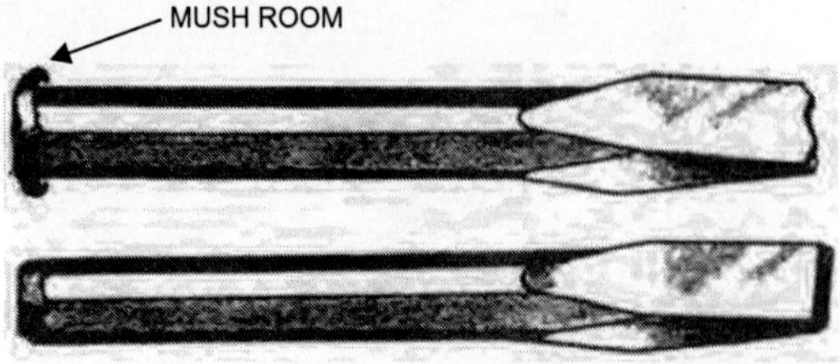

Fig. 1. 31 *(e) Chisel before and after dressing.*

When sharpening a cold chisel, the two ground surfaces should form an angle of 60°. Rivet busters are ground only at one surface only at an angle of 30°.

Metal Shears

Metal shears, Fig. 1.32, are required for cutting sheet metal. The straight blade shear is the type most frequently used; however, the curved blade shear and the scroll pivoter snips are also convenient to have available. The curved blade shears are used for making curved cuts, and the scroll pivoter snips follow an irregular line easily.

Fig. 1.32 *Different types of shears.*

Files

Files are used when you must remove metal by hand, *Fig. 1.33* to *1.36*. A file is a hardened steel tool to remove metal for polishing, smoothing, or shaping. The cutting edges of a file consist of diagonal rows of parallel teeth formed on the face of the file. Files are made in different lengths, measured from the tip to the heel. A pointed end, the tang, is shaped to fit into the handle. A handle must always be attached when filing to protect the mechanic from the sharp tang.

The handle is set tightly on the file by striking the handle on a workbench, as shown in *Fig. 1.34.* A file with single row of parallel teeth is called a *single-cut file,* Figs.1.33, 34 and 35. Files that have a row of teeth crossing another row of teeth in a crisscross pattern are known as *double-cut files, Fig. 1.35* and *Fig. 1.36.*

A file's cutting edges may be spaced close together or wide apart. The wider the cutting edges are spaced the faster the file will remove metal. A file whose cutting edges are closer together will remove less metal and can be used to smooth or polish a metal surface. Files are graded according to the spacing of their teeth. Files are classified as *coarse, bastard, second-cut* and *smooth, Fig. 1.36.* Files are also available in different shapes, *Fig. 1.37.*

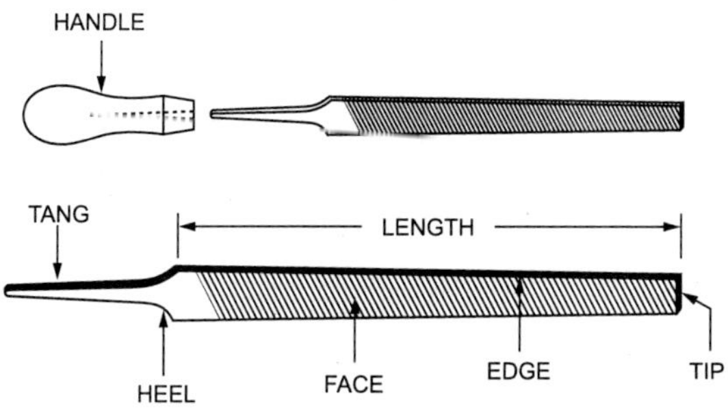

Fig. 1.33 *Parts of a file (General Motors Corp.)*

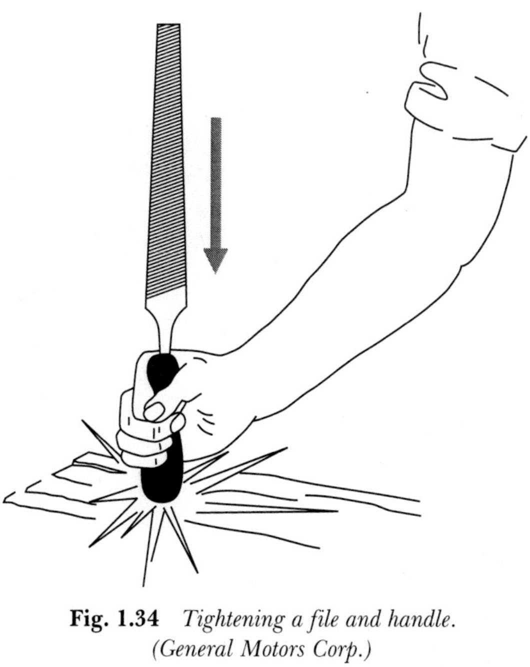

Fig. 1.34 *Tightening a file and handle. (General Motors Corp.)*

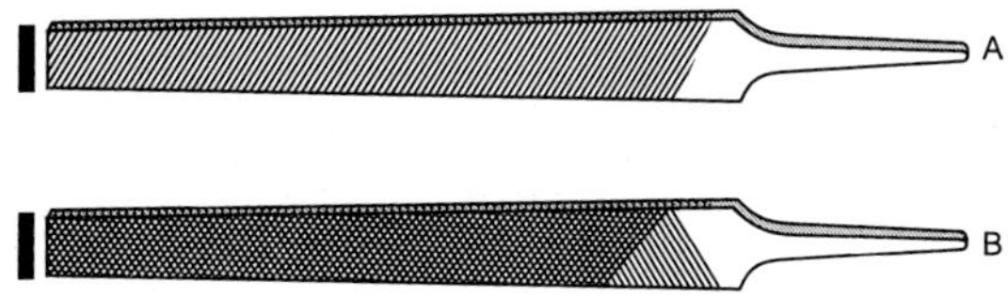

Fig. 1.35 *A-Single-cutfile, or mill file. B-Double-cut file*

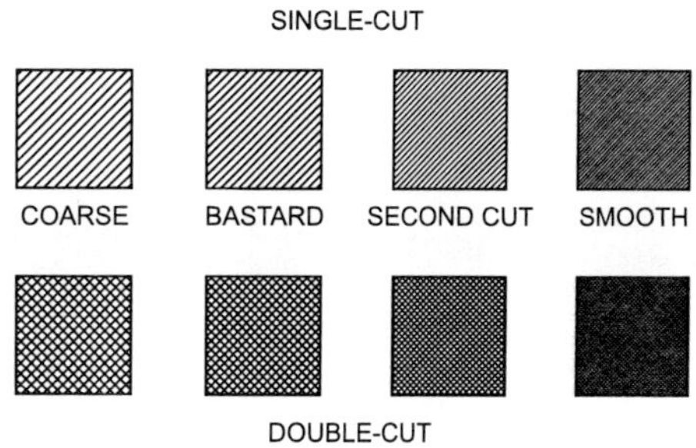

Fig. 1.36 *Files classified by cutting edges. (General Motors Corp.)*

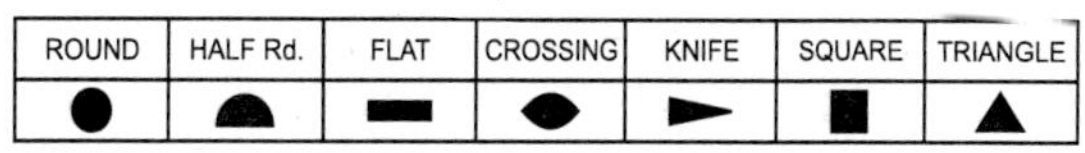

Fig. 1.37 *File shapes.*

The *mill file* is a single cut file, tapering in thickness and width for one-third of its length. It is used primarily for fine work and is available with either square or round edges.

A double-cut file that tapers in thickness and width is known as *a flat file.* It is used when a fast cut is desired. The *hand file* is single-cut and is similar in shape to a flat file, with parallel sides and a slight taper in thickness. It has square edges, one of which is a safe edge. For rough filing a *bastard file* is used.

The *round file* is tapered and usually single-cut. In larger sizes it is also available in double-cut configuration, too. The principal use of round files is to enlarge circular openings and to file concave surfaces.

A *half-round file* is a double-cut file. It tapers in thickness and width, with one flat side and one oval side. It is used mainly for rough filing on concave surfaces.

The *triangular file* is useful for filing small notches and square or cornered holes. In addition it can be used for re-cutting damaged threads on bolts.

Like all other tools, having the right file for the job makes work safer and easier. The type of metal must be considered when selecting a file. When filing cast iron, first use a bastard file, and then use a second-cut file for finishing. On soft steel, a second-cut file is used first, and a smooth-cut file is used for finishing. On hard steel, start with a smooth-cut file and finish with a dead-smooth file. On soft metals, e.g., brass or bronze, use a bastard file first and then use a second-cut file. On aluminium, Babbitt or lead, a Vixen-cut file, *Fig. 1.38*, similar to those used by automotive body repair technicians, is preferred.

Fig. 1.38 *A Vixen-cut file is used to shape soft metals and body filler.*

Whenever possible clamp the work in the jaws of a vise. If the work is soft metal, cover the jaws of the vise with *soft caps* so that the work will not be marked or damaged, *Fig. 1.39*. Never hammer on files or use them as levers. Files are brittle and will quickly break if abused.

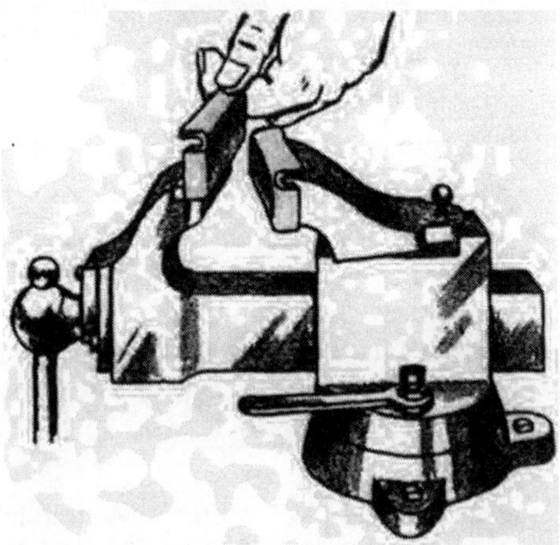

Fig. 1.39 *Soft caps can be easily added to a vise. The caps prevent marring soft materials when held in vise.*

File teeth are designed to cut only when they are pushed forward. Therefore, the preferred method of filing is to raise the file from the work before drawing it back to start the next stroke. Apply only enough pressure to the file to keep it cutting. Excessive pressure only results in increased effort being required to move the file forward. The correct way to hold a file is shown in *Fig. 1.40*.

If you are right-handed, grasp the file handle in your right hand. Hold the other end of the file in your left hand with your fingers curled over the file end. Your feet should be spread apart and your body should lean slightly forward so that your shoulder is over the work. In order to file a flat surface, the forward movement of the file must be perfectly horizontal. Any rocking of the file on a round surface, the file should be rocked as shown in *Fig. 1.41*.

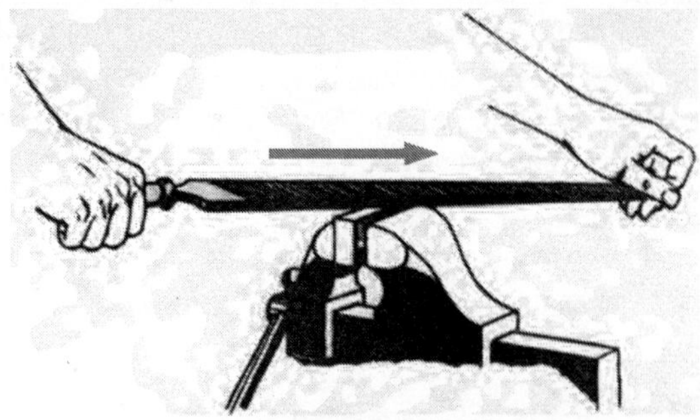

Fig. 1.40 *The correct method of using a file*

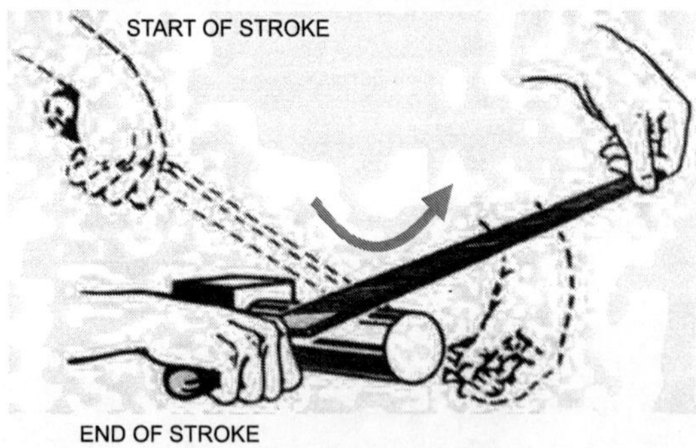

Fig. 1.41 *When filing round objects, the file should be Rocked.*

File teeth tend to become clogged, particularly when soft metals are being filed. The material clogging the file teeth will tend to scratch the metal surface being filed. Clean the file teeth by tapping the file handle or by brushing them with a file card or wire-bristled brush. Scratching can be overcome to some extent by rubbing chalk on the file before use.

1.9 HACKSAWS

There are several styles *of hacksaws* designed to cut metal. Buy one that seems sturdy. Automotive mechanics use a hacksaw to cut exhaust pipes and other metal parts that are made during a repair job. As shown in *Fig. 1.42*, a hacksaw consists of a metal frame and a detachable cutting blade. Different frames are available and some frames are made adjustable, *Fig. 1.42* (*c*), allowing various size blades to be used. The hacksaw shown in (*d*), called a low-clearance hacksaw is made to work in tight places.

(*a*)

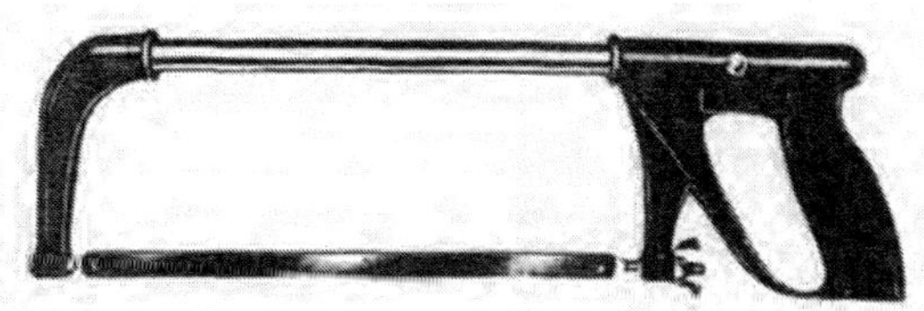

(*b*)

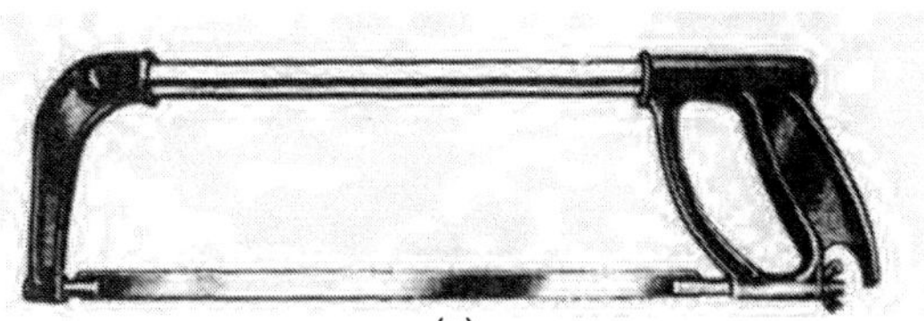

(*c*)

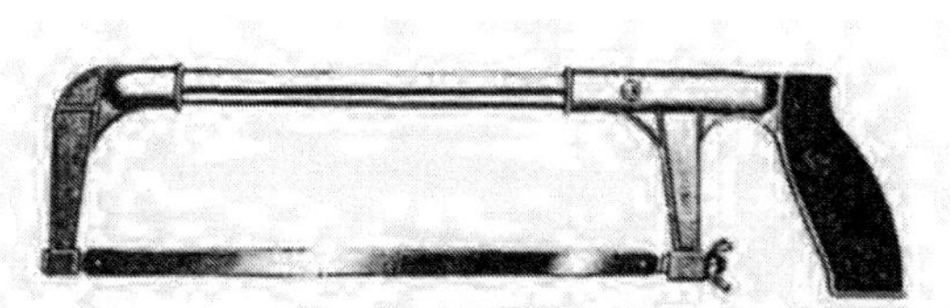

(*d*)

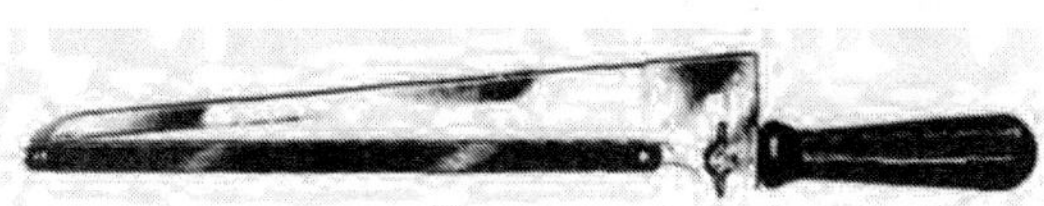

(*e*)

Fig. 1.42 *Types of hacksaws. (Proto)*

The hacksaw blade installed in the frame is the part that does the cutting. Hacksaw blades are made in different lengths and with different number of teeth on a blade, given in teeth per inch (T.P.I.). The fewer and larger the teeth, the faster the saw will cut. For cutting soft, wide materials, a blade with 14 teeth per inch is right. Harder materials require a blade with 18 teeth per inch. For cutting pipe and tubing or metal of an unusual shape, use a shape that will keep two or more teeth in contact with the wall, e.g., one with 24 or 32 teeth per inch. The general rules for hacksaw blade selection are given in *Fig. 1.43*.

A. Mild Materials in Large Sections

B. Harder Materials in Large Sections

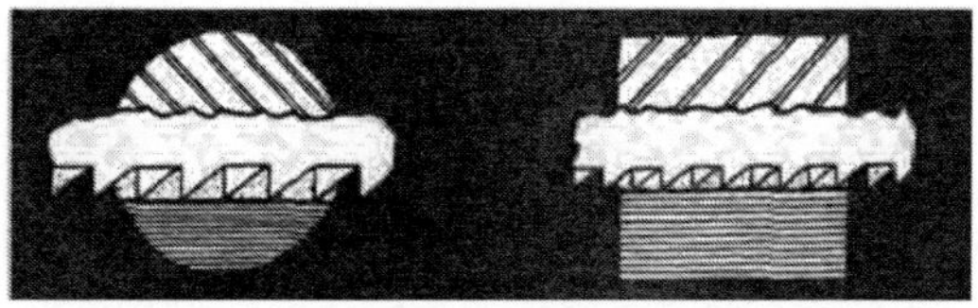

Choose coarse tooth blade to provide plenty of chip clearance, faster cutting.

Hand Blades—

14 Teeth Per Inch

Power Blades—

4 to 6 Teeth Per Inch

Choose blade with finer teeth than in A to distribute cutting load over more teeth while still maintaining good chip clearing action.

Hand Blades—

18 Teeth Per Inch

Power Blades—

6 to 10 Teeth Per Inch

C. Unusual Work Shapes

D. Pipes Tubing Conduit

Choose blade to always keep two or more teeth in contact with narrowest section. Coarse tooth blades stradle work, strip out teeth.

Hand Blades—

24 Teeth Per Inch

Power Blades—

10 to 14 Teeth Per Inch

Choose blade with finest teeth per inch to keep two or more teeth in contact with wall. Keep inside of work free of chip accumulation.

Hand Blades—

32 Teeth Per Inch

Power Blades—

14 Teeth Per Inch

Fig. 1.43 *Rules for selecting hacksaw blades. (L.S. Starrett Co.)*

Hacksaw blades are made of high-grade tool steel, which is hardened and tempered. There are two types of blades: the all-hard blade and the flexible blade. All-hard blades are hardened throughout, while only the teeth of flexible blades are hardened.

The blades are provided with holes at both ends for installation on the pins on the frame. To adjust the tension of the blade and secure it tightly in the frame, the position of one of the pins is adjustable. This adjustment is made by either turning a wing nut, *Fig. 1.42* (*a*), (*b*) and (*c*), or by turning the frame's handle, *Fig. 1.42* (*d*). The 'set' of a hacksaw refers to the amount the teeth extend out in opposite directions from the sides of the blade. The teeth of all hacksaw blades are 'set' to provide clearance for the blade.

Blades for hand-operated hacksaws are available with 14, 18, 24, and 32 teeth per inch. It is important that the number of teeth per inch be considered when selecting a hacksaw blade for a particular job, *Fig. 1.43*. Also it must be considered whether the all-hard or flexible blade is more appropriate for a particular job at hand.

Place the blade in the frame so that the teeth point forward. Mount the part to be cut tightly in a vise. Hold the saw with both hands, as in *Fig. 1.44* (*a*), (*b*), and push forward and down to cut. Release the pressure to back the saw up for the next stroke. Take about one stroke per second. If an accurate cut is to be made, mark the stock with a scribe and nick the work with a file. The nick will make it easier for the hacksaw to start cutting and assure accuracy.

Generally, an all-hard blade is considered best for sawing brass, cast-iron, steel, and other heavy stock. For cutting hollow shapes and metals of light section, e.g., channel iron, tubing, tin, copper, aluminium, or babbitt, a flexible blade is preferred.

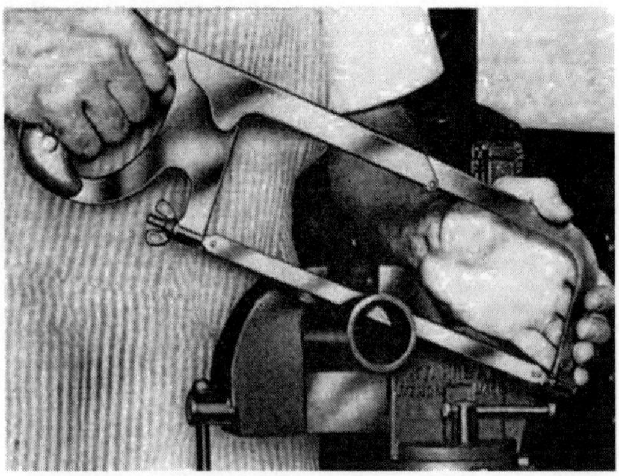

Fig. 1.44 *(a) Using a hacksaw. (L.S. Starrett Co.)*

When the material is nearly cut through, the pressure on the blade should be reduced to prevent the teeth from catching. When cutting thin stock, clamp the stock between two pieces of wood or soft metal and saw through all three pieces. This, will prevent the saw from sticking and will prevent damage to the work-piece.

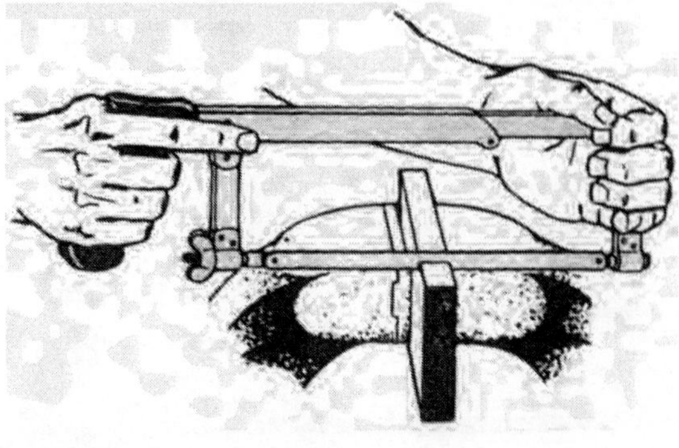

Fig. 1.44 *(b) The correct method of using a hacksaw on a flat piece.*

1.10 TWIST DRILLS

Drill bits, or *twist drills,* are tools for making holes. The material to be drilled determines the type of drill bit that must be used. Twist drills mounted or chucked in an electric drill are used to drill a hole. There are three parts to a twist drill, *Fig. 1.45*. The end of the drill is called a point; the spiral portion the body; and the part that fits in the electric drill motor is the shank. The tapered shank drill fits in a special tapered chuck in a drill press or lathe. The straight shank drill is usually used in portable drill motors.

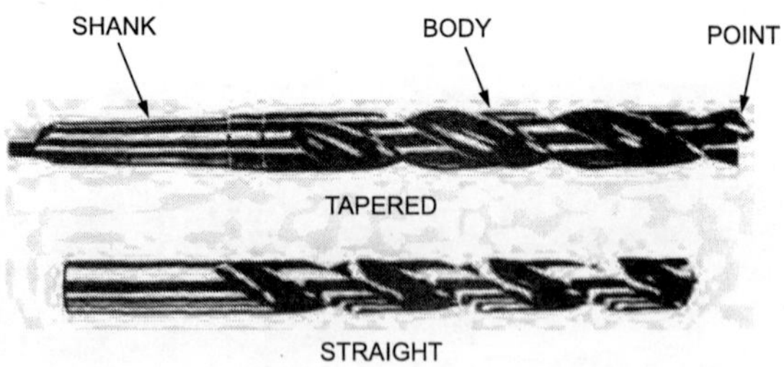

Fig. 1.45 *(a) Tapered and straight shank twist drills. (Beloit Tool of California, Inc.)*

Twist drills are made in four different size groups:

1. fractional sizes from 1/64" to 1/2" and larger in steps of 1/64",
2. letter sizes from A to Z,
3. number sizes 1 to 80, and
4. millimeter sizes.

Drill sets or indexes, *Fig. 1.45* (*b*), are sold in each size group. Drill charts are available that list each drill and give decimal and metric equivalents sizes.

Fig. 1.45 *(b) Drill set or index. (Beloit Tool of California)*

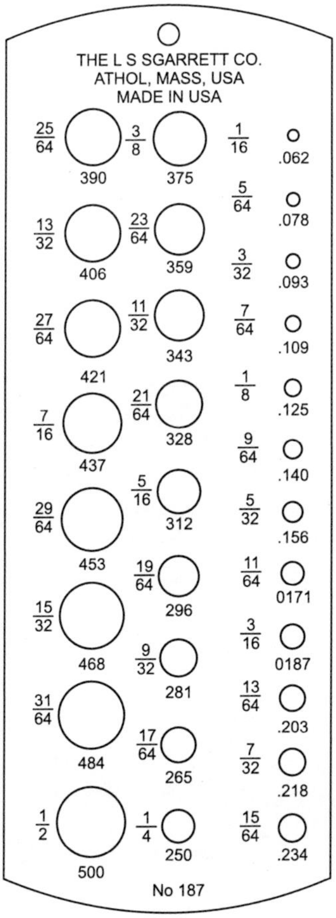

Fig. 1.45 *(c) Drill gauge. (L.S. Starrett Co.)*

The size of a twist drill is stamped on the shank, but after a lot of use, the stamp may be difficult to read. A drill may be measured with a micrometer or with a drill gauge. A drill gauge is a metal plate with holes identified by size, *Fig. 1.45 (c)*. The drill to be measured is placed in the holes until it is found which size hole best matches the drill.

Twist drill must be sharp to do a quality job of cutting. The point of the drill is sharpened with a grinding wheel, using a special fixture for this purpose. The fixture ensures that the point is ground to the correct angles, *Fig. 1.46*. Drill point angles may be checked with a special drill point gauge, *Fig. 1.47*.

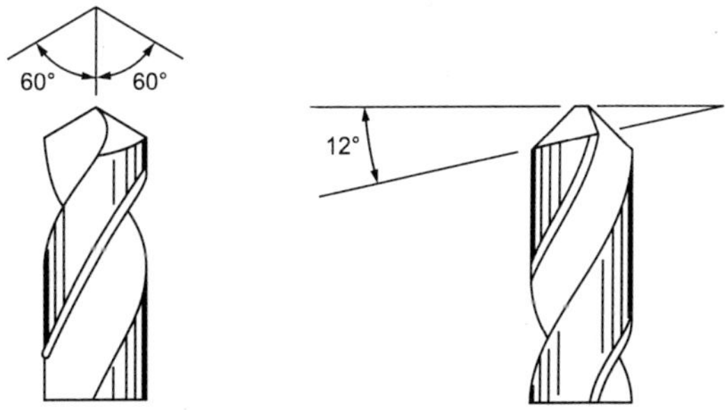

Fig. 1.46 *Drill sharpening angles.*

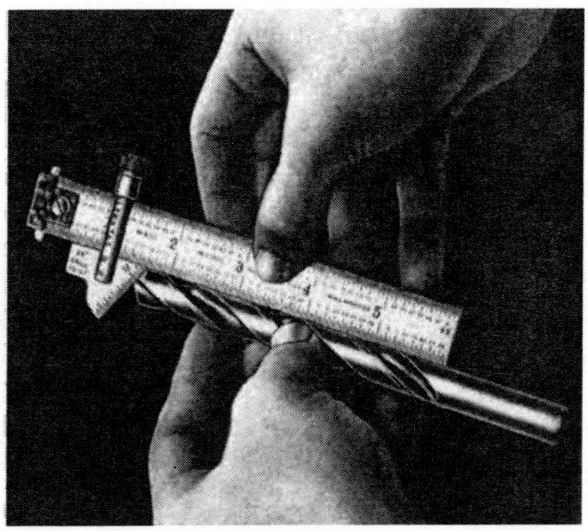

Fig. 1.47 *Drill point gauge. (LS. Starrett Co.)*

To use a twist drill, insert and tighten the drill in the electric drill chuck. Mark the centre of the hole to be drilled in the metal with a centre punch, *Fig. 1.48*. The mark will provide a place for the drill to get started and prevents it from wandering all over the metal, *Fig. 1.49*. Small pieces of metal that are to be drilled must always be adequately clamped in a vise. Always wear a face shield or protective goggles when drilling holes.

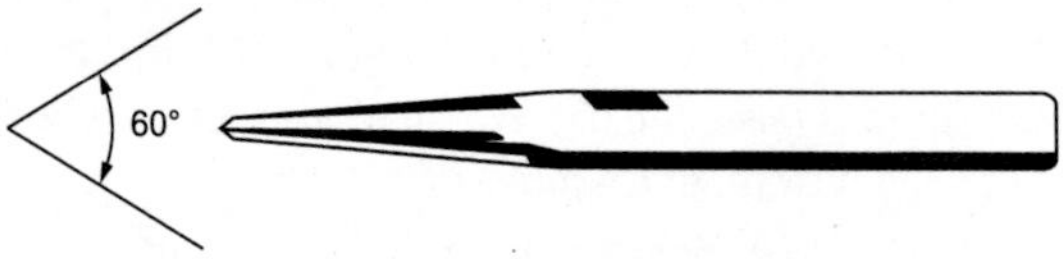

Fig. 1.48 *Centre punch used to mark the centre of a hole. (Proto)*

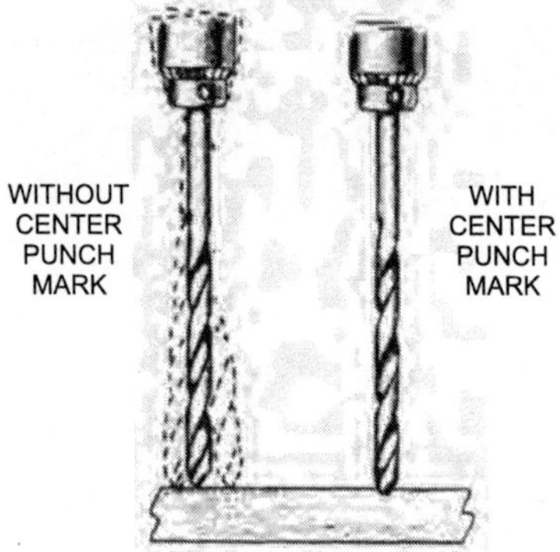

Fig. 1.49 *Centre punch mark prevents drill from wandering. (General Motors Corp.)*

1.11 REAMERS

Reamers are used to enlarge or smoothen holes and prepare them for a precision fit. They create a much smoother finish as compared to that produced by a drill. A reamer is a tool with cutting edges designed to remove a small amount of metal from a drilled hole. Many automobile components have bushings that must be finished to size with a reamer. Figures 1.50 (*a*), (*b*) and (*c*), shows two solid reamers — one a straight flute reamer and the other a spiral flute reamer — and an adjustable reamer. The spiral flute reamer is costlier than a straight flute reamer but has a less tendency to chatter in operation. Reamers are also made that adjust to many different sizes. Adjustable reamers provide more flexibility, but requires a very careful setting of its size. A micrometer should be used to for this purpose.

Solid reamers are available in standard sizes and in size variations of 0.025 mm or .001". The size is stamped on the shank. Adjustable reamers are normally available in standard sizes from 1/4" to 1" in 1/32" increment. They are designed to expand by 1/32", e.g., a 1/4" adjustable reamer will cover hole sizes from 1/4" to 9/32".

(a) Spiral flute reamer.

(b) Straight flute reamer.

(c) Adjustable reamer.

Fig. 1.50 *Reamer types.*

There are machine-driven reamers, but most automotive jobs need a hand -driven reamer. A reamer can be turned with a wrench or tap wrench in a clockwise direction only. Turning a reamer backward can quickly dull the cutting edges.

1.12 TAPS AND DIES

Taps and dies are thread-cutting tools. The mechanics must have the basic knowledge of the thread terminology when dealing with threads and threaded fasteners, *Fig. 1.51*.

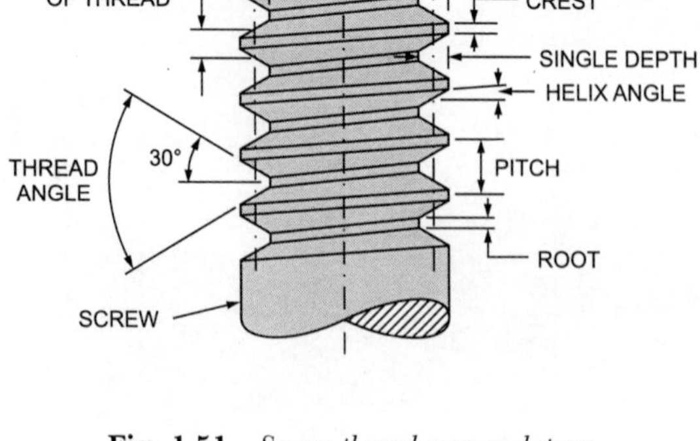

Fig. 1.51 *Screw thread nomenclature.*

Taps, *Fig. 1.52*, are internal thread-cutting tools in already made holes. Various tap shapes are provided. Some are for starting the threads. Others are for cutting the threads all the way to the bottom of a hole. If a new threaded hole is needed, a drill is first used to make the correct size hole. If you want to drill and tap a hole, then drill the hole to the correct diameter for the particular tap. The drill used to drill such a hole is called a *tap drill.* A tap drill chart provides the correct hole size for any tap in metric/English units as the case may be. Thread size, thread series, and tap drill required for a variety of threaded holes are given in *Fig. 1.53*, in English units.

Damaged internal threads must be repaired before assembling the reconditioned automotive parts. The diameter and thread pattern are marked on the shank of the tap.

Mount the tap in a tap wrench, *Fig. 1.52 (b), (c)*, and insert it in the hole. Carefully start the tap squarely in the hole, turn it clockwise into the hole to cut threads. As soon as the tap or die begins to bind, back the tool off about a quarter turn to prevent tool breakage. This will clean out metal cuttings. Then, the cut can be made another half turn deeper. Keep rotating a half turn in and a quarter turn out until the cut is complete. Forcing the tool will cause tool breakage, ruined parts, and poor quality threads.

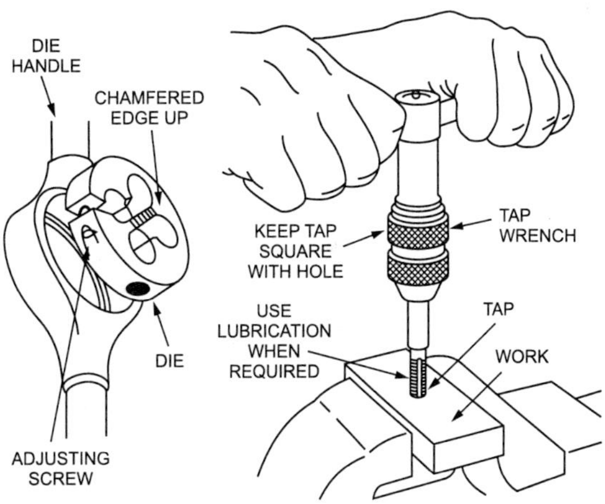

Fig. 1.52 *(b) Tap is for cutting threads in hole. Die is for cutting threads on shaft or bolt. When using tap or die, rotate clockwise until you feel a slight resistance to rotation. Then back off handle to clean threads before turning in a little deeper.*

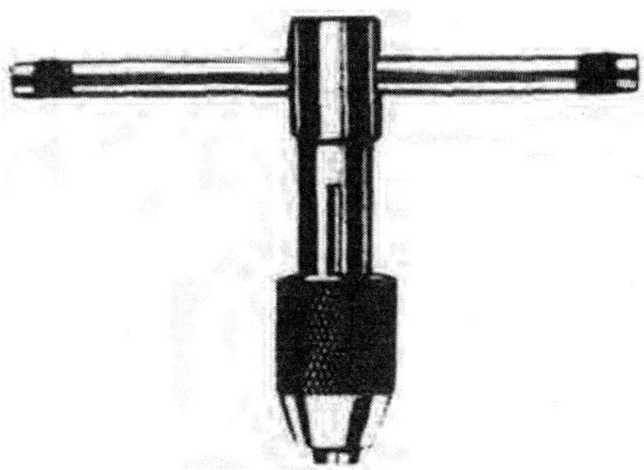

Fig. 1.52 *(c) Tap wrench. (Proto Tool Co.)*

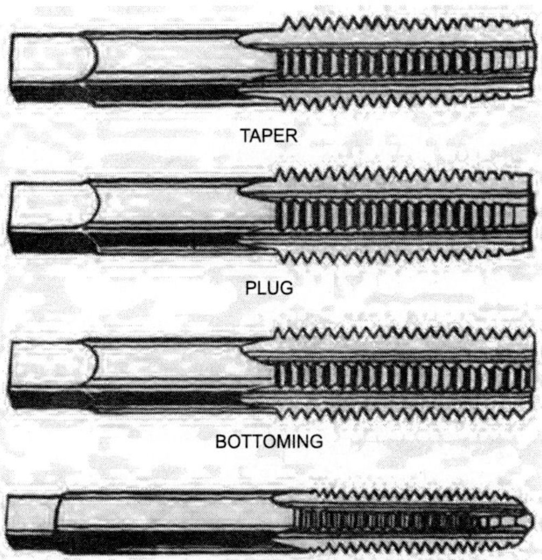

Fig. 1.52 *(a) Various types of taps are used when repairing vehicles.*

Dies, *Fig. 1.52*, are used to cut external threads, i.e., used to cut threads on rods, bolts, shafts, and pins, *Fig. 1.52*. The size and thread pattern are stamped on each die. A die stock is used to hold and turn the die during threading. Start the die squarely on the part and turn it clockwise to cut threads and move the die along the part. When threading steel parts, a lubricant such as lard oil should be used. Use of kerosene is preferred when cutting threads in aluminium. No lubricant is required for cutting threads in brass and cast iron.

When tapping a hole with a tap or cutting a thread with a die, considerable care is needed to perform a quality job. Commonly used sizes of taps and dies are sold in sets, *Fig. 1.55*, which may also include a tap wrench and a die stock.

American National Screw Thread Pitches

COARSE STANDARD THREAD (N. C.) Formerly U. S. Standard Thread				
Sizes	*Threads Per Inch*	*Outside Diameter at Screw*	*Tap Drill Sizes*	*Decimal Equivalent of Drill*
1	64	.073	53	0.0595
2	56	.086	50	0.0700
3	48	.099	47	0.0785
4	40	.112	43	0.0890
5	40	.125	38	0.1015
6	32	.138	36	0.1065
8	32	.164	29	0.1360
10	24	.190	25	0.1495
12	24	.216	16	0.1770
$^{1}/_{4}$	20	.250	7	0.2010
$^{5}/_{16}$	18	.3125	F	0.2570
$^{3}/_{8}$	16	.375	$^{5}/_{16}$	0.3125
$^{7}/_{16}$	14	.4375	U	0.3680
$^{1}/_{2}$	13	.500	$^{27}/_{64}$	0.4219
$^{9}/_{16}$	12	.5625	$^{31}/_{64}$	0.4843
$^{5}/_{8}$	11	.625	$^{17}/_{32}$	0.5312
$^{3}/_{4}$	10	.750	$^{21}/_{32}$	0.6562
$^{7}/_{8}$	9	.875	$^{49}/_{64}$	0.7656
1	8	1.000	$^{7}/_{8}$	0.875
$1^{1}/_{8}$	7	1.125	$^{63}/_{64}$	0.9843
$1^{1}/_{4}$	7	1.250	$1\ ^{7}/_{64}$	1.1093

FINE STANDARD THREAD (N. F.) Formerly S.A.E. Thread				
Sizes	*Threads Per Inch*	*Outside Diameter at Screw*	*Tap Drill Sizes*	*Decimal Equivalent of Drill*
0	80	.060	$^{3}/_{64}$	0.0469
1	72	.073	53	0.0595
2	64	.086	50	0.0700
3	56	.099	45	0.0820
4	48	.112	42	0.0935
5	44	.125	37	0.1040
6	40	.138	33	0.1130
8	36	.164	29	0.1360
10	32	.190	21	0.1590
12	28	.216	14	0.1820
$^{1}/_{4}$	28	.250	3	0.2130
$^{5}/_{4}$	24	.3125	1	0.2720
$^{3}/_{8}$	24	.375	Q	0.3320
$^{7}/_{4}$	20	.4375	$^{25}/_{84}$	0.3906
$^{1}/_{2}$	20	.500	$^{29}/_{64}$	0.4531
$^{9}/_{16}$	18	.5625	0.5062	0.5062
$^{5}/_{8}$	18	.625	0.5687	0.5687
$^{3}/_{4}$	16	.750	$^{11}/_{16}$	0.6875
$^{7}/_{8}$	14	.875	0.8020	0.8020
1	14	1.000	0.9274	0.9274
$1\ ^{1}/_{8}$	12	1.125	$1^{3}/_{64}$	1.0468
$1\ ^{1}/_{4}$	12	1.250	$1^{11}/_{64}$	1.1718

Fig. 1.53 *Tap drill chart will give size of hole to drill for each tap.*

A few rules for using taps and dies are:

1. Never force a tap handle or the tool may break. Back off the handle as described to clean out metal shavings.
2. Keep the tap and die well oiled to facilitate thread cutting.
3. Always use the right size tap in a correctly sized drilled hole.

1.13 SCREW EXTRACTORS

Sometimes a bolt or stud may break off in an automotive part. If a part of the broken bolt or stud sticks up above the part surface, a pair of pliers or vise grips may be used to unscrew it out. When a bolt or stud breaks off even or below the part surface, a screw extractor, *Fig. 1.56* (*a*), must be used to remove it.

Screw extractors are available in sets, often with drills of the same size, *Fig. 1.56* (*b*). To use a screw extractor, *Fig. 1.57* (*a*), (*b*), (*c*), drill a hole of the proper size down into the broken bolt or stud, as at (*a*). Then drive a screw extractor into the hole, as at (*b*). A wrench or tap wrench will turn out the screw extractor and the broken bolt or stud attached to it, as at (*c*).

1.14 POWER TOOLS

Power tools and equipment make a mechanic's job easier and faster. Tools powered by electricity, pneumatic – compressed air, or hydraulic pressure are called power tools. Most power tools can be purchased as either pneumatic or electric. Pneumatic tools offer the advantages of more torque, less weight,

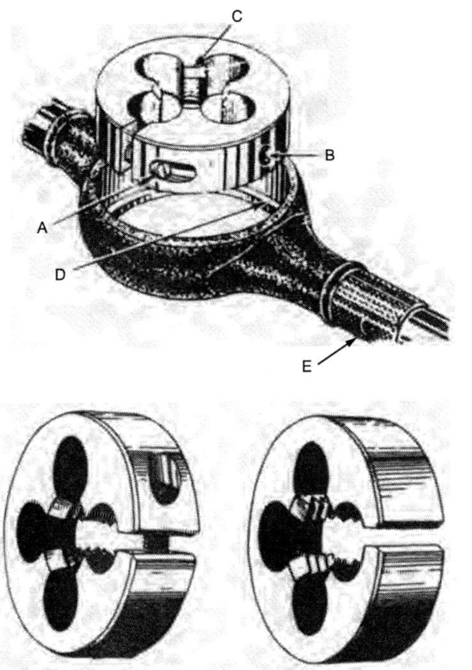

Fig. 1.54 *A die is used for cutting external threads. A -Adjusting screw. B -Drive hole. C -Cutting edge. D -Shoulder. E-Handle.*

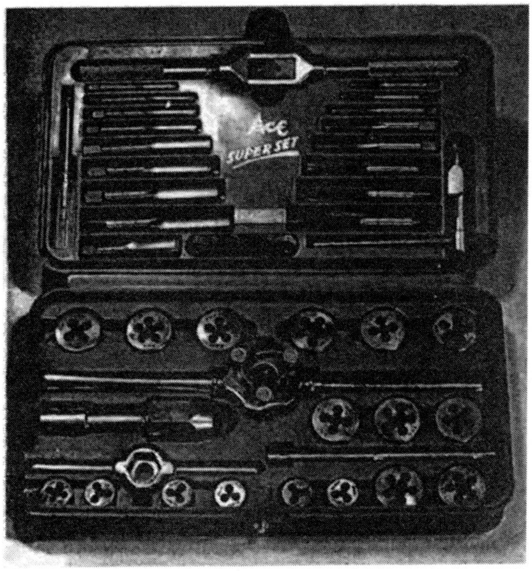

Fig. 1.55 *Tap and die set. (Beloit Tool of California, Inc.)*

Fig. 1.56 *(a) Screw extractor. (Proto Tool Co.)*

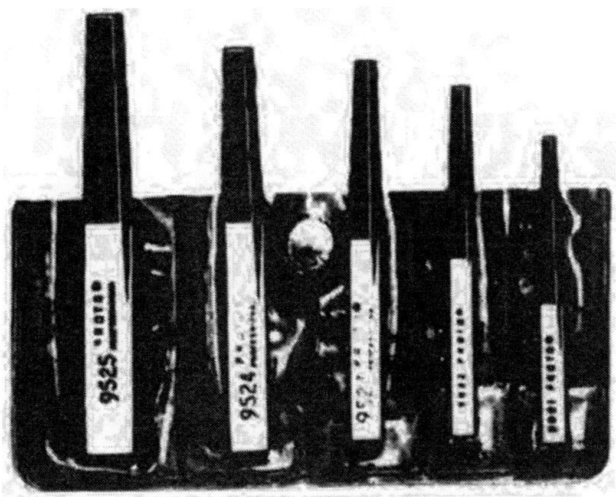

Fig. 1.56 *(b) Set of screw extractors. (Proto Tool Co.)*

and lower maintenance. They are more widely used. Their major disadvantage is their original higher cost. While electric tools can be plugged into most wall sockets, the pneumatic tool, an air compressor and an air storage tank are required. Although many power tools will become part of a mechanic's toolbox kit, many are found as part of the shop tools. Only the basic power tools are described. Specialized power tools, such as those used to work on engines, are covered in later chapters.

If not handled properly power tools can cause very serious injuries. The following rules apply to all power tools:

1. The operator must know the construction and operating principles of each tool he/she is to going to operate.
2. Check adjustments before turning on the power. Rotate the tool one revolution by hand whenever it is safe to do so.
3. No one else should be near the power tool before turning the power on. Observe rules concerning operator zones.
4. Always wear a face shield when operating a power tool.
5. Keep all machine safety guards in correct position.
6. Stay clear of power tools being operated by others.

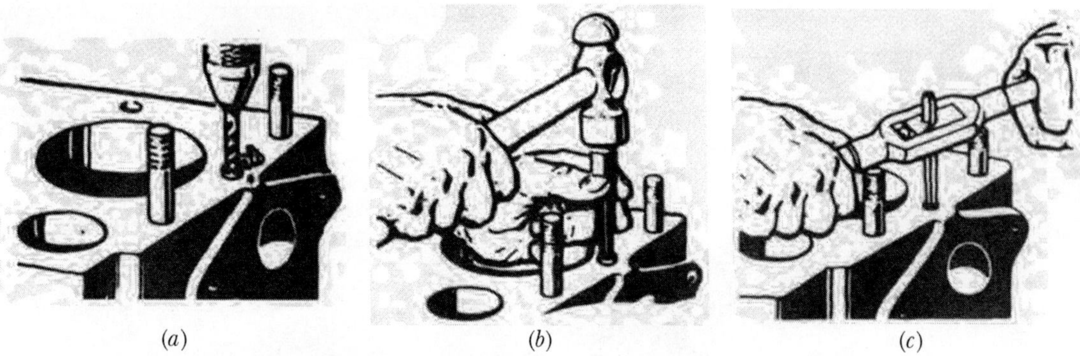

Fig. 1.57 *Using a screw extractor. (Proto Tool Co.)*

7. Start the power tool and remain with it until it is turned off and comes to a dead stop.
8. Wait for power tools to come to a dead stop before oiling, cleaning, or adjusting.
9. Wear clothes that are appropriate for shop work. Remove or fasten any loose clothing. Roll loose sleeves above elbows. Keep long hair away from equipment in operation.

1.15 TOOLS OPERATED BY ELECTRIC POWER

Electric tools, such as drills, valve grinders, cylinder hones, grinders, use electric motors. Most specialized automotive equipment such as boring bars, oscilloscopes, and wheel balancers, operate on electrical power. Common electrical power tools used in automotive repair shops are briefly discussed here.

Electric Drill

The electric drill, *Fig. 1.58*, has an electric motor (operated by a trigger on the handle) that drives a chuck. The chuck has adjustable jaws that are made to grip a drill bit, *Fig. 1.59*. A special chuck key is used to adjust the chuck jaws. If the chuck of an electric drill opens a maximum of 1/2", it is called a 1/2" drill. The common drill sizes are 1/4", 3/8", and 1/2". Important cautions that should be observed when using an electric drill are:

- Drill should be properly grounded, *Fig. 1.60*, with the third blade on the plug, or the drill is double insulated.
- Exercise proper care to always preserve the insulation of the drill cord. Do not drag the drill around by its cord. Do not kink the cord or let anyone to step on or run any other equipment over the cord. Any of these actions can result in damage to cord and result in someone getting a dangerous electric shock.
- Keep your hands and clothes away from the moving drill bit.
- Maintain a firm grip on the drill and always be ready to shut the drill off if it jams. If the drill bit jams, pull it back to free it. Do not attempt to break it free by turning it *on* and *off*. This can damage the drill.
- After use, disconnect the drill, wipe it clean and put it away in a safe location. If needed, oil the drill occasionally. Some drills are equipped with pre-lubricated bearings and need no oiling. This power tool is also used to drive engine cylinder hones, deglazers and brake hones.

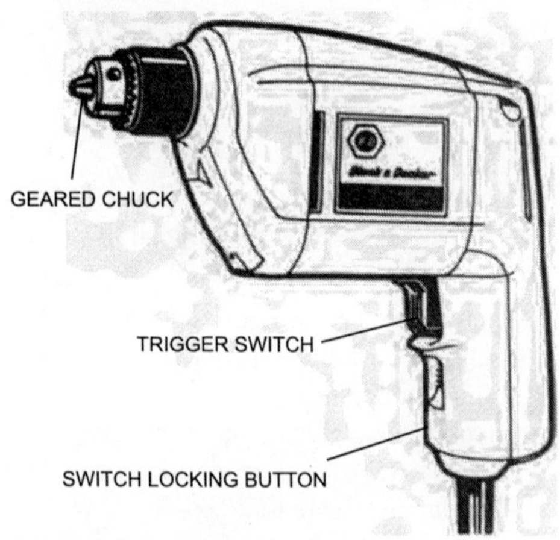

Fig. 1.58 *Electric drill. (Black and Decker Manufacturing Company)*

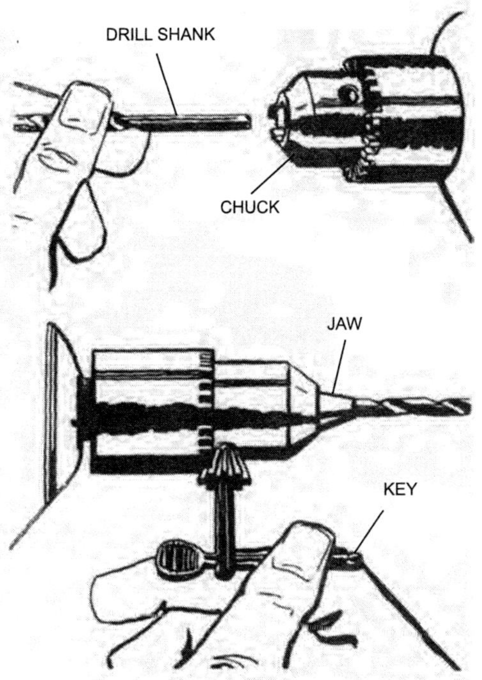

Fig. 1.59 *Installing a twist drill in a drill chuck.*

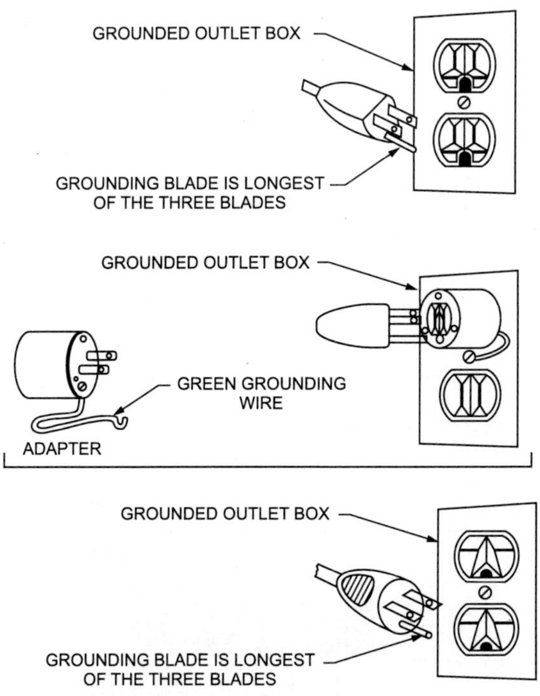

Fig. 1.60 *Various methods of grounding electric tools. (Black and Decker Manufacturing Company)*

Electric Wrench

An electric wrench, *Fig. 1.61*, has an electric motor operated by a trigger on the handle. Special heavy duty sockets are attached to a socket drive at the front of the wrench. Holding down the trigger spins the drive and socket. A reversing switch allows the mechanic to loosen or tighten bolts and nuts. The main advantage of the electric wrench is its much faster speed as compared to drive by hand. Electric wrenches are especially useful for disassembling parts that are held together with many bolts and nuts, e.g., engines and transmissions.

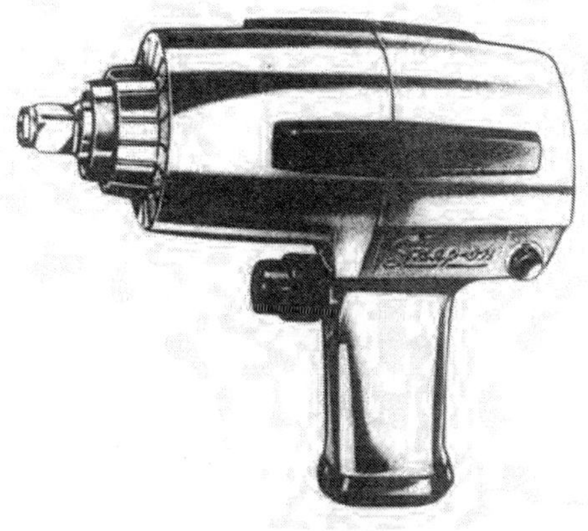

Fig. 1.61 *Impact wrench. (Snap-on Tools Corp.)*

Impact wrenches, often known as impact tools, are designed for loosening and tightening nuts and bolts quickly. Their use results in considerable time saving and a noticeable reduction in fatigue on the part of the mechanic. Designed for use with heavy-duty sockets, both electrical and pneumatic impact wrenches are available.

Grinding Wheels

Most automotive repair shops have a grinder, *Fig. 1.62*. A grinder can be either bench-mounted or a pedestal mounted. An electric motor in the grinder unit drives one or two shafts to which a grinding wheel or wire wheel may be attached. It can have two grinding wheels, with one coarse and the other fine. Grinding wheels remove metal to sharpen tools such as chisels, drills, etc. A wire wheel mounted on a grinder is used to clean parts, e.g., to remove carbon from engine valves. Some cautions when using a grinding wheel are:

1. Be careful to the possible danger of flying sparks from the grinding wheel during metal grinding. These hot sparks can set fire to your clothes or burn skin.
2. Always wear eye protection, even if the wheel is equipped with an eye shield.
3. Never hammer on the wheel or apply excessive force against it. It is brittle and can easily break.
4. Avoid any physical contact with the rotating wheel. It can hurt you.
5. Carefully adjust the light to observe the work in progress.
6. Never try to adjust the tool rest while the grinder is running. Instead shut the grinder off and let the wheel stop moving.
7. When grinding a tool, avoid overheating the tool and preserve its temper. To prevent overheating, dip the tool in water, frequently during the grinding operation.

Fig. 1.62 *Grinder (Snap-on Tools Corp.)*

Electric Sander and Polisher

An electric sander and polisher, *Fig. 1.63*, has an electric motor controlled by a trigger that drives a round rubber pad. An abrasive disk for sanding and or a cloth pad for polishing can be mounted to the rubber pad. A sanding disk is used to remove metal during automotive body work. The polishing pad is used in waxing or polishing the body finish of an automobile.

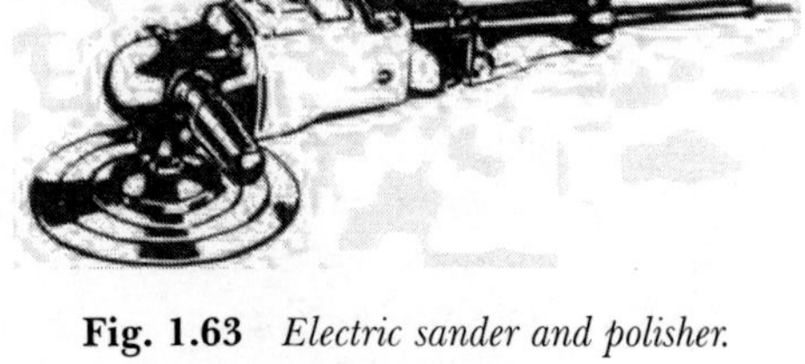

Fig. 1.63 *Electric sander and polisher. (Snap-on Tools Corp.)*

Vacuum Cleaner

Many shops have vacuum cleaners for cleaning the shop floor and also cleaning vehicle interiors after service. This is particularly important in a body shop, where dust is created during auto-body repair and refinishing. The shop vacuum cleaner can also be used to remove harmful asbestos dust from wheel-brake mechanisms during brake service.

Trouble Light

A trouble light is a light bulb mounted in a hand-held socket. The bulb is caged to avoid breakage. Some trouble lights use a reel to automatically wind up the cord. Its main purpose is to light up areas that are poorly illuminated and where it is difficult to see, *Fig. 1.65(c)*.

1.16 TOOLS OPERATED BY COMPRESSED AIR

Auto repair shops have an air compressor operated by an electric motor. It draws in air, compresses it, and stores in a tank. Air lines connected to the tank transmits air pressure to various parts of the shop. In addition to inflating tyres and blowing off components during cleaning, compressed air may be used to power pneumatic tools. Some cautions when using pneumatic tools are:

1. Check air hose connections before turning on the air.
2. Wear safety glasses when using an air hose.
3. When turning air on and off. Hold the air hose nozzle to prevent it from whipping. Do not lay the air hose down while there is pressure in it. It might whip about and hit someone.
4. Do not use compressed air to dust off hair and clothes or to sweep the floor.

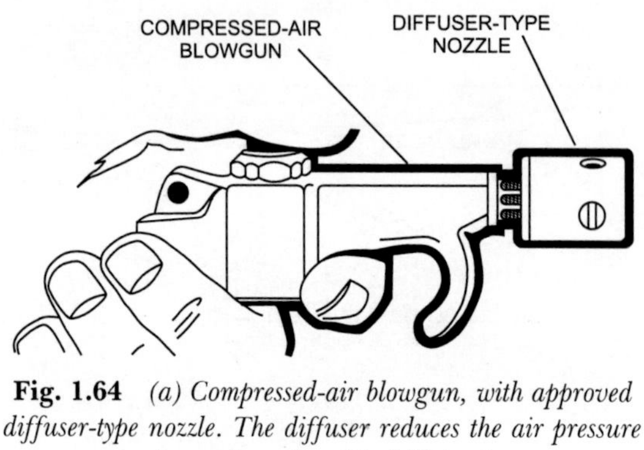

Fig. 1.64 *(a) Compressed-air blowgun, with approved diffuser-type nozzle. The diffuser reduces the air pressure to a safer value. (Ford Motors)*

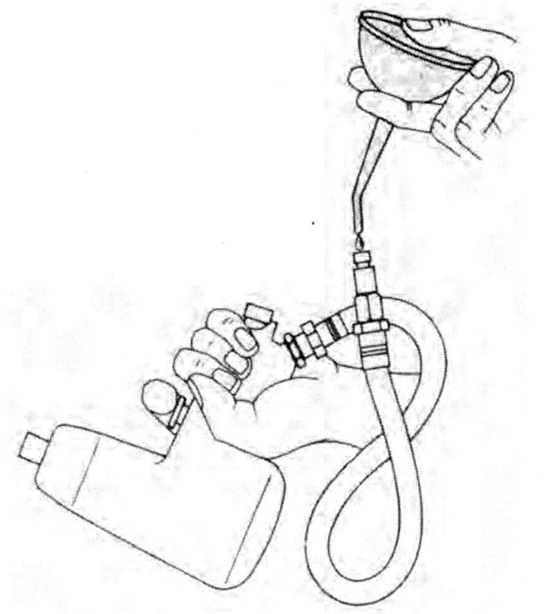

Fig. 1.64 *(b) Squirting a few drops of oil into an impact-wrench air-line. (Chicago Pneumatic Tool Company)*

5. Never point the air nozzle at anyone or use it for practical jokes.
6. Shops have air nozzles, called *blowguns,* for blowing parts dry and clean, *Fig. 1.64*. They are often equipped with a device to reduce the air pressure to a safe maximum. Never use the blowgun to blow dust off your clothes. Dirt particles are driven at high speed by the air. They can penetrate the skin or the eyes. High-pressure air directed toward an open wound, can push air into the bloodstream and could be fatal.
7. Never look into the air outlet of any pneumatic tool.
8. Every day, before using an air tool, apply three or four squirts of flushing oil to the air inlet, *Fig. 1.64* (*b*). Then connect and operate the tool. This flushes out any dirt or moisture and lubricates moving parts of the air tool.

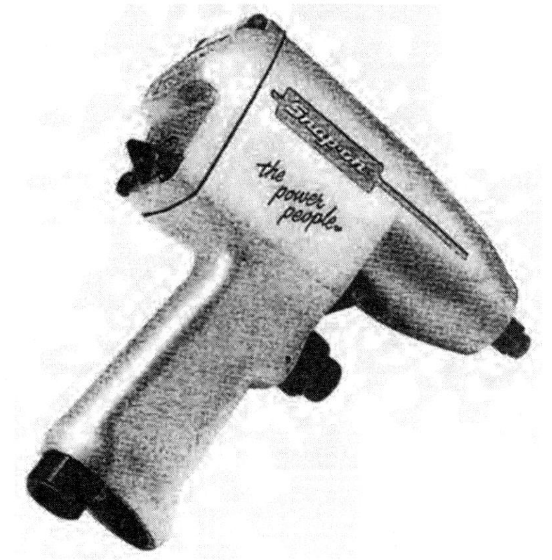

Fig. 1.65 *(a) A pneumatic impact tool is a valuable time-saving device. (Snap-on Tools Corp.)*

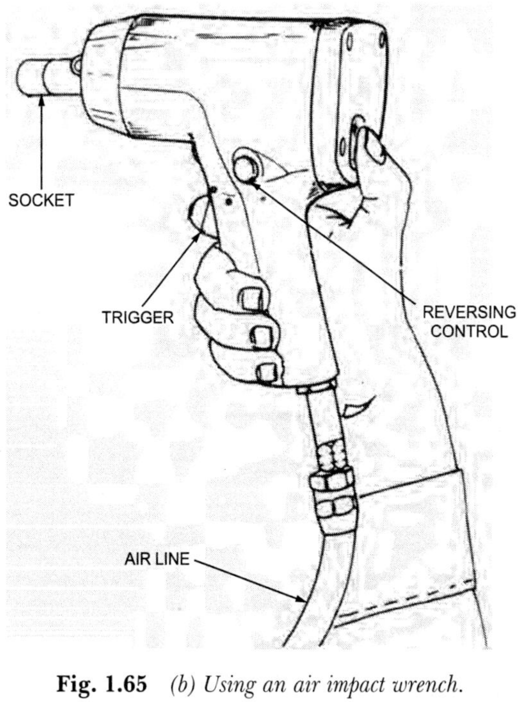

Fig. 1.65 *(b) Using an air impact wrench.*

Air Impact Wrench

Air impact wrench, *Fig. 1.65*, use an impact or pounding force to loosen or tighten bolts and nuts. They do the job very rapidly and are widely used in auto repair shops. The air-operated wrench is connected to an air line. Pulling the trigger causes the compressed air to rotate a socket attached to the drive on the wrench. A reversing switch allows the mechanic to reverse the direction of rotation, and permits both tightening and loosening of threaded

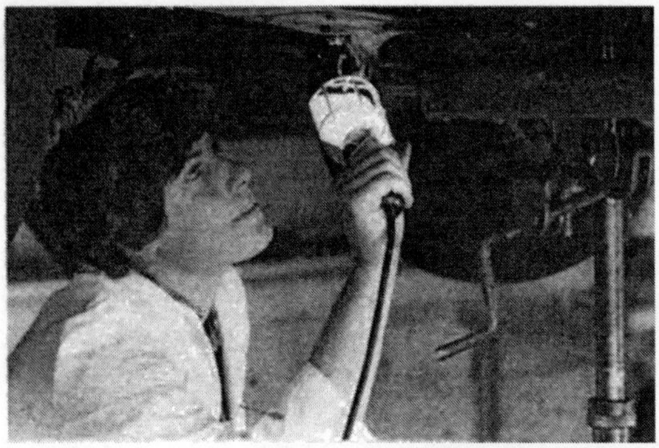

Fig. 1.65 *(c) Trouble light. (Rota-Reel)*

fasteners. Many air wrenches are designed with an impact feature. An impact wrench not only drives the socket but also vibrates or impacts it in and out. The force of impact helps to loosen a bolt or nut that is difficult to remove. Some rules for using an impact wrench are:

1. Always use *impact sockets* of the right size for the fastener.
2. Hold the wrench such that the socket fits on the nut or bolt head. Exert a slight forward force to hold the socket in place.
3. Once a nut or bolt tightens, do not impact it beyond an additional one-half turn to avoid stripping of threads or break the bolt. For accurate tightening control of a bolt or nut, use a torque wrench.
4. Soak rusty bolts or nuts with penetrating oil before impacting them.

Air Chisel

The air chisel, or air hammer, *Fig. 1.66*, uses reciprocating motion to drive a cutting or hammering tool. *Fig. 1.66* shows an air hammer driving a chisel to cut off a nut that has frozen to a stud. A few strokes will split the nut so that it can be removed. Air hammer can be used with a variety of tools— cutters, chisels, punches — to do different jobs.

Air Drill

Air drills do the same job as electric drills. *Fig. 1.67* shows the hook-up for an air drill. One advantage of the air drill is that it is lighter than a comparable electric drill. Also, repeated stalling or overloading does not damage or overheat the air drill.

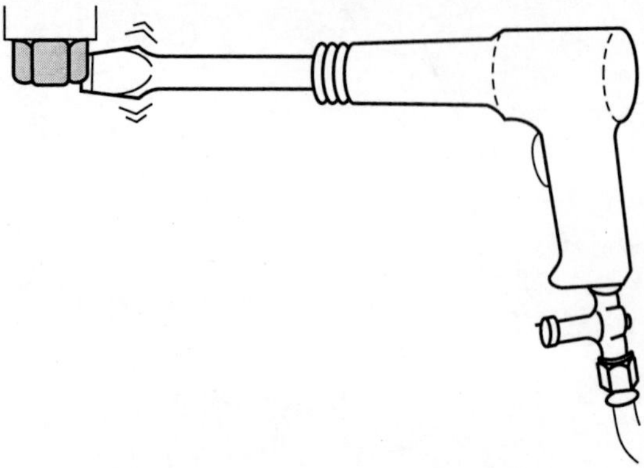

Fig. 1.66 *Using an air hammer. (Chicago Pneumatic Tool Company)*

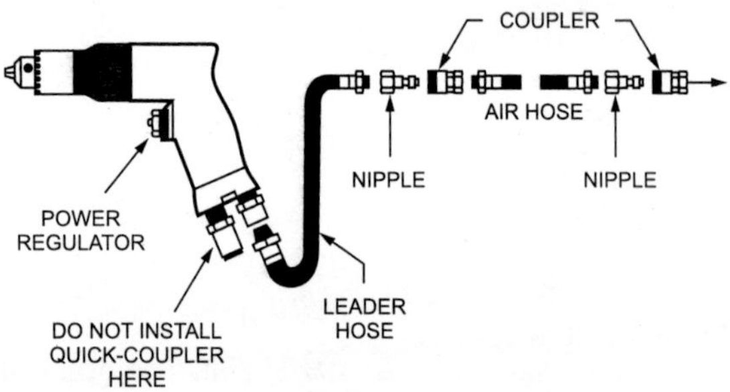

Fig. 1.67 *Air drill with hose and couplings. (Chicago Pneumatic Tool Company)*

Air Operated Jacks

A jack is a tool for raising the front or rear of an automobile. Many types of pneumatic jacks are used in repair shops. The jack shown in *Fig. 1.68* has a large cylinder filled with compressed air, which raises a part of the jack placed under the automobile. A jack stand, *Fig. 1.69 (a)*, should always be placed under the vehicle raised by a jack. Never trust a jack alone to support the automobile, *Fig. 1.69 (b)*.

Automobile Lifts or Hoists

An automobile lift is used to raise the entire automobile so that a mechanic can work underneath. Some hoists are electrical, but most use compressed air as their power source. The lift mechanism is below the shop floor level so that a vehicle may roll over it. When the mechanic operates the controls, the lift rises and raises the vehicle.

There are a number of styles of lifts. *Fig. 1.70* shows a lift with a single post that rises. The single post lift is simple to operate and may be adjusted for any size vehicle quickly by swinging the four large pads under the automobile frame. The large centre post, however, makes working on the middle of the vehicle difficult. To work on the centre of the vehicle a twin-post lift should be used. Its two posts lift the vehicle either on its ends, *Fig. 1.71,* or sides, *Fig. 1.72.*

There are special jacking and supporting places on automobiles. Always lift and support the vehicle in the correct locations, *Fig. 1.72.*

Never allow passengers to remain in a vehicle that is being lifted. Ensure that the trunk lid, hood, and all doors are closed. Various safety features prevent the lift from coming down is a seal does leak or air pressure is lost.

Fig. 1.68 *Air-operated floor jack (Hein Werner)*

Fig. 1.69 *(a) Jack stands (Hein Werner)*

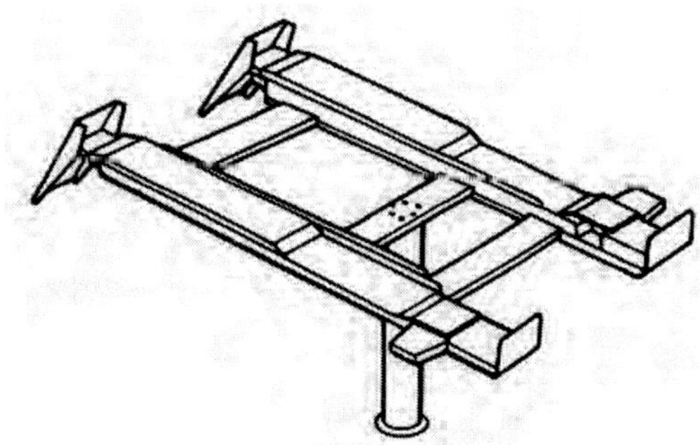

Fig. 1.70 *Single-post lift. (Lincoln St. Div of McNeil Corp.*

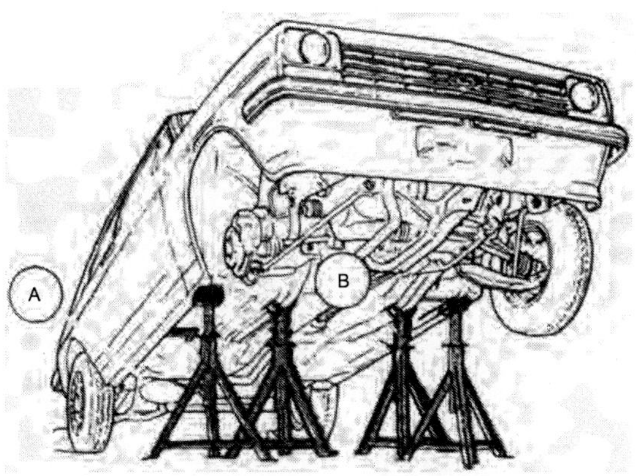

Fig. 1.69 *(b) Always place safety stands under a vehicle at the proper lift points (A or B) before going under a vehicle. (Ford Motors)*

Fig. 1.71 *Twin-post lift.(Linco!n St. Div of McNeil Corp.)*

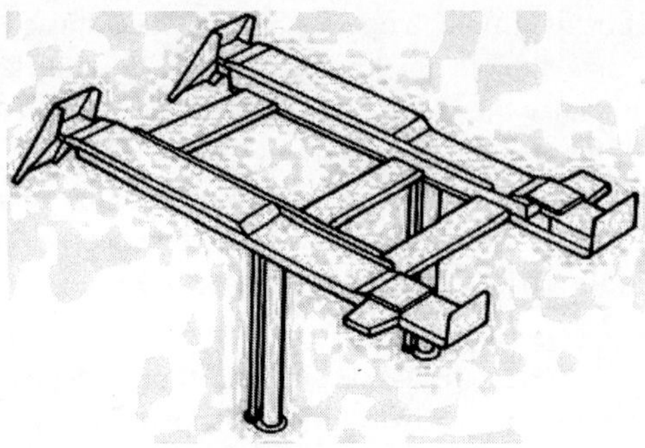

Fig. 1.72 *(a) Twin-post lift. (Lincoln St. Div of McNeil Corp.)*

1.17 TOOLS OPERATED BY HYDRAULIC POWER

Fluid pumped by a piston is a source of power. This power, called hydraulic force, is used in automotive equipment.

Hydraulic Floor Jacks

In order to work underneath an automobile it must be raised on a hoist or lifted with a hydraulic jack. It is always risky to work under a raised heavy vehicle. Great care must therefore be taken in using jacks and hoists. The hydraulic floor jack, *Fig. 1.73*, is used to raise the automobile. The handle on the jack pumps hydraulic fluid to lift the jack pad and raise the automobile. A control knob on the handle releases the fluid to lower the vehicle. These jacks are named by the weight they can lift: 1 ½, 2 and 2 ½ tons and so on.

Another style of jack, called a bumper end lift, *Fig. 1.74*, has rubber pads that are positioned under the bumper to raise the automobile but never use such a lift on vehicles that have energy absorbing units on the bumper. Never work under a vehicle with only a jack supporting it. Always use safety stands or jack stands, *Fig. 1.73*.

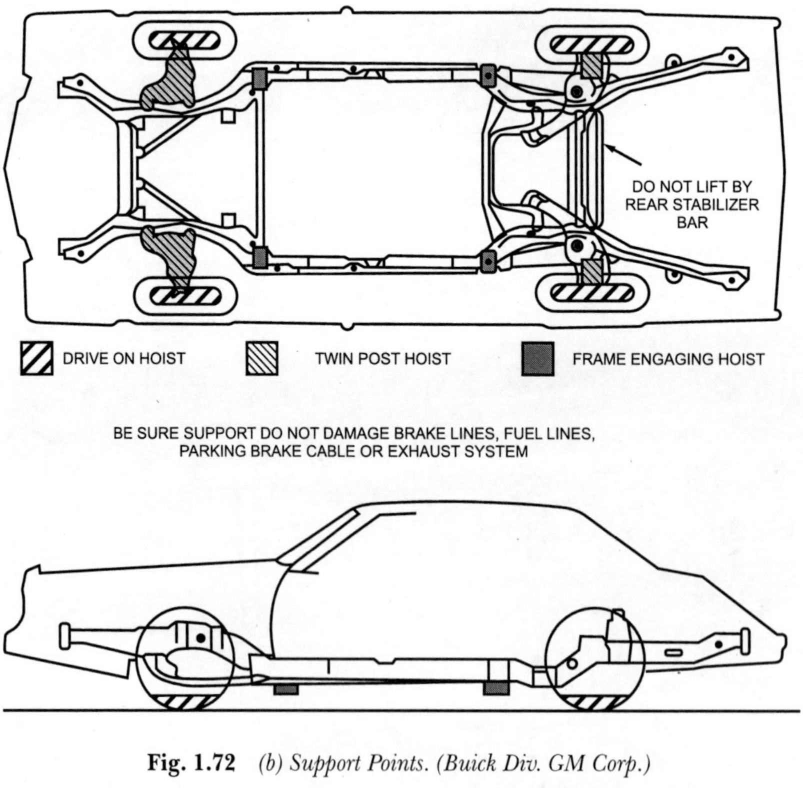

Fig. 1.72 *(b) Support Points. (Buick Div. GM Corp.)*

Fig. 1.73 *(a) Hydraulic floor jack. (Hein Werner)*

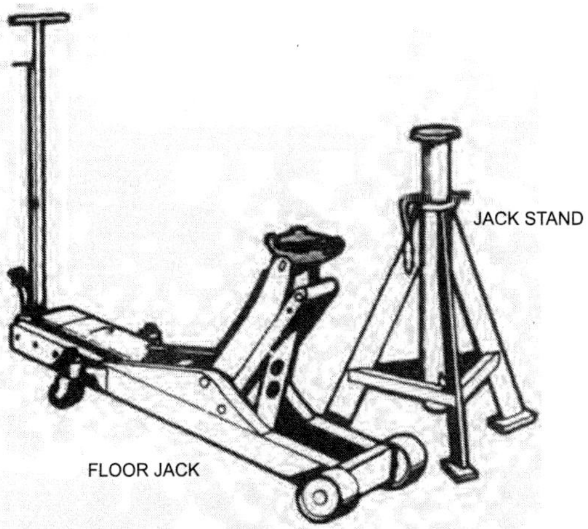

Fig. 1.73 *(b) Hydraulic floor jack and jack stand.*

Fig. 1.74 *Bumper end lift. (Hein Werner)*

Hydraulic Lift

The safest lifting is the hydraulic floor lift. The hydraulic floor lift raises the vehicle into the air high enough so that you can walk under it, *Fig. 1.75*. Various safety features prevent the hydraulic lift from coming down if a seal does leak or air pressure is lost. Therefore, you can work under the vehicle with an assurance of safety.

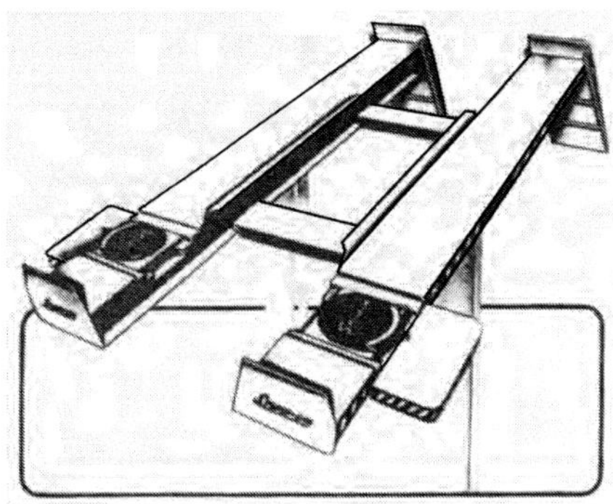

Fig. 1.75 *Hydraulic floor lift. (Snap-on Tools Corp.)*

Mobile Engine Hoist or Floor Crane

A tool capable of lifting very heavy loads is needed to remove an engine from an automobile. A mobile floor crane, also called engine hoist, *Fig. 1.76*, is designed to perform this function. Hydraulic pressure converts power to a mechanical advantage and lifts the engine from the vehicle. It is mounted on wheels so that it can be rolled into place. A lift chain or cable is attached to the engine. A mechanic pumps a hydraulic cylinder to raise the crane and engine out of the vehicle. The crane and engine may then be rolled anywhere in the shop. A control knob allows the operator to slowly lower the engine to the floor.

Hydraulic Press

A hydraulic press, *Fig. 1.77*, is a table to which a hydraulic cylinder is mounted. Pumping a handle attached to the cylinder causes a ram to lower with several tons of force. This force is used to press bearings on and off shafts or in and out of housings, or to straighten bent components. A lever or knob releases the hydraulic pressure to remove the component.

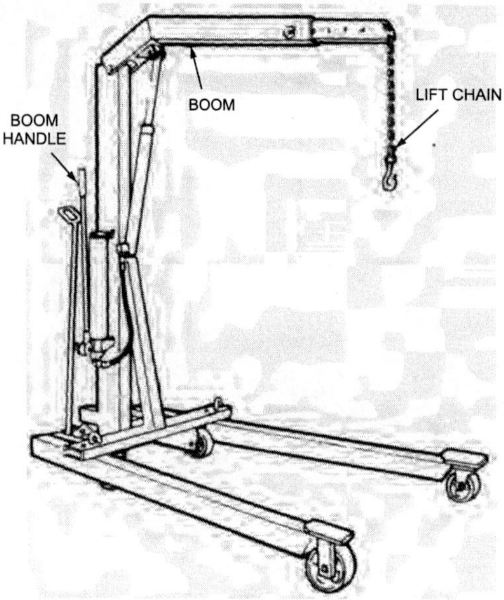

Fig. 1.76 *Hydraulic engine hoist.*

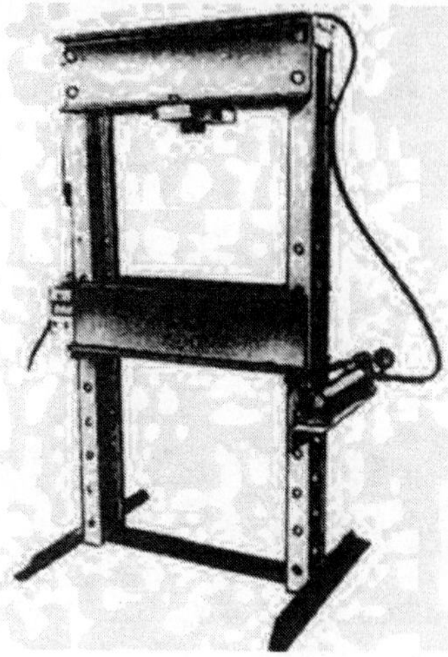

Fig. 1.77 *Hydraulic press. (Owatonna Tool Co.)*

1.18 BENCH VISE

Bench vise, *Fig. 1.78*, is one of the most common shop tools. It is used to hold a part for sawing, filing, chiseling, or otherwise being worked on. When the vise handle is turned, a screw moves the movable jaw toward or away from the stationary jaw. To protect the parts clamped in the vise, caps of soft metal are placed over the steel jaws of the vise. These are commonly called "soft jaws".

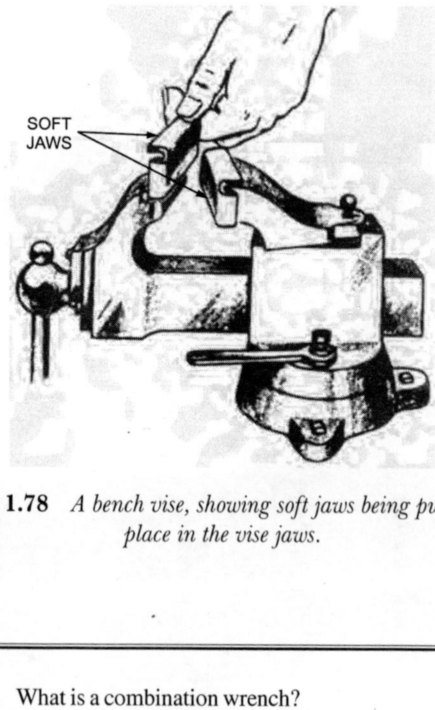

Fig. 1.78 *A bench vise, showing soft jaws being put into place in the vise jaws.*

THEORY QUESTIONS

1. What is a wrench used for?
2. What are the two ways wrench sizes are specified?
3. What are open type wrenches?
4. The size of open-end wrenches increases in:
   (*a*) 1/16" steps (*b*) 1/32" steps
   (*c*) 1/64" steps (*d*) 1/8" steps.
5. What is a socket wrench? How does a socket wrench differ from other wrench types?
6. What is the main reason for using a box wrench?
7. List three standard size drives for socket wrenches.
8. What is a combination wrench?
9. A combination wrench
   (*a*) has replaceable sockets
   (*b*) has box at one end and a open at the other
   (*c*) both ends open
   (*d*) has box shape on both ends.
10. What is a socket point?
11. The most common type of screwdriver used in a shop is the
    (*a*) double blade (*b*) Phillips head
    (*c*) flat blade (*d*) hex-head.

12. The end of a Phillips- type screwdriver is a:
(*a*) Flat blade
(*b*) Pointed end with four grooves
(*c*) Fluted end.
13. What is a wretch handle and how it works?
14. The most common box wrench has
(*a*) 6 notches, or points (*b*) 8 notches or points
(*c*) 12 notches, or points (*d*) 16 notches, or points.
15. What is a torque wrench and why it is used?
16. To tighten nuts or bolts accurately, use
(*a*) an open-end wrench (*b*) box wrench
(*c*) combination wrench (*d*) a torque wrench.
17. Describe an Allen wrench.
18. An Allen wrench has:
(*a*) Four sides (*b*) Six sides
(*c*) Eight sides.
19. What is an adjustable wrench and explain how to use it properly?
20. What is the function of a centre punch?
21. To knockout rivets and pins, the mechanic needs a
(*a*) pin punch and an aligning punch
(*b*) centre punch and a pin punch
(*c*) drill punch and a centre punch
(*d*) staring punch and a pin punch.
22. A Torx screwdriver is similar to a Phillips screwdriver. *True or False?*
23. The main difference between the box wrench and the socket is that
(*a*) the socket is bigger
(*b*) the box wrench has exchangeable handles
(*c*) the socket does not have an attached handle
(*d*) they are both combination wrenches.
24. A hammer used most often in a shop is the
(*a*) rawhide hammer
(*b*) brass hammer
(*c*) ball-peen hammer
(*d*) plastic-tipped hammer.
25. What are soft face hammers used for?
26. What is a file? Explain its common uses.
27. The most commonly used file in a shop is the
(*a*) cut file (*b*) oval file
(*c*) flat file (*d*) round file.
28. Explain the difference between a single-cut and a double-cut file.
29. When filing soft steel, which type file should be used first?
(*a*) A bastard file (*b*) A smooth-cut file
(*c*) A second-cut file.
30. A file with one row of teeth that cross another row of teeth is called a:
(*a*) crisscross file (*b*) Vixen-cut file
(*c*) double-cut file.
31. What is a hacksaw? Explain its uses.
32. The blade should be installed in the hacksaw so that it cuts on
(*a*) the forward stroke (*b*) the return stroke
(*c*) both strokes
33. Hacksaw blades are made of:
(a) high grade tool steel (*b*) chilled cast iron
(*c*) carballoy.
34. How should a blade be installed in a hacksaw?
35. List the usual lengths of hacksaw blades used.
36. Which saw blade is recommended to cut soft steel and cast iron?
(*a*) 16 tooth (*b*) 24 tooth
(*c*) 32 tooth (*d*) 14 tooth.
37. A cape chisel is used to cut:
(*a*) narrow grooves (*b*) rivet heads
(*c*) tool steel.
38. What is a chisel? Name at least four different types of chisels.
39. Why should the mushroom end of a chisel be dressed off before it is used?
40. What are the two types of shanks used for drills?
41. List the size classifications for drills.
42. If the size of a drill cannot be read on the drill, how can the drill size be determined?
43. What is reamer?
44. Describe the method of using a reamer on an automotive repair job.
45. Explain the purpose of a tap. List it different types.
46. What is a die? How does the die differ from a tap?
47. What is a screw extractor? Explain its use.
48. Two basic types of tools used in automotive repair shop are:
(*a*) power tools and machine tools
(*b*) screwdrivers and wrenches
(*c*) hand tools and power tools
(*d*) hand tools and cutting tools.
49. What are power tools and why they have developed?
50. The three types of power tools are
(*a*) mechanical, electrical and hydraulic
(*b*) hydraulic, electric, vacuum
(*c*) electric, pneumatic, hydraulic
(*d*) brake, hydraulic, electric.
51. List three ways to power a power tool.

52. Pneumatic tool include
    (*a*) air hammers, impact wrenches, and air jacks.
    (*b*) air ratchets, air jacks, and air hoses
    (*c*) air hammers, air ratchets, and air hoses
    (*d*) all of the above.
53. Hydraulic tools include
    (*a*) lifts, floor jacks, impact wrenches, and air hammers
    (*b*) lifts, portable cranes, air ratchets, and jacks
    (*c*) lifts, floor jacks, portable cranes, and hydraulic presses
    (*d*) floor jacks, hydraulic presses, air hammers, and ratchets.
54. Describe the use of an electric wrench.
55. What are three common sizes of electric drills?
56. What are the common uses of a grinder?
57. When using a grinder wheel,
    (*a*) always wear safety glasses
    (*b*) never touch the rotating grinding wheel rotating
    (*c*) never adjust the tool rest when the grinder is running
    (*d*) all of the above.
58. What is an impact wrench? What does the impact part an air impact wrench do?
59. Why mechanics should not trust a floor jack alone to support a vehicle?
60. Mechanic A says it is all right to go under a vehicle supported by a floor jack provided the saddle is properly located. Mechanic B says that never go under a vehicle unless it is supported by safety stands. Who is right?
    (*a*) Mechanic A (*b*) Mechanic B.
    (*c*) Both A and B (*d*) Neither A nor B.
61. What is the difference between a single post and a double post lift?
62. When using an air nozzle
    (*a*) never point it at anyone
    (*b*) never use it to blow dust off your clothes
    (*c*) do not direct the air stream against your face or hands
    (*d*) all of the above.
63. Explain the purpose of a mobile floor crane.
64. The portable crane is used to
    (*a*) lift the vehicle for oil change
    (*b*) lift the engine out of the vehicle
    (*c*) suspend parts to work on them
    (*d*) all of the above.
65. List two uses of a hydraulic press.
66. To remove a bolt that has broken off flush with the cylinder deck, use
    (*a*) a hacksaw and screw driver
    (*b*) a file and a wrench
    (*c*) a drill and stud extractor
    (*d*) any of the above.

❑❑❑

Fasteners, Gaskets, Seals, Hoses, and Belts

2

2.1 INTRODUCTION

Various types of hardware used on modern engines include, bolts, studs, nuts, snap rings, lock pins, machine screws, washers, keys, cotters, rivets, gaskets, seals, sealants, hoses, belts, and tubing. An automotive mechanic must know how to properly handle these devices when working on an engine. The skill in handling these elements is essential for an automotive mechanic to be proficient in repairing engines.

2.2 FASTENERS

Fasteners include all types of holding devices. A large number of fasteners are used in the construction of an automotive engine. Most common among them are shown in *Fig. 2.1.* Most types of fasteners can be removed so that the assembly can be taken apart. This type of fasteners is called ***temporary fasteners***. Another type of fasteners fastens the parts permanently together, e.g., welding soldering, brazing, and riveting. This type will allow the parts to be separated only by destroying the fasteners. The fasteners that fall under this category are called ***permanent fasteners.*** A mechanic should be able to identify the various types of fasteners and know when and how they are used.

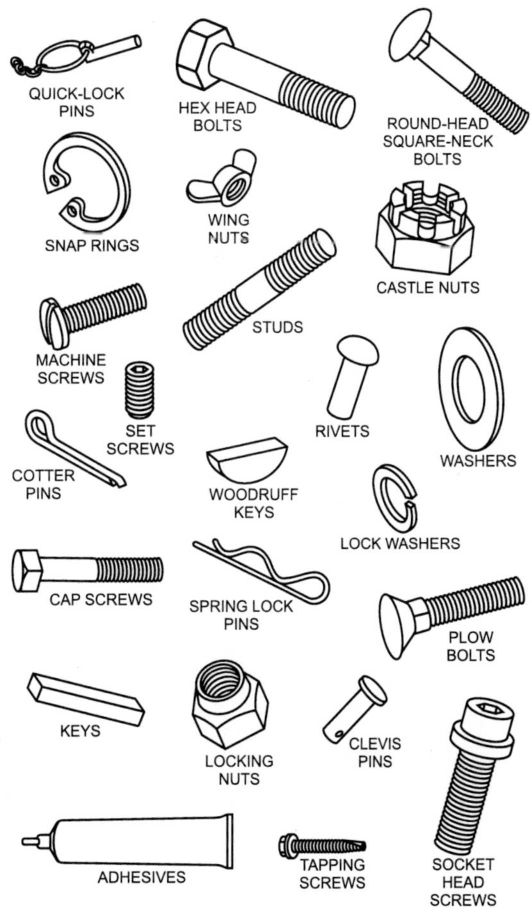

Fig. 2.1 *Basic types of fasteners used in automotive practice. (Deere & Co.)*

2.3 THREADED FASTENERS

Threaded fasteners use the wedging action of a spiral groove or thread to clamp two parts together. The common types of threaded fasteners are screws, bolts, studs, and nuts. Threaded fasteners are always used in pairs, i.e., corresponding external and internal threads cut on two mating parts,

There are several factors involved in describing threaded fasteners. You need to know the length of the fastener, the type of head, the number of threads per unit length (or pitch) of threads, and the units of measurement, e.g., mm in metric units or inches in English units.

Screws

A screw is a threaded fastener that fits in a threaded hole. The screw is turned into the threaded hole to clamp the two parts together, *Fig. 2.2.* Many types of screws are used in automotive engine assemblies.

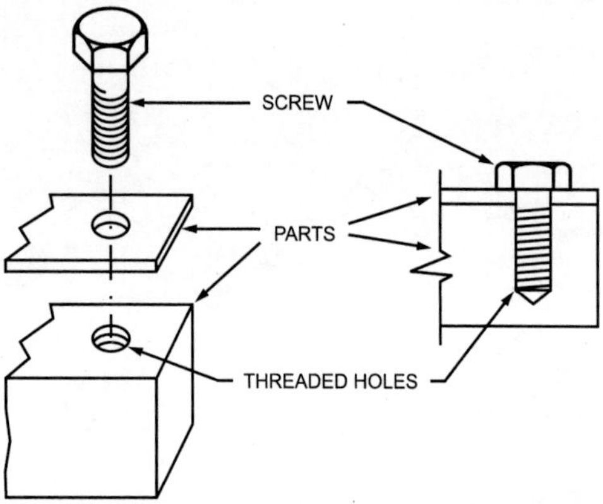

Fig. 2.2 *Screw used in threaded hole.*

The most common among them is the *cap screw,* also known as tap bolt, *Fig. 2.3*, has a hexagonal head, and is sometimes called a hex head screw. The cap screws are driven with the common automotive wrenches, such as the box, open-end, combination, and socket wrenches.

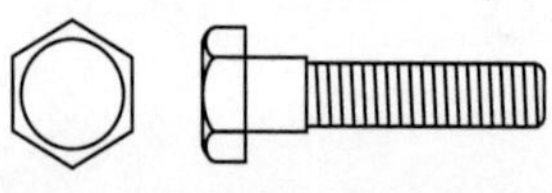

Fig. 2.3 *Hex head cap screw.*

Bolt or cap screw is usually tightened with a wrench, *Fig. 2.4* (*a*). The diameter of a cap screw normally is measured in mm or fractions of an inch.

A *hollow head* or *Allen head* screw, *Fig. 2.4* (*b*), is driven with an Allen wrench. This type of screw is generally used in aircraft assemblies than in automotive assemblies.

Sheet metal screw is a popular fastening device, *Fig. 2.5* (*a*). These sheet metal screws have a wide-spaced external thread. They do not fit into threaded holes. They use the thin metal in which they are driven for the thread. A hole is simply punched or drilled into the sheet metal, and then the screw is turned into the hole.

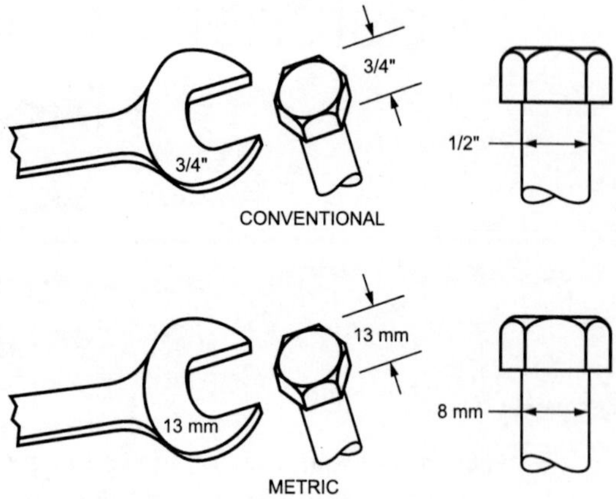

Fig. 2.4. *(a) The bolt diameters are 1/2″ and 8 mm respectively. However, the size of the hex, or wrench size, is 3/4″ and 13 mm respectively.*

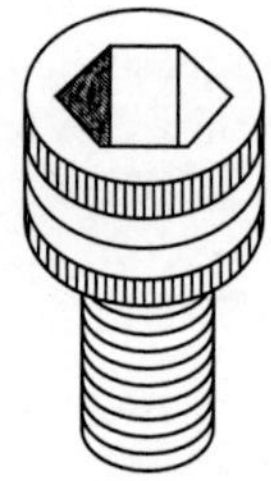

Fig. 2.4 *(b) Allen head screw.*

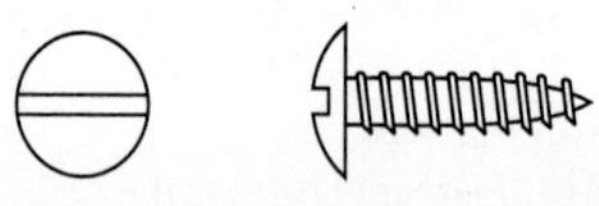

Fig. 2.5 *(a) Sheet metal screw.*

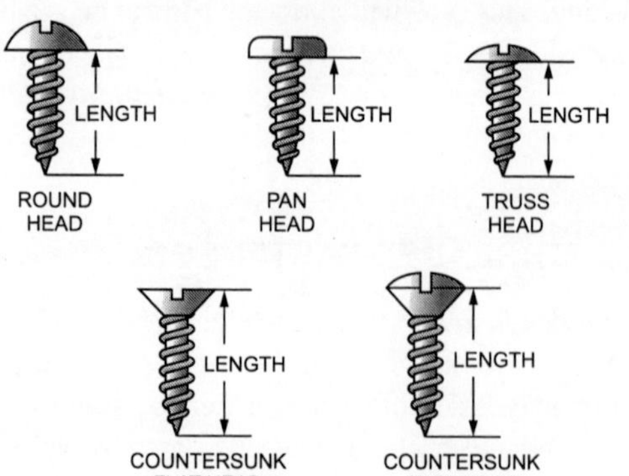

Fig. 2.5 *(b) Types of sheet metal screws.*

Another type of the sheet metal screw is a self-tapping or thread-cutting screw, *Fig. 2.5 (b)*. This screw's hard external thread will make internal threads when it is driven into a soft metal.

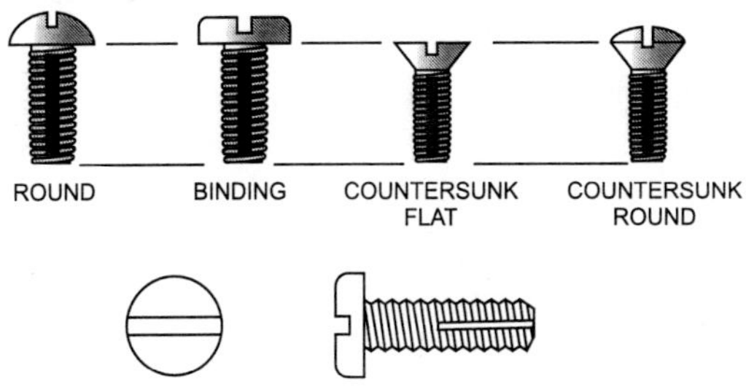

Fig. 2.5 *(c) Self-tapping screw.*

Bolts

Bolts are metal shafts with threads on one end and a head on the other end. When threaded into a tapped hole in a part (without the use of a nut) they are called tap bolts, or *cap screws*.

A bolt is a threaded fastener used with a nut instead of a threaded hole to hold automotive parts together, *Fig. 2.6*. Many bolts look like the hex head cap screw. Bolts made with a square head, *Fig. 2.7*, are called machine bolts. They are used to assemble parts that do not require strong fasteners or close tolerances.

Two wrenches are normally needed to tighten or loosen bolts, one to drive the bolt and the other to hold the nut from turning. Sometimes a bolt with a square shoulder below the head is used that fits into a corresponding square section cavity in the part adjacent to the bolt head. The arrangement stops rotation of the bolt in the bolt-hole and only one wrench is then used to tighten or loosen the nut.

Studs

A *stud* is a fastener with threads at both ends. One end of the stud fits into a threaded hole in a part. Another part, with a proper through hole, fits over the stud, and the two parts are clamped together by tightening a mating nut over the stud end, *Fig. 2.8*. Studs are often used where the positioning of a part is critical to part function. A stud may have threads on its full length, but most commonly the threads are formed on each end, as shown in *Fig. 2.8*.

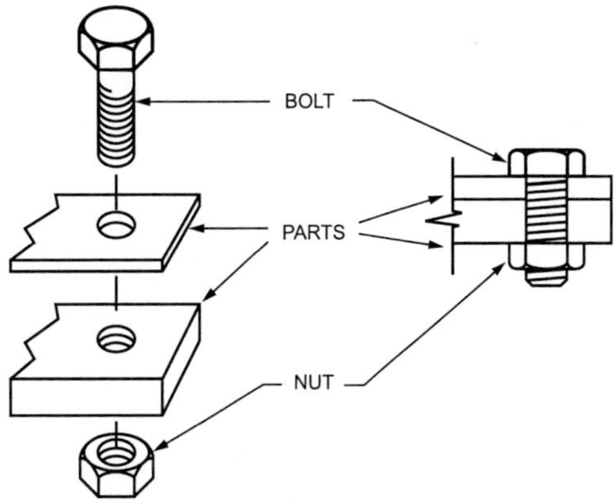

Fig. 2.6 *Bolt used with a nut to clamp two parts together.*

Fig. 2.7 *Square-head machine bolt and nut.*

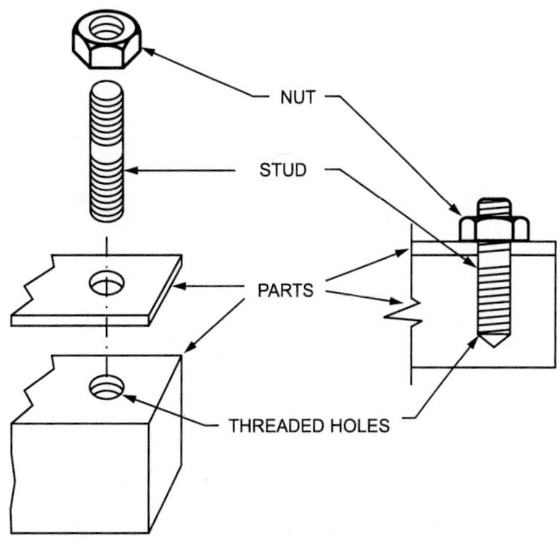

Fig. 2.8 *(a) Stud used with nut to clamp parts together.*

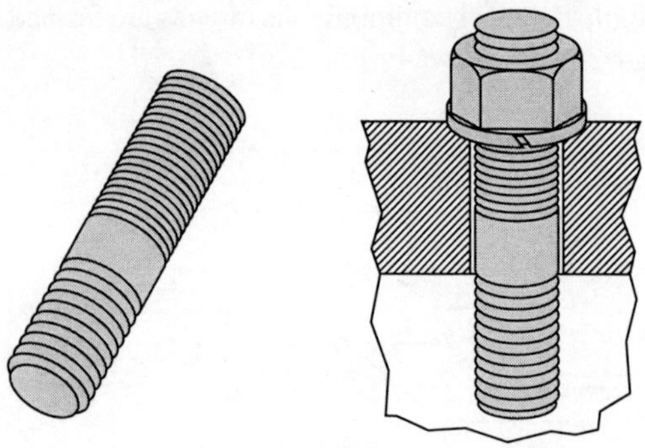

Fig. 2.8 *(b) A stud has threads at each end and has no head. (Deere and Co.)*

Nuts

Nuts have internal threads and are used with bolts and studs. Generally, they are hexagonal in shape, even though various types of nuts can be found on an engine, *Fig. 2.9 (a)*. Nuts used with square-head machine bolts are often square. Nuts may be driven with box-end, open-end, or socket wrenches.

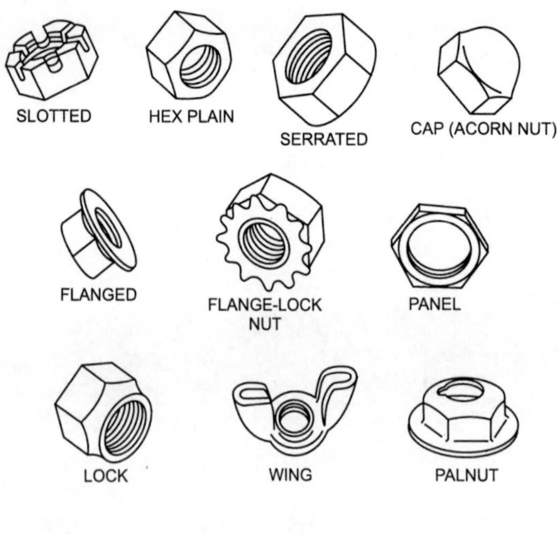

Fig. 2.9 *(a) Various types of nuts found on engines. (Deere and Co.)*

Many automotive components are subject to extreme shaking and vibration. Ordinary nuts used on these components, due to load reversals, easily loosen, causing the joints and parts to fail. In such problem areas, special nuts are used to over come the problem, by locking in place. A number of locking arrangements are available and the one most suited to the situation should be used. Use of a lock nut, locking pin, spring washer, and lock washers are a few commonly used.

The *castellated nuts, Fig. 2.9 (b)*, are generally used on a bolt or threaded shaft. Then a cotter key is inserted through the castles of the nut and through a hole in the bolt or shaft. This prevents the nut from loosening, *Fig. 2.9 (c)*.

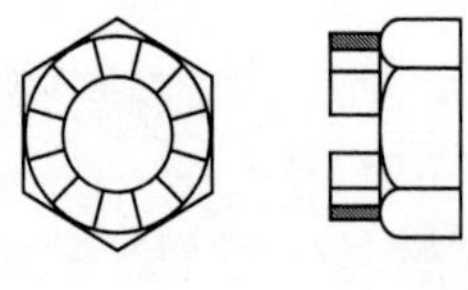

Fig. 2.9 *(b) Castellated or slotted nut.*

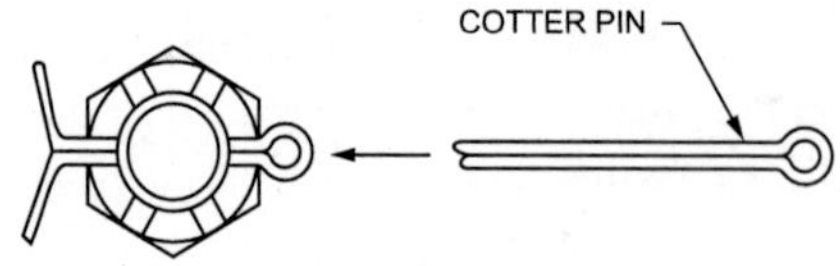

Fig. 2.9 *(c) Cotter pin used with castellated nut.*

There are several types of *self-locking nuts.* As the name implies, self-locking nuts are designed to resist loosening caused by vibration. A self-locking nut shown in *Fig. 2.9 (d)* has a composition plug, which when forced against the threads, prevents the nut from slackening by turning back.

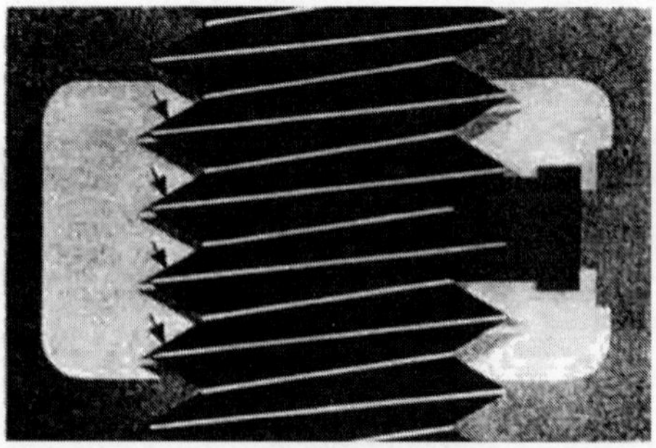

Fig. 2.9 *(d) A self-locking nut has a composition plug, which is forced against the threads to prevent the nut from turning. (Deere and Co.)*

The *palnut* is a locking device, *Fig. 2.9 (e)* that is stamped from thin sheet steel and designed to bind against the threads of the bolts when installed. The palnut is turned down to make firm contact with the regular nut, which has been torqued to the specified amount before the palnut is installed. Then the palnut is turned and is given an additional half turn.

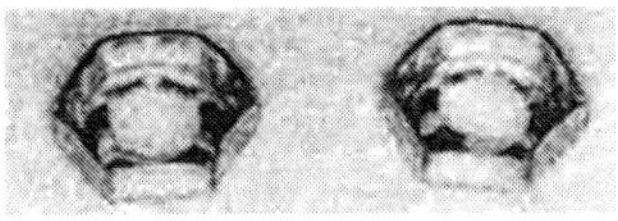

Fig. 2.9 *(e) Palnuts are made of sheet metal and have an inner prong that engages the threads of a bolt to prevent the nut from turning.*

Lock washers, in several different forms are available, that are designed to prevent nuts from getting loose, *Fig. 2.10.* A conventional flat washer is used under the nut to prevent galling of the surface contacted.

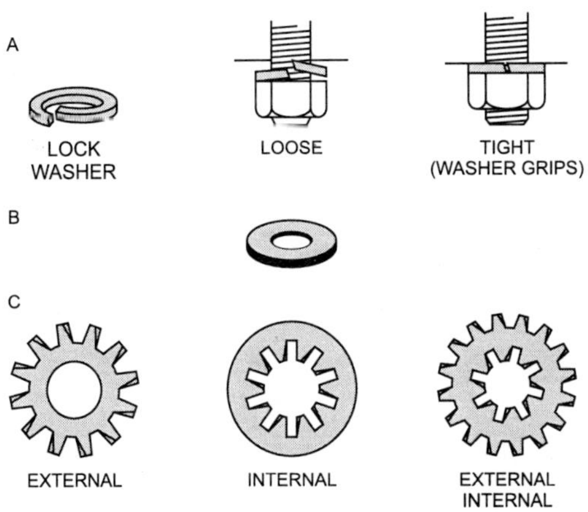

Fig. 2.10 *Various types of washers. A-A spring or lock washer. B-A typical flat washer. C-Tooth-type lock washers grip the metal. (Deere & Co.)*

Liquid Thread-lock

To prevent nuts, bolts, and screws from loosening due to vibration, a liquid thread-lock can be applied to the threaded fasteners, *Fig. 2.11.* Although this prevents the threaded fastener from vibrating loose, it allows easy removal of the fastener to allow disassembly of parts when needed. The liquid thread-lock is applied to the entire threaded portion of the fastener. The fastener is then torqued to specs. A socket and ratchet, wrench, or screwdriver may be used to remove the fastener after remaining in service for any length of time.

Fig. 2.11 *Liquid thread-lock prevents threaded fasteners from vibrating loose.*

Bolt and Nut Dimensions

The basic dimensions of a bolt are:

- *Bolt Size* – It is measurement of the outside diameter of the bolt threads, *Fig. 2.12.*
- *Bolt-Head Size* – It is the distance across the flats of the bolt head. It is the same as the Wrench Size.
- *Bolt Length* – It is measured from the bottom of the bolt head to the threaded end of the bolt.
- *Thread Pitch* – It is the distance between two corresponding points on two consecutive threads, measured parallel to the bolt axis. In English units it is the number of threads per inch.

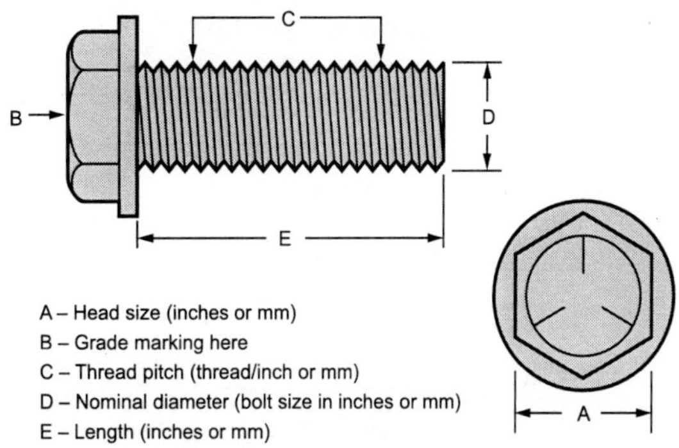

Fig. 2.12 *These are basic dimensions of bolt. These are needed when ordering bolts.*

2.4 THREAD SIZES AND DESIGNATIONS

Different types of threads are used in threaded fasteners. In a threaded pair of internal and external threads, one kind of thread cannot be used with another. A skilled mechanic must be able to differentiate one type of thread from the other.

English System Threads

Automobiles manufactured according to the Unified National System (the basic American standard) for screw threads, has two types of threads: coarse and fine. They are designated by UNC, for unified national coarse, and UNF for unified national fine, *Fig. 2.13*. The coarse thread has fewer threads per inch than the fine thread and can be easily recognized. Coarse threads are used in aluminium parts because they provide greater holding strength in soft materials. Fine threads are used in harder materials, such as cast iron and steel.

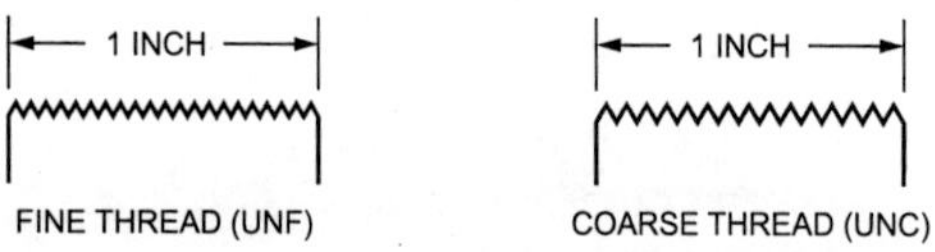

Fig. 2.13 *Unified system threads.*

English threaded fasteners are also designated by size. Different sizes of bolts, screws, studs, and nuts are given in fractions of an inch. A typical designation is 1/2-20UNF, where 1/2 represents a bolt, screw, or stud, diameter of 1/2 inch. When applied to a nut it refers to the size of the bolt that nut fits, *Fig. 2.14* (*a*). Number 20 designates the number of threads per inch length, and UNF represents Unified National Fine threads. All 1/2 Unified National Fine fasteners have the same number of threads (20) per inch. A 1/2-inch diameter coarse thread fastener is designated as 1/2-13UNC. It has only 13 threads per inch. The length of a bolt screw or stud is also given when referring to a bolt, e.g., l/2-13UNC × 1 inch long. *Fig. 2.14* (*b*) shows how to decipher the markings for an ASE bolt.

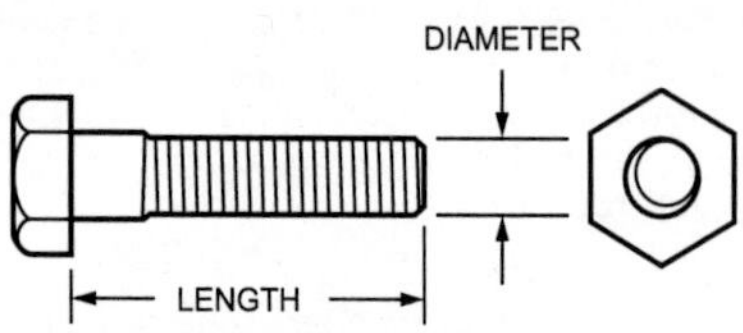

Fig. 2.14 *(a) Bolt or nut size.*

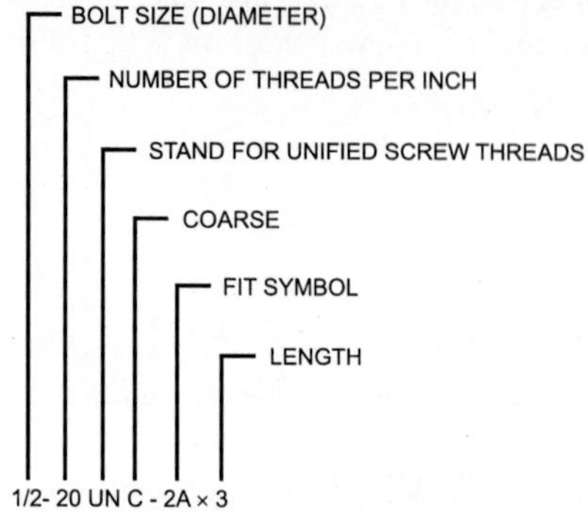

Fig. 2.14 *(b) Deciphering the markings for an SAE bolt. (Deere & Co.)*

Metric System Threads

Automobiles manufactured according to the metric system have fasteners with metric threads.

Metric and English system threads cannot be mixed. Never interchange thread types to avoid damage to threads, *Fig. 2.15*. A metric nut will not fit on an English system bolt. Metric threads could be easily confused for fine UNF threads, if not checked carefully. If a bolt, with wrong threads, is forced into a threaded hole, either the bolt or hole threads can be damaged.

A common metric system fastener might be M12 × 1.75, *Fig. 2.16* (*a*). Here M indicates that the fastener has metric threads. The first number (12 here) stands for the outside diameter of the bolt, screw, or stud or the internal diameter of a nut in millimeters Second number, after the sign ×, (1.75 here), represents the pitch of the thread, in mm. There are no abbreviations used to identify coarse and fine metric threads. Instead, the pitch number itself distinguishes between coarse and fine threads. For example M12 × 1.5 represents a fine thread metric bolt while a M 12 × 1.75 represents bolt of the same diameter with coarser threads. The larger pitch number indicates wider spacing between threads. The length of metric fasteners is measured the same as in the English system, except the measurements are given in metric units (mm) instead of inches.

Figure 2.16 (*b*) shows how to decipher the markings for a metric bolt.

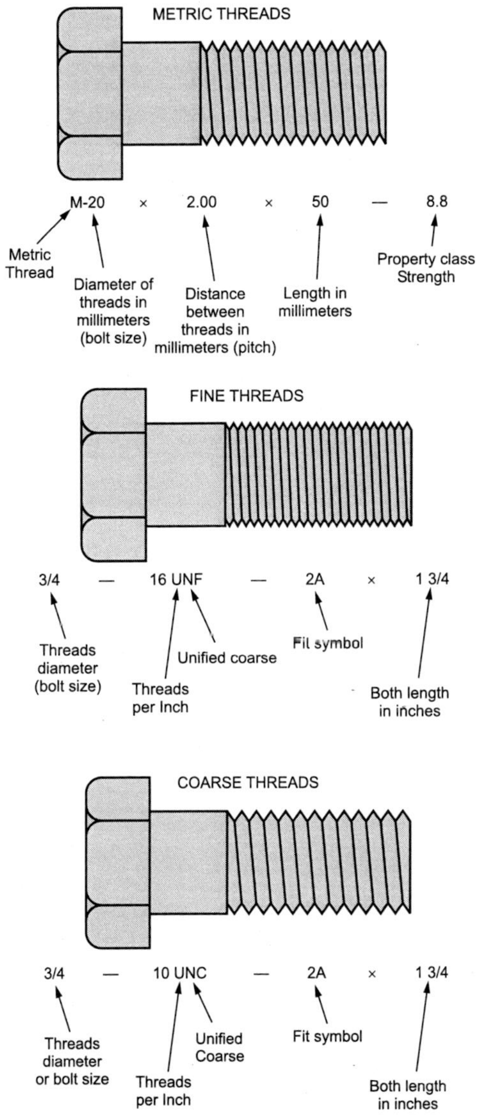

Fig. 2.15 *Coarse, fine, and metric are three thread types found on engines. Never interchange them or thread damage will result. Carefully study bolt designation numbers.*

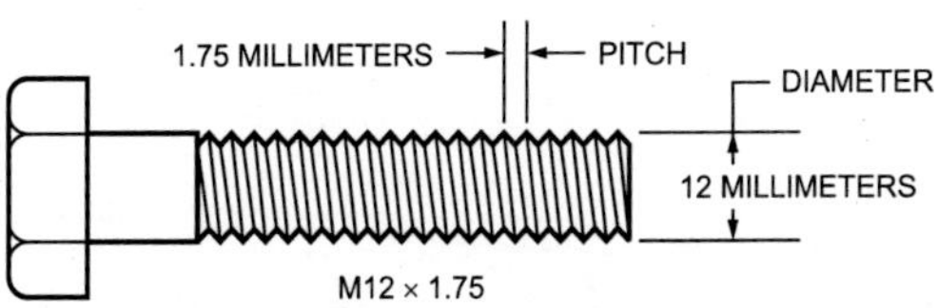

Fig. 2.16 *(a) Metric threaded bolt, M12 × 1.75.*

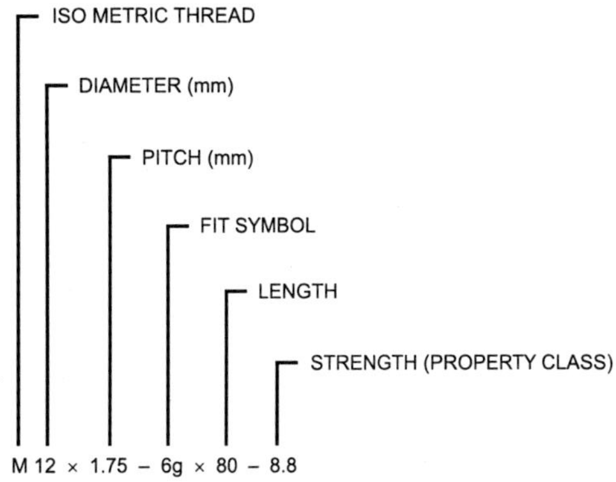

Fig. 2.16 *(b) Deciphering the markings for a metric bolt. (Deere & Co.)*

Using a Pitch Gauge

The fact that two different thread systems, each with fine and coarse threads, are in use in automotive practice is a problem for the automotive mechanic. Metric threads cannot be used with English threads. Fine threads cannot be used with coarse threads. Some vehicles use both metric and English fasteners, based upon the system their suppliers follow, and all vehicles use both fine and coarse threads. It is sometimes difficult to distinguish one thread from another.

A *thread pitch gauge, Fig. 2.17,* is used to check and determine thread type, *Fig. 2.18.* A thread gauge has a number of blades with teeth. The thread size is written on the blade. By matching the teeth on the blade with threads on a fastener, the mechanic can determine thread size. Pitch gauges are available for both metric and English threads.

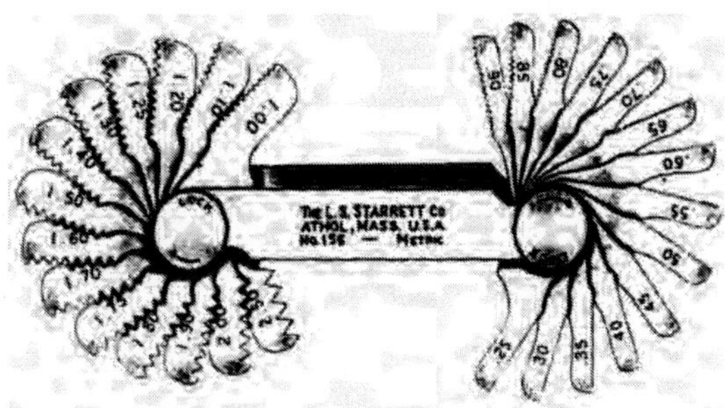

Fig. 2.17 *Pitch gauge for metric threads. (L.S. Starred Company)*

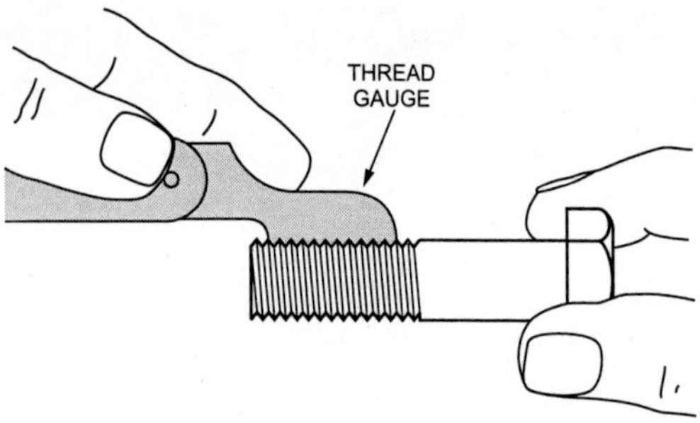

Fig. 2.18 *(a) Thread gauge will check thread type. (Deere & Co.)*

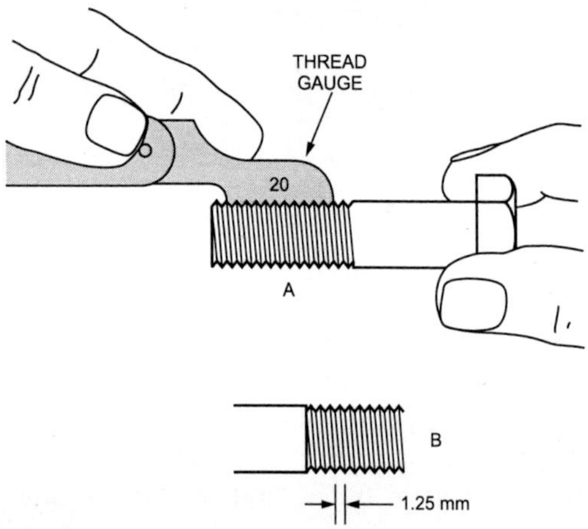

Fig. 2.18 *(b) Using a thread gauge. A-When the thread gauge seats squarely on the threads, look at the number stamped on the gauge. The number indicates the number of teeth per inch (25.4 mm). B-Pitch is the distance between the two consecutive crests. (Deere & Co.)*

Right-hand and Left-hand Threads

Bolts and nuts may have right-hand or left-hand threads. For most commonly used right-hand threads, the fastener must be turned *clockwise* to tighten. For less common left-hand threads, the fastener has to be turned anticlockwise direction to tighten. Letter "**L**" may be stamped on fasteners with left-hand threads.

Bolt Grade Markings—Head Markings

Bolts and screws used for different jobs require different metal strengths. *Bolt head markings,* also called *grade markings,* specify the tensile strength of the bolt. *Tensile strength* or *grade* refers to the strength of a bolt. It indicates the amount of torque a bolt can withstand before breaking. As the bolts are made of different materials, their tensile strengths vary and some are stronger than the others.

Conventional English bolts are marked on the head with *Lines* or *Slash Marks* — more the lines, stronger the bolt, *Fig. 2.19.* A Grade 1 or 2 has no markings on the head and is not very strong. A bolt with three marks is called a Grade 5 bolt and is much stronger than a Grade 1. The strongest bolt for automotive use has six marks and is called a Grade 8. A metric bolt is marked by a *Numbering System* — larger the number, stronger the bolt, *Fig. 2.20.*

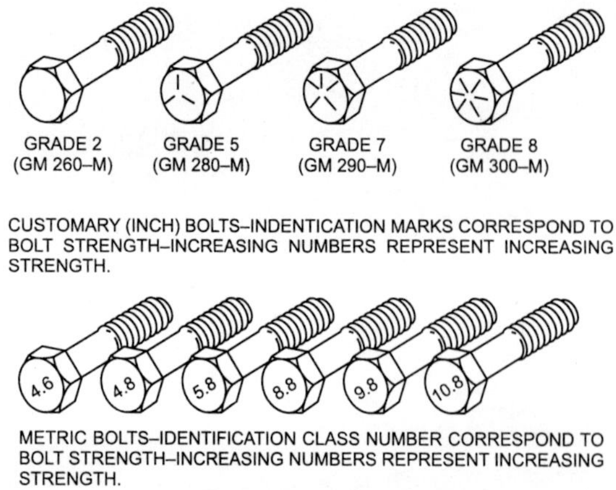

Fig. 2.19 *The strength of bolts is indicated on the top of the hex. (Cadillac Div GM Corp.)*

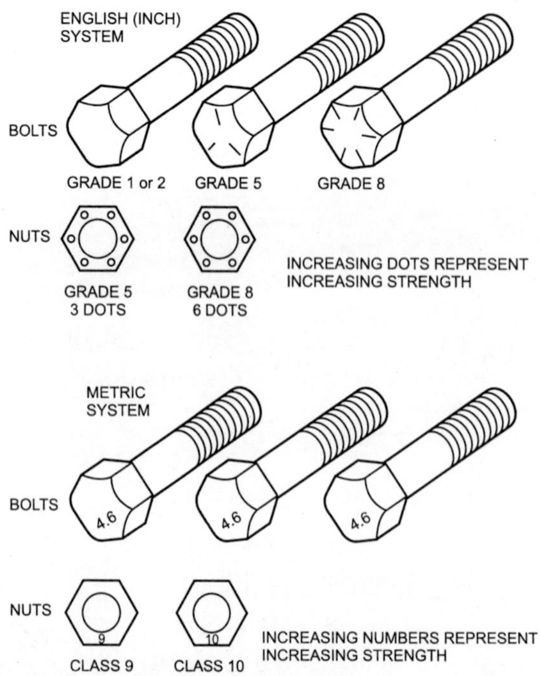

Fig. 2.20 *Strength markings for English and Metric bolts and nuts.*

CAUTION

The torque specifications listed below are approximate guidelines only and may vary depending on conditions when used such as amount and type of lubricant, type of plating on bolt, etc.

SAE STANDARD/FOOT POUNDS

Grade of Bolt	SAE 1 & 2	SAE 5	SAE 6	SAE 8		
Min. Ten Strength	64,000 P.S.I	105,000 P.S.I	133,000 P.S.I	150,000 P.S.I		
Markings on Head					Size of Socket or Wrench opening	
U.S. Standard	FOOT POINTS				U.S. Regular	
Bolt Dia.					Bolt Head	Nut
1/4	5	7	10	10.5	3/8	7/16
5/16	9	14	19	22	1/2	9/16
3/8	15	25	34	37	9/16	5/8
7/16	24	40	55	60	5/8	3/4
1/2	37	60	85	92	3/4	13/16
9/16	53	88	120	132	7/8	7/8
5/8	74	120	167	180	15/16	1.
3/4	120	200	280	296	1-1/8	1-1/18

METRIC STANDARD

Grade of Bolt		5D	8G	10K	12K	
Min. Tensile Strength		71,160 P.S.I	113,800 P.S.I	142,200 P.S.I	170,679 P.S.I	
Grade Markings on Head		5D	8G	10K	12K	Size of socket or wrench opening
Metric		POUNDS				Metric
Bolt Dia.	U.S. Dec Equiv					Bolt Head
6 mm	.2362	5	6	8	10	10 mm
8 mm	.3150	10	16	22	27	14 mm
10 mm	.3937	19	31	40	49	17 mm
12 mm	.4720	34	54	70	86	19 mm
14 mm	.5512	55	89	119	137	22 mm
16 mm	.6299	83	132	175	208	24 mm
18 mm	.709	111	182	236	283	27 mm
22 mm	.8661	182	284	394	464	32 mm

Fig. 2.21 *Basic bolt chart can be used when engine specs are not available. Size, grade, and threads and bolt determine how tight it should be torqued.*

A mechanic must always use replacement bolts and screws of similar quality. Never replace a higher-grade bolt with a lower grade bolt, because a weaker bolt could snap, possibly causing part failure and a dangerous condition.

Nuts are not always marked for quality. When marked, English nuts have dots, which indicate Grade markings. Metric nuts have numbers, which represent strength, *Fig. 2.21.*

Bolt Description

A *bolt description* is a series of numbers and letters that give specifications for the bolt as explained in *Fig. 2.15.* This information is needed when purchasing new bolts.

2.5 WASHERS

Washers are used under bolt heads and nuts. There are two basic types of washers: *flat washers* and *lock washers, Figs. 2.1* and *2.10.*

A *flat washer* or plain washer increases the clamping surface area under the fastener; it prevents the bolt or nut head from gouging or digging into the part.

A *lock washer* prevents the bolt or nut from getting loose. Under stress and vibration, a fastener could unscrew without using a lock washer.

Lock tabs or plates perform the functions of both flat and lock washers. They increase clamping surface area and secure the fastener. They are found on engine exhaust manifolds.

2.6 TORQUING BOLTS/SCREWS

It is very important that engine bolts and nuts are *torqued* or tightened correctly, *Fig. 2.22.* If *overtightened,* a bolt will stretch and possibly break or the threads could fail. If *under-tightened,* a bolt could work loose. Part movement could shear the fastener or break a gasket, causing leakage. *Torque specifications* are the fastener tightening values recommended by the vehicle manufacturer. Torque specs are normally given for all precision assemblies, e.g., for head bolts, rod bolts, oil pump bolts, etc.

Bolt Tightening Sequence

A *bolt tightening sequence* or pattern ensures that the parts clamp together evenly. An incorrect or uneven pattern could break parts, cause part warpage, and leakage.

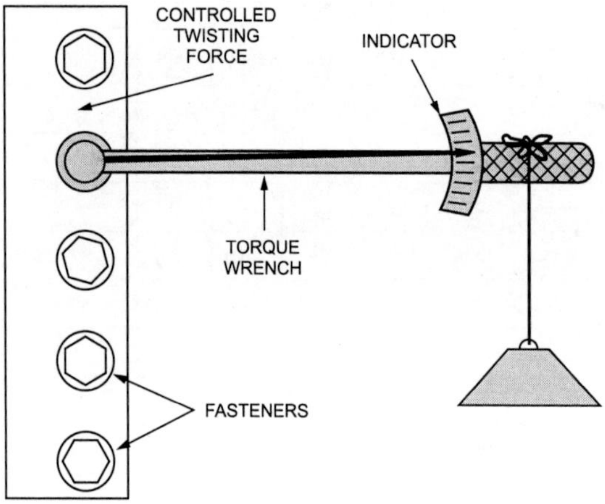

Fig. 2.22 *Torque is measured in pound-feet or Newton-meters. One pound-foot of torque is produced when one pound of weight is suspended from a one foot long lever arm.*

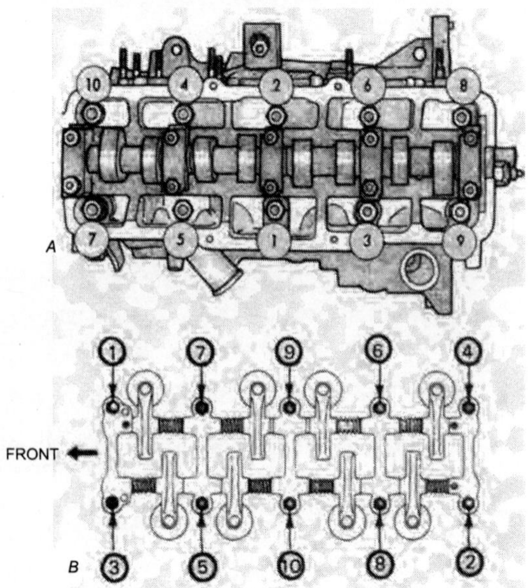

Fig. 2.23 *The figure shows a fundamental crisscross tightening sequence. Use it when several bolts hold one part. It will assure even clamping pressure between parts. Service manual will give more specific sequence for critical engine assemblies.*

A *crisscross pattern* is commonly used because it goes from side to side and end to end, starting in the middle. Each bolt is tightened a small amount at a time, jumping back and forth from the middle outward, *Fig. 2.23*. A shop service manual will show the proper sequence for critical assemblies, e.g., cylinder heads, intake manifolds, and exhaust manifolds.

A *torque wrench* is used to correctly torque the bolts and nuts. When using a torque wrench, follow the basic rules given below:

1. Exert a steady pull on the torque wrench. Never use short, jerky motions to avoid inaccurate readings.
2. The fastener threads should be clean and lightly oiled.
3. If possible, avoid using swivel joints. They can upset torque wrench accuracy.
4. Read the torque wrench carefully. Look straight down at the scale. Viewing from an angle can give an incorrect reading.
5. Always use manufacturer's specs. A general torque value chart should only be used when manufacturer's specs are ***not*** available.
6. When manufacturer's torque patterns are not available, use a general crisscross pattern.
7. Pull on the handle of the torque wrench. Do *not* let the wrench beam touch anything else.
8. Tighten bolts and nuts in at least four steps: (*i*) tighten to 1/2 recommended torque, (*ii*) tighten to 3/4th torque, (*iii*) tighten to full torque, and (*iv*) finally tighten to full torque second time to avoid any missed fastener during previous step.
9. Torque fasteners where required after use and heating.
10. Replace fasteners as recommended by the vehicle/engine manufacturer.

2.7 MACHINE SCREWS

Machine screws are similar to bolts, but they usually have screwdriver-type heads. A machine screw is torqued with a suitable screwdriver. The machine screws are smaller in diameter than the cap screws or tap bolt, and are relatively weak. Machine screws are used to secure parts when clamping loads are light. They come in various head shapes, *Fig. 2.24*. Machine screws can be found on accessory units mounted on the engine.

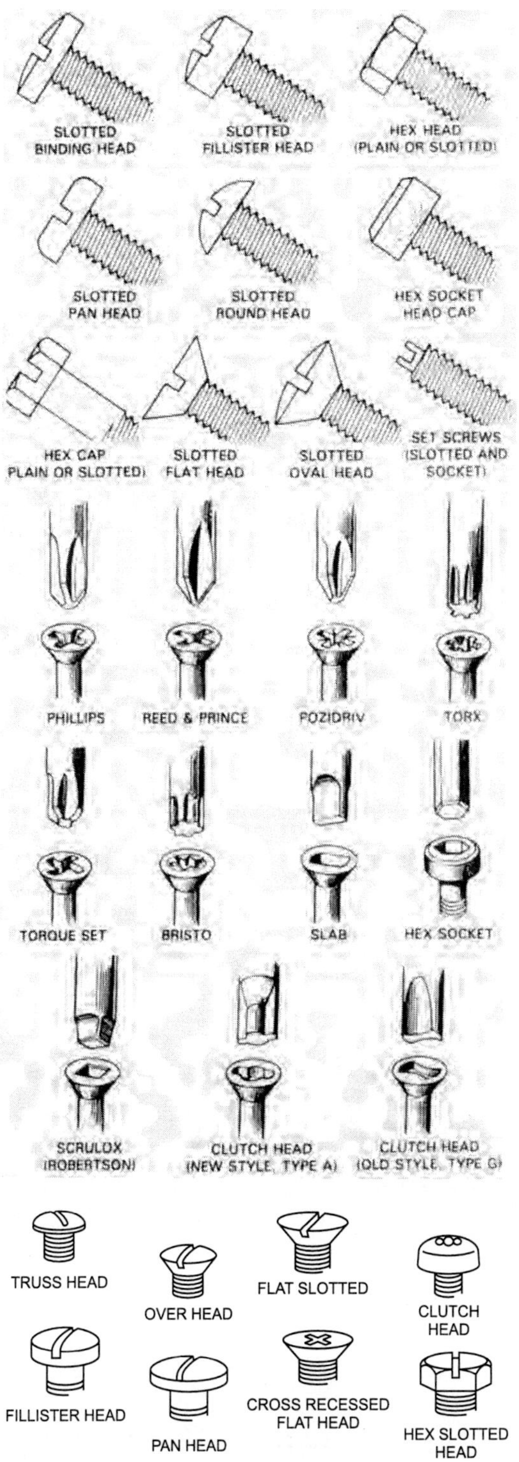

Fig. 2.24 *Different types of machine screw or head types. (Heyco and Klein Tools)*

The diameter of a machine screw is given in nominal size, which is represented by number (e.g., No.8, No. 10, etc.) or by a fraction or decimal equivalent.

2.8 REPAIRING THREADS

On extended service, internal or external threads may get stripped, but often they can be repaired with a tap or die. If the threads are severely damaged, other repair techniques can be used. A mechanic must be skilled to quickly and properly repair the damaged threads. Damaged threaded holes in parts require repairs to salvage the parts. *Fig. 2.25* shows a number of thread repair tools, used in automotive repair shops.

Minor Thread Repairs

Minor thread damage includes partial flattened threads, nicks in threads, and other minor problems. Minor thread damage can usually be repaired with a *thread chaser,* which is rethreading tool for cleaning up damaged threads. This type tool is available for fixing both internal and external threads. The chaser is run through or over the external threads to restore them, *Fig. 2.25.*

Tapping Oversize

Major thread damage generally includes badly smashed threads, stripped threads, or threads that cannot be repaired easily. When a thread chaser cannot be used to clean up damaged threads, the hole can be drilled and tapped oversize. First, drill out the hole one diameter or size larger. Then, cut new threads in the drilled hole with the correct size tap. A larger bolt can then be installed. However, in some cases, this presents a problem. If a cylinder head bolthole in the block were stripped, it would necessitate drilling out the head to accommodate the larger stud. In some cases, this could result in breaking into the water jacket.

Helicoil—Thread Repair Insert

One method of thread repair provides a patented coil of wire that forms new threads. A *Helicoil* or *thread repair insert* should be used when an oversize hole and fastener is *not* acceptable. A Helicoil will repair damaged internal threads and allow the use of the ***original size hole.***

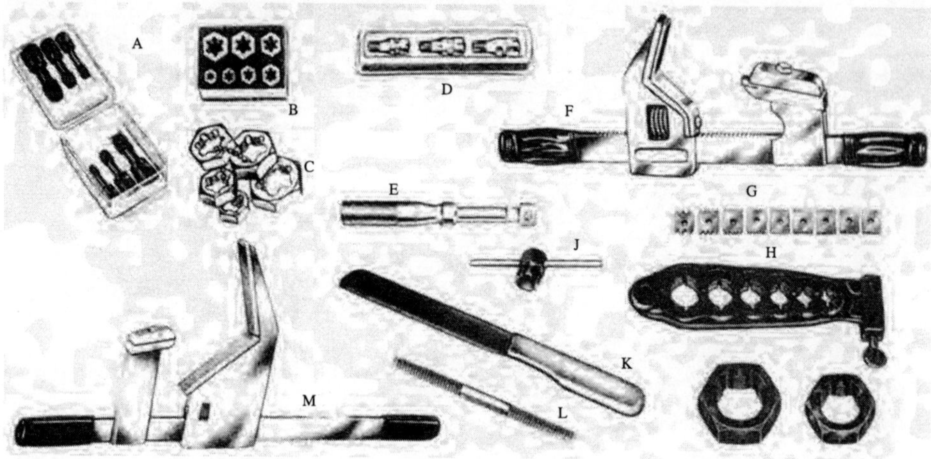

Fig. 2.25 *A-Spark plug hole thread chasers. B and C-Nut type chasers. D-Metric chasers. E to J-External chasers. K and L-Threadfiles. M-External thread chasers.*

To use a helicoil, drill the hole oversize as described in the Helicoil instructions. Tap the hole. Then, screw the Helicoil into the threaded hole. The internal threads of the helicoil will act as threads the same size as the damaged hole, *Fig. 2.26.*

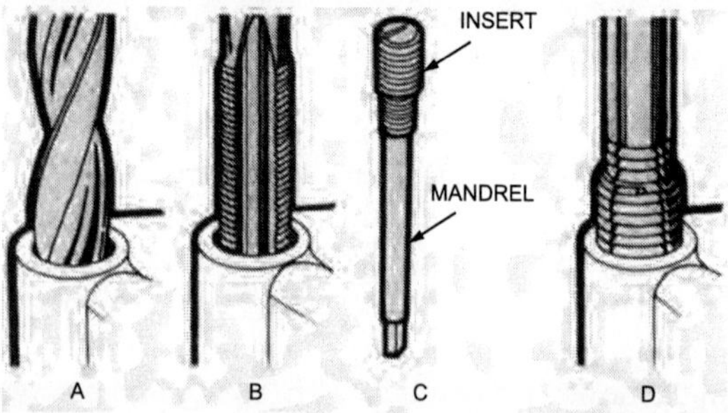

Fig. 2.26 *(a) Helicoil will let you repair stripped threads in part and use same size bolt. A-Drill hole oversize. B-Tap out hole. C and D-Use mandrel to screw helicoil into hole. (Buick Div. GM)*

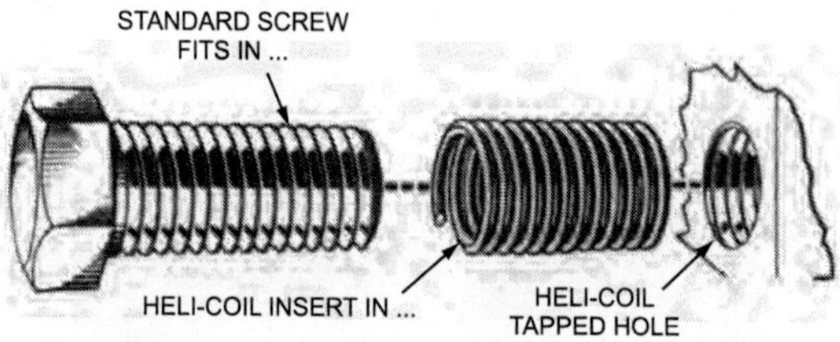

E-Original Size bolt will then thread into helicoil.

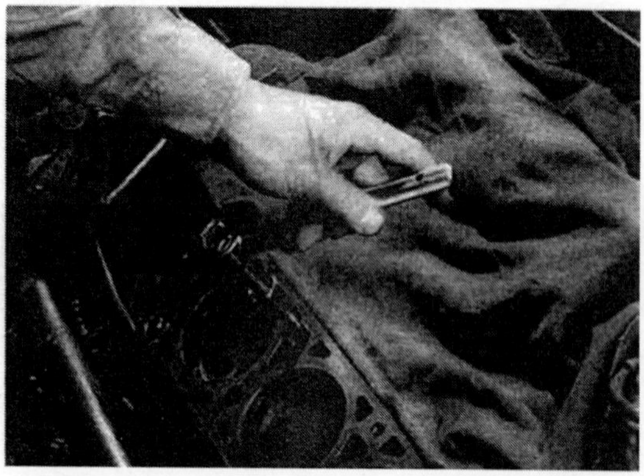

Fig. 2.26 *(b) Using a tap to repair threads in the engine block. (Fel-Pro Inc.)*

2.9 REMOVING BROKEN BOLTS

Every now and then, a bolt will break off during removal or installation. When this happens, the broken portion remaining in the hole must be removed. Also, an engine mechanic must be able to remove broken bolts and bolts with rusted or rounded off heads. This is an important skill that is not covered in service manuals.

Certain tools and methods are needed for removing problem fasteners:

1. *Vise grip pliers* may be used to remove broken bolts that are sticking out enough to be held, or bolts with rounded heads. Lock vise grip pliers tightly on bolt for removal, *Fig. 2.27 (a).*

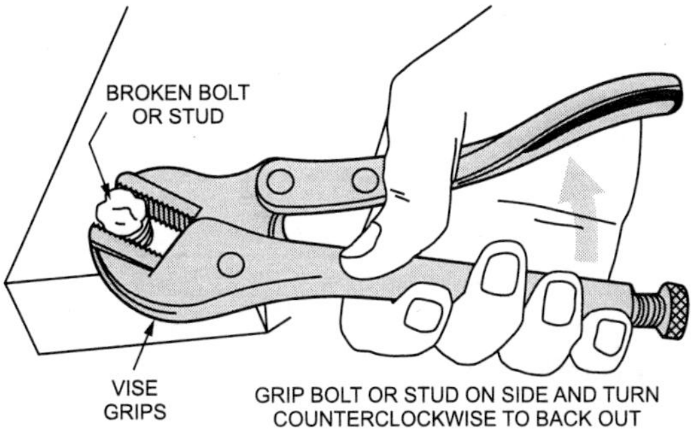

Fig. 2.27 *(a) Vise grip pliers will sometimes remove broken bolt or stud when part of fastener is sticking up above the part. Clamp vise grips tightly on first try. (Florida Voc. Ed.)*

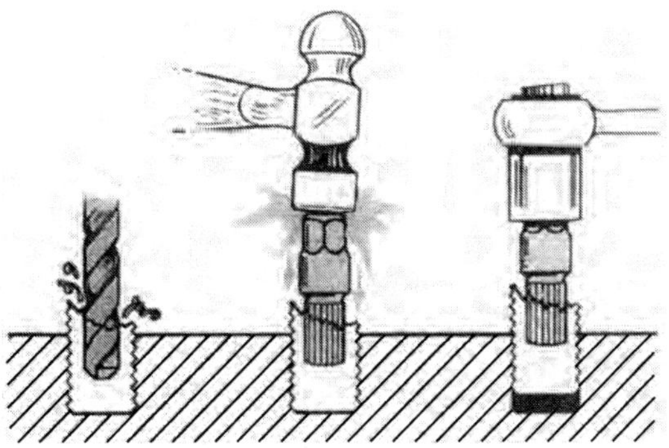

Fig. 2.27 *(b) Screw extractor is a method of removing broken bolt when below surface of part. Drill hole in centre of broken bolt. Tap extractor into hole. Use wrench to unscrew and remove broken bolt. (Florida Voc. Ed. And Lisle Tools)*

2. A *stud puller* or *stud wrench* will remove studs, bolts broken off above the surface of the part. It will also install studs. Position the stud puller so that it will not clamp on and damage the threads.
3. In some cases, broken fasteners are too short to grasp with any tool. In such a case, either cut a *screwdriver slot* in the bolt with a hacksaw or weld on another *bolt head.* Then use the screwdriver or a wrench to unscrew the broken bolt.
4. When the fastener is *broken flush* with the part surface, a *hammer and punch* can be used for its removal. Angle the punch so that blows from the hammer can drive out and unscrew the broken bolt.
5. A *screw extractor* or *"easy-out"* can also be used to remove bolts broken flush or below the part surface. To use a screw extractor, drill a hole in the exact centre of the broken fastener. Then lightly drive the extractor into the hole using a hammer. Unscrew the broken bolt using a suitable wrench, *Fig. 2.27* (*b*). Extreme care should be used to avoid breaking a tap or a screw extractor. They are case hardened and cannot be easily drilled out of a hole.
6. On some broken bolts, it might be necessary to drill a hole almost as large as the inside diameter of the threads. Then, use a tap or punch to remove the threaded *shell,* i.e., thin layer of threads remaining in hole).

2.10 ANTI-SEIZE LUBRICANT

Bolts and any connected metal part that are exposed to continuous heat may get "cold welded" together. To avoid this, an *anti-seize lubricant* should be applied to the bolt threads or the connecting metal surface(s), *Fig. 2.28.* This is particularly true of the threads on an oxygen sensor. If anti-seize lubricant is not placed on the threads, will be impossible to unscrew after a period of driving time. Anti-seize mixture can also be used on exhaust manifold bolts and other exhaust pipe connections. Keep in mind that anti-seize compound is a lubricant not a sealer.

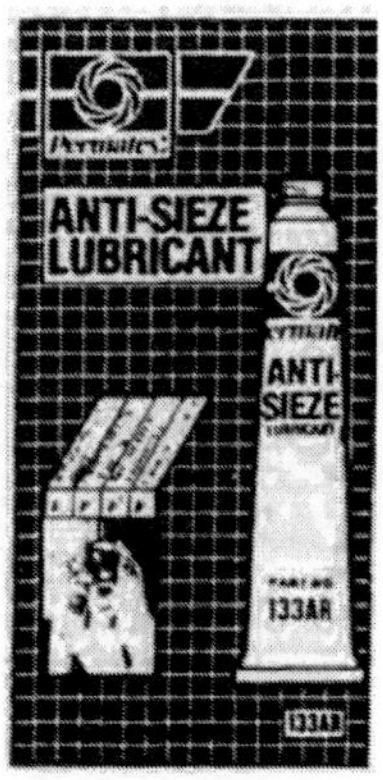

Fig. 2.28 *Anti-seize lubricant is applied to the threads of fasteners, as well as to the connections of metal pipes that are exposed to continuous heat.*

2.11 NON-THREADED FASTENERS

Many types of *non-threaded fasteners* are used in the assembly of an automobile. It is very important for an automotive mechanic to learn the most common types of non-threaded fasteners.

Pins

A *cotter pin* is a safety device normally used in a slotted nut, *Fig. 2.29*. It fits through a hole in the bolt or part. This keeps the slotted nut from turning, and possibly coming off. Cotter pins are also used with pins, shafts, and linkages. *Pins - straight, tapered, and split - are often used to secure a gear or pulley to its shaft.* Some variations of pins are shown in *Fig. 2.29.*

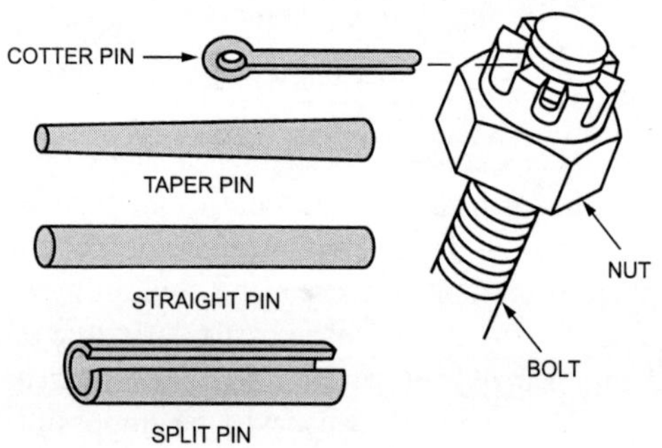

Fig. 2.29 *(a) Various types of cotter pins. (Florida Voc. Ed.)*

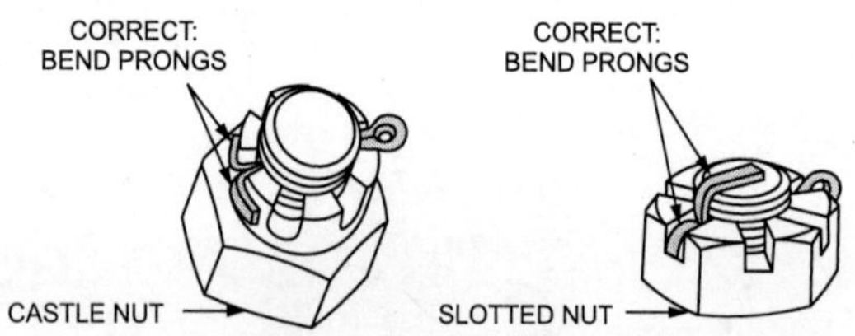

Fig. 2.29 *(b) The cotter key prevents the nut from coming loose.*

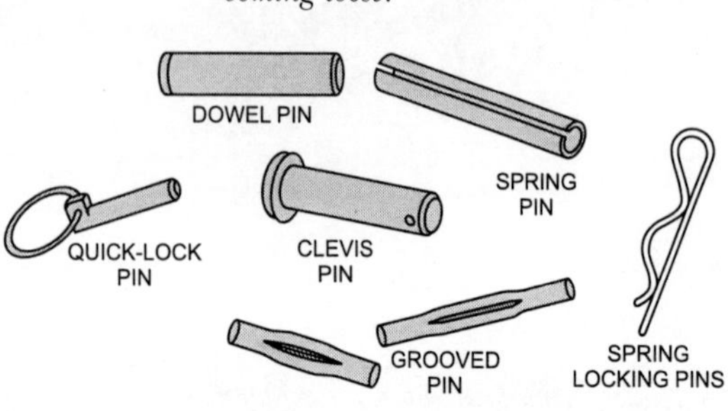

Fig. 2.29 *(c) Pins come in different shapes and sizes. (Deere & Co.)*

Pins are also used as dowel pins that fit into drilled holes to position two parts that fit together. Dowel pins, also called roll pins, require special piers or a punch and hammer for removal. Dowel pins may be straight or tapered, solid or split.

Keys

A key is a small, hardened piece of metal used with a gear or pulley to lock it to a shaft. Usually half of the key fits into a keyseat on the shaft; the other half fits into another slot called a keyway on the pulley or gear, *Fig. 2.30* (*a*). Keys used in automotive design include the Woodruff key, square key, and gib-head key, *Fig. 2.30* (*b*).

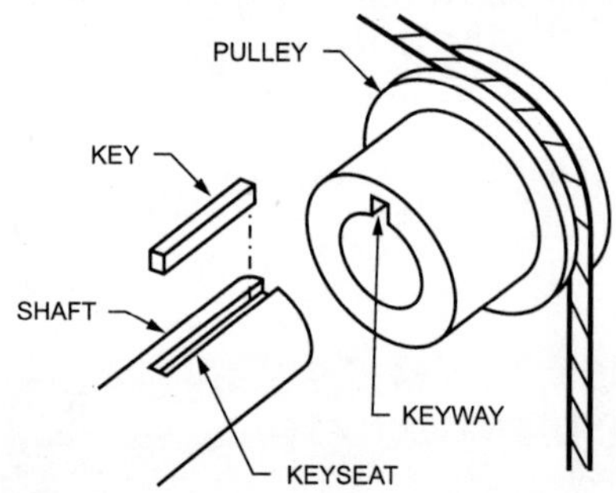

Fig. 2.30 *(a) Key used to lock pulley to a shaft.*

Snap Rings

Snap rings, Fig. 2.31 (*a*), are employed to prevent end wise movement of cylindrical parts and shafts. A snap ring fits into a groove in a part and commonly holds shafts, bearings, gears, and other similar parts in place in an assembly. There are both internal and external snap rings. An *internal snap ring* is used in a grove cut in a housing. An *external snap ring* is designed to fit in a groove cut on the outside surface on a cylindrical shape, such as a shaft. Snap rings are used extensively in transmissions, and engines.

Snap ring pliers are needed to remove and install snap rings. They have special jaws that grasp the snap ring for assembly or disassembly. An external ring is expanded with retaining pliers for installation. The spring tension from the ring holds it in the groove. An internal ring is compressed with retaining pliers to fit it in the groove, *Fig. 2.31* (*b*). Then its spring tension will hold it in position. When several gears are mounted on a shaft, a retaining ring at each end of the shaft, adjacent to the gears, will hold the gears in place.

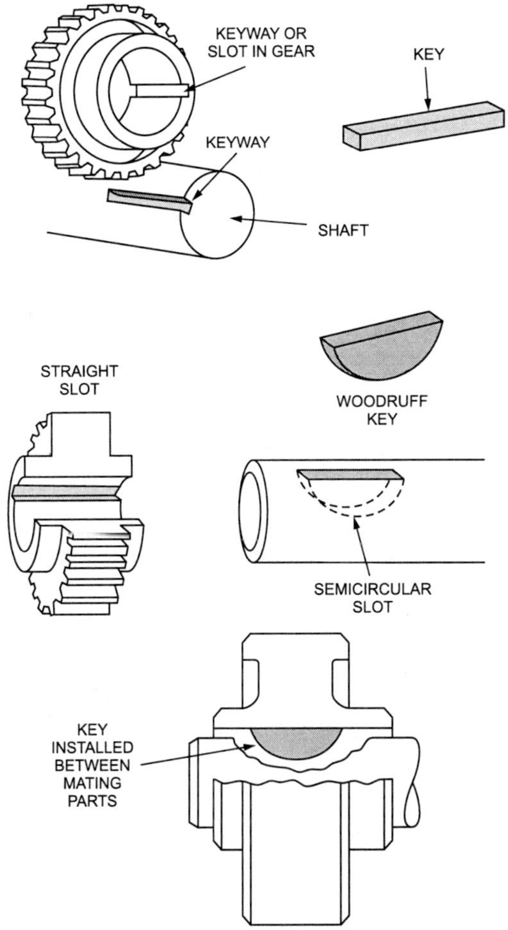

Fig. 2.30 *(b) Key and keyway lock parts to shaft. This is commonly done on front of crankshaft to hold damper and pulley. (Deere & Co.)*

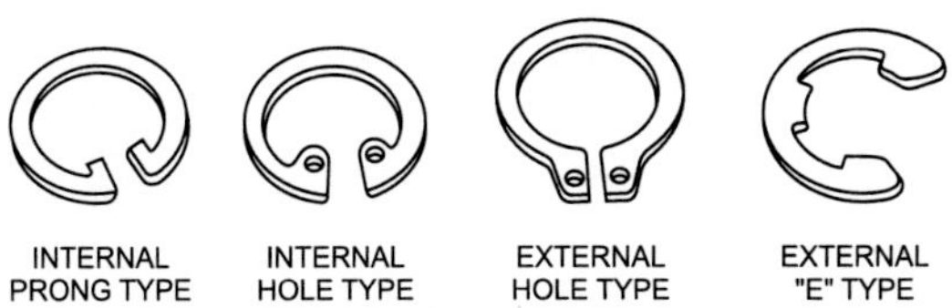

Fig. 2.31 *(a) Snap rings types.*

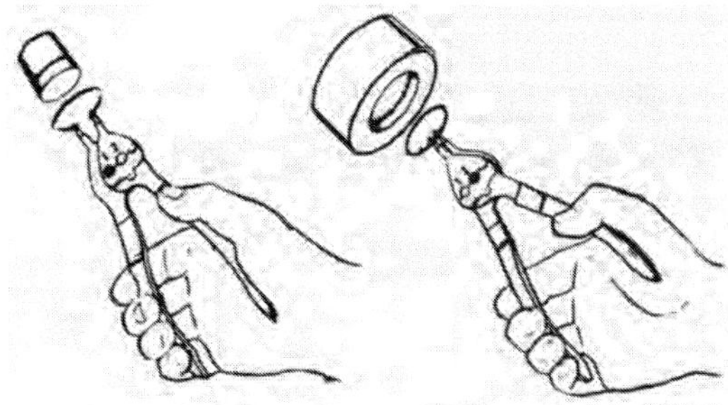

(a) External snap ring *(b) Inlemal map Hng*

Fig. 2.31 *(b) Snap ring require the use of snap ring pliers.*

Wear eye protection when working with snap rings. When flexed, the ring can fly and shoot into your face with considerable force to cause injury.

Splines

Splines are another way to lock gears on a shaft. Splines are external teeth cut on a shaft that matches up with internal teeth cut on a gear. The external and internal splines slide or mate together, allowing the shaft to spin the gear, *Fig. 2.32*. This fastening method is often used in transmission assemblies.

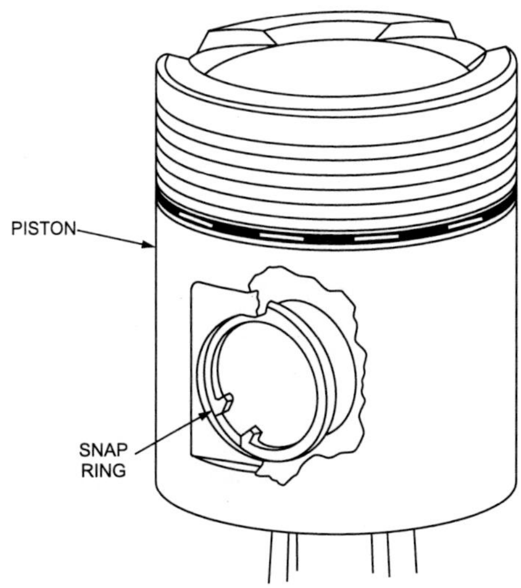

Fig. 2.31 *(c) Snap rings are frequently used to hold piston pin inside piston. They fit into small groove machined in pin or bore. (Deere & Co.)*

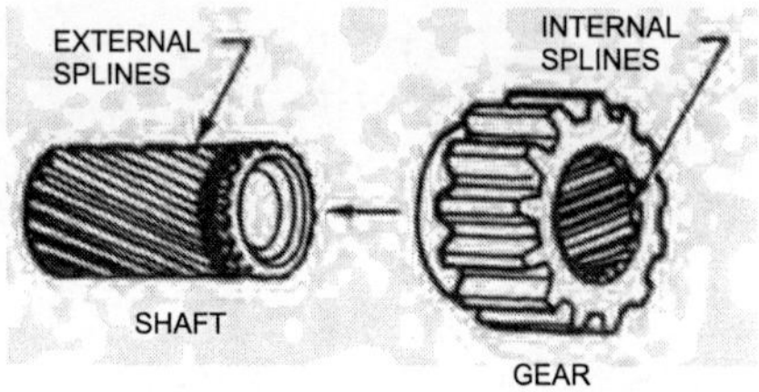

Fig. 2.32 *Internal and external splines used to hold a gear on a shaft.*

Rivets

A *rivet* is a soft metal pin with a head at one end. It is used to hold two metal parts together. A hole is drilled through the two parts together. The rivet is then placed through the hole in both pieces. The small end of the rivet is formed into a head with a rivet set or ball peen hammer. A *pop riveter* is a tool designed to install rivets when only one side of the parts to be joined can be reached, *Fig. 2.33* (*a*). Rivets used in a hand-operated blind riveter are shown in Fig, 2.33 (*b*).

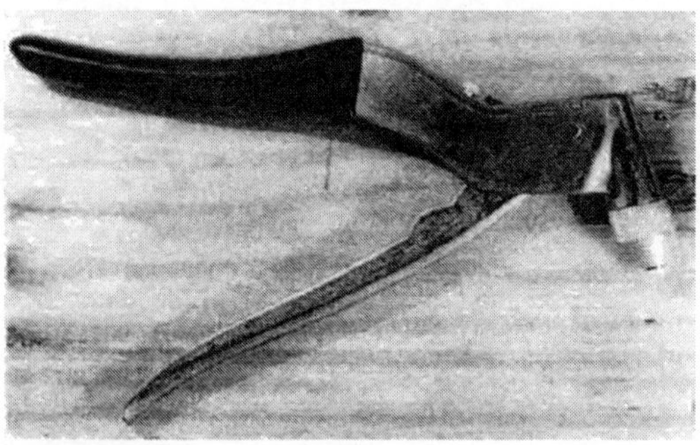

Fig. 2.33 *(a) A pop riveter installs blind rivets.*

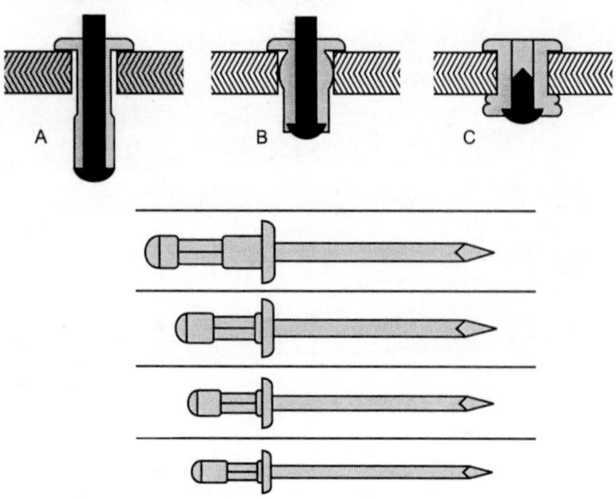

Fig. 2.33 *(b) A-Rivet is placed into hole. B-Handle of riveter is squeezed and rivet starts to compress. C-Rivet is compressed and holds the two pieces of material together.*

To install a blind rivet, drill a hole through the parts to be joined. Place the long stem of the rivet in the head of the tool and insert the short end in the hole in the parts. Squeeze the handle of the pop riveter repeatedly to set the rivet and snap off its end.

A rivet is a permanent type of fastener. Once installed, it cannot be taken out without destroying it. A rivet is taken out by first chipping off the head with a chisel and then driving it out with a punch.

2.12 GASKETS

Gaskets are used between engine parts to prevent fluid leakage. A gasket compensates for small irregularities between two flat metal surfaces, preventing oil and coolant from leaking. In case of a cylinder head gasket, combustion pressures are sealed in the cylinder. Gaskets also dampen vibration, act as spacers, and function as wear insulators. Gaskets can be made of steel, copper, pressed fiber, cork, rubber, silicon, graphite, etc. New gaskets have been developed for use in bimetal interfaces that experience more movement between cylinder block and the head, with changes in temperature. Never reuse a gasket, as it will not form a proper seal when reinstalled.

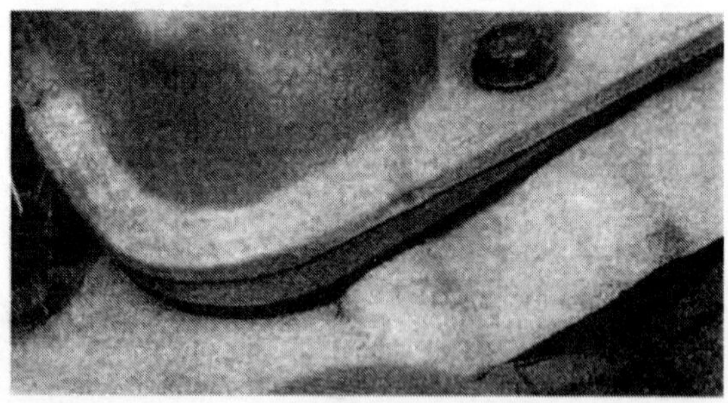

Fig. 2.34 *(a) Overtightening of this valve cover gasket caused the gasket to bulge out. Look up torque specs in service manaual and use a torque wrench.* *(Fel-Pro. Inc.)*

The most usual cause of gasket failure is over-tightening of the bolts that hold the gasket interface joint, *Fig. 2.34* (*a*). To avoid over-tightening of the gaskets, some manufacturers incorporate metal washers in the boltholes of the gasket. These metal washers limit the amount of torque that can be applied. Some manufacturers dimple the flange of the sheet metal that is bolted in place. This has the same effect as a washer placed in the bolthole of the gasket. Some gaskets must be installed in a specific manner, *Fig. 2.34 (b)*, while other gaskets do not.

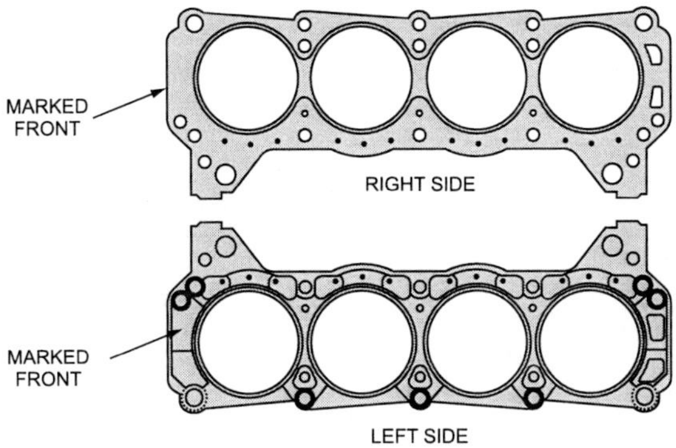

Fig. 2.34 *(b) Some gaskets must be installed in a specified manner.*

It is important to understand some basic principles about gaskets. If they are not properly serviced, serious customer complaints and mechanical failures can result.

Gasket Rules

Observe the following rules when working with gaskets:

1. Carefully inspect any leaks before engine disassembly. If an interface with a gasket is leaking, the part surface should be inspected thoroughly to determine problems.
2. Use proper care to avoid part damage during disassembly. Do not nick, gouge, or dent mating interface surfaces when removing parts. Even minor unevenness could cause leakage.
3. Clean off old engine gasket properly. All of the sticking old gasket material should be scraped or wire brushed from the parts. Be careful while scraping aluminium and other soft metal surfaces. Use a dull scraper and wire brush lightly. Soft metals are easily damaged if proper care is not given.
4. Wash and dry components thoroughly, after gasket removal, in a solvent. Blow the surfaces dry with compressed air. Then wipe the mating surfaces with a clean shop rag.
5. Check and compare the new gasket to match it to the mating interface surface. Lay it into place and see that all holes and sealing surfaces match perfectly.
6. Use sealer if necessary. Sealer is normally used where two different gaskets come together, to prevent leakage where the gaskets overlap. Check the applicable service manual for instructions. Use the sealer only as needed. Too much sealer could clog internal passages in the assembly, *Figs. 2.35* and *2.36*.

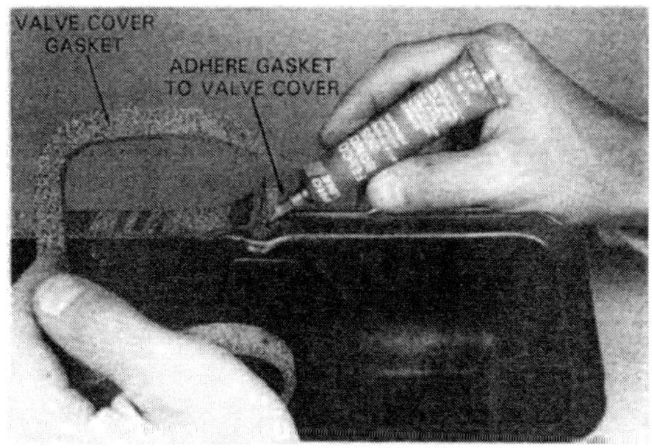

Fig. 2.35 *Mechanic is using adhesive type sealer to hold valve cover gasket in place. This will keep gasket from shifting while fitting cover on engine. (Pel-Pro Gaskets)*

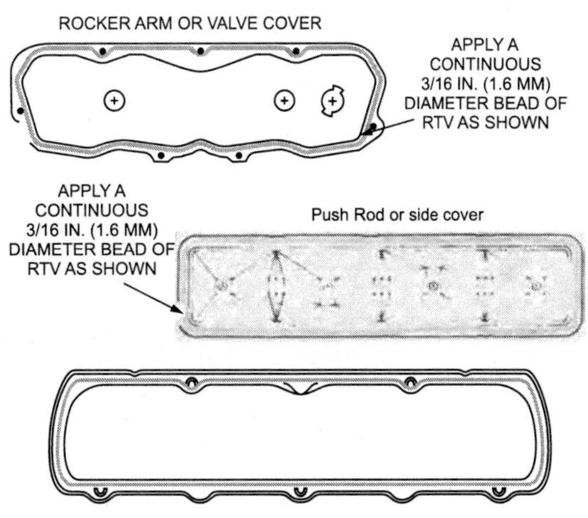

Fig. 2.36 *RTV or silicone sealer can be used instead of a cork or rubber gasket. It will seal properly if parts are clean before assembly. Note size of bead run around covers. (Chevrolet Div. General Motors)*

7. After fitting the gasket parts in place, screw all bolts by hand to ensure proper part alignment and threading of fasteners and check bolt lengths.
8. When more than one bolt is used to hold a part, tighten each bolt only a small amount at a time. Tighten one to about half of its torque spec, then the others in a crisscross manner. Tighten to 3/4th torque and to full torque. Then re-torque to full specs again.
9. Use the manufacturer recommended or a basic crisscross pattern when tightening parts having multiple fasteners. This is required to ensure even gasket compression and sealing.
10. Apply only the specified torque and do not over-tighten the fasteners. It is easy to tighten the bolts enough to dent sheet metal parts and smash or break gaskets.

2.13 ADHESIVES

Adhesives are used in many places in a modern automobile. They are used to secure external mouldings and emblems, hold weather strippings in place, and attach interior trim pieces. The properties of adhesives used in automotive practice vary, depending on the application. Some adhesives dry hard, while others remain pliable when dry.

2.14 SEALERS

A *sealer* is usually coated on a gasket to help prevent leakage, and to hold the gasket during assembly. Numerous sealers with different properties and designed for different uses. Always read the manufacturer's label and a service manual before selecting a sealer, *Fig. 2.37.*

Hardening sealers are used on permanent assemblies such as fittings and threads, and for filling uneven surfaces. They are usually resistant to most chemicals and heat.

Non-hardening sealers are for semi-permanent assemblies: cover plates, flanges, threads, hose connections, and other applications. They are also resistant to most chemicals and moderate heat, *Fig. 2.35.*

Shellac is a non-hardening sealer. It is gummy, sticky substance that remains pliable. It is frequently used on fiber gaskets as a sealer and to hold the gasket in place during assembly.

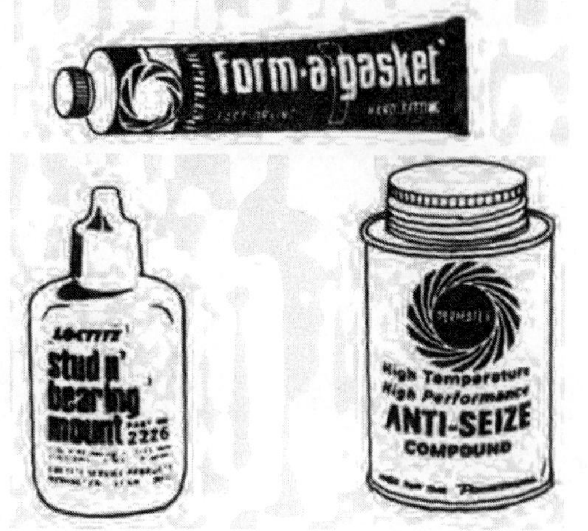

Fig. 2.37 *Hard setting sealer may be used on threaded fittings. Anti-seize compound is used when bolt rusting and corrosion could prevent removal. Locking compound can be used when fastener must not unscrew or loosen in service. (Permatex)*

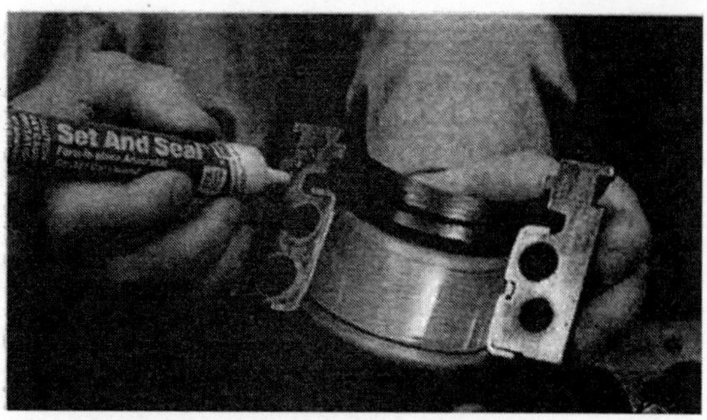

(a) Applying silicone sealer to rear main bearing cap to prevent oil leakage.

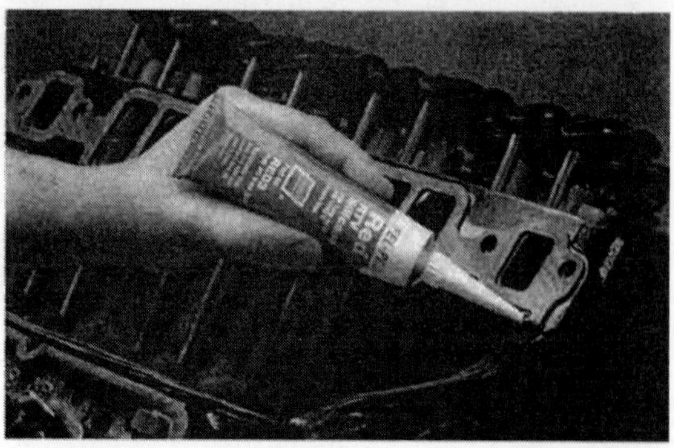

(b) Whenever two gasket ends meet, a dab of silicone sealer should be added at this point to prevent oil.

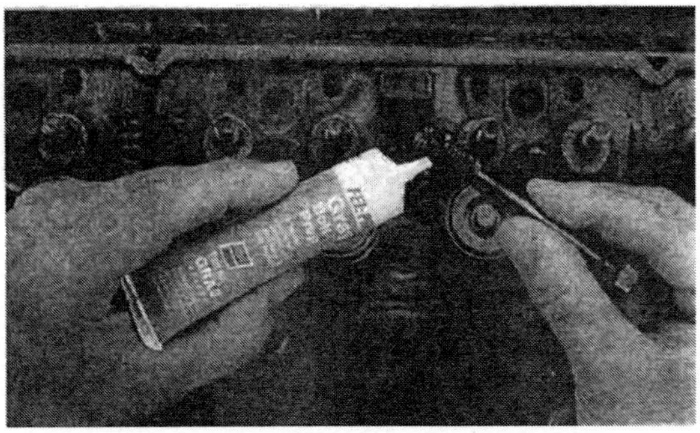

(c) *Always apply a sealer to threads of a cylinder head bolt that extends into a water jacket, or coolant will seep past threads*

Fig. 2.38 *Other uses of silicone sealer. (Pel-Pro, Inc.)*

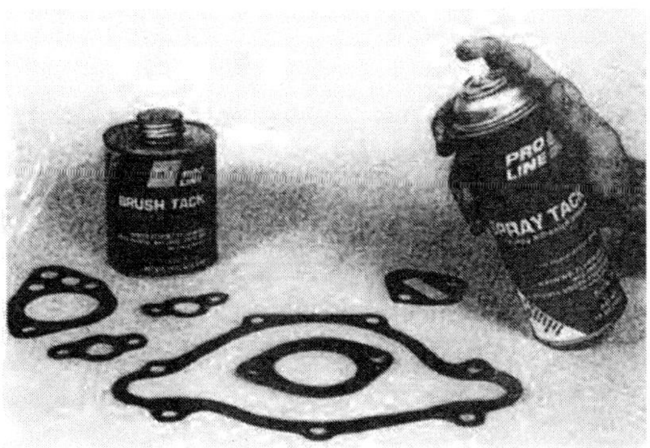

Fig. 2.38 (*d*) *Gaskets prevent leakage between parts. Various sealers and adhesives can be used with gaskets, primarily to hold the gaskets in place during part assemblies. (Pel-Pro Gaskets)*

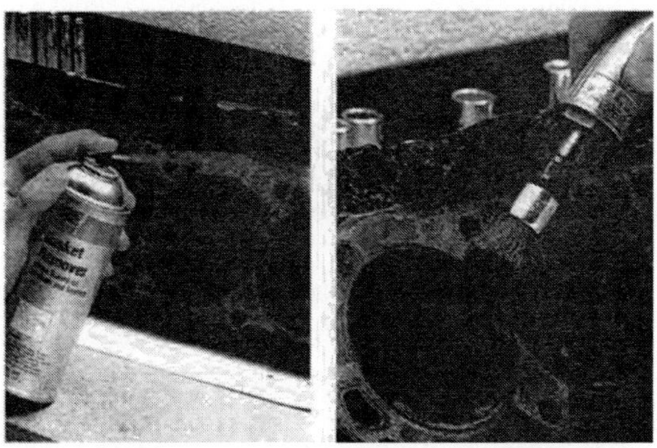

Fig. 2.38 (*e*) *Before installing a new gasket, the mating surfaces must be cleaned of all old gasket material.*

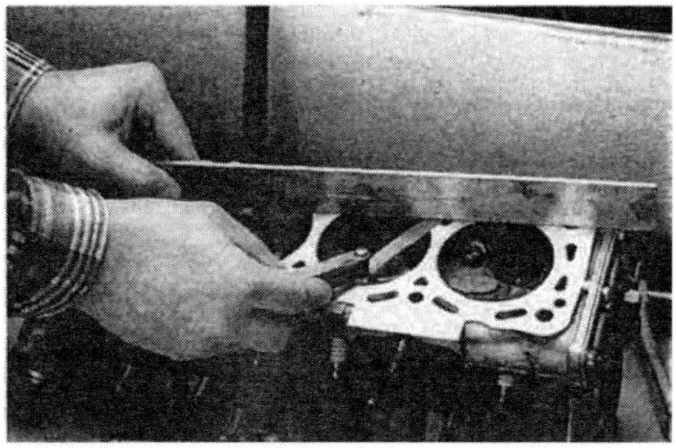

Fig. 2.38 (*f*) *After cleaning the mating surfaces of bare metal, the surface must be checked for warpage. Place a feeler gauge between head and straightedge. (Pel-Pro, Inc)*

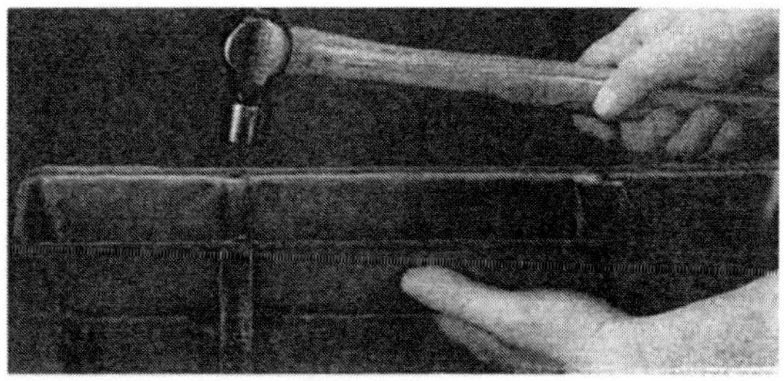

Fig. 2.38 (*g*) *If valve cover or oil pan flanges are not flat, they must also be made flat.*

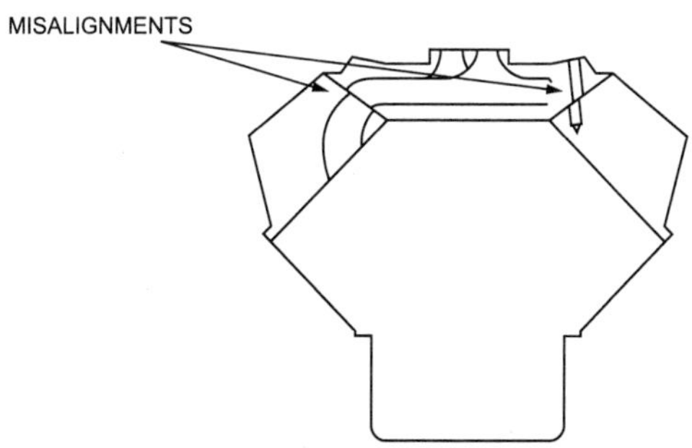

Fig. 2.38 (*h*)

Formed-in-place Sealants

Formed-in-place sealants can be used in place of some preformed conventional fiber or rubber gaskets. When selecting a form-in-place gasket, refer to a manufacturer service manual. Scrape or wire brush gasket surfaces to remove all loose material. Check that all gasket rails are flat. Using a shop rag and solvent, wipe off oil and grease. The sealing surfaces must be clean and dry before using a form-in-place sealant gasket. Two common types of form-in-place sealants are: room temperature vulcanizing (RTV) sealer, also known as silicon sealer, and anaerobic sealer.

RTV or Silicone Sealer

RTV or silicone sealer cures from moisture in the air. It is clear in its pure form. However, a dye is sometimes added to make the sealer black, red, orange, blue, or copper coloured. RTV is either an acid or an amine base. When RTV that is of the acidic formula is heated, it forms vapours that can interfere with the operation of the oxygen sensor. Use only oxygen sensor safe sealers. RTV can be used in place of paper, cork, or rubber gaskets. However, RTV cannot be used:

- In place of an intake manifold gasket.
- In place of head gasket.
- On carburetors.
- On automatic transmissions.

RTV sealer normally comes in a tube. Depending upon the brand, it can have a shelf life from one to two years. Always check the package for expiration date before use. If too old, RTV will not cure (harden) and seal properly.

RTV or silicone sealer should be applied in a continuous bead approximately 3 mm to 1.6 rnm in diameter, *Fig. 2.36*. All mounting holes must be circles. Uncured RTV may be removed with a shop rag.

To prevent oil leaks, a bead of RTV should be placed on the rear main bearing cap, *Fig. 2.38* (*a*). It can be used where two gaskets or a gasket and seal meet, *Fig. 2.38* (*b*). Some cylinder head bolts extend through a water jacket. The threads of these bolts must be coated with RTV or a suitable sealer, *Fig. 2.38* (*c*). Otherwise, coolant will seep past the threads and into the crankcase.

Components should be assembled and torqued in place while the RTV is still wet to the touch, i.e., within about 10 minutes. The use of locating dowels is often recommended to prevent the sealing bead from being smeared. If the continuous bead of silicon is broken, a leak may result.

Anaerobic Sealer

Anaerobic sealer is similar to RTV, but it can cure in the absence of air and is designed for tightly fitting, thick parts. This type of sealer can be used as a thread locking compound or a retaining compound. It can also be used between the machined surfaces of rigid metal castings. It is used between two smooth, true surfaces, not on thin, flexible flanges. Anaerobic sealer should be applied sparingly. Use 1.5 to 2 mm diameter bead on one gasket surface. Ensure that the sealer surrounds each mounting hole. Typically, torque the bolts within 15 minutes of application.

2.15 SEALS

The purpose of any seal is to prevent a liquid from flowing into areas where it is not wanted. They prevent leakage between a stationary part and a moving part. A seal allows the shaft to rotate or slide inside the stationary mating part without fluid leakage, *Fig. 2.39*. A rear axle shaft seal prevents rear axle oil from contaminating the brake linings. Valve seals prevent the oil from entering the engine combustion chambers through the valve guides. Seals are commonly made of synthetic rubber on a metal body. Seals can be made of synthetic rubber, nylon, Teflon, or steel.

There are several reasons for seal leaks. The seal may be installed backwards; it may be forced out of location, or might have become brittle on prolonged service. In some cases a bushing or bearing may be badly worn and cannot effectively control the oil flow across it. The flow of oil to the seal is increased, which the seal may be unable to handle, and some of the oil gets past the seal, even when the seal is otherwise good.

O-ring Seals

O-rings are doughnut-shaped seals that have a round cross-section. An O-ring seal is a stationary type seal that fits into a groove between two mating parts, *Fig. 2.40*. The groove holds the O-ring in place. O-rings are used where there is no rotational or axial movement of any kind. When the parts are assembled, the synthetic rubber seal is partially compressed and forms a leak-proof joint. O-ring is usually used where a connection is made with a hose or tube fitting, *Fig. 2.41*. They are generally made of neoprene.

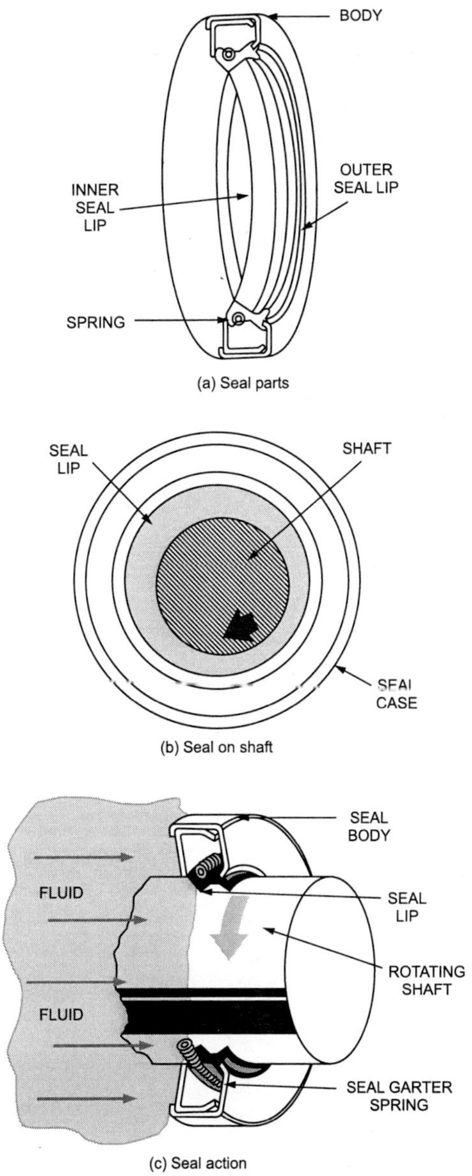

Fig. 2.39 *Seal prevents leakage between rotating shaft and stationary housing or cover. (Chrysler & Caterpillar)*

Normally, O-ring seals should be coated with system fluid (engine oil, diesel oil, or type fluid used in component). This helps the parts slide together without scuffing or cutting the seal.

Normally, sealants are not used on O-ring type seals. Check the shop manual for seal installation, when in doubt.

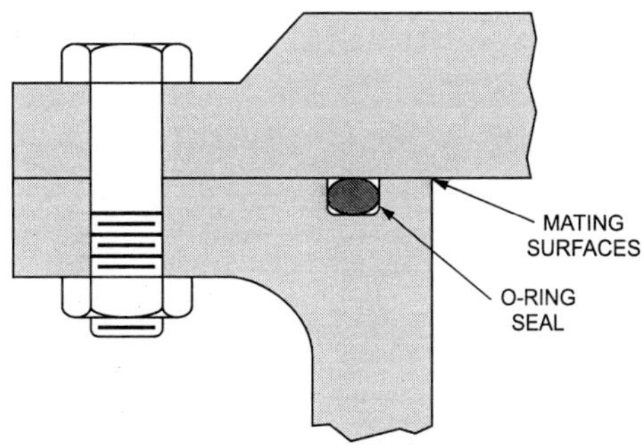

Fig. 2.40 *O-ring seal is usually stationary type seal between parts.*

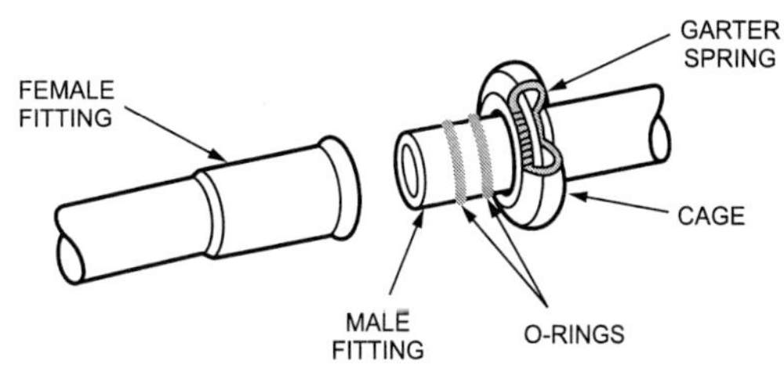

Fig. 2.41 *O-rings are used where there is no axial or rotational movement. (Ford Motors)*

Square-cut Seals

Square-cut seals are doughnut-shaped seals that have a square cross-section, *Fig. 2.42*. The square-cut seal is made of neoprene and is used where there is slight axial movement and no rotational movement between mating parts.

Sealing Rings

Like square-cut seals, *sealing rings* are doughnut shaped and have a square cross-section. However, sealing rings are either made of Teflon or steel. A Teflon sealing ring, *Fig. 2.43 (a)*, is distinguished by an angle cut. While the Teflon seal can conform to irregularities, the disadvantage is that metal particles can become embedded in the sealing surface, reducing its effectiveness. The steel sealing-ring is distinguished by the ends that are hooked together, *Fig. 2.43 (b)*. Sealing rings are used where there is axial and rotational movement.

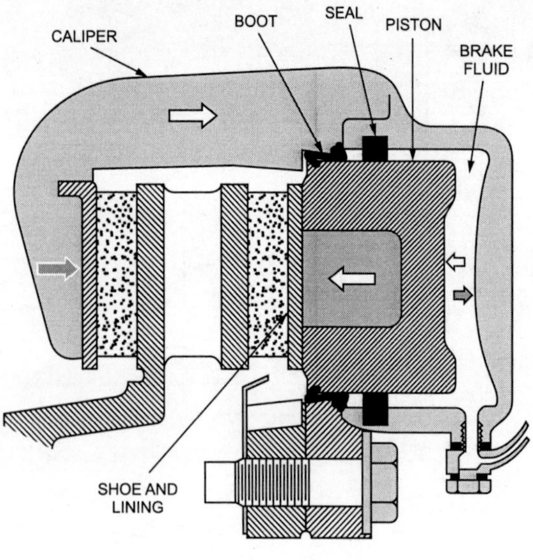

Fig. 2.42 *A square-cut seal is used where there is only axial movement, like a disc brake caliper. (Chrysler)*

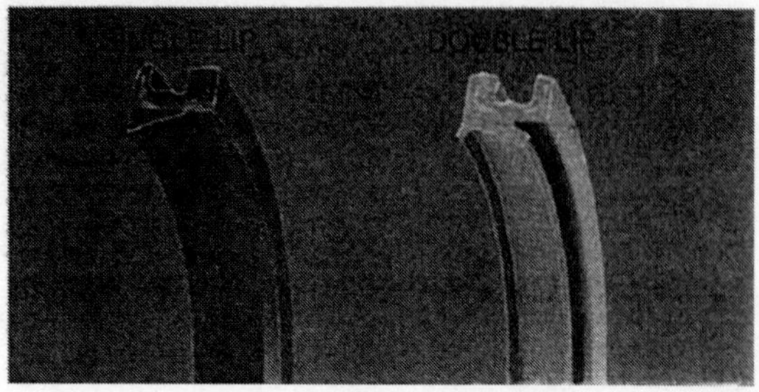

(a) Single and double lip seals. Double lip seal prevents dirt from tearing inner seal.

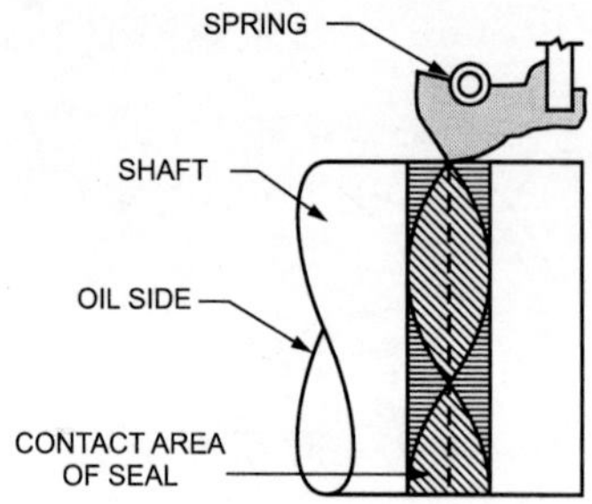

(b) Lip seal is used where there is rotational movement.

Fig. 2.44 *Shaft seals. (Pel-Pro, Inc. C-R Industries)*

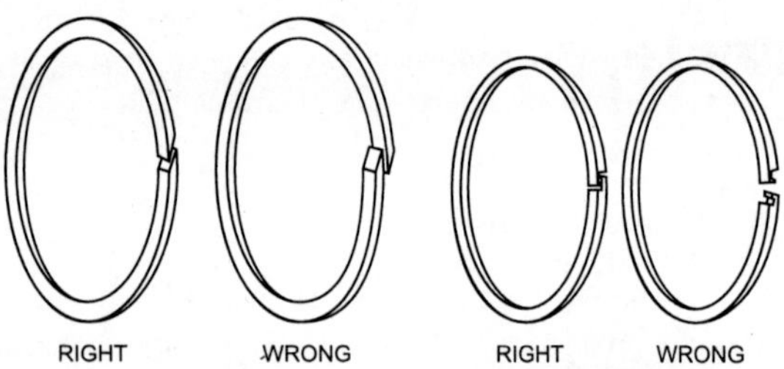

(a) Teflon sealing rings have an angle cut.

(b) Metal sealing rings have an interlocking feature.

Fig. 2.43 *Different types of sealing rings. (GM Corp.)*

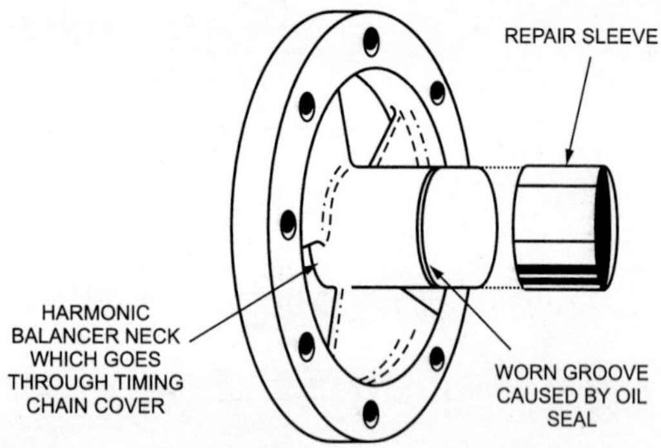

Fig. 2.45 *Groove caused by seal must be reconditioned*

Shaft Seals

Shaft seals, or *lip seals, Fig. 2.44 (a),* are normally made of neoprene. Shaft seals are used to seal against a rotating shaft, *Fig. 2.44 (b).* The shaft seal prevents the oil or lubricant from leaking out while the shaft rotates. It also prevents dirt from entering the lubricant. With prolonged service the seal can cut a groove into the shaft, as under the lip there is no lubrication, *Fig. 2.45.* When this happens the shaft must be reconditioned or replaced, as the oil will work its way under the seal at the groove.

Servicing Seals

Remember to do the following procedures when working with seals:

1. Inspect seal for any leakage before disassembly. If a seal is leaking, there may be other problems besides a defective seal, e.g., a scored shaft, misaligned seal housing, or damaged parts. Leakage needs careful inspection after disassembly.

2. Pry out the old seal without scratching the seal housing. If needed, use a special seal puller.
3. Inspect the shaft closely where it contacts the seal. It should be smooth and flat. File off any burrs that could cut the new seal. A badly worn shaft will require polishing, a shaft seal repair kit, or replacement, *Fig. 2.46*.
4. Compare the old seal to the new seal. Hold them next to each other. Both the inside diameter (ID) and outside diameter (OD) should be the same. To verify the inside diameter for accuracy, slip the seal over the shaft. It should be snug fit over its location.
5. Coat the outside of the seal housing with recommended sealer. Lubricate the inner lip of the new seal with system fluid. Then, install the seal with the sealing lip facing the inside of the part. If installed backwards, a large oil leak will result. Check that the seat is squarely and fully seated in its bore.

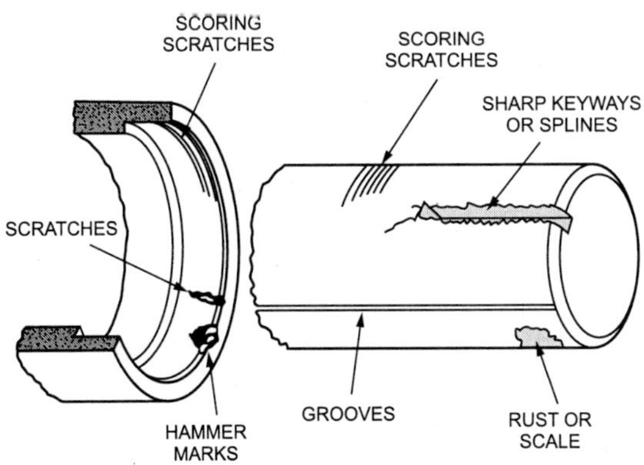

Fig. 2.46 *When servicing a seal, check for these kinds of problems. (Federal Mogul)*

2.16 HOSES AND LINES

When servicing an engine, a mechanic will have to service a number of hoses and lines. It is important to handle them properly otherwise, dangerous or damaging leaks can result. *Fig. 2.47* shows some of the types of lines and hoses found in an engine compartment.

Diesel injection lines can contain over 56,000 kPa pressure. Never attempt to remove a diesel injection component with the engine running. Fuel could squirt out to cause blindness, or puncture skin causing blood poisoning, can even be fatal. Always wear eye protection when working around a running diesel engine.

Many gasoline injection systems retain pressure after the engine has been shut off. Relieve system pressure at the pressure test fitting before disconnecting fuel lines.

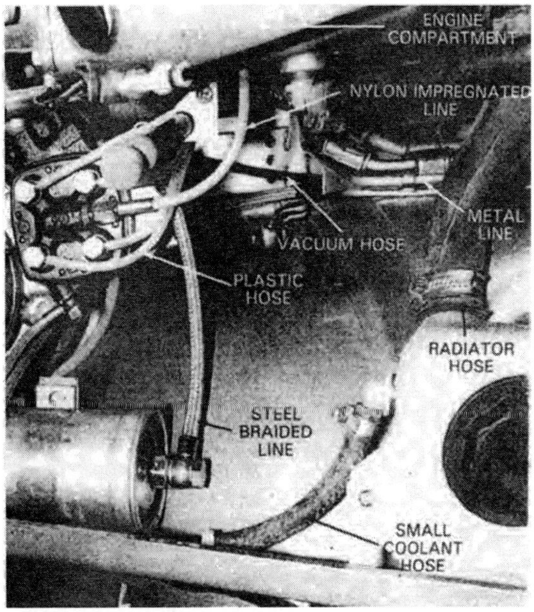

Fig. 2.47 *Few types of hoses found in engine compartment. All these hoses must be serviced properly when servicing the engine. (Saab)*

Metal Lines

Metal lines or tubing are used as fuel lines, especially from the fuel tank to the engine. They are normally double-wall steel tubing. Some of the tools needed to replace a section of metal line are shown in *Fig. 2.48.* (*a*) to (*c*). Steps followed to service a fuel line are illustrated in *Fig. 2.49* (*a*) to (*g*).

- Cut off the old line and a section of new line while progressively tightening the thumbscrew, *Fig. 2.49* (*a*).
- Use a reamer to remove metal burrs from inside the line, *Fig. 2.49* (*b*).
- Bend the line to the shape of the old line using a bender, bending spring or other type of bending tool, *Fig. 2.49* (*c*) and (*d*).

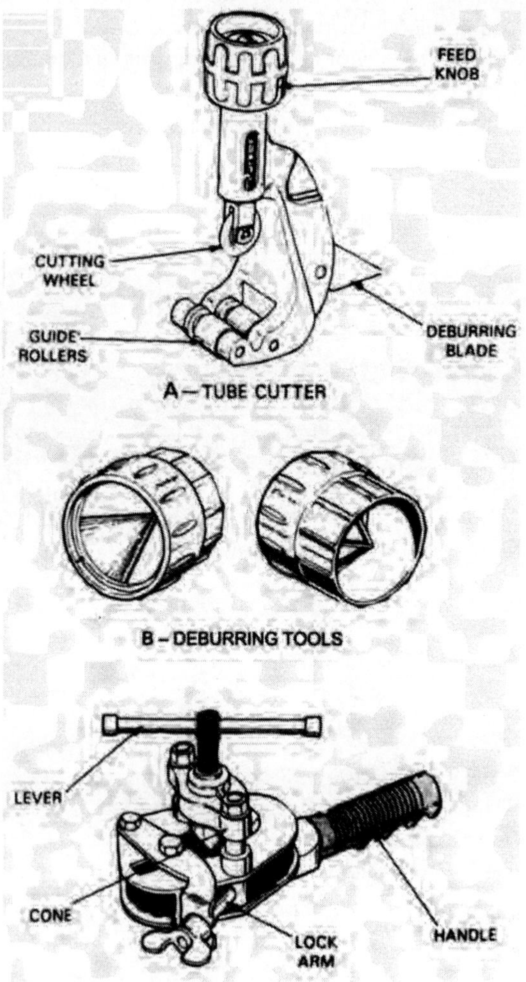

Fig. 2.48 *Basic tubing tools. (Parker)*

- Form a flare on the line, using a flaring tool, as shown in *Fig. 2.49* (*e*) and (*f*). Form the tubing inward with the flaring tool adapter.
- Then force the cone into the end of the line to produce a double-lap.
- A fitting nut can then be used on the line.
- Blow line before use.

Hoses

Hoses, Fig. 2.50, are used on an engine to make a flexible connection between two fittings. They carry fuel, coolant, oil, air, and other fluids to the engine. They must be serviced properly to avoid engine failure.

Hose clamps secure hoses tightly to their fittings. The three basic types of hose clamps are shown in *Fig. 2.51.*

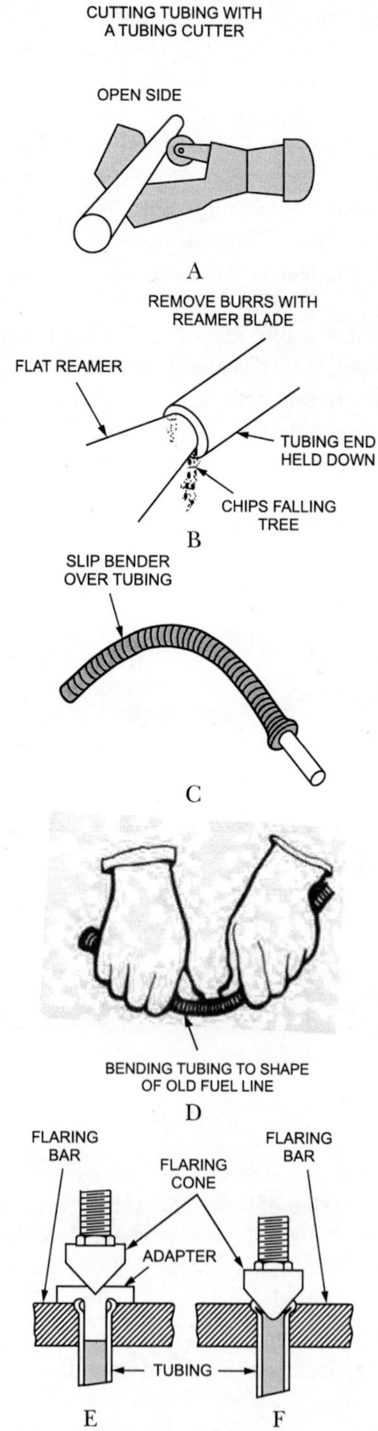

Fig. 2.49 *(Contd.)*

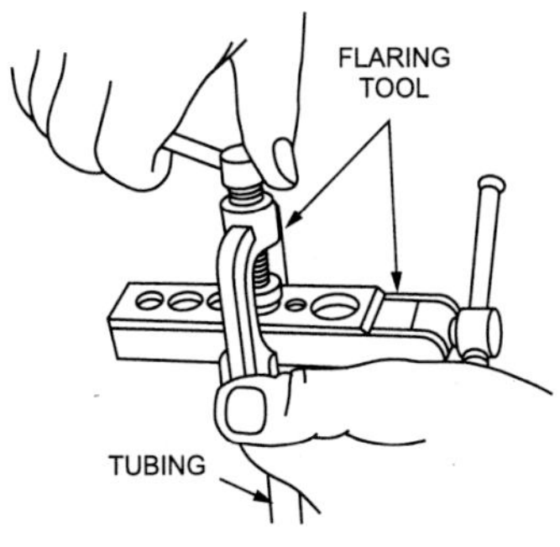

G

Fig. 2.49 *Basic steps for forming flare on end of line. This permits use of tube fitting. (Chrysler)*

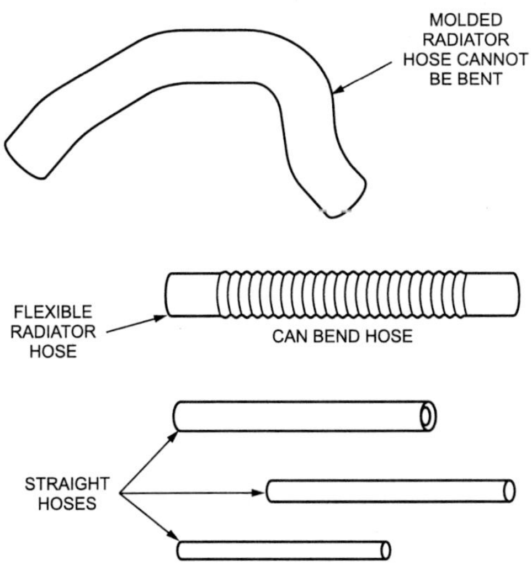

Fig. 2.50 *Hose types found in engine compartment.*

Hose Inspection

To check a hose, visually inspect and squeeze it. If you see cracking of the outer rubber layer or if the hose has hardened (aged) or softened (fluid contamination), install a new hose. Do this for the radiator hoses, heater hoses, fuel system hoses, vacuum hoses, etc.

Line and Hose Service Rules

Keep the following rules in mind when working with lines and hoses:

1. To avoid fuel spraying on you and on the hot engine, place a shop rag around the fuel line fitting during removal.

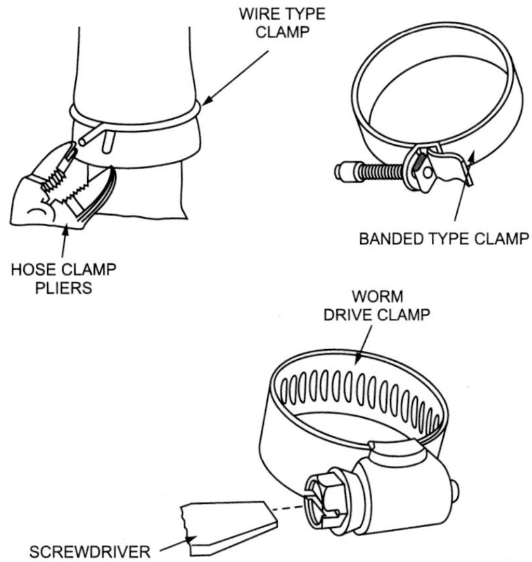

Fig. 2.51 *Clamp types for hoses. (Mopar)*

2. Use a flare nut or tubing wrench. Two wrenches may be needed where two fittings join, *Fig. 2.52*. Do not bend or kink plastic fuel lines.
3. Only use recommended double-wall steel tubing for fuel lines. Never use copper tubing.
4. Make smooth bends when forming a new line. Use a bending spring or bending tool.
5. Form double-lap flares on the ends of fuel lines. A single-lap flare is not recommended for fuel lines.
6. Reinstall all line hold-down clamps and brackets. If not properly supported, the line may vibrate and eventually fail.
7. Route all lines and hoses away from hot or moving engine parts. Always check clearance after line or pipe installation.
8. Torque fittings when recommended in the service manual, *Fig. 2.53*.
9. Only use the recommended hoses for the application. If the wrong type is accidentally used, the fluid can chemically attack and rapidly ruin the hose. A dangerous and damaging leak could result.
10. Check the condition of all hoses carefully before reinstallation, *Fig. 2.54*.
11. Cut hoses off squarely, *Fig. 2.55*.
12. Ensure that a hose fully covers its fitting or line before installing the clamps. System pressure could force the hose off if it is not properly installed.

13. Remove the vacuum hoses carefully to avoid breaking the plastic fittings, *Fig. 2.56 (a)*.
14. Label hoses if their routing is confusing, *Fig. 2.56 (b)*.
15. Finally check all the fittings to ensure there are no leaks.
16. Start the engine and carefully inspect all the connections.

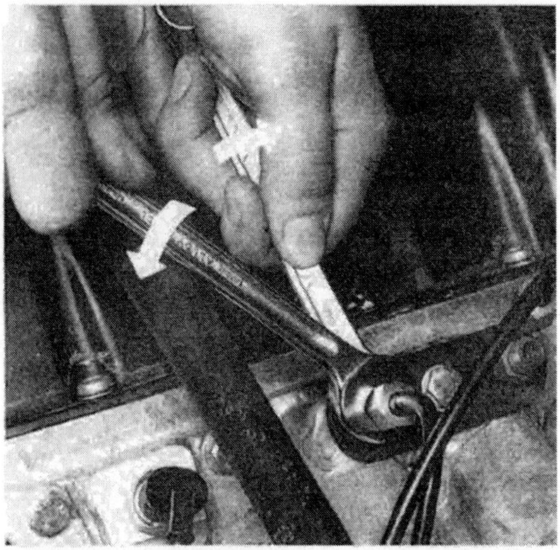

Fig. 2.52 *Often, two wrenches are needed to remove fitting when two fittings are joined. This is necessary to keep both fittings from turning and twisting off line. (Saab)*

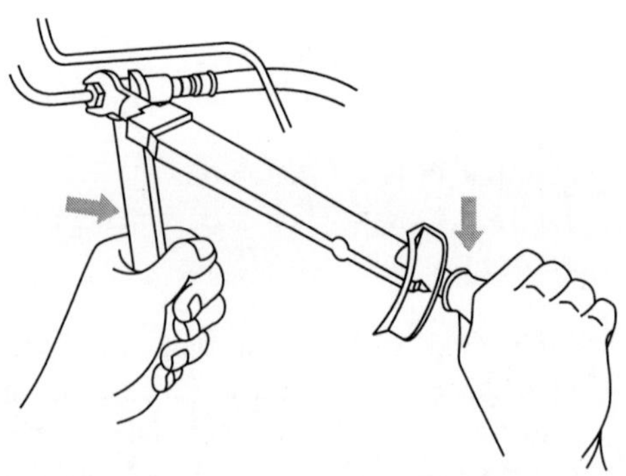

Fig. 2.53 *When the procedure described in service manual recommends, torque line fittings. This is particularly important on high pressure injection lines.*

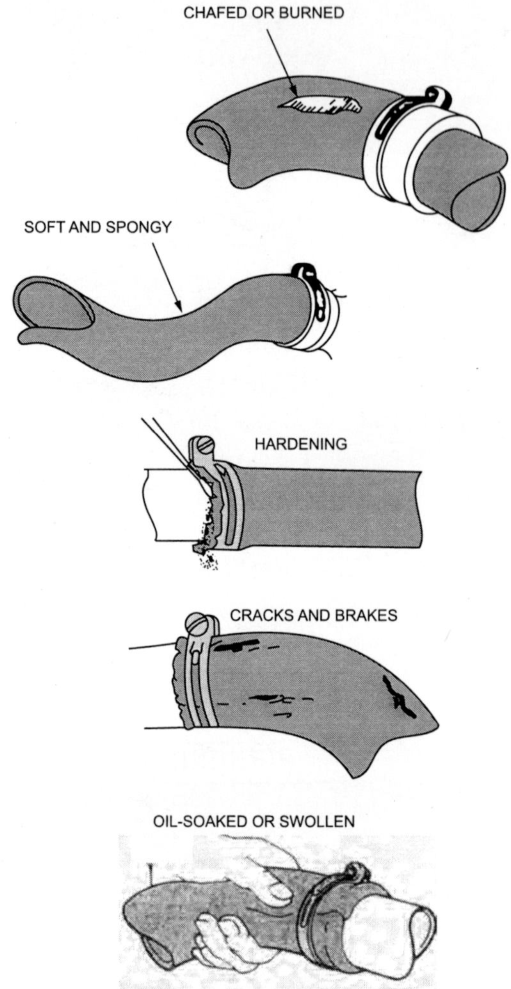

Fig. 2.54 *Inspect hoses for these problems when working on engine. (Gates Rubber Co. & Ford Motors)*

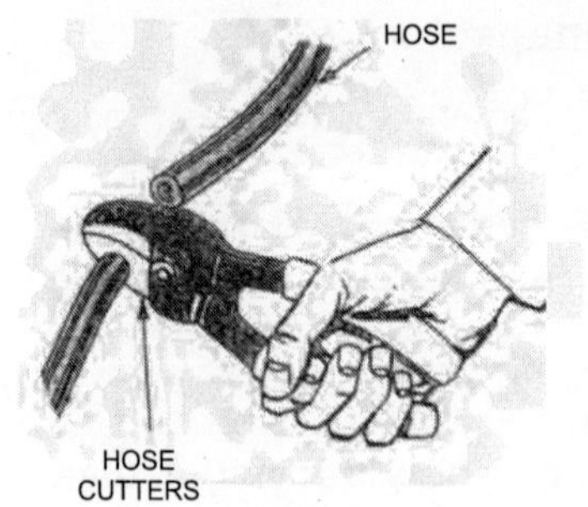

Fig. 2.55 *Use special cutter to cut hose off squarely. (Belden. NAPA)*

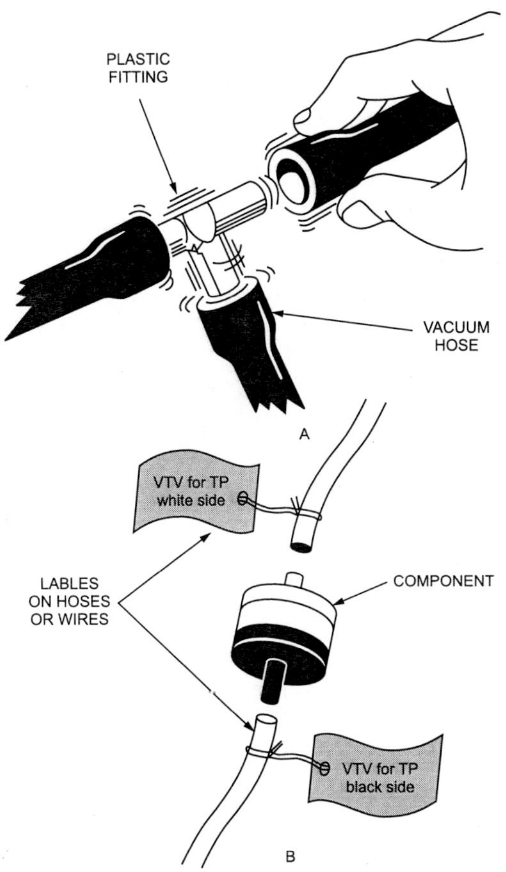

Fig. 2.56 *A - Be careful not to break plastic fittings when disconnecting vacuum hoses from engine. Grasp fitting and twist hose while pulling. B - Tag or label hoses and wires with tape on engine to avoid confusion during reassembly. (Toyota Motors)*

2.17 BELTS

Many types of belts are used to operate the water pump, alternator, power steering pump, air conditioning compressor, cooling fan, air injection pump (sometimes), and auxiliary shaft. One or more belts may be used to operate these components, *Fig. 2.57* and *2.58.*

A conventional V-belt, in cross-section, looks like the letter "V". Usually, two, three, or four V-belts are used on an engine. A *ribbed belt* is more flat in shape and has small ridges that are formed on the inside diameter of the belt, *Fig. 2.58*. This is a comparatively newer design. A ribbed belt can handle more torque than a comparable V-belt and only one or two ribbed belts are needed to operate all engine accessory units.

Fig. 2.57 *Note arrangement of drive belts on front of this engine. (Ford Motors)*

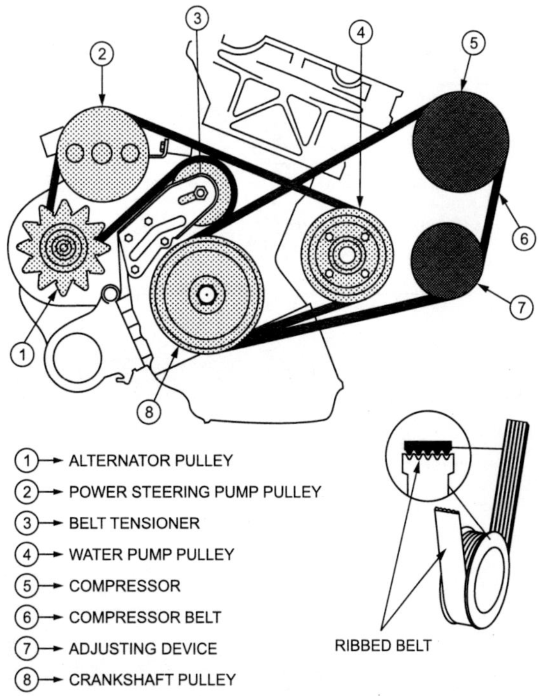

Fig. 2.58 *Ribbed belt is newer design that will handle more load. Note how only two belts are needed. (Saab)*

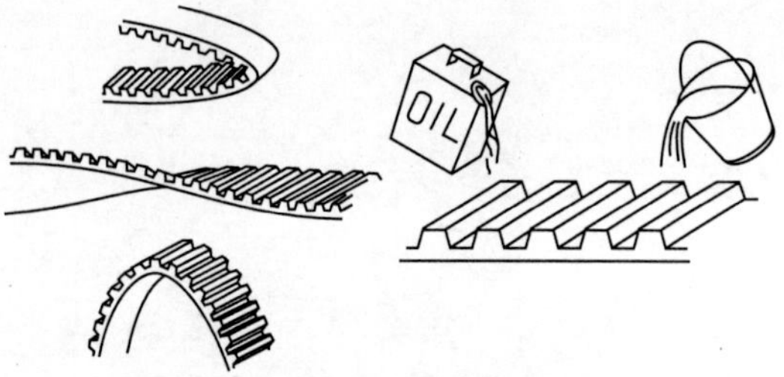

A — Do not bend, twist, or turn belt excessively. Oil and water will deteriorate belt. Fix all engine leaks.

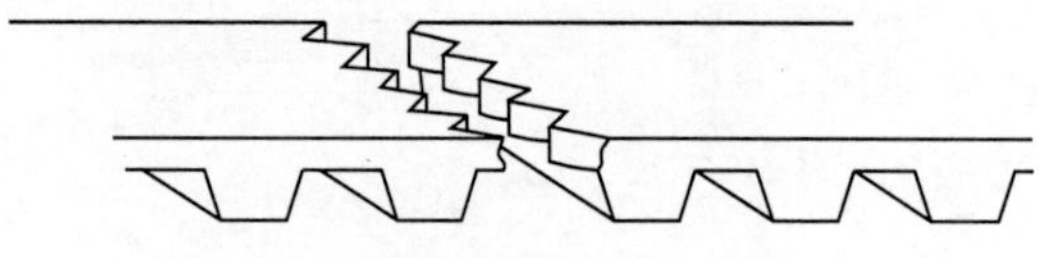

B — Belt breakage may be caused by sprocket or tensioner problem. Check these parts before installing new belt.

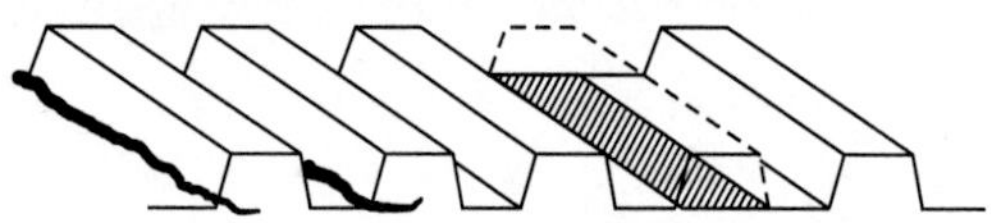

C — If timing belt teeth are missing, check for locked component. Oil pump, injection pump, etc., could be frozen.

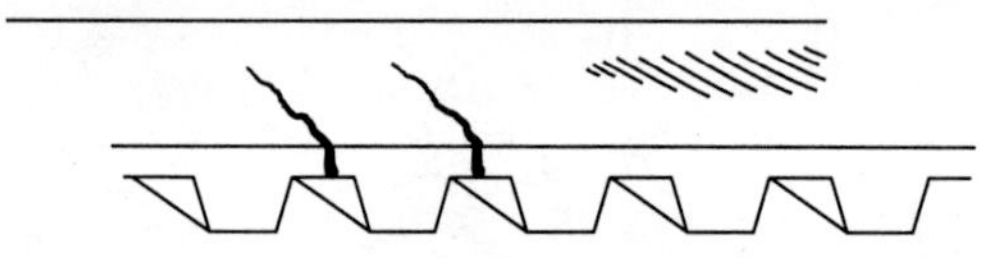

D — If there is wear or cracks on smooth side of belt, check idler or tensioner pulley.

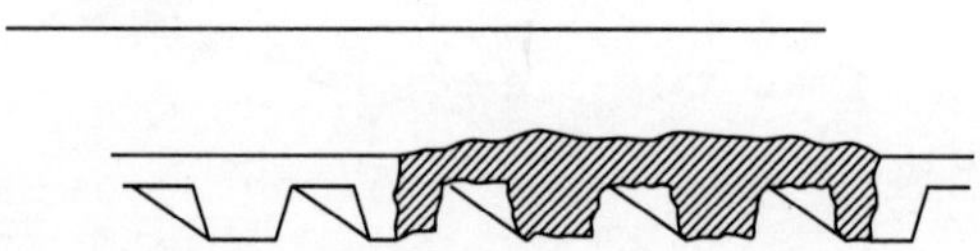

E — With damage or wear on one side of belt, check belt guide and pulley alignment.

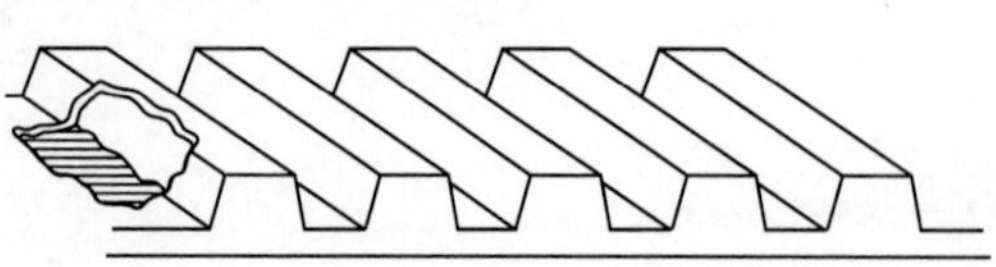

F — Wear on timing belt teeth may be caused by timing sprocket problem. Inspect sprockets carefully. Oil contamination will also cause this trouble.

Fig. 2.59 *Check cogged belts for these problems. (Toyota Motors)*

A *cogged belt* has ribs or teeth that operate in corresponding matching ribs in the belt sprockets. This type is commonly used to drive the engine camshaft and auxiliary shaft, *Fig. 2.59.*

Belt Service

Belt service usually includes periodic inspection and replacement if necessary. Inspect the belt carefully to see if any cracks or fraying is present. Twist the belt and check the underside. Look for the problems shown in *Figs. 2.59* and *2.60.*

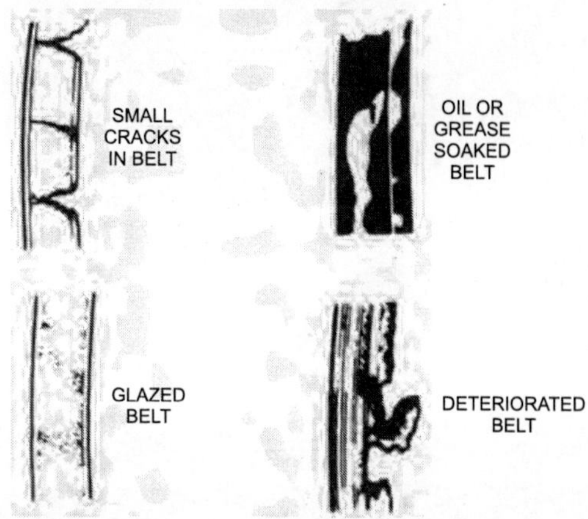

Fig. 2.60 *Carefully inspect V-belts for any problem. (Sun Electric)*

If any belt problems are found, replace the belt. Follow service manual procedures and tighten the belt to specifications. Usually, the alternator belt should be tightened just enough to avoid slippage. Overtightening could cause premature failure of the alternator bearings. Other units, e.g., power steering pump, air conditioning compressor, auxiliary drive, operate in oil and can tolerate more belt tightening, *Fig. 2.61.*

2.18 BEARINGS

There are four types of bearings used in an engine: friction bearing, roller bearing, ball bearing, and needle bearing, *Fig. 2.62.* Needle and friction bearings are commonly used in an engine. Roller and ball bearings are also used frequently in an automobile.

A *friction bearing* is a plain bearing that has two smooth, sliding surfaces. This type is commonly used as engine main, connecting rod, and camshaft bearings.

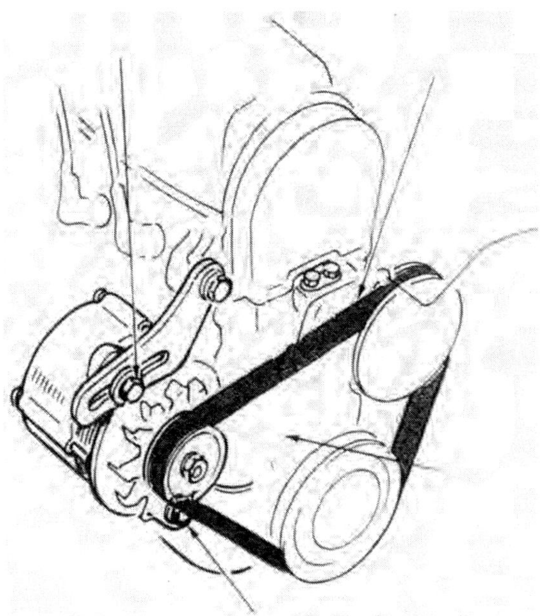

Fig. 2.61 *Tighten belts to specs. Alternator belt should not be as tight as other belts because bearings could fail due to Overtightening. (Honda Motors)*

Antifriction bearings use balls, rollers, or needle bearings between the two moving surfaces. A rolling, instead of sliding, action reduces friction in this type of bearings. This type bearing (needle type) can be used on roller lifters and sometimes the rocker arms. They are also found in engine accessory units, e.g., alternator, pumps, etc.

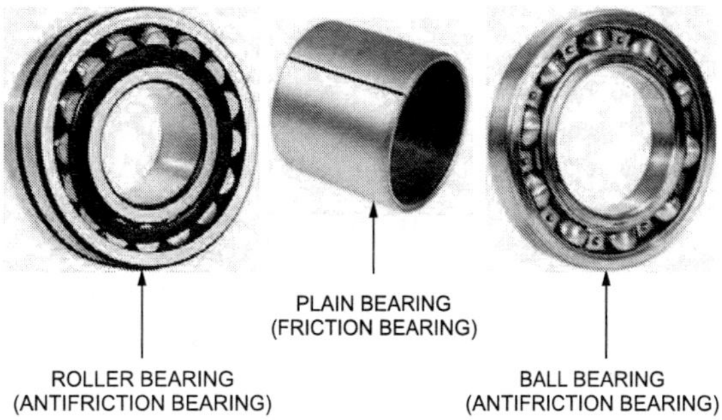

Fig. 2.62 *Basic types of bearings. Friction bearings are most common ones found in engine. A few parts, however, use needle bearings.*

THEORY QUESTIONS

1. Define the term fastener.
2. What is a cap screw?
3. Turn —————————— clockwise to tighten.
4. When a bolt threads into a part, it is called a ———.
5. Bolt size is measured across the inside diameter of threads. *True or False?*
6. List the three types of bolt threads.
7. This refers to the strength of a bolt.
   (*a*) Tensile strength or grade
   (*b*) Tensile strength and static strength
   (*c*) Static strength and grade
   (*d*) None of the above applies.
8. Largest diameter of a screw is known as the:
   (*a*) pitch diameter (*b*) major diameter
   (*c*) minor diameter (*d*) none of the above.
9. Explain the difference between a bolt and a screw.
10. Most bolts have a:
    (*a*) round head (*b*) square head
    (*c*) hexagonal head (*d*) octagonal head.
11. What is the main difference between a machine screw and a bolt?
12. Mechanic A says that metric threads can be easily confused for fine threads if not inspected carefully. Mechanic B says that metric threads can be easily confused for coarse threads if not inspected carefully. Who is right?
    (*a*) A only (*b*) B only
    (*c*) Both A and B (*d*) Neither A nor B.
13. A stud has threads on one end. *True or False?*
14. What is stud and where it is used?
15. Describe two methods used to prevent a nut from working loose.
16. How is a castellated nut and cotter pin used?
17. List the two thread designation systems used in automotive practice.
18. Before using a sheet metal screw, it is necessary to tap the hole. *True or False?*
19. Write a thread designation for a bolt with coarse threads that is one inch long and half inch in diameter.

20. Mechanic A says that the fastener's threads should be clean and lightly oiled before torquing. Mechanic B says that the torque wrench should be viewed from an angle when torquing a fastener. Who is right?
(*a*) A only (*b*) B only
(*c*) Both A and B (*d*) Neither A nor B.

21. All of the following are basic dimensions of a bolt *except*
(*a*) bolt length (*b*) thread pitch
(*c*) thread shape (*d*) bolt head size.

22. Mechanic A says that a woodruff key is a common fastener used on an automotive engine. Mechanic B says that a snap ring is a common fastener used on an automotive engine. Who is right?
(*a*) A only (*b*) B only
(c) Both A and B (*d*) Neither A nor B.

23. What is a pitch gauge and how is it used?

24. A ———— fits in a slot and prevents a part from rotating on its shaft.

25. What is a grade marking on a threaded fastener?

26. What is a dowel pin and how is it used?

27. Explain the purpose of a flat washer and a lock washer.

28. What is a helicoil?

29. A die can be used to cut threads on all of the following *except:*
(*a*) Bolts (*b*) Shafts
(*c*) Holes (*d*) Rods.

30. Silicone sealer can be used in place of a head gasket. *True or False?*

31. Stripped threads can:
(*a*) not be repaired
(*b*) be repaired with tap or die
(*c*) be repaired with a helicoil insert
(*d*) either (*b*) or (*c*).

32. A square-cut seal is used where there is:
(*a*) no axial or rotational movement
(*b*) axial movement only
(*c*) rotational movement only
(*d*) both rotational and axial movement.

33. Anti-seize lubricant should be used on the threads of an oxygen sensor. *True or False?*

34. RTV should be applied to the threads of cylinder head bolts that extend into the water jacket. *True or False?*

35. Before installing a new gasket, the mating surfaces must be:
(*a*) cleaned of all old gasket material
(*b*) straightened or surfaced if warped
(*c*) check for warpage
(*d*) All of the above.

36. The two types of snap rings are:
(*a*) split and solid
(*b*) compression and oil control
(*c*) internal and external
(*d*) treaded and press fit.

37. Slip joint pliers are needed to remove and install snap rings. *True or False?*

38. Thread inserts are used to
(*a*) repair damaged bolts
(*b*) replace broken bolts
(*c*) repair damaged internal threads
(*d*) repair damaged external threads.

39. A gasket is used to
(*a*) seal in fuel, oil, or coolant
(*b*) seal out dirt, water and air
(*c*) take up space
(*d*) all of the above.

40. When installing a new gasket, the mechanic should ————.
(*a*) tighten fasteners in crisscross pattern
(*b*) never over-tighten the fasteners
(*c*) coat the gasket with grease
(*d*) Both (*a*) and (*b*).

41. Mechanic A says that friction bearings are commonly used as engine main bearings. Mechanic B says that antifriction bearings are commonly used as engine main bearings. Who is right?
(*a*) A only
(*b*) B only
(*c*) Both A and B
(*d*) Neither A nor B.

42. All of the following are basic rules to follow when working with automotive seals *except:*
(*a*) check the size of the new seal
(*b*) check the shaft or seal housing for damage
(*c*) coat the outside of the housing with sealer before installing a new seal
(*d*) use a hammer and screwdriver to remove the old seal.

❑❑❑

Service Information, Tools and Shop Safety

3

3.1 SERVICE INFORMATION

Service information is needed by service technicians to determine specifications and service procedures and learn about any necessary special tools required.

Owner's Manuals

The owner's manual is normally kept in the vehicle's glove box. It contains non-technical information about the proper operation of the vehicle and contains data on when maintenance tasks should be performed.

Specifications on the types and levels of various fluids that are to be used are described.

1. **Starting** – describes how to start the engine under different conditions. Starting an engine with electronic fuel injection (EFI) is different than starting a carbureted engine.
2. **Towing** – specifies the maximum weight limit that the vehicle can tow behind it. In addition, it contains information on how the vehicle should be towed if it is disabled.
3. **Components, gauges, and accessories** –describes what each of the dash gauges is for and their location. Methods of setting the radio, DVD player, clock, cruise control, air conditioning, and other accessories are explained.
4. **Warranty information** – describes the terms of the manufacturer's warranty.
5. **Service assistance** – provides the addresses and telephone numbers to seek manufacturer's help for vehicle repair. This might require the following information when calling for help:
    - Vehicle identification number, which is located on the vehicle, *Figs. 3.1* (*a*) and (*b*), and the warranty papers.
    - When the vehicle was bought, as on warranty papers.
    - Current mileage.

Service Manuals

Factory and aftermarket service manuals contain specifications and service procedures. While factory service manuals cover just one year and one or more models of the same vehicle, most after market service *Fig. 2.49* manuals cover multiple years and/or models in one manual. Most manuals include:

- Capacities and recommended specifications for all fluids used
- Specifications including engine and routine maintenance items
- Testing procedures
- Service procedures including the use of special tools where needed.

Electronic Service Information

Electronic service information is normally available by subscription and provides access to an internet-site where service manual-type information is available. Most vehicle manufacturers also offer electronic service information to their dealers and to schools and colleges that offer corporate training programs, *Fig. 3.1* (*c*).

VEHICLE IDENTIFICATION NUMBER

Each 1247 Cadillac carries a 17 digit Vehicle identification Number used for license and insurance identification and in general reference to the automobile. The number is located on a plate riveted to the windshield lower frame at the driver's side of the car where it is visible through the windshield. The 1247 Cadillac vehicle identification is decoded as follows:

1 G 6 CD 5 1 8 X H 4 200001

NATION OF ORIGIN – 1-USA

MANUFACTURE – G-GENERAL MOTORS

MAKE
6-CADILLAC

CARLINE/SERIES
CB = Fleetwood D'Elegance
CD = Deville
CH = Limousine
CS = Fleetwood Sixty Special
DW = Brougham
EL = Eldorado
JG = Cimarron
KS = Seville
VR = Allante

Vin Body Code = Body Type
1 = 47 – Coupe 2 Door Notchback
1 = 57 – Coupe 2 Door Notchback
3 = 67 – Coupe 2 Door Convertible
5 = 19 – Sedan 4 Door
6 Window notchback
5 = 23 – Sedan 4 Door Limousine
5 = 23 – Sedan 4 Door Limousine
With Center Partition
5 = 69 – Sedan 4 Door Notchback

PRODUCTION SEQUENCE NUMBER
100001 – Allante
200001 – Deville
500001 – Cimarron
600001 – Eldorado
700001 – Brougham
800001 – Seville

ASSEMBLY PLANT
4 = Lake Orion Michigan
9 = Detroit, Michigan
J = Janesville Wisconsin
U = Hamtramck, Michigan

MODEL YEAR — H = 1947

CHECK DIGIT

ENGINE CODE — SEE BELOW

RESTRAINT SYSTEM
1 = MANUAL BELTS

FWD-C Commercial Body/Chassis
Vin is as Follows:
IGEDZ908XH4200001

Fig. 3.1 *(a) To make sure that you have the correct service manual for the vehicle under service, check the vehicle identification number (VIN). The VIN is a 17 digit code number that contains information about the vehicle, e.g., model year and origin.*

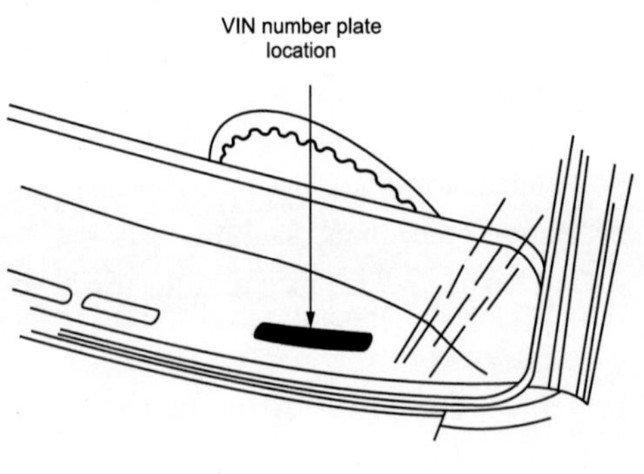

Fig. 3.1 *(b) The VIN has 17 digits and is generally located at the upper left-hand corner of the dash. It can be seen through the windshield. (CM Caddilac Div.)*

Fig. 3.1 *(c) Computerized manuals.*

Technical Service Bulletins

Vehicle manufacturers issue Technical service bulletins (TSBs) to notify service technicians of a problem and provide the necessary corrective action. Technical service bulletins are designed for dealership technicians but are republished by aftermarket companies and made available along with other service information to repair shops and vehicle repair facilities.

Internet

The Internet has opened the field for information exchange and access to technical advice. One of the most useful is the International Automotive Technician's network at www.iatn.net. This is a free site but service technicians need to register to join. A small monthly sponsor fee has to be paid; the shop or service technician can gain access to the archives, which include thousands of successful repairs, and is included in the searchable database.

3.2 COMMON STEPS IN AUTOMOTIVE ENGINE SERVICE

Engine service jobs do vary in nature and complexity. But all jobs require all or some of the following steps:

1. Testing
2. Disassembling
3. Measuring
4. Machining
5. Reinstall serviced parts or replace unusable parts with new parts
6. Reassembling
7. Adjusting.

These steps are performed with the help of hand tools, measuring tools, and power tools. Reinstall means putting back the same part that was removed from a location after servicing as needed. But part that is found to be unusable after inspection, is discarded and is replaced with a new part.

1. Testing

When you encounter a specific trouble that relates to engine, you want to follow a specific procedure to find the cause and correct it. For example, if there is a customer complaint of excessive fuel or oil consumption, then there are definite trouble-diagnosis checks that can be made to pinpoint the cause of trouble. Testing could involve listening to a running engine to detect noises, using test instruments to measure engine vacuum or compression, and analyze lubricating oil and exhaust, etc. The test results tell you what sort of job you have to do.

2. Disassembling

When an engine experiences some performance problem and the service of the defective system or part is necessary to correct the problem. You have to disassemble, or "take apart", the component or subassembly, to get to the problem. Disassembly is also called *teardown*.

3. Measuring

Before starting the service on an engine, you must first determine as to what is wrong with it. Properly cleaning and measuring the disassembled system components usually accomplish this. By various measurements, with the help of measuring tools and gauges, the parts are checked to see they are within or out of specs and whether they can be reconditioned and reused or have to be replaced to restore the various system functions and vehicle performance and solve the complaint. Linear measurements are most common kind of measurement. They are measurements you take in a straight line, in metric units.

There are other ways to measure, too. Sometimes the measurement is done by listening — as when listening to a running engine to pickup noise. When you check the oil in an engine, you measure its level in the crankcase/oil pan. Test instruments are used to measure battery conditions. You measure engine vacuum or compression to measure engine performance. The results of these measurements suggest the sort of job you have to do.

4. Machining

Sometimes you have to remove metal from a part to make it usable again. Using a machine to remove metal is called machining. Frequently, special machines are used in engine reconditioning. Boring, milling, grinding, lapping, honing, welding are some of the machining processes used in automotive repair.

5. Installing New Parts

Inspection and measurements will reveal that some parts are so worn that they need to be discarded. Then new parts must be installed in their place. For example, engine bearings that wear out of specs are replaced with new ones. Even new parts may require machining to make them fit.

6. Reassembly

After service and repair, the reconditioned parts and replaced parts have to be put back together as they were before disassembly. This is called reassembly. All the parts are put back together to make a complete and functional assembly.

7. Adjusting

As a vehicle/engine is operated, parts normally wear. This necessitates adjustments in various system parts from time to time. Adjustments might also be required after a service job. For example, after valve grinding, you install all valve train parts back together. Then you measure the different clearances and check the valve action. If it is not correct then adjustments have to be made.

3.3 SPECIFICATIONS

The term "*specifications or specs*" is frequently used in automotive service and repair shops. The vehicle manufacturers give the right measurement limits in their vehicle service/shop manuals. The specs give you the right measurements for the components and various systems of the vehicle under service. These specs include valve setting, ignition timing, piston clearances, piston-ring clearances, oil clearances for different bearings, gaskets compression, seals, and threaded fastener torques, and hundreds of other measurements.

3.4 SAFETY TIPS FOR USING HAND TOOLS

Although automotive service and repair can be very rewarding it can be dangerous too, *Fig. 3.2.* A technician must be aware of the potential hazards encountered in an automotive repair shop and must know how to minimize these hazards.

All applicable laws regarding the worker safety in a shop should be studied and carefully followed. Safety should be a part of every service procedure.

Following safety tips should be kept in mind while working with hand tools:

1. Always wear safety glasses when using power tools, *Fig. 3.3 (a).*
2. Always pull a wrench toward your body for best control and safety, *Fig. 3.3 (b).* Never *push* a wrench. If you do and a bolt or nut loosens, your entire weight is used to propel your hands

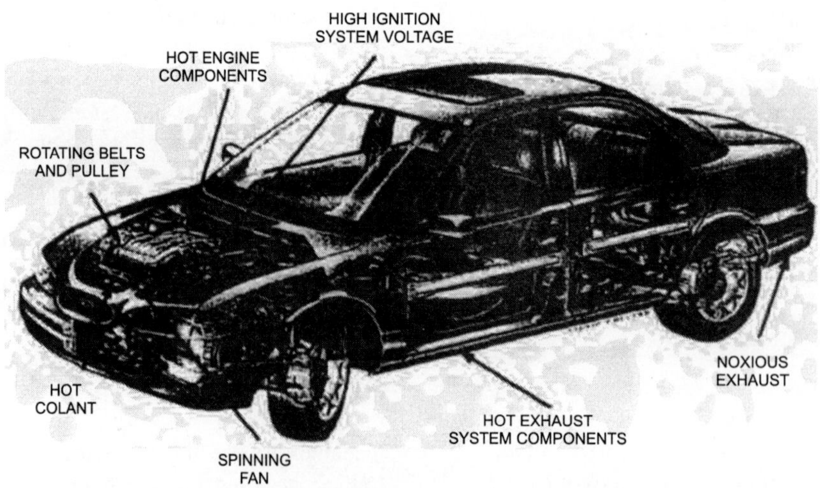

Fig. 3.2 *Automobile service and repair can be dangerous. A few of the most common hazards are identified here. (Ford Motors)*

forward. This commonly results in cuts, bruises, or other painful injury.

3. Always keep wrenches and other hand tool clean and dry to help prevent rust and allow better, firmer grip.
4. Always use a 6-point socket, *Fig. 3.4 (a)*, or box-end wrench to break loose a tight bolt or nut.
5. Use a box-end wrench for torque and an open-end wrench for speed.
6. Never use a pipe extension or other type of "*cheater bar*" on a wrench or wrench handle to increase the applied torque. If more force is required, use a larger tool or use penetrating oil and/or heat on the frozen fastener. Also, if heat is used on a bolt or nut to remove it, always replace it with a new part.
7. Always use the proper tool for the job at hand. In case a special tool is required, use the proper tool and do not attempt to use an alternate tool improperly, *Fig. 3.4 (b)*.
8. Never subject any tool to excessive heat. High temperatures can decrease the tool strength.
9. Never use a hammer on any wrench or socket handle unless you are using a special '*staking face*' wrench, designed for use with a hammer.
10. Replace any tools that are damaged or worn.

3.5 SAFETY TIPS FOR TECHNICIANS

Safe work habits, following safety precautions, can minimize the accidents and injuries, ease the workload, and keep the workers pain free. Some of the safety tips are:

Fig. 3.3 *(a) Wear safety glasses when using power tools.*

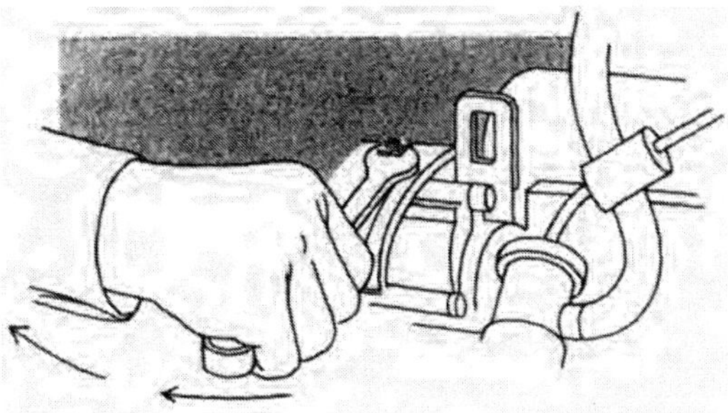

Fig. 3.3 *(b) Always pull on the wrench handle when tightening and loosening nuts and bolts.*

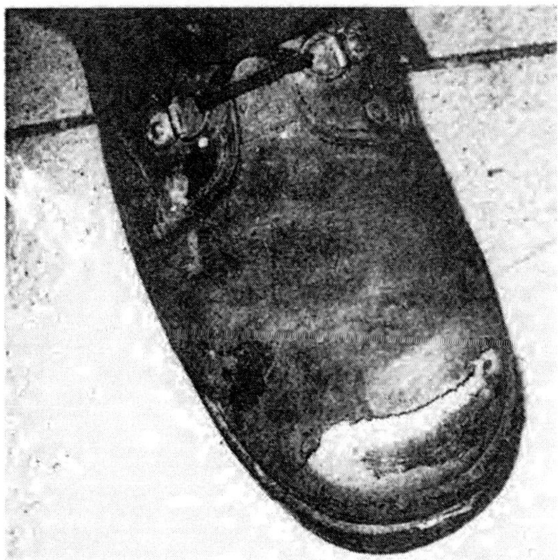

Fig. 3.3 *(c) Steel-toed shoes are a worthwhile investment to help prevent foot injury due to felling objects. Even these well-worn shoes can protect the feet of this service technician.*

Fig. 3.3 *(d) One version of a bump cap is this padded plastic insert that is worn inside a regular cloth cap.*

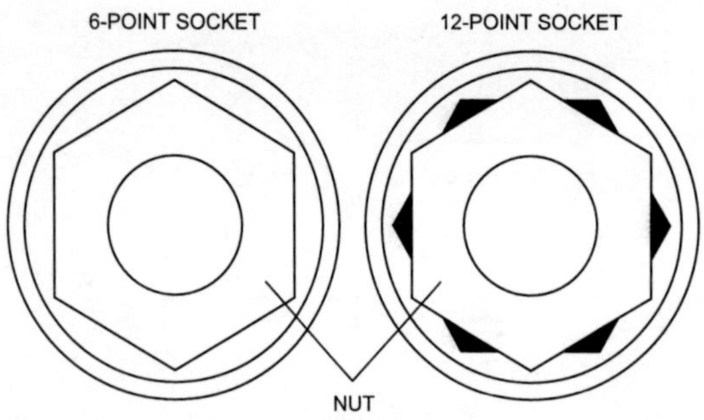

Fig. 3.4 *(a) A 6-point socket fits the head of the bolt or nut on all sides. A 12-point socket can round off the head of a bolt or nut if a lot of force is applied.*

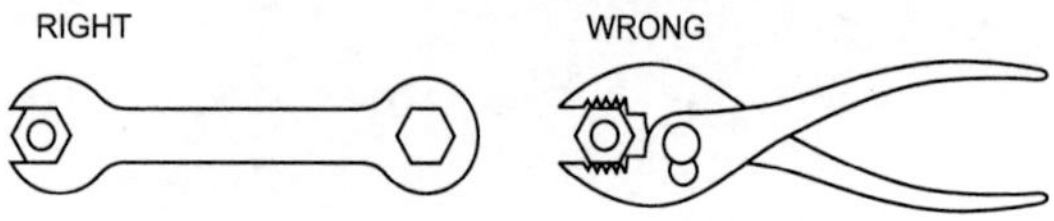

Fig. 3.4 *(b) Use the correct tool for the job.*

1. Standard safety glasses should be worn at all times while servicing any vehicle, *Fig. 3.3(a)*.
2. Wear gloves to protect hands from rough or sharp surfaces. Thin rubber glove should be worn when working around automotive liquids, e.g., engine oil, antifreeze, transmission fluid, brake fluid, refrigerants, or any other liquids that may be hazardous.
3. Technicians should always keep their toes protected with steel-toed safety shoes *Fig. 3.3 (c)*. If safety shoes are not available, then leather-topped shoes should be preferred to canvas or cloth shoes.
4. When working under a vehicle, wear a *bump-cap* or any other headgear, to protect the head against under-vehicle objects and the pads of the hoist *Fig. 3.3 (d)*.
5. Avoid wearing loose clothes, *Fig. 3.5 (a)*.
6. Ear protection should be worn if the sound in the work area requires raising voice to communicate, i.e., sound level higher than 90 dB (decibell).
7. Remove jewelry that may get caught on something or act as a conductor to an exposed electrical circuit, *Fig. 3.5 (b)*.
8. When lifting an object, get a secure grip with solid footing. Keep the load close to your body to minimize the strain. Lift with your legs and arms, not your back, *Fig. 3.5 (c)*.
9. Do not twist your body when carrying a load. Instead, pivot your feet to help prevent strain on the spine.
10. Ask for help when moving or lifting heavy objects.
11. Push a heavy object rather than pull it.
12. Always connect an exhaust hose to the tailpipe of any running vehicle to help prevent the buildup of CO inside a closed garage space. *Fig. 3.5 (d)*.
13. When standing, keep objects, parts, and tools with which you are working between chest height and waist height. When seated, work at tasks that are at elbow height.
14. Take care of your hands. Keep your hands clean by washing with soap and hot water at least 43°C.
15. Use properly grounded three-prong sockets and extension cords to operate power tools, *Fig. 3.5 (e)*. Some tools are only two-prong plugs. Make sure these are double insulated and repair or replace any electrical cords that are cut or damaged to avoid the possibility of an electric shock. When not in use, keep electrical cords off the floor to avoid tripping over them.

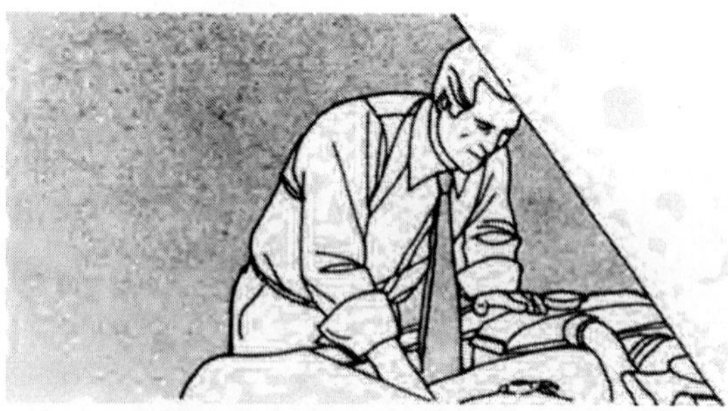

Fig. 3.5 *(a) Wear proper clothing.*

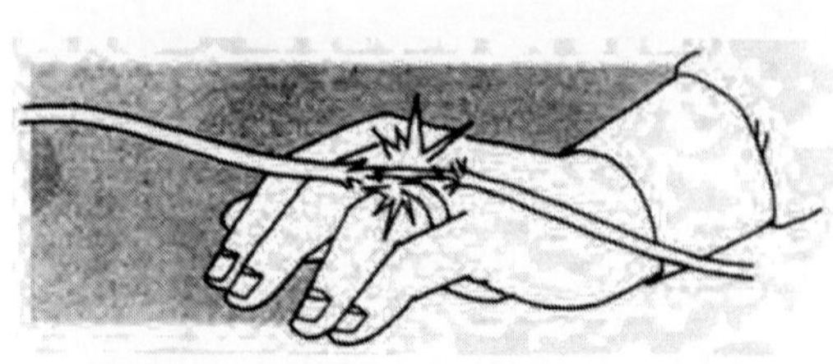

Fig. 3.5 *(b) Do not wear jewelry at-work.*

Fig. 3.5 *(c) When lifting heavy objects, keep your back straight and use your legs to lift the weight*

Fig. 3.5 *(d) Use exhaust hose when engine is running in enclose shop. Poisonous gases will then be released out of the shop. (Chrysler Corp.)*

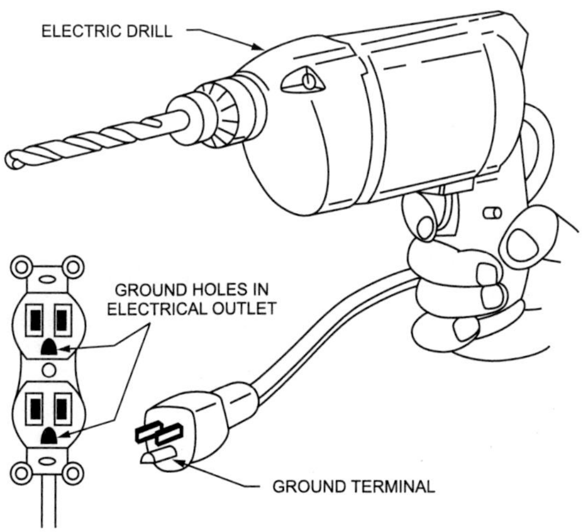

Fig. 3.5 *(e) Electric drill with three-wire cord. The third wire and terminal are to ground the electric-drill motor.*

16. Improper use of an air nozzle can cause blindness or deafness. Compressed air must be reduced to less than 206 kPa. When an air nozzle is used to dry and clean parts, make sure the air stream is directed away from anyone in the immediate area. Coil and store air hoses when they are not in use.
17. All shops must be equipped with a first aid kit and an eye wash station centrally located and kept stocked with emergency supplies.
18. Avoid anyone who does not take shop work seriously.
19. Keep the shop organized. Return all tools and equipment to their proper storage areas. Never lay tools, creepers, or parts on the floor.
20. Never carry sharp tools or parts in pockets as they can easily puncture skin.
21. Work like a professional. Avoid working over excited and too fast because they lead to overlooking a repair procedure or safety rule and result in an accident.
22. Keep equipment guards or shields in place, *Fig. 3.5 (f)*. If a power tool has a safety guard, use it.
23. Use adequate lighting. A portable shop light increases working safety and enhances work speed and precision.
24. Never stir up asbestos dust. Asbestos dust particles found in engine clutch assemblies are powerful cancer-causing agents. Do not use compressed air to blow the dust off these parts.
25. Jack up or raise a vehicle slowly and safely. Never work under a vehicle unless it is supported by jack stands, *Fig. 3.6.* It is not safe to work under a vehicle held by a floor jack. Also chock wheels when the vehicle is on stands.
26. Drive carefully and slowly in the shop floor area. It is very easy to have an accident here.
27. Keep away from engine fans. A fan is like a spinning knife and can inflict serious injuries. Also, if a part or tool is dropped into the fan, it can fly and hit someone or damage the radiator core.
28. When a vehicle engine is running, ensure that the transmission or transaxle is in park. Check that the emergency brake is set, and that the wheels are blocked. If the transmission were knocked into gear, it could run over a person.
29. No smoking in an auto repair shop. Smoking is a serious fire hazard considering fuel lines, cleaning solvents, and other flammables may be exposed.

30. Chemical pneumonia can result from inhaling oil mist. Dermatitis can result from skin contact with oil.

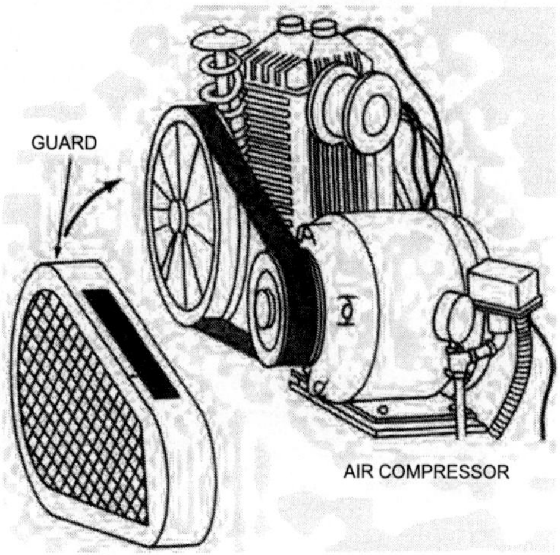

Fig. 3.5 *(f) Belts and pulleys on shop equipment should always be protected with proper guards.*

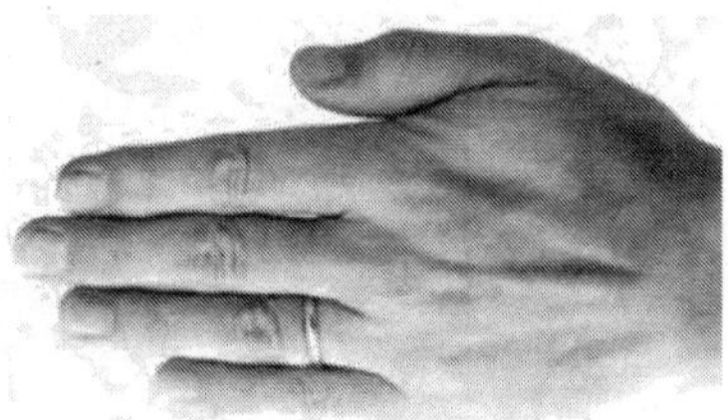

Fig. 3.5 *(g) Remove all jewelry before performing service work on any vehicle.*

Fig. 3.5 *(h) Engine exhaust connected to shop ventilation systems.*

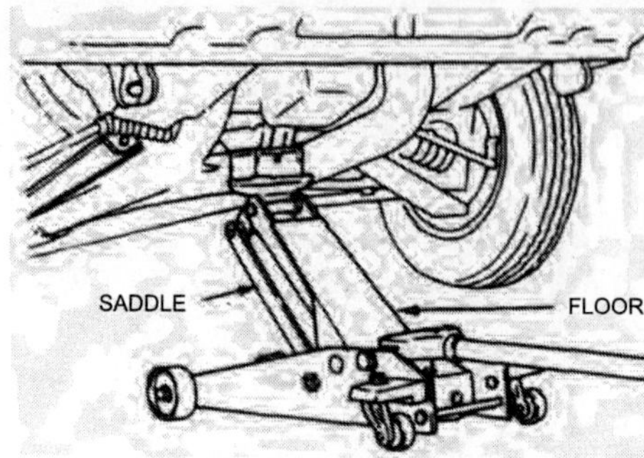

Fig. 3.6 *(a) Use floor jack to raise vehicle slowly. Place jack under an approved lift point on vehicle. Place transmission in neutral while raising.*

Fig. 3.6 *(b) Secure the vehicle on jack stands before working under it.*

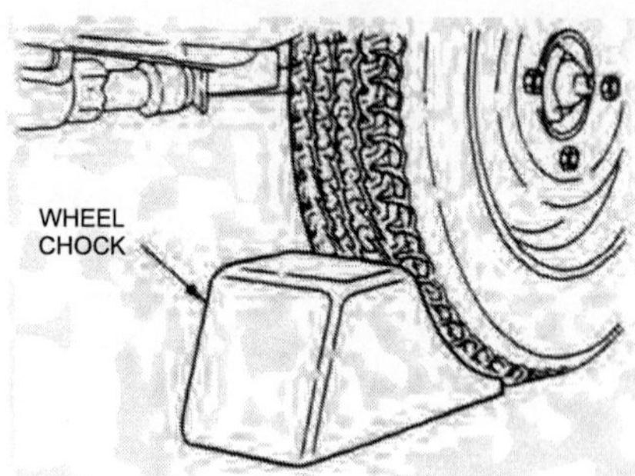

Fig. 3.6 *(c) Chock wheels to make sure that the vehicle cannot roll off of jack stands (Subaru)*

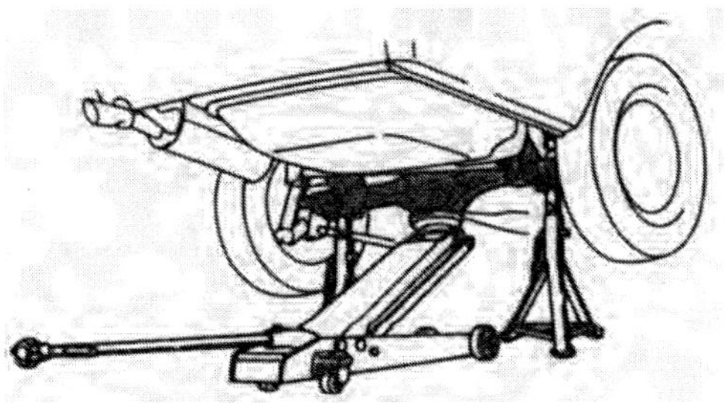

Fig. 3.6 *(d) Place floor jack under substantial part of the vehicle.*

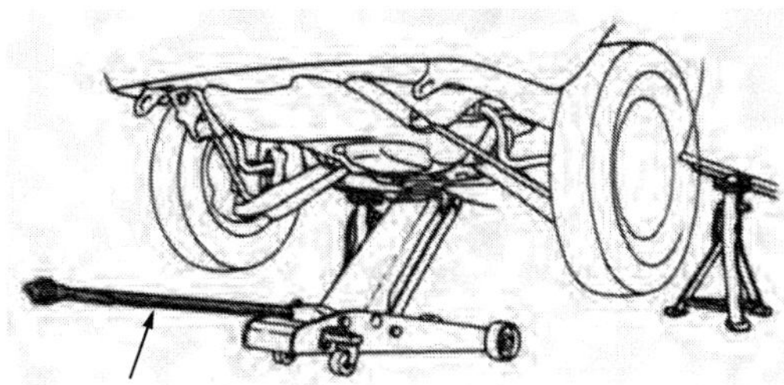

Fig. 3.6 *(e) Do not leave a floor jack handle sticking out.*

3.6 ENGINE SERVICE TOOLS AND EQUIPMENT

Late model automotive engines are supported by very complicated systems. To diagnose whether the problems an internal, mechanical problem or a problem with an engine system, demands the use of specialized testing equipment. The function of specialized engine service tools and equipment is described here.

A good engine technician knows when and how to use the various tools and equipment required to service modern automotive engines. Using the right tool for the job increases work efficiency and service quality.

Engine Stand

An *engine stand* holds an engine, at waist level, for easy disassembly during service and reassembly after service. It is in the form of metal framework on small caster wheels. The engine cylinder block is bolted to the stand. The stand allows the engine to be rotated into different positions for easy access to any side of the cylinder block. A catch tray may be mounted on the base of the engine stand to catch any dripping oil and coolant, *Fig. 3.7.*

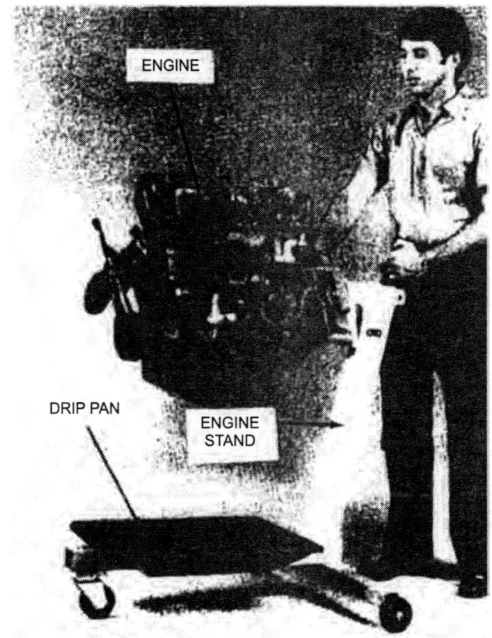

Fig. 3.7 *Engine stand is a "must" when disassembling or reassembling an engine. Catch pan will keep floor clean. Engine can be rotated into different positions for easy access to crankcase, cylinders, lifter valves, cam bore, etc. (OTC)*

Engine Crane

A portable engine crane is used to remove and reinstall the engines, for lifting heavy engine parts, transmissions, and transaxles. It has a hydraulic hand jack for raising and a pressure release valve for lowering.

Ridge Reamer

A *ridge reamer* is used to remove the metal lip formed at the top of a worn cylinder, *Fig. 3.8.* It has small cutter blades that will cut the cylinder ridge. The top portion of the cylinder that is not subjected to the rubbing, wearing action of the piston rings, will not wear, while the lower section of the cylinder, over which the rings operate, will wear to a larger diameter. If the ring ridge is not removed, the piston can be damaged while pushing it out during engine teardown.

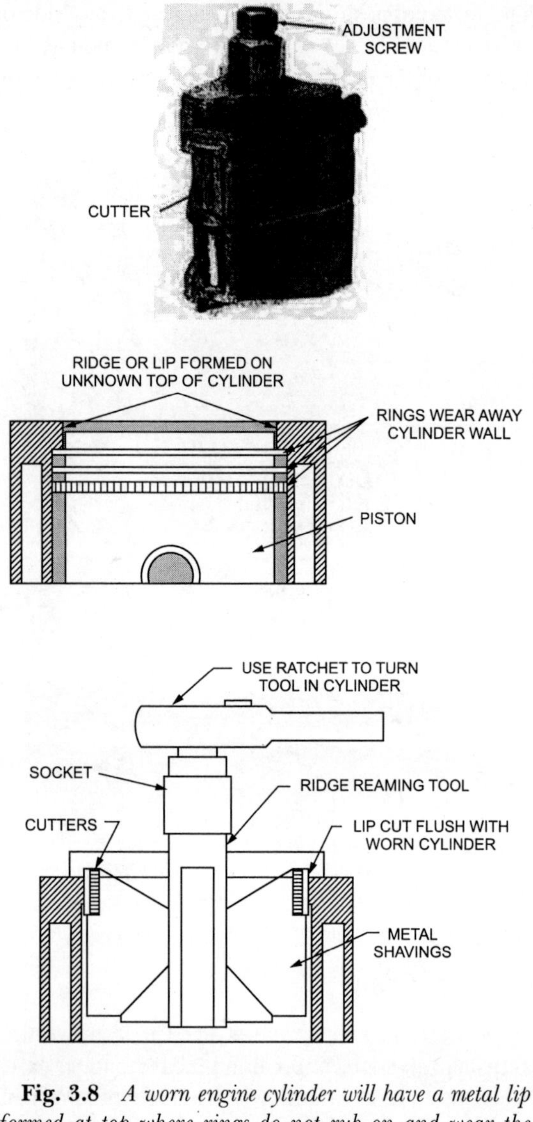

Fig. 3.8 *A worn engine cylinder will have a metal lip formed at top where rings do not rub on and wear the cylinder. Ridge reamer is used to cut this lip out and allow piston removal.*

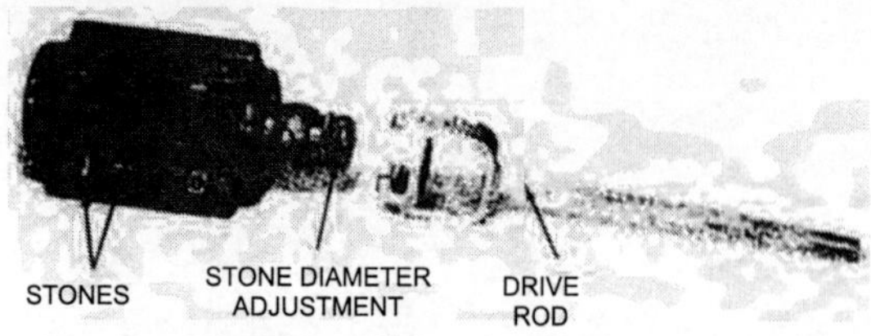

(a) Rigid hone for heavy honing.

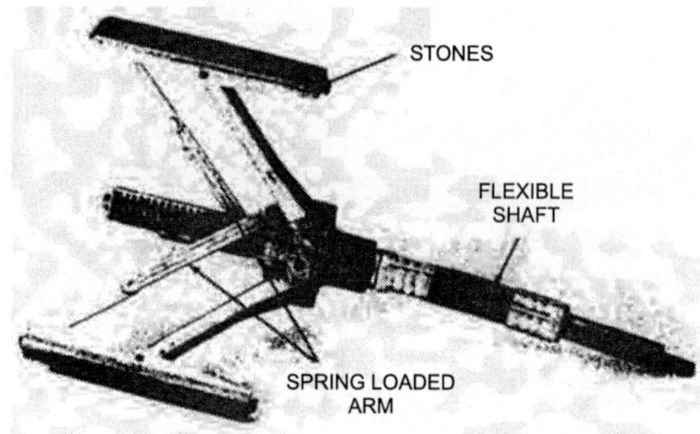

(b) Flex hone.

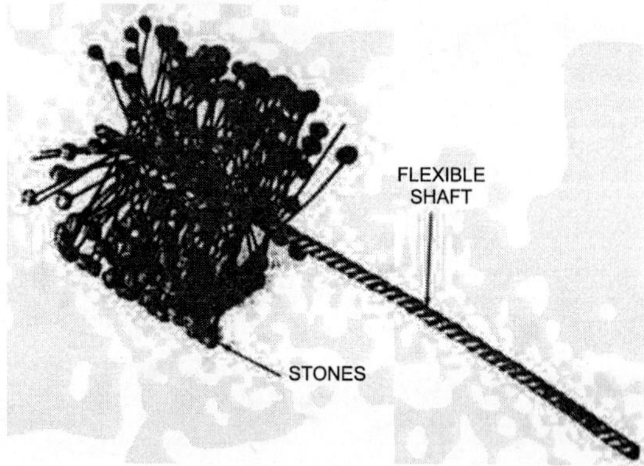

(c) Brush hone.

Fig. 3.9 *Cylinder hones are required to prepare cylinder wall surface so new rings will seal properly and engine will not consume oil or produce smoke.*

Cylinder Hone

A *cylinder hone, Fig. 3.9,* is used to texture or deglaze a cylinder wall during an engine rebuild. It has fine grit stones that are spun inside the engine cylinder using an electric drill. The drill with the hone is moved up and down to produce a crosshatched hone pattern on the cylinder wall. The installed new piston rings will wear into the cylinder wall surface and develop a good ring seal.

Ring Groove Cleaner

A *ring groove cleaner, Fig. 3.10,* is used to remove carbon deposits from inside piston ring grooves. Small cutters, of the size of ring grooves, scrape out the carbon deposits from the piston grooves as the cleaner is rotated around the piston by hand.

Piston Ring Expander

A *piston ring expander, Fig. 3.11*, is used to remove and install piston rings. Cast iron compression rings are brittle and easy to break. When installing new piston rings, extra care is needed not to over expand and break the rings. A piston ring expander helps prevent ring breakage while expediting engine assembly.

Piston Ring Compressor

A *piston ring compressor, Fig. 3.12*, will squeeze the piston rings into their grooves so as to facilitate the piston assembly installation in its cylinder. It is spring steel clamp that can be tightened around the piston rings. A wooden hammer is used to force the piston-rod assembly into the cylinder while guiding the rod over the crankshaft.

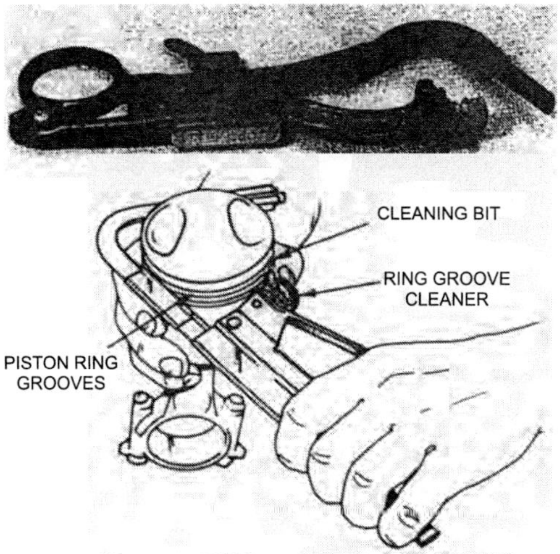

Fig. 3.10 *Used pistons can have hard carbon deposits inside ring grooves. Ring groove cleaner is required to remove these deposits from grooves so new rings fit inside groves properly. (Lisle Tools)*

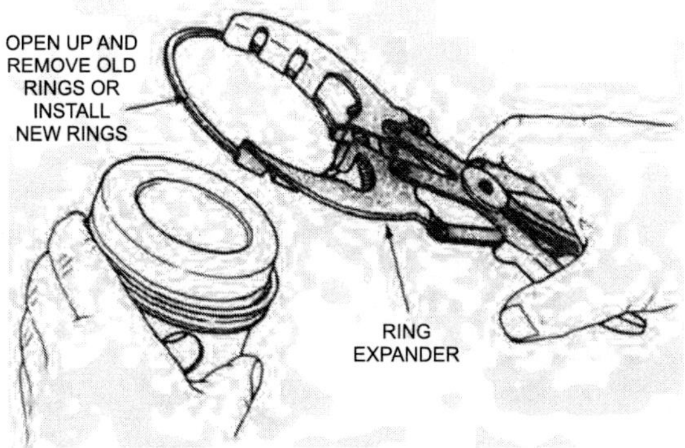

Fig. 3.11 *Compression rings are very brittle and will break easily. Ring expander will make ring installation quicker and can help prevent ring breakage. (Chrysler)*

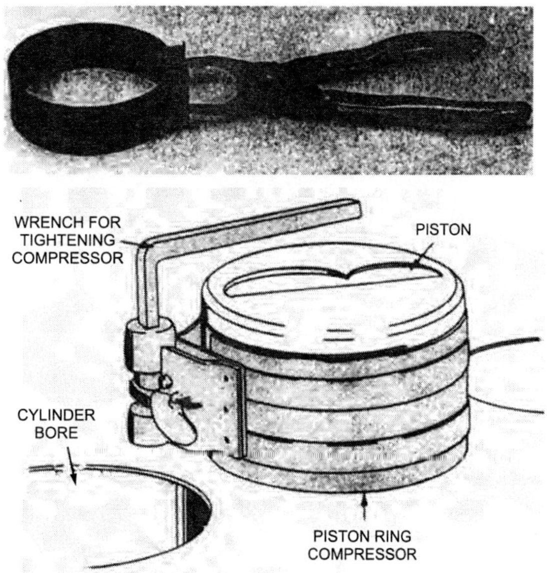

Fig. 3. 12 *(a) Ring compressor is needed to hold rings in their grooves as piston assembly is forced down into cylinder block. (General Motors Cadillac Div)*

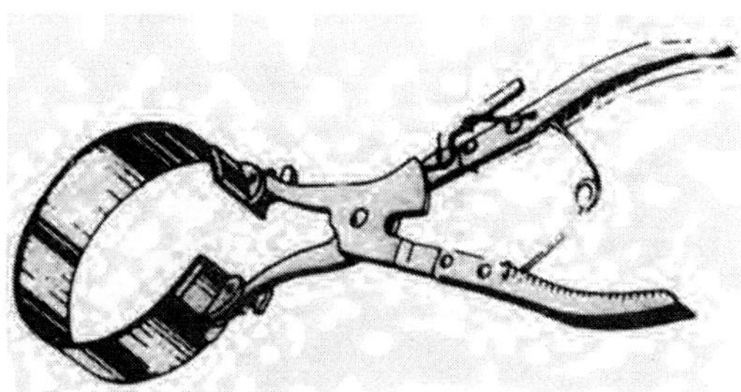

Fig. 3.12 *(b) A plier-type piston ring compressor. This style of compressor requires an assortment of bands to service different size pistons.*

Piston Knurler

A *piston knurler, Fig. 3.13,* is used to form grooves in the piston skirt to increase the outside diameter of slightly worn pistons to restore their size to within specifications. The knurler formed grooves also retain oil to avoid further piston and cylinder wear.

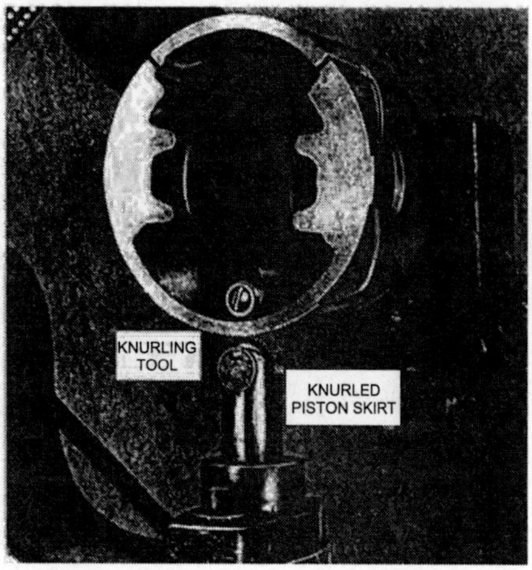

Fig. 3.13 *Knurling tool will smash grooves in piston, making its outside diameter little larger to restore the used pistons to be within specifications. (Deere & Co.)*

Flywheel Lock

A *flywheel lock, Fig. 3.14*, prevents the engine crankshaft from rotating during engine disassembly or assembly. Also, locking of the flywheel and crankshaft might be necessary when torquing flywheel bolts or the front damper bolt.

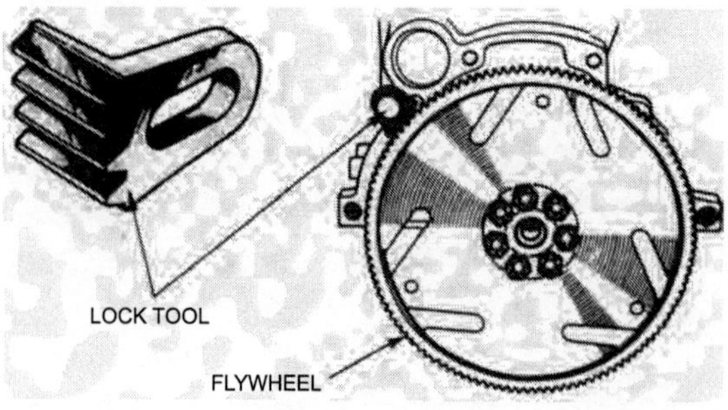

Fig. 3.14 *Flywheel lock is a handy tool used to keep crankshaft from turning while torquing flywheel bolts or front damper. (Renault Motors)*

Cylinder Head Stand

A *cylinder head stand, Fig. 3.15*, holds the cylinder head during servicing the valves and valve seats. It renders the work much easier.

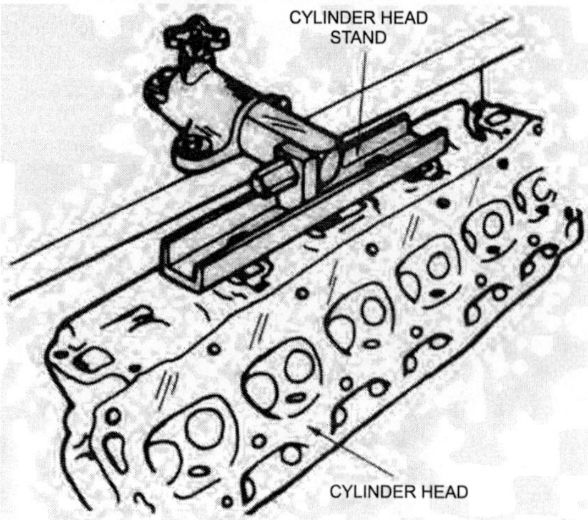

Fig. 3.15 *Cylinder head stand makes the valve service more convenient. Head can be swiveled into various positions. (K-Line Tools)*

Boring Bar

A *boring bar* is a tool for making a worn engine cylinder larger in diameter, *Fig. 3.15 (a)*. With continued use, the piston rings will wear a cylinder out-of-round and cause taper. A cylinder may be damaged by a piston assembly part, too. If not too severely worn/damaged, the cylinder may be bored oversize to install new oversize pistons during the engine overall.

Crankshaft Grinder

A *crankshaft grinder* is used to resurface worn crankshaft journals by rotating the crankshaft against a spinning stone, *Fig. 3.15 (b)*. This grinds the crank journals smaller in diameter. New undersize rod and main bearings can then be used to rebuild the engine and bring bearing clearances within specifications.

Camshaft Grinder

A *camshaft grinder* is used to resurface the cam lobes and bearing journals on a worn camshaft by rotating the camshaft against a grinding wheel, *Fig. 3.15 (c)*. This will restore the shape of the cam lobes that are prone to wear because of the high friction between the lobes and lifters or rocker arms.

Fig. 3. 15 *(a) A boring bar, like this power honing machine, will remove metal from inside of cylinder bores. This will make block cylinder like new for oversize pistons. (Sunnen)*

Fig. 3.15 *(b) A crankshaft grinder will resurface the rod and main journals smaller. Then undersize bearings can be used to produce the proper bearing-to-crank clearance. (Storm Vulcan).*

Valve Spring Compressor

A *valve spring compressor* allows a valve assembly to be removed or installed in a cylinder head, *Fig. 3.16*. It is a clamp-like device that squeezes the valve spring together so that the valve keepers can be removed. One end of the compressor fits on the valve head. The other end fits over the spring retainer.

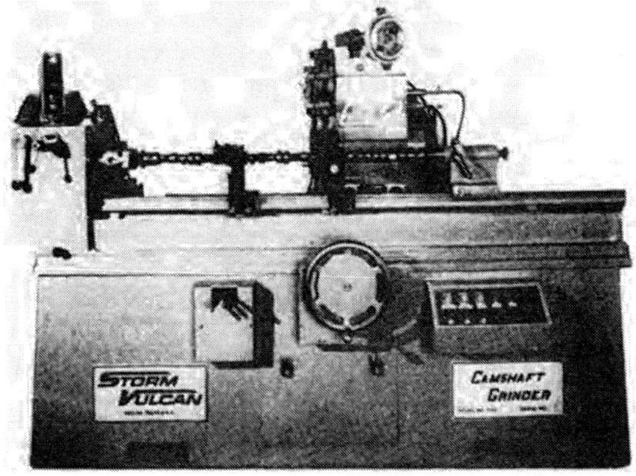

Fig. 3.15 *(c) A camshaft grinder will resurface and re-contour the cam lobes. Cam lobes are one of the highest friction and wear points in an engine. (Storm Vulcan)*

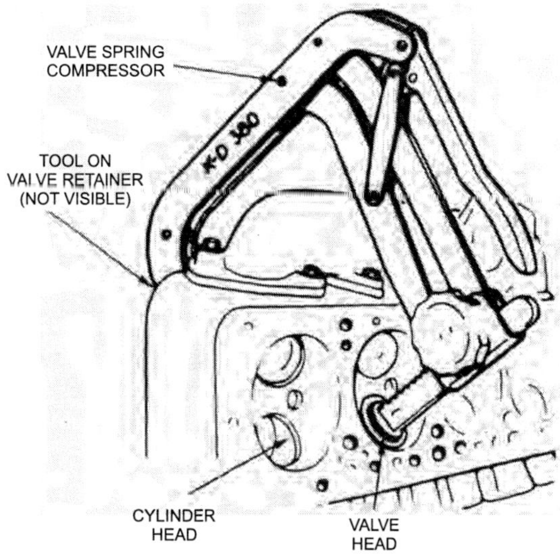

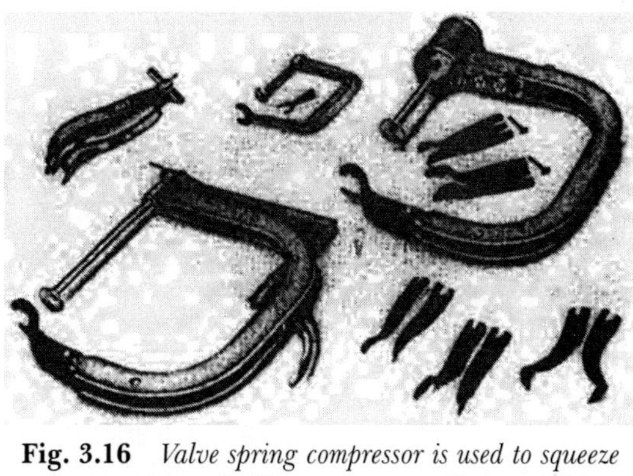

Fig. 3.16 *Valve spring compressor is used to squeeze spring together to facilitate valve assembly removal from cylinder head. (K-D Tools & Snap-on)*

Valve Adjusting Wrench

A *valve adjusting wrench, Fig. 3.17,* is a special tool for turning the rocker arm adjusting screws. It facilitates setting of rocker arm-to-valve stem clearance.

Valve Spring Tester

A *valve spring tester, Fig. 3.18,* is used to measure the tension of valve springs. In use, springs tend to set, weaken, and lose tension. The valve spring tester checks spring condition by compressing the spring to a specific height and showing spring pressure in kg. If the pressure measured is below specifications, replace the valve springs.

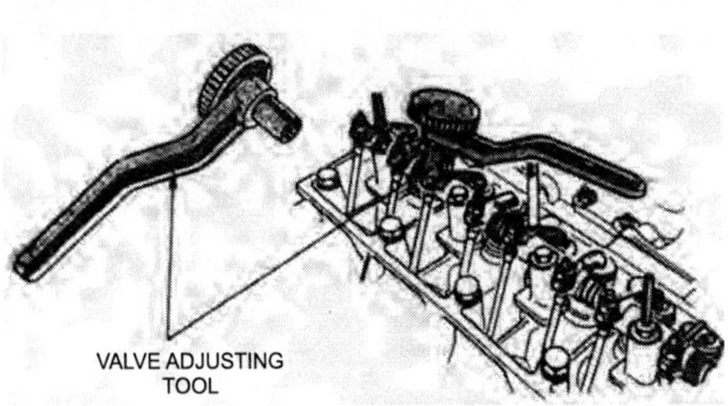

Fig. 3.17 *This tool makes valve clearance adjustment easier. (Renault Motors)*

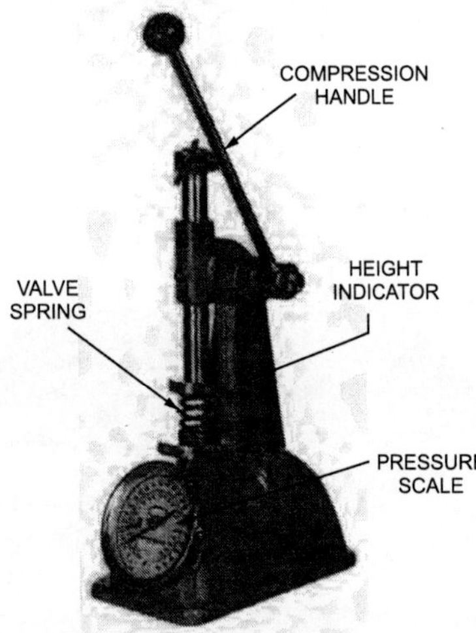

Fig. 3.18 *Valve spring tester is used to check spring pressure to make sure springs have not lost their tension. Compress spring to specified height and check pressure. Pressure should be within specifications. (K-Line Tools)*

Cam Bearing Driver

A *cam-bearing driver, Fig. 3.19,* is used to force cam bearings into or out of the cylinder block. It comprises of a long steel bar with a threaded end and special drivers. When screwed down, it produces a powerful pulling action to help bearing replacement.

Valve Guide Tools

Valve guide tools, Fig. 3.20, are required to restore the cylinder head valve guides when they are worn. With an integral guide, the old guide can be machined larger and a new sleeve or insert pressed into the cylinder head. In case of a pressed-in guide, the old guide can be driven out and a new one driven into the head.

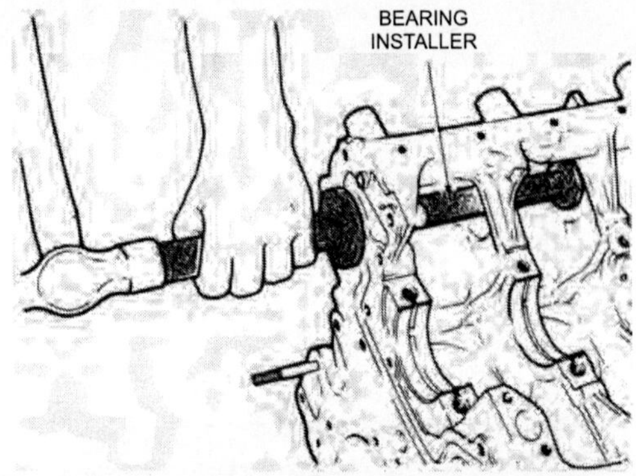

Fig. 3.19 *Cam bearing driver is used to force bearings in and out of cylinder block or head. Oil holes in bearing and part must line up. (Chrysler Corp.)*

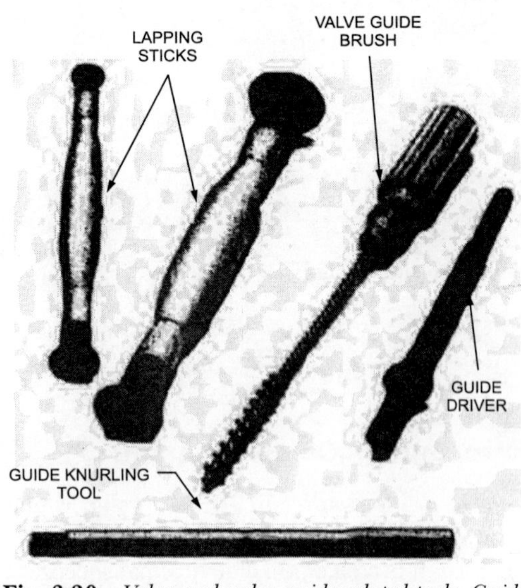

Fig. 3.20 *Valve and valve guide related tools. Guide tools are used for renewing worn valve guides in head.*

Valve Seat Grinder

A *valve seat grinder* is used to resurface the valve seats in a cylinder head. After prolonged use, the seats can become pitted and irregular. The seat grinder is used to smoothen the seat and make it concentric so that the ground valve will form a leak proof seal when closed on the seat.

Hydraulic Lifter Tester

A *hydraulic lifter tester, Fig. 3.21*, is used to measure the wear inside a lifter. It preloads the lifter plunger and produces a controlled collapse of the lifter. By timing how long it takes for the plunger to move to the bottom of its bore, the condition (good or bad) of the lifter can be determined.

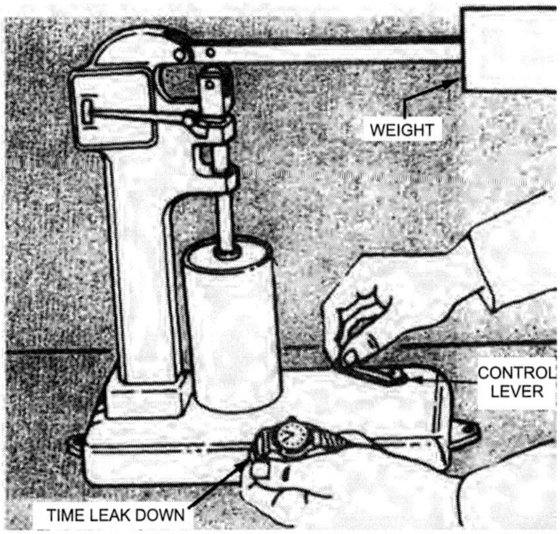

Fig. 3.21 *Hydraulic lifter tester will load lifter plunger. Amount of time that this weight takes to push plunger down can be used to determine condition of the lifter. (General Motors Buick Div.)*

Black Light

A *black light, Fig. 3.22*, is used to locate engine leaks. A special powder or fluid is added to the engine oil, and if the oil is leaking, the black light makes the fluid glow and show up distinctly.

Engine Prelubricator

An *engine prelubricator, Fig. 3.23 (b)*, can be used to fill oil passages of an engine with oil before starting. It also helps in determining worn engine bearings and other lubrication system problems during pre-teardown diagnosis.

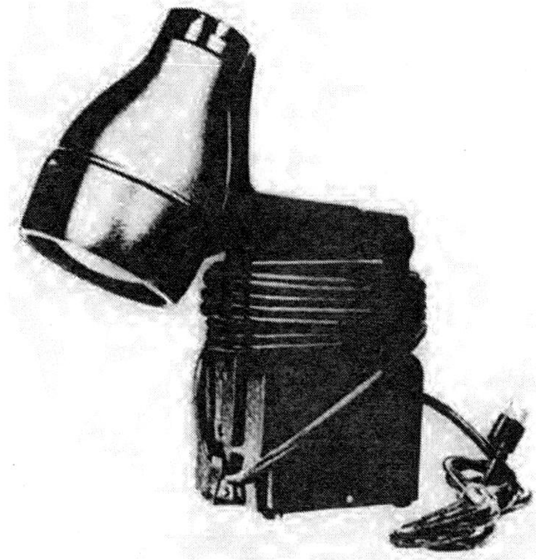

Fig. 3.22 *Black light is used to find any fluid leaks and cracks in parts. Special powders or fluids are used in conjunction with black light. (Kent Moore)*

Fig. 3.23 *(a) Engine balancer will spin crankshaft so that weight can be added to or removed from counterweights to prevent engine vibration. It is used when weight of parts is altered.*

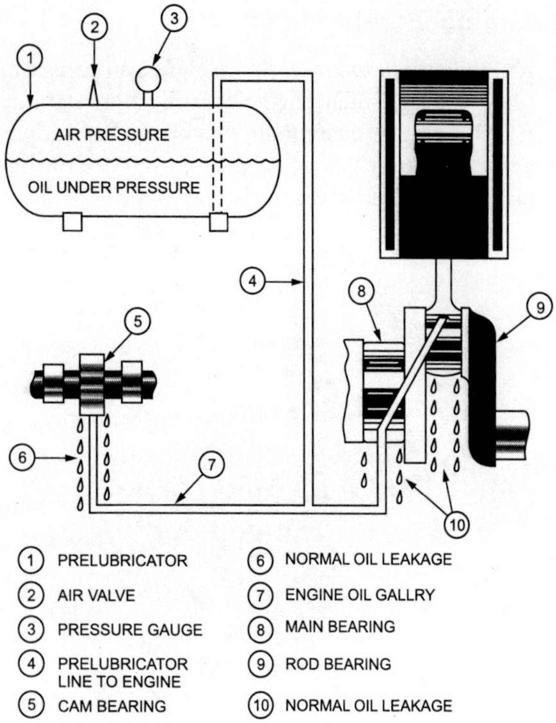

Fig. 3.23 *(b) Engine prelubricator is steel tank with air fitting and outlet hose to engine. Outlet hose is connected to engine system. Air pressure is injected into tank. Then engine oil can be forced through engine for prelubrication before starting or for finding bad bearings. Excess oil will leak out of bad bearings.*

Engine Balancer

An *engine balancer* is used to spin the engine crankshaft to determine where to add or remove weight to balance the engine, *Fig. 3.23 (a)*. A new engine is adequately balanced in the factory. If the same pistons, rods, and crankshaft are reused in their respective positions, during a rebuild, balancing is not necessary. However, if pistons, rods, or the crankshaft are changed, balancing may be needed to prevent engine vibration.

Wheel Puller

A *wheel puller, Fig. 3.24,* is used to force a hub, damper, pulley, gear, etc. off its shaft. The puller clamps or bolts on to the part. Then, when the puller screw is tightened, it will pull the component off its shaft.

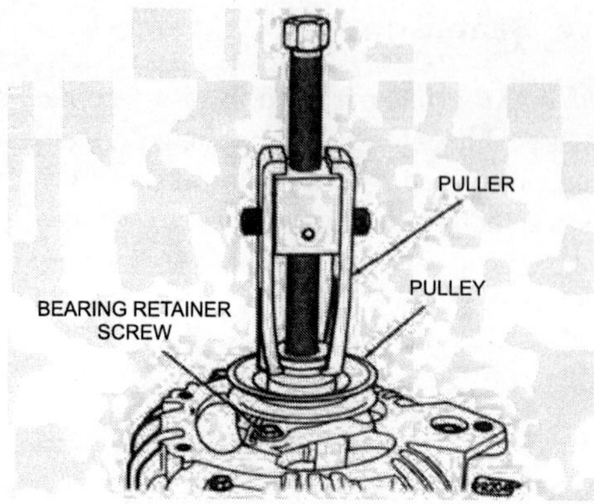

Fig. 3.24 *Wheel puller is used to force gears, pulleys, dampers, sprockets, and other parts of their shafts. (Snap-on Tools Corp.)*

Seal Drivers

Seal drivers, Fig. 3.25, are used to install and remove the front and rear crankshaft seals in an engine. A wheel-puller like tool (*a*) can be used to force out an old seal. Then, a driving tool (*b*) can be used to force the new seal into place without damage to the seal body.

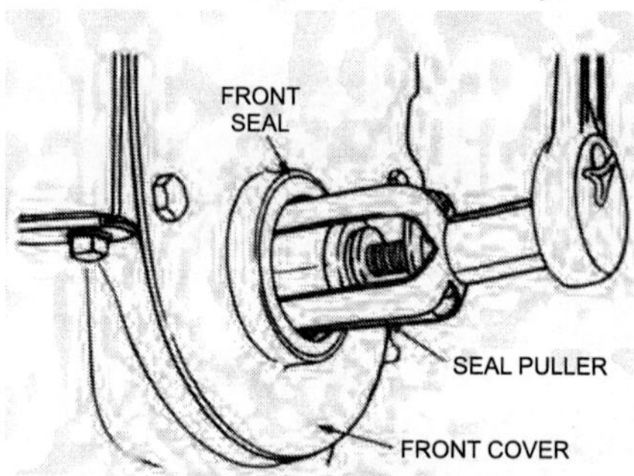

(a) Seal Puller used to remove old seal.

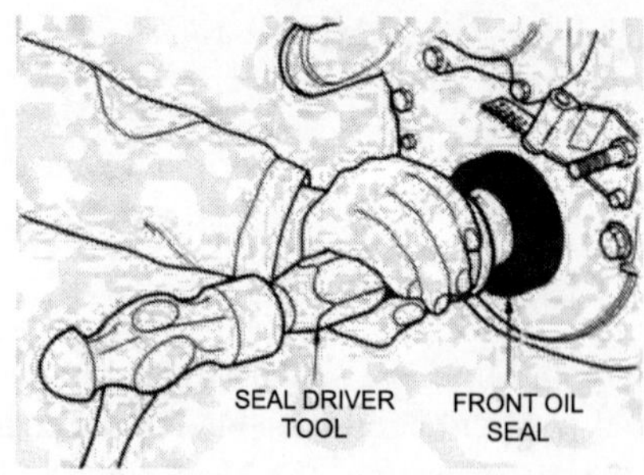

(b) Seal driver produces equal force around seal body to prevent seal damage when driving in new seal. (CMC Truck Div.)

Fig. 3.25 *Seal Tools.*

Mirror Probe

A *mirror probe, Fig. 3.26,* is used to inspect for problems when an area is blocked from view. It is small mirror hinged on a long handle. It is used to find the source of hidden leaks, cracked parts, and other problems.

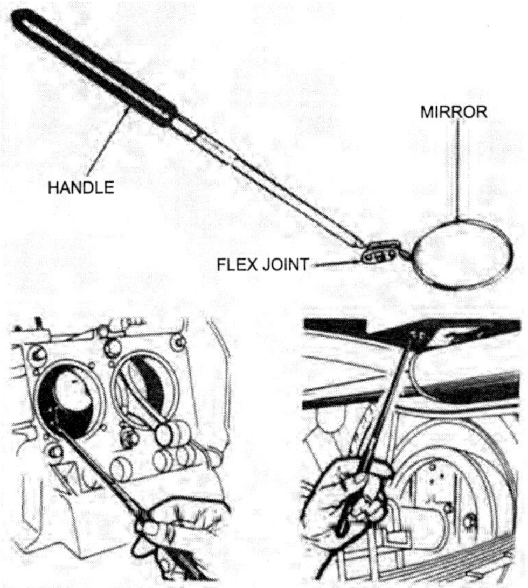

Fig. 3.26 *Mirror probe will let you look in hidden areas for finding leaks, damaged parts, etc.*

Pickup Tools

A *finger pickup tool, Fig. 3.27* (*a*), is used to retrieve dropped parts or tools. A magnetic pickup tool, *Fig. 3.27* (*b*), is used to lift the ferrous metal objects by attracting and holding due to the magnetic force.

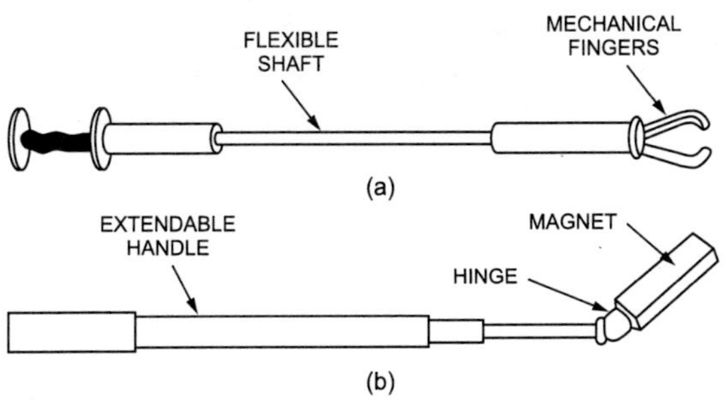

Fig. 3.27 *(a) Finger probe is used to grap and pick up small objects when the handle knob is activated; (b) Magnetic pickup is used to retrieve dropped parts and tools (Floride Voc. Ed.)*

Bench Grinder

The engine mechanic frequently uses a *bench grinder, Fig. 3.28.* The wire wheel on the grinder can be used to remove hard carbon deposits from engine valves, for example, It is also used for grinding other tools, e.g., chisels etc.

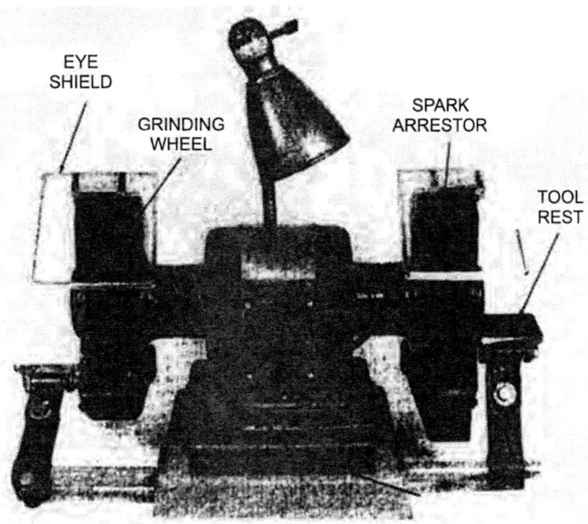

Fig. 3.28 *Bench grinder is useful tool of engine mechanic. Wire brush, for example, can be used to clean carbon off the engine valves. Keep guards and shields in place.*

Hydraulic press

A *hydraulic press, Fig. 3.29,* is used to assemble and disassemble the press-fit components, e.g., to force piston pins out of the pistons. A special fixture is required to support the piston and prevent any piston damage. It produces a lot of force for assembly and disassembly of press fit assemblies.

Hydraulic Jack

Hydraulic jacks are used to lift the vehicle, engine, and transmission during engine removal, installation and replacing motor mounts, exhaust manifolds, oil pans, etc.

3.7 ENGINE CLEANING TOOLS AND EQUIPMENT

A lot of time is spent in cleaning various parts during an engine service and reconditioning. Therefore, it is important to have a working knowledge of the most commonly used cleaning tools and equipment.

Fig. 3.29 *Hydraulic press produces powerful "driving force". It is used to push piston pins, for example. Wear eye protection during its use. (Binks)*

Blow Guns

Air powered *blow-guns, Figs. 3.30 and 3.31* are commonly used to dry and clean engine parts after being washed in solvent. It is also used to blow dust and loose dirt off an engine component before disassembly.

Fig. 3.30 *Blow guns will remove dust, dirt, and cleaning solvent from engine parts. Do not aim nozzle towards body when using a blow gun. (Binks)*

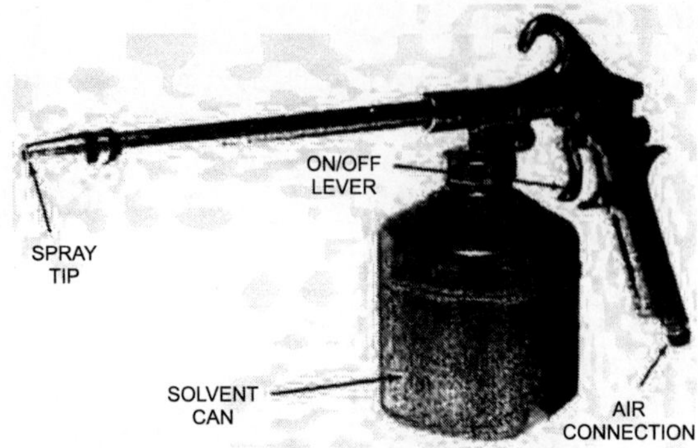

Fig. 3.31 *In this solvent cleaning gun, solvent is pulled out of Can and blown onto parts with air pressure. (Binks)*

Steam Cleaner & High Pressure Washer

A '*steam cleaner*' or a '*high pressure washer*' is used to remove heavy deposits of contaminants, e.g., dirt, grease, and oil, from the outside of engines. They provide an easy and faster method of cleaning the engine before disassembly, *Fig. 3.32.*

Cold Solvent Tank

A *cold solvent tank* is used to remove grease and oil from engine parts. After removing all old gaskets and scraping off excess grease and other contaminant deposits, the parts may be brushed clean in the solvent, *Fig. 3.33*. Then a blow gun is used to remove the solvent.

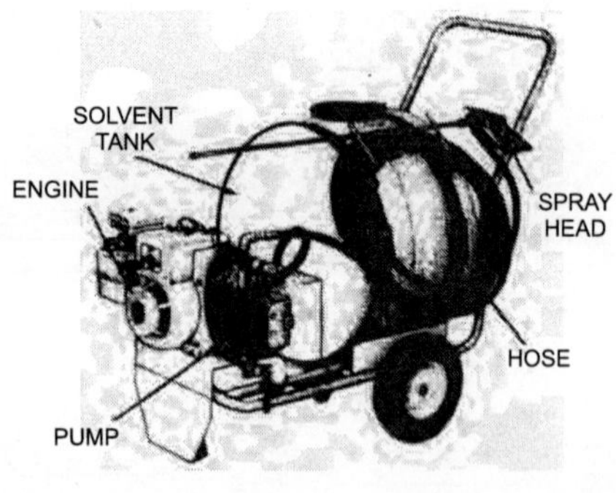

Fig. 3.32 *Pressure washers, like this one, or a steam cleaner, can be used to clean engines before teardown. (Sioux Tools)*

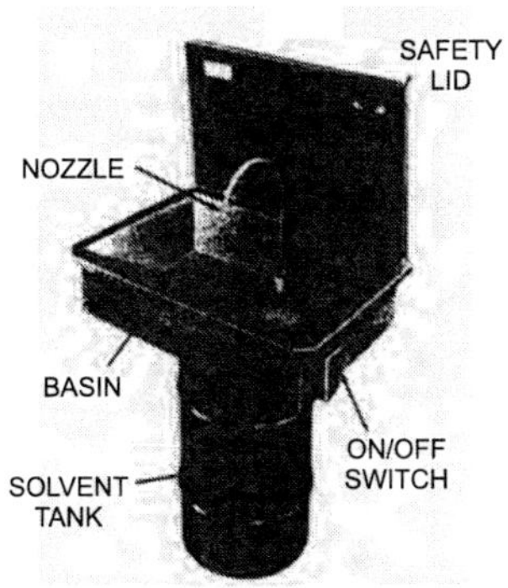

Fig. 3.33 *(a) Cold solvent tank contains mild cleaning agent for removing oil and grease from parts. (Build-All)*

Fig. 3.33 *(b) This parts cleaner can be used to remove grease and oil buildup from automotive components.*

Hot Cleaning Tank

A *hot cleaning tank*, *Fig. 3.34*, uses heat and a very powerful cleaning agent to clean engine parts. An engine block is often hot tanked during an engine overhaul. Hot tanking removes all paint, grease, oil, and mineral deposits in the water jackets. All core plugs should be removed before hot tank cleaning. All eroded bearings should be replaced after tanking.

Scrapers and Brushes

Scrapes, Fig. 3.35, are frequently used to remove old gasket material from metal surfaces and oil buildups. *Hand brushes* are used with solvent to remove oil and grease. A rotary brush, *Fig. 3.36*, is often used in a drill to remove hard carbon deposits from cylinder head combustion chambers.

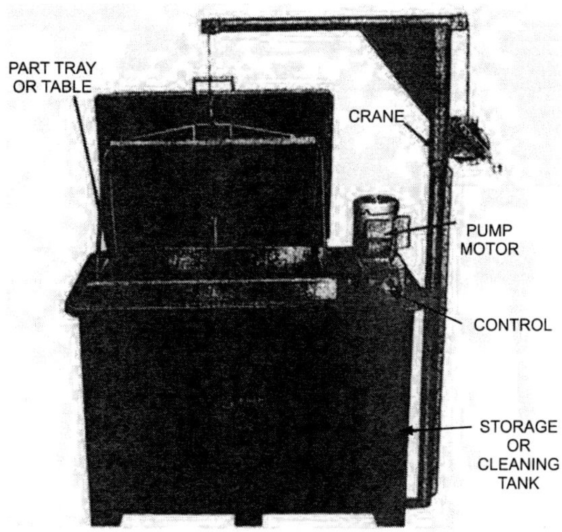

Fig. 3.34 *Hot tanks use very powerful cleaning agent for removing paint, mineral deposits, etc. Wear eye protection and gloves when working around hot tank. (BAC)*

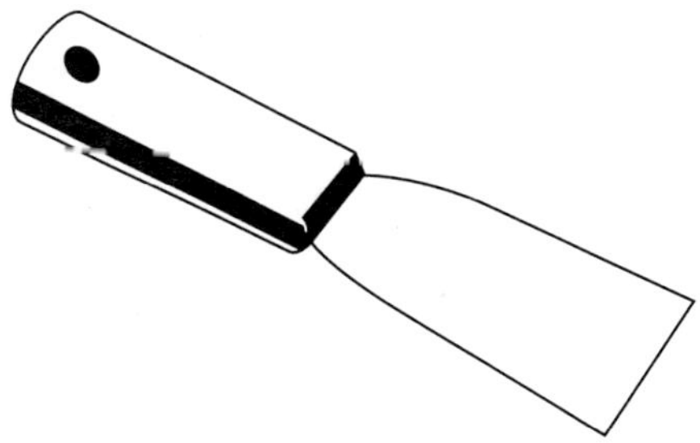

(a) Putty knife type scraper for removing old engine gaskets.

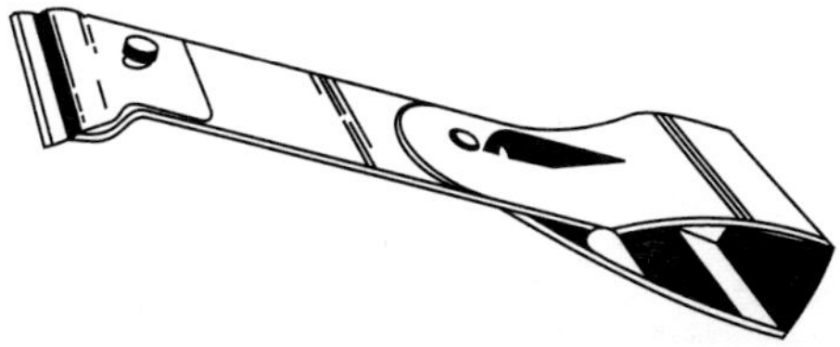

(b) Razor blade type scraper for removing paint and other thin coatings.

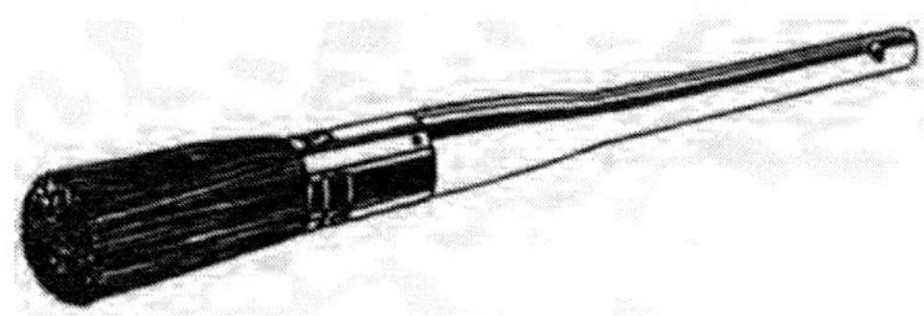

(c) Parts cleaning brush for use in cold solvent tank.

Fig. 3.35 *Scrapers. (Plew Tools Parker)*

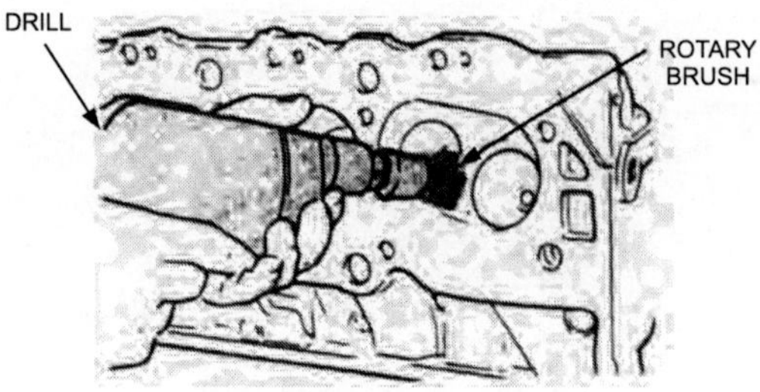

Fig. 3.36 *Rotary brush is commonly used in drill to remove hard carbon deposits form inside combustion chambers in cylinder head. Care should be exercised to avoid any damage to valve seats.*

Tubing Equipment

Tubing equipment, e.g., tube cutter, flaring tools, reamer, is commonly used to make new lines running to an engine. For example, these tools help in producing a new line to replace a damaged old fuel line, *Fig. 3.37.*

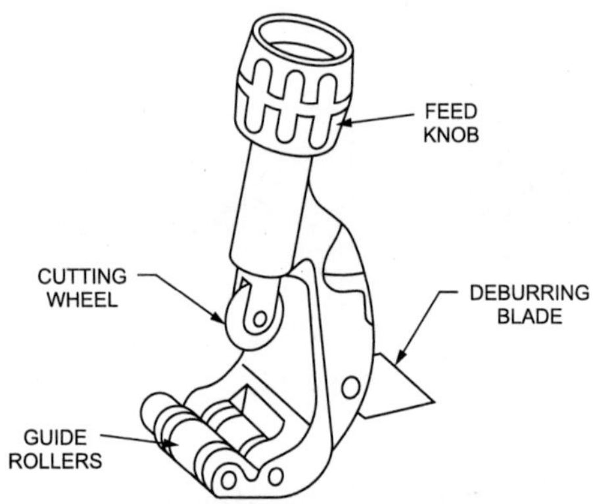

(a) Tube cutter.

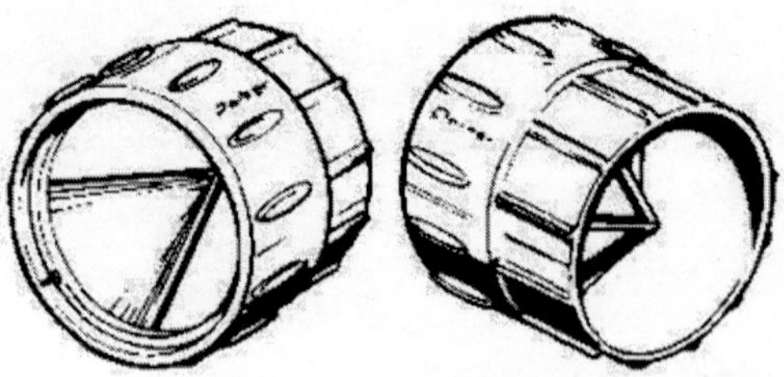

(b) Deburring tools.

Fig. 3.37 *(Contd).*

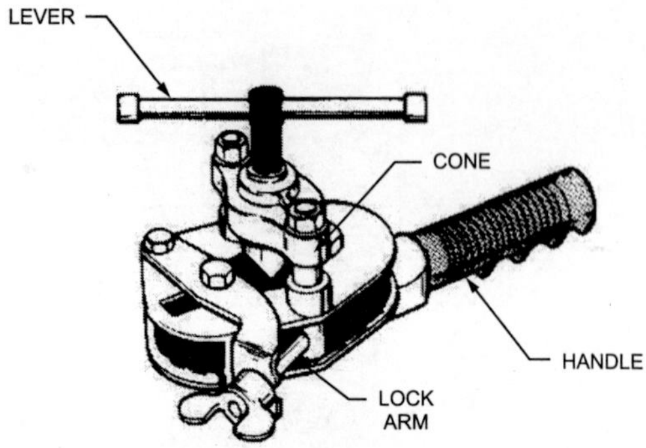

(c) Flaring tool.

Fig. 3.37 *Basic tubing tools. (Parker)*

3.8 MEASURING INSTRUMENTS

Ruler

A steel *ruler*, called scale, is frequently used to make low precision linear measurements. It is accurate to about 0.5 mm in general.

Feeler Gauges

A *feeler gauge, Fig. 3.38,* is used to measure small clearances or gaps between adjacent parts.

A *flat feeler gauge* has precision ground, steel blades of various thicknesses. Thickness of blade is written on each blade in hundredths of millimeter, i.e., 0.01, 0.06, 0.10, 0.1, 0.2, 0.23 mm. This type feeler gauge is generally used to measure small distances between parallel surfaces.

A *wire feeler gauge* has precision sized wires that are also labelled by diameter or thickness. It is usually used to measure slightly larger clearance spaces or gaps than a flat feeler gauge. The wire gauge also makes it more accurate for measuring unparallel or curved surfaces.

To measure with either type feeler gauge, by trial find the gauge thickness that just fits between the two parts being measured. The gauge should drag slightly when pulled between the two surfaces. The size given on the side of the gauge gives the clearance between the two parts.

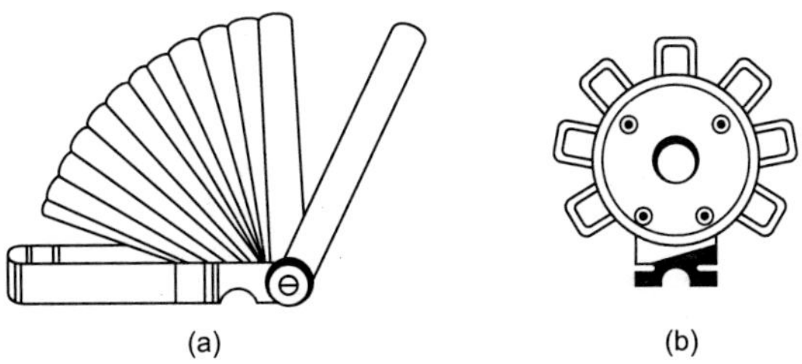

Fig. 3.38 *(a) Flat feeler gauge is for checking clearances between parallel surfaces, (b) Wire feeler gauge is for checking clearance between unparallel surfaces. (Plew Tools Parker)*

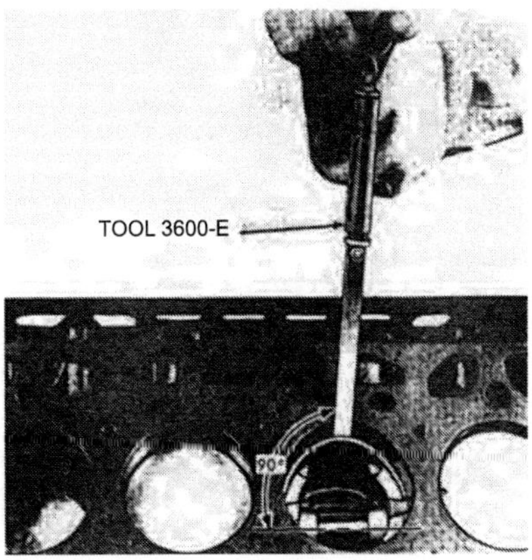

Fig. 3.38 *(c) The feeler gauge is attached to a spring scale to check the clearance between a piston and a cylinder.*

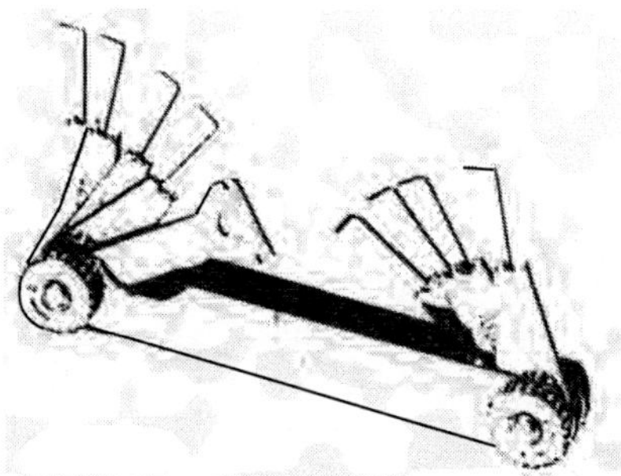

Fig. 3.38 *(d) Wire-type feeler gauge is used for measuring gap on spark plugs. (Snap-on Tools Corp.)*

Straightedge

A *straightedge* is normally used with a flat feeler gauge to check the flatness of engine parts. Cylinder heads, blocks, and manifolds are the most common engine parts checked with a straightedge, *Fig. 3.39*.

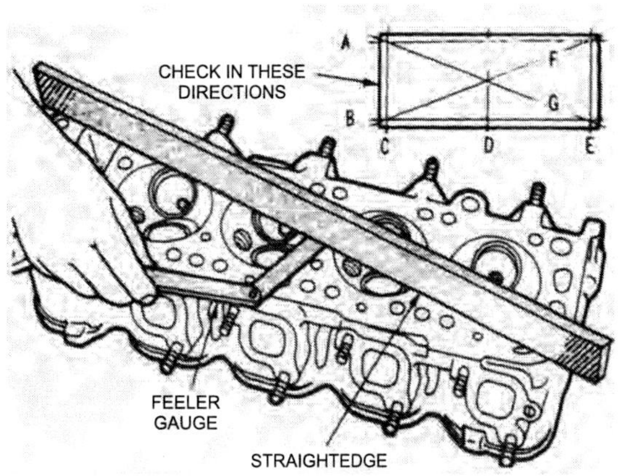

Fig. 3.39 *Straightedge will show if the part is warped. Feeler gauge that fits between part and straightedge equals amount of warpage. Check at different angles. (Chrysler)*

Combination Square

A *combination square* has a sliding square mounted on a steel ruler. It is required when the ruler must be held perfectly square against the part being measured, *Fig. 3.40*.

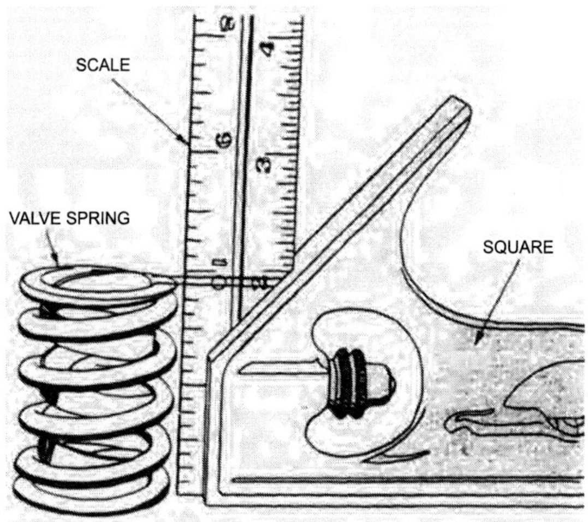

Fig. 3.40 *Combination square is used to check valve spring. If spring is not square, it should be replaced. (GM Cadillac Div.)*

Calipers and Dividers

The outside caliper may be used for making external measurements where about 0.04 mm measurement accuracy is acceptable, *Fig. 3.41 (a)*. The inside caliper is used for internal measurements in holes and other internal surface features, *Fig. 3.41 (b)*.

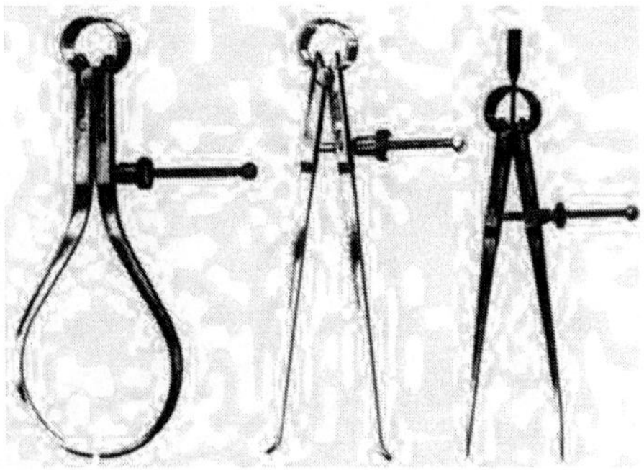

(a) Outside caliper (b) Inside caliper (c) Dividers

Fig. 3.41 *Calipers and dividers.*

Dividers are similar to calipers but have straight, sharply pointed tips, *Fig. 3.41 (c)*. These are usually used for layout work on sheet metal parts. These also are used to transfer and make surface measurements

Dial Caliper

A *dial caliper* has measurement accuracy of 0.01 mm and is used to make inside, outside, and depth measurements quickly and precisely, *Fig. 3.42*.

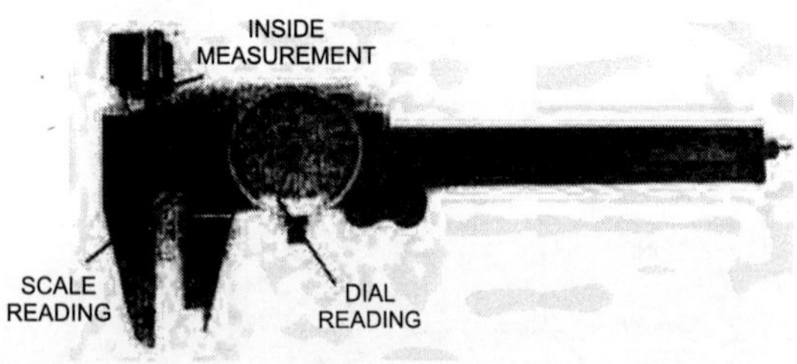

(a) Inside measurement

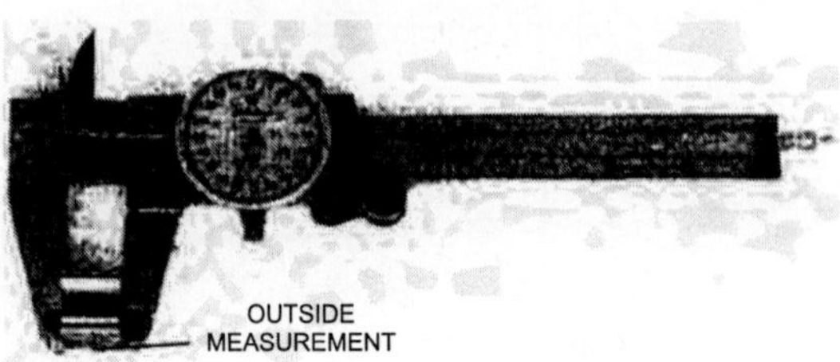

(b) Outside measurement
Fig. 3.42 *(Cond).*

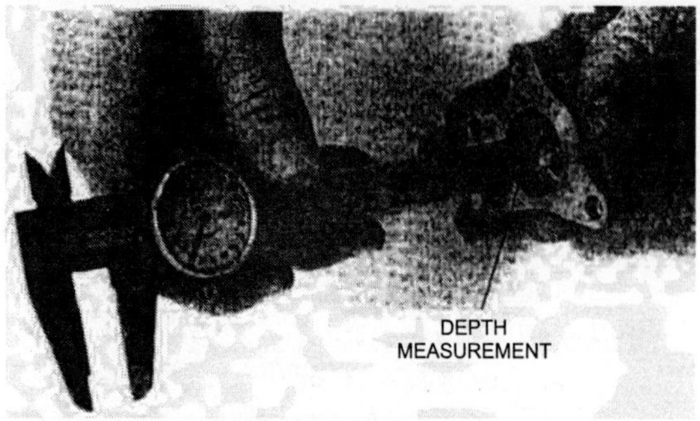

(c) Depth measurement

Fig. 3.42 *Dial caliper is very useful because it will measure inside, outside, and depth with an accuracy 0.001 in. Read ruler scale and then add dial reading in thousandths of an inch.*

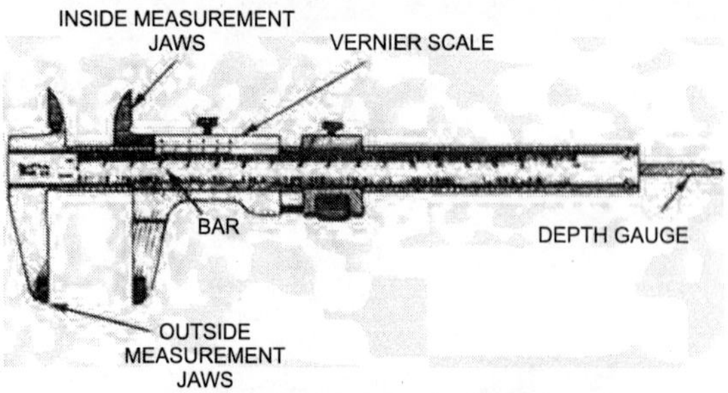

Fig. 3.42 *(d) Nomenclature of a Vernier caliper. (Central Tools)*

Micrometers

A *micrometer* is commonly used when making very accurate engine measurements. It can easily measure to 0.01 mm. An *outside micrometer, Fig. 3.43*, is used for measuring external dimensions, diameters, or thicknesses. It is fitted around the outside of the part. Then the thimble is turned until the part is lightly touching both the spindle and anvil. Measurement is obtained by reading graduations on the hub and thimble.

Reading a Metric Micrometer

One revolution of the thimble = 0.5 mm. To read a metric micrometer, follow the four steps listed below:

1. Read the *sleeve number*. Each sleeve number = 1 mm, *Fig. 3.44*. Here 5 = 5.00 mm.
2. Count and record the *sleeve graduations* visible to the right of the sleeve number. Each line = 0.5 mm

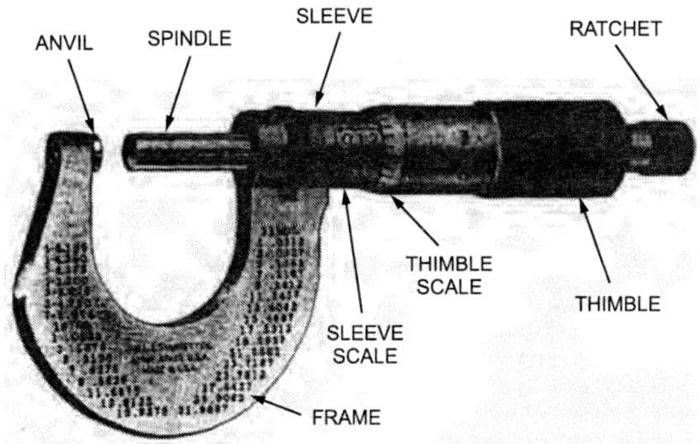

Fig. 3.43 *(a) Nomenclature of a micrometer. (Starret)*

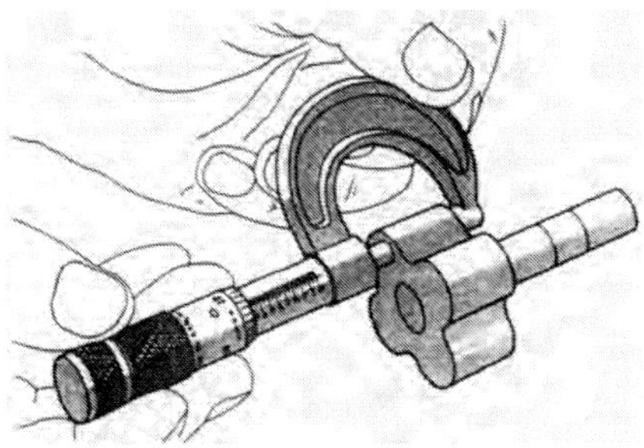

Fig. 3.43 *(b) Measuring with a micrometer.*

3. Read the *thimble graduations* lined up with the horizontal sleeve line. Each thimble graduation = 0.01 mm.
4. Add the values obtained in the previous three steps. This will give the metric micrometer reading, e.g., 5.5 mm in *Fig. 3.44.*

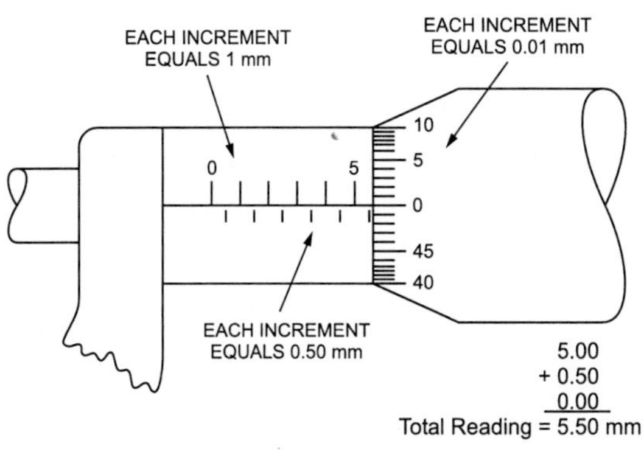

Fig. 3.44 *Note scale divisions on metric micrometer. These are in millimeters. It is read like a conventional micrometer.*

Micrometer Rules

Remember the following micrometer rules:

1. Never drop or over tighten a micrometer. It is a very delicate and its accuracy can be thrown off easily by mishandling.
2. Store micrometers to avoid damage by large, heavy tools. Keep them in their storage boxes.
3. Grasp the micrometer frame in palm and turn the thimble with thumb and finger. It should just grab on the part being measured.
4. Hold the micrometer squarely with the work otherwise false readings can result.
5. Rock or swivel the micrometer as it is touched on round part surfaces, in order to ensure that most accurate diameter measurement is obtained.
6. To avoid any rusting of the micrometer during storage, place a thin film of oil on it.
7. Always check the accuracy of a micrometer if it is dropped, struck, or after long period of use.

Special Micrometers and Gauges

Various special micrometers and gauges are used by auto mechanics.

A telescoping gauge measures internal part bores or openings. It is commonly used to measure cylinder block cylinder bores. The spring-loaded gauge is expanded to the size of the measured feature, *Fig. 3.45* (*a*). Then, it is locked to size and measured with an outside micrometer, *Fig. 3.45* (*b*). A hole gauge is used to measure very small holes in parts. The hole gauge is inserted and adjusted to fit the hole. Then, it is removed and measured with an outside micrometer.

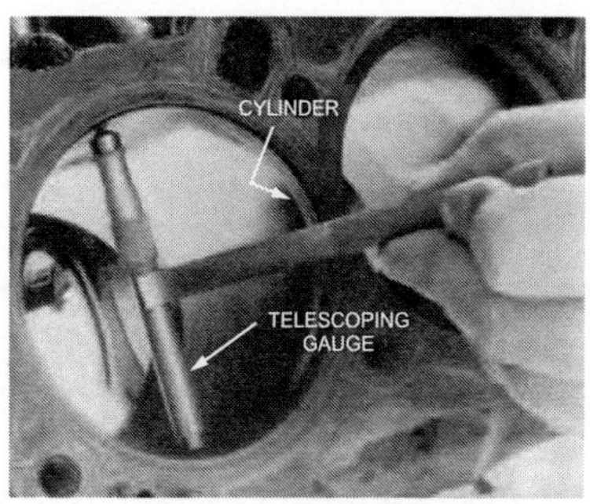

(a) Telescoping gauge is adjusted to fit inside bore.

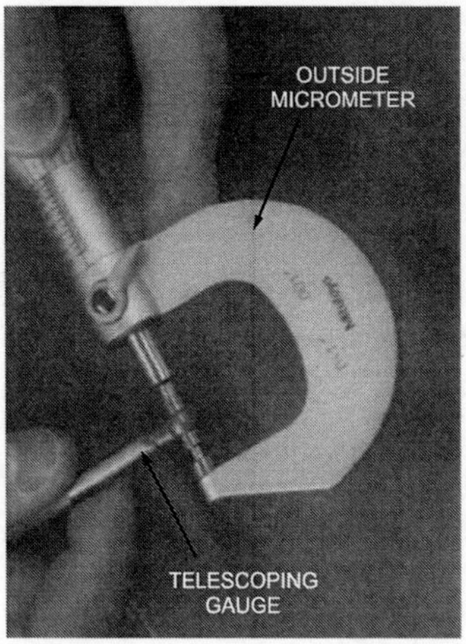

(b) Then, outside micrometer is used to measure gauge and determine diameter of bore.

Fig. 3.45 *Telescoping gauge.*

An *inside micrometer* is a special measuring tool used for internal measurements in large holes, cylinders, or other part openings. It is read in the same manner as an outside micrometer, *Fig. 3.46.*

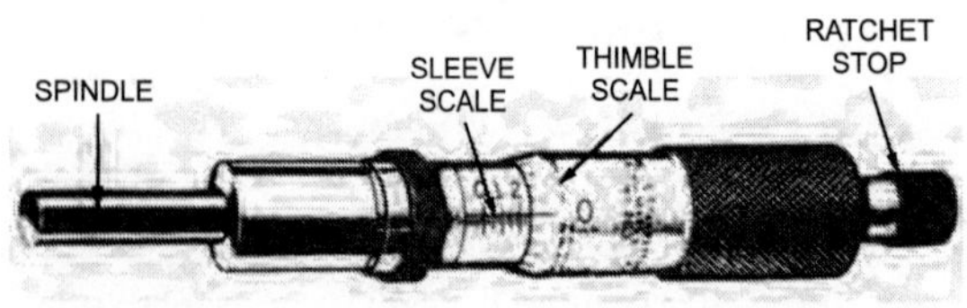

Fig. 3.46 *An inside micrometer does the same thing as a telescoping gauge but will not measure small diameters. (Starret)*

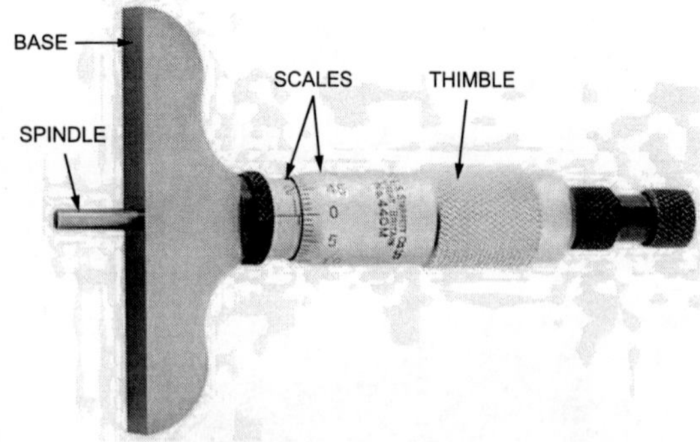

Fig. 3.47 *Depth gauge is more specialized tool for depth measurements. (Starret)*

A *depth micrometer* is useful when precisely measuring the depth of an opening. The base of the micrometer is positioned squarely in the part. Then, the thimble is turned until the spindle contacts the bottom of the opening. The depth micrometer is read in the same manner as an outside micrometer. However, the hub markings are reversed, *Fig. 3.47.*

Dial Indicator

A *dial indicator* measures part movement in hundredths of a millimeter (0.01 mm). The needle on the indicator face registers the amount of plunger movement, *Fig. 3.48.*

In auto mechanics practice a dial indicator is used to check gear teeth backlash (clearance), shaft endplay, cam lobe lift, and many other similar kinds of part movements. A magnetic mounting base or clamp mechanism is normally used to secure the dial indicator to the work. *Fig. 3.49 (a)* shows a special fixture for using a dial indicator to measure valve spring installed height. *Fig. 3.49 (b)* shows a dial bore gauge used for cylinder wear.

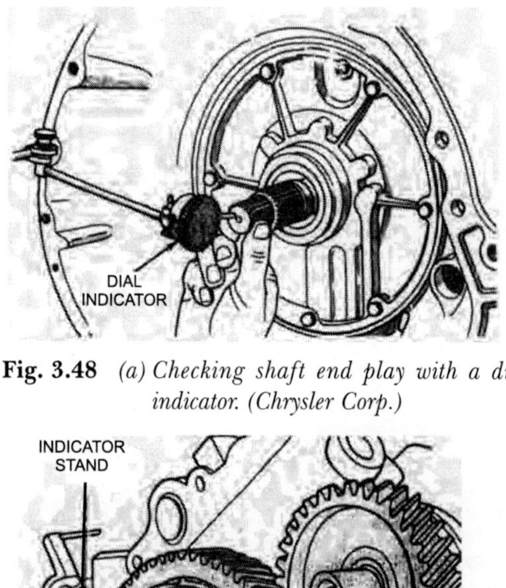

Fig. 3.48 *(a) Checking shaft end play with a dial indicator. (Chrysler Corp.)*

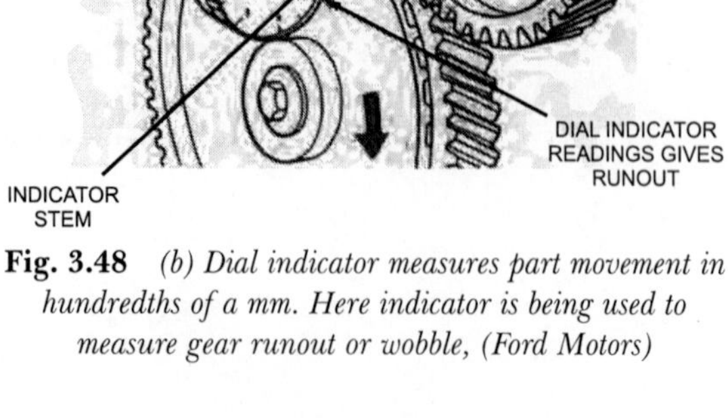

Fig. 3.48 *(b) Dial indicator measures part movement in hundredths of a mm. Here indicator is being used to measure gear runout or wobble, (Ford Motors)*

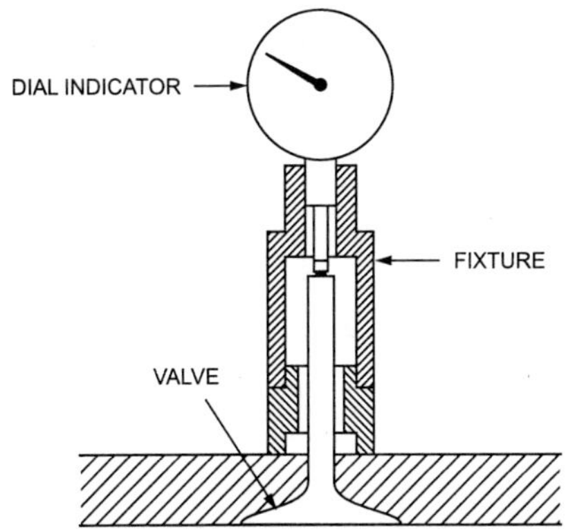

Fig. 3.49 *(a) A special fixture made to check valve spring installed height with dial indicator. Such fixtures are for specific engine makes and will not be useful for others (K.D. Tools)*

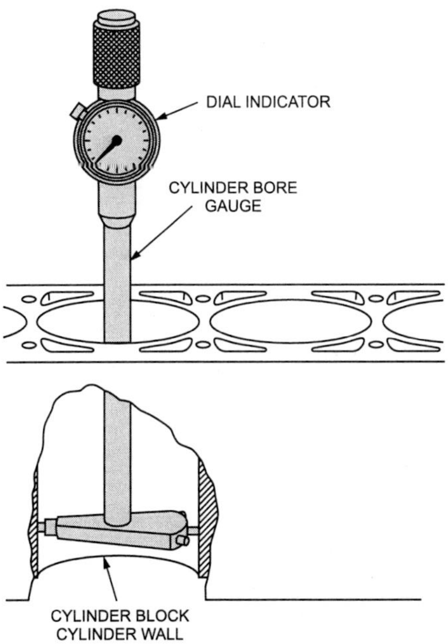

Fig. 3.49 *(b) Dial bore gauge is used to check condition of engine cylinders. It is slid up and down in cylinder. Indicator movement gives amount of cylinder taper and out-of-roundness. (American Motors)*

Using a dial indicator

To take measurements using a dial indicator:

1. Mount the indicator securely and position the dial plunger parallel to the movement to be measured.
2. Preload or partially compress the indicator plunger before locking the indicator into place. Part movement in either direction should cause dial pointer movement.
3. Move the part back and forth or rotate the part while reading the indicator. Pointer movement equals part movement, clearance, or run out.
4. Dial indicator is a very delicate tool, be careful not to damage it.

Plastigauge

Plastigauge is a special measuring device normally used to check clearances between internal surfaces, *Fig. 3.50*. It is commonly used to measure clearances in engine connecting rod bearings, crankshaft main bearings, and oil pumps.

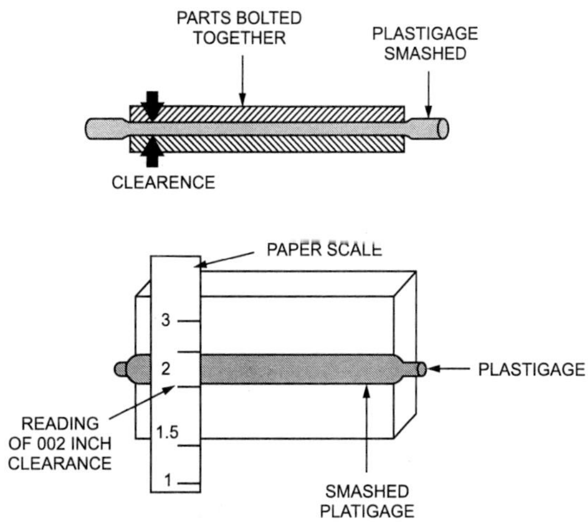

Fig. 3.50 *Platigage is placed between two parts to measure clearance between them. Parts will smash Plastigage to determine clearance. Then, paper scale can be compared to with a smashed Platigage to determine clearance – the more the wire is smashed, the less the clearance.*

To measure with Plastigage, place a small piece of the round plastic–like thread on a clean, unoiled part surface. Bolt the parts together, torque to specifications and disassemble. Compare the paper covering's scale to the smashed Plastigage strip. The corresponding paper scale width indicates part clearance.

Torque Wrenches

A *torque wrench* measures the amount of turning force being applied to a fastener. Conventional metric system torque wrench scales read in Newton-meters (N-m). Three general types of torque wrenches are the ***flex bar, dial indicator***, and ***sound indicating types***, *Fig. 3.51*.

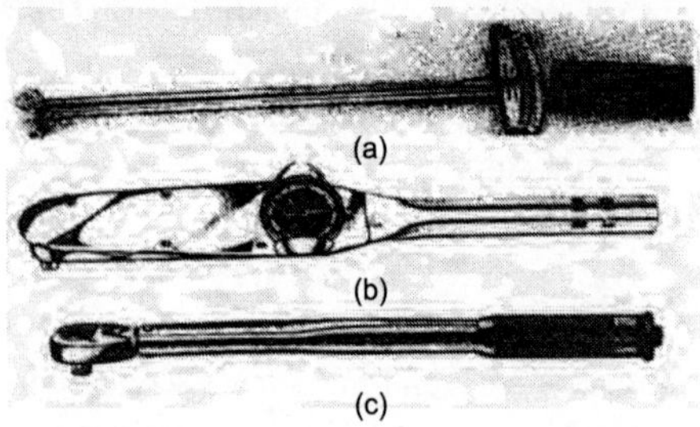

Fig. 3.51 *Some commonly used torque wrenches: (a) Flex bar torque wrench, (b) Dial indicator torque wrench, (c) Click on sound indication torque wrench (Snap-on Tools)*

The flex bar torque wrench is cheaper and adequate. The dial indicator torque wrench is very accurate but is hard to use in tight quarter. The sound indicating type torque wrench is very fast and easy to use. It makes a "pop" or "click" sound when a preset torque value is reached and the technician does not have to watch an indicating needle while torquing.

3.9 METERS, TESTERS AND ANALYZERS

It is important for a qualified mechanic /technician to be familiar with test equipment and the theory behind them otherwise he/she will be handicapped when attempting to service an automotive engine, tune-up an engine, test a particular circuit, determine the cause of the engine miss under load, or determine the cause(s) of excessive fuel or oil consumption. These are only some of the problems that mechanic can diagnose by using diagnostic test equipment.

Most late model engines are controlled by electronics. Sensors and a computer monitor various engine operating conditions and then use electrical signals to control various engine functions like ignition timing, fuel injection, idle speed, turbocharger boost, etc.

One particular demonstration of automotive knowledge that makes an engine service technician / mechanic stand out as an expert is the ability to systematically test, evaluate, and correct engine performance problems in all five major engine systems: starting, charging, fuel distribution, compression, and ignition.

A technician/mechanic, however, performs as an expert with the help of a range of very accurate meters, testers, and analyzers. For electrical systems testing, basic instruments include the *ammeter* for measuring the current in amperes, the *voltmeter* for measuring the voltage, and the *ohmmeter* for measuring resistance.

A mechanic/technician also frequently uses a *battery-starter tester, an alternator-regulator tester, a battery charger, a Multimeter, an infrared exhaust analyzer,* and *an engine systems analyzer.* For an 1C engine troubleshooting, a *compression tester*, *vacuum gauge*, and *cylinder leakage tester* are commonly used.

To be a good technician in a modern auto repair facility, one should be skilled in making automotive electrical tests. A review of common electrical test equipment and their use is given here.

Each electrical function, e.g., volts, ohms, and amps, has a special tester. These special testers usually test only one or two functions. Each type of meter must be set for the proper measurement and then connected properly. Wrong settings or connections could damage the meter or obtain an incorrect measurement.

Meter Design

Basically, two types of meters are used in automotive service work: *analog meters* and *digital meters*. These sensitive instruments are designed to measure and indicate electrical values. An analog meter has a calibrated dial (face with marked values) and a pointer that indicates the value of the electrical unit under test, *Figs. 3.62, 3.63*. *Digital meters* "spell" the value of the unit under test, *Figs. 3.63, 3.64.* This type of meter is especially well suited for use in computerized engine systems analyzers, on-board computer testers, and exhaust emissions analyzers.

Most analog meters (ammeters, voltmeters, and ohmmeters) used in the automotive service field are of the moving coil type, *Figs. 3.52* and *3.53.* These instruments consist of a horseshoe-shaped or hoop-shaped permanent magnet and a movable coil. In addition, a damping spring (similar to hairspring of a watch) is attached to the pointer of the meter. The damping spring is used to prevent the pointer from going full scale too quickly when current is applied and removed. This long established meter design is known as the **D'Arsonval movement.** Usually the D'Arsonval movement is "jeweled" to cut down on friction to ensure the accuracy of its readings. Therefore, this type meters are reliably accurate and sensitive. Therefore, these must be handled carefully. If subjected to shock the meter can damage the movement and ruin the meter.

When using any analog meter with multiple ranges, always start by adjusting the range selector switch to the highest range. Then reduce the switch setting until the meter is able to 'read' the electrical value. Otherwise, a low range setting could result in "pegging the meter," which happens when the pointer strikes hard against the high end of the meter scale.

Ammeter/Voltmeter Operation

The basic design of an ammeter and a voltmeter is the same. However, the ammeter is provided with a heavy *shunt* (alternate path for current) of low resistance connected across the movable coil, *Fig. 3.53*. Another resistance element is connected in series with one end of the shunt and the coil. This resistance is used for calibrating the instrument.

The voltmeter does not have a resistance shunted across the coil, but it has the calibrating resistance in series with the movable coil, *Fig. 3.52*. When hooking up a voltmeter leads, always connect them in parallel to the terminals of the device or circuit, *Fig. 3.54*. But always connect ammeter leads in series with the circuit, *Fig. 3.55*.

When the circuit to be tested is closed, current flowing through the movable coil of the ammeter or voltmeter reacts with the magnetic field, causing the coil to rotate against the tension of the damping spring. Relative movement of the coil is directly proportional to the current flowing through it. A pointer attached to the coil moves across a calibrated scale on the meter face to indicate the amount of current or voltage flowing through the coil.

Ammeters can be designed to cover different ranges of current (0-10, 0-100, etc.) by providing different shunts. When working with an ammeter, ensure to use the leads provided by the meter manufacturer. Leads having a different resistance than that of the original ones will seriously affect the accuracy of the instrument.

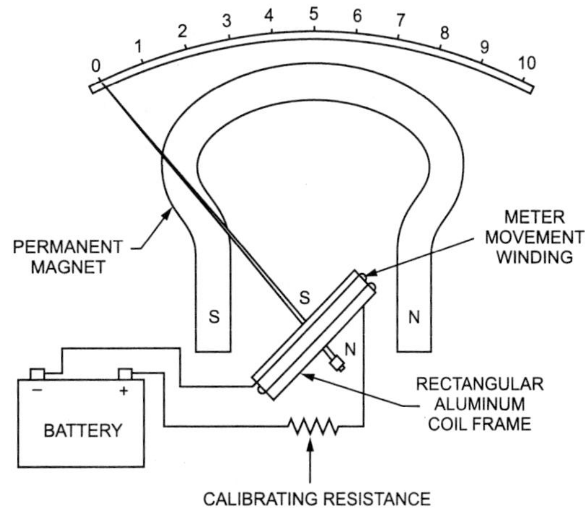

Fig. 3.52 *Schematic drawing of the makeup of a typical voltmeter. Note calibrating resistance in series with moving coil.*

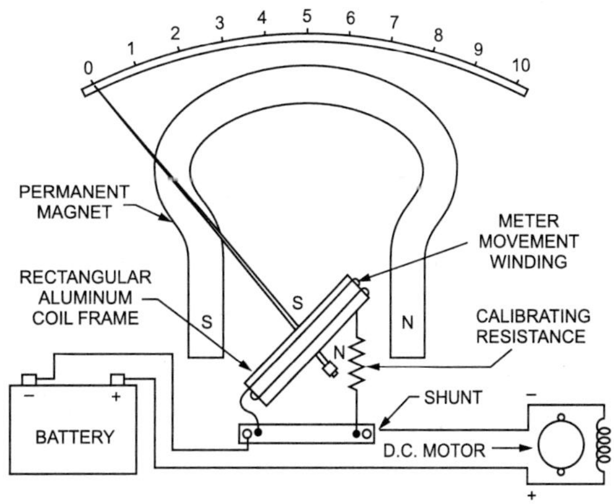

Fig. 3.53 *Note that the ammeter has a heavy resistance (shunt) connected across the moving coil and another resistance connected in series with the moving coil.*

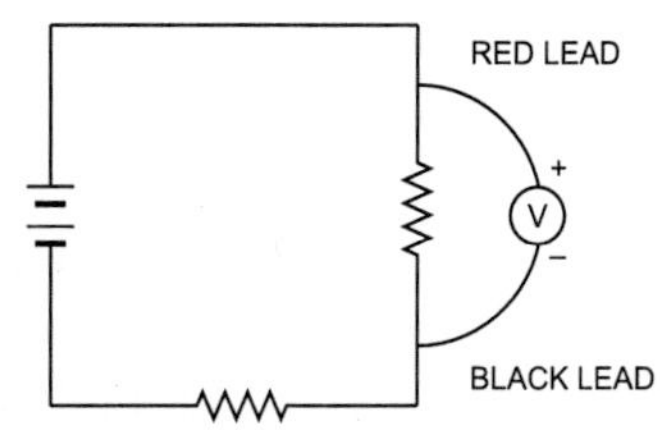

Fig. 3.54 *Schematic view showing hook up of voltmeter across a circuit. Always connect the test leads so that the meter is in parallel to the circuit.*

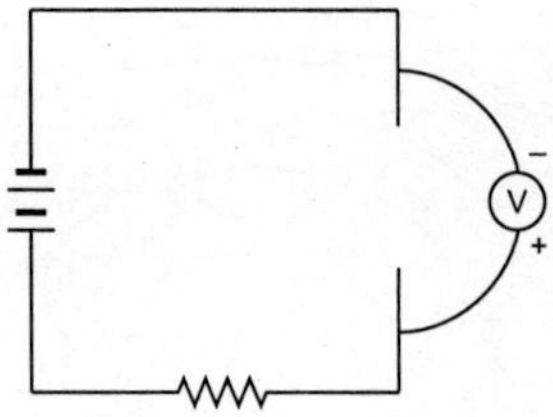

Fig. 3.55 *Schematic view showing hook up of an ammeter in series with circuit; connect meter test leads; read meter.*

A single voltmeter can also be designed to cover different ranges of voltage (0-2, 0-10, etc.). Varying the resistance in series with the movable coil does this. A separate resistance is provided for each range of the instrument. The resistance value – as well as the voltage range – is changed by means of a switch located on the instrument.

Usually, ammeters and voltmeters used for automotive service work should have an accuracy of 1% of full-scale deflection. These instruments should also be compensated for changes in temperature.

Voltmeter

Hook up ***voltmeter*** in parallel where the voltage is to be measured. Power must be on and the circuit must be complete.

Voltage Drop Test

Place the positive (red) lead of the voltmeter on the side of the circuit coming from the battery positive terminal. Place the negative (black) lead on the other side of the part being tested. The wire from this side should lead to a ground or other parts. Check that the meter is on the correct scale before connecting the leads.

In this test the voltage used to operate that part of the circuit is measured. If it is the only part, you should read battery voltage. If there are other parts in that circuit, you will read less than battery voltage, *Fig. 3.56*.

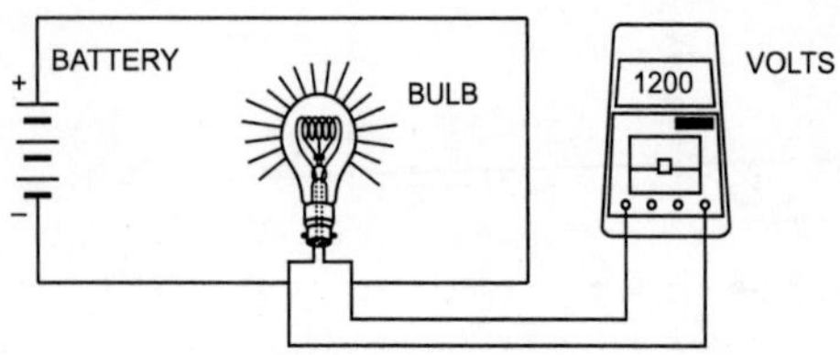

Fig. 3.56 *Measuring a voltage drop across a part.*

Voltmeter Starting Circuit Test

To make a starting circuit test, connect the positive (red) voltmeter lead to the positive battery terminal. Connect the negative voltmeter lead to the positive coil terminal, *Fig. 3.57*. With the ignition switch off, no voltage is supplied to the ignition coil, and no reading will occur on the voltmeter. During starting, the starting controls of the ignition switch bypass the ballast resistor, so its resistance is nearly zero. Because the resistance is nearly zero, the voltage drop across it is also nearly zero. With the ignition switch on and the engine running, the voltmeter reads the voltage drop across the ballast resistor. The meter should read 5-8 volts.

If any voltmeter reading is not as specified in the service manual, check and repair, if necessary, the battery, battery cables, connections, or starter until the specified readings are restored.

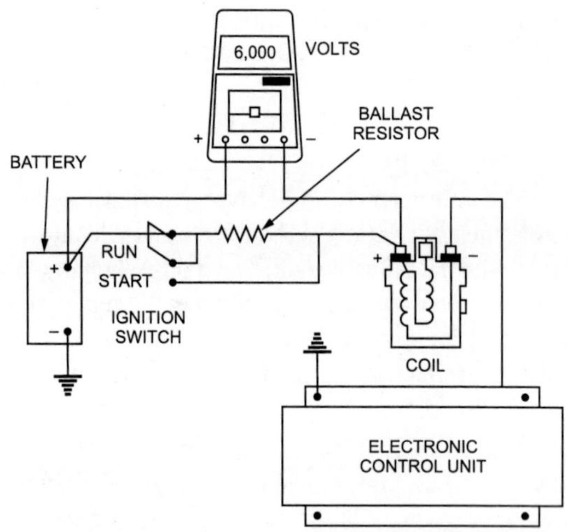

Fig. 3.57 *Voltmeter starting circuit test.*

Ammeter

Ammeters measure current flowing through the circuit. Always connect ammeter in series with the circuit you are testing. If you are measuring a parallel circuit, connect the meter before the branches split, or you will only be measuring the current in the branch where the meter is, *Fig. 3.58*.

To connect an ammeter, turn off power and disconnect the circuit at the proper point. Place the positive ammeter lead to the side of the circuit coming from the battery positive terminal. Place the negative lead on the side going toward ground. Check that the meter is on the proper setting. Turn on power and measure amps. Be careful: If the meter needle tries to move to the left, quickly disconnect the power. Reverse the leads and retest.

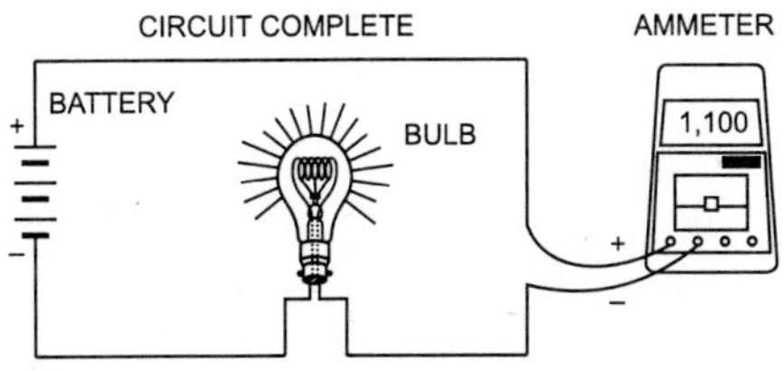

Fig. 3.58 *Measuring current flow with an ammeter.*

Ohmmeter Operation

The ohmmeter is designed to measure resistance of an electrical circuit or unit in ohms. A typical automotive application of an ohmmeter is testing the resistance of high-tension wiring. It is also used to test circuit continuity.

Generally, the ohmmeter utilizes a D'Arsonval movement. It has a calibrated resistance, a variable resistance, and a self-contained dry cell, *Fig. 3.59.* All ohmmeters need dry cells to serve as the voltage supply needed to force current through the circuit or unit being tested. The power must be off in the circuit being tested.

A special low-resistance wire is used to make leads that connect the ohmmeter to the circuit. Since they are used for calibrating the instrument, these same leads should be used when taking measurements. To calibrate the ohmmeter, it is necessary to "zero the meter." Joining the two leads together to establish zero resistance does this. The variable resistance is then adjusted to bring the pointer to the zero mark on the calibrated dial.

To measure resistance, just connect the ohmmeter leads (or connect the probes) to the terminals of the unit to be tested. Then read the resistance in ohms on the dial of the meter. Generally, ohmmeters used in automotive service application have two ranges: 0-500 ohms and 0-50,000 ohms.

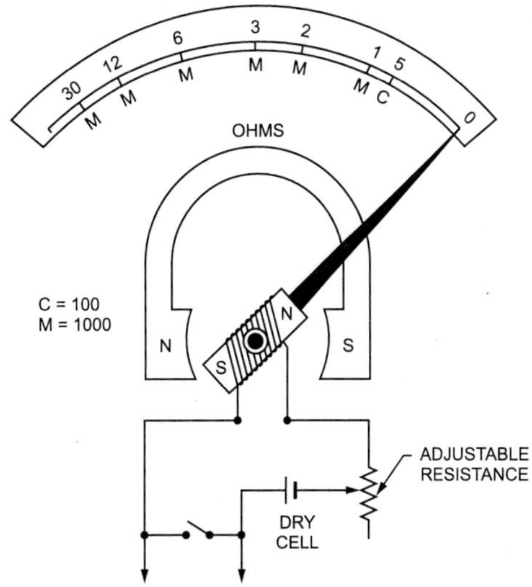

Fig. 3.59 *Schematic view of a typical ohmmeter.*

Ohmmeter

Ohmmeters measure the resistance in the circuit. You can either test a single part or the entire circuit, though usually only a part is measured in ignition testing. Before using an ohmmeter, you must zero the meter to get an accurate reading. The meter is zeroed by simply putting the two leads together and adjusting the meter to read 0 Ω. Remember that when using an ohmmeter that the power must be off. Also, the part being tested must be disconnected from the circuit on at least one side.

Resistance Test

To measure resistance, disconnect the circuit on either side of the part under test. Check that the power is off. Then, connect one lead of the ohmmeter to one side of the part and the other lead to the other side. Read the indicated on the meter, *Fig. 3.60.*

Multimeter or VOM

A ***VOM (volt-ohm-milliammeter)***, also called multimeter, is a voltmeter, ohmmeter, and ammeter combined into one case. A ***Multimeter*** tests a variety of electrical functions, *Fig. 3.61*. It identifies problems in the electrical system. Measurements often require a different meter setting, or a different meter, and a different way of connecting the meter's test leads or wires.

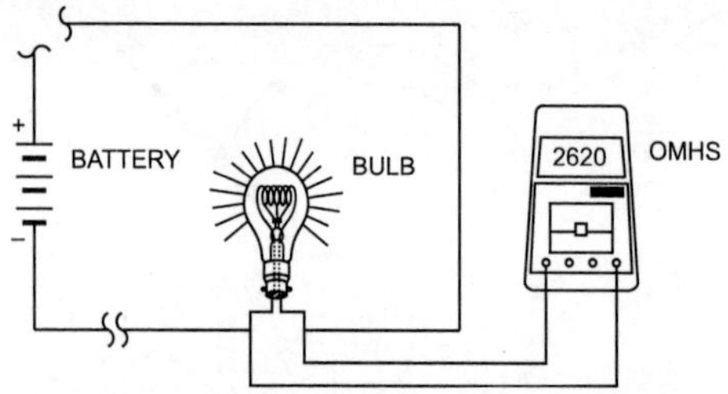

Fig. 3.60 *Measuring resistance with an ohmmeter.*

Fig. 3.61 *Multimeters. (Snap-on Tools Corp.)*

It will measure voltage, current, and resistance. It is used to check a faulty electrical component on an engine. If the current, voltage, or ohms test value is not within specs, the circuit or component is defective. As shown in *Fig. 3.62*, a ***function knob*** or ***control knob*** can be turned to select the type of measurement to be made (volts, amps, or ohms).

The measurements made by the meter can be displayed in different ways. The most common type of meter is the dial and needle or ***analog*** type. An ***analog meter*** uses a needle to register the electrical value, *Fig. 3.63 (a)*. Several curved scales and a moving needle indicate measurements. The disadvantage of this type of meter is the difficulty of reading the right scale, especially if the needle is vibrating.

Fig. 3.62 *Analog meter uses a needle to indicate reading. It is handy to detect a bad connection. Wiggle wires while watching the meter needle. If needle fluctuates, there may be a poor electrical connection. This type of meter should not be used to check some engine sensors. It can draw too much current and burn the sensors. (Saab)*

Digital multimeters are becoming more common. The measurements are displayed as numbers. The advantages of these meters are: they are more accurate, and easy to read. Some digital meters are specifically designed for ignition system service and provide settings marked for specific ignition system tests.

A ***digital meter*** uses a number display to show the electrical values, *Figs. 3.63 (b)* and *3.64*. It is a more modern type that is normally recommended when checking engine sensors. A digital meter draws less current than an analog meter. It is less likely to damage delicate electronic components.

Rules to remember when using a multimeter are:

- Do not connect the meter, when set on ohms, to a source of voltage. This could damage the meter or blow its fuse. Disconnect the battery from the circuit before measuring resistance.
- Follow applicable service manual recommended procedures when making electrical tests. A wrong meter connection could cause serious damage to computer systems.
- Disconnect electrical terminals carefully to avoid any damage to the terminal and cause high circuit resistance and an engine malfunction.

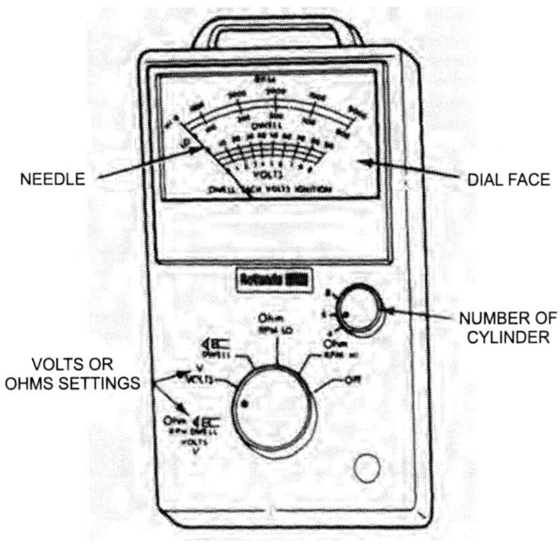

(a) Analog meter uses needle that swings across meter face.

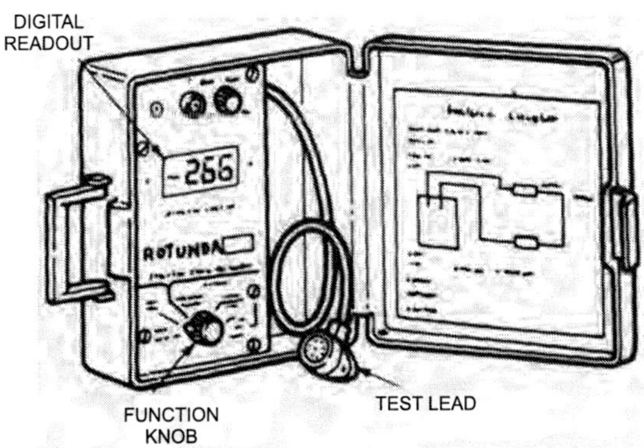

(b) Digital meter uses number output

Fig. 3.63 *Both analog and digital meters are needed in engine repair with modern computer controlled engines. (Ford Motors)*

- Preset the meter to a high range before measuring an unknown electrical quantity, e.g., when measuring current, set the meter to a high range. This will prevent excess needle deflection or current flow through the meter that could damage the meter.
- Use a high impedance (high resistance) meter when checking electronic components, particularly in an engine computer control system. Some meters with low impedence will draw too much current through the electronic device, and ruin it.

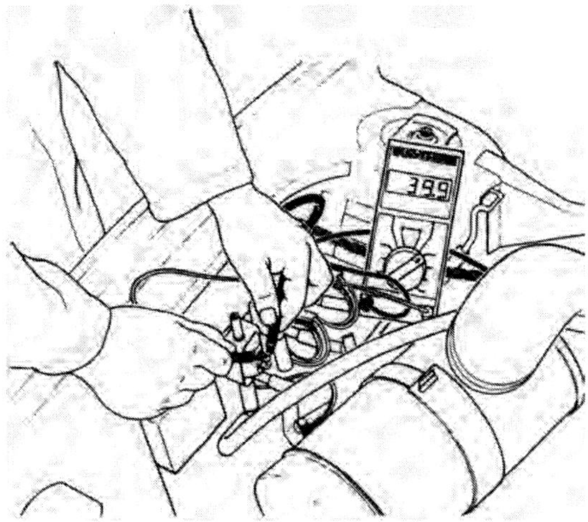

Fig. 3.64 *Digital meter has number display instead of needle. This is more modern type that is usually recommended for testing sensors. Here mechanic is shown measuring resistance of vacuum valve winding to check for open or shut. Readings must be within specs. (Peugeot)*

- When using a pointed test probe, be careful not to stick the sharp probe into your hand. The tip can easily slip off of a wire and prick through the skin. Do not prick holes in wires going to some engine sensors, e.g., oxygen sensors. They produce such a small voltage that moisture entering through the hole in the wire insulation could affect sensor signals to the engine control computer.
- Always read the meter's operating instructions before use. Meter ranges, specifications, and capabilities do vary.
- Position the meter so that it cannot be damaged from a fall or hot or rotating engine parts. Keep the leads away from the hot exhaust manifolds and rotating engine fan.

Normally, hook up voltmeters in parallel with the power on - the positive lead toward the battery positive side. Hook up ohmmeters with the power off, across a component that is disconnected from the circuit. Always connect ammeters in series with the circuit and before branches split from the circuit. Always connect the ammeter before turning the ignition switch on.

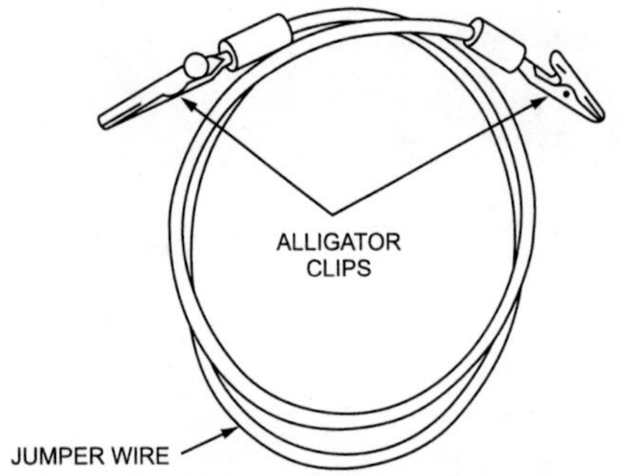

Fig. 3.65 *(a) Jumper wire for bypassing component or connecting voltage directly to circuit.*

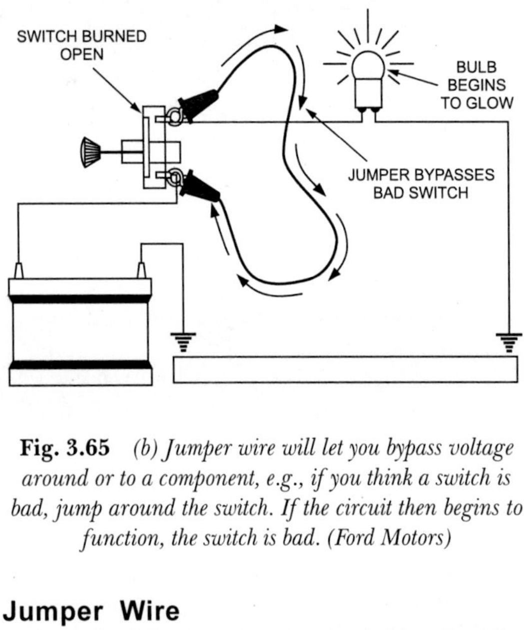

Fig. 3.65 *(b) Jumper wire will let you bypass voltage around or to a component, e.g., if you think a switch is bad, jump around the switch. If the circuit then begins to function, the switch is bad. (Ford Motors)*

Jumper Wire

A ***jumper wire*** is handy for testing switches, relays, solenoids, wires, and other non-resistive components. The "***jumper***" can be substituted for the component, *Fig. 3.65(a)*. If the circuit begins to function with the jumper in place, then the component being bypassed is defective. For example, a jumper wire is used to bypass a component in a circuit, *Fig. 3.65 (b)*, e.g., the jumper wire could be used to bypass a switch. If the circuit begins to function with the switch bypassed with jumper connected across switch terminals, then the switch is open and should be replaced. Jumper wires can also be used to connect a meter to various engine components for testing.

Jumper Cables

Jumper cables are used to start a vehicle with a dead (discharged) battery. The cables are connected between the dead battery and another vehicle's battery, *Fig. 3.66*. This will allow you crank and start the engine.

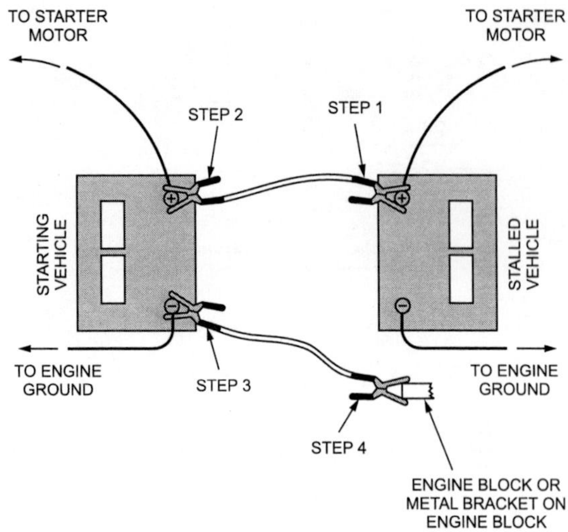

Fig. 3.66 *Jumper cable usage guide.*

When connecting jumper cables, connect directly to both battery terminals on the good battery. Connect the positive cable to the positive battery terminal on the discharged battery. Finally, connect the negative terminal to a good vehicle ground so as to prevent sparks from occurring around the discharged battery.

Remote Starter Switch

A ***remote starter*** switch is handy when it is required to rotate the crankshaft, *Fig. 3.67*. It is connected to the starter solenoid and to the battery. When the remote starter switch is pressed, the starter is energized to crank the engine. It is usually used when adjusting engine valves, during engine removal for flywheel-torque converter bolt removal, and to make compression tests.

Special Electrical Testers

Various types of specialized electrical testers designed to perform specific tests are available. Some of the most commonly used include:

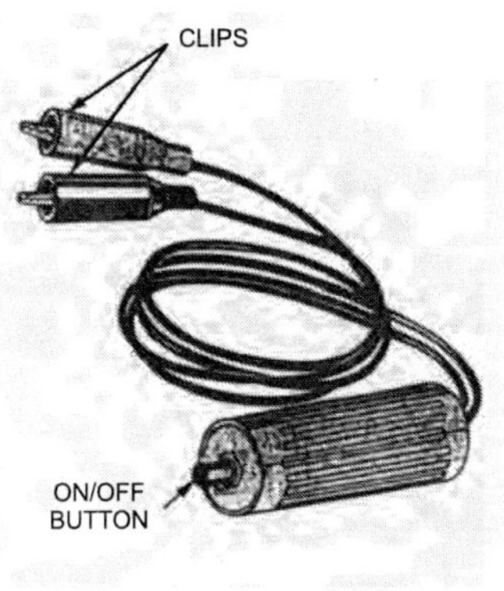

Fig. 3.67 *Remote starter switch is connected to starter solenoid so engine can be cranked from engine compartment or under engine. (Ford and Parker)*

A ***charging system*** tester is an electronic device for quickly checking the voltage output of the alternator, *Fig. 3.68*. It is connected to the electrical system. Its indicator lights show whether the charging system is working properly or not.

An ***ignition system tester*** is used to check the major components of an electronic ignition system, *Fig. 3.69*. It uses indicator lights to show whether the component is working properly or not.

A ***load tester*** is used to check batteries, the charging system, the starting system, and other electrical tests. It is very fast and easy to use. A load tester is shown in *Fig. 3.70*.

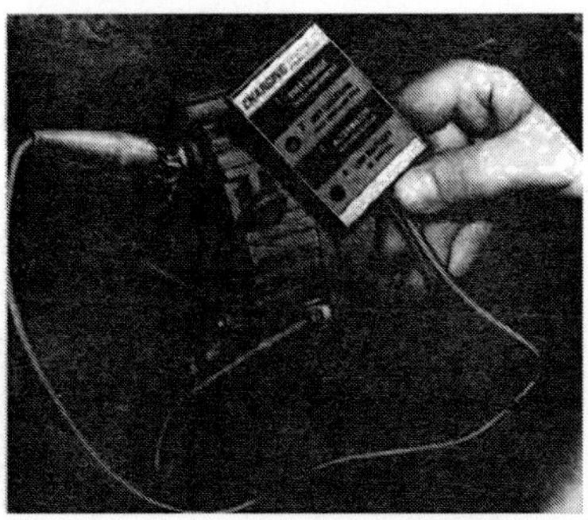

Fig. 3.68 *Special charging system tester will quickly show condition of alternator and regulator with indicator lights. (Belden)*

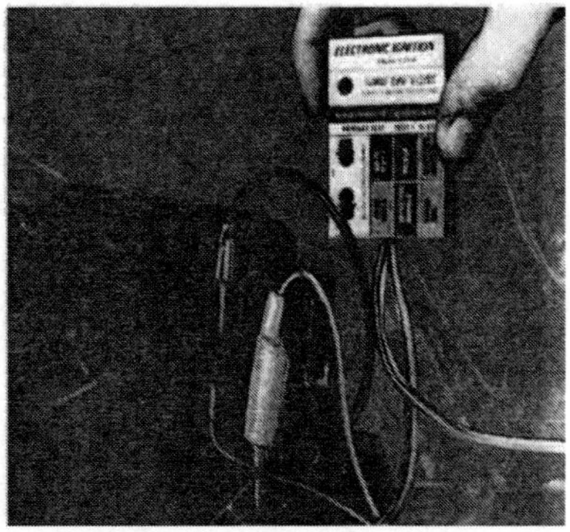

Fig. 3.69 *Electronic ignition system tester will also show system condition with indicator lights. Testers vary and it becomes important to read operating instructions before using the tester. (Be/den)*

Fig. 3.70 *Load tester will quickly check condition of battery, charging system, and starting system. (Lincoln)*

A ***computer system tester*** is used to analyze the condition of the on-board computer and its related components. It is usually connected to a specified test point on the system. Then, a digital code or engine trouble code in the form of number output, flashing lights, etc., can be used to isolate the location of any defective parts. The service manual will have a chart describing what each number code or value means, *Fig. 3.71*.

Many other electrical-electronic testers are used by an engine technician and are described in other chapters.

Fig. 3.71 *This modern tester will check computer control system. It will perform numerous tests. Variations of this tester are available and it is important to follow the operating manual instructions.*

Test Light

A ***test light*** is a fast method of checking an electrical circuit for power or voltage, *Fig. 3.72(a)*. It has an alligator clip that connects to ground. Then the pointed tip can be touched to the circuit to check for power. If there is no voltage, the light will glow. If it does not glow, there is an open or break between the power source and the test point. A ***self-powered test light*** is similar to a flashlight with a lead attached. It utilizes batteries and is used to check for ***circuit continuity***. To use this type of test light, the normal source of power (battery or feed wire) must be disconnected. If the light glows, the circuit or part has continuity (low ohms). If it does not glow, there is an open or break (high ohms) between the two points. A test light is used to check the voltage in different sections of a circuit, *Fig. 3.72 (b)*. For example, if the engine does not start and there are no sparks at the spark plugs, you could use test light to make sure voltage is reaching the ignition coil. It will quickly check for ***circuitry continuity***.

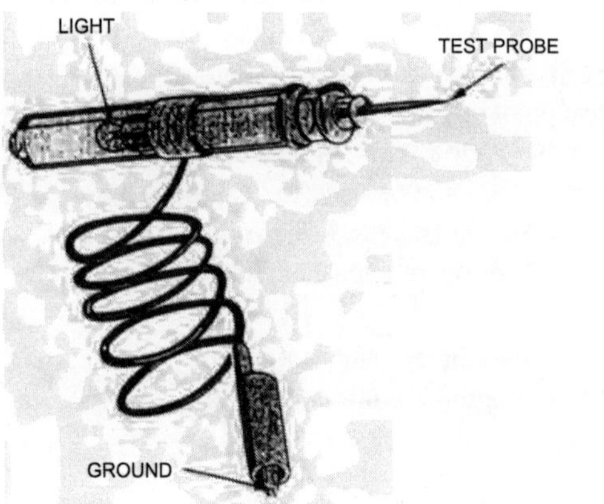

Fig. 3.72 *(a) Test light for making sure circuit or component is getting electricity.*

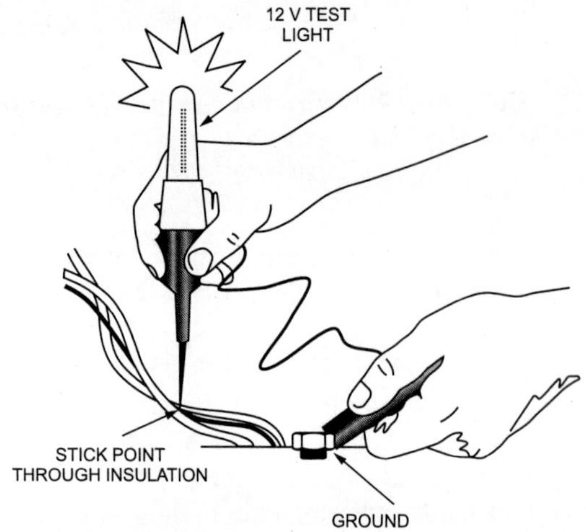

Fig. 3.72 *(b) Test light is fast way of making sure power is being fed to section of circuit. Sharp probe can be touched on metal connections or pierced through insulation to check for voltage. Do not pierce insulation on sensor wires carrying low voltage output. (Lisle Tools)*

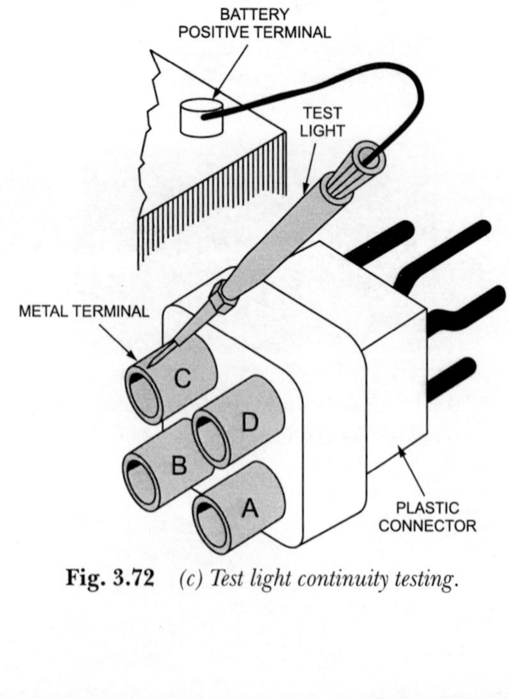

Fig. 3.72 *(c) Test light continuity testing.*

A test light, has two wires hooked to a filament or bulb. If you touch one wire to one terminal of the battery and the other wire to the other terminal, current will flow through the wires to light up the bulb. The test light will also light when you touch one wire to the battery terminal (positive) that is not connected to the engine or other metal parts of the vehicle, and touch the other wire to any metal part that is connected to the battery, *Fig. 3.72 (c)*. All these metal parts hooked to one side of the battery are called ground (negative).

Battery Charger

A ***Battery charger*** is used to recharge a discharged vehicle battery. Normally, the red charger lead connects to the battery positive (+) terminal. The black charger lead connects the negative (–) terminal.

Always connect the battery charger leads to the battery ***before*** *turning the charger* ***ON****. This is required to prevent sparks that could ignite any battery gas. The gases present around the top of a battery can explode violently.*

Timing Light

The ***timing light*** is a strobe light that freezes crank pulley rotation, *Fig. 3.73*. If oscilloscope were set up correctly, the crank pulley will appear not to rotate. This freezing of rotation is a result of the pulse of light emitted by the timing light happening at the same spot on the crank pulley. Correct ignition timing causes the spark plug to fire when the pistons are a specified number of degrees *before top dead centre* (BTDC).

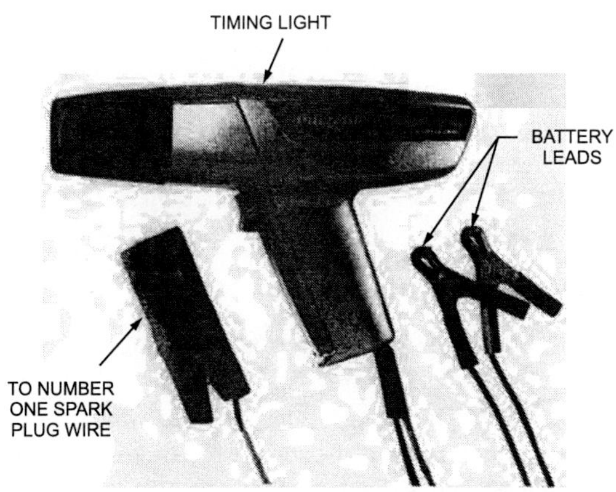

Fig. 3.73 *Timing light is strobe light. It is aimed on timing marks and spinning damper or flywheel to adjust when spark plugs fire in cylinders in relation fo piston position. Note lead connection points. (Peerless)*

Using the Timing Light

The timing light hookup for most vehicles, as shown in *Fig. 3.74*, is simple. The black lead clamps to ground, the red lead clamps to the distributor side of the coil and the induction pickup snaps on the # 1 spark plug wire.

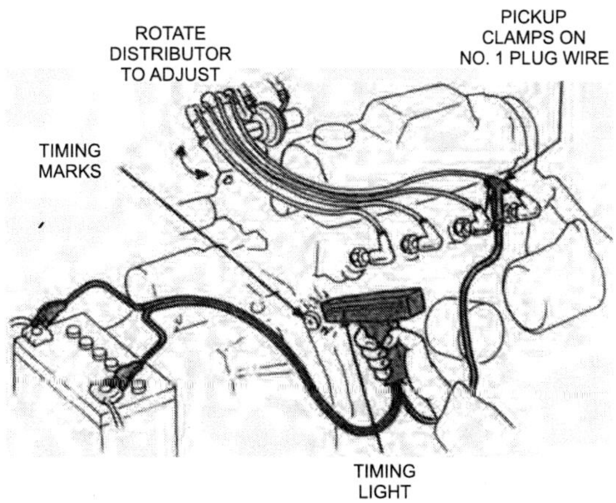

Fig. 3.74 *Timing light hookup. (American Motors)*

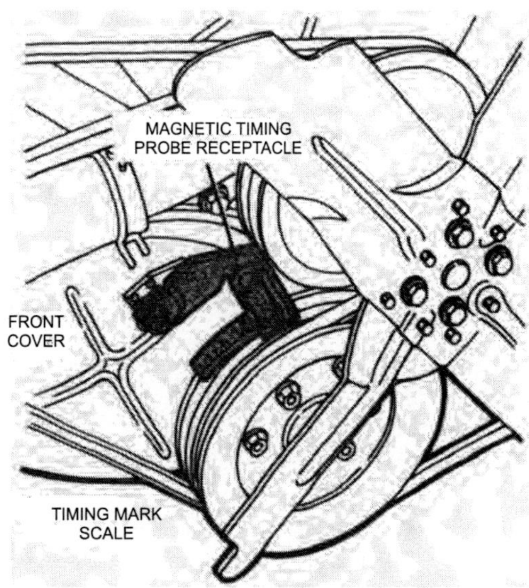

(A) Six Cylinder

Fig. 3.75 *(Contd).*

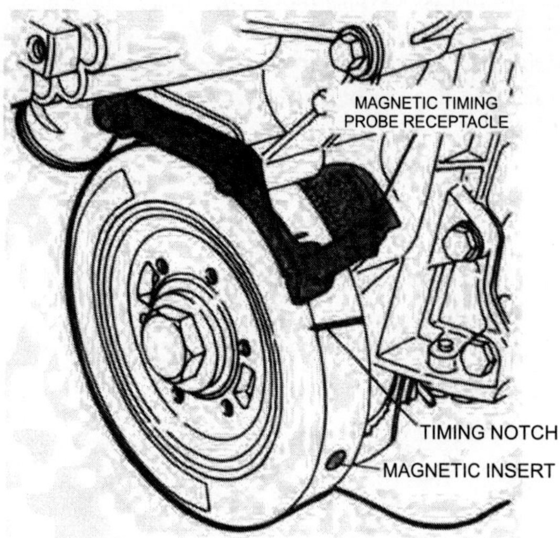

(B) Eight Cylinder

Fig. 3.75 *Index marks on the crank pulley and timing cover. (Ford Motors)*

The steps in timing most vehicles are the following:

1. Find the ignition specifications in the applicable service manual.
2. Remove and plug the distributor vacuum hose.
3. Loosen the distributor hold-down bolt.
4. Hook up the timing light.
5. Start the engine.
6. Adjust timing to the specs by turning the distributor, *Fig. 3.75.*
7. Tighten the distributor hold-down bolt.
8. Check timing again.
9. Shut off the engine, remove the timing light, and reconnect the distributor timing hose.

Oscilloscopes

The *oscilloscope*, *Fig. 3.76*, is an electronic device that allows mechanics/technicians to visually observe and measure the instantaneous voltage in an electrical circuit. The *oscilloscope* is probably the most efficient and complete diagnostic piece of equipment used in modern automotive repair shop. With it, you can "see" inside the engine. Basically, the oscilloscope consists of a cathode ray tube (CRT) and operational circuitry. It operates much the same manner as a television set.

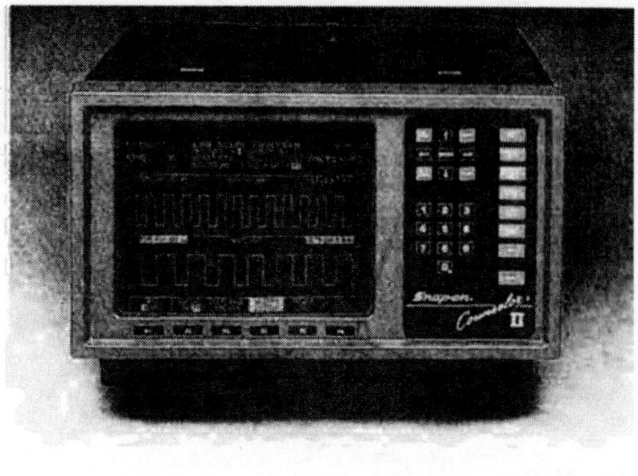

Fig. 3.76 *This digital oscilloscope can be used as a module in a console or as a stand-alone unit. (Snap-on Tools Corp.)*

Oscilloscope can be used to test conventional and electronic ignition systems, alternator diode condition, and cylinder balance. Oscilloscopes can be used to test ignition coil performance, distributor component condition, and spark plug circuit firing voltages.

The oscilloscope produces a graph-like picture showing voltage values with respect to time. The picture usually is referred to as the pattern or waveform. This pattern on the oscilloscope face or screen provides a visual means for comparing the performance of the vehicle's ignition system or charging system to a normal pattern for the system being tested.

Technically, the pattern is produced by the cathode ray, or electron beam, striking a phosphor coating on the inside of the CTR screen. When the electrons strike this material, it gives off a brilliant glow, making it possible to see the path of the beam as it moves across the screen. The electron beam is moved or deflected by voltage applied to metal plates within the tube. Horizontal deflection, or sweep, moves the beam from left to right. Vertical deflection, or sweep, moves the beam up and down on the screen of the CRT. The screen, then, shows "time" on the horizontal axis called the "*x*-axis." The vertical axis is called "*y*-axis."

It displays voltage and time, *Fig. 3.77*, indicating the condition of the alternator, coil, condenser, and all electronic ignition system components. It can also tell cylinder balance, carburetor balance, distributor faults, and resistance or leakage within the spark plugs.

The primary use of the oscilloscope is to study the primary and secondary ignition circuits, *Fig. 3.78 and 3.79*. Finally, it can do all of the jobs of the meters already described.

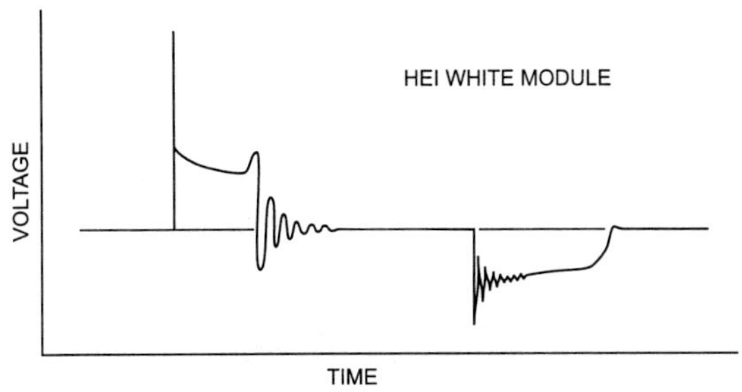

Fig. 3.77 *On this particular oscilloscope, the pattern, or waveform, is normal for a single cylinder in GM High Energy Ignition System (HEI). (Peerless Instrument Co.)*

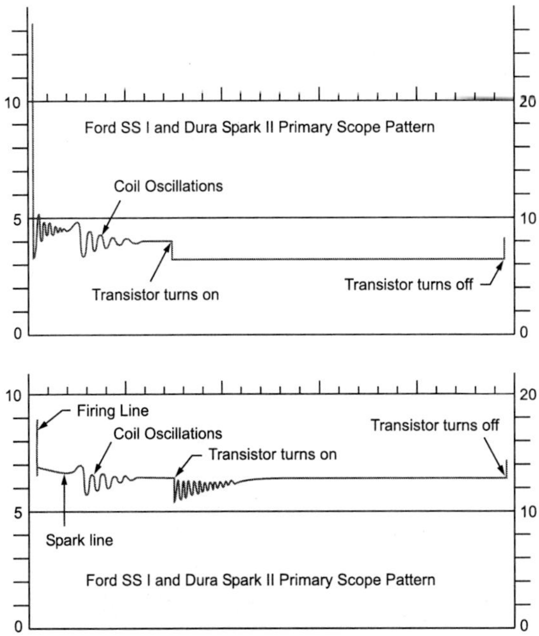

Fig. 3.78 *Typical Dura-Spark ignition patterns. (Sun Electric Corp.)*

To fully understand and learn the use of the oscilloscope, you must know the ignition system you are working on and what the patterns mean. The specific information is provided in service and the oscilloscope's manufacturer manuals.

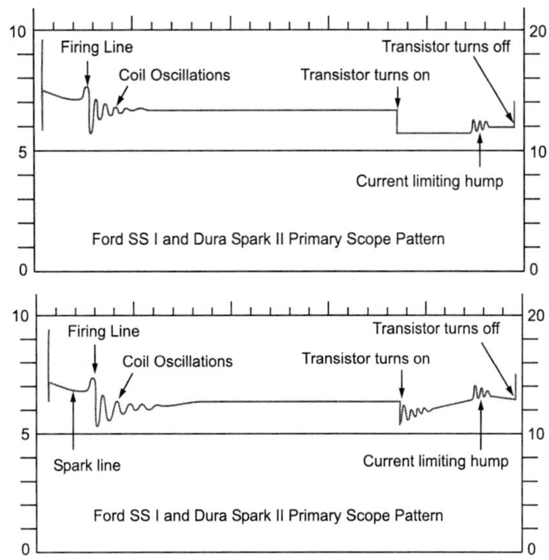

Fig. 3.79 *GM high-energy ignition oscilloscope patterns. (Sun Electric Corp.)*

The hookup of an oscilloscope uses a variety of clamp-type connectors, *Fig. 3.80 (a)*. When the engine is running, hooked up oscilloscope shows patterns that a trained technician can interpret to "see" what is going on with the specific part or parts being diagnosed.

In addition to being able to measure voltage in an electrical circuit, the scope can be used to determine the polarity of the voltage. The vertical movement of the electron beam indicates polarity. When the scope leads are properly connected to the secondary circuit of the ignition system, the voltage (normally negative) will appear above the zero reference line, *Fig. 3.80 (b)*.

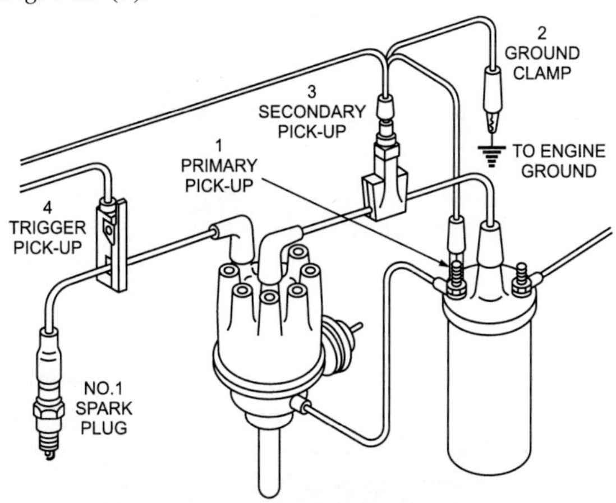

Fig. 3.80 *(a) Oscilloscope hookup.*

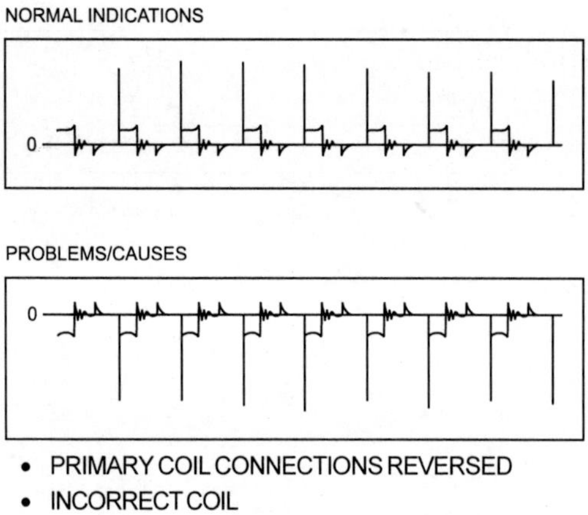

Fig. 3.80 *(b) When testing spark polarity on an oscilloscope, the normal pattern is above the zero reference line (top). Reversed polarity (bottom) means the incorrect coil has been installed or the coil connections have been reversed. (Peerless Instruments Co.)*

The voltage on the electrical system to which it is connected controls the oscilloscope pattern. Oscilloscopes can also be connected to nonelectrical components by means of special pickups known as *transducers*. These devices, which convert other forms of energy to electrical impulses, are used to observe engine compression and valve action, and to locate noises and vibrations.

Tachometers

A *tachometer* is designed to measure the speed of a rotating part in rpm. In automotive service applications the tachometer is used to measure the engine speed. Modern tachometers normally consist of a jeweled D'Arsonval movement, zero adjuster, and solid-state circuitry with low and high ranges provided. The low range may be 0-1200 rpm, while the high range could be 0-6000 rpm, for example.

The two types of tachometers are used with the automotive engines: *Mechanical* and *Electronic*. Mechanical tachs are not as popular as they once were because they are noisy, expensive, and need lubrication.

The electronic tachometer is the most versatile. Tachometers measure engine rpm in different ways. The standard tachometer for gasoline engines senses the firing impulses sent to the spark plugs by the distributor. Other tachometers for gasoline engines have a photoelectric pickup or a magnetic pickup. Diesel engines use a special pickup on the alternator, photoelectric cell, magnetic pickup, or fuel line sensor to drive the tachometer.

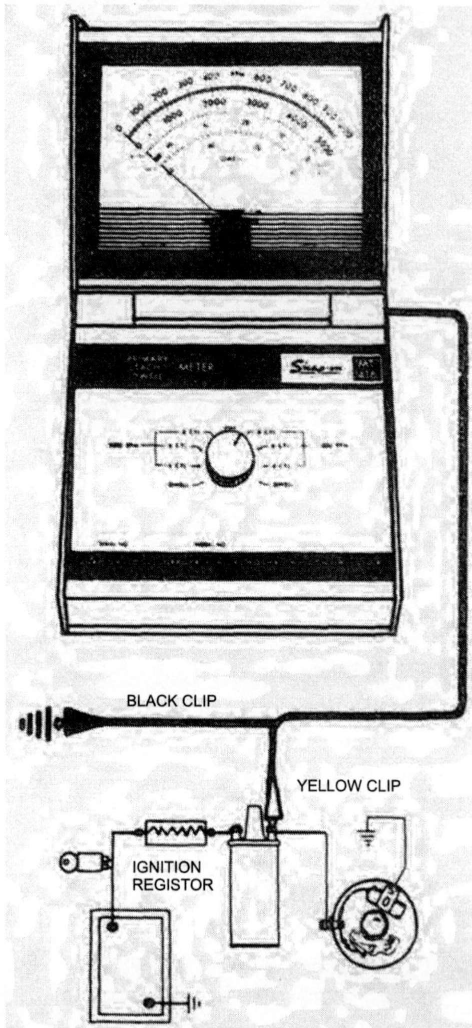

Fig. 3.81 *Tachometer connected to a contact-point ignition system. (Snap-on Tools Corp.)*

The tachometer is usually connected to the engine ignition system and operates electrically, *Fig. 3.81*. To test the engine speed, connect one lead of the tachometer to the primary terminal of the distributor. Connect the other lead to ground. It measures the number of times the primary circuit is interrupted (on SI engines). The tachometer is essential instrument because the engine idle speed must be adjusted to a specific rpm. Also, many engine tests must be made at specified engine speeds. The tachometer has a selector knob or buttons that can be set to the number of cylinders in the engine under test. Set the control for low range to test curb idle and fast idle. Set the control for high range to make tests at speeds above the low range rpm.

Some vehicles have tachometers mounted on the instrument panel, *Fig. 3.82*. They indicate how fast the engine is turning. This helps the driver to keep the rpm within the range at which the engine develops maximum torque.

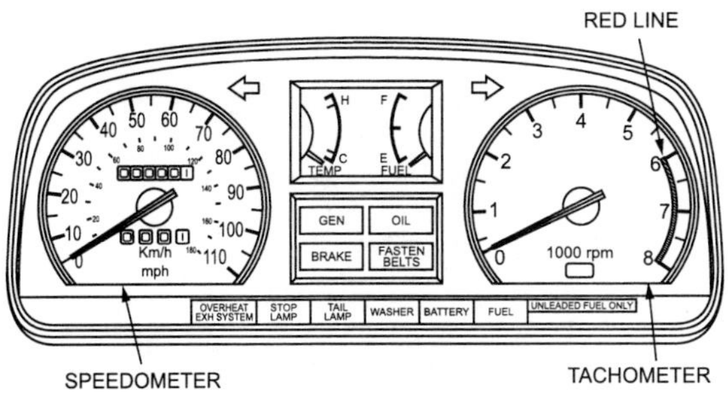

Fig. 3.82 *Engine tachometer mounted in the vehicle instrument panel.*

Using a Tachometer

To hookup a standard tachometer on a breaker-point type ignition system, connect the two probes, a red lead and a black lead. Connect the red lead to the negative side, or point side, of the coil. Connect the black lead to any ground, *Fig. 3.83*. Read the rpm on the scale.

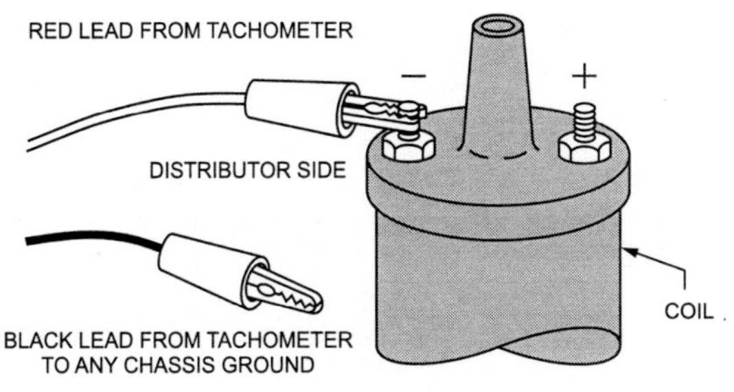

Fig. 3.83 *Tack/Dwell meter connections.*

Tack/Dwell Meters

Often, a dwell meter is included in the same case as a tachometer. Usually these are hand-held units called "tach-dwell meters", *Fig. 3.84*. The *dwell meter* is used usually for measuring the angle through which the ignition distributor shaft turns while the breaker points are kept closed, known as *point dwell*, or *cam angle*, *Fig. 3.85*. Vehicle manufacturers specify the cam angle for different engine applications in their product lines. Since the cam angle relates to opening and closing of the breaker points, adjusting the point gap also adjusts the cam angle. After making the point gap setting, it can be checked for accuracy, and readjusted if necessary, by hooking up the dwell meter and measuring the cam angle.

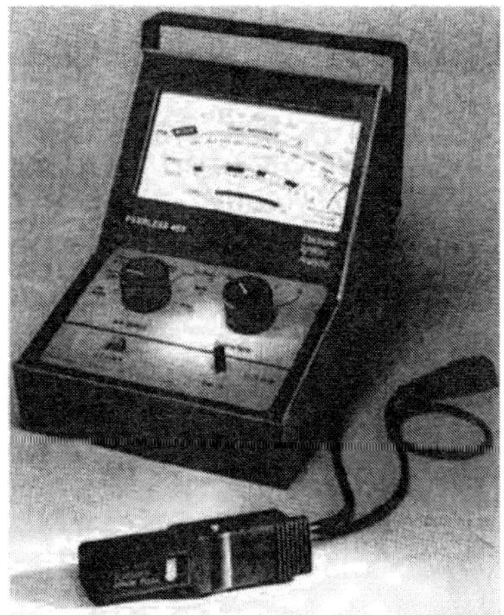

Fig. 3.84 *(a) Tack-dwell meter will measure engine speed for throttle adjustment. Dwell meter will measure in degrees for contact point and computer-controlled carburetor service. (Peerless)*

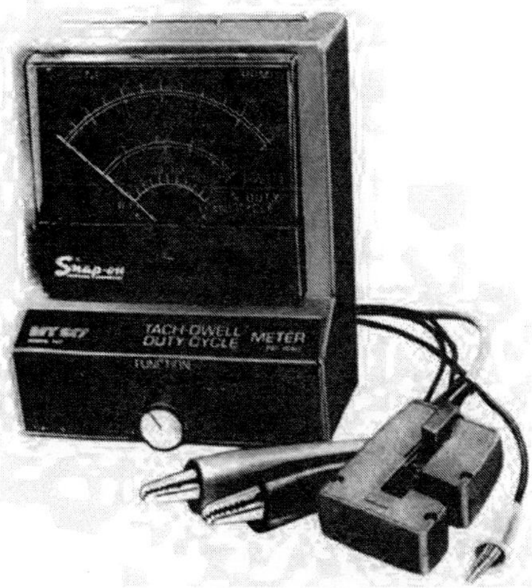

Fig. 3.84 *(b) Tack/dwell meter. (Snap-on Tools Corp.)*

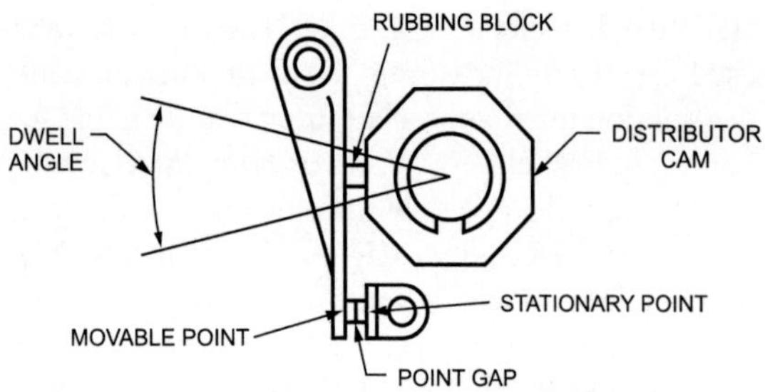

Any change in point gap will change the dwell angle

Fig. 3.85 *(a) Dwell angle.*

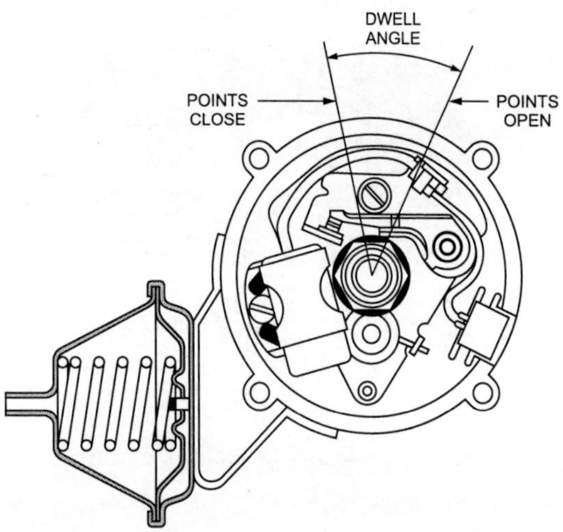

Fig. 3.85 *(b) Dwell angle.*

Using the Tach/Dwell Meter

Because the dwell meter and tachometer are essentially one unit, the dwell meter connects the same way as the tachometer does. Connect the red lead to the negative side of the coil and the black lead to ground, *Fig. 3.83*. Then read the dwell from the scale. To change the dwell, adjust the point gap closer to or farther away from the stationary point as necessary to set the angle to specs.

Electronic ignition systems do not need to be set with a dwell meter. In these systems solid-state electronic devices are used in place of points. These electronic devices last longer and are more accurate than point ignition systems.

Battery/Starter Tester

Battery testing equipment includes the hydrometer, the voltmeter, and the multi-purpose testing unit, which comprises voltmeter, ammeter, and load control, *Figs. 3.86 and 3.87*. Housed in a single portable, cabinet, some of these combination units are used to test batteries, starters, alternators, voltage regulators, and related circuitry. The tester shown in *Fig. 3.86* has a 0-750 amperes current range, zero amps control, and light emitting (LED) displays. It also includes a "ripple" indicator that analyzes the quality of an alternator's DC output. This combination unit is used to perform: battery load, starter current load, cranking voltage, engine cranking speed, diode-stator condition, alternator current output, and voltage regulator operation.

Fig. 3.86 *Starting and charging systems tester shows amperes and volts in digital display. It has an automatic return load control knob and a ripple indicator that reveals alternator defects. (Snap-on Tools Corp.)*

Fig. 3.87 *Portable battery tester also checks condition of alternator, regulator, and primary circuit cables. Tester has a built-in discharge load comparable to vehicle's starter current draw. (Christie Electric Corp.)*

The portable battery tester shown in *Fig. 3.87* is designed to test all recognized battery ratings, including cold cranking performance and ampere-hours. It has a compensation adjustment for battery temperatures from –18° to 52°C. A colour-coded scale shows condition of battery, cables, alternator, and regulator.

Battery Fast Chargers

Battery fast chargers are automatic and foolproof. Most have alternator protection to guard against reverse hookup. One type of fast battery charger uses a light emitting diode (LED) indicator that shows "green" for a correct connection or "red" for a reverse connection.

Charging rate output selection is provided on fast chargers. Some of them have an electric timer with 120-minute "hold" position for slow charging. Maximum charging rates are in the 60-100 amp range; cranking power is in the 200-300 amp range, *Fig. 3.88*.

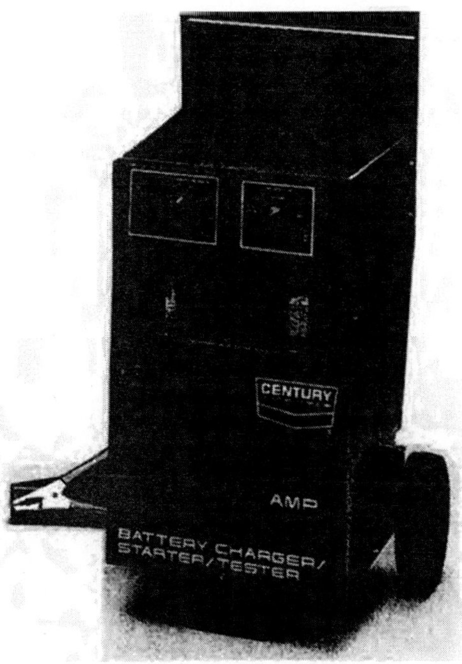

Fig. 3.88 *Battery fast charger offers low, medium, and' high charging rates. (Century Mfg. Co.)*

An enhancement of the fast charger is the fast charger/tester. In addition to alternator protection and slow charge timer, the charger/tester provides a "start charge" for completely discharged batteries. It also has a thermal overload device and battery type selector (conventional or maintenance-free). A battery rating control compensates for capacity of the battery in cold cranking amps or ampere-hours.

The battery fast charger/tester can analyze the condition, state of charge, and operating performance of the battery. It can also test the charging system, starter, and cables. It will reveal any short circuits in the electrical system, too.

Exhaust Gas Analyzers

With legal limits imposed on the exhaust gas emissions, service technicians need suitable test equipment to verify, through various modes of engine operation, whether or not the engine is emitting excessive amounts of air pollutants. *Infrared exhaust gas analyzers* are commonly used to detect the amounts of hydrocarbons (HC) and carbon monoxide (CO) in the exhaust gas, *Fig. 3.89*.

Figure 3.90 shows the infrared type of exhaust-gas analyzer. To use it, you insert a probe into the tall pipe of the vehicle exhaust system. The probe draws out some of the exhaust gas and carries it through the analyzer.

Infrared exhaust gas analyzers measure HC and CO by determining how much infrared is absorbed by a sample of the vehicle's exhaust gas. In the infrared block diagram shown in *Fig. 3.91*, infrared energy generated by an electrically heated element is projected through a sample tube (cell) that has the vehicle's exhaust gas flowing through it. Infrared radiation not absorbed by the exhaust gases hits on infrared detector and is converted to an electronic signal that causes the digital display of the test result to change. With this infrared gas analyzer, digital readouts have a 0-1999 parts per million (ppm) HC range, and a 0-9.99 % CO range.

A 4-gas analyzer is an advanced type of exhaust gas analyzer. This unit can test for HC, CO, CO_2 and O_2. Some 4-gas analyzers use emissions analysis to determine whether the cause of an engine performance problem is in the fuel system or in the ignition system, *Fig. 3.92*.

In addition to HC and CO, some exhaust-gas analyzers also measure the amount of CO_2 and O_2 in the exhaust gas. A tester that measures HC, CO, and CO_2 is called a three-gas analyzer. Some equipment manufacturers couple the 4-gas analyzer module with an engine analyzer said to provide "all systems" engine diagnosis. This computer test centre is capable of identifying a variety of problems, e.g., rich, lean, or unbalanced air-fuel mixtures; closed loop system malfunctions; air pump or catalytic converter malfunction; EGR valve leaks; blown head gaskets; excessive misfire; excessive spark advance; and leaks or restrictions in exhaust systems.

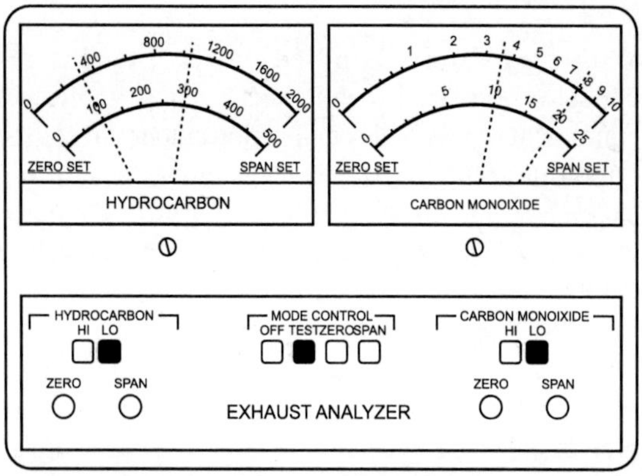

Fig. 3.89 *Diagram of a typical infrared exhaust analyzer. Dashed lines show normal range of readings on low scale of HC and CO meters.*

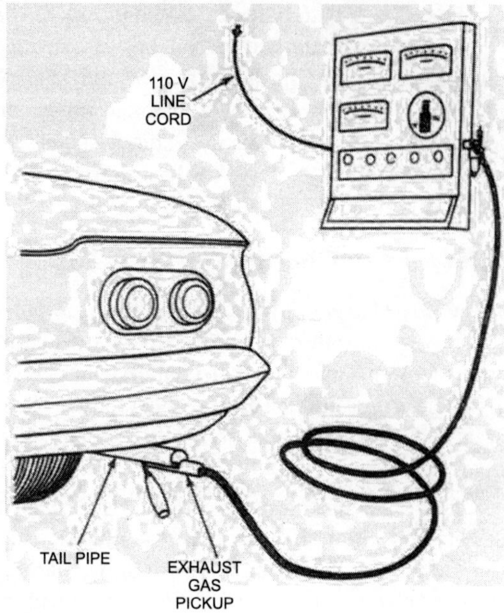

Fig. 3.90 *Exhaust gas analyzer connected for test.*

Other equipment manufacturers produce a digital performance gas analyzer that combines digital display with computer electronics. The unit is microcomputer controlled with an automatically timed warm up period and an automatically timed gas calibration. The digital display reads 2-2000 ppm HC in 10-ppm increments, and 0-10 % CO in 0.05% increments.

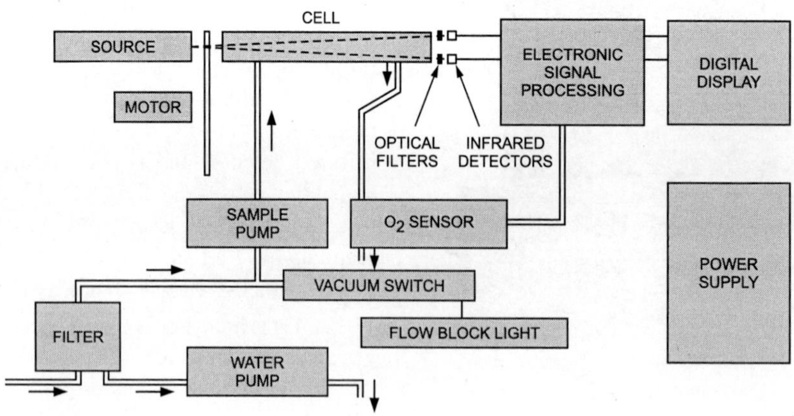

Fig. 3.91 *Infrared exhaust analyzer block diagram traces flow of exhaust gas sample through the cell. The sample absorbs infrared energy from the source. The digital display reads in HC and CO values. (Peerless Instrument Co.)*

A different kind of tester is required for NO_x, but it works in the same general way. It draws exhaust gas from the tail pipe and runs the gas through the analyzer. The reading is reported in grams per km. Generally, NO_x testers are available only in the testing laboratories. They are not much used in automotive-service shops.

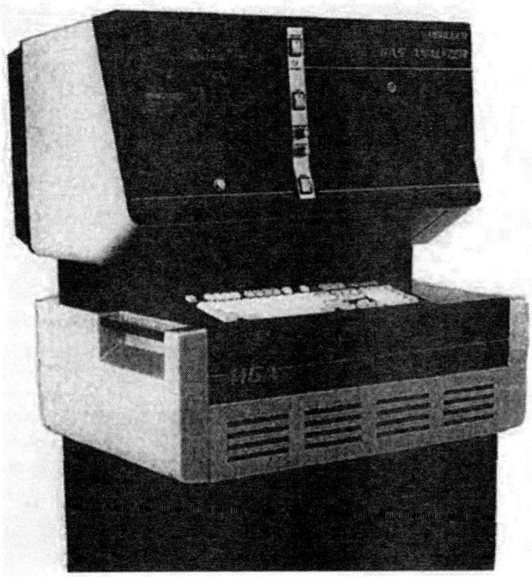

Fig. 3.92 *To meet challenge of diagnosing computer controlled engine systems, one test equipment manufacturer couples a four-gas analyzer with an engine analyzer. Note the video monitor and the printout. (Sun Electric Corp.)*

Engine Analyzers

Engine analyzer is a generic term used to describe a host of testers, ranging from a hand-held, ten-function tune-up analyzer to a massive console of equipment modules that includes a large screen oscilloscope, an automatic computer analyzer, and banks of large scale test meters, *Fig. 3.93*. In between, a popular group of compact digital engine analyzers are available for tying in with on-board computer systems, *Fig. 3.94*. These special testers interpret the signals that indicate malfunctions in various engine systems. Data is entered into the tester and specific tests are requested. Trouble codes are digitally displayed.

Console-type engine analyzers come in many packages. Some can be built by adding modules to a basic test stand to suit the individual needs.

Others are unitized and put on wheels to provide mobility. Still others are suspended from an overhead track and travel from service bay to service bay. Some consoles are used in conjunction with a dynamometer to simulate road testing.

Fig. 3.93 *This modular analyzer can be expanded as necessary. (Sun Electric Corp.)*

Fig. 3.94 *This type of compact analyzer is often called a scan tool. (OTC Div. of SPX Corp.)*

The most sophisticated of engine analyzers is the automatic computer analyzer. Factory approved specifications are programmed into the computer. Data is then collected, based on the vehicle's actual operating characteristics. The computer analyzes this information and deduces which components are faulty in any of the major engine systems i.e., starting, charging, fuel distribution, compression, timing and ignition. Test results are displayed automatically on the computer terminal. In addition a multi-copy printer provides a hard-copy printout. The printout contains both "specified" and "actual" readings for each test for the mechanic/technician and customer's verification.

Dynamometer

The largest piece of test equipment is the ***dynamometer***. As shown in *Fig. 3.95*, the vehicle should be driven onto rollers. Running the drive wheels against the rollers shows the relative power of the vehicle. Adjustments can be made to the engine, then it is possible to see the change in relative power and rpm. This is probably the most efficient way to tune an engine. Every adjustment in carburetion emission and timing is immediately seen as a change in relative power?

Fig. 3.95 *Dynamometer.*

3.10 TESTING SAFETY RULES

When using test equipment, remember the following safety rules:

- Carefully read and follow the operating instructions for the test equipment. Failure to follow directions can cause bodily injury and damage to the part or the instrument.
- If the engine is required to be running during the tests, set parking brake. Block the wheels. Place an automatic transmission in park or manual transmission in neutral, as the case may be. Connect an exhaust vent hose to the tailpipe when working in an enclosed auto repair shop.
- Refer to the auto manufacturer's service manual for specific testing procedures.
- Wear eye protection when working around an engine fan. Remove jewelry and secure long hair.
- Keep test equipment leads or hoses away from engine fan, belts, and hot engine parts. They can be damaged easily.
- Never look into a carburetor or throttle body when cranking or running the engine. Do not cover the air inlet with hand. Because, any possible engine backfire would cause a serious burn.

3.11 USING WELDING EQUIPMENT

Both oxygen-acetylene and electric arc welding equipment are often used in automotive repair shops. Careless or improper use of welders can cause burn injuries and start fires. To use ***oxygen-acetylene welding equipment*** safely the following rules must be followed:

1. Only a qualified welder should operate the welding equipment or supervise the process.
2. Fasten cylinders with a chain or by some other dependable device so that they will not roll or fall.
3. Close cylinder valves and replace protective cover on oxygen tank before moving cylinder.
4. Keep welding equipment clean and free of oil and grease. Use only clean shop rags for wiping equipment. Oil combined with oxygen can cause an explosion.
5. Inspect hoses before use. Make sure that the hoses are properly connected and that all connections are fairly tight. Service or replace any leaking hoses immediately.
6. The workspace should be properly ventilated.
7. Keep all flammable material away from work area.
8. When welding, always wear welding goggles.
9. Release regulator pressure screw. Open cylinder valves gradually.
10. Open acetylene cylinder valve 1¼ turns or less. Keep wrench at hand so that the valve may be shut off quickly if needed.
11. Keep acetylene pressure in the hose 103.4 kPa.
12. Use a friction torch lighter to ignite torch. Never use an open flame.
13. Close the acetylene valve first if the torch backfires.

14. Be careful to keep the lighted torch pointing away from your body and anyone near by.
15. Keep sparks and flame away from cylinders and hoses.
16. Close the cylinder valve when done welding.
17. Before leaving your work, either cool the section of metal that has been welded or mark "hot" on the metal with chalk or soapstone.

The safety rules to be followed when using the ***electric arc welder***:

1. Only a qualified arc welder should use welding equipment or the supervisor should be present for guidance.
2. Wear helmet with proper observation window, treated gauntlet gloves, and treated leather apron. All assistants and observers should also wear this equipment.
3. Keep shirt sleeves and pants cuffs rolled down. Wear leather jacket.
4. Do electric welding only in a correctly constructed booth or room or behind proper screens.
5. The work area should be adequately ventilated.
6. Keep all flammable material away from work area.
7. Keep the work floor area clear of all obstructions.
8. Take immediate precaution to check for any problem in the system if electrode holder, holder cable connection, cable, or cable terminals at the welding machine, ground clamp, lugs, or cable get hot.
9. Hang up electrode holder and turn off welder when work is being changed or is completed.

3.12 SHOP SAFETY

Automotive repair shop-work is varied and as mechanic you will learn to do many automotive jobs in the shop. All jobs can be completed easily and safely if you are properly trained and follow the shop safety rules. Safety means protecting yourself and others from possible injury. You do not to get hurt yourself or others around you if you become careless and absentminded.

Safety is your job. In the shop, you are *safe* when you adequately protect your eyes, your fingers, your hands and your body from danger of getting injured and you remain constantly careful about the safety of others you work with.

Shop Layouts

The term ***shop layout*** means the locations of workbenches, vehicle hoists, machine tools, work areas, washing and cleaning areas, tool racks, etc. A mechanic should always find out where every thing is located in the shop. Many shops have painted lines on the floor to mark off work areas. These lines guide customers and workers away from danger zones where machines are being operated. These lines also remind the workers to keep their tools and equipment inside work-area lines. Many shops have warning signs posted around machinery. The signs are posted to remind everyone about safety, and about how to use machines safely. Follow the posted instructions at all times when in the shop. The most common cause of accidents in the shop is failure to follow the posted instructions.

An engine technician will be exposed to many potentially dangerous situations.

Types of Accidents

Each of the six kinds of life threatening accidents that must be prevented are *fires, explosions, asphyxiation, chemical burns, electric shock, and physical injuries.*

Fires

Fires can cause horrible destruction and death. When servicing IC engines, gasoline, diesel, cleaning solvents, and oily rags are some of the many possible causes of a shop fire.

Gasoline is very dangerous flammable in the shop. Even a small amount of gasoline when set on fire may consume the whole vehicle or even the complete shop. To avoid starting a gasoline fire, some rules for handling gasoline are:

1. Never use gasoline as a cleaning agent. Cleaning solvents are not as flammable as gasoline.
2. Keep sources of heat like welding and cutting equipment, away from the engine's fuel system.
3. Wipe up gasoline spills immediately. Do not spread "quick dry" or oil absorbent on spilled

gasoline because the absorbent will become even more flammable.

4. Disconnect the vehicle battery before disconnecting a fuel line on an engine.
5. Wrap a rag around any fitting when disconnecting a fuel line. This will prevent fuel from leaking or spraying out and possibly initiating a fire when striking hot engine parts or electrical system components that could produce a spark.
6. Store gasoline and other flammable materials in approved containers with a tight lid, *Fig. 3.96.*

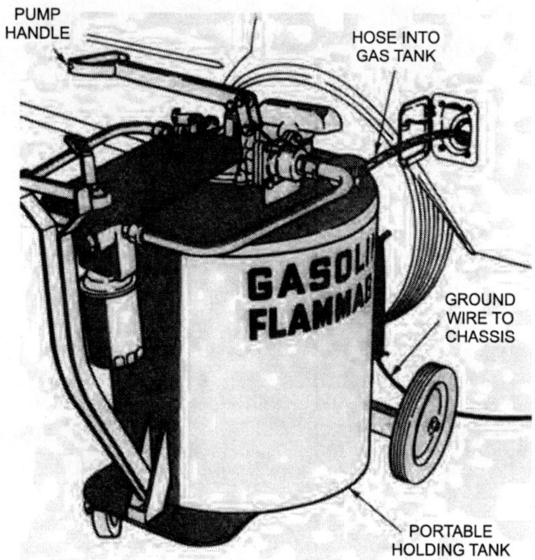

Fig. 3.96 *(a) When gasoline is removed from a vehicle, it must be stored properly. (Chrysler Corp.)*

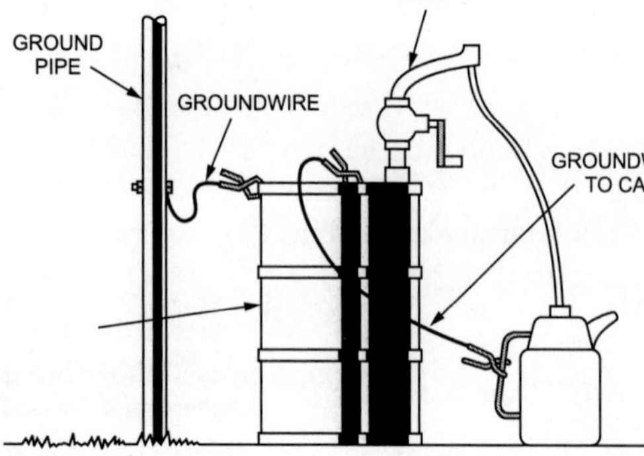

Fig. 3.96 *(b) Outside fuel and solvent storage.*

Electrical fires occur when a current carrying conductor "shorts" to ground. The excess current carrying conductor then becomes "red hot", melting its insulation. The insulation can then catch fire and result in other wires getting shorted and begin to burn, *Fig. 3.97.*

Always disconnect the vehicle battery to prevent electrical wires by accidentally shorting a hot wire to ground.

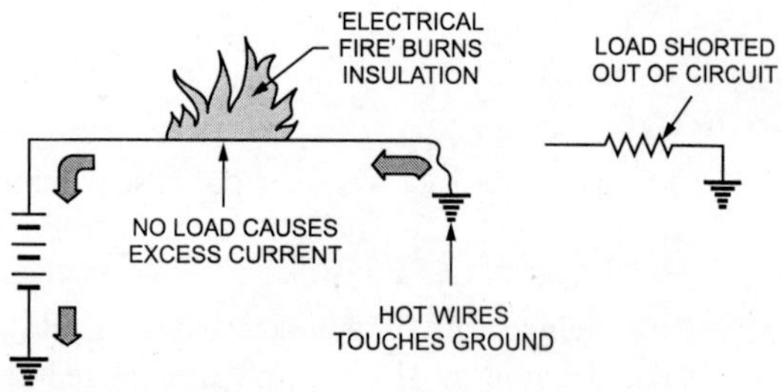

Fig. 3.97 *Electrical fire can cause tremendous damage to a vehicle in a matter of seconds. Hot wire touches ground and high current causes enough heat to make wire insulation burn. Always disconnect battery when wiring is disconnected to service automotive engine.*

Explosions

Different types of explosions are possible when servicing an engine or its systems. Vehicle batteries can explode, *Fig. 3.98.* Hydrogen gas can surround the top of a storage battery being charged or discharged. This gas is highly explosive. A slightest spark or flame can ignite the gas and cause the battery to explode. Chunks of battery case and acid can blow into eyes and face of any person nearby. Blindness, facial cuts, acid burns, and scars can result.

Various other sources, e.g., special sodium filled engine valves, welding tanks, propane filled bottles, and fuel tanks can all explode if mishandled.

Asphyxiation

It is caused by breathing toxic or poisonous substances in the air. Mild cases of asphyxiation will cause dizziness, headaches, and vomiting. Severe asphyxiation can cause death.

Most dangerous source of asphyxiation in an auto repair shop is an 1C engine. An engine's exhaust gases are deadly poison. Connect an exhaust hose to the tailpipe of any vehicle being operated in the repair shop, and turn on the exhaust system, *Fig. 3.99.* Other shop substances that are harmful when inhaled are discussed in their related chapters.

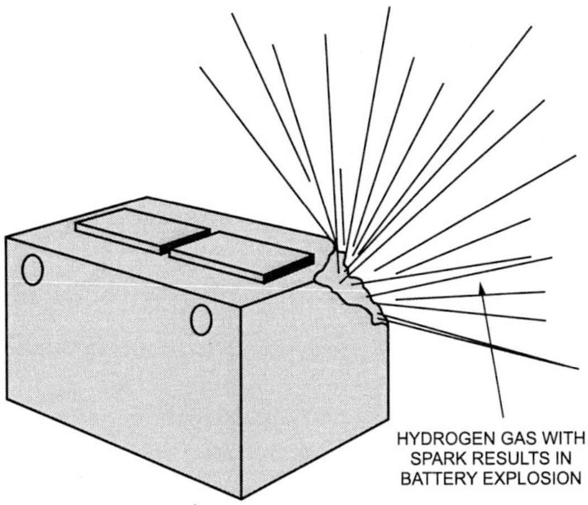

Fig. 3.98 *Vehicle battery can explode with blinding or lethal force. Hydrogen gas can collect around top of battery. Spark or flame can ignite gases. Then, gases inside the battery can ignite, blowing chunks of battery case and battery acid into shop.*

Fig. 3.99 *Use exhaust hose when engine is running in enclose shop. Poisonous gases will then be released out of the shop. (Chrysler Corp.)*

Chemical burns

Various solvents and part cleaners, battery acid, and some other shop materials can cause chemical burns to the skin. Always carefully read the directions and warnings on any chemicals that are used. Carburetor cleaner, used to decarbonize, is a super powerful and can severely burn hands in a matter of seconds. Wear

FIRES	TYPE	USE	OPERATION
A CLASS 'A' FIRES Ordinary combustible materials e.g., wood paper, textiles, and so forth Requires .... Cooling – Quenching	**FOAM** Solution of Aluminium Sulphate and Bicarbonate of Soda	OK FOR A B NOT FOR C	**FROM:** Direct steam into the burning liquid. Allow foam to fall lightly on fire.
B CLASS 'B' FIRES Flammable liquids, greases, gasoline, oils, paints, and so forth. Requires .... Blanketing or Smoothering	**CARBON DIOXIDE** Carbon Dioxide Gas under Pressure	NOT FOR A OK FOR B C	**CARBON DIOXIDE:** Direct discharge as close to fire as possible. First at edge or flames and gradually forward and upward.
	DRY Chemical	MULTI-PURPOSE TYPE: OK FOR A B C ORDINARY BC TYPE: NOT FOR A; OK FOR B C	**DRY CHEMICAL:** Direct steam at base of flames, use repaid left-to right motion toward flames
C CLASS 'C' FIRES Electrical equipment, Motors, switches, and so forth. Requires .... A nonconducting agent	**SODA-ACID** Bicarbinate of soda solution and sulphuric acid	OK FOR A NOT FOR B C	**SODA-ACID:** Direct steam at base of flame.

Fig. 3.100 *This chart illustrates the various fire classifications and fire extinguisher types. Always use an extinguisher designed for electrical and chemical fires (types B and C) in an automotive repair shop.*

rubber gloves and eye protection when using carburetor cleaner. If a skin burn occurs, follow the label directions on the container.

Electric Shock

It can occur when using improperly grounded electric power tools. Never use an electric tool unless it has a functional ground prong (third, round prong on plug socket). This prevents current from accidentally passing through body. Never use an electric tool on a wet shop floor.

Physical Injury

Physical injuries, i.e., cuts, broken bones, strained backs etc., can result from a number of accidents. Every repair technique needs to be carefully considered and evaluated to determine whether a particular operation is safe or dangerous and take precaution as required.

3.13 FIRE PREVENTION

Because there are many *combustibles*, such as gasoline and solvents, used in auto repair shops, special precautions are required to prevent fire. Fuel, thinners, and other combustibles should always be kept in closed containers that are designed for this purpose.

Unshielded flames should never be allowed. All shops should be provided with an ample number of appropriate *fire extinguishers*, *Fig. 3.100*. Everyone should be familiar with the location and use of the extinguishers, *Fig. 3.101*.

As an additional precaution against fire, rags that are soaked with fuel, oil, or solvents should be kept in an *oily waste container, Fig. 3.102*. Care must be exercised so that spontaneous combustion does not occur. Flame and sparks should be kept away from batteries. The fumes produces by a battery are highly explosive.

Ventilation

One of the most important safety precautions to be followed in an auto repair shop is to ensure enough *ventilation*. If it is necessary to operate an engine, the vehicle should be driven outside. Exhaust gases contain CO, which is a deadly poison. In small quantities, it produces drowsiness and headaches. In large quantities, it will cause death.

3.14 SAFETY IN HOISTING A VEHICLE

Many chassis and underbody service procedures require the vehicle to be hoisted or lifted off the ground. The simplest methods involve the use of drive-on ramps or a floor jack and safety (jack) stands, whereas in-ground or surface mounted lifts provide greater access.

Setting the Pads is a critical part of this procedure. All vehicle service manuals include recommended locations to be used when hoisting a vehicle. Late model vehicles have a triangle decal on the driver's door indicating the recommended lift points. The recommended standards for the lift points typically include the following points, *Fig. 3.103*:

1. The vehicle should be centred on the lift or hoist to avoid overload or put too much force either forward or rearward.
2. The pads of the lift should be spread as far apart as possible to provide a stable platform.
3. Each pad should be placed under a portion of the vehicle that is strong and capable of supporting the weight of the vehicle. In this regard
   - Pitch welds at the bottom edge of the body are generally considered to be strong.
   - Boxed areas of the body are the best places to position the pads on a vehicle without a frame. Use care to check whether the arms of the lift might come into contact with other parts of the vehicle before the pad touches the intended location. Commonly damaged areas include the following:

1. Rocker panel mouldings.
2. Exhaust system, including catalytic converter.
3. Tyres or body panels.
4. Raise the vehicle about 30 cm off the floor, then stop and shake the vehicle to check for stability. If the vehicle seems to be stable, then continue to raise the vehicle to the desired height. Most hoists can be safely placed at any desired height. For ease of working, the area in which to work should be at the technician's chest level.
5. Before lowering the hoist, the safety latches must be released and the direction of the controls reversed, from ↑ to ↓. The speed of lowering is often adjusted to be as slow as possible for

additional safety. Never take off eyes off the vehicle while hoisting or lowering on a hoist to avoid any possible accidents.

Fig. 3.101 *Make sure fire extinguishers are displayed prominently. This particular extinguisher can be used for type A, B, or C fires.*

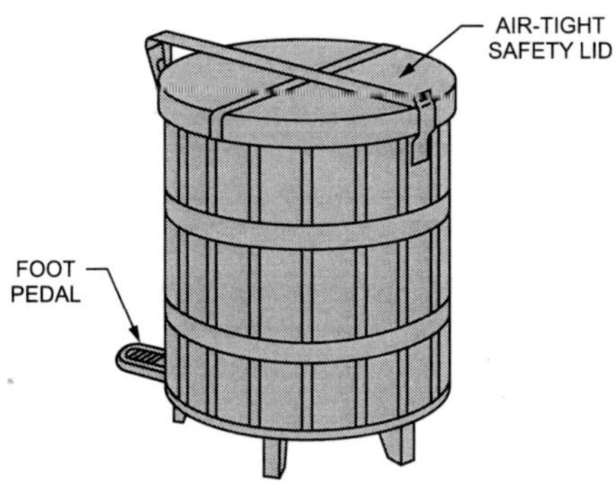

Fig. 3.102 *Always place rags in an appropriate container*

3.15 DISPOSING OF AUTO SHOP WASTES

Automotive service and maintenance facilities frequently generate hazardous wastes, which are usually regulated by the Resource Conservation and Recovery Act (in USA) and similar acts in most other countries. Any business that maintains or repairs vehicles, heavy equipment, or farm equipment should comply with the regulations of the act.

Draining automotive fluids and replacing non-repairable parts are the most common automotive repair activities that produce hazardous wastes. Automotive fluid and solid wastes include:

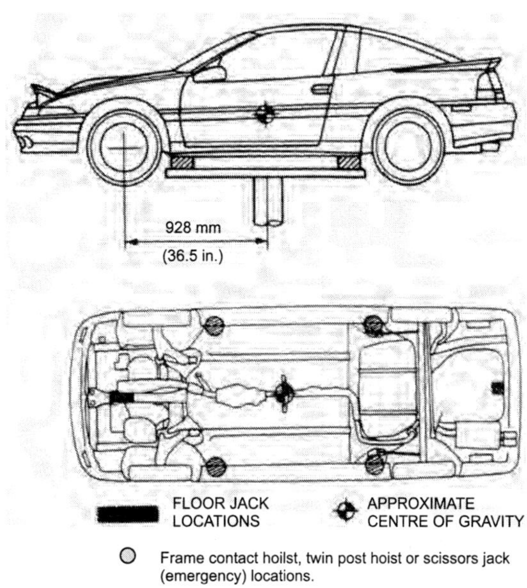

Fig. 3.103 *Typical points that should be supported when servicing a vehicle. (Chrysler Corp.)*

- Used motor oil, which is combustible and may contain toxic chemicals.
- Other discarded lubricants, e.g., transmission and differential fluids that may contain toxic chemicals.
- Used part cleaners and degreasers that are combustible and may contain toxic chemicals.
- Carburetor cleaners that contain flammable or combustible liquids.
- Worn out batteries that contain lead and toxic chemicals.
- Fans belts, mufflers, and catalytic converters.
- Paint and body repair products waste.
- Engine oil filters.
- Air-conditioning refrigerants.

One method to deal with hazardous wastes is to minimize the quantity of wastes produced. This can be accomplished by practicing good housekeeping, improving inventory control, and following proper spill containment techniques.

When the hazardous wastes are produced, they must be disposed off properly. Legal regulations require that a registered hazardous waste hauler should collect these wastes. Several companies offer pickup and recycling services. Repair or maintenance facilities that generate 100 kg or more of hazardous waste monthly are required to fill a Hazardous Waste Manifest before shipping the wastes to a disposal or recycling site. This manifest is a tracking document that must accompany hazardous wastes when shipped from the facility. It contains information about the origin, character, and destination of the wastes. Always follow the state and federal laws and procedures applicable when disposing the facility wastes. Used oil and lead-acid batteries when sent off for recycling may not be regarded as hazardous. Check with the state hazardous waste management agency for all legal requirements that need to be fulfilled.

Unless recycled for scrap metal, used oil filters are considered hazardous waste. If not recycled, they must be listed on the manifest as hazardous. Before disposal, oil filters should be gravity drained for any free-flowing oil. Store them in a closed, labelled container for pickup by a recycler.

Antifreeze

It might pick up heavy metal and chlorinated solvents when circulating through an engine's cooling system and it may then become a hazardous waste. Used antifreeze should not be mixed with used oil. If it occurs, the entire mixture should be classified as a hazardous waste even though the oil may not be hazardous.

Refrigerants

Refrigerants removed from automotive air conditioning systems during service must not be vented to the atmosphere. State regulations require that they be recovered and recycled. Recovery systems are available for this purpose. Technicians servicing automotive air conditioning systems must be certified in refrigerant recycling.

Material Safety Data Sheets

All chemical manufacturers are required to provide a *Material Safety Data Sheet (MSDS)* for each chemical they produce. These sheets list all known dangers and treatment procedures for a specific chemical. There should be a MSDS for every chemical used in the auto repair shop. Read the MSDS for any chemical that you are not familiar with.

Hazardous Waste

Hazardous waste is a solid, liquid, or gas that can harm public and the environment. Several criteria to determine if a substance is hazardous:

1. A material is considered an *ignitable hazard* if it will ignite and burn easily. Gasoline, diesel oil, solvents, and other chemicals used in an auto repair shop fall in this category.
2. A material or waste that dissolves metals and other materials or burns human skin is a *corrosive hazard.* Battery acid and many parts cleaners are considered a corrosive hazard.
3. Materials that generate toxic mists, fumes, vapours, and flammable gases are a *reactive hazard.* Any material that reacts violently with other materials or releases poisonous gases is considered reactivity hazard.
4. Materials like lead, cadmium, chromium, arsenic, and other heavy metals that can pollute and render water and soil harmful to life are considered a *toxicity hazard.* Used motor oil, solvents, and other chemicals in the auto repair shop are a *toxic hazard* and must be disposed off properly.

Used motor oil, cleaning solvent, and other waste materials should be sent to an established waste disposal facility where these materials are treated and recycled to help prevent environmental pollution. Never pour used oil or other wastes down a drain because this will contaminate water, soil, and air.

Dangers of Exposure to Asbestos

Friction materials, e.g., brake and clutch linings often contain asbestos. While asbestos has been eliminated from most original equipment friction materials, the automotive service technician would not know whether the vehicle being serviced is or is not equipped with friction materials containing asbestos. It is safe to handle all friction materials as if they do contain asbestos.

Asbestos exposure can cause scar tissue to form in the lungs. This condition is known as *asbestosis*. It gradually causes increasing shortness of breath, and the scaring to the lungs is permanent.

Even low exposures to asbestos can cause *mesothelioma*, a type of fatal cancer of the lining of the chest or abdominal cavity. Asbestos exposure can also increase the risk of *lung cancer* as well as cancer of the voice box, stomach, and large intestine. It usually takes 15 to 30 years or more for cancer or asbestos lung scarring to show up after exposure.

The federal Environmental Protection Agency (EPA) has established procedures for the removal and disposal of asbestos. The EPA procedures require that products containing asbestos be "wetted" to prevent the asbestos fibers from becoming airborne. According to the EPA, asbestos-containing materials can be disposed off as regular waste — only when asbestos becomes airborne it is considered to be hazardous.

Asbestos handling guidelines

Asbestos levels in air in the shop area can be kept below the recommended levels by using a solvent or a special vacuum.

HEPA Vacuum

A special high efficiency particulate air (HEPA) vacuum system has proven to be effective in keeping asbestos exposure levels below 0.1 fibers / cm^3.

Solvent Spray

Many technicians use an aerosol can of brake cleaning solvent to wet the brake dust and prevent it from becoming airborne. Commercial brake cleaners are available that use a concentrated cleaner that is mixed with water. The waste liquid is filtered, and when dry, the filter can be disposed as solid waste.

Never use compressed air to blow brake dust. The fine powder like brake dust can create a health hazard even if asbestos is not present or is present in dust rather than fiber form.

Used Brake Fluid

Generally, brake fluid is made from polyglycol, is water soluble, and is considered hazardous if it has absorbed metals from the brake system.

- Collect brake fluid in containers clearly marked to indicate that it is dedicated for that purpose.
- A hazardous brake fluid has to be managed appropriately using only an authorized waste receiver for its disposal.
- Do not mix brake fluid with used engine oil.
- Do not pour brake fluid down drains or onto ground.
- Recycle brake fluid through a registered recycler.

Used Oil

Used oil is any based-based or synthetic oil that has been used. During normal use, impurities such as dirt, metal scrapings, water, blow by carbon or chemicals can get mixed in with the oil. Eventually, this used oil has to be replaced with virgin or re-refined oil.

If used oil is dumped down the drain will enter a sewage treatment plant, concentrations as small as 50 to 100 parts per million (PPM) in the waste water can foul sewage treatment processes. Never mix a listed hazardous waste, gasoline, waste water, halogenated solvent, antifreeze, or an unknown waste material with used oil. Adding any of these substances will cause the used oil to become contaminated, which classifies it as hazardous waste.

Used oil may also be stored in units that are permitted to store regulated hazardous waste.

Disposal of Used Oil

Once oil has been used, it can be collected, recycled, and used over and over again. Recycled used oil can sometimes be used again for the same job or can take on a completely different task, e.g., used oil can be refined and sold as an engine oil or processed for furnace fuel oil. After collecting used oil in an appropriate container, the material must be disposed off in one of the two ways:

- Shipped offsite for recycling
- Burned in an onsite or offsite EPA approved heater for energy recovery.

Used Oil Filter Disposal Regulations

Used oil filters contain used oil that may be hazardous. Before an oil filter is placed into the trash or sent for recycling, it must be drained, using one of the following hot draining methods approved by EPA.

- Puncture the filter antidrain back valve or filter dome end and hot drain for at least 12 hours.
- Hot drain and crushing.
- Dismantling and hot draining.

- Any other hot draining method, which will remove all the oil from the used filter. After the oil has been drained from the oil filter, the filter housing can be disposed off in any of the following ways:
- Sent for recycling
- Pickup by a service contract company
- Disposing off in regular trash.

Solvents

The major sources of chemical danger are liquid and aerosol brake cleaning fluids that contain chlorinated hydrocarbon solvents. The most common such solvents are 1, 1, 1-trichloroethane, trichlorethylene, and tetra-chloroethylene, which is also known as perchloro-ethylene. These solvents are all members of the same chemical family, and share the same basic characteristics. They are colourless liquids with an odour of chloroform or ether. They are narcotic, and in large quantities can dull the senses, induce sleep, or cause a stupor. Very high levels of exposure over even a short period of time can be fatal. Repeated exposure to these solvents, in high concentrations and over a long period of time, can result in liver, kidney, and lung damage, and may potentially cause cancer. Also, if these solvents are exposed to high heat or open flame, they decompose into deadly gases such as hydrogen chloride, phosgene, and carbon monoxide. Because 1, 1, 1-trichloroethane and trichlorethylene are known to be ozone depleters, their manufacture and use is prohibited. Perchloroethylenes do not deplete ozone and their use continues. Several other chemicals that do not deplete the ozone, e.g., heptanes, hexane, and xylene, are now being used in non-chlorinated brake cleaning solvents.

Sources of Chemical Poisoning

The health hazards presented by brake cleaning solvents occur from three forms of exposure: *ingestion, inhalation*, or *physical contact*. Swallowing brake-cleaning solvent is harmful. Brake cleaning solvents should always be handled and stored properly, and kept out of the reach of children. The dangers of inhalation are perhaps the most serious problem, because even very low levels of solvent vapours are hazardous. Allowing brake cleaning solvents to come into contact with the skin presents a danger and cause irritation of the tissues, plus they can be absorbed through the skin directly into the bloodstream. The transfer begins immediately upon contact, and continues until the liquid is wiped or washed away.

There is no specific standard for physical contact with chlorinated hydrocarbon solvents or the chemicals replacing them. All contact should be avoided whenever possible. The law requires the employer to provide appropriate protective equipment and ensure proper work practices by an employee handling these chemicals.

Effects of Chemical Poisoning

The effects of exposure to chlorinated hydrocarbon and other types of solvents can take many forms. Short-term exposure at low levels can cause headache, nausea, drowsiness, dizziness, lack of coordination, or unconsciousness. It may also cause irritation of the eyes, nose, and throat, and flushing of the face and neck. Short-term exposure to higher concentrations can cause liver damage with symptoms such as yellow jaundice or dark urine. Liver damage not becomes evident until several weeks after the exposure.

In addition to the symptoms already mentioned, long term or repeated exposure to Perchloro-ethylenes may cause irritation or burning of the skin. It also increases the risk of damage of the liver or kidneys. In case you experience any of these symptoms, and suspect that they may be a result of exposure to brake cleaning solvents, seek medical treatment immediately.

Used Solvents

Used or spent solvents are liquid materials that have been generated as waste and may contain xylene, methanol, ethyl ether, and methyl isobutyl ketone (MIBK). These materials must be stored in approved safety containers with the lids or caps closed tightly. The container should not leak and be damaged due to dents and rust. The containers should be stored in a protected area equipped with secondary containment or spill protector, such as a spill pallet. Additional requirements include:

- Containers should be clearly labelled "*Hazardous Waste*" and the date the material was first placed into the storage receptacle should be noted.

- Labelling is not required for solvents being used in a parts washer.
- Used solvents will not be counted toward a facility's monthly output of hazardous waste if the vendor under contract removes the material.

Coolant Disposal

Coolant is a mixture of antifreeze and water. New antifreeze is not considered to be hazardous even though it can cause death if ingested. Used antifreeze may be hazardous due to dissolved metals from the engine and other components of the cooling system. These metals may include iron, steel, aluminium, copper, brass, and lead (from older radiators and heater cores).

- Coolant should be recycled either onsite or off site.
- Used coolant should be stored in a sealed and labelled container.
- Often used coolant may be disposed off into municipal sewerage with a permit. Check with authorities and obtain a permit before discharging used coolant into sanitary sewers.

Airbag Handling

Airbag modules are pyrometer devices that can be ignited if exposed to an electrical charge or the body of the vehicle is subjected to a shock. Precautions for airbag safety include:

1. Disarm the airbag(s) in case work is needed to be done in the area where a discharged bag could make contact with any part of the technician's body. Consult service information for the exact procedure to follow for the vehicle under service.
2. Do not expose an airbag to extreme heat or fire.
3. Always carry an airbag pointing away from your body.
4. Place an airbag module facing upward.
5. Always follow the manufacturer's recommended procedure for airbag disposal or recycling, including the method of packaging to use during shipment.
6. Always wash hands or body well if exposed to a deployed airbag. The chemicals involved may cause skin irritation and a possible rash to develop.
7. Wear protective gloves if handling a deployed airbag.

Used Tyre Disposal

Used tyres are an environmental concern due to the following reasons:

1. In a landfill, they tend to "float" up through the other trash and rise to the surface.
2. The inside of tyres traps and holds rain water, which is a breeding ground for mosquitoes, Mosquito-borne diseases include *encephalitis* and *dengue fever*
3. Used tyres are a fire hazard and, when burned, create a large amount of black smoke that contaminates the air.

Used tyres should be disposed off in one of the flowing ways:

1. Used tyres can be reused until the end of their useful life.
2. Tyres can be recycled or shred for use in asphalt.
3. Tyres can be retreaded.
4. Shredded tyres may be sent to a landfill.
5. Tyres may be burned in cement kilns or other power plants where the smoke can be controlled.
6. Use only a registered scrap tyre handler to transport tyres for disposal or recycling.

THEORY QUESTIONS

1. An ———— ———— allows for easy engine service because the cylinder block can be rotated into any position desired. To carry out repair and maintenance work.
2. A ridge reamer is used to remove the metal lip formed at the top of a worn ————
   (*a*) Valve guide (*b*) Engine cylinder
   (*c*) Piston (*d*) Valve.
3. The tool used to cut the metal lip out of the top of a worn cylinder to facilitate piston removal.
   (*a*) Boring bar (*b*) Ridge reamer
   (*c*) Hone (*d*) Ring groove cleaner.

4. A cylinder hone is used to ————— a cylinder wall to let the ————— seal properly.

5. A ———— ———— is a machine shop tool used for making the cylinders larger in diameter when they are worn or damaged due to some part breakage.

6. Technician A says a cylinder hone is used to increase the diameter of an engine's cylinder wall. Technician B says a cylinder hone is used to deglaze an engine's cylinder wall. Who is right?
   (*a*) A only (*b*) B only
   (*c*) Both A and B (*d*) Neither A nor B

7. Explain the difference between a piston ring expander and a piston ring compressor.

8. Why it is required to remove the carbon from the piston ring grooves on a used piston?

9. A technician A says piston knurling is used to restore slightly worn piston pins to within specifications. Technician B says piston knurling is used to restore slightly worn pistons to within specifications. Who is right?
   (*a*) A only (*b*) B only
   (*c*) Both A and B (*d*) Neither A nor B.

10. A valve spring compressor fits onto the cylinder head deck and the valve spring retainer. *True or False*?

11. You should wear eye protection when using a wheel puller since it can exert a lot of force. *True or False*?

12. How a technician can determine if the valve springs should be replaced?

13. When it becomes necessary to balance an engine?

14. A ———— ———— is normally used for installing and removing piston pins.

15. What is the difference between a cold solvent tank and a hot tank?

16. Describe the use of a flat feeler gauge and a straight edge.

17. A dial caliper measures to an accuracy of——mm.

18. Describe the basic steps for using a conventional outside micrometer.

19. Technician A says that a crankshaft grinder is used to resurface worn crankshaft journals. Technician B says a crankshaft grinder is used to resurface cylinder head? Who is right?
   (*a*) A only (*b*) B only
   (*c*) Both A and B (*d*) Neither A nor B.

20. Technician A says that an outside micrometer is used for measuring external dimensions, diameters, or thicknesses. Technician B says that an outside micrometer is used to measure internal depths, lengths, and diameters. Who is right?
   (*a*) A only (*b*) B only
   (*c*) Both A and B (*d*) Neither A nor B.

21. A ———— ————should be used to clean a cast iron engine block during an engine overhaul.
   (*a*) Cold solvent tank (*b*) Acid dip
   (*c*) Hot tank (*d*) Steam cleaner.

22. All of the following cleaning equipment should be used when rebuilding an automotive engine except:
   (*a*) Bench grinder (*b*) Blowgun
   (*c*) Rotary brush (*d*) Plastigage.

23. List six types of accidents that can happen in an auto repair shop.

24. How does an electrical fire occur?

25. ————— is the most common cause of shop fires.

26. Apply 'quick dry' or absorbent to gasoline spills. *True or False*?

27. Why is a ground prong provided on auto shop electrical equipment?

28. What should be done when disconnecting a fuel line?

29. Storage batteries can explode! *True or False*?

30. ———— can cause severe chemical buns.
   (*a*) Diesel fuel (*b*) Gasoline
   (*c*) Carburetor cleaner (*d*) All of the above
   (*e*) None of the above.

31. Technician A says that a battery can produce gases that can cause an explosion. Technician B says that special sodium-filled engine valves, if mishandled, can cause an explosion. Who is right?
   (*a*) A only (*b*) B only
   (*c*) Both A and B (*d*) Neither A nor B.

32. Technician A says that battery should be disconnected before removing a fuel line from an engine. Technician B says that a rag should be wrapped around the fitting before disconnecting a fuel line. Who is right?
   (*a*) A only (*b*) B only
   (*c*) BothA and B (*d*) Neither A nor B.

33. The most dangerous source of asphyxiation in an auto repair shop is a(n) —————.
   (*a*) Oily shop rag (*b*) Gasoline can
   (*c*) Engine (*d*) All of the above.
   (*e*) None of the above.

34. Technician A says that battery acid can cause chemical burns to the skin. Technician B says that some carburetor cleaners can cause chemical burns to the skin. Who is right?
   (*a*) A only
   (*b*) B only
   (*c*) Both A and B
   (*d*) Neither A nor B.

35. All of the following are safety rules to follow when working in an auto shop, except:
   (*a*) Keep the auto shop organized
   (*b*) Never carry sharp parts and tools in pocket
   (*c*) Always keep equipment guards or shields in place
   (*d*) Always lift heavy engine parts with your back, not with your legs.

36. Hazardous materials include all of the following except ———.
   (*a*) Engine oil (*b*) Asbestos
   (*c*) Brake cleaner (*d*) Water.

37. To determine if a product or substance being used is hazardous, consult ———.
   (*a*) EPA guidelines
   (*b*) SAE standards
   (*c*) MSDS
   (*d*) A dictionary.

38. Technician A says that used batteries are considered a hazardous waste. Technician B says that used carburetor cleaner is considered a hazardous waste. Who is right?
   (*a*) A only (*b*) B only
   (*c*) Both A and B (*d*) Neither A nor B.

39. Exposure to asbestos dust can cause what condition?
   (*a*) Asbestosis (*b*) Mesothelioma
   (*c*) Lung cancer (*d*) All of the above.

40. Wetted asbestos dust is considered to be ———
   (*a*) Solid waste (*b*) Hazardous waste
   (*c*) Toxic (*d*) Poisonous.

41. Technician A says that refrigerants can be vented into the atmosphere as long as proper filtering systems are utilized. Technician B says that refrigerants must never be vented into the atmosphere during air conditioning system service. Who is right?
   (*a*) A only (*b*) B only
   (*c*) Both A and B (*d*) Neither A nor B.

42. An oil filter should be hot drained for how long before disposing it off?
   (*a*) 30 to 60 minutes (*b*) 4 hours
   (*c*) 12 hours (*d*) 24 hours.

43. All except the following methods should dispose off used oil.
   (*a*) Shipped offsite for recycling
   (*b*) Disposed off in regular trash
   (*c*) Burned offsite in a waste oil-approved heater
   (*d*) Burned onsite in a waste oil-approved heater.

44. All of the following are proper ways to dispose off a drained oil filter except ———.
   (*a*) Sent for recycling
   (*b*) Disposed in regular trash
   (*c*) Picked up by a service contract company
   (*d*) Considered to be hazardous waste and disposed off accordingly.

45. Used coolant is considered to be hazardous waste because it contains ———.
   (*a*) Antifreeze (*b*) Metals
   (*c*) Water (*d*) Additives.

46. Technician A says that for hoisting a vehicle the pads of a lift should be put under a notch at the pinch weld seams of a unibody vehicle. Technician B says that the pads be placed on the four corners of the frame of a full-frame vehicle. Who is right?
   (*a*) A only (*b*) B only
   (*c*) Both A and B (*d*) Neither A nor B.

47. The correct location for the pads when hoisting the vehicle can often be found in the ———
   (*a*) Service manual (*b*) Shop manual
   (*c*) Owner's manual (*d*) All of the above.

48. For the optimum working position, the work should be ———
   (*a*) At shoulder level (*b*) At knee level
   (*c*) At eye level
   (*d*) At chest or elbow level.

49. When working with hand tools, always pull a wrench. *True or False*?

50. When working with hands tools, always push a wrench. *True or False*?

51. What is not considered to be personal safety equipment?
   (*a*) Safety glasses (*b*) Rubber gloves
   (*c*) Air impact wrench (*d*) Ear protection.

52. A distributor tester will check the ———
   (*a*) Pickup coil
   (*b*) Distributor shaft for wear
   (*c*) Rotor for wear (*d*) All of the above.

❑❑❑

Cooling and Lubrication Systems Diagnosis and Service

4

4.1 INTRODUCTION

The cooling and lubrication systems are the key to engine performance and service life. Failure of any of these systems will cause engine destruction in matter of minutes. An engine technician must be capable of understanding the operation of these systems. Must understand the construction and function of each of the components of these systems, and be properly trained to quickly and correctly diagnose and repair any problems with these systems. These systems are common causes of engine failure and sources of trouble.

4.2 COOLING SYSTEM DIAGNOSIS AND SERVICE

The cooling system must allow the engine to warm up to the required operating temperature as quickly as possible and then maintain that temperature during operation. It must be capable of doing so under all possible operating environment conditions, e.g., air temperatures as low as – 40°C and as high as 46°C.

Satisfactory cooling system operation depends upon the design and operating conditions of the system. The design is based on heat output of the engine, radiator size, type of coolant (cooling medium), size of coolant pump, type of the system fan, thermostat, and system pressure. Unfortunately, the cooling system is usually neglected until there is a problem. Specified routine care and maintenance can avoid cooling system problems.

A faulty cooling system is one of the major causes of engine failure. As part of the engine rebuild, cooling system components and passages must be restored to specifications. Routinely check and, if required, replace the water (coolant) pump and hoses after engine overhaul or repair. The radiator should also be checked for any leaks and any clogging of the radiator core and coolant flow passages. Overheating is one of the most common causes of engine failure.

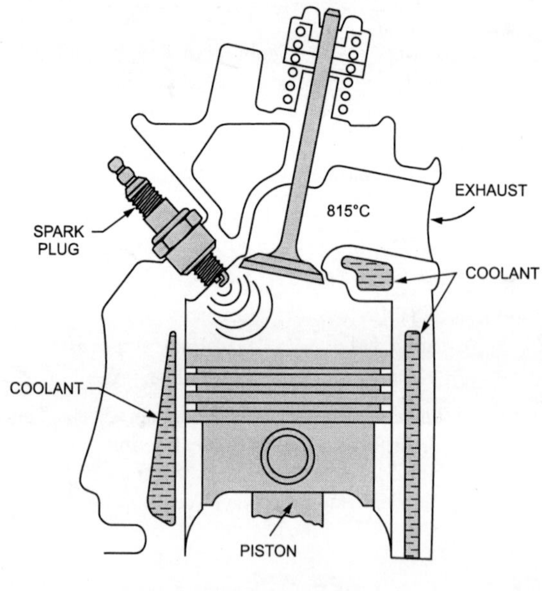

Fig. 4.1 *Typical combustion and exhaust temperatures.*

Peak combustion temperatures in an 1C engine cycle vary from 2200°C to 3300°C. The average combustion temperatures will be 650°C to 950°C.

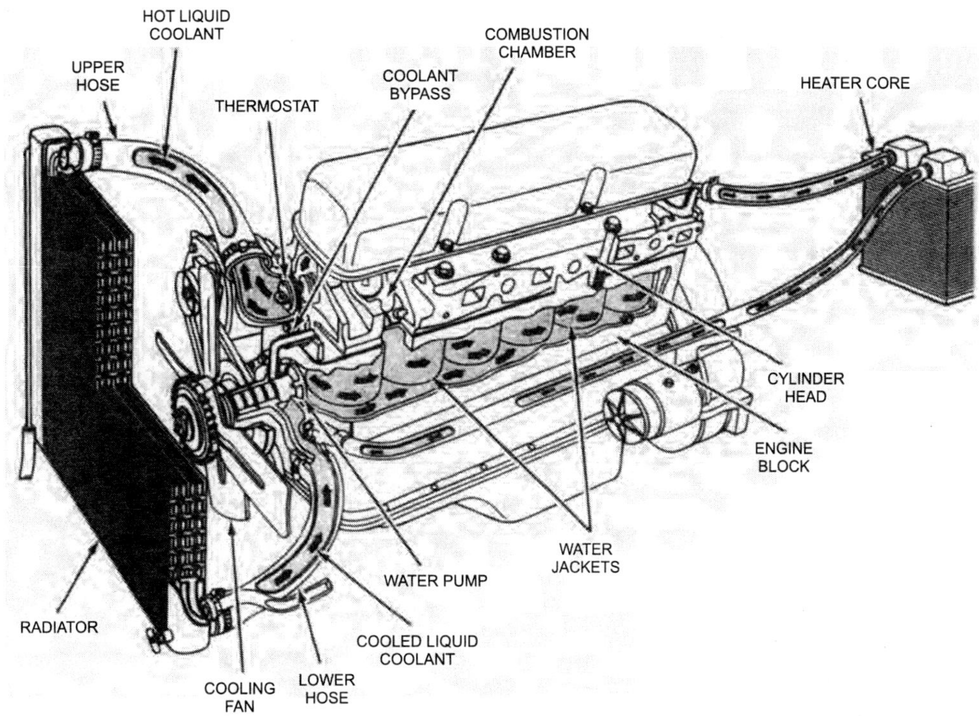

Fig. 4.2 *Coolant flow through a typical engine cooling System.*

Continuously high temperatures tend to weaken the engine parts. Therefore, heat must be removed from the engine with the help of the cooling system to keep the head and cylinder walls at a temperature that is within the range for maximum efficiency, *Fig. 4.1.*

The coolant flow rate may be as high as 4 litres per minute per horsepower produced by the engine. Much of the cooling capacity of the cooling system depends upon the functioning of the radiator. Cooling airflow through the radiator is augmented by a belt- or electric motor — driven cooling fan. Radiators are designed for maximum rate of heat transfer using minimum space.

Figures 4.2 and 4.3 show the coolant flow through a typical engine cooling system. There is a normal operating temperature range between low-temperature and high-temperature extremes. The thermostat controls the minimum normal temperature. The rated temperature of the thermostat indicates the temperature at which the thermostat starts to open. The thermostat is fully open at about 11°C higher than its opening temperature. For example, a

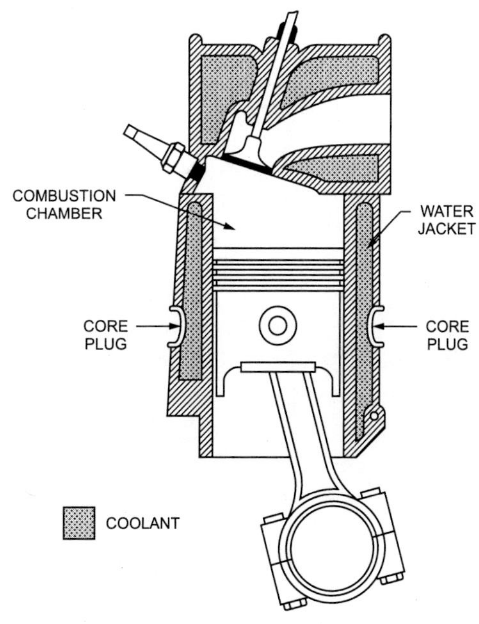

Fig. 4.3 *Coolant circulates through the jackets in the Engine block and cylinder head.*

thermostat with temperature rating of 82°C starts to open at 82°C and will be fully open at 93°C.

If the radiator, water (coolant) pump, and coolant passages are functioning correctly, the engine should always be operating within the opening and fully open temperature range of the thermostat, *Fig. 4.4*.

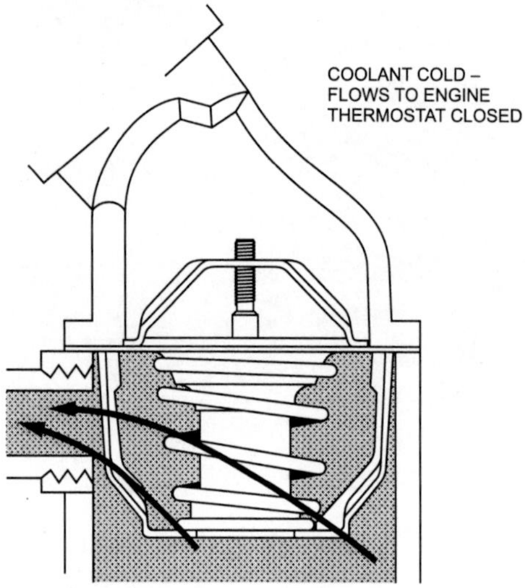

(a) When the engine is cold, the coolant flaws through the bypass.

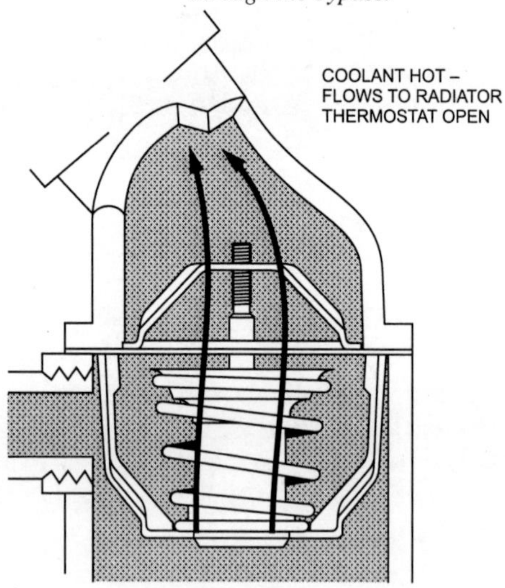

(b) When the thermostat opens, the coolant start flow to the radiator.

Fig. 4.4 *Thermostat action.*

A bypass around the closed thermostat allows a small part of the coolant to circulate within the engine during warm-up. It is a small passage that leads from the engine side of the thermostat to the inlet side of the coolant pump. It allows some coolant to bypass the thermostat even when the thermostat is open. The bypass may be cast or drilled into the engine and pump parts, *Fig. 4.6.* The bypass allows uniform engine warm-up. Its operation eliminates hot spots and prevents the building of excessive coolant pressure in the engine when the thermostat is in closed position.

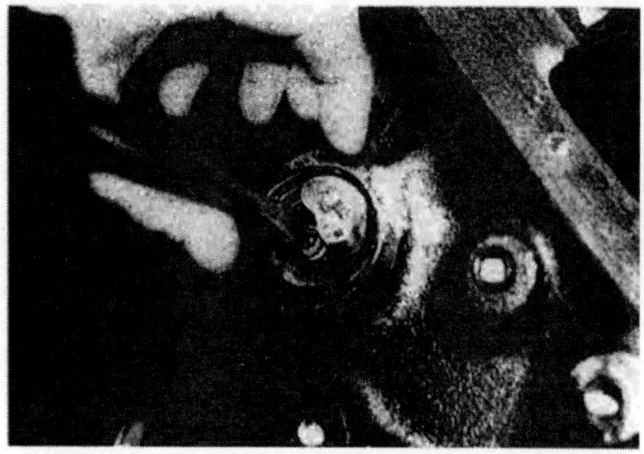

(a) Losen the screw that tightens the block heater element into the core plug opening in the side of the block.

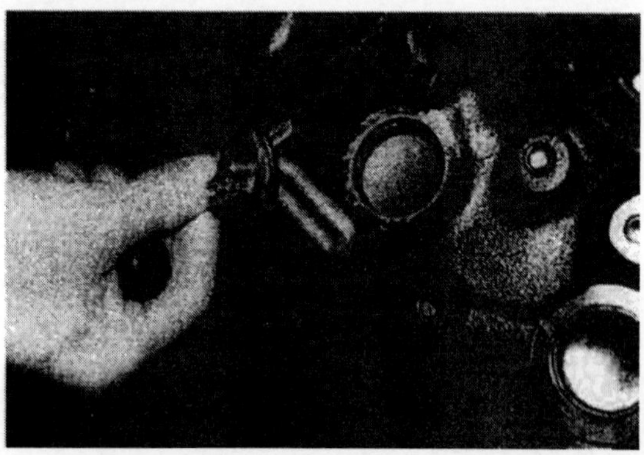

(b) Block heater element removed from block.

Fig. 4.5 *The block heater warms the coolant around the element, and the warm coolant rises, drawing cooler coolant up. As a result of this thermal circulation, all coolant surrounding the entire engine is warmed.*

To reduce cold-engine problems and to help start engines in cold climatic conditions, many automotive manufacturers offer cylinder block heaters as an option. These block heaters can be plugged into domestic electrical supply to make the heating element to warm up the coolant in cylinder jackets, *Fig. 4.5.* The heater warms the coolant around the element, and the warm coolant rises up and is replaced by cooler coolant. As a result of this thermal circulation, all coolant in the engine is heated.

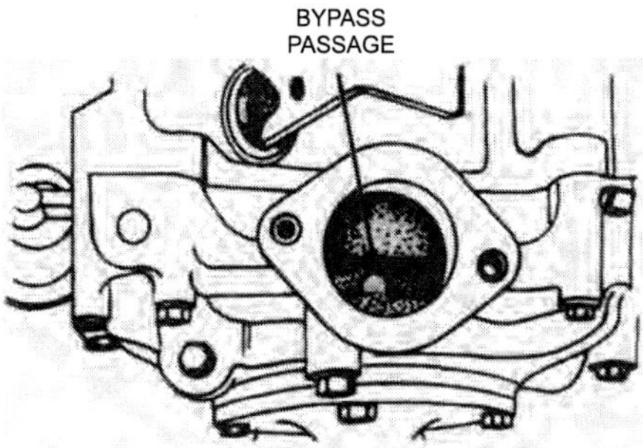

Fig. 4.6 *This internal bypass passage in the thermostat housing directs cold coolant to the water pump.*

4.3 COOLING SYSTEM SERVICE

In the following sections will review the most common cooling system service procedures done by an engine technician. Cooling system failures can crack cylinder blocks, warp cylinder heads, "blow" head gaskets, burn valves, melt pistons, and do other expensive damage to engines. An engine technician must know how to correctly test, service, and repair this vital system.

Cooling System Troubleshooting

A carefully done inspection will frequently reveal the source of the cooling system problem. Obvious troubles include:

1. Coolant leaks, *Fig. 4.7.*
2. Loose or missing fan belts, *Fig. 4.8.*
3. Low coolant level.
4. Abnormal water pump noises.
5. Radiator outside surface blocked by leaves and debris.
6. Coolant mixed in oil that make the oil look like milk.
7. Combustion leakage into coolant — air bubbles in coolant.

Cooling system problems can be broadly grouped into three general categories: ***leaks, overheating, and overcooling.***

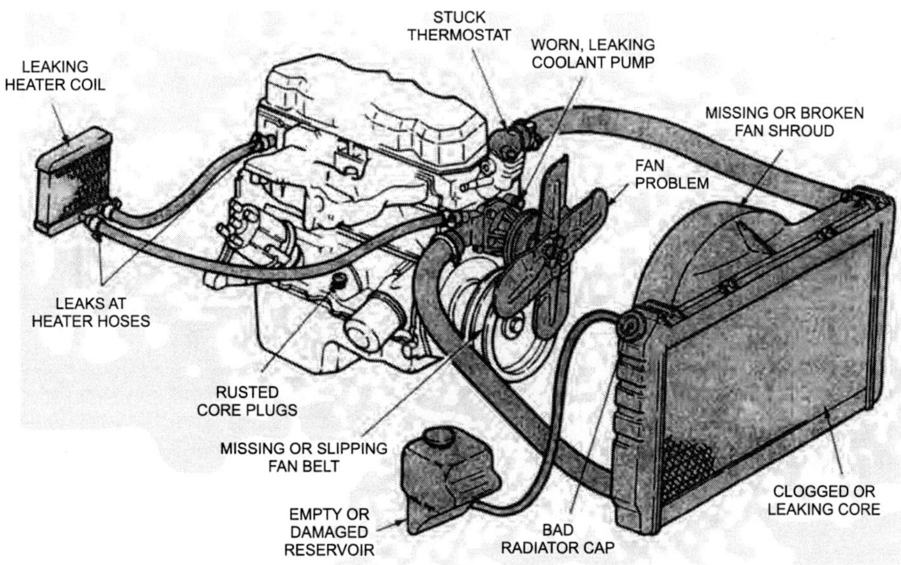

Fig. 4.7 *Some common types of troubles found on a forced circulation cooling system. (GM - Pontiac Div.)*

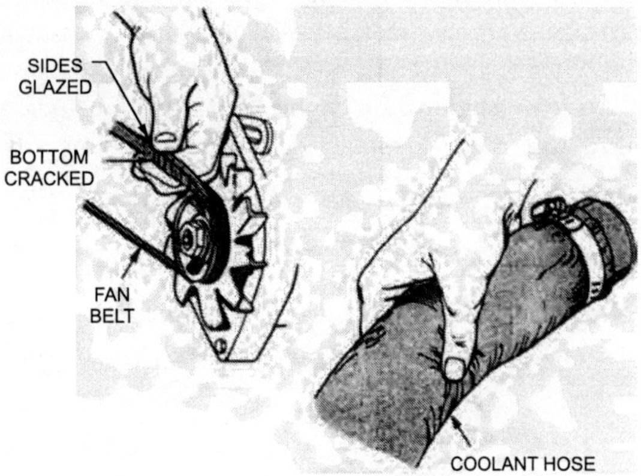

Fig. 4.8 *Inspect cooling system and check condition of belts and hoses. Check for loose or deteriorated belts. Check hoses for leaks, hardening, or softening. (Gates Rubber Co.)*

Coolant Leaks

Coolant leaks show up as a wet, discoloured i.e., darkened green, orange, or rust coloured, areas in the engine compartment or on the ground. The leaking fluid will smell like antifreeze and will have the same general colour. Leaks can occur almost anywhere in the cooling system.

A low coolant level may indicate a leak. If not visible, the leak may be internal due to a cracked engine part or blown head gasket.

Engine Overheating

High temperatures oxidize the engine oil and this breaks the oil down, producing hard carbon and varnish. Continued high temperatures produce carbon that will plug piston rings. The produced varnish will cause the hydraulic valve lifter plungers to stick. Also, high temperatures always lower the oil viscosity to thin the oil, resulting in too thin a lubricant oil film between mating surfaces. Metal-to-metal contact within the engine will occur resulting in high friction, losses of power, and rapid wear of engine parts. Thin oil finds its way up the piston rings into the combustion chamber and through the valve guides to cause excessive oil consumption.

The combustion process is sensitive to engine temperature. High coolant temperatures raise the combustion temperatures to a point that may result in onset of detonation and preignition. Continuous abnormal combustion will damage the engine.

Engine overheating is a serious problem that can cause major engine damage. The driver may notice the engine temperature light turned on, temperature gauge reading high i.e., the gauge needle in red zone, or the coolant boiling.

Common causes of engine overheating are:

- ***Low coolant level*** — Coolant leak or lack of maintenance has allowed coolant level to drop too low.
- ***Rust or Scale*** — Mineral accumulations in system have clogged radiator core or built up in water jackets.
- ***Stuck Thermostat*** — Thermostat is stuck and fails to open normally and restricts coolant flow.
- ***Retarded Ignition Timing*** — Late ignition timing causes too much heat transfer into exhaust valves, ports, and exhaust manifold.
- ***Loose Fan Belt*** — Coolant pump drive belt slips under load and results in reduced coolant circulation, *Fig. 4.9.*
- ***Defective Coolant Pump*** — A leaking seal, broken pump shaft, or damaged impeller blades preventing normal pumping action.

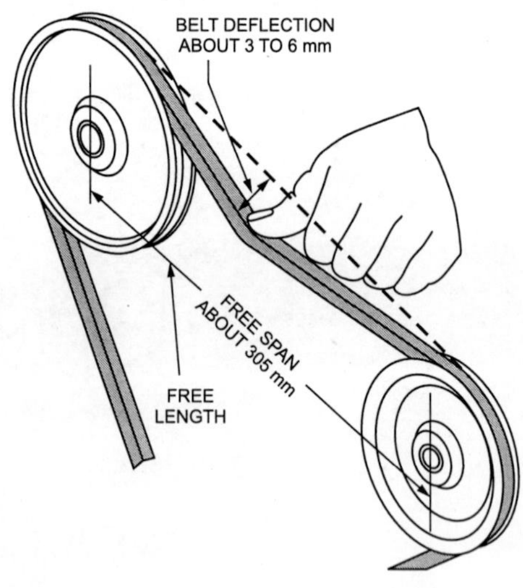

Fig. 4.9 *An easy way to check belt tension is to push in on belt with your thumb. Belt should not deflect over about 3 to 6 mm. (GM Chevrolet Div.)*

- ***Missing Fan Shroud*** — Air circulates between fan and back of radiator, reducing airflow through the radiator.
- ***Collapsed Lower Hose*** — Suction applied by coolant pump may flatten hose if spring is missing or hose is badly deteriorated.
- ***Ice in Coolant*** — Coolant frozen due to lack of antifreeze can block circulation and cause overheating.
- ***Engine Fan Problems*** — Fan clutch or electric fan troubles can prevent enough airflow through the radiator.

Common Causes of Overheating

Overheating can be caused by various defects occurring in the cooling system. Some of the common causes are:

1. Low coolant level
2. Plugged, dirty, or blocked radiator
3. Defective fan clutch or electric fan
4. Incorrect ignition timing
5. Low engine oil level
6. Broken or slipping fan belt
7. Defective radiator cap
8. Dragging brakes
9. Leaking head gasket
10. Frozen coolant in severe cold weather
11. Defective thermostat
12. Defective coolant pump, e.g., damaged impeller, and impeller slipping on the shaft internally.

Some late model engines are protected from overheating damage, due to loss of coolant by the Engine Control Unit (ECU). The ECU monitors the coolant temperature sensor to detect overheating. If overheating is detected, the computer cuts off spark to one cylinder at a time in a controlled sequence and retards ignition timing to reduce top speed. The outside air pulled into the "dead cylinders" cools the engine and prevents overheating damage for a time.

Engine Overcooling

For proper engine operation its operating temperatures must be above a minimum temperature condition. When the engine temperature is too low, there is not sufficient heat available to adequately vaporize the fuel in the inducted charge. This requires additional fuel to make the mixture combustible. The heavy, less volatile part of the fuel does not vaporize, and it remains as unburned liquid fuel. The cool engine surfaces quench part of the combustion gases, allowing partially burned fuel as soot in exhaust.

Combustion is a rapid oxidation process that releases heat as the hydrocarbon fuel chemically combines with oxygen from the air. For each litre of fuel consumed, moisture equal to a litre of water is produced during combustion. A part of this moisture condenses and finds its way into the oil pan, along with unburned fuel and soot, as blowby, and causes sludge formation. The condensed moisture combines with unburned hydrocarbons and additives to form carbonic acid, sulphuric acid, nitric acid, hydrobromic acid, and hydrochloric acid. These acids are responsible for engine wear, by causing corrosion and rust within the engine. Rust occurs rapidly when the coolant temperature is below 55°C. Below 45°C, water from the combustion process will actually accumulate in the oil. High cylinder wall wear rates occur whenever the coolant temperature is below 65°C.

Overcooling causes slow engine warm-up, insufficient warmth from the heater, and sluggish engine performance. Overcooling also reduces fuel economy due to larger amounts of combustion heat being transferred into the metal engine parts. Overcooling may be caused by:

- ***Stuck Thermostat*** — Thermostat stuck open, allowing too much circulation of coolant in the cooling system.
- ***Locked Fan Clutch*** — The fan operates all the time to cause excess airflow through radiator.
- ***Shorted Fan Switch*** — The fan operates all the time to cause excess airflow through radiator, increasing warm-up time.

4.4 THERMOSTAT SERVICE

A ***stuck thermostat*** can either cause engine overheating or engine overcooling. If the thermostat is stuck shut, coolant will ***not*** circulate through the radiator. This could cause overheating and make the coolant boil at all engine speeds. In case the thermostat is stuck open, too much coolant may circulate through the radiator. The engine may not attain normal operating temperature. This can cause the engine to run poorly for extended periods in cold weather.

Engine efficiency i.e., power, mileage per litre of fuel, and drivability, will be reduced.

Testing the Thermostat

To check thermostat action, watch the coolant through the radiator filler neck. When the engine is cold, coolant should not flow through the radiator. When the engine warns up, the thermostat should open to allow coolant circulation through radiator. If this does not happen, the thermostat has a problem and may have to be removed from the engine for testing.

Three basic methods can be used to check the operation of a cooling system thermostat.

1. Hot water method

If the thermostat is removed from the vehicle and is closed, insert a 0.4 mm feeler gauge in the opening so that the thermostat will hang on the feeler gauge. The thermostat should then be suspended, by the feeler gauge, in a bath, along with a thermometer. Heat up the bath until the thermostat opens enough to release and fall from the feeler gauge. The temperature of the bath when the thermostat falls gives the opening temperature of the thermostat. If this temperature is within 4°C of the temperature stamped on the thermostat, the thermostat is good for use. If the difference is greater, the thermostat should be replaced.

2. Infrared pyrometer method

An infrared pyrometer can be used to measure the temperature of the coolant near the thermostat. The area on the engine side of the thermostat should be at the highest temperature that exists in the engine. A properly operating cooling system should cause the pyrometer to read as follows:

- As the engine warms up, the temperature reaches near thermostat opening temperature.
- As the thermostat opens, the temperature drops, sending coolant to the radiator.
- As the thermostat cycles, the temperature should range between the opening temperature of the thermostat and 11°C above the opening temperature.

If the temperature rises higher than 11°C above the opening temperature of the thermostat, inspect the cooling system for any restriction or low coolant flow. A clogged radiator could also cause the excessive temperature rise.

3. Scan tool method

A scan tool can be used on many vehicles to read the actual temperature of the coolant as detected by the engine coolant temperature (ECT) sensor. Although the sensor or the wiring to and from the sensor may be defective, at least the scan tool can indicate what the computer 'thinks' the engine coolant temperature is.

Thermostat Replacement

The thermostat is normally located under the ***thermostat housing,*** which is a fitting for upper radiator hose. The thermostat housing can be located at either the coolant inlet or outlet on the engine.

Fig. 4.10 *Some thermostats are an integral part of the housing. This thermostat and radiator hose housing is serviced as an assembly. Some thermostats simply snap into the engine radiator fill tube underneath the pressure cap.*

A faulty thermostat may cause an overheating engine. Also an engine that does not get warm enough always indicates a defective thermostat. To replace a faulty thermostat, drain the coolant from the radiator petcock to lower the coolant level below the thermostat. It is not necessary to completely drain the system. Remove the upper hose from the thermostat housing neck; then remove the housing to expose the thermostat, *Fig. 4.10.* Tap the housing free with a rubber hammer. Lift off the housing and thermostat. Scrape all of the old gasket material off the thermostat housing and the sealing surface of the engine. The gasket flanges of the engine and thermostat housing should be cleaned, and the gasket surface of the housing must be flat and free from any warpage. Place it on a flat surface and check for gaps between the housing and the flat surface. If warped, file the surface flat. This will prevent coolant leakage. Place the thermostat in the engine with its sensing pellet towards the engine. Ensure that the thermostat position is correct, and install the thermostat housing with a new gasket. Make sure the temperature rating is correct. Place the new thermostat into the engine. Normally, the rod (pointed end) on the thermostat should face the radiator hose. The pellet chamber should face the inside of the engine.

Position the new gasket with approved sealer, *Fig. 4.11*. Start the fasteners by hand. Then torque them to specifications in a crisscross pattern. Do not over-tighten the thermostat housing bolts or warpage might be caused. Most of the housings are made of soft aluminium or "pot metal".

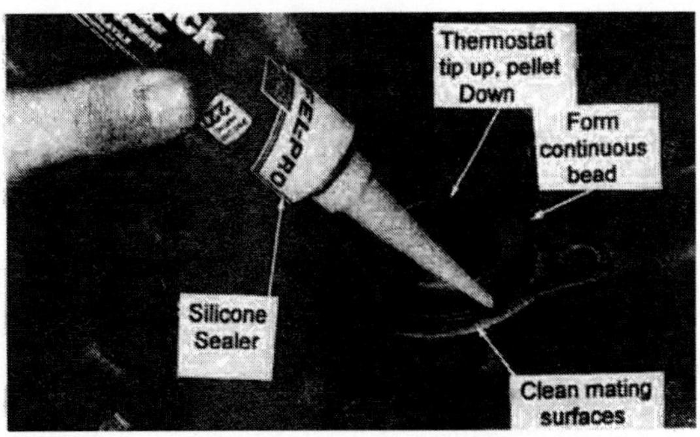

Fig. 4.11 *When installing thermostat, clean housing and engine gasket surfaces. Make sure thermostat is pointing in right direction, pellet towards engine. Use fiber gasket or silicone sealer. Tighten bolts equally to specs. (Fel-Pro)*

Some thermostat housing designs use O-ring instead of a gasket. In many cases, this O-ring can be reused if it is not damaged. If it shows any signs of damage replace it with a new one.

Then install the upper hose and refill the system with coolant. Install the proper size of radiator hose clamp.

Failure to set the thermostat into the recessed groove for it will cause the housing to become tilted when the bolts are tightened and may cause the housing to develop a crack leading to a leak.

4.5 COOLING SYSTEM BLEEDER SCREW

A ***cooling system bleed screw*** is sometimes installed to help remove trapped air when refilling the system. Some systems have more than one bleed screw or valve. Many late model vehicles with low hood lines require a bleed valve to empty out air pockets formed in areas of the system. Air trapped in the cooling system can quickly cause engine overheating or damage to the part near the air pocket and hot spot. Part cracking or warpage could result.

To bleed a cooling system, fill the system with recommended coolant. Start and warm the engine to full operating temperature. Open the bleed screw until all the air is purged out and coolant leaks from the valve. Refill the reservoir as necessary after bleeding.

4.6 ANTIFREEZE / COOLANT

Coolant is a mixture of antifreeze and water. Water is able to absorb more heat / gallon than any other liquid coolant. Under standard conditions, water boils at 100 °C and freezes at 0°C. On freezing, water increases in volume by ≈ 9 %. This expansion of the freezing water can easily crack engine blocks, cylinder heads, and radiators. All automotive manufacturers recommend the use of *ethylene glycol-based antifreeze* mixtures to protect the engines against this expansion problem.

Figure 4.12 shows a curve depicting freezing point as compared with the percentage of antifreeze mixture. It can be noted here that the freezing point increases as the antifreeze concentration is increased above 60%. The normal mixture is 50% antifreeze and 50% water. Ethylene glycol antifreezes contain anticorrosion additives, rust inhibitors, and water pump lubricants. At maximum level of protection, an ethylene glycol concentration of 60% will absorb about 85% as much heat as will water. Ethylene glycol-based antifreeze also has a higher boiling point than water, *Fig. 4.13*. If the coolant boils, it vaporizes and does not act as a cooling agent because it is not in liquid form and in contact with the cooling surfaces.

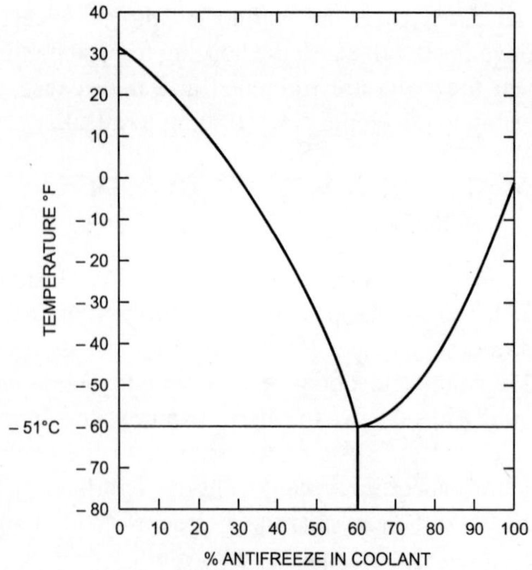

Fig. 4.12 *Graph showing the relationship of the freezing point of the coolant to the %age of antifreeze used in the coolant.*

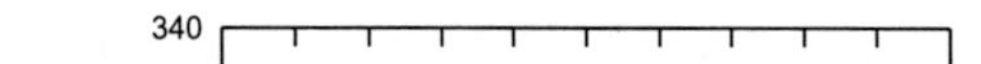

Fig. 4.13 *Graph showing how the boiling point of the coolant increases as the %age of antifreeze in the colant increases.*

An antifreeze and water mixture is an example wherein the freezing point differs from the freezing point of either pure antifreeze or pure water. Pure antifreeze is usually 95% ethylene glycol, 2% to 3% water, and 2% to 3% additives, and it freezes at –4°C. The boiling point of antifreeze and water is also a factor of mixture concentrations. 50/50 mixture boils at 103°C at Sea level, and boils at 130°C with a 100-kPa-radiator pressure cap, *Fig. 4.14*.

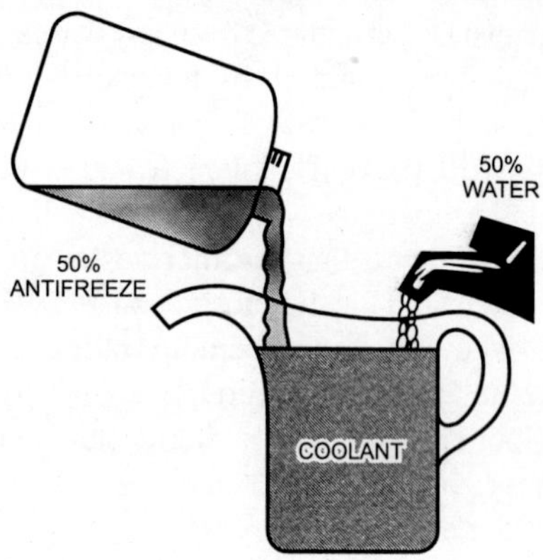

Fig. 4.14 *Coolant is commonly mix of 50% water and 50% antifreeze. (Volvo)*

4.7 ANTIFREEZE SERVICE

Antifreeze should be checked and changed at recommended time intervals. On prolonged use, antifreeze will break down and become very corrosive. It can lose its rust inhibiting characteristics and as a consequence the cooling system can rapidly accumulate rust. A visual inspection of the antifreeze will reveal its condition. Rub your finger inside the radiator filler neck. Check for the presence of rust, oil due to internal engine leak, scale or transmission fluid from a leaking oil cooler. Also, find out the service life of the antifreeze. If contaminated or too old, replace the antifreeze. If the cooling system is badly rusted, it is required to flush (clean) the system.

Changing Antifreeze

Antifreeze should be changed when contaminated or after every two years. Some coolants can even last for five years between services. Refer the applicable service manual for exact coolant change schedules. Open the petcock on the bottom of the radiator or remove one end of the lower radiator hose and allow the old coolant to drain. If the antifreeze is not contaminated with rust, you may refill the system. Tighten the petcock or hose. Pour in the specified amount of antifreeze, usually about 7 to 8 litres.

Start the engine to warm up. The coolant level may ***drop*** when the thermostat opens. Add more water, if needed. Then install the radiator cap.

Aluminium Components

Many late model vehicles use aluminium cooling system and engine parts. In such vehicle cooling system, only use antifreeze designed for aluminium components. Some types of antifreeze can corrode aluminium. Refer the applicable vehicle service manual or the antifreeze label for details.

Testing Antifreeze Strength

Antifreeze strength is a measurement of the concentration of antifreeze compared to water. It determines the freeze up protection of the solution. State of the coolant can be checked using a coolant hydrometer, *Fig. 4.15*. The hydrometer measures the density of the coolant. The higher the density is, the more concentration of antifreeze in the water. Most coolant hydrometers read the freezing point and boiling point of the coolant. In case the engine is overheating and the hydrometer reading is near – 60°C, suspect that pure (100%) antifreeze is present. For best results, the coolant should have a freezing point lower than – 29°C and boiling point above 112°C. Ignore the chill effect, as it does not affect the engine coolant temperature.

Recycling Coolant

Used coolant (mixture of antifreeze and water) may contain heavy metals, e.g., lead, aluminium, and iron absorbed during its use in the engine cooling system. Dispose the used antifreeze properly by storing and sending it to a recycling centre.

Recycle machines filter out these metals and other contaminants and reinstall the depleted additives and restore the coolant like the new one and reused into the vehicle.

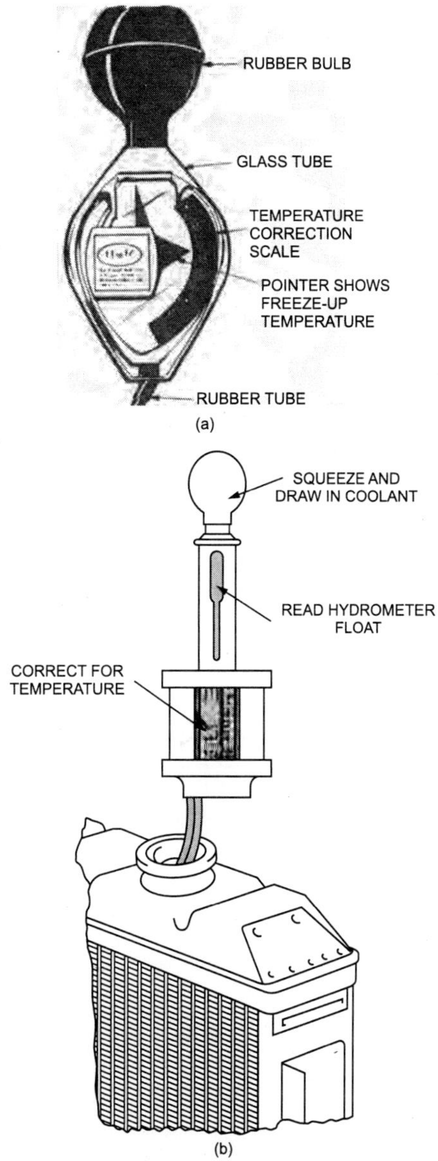

Fig. 4.15 *(a) Hydrometer will measure antifreeze strength, (b) Draw coolant into hydrometer. Read pointer or float. Then correct for temperature. Antifreeze must be strong enough to prevent winter ice formation. (Florida Dept. of Voc. Ed.)*

4.8 COOLING SYSTEM HOSES AND HOSE SERVICE

Coolant system hoses are critical to engine cooling. With aging, the hoses become either soft or brittle and may swell in diameter. Their condition depends upon their material and on the engine service conditions. If a hose breaks while the engine is running, all coolant will be lost. A hose should be replaced anytime it appears to be abnormal, *Fig. 4.16*.

Old radiator *hoses* and *heater hoses* are frequent sources of cooling system problems. After few years of service, hoses deteriorate. They may become soft and mushy or hard and brittle. Cooling system pressure can rupture the hoses and result in coolant loss.

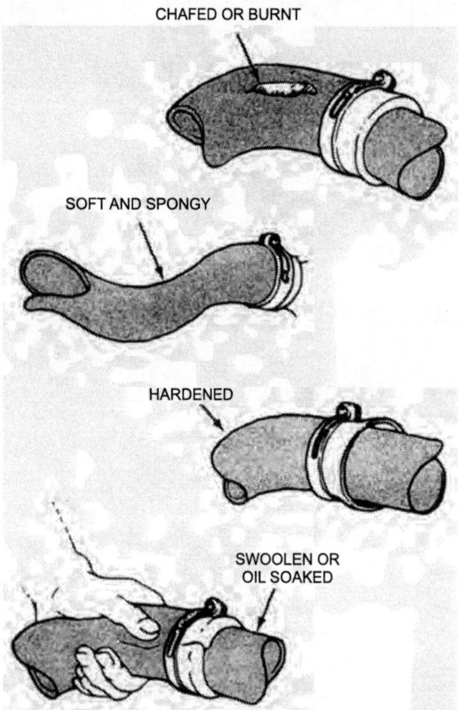

Fig. 4.16 *All cooling system hoses should be checked for wear or damage.*

The coolant pump suction can collapse a softened lower radiator hose. The collapsed hose will restrict coolant circulation and cause engine overheating. The spring inside the lower radiator hose normally prevents hose collapse and it should never be removed from the hose.

Checking Cooling System Hoses

Regularly check the radiator and heater hoses for cracks, budges, cuts, or any other sign of failure. ***Hand-squeeze*** the hoses to check their condition, i.e., hardened or softened and defective. Flex (bend) the heater hoses and look for surface cracks. If any defect is found, the affected hose(s) should be replaced.

Hose Replacement

To remove a hose, loosen the hose clamps. Twist the hose while pulling out. If a new hose is to be installed, for easier hose removal, and to avoid possible damage to the radiator, use a utility knife and slit the hose lengthwise. Then simply peel the hose off.

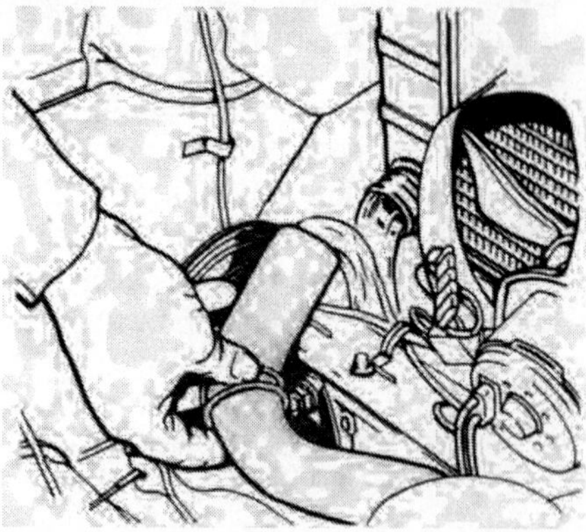

(a) Loosen clamp. Twist hose and pull off.

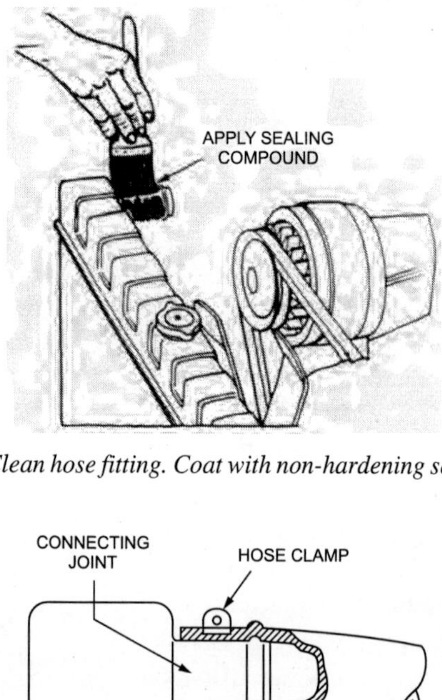

(b) Clean hose fitting. Coat with non-hardening sealer.

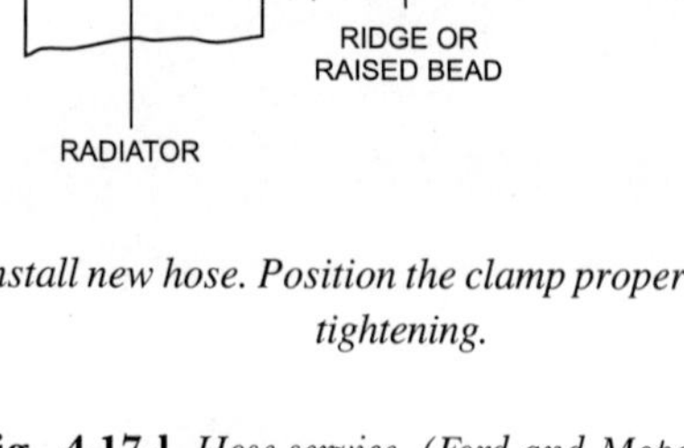

(c) Install new hose. Position the clamp properly before tightening.

Fig. 4.17-1 *Hose service. (Ford and Mopar)*

Clean the metal hose fittings. Coat them with a non-hardening sealer. Install the new hose. Position the hose clamps such that they are fully over the metal hose fitting. Then tighten the clamps.

Take care to avoid bending the soft metal hose neck on the radiator. Clean the hose neck before a new hose is slipped in place. The clamp is placed on the hose, and then the hose is pushed fully over the neck. The hose should be cut so that the clamp is close to the bead on the neck. This is especially important on aluminium hose necks to avoid corrosion. When the hoses are in place and the drain petcock is closed. Install coolant and pressure test the system. Check all fittings for any leaks, *Fig. 4.17-1* and *4.17-2*.

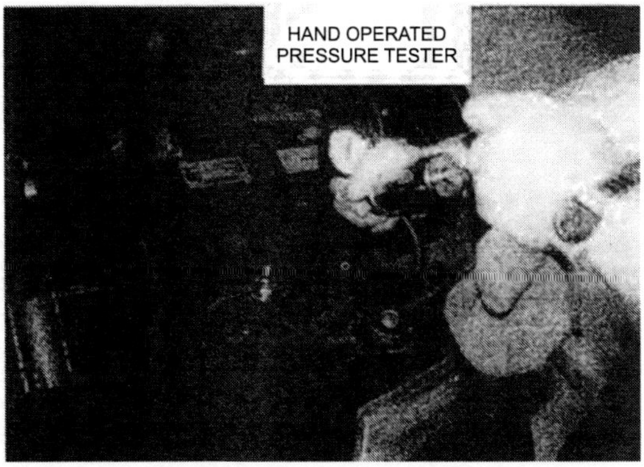

(a) Using a hand-operated pressure tester to pressurize the entire cooling system.

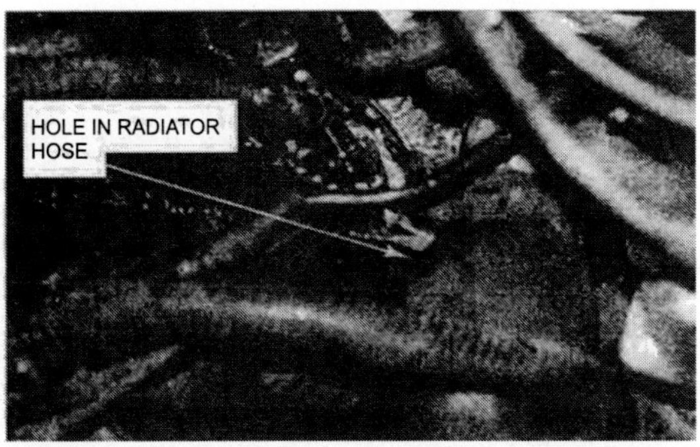

(b) Notice the coolant leaking out of a hole in the radiator hose. The "hot coolant" smell made the driver to look, for the problem.

Fig. 4.17-2

4.9 COOLING SYSTEM PRESSURE TEST

A ***cooling system pressure test*** is used to quickly locate leaks. Pressure testing using a hand-operated pressure tester is a quick and easy cooling system test. The radiator cap is removed, with engine in cold condition, and the tester is attached in place of radiator cap. By operating the plunger on the pump, the entire cooling system is pressurized, as specified by the vehicle manufacturer. Most systems specify this pressure limit as 95 kPa, *Fig. 4.18.* With pressure in the system, inspect all parts for leakage. Check at all hose fittings, at gaskets, under the coolant (water) pump, around the radiator, and at the engine core plugs. If a leak is found, tighten, repair, or replace parts as required. An excess pressure applied, may cause the water pump, radiator, radiator core, or hoses to fail. If the cooling system is free from leaks, the pressure should stay and not drop. If the pressure drops, inspect for any leaks anywhere in the cooling system including:

1. Heater hoses
2. Radiator hoses
3. Radiator
4. Heater core
5. Cylinder head
6. Core plugs in the side of the cylinder block or cylinder head.

Pressure testing should be performed when there is a leak or suspected leak. The pressure tester can also be used to test the radiator cap. An adapter is used to connect the pressure tester to the radiator cap. Replace any cap that does not hold pressure, *Fig. 4.19.*

4.10 COOLANT DYE LEAK TESTING

One of the best methods to check for a coolant leak is to use a designated fluorescent dye in the coolant. Operate the vehicle with the dye in the coolant until the engine reaches normal operating temperature. Use a black light to inspect all areas of the cooling system. In case there is any leak, it will be easily spotted because the dye in the coolant will be seen as bright green.

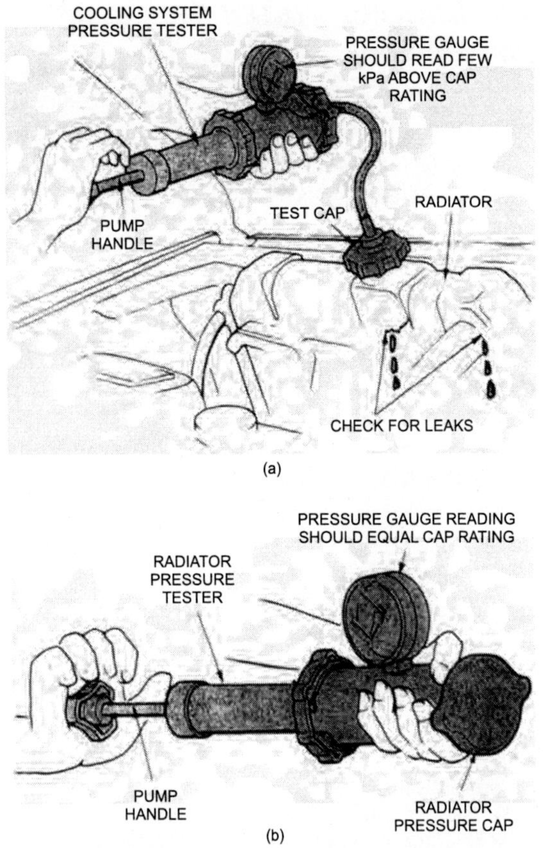

Fig. 4.18 *(a) To pressure test cooling system, install tester on radiator filler neck. Pump handle until gauge reads a few kPa above cap pressure. Then check the leaks in system, (b) To test pressure cap, install cap on tester using adapter. Pumphandle until gauage stops increasing. Reading should stop going up at cap rated pressure. It should also hold pressure without leaking. (Honda)*

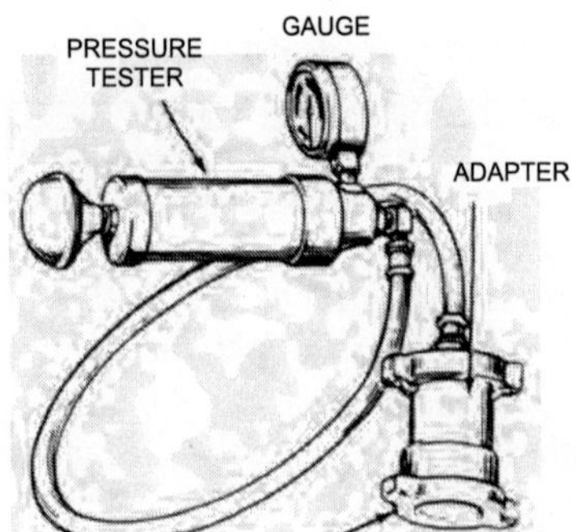

Fig. 4.19 *The pressure cap should be checked for proper operation using a pressure tester as part of cooling system diagnosis.*

4.11 RADIATOR AND PRESSURE CAP SERVICE

When overheating problems occur and the system has ***no*** leaks, check the radiator and the pressure cap. They are frequent sources of overheating. The pressure cap could have defective seals, allowing pressure loss. The radiator may be clogged and preventing adequate amount of air or coolant flow for the system.

Checking Radiator and Pressure Cap

Inspect the outside of the radiator for leaves, mud, road dust, etc. Also, check the radiator shroud for its position and physical condition, because these could limit air circulation through the radiator core.

If necessary, use a water hose to push debris out of the radiator core. Spray water from the rear to push debris out the front of the radiator. Low pressure compressed air, not high enough to damage the radiator core, may be used for this purpose.

Check the radiator cap and filler neck. Look for any cracks or tears in the cap seal. Check the filler neck sealing surfaces for nicks and dents. Replace the cap or get the neck repaired as necessary.

The radiator filler cap is fitted with a pressure cap. The cap has a spring-loaded valve that closes the cooling system vent. This causes cooling pressure to build up to the pressure setting of the cap, as the system heats up to operating temperatures. At this set point the valve will open and release the excess pressure to prevent cooling system damage, *Fig. 4.20.*

Engine cooling systems are pressurized to raise the boiling temperature of the coolant. The cooling temperature increases by approximately 1.6°C for each 6.67 kPa increase in pressure. At standard atmospheric pressure, water boils at 100°C. With a 100-kPa-pressure cap, water boils at 125°C, which is a maximum operating temperature for an engine. For proper cooling, the system must have the right pressure cap correctly installed. The high coolant system temperature serves two functions:

1. It allows the engine to run at an efficient temperature, about 93°C, with no danger of boiling the coolant.
2. The higher the coolant temperature, the more heat the cooling system can transfer from the engine to the atmospheric air. The heat transferred by the engine cooling system is

proportional to the temperature difference between the coolant and the atmospheric air. This characteristic has led to the design of small, high-pressure radiators that are capable of handling large quantities of heat.

Proper operation of the pressure cap is especially important at high altitudes. Boiling point of water lowers by 0.555°C for every 165 m increase in altitude.

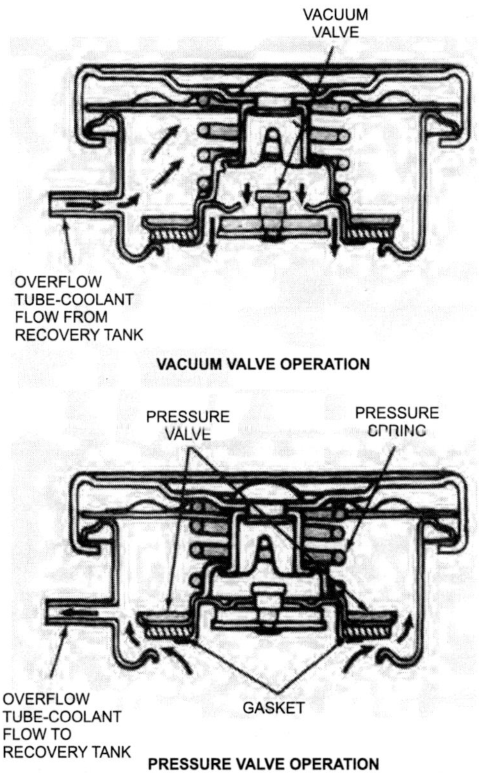

Fig. 4.20 *The pressure valve maintains the system pressure and allows excess pressure to vent. The vacuum valve allows coolant to return to the system from the recovery tank.*

Pressure Testing Radiator Cap

A ***radiator cap pressure test*** is used to measure cap-opening pressure and checks the condition of the sealing washer. The cap is installed on a cooling system pressure tester, *Fig. 4.19.* Pump the tester to pressurize the cap and keep watching the pressure gauge. The cap should release air at its rated pressure, stamped on cap. It should hold that pressure for at least one minute.

Cleaning the Radiator Exterior

Engine overheating can result from exterior radiator clogging as well as internal clogging. Dirt and insects cause external clogging. This type of clogging can be observed when looking straight through the radiator while a light is held behind the radiator. It is most likely to occur on off-road vehicles. The exterior clogging can usually be cleaned with water from a hose. The water is aimed at the engine side of the radiator. The water should flow freely through the core at all locations. It this does not clean the core; the radiator should be removed for cleaning at a radiator shop.

Radiator Repair

A dedicated ***radiator repair shop*** specializes in radiator repair. It is equipped with facilities to properly disassemble, rod out i.e., clean, solder to reassemble or repair, and pressure test a radiator. Some late model radiators have plastic components. Special sealers, seals, and repair techniques are needed on these radiators. Refer to the applicable service manual for details.

4.12 COOLANT PUMP SERVICE

A worn and corroded impeller on a coolant pump can reduce the amount of coolant flow through the engine, *Fig. 4.21.* A ***bad water pump*** may leak coolant due to a worn seal, fail to circulate coolant due to broken shaft or damaged impeller, or it may produce a grinding sound due to defective bearings.

Rust in the cooling system or lack of antifreeze are common causes for coolant pump failure. These conditions could speed up seal, shaft, and bearing wear. A too tight fan belt is another common cause of premature coolant pump failure.

If the seal of the pump fails, coolant will leak out of the bleed weep hole, *Fig. 4.22.* The hole allows coolant to escape without getting trapped and forced into the coolant pump bearing assembly. In case the pump bearing is damaged, the pump will usually be noisy and will have to be replaced. Before replacing a failed coolant pump due to a loose or noisy bearing, check the following:

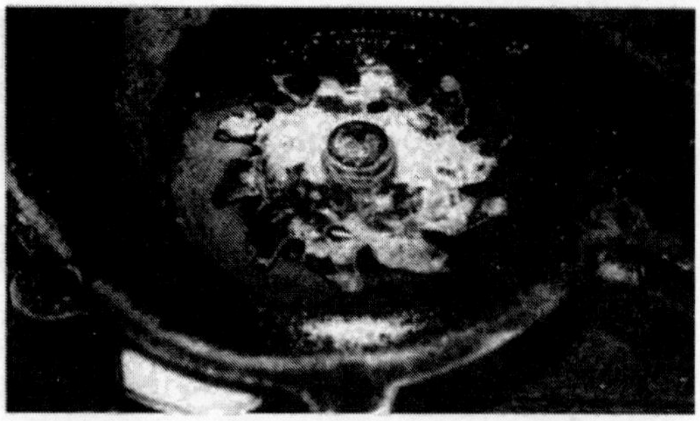

Fig. 4.21 *This severely corroded water pump could not circulate enough coolant to keep the engine cool. As a result the engine overheated and blew a head gasket.*

Fig. 4.22 *The bleed weep hole in the water pump allows coolant to leak out of the pump and not to be forced into the bearing. If the bearing failed, more serious damage could result.*

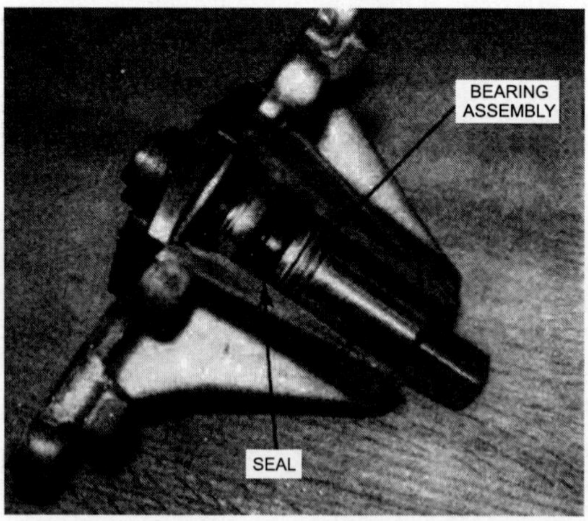

Fig. 4.23 *A cutaway of a typical water pump showing the long bearing assembly and the seal. The weep hole is located between the seal and the bearing. If the seal fails, then coolant flows out of the weep hole to prevent the coolant from damaging the bearing.*

1. Check drive belt tension
2. Check for bent fan
3. Check fan for balance

A too tight drive belt exerts excessive force against the coolant pump bearing. Also, a bent or out of balance impeller, will develop vibration that can damage the coolant pump bearing, *Fig. 4.23.*

4.13 CHECKING COOLANT PUMP

To check for a ***defective pump seal,*** pressure test the system and watch for coolant leakage at the pump. Coolant might leak out of the small drain hole at the bottom of the pump or at the end of the pump shaft. Replace or repair the pump if a leak is found.

To check for ***worn pump bearings,*** attempt to wiggle the fan or pump pulley up and down. If the pump shaft is loose in it's housing, the pump bearings are badly worn. Replace the pump. A stethoscope can also be used to listen for worn, noisy water pump bearings.

To check for ***water pump operation*** run the engine to warm up to a temperature enough to open the thermostat. Stop the engine. Squeeze the top radiator hose while someone starts the engine. You should feel a pressure surge through the hose if the pump is functioning and circulating coolant. If not, pump shaft or impeller problems exist. You can also watch for coolant circulation in the radiator with the engine at normal operating temperature.

Water Pump Removal

To remove the water pump, unbolt all brackets and other components like air conditioning compressor, power steering pump, alternator, etc. preventing pump removal. Then, remove the bolts holding the pump to the engine. Use care to mark/organize to facilitate reassembly.

Scrape off old gasket or sealer material from the engine-to-pump mating surfaces to perfectly clean them to prevent coolant leakage. On soft aluminium parts, be careful not to gouge or scratch the mating sealing surfaces.

Installing Water Pump

To install a ***water pump gasket,*** use specified sealer to stick the new gasket, to the pump sealing surface, in right location. This will keep the gasket in alignment over the bolt holes during pump installation, *Fig. 4.24*.

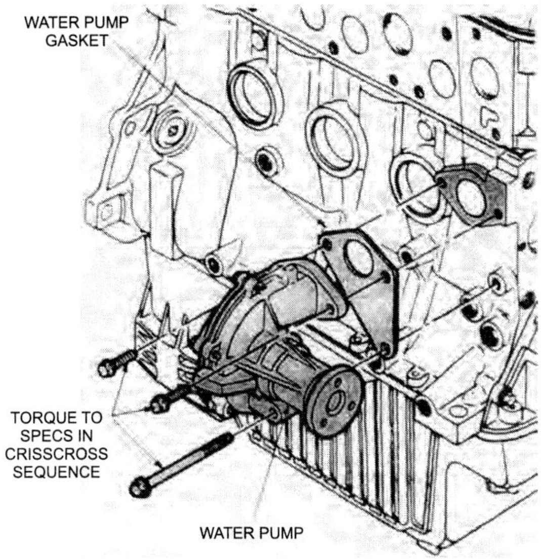

Fig. 4.24 *When installing water pump, make sure all old gasket material is removed from the mating surfaces. Use approved sealer to hold gasket in place. Start bolts by hand and check lengths. Then torque to specs in crisscross pattern. (Ford Motors Co.)*

To use a chemical gasket, using a sealer in place of fiber gasket, squeeze out a bead of approved sealer around the pump sealing surface. Form a continuous bead of uniform width of about 3 mm. Fit the pump onto the engine, Move it straight into place, keeping the mating surface on the pump parallel to its counter part on block. Do not shift the gasket or break the sealant bead. Start all of the attaching bolts by hand. Screw each in about two turns. Verify that all bolt lengths are correct per design.

Torque all bolts in small steps at a time in a crisscross pattern. Go over the bolts several times to assure correct tightening. Install the other components in the reverse order they were removed and tighten the belt as per specifications.

If necessary, refer to an applicable service manual. It will give detailed directions on pump service for the exact make and model of the vehicle.

Water Pump Rebuild

A ***water pump rebuild*** involves pump disassembly, cleaning, part inspection, worn or defective part replacement, and reassembly. Usually the technicians purchase new or factory rebuilt pumps. Rebuilding in the repair shop takes too much time and would not usually be cost effective.

4.14 COOLING FANS

Air is forced across the radiator core by a cooling fan. The fan is designed to move enough air at the lowest fan speed to cool the engine when it is at its highest coolant temperature. The fan shroud is used to increase the cooling system efficiency. The horsepower required to drive the fan increases at a much faster rate than the increase in fan speed. Higher fan speeds also increase fan noise. Fans with flexible plastic or flexible speed blades have been used. These fans have high blade angles that pull a high volume of air when turning at low speeds. As the fan speed increases, the fan blade angle flattens, reducing the horsepower required to rotate the blade at high speeds, *Fig. 4.25*.

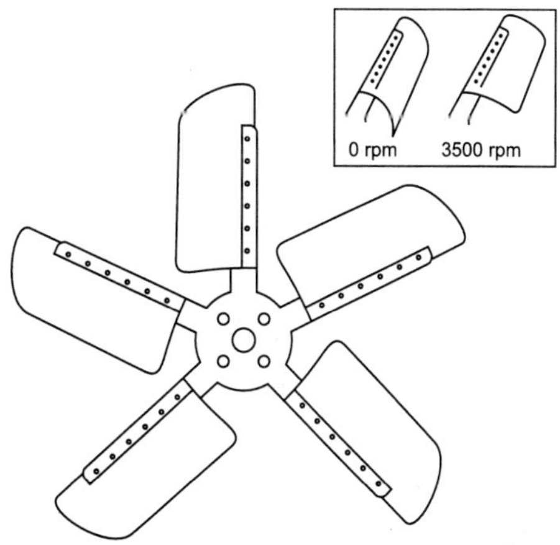

Fig. 4.25 *Flexible cooling fan blades change shape as the engine speed changes.*

Engine Fan Service

A ***faulty engine fan*** can cause engine overheating, overcooling, vibration, and coolant pump wear or damage. Always check for bent blades, cracks, and other problems. A flexible fan is particularly prone to such problems. If any problems are found, replace the fan. Remember that a fan with cracked or bent blades is extremely dangerous. Broken blade may be thrown out with great force resulting in severe lacerations or death.

4.15 THERMOSTATIC FANS

Most cooling fans have been computer-controlled electric motor units. On some rear-wheel-drive vehicles, a belt drives a thermostatic cooling fan from the crankshaft. It turns faster as the engine speed increases. Generally, the engine is required to produce more power at higher speeds. Therefore, the cooling system will also transfer more heat. Increased fan speed aids in the required cooling. Engine speed also becomes critical at low engine speeds in city traffic where the vehicle moves slowly.

The thermal fan is designed so that it uses little power at high engine speeds and minimizes noise. The thermal fan has a ***silicone coupling fan drive*** mounted between the drive pulley and the fan.

Another type of thermal fan has a *thermostatic spring* added to the silicone coupling fan drive. The thermostatic spring operates a valve that allows the fan to freewheel when the radiator is cold. As the radiator warms to about 65 °C, the air hitting the thermostatic spring will cause the spring to change its shape. The new shape of the spring opens a valve that allows the drive to operate like the silicone coupling drive. When the engine is very cold, the fan may operate at high speeds for a short time until the drive fluid warms slightly. The silicone fluid will then flow into a reservoir to let the fan speed drop to idle, *Fig. 4.26.*

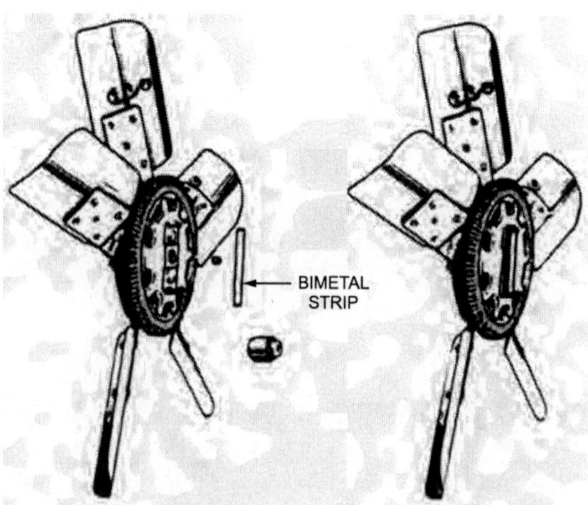

Fig. 4.26 *The bimetallic temperature sensor spring controls the amount of silicone that is allowed into the drive unit, which controls the speed of the fan.*

A common cause of overheating is an inoperative cooling fan. Most front-wheel-drive vehicles and many rear-wheel-drive vehicles use electric motor driven cooling fans. A fault in the cooling fan circuit often causes overheating during slow city driving conditions.

Even slight overheating can soften or destroy rubber vacuum hoses and gaskets. The gaskets most prone to overheating damage are rocker cover (valve cover) and induction manifold gaskets. Gasket and / or vacuum hose failure often results in an air (vacuum) leak that leans the air-fuel mixture. The resulting lean mixture burns hotter in the cylinders and translates into engine overheating.

Most electric cooling fans are computer controlled. To save energy, most cooling fans are turned off whenever the vehicle is travelling faster than 55 km/h. The ram air from the vehicle's travelling at that speed should be enough to keep the radiator cool. Of course, if the computer senses that the temperature is still too high, the computer will turn on the cooling fan, to 'high' speed in order to cool the engine to avoid severe engine damage.

4.16 TESTING A FAN CLUTCH

To test a thermostatic fan clutch, start the engine. The fan should slip when cold. When the engine warms, the clutch disengages. Air thrown by the fan then starts to flow through the radiator and over the engine. You can hear and feel the air when the fan clutch locks up.

If the fan clutch is locked at all times during engine operation, it is defective and must be replaced. Excessive play or oil leakage also indicates the fan clutch failure.

4.17 FAN BELT SERVICE

A ***loose fan belt*** may slip and not rotate the coolant pump and fan at correct speed. The engine may even overheat. Always inspect the condition and tightness (tension) of fan belts when servicing a cooling system, *Fig. 4.9.* If a fan belt is cracked, frayed, glazed (hard, shiny surface), or oil soaked, it should be replaced.

Use care to keep your hands from running engine belts. A belt can pull your fingers into the pulleys, resulting in severe hand injuries.

Electric Cooling Fan Service

Most ***electric cooling fans*** are controlled by a heat sensitive switch located somewhere in the cooling system (radiator, engine block, thermostatic housing.) When the engine is cold, the switch keeps the electric fan motor OFF to speed up engine warm-up. When the cooling system reaches a predetermined temperature, the switch closes and the fan begins to throw air on to the radiator and the engine for cooling purposes.

4.18 TESTING AN ELECTRIC COOLING FAN

To test an electric cooling fan, observe whether the fan turns ON when the engine is warm. Make sure the fan motor is spinning at a ***normal speed*** and is forcing enough air through the radiator.

If the fan does ***not*** function, check the fuse, electrical connections, and supply voltage to the motor. If voltage is applied to the fan, and still the motor does not operate, replace the fan motor.

If the engine is warm and voltage is not supplied to the fan motor, check the action of the fan switch. Use either a VOM (multimeter) or test light. The switch should have almost zero resistance (i.e., allows current and voltage to pass) when engine is warm. Resistance should be infinite (i.e., stop current and voltage) when the engine is cold.

In case the above tests do not isolate the trouble, reference should be made to the applicable service manual for instructions. There may be a defective relay, connection, or some other problem.

4.19 HEATER CORE

Most of the heat absorbed from the engine by the cooling system is wasted. The vehicle heater, however, recovers some of this heat. Heated coolant is passed through tubes in the small core of the heater. Air is passed across the heater fins and is then sent to the passenger compartment. In some vehicles, the heater and air conditioning work in series to maintain vehicle compartment temperature, *Fig. 4.27.*

Heater Problem Diagnosis

When a vehicle's heater does not provide the desired amount of heat, generally the trouble can be traced to a defective thermostat and the engine not reaching the normal operating temperatures. Many other causes besides a defective thermostat can result in lack of heat from the heater. To determine the exact cause, follow this procedure:

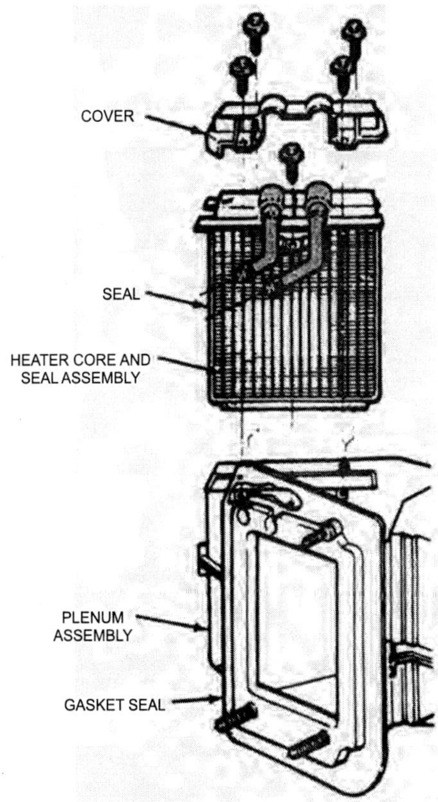

Fig. 4.27 *A heater core is similar to a small radiator but is mounted within the dash.*

1. After operating the engine for some time, feel the upper radiator hose. If the engine is up to proper operating temperature, the upper radiator hose should be too hot to keep hand on it. The hose should be pressurized too.
   - If the hose is not hot enough, replace the thermostat.
   - If the hose is not pressurized, test it. Replace the radiator pressure cap if it will not hold the specified pressure.
   - If every thing is fine, follow next step.
2. With the engine running, feel both heater hoses, with the heater set to the maximum heat position. Both hoses should be too hot to hold. If both hoses are warm or cool, check the heater control valve for proper operation. If one hose

is hot and the other (return) is just warm or cool, remove both hoses from the heater core or engine and flush the heater core with water from a garden hose.

3. Heat from the heater that "comes and goes" is most likely the result of low coolant level in the system.

4.20 COOLANT TEMPERATURE WARNING LIGHT

Most vehicles are equipped with a heat sensor for the engine operating temperature. In case the "hot" light comes on during driving or the temperature gauge goes into the red zone, then the coolant temperature is about 120°C to 126°C, which is still below the boiling point of the coolant (assuming a properly operating system and radiator cap). If this happens, follow the steps given below:

1. Shut off the air conditioning and turn on the heater. The heater will help rid the engine of some extra heat. Set the blower speed to high.
2. If possible, take a break, shut the engine off, and let it cool.
3. Never remove the radiator cap when the engine is hot. There may be steam in the radiator.
4. Do not continue to drive with the hot light on; this can cause a serious damage to the engine.
5. If the engine does not feel or smell hot, it is possible that the problem may be a faulty hot light sensor or gauge. Continue to drive, but to be safe, stop occasionally and check for any evidence of overheating or coolant loss.

4.21 COOLING SYSTEM INSPECTION CHECKLIST

Drive Belts – Look for

- Glazing, deterioration, cracks, fraying, or any other damage
- Age, replace every 4 years
- Pulley alignment
- Tension

Hoses – Look for

- Hardness, cracks, brittleness, cuts, or any other damage
- Sponginess or interior damage
- Loose connections or leakage
- Age, replace every 4 years

Coolant Leakage – Look for leakage at

- Core plugs
- Water pump shaft seal
- Water pump gaskets
- Thermostat gasket
- Hose connections

Radiator Cap – Check for

- Proper filler neck fit
- Proper spring action
- Specified pressure rating
- Damaged or brittle seal
- Pressure test

Coolant Pump – Check for

- Coolant leakage at the gasket
- Coolant leakage at the shaft
- Proper hose connections
- Pulley and belt alignment
- Excessive play

Radiator – Check for

- Leaks and/or corrosion at the tank seams or on the radiator tubes
- Clogged or bent radiator core fins
- Loose or missing mounting bolts
- Blocked air intake paths
- Coolant recovery tank installation
- Coolant level and concentration
- Rust, or any other coolant contamination
- Coolant age, replace every 2 years
- Pressure test

Radiator Fan – Check for

- Bent or cracked fan blades
- Any binding between the fan and shaft
- Fan clutch or electric fan operation

Fan Shrouds and Air Deflectors – Check for

- Location and mounting
- General condition
- Clearance

Heater Assembly – Check for

- Rust and/or corrosion on the core
- Hose connections
- Hose condition
- Heater control valve operation, leakage, and mounting

- Proper core assembly mounting
- Coolant leakage
- Condition of air ducts

Abnormal Noises – Check for

- Engine thump – coolant flow restriction
- Screeching – loose drive belt or defective coolant pump bearing
- Buzz or whistle – poor radiator cap seal or vibrating fan shroud
- Ringing or grinding – coolant pump shaft bearing, loose pulley, or tensioner bearing
- Gurgling radiator – plugged radiator or air in the coolant

4.22 COOLING SYSTEM MAINTENANCE

The cooling system is one of most maintenance-free systems in an automotive engine. Normal maintenance involves an occasional check on the coolant level, and a visual inspection for signs of coolant system leaks, and for the condition of the coolant hoses and fan drive belts.

The coolant level should only be checked when the engine is cool. Removing the radiator pressure cap from a hot engine may release the cooling system pressure while the coolant temperature is still above its atmospheric boiling temperature. When the cap is removed, the pressure in the system will instantly drop to atmospheric pressure level, causing the coolant to boil immediately. Vapours from the boiling coolant will blow coolant from the system. This results in coolant loss and someone may be injured or burned by the high temperature coolant that is blown out of the filler opening.

Drive belt condition and proper installation are important for the proper operation of the cooling system. If a wrong application of a fan is installed, the blades of the fan may not be angled correctly to provide adequate airflow through the radiator.

Flush and Refill

The coolant-antifreeze mixture is renewed at periodic intervals. Manufacturers recommend that a cooling system be flushed and that the antifreeze be replaced at specified intervals. Draining coolant when the engine is cool eliminates the danger of being injured by hot coolant. Opening a petcock in the bottom tank drains the radiator, and the coolant in the block is drained into a suitable container by opening plugs located in the lower part of the cooling passage.

Water should be run into the filler opening while the drains remain open. This flushing should continue until only clear water comes out of the system.

The volume of the cooling system should be determined. It is specified in the owner's manual and in the engine service manual. The antifreeze quantity required for the protection desired is shown on a chart that comes with the antifreeze. Open the bleeder valves and add the correct amount of the specified type of antifreeze followed by enough water to completely fill the system. The coolant recovery reservoir should be filled to the "level-cold" mark with the correct antifreeze mixture.

Burping the Cooling System

In cooling systems, small air pockets can occur. The engine must be thoroughly warmed to open the thermostat. This allows full coolant flow to remove the air pockets. The heater must also be turned to full heat.

The cooling system will not function correctly if air is not burped from the system after a refill. An easy method of burping is by replacing the radiator cap after the refill, but only to the first locked position. Drive the vehicle for several minutes and check the coolant level in radiator. Without the radiator cap tightly sealed, no pressure will build in the cooling system. Driving the vehicle helps circulate the coolant enough to force all air pockets up and out of the radiator filler. Top off the radiator after burping and replace the radiator cap to the fully locked position. Failure to burp the cooling system to remove all the air will often result in lack of heat from the heater and may result in engine overheating.

Flushing a Cooling System

Deposits that restrict coolant flow in the radiator may cause engine overheat problems.

Flushing (cleaning) of a cooling system should be done when rust or scale is found in the system. Flushing involves running water or a cleaning chemical through the cooling system in order to wash out contaminants.

Fast flushing is a common method of cleaning a cooling system because the thermostat does not have to be removed from the engine. A water hose is connected to a heater hose fitting. The radiator cap is removed and the drain cock is opened. When the water hose is ON and water flows into the system, rust and loose scale are removed.

Reverse flushing of a radiator requires a special adapter that is connected to the radiator outlet tank by a piece of hose. Another hose is attached to the inlet tank. Reverse flushing can often loosen these deposits. Back flushing requires the use of a special gun that mixes air with water. Low-pressure air is used to avoid damage to the cooling system, *Fig. 4.28*. Deposits will come out of the filler opening and out of the hose connected to the upper hose neck. This can be done on the engine block as well.

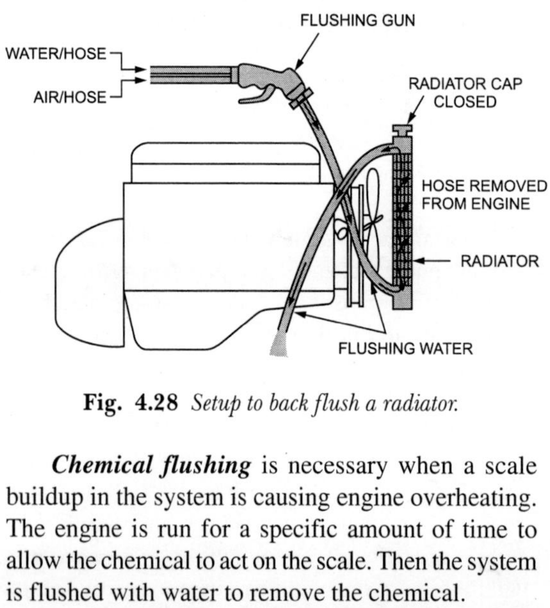

Fig. 4.28 *Setup to back flush a radiator.*

Chemical flushing is necessary when a scale buildup in the system is causing engine overheating. The engine is run for a specific amount of time to allow the chemical to act on the scale. Then the system is flushed with water to remove the chemical.

Always follow manufacturer's instructions when using a cooling system cleaning agent. The chemical may cause eye and skin burn. In case, after back flushing, some deposits still block the radiator core, then the radiator will have to be removed and sent to a dedicated radiator repair shop for cleaning.

4.23 LUBRICATION SYSTEM SERVICE

This is very important system that affects engine service life. The following sections review the types of service tasks that an automotive engine technician might be called upon to do when working on an engine lubrication system.

4.24 LUBRICATION SYSTEM TROUBLES

Problems found in a lubrication system are limited in number. Normally they include:

1. ***High oil consumption*** – Requires oil must be added to engine often.
2. ***Low oil pressure*** – Oil gauge reading is low, oil pressure indicator light glows, or abnormal engine noises.
3. ***High oil pressure*** – Gauge reading is high, oil filter swelled.
4. ***Defective indicator or gauge circuit*** – Inaccurate operation or readings.

When diagnosing these problems, make a visual inspection of the engine for obvious problems. Check for oil leakage, disconnected sending unit wire, low oil level, smashed oil pan, or other trouble that would relate to the *symptoms*, *Fig. 4.29*.

High Oil Consumption

High oil consumption is caused by external oil leakage out of the engine or by internal leakage of oil into the combustion chambers.

External oil leakage is easily detected as darkened, oil-wet areas on or around the engine. Oil will be dark brown or black if used for an extended period of time. Oil may be spotted in small puddles, under the engine compartment, in the garage or other parking place. Leaking gaskets or seals are usually the source of external engine oil leakage.

To locate oil leakage, it may be necessary to raise the vehicle on a lift and inspect for leaks under the engine. Trace the oil leakage to its highest point. The parts around the point of leakage will be ***washed clean*** by the constant seepage of oil.

Internal leakage appears as ***blue smoke*** coming out of the vehicle's exhaust system tailpipe. For example, oil can enter the engine combustion chambers and be burned if the engine piston rings and cylinders are badly worn.

Be careful not to confuse black smoke caused by excess fuel in cylinder, and white smoke caused by coolant leaking into gasoline engine cylinder with the blue smoke caused by engine lubricating oil.

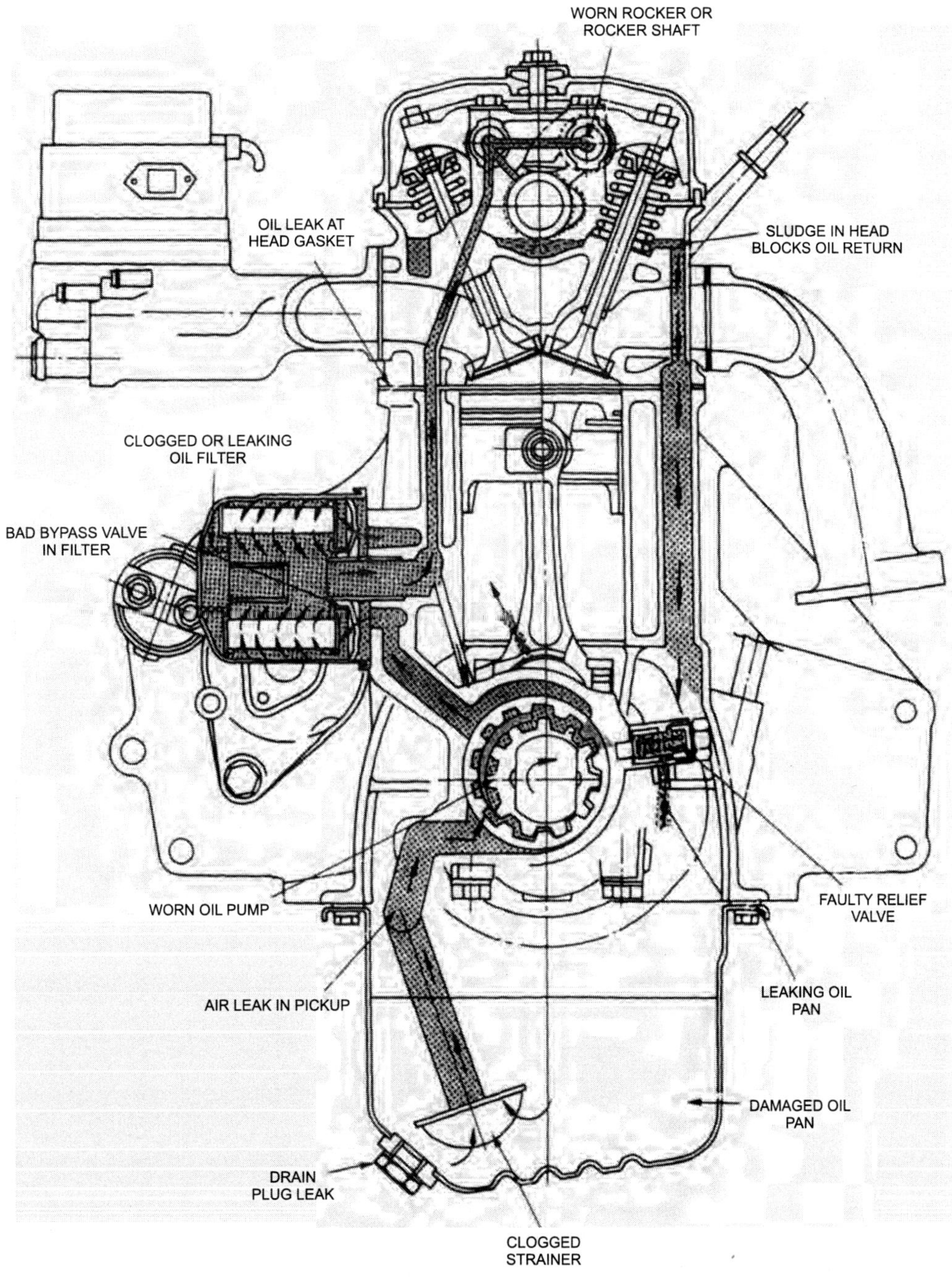

Fig. 4.29 *Some of the common problems that can develop with an engine lubrication system. You must check for these troubles during diagnosis.*

4.25 OIL PRESSURE PROBLEMS

Low oil pressure is indicated when the oil pressure indicator dash light glows, oil pressure gauge reading is low, or when the hydraulic valve lifters or bearings rattle.

Common causes of low oil pressure include:

- Low oil level – Oil level in oil pan not high enough to immerse oil pickup or oil strainer.
- Worn connecting rod or main bearings – Pump can not provide enough oil volume to keep up demand to maintain the required oil pressure.
- Thin or diluted oil – Low viscosity or gasoline in oil.
- Weak or broken pressure relief valve spring – Relief valve opening too early to prevent needed oil pressure.
- Cracked or loose oil pump pickup tube – Air being induced into oil pump.
- Worn oil pump (excess clearance between rotors or gears and housing.
- Clogged oil pickup screen – Reduced amount of oil entering the pump.

High oil pressure is rarely a problem. The most frequent causes of high oil pressure are:

- Pressure relief valve stuck in closed position – not opening at specified pressure.
- High relief valve spring tension – wrong spring or improperly shimmed spring.
- High oil viscosity – excessively thick oil or use of oil additive that causes viscosity increase.
- Restricted oil gallery – Due to defective cylinder block casting or debris in oil passage. A ***defective indicator*** or ***gauge*** may look to be a low or high oil pressure problem. Instead, the sending unit, circuit wiring or gauge may be at fault.

4.26 OIL PRESSURE TEST

An ***oil pressure test*** uses a gauge to measure actual pressure in the engine. A pressure gauge is screwed into the hole for the oil pressure sending unit, *Fig. 4.30*.

Start and run the engine to warm it up. Then run the engine at the service manual recommended rpm. Read the test pressure gauge and compare to specs. If oil pressure is too low or high, carry out the necessary repairs.

Depending upon the number of kilometers on the type of engine, oil pressure should be at least 138 to 207 kPa at cruising speeds. Refer to the applicable service manual for specs when testing the oil pressure.

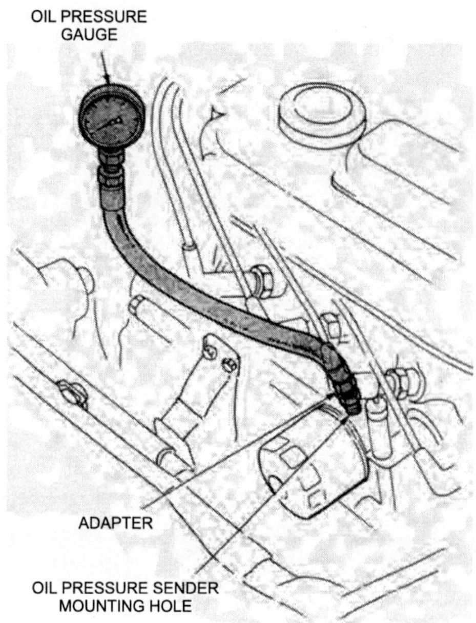

Fig. 4.30 *If symptoms indicate a low oil pressure, measure actual pressure. Remove oil-sending unit and install test gauge. Start the engine and warm-up engine to the operating temperature. Test gauge reading should be up to specs at idle and at cruising rpm. (Honda)*

4.27 ENGINE OIL AND FILTER SERVICE

It is very important that the oil and oil filter of an engine be serviced regularly. Lack of maintenance can greatly shorten engine service life, *Fig. 4.31*. Used oil will be contaminated with dirt, carbon, wear metal particles, gasoline, acids, and other harmful substances. Some of the smaller particles and corrosive chemicals are not trapped in the oil filter. These particles continue to circulate through the engine, and accelerate part wear and corrosion.

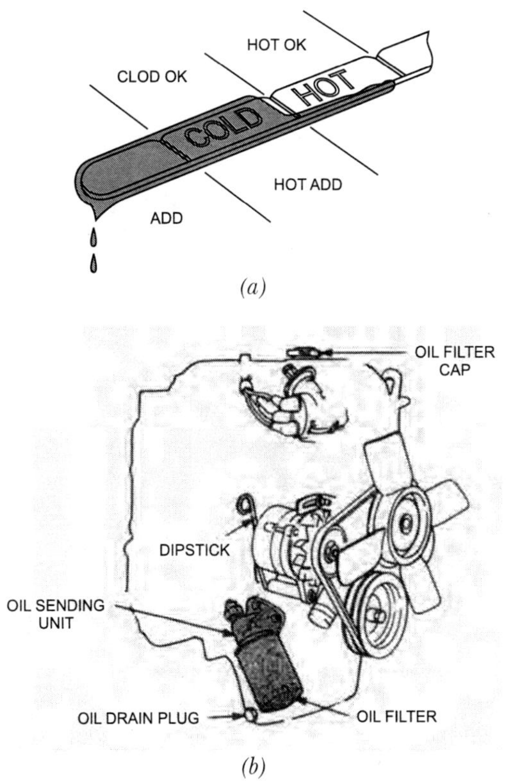

Fig. 4.31 *(a) When checking oil level, make sure that you use correct markings on dipstick, (b) Typical locations for filler cap, dipstick, and oil filter. Also note the oil-sending unit. (Mazda)*

4.28 OIL AND FILTER CHANGE INTERVALS

Automotive vehicle manufacturers specify a maximum number of kilometers (miles) a vehicle can be driven between successive oil changes. If the oil is not changed at the specified interval, the vehicle's warranty will become void.

New passenger cars can generally be driven about 10000 km between oil changes. Older model cars should have their oil changed at much lesser kilometers interval. An older, and worn engine will contaminate the oil more quickly as compared to a new engine. More product of combustion will blow past the rings and enter the crankcase and mix with oil. Also, larger engine bearing clearance will require more oil from the lubrication system.

A vehicle regularly driven for shorter distances and parked will require frequent oil changes. This is due to the fact that the engine may not be reaching full operating temperature, and the oil gets contaminated with fuel, moisture, and other substances more quickly.

Diesel engines and turbocharged engines normally require more frequent oil and filter service than naturally aspirated engines.

4.29 CHANGING ENGINE OIL AND FILTER

To change the engine oil, run the engine to allow it to warm-up to normal operating temperature. This will help suspend debris and contaminants in the oil and allow the oil to drain quickly and thoroughly. Then, follow the basic steps, shown in *Fig. 4.32*, to change engine oil and the oil filter.

- Park the vehicle on level ground so that all oil drains from the pan.
- Be careful not to let hot oil pour on hand or arm.
- Check the condition of the drain plug threads and the O-ring washer. If found defective change them.
- Do not over tighten the oil pan drain plug, as it will strip easily.
- Wipe any oil on the filter O-ring before assembly.
- Only hand-tighten the filter and avoid distortion of the filter canister and prevent leak.
- Fill the engine with the recommended type and amount of oil.
- Inspect for any oil leaks with the engine running before allowing the customer to go.

Used Oil Disposal

Used motor oil is a hazardous waste. The old oil should be stored in an approved container for recycling. Recycling old oil not only saves environment from being polluted, it also helps conserve natural resources. One gallon of used oil can be refined into 2.5 quarts of high quality like-new oil. It takes about 40 gallons of crude oil to produce this much oil. Used oil may be burned as a fuel by heaters in shop.

Always send used oil to a recycling facility. Some recycling companies provide a pick-up service while some others require the oil to be delivered to their facility.

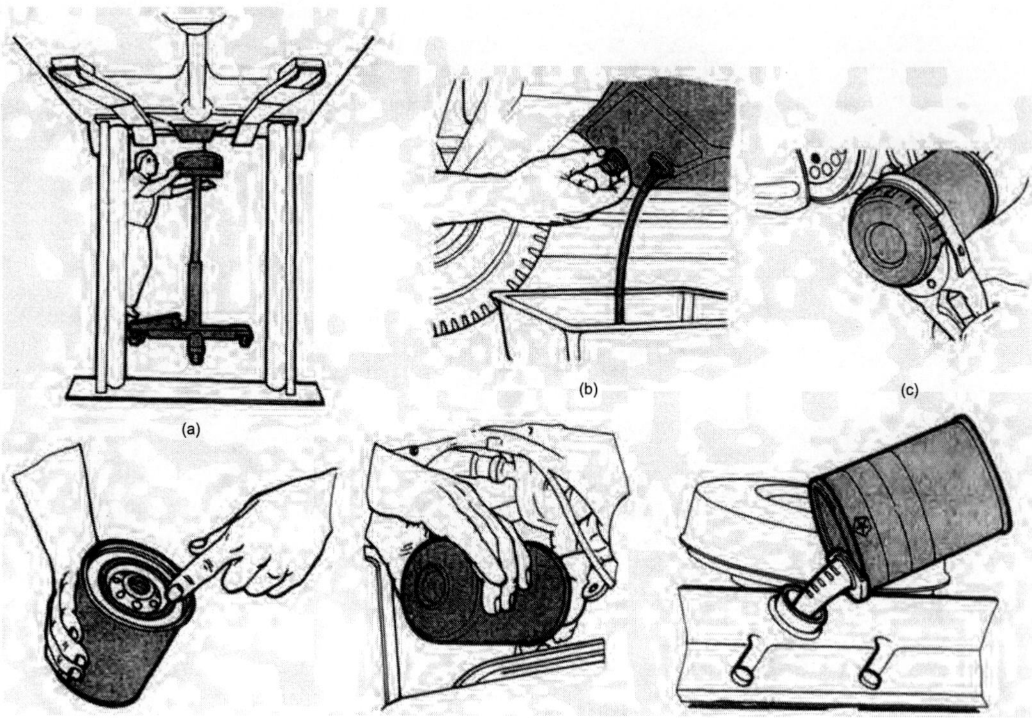

Fig. 4.32 *Changing engine oil and filter, (a) Use lift or floor jack and jack stands to hoist vehicle in level position. Place catch pan under drain plug, (b) Unscrew drain plug and allow oil to pour into pan. Use caution as hot oil can cause painful burns, (c) Use filter wrench to remove old filter. Turn counterclockwise, (d) Wipe clean oil on new oil filter O-ring. This will assure proper tightening, (e) Install and tighten oil filter by hand. Hands and filter should be clean and dry. Use a rag if needed, (f) Install right type and quantity of oil. Pour oil into filter or breather opening in valve cover or induction manifold. (Mopar)*

4.30 OIL PUMP SERVICE

A ***bad oil pump*** that creates no or low oil pressure can end up in severe engine damage. With worn internal pump parts, the pump will have reduced output, *Fig. 4.33*. The pump drive shaft can strip in the pump or distributor, disrupting the pump operation. If tests indicate a faulty pump, remove and replace or rebuild the pump to restore its intended function.

Oil Pump Removal

Since the oil pump location can vary form model and make of vehicle, the removal procedures will vary too. Refer to the applicable service manual for pump removal procedure details.

Oil Pump Rebuild

In general, technicians/mechanics install a new or factory rebuilt oil pump when replacing the old one. It is usually costly to completely rebuild an old oil pump in an automobile repair shop. However, to understand the procedure to overhaul old pumps, *Figs. 4.34* through *4.36*, summarize the service procedure for typical oil pump designs. Install the pump in reverse order of its removal. Assure the correct position of the gasket. Torque all attaching bolts as specified. *Figs. 4.37* and *4.38* show installation methods of oil pumps mounted on the front of an engine.

If an oil pickup is used, make sure it is properly installed and tightened in place, *Fig. 4.39*.

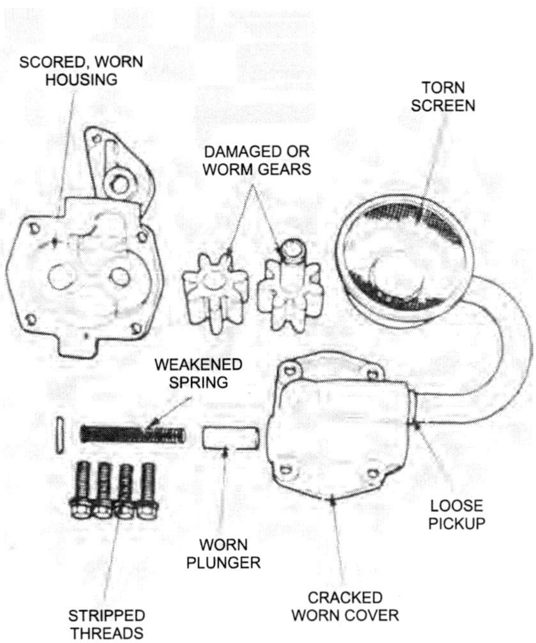

Fig. 4.33 *When servicing oil pump, check conditions of gears or rotors and inside of housing. Slightest signs of wear or damage can upset pump delivery and require pump replacement. (GM Pontiac Div.)*

4.31 PRESSURE RELIEF VALVE SERVICE

A ***bad pressure relief valve*** can be a source of oil pressure problems. The valve may be located in the oil pump itself, oil filter housing, or engine block.

If system test symptoms indicate a faulty pressure relief valve, disassemble it for service. It is also serviced during an engine overhaul. Remove the cup plug or cap holding the pressure relief valve. Then slide the spring and piston out of their bore.

Measure the ***spring free length*** and compare to specs. If the spring length does not match the specs, replace it. The spring tension may be tested on a spring tester to assure the spring meets its design specs.

Check the relief valve bore wear with a micrometer and small hole gauge. Also, check the valve for scratches and scoring. Slide the plunger in and out of the bore. If any problems are found, replace the parts.

4.32 OIL PAN SERVICE

An engine oil pan may need to be removed for different reasons: to service engine bearings, oil pump, a damaged pan, a stripped pan oil drain plug, or during an engine overhaul.

Some vehicle designs allow engine oil pans to be removed with the engine in the vehicle's engine compartment. After removing any bolt-on cross members, steering system components that can cause interference, and other obstructions, raise the engine off its mounts. This may be required to allow enough clearance to unbolt and slide off the oil pan for removal.

Some other vehicle designs do not allow in-vehicle oil pan removal. The engine has to be lifted from the vehicle engine compartment before the oil pan will come off. For exact details refer to the applicable service manual.

Oil Pan Removal

Drain the oil before removing an oil pan by removing the drain plug. Reinstall the drain plug after oil removal. Remove the bolts around the outside of oil pan flange. Then, to free the pan from the cylinder block, tap on it lightly with a rubber hammer while supporting it with a hand. If stuck tight, gently pry between the pan flange and the block avoiding any bending of oil pan flange. This is necessary to avoid any leakage later upon assembly.

Carefully remove all old gasket or silicone adhesive material, from the pan and engine block mating surfaces, using a gasket scraper. With an aluminium pan, do not gouge the sealing surface. Inspect the inside of the pan for any debris. Metal bits from bearings, plastic bits from timing gear teeth, rubber particles from valve stem seals indicate engine mechanical problems.

Oil Pan Installation

Wash the oil pan thoroughly in cold soak cleaner before installing it. Inspect the drain plug hole for stripped threads. Lay the pan upside down on a flat worktable. Verify the flatness of the pan flange sealing surface. Restore the flange shape, if needed, with light hammer blows.

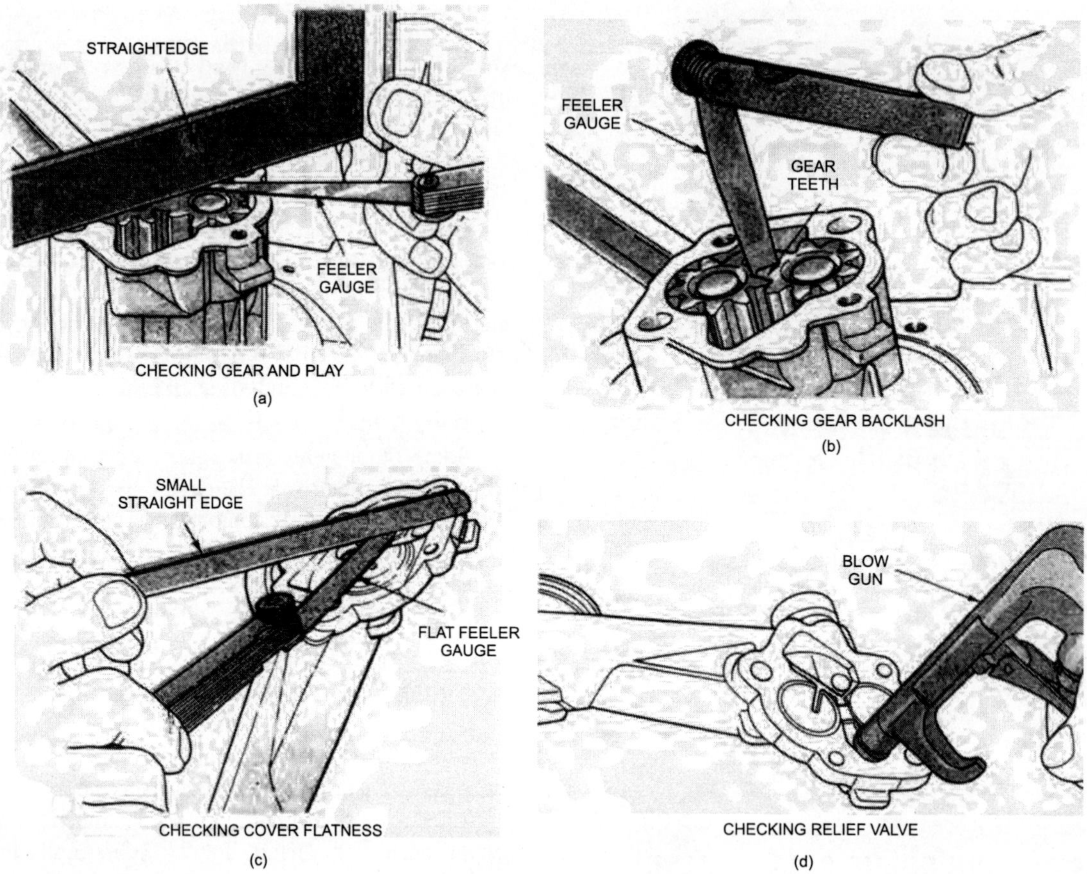

Fig. 4.34 *Rebuilding in-pan type oil pump, (a) Use straight and flat feeler gauge to check gear end play. Sand parts with 400 grit paper on perfectly flat surface if required, (b) Check gear backlash and replace parts if not within specs, (c) Check the cover flatness with straightedge and feeler gauge. If only slightly warped, true by rubbing on 400 grit laying on flat surface, (d) Relief valve action can be checked with air nozzle. (Dodge)*

Either a gasket or chemical sealer may be used for the oil pan, *Fig. 4.40*. Install a new gasket with an approved gasket adhesive to position it at the correct position. If rubber seals are used on each end, press them into their grooves. Apply silicone sealer where the gaskets meet the copper seals.

To install a chemical or silicone gasket, clean the pan and block mating surfaces with a suitable solvent. Then run a uniform bead, about 3 mm wide, of sealer around the pan or block flange as recommended. Form a continuous bead and place extra sealer at part or gasket seal joints.

Assemble the pan carefully into correct place on the engine block. Start all mounting bolts by hand. Check that the gasket or sealer does not shift or smear, *Fig. 4.41*. Then, tighten all bolts in small steps at a time in a crisscross pattern. Torque each bolt to recommended specs. Carefully check for correct drain plug torque specification. Fill the engine with recommended type and grade oil. Check for any oil leaks with the engine running.

4.33 OIL PRESSURE INDICATOR AND GAUGE SERVICE

A ***bad oil pressure indicator*** or ***gauge*** *may* raise a false flag to scare the driver into thinking that there are major engine problems. The indicator light may flicker or stay on, indicating a low oil pressure problem. The gauge may read high or low, suggesting a lubrication system problem.

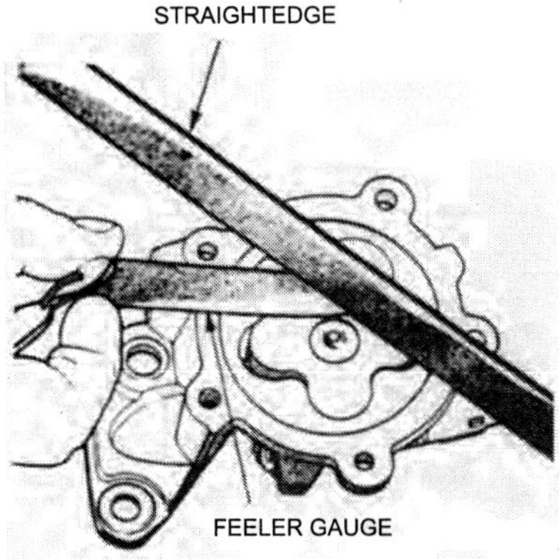

(a) Checking rotor end play.

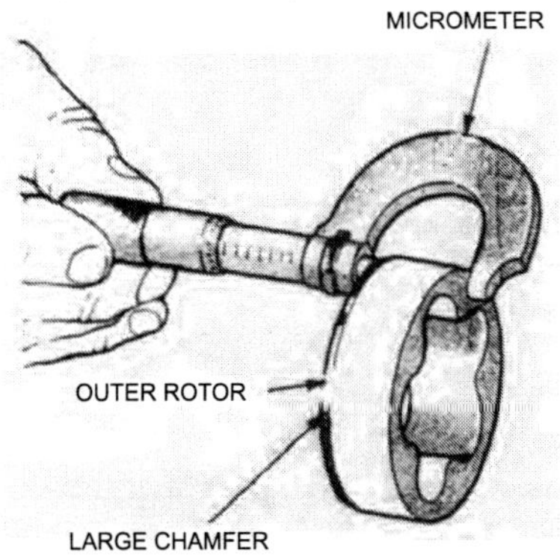

(b) Measuring outer rotor thickness.

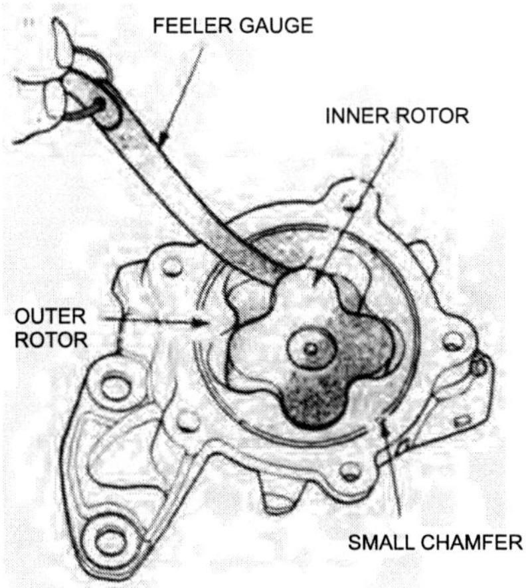

(c) Measuring clearance between rotors.

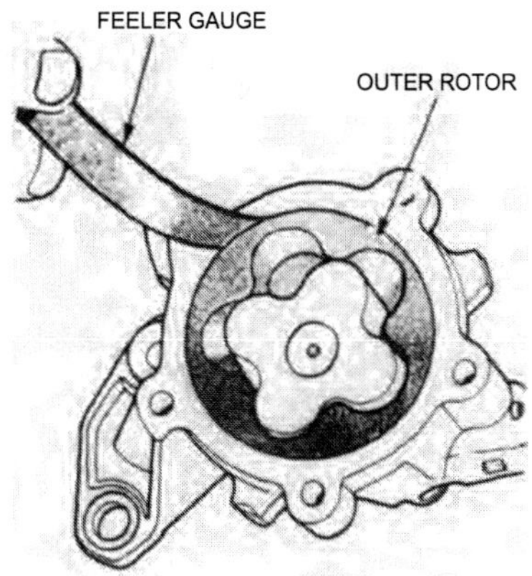

(d) Checking outer rotor clearance.

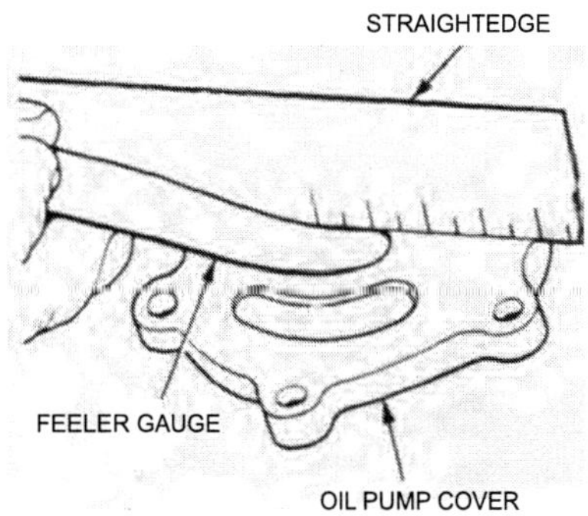

(e) Using straightedge and feeler gauge to check cover flatness.

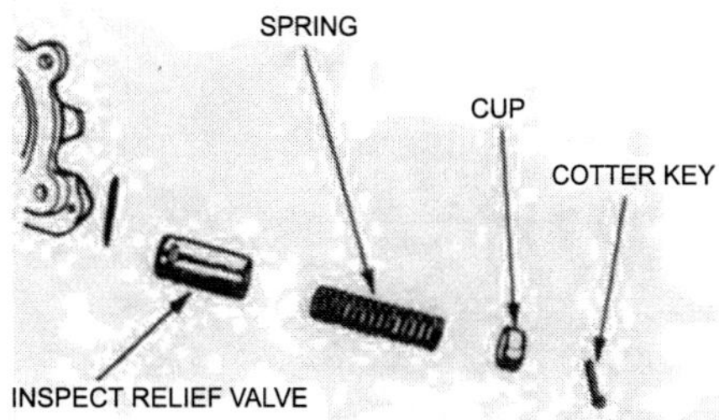

(f) Inspect parts of oil Pressure relief valve.

Fig. 4.35 *Rebuilding a rotor gear oil pump. (Chrysler)*

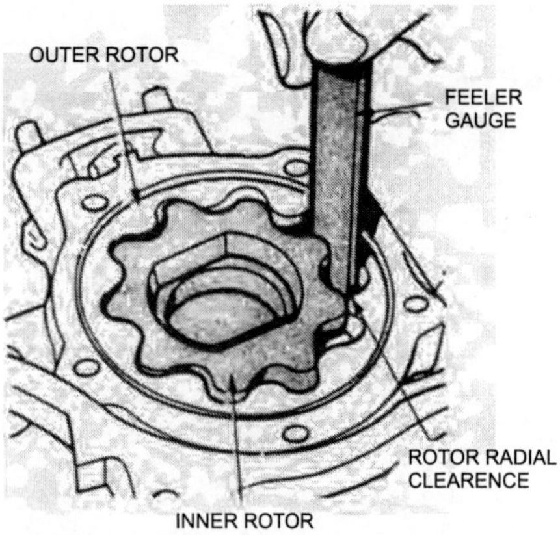

(a) Measuring inner rotor radial clearance.

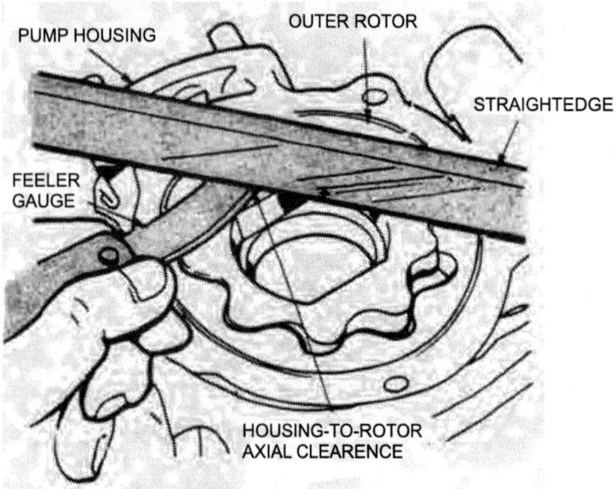

(b) Measuring housing-to-rotor axial clearance.

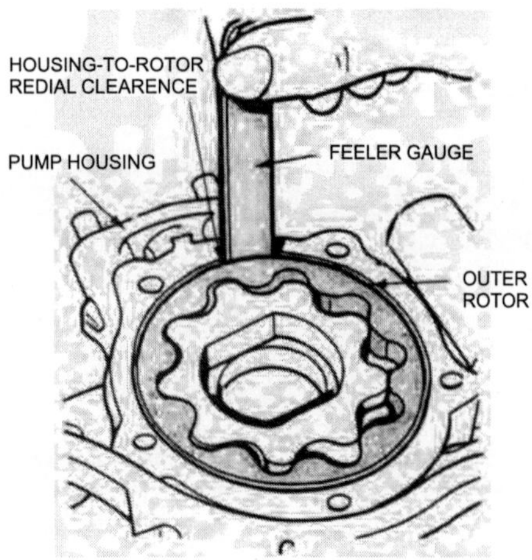

(c) Measuring housing-to-rotor radial clearance.

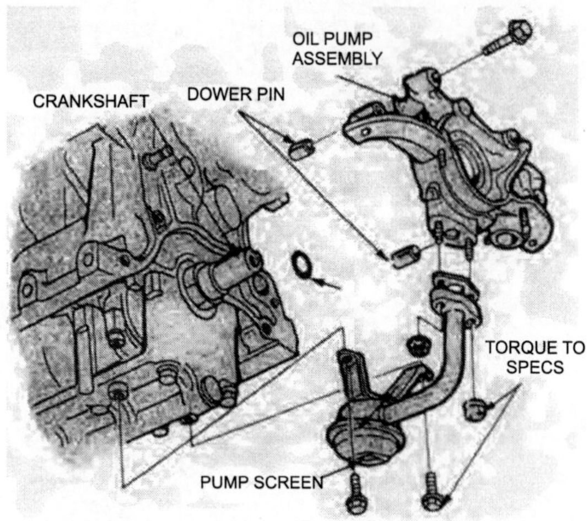

(d) Install assembled pump on front of engine using approved service manual methods. Dowel pins assure that pump aligns around crankshaft properly.

Fig. 4.36 *Servicing crank-mounted oil pump. (Honda)*

Check the indicator or gauge circuit for any problems. The wire going to the sending unit could be disengaged causing no light, flickering light, or improper gauge readings. The sending unit wire may also be shorted to ground to cause the light to stay on or gauge always reading high.

To check the action of the indicator or gauge, remove the wire from the sending unit, *Fig. 4.42*. Touch it on a metal part of the engine. This should cause the indicator light to glow or the oil pressure gauge read maximum. If it does the sending unit may be defective. If it does not, then the circuit, indicator, or gauge may be faulty.

Always refer the service manual before testing an indicator or gauge circuit. Vehicle manufacturers might have recommended a special gauge tester. This is particularly important with some computer-controlled systems. The tester will place a specific resistance in the circuit to avoid circuit damage. Refer to the applicable service manual for instructions.

Usually a low-pressure light comes on when oil pressure drops below 34 to 69 kPa.

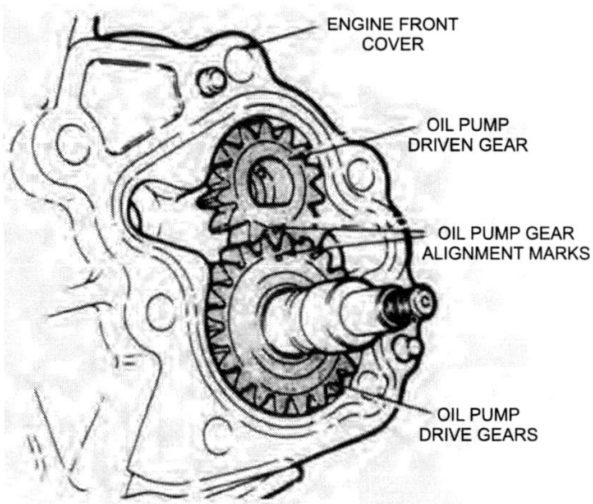

Fig. 4.37 *This gear pump, mounted on front of the engine, has alignment marks. Gear teeth have been lapped together at factory and must be timed together for proper meshing. (Chrysler)*

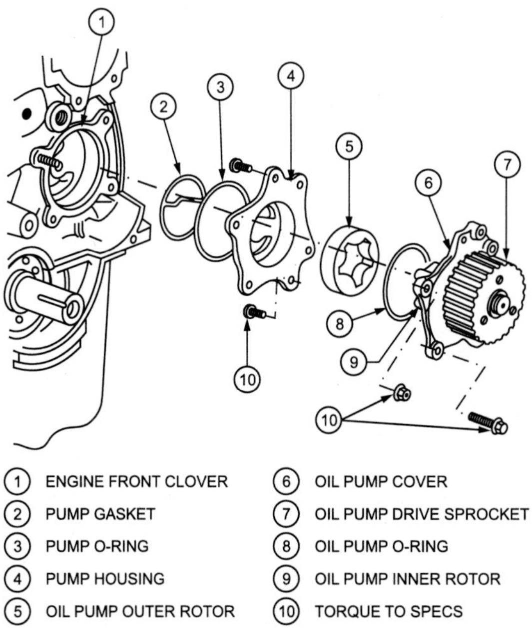

Fig. 4.38 *Note how seals install on this front-mounted oil pump. Tighten bolts to specs to prevent part warpage. (Honda)*

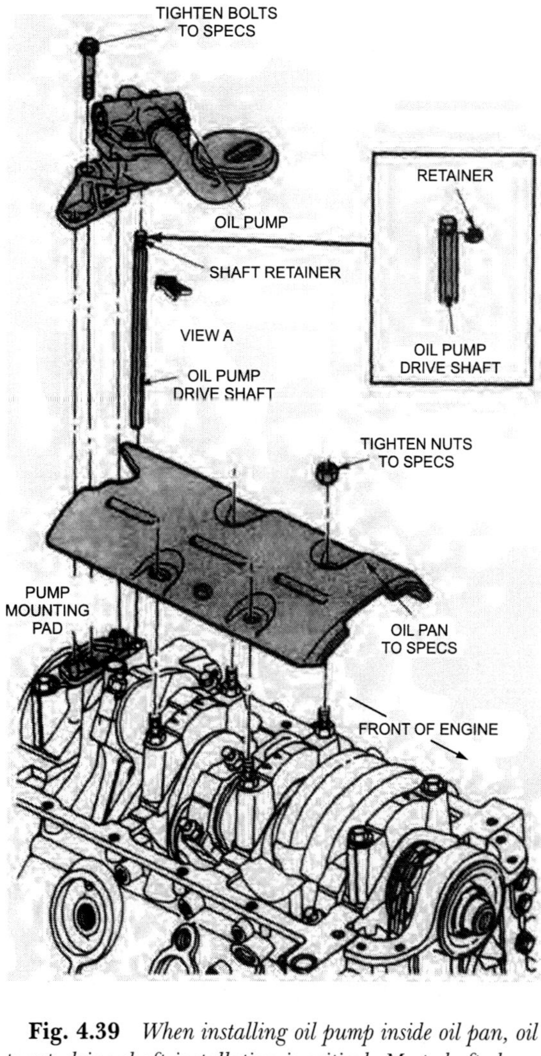

Fig. 4.39 *When installing oil pump inside oil pan, oil pump drive shaft installation is critical. Most shafts have small clip that keeps shaft from pulling or falling out during distributor removal. Make sure this clip is positioned correctly. Tighten all oil pump bolts in exact specs. Double-check pickup tube. (Ford Motors)*

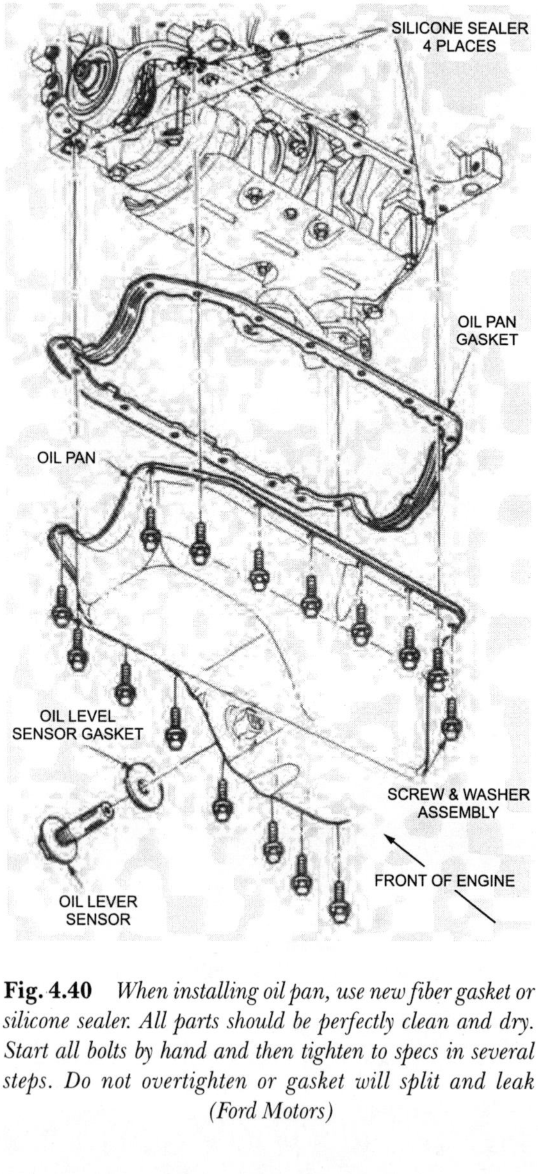

Fig. 4.40 *When installing oil pan, use new fiber gasket or silicone sealer. All parts should be perfectly clean and dry. Start all bolts by hand and then tighten to specs in several steps. Do not overtighten or gasket will split and leak (Ford Motors)*

Fig. 4.41 *To install oil pan. adhere gasket to cylinder block. Then place bead of silicon sealer where seal touches gasket. (Fel-Pro)*

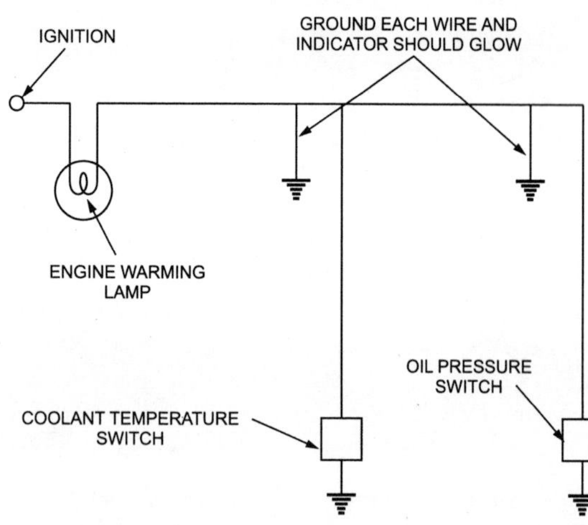

Fig. 4.42 *Usually, we can check oil sending unit or temperature sending unit by grounding wire going to unit. When lead is grounded, indicator light should glow. With a gauge, do not ground too long or damage could result. (Ford Motors)*

THEORY QUESTIONS

1. This is not the purpose of the radiator cap:
   (*a*) Relieve vacuum in system
   (*b*) Lower boiling point of coolant,
   (*c*) Relieve excess pressure
   (*d*) Seal top of the filler neck.
2. Name the two ways of checking antifreeze.
3. A driver complains of engine overheating only at idle. The cooling system is full of coolant and belts are properly adjusted. Technician A notices a missing fan shroud and says that could be the cause of overheating. Technician B says a thermostat stuck closed is a more likely cause of overheating. Who is right?
   (*a*) Technician A (*b*) Technician B
   (*c*) Both A and B (*d*) Neither A nor B.
4. A missing lower hose spring can cause engine overheating. *True* or *False?*
5. Permanent antifreeze is usually ——————.
   (*a*) Glycerin (*b*) Methanol
   (*c*) Kerosine (*d*) Ethylene glycol
6. As the %age of antifreeze in the coolant increases, ——————.
   (*a*) The freeze point decreases up to a point
   (*b*) The boiling point decreases
   (*c*) The heat transfer increases
   (*d*) All of the above occur
7. Describe the method of checking worn coolant pump bearings.
8. An electric engine cooling system fan fails to turn on when the engine warms up. Technician A says that the first step is to remove and test the fan motor. Technician B says that apply battery voltage directly to the motor at the fan switch. Who is right?
   (*a*) Technician A (*b*) Technician B
   (*c*) Both A and B (*d*) Neither A nor B.
9. How and why you perform an engine cooling system pressure test?
10. It is very difficult to break a thermostat housing. *True or False*?
11. Heat transfer is improved from the coolant to the outside air when ——————.
    (*a*) The temperature differential is higher
    (*b*) The temperature differential is smaller
    (*c*) The coolant is 95% antifreeze
    (*d*) Both (*a*) and (*c*).
12. A coolant pump is a positive displacement-type of pump. *True or False*?
13. The procedure that should be used when refilling an empty cooling system is to:
    (*a*) Determine capacity of the cooling system, then fill the system halfway with antifreeze and the rest with water
    (*b*) Fill completely with antifreeze, but mix 50/50 solution for the overflow bottle
    (*c*) Fill the cylinder block and one-half of the radiator with 100% antifreeze and fill the rest of the radiator with water
    (*d*) Fill the radiator with antifreeze, start the engine, drain the radiator, and fill with 50/50 mixture of antifreeze and water.
14. Which statement is true about thermostat?
    (*a*) The temperature marked on the thermostat is the temperature at which the thermostat should be fully open
    (*b*) The temperature marked on the thermostat is the temperature at which the thermostat should start to open
    (*c*) Both (*a*) and (*b*)
    (*d*) None of the above.
15. Technician A says that the radiator should be inspected for leaks and proper flow before installing a rebuilt engine. Technician B says that overheating during slow city driving can only be due to a defective electric fan. Who is right?
    (*a*) Technician A (*b*) Technician B
    (*c*) Both A and B (*d*) Neither A nor B
16. The normal operating temperature (coolant temperature) of an engine equipped with a 195°F thermostat is
    (*a*) 175° to 195°F (*b*) 45° to 205°F
    (*c*) 195° to 215°F (*d*) 175° to 215°F
17. Technician A says that an automotive engine normally operates between 104°–121°C. Technician B says that an engine operates between 82.2° to 90.5°C. Who is right?
    (*a*) A only (*b*) B only
    (*c*) Both A and B (*d*) Neither A nor B.

18. Technician A says that a liquid cooling system is better than an air cooling system because it improves the efficiency of the heater. Technician B says that a liquid cooling system is better than an air cooling system because engine temperature is more precisely controlled. Who is right?
    (*a*) A only (*b*) B only
    (*c*) Both A and B (*d*) Neither A nor B.

19. All of the following will cause an automotive engine to overheat except:
    (*a*) a loose fan belt
    (*b*) An inoperative electric cooling fan
    (*c*) A locked fan clutch
    (*d*) Retarded ignition timing.

20. Technician A says that when the engine is cold, coolant. should circulate through the radiator. Technician B says that when the engine is cold, coolant should not circulate through the radiator. Who is right?
    (*a*) A only (*b*) B only
    (*c*) Both A and B (*d*) Neither A nor B.

21. A driver complains that the heater works sometimes, but sometimes only cold air comes out while driving. Technician A says that the water pump is defective. Technician B says that the cooling system could be low on coolant. Who is right?
    (*a*) Technician A only (*b*) Technician B only
    (*c*) Both A and B (*d*) Neither A nor B.

22. Normal oil pump pressure in an engine is _____.
    (*a*) 3 to 7 psi (*b*) 10 to 60 psi
    (*c*) 100 to 150 psi (*d*) 40 to 250 psi

23. As oil is used in an engine _____________ .
    (*a*) It becomes thinner as a result of chemical breakdown that occurs with age
    (*b*) It becomes thicker because of oxidation, wear metals, and combustion byproducts
    (*c*) It becomes thicker because of temperature changes
    (*d*) It becomes thinner because of oxidation, wear metals, and combustion by-products.

24. Technician A says that the same brand engine oil should be used throughout an engine's entire service life. Technician B says that any engine oil of the correct API and ASE rating can be used because all engine oils are compatible. Who is right?
    (*a*) Technician A (*b*) Technician B
    (*c*) Both A and B (*d*) Neither A nor B

25. Technician A says that the oil will remain perfectly clean if just the oil filter is changed regularly. Technician B says that oil filters can filter particles smaller than the human eye can see. Who is right?
    (*a*) Technician A (*b*) Technician B
    (*c*) Both A and B (*d*) Neither A nor B

26. In typical engine lubrication systems, what components are the last to receive oil and the first to suffer from a lack of oil or oil pressure?
    (*a*) Main bearings (*b*) Oil filters
    (*c*) Rod bearings (*d*) Valve trains

27. A ______ _______ _______ limits maximum oil pressure in an engine lubrication system.

28. Diesel and turbocharged engines should have their oil changed more often than naturally aspirated gasoline engines. *True or False*?

29. Internal oil leakage shows up as _____ ____ coming out of the vehicle exhaust tailpipe.

30. Give some of the reasons you may have to remove the engine oil pan.

31. Technician A says that a low viscosity oil is very thick and resists flow. Technician B says that a low viscosity oil is thin and flows easily. Who is right?
    (*a*) A only (*b*) B only
    (*c*) Both A and B (*d*) Neither A nor B.

32. All of the following can cause low oil pressure except:
    (*a*) A clogged oil pump oil pickup
    (*b*) Worn connecting rod bearings
    (*c*) Main bearings
    (*d*) Restricted oil gallery.

33. When performing an oil pressure test, the pressure gauge should be attached to ___________.
    (*a*) The oil filter fitting
    (*b*) The oil pressure sending unit
    (*c*) The oil pump
    (*d*) None of the above.

34. All of the following are basic rules to follow when changing the engine's oil and oil filter except:
    (*a*) Park the vehicle on a level groud when draining the oil
    (*b*) Apply oil to the oil O-ring before installing filter
    (*c*) Use a filter wrench to tighten engine oil filter
    (*d*) Do not overtighten the oil pan drain plug.

❑❑❑

Starting and Charging Systems Diagnosis and Service

5

5.1 INTRODUCTION

In this chapter we will review basic testing, service, and repair of the starting and charging system components. However, keep in mind that the battery, charging system, and starting system are interacting. A malfunction in one component can adversely affect the operation of another component.

Figure 5.1 shows a diagram for a modern electrical diagnostic connector circuit. The *diagnostic connector* can be used to measure voltages in various parts of the circuit without disconnecting wires.

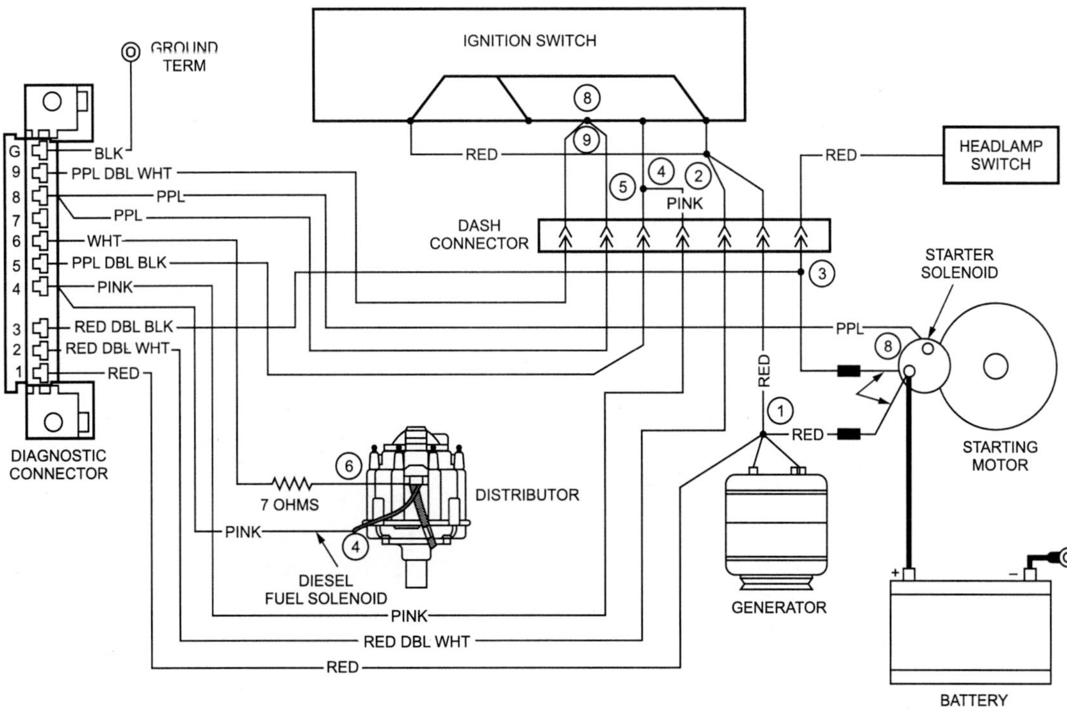

Fig. 5.1 *The wiring diagram shows how special test connector is wired. Special tester or voltmeter can be used to quickly determine the circuit voltage at connector, without disconnecting any wires. (Buick Div. General Motors)*

Special testers can be plugged into the diagnostic connector to analyze troubles.

5.2 CRANKING CIRCUIT

The cranking circuit includes those mechanical and electrical components required to crank the engine for starting. A modern cranking circuits includes:

1. **Starter motor** – The starter motor is usually a 0.4 to 2 kilowatt electric motor, *Fig. 5.2.*
2. **Battery** – The battery must be of the correct capacity and at least 75% charged to provide the necessary current and voltage for correct operation of the starter.
3. **Starter solenoid or relay** – The high current required by starter must be able to be turned ***on*** and ***off***. A large switch would be required if the current were controlled by the driver directly. Instead, a small current ignition switch operates a solenoid or relay that controls the high starter current.
4. **Starter drive** – The starter drive uses a small gear that engages the engine flywheel ring gear and transmits starter motor power to crank the engine.
5. **Ignition switch** – The ignition switch and safety control switches control the starter motor operation.
6. **Neutral safety (clutch switch)** – This switch prevents the operation of the starter unless the gear selector is in park or neutral or the clutch pedal is depressed. *Fig. 5.3.*

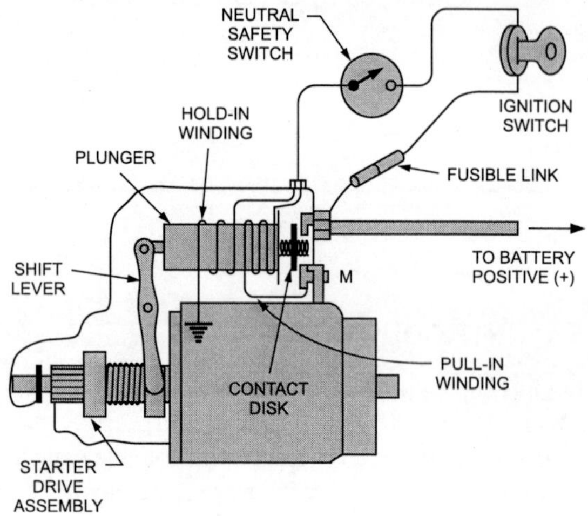

Fig. 5.3 *Wiring diagram of a typical starter motor solenoid. Note that both the pull-in winding and hold-in winding are energized when the ignition switch is first turned to the "start" position. As soon as the solenoid contact disc makes electrical contact with both B and M terminals, the battery current is conducted to the starter motor.*

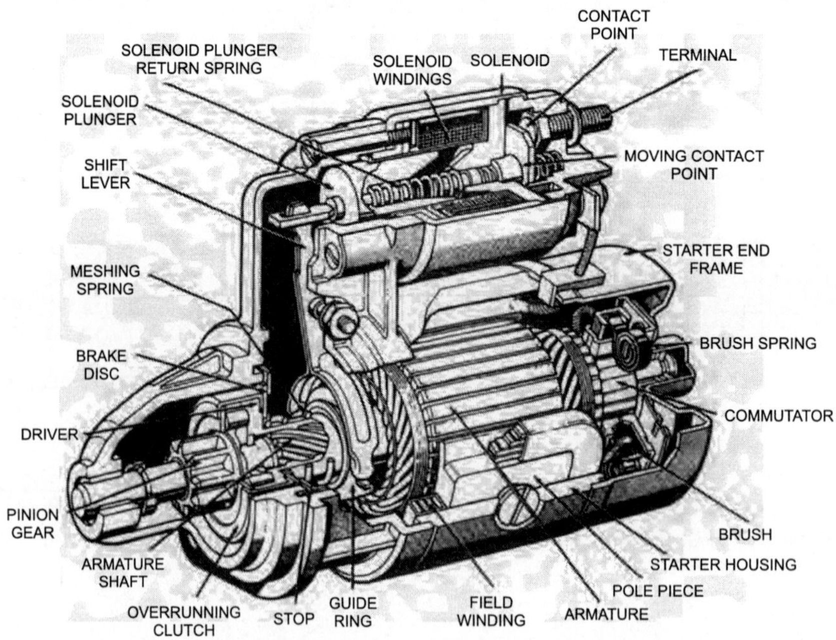

Fig. 5.2 *A cutaway view of a typical starter motor.*

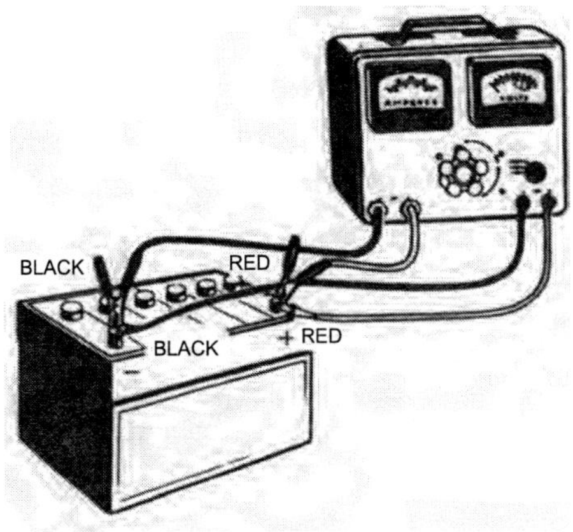

Fig. 5.4 *(a) Battery capacity test. (Chrysler Corp.)*

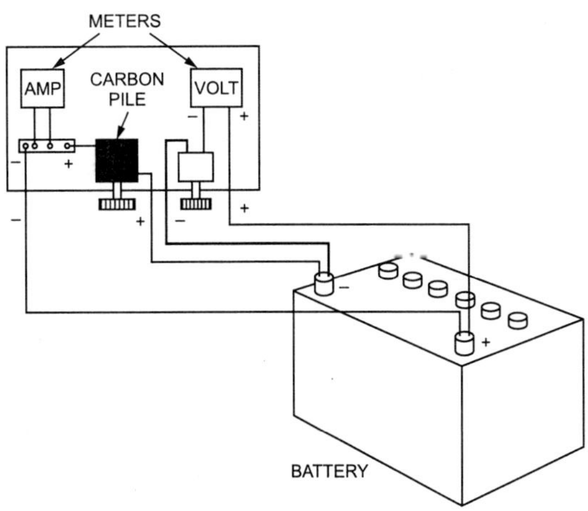

Fig. 5.4 *(b)*

5.3 BATTERY LIFE AND MAINTENANCE

If a battery is not maintained properly, its service life will be decreased. Battery maintenance should be done periodically during tune-ups; occasional oil change or anytime symptoms indicate battery problems.

Batteries wear out and require replacement. The life of a battery depends on the design, materials and workmanship used in its construction and also on the *care* it receives in service. There are certain types of abnormal operation that will cause serious trouble to the battery unless they are found and eliminated. When the battery is given casual service, these conditions may not be detected in time to prevent trouble. But if the battery is inspected and checked periodically these troubles may be found.

Battery maintenance includes making certain that the battery case is kept clean and adding clean distilled water, if needed. If distilled water is not available, clean drinking water, low in mineral content, may be used. Because water is the only thing in a battery that is consumed, acid should never be added to a battery.

Lack of water produces too low a level of electrolyte in the cell and thus causes disintegration of the exposed part of the plates. The acid in electrolyte becomes too concentrated and detrimental to separators and plates alike. Water added to the battery to compensate for losses must be distilled to prevent the addition of dissolved materials. In freezing weather it should be added just before the automobile is operated to avoid it freeze before combining with the acid.

Battery life is also shortened by mechanical breakage, caused by loose battery holders, which allow the battery to be jarred and bounced when the vehicle travels on rough roads.

Check the battery for any physical damage to case, case terminals, and replace defective cables.

Inspect the battery anytime the hood is opened. Look for a buildup on the battery top, case damage, loose or corroded connections, or any other problem that could upset battery operation, *Fig. 5.5.*

Due to normal release of gases in the battery electrolyte, loss of active material from the plates does take place and thus constitutes the wearing of the battery.

If the batteries are allowed to rundown in cold weather they may freeze and be completely destroyed. A battery that remains continuously uncharged becomes sulphated thus causing loss of its life. Overcharging of a battery shortens its life by overheating it, by producing excessive gas bubbling, and by having too concentrated an electrolyte.

Some important service checks are discussed below:

1. Adding Water

Electrolyte level in the battery must be above the plates. Distilled water has to be added, at periodic intervals to replace the water lost from the battery during the battery action and to prevent the electrolyte level from falling below to expose the tops of battery plates and separators. Elements exposed to air are liable to be ruined.

Fig. 5.5 *(a) Cleaning the battery with baking soda and water. (Chrysler Corp.)*

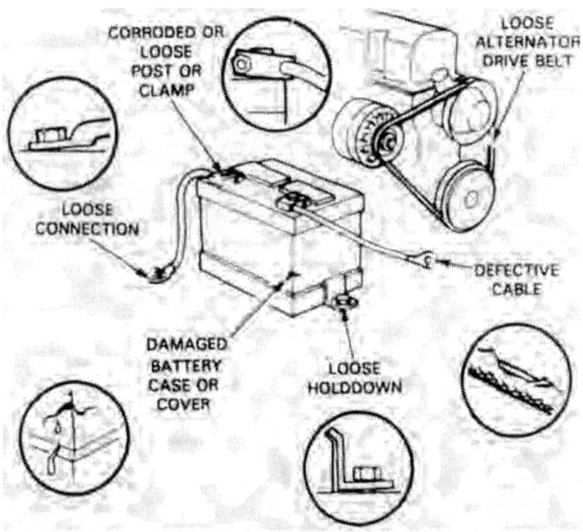

Fig. 5.5 *(b) Various types of problems that make an engine difficult to start. (General Motors Corp.)*

Do not overfill a battery, because normal gassing of the electrolyte will cause the electrolyte to escape and initiate corrosion of the battery terminals, hold-down brackets, and battery tray. Fill battery to the indicator that is approximately 38 mm from the top of the filler tube.

Just enough water should be added, so that the electrolyte does not overflow. Electrolyte that escapes and results in loss of battery capacity and causes corrosion of any metal or fabric with which it comes into contact.

If too much water is required for topping up and if it happens consistently, investigate the reason for this. This may be due to overcharging and may be avoided by adjusting the regulator setting. Leakage due to a cracked container may be another reason.

2. Cleaning the Battery

The powdery residue that builds up on the battery posts and terminals is corrosive and can burn skin and damage eyes. Also, some sulphuric acid may have overflowed onto the top of the battery if one or more of the cells are overfilled.

Carefully brush off loose corrosion and dirt. Gently hose off the battery and surrounding area. Now mix up a batch of corrosion eater, i.e., baking soda mixed in water. Add about two tablespoons of baking soda to 2 cups of water in a jar or can and mix thoroughly. Brush this solution or slowly pour it over the corroded terminals and hold-down brackets. Be very careful not to get any in the cells. The baking soda will froth with a great deal of energy as it neutralizes the acid in the corrosion. When the foaming stops, gently hose off the affected areas until the entire residue is rinsed away. Repeat the process, if necessary, until all corrosion is gone. Be sure to flush the floor under the vehicle, or it will be permanently stained.

The moisture, dust and grime collected on top of battery should be cleaned off, as it forms a path of flow of current, and permit the battery to discharge slowly across the battery top. To clean the battery top, tighten the plugs, sprinkle or brush baking soda solution on it, wait for the foaming to stop, and then flush off with clean water and dry it.

3. Cleaning Terminals and Cable Clamps

Battery cable connections should be checked and cleaned to prevent voltage drop at the connections. One common reason for an engine not starting is loose or corroded battery cable connections.

The battery terminals and cable clamps should be clean and tight. A loose connection at the terminals will result in a faulty starting and discharged battery. Accumulation of corrosion around the cable clamps and terminals is more or less a natural condition. The electrolyte may escape from the battery cells, due to overfilling or overcharging and corrosion is accelerated. The periodic cleaning of the battery and the terminals and cable clamps will prevent the corrosion from accumulating and cause a bad connection.

To remove corrosion to ensure clean and tight connections, first detach the clamp from the battery terminal posts. Detach the battery ground cable first to avoid accidentally grounding the insulated terminal with a tool. If this happens with the ground cable still connected, a very heavy current would flow and might cause considerable damage.

With the cable clamps off, use special wire brush or steel wool to clean the clamps and the terminal posts to bright metal. Then replace the cable clamps such that they seat properly on the posts. Apply a coating of Vaseline or petroleum jelly, or special battery anti-corrosion paste on the clamps and posts to provide protection against corrosion.

Battery cables with damaged insulation, loose or broken strands, or with excessive corrosion should be replaced by new cables of the same size. Using cables of smaller size introduces excessive resistance in the starting circuit and causes starting trouble.

4. Tightening Battery Hold-down Clamps

The battery holder should be kept properly clean and painted to avoid its corrosion.

The hold-down clamps should be kept tight enough to avoid bounce or relative movement of the battery, in its carrier but not so tight as to cause the battery case to warp, bulge, or break.

5. Checking State of Battery

Remove vent plugs, taking care that no flame is brought near the vents, as the gas inside is highly inflammable. The details of taking the specific gravity reading of the battery electrolyte with a hydrometer and the significance of various readings, to ascertain the state of charge of the battery, have already been discussed. If the battery is overcharged, adjust the regulator setting and if the battery is found discharged, remove it, recharge and put in service again.

Some battery manufacturers recommend the use of the light-load test as the basic test of battery condition. Then the use of hydrometer to check specific gravity only on fully charged batteries.

Cable Removal

Even if it is not visible, a thin film of corrosion builds up between the battery post and cable clamp. Clean this area by removing the cable clamps from the posts, and using a terminal cleaner, *Fig. 5.6.*

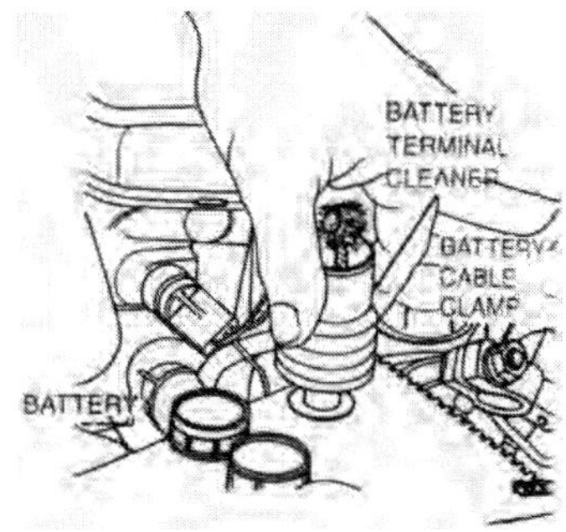

Fig. 5.6 *Battery post cleaners (Ford Motors)*

First, remove the cable that is grounded to the engine block and body. Removing this cable first will prevent possible shorts and sparking, when disconnecting the other cable. Most vehicles have the negative side of the battery grounded, but some may have the positive side grounded. Always check carefully to be sure.

Use a box wrench or battery pliers to loosen the bolts on cable clamps. Regular pliers and open-end wrenches are too thick and may damage the battery top when turned. Do not pry or twist the cable clamps to remove them because this can break the post loose from the plate connector. In particular, do not pry with a screwdriver, as it is a sure way to puncture the battery case. Use a battery clamp puller, *Fig. 5.7.*

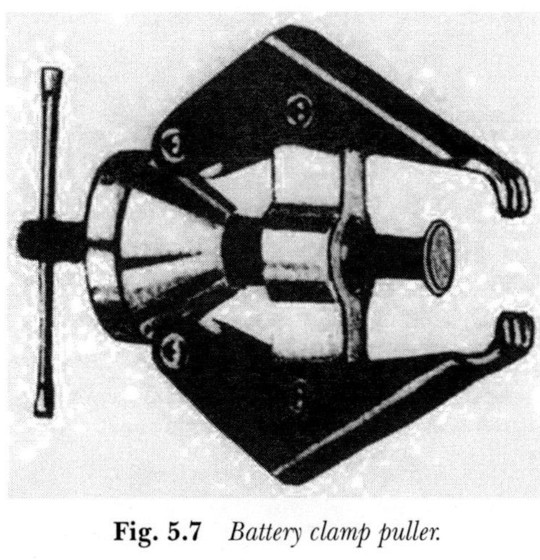

Fig. 5.7 *Battery clamp puller.*

Cleaning Battery Terminals

Slip the cup of the wire brush battery post cleaner over each post, in turn, and twist a number of times. Use the cone-shaped brush on the inside surface of the cable clamps and twist, and continue until the contact surfaces are bright *Fig. 5.8.* Reinstall the cable clamps maintaining the right polarity. The negative post is smaller than the positive post and takes the smaller clamp. Clamps should be made tight enough to ensure good electrical contact, but do not over-tighten. It is easy to distort or break the cable clamps, pull the threads out of the side of the case or strip the threads on the bolt. Coat the posts and clamps with a commercial corrosion inhibitor.

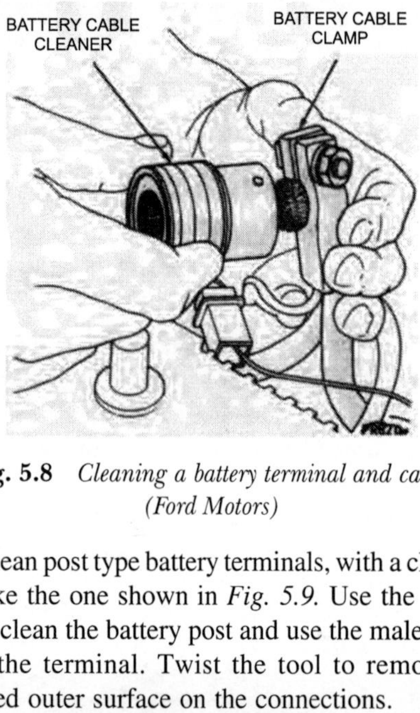

Fig. 5.8 *Cleaning a battery terminal and cable. (Ford Motors)*

Clean post type battery terminals, with a cleaning tool like the one shown in *Fig. 5.9.* Use the female end to clean the battery post and use the male end to clean the terminal. Twist the tool to remove the oxidized outer surface on the connections.

Use a small wire brush to clean side terminals. Polish both the cable end and the mating surface on the battery terminal.

Never use a knife or scraper to clean battery terminals. This removes too much metal and can ruin the connection caused by surface irregularities.

Coat the terminals with petroleum jelly or with grease before reinstalling the cables on terminals. This will help keep acid fumes away from the connections and avoid corroding again. Tighten the fasteners just enough to secure the connection.

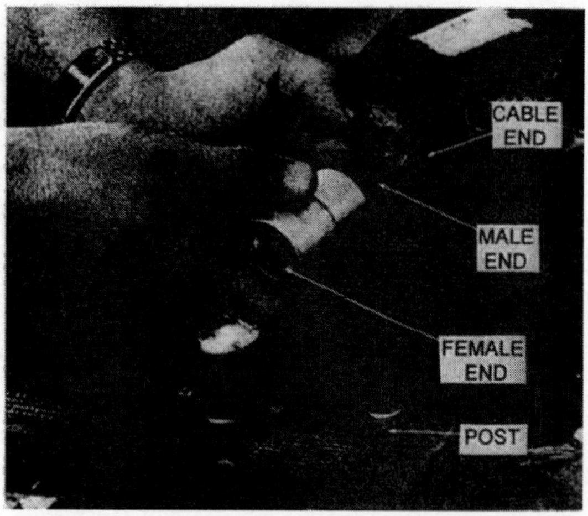

Fig. 5.9 *Mechanic is using special post type cleaner to remove deposits from surface of metal on post and cable end. (Peerless)*

Safety Considerations

Batteries contain acid and release explosive, hydrogen and oxygen, gases during normal charging and discharging cycles. To prevent physical injury or damage to the vehicle, always follow the following safety rules:

1. When working on any electrical component of a vehicle, disconnect the negative battery cable from the battery, to cause all electrical circuits in the vehicle to open, which will prevent accidental electrical contact between an electrical component and ground. An electrical spark has the potential to cause explosion and personal injury.
2. Wear eye protection when working around a battery.
3. Wear protective clothing to avoid skin contact with battery acid.
4. Never smoke or use an open flame around a battery.
5. Follow all safety precautions as stated in the service procedures for the equipment used in battery service and testing.

Battery Problems

A ***dead*** or ***discharged battery*** is a very common problem. The engine will fail to crank and start. Even though the lights and horn may work, there will not be enough current in the battery to operate the starting motor at the needed speed. A discharged battery could be caused by:

1. Defective battery, e.g., sulfated plates.
2. Charging system problem, e.g., loose alternator belt.
3. Starting system problem, e.g., defective solenoid or starter motor.
4. Poor cable connections.
5. Engine performance problem requiring excessive cranking time.
6. Electrical problem drawing current out of battery with the ignition key ***off.***

5.4 BATTERY SERVICE

Battery Terminal Test

A ***battery terminal test*** quickly checks for a poor electrical connection between battery cables and terminals. A voltmeter is used to measure voltage drop across the cables and terminals when cranking the engine, *Fig. 5.10.* If the voltmeter reads ***over*** 0.5 volt, there is a high resistance at the cable connection and requires cleaning.

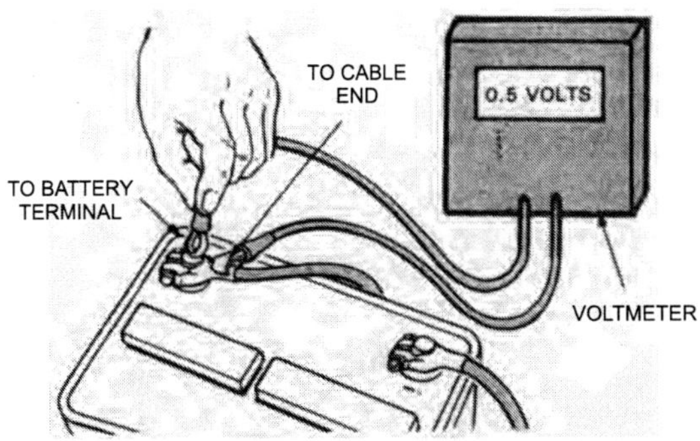

Fig. 5.10 *Voltmeter can be used to quickly check for poor battery cable connections. Measure voltage drop across cable and terminal while cranking engine. Voltage drop should be less than 0.5 V. (NAPA)*

Jump Starting

In emergency situations, it will be necessary to ***jump-start*** a vehicle by connecting another battery to the discharged battery. To jump start a vehicle, use the battery from another vehicle. This is called the ***booster*** battery. You must take certain precautions when jump starting, because any shorting of the jumper cables together or connecting them backwards can cause serious damage to the charging, computer and other electrical systems. Or even worse, the battery could explode and cause physical injury. Observe the following precautions:

1. Wear eye protection to protect your eyes from electrolyte getting in to eyes if the battery erupts or explodes.
2. If the dead battery is the maintenance-free type, check the charge indicator. If it shows light yellow, ***do not try to jump-start.*** The battery could explode!
3. In cold weather, the electrolyte in the battery could freeze – especially if the battery was run down to begin with. ***Never attempt to jump-start a frozen battery.*** It could explode!
4. The battery could explode if the booster battery is connected backward, i.e., positive to negative, negative to positive. Even if a battery does not explode, both batteries could be ruined.

The jump-start procedure requires two jumper cables to execute it. Follow the steps given below:

1. Wear eye protection.
2. If either battery has vent plugs, *Fig. 5.11*, remove them. Cover the holes with cloths to stop splashing of the electrolyte in case of an explosion.
3. The two vehicles should not touch each other as it could damage electrical equipment, if the two electrical systems have different voltage, or different ground polarity.
4. Check that all electrical equipment except the ignition system is turned off on the vehicle being jump-started.
5. Connect the end of one jumper cable to the positive (+) terminal of the booster battery. Connect the other end of this cable to the positive (+) terminal of the dead battery.
6. Connect one end of the second cable to the negative (–) terminal of the booster battery.
7. Connect the other end of the second cable to the engine block of the vehicle to be jump-started. Do not connect it to the negative (–) terminal of the vehicle with the dead battery! A spark from this connection could cause the

battery to explode. Do not lean over the battery when making this connection!

8. Start the vehicle with the booster battery. Then start the vehicle with the dead battery. After the dead battery engine starts, disconnect the booster cables. First disconnect the cable from the engine block. Then disconnect the other end of this cable. Finally disconnect the positive cable.
9. Discard the cloths used to cover the vent holes. They may have acid on them. Install vent plug if removed.

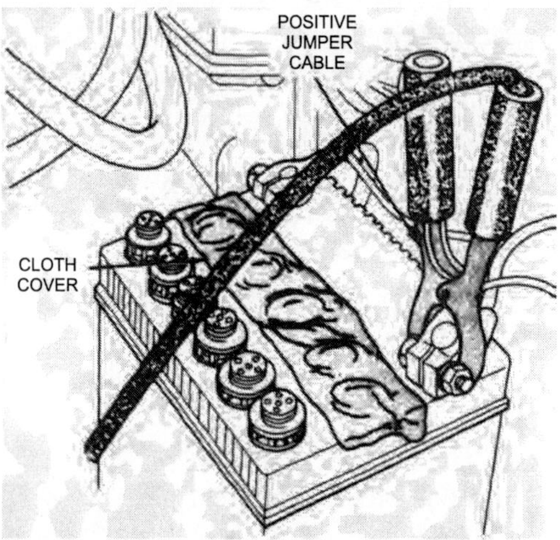

Fig. 5.11 *Connecting jumper cable and covering the openings in the battery cover with a clean cloth. The vent plugs are shown removed here. These may be left in after making sure that the vents in the plugs are open. (Chrysler Corporation)*

Battery Voltage Testing

Testing the battery voltage with a voltmeter is a simple method for determining the state of charge of any battery. A battery that "looks good" may not be good. This test is commonly called ***open-circuit battery voltage test,*** because it is conducted with an open circuit — with no current flowing and no load applied to the battery.

1. If the battery has just been charged or the vehicle has recently been driven, it is necessary to remove the surface charge from the battery before testing. A surface charge is a charge of higher-than-normal voltage that is only on the surface of the battery plates. The surface charge is quickly removed when the battery is loaded and therefore does not accurately represent the true state of the battery.
2. To remove the surface charge, turn the headlights on high beam for one minute, then turn the headlights off and wait for two minutes.
3. With the engine and all electrical accessories off, and vehicle doors shut so as to turn off the interior lights, connect a voltmeter to the battery posts. Connect the red positive lead to the positive post and the black negative lead to the negative post.

   If the meter reads negative, the battery has been reverse charged (has reverse polarity) and should be replaced, or the meter has been connected incorrectly.
4. Read the voltmeter and the reading with the following state-of-charge chart.

Open Circuit Test	*Percent Charge*
11.7 volts or less	0%
12.0	25%
12.2	50%
12.4	75%
12.6 or more	100%

Battery Ratings

Batteries are rated according to the amount of current they can produce under specific conditions.

Every automotive battery must be able to supply electrical power to crank the engine in cold weather and still provide voltage high enough to operate the ignition system for starting. ***Cold-Cranking Power*** of a battery is the number of amperes that can be supplied by a battery at –18°C for 30 seconds while the battery still maintains a voltage of 1.2 volts per cell or higher, i.e., the battery voltage would be 7.2 volts for a 12-volt battery. The cold cranking performance rating is called ***Cold-Cranking Amperes (CCA).*** Try to buy a battery with the highest CCA for the money you spend.

Battery Chargers

When tests show that the battery is discharged, or the state of the charge of a battery is low, it must be recharged. It is best to slow-charge any battery to prevent possible overheating damage to the battery. Remember, it may take 8 hours or more to charge a fully discharged battery. The initial charge rate should be about 35 amperes for 30 minutes to help start the charging process. Fast charging a battery increases the temperature of the battery and can cause warping of the plates inside the battery. It also increases the amount of release of hydrogen and oxygen that can create a health hazard. The battery temperature should not exceed 52°C. ***Most batteries should be charged at a rate equal to 1% of the battery cold-cranking amperes (CCA) rating.***

Fast charge:	15 amperes maximum
Slow charge	5 amperes maximum

A ***battery charger*** may be used to recharge the battery. The battery charger will force current back into the battery to restore the charge on the battery plates.

Before connecting a battery charger, to the battery, make sure that the charger is turned ***off.*** Wear eye protection. Also ensure that the work area is well ventilated. If a spark ignites any battery gas, the battery could ***explode.***

Connect the ***red*** charger lead to the positive terminal. Connect the ***black*** charger lead to the negative terminal of the battery. With side terminal batteries, use adapters. Make sure that you do not reverse the charger connections or the charging system could be damaged. Set the charger controls as recommended and turn on the power.

Battery Load Test

One of the most accurate tests to determine the condition of any battery is the ***load test.*** It tests the battery under full current load. The battery load test actually measures the current output and performance of the battery. It is one of the most common and informative battery tests used in modern automotive service shops, *Fig. 5.12.*

Most automotive charging and starting testers use a carbon pile to create an electrical load on the battery. The amount of load is determined by the original capacity of the battery being tested. The capacity is measured in CCA. An older type of battery rating is called the ***ampere-hour rating. The proper electrical load to be used to test a battery is one-half the CCA rating with a minimum 250-ampere load.*** After the battery has been tested to be at least 75% charged by observing the built-in hydrometer or by performing an open-circuit voltage test, a load test can be performed.

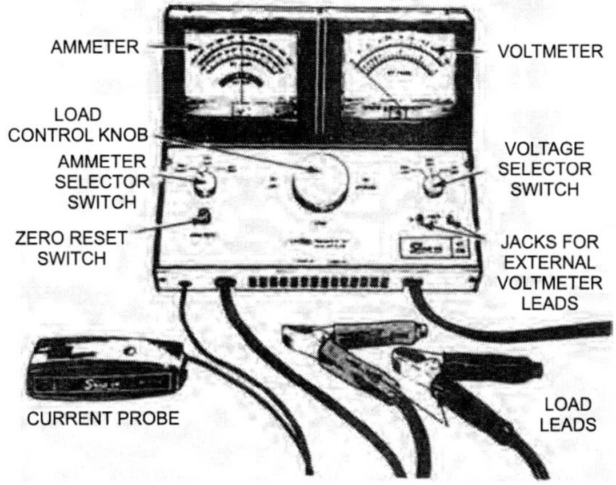

Fig. 5.12 *Note controls on this load tester. (Snap-On Tools)*

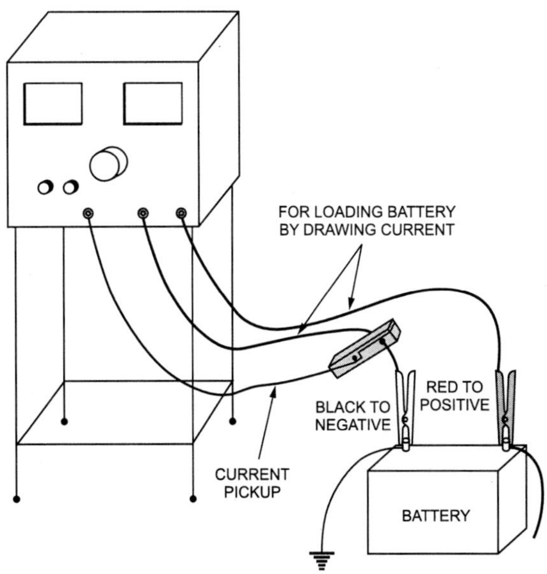

Fig. 5.13 *To load test battery, connect tester as shown here or follow equipment operating instructions supplied by the manufacturer. Basically, use tester to load battery and draw specified current depending upon the battery rating. Battery voltage must stay above 9.5 volts for battery condition to be acceptable.*

Connecting Load Tester

Connect the load tester to the battery terminals. If the tester is an ***inductive type,*** i.e., clip-on ammeter lead senses field around outside of cables, use the connections shown in *Fig. 5.13*. If the tester is *not* inductive, connect the ammeter in series. Control settings and exact procedures vary with equipment. Always follow the manufacturer's directions provided with the testing equipment.

Determine the Battery Load

Before loading a battery, calculate how much current draw should be applied to the battery. If the ***ampere-hour rating*** is given, load the battery to ***three times*** its amp-hour rating. For example, if the battery is rated at 60 amp-hours, test the battery at (60 × 3 = 180) 180 amps. *Fig. 5.14* is a chart showing load test amps. A load conversion chart will normally be provided with the load testing equipment. Refer to this material when in doubt.

BATTERY RATINGS			*LOAD TEST AMPS*
Cold Cranking Current	*Amp-Hours (Approx.)*	*Watts*	
200	35-40	1800	100 amps
250	41-48	2100	125 amps
300	49-62	2500	150 amps
350	63-70	2900	175 amps
400	71-76	3250	200 amps
450	77-86	3600	225 amps
500	87-92	3900	250 amps
550	93-110	4200	275 amps

Fig. 5.14 *This chart shows how different ratings affect the amount of current draw during battery load test*

Loading the Battery

Check that the tester is connected correctly. Then turn the load control knob until the ammeter reads the correct load for the battery under test. Apply the load for a full 15 seconds and observe the voltmeter at the end of the 15-second period while the battery is still under load. A good battery should indicate above 9.5 V at room temperature. This voltage is based on the battery temperature 21°C. A cold battery may show a lower voltage. A temperature compensation chart like the one shown in *Fig. 5.15* may be used to compensate for temperature. It allows for any reduced battery performance caused by a low temperature. Turn the load completely ***OFF*** so the battery will not be discharged.

Approximate Electrolyte Temperature	*Minimum Acceptable Voltage Under Load For Good Battery*
16°C	9.5 V
10°C	9.4 V
4°C	9.3 V
–1°C	9.1 V
–7°C	8.9 V
–12°C	8.7 V
–18°C	8.5 V

Fig. 5.15 *Consider the temperature too, when testing battery condition. Note how temperature affects voltage reading during load test.*

Many battery manufacturers recommend performing the load test twice. The first test the first load period is used to remove the surface charge on the battery. Wait for minimum 30 seconds to allow the battery to recover and then run the second test to obtain a truer indication of the battery condition. If the battery fails the test, recharge the battery and retest. If a second failure occurs, replace the battery.

5.5 BATTERY TROUBLES

Loss of Changeability From Age

With use, a battery gradually loses its ability to receive full charge. Its plates and separators undergo deterioration. Some of the lead sulphate, formed on plates during discharge, sheds and drops to the bottom of the battery case. Thus, the battery suffers constant decrease in its capacity and its ability to crank an engine and has to be discarded.

Battery Self-Discharge

A battery when left standing, gets discharged by itself. This happens due to constant chemical activity taking place in it. This phenomenon, not productive of current, is termed *self-discharge*. The causes of self-discharge are higher temperatures, strength and contamination of electrolyte, damaged separators and long periods of storage.

Self-discharge is faster at higher temperatures as the chemical activity is stimulated by increased temperature, e.g., a battery kept at 100°F loses half its charge in 30 days while the battery kept at 0°F suffers no loss of charge.

A strong electrolyte, with higher percentage of sulphuric acid also accelerates self-discharge. Pure sulphuric acid (not commercial) and distilled water should be used for preparing electrolyte. Every care should be taken to avoid dirt or dust entering the cells during inspection and testing of battery.

Long-term storage of the battery causes stratification of the electrolyte, i.e., the bottom layers become dense than the top layers. This sets up local equalizing currents resulting in increased self-discharge rate.

Overcharging and its Effects

When a battery comes to its fully charged condition, whole of the lead sulphate on plates has been converted to active materials. If the charging of the battery is continued after it, i.e., the battery is over-charged, the oxygen and hydrogen (water broken into its constituents) continue to accumulate at the plates and has nothing to combine with. These gases bubble up through the electrolyte and escape to the atmosphere through the cap vents. Actually this gassing starts well before the battery reaches full charge. At full charge the battery gases freely and water is rapidly lost from the battery cells. At the same time, overcharging develops considerable heat and water is lost due to evaporation. The level of electrolyte will drop to expose the tops of plates to air. The plates will dry out and harden, and loose their ability to receive charge. The remaining electrolyte gets concentrated due to loss of water and the lower parts of the plates are worked harder and deteriorate more rapidly. Also the separators get damaged.

High temperatures developed due to overcharge cause the plates to buckle and crush the separators and permit the positive and negative plates to short circuit. This ruins the plates completely.

Also the oxygen accumulating at the positive plates attacks the grid structures of the plates, causing them to oxidize or turn into lead oxide. This weakens the plate grid and also causes the plates to swell. This swelling pushes the positive sides of the cell covers and the upper edges of the positive plates to be pushed up against the negative plate straps. The short circuit causes battery failure.

Detection of Overcharge

1. A battery that is overcharged requires excessive amounts of water.
2. Also, the raised positive sides of the cell covers are a proof of overcharge occurring.
3. An overcharged battery is apt to gas heavily and spray electrolyte. Battery top wet with electrolyte and acid corrosion of surrounding metal parts indicates either overcharging or over filling.

Correction of Overcharge

Overcharging can take place if the generator-regulator system is faulty or the voltage regulator setting is high. The generator-regulator system setting should be adjusted.

High temperatures reduce battery resistance to charging current. Therefore, as temperature goes up, current input from a constant voltage source goes up. Voltage regulators are temperature compensated, so their voltage goes down as temperature goes up. This provides some protection against overcharge. Reducing the voltage-regulator setting and also reducing the battery electrolyte gravity will provide additional protection against battery overcharge in hot climates.

Run-Down Battery

A battery may run-down due to the faulty generator-regulator system, excessive current demands on it, due to its extra-high self-discharge rate, or it being defective or old. A discharged battery looses is ability to crank the engine at normal cranking rate or it may fail to crank the engine at all. Periodic battery checks will help to keep track of its state of charge and any abnormal condition arising may be avoided. A partly discharged battery operated for long periods gets permanently damaged. The sulphate on plates gets changed to a hard, crystalline substance, which even on giving the battery its full charge, is not reconverted to active material. The battery capacity is thus reduced. In addition, the soleplate causes the battery plates to swell, which damages the plate grids and weaken the plates.

Correcting Cause of Run-Down Battery

Check setting of the generator-regulator system and wiring circuit, to determine whether the trouble lies in the generator, regulator, wiring or battery. If the battery is suspected, it should be charged and then tested. This will reveal whether the battery is capable of giving more service or should be discharged.

Cycling Failure

A battery is repeatedly charged and discharged during its normal life. The swelling and contraction of the active material in the plates during the discharge — charge cycle causes the active material to loosen and gradually shed from the plate.

In a cycling failure, the positive plate material becomes soft, muddy, and brown in colour. It is so soft that any gassing or jarring action tends to cause the material to shed off badly and fall to the sediment chambers of the cells. On accumulation it produces short circuit between the bottoms of the negative and the positive plates.

If this loose material be kept floating in the electrolyte by jarring or by heavy charging currents, it may built up on top of the negative plates until it finally touches the positive plate strap to cause the short circuit.

Sulphation

The active materials in the negative and positive plates are converted into lead sulphate during battery discharge. If the battery is recharged within a reasonable time, the lead sulphate will reconvert into active materials without trouble. However, if the battery is allowed to stand for any length of time in a discharged condition, the lead sulphate will stabilize into a hard substance that resists reconversion to active materials. During this process, the negative plates turn grayish white. The positive plates tend to turn milky white but appear less affected than negative plates. In addition, the plates tend to swell and break the grids. Over a period of time, a sulphated battery may become so badly damaged that it will be ruined. The other prominent causes of Sulphation are very high specific gravity of electrolyte and insufficient electrolyte in the battery.

To avoid Sulphation the battery should be inspected regularly and it should be topped up when required to keep the electrolyte level and gravity at proper values. Do not use heavily discharged batteries. Instead recharge them immediately.

Internal Short Circuits

Internal short circuits result from shedding of material from the plates and from mechanical damage. The shredded loose material gradually builds up along the plate edges until it shorts between the positive and negative plates or between the negative plates and the positive plate strap. Failure of separators also causes short-circuiting. Short circuits can be found by looking for small orange-coloured spots either on the plates or on the separators. The orange-coloured spots are produced by heat resulting from short-circuits.

The remedy is to clean the battery and replace the damaged separators.

Cracked Case

The container develops cracks generally due to mechanical shocks caused by rough roads or rough handling. If the cracks are on the outer walls the electrolyte will leak while the cracks on the interior walls cause internal short-circuiting of the plates of the two dying cells.

If the *hold-down clamps* are *too loose,* the battery may bounce around in the battery holder. This may strain the case to such an extent that it cracks. Also, if the battery freezes, the expanding electrolyte is liable to break the case and break the plates and separators and ruin the battery and it has to be replaced.

Bulged Case

A battery that is overcharged for long periods, particularly in hot climates, may get so overheated that the case softens and begins to bulge. Too tight hold-down clamps can also cause a case to bulge. If it be noted right at its beginning, the battery may be saved by removing the cause of bulging. If overcharge is the cause, the generator-regulator system must be checked and proper adjustments made. If the clamps are too tight, they should be loosened and then retightened correctly.

Corroded Battery Holder

When a battery is being charged, it starts gassing as it approaches full charge. If the battery is overfilled, gassing will spray electrolyte over the battery. Even otherwise, electrolyte will spray from the plug vents when the battery gasses violently from overcharging. In either case, the acid in the electrolyte will attack any metal it touches, including the holder. A badly corroded holder might have to be replaced. But if corrosion is not too far advanced, it should be cleaned off with a wire brush and soda solution and then painted with acid proof paint. The generator-regulator system should be checked and adjustments made if overcharging is taking place. Overcharging has to be avoided always.

Low Specific Gravity

If a hydrometer check discloses a low reading, indicating a run-down battery. If the battery is old or is suspected of having internal defects, it should be recharged and then given a high-discharge test.

Cell Readings More Than 25 Gravity Points Apart

A gravity variation of more than 25 gravity points between cells could be due to any of several conditions.

1. Partial short in the low cell, which prevents normal recharging activities in the cell, so that the cell does not come up to charge with other cells. A partial short might be due to a defective separator, buckled plates etc.
2. Due to loss of electrolyte from the low cell, which would then be filled with water so that the gravity reading would be lowered. Electrolyte loss can result from a cranked case, a poor cover seal, a loose vent plug, or excessive electrolyte spraying from overfilling or overcharging.
3. Improper activation of the battery.

5.6 CHOOSING A CORRECT BATTERY

Two conditions are important in choosing the correct battery for an automotive application:

1. Its ability to power the starter to enable minimum starting speed under very cold conditions.
2. The expected use of the battery for running accessories when the engine is not running. The first criterion is usually the deciding factor. *Fig. 5.16* shows a graph comparing the power required by the starter motor and the power available from the battery, plotted against temperature. The point at which the lines cross is the cold start limit of the system. European standards generally use the figure of –18°C as the cold start limit, and a battery to meet this requirement is selected.

5.7 STARTING SYSTEM SERVICE

A starting system has only about five major components that can cause problems in starting the engine. *Fig. 5.17* shows the most common starting system problems. In case any of the components shown has high resistance, low resistance, wear, or damage, the engine may not start normally. A few simple tests can be used to help determine the source of starting system problem.

5.8 COMMON STARTING SYSTEM PROBLEMS

1. A ***dead*** or ***discharged battery*** is one of the most common causes a starting system fails to start an engine. A starting motor draws about 200 amps and it will not function properly without a fully charged and well-connected battery of the starting system. If needed, load test the battery as already described.
2. If the ***starting motor does not crank*** the engine properly, turn on the headlights and try to start the engine — a helper may be needed. If the starting motor does not crank the engine and ***headlights do not light,*** the system battery may be dead or the electrical connections to the battery and/or starting motor may be faulty. Charge or replace battery and retest starter.
3. If the ***lights turn on but become dim*** when the starter motor is cranked, the starting motor may be short-circuiting, the battery may be weak, or the engine may be dragging due to mechanical problems. Charge or replace battery and retest starter.

4. If the ***starter does not crank properly but the headlights remain bright,*** there may *be* excessive resistance or a short in the starting system.
5. A ***"no cranking problem"*** results when the engine crankshaft does not rotate with the ignition key at start. The most common causes are a dead battery, a bad electrical connection, or a faulty system component.
6. **A *"slow cranking problem"*** occurs when the engine crankshaft rotates at lower than normal speed. The common causes are a weak battery, high resistance in cables or defective starter motor.
   (*i*) Feel both battery cables and connections. If either is ***hot to the touch*** after cranking the engine, replace the cables or clean the connections.
   (*ii*) **Weak battery** – Test the battery or charge the battery and try to crank the engine again.

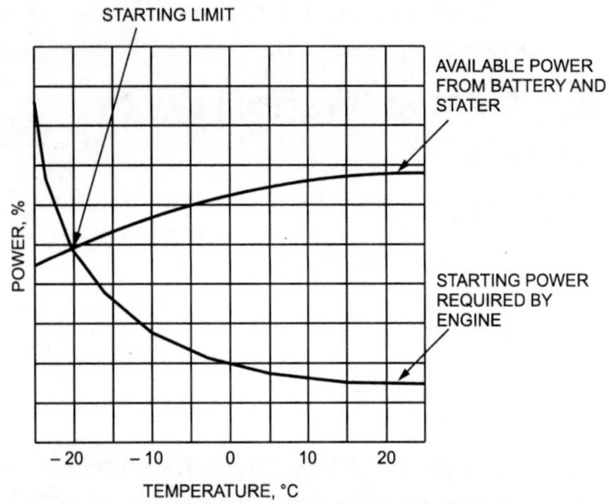

Fig. 5.16 *Comparison of power required by the starter motor and the power available from the battery plotted against temperature.*

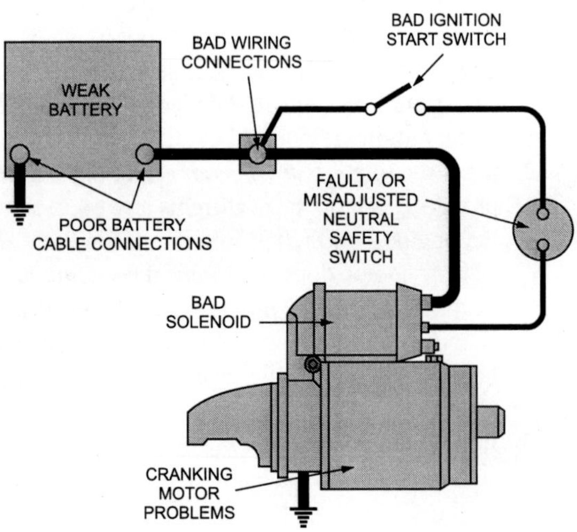

Fig. 5.17 *Types of problems that can be found in an automotive starting system. (Echlin)*

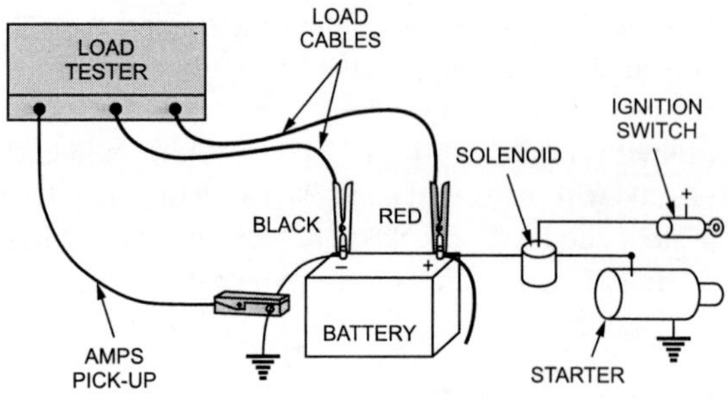

Fig. 5.17 *(a) Load tester will also measure current draw of starting motor. High or low current draw when cranking will indicate possible problems.*

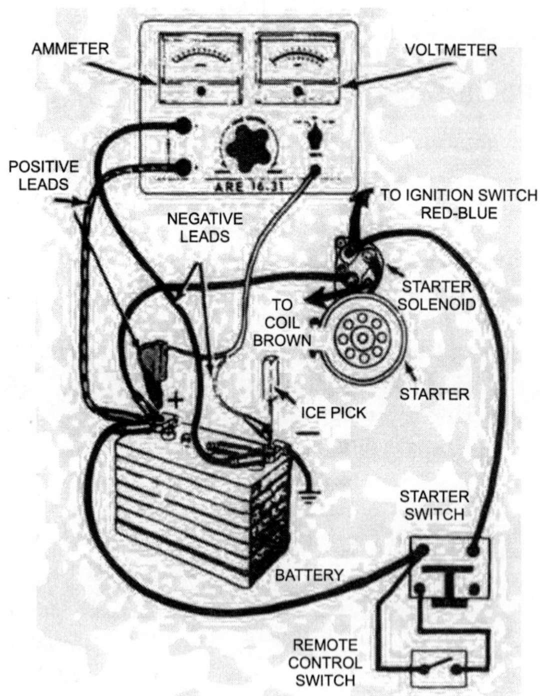

Fig. 5.17 *(b) Starter motor load test connections.*

(*iii*) **Disable ignition** – Remove coil wire from distributor and ground wire or disconnect power load to GM HEI System. Check if it cranks normally. ***If it cranks normally*** – Problem is advanced ignition timing – Check timing and distributor (if equipped) for defective parts such as a broken rotor or cracked distributor cap.

(*iv*) ***If it does not crank normally*** – Perform starter amperage test. If amperage is too high, check if the engine can be rotated by hand with a wrench on engine. ***If engine turns freely*** – Replace or overhaul solenoid/starter. ***If engine does not rotate freely*** – Find and correct engine mechanical problem.

7. It is important to ***listen to the starting motor operate*** when a starting problem exists. A ***buzzing*** or ***clicking sound*** from the solenoid, without cranking, is commonly due to discharged battery, or bad battery cable connections. Low current flow causes the solenoid plunger to rapidly kick in and out, making a clattering sound. Clean battery cables at battery terminals and at starter solenoid and block (ground) and charge battery.
8. A ***single click sound,*** without cranking, may result from the bad starting motor, burned solenoid contacts, dead battery, or engine mechanical problems. The click sound is probably due to the solenoid closing or the pinion gear contacting the flywheel ring gear.
9. If the starting motor does not crank the engine but a ***consistent clicking or buzzing sound*** is heard, low current to the starter motor is the most likely reason. Check the battery voltage and the connections between the battery and the starting motor.
10. A ***grinding noise*** during starting motor operation is normally a sign of broken flywheel ring gear teeth, or starting motor pinion gear teeth. The grinding may be due to the gears clashing over each other.
11. A ***humming sound,*** after momentary engine cranking, may be due to a bad starting motor overrunning clutch or pinion gear unit. Pinion gear wear may make the pinion gear disengage from the flywheel ring gear too soon. This can allow the motor armature spin faster making the humming sound.
12. ***Normal cranking*** without the engine starting is normally not caused by the starting system. This could be due to a problem in the fuel supply or ignition systems.
13. For a diesel engine normal cranking without starting may be due to low cranking rpm.

    Some other common starting system problems are:

Inoperative Starter

1. Loose or corroded system battery terminals.
2. Discharged system battery.
3. Dead battery.
4. Open cranking circuit.
5. Faulty ignition switch.
6. Defective starting motor.
7. Inoperative solenoid or relay.
8. Inoperative neutral safety switch.

Inoperative Starter and Light Dim

1. Weak battery or dead cell.
2. Loose or corroded battery connections.
3. Armature rubbing on pole shoes.
4. Grounded starter motor field.
5. Internal ground on starting motor windings.

Starter Turns But Drive Does Not Engage

1. Broken teeth in flywheel ring gear.
2. Rusted starting motor drive shaft.
3. Defective starting motor drive.
4. Slipping overrunning clutch.

Slow Cranking Speed

1. Discharged battery or defective cell.
2. Excessive resistance in the starter.
3. Excessive resistance in the starting circuit.
4. Engine oil too heavy for environment conditions.
5. Excessive engine friction.
6. Burnt solenoid contacts.
7. Bent armature.
8. Loose pole show screws.
9. Worn armature shaft bearings.

Starter Does Not Disengage

1. Defective ignition switch.
2. Short circuit in solenoid.
3. Stuck solenoid contact switch plunger.
4. Broken solenoid plunger spring.
5. Defective starting motor relay.

6. Loose starting motor mounting bolts.
7. Worn drive end bushing.
8. Broken drive yoke return spring.
9. Defective overrunning clutch.

5.9 STARTER CURRENT DRAW TEST

A ***starter current draw test*** measures the number of amperes used by the starting system. It tells about the condition of the starting motor and other starting system components. If current draw is higher or lower than the specified value, there is a problem in the starting system.

To perform a starting motor current draw test, connect a load tester to measure battery voltage and current flow out of the battery, *Fig. 5.17(a)*. To keep the engine from starting during the test, disconnect the ignition coil supply wire or ground the coil wire. For a diesel engine, disable the fuel injection system. See shop manual for details. Crank the engine and note the current readings. If the readings do not fall within specs, something is wrong in the starting system. Further tests would be necessary.

Engine Displacement	*12 V System Current*
Most 4-6 Cylinders	125-175 Amps Max.
Under 300 C.I.D.	150-200 Amps Max
300 C.I.D. or Over	175-250 Amps Max.

Cranking Circuit Troubleshooting Chart

Cranking Voltage	*Cranking Amps*	*Possible Problem*
Voltage Within Specs	Current Within Specs	System OK
Voltage OK	Current Low Engine Cranks Slowly	Starter Circuit Connections Faulty
Voltage Low	Current Low Engine Cranks Slowly	Battery Low
Voltage Low	Current High	Starter Motor Faulty

Fig. 5.18 *Chart shows how test voltages and current values may indicate potential problems in starting system.*

Do not crank the engine for more than 15-25 seconds or starting motor damage may result due to starter overheating. Allow the starting motor to cool for a few minutes if more cranking time is needed. Table in *Fig. 5.18* below gives the average current draw values for various engine sizes and cranking circuit troubleshooting.

5.10 STARTER MAINTENANCE AND SERVICE

Regular periodic check of the condition of the starting motor and engine cranking system reduces the possibility of system failures on the road. The frequency of system inspection depends on the type of automotive operational environment. However, a visual inspection of the cranking system, an operational test of starting motor performance, and a test of the neutral safety switch operation should be done annually or as recommended by the manufacturer in service manuals or as part of the engine tune-up.

A faulty starting motor can cause a wide range of slow cranking, no cranking, overheating of starter cables, and abnormal noises while cranking. If the battery, cables, solenoid/relay, and other starting system components are in good condition but the engine does not crank properly, the starter may be bad.

Starting Motor Removal

Before deciding to remove the starting motor, inspect it closely for problems. Check that starter-to-engine bolts are tight to specs. Loose bolts can upset starting motor operations by causing a poor ground or inaccurate pinion gear to flywheel ring gear meshing. Also ensure that all electrical connections to the motor and solenoid are tight. If necessary, remove the starter for cleaning and thorough inspection. Steps for starting motor removal commonly include:

1. Disconnect the battery and then unbolt the battery cable. Remove the wires from the solenoid (starter-mounted solenoid type) and any braces on the motor, *Fig. 5.19 (a)*.
2. ***Starter shims*** may be used to adjust the distance between the starting motor pinion gear and the flywheel ring gear, *Fig. 5.19 (b)*. Shims help position the starting motor farther away from or closer to the ring gear. Always check for shims, during starter removal. If used, they must be

put in the same place during starting motor reassembly to the engine. Otherwise the pinion and flywheel ring gear may not mesh and a grinding noise will be produced.

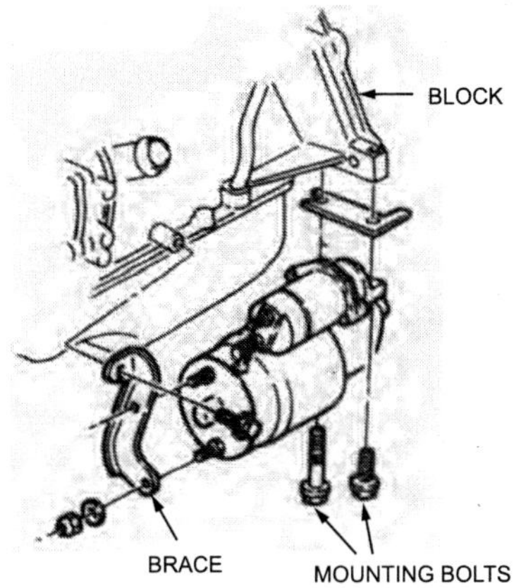

Fig. 5.19 *(a) Removing the starter motor. (Chevrolet Div. CM)*

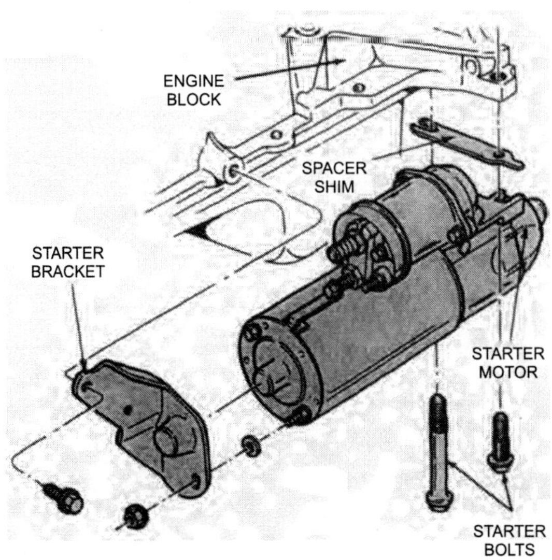

Fig. 5.19 *(b) During starting motor installation, connect the wires correctly as per instructions, replace any shims and secure all brackets. When removing starting motor hold on to it securely because it is heavy for its small size. (General Motors Corp.)*

3. When the starting motor is located inside the block valley, the intake manifold might have to be removed to access the motor. For this, first disconnect the battery and remove wires, fuel lines, and cables between the manifold and vehicle chassis. Then, remove the intake manifold bolts, brackets, and other parts interferring with the manifold removal. Remove the manifold, to allow access to the bolts that attach the starting motor to the cylinder block.

Starter Motor Disassembly

1. The solenoid should to be removed from the starter motor. Remove the solenoid terminal nut or screw and disconnect the wire before removing the solenoid from the starter motor, *Fig. 5.20 (a)*.
2. Use a screwdriver to remove the mounting screws from the solenoid. *Fig. 5.20 (b)*.

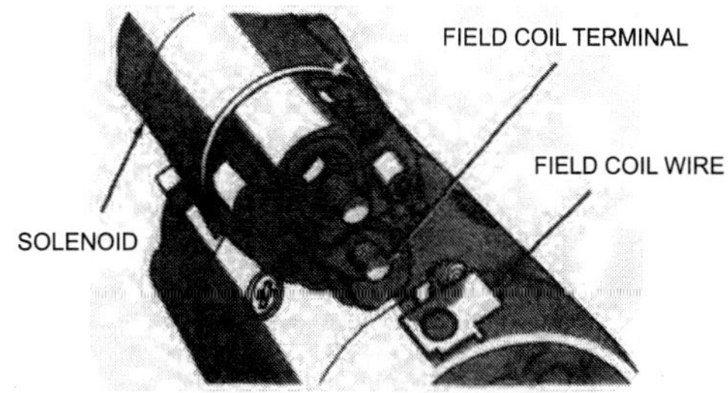

Fig. 5.20 *(a) Disconnecting the field coil wire. (Chrysler Corp.)*

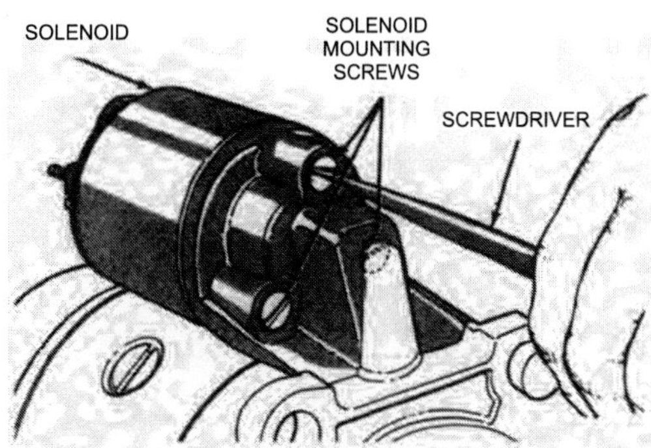

Fig. 5.20 *(b) Removing the solenoid mounting screws. (Chrysler Corp.)*

3. Rotate the solenoid through about 90°, rotation in either direction. Disconnect the solenoid from the starter motor by working the solenoid hook off the starter shift lever.
4. Remove the two screws and remove the end frame bearing cap, if so equipped.

5. Then remove the end frame bearing washer or washers.
6. Remove two through bolts for the starter and then pull off the end frame, *Fig. 5.20 (c)*.
7. Locate the two brushes that are connected to the field windings. Disengage the brush springs, then pull the two brushes out of their holders.
8. Slide the field frame assembly (centre housing) off the starter motor armature as shown in *Fig. 5.20 (d)*.
9. Remove the clutch shift lever pivot bolt from the drive end housing, *Fig. 5.20 (e)*.
10. Remove the overrunning clutch drive from the armature shaft as shown in *Fig. 5.20 (f)*.
11. Pull the drive end housing off the armature.

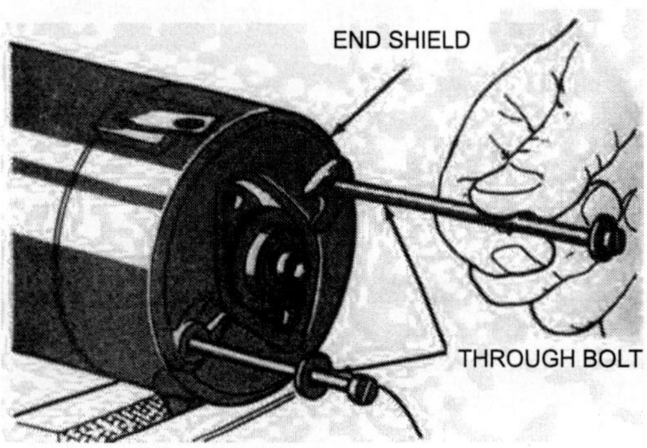

Fig. 5.20 *(c) Removing the through bolts. (Chrysler Corp.)*

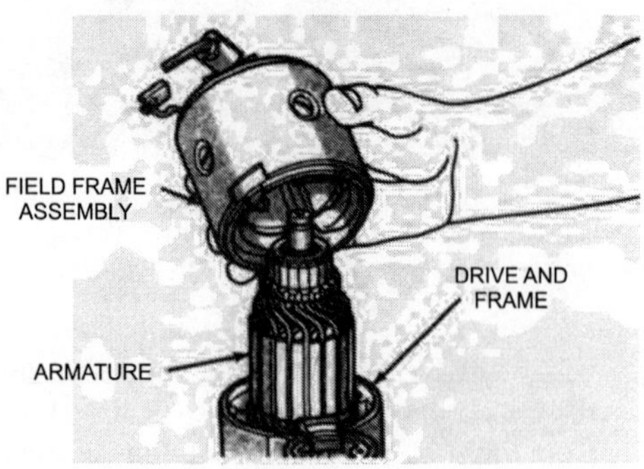

Fig. 5.20 *(d) Removing the field frame assembly. (Chrysler Corp.)*

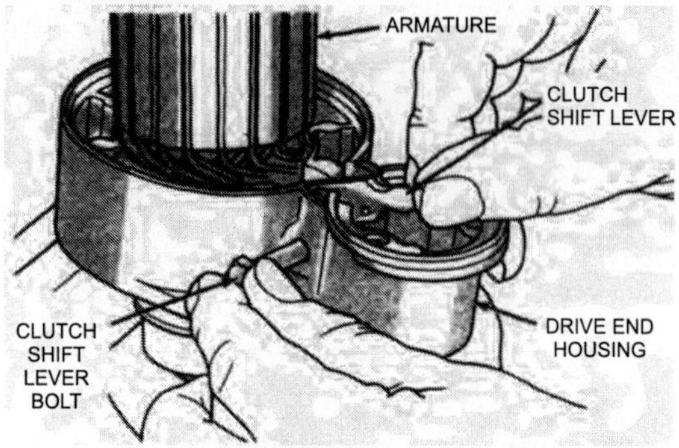

Fig. 5.20 *(e) Removing the shift lever bolt. (Chrysler Corp.)*

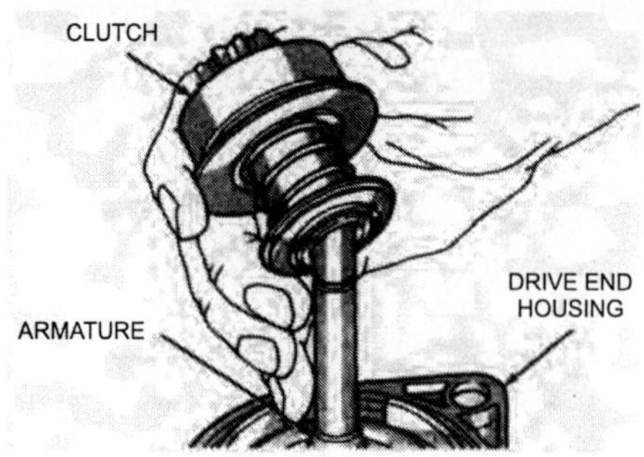

Fig. 5.20 *(f) Removing the overrunning clutch drive. (Chrysler Corp.)*

Disassemble the starting motor and examine the commutator and brushes. If the commutator is dirty, clean it with a strip of # 00 sandpaper. In case the commutator is rough, pitted, or out-of-round, or if the mica insulation between commutator bars is high, chuck the armature in a lathe and recondition the commutator. If necessary, undercut the mica 0.8 mm.

Brushes should be at least half of their full length. If not, replace them. The brushes should have free movement in the brush holders. In addition, they should have the specified spring tension and make full contact with the commutator.

Never use a grease-dissolving chemical solution or high-temperature cleaning solution to clean the commutator and field windings as it could damage the insulation. This also applies to overrunning clutch drive unit.

Overrunning Clutch (Pinion Gear) Service

Normally, the overrunning clutch or pinion gear assembly is replaced anytime the starting motor is disassembled. Test the operation of the overrunning clutch drive unit. It should spin freely in the overrunning direction, and should not slip in the driving direction. The pinion gear is subject to tremendous wear and tear when engaged and disengaged from the engine flywheel ring gear. It is preferred to replace the pinion gear during starting motor service, *Fig. 5.20* (*g*).

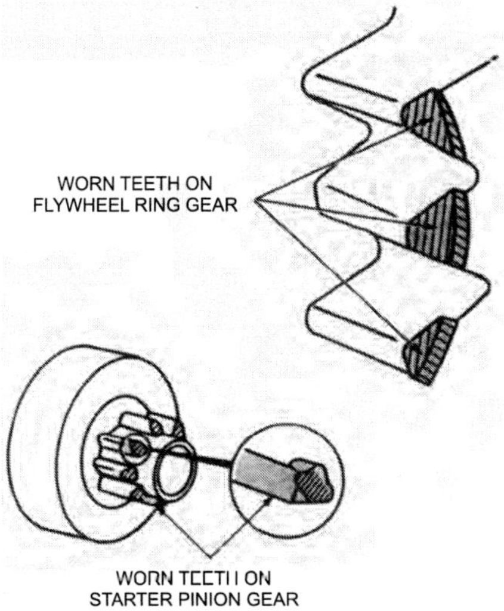

Fig. 5.20 *(g) Worn teeth is a common problem on starter motor pinion gear and sometimes flywheel ring gear. Worn teeth can cause starter motor disengagement prematurely. (Ford Motors)*

Installing Starting Motor

Install the starting motor in the reverse order of its removal. Ensure that any spacer shims are replaced between the starting motor and engine block. Use a special lubricant to coat the armature shaft, drive end bushing, and commutator end bushing. Make sure the brushes are fully seated. Properly align the motor housing and end frame, and install the through bolts to secure it, *Fig. 5.20* (*h*). If the starting motor has a solenoid on it, connect the electrical wires on the solenoid before bolting the starting motor to the engine. Tighten the starting motor bolts to the specified torque, Replace any bracket or shield and reconnect the battery cables to clean terminals. Crank the engine several times to check starting motor operation.

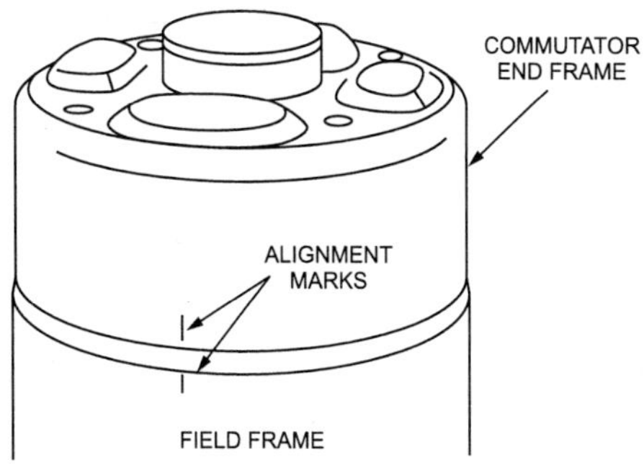

Fig. 5.20 *(h) Before disassembling a starter motor, scratch alignment marks on the field frame and the end frame to facilitate reassembly after service. (General Motors Corporation)*

5.11 STARTING MOTOR TESTS

There are several ways of testing a starting motor to determine its operating condition. Begin by making on-vehicle tests, followed by no-load test, and isolate the cause of the problem with bench tests.

5.12 IN-VEHICLE TESTS

Testing the starting system to isolate problems begins in the vehicle. Many problems can be repaired without removing the starter from the vehicle, saving time and effort. In fact, the problem may not be in the starter but may be caused by a discharged battery or corroded connections.

System Quick Check

1. Test the battery, and, if necessary, recharge it.
2. Clean and tighten battery connections. Check all starter system connections and clean as needed. Carefully check the large-diameter cable that runs from the battery to the solenoid and starter motor, as this cable supplies the heavy current required for cranking the engine.
3. Check that the ground strap is securely fastened between the engine block and the frame. The connections must be clean and tight. Losses due to corrosion or a loose connection can render the starter motor inoperative.
4. Try staring the engine. If the starter cranks the engine normally but the engine fails to start, then check the ignition and fuel systems. Starter

is not the problem. But if the starter does not turn over the engine at all or cranks very slowly, then troubleshoot the starting system using Table in *Fig. 5.21*.

Symptom	*Indication*	*Probable Cause*
No cranking	No lights	Open fusible link
No cranking	Light bright and no sound from solenoid	Open circuit in Solenoid
No cranking	Light bright and solenoid clicks	Open circuit in starter motor
No cranking	Headlights dim	Jammed drive assembly
Slow cranking	Headlights dim	Pitted contacts in solenoid. Worn brushes in starter motor. Pitted commutotor. Shorted commutator. Engine problem.

Fig. 5.21 *Troubleshooting chart for cranking problems.*

If the quick check fails to clearly isolate the problem area, then more detailed tests are necessary.

Cranking Circuit Tests

The following tests will reveal *excessive resistance* in the cranking circuit.

I. Use an expanded-scale voltmeter to test the voltage drop across the various points shown in *Fig. 5.22.* Remove the primary lead from the ignition coil and crank the engine (without starting it). Maximum allowable voltage drop, typically, should be as follows:

1. With the voltmeter leads connected to the positive post of the system battery and to the starting motor terminal 0.5 volt.
2. With the voltmeter connected to the positive post of the system battery and to the battery terminal of the starter relay 0.1 volt.
3. With the voltmeter connected to the positive post of the system battery and to the starting motor terminal of the starter relay 0.3 volt.
4. With the voltmeter connected to the negative terminal of the system battery and to ground 0.3 volt.

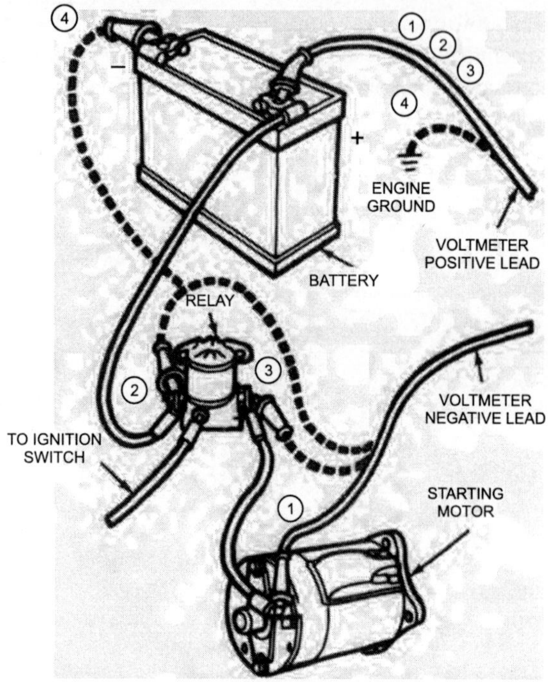

Fig. 5.22 *(a) To test for high resistance in automotive engine cranking system, use an expanded scale voltmeter (0.1 V calibrations) to test voltage drop between four sets of points indicated by numbers, i.e., 1-1. 2-2. 3-3. 4-4. (Ford Motors Corp.)*

II. For this test the engine should be brought to normal operating temperature. A cold engine will cause the starter to draw more current and make the results inconclusive.

To prevent the engine from starting during testing, disconnect and ground the high-voltage coil wire in a standard ignition system. On most electronic ignition systems, disconnect the wiring harness at the distributor. Follow the instruction in the applicable service manual to disconnect the wiring. The tester is a standard starter system tester containing a voltmeter, an ammeter, and a carbon pile rheostat.

1. Connect the voltmeter's negative lead to the solenoid starter terminal. Connect the meter's positive lead to the battery positive terminal, *Fig. 5.22* (*b*). Select the low voltage scale, and

do not bridge the starter relay except when cranking. Note the voltage reading while cranking the engine. This is the voltage drop while cranking. The maximum allowable voltage drop is 0.5 V.

2. Change the voltmeter's positive lead to the solenoid battery terminal. The voltage drop while cranking should not be more than 0.3 V.
3. Connect the voltmeter's negative lead to the solenoid battery terminal. Connect the meter's positive lead to the battery positive terminal. Crank the engine and observe the voltage drop reading. It should not be more than 0.2 V.
4. Connect the voltmeter's negative lead to the battery negative terminal. Connect the meter's positive lead to the engine block (ground). The voltage drop during cranking should not exceed 0.1 V.

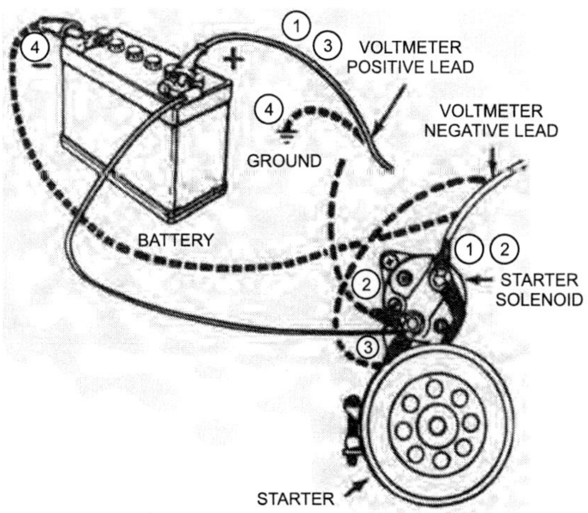

Fig. 5.22 *(b) Cranking Circuit test connections.*

Amperage Draw

The following test, with the engine at normal operating temperature, of the starting motor on the vehicle under load gives *amperage draw.*

1. Disconnect and ground ignition coil secondary wire.
2. Connect a remote control switch between the battery positive terminal and S terminal on the starter solenoid.
3. Connect the load tester as shown in *Fig. 5.17 (b)* and place the tester load control knob in the full counterclockwise position (decease).
4. Close the remote control switch and note the voltage indicated on the voltmeter after the starting motor has reached maximum specified rpm.
5. Do not crank engine more than 15 seconds, then open the remote control switch.
6. Turn the tester load control knob clockwise (increase) until the voltmeter indicates the same voltage as noted when the starting motor cranked the engine (step # 4).
7. Note the amperage draw on the ammeter scale and compare it with the manufacturer's specification (between 150-250 amps)

No-Load Test

The *'no-load test'* determines how fast the armature spins and the amount of current draw at a specified voltage.

1. To perform this test, connect the test equipment leads as shown in *Fig. 5.23*, attaching a tachometer to the end of the armature shaft. Connect a remote starter switch between the battery and starter terminals on the solenoid

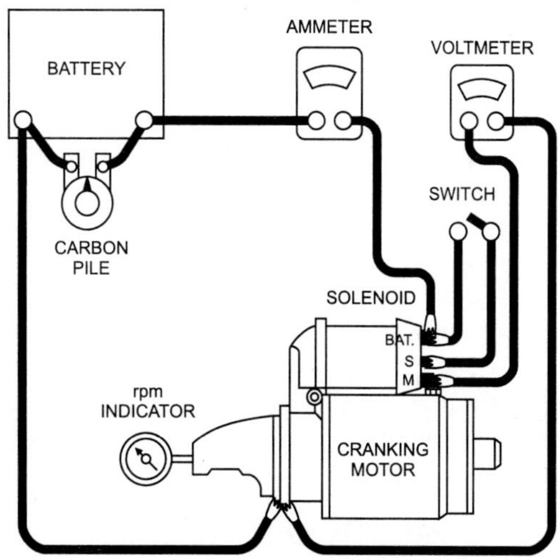

Fig. 5.23 *Schematic diagram shows ammeter and voltmeter lead connections and the attachment of the tachometer for performing a no-load test on a starter motor.*

2. Adjust the carbon pile to obtain a reading on the voltmeter of 10 V. Do not set for a higher voltage; excessive speed under a no-load condition could damage the starter.

3. Close the remote starter switch and record rpm, amperage, and voltage readings. The starting motor will run at no-load speed. Open the remote starter switch. Compare the readings with the factory specifications for the unit under test. Typical good values are in the range of 50-85 A and 6000-12000 rpm at 10 V.
4. If the current and rpm readings are within specs, the starter is good.

Analysis of No-Load Test Results

If under no-load conditions, the starting motor rotates at normal rpm at specified voltage and amperage draw, the starting motor is in good operating condition. However, if problem is indicated, consider the following:

1. If the starting motor fails to spin under no-load tests and shows high amperage draw, there may be a direct ground in the armature or field windings or *"frozen"* armature shaft bearings.
2. Low no-load speed and a high amperage draw indicate a dragging armature, worn bearings, or tight and dirty bearings, a shorted or grounded armature, or shorted field winding.
3. Low no-load speed and a low amperage draw indicate high resistance in the starting motor. One of the field windings may be *"open."* Other possible causes may be broken brush springs, excessively worn brushes, high insulation between the commutator bars, or a very dirty or oily commutator. Also, Check for poor connections, defective leads.
4. If the armature does not spin and there is little or no current draw, check for an open field winding, an open armature winding, or a broken brush or spring.
5. High no-load speed and high amperage draw are indications of *"shorted"* field windings.

5.13 STARTING MOTOR BENCH TESTS

When the no-load test results analysis indicate the trouble in the starting motor, perform the following *bench tests* to pinpoint the cause of the problem.

Disconnect both the battery cables at the battery to avoid short circuits during starter motor removal. These can cause burns to the hands or damage to components. Disconnect all wires from the solenoid. Remove the mounting bolts securing the starter motor to the flywheel housing and carefully remove the starter motor.

Armature and Field Open Circuit Tests

An open circuit armature can be detected by examining the commutator for evidence of any burning. An arc formed every time the commutator segment connected to the open circuit winding passes under a brush causes the spot burned on the commutator. An open circuit test of the field can be made with a 120 volts test lamp, *Fig. 5.24*, or by connecting a jumper lead from the positive post of a battery to the starting motor terminal, the voltmeter negative lead to the negative post of the battery, and the voltmeter positive lead to each field brush in turn. If the lamp fails to light or if no reading is obtained on the voltmeter, the field winding is open.

Armature and Field Grounded Circuit Tests

In case the armature or field coil insulation has failed, it would allow a conductor to touch the starting motor frame or armature core. This test can be made with a 120-volt test lamp or by using a voltmeter.

To test the armature with a test lamp, touch one test prod to the armature shaft and touch the other prod to each commutator bar in turn. In case the lamp lights, the armature winding is grounded.

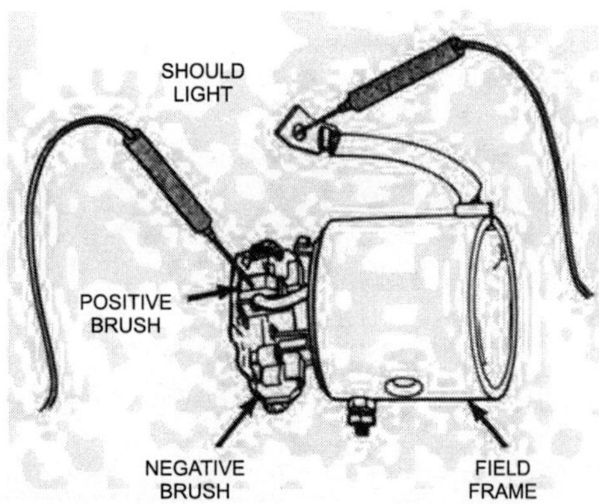

Fig. 5.24. *Bench tests of starting motor circuitry include testing the field windings for an open circuit by means of a 120 V test lamp. If the lamp fails to light, the field winding is open. (General Motors Corp.)*

To perform the test with a voltmeter, connect a jumper lead from the positive post of a battery to the armature shaft. Connect the voltmeter leads to the negative post of the battery and to each commutator bar in turn. In case any voltage is indicated, the winding is grounded.

To perform the test for grounded field circuit windings, connect a jumper lead from one terminal of the starting motor to one post of the battery. Connect the other post of the battery with a test prod while keeping the brushes away from the frame of the starting motor. Then touch the other prod to the field frame of the starting motor. In case the lamp lights or if any voltage is indicated on the meter, the field windings are grounded.

Component Tests

Disassemble the starter assembly by removing the two long bolts that hold the drive gear housing, field frame, and end frame together. Some designs require that the solenoid be removed before the starter can be disassembled. Next remove the shift lever mechanism to free the drive assembly and armature. Separate the housing and withdraw the armature, being careful not to damage the brushes.

Armature Test

A special tester called a growler is used for checking the armature, *Figs. 5.25* (*a*) and (*b*). A growler is a test box containing a cradle for the armature, a meter, a test light, and test probes. The cradle contains a powerful electromagnet powered from the AC power line. Because it is an AC magnet, the magnetic field alternates at a 60-cycle rate, in response to the frequency of the power supply. The armature is set in the cradle and the electromagnet energized. A thin steel blade as a hacksaw blade is held lengthwise against the armature core as the armature is slowly rotated in the cradle. The blade will vibrate when held against a core containing a shorted winding. This test is necessary because of the construction of the armature. The bindings are buried in layers around the core and are connected in series. It is impossible to use a simple continuity test to tell whether all windings are free from shorts to each other or to ground. Replace the armature if there is a short found.

Grounded windings are detected using the test light and probes on the tester. Touch one probe to the armature shaft and the other to each commutator bar, one at a time. If the test lamp lights, the armature is grounded and must be replaced.

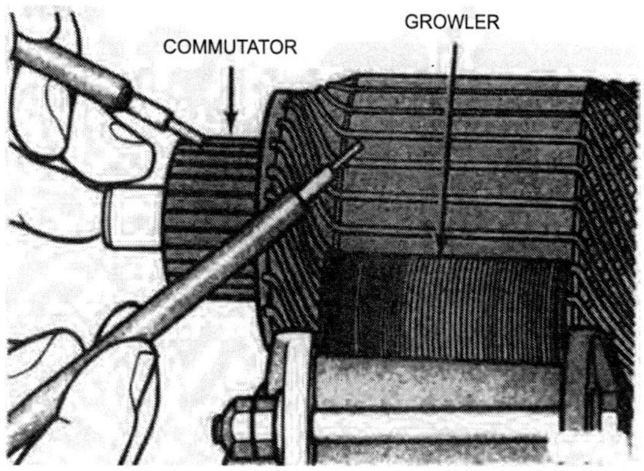

Fig. 5.25 *(a) Testing a starter motor armature with growler.*

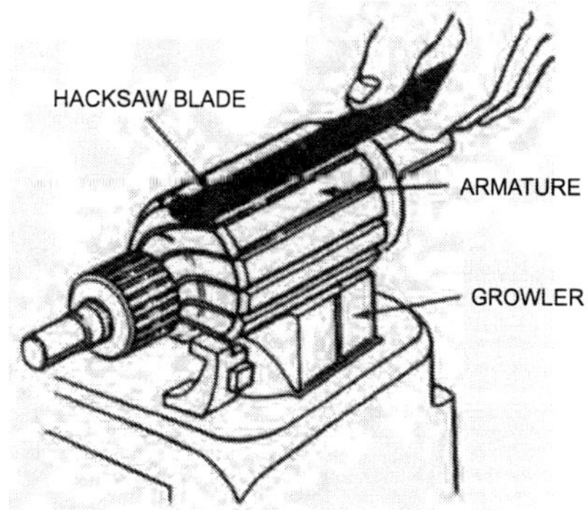

Fig. 5.25 *(b) Checking armature for short circuits on a growler.*

Field Coil Tests

To check for field coil opens or ground, a 12 V test lamp is used, *Figs. 5.26* (*a*) and (*b*).

1. ***Open coil test:*** Touch one test lamp lead to the series field coil terminal and the other to one of the insulated brushes. The lamp should light up; if it does not, the field coil is open and must be repaired or replaced. Repeat the process for each insulated brush, one at a time.
2. ***Grounder coil test:*** Check that brushes, brush holders, and leads do not touch each other or the frame. Touch one lamp lead to one of the grounded brush holders. Touch the other test lead to either insulated brush. If the lamp lights,

one of the field coils is grounded and must be repaired or replaced.

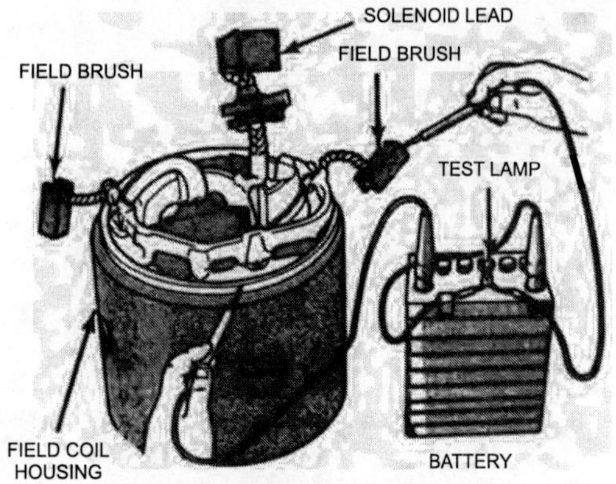

Fig. 5.26 *(a) Field coil test.*

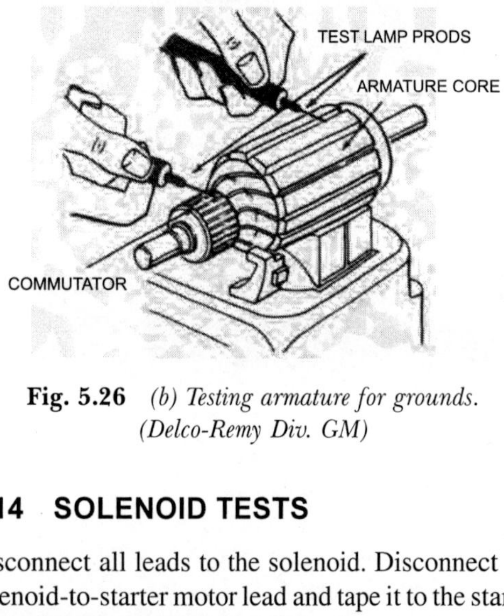

Fig. 5.26 *(b) Testing armature for grounds. (Delco-Remy Div. GM)*

5.14 SOLENOID TESTS

Disconnect all leads to the solenoid. Disconnect the solenoid-to-starter motor lead and tape it to the starter motor housing to avoid accidental energizing of the motor during testing.

Pull-in Winding Test

1. Connect a remote starter switch between the positive post of the battery and the ignition switch terminal on the solenoid, *Fig. 5.27*. The terminal is usually marked with an S. Check that the ground connection from the solenoid housing to the battery negative post is good.
2. Close the remote starter switch and momentarily apply jumper to the solenoid motor terminal (usually marked with an M) to the starter motor housing. The solenoid should engage and remain engaged. If not, the pull-in winding is defective. Do not allow the solenoid to remain energized for long periods; the solenoid winding could overheat and be damaged.

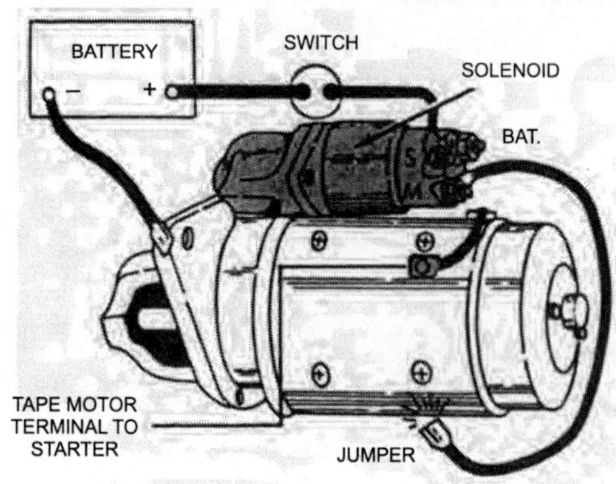

Fig. 5.27 *Solenoid pull-in winding test.*

Hold-in Winding Test

This test is performed with the same hookup used in the pull-in winding test, but without the jumper, *Fig. 5.28*.

1. Close the remote starter switch, but do not use the jumper. Manually pull out the starter drive assembly. The drive assembly should remain engaged. If it does not, the hold-in winding is defective.
2. Open the remote starter switch. The drive assembly should return without hesitation. If it does not, the return spring is defective.

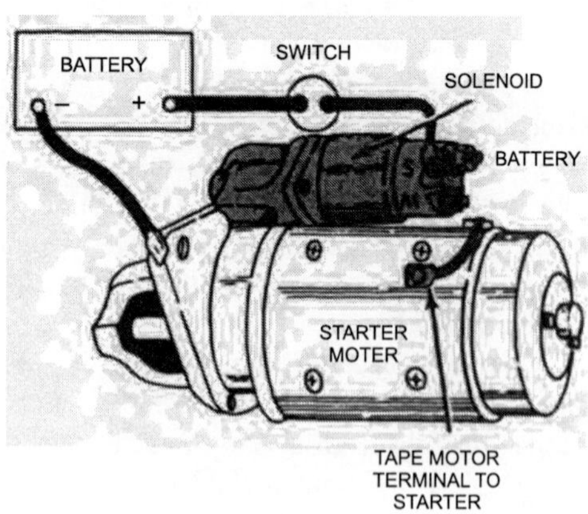

Fig. 5.28 *Solenoid hold-in winding test.*

Pinion Clearance Check

After removing the starter from the vehicle and before disassembly, check pinion end-play as described below:

1. Clamp the starter in a vise using pads or a shop rag to prevent damage to the housing. Clamp the starter just tight enough to hold the starter securely; do not over-tighten to avoid damage to the starter.
2. Connect leads from a 12 V battery to terminal S on the solenoid and to the starter housing. Connect a jumper from terminal M on the solenoid to the starter frame. Hold the pinion forward in the starting position until the battery is disconnected.
3. Push the pinion gear as far back from the stop retainer as possible and measure the clearance between the stop and the end of the pinion gear, using a feeler gauge, *Fig. 5.29*. Clearance should be in the range of 0.25-3.6 mm.
4. If pinion gear clearance is not as specified, disassemble the starter and replace worn parts.

Brushes, Springs, and Holders

Replace brushes if they are worn to ½ or more of their original length or if they are oil soaked or pitted. Check brush spring tension. If springs are weak or distorted, replace them. Replace deformed or damaged brush holders.

Overrunning Clutch

The pinion gear should turn freely and in one direction only. Check for chipped or broken teeth on the pinion gear it shows excessive wear or chipping. Check for cracks in the gear or clutch housing.

Check the overrunning clutch for slippage by clamping the armature in a vise, and use care to protect the armature from damage by using pads or cushioning it with a shop towel. Use a torque wrench and a socket that fits the pinion gear to apply a force of 67.8 Nm to the clutch assembly. If the clutch slips, replace it.

Armature Bushings

Check armature shaft end bearings for wear by installing in an armature in the bushings and observing if any side play exists. If side play exists, replace the bushings. Lubricate the shaft ends with a silicone-base lubricant before reassembling.

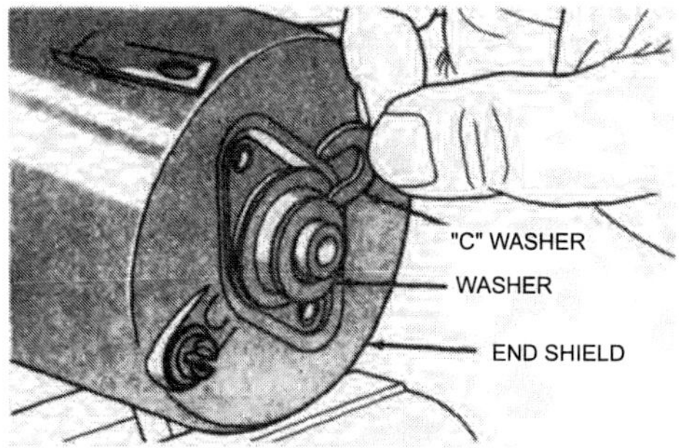

Fig. 5.29 *(a) Installing the new "C" washer. (Chrysler Corp.)*

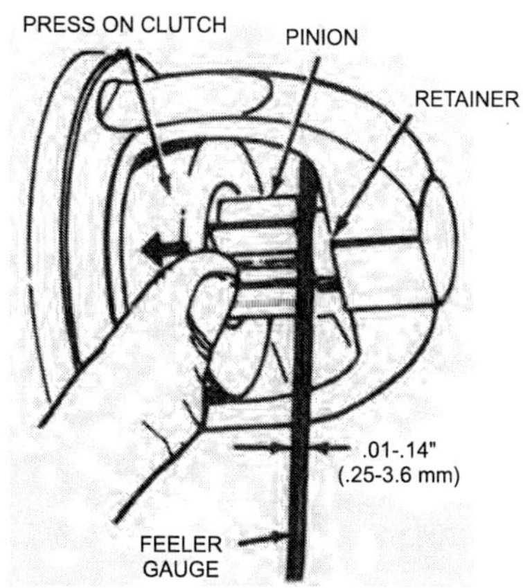

Fig. 5.29 *(b) Checking starter motor pinion end play.*

5.15 CHARGING SYSTEM SERVICE

It is important for mechanics and technicians to understand the charging system and the application of diagnostic equipment, required for testing, so as to be able to intelligently test and diagnose the charging system problems. This helps in pinpointing the faulty unit, and determining the specific part or junction producing the problem.

Automobile battery is a vital part of the charging system. It acts as an electrical reservoir. It supplies electrical power when the alternator is providing insufficient output, to meet the vehicle demands. The battery, to charge itself, absorbs any excess alternator output over the vehicle demand. With continuous operation of the alternator, the battery becomes charged, its voltage increases and resists its charging. This results in decreased charging rate. The resistance to current flow in the battery is measured in volts and is called the *battery counter voltage.*

Maximum output of an alternator is based on the alternator size and its rotating speed, which in turn depends upon the engine operating speed. The alternator design limits maximum output. A voltage regulator in the charging system limits the maximum charging system voltage. The voltage regulator does not affect the output until the voltage reaches the voltage regulator setting. Voltage regulator setting has no effect on maximum alternator output amperage. Its purpose is to keep voltage below a safe limit so as to protect electrical operating units of the vehicle, especially the battery. Any excess voltage will shorten the life of the battery, ignition points, light bulbs, radio, staring motor, etc. The voltage regulator is set just high enough to fully charge the battery without overcharging it.

The battery counter voltage controls voltage of the charging system whenever the system voltage is below the regulator voltage setting. A charging system operating with a partly charged battery will have low voltage. As the battery becomes fully charged the charging system voltage will rise until it reaches regulated voltage, at which time the regulator will take control and prevent any further increase in charging system voltage.

Even though a charging system has two major parts (alternator and regulator), at times problems come from other faulty components, e.g., bad starting motor, battery, or wiring. There are four common types of trouble symptoms caused by the charging system:

1. Dead battery with slow or no cranking.
2. Overcharged battery that requires frequent water to be added to battery.
3. Abnormal noises, e.g., grinding, squealing, buzzing, etc.
4. Indicator shows problem, i.e., light glows all the time or incorrect indicator reading.

5.16 CHARGING SYSTEM INSPECTION

Inspect the charging system alternator at recommended intervals. Before performing charging system electrical tests, test the state of charge of the battery and check the condition of starting and charging system cables, wires, and electrical connections. Open the hood and visually inspect the parts of the charging system for the troubles like the ones shown in *Fig. 5.30.*

Make sure that starting motor, alternator, and regulator *mounting bolts* are adequately tight to allow good ground circuits. Check for *alternator belt problems.* Look over the alternator drive belt for signs of wear or slippage. Make sure the belt is adjusted properly. A loose belt may squeal or flap up and down. Also inspect the condition of the alternator belt for cracks, glazing (hard shiny surface), grease or oil contamination, and general wear. See that the belt tension adjustment is correct. The belt should deflect 6-7 mm in the centre of its long span. Belt tension testers normally show colour-coded *good* and *bad* ranges. If not, check the tester reading against manufacturer's specifications.

If needed, adjust alternator belt tension, *Figs. 5.31 (a) and (b).* Loosen the alternator mounting and adjusting bolts. Then pry on a stiff surface of the end frame. Pull hard enough to produce proper tension. Tighten the adjusting bolt while holding the pry bar. Then tighten the other mounting bolts and recheck tension. Remember that, over tightening the belt will result in quick failure of alternator bearings because of excessive load imposed by the belt.

Two components of an alternator assembly wear in operation; the *field brushes* and the *bearings.* Excessively worn brushes normally appear as intermittent faulty operation of the alternator and sometimes as noise in the vehicle radio that changes frequency as the engine speed is changed. The worn field brushes will have to be replaced with new ones. The alternator has to be removed from the engine to replace the brushes. Depending upon the alternator design it may or may not be necessary to disassemble the alternator for replacing the field brushes.

Defective alternator bearings either *rumble* or *whine.* The pitch of the noise changes as the engine speed changes and the sound persists whether the alternator is charging or not. To isolate an alternator-bearing problem, removing the connecting wire junction from the regulator can easily stop alternator charge. It is necessary to disassemble the alternator to replace the bearings. Maintenance is minimized by the use of pre-lubricated rotor bearings and long brushes in most modern alternators.

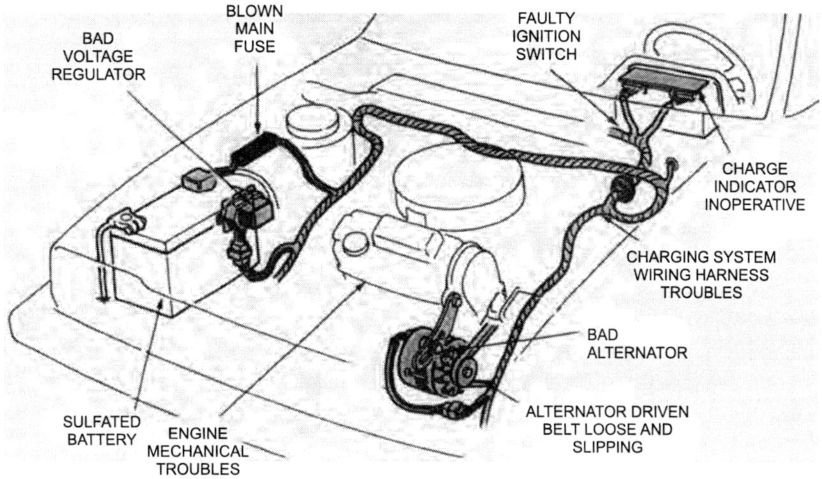

Fig. 5.30 *Types of problems that can be found in an automotive vehicle charging system.*

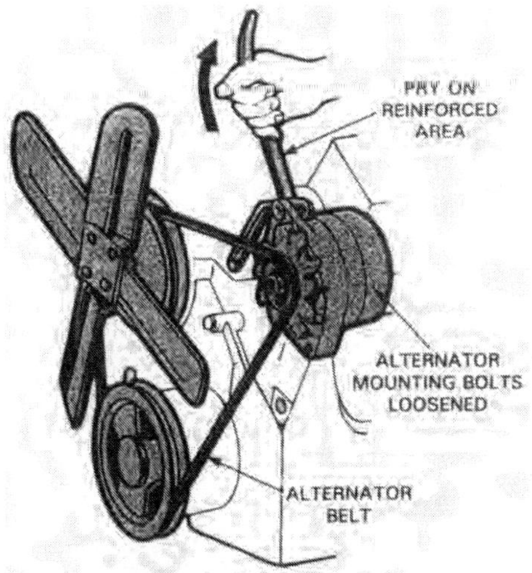

Fig. 5.31 *(a) Adjust alternator belt by prying on reinforced area housing. Do not tighten belt too much to avoid alternator bearing failure (Florida Dept. of Voc. Ed.)*

5.17 CHARGING SYSTEM PRECAUTIONS

Some special precautions need to be taken while testing and servicing the alternator because the alternator output terminal is connected to the battery at all times.

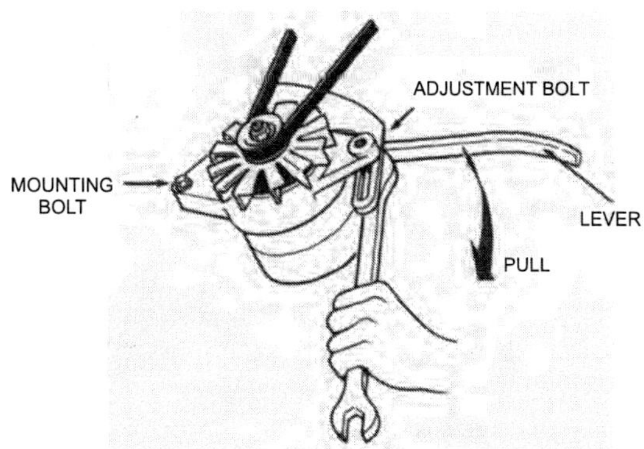

Fig. 5.31 *(b) Adjusting the alternator belt.*

- Be careful to avoid reverse polarity while performing battery service of any kind. A surge of current in opposite direction could burn alternator diodes and damage vehicle electrical wiring.
- Never 'short' or "ground" the system when disconnecting wires or connecting test equipment leads to the terminals of the alternator or regulator. Grounding of field terminal at either the alternator or regulator will damage the regulator. Grounding of alternator output terminal will damage the alternator and/ or the charging circuit.

- Never operate an alternator on an open circuit. With no battery or electric load in circuit the regulator may build voltage 50 to over 110 volts, which may damage rectifier diodes and be dangerous to anyone who might touch the alternator output terminal.

5.18 QUICK CHARGING SYSTEM CHECKS

If a problem exists, e.g., low output or overcharging, *check for complete field circuit* (rotor) by placing a long screwdriver on the alternator rear-bearing surface. When the field circuit is complete, there will be a strong pull on the blade of the screwdriver, indicating that the field winding is energized. In case there is no field circuit, an alternator will not charge because the field windings must be "excited" by the battery voltage.

Normally, certain other *quick checks* for possible charging system problems can be made before carrying out extensive electrical tests.

In case the charging system incorporates a fuse link, *Fig. 5.32*, check it for continuity and appearance. A fuse link is designed to "*blow*" if heavy current flows in the circuit. If the fuse link is blackened or is "*open*", replace it with a "*matching*" fuse link. Do not bypass it or replace it with a wire.

Check the *charge indicator lamp* on the vehicle dashboard. In a normal charging system, the only time the lamp should light is when the ignition key is in the "run" position and the engine is not running. Once the engine starts and the alternator begins to produce voltage, the charge indicator lamp should turn off. But if it stays on even after the engine starts, there is a malfunction in the charging system. If the indicator lamp fails to light in any ignition key position, the indicator valve lamp may be defective.

On alternators with an integral voltage regulator, and charging systems having an indicator lamp outage, make the following checks. Unplug # 1-2 connector and ground the #1 wire. If the indicator lamp fails to light, check for an "open" in the #1 wire. If the lamp lights, reconnect the #1-2 connector and "ground" the tab in the alternator end frame access hole, *Fig. 5.33*. If the light stays on, replace the regulator. But if the light turns off when the tab is "grounded", check the rotor, brushes, and slip rings for an "open".

Before condemning any particular component or assembly in the charging system under inspection, make further tests of *current output* and *circuit resistance.*

When making tests or performing service operations on the charging system, consult the manufacture's service manuals for step-by-step instructions and acceptable values or ranges for a comparison with test results.

5.19 CHARGING SYSTEM TESTS

To determine the operating condition of the charging system, charging system tests should be performed when symptoms indicate low alternator voltage and current. The following four tests are made on a charging system:

1. **Charging System Output Test** – It measures the current and voltage output of charging system under a load.
2. **Regulator Voltage Test** – It measures the charging system voltage under low output, low load conditions.
3. **Regulator Bypass Test** – It connects full battery voltage to alternator field, leaving regulator out of circuit.
4. **Circuit Resistance Tests** – It measures resistance in insulated and grounded circuits of the system.

5.20 IN-VEHICLE CHARGING SYSTEM TESTS

In-vehicle testing identifies the problems in a replaceable unit or repairable element so the mechanic can correct the problem. Carefully examining the symptoms and logically analyzing the problem can help save a lot of time consuming troubleshooting. The principles that apply to tests for systems with externally mounted voltage regulators are applicable to systems with internally mounted regulators. However, the ***specific test methods will vary*** and to be certain always refer to the manufacturer's service manual.

Pretest Checks

Check the specific gravity of the battery's electrolyte and recharge the battery, if needed. Clean and tighten the battery connectors, as necessary. Check and adjust the drive belt tension to be within specs.

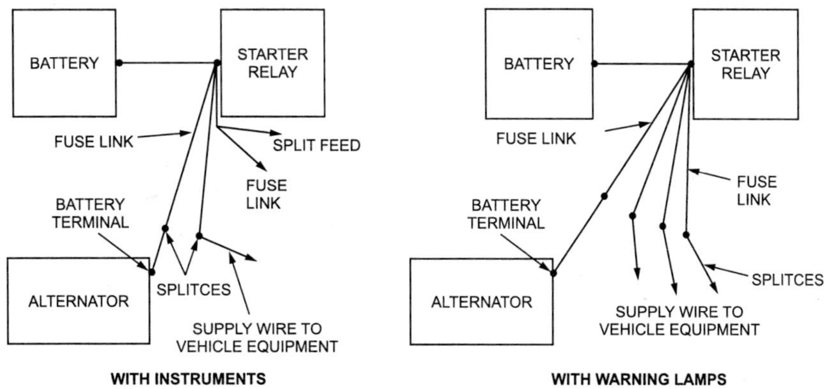

Fig. 5.32 *A fuse link is installed between the starter relay terminal and the alternator battery terminal. An open circuit usually means that the fuse link has "blown". (Ford Motors)*

Charging Circuit Resistance Test

1. Disconnect the ground cable from the battery negative post before making test connections. This will prevent any accidental short-circuiting.
2. Disconnect the lead at the BAT terminal on the alternator. Connect a 0-100 A direct current ammeter in series between the BAT terminal and the lead disconnected from it, *Fig. 5.34*.
3. Connect the positive lead of a dc voltmeter to the disconnected BAT lead. Connect the negative lead of the voltmeter to the positive terminal of the battery to measure voltage drop during the test.
4. Disconnect the regulator field wire (usually green) from the alternator field terminal (F). Connect a wire from the alternator field terminal (F) to ground.
5. Connect a tachometer for measuring engine speed following the tachometer manufacturer's instructions. Typically two connections are required: the +12 V connection at the coil and a good ground connection.
6. Reconnect the ground cable to the battery negative post.
7. Connect a carbon pile rheostat (load) across the battery, ensuring that it is set to **off** or **open** to avoid severe sparking and the danger of an explosion.
8. Start the engine and quickly reduce the speed to idle. Racing the engine with the field coil grounded can cause damage to electrical components. Adjust the engine speed and the carbon pile to maintain a current of 20 amperes flowing in the circuit.

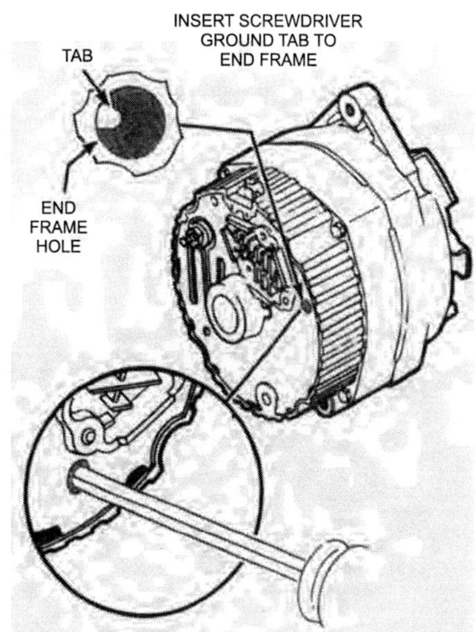

Fig. 5.33 *Some alternators have a test hole that provides a point of access for using a screwdriver to ground the field winding. (Chevrolet Div. General Motors)*

9. The voltmeter should indicate a voltage drop of 0.7 V or less. If the voltage drop is higher, move the voltmeter positive lead from point to point in the circuit. Start at the starter motor solenoid main battery connection and move down the circuit until you locate the connection with the greatest voltage drop.

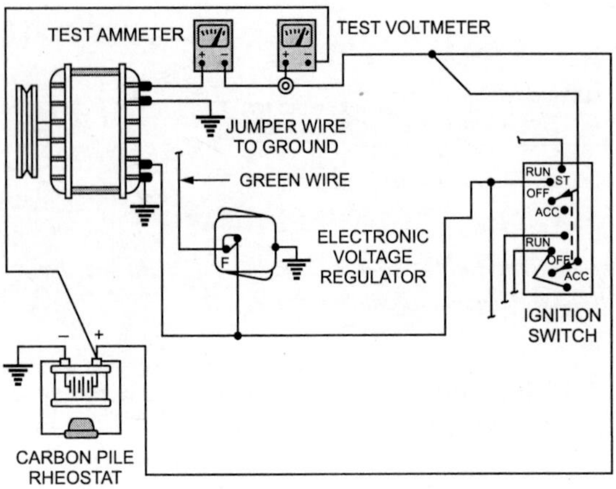

Fig. 5.34 *Charging circuit resistance test.*

Current Output Test

1. Make sure the battery negative cable is disconnected before making the test connections. All test connections are the same as for the resistance test, except for the voltmeter, *Fig. 5.35.*

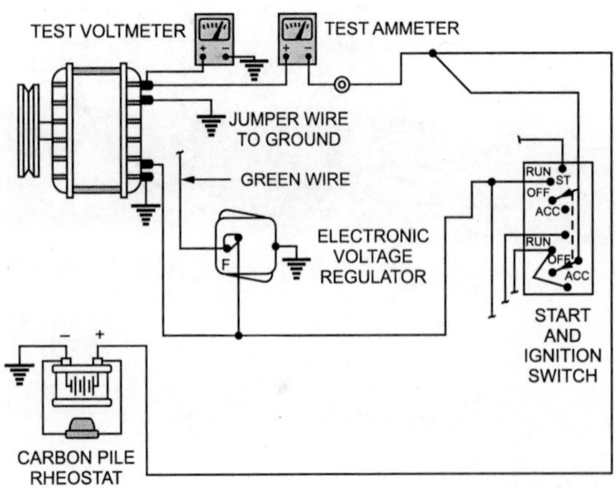

Fig. 5.35 *Alternator current output test hookup.*

2. Connect the negative lead of the voltmeter to ground and the positive lead to the BAT terminal on the alternator.
3. Start the engine and quickly reduce the speed to idle. Racing the engine with the test connections can damage other electrical components.
4. Adjust the engine speed and the carbon pile for an indication of 1250 rpm on the tachometer and 15 V on the voltmeter, respectively. Be careful not to allow the voltage to exceed 16 V, to avoid component damage.
5. The current output, as indicated on the ammeter, should be within specs for the vehicle under test (typically a minimum of 47 to 50 A).

Voltage Output Test

The specific gravity of the battery should be more than 1200 for a proper regulated voltage check. Charge the battery or use a good test battery before testing the regulator.

1. Connect the positive lead of the voltmeter to the positive post on the battery. Connect the negative lead from the voltmeter to a good vehicle body ground, *Fig. 5.36.*
2. Check the ambient temperature of the voltage regulator. The temperature affects the voltage output as shown in Table 5.1

Table 5.1 Regulator Ambient Temperature and Voltage Range

Ambient Temperature	*Voltage Range*
–20° F	14.9-15.9 V
80° F	13.9-14.6 V
140° F	13.3-13.9 V

3. Start and run the engine at 1250 rpm with all lights and accessories turned off. Check the voltmeter. The regulator is working properly if voltage readings are within specs.
4. If the voltage is not within limits or is fluctuating, check that the regulator has a good ground. With the engine off, disconnect the regulator terminals. Turn the ignition on, but do not start the engine. The battery voltage should appear at both the blue and green leads of the connector. If all the tests are satisfactory. Replace the regulator and repeat the tests.

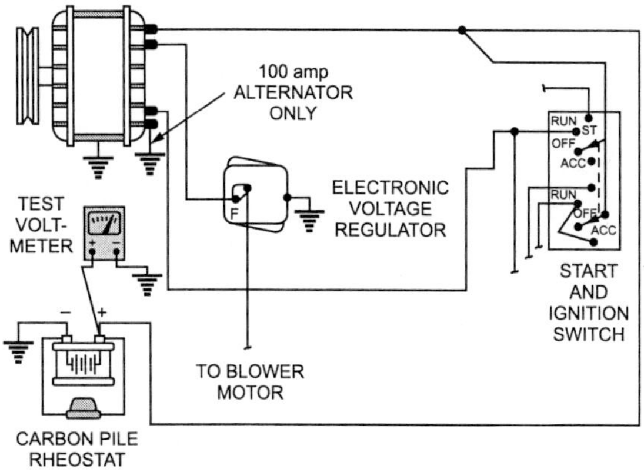

Fig. 5.36 *Voltmeter connections for voltage output test.*

5.21 CHARGING SYSTEM OUTPUT TEST

A *charging system output test* is made to ensure that the alternator can produce sufficient current to operate the vehicle's electrical devices and still have enough current to recharge the battery. This test can be done with a load tester or an accurate voltmeter.

To use a *load tester,* connect the tester to the battery as described in the vehicle's operating manual. Start the engine and load the system to the current specifications of the alternator. The system should be able to produce specified current and maintain specified voltage. If not, there is some defect in the system.

To use a *voltmeter* to check the charging system, connect the voltmeter to the battery cables. Then follow three basic steps:

1. Measure, and note the battery voltage with the engine off, *Fig. 5.37 (a)*. The battery voltage should be around 12.3 to 12.5 V.
2. Start and fast idle the engine while all electrical system devices are off. Again, read and note the voltmeter voltage value. This voltage should increase above the battery voltage by 2 -3 volts. This checks regulator no-load voltage, *Fig. 5.37 (b)*.
3. Turn on all electrical system accessories, e.g., headlights, wipers, radio, air-conditioner, etc., with the engine running. Read and record the voltmeter reading, *Fig. 5.37 (c)*. Voltage will drop, but it must not drop below battery voltage reading in step # 1, i.e., reading remains above 12.5 V.

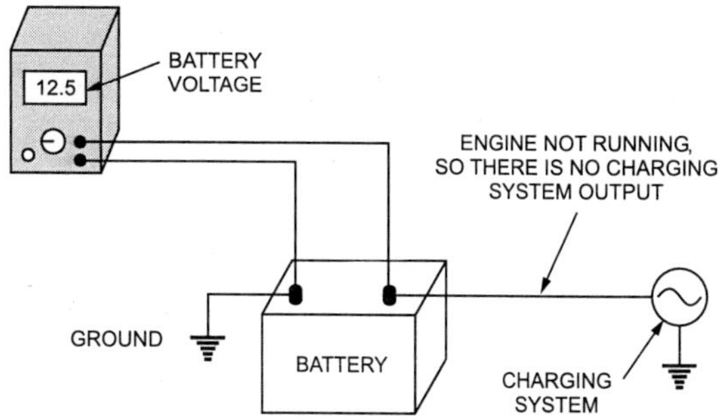

(a) Battery voltage is measured to check battery condition and allow for meter accuracy.

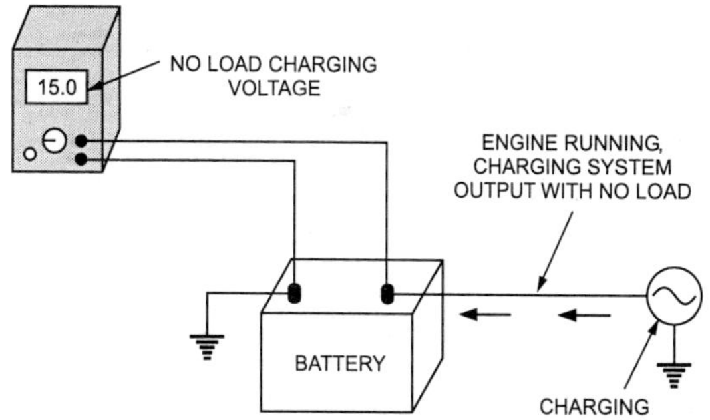

(b) No-load test checks action of voltage regulator and alternator. Voltage should increase over battery voltage by only as specified.

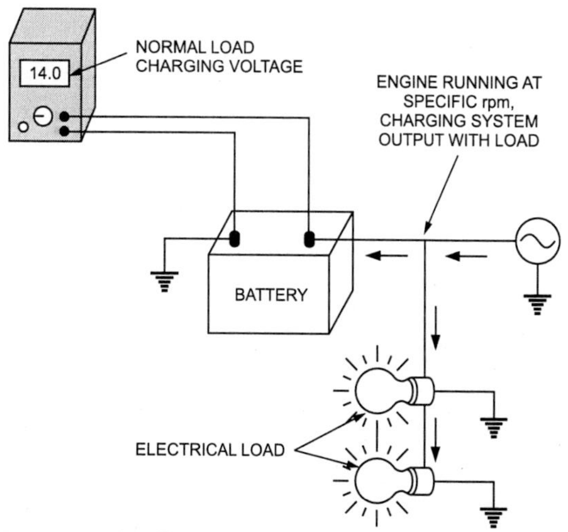

(c) Turn on headlight high beam and taillights. The voltage will drop but should stay above battery voltage. If voltage is higher than in (a), charging system should be recharging battery.

Fig. 5.37 *A simple method of using a voltmeter to check charging system output.*

If voltage in step # 3 remains above battery voltage, there is sufficient charging output to run the electrical system accessories and recharge the battery. But if the voltage across the battery is equal to or below the battery voltage across the battery voltage, the charging system is *not* working properly. Other tests are then needed to pinpoint the charging system problem.

5.22 REGULATOR BY-PASS TEST

If the problem detected is that of *"undercharging"* preferred way to initiate electrical tests on the charging system is to bypass the voltage regulator. This is called *"full fielding"* because the alternator field winding is receiving full current input. *Full fielding* can be achieved by the use of a plastic *'full fielder,* " a piece of *"jumper* " wire, or a screwdriver, *Fig. 5.33.* If with full fielding the alternator produces at least a normal charging voltage (13.8-14.9 volts), the voltage regulator is defective. But if the alternator is still "undercharging," then there is a problem in the alternator or wiring.

Other tests, like circuit resistance tests, would be needed to find the exact problem. Refer to the service manual for details.

5.23 ALTERNATOR REMOVAL

If the charging system fails to meet current output specifications, and *"full fielding"* (bypassing voltage regulator or computer modules) shows that the alternator is defective, remove the alternator from the engine, for disassembly and bench test. Before unbolting the alternator, first *disconnect the battery* to avoid damage to parts if wires are shorted. Shown in *Figs. 5.38* (*a*) and (*b*), most alternators are held to the front of the engine with two bolts.

- Place the ignition key in the "off" position.
- Remove the ground cable from the negative post of the battery.
- Disconnect the leads from the alternator output terminal (BAT) and from the two field terminals (FLD).
- Disconnect the ground lead from the alternator.
- Remove the wiring retainer nut and disengage the retainer from the alternator.
- Unscrew the alternator mounting bolts and the adjusting arm bolts.
- Slip off the drive belt, or belts, and remove the alternator from the engine.

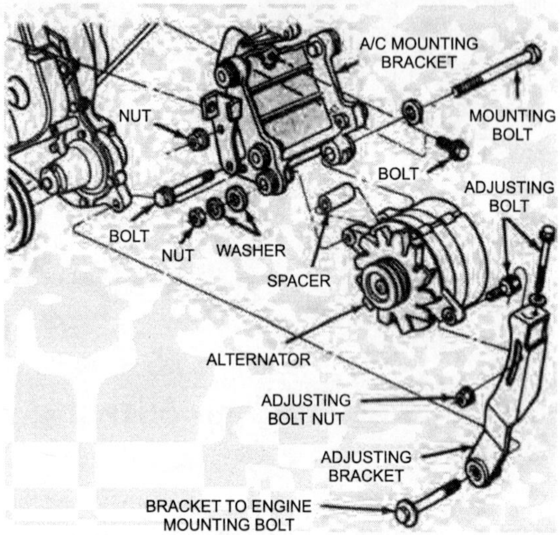

Fig. 5.38 *(a) Exploded view of a typical alternator installation illustrates arrangement of brackets, mounting bracket, adjusting bracket, bolts, and nuts. (Chrysler Corp.)*

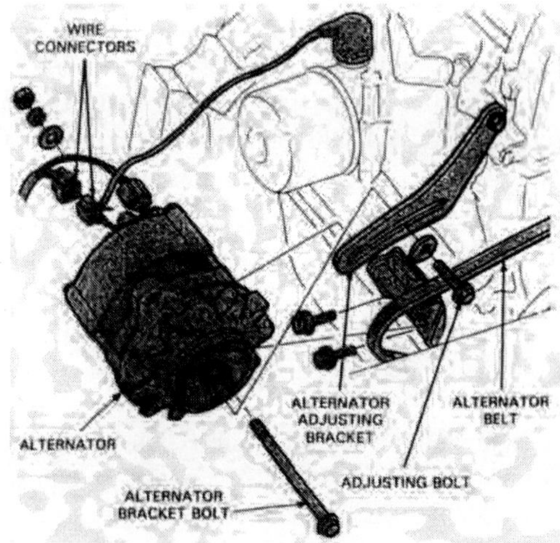

Fig. 5.38 *(b) Two or three bolts normally hold alternator on engine. Disconnect battery to avoid shorts that could damage electrical system when replacing the alternator. (Honda Motors)*

When removing the wires from the back of the alternator, note their location and whether special insulating washers are used. This will facilitate reassembly of the alternator to the engine, after service.

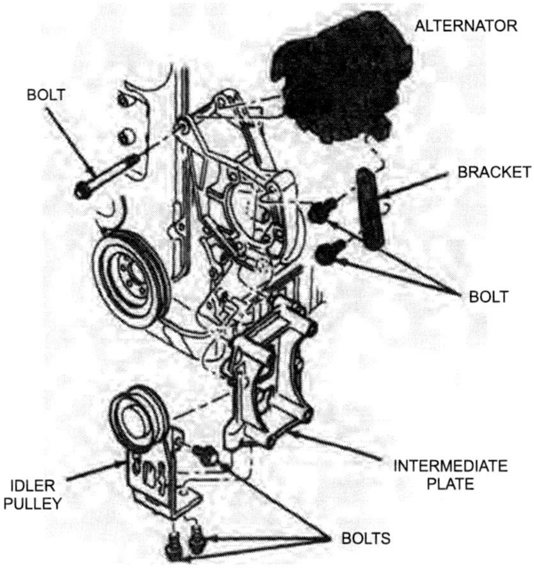

Fig. 5.38 *(c) Removing an alternator. (Ford Motors)*

5.24 ALTERNATOR DISSASSEMBLY

To disassemble the alternator:

- Scribe marks on the front and rear bushings or end shields and the stator frame to facilitate reassembly.
- Remove the dust cover, *Fig. 5.39*.

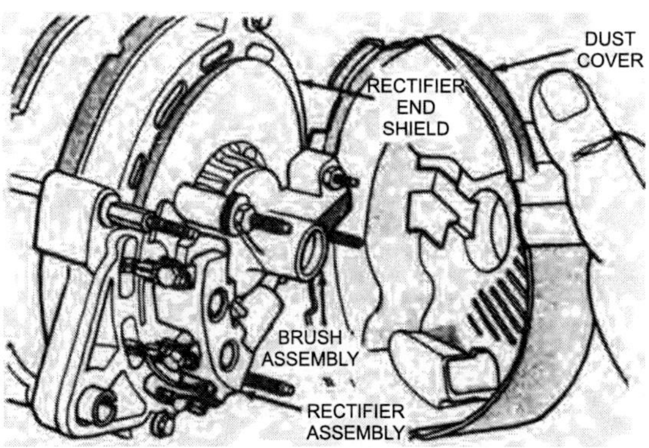

Fig. 5.39 *Disassembly of this Chrysler alternator with voltage regulated by electronics begins with removal of the dust cover.*

- Remove the brush holder mounting screws and the brush holder assembly, *Fig. 5.40*.

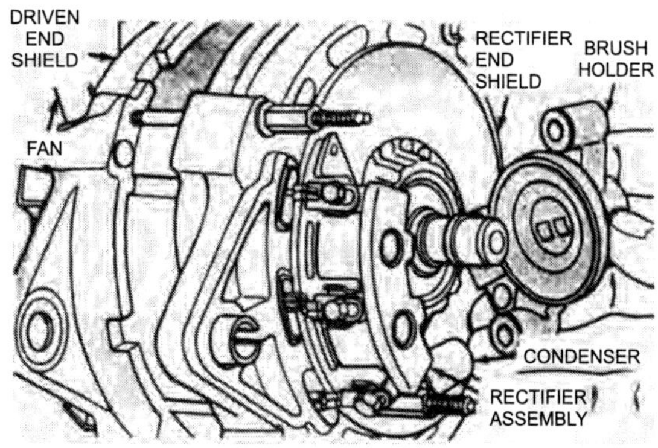

Fig. 5.40 *Brush holder assembly, when removed, allows access to the rectifier assembly and the condenser. (Chrysler Corp.)*

- Remove the (three here) stator-to-rectifier attaching screws.
- Remove the two, rectifier assembly mounting screws.
- Remove the rectifier insulator.
- Remove the condenser (capacitor) mounting screw and remove the rectifier assembly.
- Remove the through bolts, then separate the end shields by using two screw drivers to pry gently between the drive end shield and the stator, *Fig. 5.41*. The stator should remain with the rear end shield.

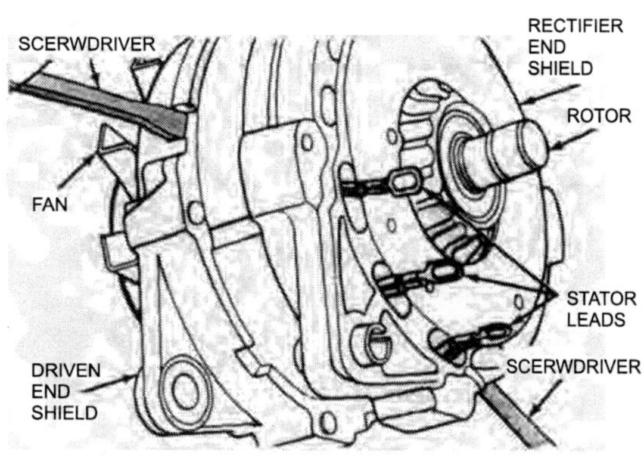

Fig. 5.41 *To separate end shields, use two screwdrivers to pry between the stator and the drive end shield. (Chrysler Corp.)*

- Remove the pulley nut, pulley washer, pulley, and fan.
- Remove the bearing spacer and press the rotor out of the drive end shield, *Fig. 5.42*.
- Remove the stator form the rear end shield, *Fig. 5.43*.

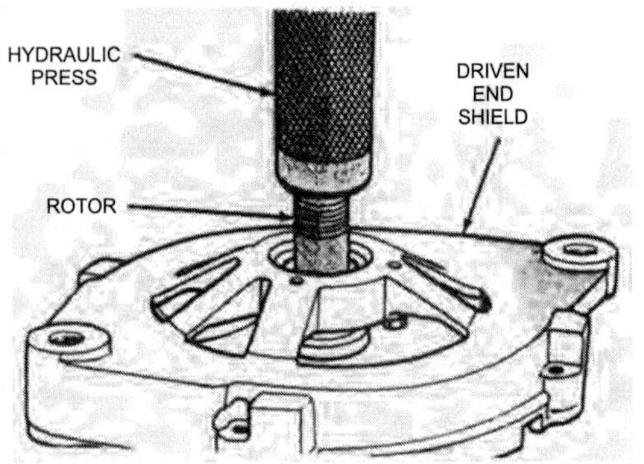

Fig. 5.42 *In case rotor removal is necessary, support the drive end shield on the bed of a hydraulic press and press out the rotor. (Chrysler Corp.)*

Fig. 5.43 *Remove the stator from the rectifier end shield for carrying out bench tests. (Chrysler Corp.)*

5.25 BENCH TESTS

Removing the alternator for bench testing detaches the unit from the rest of the charging system circuit and allows it to be disassembled. The individual components of the alternator can then be tested and repaired. The procedures for removing the alternator and disassembling it vary from vehicle to vehicle, *Fig. 5.38*. Two bolts generally hold the alternator in place. Remove the locking bolt and loosen the pivot bolt to allow the alternator to rotate, releasing the belt tension. Slip the belt off the pulley. Disconnect the wiring harness and the battery cable and remove the remaining bolt. To disassemble the alternator, just remove the screws that hold together the front and rear housings. Carefully separate the individual assemblies for testing. The following tests are for typical models. Use a shop manual for applicable specific testing procedures.

When all the major parts of an alternator are disassembled, they can now be tested individually by making various *bench tests*.

Field Coil Current Draw Test – Internally Grounded Rotor

1. Place the insulating pad on the workbench and put the alternator on the pad. Connect a jumper wire between the housing of the alternator and the negative terminal of a fully charged battery.
2. Connect the positive lead of an ammeter to the field terminal of the alternator. Connect the ammeter's negative lead to the positive battery terminal. Connect a voltmeter across the battery's terminals.
3. Rotate the alternator rotor slowly by hand. The ammeter should indicate a current of 4.5-6.5 A, and the voltmeter should read 14 V.
4. A low current reading indicates a high resistance in the rotor/field coil circuit. Worn brushes that make uneven contact, corroded slip rings, poor connections, or a problem in the rotor coils themselves can cause high resistance.
5. A high current reading indicates a shorted rotor coil or a grounded rotor.
6. No reading on the ammeter indicates an open rotor winding (coil) or effective brushes.

Rotor Tests

1. An ohmmeter is used to bench-test the rotor.
2. Touch the ohmmeter leads to the slip rings (one at a time) and to the rotor shaft to check for shorts to ground, *Fig. 5. 44*. There should be no continuity indicated. If the ohmmeter shows any continuity, replace the rotor.
3. Touch the ohmmeter leads to the slip rings, *Fig. 5.45*. The ohmmeter should read between 1.5-2.1 Ω at room temperature. If the rotor is

still warm from recent operation, the reading will be slightly higher.

4. A reading above 3.5 Ω means that the rotor must be replaced. A reading below 1.5 Ω indicates that the field coil is shorted and the rotor must be replaced.

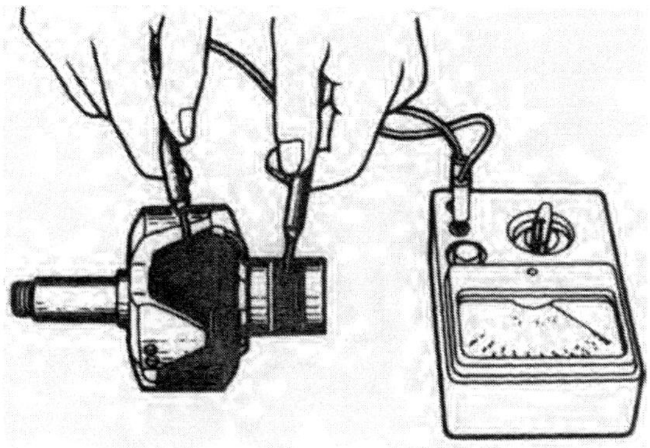

Fig. 5.44 *Rotor coil grounding test.*

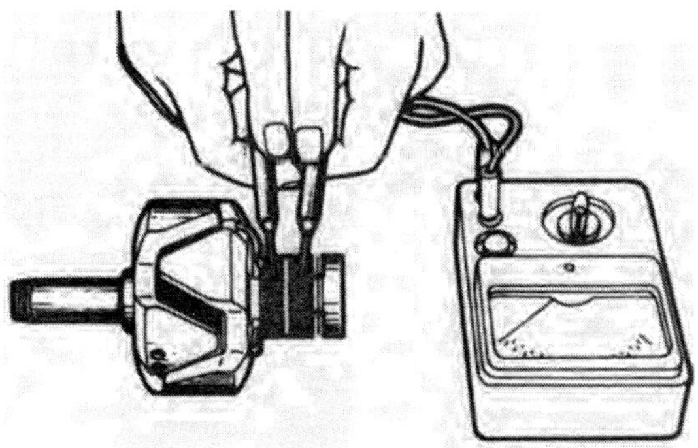

Fig. 5.45 *Rotor coil continuity test.*

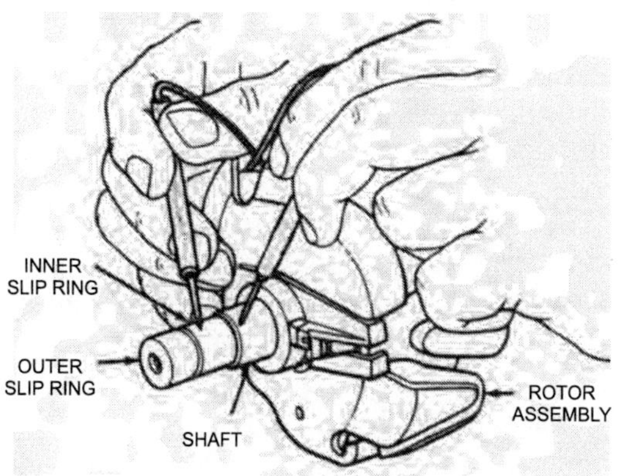

Fig. 5.46 *To perform test for a "short to ground" touch the test tamp probes to the rotor shaft and slip ring. If the lamp lights, the rotor is grounded. (Chrysler Corp.)*

Open Circuit Test – To test the rotor for *'open circuit'* connect the test lamp leads to each slip ring. If the lamp fails to light, the circuit is incomplete or "open".

Short Circuit Test – To test for a *short circuit* connect one test lamp lead to the rotor shaft. Connect the other lead to one slip ring, *Fig. 5.46.* If the lamp lights, there is a "short to the ground" between windings or slip rings and the rotor shaft. If the rotor fails either of the above two tests, replace it.

Stator Test – Y-Connected Models

An example stator test procedure can be as follows:

1. Use the ohmmeter or the test lamp to test the stator windings for continuity and shorts to ground.
2. Remove the varnish from a small area of the stator frame using, sandpaper or sharp knife to use the area for the ground connection.
3. Press one test probe firmly against the cleaned area on the stator frame. Press the other test probe to each of the three winding leads, one at a time, *Fig. 5.47.*

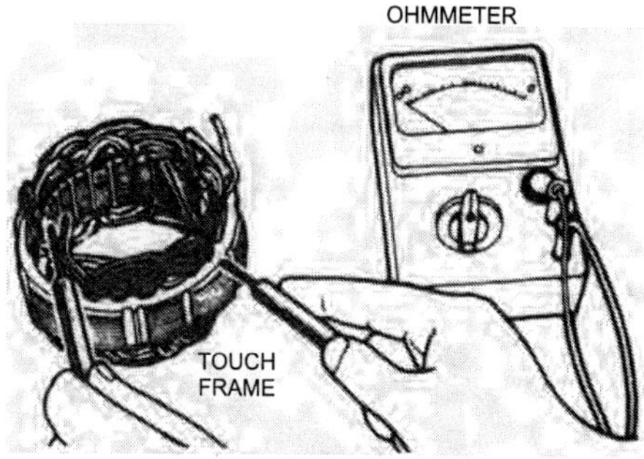

Fig. 5.47 *Stator coil ground test.*

4. If the test lamp lights or the ohmmeter indicates no resistance, the stator winding is grounded and the stator must be replaced.
5. Press one test probe to one winding lead and touch the second probe to each of the other two stator-coil leads, one at a time. The test lamp should light or the ohmmeter should indicate continuity in each case, *Fig. 5.48.* If not, windings are open, and the stator must be replaced.

Grounder Circuit Test – To test the stator for a *grounded circuit,* connect one lead of the test lamp to the stator core, after removing varnish from its surface contact area. With the other test lamp lead prod each of three stator leads. If the test lamp lights up, the stator lead is "grounded".

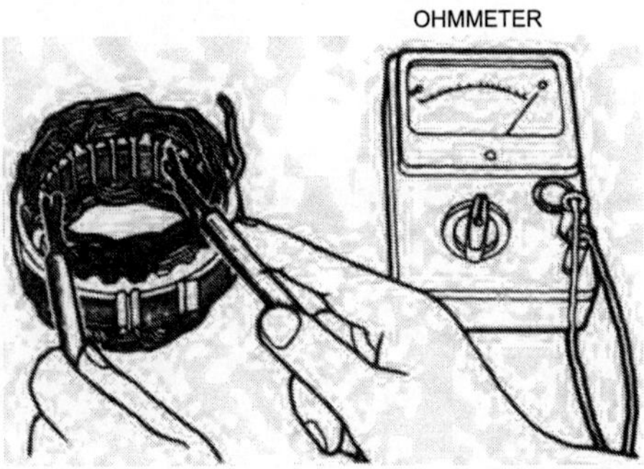

Fig. 5.48 *Stator coil continuity test.*

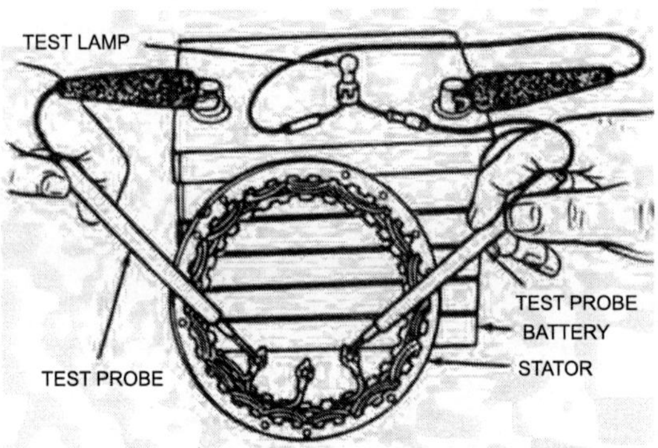

Fig. 5.49 *To test the stator for "open" touch the test lamp probes to two stator leads at a time. If the lamp does not light up, the stator has an open circuit. (Chrysler corp.)*

Open Circuit Test – To test the stator windings for an open circuit, use a test lamp to contact each one of three stator-leads in turn, two at a time, *Fig. 5.49*. If the lamp fails to light up, there is an "open" in the stator coil.

If the stator fails either of the above two tests, replace it.

Diode Tests

A number of testers on the market allow diode testing without removing the stator leads: others require that the leads be disconnected. To make a ***positive diode test,*** clip one test lead to the output (BAT) terminal of the alternator and clip the other test lead to the metal strap or pin of each positive diode in turn, *Fig. 5.50*. Take the meter readings. They should fall in the good band and within close range of each other.

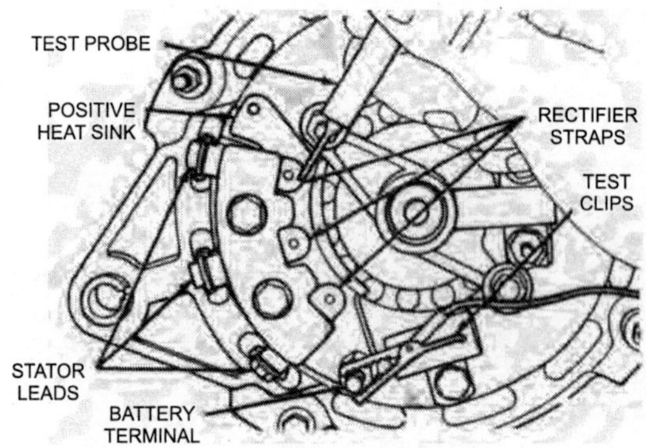

Fig. 5.50. *To test positive rectifier (diode) condition, clip one text meter lead to the alternator battery terminal. Touch the other lead to each rectifier strap in turn. The meter readings should be in the good range and relatively close to each other. (Chrysler Corp.)*

To conduct ***test on negative diodes,*** move the test lead from the output terminal of the alternator to the rectifier end shield. Touch the other test lead to each negative diode in turn. Take the meter readings. The meter readings should be in the *good band* and be close to each other.

To test individual diodes, connect one test lamp clip to the diode base and the other test lamp lead to the diode lead. Note whether or not the lamp lights up. Then reverse the test lamp connections. The lamp should light only once, and in the same direction for each diode. In case the diode lights both times, the diode is *"shorted,"* Whereas, if the lamp does not light up at all, the diode is *"open"*.

To replace the diode follow the instructions in the manufacturer's service manual. An *example diode test procedure* **can be as follows:**

1. Remove the nuts securing the stator leads and the positive and negative rectifier connectors.
2. Use a 12 V battery and a # 67 bulb for testing. Connect one test lamp to the pressure battery post. Connect the other lamp lead to the end of a test probe.

3. Connect a second test probe to the battery negative terminal.
4. Test diode continuity by touching one probe to the heat sink and the other probe to the rectifier pins. *Fig. 5.51*. Now reverse the probes.

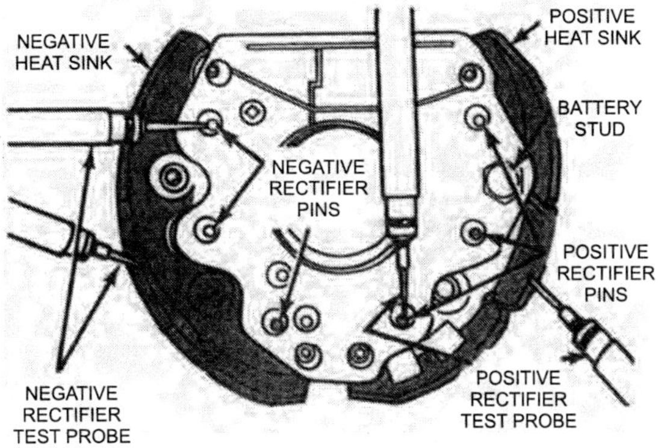

Fig. 5.51 *Rectifier testing points.*

5. If the test lamp lights with the probes in one position but not the other, the diode is good. If the lamp lights with the probes in both positions, the diode is shorted, and must be replaced. Also, if the lamp does not light with the probes in either position, the diode is open and must be replaced.
6. Test each rectifier in both assemblies. The lamp should light with the probe in the same relative position for each diode on the same rectifier assembly.

5.26 ALTERNATOR REASSEMBLY

After completing all the bench tests, and replacing the pinpointed defective parts, and connecting the diodes to the stator leads as required, the parts can be reassembled in the reverse order of disassembly, *Fig. 5.52*.

5.27 ALTERNATOR REINSTALLATION ON ENGINE

To install an alternator, position the assembly on the engine and install the pivot bolt and nut and the outer mounting bracket bolts, *Fig. 5.40*. Install the alternator drive belt and the adjusting bracket bolt. Adjust the alternator drive belt to the specified tension, *Fig. 5.31*, but take care not to pry against the stator section of the alternator. Then tighten all alternator mounting bolts and nuts. Connect the lead wires to the alternator field (FLD) terminals, the output battery terminal, and the ground terminal, *Fig. 5.53*. Reconnect the ground cable to the negative post of the battery. The alternator will be polarized when the ignition key is turned "on". Start and operate the engine. Check alternator operation. Test the current output and compare the test result with the manufacturers specifications.

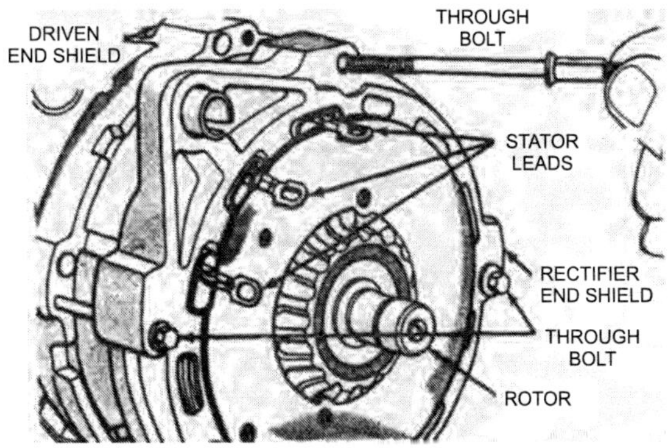

Fig. 5.52 *Reassemble the alternator in the reverse order of disassembly. Manually press the stator and the end shields, and install the through bolts. Ensure that the rotor turns freely and the fan does not hit the stator winding leads. (Chrysler Corp.)*

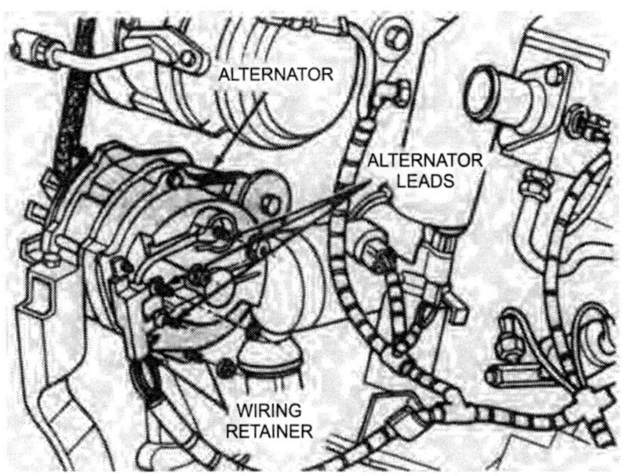

Fig. 5.53 *With the slternator securely mounted and the drive belt adjusted to proper tension, connect the leads to the alternator battery terminal, field terminals, and ground terminal. (Chrysler Corp.)*

5.28 CHARGING SYSTEM DIAGNOSIS

Whenever the charging system is not functioning properly, start troubleshooting using the information provided in the manufacturer's service manual. Most manuals have a system diagnosis chart in them, *Fig. 5.54*. From the symptoms, these charts identify the possible sources of faulty operation and the actions that must be taken to repair the charging system.

5.29 TROUBLE SHOOTING CHARGING SYSTEM

Common charging system problems and their possible causes are as given below.

Charge Indicator Light Flickers

- Loose or worn alternator belt.
- Loose or corroded electrical connections.
- Defective alternator.
- Defective voltage regulator.
- Poor ground at voltage regulator.
- Loose alternator ground wire or strap.
- Corroded or loose battery cable clamps or terminals.

Charge Indicator Light Stays On

- Loose, worn or broken alternator drive belt.
- Open or grounded wiring from battery to alternator.
- Defective alternator component(s).
- Defective voltage regulator.
- Grounded field circuit.
- Corroded or loose battery cable clamps or terminals.
- Malfunction in other electrical systems.

Low or Unsteady Charging

- Excessive charging circuit resistance.
- Corroded or shorted cables.
- Open stator winding.
- High resistance across fuse link.
- Defective alternator diodes.
- Excessive carbon on slip rings.

Chart – ***Charging system diagnosis. (Ford Motors)***

Condition	*Possible Source*	*Corrective Action*
Battery does not stay charged — Engine starts OK	- Battery - Loose or worn alternator belt - Wiring/cables - Alternator - Regulator - Other vehicle electrical systems	- Test battery, replace if needed. - Adjust or replace belt. - Service as needed. - Test/replace components as needed. - Test/replace if necessary. - Check other systems for current draw. Service as necessary.
Alternator noisy	- Loose or worn alternator belt - Bent pulley flanges. - Alternator	- Adjust tension or replace belt. - Replace pulley - Service or repair alternator
Lights and/or Fuses burn out frequently.	-Wiring. - Alternator / Regulator - Battery	- Service as needed. - Test, service, replace if needed. - Test, replace if necessary.
Charge indicator light flickers after engine starts or comes on while vehicle is being driven	- Loose or worn alternator belt. - Alternator - Field circuit ground. - Regulator - Lamp circuit wiring and connector.	- Adjust tension or replace belt. - Service or replace. - Service or replace defective wiring. - Test, replace if necessary. - Service as needed.
Charge indicator light flickers while vehicle is being driven.	- Loose or worn alternator belt. - Loose or wrong wiring connections. - Alternator - Regulator.	- Adjust tension or replace belt. - Service as needed - Sen/ice as needed. - Test, replace if necessary.
Charge indicator gauge shows discharge.	- Loose or worn alternator belt. - Wiring (battery to alternator for ground or open - Field circuit ground. - Alternator - Regulator - Charge indicator gauge wiring and connections - Gauge - Other vehicle electrical systems malfunction.	- Adjust tension or replace belt. - Service or replace wiring. - Service or replace damaged wiring. - Service or replace - Test, replace if necessary. - Service as needed. - Replace gauge. - Service as needed.

Fig. 5.54 *Most manuals have a system diagnosis chart in them*

Over-charging

- Defective voltage regulator.
- Internally grounded alternator field.
- Grounded alternator field wire, field terminal, or connections.

Lights and Fuses Burn Out

- Too high alternator output.
- Defective voltage regulator.
- Defective charging circuit wiring.
- Grounded alternator circuit field wire, field terminal, or electrical connection.
- Internally grounded alternator field.

Alternator Noise

- Loose or worn alternator drive belt.
- Defective drive belt pulley.
- Loose alternator mounting bolts.
- Worn alternator bearings.
- Interference between rotor and stator leads.
- Defective diodes.
- Defective wiring in stator.

THEORY QUESTIONS

1. A —— —— —— prevents engine cranking unless the automatic transmission shift selector is in park or neutral.
2. Discuss common problems occurring in the starting system of an automotive engine. Suggest suitable remedies in each case.
3. A vehicle owner complains of a rapid clicking or buzzing sound and no engine cranking when the key is turned to start. Mechanic A says that the cranking motor must be bad, high current draw is overloading the solenoid. Mechanic B says that a battery load test should be performed because a weak battery or poor cable connections could let the solenoid kick in and out. Which mechanic is right?

   (*a*) A only (*b*) B only
   (*c*) Both A and B (*d*) Neither A nor B.
4. A humming sound, after momentary cranking, is normally caused by:

   (*a*) Faulty solenoid
   (*b*) Bad armature
   (*c*) Worn pinion gear unit
   (*d*) Loose flywheel bolts
   (*e*) All of the above
   (*f*) None of the above.
5. Mechanic A says that the starter pole pieces are located next to the commutator. Technician B says that the starter pole pieces are mounted inside the field windings. Who is right?

   (*a*) A only (*b*) B only
   (*c*) Both A and B (*d*) Neither A nor B.
6. The starter solenoid is used on some automobile engine starting systems to ———.

   (*a*) Push the starter pinion gear into mesh with the flywheel ring gear
   (*b*) Bypass the resistance wire in the ignition circuit.
   (*c*) Both (*a*) and (*b*).
   (*d*) None of the above.
7. In which direction does a current carrying conductor tend to move when placed in a magnetic field?

   (*a*) From a strong magnetic field to a weak magnetic field
   (*b*) From a weak magnetic field to a strong magnetic field
8. The rotating part of a starting motor is known as a(n) —————.
9. To what are the ends of an armature coil connected?
10. In an electric motor, the rotating coils are cutting the magnetic lines of force, thereby generating voltage. Relative to the applied voltage, in which direction does the induced voltage cause the current to flow?

    (*a*) In the same direction as the applied voltage.
    (*b*) In opposite direction to that of the applied voltage.
11. The speed of a series-wound motor will vary with the load. *True* or *False?*

12. When starting an engine or when testing starting motor performance, do not operate the starting motor for more than ———— seconds at a time.
    (*a*) 30 (*b*) 20
    (*c*) 25 (*d*) 15
13. Give three starting motor and cranking system maintenance procedures that should be performed annually or as part of an engine tune-up.
14. What is the purpose of the overrunning clutch in a positive engagement starting motor drive mechanism?
15. Why is a series-mound motor used for cranking an internal combustion engine?
    (*a*) Low torque (*b*) Low speed
    (*c*) High torque (*d*) High speed.
16. When checking internal parts of a starting motor, brushes should be at least half of their full length. *True* or *False?*
17. A vehicle when tested during service, it was found that the starter turned but the starter drive did not engage. Mechanic A said it could be caused by broken teeth in the flywheel ring gear, Mechanic B said it could be caused by a slipping overrunning clutch. Who is right?
    (*a*) A only (*b*) B only
    (*c*) Both A and B. (*d*) Neither A nor B.
18. A vehicle owner complained that the starter did not disengage when the engine started. Mechanic A said a faulty ignition switch could cause it. Mechanic B said it could be caused by loose starting motor mounting bolts. Who is right?
    (*a*) A only (*b*) B only
    (*c*) Both A and B (*d*) Neither A nor B.
19. While testing a stalled vehicle, it was found that the starter would not crank the engine and the headlights dimmed when the ignition switch was turned to "start" position. Mechanic A said it could be caused by a defective starter drive mechanism. Mechanic B said grounded starting motor field could cause it. Who is right?
    (*a*) A only (*b*) B only
    (*c*) Both A and B (*d*) Neither A nor B.
20. The no-load starting motor test will determine how fast the ———————will spin.
21. The no-load starting motor test will also determine the amount of ———————— at a specified ————.
22. Using a 120 -volt test lamp, describe the method of testing for a grounded armature?
23. An open circuit armature may be detected by examining the ——————————— for evidence of burning.
24. Why should the technician avoid soaking alternator parts in clean solvent?
25. What holds the starter parts together?
26. What checks are run after starter assembly?
27. What part of the starter drive mechanism is most likely to fail?
28. What problem exists when a starter turns slowly with a high cranking voltage?
29. What is the general rule regarding starter system cable voltage drop?
30. When checking alternator drive belt tension, how much the alternator drive-belt deflects in the middle of its long span?
    (*a*) No deflection (*b*) 6 to 7 mm
    (*c*) 12.5 mm (*d*) 25.4 mm.
31. If there is no field current, an alternator will not charge because the field windings must be ————by battery voltage.
    (*a*) Regulated (*b*) Charged
    (*c*) Excited (*d*) Controlled
32. How can a voltmeter be used to check charging system operation?
33. Mechanic A says that one of the functions of an automotive charging system is to supply all the vehicle electrical systems electricity when the engine is running, Mechanic B says that one of the functions of an automotive charging system is to provide voltage output slightly higher than battery voltage, Who is right?
    (*a*) A only (*b*) B only
    (*c*) Both A and B (*d*) Neither A nor B.
34. A charging system voltage regulator normally maintains alternator output at approximately —
    (*a*) 11-13 volts (*b*) 16-18 volts
    (*c*) 13-15 volts (*d*) 13-14 volts.
35. If a charging system has fusible link, check it for appearance and ————————.
    (*a*) voltage (*b*) resistance
    (*c*) continuity (*d*) amperage draw.
36. If a fuse link has "blown" replace it with a ——— link.
37. What is the term used to describe a charging system circuit that is "incomplete"?

(*a*) Grounded (*b*) Shorted
(*c*) Closed (*d*) Open.

38. The charge lamp should light up when the ignition key is in the "run" position and the engine is off. *True or False?*

39. What is the term used for bypassing the voltage regulator or computer modules in order to test unregulated alternator voltage output?

40. Which one of the following charging system components is not a bench test?
(*a*) Stator test (*b*) Rotor test
(*c*) Voltage regulator test
(*d*) Diode or rectifier test.

41. A vehicle in for service has a charge indicator light that flickers. Mechanic A says it could be caused by a worn alternator drive-belt. Mechanic B says it could be caused by open or grounded wiring from battery to alternator. Who is right?
(*a*) A only (*b*) B only
(*c*) Both A and B (*d*) Neither A nor B.

42. Tests show that a vehicle's charging system has an excessive charging rate. Mechanic A says that a defective voltage regulator could cause it. Mechanic B says that a grounded alternator field could cause it. Who is right?
(*a*) A only (*b*) B only
(*c*) Both A and B (*d*) Neither A nor B.

❑❑❑

Carbureted-Fuel-System Diagnosis and Service

6

6.1 INTRODUCTION

The engine must have a continuous supply of combustible air-fuel mixture to keep running. In many engines, a carburetor mounted on the intake manifold supplies the mixture. The mixture then flows to the engine cylinders when the intake valves open. The amount of mixture that enters intake manifold is controlled by the position (from fully closed to wide open) of the throttle valves. Their position determines the amount of air-fuel mixture that can flow through into the intake manifold.

An automotive engine relies on clean air and fuel. Unfiltered air can cause rapid part wear. Unfiltered fuel can clog injectors, carburetor passages, and cause numerous system performance problems. Double-wall steel fuel lines and synthetic rubber hoses carry fuel from the fuel tank to the engine. Fuel filters can be located in fuel tank, inline, at the carburetor, or strainers may be inside injectors.

Either a mechanical or electrical fuel pump can force fuel through lines and to the engine. A carburetor is a mixing device for air and fuel. It must alter the mixture ratio with changes in engine operation. The seven carburetor systems are the float, idle, off-idle, acceleration, high speed, full power, and choke systems. An automotive mechanic must be able to visualize the operation of each system to be good at carburetor troubleshooting and repair.

A computer-controlled carburetor uses engine sensors, an electronic control unit (ECU), and an electro-mechanical carburetor. When trouble occurs, always check the carburetor for problems: fuel leaks, vacuum leaks, missing wires, etc.

Most new engines come equipped with electronic fuel injection. An automotive mechanic must know how to do the more common tests and repairs done on fuel injection systems. Electronic fuel injection (EFI) systems use engine sensors, a computer, and electrically controlled injectors. The sensors monitor engine temperature, air-fuel ratio, airflow, throttle position, and other variables. The sensors produce different voltage or current values with changes in operating conditions. The computer uses these signals from the sensors to control engine functions.

A throttle body injection system has the injector(s) inside the throttle body assembly. Multi-port injection has one injector for each cylinder. Multi-port is replacing single-point injection because it is more efficient.

A diesel injection is a high-pressure system that forces fuel into the engine combustion chambers. An injection pump produces this high pressure and times injection with the engine's power strokes. A throttle lever on the injection pump connects to the driver's accelerator pedal. Movement of the pedal rotates the lever to change injection quantity and engine power output. A fuel shutoff solenoid is used to prevent fuel flow into the pumping plunger area of the injection pump. This allows the driver turn the engine off with the ignition key switch.

Some late model systems use a computer to control injection pump operation. Engine sensors report various conditions to the computer. The computer can then send electrical signals to actuators on the pump. The actuators are small electric motors that operate screw mechanisms. In this way, the computer signals are converted into mechanical movement for controlling the injection timing and injection quantity.

A trained and qualified automotive mechanic/technician must know all these fuel supply system variations to be efficient and reliable in automotive service. Here, in this chapter only the carbureted-fuel-system has been described. Fuel Injection Systems have been described in the next chapter.

6.2 IMPORTANT TERMS

Some terms that are frequently used to describe engine performance problems are defined here. These problems can be caused by conditions in the fuel supply system, or by conditions in the other engine systems, or the engine itself.

1. **Detonation/Spark Knock** – It is a secondary explosion that occurs after the spark at the spark plug. It produces too rapid burning of the air-fuel mixture and a pinging or knocking. Usually, it is worse during acceleration and can damage the engine if severe.
2. **Dieseling** – This is the condition when the engine continues to run after the ignition is tuned off. The engine runs unevenly and may knock. Ignition is caused by hot spots in the combustion chambers, e.g., carbon, exhaust valve, and spark plugs. It is supported by air and fuel leaks past the throttle valve and through the idle circuit of carburetion system.
3. **Hesitation / Sag** – It is the momentary lack of response from the engine as the throttle valve is opened. It can occur at all vehicle speeds and is more severe when the engine is first starting out. It may cause the engine to stall.
4. **Miss** – It is the failure of a cylinder to fire. It causes steady pulsation or jerking. Usually, it is more noticeable, when the load increases. It is not normally noticeable at higher engine speeds. Engine exhaust can have a steady spitting sound at low speeds.
5. **Roll** – This is a condition when the engine speed varies under steady throttle or cruises. It feels as if the vehicle speeds up and slows down with no change in throttle-valve position.
6. **Preignition** – it is the condition when the ignition starts before the spark occurs, due to hot spots in the combustion chamber. It can cause a rumbling sound and rough engine operation.
7. **Rough Idle** – It is the condition when the engine runs roughly at idle, sometimes severe enough to cause the vehicle to shake.
8. **Sluggish** – This is the condition under which engine power is limited under load or when climbing steep gradients. The engine slows on acceleration and loses too much speed going up hill.
9. **Spongy** – In this condition the engine response is less than anticipated from it when throttle is opened. There is no speed increase when throttle is opened a small amount. Further increase in throttle opening will eventually get some response.
10. **Stalls** – This is the condition when the engine quits running. This could occur when idling or when the vehicle is moving, e.g., when the throttle is suddenly opened.
11. **Stumbles** – This is the condition when opening the throttle causes the engine power to drop off. It can even cause the engine to stall. It is like hesitation but is more severe. Engine will recover if throttle is closed and then tapped a couple of times.

6.3 GENERAL CARBURETED-FUEL TROUBLE SHOOTING

Fuel-system troubles usually show up as faulty engine operation or poor vehicle ***driveability.*** The vehicle with problems as poor acceleration, hard starting, missing, loss of power, stumble, hesitation, and stalling has poor driveability. The driveability problems that could be caused by the fuel system problems are discussed here. But the conditions discussed here could also be caused by troubles outside the fuel system.

Some modern vehicles have built-in electronic diagnostic systems. When a fault occurs, the system records the fault in its memory. Either a ***check engine*** light comes on, or the electronic system gives the driver verbal warning that something is wrong. Then, when the mechanic checks the system, during service, all that is needed is to activate the diagnostic system. One or more code numbers then appear on the instrument panel, called from the memory. The code numbers tell the mechanic about what the problem is.

1. The initial step in troubleshooting the fuel system is to check complete system to determine whether the problem is too much or too little fuel and to eliminate any obvious problem. Too much gasoline is known as **flooding.** Pumping the accelerator pedal too many times while starting an engine will flood the engine. Open the hood and smell. A flooded engine normally gives a strong odour of gasoline under the hood. Allow a flooded engine a short time to dry out, or crank it with the accelerator pedal all the way to the floor. This allows lot of air to enter the engine cylinders and dry them out.
2. An engine may be flooded because of a problem in the carburetor. This too will give a strong smell of gasoline under the hood.
3. If the cause is a problem with the float system in the carburetor, the outside of the carburetor may be wet with gasoline. Keep in mind that an overflowing carburetor is a serious fire hazard.
4. Too little fuel supply will cause the engine to stop or fail to start. Frequently the driver fails to note that the fuel gauge reads empty. Also, engine can get too little fuel if something is wrong with the fuel system. A few simple checks may identify the problem. If there is enough fuel in the fuel tank, no gasoline smell is found under the hood, the carburetor is dry from outside, and there is no ignition system problem, then the rest of the fuel supply system should be carefully checked for any problem.
5. Remove the air cleaner element. Look down the carburetor and pump the accelerator pedal linkage. Each time the accelerator pedal is depressed, gasoline should squirt into the venturi from the accelerator pump jet. If no gasoline comes out, the carburetor may be dry.
6. Removing and inspecting the spark plugs is a good way to determine if the engine is getting the correct air fuel mixture. The spark plug firing end should be dry and light tan coloured or almost white. If the spark plug is wet with gasoline or black soot, the fuel system is providing too rich a mixture.

These quick checks should enable the mechanic to identify the fuel system component that needs service. But how well the component is operating will still be unknown. During a major tune-up, the fuel system is checked and serviced to bring engine operation to restore its performance efficiency. More complex troubleshooting procedures that require specialized testing equipment are used to correctly recondition the fuel supply system.

6.4 FUEL SUPPLY SYSTEM SERVICE

The fuel supply system is fairly reliable and needs no preventive service beyond periodic cleaning or replacement of filters. Always follow the manufacturer's service manual recommendations. Always keep a fire extinguisher handy when working on a vehicle's fuel system.

Fuel Tank, Fuel Lines and Hoses Service

Specialized repair is necessary to recondition a ruptured or leaking fuel tank. Welding the tank requires extreme caution and special techniques to prevent possible explosion. The most common practice is to replace defective fuel tank.

During engine operation, the fuel pump lowers the pressure on its inlet or suction side. The fuel tank and lines are often described as the suction side of the fuel system. If the right amount of fuel does not reach the pump from the tank, the carburetor will not have sufficient fuel for highspeed operation of the engine.

Common causes of system failures in the suction side of the system are shown in *Fig. 6.1.* The air vent in the fuel tank must be open to allow atmospheric pressure to push on the fuel in the tank. If a plugged vent is suspected, operate the engine without the fuel tank cap or with a different cap.

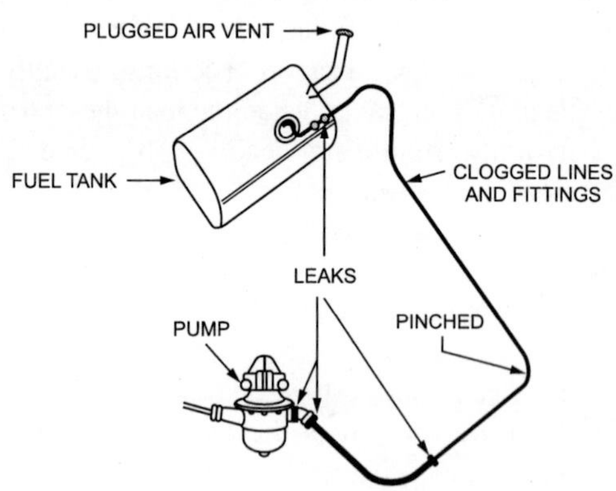

Fig. 6.1 *Common problems in the suction side of the fuel system.*

Faulty fuel lines and hoses are a common source of fuel leaks. Metal fuel lines seldom cause problems. However, clogged, pinched, kinked, rusted, or leaking fuel lines can also stop the flow of fuel in the suction side. Clogged fuel lines should be disconnected and blown clear with compressed air. Pinched lines should be replaced. Leaking lines or fittings should be tightened or replaced.

Fuel hoses can become hard and brittle after being exposed to engine heat. Engine oil can soften and swell fuel hoses. Always inspect hoses closely and any in poor condition should be replaced.

Do not bend or mishandle plastic or nylon fuel lines. If the line is kinked, it will cause a permanent restriction. If a plastic or nylon fuel line is damaged in any manner, do not try to repair, just replace it.

Some new designs of fuel lines have a special snap type fitting, called a ***push-on fitting.*** Do not pry the fitting apart to avoid any damage to it. In some cases, a special tool, available at most auto parts stores or tool suppliers, may be needed to release the fuel line fitting for service.

When servicing push-on fuel line fitting, also replace the O-ring seal to prevent any fuel leakage. Only use the right size, shape and material seal for the fuel line and make/model of the vehicle under service. Installing a wrong seal can result in a fire hazard.

Certain rules that should be observed when working with fuel lines and hoses are:

1. Wrap a rag around the fuel line fitting during removal. This will prevent fuel from spraying on the mechanic or the hot engine.

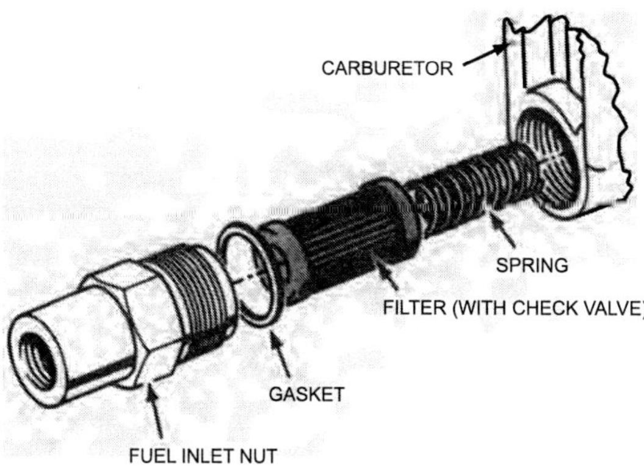

(a) Fuel inlet filter. (Cadillac Div. GM)

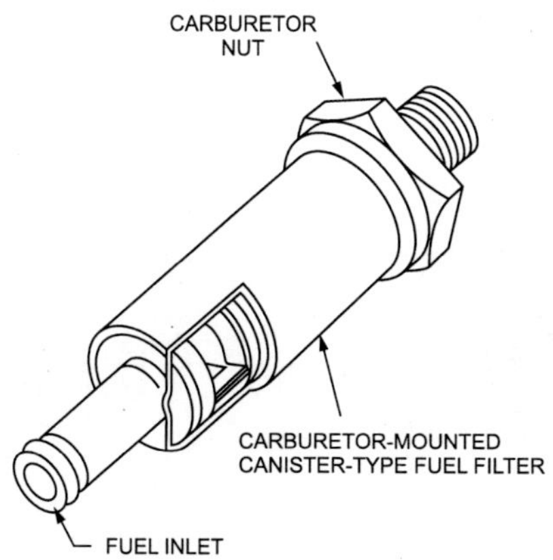

(b) Carburetor mounted canister-type fuel filter. (Chrysler Corporation)

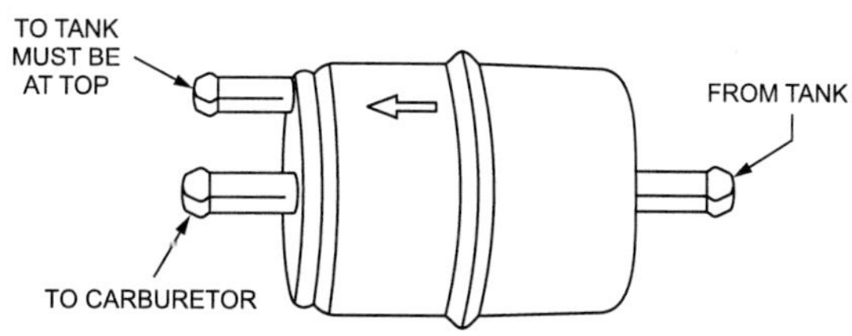

(c) In-line fuel filter is replaced at regular time intervals. (American Motors Corp.)

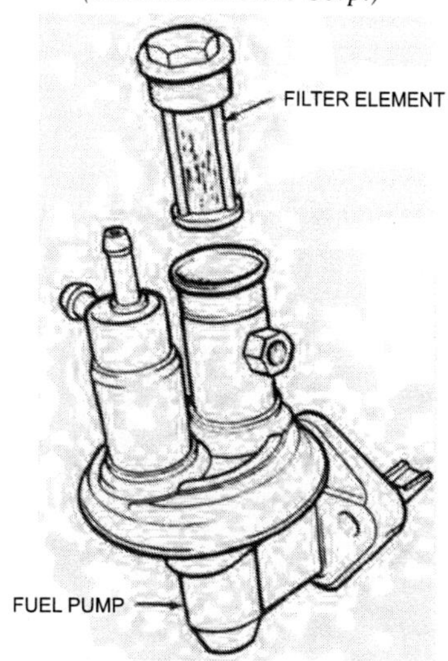

(d) Fuel pump mounted fuel filter.

Fig. 6.2 *Various types of fuel filters. (Chrysler Corp.)*

2. Use a tubing wrench on fuel system fittings.
3. Only use recommended double-wall steel tubing for fuel lines. Do not use copper tubing.
4. When installing a new fuel line, make smooth bends. Use a bending spring or a bending tool.
5. Form only double-lap or ISO flares on the ends of the fuel line.
6. Route all fuel lines and hoses away from hot or moving parts. Check the various clearance distances from adjacent parts after installation.
7. Reinstall fuel line hold-down clamps and brackets. If not adequately supported, the fuel line can vibrate and fail.
8. Only use recommended synthetic rubber hoses in a fuel system. If a wrong type rubber hose is used, the fuel can chemically attack the hose and a dangerous fuel leak could be caused.
9. Make sure a fuel hose slides fully over it's fitting before installing the clamps. Pressure in the fuel system could force the hose off the fitting.
10. Recheck the fittings for any leaks. Start the engine and carefully inspect all connections.

Fuel Filter Service

Fuel filter service involves periodic replacement or cleaning of system filters, *Fig. 6.2.* It also includes locating clogged fuel filters. Clogged fuel filters on either side of the fuel pump can restrict the flow of fuel to the carburetor, gasoline injectors, or diesel injection fuel pump. Engine performance problems, e.g., a loss of engine power or a rough, pulsating feel, especially at high cruising speeds will usually show up.

Most modern vehicles use an in-line throwaway paper filter between the fuel pump and carburetor. Some use a bronze filter element mounted at the carburetor inlet. Check the bronze element by blowing on one end. The element should allow air to pass freely. Check the paper element by blowing on the fuel inlet end. If the filter does not allow air to pass freely replace the element.

Fuel filters can be located:

1. In the fuel line before the carburetor, fuel injectors, or diesel injection pump.
2. Inside the fuel pump.
3. In the fuel line right after the electric fuel pump.
4. Under the fuel line fitting in the carburetor.
5. A fuel strainer is also located in the fuel tank on the end of the fuel pickup tube.
6. Inside the injector body (strainer type).

When in doubt, refer to the applicable service manual.

6.5 FUEL PUMP SERVICE

A faulty fuel pump may cause a poor high-speed engine performance. If pump pressure or volume are too low, lack of fuel to the carburetor will cause high speed miss. If pressure is too high, the carburetor will flood. ***Fuel pump*** problems usually show up as low fuel pressure, inadequate fuel flow, abnormal fuel pump noise, or fuel leakage from the pump. Both mechanical and electrical fuel pumps can fail after prolonged operating life.

Measuring Fuel Pump Pressure

A fuel pump should be checked for: ***fuel pump volume*** and ***fuel pump pressure.***

Inspect fuel lines for kinks and bends and check connections for leaks. The fuel pump volume and pressure may be checked with a fuel pump tester, *Fig. 6.3.* Remove the air cleaner and disconnect the main fuel line at the carburetor, or at the "*T*" or junction if there is more than one carburetor. Attach the tester to the carburetor inlet with the gauge vertical and facing you. Use adapters provided with the tester. Connect the fuel line to the fitting on the fuel pump tester hose. Close the shutoff valve on the fuel discharge hose of the tester.

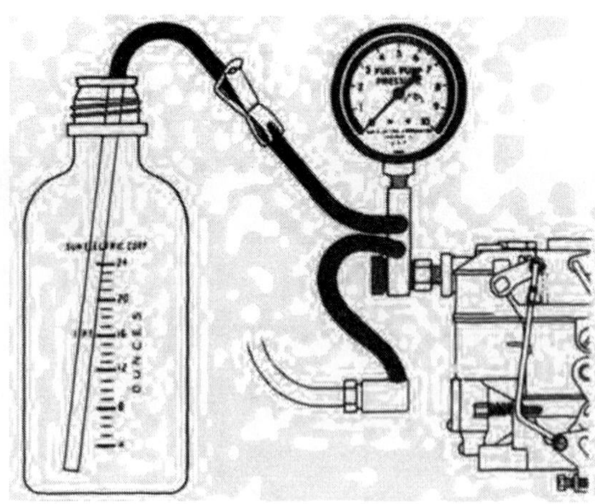

Fig. 6.3 *Fuel pump tester connected between pump fuel-line and carburetor. (Sun Electric Corp.)*

Start the engine and adjust its speed to about 500 rpm, or as specified. Insert the volume test hose into a graduated container and open the shutoff clamp. When fuel reaches the 136 mm^3 level in the container, put the end of the hose in the fuel and look for bubbles. Be careful to prevent fuel spillage. Note the time needed to pump one 0.47 litre of fuel. Then close the shutoff clamp. With the engine still running at test speed, note the gauge pressure reading on the fuel pump tester. Compare volume and pressure test readings with the specifications. Carefully dispose off fuel in the graduated container to prevent fire hazard.

If pressure is too low or too high or if it changes substantially at different speeds, the pump should be replaced. Remove the gauge and reconnect the fuel line to the carburetor.

The carburetor bowl holds enough gasoline to let the engine run for a short time. Start the engine, and let it idle at specified speed with a mechanical fuel pump, and collect the fuel. With an electric fuel pump, you may have to connect some jumper wires to run the pump. Refer to the service manual for detail..

To check fuel pressure, connect a pressure gauge on the outlet side of the fuel pump, *Figs. 6.4 and 6.5*. Compare the pressure readings to specs.

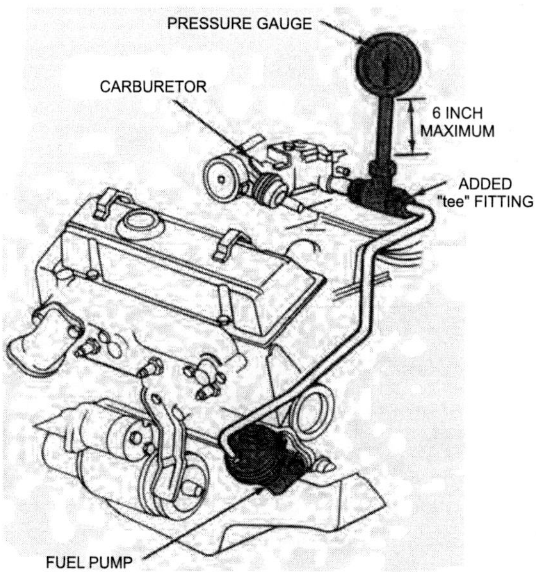

Fig. 6.4 *Checking fuel pump pressure. (Chrysler Corp.)*

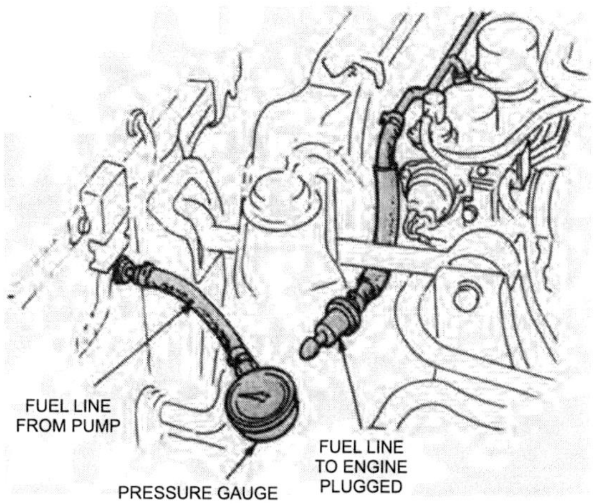

(a) To check pump pressure, disconnect fuel line after pump and install pressure gauge. Start engine or activate electric pump. Read gauge and compare the reading to specs.

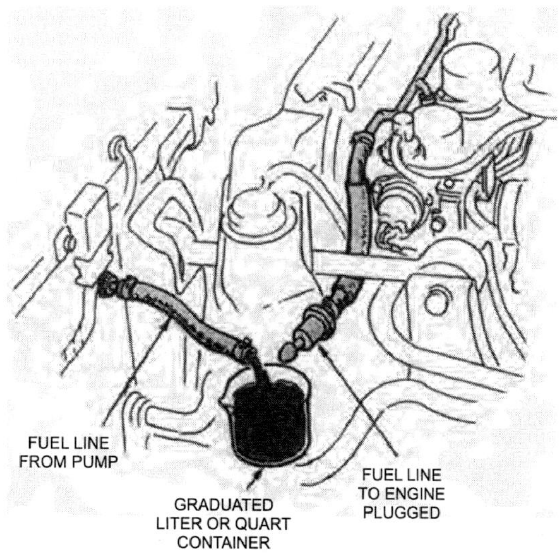

(b) To check pump volume, insert fuel line hose into graduated container. Start engine or activate electric fuel pump and allow fuel to flow into container for specific time period. Compare output to specs.

Fig. 6.5 *Performance testing fuel pump. Always check for clogged in-tank strainer or fuel filter before discarding pump. (Honda Motors)*

Depending upon the type of fuel system, the ***fuel pressure test*** should show a:

1. Gauge reading of 30 to 40 kPa for *carburetor type fuel system.*

2. Gauge reading of 100 to 275 kPa for *gasoline injection* type fuel system.
3. Gauge reading of 40 to 70 kPa for *Diesel engine* fuel supply system.

Always use specified values when determining fuel pump condition. Pressures vary for each make and model vehicle. If fuel pump pressure is ***not*** within specs, check pump volume and the lines and filters before replacing the fuel pump.

Insufficient fuel pressure may not open the carburetor float valve enough to keep the carburetor bowl full. Low fuel pressure indicates a faulty pump.

Low fuel pump pressure can be caused by a weak diaphragm spring, ruptured diaphragm, leaking check valves, or physical wear of relatively moving parts.

Measuring Fuel Pump Volume

Fuel pump volume, also called ***capacity,*** is the amount of fuel the pump can deliver in a specific amount of time.

A properly operating fuel pump provides a certain volume of fuel per minute at a specified pressure. Check the service manual for the engine under service. To check the volume of fuel, remove the fuel line at the engine and place the fuel line into a graduated container of known volume, *Fig. 6.6.*

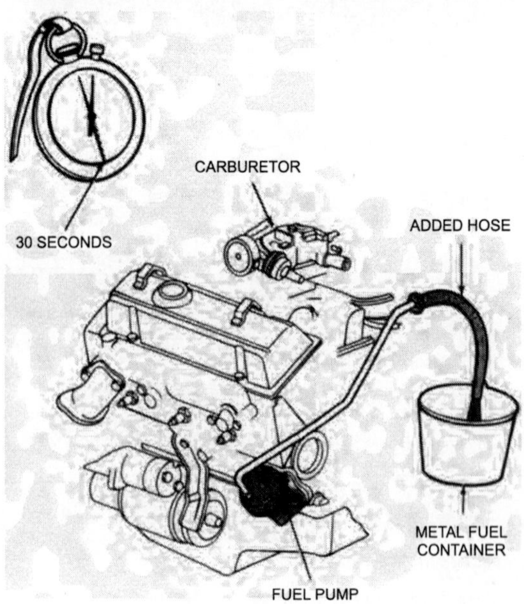

Fig. 6.6 *Measuring volume of fuel from fuel-pump. (Chrysler Corp.)*

Measure the volume of fuel pumped into the container for the prescribed time, usually 30 seconds. Multiply this by 2 to determine the fuel volume delivered per minute. ***Fuel pump volume output*** should be minimum of about 0.47 litres in 30 seconds for carburetor systems. Fuel injection systems typically have a slightly higher volume output from the supply pump. Refer to the factory values for the specific fuel pump and vehicle make and model.

If the measured volume is below specifications, the fuel pump could be faulty, or the fuel lines might be plugged. Replace the faulty pump as a unit. Clear plugged lines with short blasts of high pressure air.

Mechanical fuel pump noise from inside the pump is caused by a weak or broken rocker arm return spring or by wear of the rocker arm pin or arm itself. Mechanical fuel pump noise can easily be confused with valve or tappet clatter. Use a stethoscope to verify mechanical fuel pump noise.

Most electric fuel pumps make some buzz or humming sound when running. Only when the pump noise is abnormally loud should an electric fuel pump be considered faulty.

A ***clogged fuel tank strainer*** can cause excess electric fuel pump noise. Pump speed can increase noise because fuel is not entering the pump normally.

Fuel pump leaks are caused by physical damage to the fuel pump body or deterioration of the diaphragm or gaskets. Most mechanical fuel pumps have a small vent hole in the pump body. When the diaphragm is ruptured the fuel will leak out of this hole.

A ***ruptured mechanical fuel pump diaphragm*** can contaminate the engine oil with gasoline. Fuel can leak through the diaphragm, pump, into the side of the block, and down into the oil pan.

Measuring Fuel Pump Vacuum

Fuel pump vacuum should be checked when a fuel pump fails pressure and volume tests. A vacuum test will eliminate possible problems in the fuel lines, hoses, filters, and pickup screen in the tank.

Typically, fuel vacuum should be about 178 to 254 mm of mercury. A good vacuum reading indicates a good fuel pump. If the pump fails the pressure or volume test but passes the vacuum test, the fuel supply lines and filters may be at fault.

Fuel Pump Replacement

Replacing a fuel pump in a carbureted fuel supply system is a relatively simple task. You should remember to:

1. Relieve the system fuel pressure if needed.
2. Place a shop rag around the pump fittings when loosening them, *Fig. 6.7.*

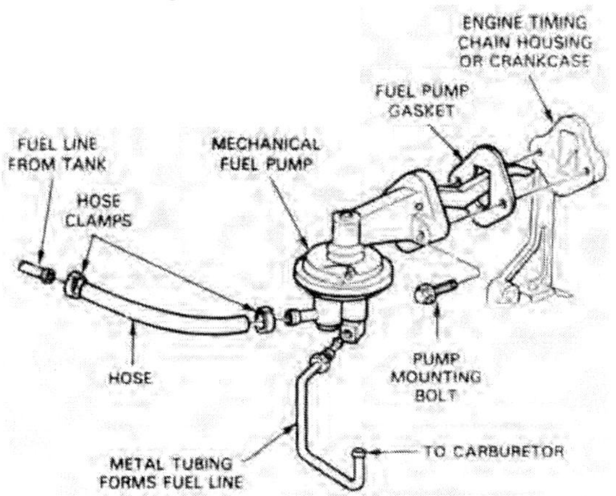

Fig. 6.7 *When replacing mechanical fuel pump, clean gasket surfaces. Position eccentric away from pump arm to ease starting of bolts. Torque bolts and tighten clamps properly. Check for leaks after replacement. (Chrysler Corp.)*

3. Seal off fuel lines when pump is removed to prevent leakage and a potential fire hazard.
4. Clean mating surfaces on engine properly before installing new mechanical fuel pump. Use gasket sealer to hold the mechanical pump gasket in position during assembly.
5. Make sure to install the correct replacement pump.
6. Position mechanical pump eccentric in the released position by cranking the engine. This will allow you to start bolts in their threaded holes in engine block more easily.
7. Check supply voltage to electric fuel pump before discarding and replacing pump.
8. Check fuel filter for clogging before discarding pump.
9. Use service manual as needed for specific instructions on specific pump type.
10. Torque all fasteners to specs.
11. Tighten hose clamps, fittings, and fasteners properly.
12. Check for fuel leaks before releasing the serviced vehicle to customer.

6.6 AIR CLEANER

The air cleaner filters dirt or dust particles out of the air that enter the carburetor, *Fig. 6.8.* Unfiltered air would rapidly cause engine damage. A clogged air cleaner reduces the amount of air going into the engine. The air cleaner handles the most air when the engine is running under load at full throttle high speed. Under these conditions the air restrictions are most evident.

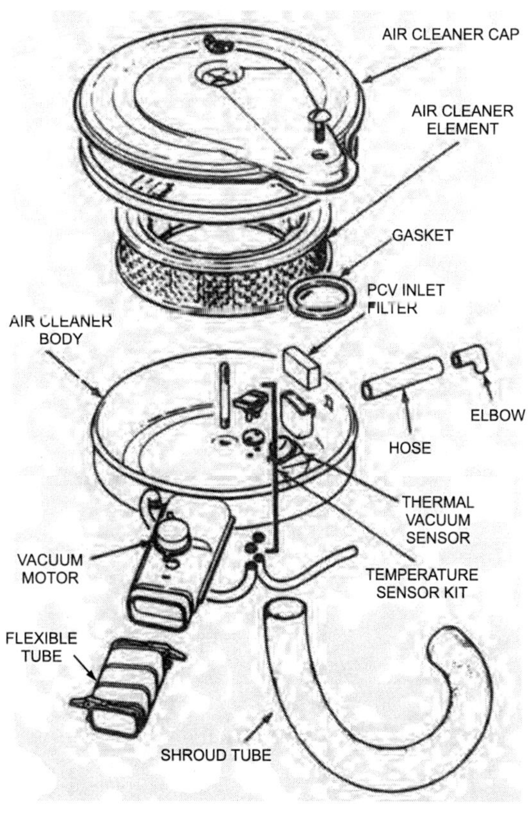

Fig. 6.8 *(a) Disassembled carburetor air cleaner and related parts for an in-line engine. (American Motors Corp.)*

1. Paper Element

Most air cleaners have paper elements. The air cleaner may be checked for clogging after removing the filter element. Automobile manufacturers recommend replacing the paper elements at specific intervals.

Check the service manual for these intervals. Remove the cap or cover from the air cleaner and lift out the paper filter element. Examine it for oil. If the element has oil on it for more than half its circumference, check the crankcase ventilating system. It is carrying oil up to the air cleaner.

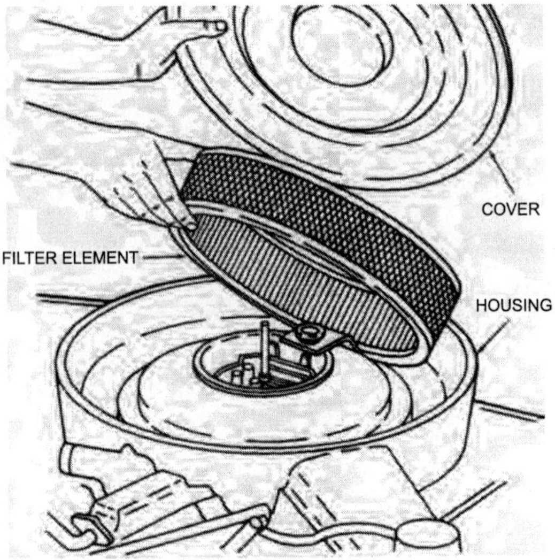

Fig. 6.8 *Air cleaner and cover. (Chrysler Corporation)*

Before replacing the filter element, clean all dust and oil out of the air-cleaner body. Make sure that the plastic rings on both sides of the element are smooth and seal properly at the top and bottom.

2. Polyurethane Element

Some air cleaners have a plastic wraparound, or polyurethane element, outside the paper element, *Fig. 6.9.* When servicing this air cleaner, remove the polyurethane element. Inspect it carefully for rips or other damage. If it is serviceable, wash it in an approved filter cleaner. Then gently fold it and squeeze it to remove the excess cleaner. Clean the cover and other parts of the air-cleaner assembly. Then dip the cleaned element in SAE 10W-30 engine oil. Squeeze out the excess oil. Install the element and its support in the air cleaner. Make sure that the element is not folded or creased. It must seal all the way around. Use a new gasket when installing the air cleaner on the carburetor.

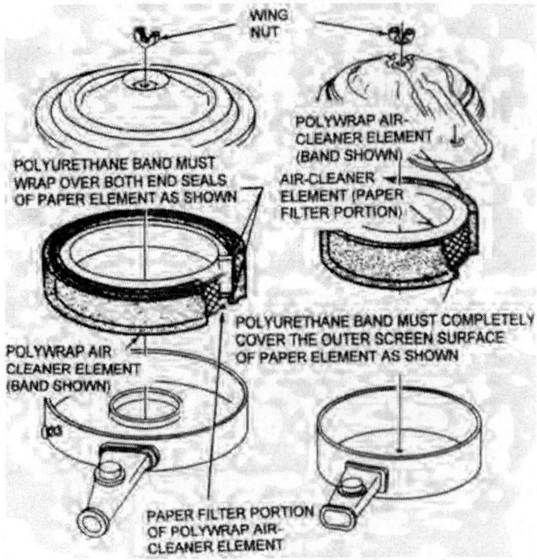

Fig. 6.9 *Exploded view of air cleaners that use a polyurethane band, showing locations of the band. (Chevrolet Div. GM)*

6.7 CARBURETOR SERVICE

A carburetor is vital to efficient automotive engine operation. Here is the information needed to service carburetors.

Do not service or adjust a carburetor right away when you suspect it of giving poor service. Many times the problem lies elsewhere. Before you can do any meaningful work on the carburetor, examine the ignition system, compression, and valve adjustment. Unless these systems are working correctly, there is no sense in adjusting the carburetor.

After making sure that the rest of the engine is working correctly, turn to the fuel system. Make sure there are no leaking fuel lines, clogged fuel lines or filters, or worn pump before going to the carburetor. The chart in *Fig. 6.10* will help diagnose the problem.

Once you have isolated the carburetor as the source, try outside adjustments. Use the shop manual to ensure that the throttle linkage is correctly adjusted. This linkage adjustment is especially important on multi-carburetor installations. Remove and inspect the mixture and idle speed adjusting screws. These screws have sharp, pointed tips. If the tips are deformed they will disrupt that circuit.

Condition	*Possible Cause*	*Remedy*
Engine would not start	Choke not closing. Choke linkage bent.	Check choke operation. Check linkage.
Engine Starts, Then Dies	Choke vacuum kick setting too wide. Fast idle rpm too low. Fast idle cam index incorrect. Vacuum leak. Low fuel pump outlet. Low carburetor fuel level.	Check setting and adjust. Reset rpm to specs. Reset fast idle cam index. Inspect vacuum system for leaks. Repair or replace pump. Check float setting.
Engine Quits Under Load.	Choke vacuum kick setting incorrect. Fast idle cam index incorrect. Incorrect hot fast idle speed rpm.	Reset vacuum kick setting. Reset fast idle cam index. Reset fast idle rpm.
Engine Starts, Runs Up, Then idles Slowly with Black Smoke	Choke vacuum kick set too narrow Fast idle cam index incorrect. Hot fast idle too low.	Reset vacuum kick. Reset fast idle cam index. Reset fast idle rpm.
	HOT STARTING SYMPTOMS	
Engine Would Not Start	Engine flooded.	Allow fuel to evaporate.
	COLD ENGINE DRIVEABILITY SYMPTOMS	
Engine Starts In Gear	Choke vacuum kick setting incorrect. Fast idle rpm incorrect. Fast idle cam index incorrect.	Reset choke vacuum kick. Reset fast idle rpm. Reset fast idle cam index.
Acceleration Sag or Stall	Defective choke control switch. Choke vacuum kick setting incorrect. Float level incorrect–too low. Accelerator pump defective. Secondary throttles not closed.	Replace choke control switch. Reset choke vacuum kick. Adjust float level. Repair or replace pump. Check lockout adjustment.
Sag or Stall After Warm-up	Defective choke control switch. Defective accelerator pump–low output Float level incorrect–too low.	Replace choke control switch. Replace pump. Adjust float level.
Backfires and Black Smoke	Plugged heat crossover system.	Remove restriction.
	WARM ENGINE DRIVEABILITY SYSMPTOMS	
Hesitation With Small Amount of Gas Pedal Movement	Vacuum leak. Accelerator pump weak or inoperable. Float level setting too low. Metering rods sticking or binding. Carburetor idle or transfer system plugged. Frozen or binding heated air inlet.	Inspect vacuum lines. Replace pump. Reset float level. Inspect and/or replace rods. Inspect system and remove restrictions. Inspect heated air door for binding.
Hesitation With Heavy Gas Pedal Movement	Defective accelerator pump. Metering rod carrier sticking or binding. Large vacuum leak Float level setting too low. Defective fuel pump, lines or filter. Air door setting incorrect.	Replace pump. Remove restriction. Inspect vacuum system and repair leak Reset float level. Inspect pump, lines and filter. Adjust air door setting.

Fig. 6.10 *Carburetor troubleshooting.*

Other external checks you should make include tightening the carburetor, mounting nuts, inspecting all vacuum connections, and replacing the fuel inlet filter, *Fig. 6.11*.

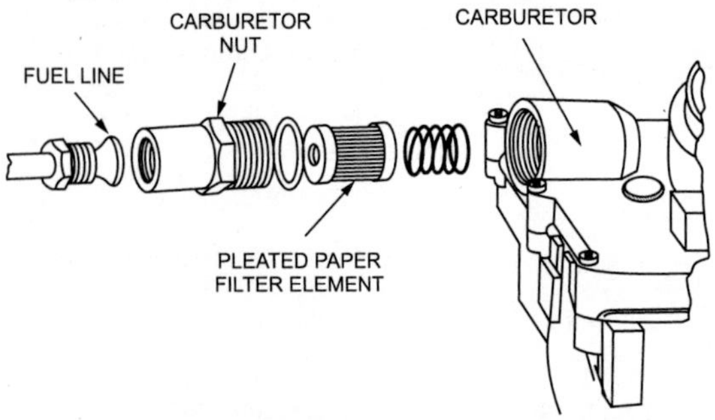

Fig. 6.11 *Fuel inlet filter to carburetor. (Chrysler Corp.)*

How well the carburetor is working may be determined by analyzing the engine's exhaust. Most exhaust analyzers use an infrared light beam to measure an exhaust sample, *Fig. 6.12*.

Fig. 6.12 *Analyzer connected to the vehicle exhaust. (Marquette)*

Exhaust gas is collected at the tailpipe and drawn into a water trap where the collected water is removed. The gas continues through a selector valve and a particle filter, which removes solid particles. The gas then flows into the infrared sampling tubes in the tester. An infrared light beam is directed through the sample tubes and focused on a detector. The amount of energy that makes its way through the exhaust sample to the detector depends on the amount of carbon monoxide gas.

The amount of CO in the exhaust depends on the air-fuel ratio. CO instruments read in % – the higher the reading on the gauge, *Fig. 6.13*, the richer the mixture. Some automotive manufacturers specify carburetion settings in CO, and many states check it for compliance.

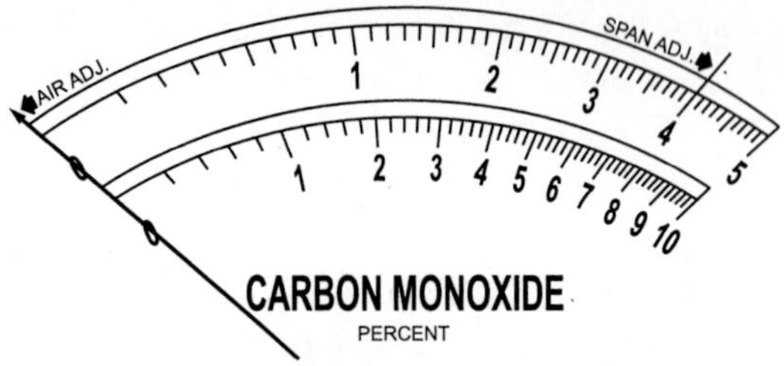

Fig. 6.13 *Carbon monoxide gauge on infrared analyzer. (Marquette)*

A second instrument found on the infrared analyzer measures unburned hydrocarbons (HC) in the exhaust. This reading shows how much gasoline does not get burned in the engine cylinders. The readings on the gauge, *Fig. 6.14*, are in parts per million (PPM). High HC levels are caused by leaking exhaust valves, ignition misfires, or intake manifold leaks that make the mixture too lean to ignite.

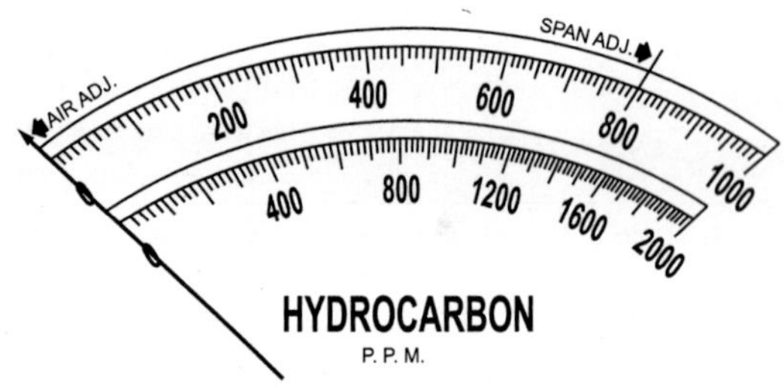

Fig. 6.14 *Hydrocarbon gauge on infrared tester. (Marquette)*

After the tester is connected to the vehicle and calibrated, a series of tests determine whether CO and HC levels are within specifications. Tests are normally conducted at idle speed, cruising rpm, and during acceleration and deceleration. The test is often used with a chassis dynamometer so that actual road conditions can be duplicated.

High CO and HC readings at idle with rough idle and black smoke from the tailpipe indicate one of the following:

1. idle mix too rich;
2. leaking needles and seats;

3. leaky power valve;
4. wrong float setting;
5. faulty choke action.

Low CO readings and high HC readings during cruising operation with surging indicate a carburetion mixture that is too lean.

Normal CO readings and high HC readings at cruising speeds with black smoke indicate one of the following:

1. float not properly set;
2. cruise circuit too rich;
3. incorrect power circuit action.

Low CO readings and normal HC readings at high speed with engine detonation indicate a mixture that is too lean or a power circuit that is not operating.

The manufacturers of infrared testers normally provide a test procedure for the mechanic to follow and a diagnostic chart to help in understanding the results. This tester is useful for testing much more than the carburetion system. Any thing that affects combustion, e.g., ignition and compression, will be shown in the tests. This equipment helps find out whether exhaust emission control equipment is working properly.

Inspecting Carburetors

A visual inspection of the carburetor may provide clues to the carburetor problem. Remove the air cleaner. Look for fuel leakage, sticking choke, binding linkage, missing or disconnected vacuum hoses, or any other troubles. While inspecting the carburetor, also check the rest of the engine compartment. Look for any disconnected wires and hoses. Listen for the ***hissing sound*** of a vacuum leak. Try to locate any event that could upset engine operation.

A heavy covering of road dirt usually indicates that the carburetor has been in service for a long time. Adjustments or repairs may be necessary.

Incorrect Air-Fuel Mixture

Many internal carburetor problems show up as an air-fuel mixture that is too rich or too lean.

A ***lean air-fuel mixture*** is caused by any condition that allows too much air and too little fuel to enter the engine cylinders. A lean mixture will cause the engine to ***miss erratically*** at random. The cause may be a vacuum leak, incorrect mixture screw adjustment, clogged fuel passage, low float level, or other problems.

A ***rich fuel-air mixture*** results from too much fuel entering the engine cylinders. A very rich mixture will make the engine roll, lope, and emit black smoke. A clogged air-bleed, high float level, incorrect choke setting, or other carburetor troubles can produce a rich mixture.

Carburetor Problem Diagnosis

When diagnosing carburetor problems, try to determine which carburetor circuit is at fault. For example, if the engine only runs poorly when cold, check the choke and fast idle circuits.

If the engine only misses at idle, suspect the idle or low speed carburetor circuits and check them first. Use this type of logic to narrow down the possible problem sources.

Use a factory or service manual diagnosis chart when you have difficulty locating a carburetor problem. Such charts are designed for each type of carburetor.

Carburetor Overhaul – Rebuild

A carburetor rebuild is necessary when carburetor passages become clogged, gaskets or seals leak, rubber parts deteriorate, or components wear and fail.

A ***carburetor rebuild/overhaul*** includes:

1. Disassembly and cleaning of major parts.
2. Inspection of parts to determine wear and/or damage.
3. Installation of carburetor ***rebuild kit*** or new gaskets, seals, needle valve and seat, pump diaphragm or cup, and other nonmetal parts in the carburetor assembly.
4. Reassembly of the carburetor.
5. Major adjustments.

When internal carburetor work is to be done, you will need a carburetor overhaul kit. The kit will include all gaskets and small parts needed to disassemble, clean, and reassemble the carburetor.

Carburetor cleaning is done in a cold, non-caustic solvent called carburetor cleaner. Carburetor cleaner is very dangerous if you splash it on you or breathe its fumes. Always use a carburetor cleaner in a well-ventilated place and be careful not to spill it around. Use the wire basket provided to slowly lower the parts into the solution, *Fig. 6.15*.

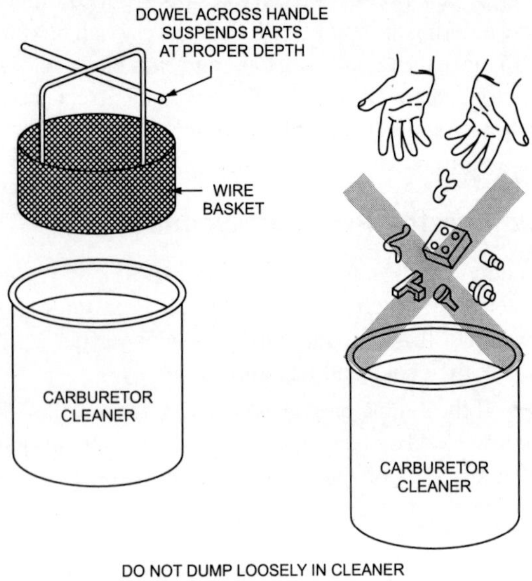

Fig. 6.15 *Carburetor cleaner.*

Because a carburetor does not have many heavy, fast-moving parts, it does not wear out as do internal engine parts. Usually a thorough cleaning and installing a carburetor kit will restore carburetor efficiency.

Carburetor floats are either mounted in the fuel bowl, *Fig. 6.16*, or attached to the air horn assembly, *Fig. 6.17*. The float assembly should also be replaced during a carburetor rebuild. This is an inexpensive insurance, since the floats can absorb gasoline as they age. This makes the float heavier. Consequently, it rides lower in the fuel bowl and allows more fuel into the bowl than specified. This causes an overly rich air-fuel mixture. A new fuel filter and a new air filter completes the carburetor rebuild.

Carburetor Removal

Carburetor removal is necessary during engine repairs or carburetor service. Begin by removing the air cleaner and all wires, hoses, and linkages that would hinder removal of the carburetor. When removing the carburetor, mark all disconnections to facilitate reinstallation. You may want to label the vacuum hoses and wires that connect to the carburetor. Late model vehicles have a large number of hoses and wires, *Fig. 6.18*. Block the manifold entrance with a rag to keep dirt and foreign objects out. If any thing is dropped down in the intake manifold, e.g., a nut, make sure you take it out before starting the engine. Otherwise severe engine damage will be caused. Remove the four nuts or bolts that secure the carburetor to the engine intake manifold, *Fig. 6.19*. Carefully lift the carburetor off the engine while holding it level. Do not splash the fuel in the bowl. This would stir up any dirt in the bottom of the bowl. To prevent any damage to the throttle plates, mount the carburetor on a stand.

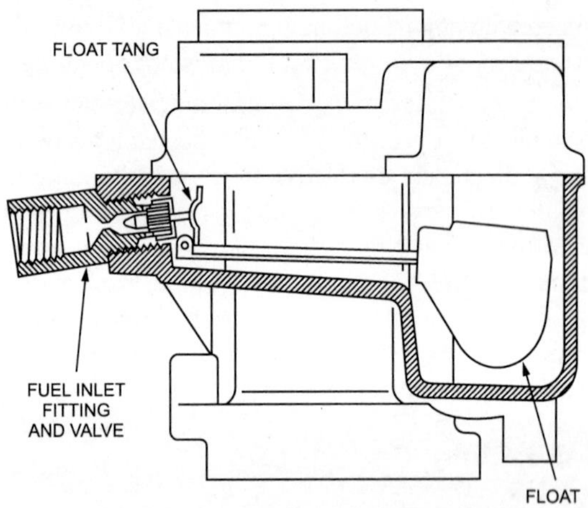

(a) Float is mounted in the fuel bowl.

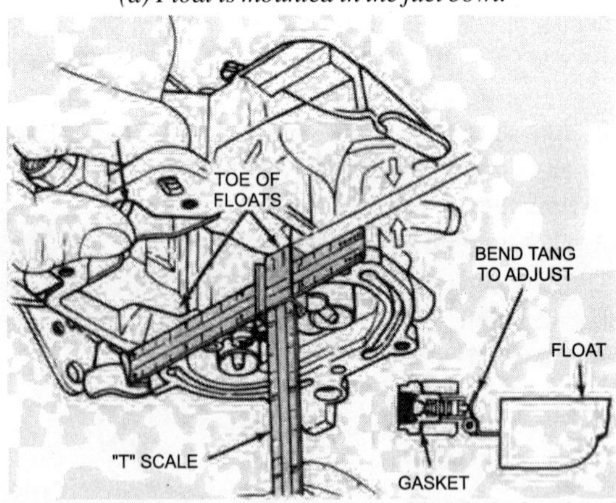

(b) When a float is mounted in the fuel bowl, the bowl and float must be inverted to take the float measurement. Place the carburetor upright and then bend the tang to adjust the float level. (Chrysler Corporation)

Fig. 6.16 *Float mountings (Chrysler Corporation)*

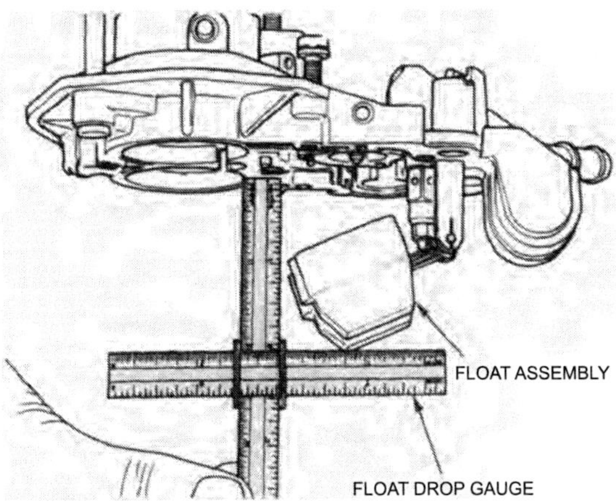

(a) Floats mounted to the air horn must have the float drop measures.

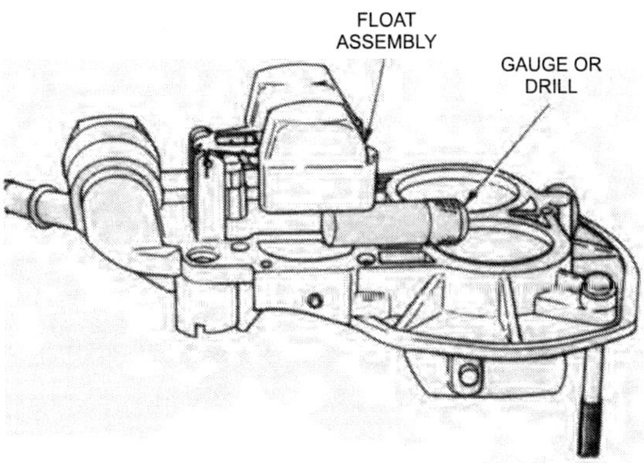

(b) Measuring float level.

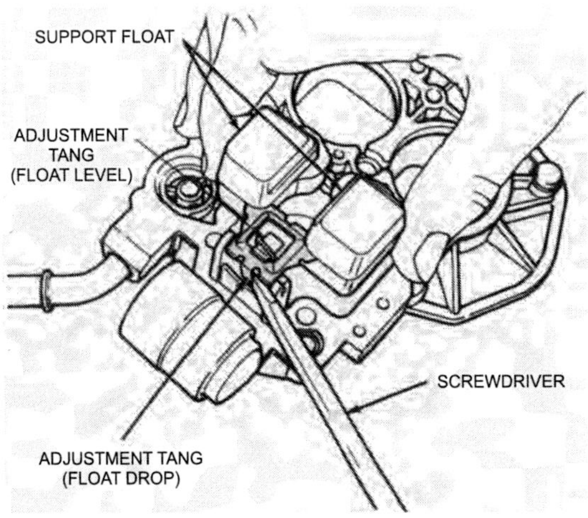

(c) Adjustment points for float drop and float level.

Fig. 6.17 *Float mountings (Chrysler Corporation)*

Carburetor Disassembly

Use the detailed instructions and exploded view drawings of the carburetor provided in the service (shop) manual. Do not attempt carburetor disassembly without the shop manual use. Generally, ***all plastic*** and ***rubber*** parts should be removed. Also remove any part that will prevent thorough cleaning of the carburetor passages. If not worn or damaged, the throttle plates and shaft can be left in place.

To prevent part damage, always use proper disassembly techniques. When unscrewing jets, use a special jet tool or the correct size screwdriver. Disassemble the carburetor over a clean cloth, laying each piece on the rag in the order it is removed. Keep parts organized and note the location of each part. Jet sizes are sometimes different and must be reinstalled in their original locations.

Carburetor Parts cleaning

Clean all metal parts in carburetor cleaner, also known as de-carbonizing cleaner as it removes deposits from a carburetor. A very powerful cleaning agent, it will remove gum, carbon, oil, grease, and other deposits inside air and fuel passages and on external part surfaces. Carburetor cleaner will destroy all rubber, paper, or cork gaskets and diaphragms, and any electrical parts. Make sure these vulnerable parts are removed before cleaning. All gaskets should be replaced, but dipping them in carb cleaner will only gum up the cleaner.

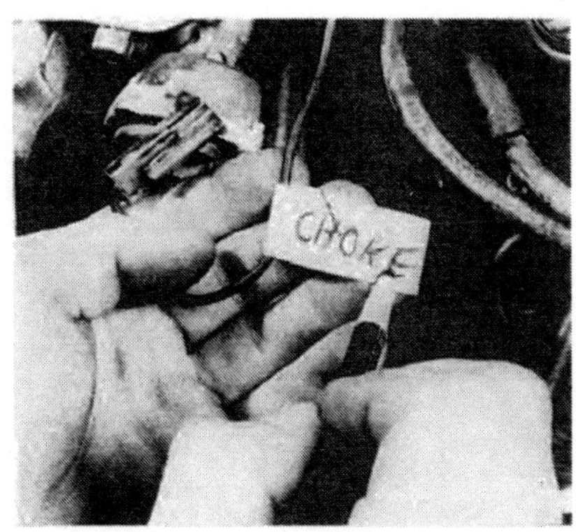

Fig. 6.18 *When removing carburetor, it is important to label all wires and vacuum hoses. This will facilitate reinstalling the carburetor. (Fel-Pro)*

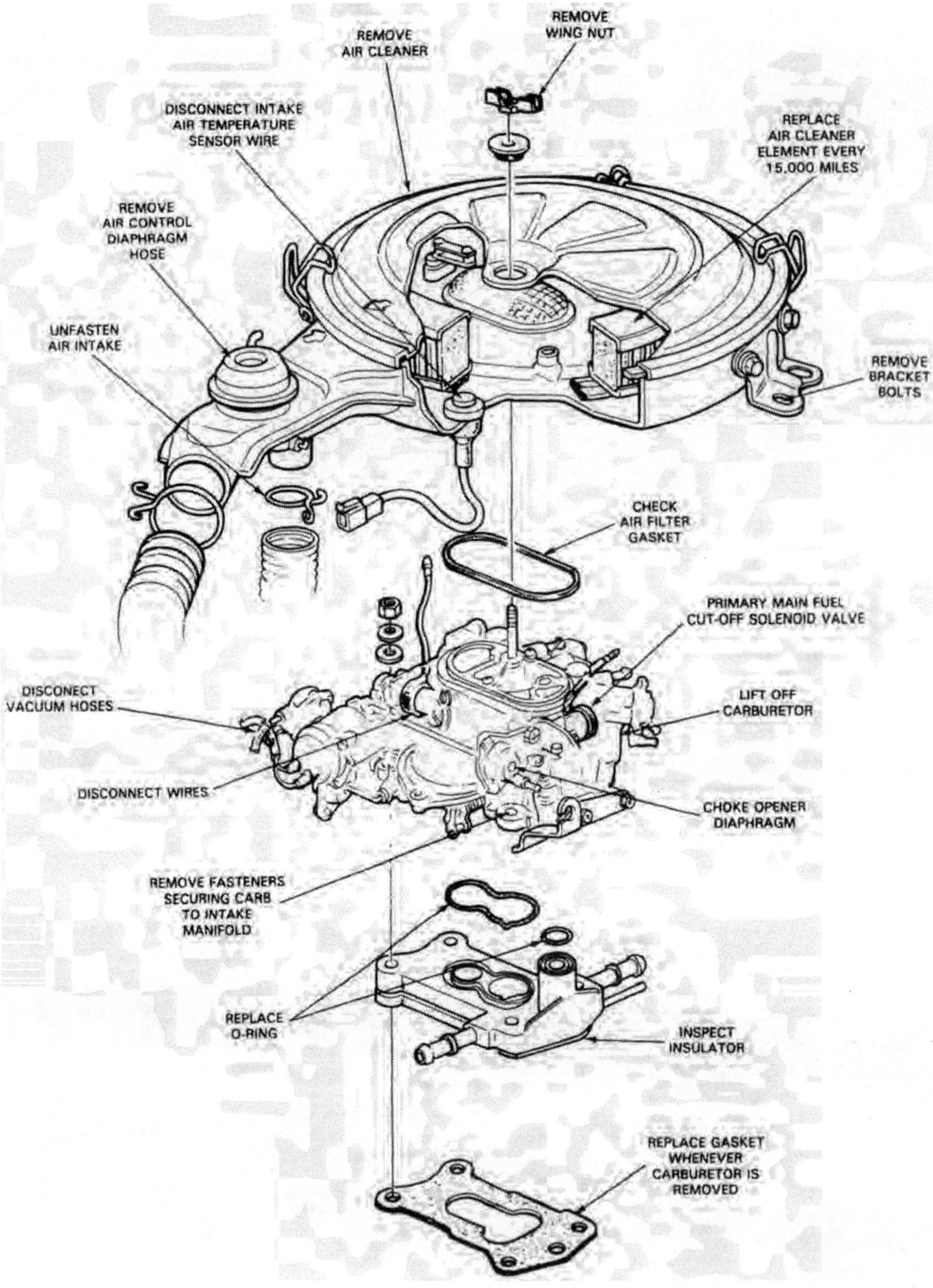

Fig. 6.19 *Carefully note the parts that must be serviced when removing or installing carburetor, (Honda Motors)*

Place metal components in a tray and lower them into the cleaner. Allow the carburetor parts to soak for the specified time, *Fig. 6.20.*

Carburetor cleaner can cause serious chemical burns to the skin and eyes. Wear rubber gloves and eye protection when working with carburetor cleaning solution.

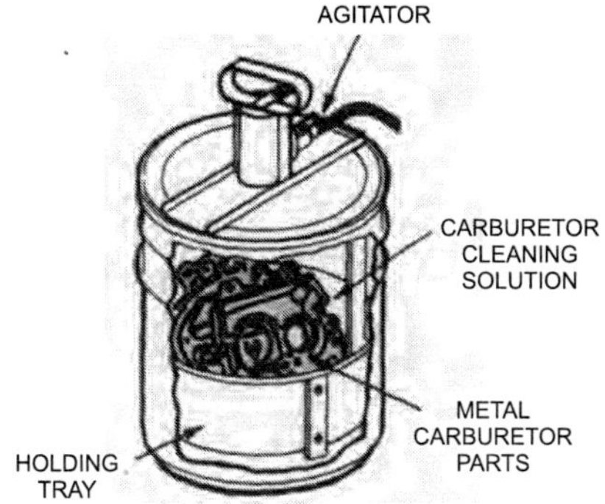

(a) Submerse metal parts in de-carbonizing cleaner, often called carburetor cleaner. It is very powerful so do not let cleaner splash into your eyes or get on your hands. Soak parts for a period recommended by solution manufacturer.

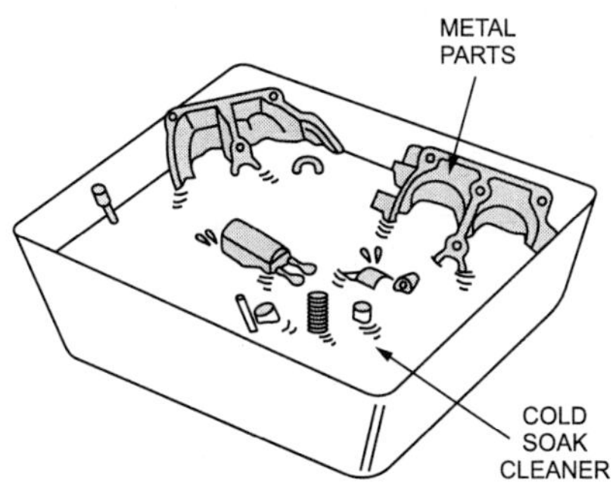

(b) After soaking in carburetor cleaner, wash parts in cold soak cleaner and blow dry.

Fig. 6.20 *After completely disassembled, all metal parts of carburetor should be cleaned. (Toyota Motors)*

Rinse cleaned parts with water, clean kerosene, or clean cold soak cleaning solution and then blow-dry with compressed air. Use the compressed air to blast out all the small passages in the carburetor body. Use eye protection when blasting with compressed air. Do not use wire or a drill to clean out carburetor passages. This could scratch and enlarge the passages, upsetting carburetor operation.

Inspecting Carburetor Parts

1. Check for wear and excessive play between the throttle shaft and throttle body.
2. Check for binding of the choke plate and linkage.
3. Inspect carburetor bodies for warpage, cracks, and other damages.
4. Check float for leakage or bent hinge arm.
5. Check all gasket-mating surfaces of the carburetor for nicks, burrs, and cleanliness.
6. Inspect for damaged or weakened springs and stripped fasteners.
7. Check pointed ends of idle mixture screws for damage.

Replace parts that show wear or damage. Also, discard all of the old parts included in the new carburetor rebuild kit.

Carburetor Reassembly

Follow the directions in the applicable service manual to reassemble a modern carburetor. There can be a number of different carburetors for just one model year. This results in thousands of different reassembly procedures and adjustments. Each is critical to proper engine performance.

Basically, the carburetor is assembled in reverse order of its disassembly. The manual will give procedures for installing each part. An exploded view of the carburetor shows how parts fit together. One such carburetor exploded view is shown in *Fig. 6.21.*

Reassemble the carburetor, using the new gaskets from the new build kit. Check the fit of each gasket before installing. Besides the fit, check the new gasket for the correct number of holes and their position for proper alignment.

Carburetor Preinstallation Adjustments

As you assemble the carburetor, several adjustments should be made. These adjustments include:

1. Float drop adjustment, *Fig. 6.22.*
2. Float level adjustment, *Fig. 6.22.*
3. Idle mixture screw "rough" adjustment, *Fig. 6.23.*
4. Choke linkage and spring adjustments, *Fig. 6.24.*

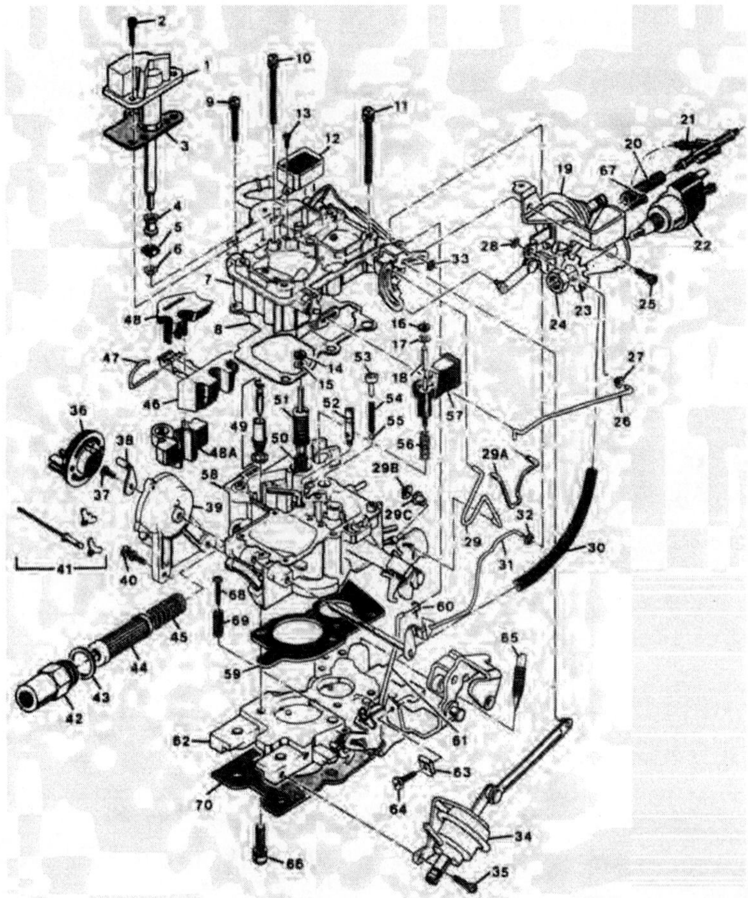

Air Horn Parts

1. Mixture Control (M/C) Solenoid
2. Screw Assembly-Solenoid Attaching
3. Gasket-M/C Solenoid To Air Horn
4. Spacer-M/C Solenoid
5. Seal-M/C Solenoid To Float Bowl
6. Retainer M/C Solenoid Seal
7. Air Horn Assembly
8. Gasket-air Horn To Float Bowl
9. Screw-air Horn To Float Bowl (Short)
10. Screw-air Horn To Float Bowl (Long)
11. Screw-Air Horn To Float Bowl (Large)
12. Vent Stack And Screen Assembly
13. Screw-vent Stack Attaching
14. Seal-pump Stem
15. Retainer-pump Stem Seal
16. Seal-T.P.S. Plunger
17. Retainer-T.P.S. Plunger Seal
18. Plunger-T.P.S. Actuator

Choke Parts

19. Vacuum Break And Bracket Assembly- Primary
20. Hose-vacuum Break Primary
21. Tee-vacuum Break
22. Solenoid-idle Speed
23. Retainer-Idle Speed Solenoid
24. Nut-idle Speed Solenoid Attaching
25. Screw-vacuum Break Bracket Attaching
26. Link-air Valve
27. Bushing-Air Valve Link
28. Retainer-Air Valve Link
29. Link-fast Idle Cam

29a. Link-fast Idle Cam

29b. Retainer-link

29c. Bushing-link

30. Hose-vacuum Break
31. Intermediate Choke Shaft/Lever/Link Assembly
32. Bushing-intermediate Choke Link
33. Retainer Intermediate Choke Link
34. Vacuum Break And Link Assembly-Secondary
35. Screw Vacuum Break Attaching
36. Electric Choke-cover And Coil Assembly
37. Screw-choke Lever Attaching
38. Choke Coil Lever Assembly
39. Choke Housing
40. Screw-choke Housing Attaching
41. Choke Cover Retainer Kit

Float Bowl Parts

42. Nut Fuel Inlet
43. Gasket-fuel Inlet Nut
44. Filter-fuel Inlet
45. Spring-fuel Filter
46. Float And Lever Assembly
48. Upper Insert-Float Bowl

48a. Lower Insert-Float Bowl

49. Needle And Seat Assembly
50. Spring-pump Return
51. Pump Plunger Assembly
52. Primary Metering Jet Assembly
53. Retainer-pump Discharge Ball
54. Spring-pump Discharge
55. Ball-pump Discharge
56. Spring-T.P.S. Adjusting
57. Sensor-throttle Position (TPS)
58. Float Bowl Assembly
59. Gasket-float Bowl

Throttle Body Parts

60. Retainer-pump Link
61. Link-pump
62. Throttle Body Assembly
63. Clip-cam Screw
64. Screw-fast Idle Cam
65. Idle Needle And Spring Assembly
66. Screw-throttle Body To Float Bowl
67. Screw-vacuum Break Bracket Attaching
68. Screw-idle Stop
69. Spring-idle Stop Screw
70. Gasket-insulator Flange

Fig. 6.21 *Service Manual Will Give Exploded View Of A Carburetor As Shown Here. It Will Help Ensure That every Part Is Installed Properly. (Buick Div. GM)*

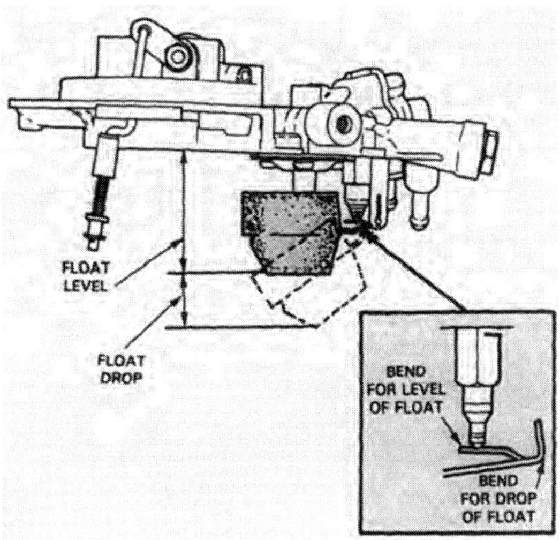

Fig. 6.22 *During assembly, adjust float drop by measuring from float to specific location of carburetor. Bend float hinge tang if needed. Float level should also be adjusted. Float should be in up position when measuring. Bend correct tang on hinge as needed. (Renault Motors)*

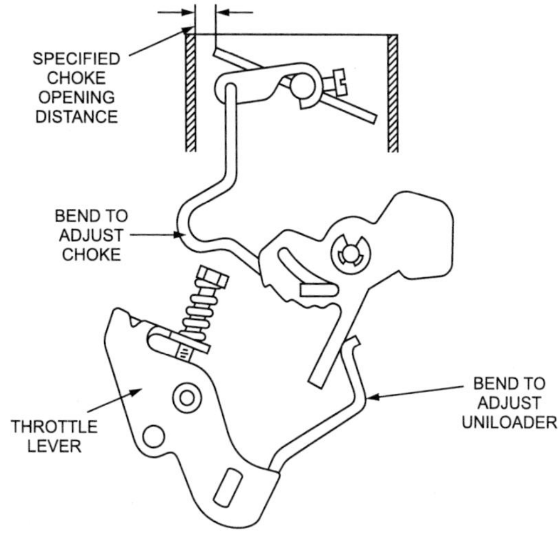

Fig. 6.24 *Mechanical choke unloaded cracks choke plate open when throttle valve opens. This prevents engine flooding with fuel. It should be checked during carburetor service. (Ford Motors)*

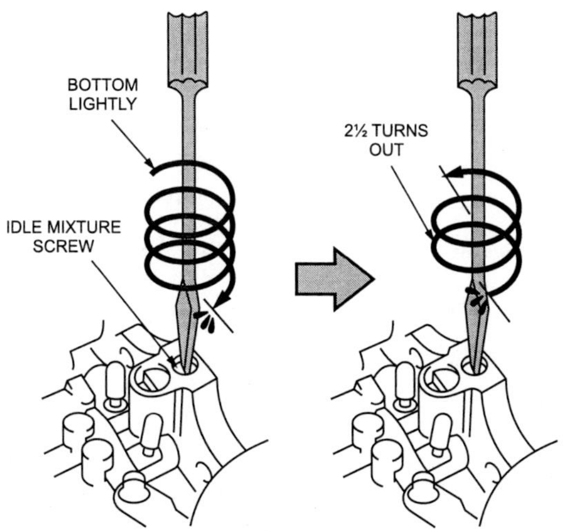

Fig. 6.23 *Rough idle mixture adjustment is done before installing carburetor. Bottom screws and then back them out about 2 ½ turns. Readjust mixture after installation. (Toyota Motors)*

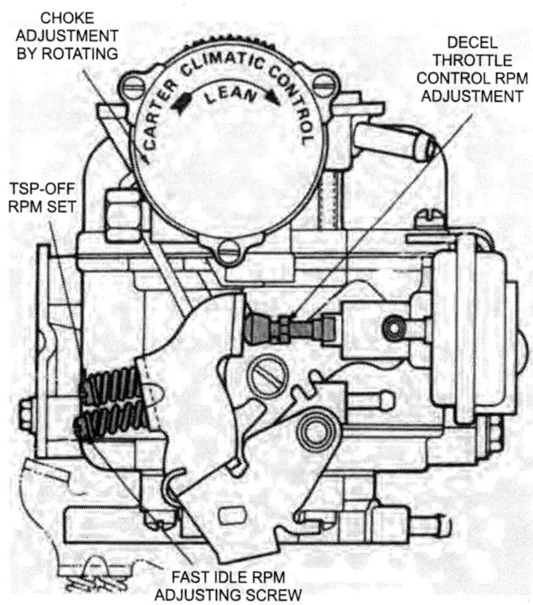

Fig. 6.25 *Note speed adjustment screws and choke adjustment. Locations will vary with different carburetors. (Ford Motors)*

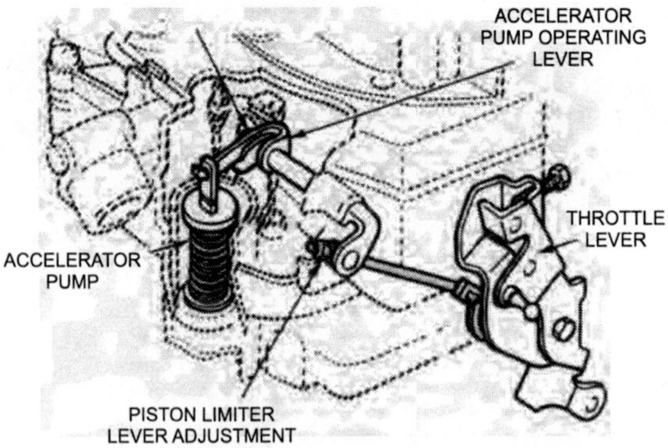

Fig. 6.26 *Accelerator pump must be adjusted to spray fuel into air horn as soon as throttle begins to open. Adjust by threaded nut, bending linkage, moving linkage to different hole, etc. (Ford Motors)*

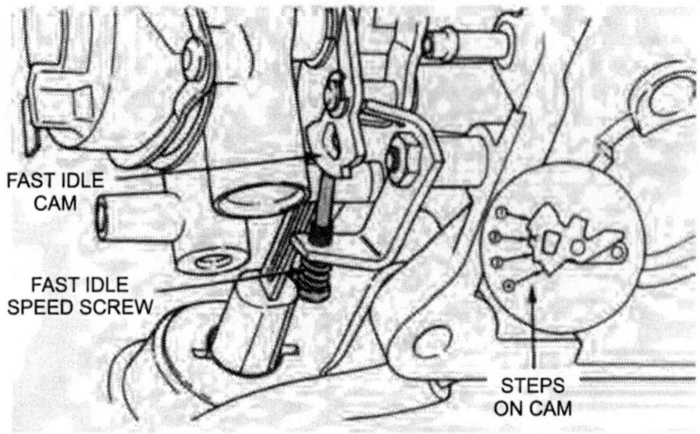

Fig. 6.27 *Fast idle adjustment is sometimes done with cam mechanism. Position fast idle screw on correct cam step. Then use tachometer to measure engine rpm while turning screw. (Ford Motors)*

5. Choke spring adjustment, *Fig. 6.25*.
6. Accelerator pump adjustment, *Fig. 6.26*.

There are several other adjustments that must be made ***after*** installing the carburetor on the engine.

6.8 CARBURETOR INSTALLATION

1. Place a new base plate gasket on the intake manifold in proper alignment of all features.
2. Fit the carburetor over the mounting studs if so equipped.
3. Install and carefully torque the fasteners that secure the carburetor to the intake manifold, to specs following the correct procedure, *Fig. 6.19*.

   Do not overtighten the carburetor hold-down nuts or bolts. Overtightening can snap off the carburetor base flange.
4. Reinstall all of the hoses, lines, wires, and linkage rods or cables.
5. Recheck that all of the vacuum hoses have been installed correctly.
6. Hand operate the throttle to make sure the throttle plates are not binding or hitting on the base plate gasket.

Install the carburetor and then crank the engine to fill the float bowl. Check for leaks and then start the engine. If the carburetor leaks, shut the engine off immediately, locate the leak, and correct it.

6.9 CARBURETOR ADJUSTMENTS

Before making carburetor adjustments, make sure that the ignition system is in good condition and that compression is within 10% of specs for all cylinders. There must be no leaks in the intake manifold. The carburetor float level must be within specs. The engine must be at operating temperature so that the idle speed can be set.

Cold Idle Speed Adjustment

1. To set the cold fast idle or fast idle cam, connect a tachometer to the engine, according to the manufacturer's recommendations.
2. Warm the engine to operating temperature. Set the emergency brake and block the wheels of the vehicle.
3. Following the service manual directions, set the fast idle cam to hold the throttle open for fast idle. The fast idle screw must contact a specified step on the fast idle cam, *Fig. 6.27*.
4. Turn the fast idle screw until the tachometer reads speed within specs. Cold fast idle speeds vary from approximately 750 to 950 rpm. Some specs are given with the automatic transmission in drive. This will reduce the cold fast idle speed value.

Hot Idle Speed Adjustment

- Many carburetors use a solenoid to control the hot idle speed at the engine. When the engine is running, the solenoid acts on the throttle lever to hold the throttle open for the correct idle speed.
- Sometimes, a solenoid is used to maintain idle speed when the air conditioner is turned on. In any case, follow the manufacturer's recommendations, as laid down in the service manual, to adjust the solenoid.
- To set hot idle speed, turn the solenoid adjusting screw until the tachometer reads the correct rpm, Fig 6.28. Sometimes the adjustment is on the solenoid mounting bracket or it may be on the solenoid plunger itself.
- Typically, hot idle speed is from 650 to 850 rpm. It is lower than cold idle speed but higher than curb idle speed.

The choke spring coil forces the choke valve closed when it is cold, after depressing the accelerator pedal. At the same time, the fast idle cam is positioned by linkage connected to the choke valve shaft, *Fig. 6.29*. Therefore, the fast idle screw is adjusted after it touches the fast idle cam with the engine running. When the choke valve is wide open, the fast idle screw will not touch the fast idle cam. The engine will then be at a "curb" idle speed.

On some computerized engines, the computer controls the idle and fast idle speed. No attempt should be made to adjust the idle or fast idle speed on those engines. Always consult the applicable service manual.

Curb Idle Speed Adjustment

Curb idle speed is the lowest idle speed setting, approximately 500 to 750 rpm. On old carburetors, it is the idle speed setting that controls engine speed under normal, warm engine conditions, *Fig. 6.19*.

On late model carburetors using a fast idle solenoid, curb idle speed can also be termed ***idle drop*** or ***low idle speed adjustment.*** In this case, it is very low idle speed that only occurs when the engine is shut off. When the solenoid is de-energized, the throttle plates drop to the curb idle setting to keep the engine from dieseling – engine running with ignition key off.

Fig. 6.28 *Note how solenoid is moved into or away from throttle lever to change hot idle speed. Curb idle screw is set with solenoid de-energized. (Ford Motors)*

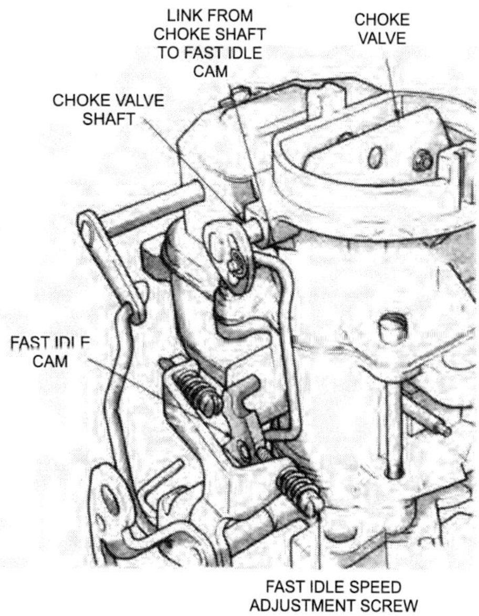

Fig. 6.29 *The fuel mixture screws are located behind concealment plugs. (Chrysler Corporation)*

To adjust curb idle speed, unplug the wires going to the idle speed solenoid. Then, turn the curb idle speed screw until the tachometer reads speed within specs.

Idle Mixture Check

An infrared exhaust gas analyzer is used to check the ***idle mixture.*** Always consult the correct service manual for specification on the carburetor under service. The following procedure is just an example:

1. Set parking brake and place transmission or transaxle in neutral. Turn off all lights and accessories. Connect a tachometer as recommended. Start the engine until it reaches the normal operating temperature.

2. On feedback-equipped vehicles, turn engine off and then disconnect negative battery cable for 10 seconds before reconnecting the cable. Disconnect oxygen sensor electrical connection. Restart the engine and run at 2500 rpm for 10 seconds before returning to curb idle speed.
3. Disconnect the electrical connection at the radiator fan, if so equipped. Allow engine to idle for two minutes.
4. Insert probe from exhaust gas analyzer into exhaust tailpipe.
5. Adjust the idle speed, if possible, to specified idle rpm.
6. Check reading of CO at the exhaust gas analyzer. The reading should be between 0.1 to 0.3%. If it is not the idle mixture may need adjustment. It is possible that a high float level, dirty air filter, etc., may also cause a high percentage of CO in the exhaust.
7. If the reading is within specs, turn the engine off and reconnect the fan motor and the oxygen sensor wire. Finally disconnect tachometer.

Idle Mixture Adjustment

If the CO reading is not within specifications, the air-fuel mixture may have to be adjusted. Prior to adjustment, the concealment plugs must be removed to gain access to the mixture screws, *Figs. 6.30 (a) and 6.30 (b)*. This requires removing the carburetor from the intake manifold. After the plugs have been removed, the carburetor must be reinstalled on the engine. With the engine running and test equipment hooked up, the fuel mixture screws can be adjusted.

To lean the mixture, turn the mixture screws clockwise (drive forward). To enrich the fuel mixture, turn the mixture screws counterclockwise (unscrew backward). The mixture screws should be turned only 1/16 of a turn at a time, to permit precise adjustment.

On two- and four-barrel carburetors, both fuel mixture screws must be turned the same amount and in the same direction. Alternate between fuel screws after each adjustment. For example, if you turn one fuel mixture screw 1/16 of a turn, turn the other screw 1/16 of a turn in the same direction. This will make the engine run smoothly at idle.

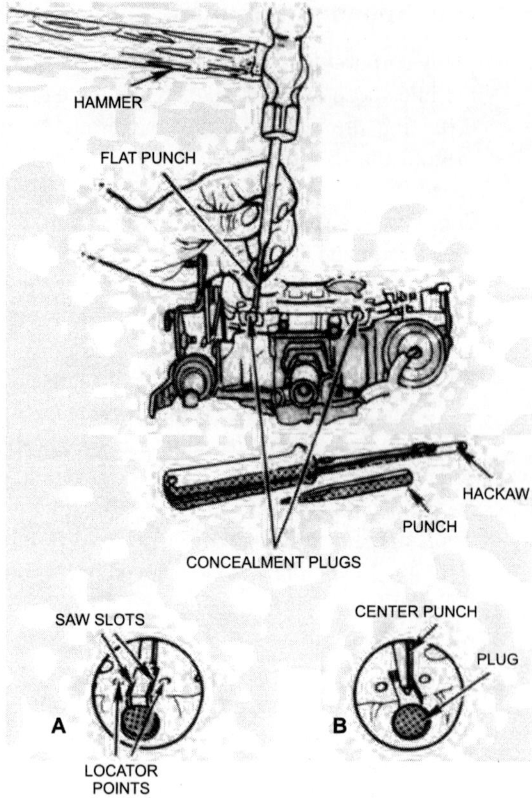

Fig. 6.30 *Gaining access to the fuel mixture screws. (Rochester)*

Idle mixture screws are sealed because the Federal laws prohibit the messing with these screws unless an infrared test equipment is used while adjusting the fuel mixture. To otherwise temper with these is a punishable offence.

6.10 COMPUTER CONTROLLED CARBURETOR SERVICE

A computer controlled carburetor system requires specialized service techniques. Most systems have a self-diagnostic mode that allows easy location of system problems. Some computer systems flash light on and off in a Morse (number) type code. The mechanic can use the code chart in the service manual to pinpoint the system trouble.

Computer Controlled Carburetor Problems

Computer controlled carburetors can suffer from the same problems as conventional carburetors. Gaskets can rupture and leak. Seals can harden and split. Passages can become clogged. Vacuum hoses can become defective or loosen and come off.

Before suspecting the computer system or any of its components, check all other possible causes of the problem.

1. Check the ignition system components – ignition timing, spark plug wires, spark plugs, pick-up coil, etc.
2. Check the operation of the air cleaner door.
3. Check all emission controls – EGR system, evaporative system, PCV system, EFE system etc.
4. Check the engine compression – piston rings and valves.
5. Check vacuum leaks at the intake manifold gasket, vacuum hoses, and carburetor. A malfunction of any one of these parts could upset engine operation and appear to be a computer system problem. If none of these checks reveal the cause of the performance problem, the computer control system may be defective and should be tested. The following symptoms could indicate a possible computer system problem:
    1. Detonation.
    2. Poor idling.
    3. Missing.
    4. Hesitation.
    5. Surging.
    6. Poor fuel economy.
    7. Sluggish engine.
    8. Hard starting.
    9. Exhaust odour or smoke.
    10. Engine cut-out.
    11. Engine computer light flashing.

Computer Controlled Carburetor System Testing

In most cases, a built-in diagnostic system catches and indicates computer system problems. When the fault exists in the computer network, a ***"check engine"*** light will glow in the instrument panel, to warn the driver and mechanic of a possible problem.

Reading Trouble Codes

To isolate which part of the computer system is defective, connect a scan tool to, or ground the proper

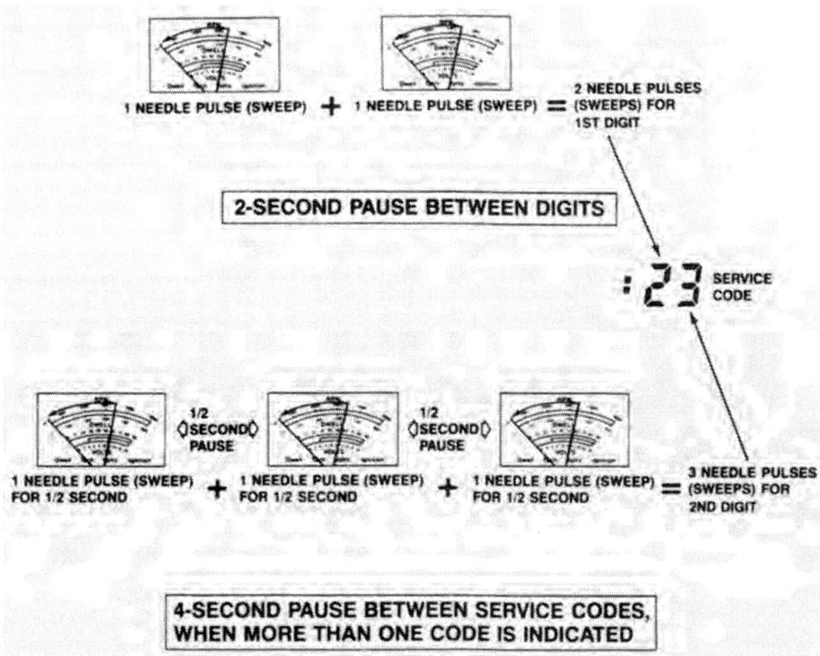

Fig. 6.31 *Analog or needle type voltmeter is needed to read the self-diagnosis code of many computer controlled carburetor systems. When activated by jumper wire, computer will send voltage pulses to test socket and meter. Two meter deflections, a short pause, and three more deflections would represent a 23 code. Your manual will tell what problem 23 represents. (Ford Motors)*

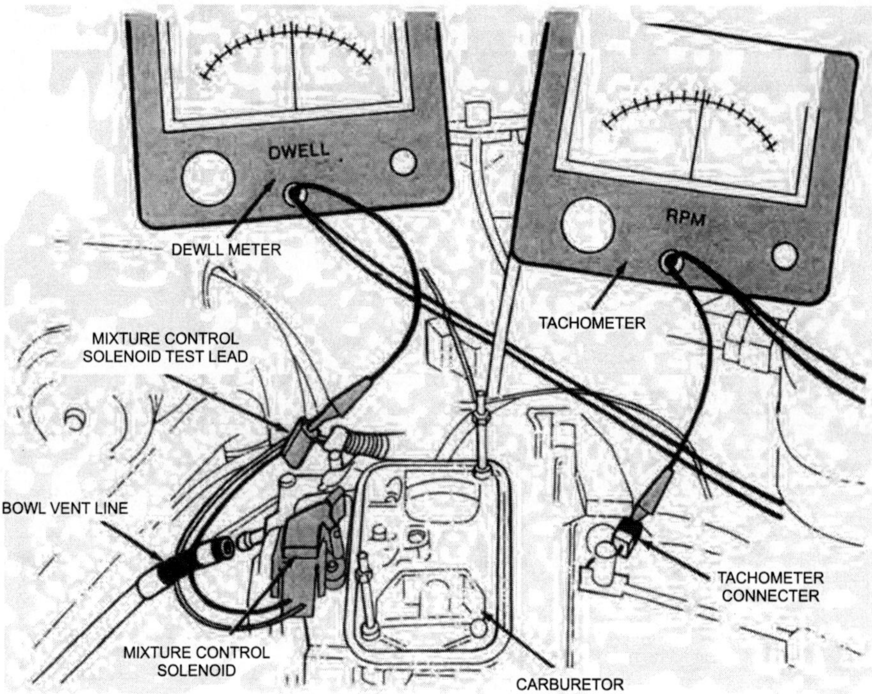

Fig. 6.32 *Note basic connections for using a dwell meter to check computer signals to carburetor. Dwell meter will measure pulse width going to mixture control solenoid. This will help you determine whether there is a problem in the control system, sensor system, or carburetor itself. (General Motors)*

Relationship of Dwell Meter Readings to Mixture Control Solenoid Cycling

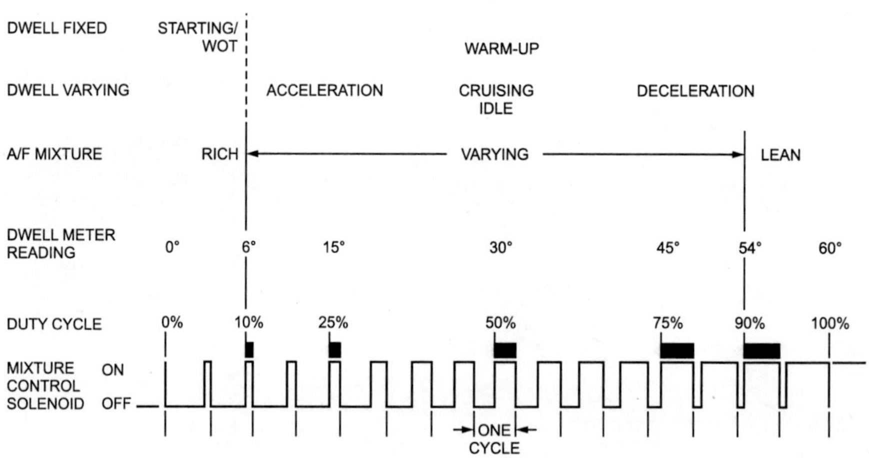

Fig. 6.33 *Chart shows relationship between dwell and mixture control solenoid operation. Note how low dwell richens mixture and high dwell leans mixture. (Buick Div. General Motors)*

terminals on, the data link connector. Grounding the connector will cause the "check engine" "light to flash" a code. On other systems, an analog (needle type) voltmeter can be connected to a specified point. Each pulse (sweep) of the voltmeter needle would signal the numbers of the code.

For example, the light or voltmeter may flash or sweep two times and then pause, indicating the number two. Then, after a few seconds, it may flash or sweep three times to indicate the number three. This would mean a trouble code 23 (2, pause, 3) as in *Fig. 6.31*.

The mechanic can then pinpoint the trouble by referring to a code chart in the service manual. For detailed information refer to a service manual for the computer system under service. Systems codes and test procedures do vary with applications.

Other Tests

A system performance test is used to find problems not included in the system code or diagnostic warning system. The check is needed when the warning light does not indicate a system problem and no cause for a problem can be found. It is the last test made when an engine performance problem exists. The applicable service manual will give step-by-step directions to make this test.

Measuring Dwell Signal to Carburetor

Some computer controlled carburetor systems require a dwell meter and tachometer to check the computer output. The dwell meter measures the duration of the electrical pulses to the carburetor mixture control solenoid. The tachometer measures engine rpm.

Usually, the dwell meter and tachometer are connected as shown in *Fig. 6.32*, The dwell meter connects to a specified connector in the wiring harness. The tachometer connects to the ignition system.

Figure 6.33 shows the relationship between the dwell meter reading and the air-fuel mixture of the carburetor.

When testing the system, compare dwell meter readings to specs. If the dwell is too high or low for engine conditions, there is a problem in the computer system. Normally, the dwell reading will rise and fall within a certain range-around 35° ± 5° with the engine at full operating temperature and at a constant speed.

If the engine is cold and/or has just been started, the dwell should be fixed at 6° for about 2 minutes. After this time dwell will be around 30° at idle and will vary as the engine is accelerated or decelerated.

The mechanic should be able to hear the mixture control solenoid "click" after starting the engine. If not, the mixture control solenoid may be defective. After the engine has reached operating temperature, remove a vacuum hose. The dwell should decrease. Then, reconnect the vacuum hose and choke the engine. The dwell should increase. If not, consult the applicable service manual for test and repair procedures.

If the computer dwell signal is not within specs, use common sense, basic testing methods, and the service manual to isolate the problem source. The service manual usually gives a step-by-step procedure for isolating specific problems using a digital VOM or special testers, *Figs. 6.33 and 6.34*.

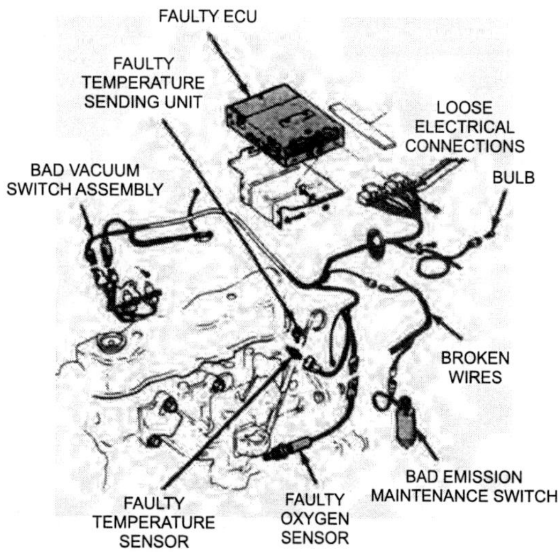

Fig. 6.34 *Study types of problems that can occur in a computer controlled carburetor system.*

6.11 CARBURETOR PROBLEMS

Engine troubles can be caused by many causes besides problems in fuel system. Troubles caused by conditions inside the carburetor itself are listed and described below:

1. Poor Engine Performance

- Air leak at carburetor or manifold
- Air leak in fuel line.
- Clogged or dirty carburetor air filter.
- Clogged fuel lines or fuel filter.
- Defective fuel pump.
- A stuck power piston or valve.
- Dirt or gum clogging the fuel nozzle or jets.
- Incorrect fuel level in fuel bowl.
- Automatic choke setting incorrect.
- The choke stuck or not operating.
- The throttle valve not fully opening.
- Dirt in carburetor jets and passages.
- Worn or inoperative accelerator pump.
- Wrong or incorrectly set metering rod
- Inoperative power valve, economizer, or jet
- Damaged or wrong main metering jet
- Worn idle needle valve and seat
- Loose jets
- Defective gaskets
- Worn throttle valve shaft
- Clogged exhaust system
- Defective manifold heat control valve
- Leaking vacuum lines
- Defective sensors
- Defective computer.

2. Poor Idling

- Incorrect adjustment of needle valve.
- Incorrect float level.
- Sticking float needle valve.
- Defective gasket between carburetor and intake manifold
- Stuck PCV valve
- A leaky vacuum hose.
- Defective gaskets in carburetor.
- Loose carburetor-to-manifold attaching fasteners.
- Loose intake manifold mounting bolts.
- Idle discharge holes partially clogged.
- Defective automatic choke.
- Loose jets in carburetor.
- Leaking vacuum lines to accessory equipment.
- Worn main metering jet.
- Restricted or clogged air cleaner.
- High float level.
- Defective sensors.
- Defective computer.
- Defective O_2 feedback solenoid.
- Vacuum leak that is partly compensated for by a rich mixture idle adjustment.

3. Failure of Engine to Start Unless Primed

- No gasoline in the fuel tank.
- Wrong fuel tank cap.
- Plugged tank or cap vent.
- Holes in the fuel-pump flex line.
- Carburetor jets or lines clogged.
- A defective choke.
- A clogged fuel filter.
- Air leaks into the manifold.

5. Hard Starting

- Incorrect choke adjustment.
- Defective choke.
- Incorrect fuel level.
- Incorrect fuel pump pressure.
- Sticking fuel inlet needle.
- Improper starting procedure.

6. Poor Acceleration

- Accelerator pump incorrectly adjusted.
- Accelerator pump inoperative.
- Corroded or bad seat on accelerator bypass jet.
- Accelerator pump leather hard or worn.
- Clogged accelerator jets or passages.
- Defective ball checks in accelerator system.
- Incorrect fuel level.
- Misadjusted throttle position sensor.

7. Carburetor Floods

- Float level too high.
- Stuck float needle valve.
- Defective gasket in carburetor.
- Cracked carburetor body.
- Excessive fuel pump pressure.

8. Excessive Fuel Consumption

- A high float level.
- A leaky or saturated float.
- A sticking or dirty float needle valve.
- Worn jets or nozzles.
- A stuck metering rod or power piston.
- Idle too rich or too fast.
- A stuck accelerator-pump check valve.
- A leaky carburetor or lines.
- A dirty air cleaner.

- Defective fuel economizer.
- Defective manifold heat control valve
- Turbo waste gate stuck closed.
- Excessive fuel pressure.

9. Slow Engine Warm-up

- Defective choke.
- Defective manifold heat control valve.

10. Smoky, Black Exhaust

- Due to an over rich mixture
- All items listed in # 8 above.

11. Engine Stalls as it Warms-up

- Defective choke.

12. Engine Stalls after high-speed Driving

- Malfunctioning anti-percolator.

13. Engine Backfires

- Excessively rich or lean air-fuel mixture.
- Defective air-injection-system antibackfire valve.

14. Engine Runs but Misses

- Vacuum leak caused by vacuum hose or intake-manifold leak.
- Improper amount and ratio of air-fuel mixture are not reaching the engine due to clogged jets or worn carburetor jets.
- Incorrect fuel level in the float bowl.

CHOKE TROUBLE SHOOTING

The basic check for choke operation is to ensure that the choke valve is closed when the engine is cold. When the engine is hot, the choke valve should be open.

Choke troubles may be the cause of:

- ***Hard or no Starting Cold.*** This could be caused by the choke valve remaining in the open position. If the temperature is mild, the engine may start but will stall as soon as it is started. If the temperature is extremely cold, the engine may not even start if the choke valve remains open. This could be due to a defective choke coil spring.
- ***No Start Cold.*** This could be caused by the choke valve remaining closed, which would not allow air to enter the carburetor. This may be caused by the choke coil spring being set too tight. When the choke coil is set too tight, the spring coil exerts excessive pressure on the choke valve, keeping it closed. To determine if this is the cause, wedge the choke valve open with the shank of a screwdriver. The vehicle will start if this is the cause and the engine is not flooded.

THEORY QUESTIONS

1. The rate of flow through a carburetor stays same under all operating conditions. *True or False?*
2. Wear eye protection and rubber gloves when using carburetor cleaner. *True or False?*
3. Explain the five steps for a carburetor rebuild.
4. Why is the air filter important to engine service life?
5. What are the symptoms of a clogged air filter?
6. List typical fuel pressures for carburetor, gasoline injection and diesel injection systems.
7. A very rich fuel mixture will make the engine emit ....
8. How does a computerized carburetor system works?
9. If the engine cranks normally but does not start, the trouble could be
   (*a*) an ignition problem (*b*) overchoking
   (*c*) lack of fuel (*d*) all of the above.
10. If the engine starts, speeds up, and then dies, the problem could be
    (*a*) low battery (*b*) defective spark plugs
    (*c*) stuck EGR valve (*d*) defective ignition coil.
11. When starting an internal combustion engine, the fuel mixture should be:
    (*a*) rich (*b*) lean
    (*c*) average (*d*) none of the above.
12. The best way to check a customer complaint that the engine lacks power is with
    (*a*) a road test
    (*b*) an exhaust gas analyzer

(*c*) a dynamometer and oscilloscope
(*d*) a fuel pump tester

13. The purpose of a venturi in a carburetor is to:
(*a*) increase speed of air passing through carburetor
(*b*) maintain correct air/fuel ratio
(*c*) provide extra fuel for acceleration
(*d*) all of the above.

14. A vehicle with poor driveability might have
(*a*) stalling (*b*) hard starting
(*c*) poor acceleration (*d*) all of the above

15. In a carburetor venturi, which point has the highest vacuum?
(*a*) Entrance to the venturi
(*b*) Narrowest point of the venturi
(*c*) Point 25 mm below the venturi
(*d*) None of the above.

16. When the idle system is no longer supplying fuel to the engine, which system then supplies fuel?
(*a*) Air bleed system (*b*) Main system
(*c*) Vaporizing system (*d*) None of the above.

17. Which should be adjusted first?
(*a*) Ignition system
(*b*) Carburetor

18. Dieseling, or run-on, can be caused by
(*a*) defective air pump
(*b*) incorrectly adjusted idle-stop solenoid
(*c*) throttle valve stuck closed
(*d*) EGR valve stuck open.

19. Briefly describe the procedure to adjust a single-barrel carburetor.

20. Briefly describe the procedure to adjust a two-barrel carburetor.

21. What is the advantage of analyzing the exhaust gas?

22. Which of the following parts should not be washed in carburetor-cleaning solution?
(*a*) Carburetor float
(*b*) Anti-stall dashpot
(*c*) Idle needle valve
(*d*) Accelerator pump plunger
(*e*) Accelerator pump diaphragm
(*f*) Main jets
(*g*) Throttle valve.

23. List three causes of carburetor flooding.

24. How do you adjust the various engine idle speeds?

25. List five causes of excessive fuel consumption that start in the fuel system.

26. A dwell meter is used to determine if a problem exists in the carburetor. Computer, or oxygen sensor. *True or False?*

27. Mechanic A says that the "check engine" light tells the driver of a problem. Mechanic B says that by the mechanic uses "check engine" light to pull trouble codes from the computer. Who is right?
(*a*) A only (*b*) B only
(*c*) Both A and B. (*d*) Neither A nor B.

28. A computer controlled carburetor system normally functions in ——— mode:
(*a*) cycling loop (*b*) repeating loop
(*c*) closed loop (*d*) open loop.

29. Mechanic A says that an automotive engine's fuel return system is used to regulate fuel pressure in a gasoline fuel injection system. Mechanic B says that an automotive engine's fuel return system is used to cool the fuel in order to prevent vapour lock. Who is right?
(*a*) A only (*b*) B only
(*c*) Both A and B (*d*) Neither A nor B.

30. Mechanic A says that a fuel filter can be located inside the vehicle's fuel pump. Mechanic B says that a fuel filter can be located inside an injector body. Who is right?
(*a*) A only (*b*) B only
(*c*) Both A and B (*d*) Neither A nor B.

31. Mechanic A says that a carburetor's main discharge tube is a fuel passage between the fuel bowl and air horn. Mechanic B says that a carburetor's main discharge tube is a fuel passage between the throttle valve and intake manifold. Who is right?
(*a*) A only (*b*) B only
(*c*) Both A and B (*d*) Neither A nor B.

32. Which of the following engine components can be operated off carburetor Vacuum?
(*a*) Choke break. (*b*) Charcoal canister
(*c*) EGR valve (*d*) All of the above.

33. Mechanic A says that a carburetor normally provides an 18:1 air-fuel ratio for cold engine starting, Mechanic B says that a carburetor normally provides an 8:1 air-fuel ratio for cold engine starting. Who is right?
(*a*) A only (*b*) B only
(*c*) Both A and B (*d*) Neither A nor B.

34. The acceleration system of a carburetor ————.
    (*a*) supplies a lean air-fuel mixture at cruising speeds
    (*b*) provides correct air-fuel mixture slightly above idle speed.
    (*c*) enriches fuel mixture slightly when engine power demands are high.
    (*d*) squirts fuel into air horn when the throttle valve opens

35. Mechanic A says that several computer-controlled feedback carburetors require the use of a dwell meter to check computer output. Mechanic B says that several computer-controlled feedback carburetors require the use of a tachometer to help test computer output. Who is right?
    (*a*) A only (*b*) B only
    (*c*) Both A and B (*d*) Neither A nor B.

❑❑❑

Fuel-Injection Systems Diagnosis and Service

7

7.1 INTRODUCTION

An engine must have a continuous supply of combustible air-fuel mixture to keep running. Automotive vehicle manufacturers have converted over to gasoline injection in order to achieve government specified fuel economy to avoid a "gas guzzler" charge being added to the cost of the vehicles. Also it helps control the exhaust emissions to specs as laid down by government. For this reason, modern engines are equipped with electronic fuel injection (EFI) because it is more efficient than carburetion. Modern ***gasoline injection systems*** use engine sensors, a computer, and solenoid-operated fuel injectors to meter the right amount of fuel into the engine. These systems use electric-electronic devices to monitor and control engine operation.

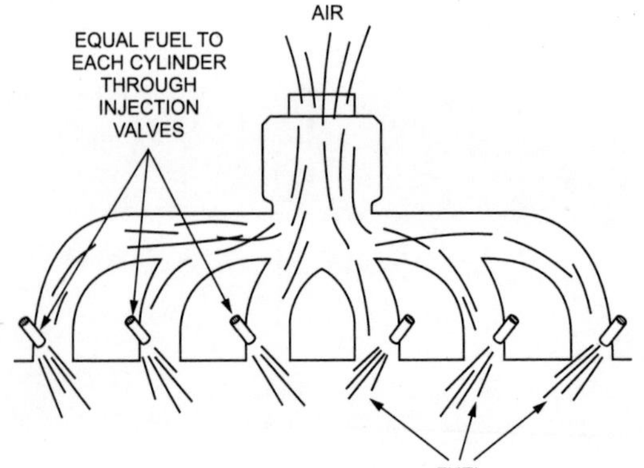

Fig. 7.1 *(a) Port injection. An injection valve in each intake port sprays fuel into the intake air.*

An electronic control unit (ECU) or computer receives electrical signals in the form of different current or voltage values, from the various sensors. It then uses stored data to operate the injectors, ignition systems, and other engine related devices. The system controls the exhaust emissions to meet specs and gets better fuel economy.

In the gasoline fuel-injection system, the fuel is injected into the intake manifold through fuel-

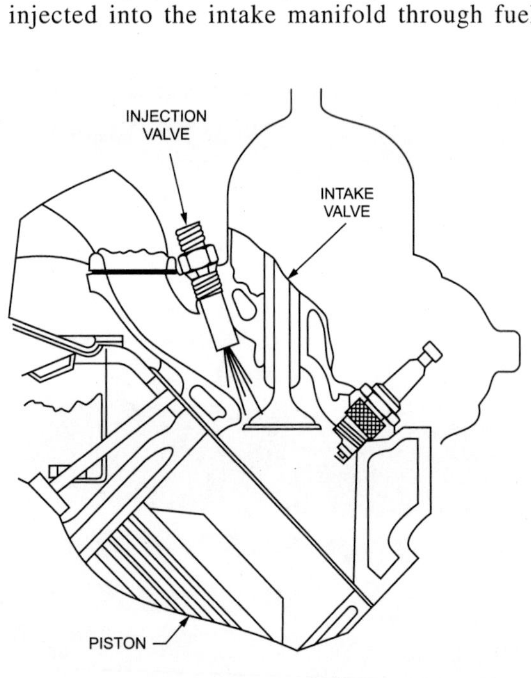

Fig. 7.1 *(b) The injection valve sprays fuel into the intake air just before it passes the intake valve and enters the cylinder.*

injection valves (injectors). Two common types of electronic fuel injection are ***multiport*** and ***throttle body injection,*** *Figs. 7.1 and 7.2.*

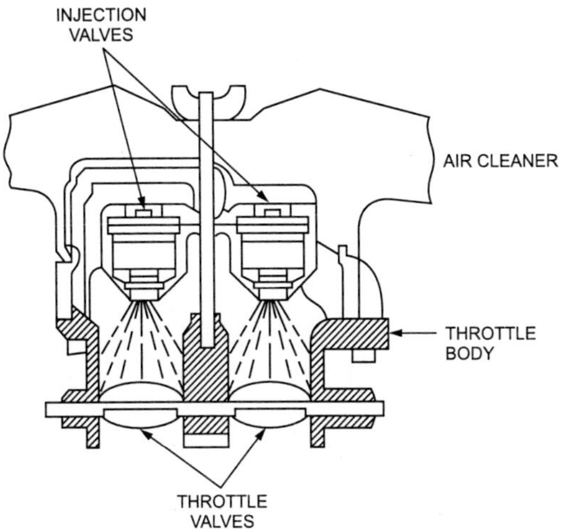

Fig. 7.2. *The throttle body for the TBI system is similar to the carburetor throttle body. Throttle valves in both systems control the amount of air entering the intake manifold.*

Multiport injection has one injector for each engine cylinder. Fuel is injected in more than one location. This is the most common type and is often called ***port injection.***

In a ***throttle body injection (TBI),*** a throttle body is mounted on the intake manifold. The throttle body is quite similar to the carburetor throttle body for one- or two-barrel carburetors. It has either one or two throttle valves. As in the carburetor, the position of the throttle valves determines the amount of air that flows through into the intake manifold. One or two injectors are mounted inside the throttle body assembly. Fuel is sprayed into one point or locaton at the centre inlet to the engine intake manifold.

An automotive engine mechanic must understand the operation, construction, and basic service of gasoline injection systems and be able to quickly and accurately find the problem sources in a fuel injection system.

7.2 SERVICING FUEL INJECTION SYSTEMS

Fuel injection systems do not need much service, but the service they do require should be given regularly. All regular injection system is limited to the fuel supply system, on both diesel and gasoline systems. Under normal conditions, the computer, injection pump, or injectors do not need any service. Fuel injection problems cause the same general symptoms as those caused discussed for carburetion systems.

Fuel Supply

A major injection problem is fuel contamination. Either dirt or water gets into the fuel supply and clogs or damages the system. Dirt can cause damage to injectors if it is not filtered out. Large amounts of dirt can quickly clog the fuel filter. This can keep the engine from developing full power or even running at all. Water corrodes or rusts the many metal parts in the system. The presence of water in substantial amounts can cause the engine to stumble or even will not run at all.

Cleanliness is extra important with injection systems.

Fuel Source

When fuel problems arise, fuel source should also be considered. The most common source of bad fuel is the fuel service station's fuel tanks. Water and dirt collected in these, normally underground, tanks, if they start a leak or if they are not properly maintained.

Fuel Filters

Fuel filters must be replaced at the recommended intervals. On diesel engines, a water separator is beneficial. The water separator swirls the fuel in a cylinder and allows the water to settle to the bottom. Periodic draining of the separator is necessary. A pet cock is provided for this purpose.

Fuel Hazards

Whenever working around injection systems, keep the fire hazard in mind. Loosening or removing fuel lines on a running engine may cause high pressure fuel to squirt over a large area. This can start a very large fire quickly. Also, the high-pressure spray emitted from a diesel injector will force diesel fuel directly through your skin. Diesel fuel injection will cause severe headaches, blood poisoning, dizziness, and other health problems. See a doctor immediately if you accidentally get sprayed from a diesel injector.

7.3 TROUBLE SHOOTING FUEL INJECTION SYSTEMS

Before starting to troubleshoot fuel injection systems, you should obtain the correct service manual first. Although most fuel injection systems operate in the same way and have the same common components, normal pressures and sensor resistances may vary. The biggest mistake a mechanic can make when troubleshooting is to start replacing components before checking them thoroughly.

Inspection

Fuel Supply: Use of logic and common sense can solve many fuel injection problems. If the fuel does not reach the fuel injection system, the engine would not start. In any injection system check the supply of fuel in the fuel tank, condition of fuel filters, pump operation, fuel pressure, and fuel flow to begin with. Only after these items have been checked, the injectors and other fuel injection system parts be suspected.

Ignition: Make sure the ignition system and all filters are in good condition before digging into the ignition system. Valve adjustment and engine compression should also be checked.

Electrical: A visual check of the injection system wiring will locate most electrical problems. The most common ones are corroded connections and broken wires. Other frequent causes are bad ground connections, shorted or open sensors, and physically damaged airflow sensors. Sensor resistance can be checked with an ohmmeter. The ohmmeter can also be used to check the condition of wiring in the fuel injection or glow plug systems. If the fuel injection system is on a computer-controlled engine, the computer may be able to diagnose some of its own problems. A flashing code light indicates the most likely trouble. By checking the code in the applicable service manual, the mechanic can find the most effective troubleshooting procedure.

7.4 GASOLINE INJECTION SYSTEM SERVICE

When a customer brings the vehicle for service and complains of an engine performance problem, the engine mechanics may have to diagnose and repair the problem. He/she must find out whether the problem is a mechanical problem, or a problem in an engine support system. Here we will focus on finding and correcting common problems in EFI systems.

A *no start condition* is often caused by a malfunction of the electric fuel pump. If the electric fuel pump does not buzz for a second or two when the ignition is turned on, check the pump fuse. All EFI electric fuel pumps are fused someplace in the vehicle's electrical circuit.

A common cause of ***hard starting, poor idle, low power, and engine misfire*** is a plugged fuel injector or cold start valve.

EFI system problems often occur in the connections between the components. Make a quick check of all harnesses and connections for looseness or corrosion. All the harnesses and connectors should be disconnected and inspected carefully, then reconnected, *Fig. 7.3.*

Diagnostic procedures are provided in different forms by different manufactures. These may be in the form of trouble diagnosis chart for each type of fuel injection system, *Fig.7.4,* or in the form of performance diagnosis procedures that list various problems and the possible causes of each.

Diagnosing EFI Problems

To diagnose problems in a gasoline injection system will require:

- Complete understanding of system operation.
- Basic troubleshooting expertise.
- Service manual of the vehicle/engine under service.

When locating problems, visualize the operation of the system. Relate the function of each component of the system to the problem complaint. This will allow you to eliminate several possible sources and concentrate on other potential sources the problem, *Fig. 7.5.*

Careful checking of the engine compartment will sometimes locate gasoline injection problems. Check the condition of all hoses, wires, lines, fuel pump, and other parts. Look for fuel leaks, vacuum leaks, linked lines, loose electrical connections, fasteners, and other visible troubles.

With EFI, you might disconnect and check the terminals of the wiring harness for rust, corrosion, or burning. High resistance at terminal connections is a frequent cause of problems.

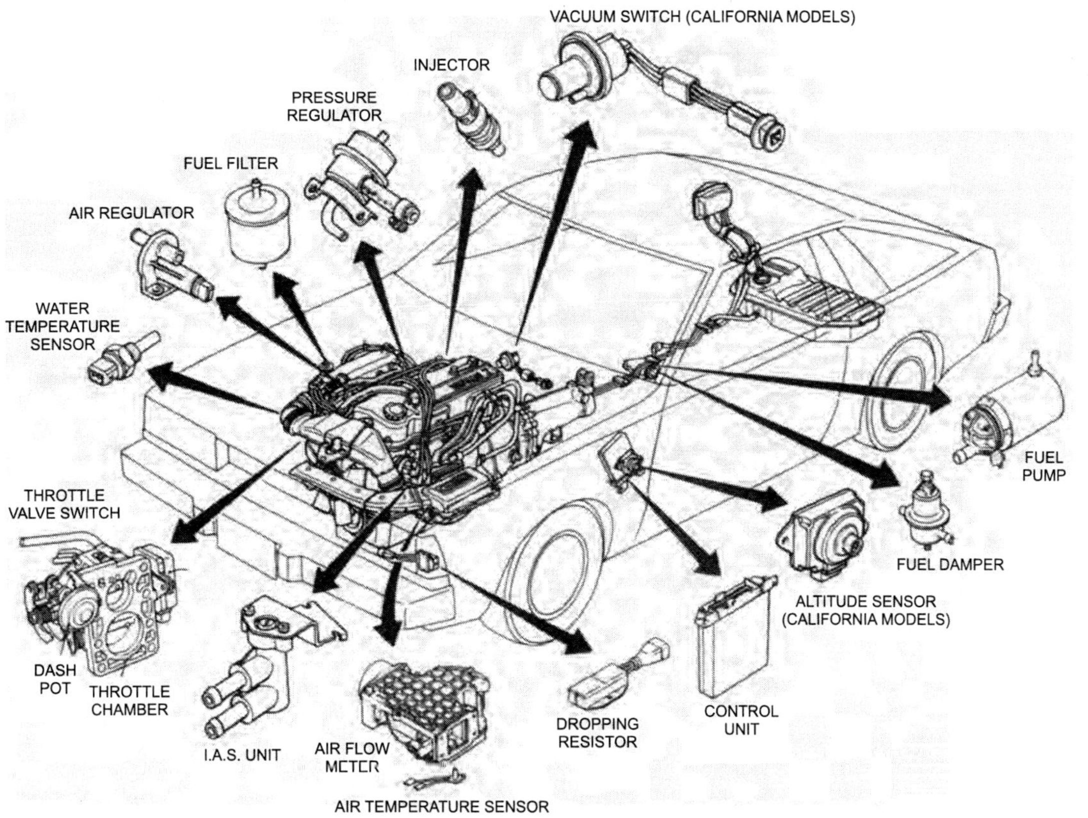

Fig. 7.3 *Parts and location of airflow-controlled EFI. (Nissan Motor Corporation)*

Electronic Control Unit Self-Diagnosis

Modern computer control systes have "onboard" electronic self-diagnosis systems that test themselves and can indicate the source of a problem. There are two types of computer control systems in use at present. The first has been in use for many years and is known as ***on-board diagnostics generation one*** or ***OBD I.***

The second system, known as ***on-board diagnostics generation two (OBD II),*** was implemented on all new vehicles starting with the 1996 model year. This system can detect sensor malfunctions even before they become noticeable to the driver or the mechanic.

Even though each manufacturer's systems have their differences, they are alike in many ways. Both diagnostic systems work as given below:

1. ECU illuminates a ***malfunction indicator lamp (MIL)*** or ***check engine light*** on the dash.
2. ECU displays a digital number in the dash climate control or driver information centre.
3. ECU lights LEDs (light emitting diodes) on the computer itself to indicate problems.

OBD II systems will flash the MIL if a problem exists that could cause damage to the vehicle's emission system. When the MIL light illuminates, it tells the driver and mechanic that a problem exists, *Fig. 7.6.*

Data Link Connector

A test terminal, also called a ***data link connector,*** is used to access he computer system. Each terminal on the connector is wired to an ECU, battery power, or ground. OBD I data link connectors can be accessed by grounding terminals or with an analog meter or scan tool. OBD II data link connectors must be accessed using a scan tool.

The data link connector can be located almost anywhere on OBD I systems, therefore, check the

COMPLAINT

ENGINE DOES NOT START IN COLD CONDITION	ENGINE DOES NOT START IN WARM CONDITION	ENGINE STARTS POORLY IN COLD CONDITION	ENGINE STARTS POORLY IN WARM CONDITION	IRREGULAR IDLE DURING WARM-UP	IRREGULAR IDLE WITH WARM ENGINE	ENGINE BACKFIRES INTO INTAKE MANIFOLD	ENGINE BACKFIRES INTO EXHAUST SYSTEM	ENGINE MISSES WHEN DRIVEN ON ROAD	DRIVING PERFORMANCE UNSATISFACTORY	ENGINE RUNS ON	FUEL CONSUMPTION AT IDLE TOO HIGH	CO CONCENTRATION AT IDLE TO HIGH	CO CONCENTRATION AT IDLE TO LOW	IDLE SPEED CANNOT BE ADJUSTED	POSSIBLE CAUSE	CHECK OR CORRECTION
●	●														ELECTRIC FUEL PUMP NOT OPERATING	CHECK PUMP FUSE, PUMP RELAY AND PUMP
●	●							●							LOOSE CONTACT AT ELECTRIC FUEL PUMP	CHECK PUMP WIRING
●		●		●											"COLD" CONTROL PRESSURE OUTSIDE TOLERANCE	TEST PRESSURE
	●		●		●	●			●				●		"WARM" CONTROL PRESSURE TOO HIGH	TEST PRESSURE
	●		●		●		●		●		●	●			"WARM" CONTROL PRESSURE TOO LOW	TEST PRESSURE
			●	●	●						●			●	AUXILIARY AIR VALVE DOES NOT CLOSE	CHECK VALVE FOR CORRECT FUNCTION
●		●		●											AUXILIARY AIR VALVE DOES NOT OPEN	CHECK VALVE FOR CORRECT FUNCTION
●		●													COLD START VALVE DOES NOT OPEN	CHECK COLD START VALVE
●		●	●	●	●		●		●	●	●	●			COLD START VALVE LEAKING	CHECK COLD START VALVE
								●							PRIMARY SYSTEM PRESSURE OUT OF TOLERANCE	TEST PRESSURE AND ADJUST WITH SHIMS
●	●	●	●												AIR FLOW SENSOR PLATE STOP INCORRECTLY SET	CHECK AND RESET
●	●	●	●		●				●	●		●			SENSOR PLATE AND/OR PLUNGER NOT MOVING FREELY	CHECK FOR FREE MOVEMENT
●	●	●	●	●	●	●			●				●		LEAKS IN AIR INTAKE SYSTEM (FALSE AIR)	CHECK AIR SYSTEM FOR LEAKS
●	●	●	●	●	●		●				●	●			FUEL SYSTEM LEAKAGE	INSPECT FUEL SYSTEM FOR LEAKS
	●		●	●	●					●					INJECTOR LEAKING, OPENING PRESSURE LOW	CHECK INJECTORS ON TESTER
	●		●		●		●		●		●	●			IDLE MIXTURE TOO RICH	CHECK AND ADJUST CO LEVEL
	●		●		●	●			●				●		IDLE MIXTURE TOO LEAN	CHECK AND ADJUST CO LEVEL
									●						THROTTLE BUTTERFLY DOES NOT OPEN COMPLETELY	CHECK BUTTERFLY AND STOPS IN THROTTLE VENTURI
●		●													THERMO-TIME SWITCH DEFECTIVE	TEST FOR RESISTANCE READINGS vs. TEMPERATURE

Fig.7.4 *Fig.7.4. Trouble-diagnosis chart for the Bosch K-type fuel-injection system. (Robert Bosch Corp.)*

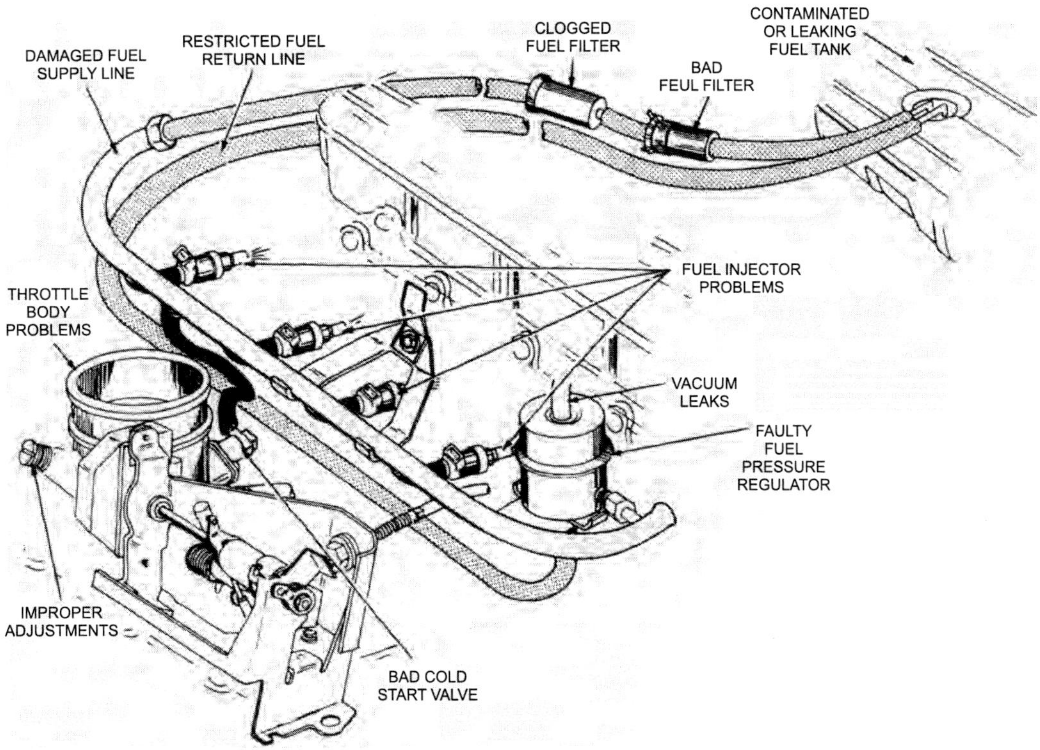

Fig. 7.5 *Various types of fuel injection system problems.*

service manuals for individual applications. OBD II systems mount the data link connector in a location accessible from the driver's seat.

Activating Self-Diagnosis

Warm the engine to operating temperature. This will ensure that the system has the proper conditions to enter closed loop operation. Shut off the engine and turn the ignition key to on or run.

The recommended method of reading codes on OBD I and OBD II equipped vehicles is to use a scan tool.

Reading Codes

If the output is a digital display, simply read the number and compare it with the chart in the service manual.

If the output is flashed from the dashboard MIL or ECU mounted LED lamp, count how many flashes occur between each pause. Write down the number flashed and compare it to the service manual.

If an analog meter is used, count the number of needle deflections between pauses. Write down the number and compare it to the chart in the service manual.

Each system may vary. The service manual will provide the instructions for each make and model of vehicle.

Interpreting Trouble Codes

When looking up trouble codes in the service manual, start from the lowest number stored and work up. If multiple codes were stored, you can often eliminate one or more codes by using this method. If you had codes 15 and 45 stored in the ECU, look up code 15 first and then 45.

OBD I codes will be two digit numbers and each commands with a particular problem listed in the service manual. OBD II codes are three digit numbers with an ***alphanumeric designator.*** The alphanumeric designator indicates in what system the problem is located and whether the code is SAE or manufacturer specific. The first digit in the three number codes indicates the nature of the code. The last two digits of the number indicate the specific fault, *Fig. 7.6.*

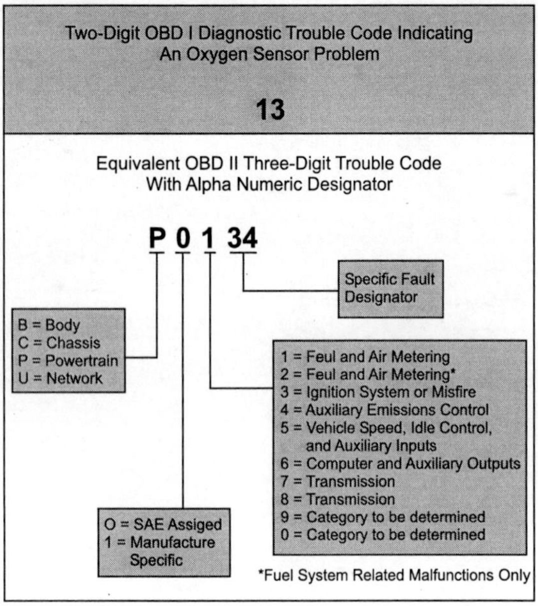

Fig. 7.6 *Most modern vehicle computer systems will test themselves. Codes may be read on a vehicle's digital display, grounding the DTC to cause the MIL light to flash, or via a scan tool. OBD I ECU's will produce a two number code. OBD II systems will produce a two number code with an alpha numeric designator. Do not try to correlate the older fault codes with OBD II fault*

Do not attempt to correlate OBD I codes with the OBD II fault designator. In most OBD I systems, only one or two codes are used to indicate a problem with a specific component. OBD II systems can use up to 20 or more SAE and manufacturer specific codes for problems with the same sensor.

All codes, no matter which system is used, should be looked up in a ***trouble code chart.*** These charts give the meaning for the codes present in the ECU.

Clearing Trouble Codes

To ***clear*** or remove the trouble codes from an OBD I computer system, you must usually disconnect the battery or the fuse to the computer. Disconnect them for about 10 seconds. To clear the codes on vehicles with OBD II system, use a scan tool.

After clearing the codes, check to see if the engine light comes back on after engine operation. If it does, it would indicate there is still a problem in the system and further tests are needed.

7.5 RELIEVING EFI SYSTEM PRESSURE

High pressure in the lines of a fuel injection system can cause serious harm if opened without first releasing pressure. Always relieve fuel pressure before disconnecting any EFI fuel line. Many systems retain fuel pressure as high as 345 kPa. Even when the engine is not running.

Most EFI systems have a port to test and relieve fuel pressure. *Fig. 7.7.* A gauge is connected to the port, *Fig. 7.8.* Fuel is discharged through drain tube after opening the drain valve. Make sure that the pump is not energized and that the ignition is at the "off" position when relieving pressure. Or, the pressure regulator may allow pressure relief for bleeding pressure back to the fuel tank.

On systems that do not have a port to relieve fuel pressure (Chrysler), the electrical connection to the injector must be removed, *Fig.7.9.* Attach two jumper wires to the electrical connections of the injector, *Fig.7.10.* Attach the other end of one jumper wire to a good ground. Touch the other end of the other jumper wire to the battery positive post for no longer than 10 seconds to avoid shorting out the injector, *Fig.7.11.* The energized injector will discharge the fuel through the injector. The system will then be safe to open and service. To test fuel pressure on this system, a "T" must be inserted in the fuel return line.

7.6 FUEL PRESSURE REGULATOR SERVICE

A ***faulty fuel pressure regulator*** can cause a too rich or too lean mixture. If the output pressure is high, too much fuel will spray out of the injectors, resulting in a rich mixture. If the regulator bypasses too much fuel to the fuel tank, not enough fuel will spray out of the injectors resulting in a lean mixture.

To test a fuel pressure regulator, install a pressure gauge at the test fitting on the fuel rail or before the regulator. Start and idle the engine while reading the gauge. Then, apply the recommended vacuum to the pressure regulator

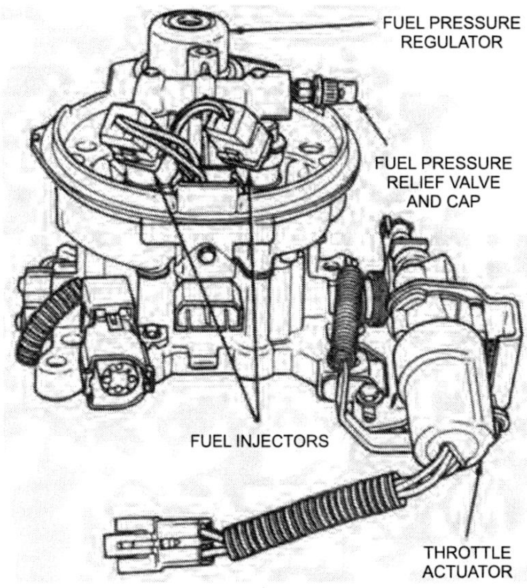

Fig. 7.7 *With a fuel pressure relief valve, a gauge can be easily connected to this point. Fuel is then discharged through the gauge. (Ford Motors)*

and note the pressure drop, *Fig.7.12.* If either reading is not within specs, the regulator could be faulty. However, check other possible causes like bad fuel pump, clogged filter, smashed return line, etc,. before condemning the regulator. If fuel is found in the vacuum hose connection of pressure regulator, the regulator is bad.

Low Fuel Pressure

Low fuel pressure causes a lean air-fuel mixture. This can cause a surging while driving at a steady speed. Causes for low fuel pressure include:

- Plugged fuel filter.
- Defective fuel pump.
- Leaking fuel pump hose coupling (inside fuel tank).
- Defective fuel pressure regulator-stuck open.
- Defective fuel pump check valve or accumulator.
- Leaking fuel line.

Testing Fuel Pressure

Always follow the individual service manual for test procedures and specifications.

To test fuel pressure, connect a fuel pressure gauge to the test port, *Fig. 7.8.* Fuel pressure should read as soon as the ignition is turned **on**, with engine not running. Note the pressure and compare it to the specs. If the pressure is within specs, start the engine and read the operating pressure. Fuel pressure should drop about 34.5 to 69 kPa after engine has started and is idling. If fuel pressure does not drop, remove the vacuum hose from the pressure regulator. Check that suction can be felt at the end of the vacuum hose. If not, locate and repair the leak. If suction is felt, reconnect the vacuum hose to the pressure regulator. Fuel pressure should drop 21 to 42 kPa after reconnecting the vacuum hose. If the pressure does not drop, replace the pressure regulator.

If the fuel pressure starts to drop shortly after connecting the gauge with the ignition on and the engine not running, the fuel pump check valve or accumulator is bad, the fuel pump coupling hose is leaking, or the fuel injector is leaking. To isolate the source of leak, energize the fuel pump, *Fig. 7.13.* Then, pinch shut the fuel return line or flex hose located in the fuel return line, *Fig.7.14.* Fuel pressure should be around 517 to 689.5 kPa with the pump energized and the fuel line pinched. If not, the fuel pump or filter is defective. However, if fuel pressure is good and does not fall, the fuel pump check valve or accumulator is bad or the pump coupling hose is leaking. On the other hand, if fuel pressure continues to fall, one or more of the injectors are leaking.

High Fuel Pressure

If fuel pressure is above specs, the fuel return line is restricted or the pressure regulator is defective. If there no restrictions in the fuel return line, replace the regulator.

Hard Starting

A ***hard-start condition*** experienced on fuel injection engines can be caused by many events. If a fuel injector is leaking, it will flood the engine. A fuel pump check valve that is bad will permit fuel to drain back into the fuel tank. Other systems may have a fuel ***pressure accumulator*** that prevents fuel from draining back into the tank. If the accumulator or check valve is bad, it will take time for the fuel to pressurize in the lines.

(a) Remove dust cap to access the Schrader valve.

(b) Screw the threaded adapter to the Schrader valve. The lever is then lowered to allow pressurized fuel to the gauge.

(c) If fuel pressure is at the specified level, no further testing is needed. Note the drain valve and drain tube. To relieve fuel pressure, open the drain valve with the ignition in the "off position.

Fig. 7.8 *Testing fuel pressure. (Rochester)*

Fig.7.9 *On systems that do not provide a port to test or relieve pressure, disconnect the electrical connection at the injector.*

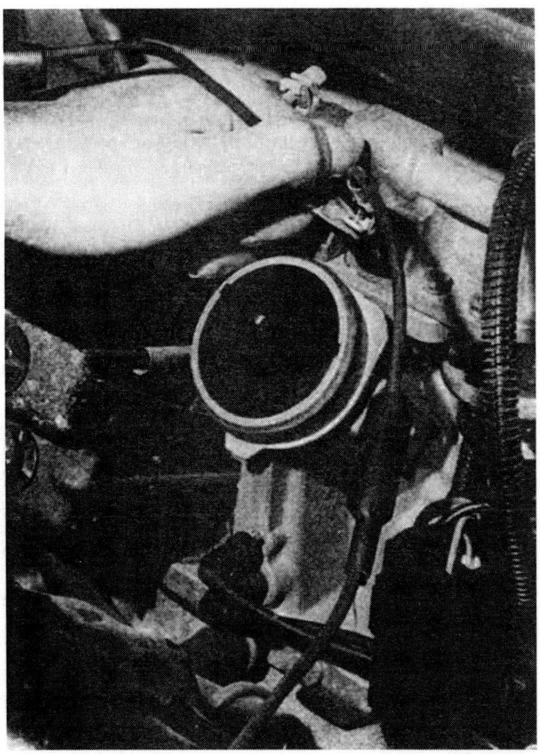

Fig. 7.10 *Connect one end of each jumper wire to the fuel injector electrical connection.*

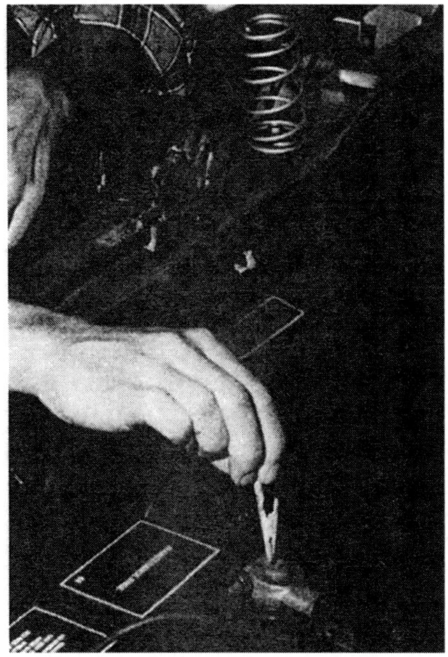

Fig. 7.11 *Connect the free end of one jumper wire to a good ground. With the remaining jumper wire end, touch it to the battery positive post. Fuel will be discharged through injector into the intake manifold.*

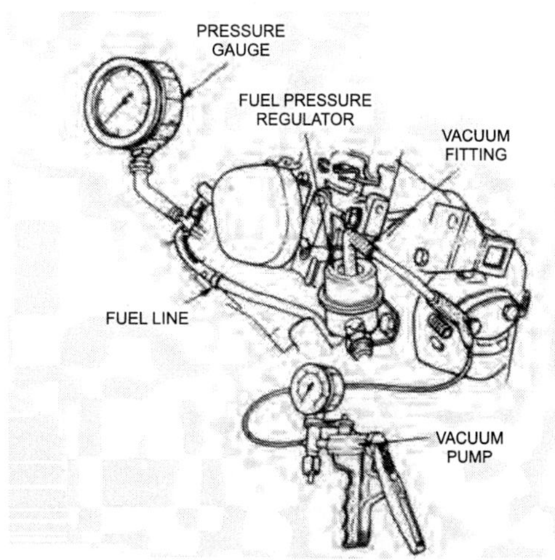

Fig. 7.12 *Fuel pressure regulator is common source of engine performance problems. To test regulator, connect pressure gauge to system and vacuum pump to regulator diaphragm. Check pressure with and without vacuum applied to diaphragm. Compare to specs. Check fuel return line, filters, and pump before discarding regulator. (Renault Motors)*

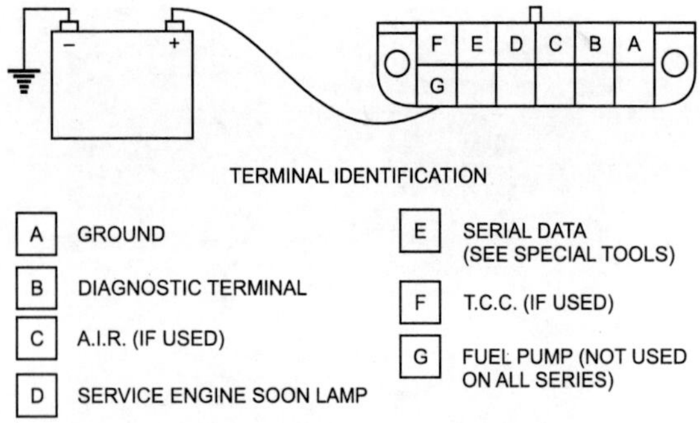

Fig. 7.13 *To energize the fuel pump on some computerized GM engines. Connect a jumper wire from a hot source to terminal G of the diagnostic connector. (Cadillac Div. GM)*

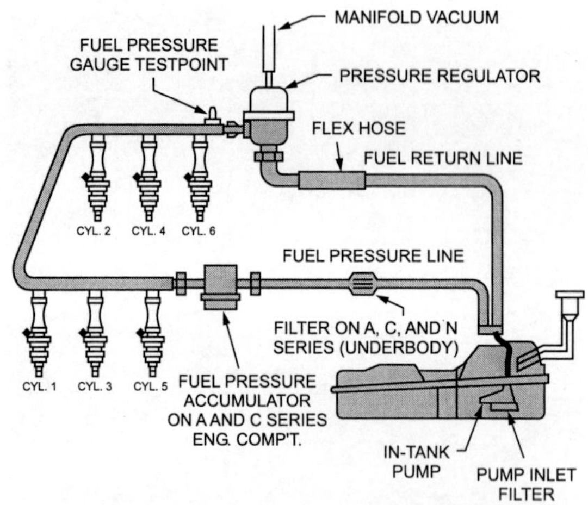

Fig. 7.14 *Pinch the flex hose or return line closed. With pump energized, fuel pressure should be at least 517 kPa. Note the fuel pressure accumulator. (General Motors)*

To pinpoint if a fuel injector or check valve is causing a hard start, connect a fuel pressure gauge and start the engine. As soon as maximum fuel pressure is reached, turn the engine off. Watch the gauge pressure. If pressure is maintained for at least 15 minutes, all is well. However, if pressure drops during that time, the injector, check valve, or accumulator is defective. A leaky fuel injector will show up in an injector balance test as shown in *Fig. 7.15. A* defective pressure regulator may also cause a hard-start condition.

A defective fuel pump relay may also cause hard starting. The fuel pump is wired in parallel with the oil pressure switch. This acts as a back up for the relay. If the relay is defective, it will take a lot of cranking for the engine to build up enough oil pressure to close the switch. A closed switch then allows current to the fuel pump motor.

7.7 FUEL INJECTOR SERVICE

A fuel injector has only one moving part, but if defective it can develop more than one symptom.

A ***bad injector*** can cause a wide range of problems: rough idle, hard starting, poor fuel economy, and engine miss. It is important that each fuel injector provides the correct fuel spray pattern.

Another common problem is a ***leaky fuel injector.*** A leaky injector will flood the engine and may cause the engine to run on after the ignition key has been turned to the "***off***" position.

Sometimes, a leaky fuel injector can be detected by removing all the spark plugs from the engine. Check the condition of porcelain insulators. If any are wet, the injector for the cylinder is leaking. In some cases, however, an injector balance test may be necessary to detect a leaky injector.

A leaking injector richens the fuel mixture by allowing extra fuel to drip out the closed injector valve. The injector valve may be worn or dirty, or the return spring may be weak or broken.

The most common problem with a fuel injector is that the orifices in the nozzle becomes dirty and restrict the flow of fuel and make the air-fuel mixture too lean. A ***dirty injector*** can cause stalling, hesitation on acceleration, and hard starting. If foreign matter collects in the valve, a poor spray pattern and inadequate fuel delivery may result.

Sometimes, adding fuel injector cleaner to full gasoline tank can clean the fuel injector. The vehicle should be driven until the tank is almost empty. If this does not correct the problem, the injector should be cleaned with a cleaning system recommended by the manufacturer.

In many of these systems, the injectors can be cleaned when installed on the engine. In other cases, the injectors must be removed for cleaning. Follow the manufacturer's recommendations when cleaning

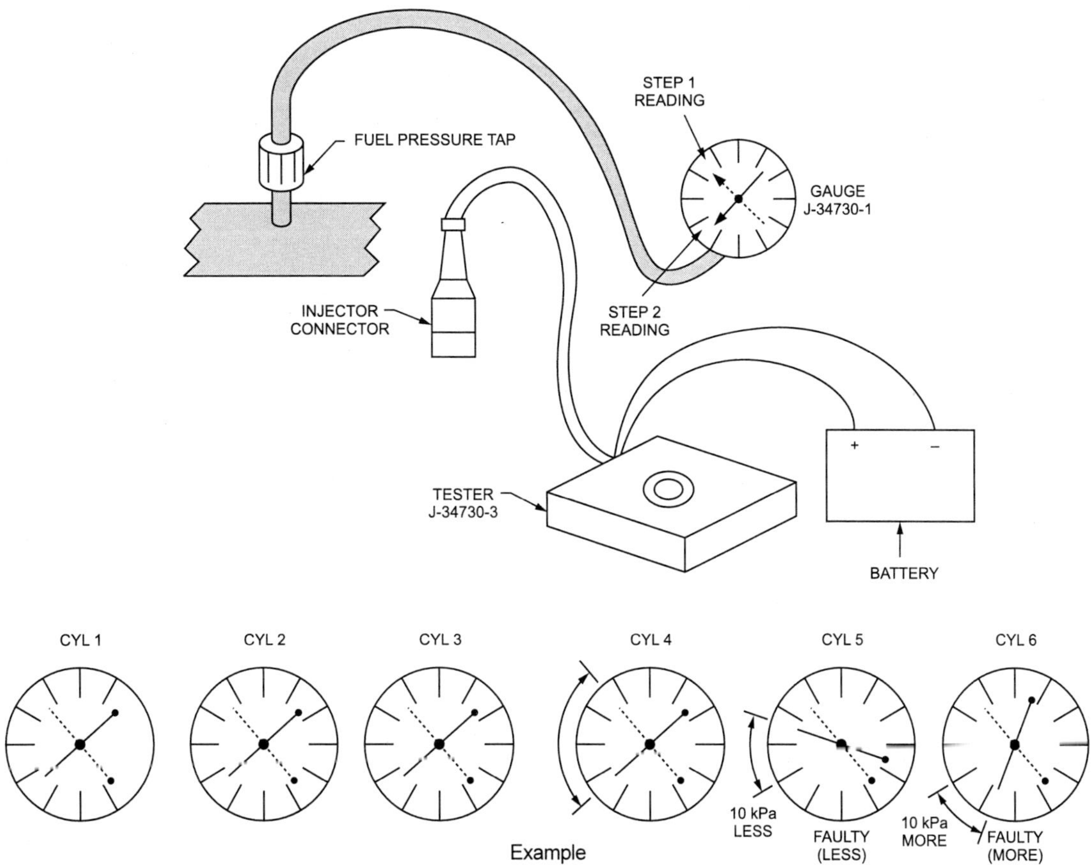

Fig. 7.15 *With gauge and tester installed, record the drop in pressure after energizing the injector with the tester. Before moving the tester to the electrical connection at the next injector, turn the ignition to the "off position. Once the tester has been connected, the ignition is again turned to the "on" position. After testing all injectors, compare the readings. If any reading is more or less than 10 kPa, the corresponding injector is defective. Replace it. (General Motors Corp.)*

injectors. If cleaning fails to solve the problem, the injector should be replaced.

An ***inoperative EFI injector*** normally has shorted or opened coil windings. Current is reaching the injector, but due to bad coil, a magnetic field cannot form and open the injector valve to allow fuel spray.

Checking Injector Operation

Checking the fuel injectors or a multi-point system needs time and special tools for testing. A defective fuel injector on a multi-point system may cause an engine miss, dieseling, and flooding. An ***oscilloscope*** can detect an injector that is only partially defective. An injector balance test is another method for detecting bad fuel injector, *Fig. 7.15.*

To quickly check to see if each EFI injector is opening and closing, place a stethoscope i.e., a listening device, against each injector. A clicking sound indicates the injector is opening and closing. If no clicking sound is heard, the injector is not working. The injector solenoid, wiring harness, or computer control circuit may be bad, *Fig. 7.16 (a).*

With the engine not running, check the condition of the coils on the inoperative injectors. Use an ohmmeter to measure the resistance across the injector coil and check for shorts to ground. If the coil is open (infinite resistance) or shorted (zero resistance to ground), replace the injector, *Fig. 7.16 (b).*

If the injector tests well, check the wiring going to that injector. Following the service manual, check supply voltage to the inoperative injector. Also, measure the resistance in the circuit between the injector solenoid and the computer. A high resistance would indicate a frayed wire, broken wire, or poor electrical connection.

Always refer to a wiring diagram when correcting complex fuel injection electrical problems. The diagram will show all electrical connections and components.

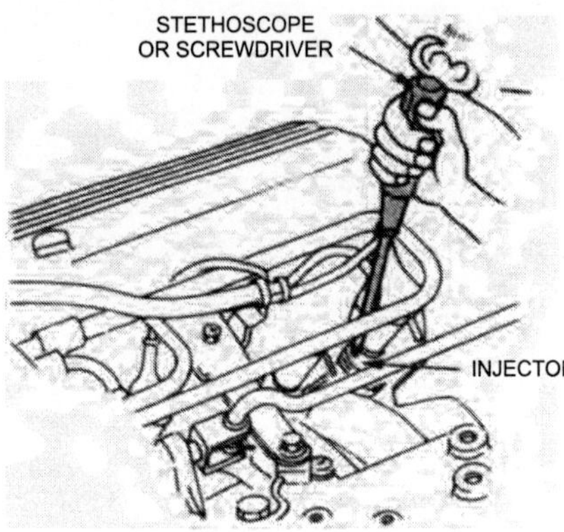

(a) Use stethoscope or screwdriver to listen for clicking noise that shows injector is working.

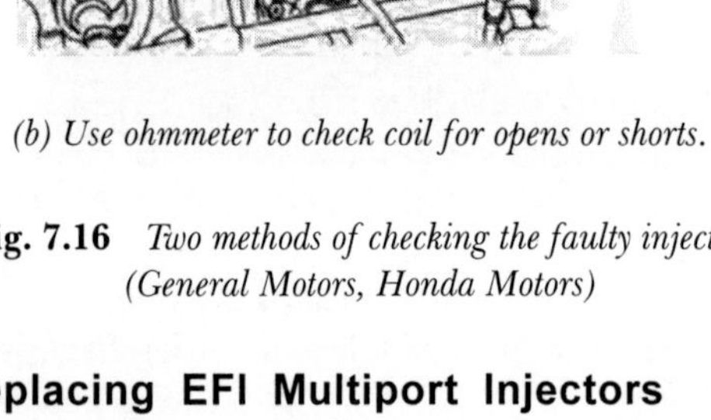

(b) Use ohmmeter to check coil for opens or shorts.

Fig. 7.16 *Two methods of checking the faulty injector. (General Motors, Honda Motors)*

Replacing EFI Multiport Injectors

An EFI multiport injector is easy to replace. Bleed off the fuel pressure and then just remove the hose from the injector and fuel manifold. Unplug the electrical connection and remove any fasteners holding the injector. Pull the injector out of the engine.

Inspect the boot and other rubber parts carefully for any defects. Some manufacturers recommend that the boot, seals, and hose be replaced if the injector is removed for service.

Install the new or serviced injector in reverse order. Always follow individual shop manual instructions when servicing a throttle body injection unit or multiport injector.

Fuel Injection Hoses/Clamps

The fuel hose on fuel injection systems is of a special type. Use only a fuel hose that is marked "EFM/EFI." Regular fuel hose cannot be used on a fuel injection system. Also, use only fuel hose clamps that have a "rolled edge." If regular worm-type clamps are used, the sharp edges can cut into the fuel hose. The fuel in a fuel injection system is under greater pressure than the fuel in a carbureted system. A small nick or cut in the fuel line will have a serious effect in a fuel injection system.

If gasoline drips out of the vacuum hose when removed from the fuel pressure regulator, the regulator is defective and should be replaced.

7.8 PORT FUEL-INJECTION SYSTEM DIAGNOSIS

To determine if a port fuel-injection system, including the fuel pump, injectors, and fuel-pressure regulator, are operating correctly, follow the steps given below:

1. Attach the fuel-pressure gauge to the Schrader valve on the fuel rail.
2. Turn the ignition key on or start the engine to build up the fuel pump pressure – about 241 to 310 kPa.
3. Wait 20 minutes, observe the fuel pressure retained in the fuel rail, and note the kPa reading. If the pressure drops less than 140 kPa in 20 minutes, everything is fine. If the pressure drop is more than 140 kPa in 20 minutes, there is a possible problem with:
   - The check valve in the fuel pump.
   - Leaking injectors.
   - A defective or leaking fuel-pressure regulator. To determine which unit is causing problem, perform the following procedure:
   - Reenergize the electric fuel pump.

- Clamp the fuel supply line and wait 10 minutes (Do not clamp plastic fuel lines. Instead, connect shut-off valves to the fuel system to shut off supply and return lines using the proper disconnect tool). If the pressure drop does not occur, replace the fuel pump. If the pressure drop still occurs, continue with the next step.
- Repeat the pressure buildup of the electric pump and clamp the fuel return line. If the pressure drop time is now fine, replace the fuel pressure fuel pressure regulator.
- If the pressure drop still occurs, one or more of the injectors is leaking. Remove the injectors with the fuel rail and hold over paper. Replace those injectors that drip one or more drops after 10 minutes with pressurized fuel.

7.9 ENGINE SENSOR SERVICE

Most EFI engine sensors can be tested with self-diagnosis, a scan tool, or a special tester/analyzer. However, an engine sensor can also be tested with a digital meter or sometimes a test light. Refer to the service manual for specific details.

Throttle Position Sensor Service

A throttle position sensor should produce a correct amount of resistance for different throttle openings. For example, the sensor might have higher resistance with the throttle plates closed and a lower resistance when the throttle plates are open. Compare the ohmmeter readings taken to specs to determine the condition of a throttle position sensor.

To replace a throttle position sensor, usually you have to un-stake the attaching screws. They may be soldered in place to prevent tempering. In this case, use a file to grind off the stake and remove the sensor.

After installing the new throttle position sensor, adjust the sensor. Refer to the service manual for detailed procedures. For most systems, the sensors should read a specified resistance with the throttle plates in specified positions.

Oxygen Sensor Service

Usually vehicle manufacturers recommend that you test the oxygen sensor with a specialized system tester or by the self-diagnosis mode. A digital voltmeter can also be used to check oxygen sensor. Normally, oxygen sensor output voltage should be about 0.5 volts or higher.

To replace an oxygen sensor, unplug the wire connection at the sensor, unscrew and remove the old sensor.

Coat the threads with anti-seize compound. Start the sensor by hand. Then screw in and tighten the sensor with a wrench. Do not overtighten the sensor or it might be damaged. Check the fuel injection system operation after sensor installation.

Temperature Sensor Service

Electronic fuel injection systems can use both a coolant temperature sensor and an intake air temperature sensor. If these sensors are defective, they cause the air-fuel mixture to be rich or lean. The internal resistances of these sensors change with temperature.

A digital ohmmeter is frequently used for testing a temperature sensor. Check the service manual for resistance values for various temperatures. If the ohmmeter test readings are out of specs for each temperature value, the sensor is bad and should be replaced.

Servicing Other EFI Sensors

Other sensors in an EFI system are tested using the same general procedures described above. You might use the self-diagnosis mode, a specialized analyzer, or a digital ohmmeter to check each sensor. Always refer to the individual service manual for exact procedures to be followed.

7.10 COMPUTER (ECU) SERVICE

Always follow manufacturer recommended procedure to check and service an electronic control unit. If the tests find that the ECU is bad, replace it.

A programmable read only memory (PROM) is a computer chip used in some on-board computers. It is a computer chip calibrated for the particular engine. When a computer tests faulty, you should normally reuse the old PROM in the new computer. Some ECU applications use an erasable PROM that is affixed to the board. Depending on the system, it may be required to program or "burn in" the vehicle operating information from a scan tool or computerized analyzer. In some vehicles, the ECU is in the engine compartment.

7.11 GASOLINE INJECTION ADJUSTMENTS

Not all vehicles have provisions for adjustments; therefore, refer to the right service manual, for adjustment procedures. As with a carburetor, there are a number of adjustments on some gasoline injection systems. These include:

1. Engine idle speed adjustment *Fig. 7.17.*
2. Throttle plate stop adjustment.
3. Minimum idle air rate adjustment.
4. Throttle cable adjustment, *Fig. 7.17.*

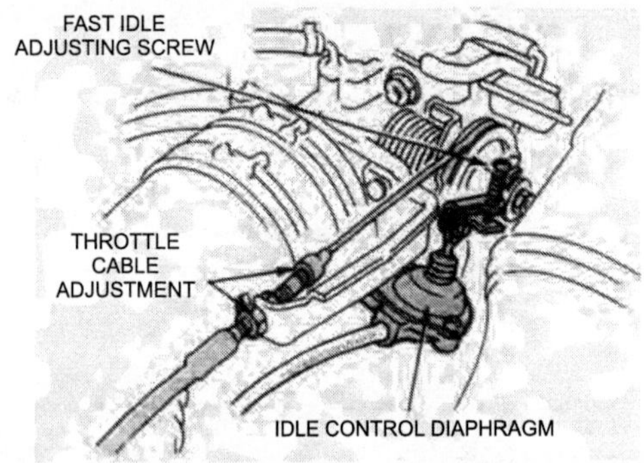

(a) Throttle cable and fast idle adjustment.

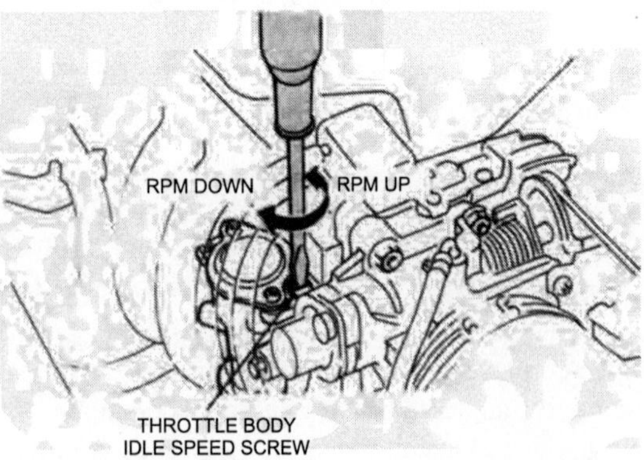

(b) Curb idle speed adjustment.

Fig. 7.17 *Refer to service manual when carrying out adjustments on the fuel injection system. (Honda Motors)*

7.12 STARTING AN EFI ENGINE

When starting carbureted engines, the driver usually has to depress the accelerator once or twice before turning the ignition key. This action sets the choke and squirts a shot of fuel into the intake manifold.

On electronic fuel injection systems, the accelerator pedal is ***never*** depressed before or during engine starting. As there is no choke to set. When the engine is not being cranked, the driver can depress the accelerator pedal any amount, fuel will ***not*** eject out of the injector. The fuel injector is energized by the ECU when it sees a signal from the crankshaft position sensor.

If the accelerator is depressed to the vehicle floor while cranking the engine, the ECU interprets this as a flooded engine. And shuts off the supply of fuel. Therefore, no fuel is supplied during cranking and the engine will not start.

If the engine does not start by just turning the ignition key to start, there is something wrong with the system. A maladjusted or binding throttle position sensor could also cause a no-start condition.

7.13 DIESEL INJECTION SERVICE

Many diesel-engine problems can be traced to the fuel injection system. The diagnostic chart lists various engine problems that could be caused by the fuel system, along with checks or corrective actions. An automotive mechanic should be trained and skilled to diagnose and repair diesel engine complaints by customers during vehicle service. For example, if a customer complaints a "miss" in the diesel engine in his/her vehicle, the mechanic must be able to pinpoint the source of problem. The problem could be an internal engine mechanical problem or a problem with the engine fuel injection system. The following information will help you "build" your diesel engine fuel injection system service skills.

Diesel Injection Diagnosis

Start diagnosis by checking the operation of the engine. Check for:

1. Abnormal exhaust smoke.
2. Excessive knock.
3. Engine "missing".
4. "No start" condition for engine.

5. Lack of engine power.
6. Poor fuel economy.

Refer to the troubleshooting chart provided in the vehicle's service manual for details. The chart will list the possible causes for each condition. The chart information will be accurate because it is designed for the particular type of injection system in the engine under service.

Figure 7.18 shows some typical diesel injection system problems.

Engine Cranks Normally But Will Not Start

If the fuel is dirty, it may have clogged the system. This requires cleaning of the fuel system components. A wrong fuel can also cause engine failure to start. Only use the recommended fuel in the engine.

Check that the injection pump timing is correct and see that the injection-pump timing mark aligns with the mark on the adapter.

To check fuel nozzles for fuel flow, loosen the injection line at a nozzle. Do not disconnect it. Wipe the connection dry. Crank engine for 5 seconds. Fuel should flow from the loose connection. If it does not, check the fuel solenoid and the fuel-supply line to the injection pump. Disconnect the fuel line at the fuel inlet at the injection pump. Connect a hose from this line to a metal container. Crank the engine to check the fuel flow. If no fuel flows, the trouble is in the fuel supply line or the fuel supply pump. If fuel flows, the problem is in the injection-pump fuel filter or the pump itself. Replace the filter first. If this does not correct the problem, replace the injection pump.

To check for a plugged fuel-return line, disconnect the line at the injection pump and connect this line to a metal container. Connect a hose from the injection pump connection to the metal container. Crank the engine. If it starts and runs, the trouble is in the fuel-return line.

The possible causes and corrective actions are:

1. **Incorrect or dirty fuel:** Flush system and use correct fuel.
2. **No fuel to nozzles or injection pump:** Check for fuel to nozzles.
3. **Restricted air or fuel flow:** Clear the causes of restriction.
4. **Bad fuel shutoff solenoid:** Replace the solenoid.
5. **Plugged fuel return:** Check return and clean if necessary.
6. **Pump timing off:** Reset fuel injection timing.
7. **Injection pump problems:** Isolate and repair the pump problem or replace the pump if necessary.
8. **Inoperative glow-plugs, incorrect starting procedure, or internal engine problems:** Replace glow-plugs, follow the correct method of engine starting and make additional tests to determine internal engine problems.

Engine Starts But Stalls on Idle

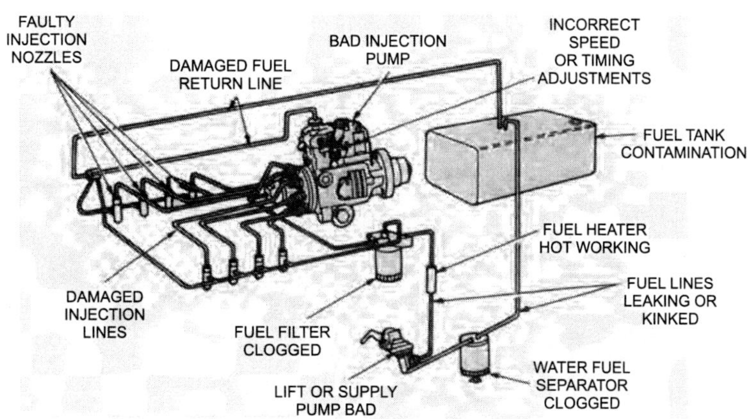

Fig. 7.18 *Various types of problems found in a diesel fuel injection system. (General Motors)*

This could be caused by any of the troubles discussed in the previous case. These include incorrect or dirty fuel, limited fuel to nozzles or injection pump, a restricted fuel return, incorrect pump timing, or defects in the injection pump. Also, stalling on idle after starting could be caused by an incorrectly set idle, or by low fuel in the tank.

The possible causes and corrective actions are:

1. **Fuel low in tank:** Fill tank.
2. **Incorrect or dirty fuel:** Flush system and use correct fuel.
3. **Limited fuel to nozzles or injection pump:** Check for fuel supply continuity to nozzles and to pump.
4. **Restricted fuel return:** Check fuel return line and clean as necessary.
5. **Idle incorrectly set:** Reset idle.
6. **Pump timing off:** Retime the pump.
7. **Injection pump trouble:** Replace pump.
8. **Internal engine problems:** Make tests to pinpoint the problem source in engine.

Rough Idle–No Abnormal Noise or Smoke

First check if the low-idle speed is correctly set. Then look for injection-line leaks. Wipe off the lines and run the engine. If there are leaks, tighten connections or replace lines as necessary to correct the problem.

Check the return fuel line for any restriction. Disconnect the fuel-return line and see if the engine runs normally.

To check for a defective nozzle, start the engine and then loosen the injection-line joint at each nozzle one at a time, taking proper care not to allow spraying of leaking fuel onto you and hot engine parts. This relieves the pressure and prevents normal action of the injection nozzle. A good nozzle is prevented from injecting fuel into its cylinder the rhythm of the engine will change. The engine will run more roughly than before. If loosening the connection at the nozzle does not change the engine idle, then that nozzle was not working before loosening the connection and should be replaced.

To check for internal fuel leaks at the fuel nozzle, disconnect the fuel-return system from the nozzles on one bank at a time. With the engine running, observe the fuel leakage at the nozzles. Replace any nozzle that has excessive fuel leakage.

Rough idle can also be caused by a fuel-supply problem. The fuel-supply pump, line, and fuel filter should be inspected carefully for any problems. Dirty fuel or the wrong fuel can cause rough idle, too.

The possible causes and corrective actions are:

1. **Low idle speed setting incorrect:** Reset the low idle speed.
2. **Injection line leaks:** Locate and fix leaks.
3. **Restricted fuel return:** Clean and remove restriction.
4. **Nozzle(s) may have problem:** Check, repair or replace as needed.
5. **Fuel-supply-pump problem:** Check the pump, repair or replace as needed.
6. **Non-uniform fuel distribution to nozzles:** Selectively replace nozzles until the problem is eliminated.
7. **Dirty or wrong fuel in system:** Flush system, and use recommended fuel.

Rough Idle With Abnormal Noise and Smoke

Check the fuel injection-pump timing. Then, disable the nozzles by loosening the injection-line connection at the nozzle, one at a time, and check their operation. A disabled nozzle that does not change the noise or smoke is the bad nozzle. Replace it.

The possible causes and corrective actions are:

1. **Injection-pump timing off:** Reset timing.
2. **Nozzle problem:** Check the nozzles in a sequence and replace the bad ones.

Diesel Engine Miss

A ***diesel engine miss*** is caused when one or more cylinders do not fire (burn fuel) properly. A diesel engine can also run roughly because of injection system problems.

A miss in a diesel engine can be caused by: faulty injectors, clogged fuel filters, incorrect injection timing, wrong or dirty fuel, low engine compression, injection system leak, air leak, or faulty injection pump.

The possible causes and corrective actions are:

1. **Plugged fuel filter:** Replace filter.
2. **Injection-pump timing off:** Reset injection–pump timing.
3. **Wrong or dirty fuel:** Flush fuel system – Fill correct fuel.

Lack of Engine Power

This is a common customer complaint that could result from many conditions in the engine systems or the engine itself as well as outside the engine. For example, dragging vehicle brakes, excessive resistance in the powertrain, or underinflated tyres can give an impression of low power. In the fuel system, loss of power could be caused by use of wrong or dirty fuel, restricted fuel return, a restricted fuel supply, or plugged nozzles.

When a diesel engine ***lacks power,*** check the throttle cable adjustment, governor setting, fuel filters, air filter, engine compression, and other factors affecting combustion. A restricted fuel tank vent can prevent normal fuel flow to the injection pump and engine so that the engine could not produce full power. An overheated engine will lose power, too.

Keep in mind that a diesel engine does not produce as much power as a gasoline engine of equal size.

The possible causes and corrective actions are:

1. **Wrong or dirty fuel:** Flush system – Use right fuel.
2. **Restricted fuel return:** Clear any restriction found.
3. **Plugged fuel tank vent:** Clear the vent.
4. **Restricted fuel supply:** Check fuel-lines, fuel-supply pump, injection pump, recondition or replace parts as needed.
5. **Plugged fuel filter:** Replace filter.
6. **Plugged nozzles:** Selectively test nozzles, replace as needed.
7. **Internal engine problems, loss of compression, compression leaks, etc.:** Make needed tests and repair or replace parts as necessary.

Poor Fuel Economy

Poor fuel economy due to problems in fuel system may be due to:

1. **A fuel leak:** Correct the leak.
2. **Clogged air filter:** Replace filter.
3. **Incorrect injection timing:** Reset injection-timing.
4. **Leaking injectors:** Check and replace as necessary. Normally, a diesel engine will get better fuel economy than an equivalent gasoline engine.

Noise or Rap From One or More Cylinders

A possible cause of noise from one or more cylinders could be presence of air in the fuel system. Air could cause a very uneven flow of fuel to the nozzles. The air present in fuel expands and contracts with changing pressure, causing too much or too little fuel to feed.

To correct the problem, loosen the injection line at each nozzle to allow the air to bleed from the system.

Another cause of the noise or rap could be a nozzle sticking open or having a very low opening pressure. The defective nozzle is located by opening the injection lines at each nozzle in turn.

The possible causes and corrective actions are:

1. **Air in fuel system:** Check the system to determine the cause and correct it.
2. **Gasoline in fuel system:** Flush the fuel and fill the right fuel.
3. **Air in high-pressure line:** Bleed fuel system to remove the air.
4. **Nozzle sticking open or with low opening pressure:** Replace defective nozzle.
5. **Engine problems:** Test, isolate the problem source, and correct or replace any defective parts as needed.

Combustion Noise With Black Smoke

Incorrect injection-pump timing, or injection-pump troubles could cause this. The cylinders receive too much fuel, or receive fuel at the wrong time. It could also be due to an injection nozzle sticking open and internal engine problems.

The possible causes and corrective actions are:

1. **Injection pump timing off:** Reset timing.
2. **Injection pump problem:** Replace injection pump.
3. **Nozzle sticking open:** Clean, replace if necessary.
4. **Internal engine problems:** Test, isolate the

problem source, and correct or replace any defective parts as needed.

7.14 DIESEL INJECTION SERVICE RULES

Basic rules to keep in mind when servicing diesel injection systems are:

1. Always wear safety glasses when working on a diesel injection system.
2. Cap lines and fittings to prevent foreign matter from entering the fuel system.
3. Handle the diesel injector and injection pump carefully. Do not pry on them or let them fall.
4. Use caution when dealing with high pressure inside a diesel injection system as it can cause serious injury.
5. Always bleed a fuel injection system after repairs.
6. Clean dirt around fittings before disconnecting them.
7. Replace bent, kinked, or damaged fuel injection lines.
8. Tighten all fasteners and connections to torque specs. It is very important in case of a diesel engine.
9. Cover the air inlet with a piece of screen mesh when the engine is to be operated with the air cleaner removed. This is needed to keep other objects from being sucked into the engine.
10. Never use your hand to cover the air inlet while the engine is running. You may get hurt.
11. Check fuel filters and water separators periodically. Presence of water can cause corrosion of costly injection system parts.
12. Observe all safety precautions concerning ventilation and fire hazards when servicing a diesel injection system.
13. Avoid mistakes that could upset engine performance or cause engine damage. When in doubt, follow instructions in the service manual for the make and model of vehicle under service.
14. Always use the procedure given in the service manual when working on an electronic or computer controlled diesel injection system. Even a small mistake made could ruin expensive system components, e.g., computer, sensors, actuators, etc.

7.15 DIESEL INJECTION SYSTEM MAINTENANCE

Follow the instructions given in the service manual on periodic maintenance of a diesel injection system. Clean or change the filters periodically. If fuel leaks are detected, use a piece of cardboard around each fitting. If there is a serious leak, it will strike the cardboard and not your hand or other body part. Replace any fuel injection line or return hose that is defective.

Fuel filters are normally located in the fuel tank (soak filter), in the fuel line (main filter), and sometimes in the injector assemblies (final filter screens). These filters should be kept in good clean condition.

The main fuel filters may have a drain. The drain can be used to bleed off trapped water. Water mixed with diesel fuel causes rapid corrosion and pitting of injection system components.

7.16 TESTING DIESEL INJECTION SYSTEM OPERATION

Operation of a diesel injection system can be checked in several ways. Most common methods are explained here.

Cylinder Balance Test

A ***cylinder balance test*** made on a diesel engine involves disabling one injector at a time to check the firing of that cylinder. Loosening the injector line to it, *Fig. 7.19 (a),* disables the cylinder.

To perform a cylinder balance test on a diesel engine, wrap a shop rag around the injector and loosen one injector fitting just enough to allow fuel to drip from the connection. When the fitting is loosened, fuel should slowly leak out of the connection. This fuel leakage will prevent the injector from opening. The inoperative injector should cause the engine rpm to drop and end up in rough idle. But, if "killing" the cylinder does not affect engine operation, that cylinder has not been firing. The injector could be bad, compression low, or a problem in injection pump. To pinpoint the problem source, some additional tests will be required. Check all injectors, one at a time.

Glow Plug Resistance-Balance Test

A ***glow plug resistance-balance test*** is a safe way of determining if each cylinder is firing. When combustion is occurring in a cylinder, it will raise the temperature and internal resistance of the glow plug in that cylinder.

To perform this test, allow the engine to cool. Unplug the wires to all of the glow plugs. Connect a digital ohmmeter across each glow plug and ground. Record the ohms reading of each glow plug. Start the engine and let run for a few minutes. Shut the engine off and recheck and record each glow plug resistance.

If a cylinder is not firing, the resistance in ohms of its glow plug will not increase as much as the others. *Fig. 7.19 (b).*

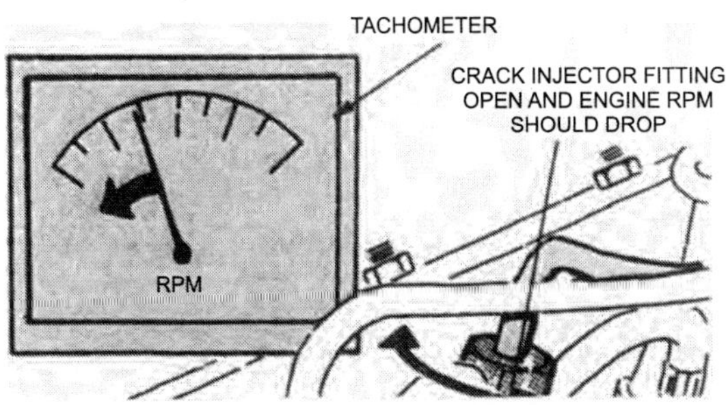

(a) Cracking open injector fitting will help isolate a dead cylinder. When the line is cracked open, injector will stop operating. This will make engine rpm to drop and the engine will idle more roughly. But if idle stays same, then that cylinder is not producing power.

Digital Pyrometer Balance Test

A ***digital pyrometer*** is an electronic device for making very accurate temperature measurements. It may occasionally be used to check the operation of a diesel engine.

Touch the digital pyrometer on the exhaust manifold at each exhaust port. The temperature of the exhaust manifold with the engine running should be almost same. If the manifold is cooler next to one exhaust port, that cylinder is not firing, *Fig. 7.19 (c).*

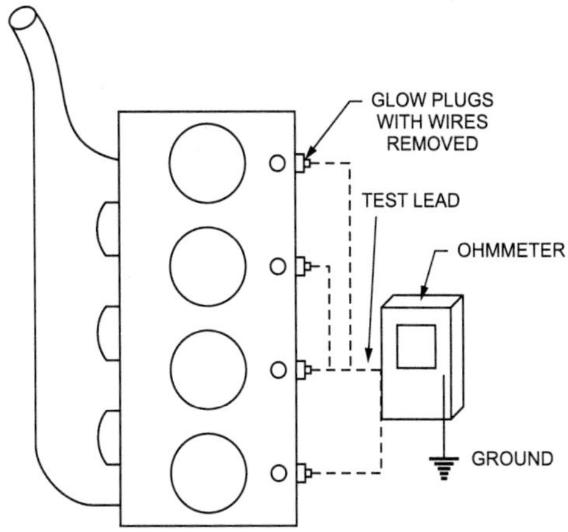

(b) Glow plug resistance-balance test will also check for dead cylinder. Measure resistance at each glow plug with engine cold. Start engine and let it warm up. Then recheck glow plug resistance. If all glow plugs do not have same resistance change, something is wrong in cylinder that has different resistance change from others.

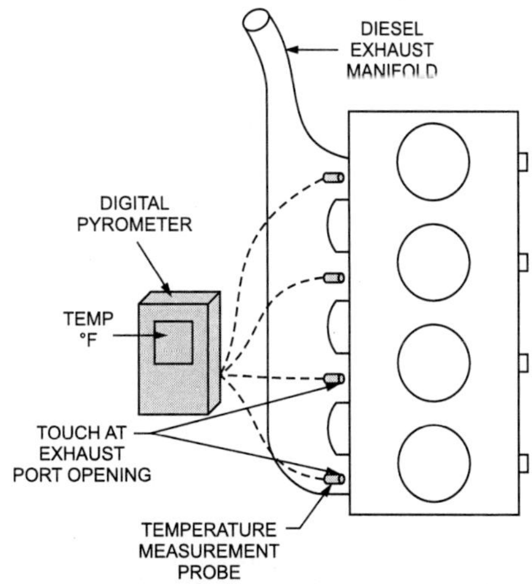

(c) Digital pyrometer will check for dead cylinder by measuring temperature of outside of exhaust manifold. If area next to one cylinder head port is cooler. That cylinder is not producing combustion power.

Fig. 7.19 *Three ways to check whether each cylinder in diesel engine is producing power or combustion. Wear gloves and eye protection. (Volvo)*

7.17 DIESEL FUEL-SYSTEM SERVICE PROCEDURES

After a problem is determined, it must be corrected. Specific servicing operations on the diesel-engine fuel-injection system that include servicing the fuel-injection lines, fuel filter, injectors, and injection and fuel-supply pumps are given here. *Fig.7.20* shows the basic parts of a diesel fuel injection system on a six-cylinder in-line engine.

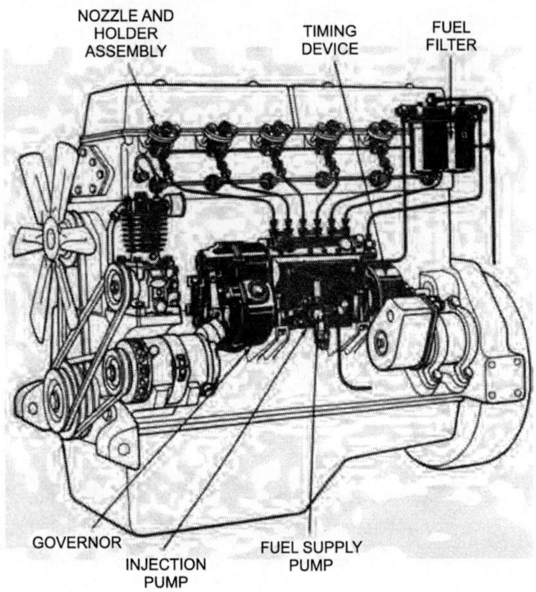

Fig. 7.20 *Fuel system for a six-cylinder diesel engine using an in-line fuel-injection pump. (Robert Bosch Corp.)*

7.18 INJECTION LINES AND FITTINGS

Fuel injection lines in diesel fuel-injection systems are also known as ***high-pressure fuel lines, injection tubes, and injection pipes.*** These lines carry the fuel, under high pressure from the injection pump to the injection nozzles in the cylinders. These pipes should be able to withstand the pressures that they are designed to withstand safely and should be non-corrosive.

A defective fuel injection line should be replaced with a recommended or approved by the engine manufacturer. Fuel-line connections, or couplings, take various forms, *Fig. 7.21*. When loosening a coupling of the type having two nuts, use two wrenches, *Fig. 7.22*. Hold the fitting with one wrench and turn the coupling nut. If you try to loosen the nut without using the backup wrench, you can twist the line and damage it. It is not necessary to use the backup wrench when disconnecting the lines from the fuel-injection pump.

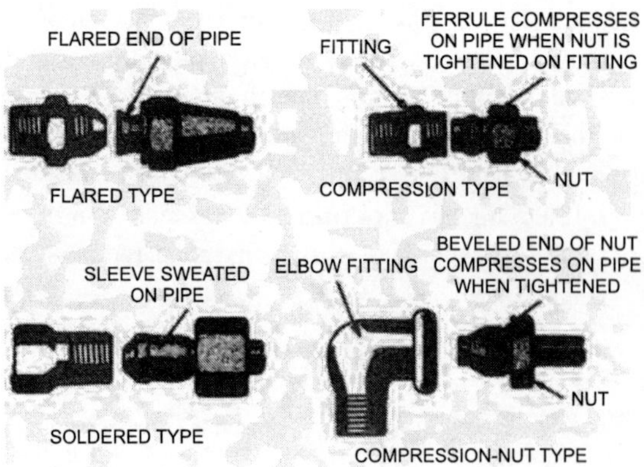

Fig. 7.21 *Types of fuel-line couplings, or connectors.*

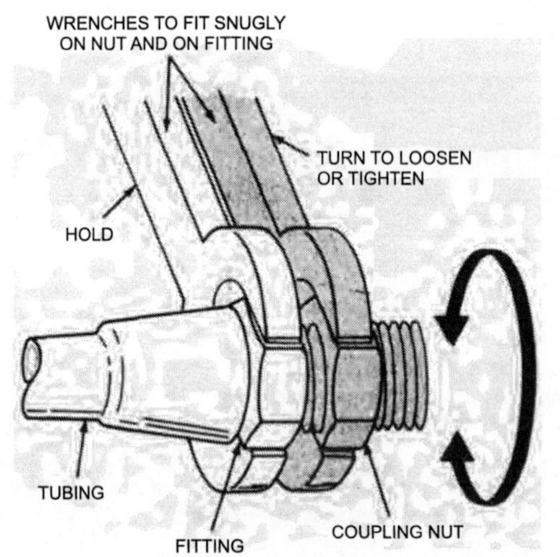

Fig. 7.22 *Using two wrenches to loosen or tighten coupling nuts to avoid twisting and damaging the line.*

New injection lines are preformed. Install them carefully to avoid twisting or bending them out of shape. When the line to be replaced is under other lines, remove the upper lines before you remove the actual line, *Fig. 7.20.*

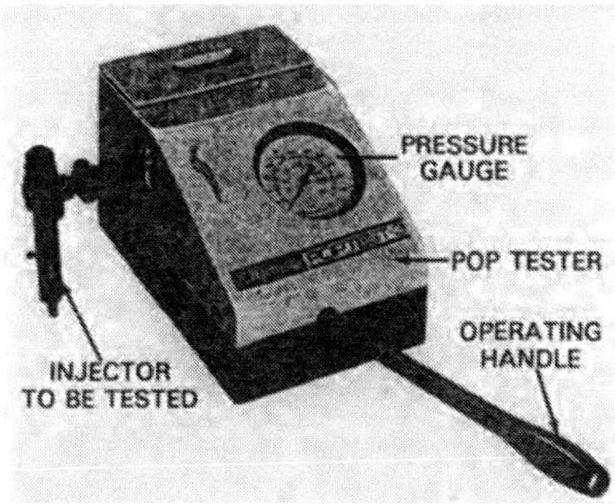

(a) Pop tester will allow measurement of injector performance.

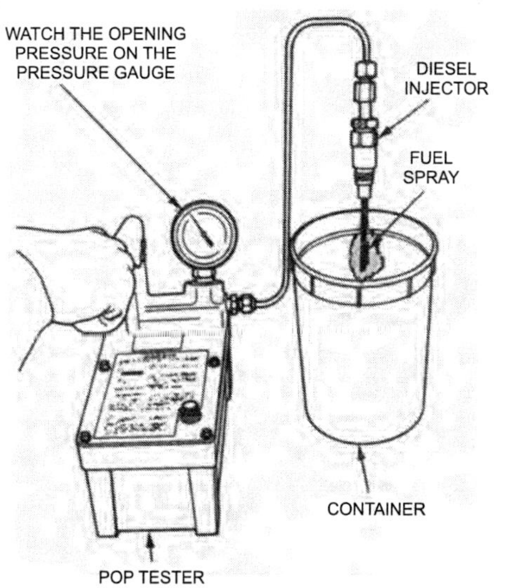

(b) Mount nozzle on pop tester. Pump handle while watching pressure gauge and spray pattern.

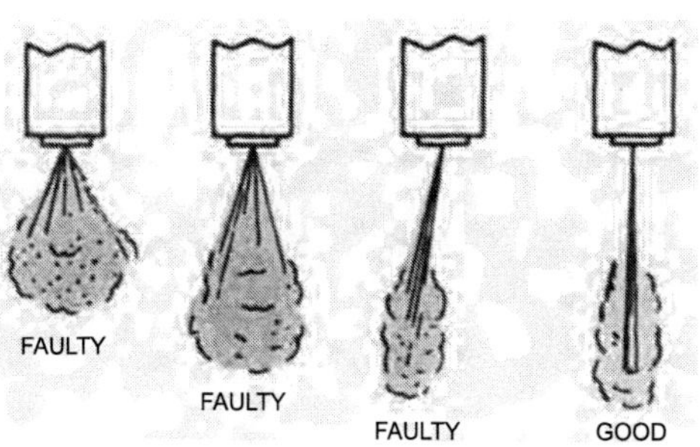

(c) Good and bad spray patterns.

Fig. 7.23 *Testing diesel fuel injectors. (Toyota, Ford, and Hartridge)*

7.19 FUEL FILTER

The fuel for a diesel engine must be clean and free of contaminants and water. Even tiny specks of dirt can clog the fuel nozzles or damage the injection-pump parts. Therefore, a fuel filter forms an essential part of the system. The filter contains a cartridge of filtering material, in the form of a special pleated paper or fiber material, through which the fuel must pass. The filter traps any particles and keeps them from entering the fuel system. The filter should be replaced at regular time intervals as recommended by the engine manufacturer.

7.20 DIESEL INJECTOR SERVICE

Injectors or ***injection nozzles*** should not be removed unless there is positive evidence that they need servicing or replacement. A bad diesel injector can cause the engine to miss. It may also reduce engine power or cause exhaust smoking and knocking.

One way to check for a faulty injection nozzle is to run the engine at fast idle. Wrap a piece of cloth around the connection to keep fuel from squirting and then loosen the connector at each nozzle in turn, one at a time. If loosening a connector causes the engine speed to drop off, the nozzle is probably working correctly. If the engine speed remains the same, then the nozzle is probably not performing its function. It could be clogged so that no fuel flows through, or the holes could be partly clogged so that the fuel spray is insufficient or does not produce the required spray pattern.

The injector nozzles are exposed to the direct heat and by-products of combustion. They can wear, become clogged with carbon, or be damaged. This can cause incorrect opening pressure, incorrect spray pattern, fuel leakage, other usual indications of trouble that include:

- One or more cylinders knocking
- Loss of engine power
- Smoky black exhaust
- Engine overheating
- Excessive fuel consumption.

A variety of nozzles are used, but all have a check valve. The check valve opens when spray pressure is applied so that the fuel can flow through. When the pressure drops, the check valve closes to shut off the flow quickly and completely.

Injector Substitution

An easy way to verify an injector problem is to ***substitute*** a good injector for the possible bad injector. If the cylinder fires with the good injector, then the old injector is bad. If the cylinder still misses with the good injector, then other engine or injection problems exist.

Injector Removal

If tests indicate a faulty injector, remove the injector for service. Following the service manual, disconnect the battery to prevent engine cranking. With proper tools, remove the injection line. Use adequate care not to bend or kink the high pressure fuel-line. Unscrew or unclamp and remove the injector.

Pop Testing Diesel Injector

A ***pop tester*** is a device used for checking a diesel injection nozzle removed from the engine. One type of tester is shown *in Fig. 7.23 (a).*

To use the pop tester, fill the tester reservoir with the correct ***calibration*** or ***test fluid. Do not*** use diesel fuel because it is highly inflammable and test results may not be reliable.

Open the tester valve and operate the tester handle. Until only the fluid, no air bubbles, sprays out the tester, close the valve. Connect the nozzle to the tester, *Fig. 7.23 (b).* Wear eye protection and use caution and keep your hands away from the fuel spray to avoid injury.

To check ***injector opening pressure,*** purge any remaining air from the nozzle by pumping the tester lever up and down. Then pump the handle slowly while watching the pressure gauge. Note the reading when the injection nozzle opens. Repeat the test until you are sure you have an accurate reading.

Typical diesel ***injector opening pressure*** is approximately 11713 to 15158 kPa. If the opening pressure is ***not*** within service manual specs, rebuild or replace the injector.

To check the ***injector's spray pattern,*** operate the pop tester handle while watching the fluid spray out of the nozzle. As shown in *Fig. 7.23 (c)*, there should be a narrow, cone-shaped mist of fluid. A solid stream of fuel, uneven spray, excessively wide spray, a spray filled with liquid droplets indicates that the injector requires service or be replaced.

To check ***injector leakage,*** slowly operate the tester handle to maintain a pressure ***lower*** than the nozzle opening pressure. Most vehicle manufacturers recommend a pressure of about 2060 kPa below opening pressure. With this pressure, the diesel injector should ***not*** leak or drip for 10 seconds. Leakage would indicate a dirty injector nozzle or worn components.

Diesel Injector Rebuild

An ***injector nozzle rebuild*** involves disassembly, cleaning, inspecting, replacing defective or worn parts, reassembly, and then testing the injector to verify its performance. As the injector designs vary, always follow the service manual procedures when servicing injectors.

Installing Diesel Injectors

Coat the threads of the injector with anti-seize compound before installing the injector. Use a new heat shield or seal to prevent compression leakage. First screw the injector into the cylinder head by hand. Then, torque it to specs. Reconnect the injector line, using care not to bend or kink it. Torque the connection properly, *Fig. 7.24.*

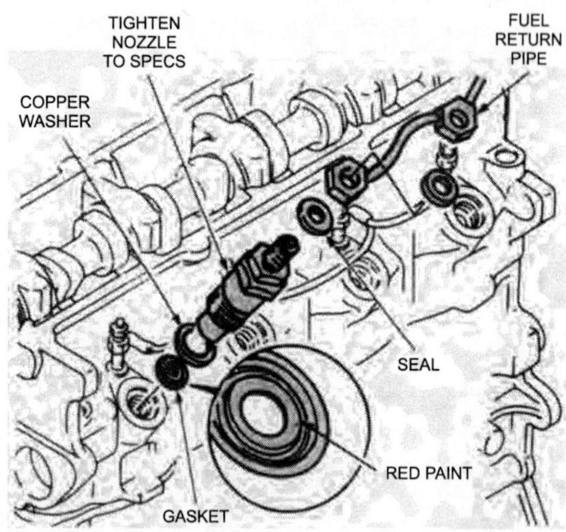

Fig. 7.24 *When replacing injector nozzle, place anti-seize compound on threads. Use new gaskets or seals if necessary. Torque fasteners to specs. Avoid kinking or bending injection line when disconnecting from or connecting to nozzle. (Ford Motors)*

Reconnect the wire on the fuel shutoff solenoid and start the engine. Check the fuel system for any leaks and proper operation. If necessary, ***bleed air out of*** the system.

7.21 GLOW PLUG SERVICE

Inoperative glow plugs cause a diesel engine hard to start especially when cold. If only one or two glow plugs are inoperative, the engine could miss while still cold.

If glow plug problems are indicated, use a test light or VOM to check for voltage to the glow plugs. Touch the probe on the feed wires to the glow plugs. A clip-on ammeter can be used to check current flow to the glow plugs.

An ohmmeter is used to determine the condition of each glow plug. Connect the ohmmeter across each glow plug terminal and ground. The glow plug resistance should be within specs. If the resistance is outside specs, replace the glow plug, *Fig. 7.25.*

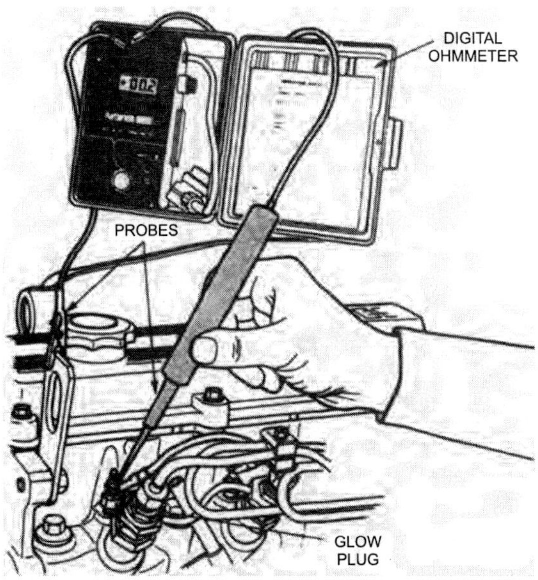

Fig. 7.25 *Ohmmeter will check for opened or shorted glow plug. Resistance must be within specs. Plugs can be checked while in or out of engine. (Ford Motors)*

Glow Plug Replacement

To remove glow plugs, disconnect the wires going to them. Then, use a deep well socket and ratchet to unscrew the glow plugs. Be careful, a glow plug can reach temperatures above 538°C. If touched, it can seriously burn your hand.

Remove the glow plugs and check each one carefully to find any damage or a heavy coating of carbon. Carbon build up can insulate the plug and cause the engine hard to start. Clean the glow plugs that are good to be reused and replace the ones that are defective.

When installing the glow plugs, coat their threads with anti-seize compound. First start the plugs by hand and then torque to specs.

7.22 INJECTION PUMP SERVICE

A ***bad fuel injection pump*** may keep the engine from running, or cause engine miss firing, smoky exhaust, and other troubles. Generally, a fuel injection pump is quite trouble free unit. However, contamination by water, prolonged service, leaking seals, and physical damage from accidents could require the pump replacement, repairs, or adjustment.

Most mechanics and repair shops replace the fuel injection pump when internal parts of the old pump are defective. A diesel injection pump is a precise mechanism and specialized tools and equipment are necessary for its repairs.

Injection Pump Timing

Injection timing is very critical to diesel engine performance. Injection pump timing must be set whenever the pump is removed from the engine or when an incorrect adjustment is pinpointed.

Diesel ***injection pump timing*** is adjusted by rotating the injection pump on its mounting. To advance injection timing, turn the pump ***opposite*** to drive shaft rotation. To retard injection timing, turn the pump housing ***in*** the direction of the drive shaft rotation.

There are a number of methods used for adjusting injection timing. Refer to the service manual for details on and the adjustment method to be used. Also, follow the operating instructions for the specific type of timing device.

Basically, injection pump timing can be adjusted by:

1. Aligning timing marks on engine and injection pump with # 1 piston at TDC – for rough adjustment.
2. Using a dial indicator, measure the injection pump plunger stroke in relation to engine piston position, *Fig. 7.26.*

3. Using a luminicity device to detect combustion flame in # 1 cylinder, *Fig. 7.27*.
4. Using a fuel pressure (injection) detector.

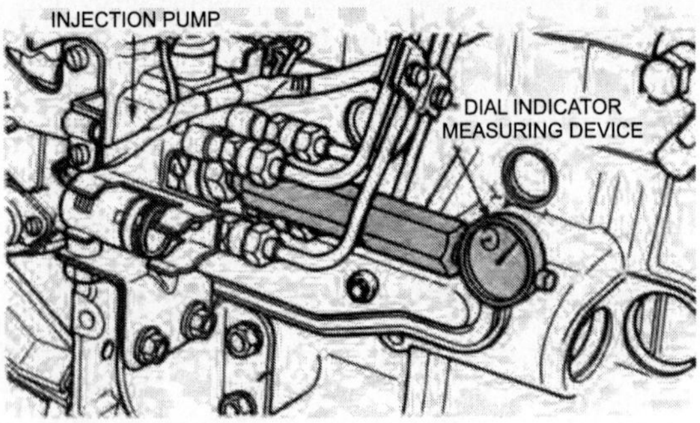

Fig. 7.26 *Some manufacturers recommend dial indicator for measuring pumping plunger movement to set injection timing. Plunger movement is compared to timing marks on crank damper so injection can be timed with engine piston position. Follow the instructions given in service manual. (Ford Motors)*

Injection Pump Throttle Cable Adjustment

Injection pump ***throttle cable*** or ***linkage adjustment*** is done to ensure that the pump's throttle lever opens and closes freely. As parts wear, the cable or linkage may require adjustment. During adjustment, check for any kind of binding experienced. Lubricate if necessary, *Fig. 7.28 (a)*.

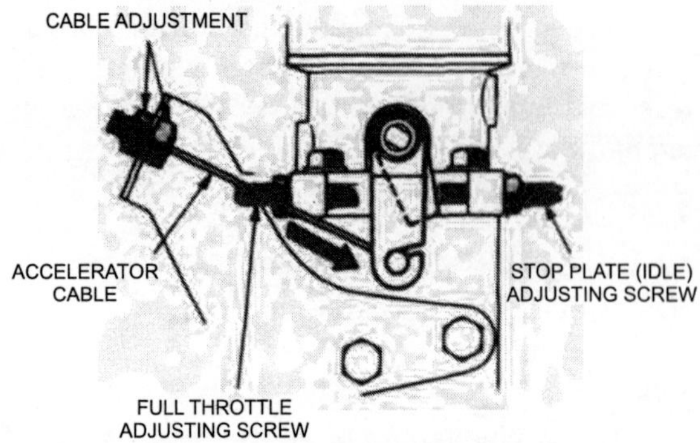

(a) Cable adjustment, idle adjustment, and maximum speed adjustment.

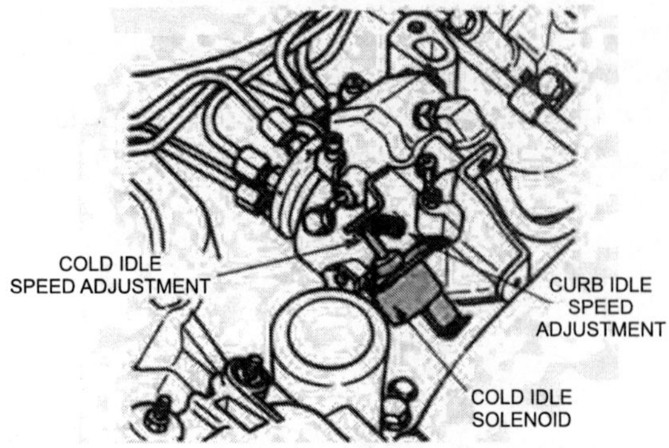

(b) The view shows cold idle solenoid adjustment and curb idle screw. Follow service manual procedures when adjusting.

Fig. 7.28 *Speed adjustments on diesel are fairly simple. Turn adjustment screw until tachometer reads with in specs. (Ford Motors).*

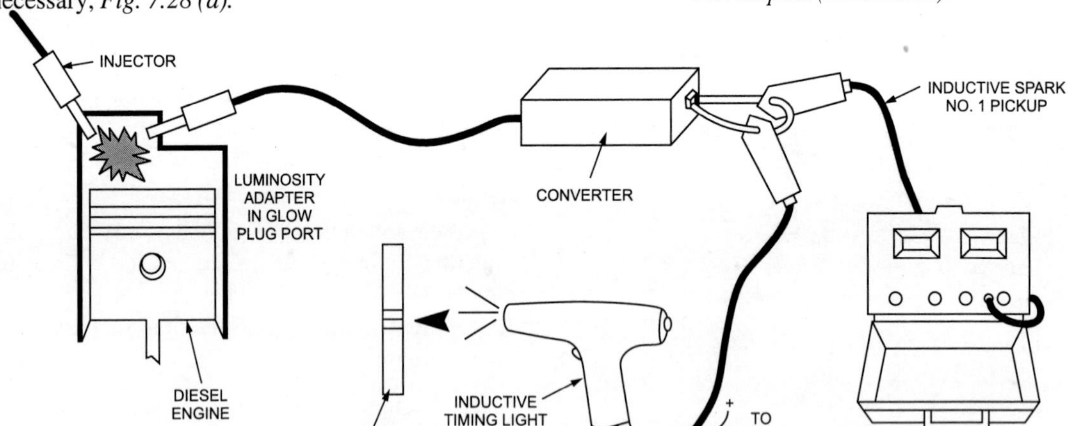

Fig. 7.27 *Luminosity probe is used to detect combustion flame. Converter amplifies signal from probe so conventional timing light or tachometer can be used on diesel. A diesel does not have an electrical ignition system for powering a timing light or tachometer.*

7.23 INJECTION PUMP SPEED ADJUSTMENTS

There are three speed adjustments commonly used on a diesel injection pump: ***curb idle speed, maximum speed,*** and ***fast (cold) idle speed.***

Diesel Idle Speed Adjustments

Diesel idle speed is usually set using a special diesel tachometer. An idle stop screw on the injection pump is turned to adjust engine speed.

A ***diesel tachometer*** may be similar to or part of the injection timing device. With all designs, the diesel tack will read engine speed in rpm.

Diesel Curb Idle Speed

To adjust diesel ***curb idle speed,*** place the transmission in neutral or park. Start the engine. Run the engine until it reaches normal operating temperature. Connect the tachometer to the engine.

Compare the tack reading with specs. Curb idle speed is usually given on the engine compartment emission control decal or in the service manual.

If necessary, turn the curb idle speed adjusting screw on the injection pump to increase or decrease idle rpm, *Fig. 7.28 (b).* Loosen the locknut before turning the adjusting screw to allow it to turn and make the adjustment. Hold the adjusting screw and retighten the locknut when the tachometer reads correctly.

Diesel Cold Idle Speed Adjustment

To set the ***cold idle speed*** of a diesel engine, connect a jumper wire from the battery positive terminal and the fast idle solenoid terminal. This will activate the fast idle, even when the engine is warm.

Increase the engine speed momentarily to release the solenoid plunger. Again compare the tachometer readings to specs. Adjust the solenoid if necessary.

In some applications a starter lever replaces a cold idle solenoid. Refer to manufacturer service manuals in each case for procedures on adjustment.

Diesel Maximum Speed Adjustment

A diesel engine ***maximum speed adjustment*** is set to limit the maximum possible diesel engine rpm. If maximum governor rpm is set too high, it could cause internal engine damage. If the diesel engine maximum rpm is too low, the engine will not produce necessary maximum power.

To set the maximum speed of a diesel engine, position the tachometer such that it can be read from within the driver's compartment. Put the transmission in neutral or park, apply brake with your left foot and press the accelerator pedal slowly to the vehicle floor.

Once the tachometer reads maximum specified speed, engine speed should no longer increase. If the maximum speed is not within specs, loosen the locknut, turn the maximum speed adjusting screw on the injection pump. Tighten the locknut on the adjusting screw after adjustment. Again check the maximum rpm.

7.24 FUEL CUTOFF SOLENOID SERVICE

A ***defective fuel cutoff solenoid*** usually will keep the diesel engine from running Solenoid windings may fail and it can not open to allow fuel to enter the pumping chambers in the fuel pump. The solenoid may also fail in the open position, keeping the diesel engine from shutting off.

Always check the supply circuit to the fuel cutoff solenoid. Wires feeding electric current to the solenoid could be open or short circuited, causing faulty operation of the solenoid.

Testing Fuel Cutoff Solenoid

Use a test light to check the power to the solenoid first. The test light should glow with the ignition key on or the engine cranking. If it does not, disconnect the wire from the solenoid. Check for power on the incoming wire alone. If there is no voltage, the circuit is bad. If there is voltage, the solenoid is probably faulty.

An ammeter can also be used to measure the resistance of the windings in the solenoid. If open or shorted, replace the solenoid.

Replacing Fuel Cutoff Solenoid

To replace the fuel cutoff solenoid, just unscrew the old solenoid. Be very careful not to allow any dirt or debris fall into the opening in the injection pump.

Make sure it is perfectly clean, install the new solenoid into the pump. Torque the fasteners to specs. Connect its feed wires and check the engine operation.

7.25 COMPUTER CONTROLLED DIESEL INJECTION SYSTEM SERVICE

Most computer controlled diesel injection systems have self-diagnosis feature. The computer will produce a number code indicating possible troubles with components in the system. This is similar to the self-diagnosis described for gasoline injection.

Since the procedures vary for different applications and a small mistake made could cause part damage. Always refer to the applicable service manual for detailed procedures of activating and testing a computer controlled diesel injection system.

Sensor and Actuator Problems

As discussed in computer controlled gasoline injection system, engine sensors and actuators on computer controlled diesel injection systems can fail and upset engine operation. Use same basic techniques to check the internal resistances of components, current levels, and voltage output at various wires. If a resistance, current, or voltage value is found to be out of specs, you must take a corrective action to restore the system function.

THEORY QUESTIONS

1. Explain the difference between throttle body and multiport injection in a gasoline injection system.
2. List the seven basic components of an electronic gasoline fuel injection system.
3. How does an electronic gasoline injector open?
4. This device is used to control gasoline engine idle speed on a gasoline fuel injection engine:
   (*a*) Aneroid bellows (*b*) Dashpot
   (*c*) Pressure regulator (*d*) Air bypass valve.
5. A TBI assembly and a multiport injection throttle body perform the same functions of controlling air and fuel flow. *True* or *False?*
6. Before replacing a fuel filter, the pressure in the fuel system must be released. *True* or *False?*
7. Any type of fuel hose and clamps can be used on a fuel injection system. *True* or *False?*
8. Mechanic A says that low fuel pressure can cause a surging condition. Mechanic B says that low fuel pressure can be caused by a bad fuel pump or a bad pressure regulator. Who is right?
   (*a*) A only (*b*) B only
   (*c*) Both A and B (*d*) Neither A nor B.
9. Mechanic A says that removing the vacuum hose from the pressure regulator should cause a drop in fuel pressure. Mechanic B says that removing the vacuum hose from the pressure regulator should cause fuel pressure to increase. Who is right?
   (*a*) A only (*b*) B only
   (*c*) Both A and B (*d*) Neither A nor B.
10. In a gasoline fuel-injection system, gasoline fuel is sprayed into the air
    (*a*) in the combustion chamber
    (*b*) before it reaches the intake valve
    (*c*) after it passes the intake valve
    (*d*) all of the above.
11. Gasoline fuel-injection systems can be classified in two ways, according to whether they are:
    (*a*) continuous or controlled
    (*b*) timed or pulsed
    (*c*) timed or continuous
    (*d*) none of the above.
12. An engine ———— is a device that can change resistance or voltage output with a change in operating condition.
13. List and explain eight engine sensors.
14. A typical gasoline fuel-injection system for an eight-cylinder engine has the injection valves grouped into
    (*a*) two groups (*b*) three groups
    (*c*) four groups (*d*) none of the above
15. The amount of fuel injected by the timed injection system depends on
    (*a*) how far the injection valves open
    (*b*) when the injection valves open
    (*c*) how long the injection valves stay open
    (*d*) all of the above.
16. Throttle-body injection has
    (*a*) one injection valve for each cylinder
    (*b*) one or two injection valves for the engine

(*c*) one injection valve for each spark plug
(*d*) none of the above.

17. In the digital fuel-injection system, the injectors are mounted
(*a*) in the valve port
(*b*) in the intake manifold
(*c*) above the throttle valves
(*d*) below the throttle valves.

18. Mechanic A says that gasoline injection systems use a computer to help regulate the correct amount of fuel sprayed into the engine. Mechanic B says that gasoline injection systems use engine sensors to help regulate the correct amount of fuel sprayed into the engine. Who is right?
(*a*) Mechanic A (*b*) Mechanic B
(*c*) Both A and B (*d*) Neither A nor B.

19. All of the following are basic components of a multiport fuel injection system *except:*
(*a*) a fuel rail
(*b*) a throttle body
(*c*) a pressure regulator
(*d*) a centrally located fuel injector.

20. Mechanic A says that a throttle body for a TBI system normally contains the system's injector(s) and fuel pressure regulator. Mechanic B says that throttle body for a TBI system normally contains only the injector(s). Who is right?
(*a*) A only (*b*) B only
(*c*) Both A and B (*d*) Neither A nor B.

21. Mechanic A says that to clear the trouble codes from most computer systems, you should disconnect the ECU from battery power. Mechanic B says that most computer control systems activate a light in the dash to indicate a diagnostic trouble code. Who is right?
(*a*) A only (*b*) B only
(*c*) Both A and B (*d*) Neither A nor B.

22. Mechanic A says that the computer controls injector pulse width in a multiport fuel injection system. Mechanic B says that the electronic fuel pump controls injector pulse width in a multiport fuel injection system. Who is right?
(a) A only (b) B only
(c) Both A and B (d) Neither A nor B.

23 Mechanic A says that a throttle body assembly for a multiport injection system is primarily used to control fuel flow into the engine. Mechanic B says that a throttle body assembly for a multiport injection system is primarily used to control airflow into the engine. Who is right?
(*a*) A only (*b*) B only
(*c*) Both A and B (*d*) Neither A nor B.

24. All of the following are typical sensors used in a throttle body fuel injection system *except:*
(*a*) oxygen sensor
(*b*) knock sensor
(*c*) Engine coolant temperature sensor
(*d*) Mass airflow sensor.

25. Low fuel pressure can be caused by a:
(*a*) Pressure regulator valve that is stuck open
(*b*) Defective fuel pump
(*c*) Both (*a*) and (*b*)
(*d*) Neither (*a*) nor (*b*).

26. When the electronic control unit (ECU) fails, the normal procedure is to
(*a*) adjust the ECU
(*b*) replace the ECU
(*c*) send ECU to a rebuilders
(*d*) rewire the ECU.

27. What is a service manual trouble code chart?
Mechanic A says that a leaking injector may cause hard starting or dieseling. Mechanic B states that a defective fuel pump relay can cause hard starting. Who is right?
(*a*) A only (*b*) B only
(*c*) Both A and B (*d*) Neither A nor B.

28. Mechanic A says that the accelerator pedal must be pumped several times prior to and during the starting of an EFI engine. Mechanic B says that only the ignition key has to be turned to start an EFI engine. Who is right?
(*a*) A only (*b*) B only
(*c*) Both A and B (*d*) Neither A nor B.

29. A customer vehicle equipped with a computer controlled fuel injection system has the "check engine light" glowing. Mechanic A says that a scan tool should be connected to the data link connector for reading the code. Mechanic B says that only a digital meter should be used on a computer system to avoid part damage. Who is right?
(*a*) Mechanic A (*b*) Mechanic B
(*c*) Both A and B (*d*) Neither A nor B.

30. How do you use voltage pulses or light flashes to determine a computer self-diagnosis code?

31. During self-diagnosis, the engine light flashes twice, pauses, and flashes four more times. This would represent a trouble code ————.

32. Describe the procedure used to check the operation of a fuel pressure regulator?

33. The opening of the solenoid-operated injection valve can be checked with
(*a*) a pressure gauge (*b*) a voltmeter
(*c*) an oscilloscope (*d*) a dynamometer.

34. This component controls injection quantity.
(*a*) Delivery valves (*b*) Transfer pump
(*c*) Solenoid (*d*) Metering sleeve.

35. Glow plugs are ——————— ——————— that warm the air in the prechamber and aid ——————— starting of a diesel engine.

36. What is the most common symptom of an inoperative glow plug?

37. In the diesel engine, the fuel is injected into the engine cylinders
(*a*) at the start of the power stroke
(*b*) at the end of exhaust stroke
(*c*) toward the end of the compression stroke
(*d*) none of the above

38. A customer comes into the shop with a vehicle equipped with a diesel engine and complains of a miss at idle speed. Mechanic A says that the shop's digital pyrometer should be used to check for a cold section next to a port on the exhaust manifolds. Mechanic B says that cranking open the injection lines is a safer and faster way of finding a dead cylinder. Who is right?
(*a*) Mechanic A (*b*) Mechanic B
(*c*) Both A and B (*d*) Neither A nor B.

39. Diesel oil is commonly used inside a pop tester. *True* or *False?*

40. Typical diesel injector opening pressure is about to ——————— kPa.

41. In a diesel-engine fuel system, the
(*a*) fuel is delivered to the injection nozzles at high pressure
(*b*) timing of fuel delivery changes with engine speed
(*c*) amount of fuel delivered changes with operating requirements
(*d*) all of the above.

42. Tests indicate a faulty diesel injection pump. Mechanic A says the pump should be disassembled to find and repair the source of the problem. Mechanic B says it is better to install a new or rebuilt injection pump. Who is right?
(*a*) Mechanic A (*b*) Mechanic B
(*c*) Both A and B (*d*) Neither A nor B.

43. All of the following are basic parts of a diesel injection system *except:*
(*a*) Fuel rail
(*b*) Glow plug controller
(*c*) Injector nozzles
(*d*) Actuators.

44. Mechanic A says that a diesel injection pump's governor is used to control engine idle speed. Mechanic B says that a governor is used on a diesel injection pump to limit maximum engine speed. Who is right?
(*a*) A only (*b*) B only
(*c*) Both A and B (*d*) Neither A nor B.

45. Mechanic A says that an inline injection pump uses one or two pumping plungers to provide fuel for all the engine cylinders. Mechanic B says that an inline injection pump has one pumping plunger for each engine cylinder. Who is right?
(*a*) A only (*b*) B only
(*c*) Both A and B (*d*) Neither A nor B.

46. Mechanic A says that during injection system diagnosis, you should check the engine for excessive knocking. Mechanic B says that during diesel injection system diagnosis, you should check for abnormal exhaust smoke. Who is right?
(*a*) A only (*b*) B only
(*c*) Both A and B (*d*) Neither A nor B.

47. If a diesel engine's cylinder is not firing, the resistance of its glow plug will not ——————— ——————— ——————— the other glow plugs.
(*a*) decrease as much as
(*b*) increase as much as
(*c*) Increase more than
(*d*) decrease more than

48. All of the following are various methods used to check the operation of a diesel injection system *except:*
(*a*) digital pyrometer balance test
(*b*) cylinder balance test
(*c*) glow plug resistance-balance test
(*d*) spark test.

49. Mechanic A says that a dial indicator can be used to adjust diesel injection pump timing. Mechanic B says that a luminosity device can be used to adjust diesel injection pump timing. Who is right?
(*a*) A only (*b*) B only
(*c*) Both A and B (*d*) Neither A nor B.

❑❑❑

Ignition Systems Diagnosis and Service

8

8.1 INTRODUCTION

Ignition systems are fairly simple collections of electrical and mechanical parts, yet ignition failures are a major cause of no-start or engine performance problems. Because proper ignition system performance is critical to engine operation, some regular maintenance is necessary.

Electronic ignition systems were designed to reduce or eliminate contact point-type maintenance needs but have introduced some problems of their own. Instead of burned points or defective condensers, we now have easily damaged high tension spark plug wires, low voltages travelling in the wiring harness from pickup to module, and mysterious electronic ailments.

This chapter will introduce the normal adjustments and maintenance needed on both point and electronic ignition systems. Here the most common ignition system tests, adjustments, and repairs done by an automotive mechanic will be explained.

8.2 CONTACT-POINT IGNITION-SYSTEM SERVICE

The following sections describe the servicing of contact-point ignition systems.

Causes of Ignition System Failure

Ignition-system failures in contact-point systems can be grouped into three categories.

1. *Loss of energy in primary circuit.* This may be caused by several conditions.

- Resistance in the primary circuit due to defective leads, bad connections, burnt distributor contact points or switch, or open primary winding.
- Improper point setting.
- Discharged battery or defective alternator.
- Defective condenser – shorted, low insulation resistance, high series resistance.
- Grounded primary circuit in coil, wiring, or distributor.

2. *Loss of energy in the secondary circuit.*

- Spark plugs fouled, broken or improperly gapped.
- Defective high-voltage wiring that allows high-voltage leaks.
- High-voltage leaks across coil head, distributor cap, or rotor, *Figs. 8.1* and *8.2*.

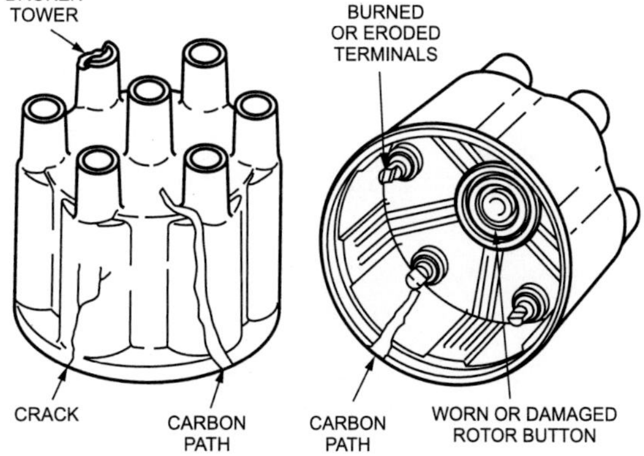

Fig. 8.1 *Defects in distributor caps that require them to be replaced. Carbon paths allow voltage leakage to ground, so the sparkplugs do not fire. (American Motors Corporation)*

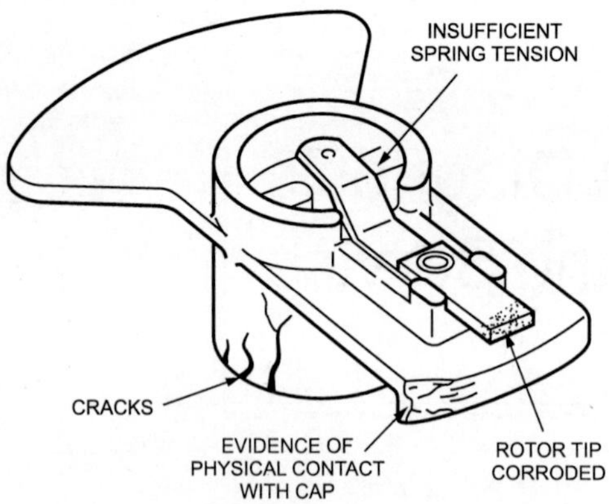

Fig. 8.2 *Rotor defects that require the rotor to be replaced. (American Motors Corporation)*

- Defective ignition coil.
- Defective connections in high-voltage circuits.

3. *Off-set ignition timing.*
- Improper set timing.
- Defective vacuum advance mechanism.
- Defective centrifugal advance mechanism.
- Preignition caused by plugs of wrong heat range, fouled plugs, etc. *Fig. 8.3.*

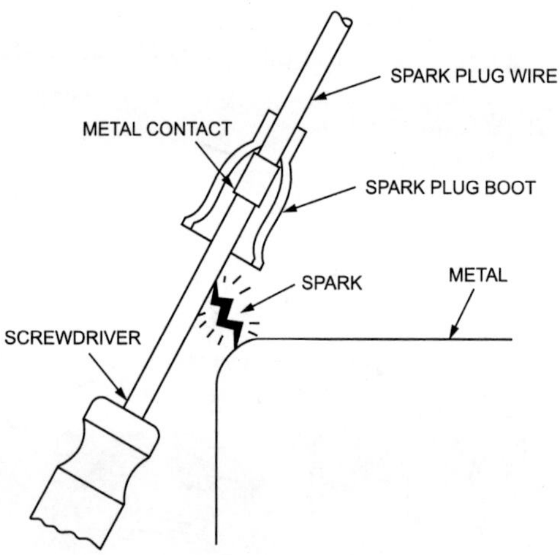

Fig. 8.3 *Checking for spark at spark plug wire.*

8.3 CONTACT- POINT IGNITION SYSTEM TROUBLESHOOTING

The ignition system provides a high-voltage spark in each of the engine's cylinders at the correct time to ignite the air-fuel mixture. If the engine does not operate normally, several quick checks can be made to determine whether the ignition system is at fault. The following applies to the contact-point ignition systems.

Engine Does Not Start

- An engine that cranks over fast enough to run, but will ***not start,*** the trouble could be either in the ignition system or the fuel system. An automobile that loses power and quits on the road is likely to have ignition trouble.
- When either of these problems occurs, first eliminate the obvious. Check the fuel gauge to make sure that the vehicle is not out of gasoline. Open the hood, smell for gasoline, and have a complete visual inspection of the engine. Check that the carburetor and fuel pump are in satisfactory condition. If every thing looks good, begin a step-by-step inspection of the ignition system.
- Check all of the ***ignition system wiring*** very carefully. A coil wire may have fallen out or one of the primary leads to the coil or distributor may have come loose.
- If the wiring looks good, pull off one of the spark plug wires at the spark plug. Insert the blade of a small screwdriver into the boot that covers the spark plug so that it touches the metal connector. Prop the screwdriver up so that the rest of the blade is about 6-10 mm away from any metal part of the engine, *Fig. 8.3.* Ask your assistant to crank the engine. If the ignition system is good, a strong spark should jump from the screwdriver to the adjacent metal surface. If the ignition system is fine, proceed to the fuel system checks.
- ***No spark*** at the spark plug wire, however, means the ignition system is at fault. Check the ignition system further.
- Connect a test ammeter into the ignition-coil primary circuit. Watch the ammeter while cranking the engine. If it gives a small, steady reading that fluctuates slightly, the primary circuit is probably good. The trouble is probably a defective coil secondary winding or secondary leads, a defective condenser, or high voltage leakage across the cap, rotor, or coil head. If the ammeter shows a fairly high and steady reading:
  (*a*) The contact points are out of adjustment,

(*b*) The condenser is shorted,

(*c*) The coil primary circuit is grounded.

- If there is no ammeter reading, the primary circuit is open. This could be due to out-of-adjustment contact points, a loose connection, defective wiring or switch, or an open coil primary winding.
- Pull the ***secondary coil wire*** out of the middle of the distributor cap or the lead from one spark plug. Use insulated pliers to hold it to avoid getting a shock. Position this wire about 6-10 mm away from a metal part of the engine, *Fig. 8.4.* Ask your assistant to crank the engine. Look for a spark to jump from the coil wire to the metal. If several strong sparks jump every time the engine cranks over, the primary system and the coil are functioning properly. The problem must be in the ignition secondary system, i.e., the rotor, or the distributor cap.

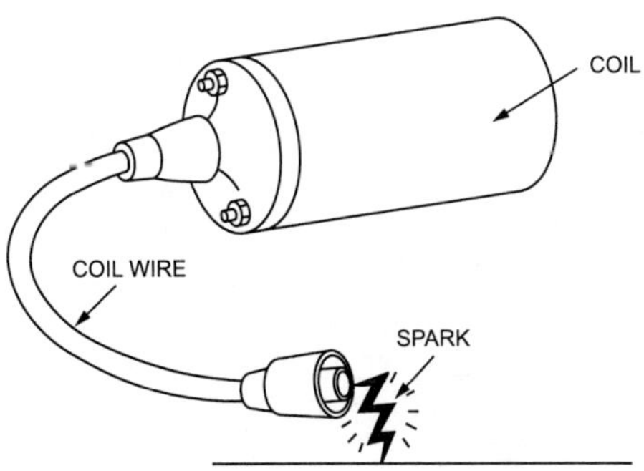

Fig. 8.4 *Checking for spark at coil.*

- Remove the ***distributor cap*** from the distributor and look inside the cap. It may be cracked or could have water or dirt in it that would interfere with current flow. Wipe the cap with a clean shop rag. Check the centre contact in the cap that touches the rotor. These sometimes break off. Replace cap if damaged. If the cap is clean and the contacts are fine, check out the rotor.
- Try to turn the ***rotor*** on the distributor shaft. It should not turn. If it does, the key — the small projection inside the rotor hole that fits into a groove on the shaft and holds the rotor stationary with respect to the shaft — is broken. Check out the contact on the rotor to ensure that it is not broken. If the rotor is damaged, replace it.
- ***Ignition primary problems*** are common on older vehicles with mechanical contact-breaker points. In their case, if there is no spark condition at the coil wire, check the breaker points. Leave the coil wire in position to check for a spark. A common problem is too dirty or burnt contact point surfaces. Have your assistant crank the engine one crank at a time until the contact points are closed. Turn on the ignition key. Touch a screwdriver blade between the contact point wire terminal and a metal part of the distributor. There should be a small spark that indicates that current is reaching the points. Use a screwdriver to open and close the contact points by pushing on the movable contact point, *Fig. 8.5.* Be careful not to touch any other metal with the screwdriver when opening and closing the contact points. If the contact points are in good condition, each time they are opened a spark will jump from the coil wire to the metal.

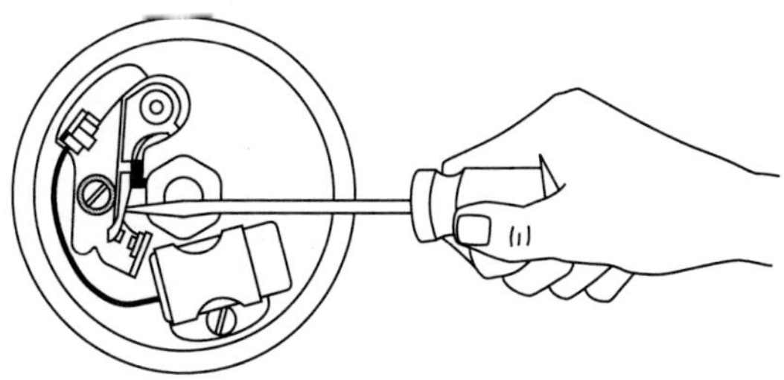

Fig. 8.5 *Checking for spark at contact points.*

- If there is no spark at the coil wire when the contacts are opened and closed with a screwdriver, the contact points are probably bad. Clean their surfaces with a small ignition point file. If the file is not available, the points can be cleaned with a fingernail file or simply scraped with a screwdriver tip. Clean both contact point surfaces and make the contact point check again with the screwdriver. If they work now the engine should be able to start and run.

Engine Misses

Engine missing, as caused by defects in ignition system, could be due to:

- Worn or out-of-adjustment contact points.
- Defective condenser.
- Centrifugal advance or vacuum-advance mechanism malfunction.
- Defective secondary wiring.
- Defective ignition coil.
- Poor electrical connections.
- High-voltage leakage across ignition-coil head, cap, or rotor, *Figs. 8.1 and 8.2.*
- Defective spark plugs. If no change in engine speed takes place when a spark plug is shorted out, then the cylinder with that plug is not firing.
- Using a wrong ignition coil for the engine, or reversed connections to the ignition coil can cause missing.
- Installing a battery backwards can cause missing. This reverses the polarity of the coil. In addition, it can cause damage to the charging system and other electronic devices on the vehicle. Coil terminals are usually marked to prevent wrong connections.

With reversed coil polarity, the electrons have to jump from the relatively cool, outer spark plug electrode to the hotter centre electrode. This requires a higher secondary voltage and enhances the possibility of the engine missing, particularly at high speed. Normally the coil and battery are connected so that electrons jump from the hot centre electrode to the outer electrode. With the emitting electrode hot, the electrons can jump the spark plug gap more easily. Voltage requirements are considerably lower.

Reversed polarity can be checked by looking at the coil primary connections. The negative terminal should be connected to the wire that goes to the distributor. If the polarity is reversed, the oscilloscope pattern will be upside-down.

Engine Overheating

Improper ignition timing can cause engine overheating. Retarded or late ignition timing the air-fuel mixture burn later in the power stroke and results in more heat passing through the cylinder walls into the coolant.

8.4 USING AN OSCILLOSCOPE FOR TROUBLESHOOTING

The automotive oscilloscope is the standard troubleshooting tool for ignition systems. A cathode ray oscilloscope will show a graph of the changing voltage in the system as pattern on a television like screen. A qualified and skilled mechanic can interpret the pattern and determine if any ignition system component is working incorrectly *Fig. 8.6.*

Fig. 8.6 *An oscilloscope. (Sun Electric Corporation)*

Oscilloscope Pattern

The pattern of light traced on the oscilloscope screen is called a pattern, a trace, or a display. It is a graph that shows voltage in relation to time. The vertical height of the pattern from the zero line shows voltage at any instant along the zero line, which represents time. Most oscilloscopes are designed to display a primary pattern from the distributor primary terminal or a secondary pattern from the secondary terminal. The secondary pattern, *Fig. 8.7,* is more generally used for showing overall ignition system operation. Each section of the pattern represents a specific part of ignition system operation: the firing section, the intermediate section, and the dwell section.

The firing section shows a voltage picture of the firing of the spark plug. It is made up of two lines, the firing line and the spark line. The vertical firing line shows the voltage required to overcome the rotor and spark plug gaps. The horizontal spark line shows the voltage required to maintain the spark. The firing of the spark plug involves two separate events. The first one is the creation of the peak, or ionization voltage. The second event is the arc-maintaining voltage or spark line. Once the plug fires there is a drop in the secondary voltage. As the spark continues to cross the electrode gap, the spark voltage remains fairly constant until the spark quits at the beginning of the intermediate section.

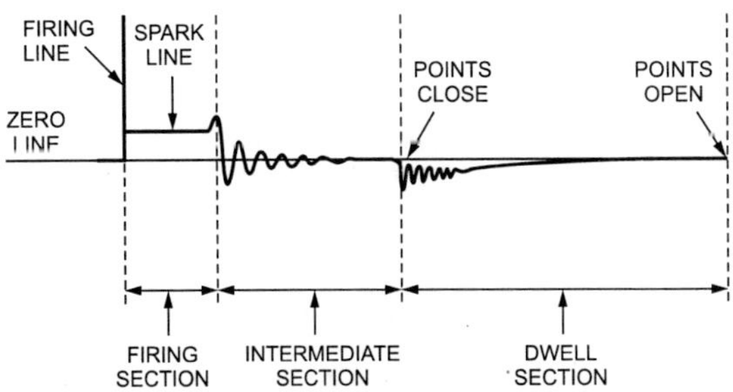

Fig. 8.7 *Basic secondary oscilloscope pattern. (Chevrolet Div. General Motors)*

The intermediate section is a series of wavy lines or oscillations, which disappear at the beginning of the dwell section. It shows the dissipation of the energy remaining in the coil and condenser after the spark plug has stopped firing. The number of oscillations depends on dwell angle, engine speed, duration of spark, degree of coil saturation, and coil and condenser condition.

The dwell section covers the period from the time the distributor contact points close until the contacts open to fire the next cylinder. Closing the points causes a short downward line followed by a series of small oscillations showing condenser operation.

The oscilloscope pattern for a solid state ignition is compared to a conventional pattern in *Fig. 8.8.* The two patterns are similar except that the solid state system gives no signal corresponding to condenser oscillations. The firing and spark lines are the same in both patterns. An oscillation after the spark line shows the transistor turning on, and the end of the pattern shows the transistor turning off.

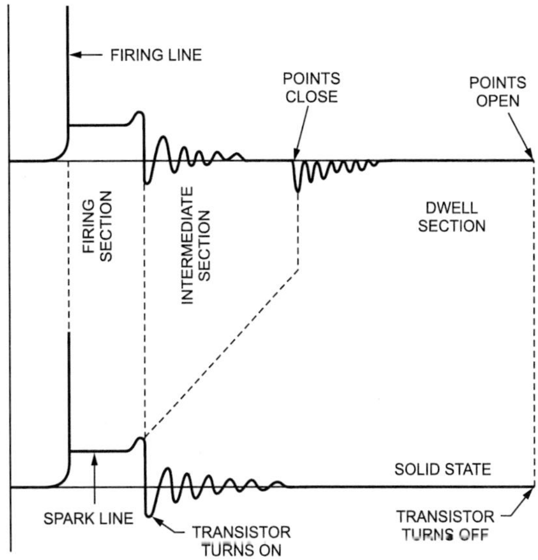

Fig. 8.8 *Comparison of solid state and conventional pattern.*

8.5 SERVICING THE DISTRIBUTOR PRIMARY CIRCUIT

No ignition system can work correctly if it does not have a good ground. Start with the battery. Keep battery terminals clean and free of corrosion. Concerning the primary circuit, standard ignition needs the most service. Little servicing is possible in electronic primary ignition circuits, except for testing and parts replacement.

Contact Point Service

Bad contact points (points having burnt, pitted, misaligned contacts, or worn rubbing block) causes a wide range of engine performance problems: high speed missing, no-start problem, and many other ignition related problems.

A ***faulty condenser*** could leak (allow some DC current to flow to ground), be shorted (direct electrical connection to ground), or be opened (broken lead wire to condenser foils). If leaking or open, a condenser will cause ***point arcing*** and ***burning.*** If the condenser is shorted, primary current will flow to ground and the engine will **not** start. The distributor and condenser are held in place by small screw. To prevent dropping the screws, use a magnetic or clip-type screwdriver that firmly holds the screws. Use a small wrench to disconnect the primary wires from the points. Lift the points and condenser out of the distributor.

Wipe the distributor cam and breaker plate clean. If recommended, apply a small amount of lubrication to points on the distributor (wick in centre of shaft, cam wick, or oil hole in side of distributor housing).

To prolong service life, place a small amount of grease on the side of the breaker rubbing-block. This will reduce friction between the distributor cam and fiber block.

Fit the points and condenser into the distributor. Then, install their screws. If the distributor has a window (square metal plate) in the distributor cap, tighten the point hold-down screws. If the distributor does *not* have a window, only partially tighten the screws so that the points can be adjusted.

Inside the distributor, the most common service is changing the contact points. Points can be burned, pitted, or oil coated. Use a small screwdriver to pry open the points and then inspect the contact surfaces. If they are blackened or pitted, replace them. Filing points is a temporary repair but not the best way to ensure trouble-free driving. Besides, the labour cost can sometimes exceed the slight cost of a new set of points. The condenser is usually replaced whenever the points are replaced.

On vehicles with mechanical breaker points, the contact points act as the switch that triggers the high voltage buildup in the coil. After prolonged use the current flow across the contact points breaks down the contact surfaces. It becomes severe enough to affect current buildup in the coil. The engine begins to lose power and waste gasoline. If the contact points are not serviced, they may get bad enough so the vehicle stops running. Therefore, contact points have to be changed regularly.

- *Burned or worn contact points* should be replaced. They are supplied in sets; one stationary point, one moveable point on the contact arm. Some sets are supplied assembled, with an attached condenser.
- To ***change the contact points and condenser,*** the distributor cap must be removed. Most distributors have spring clips on each side of the cap, which are unsnapped to free the cap. Some distributors use spring loaded hold-down screws. Lift the cap off far enough to work inside the distributor.
- Then remove the ***rotor,*** which turns with the distributor shaft and distributes high voltage to each of the contacts in the distributor cap. The rotor can be removed from the distributor shaft by pulling it straight up, *Fig. 8.9.* On GM distributors the rotor screws must be removed first. Examine the contact surface on the end of the rotor. If there is any evidence of burning, it should be replaced.

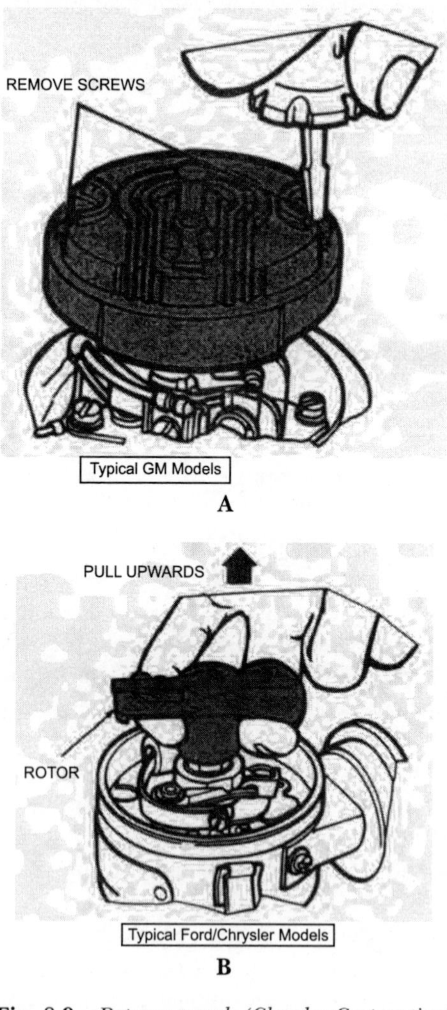

Fig. 8.9 *Rotor removal. (Chrysler Corporation)*

- With the distributor cap and rotor removed, the contact points should be visible. Some vehicles have a metal shield covering the contact points. Remove it by loosening the screws, which mount it to the distributor. Check the contact points. Make sure the rubbing block on the moveable contact is resting on the top of a cam lobe and *the points are in the open position for servicing, Fig.8.10.* If the contact points are not open, ask your assistant to crank the engine a little at a time until the points are open.

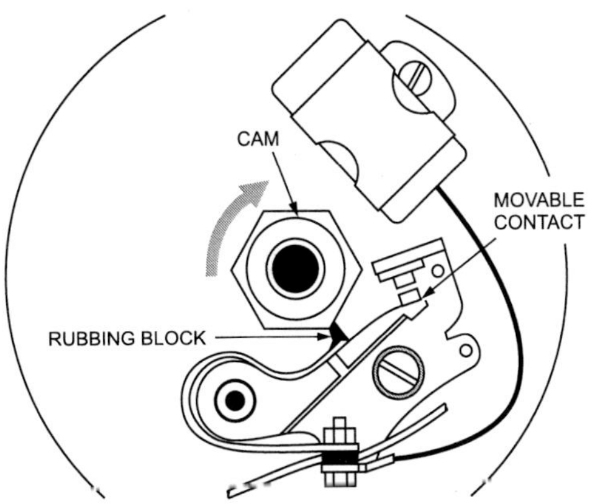

Fig. 8.10 *Contact points in open position.*

- When the points are open, they can be inspected. Use a small screwdriver to gently pry the contact points apart enough to inspect the contact surfaces. If they show ***any sign of pitting or burning,*** they should be replaced.
- Before disassembly, carefully examine the ***wires*** connected to the contact points. One wire connected from the distributor terminal delivers primary current into the distributor from the coil. Another wire is connected from the condenser to the contact points. Both these wires are protected from touching any metal that is grounded. If either of these wires is grounded the vehicle will not run. Make a descriptive sketch of the manner the wires are connected so they can be reinstalled correctly.
- Remove the small screws that hold the contact points down. Also, remove the screws that hold down the condenser. Lift the points assembly out of the distributor. Wipe any dirt or oil out of the distributor with clean shop rag.
- Most new contact point sets come with a small tube of distributor cam lubricant. This is special high temperature grease. Wipe a small amount of this grease on the distributor cam to prevent wear to the rubbing block on the contact points. Do not use any other type of grease for this purpose.
- Remove the new contact point set from the package. Do not touch the point surfaces. Skin oil can coat the points and cause quick burning of the points.
- Position the new contact point set in the distributor. Start the screws but do not tighten them. Mount the new condenser. Reconnect the wires just as they were. Check that the rubbing block of the new contact point set is positioned on top of the cam lobe. Refer to the service manual to find the correct spacing of the points. The dimension will be given in hundredths of mm. Choose a flat feeler gauge of the correct size and see that it is perfectly clean. Slide it between the contact points. When the gap is proper, there will be a slight drag felt by the gauge when pulled gently, *Fig 8.11.*

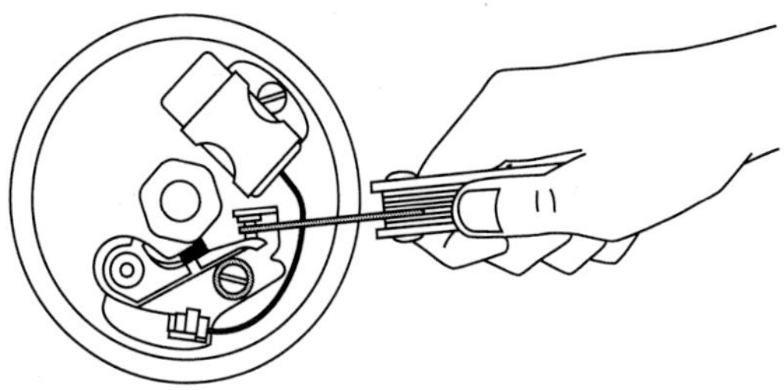

Fig. 8.11 *Checking contact point spacing with feeler gauge.*

- If the gap is too wide or too narrow, adjust the points by sliding the contact set around. When the point adjustment is correct, tighten the contact point mounting screws. Again check the points spacing with the feeler gauge. Also recheck the wiring. If everything is good, the shield, rotor, and cap are put in place. The rotor is keyed to the distributor shaft so that it will fit in only one position. Look at the underside of the rotor for the key or groove. Push the rotor on the distributor shaft and turn it until it engages the key. Install the distributor cap. It is also keyed or grooved so that it fits in the proper position on the distributor. Move it back and forth until it drops into position. Reinstall the spring clips or hold-down screws. Install the coil wire. Check all the plug wires to make sure none have pulled out when the distributor cap was moved around. If everything else is fine the engine is ready to be started.

8.6 TIMING ADJUSTMENT

The initial timing determines when the spark plugs receive the spark from the distributor. Changing the contact points also changes the relationship between the rubbing block and the distributor cam. Which affects ignition timing. Every time the contact points are changed and adjusted, ignition timing must be checked and adjusted if needed. The timing should also be checked and if necessary adjusted when you are doing a tune-up on a vehicle with electronic ignition.

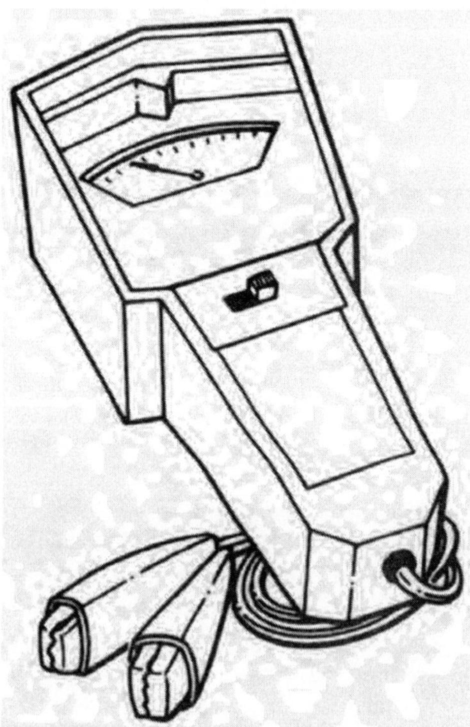

Fig. 8.12 *Dwell meter.*

After adjusting the contact points, run the engine until it is at operating temperature. Check the timing specifications in the owner's manual, or service manual. The specifications give the degrees the spark should be set before or after top dead centre at the engine idle speed. The degrees refer to the degrees of crankshaft rotation. Top dead centre means that the piston in the #1 cylinder is at the top of its compression stroke. For example, the specification 10° before TDC means that the spark should enter the cylinder 10° before the piston reaches TDC on its compression stroke.

Setting the contact point spacing with a feeler gauge gives satisfactory performance from most engines. A more accurate contact point adjustment can be done with a dwell meter, which electronically measures the time that the contact points are closed for coil saturation. It makes possible very precise adjustments, *Fig. 8.12.*

If a dwell meter is used, the contact points are still installed and adjusted with a feeler gauge. The distributor is put back together so the engine will run. The dwell meter is connected to the engine according to the directions provided with the meter.

The dwell meter must be calibrated and set for the proper number of engine cylinders. This procedure is outlined in the dwell meter instruction booklet. The engine is then started. The needle points to degree numbers on the face of the dwell meter. The proper dwell for the vehicle can usually be found in the owner's manual or service manual.

Both point and electronic systems must have initial timing adjustments made occasionally. This is normally done by rotating the distributor and needs a timing light. Modern emission controls require the ignition timing setting to be very precise.

Adjusting Distributor Points

Distributor points can be adjusted using either a ***feeler gauge*** or ***a dwell*** meter (meter that electrically measures point setting in degrees of distributor rotation).

To use a feeler gauge to set the gap of a distributor points, crank the engine until the points are ***fully open.*** The point-rubbing block should be on top of a distributor cam lobe, *Fig. 8.13.*

Distributor point gap is the recommended distance between the contacts in the fully open position. Refer the service manual for this spec. It could also be given in the sticker in the engine compartment.

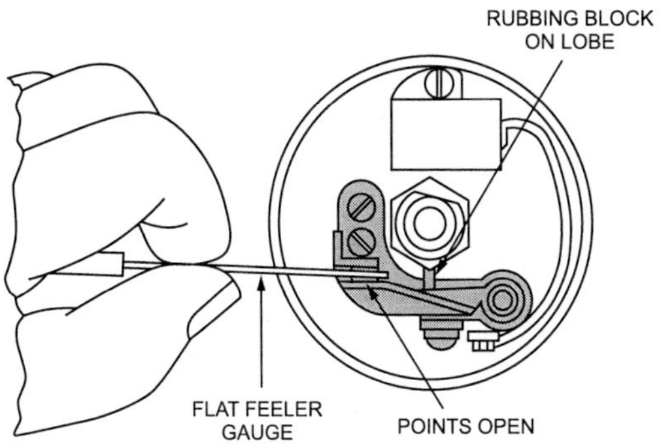

Fig. 8.13 *Older contact points are also adjusted with feeler gauge. Crank engine until distributor cam lobe is touching rubbing block on points. Points should be fully open. Adjust points so they just touch correct thickness gauge. (Chrysler Corporation)*

Typical point gap settings average around 0.38 mm for eight cylinder engines, 0.53 mm for 6-cylinder engines, and 4-cylinder engines.

With the distributor points open, slide the specified thickness feeler gauge between the points. Adjust the points so that just they touch the blade. If needed, tighten the hold-down screws and recheck point gap.

Point-System Dwell

On point-type systems, check the dwell i.e., point gap, first. Remove the distributor cap and rotor. Crank the engine with the help of starter until the points are fully open, with the points rubbing block on the tip of the distributor cam lobe, *Fig. 8.14*. Measure the gap with a feeler gauge. Loosen the point hold-down screw and move the points if adjustment is needed. Tighten the screw and again check the point gap. Wipe a small amount of cam grease on the distributor cam to minimize rubbing block wear and then reinstall the rotor and cap.

Using a Dwell Meter

To use a dwell meter to adjust distributor points, follow the directions provided with the meter. Typically, connect the red lead to the distributor side of the coil (wire going to contact points). Connect the black lead to ground (any metal part on engine), *Fig. 8.15*.

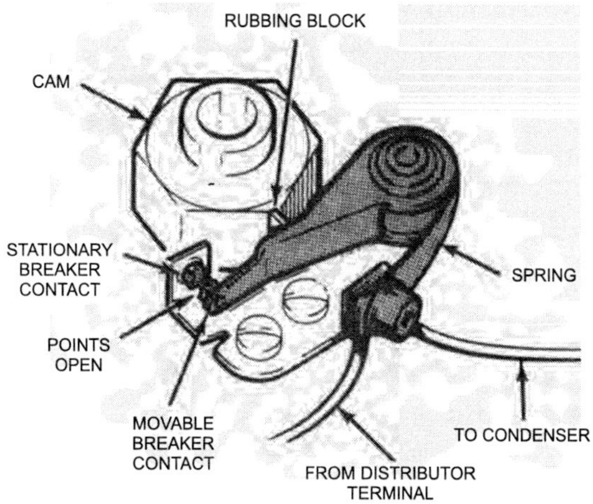

Fig. 8.14 *Points replacements.*

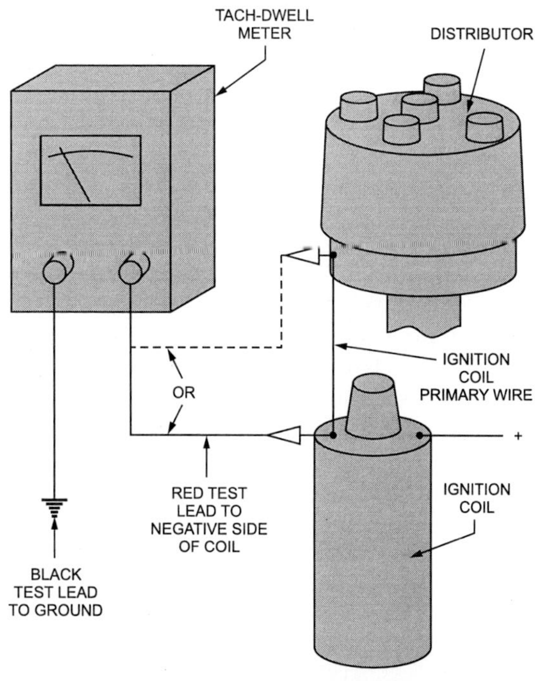

Fig. 8.15 *Note basic hookup for tack-dwell meter. Tachometer is used for engine speed adjustment. Dwell meter is for setting contact points.*

If an opening is provided in the distribute cap, the points should be set with the engine running. Install the distributor cap and rotor. Start the engine. Set the meter controls correctly. Adjust the points using an Allen wrench or special screwdriver type tool. Turn the point adjustment screw until the dwell meter reads within specs.

If the distributor cap does ***not*** have an adjustment window, set the points with the cap removed. Instead of starting the engine, ground the coil wire, i.e., connect output end on engine. Crank the engine with the starting motor. This will simulate engine operation and allow point adjustment with the dwell meter.

Dwell specifications – Recommended point settings in degrees, vary with the number of cylinders in the engine. An 8-cylinder engine will normally need 30° of dwell. An engine with fewer cylinders will normally require more dwell time in degrees. Always consult correct service manual to determine the exact dwell value for the engine under service.

Dwell variation — Change in dwell meter reading from the original set value, indicates distributor wear problems. Dwell should remain constant as engine speed is changed. If dwell varies with speed change more than 3°, the distributor should be rebuilt or replaced. The distributor shaft, bushings, or advance plate might have worn and become loose, causing a change in dwell reading. Any change in dwell or point gap will change ignition timing. For this reason, the points should always be adjusted *before* setting the ignition timing.

Points can also be better set with a dwell meter. Dwell is the amount of time the points are closed. A dwell meter measures this time in degrees of camshaft rotation. Dwell meters are hooked between the coil power (negative) lead and ground. Dwell can be read while cranking or while the engine is running. Dwell meter has the advantage of checking dwell with the engine running. It is more accurate than the feeler-gauge method because dwell changes slightly with engine rpm.

Adjust the dwell just as you adjust the point gap, because dwell and point gap are basically same. *Fig. 8.16* shows how point gap and dwell are related.

After setting the dwell, set the timing. Dwell adjustments affect the ignition timing, but timing adjustments do not affect dwell. Dwell affects timing because a narrow point gap lengthens the time between point openings, thus retarding the spark. A wide point gap (small dwell) advances the spark because the point openings occur earlier each time.

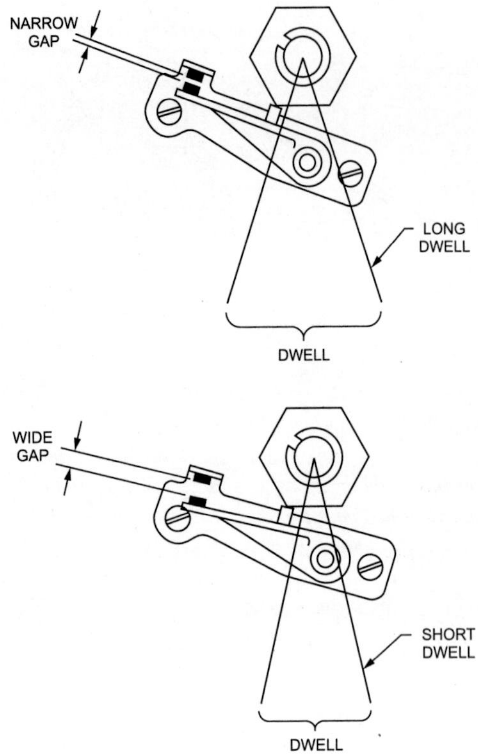

Fig. 8.16 *Point gap affects dwell.*

Ignition Timing

The ignition system must be timed so that the sparks jump across the spark-plug gaps at exactly the right time. Adjusting the distributor on the engine so that the spark occurs at this correct time is called ***setting the ignition timing. Initial ignition timing*** is the spark timing set by the mechanic with the engine idling (no centrifugal, vacuum, or electronic advance). The ignition timing is normally set at idle, or at a speed specified by the engine manufacturer. On engines with idle-speed solenoids, the solenoid must be energized. Also, the vacuum hose from the intake manifold has to be disconnected so that the vacuum advance does not work. Adequately plug the vacuum hose.

Initial ignition timing is commonly changed by turning the distributor housing in its mounting in the engine. This makes the pickup coil and electronic control unit or breaker points fire the ignition coil sooner or later. The contact point opening or dwell should be adjusted before the ignition is timed.

Turning the distributor in the direction opposite to distributor shaft rotation advances the spark timing. Turning the distributor in the direction of the shaft rotation retards the spark timing, *Fig. 8.17.*

Fig. 8.17 *Engine ignition timing is changed by turning distributor housing in block. A few computer controlled systems cannot be adjusted or provision for adjustment is made on computer itself. (Peerless Instruments)*

Some computer controlled ignition systems require moving the mounting for crankshaft position sensor or adjusting a screw or lever on the computer to change timing. In some cases, timing cannot be adjusted and the computer must be replaced if the timing is not within specs.

Each ignition system has special details that must be checked. Always check the correct shop manual or engine decal, or owner's manual for the timing specifications and the correct procedure to be followed to adjust ignition timing. On ignition systems with a vacuum advance, the vacuum hose is usually disconnected and plugged. On computer-controlled systems, sometimes a jumper wire has to be connected or a special connector has to be unplugged. Follow the directions provided by the service manual.

When the ignition timing is ***too advanced,*** the engine can suffer from spark knock or ping. A light, tapping sound may be caused when the engine is accelerating or is under a load. The ***ping*** — abnormal combustion will sound like a small hammer tapping on the engine.

When ignition timing is ***too retarded,*** the engine will have poor fuel economy, low power, and will be very sluggish during acceleration. If excessively retarded, combustion flames blowing out of the opened exhaust valve can cause overheating of the engine and crack the exhaust manifolds.

Measuring Ignition Timing

A *timing light* is used to measure ignition timing. A timing light normally has three leads, *Fig. 8.18.* Two small leads connect to the battery. The third, larger lead connects to the #1 ***spark plug wire.***

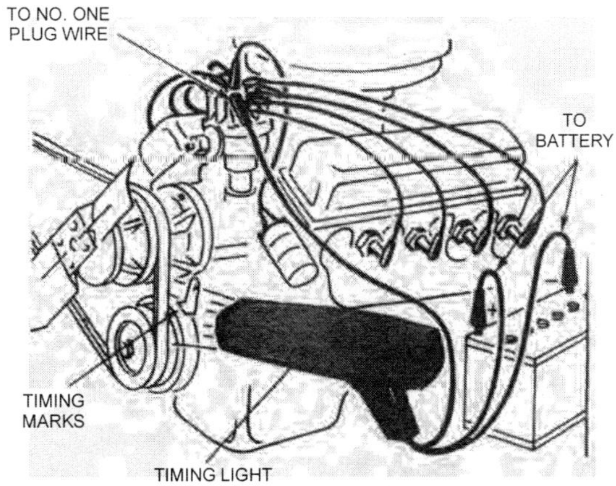

Fig. 8.18 *Timing light is connected as shown. Two small leads go to battery. Large lead connects to # 1 spark plug wire. Shine light on engine timing marks following recommended procedure for a particular engine.*

Depending upon the type of timing light, the large lead may clip around the plug wire (inductive type), or it may also need to be connected directly to the metal terminal of the plug wire (conventional Type).

When the engine is running, the timing light will flash ***on*** and ***off*** like a strobe light. This action can be used to make a moving object appear stationary.

Before measuring engine timing, disconnect and plug the vacuum advance hose going to the distributor. This will prevent the vacuum advance from functioning and upsetting the readings. Check the emission sticker or service manual to determine if any other devices must also be disconnected.

When adjusting timing on vehicles with computer-controlled ignitions, the base timing must be triggered. Base timing is simply the ignition timing without computer control. Procedures for triggering base timing vary. Refer the service manual for specific instructions that need to be followed.

Start the engine and point the timing light on the timing marks. The timing marks may be on the front cover and harmonic balancer of the engine, *Fig. 8.19.* The timing marks may also be on the engine flywheel.

The flashing timing light will make the mark or marks on the harmonic balancer or flywheel appear to stand still. This will let you determine whether the engine is timed properly. For example, if there is only one reference line on the harmonic balancer, just read initial timing by noting the degree marks lined up with the reference line, *Fig. 8.19.*

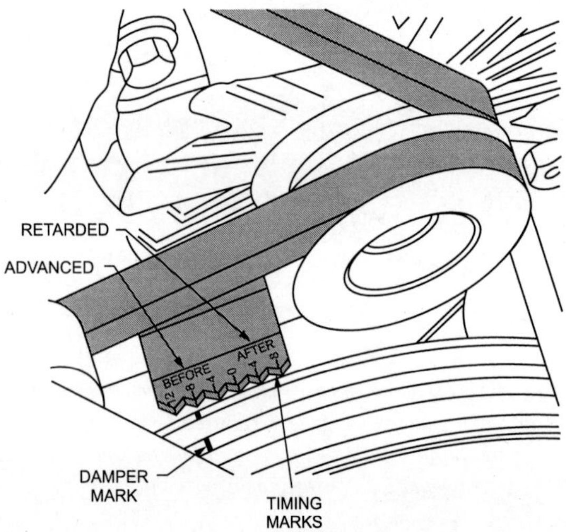

Fig. 8.19 *Timing marks are frequently on front of engine, near front damper or pulley. However, they can sometimes be on flywheel at rear of engine. Before means advanced and after means retarded. (Chevrolet Div. General Motors)*

Ignition Timing Specs

Ignition timing is very critical to the performance of an engine. If the ignition timing is off even two or three degrees, engine fuel economy and power can drop significantly.

Adjusting Ignition Timing

Adjust the ignition timing if the timing marks are not lined up correctly. Loosen the distributor hold-down bolt.

A ***distributor wrench,*** which is a long, special shaped wrench for reaching under distributor housing, is handy. Loosen the distributor bolt just enough to allow distributor rotation. Do *not* remove the bolt, *Fig. 8.17.*

With the distributor hold-down loosened, shine the timing light on the engine timing marks. Turn the distributor one way or the other until the correct timing marks line up. Tighten the hold-down bolt and again check the ignition timing. Reconnect the distributor vacuum hose and disconnect the timing light.

Timing Marks: Warm the engine to normal temperature and then stop the engine. Locate the timing marks on the crankshaft pulley or flywheel and highlight the timing marks with chalk or paint. Connect the power wires of the timing light to the battery and clip the pickup lead around the # 1 spark plug wire, *Fig. 8.20.* The timing light is simply a strobe light triggered by the firing of the number one spark plug.

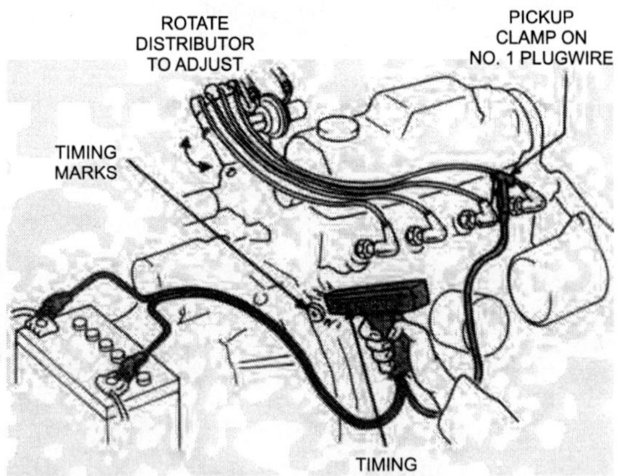

Fig. 8.20 *Ignition timing adjustment.*

Adjustment: Start the engine and let it idle at the specified speed. Aim the timing light at the mark on the crankshaft pulley or flywheel. It will **freeze** the motion of the mark, so you can turn the distributor body until the marks on the pulley and the block pointer line up. Tighten the bolt and again check the timing. If it is correct, reconnect the vacuum advance hose and remove the timing light.

Timing is checked with a timing light, which comes with instructions that explain how it is to be hooked up. The light is connected to a power source and to the engine's #1 spark plug wire (identified in the service manual). When the engine is running, the high voltage in the #1 spark plug wire makes the timing light flash, producing a strobe light effect in time with the ignition firing.

The timing light is aimed at timing marks on the crankshaft front pulley. A stationary pointer is mounted on the engine block near the pulley. The timing marks may be on the pulley or there may be a notch on the pulley and timing marks on the stationary pointer, *Fig. 8.21.* When the timing light is flashing, the timing marks and pointer will be visible. The shop service manual will explain what the marks on the crank pulley mean. Typically, there is a centre mark for top dead centre and a number of marks for degrees before and after top dead centre.

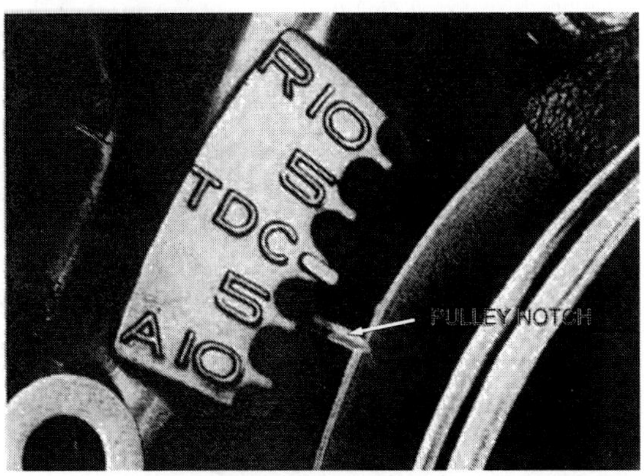

Fig. 8.21 *Timing marks on the pointer are lined up with a notch on the pulley. (Chevrolet Div. General Motors)*

Many modern vehicles are equipped with a magnetic probe receptacle, *Fig. 8.22,* to accommodate a magnetic timing tester, which senses engine crankshaft position. The magnetic probe timing system is much more accurate as the tester displays a digital reading of ignition timing.

When the engine attains operating temperature, turn it off and connect the timing light. Wipe off the timing marks on the crankshaft pulley so they are visible. Change the timing by moving the distributor housing back and forth, after loosening the hold-down clamp by turning the clamp bolt counterclockwise.

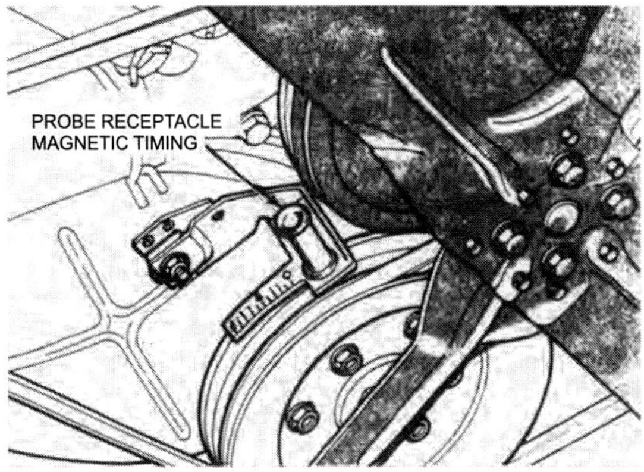

Fig. 8.22 *Receptacle for a magnetic probe timing tester. (Chrysler Corporation)*

Again start the engine. On older vehicles specs will indicate that the timing is checked with the vacuum advance disconnected. If necessary, locate the vacuum advance on the side of the distributor and pull off the vacuum line connected to it. Aim the timing light down on the pointer, taking care not to get the timing light wires caught in the fan or belts. The spinning fan and belts can damage the light or cause serious injury. One of the marks on the moving crank pulley will line up with the stationary pointer. If the pointer and crank pulley scale are not aligned to the proper mark, move the distributor a small distance to get them aligned. When they are aligned, tighten the distributor hold-down clamp by turning the bolt clockwise. Check the timing again. If it is in alignment, stop the engine and disconnect the timing light. If it was disconnected, reconnect the vacuum line.

Testing Centrifugal Advance – In-vehicle

A timing light can be used to test the operation of centrifugal advance. Remove the vacuum hose going to the distributor. Start the engine and let it idle. Then, while shining the timing light on the engine timing marks, gradually increase engine speed to about 3500 rpm.

If the centrifugal advance is working, the timing mark should steadily move to a more advanced position with the increase in speed, *Fig. 8.23.*

If the timing mark jumps around or ***does not move*** to advance smoothly, the centrifugal advance is faulty. It could be worn, rusted, have weak springs, or other mechanical problems.

Testing Vacuum Advance – In-vehicle

To test the vacuum advance, remove the vacuum advance hose from the distributor. Start the engine and increase its speed to about 1500 rpm. Note the location of the timing marks. Then, reconnect the vacuum hose on the distributor to communicate with diaphragm.

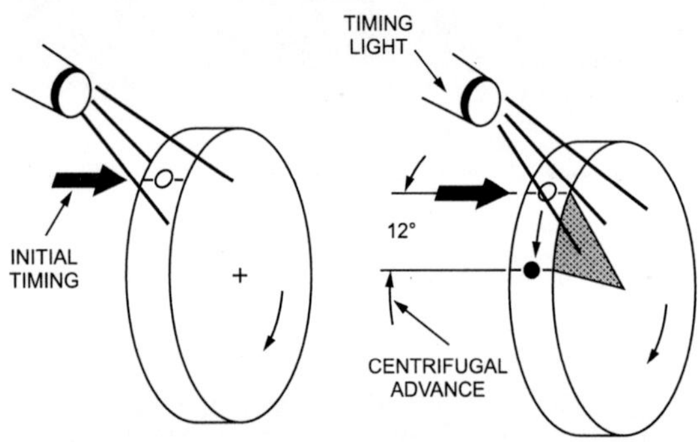

(a) Initial timing would be timing set with no distributor advance.

(b) Centrifugal advance would come in as engine speed increases.

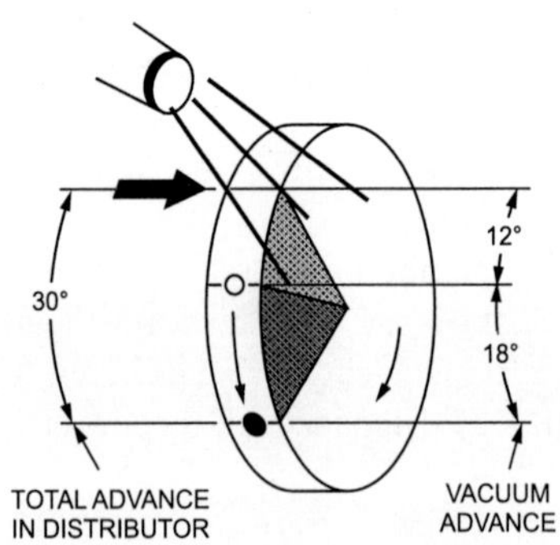

(c) Vacuum advance would be additional advance when the engine is cruising at low load.

Fig. 8.23 *Ignition timing. (Bosch)*

As soon as the vacuum is reconnected, the force acting on the diaphragm causes the engine speed to increase and the timing mark should advance, *Fig. 8.23.* If it does not, the vacuum advance mechanism is not working. Check the vacuum advance diaphragm and the supply vacuum to the distributor.

Testing a Vacuum Advance Diaphragm

To check the vacuum advance diaphragm, apply a vacuum to the unit using a vacuum pump or you mouth suction. When suction is applied to the diaphragm, the advance plate in the distributor should swing around. When the vacuum is released, the advance plate should snap back into its normal position, *Fig. 8.24.* If the ***advance diaphragm leaks,*** it will not hold vacuum and must be replaced.

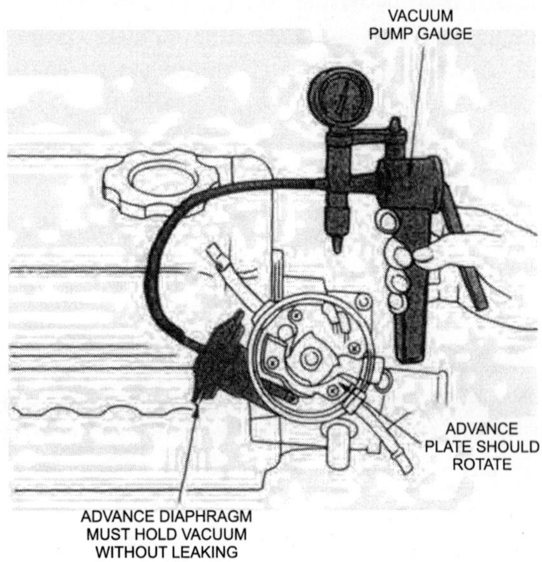

Fig. 8.24 *Test vacuum advance diaphragm for operation and leakage with hand vacuum pump. If diaphragm leaks or fails to rotate advance plate, replace diaphragm (Honda Motors)*

Measuring Total Advance – In-Vehicle

Special timing lights are available that are capable of measuring total advance with the distributor installed in the engine. The timing light has a ***degree meter*** built into the back of its case. The meter will register advance quickly and accurately, *Fig. 8.23.*

A *distributor tester* may also be used to check distributor operation. The distributor is removed and mounted in the tester. The tester can be used to check all distributor functions.

8.7 SERVICING THE DISTRIBUTOR SECONDARY CIRCUIT COMPONENTS

A distributor is critical to the operation of an ignition system. It senses engine speed, alters ignition timing, and distributes high voltage to each spark plug wire. If any part of the distributor is faulty or defective, engine performance suffers.

Ignition system service is similar for both standard and electronic ignition systems when dealing with the secondary circuit. Secondary circuit service covers the distributor rotor, cap, wires, spark plugs, and coil. Much of the service is the same for standard and electronic ignitions.

Rotor and Cap Service

A ***bad distributor cap*** or ***rotor*** can cause engine missing, ***backfiring*** (popping noise in intake manifold), and other engine performance problems. In some cases the conditions can cause no-start situation.

A common trouble arises when a ***carbon trace*** (a small line of carbon-like substance that conducts electricity forms on the distributor cap or rotor). The carbon trace will short coil voltage to ground or to a wrong terminal in the cap. A carbon trace can cause the spark plugs to fire poorly, out of sequence, or not at all.

When problems point to possible distributor cap or rotor troubles, remove and inspect them. Using a drop light, check the inside of the cap for cracks and carbon traces. A carbon trace is black, making it difficult to see on a black coloured distributor cap.

If a crack or carbon trace is found, replace the cap or rotor. Also check the rotor tip for excessive burning, damage, or looseness. Make sure the rotor fits snugly on the distributor shaft.

Remove the distributor cap by unsnapping the spring arms or disengaging the screws. The cap that is greasy or dirty on the outside creates a path that can allow the high voltage current to escape. The current may travel or "flashover" to another terminal or to ground. Flashover most often happens when the wire connections are loose. Also check the distributor cap for cracks. Rotors and caps should be wiped off with dry shop cloth whenever they are removed. When the cap and terminals become crusty with burned deposits, they should be replaced. These parts can also get ***carbon tracking,*** *Fig. 8.25.* Grease and dirt inside the cap can also cause flashover to another terminal. Carbon tracking occurs when conductive deposits form inside the cap. The carbon track leads to another spark plug terminal or to ground (the distributor body). This causes the ignition to misfire on that spark plug. Wiping off carbon tracks may help, but with powerful electronic ignitions, replacing the cap may prove the best remedy, because erosion and damage to the cap has started.

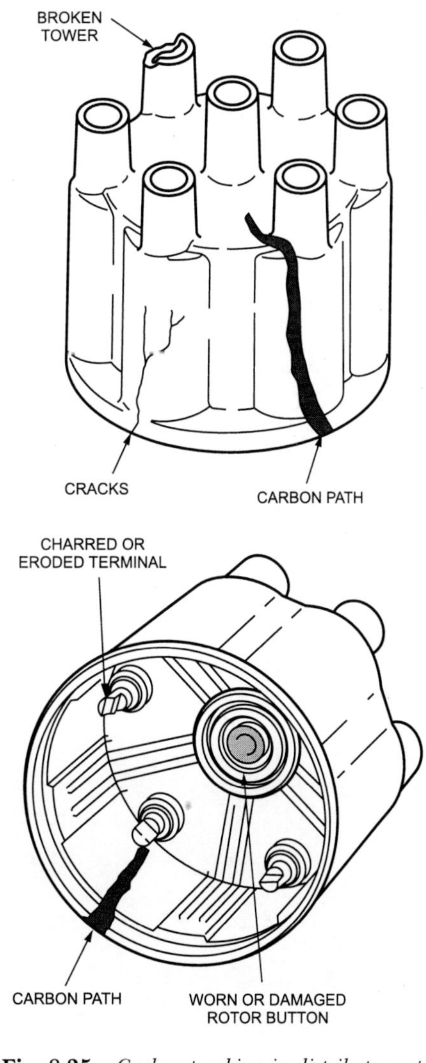

Fig. 8.25 *Carbon tracking in distributor cap. (American Motors Corp.)*

Some systems have silicone grease on the distributor cap terminals. The grease reduces corrosion and radio interference. Do not remove this grease when checking, or be sure to replace it when necessary.

Distributor Cap and Rotor Replacement

Distributor caps may be secured by either screws or spring type metal clips. Normally, turn the screws counterclockwise for removal. With clips, pry on the top of the spring clips, being careful not to crack the cap. The clip should pop free. Wiggle and pull upward to remove the cap from the distributor body.

Rotors may be held by screws, or they may be force fitted around the distributor shaft. Pulling by hand will usually free a press-fit rotor. However, if stuck, carefully pry under the rotor.

With some ignition systems, the ignition coil is housed inside the distributor cap. In this case, the coil must be taken out of the old cap and installed in the new one.

If a distributor cap is ***not*** installed correctly, rotor and distributor damage can result. The whirling rotor can strike the sides of the cap.

If the rotor contact at the centre of the distributor cap is burned away, the cap should be replaced. If the terminal posts are burnt or badly grooved, the cap should be replaced. Check the rotor resistance rod, if there is one, to see if it is broken or has excessive burning on the tip, *Fig. 8.26.* If any of these defects are found, the rotor should be replaced. Install the rotor on the shaft. Be sure the rotor groove, flats, or tabs match the distributor shaft groove or the advance mechanism openings. Mark the location of #1 spark plug cable on the old distributor cap.

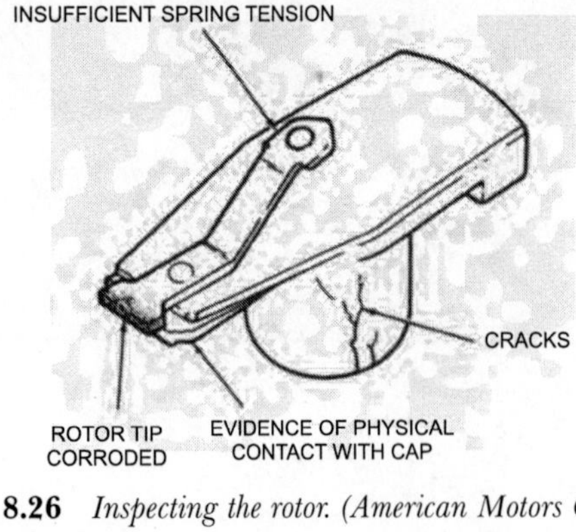

Fig. 8.26 *Inspecting the rotor. (American Motors Corp.)*

Remove the old cap from the distributor and move it to one side. Do not remove the spark plug cables at this time. Install the new distributor cap onto the distributor. Ensure that the distributor cap is centred over the rotor and fits squarely down on the distributor. Transfer one spark plug cable at a time to the new distributor cap, starting with the #1 spark plug cable. Check the installed cables against the firing order.

Start the engine and check for proper operation. If the engine misfires you may have installed the spark plug cables incorrectly. Recheck the firing order.

Pickup Coil Service

Most electronic ignition distributors use a pickup coil to sense trigger wheel rotation and engine speed. If the distributor fails to produce these tiny signals, the complete ignition system and engine can stop functioning.

A ***bad pickup coil*** can produce a wide range of engine problems: Stalling, missing, no-start troubles, and loss of power at specific speeds. If the tiny windings in the pickup coil break, they can cause intermittent problems that only occur under certain conditions. Also, because of vacuum advance movement, the thin wire leads going to be pickup coil can break internally.

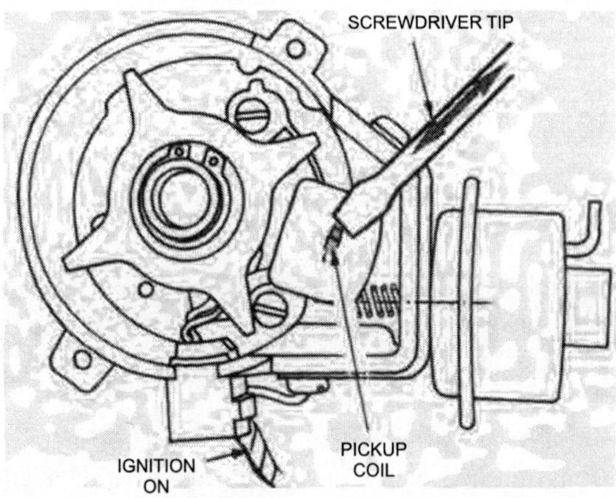

Fig. 8.27 *With coil wire removed and held next to ground and with ignition on, pass screwdriver tip over pickup coil. This should make ignition coil fire and spark jump out of coil wire. Electric soldering gun will also fire ignition when held near pickup. This will indicate that the module, ignition coil, etc., are working properly.*

With the ignition key on, try moving a screwdriver or electric soldering gun over the pickup coil. This should make the ignition coil fire, *Fig. 8.27.*

The ***pickup coil ohmmeter test*** compares actual pickup coil resistance with specs. If resistance is too high or low, the unit is bad. Connect an ohmmeter across the pickup coil output leads, *Fig. 8.28.* Observe the meter reading. ***Wiggle the wires*** to the pickup coil while watching the meter. This will help locate a break in the leads to the pickup. Also, lightly tap on the coil with the handle of a screwdriver. This could uncover any break in the coil windings.

Pickup coil resistance will usually vary between 250 and 1500 ohms. Check the applicable service manual for specs. If the meter reading changes when the wire leads are moved or when the coil is tapped, replace the pickup coil.

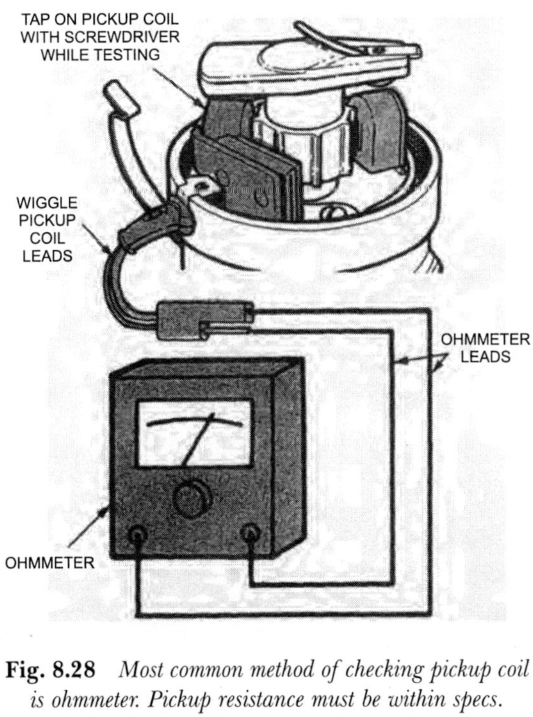

Fig. 8.28 *Most common method of checking pickup coil is ohmmeter. Pickup resistance must be within specs. Wiggle wires to pickup because they can be broken internally. (Chrysler Corporation)*

A low reading AC voltmeter can also be used to test a pickup coil. With the engine cranking, a small AC voltage of about 3 to 8 volts should be produced by the pickup coil. Refer to specs for exact voltage values and procedures,

A distributor pickup coil can usually be replaced by removing the distributor cap, rotor, and the screws on the advance plate. Often the pickup coil is mounted around the distributor shaft. Because the procedures vary for various applications, follow the procedure detailed in the service manual.

The ***pickup coil air gap*** is the space between the pickup coil and a trigger wheel tooth. With some designs, it must be set after installing the pickup coil, *Fig. 8.29.*

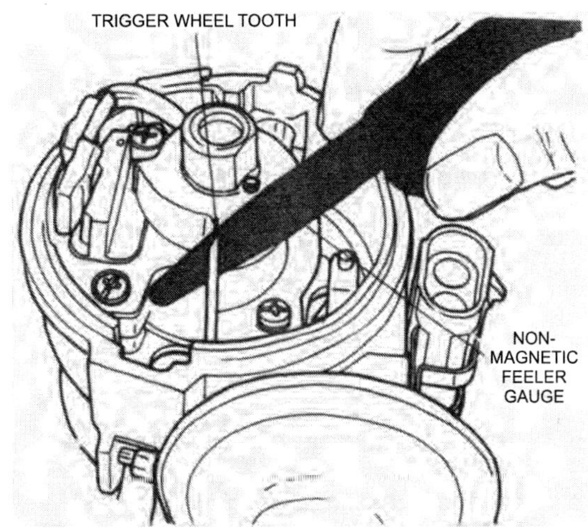

Fig. 8.29 *Some pickup coil gaps must be adjusted. Use a non-magnetic feeler gauge for accurate reading. (General Motors Corporation)*

To obtain an accurate reading, slide a ***nonmagnetic feeler gauge*** (plastic or brass gauge) between the pickup coil and trigger wheel. One of the trigger wheel teeth must point at the pickup coil. Move the pickup coil in or out until the correct size gauge fits in the gap. Tighten the pickup screws and double-check the air gap setting.

Inspecting and Servicing Spark Plug Cables

Ignition wires are not wires at all but carbon-coated threads covered with insulation. Bending and other rough handling will break the threads and cause high resistance. Many plug wires are damaged when they are removed from the spark plugs. Always twist the boot that fits over the plug before attempting to pull it off. This will break the suction, which holds the boot tightly to the plug. Special tools are available for plug boot removal, *Fig. 8.30.*

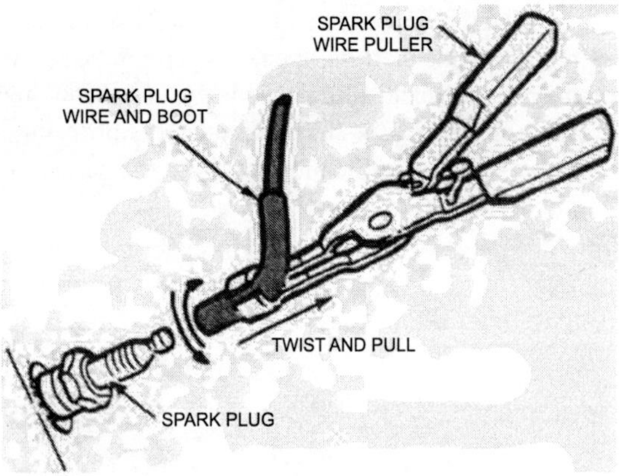

Fig. 8.30 *(a) Spark plug boot removal tool. (Chrysler Corporation)*

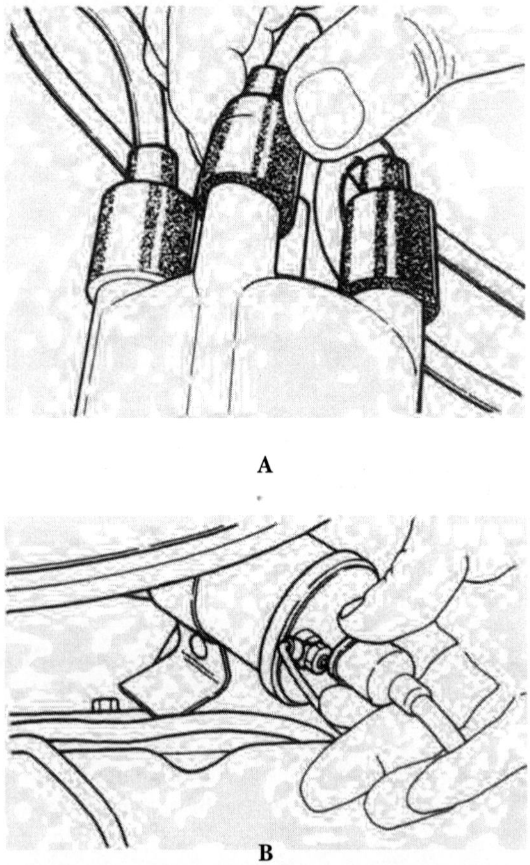

Fig. 8.30 *(B) Installing a cable and boot on (A) a distributor cap and (B) a coil tower. (Chrysler Corp.)*

A ***faulty spark plug wire*** can have a burnt or broken conductor or it could have a deteriorated insulation. Most spark plug wires have a resistance conductor that is easily separated. If the conductor is broken, voltage and current cannot reach the spark plug.

If the spark plug insulation is defective, sparks may leak through the insulation to ground or to another wire, instead of reaching the spark plug.

Inspect the wires carefully. Small burn marks, especially where two wires cross, indicate the wires are arcing or shorting out. Look for cuts or tears in the insulation or scorch marks from a hot manifold. One way to inspect wires is to check them in the dark. Lift the hood with the engine running and look for small sparks coming from the wires. If the wires have any of the above problems, replace them.

The spark plug cables carry the high secondary voltage from the distributor cap to the spark plugs. High under hood temperatures eventually break down the cable insulation. The cables must be inspected and, if necessary, replaced at regular intervals.

Using a pair of insulated spark plug cable pliers, remove the spark plug end of one spark plug cable by first twisting the rubber-like boot back and forth. This breaks the seal that the protective boot forms around the top of the spark plug. Be sure you twist the boot not the wire. Then, gripping the end of the boot, pull or snap the cable straight out away from the spark plug, *Fig. 8.31.* Repeat this procedure for each of the spark plug cables. Before removing any cables from the distributor cap, mark the location of the #1 cable distributor cap terminal with a piece of chalk.

Remove the same cable that you disconnected at the spark plug. Again, break loose the rubber-like nipple from the distributor cap tower. To do this, peel the nipple back from around the tower of the distributor cap. Then grip the upper part of the nipple and the cable. Pull the spark plug cable straight out, away from the distributor cap. Visually check each spark plug cable for dry, cracked insulation, cuts, and breaks in the wires. Wires that are oil or grease soaked, dirty, or salt-encrusted should be replaced unless they can be cleaned satisfactorily with a damp cloth and a mild detergent.

When the thread strands break, resistance increases. ***Test each cable for resistance*** with an ohmmeter. Calibrate the ohmmeter to the × 1000 scale. Connect one tester lead to the spark plug end of the cable. Touch the other tester lead to the terminal on the distributor cap end of the cable. Check the readings against service manual specs to determine if the cables should be replaced. Twist the wire while watching the reading on the ohmmeter. If the needle fluctuates, replace the wire because the fluctuation indicates a broken or partially separated conductor within the wire. This test can also be done without removing the wires from the cap, as shown *in Fig. 8.32.*

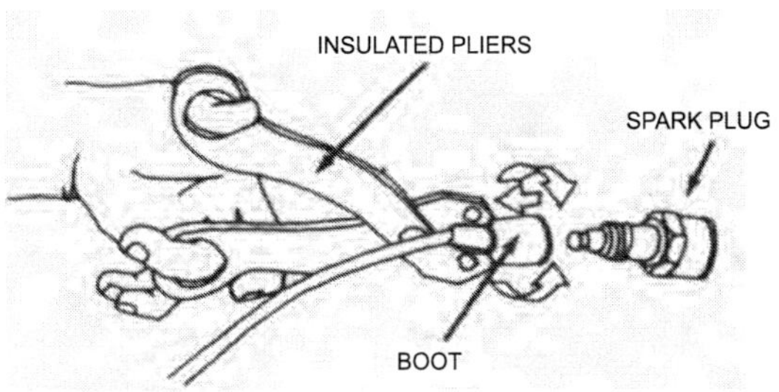

Fig. 8.31 *(a) Use insulated pliers to remove a spark plug cable from the spark plug. (General Motors)*

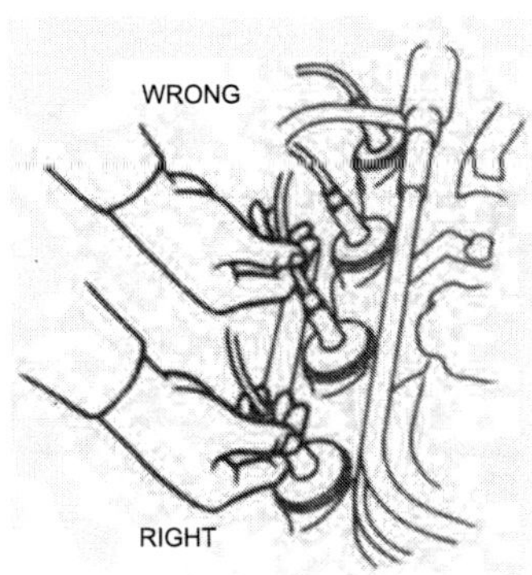

Fig. 8.31 *(b) Wrong and right ways of removing cables from spark plugs. Pull on the rubber boot, not on the cable. (Toyota Motors)*

To ***replace the cables,*** start with the number 1 spark plug wire of the firing order. Lay the old wire alongside the new one; match the cable lengths to be sure you have the correct length for the cylinder you are working on. Route the new cable on the engine in the place of the old. In some cases, the wire must be routed through small, tight passages. Do not force, yank, or pull the cable because this will damage it.

Set the cable into the clips or "looms" and support the wire and keep it apart from the other cables. Always make sure that the cables are as short as possible. Do not make any sharp bends in them or let them come into contact with heated, vibrating or moving parts.

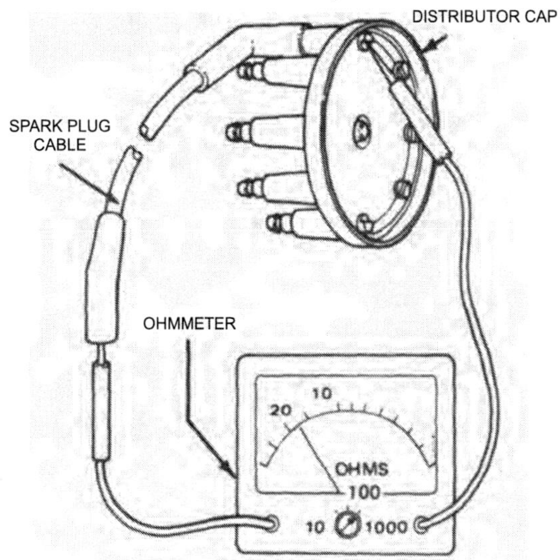

Fig. 8.32 *Checking spark plug cable resistance with an ohmmeter. (Ford Motors)*

Spark plug cables should be routed properly to prevent inductive cross firing. Cross firing occurs when the spark from one spark plug wire induces a current in the wire lying next to it. It causes the wrong plug to fire in the wrong cylinder at the wrong time. Cross firing happens when the spark plug cables are parallel over long distances, particularly those cables that connect to the spark plugs or cylinders that are side-by-side. Push the spark plug boots firmly over each spark plug and push each distributor cap terminal and boot firmly into place.

Check the cable installation with the firing order to make sure no cables are out of order. Start the engine and check for proper operation.

Secondary Wire Resistance Test

A ***secondary wire resistance test*** will check the condition of a spark plug or coil wire conductor. To do this test, connect an ohmmeter across each end of the spark plug or coil wire, as shown in *Fig. 8.33.* The meter will read internal resistance in ohms. Compare the meter reading to service manual specs.

Typically, spark plug wire resistance should not be over approximately 5000 Ω per inch or 100 KΩ (100,000 Ω) total. Since specs vary, always check in the correct service manual for an exact value. A bad spark plug will have almost infinite (maximum) resistance.

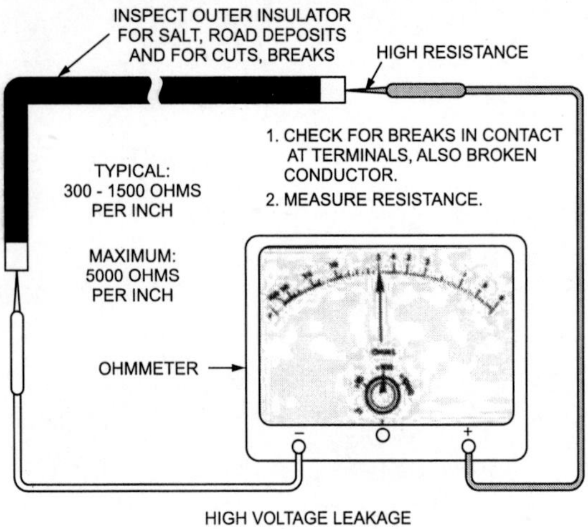

Fig. 8.33 *Ohmmeter can be used to check for internal connector breakage in coil and spark plug wires. If resistance is too high, the wire is defective. Replace it. (Ford Motors)*

Secondary Wire Insulation Test

A ***secondary wire insulation test*** checks the sparks arcing through the insulation to ground. An ohmmeter test will ***not*** detect bad insulation.

To check spark plug and coil wire insulation, place fender covers over the sides of the vehicle hood to block out light. Start the engine. Carefully check each wire for leakage.

A special test light or grounded screwdriver can be moved next to the wire insulation. If sparks jump through the insulation and onto the tool, the wire insulation is bad.

Spark plug wire leakage is a condition in which electric arcs pass through the wire insulation,. If bad wire insulation is detected, replace the wire.

Replacing Spark Plug Wires

Installing new spark plug wires is simple if one wire at a time is replaced. Compare old and new wire lengths. Replace each wire with one of equal length. Make sure that the wire is fully attached on plug and in the distributor or coil module.

Spark plug wire replacement is more complicated if all of the wires are removed at once. Then, you must use the engine firing order and cylinder numbers to connect each wire correctly.

Figure 8.34 gives ***firing order illustrations*** similar to the types found in a service manual. Each can be used to trace wires from a distributor cap tower to the correct spark plug or cylinder.

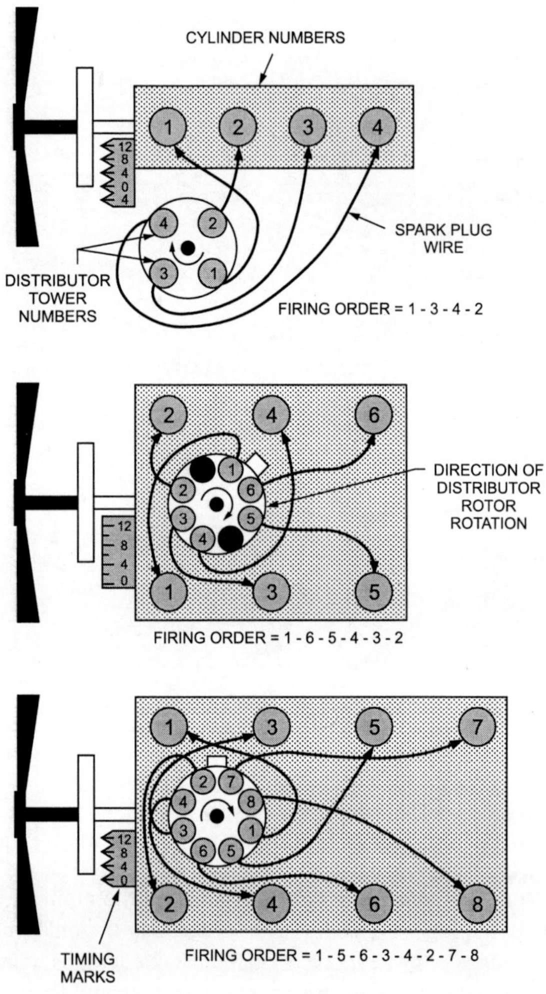

Fig. 8.34 *The figure shows service manual illustrations that depict firing order, distributor rotor rotation and spark plug wire routing. Such illustrations are needed during many engine tune-up type jobs.*

Engine Firing Order

Engine firing order refers to the sequence in which the spark plugs fire to cause combustion in each cylinder. A four-cylinder engine may have one of two firing orders: 1-3-4-2 or 1-2-4-3. The cylinders are numbered 1-2-3-4 starting at the front of the engine. In this way, you can tell which cylinders will fire in sequence. Firing orders and cylinder numbers for V-6 and V-8 engines vary.

The engine firing order is often cast into the top of the intake manifold. When not on the manifold, refer to the service manual to know the firing order, *Fig.8.34*. Engine firing order is commonly used when installing spark plug wires and when performing engine tune-up.

8.8 SPARK PLUG SERVICE

Bad spark plugs can cause a wide range of problems: ***misfiring, lack of power, poor fuel economy,*** and ***hard starting.*** After prolonged use, the spark plug tip can become coated with ash, oil, and other contaminants. Also, the plug electrodes can burn and widen the spark plug point gap. This can make it more difficult for the ignition system to produce an arc between the electrodes.

Spark plug removal and service is an important part in every electrical tune-up. Spark plug firing voltages are best checked under actual engine operating condition with an oscilloscope.

Remove the spark plugs and check them carefully to determine their condition. A close study of the spark plugs provides the mechanic with valuable clues to the condition of the engine and its systems. Spark plugs will foul or wear rapidly if they are not of the correct heat range.

Platinum Spark Plugs

Platinum spark plugs use precious metal on their electrodes to make them more resistant to electric arc wear or corrosion.. This is especially important with modern distributorless ignition systems. Platinum plugs are now commonly used to increase spark plug service life up to 160,000 km before replacement is necessary.

Spark Plug Removal

To remove the spark plugs, first check that the spark plug wires are numbered or located correctly in their clips. Grasping the spark plug wire boot, pull the wire off. Twist the boot back and forth, *Fig. 8.35 (a)*. Do not remove a spark plug wire by pulling on the wire. Pulling on the wire can break the strand type conductor.

Blow debris away from the spark plug holes with compressed air. Then, using a spark plug socket, extension, swivel, and ratchet, as needed, unscrew each plug, *Fig. 8.35 (b)*.

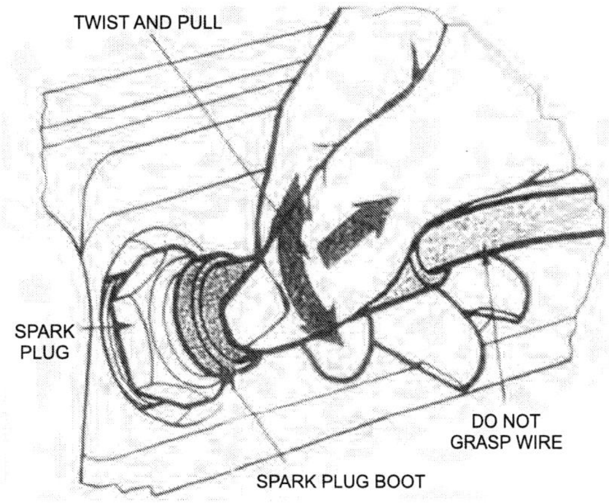

Fig. 8.35 *(a) Remove spark plug wire by pulling and twisting on boot, not wire itself to avoid conductor breakage. (Chrysler Corp.)*

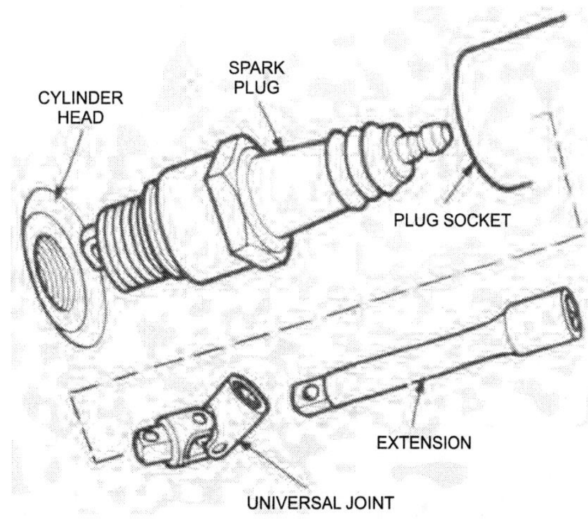

Fig. 8.35 *(b) Spark plug socket and ratchet are used to unscrew spark plugs. (Chrysler Corp.)*

As you remove each spark plug, lay it in order on the fender cover or workbench. Do not mix up the plugs. After all of the plugs are out, check them to diagnose the condition of the engine. To prevent thread damage, it is preferable to remove spark plugs from aluminium heads with the engine cold.

Spark plug removal is sometimes difficult because of their location on the engine and the engine accessories that are sometimes in the way. Special tools, such as a spark plug socket with a hex at the drive end and a flexible head ratchet, make many difficult jobs easy. Use a 6-point deep spark plug firmly so that it may be placed in the socket and started into the cylinder head by hand.

Because the condition of a spark plug provides information about the cylinder it was removed from, it is advantageous to keep the plugs in order using a spark plug holder made from a block of wood. Placing the spark plugs in the holder with the terminal end down permits their inspection without removing them from the holder.

Check the appearance and condition of both ends of the insulator, the electrodes, seat gasket, and the shell of the plug. This will throw light on how efficiently the engine has been operating and how suitable the spark plug is for the type of operation. Frequently the same type of spark plug used in two engines of the same make and model show a significant difference in appearance. The causes of these differences are the condition of the engine, piston rings, carburetor setting, kind of fuel used, and the engine operating conditions.

Gasket Inspection

Examine the engine seat gasket of each plug after removing the plug. The gasket performs two important functions:

1. It conducts away much of the heat absorbed by the spark plug insulator tip from the burning fuel in the combustion chamber.
2. It maintains a gas-tight seal between the plug and its seat in the cylinder head. If a spark plug gasket is not tightly seated, leaking combustion gases will cause overheating of the plug. This not only shortens the life of the plug caused by excessive wear of the overheated electrodes, but it may also lead to preignition.

If the gasket is flattened too much, the spark plug shell may be distorted or cracked and the plug gap may be changed due to the excessive applied torque.

A new spark plug gasket should be used every time a new or cleaned spark plug is installed for better performance and longer spark plug life.

Spark Plug Inspection

To ***read spark plugs,*** closely inspect and analyze the condition of each spark plug tip and insulator. This will give valuable information on the condition of the engine, fuel system, and ignition system.

Each removed spark plug should be compared to an analysis chart like the one shown in *Fig. 8.36.* Under normal engine operating conditions, spark plugs wear out from intense heat, from the action of sulphur and lead compounds in the fuel, and from the erosion caused by the electric spark on the electrodes. Plugs that have been in operation for a long period of service should be replaced. Worn plugs cause loss of power, loss of speed, hard starting, and sluggish performance.

- ***Brown to grayish-tan deposits and slight – electrode wear*** are signs of spark plug wear, correct spark plug heat range, and mixed periods of high speed and low speed driving. Spark plugs that look like this may be cleaned, re-gapped, and reinstalled.
- A ***black* or *wet plug*** indicates that the plug is ***not*** firing or that there is an engine problem, e.g., worn piston rings, leaking valve stem seals, low engine compression, or rich fuel mixture, in that cylinder.
- A frequent cause of spark plug misfiring is ***dirt on the top of the insulator.*** When combined with moisture, the dirt allows high voltage to go across the insulator top to the shell and away from the firing end. The dirt should be cleaned off with a suitable solvent.
- ***Soft, fluffy, dry, black carbon deposits*** mean too much fuel in the combustion chamber. This can be caused by engine operation at low speeds or by too-rich air-fuel mixture., excessive idling, stuck manifold heat riser, restricted air cleaner, or improperly operating automatic choke. Using ***"too cold"*** a plug will allow excessive carbon buildup. If the fouling continues, a change to a spark plug one or two steps, hotter might be necessary.

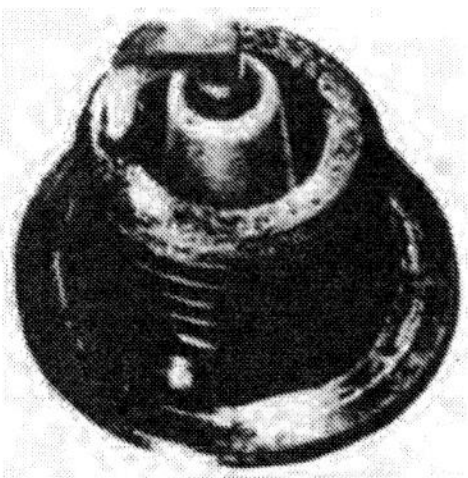

NORMAL

Brown to grayish tan colour and slight electrode wear. Correct heat range for engine and operating conditions.

RECOMMENDATION. Properly service and reinstall. Replace if over 10,000 miles of service.

SPLASHED DEPOSITS

Spotted deposits. Occurs shortly after long delayed tune-up. After a long period of misfiring, deposits may be loosened when normal combustion temperatures are restored by tune-up. During a high-speed run, these materials shed off the piston and head and are thrown against the hot insulator.

RECOMMENDATION. Clean and service the plugs properly and reinstall.

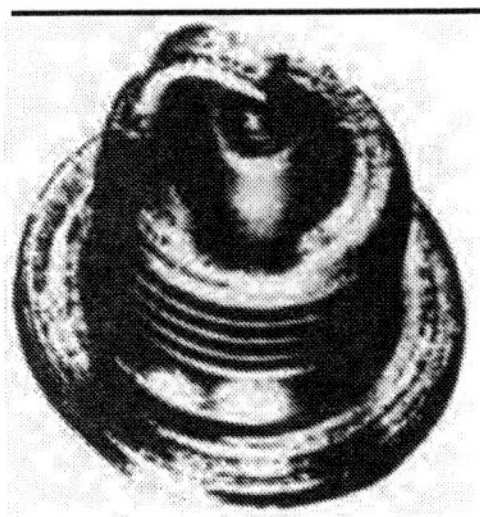

CARBON DEPOSITS

Dry soot.

RECOMMENDATION. Dry deposits indicate rich mixture or weak ignition. Check for clogged air cleaner, high float level, sticky choke or worn breaker contacts. Hotter plugs will temporarily provide additional fouling protection.

HIGH SPEED GLAZING

Insulator has yellowish, varnish-like color. Indicates combustion chamber temperatures have risen during hard, fast acceleration. Normal deposits do not get a chance to blow off, instead they melt to form a conductive coating.

RECOMMENDATION. If condition recurs, use plug type one step colder.

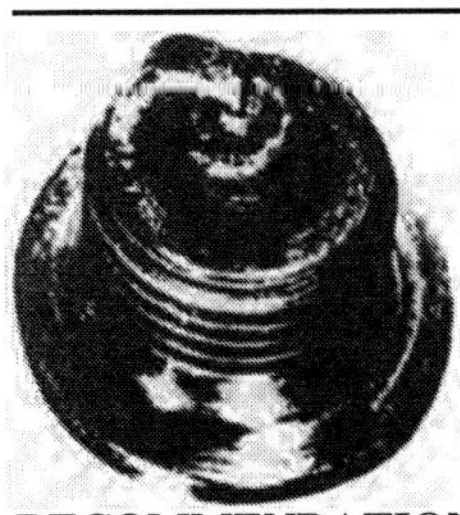

OIL DEPOSITS

Oily coating.

RECOMMENDATION. Caused by poor oil control. Oil is leaking past worn valve guides or piston rings into the combustion chamber. Hotter spark plug may temporarily relieve problem, but positive cure is to correct the condition with necessary repairs.

ASH DEPOSITS

Light brown to white coloured deposits encrusted on the side or center electrodes or both. Derived from oil and/ or fuel additives. While non-conductive, excessive amounts may mask the spark, causing misfire.

RECOMMENDATION. If excessive deposits accumulate in short mileage, corrective measures may include installation of valve guide seals to prevent seepage of oil into combustion chamber.

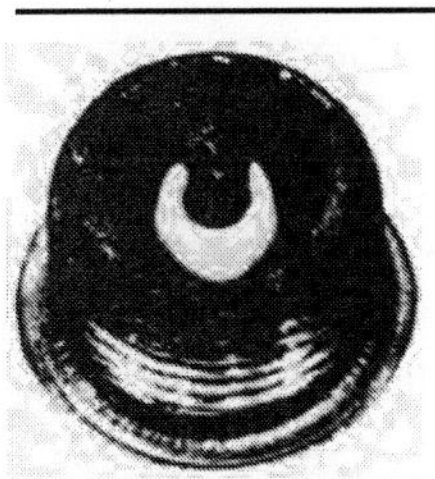

TOO HOT

Blistered, white insulator, eroded electrodes and absence of deposits.

RECOMMENDATION. Check for correct plug heat range, over advanced ignition timing, cooling system level and/or stoppages, lean fuel/air mixtures, leaking intake manifold, sticking valves, and if car is driven at high speeds most of the time.

MECHANICAL DAMAGE

Mechanical Damage to the plug's firing end is caused by some foreign object in the combustion chamber. It may also be due to the piston striking the firing tip of improper reach plugs. When working on an engine, be sure to keep the carburetor throat and any open plug holes covered. Consult the catalog for proper reach plugs.

Fig. 8.36 *Spark plug analysis chart (Champion Spark Plug Co.)*

- ***Red, brown, yellow, and white coatings*** on the insulator are by products of combustion from the fuel and lubricating oil. Under severe operating conditions, these deposits may cause missing, particularly at high speeds and on hard pulls. If the insulator is not too heavily coated, these deposits can be removed by abrasive cleaning. If deposits are packed between the insulator and the shell of the plug, they cannot be completely removed. In this case, replace the spark plug.
- The combined action of intense heat, pressure, and corrosive gases within the combustion chamber together with the spark discharges, causes the ***spark plug gap to widen.*** An excessive gap wear, at low kilometers means that the engine is operating under very high loads or that the wrong plug (too hot) is installed. Before changing to a cooler plug, ***check the heat range.*** If the plug showing rapid electrode wear is the correct type, check the seat gasket. If the gasket is in good condition, install spark plugs at least one step colder.
- The spark plug conducts electrical energy of several thousand volts to the spark plug in the combustion chamber. This electrical energy will always take the path of least resistance. Many things can cause this electrical energy to take a path away from the spark plug. A dirty or oily upper insulator, *Fig. 8.37,* causes a shorted upper insulator, resulting in flashover. A gap that is too wide as a result of wear or improper setting at the time of plug installation is another reason for ***flashover.***

Fig. 8.37 *Flashover is caused by dirty insulator or wide electrode gap. (Chevrolet Div. General Motors)*

- Sparking through the insulator to the shell results from a ***break in the insulator.*** This is caused by careless removal or installation of the spark plug. A ***fractured plug*** must be replaced.
- A ***broken lower insulator*** is generally visible, especially after the spark plug has been cleaned. One cause of this is carelessness in re-gapping: Never bend the centre electrode to adjust the gap. Never allow the gapping tool to push too hard against the tip of the centre electrode or insulator when bending the side electrode to adjust the gap. If the lower insulator is broken, replace the spark plug.
- A break in the lower insulator may also occur if the engine has been operated for long periods with heavy detonation or preignition. Over-advanced ignition timing, and the use of low-octane fuel usually cause this "heat shock failure". Rapid increase in tip temperature under severe operating conditions causes the heat shock and the break. Installing a colder spark plug may correct the engine operation. Check the spark plug type being used against the specs chart. A spark plug that is hotter than recommended might cause this problem. If the recommended type is being used, install a step colder type.

Cleaning and Gapping Spark Plugs

To save the labour cost, it is better to remove and replace spark plugs with new ones. There are times when it is necessary to clean and set the gap spark plugs and reuse them.

Spark plugs coated with carbon or oxide deposits can be cleaned in a blast cleaner. Follow the instructions of the cleaner manufacturer carefully. Plugs should be cleaned until the inside of the shell, the entire insulator, and the electrodes are clean. However, excessive blast cleaning will wear down the insulation and damage the spark plug. Spark plugs that cannot be cleaned with a reasonable amount of blasting should be replaced.

Plugs with an oily or wet deposit should be cleaned in a non-flammable degreasing solution and thoroughly dried before blast cleaning to prevent the cleaning compound from packing into the shell. Spark plug service also includes cleaning the top insulator and terminal with a cleaning solvent to remove all oil and dirt.

After cleaning check the spark plug carefully for cracks or other defects that might not have been visible before cleaning. If the plug appears to be in good condition, the end of the centre electrode should be filed lightly to provide a flat, square surface, *Fig. 8.38.* If blast cleaning has not removed all deposits from the electrodes, they should be cleaned with several strokes of fine abrasive or a file.

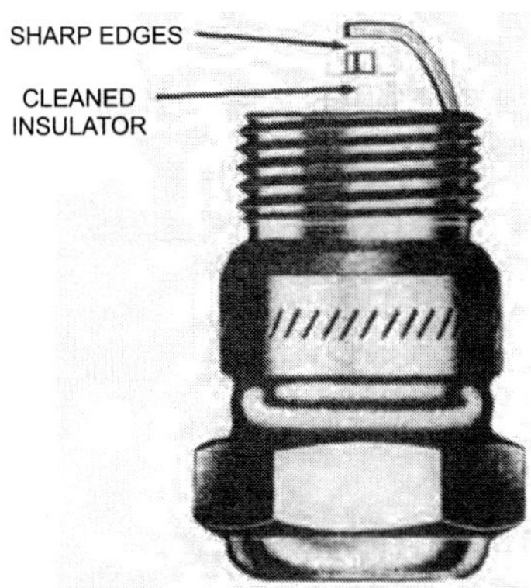

Fig. 8.38 *Cleaned, filed, and re-gapped spark plug. (Champion Spark Plug Co.)*

Also, look at the threads for carbon deposits. Clean them with a hand wire brush, using care not to injure the electrodes or insulator tip. Clean threads permit easy installation and proper seating of spark plugs when being installed back in the engine. Spark plugs with nicked or damaged threads should be replaced. Obtain the correct replacement plug recommended by the manufacturer. Spark plug gaps vary from approximately 0.76 mm on contact point ignitions to over 1.52 mm on electronic ignition systems.

The gap of each spark plug must be checked and adjusted if needed before installing the plugs. The gap should be checked even when installing the new spark plugs. Refer the service manual for gap specs. Select a wire feeler gauge of the correct thickness to measure the gap. Do not use a flat feeler gauge for this job as it will be inaccurate for measuring used spark plug gaps.

Slide the wire feeler gauge between the spark plug electrodes as shown in *Fig. 8.39 (a).* The wire should fit between the spark plug electrodes with a light drag. A too wide or too narrow gap should be adjusted.

An electrode-adjusting tool is included in most spark plug wire feeler gauge sets. The notch on the adjusting tool is inserted around the spark plug ground electrode. The tool is then used to bend the ground electrode closer to or away from the centre electrode, *Fig. 8.39 (b).* Never try to bend the centre electrode.

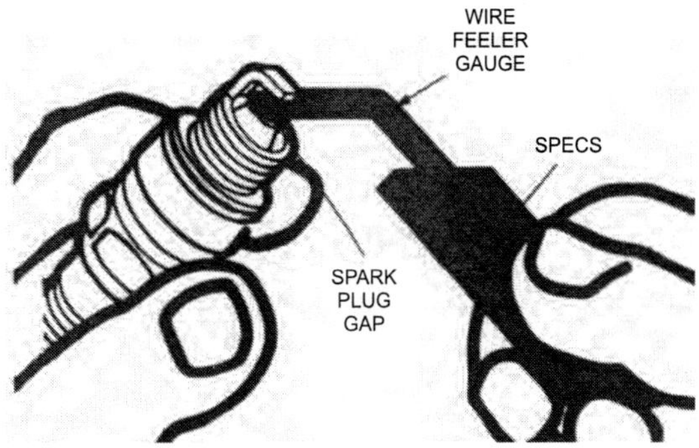

(a) Use wire feeler gauge to measure gap between electrodes. It should just drag when pulled through gap.

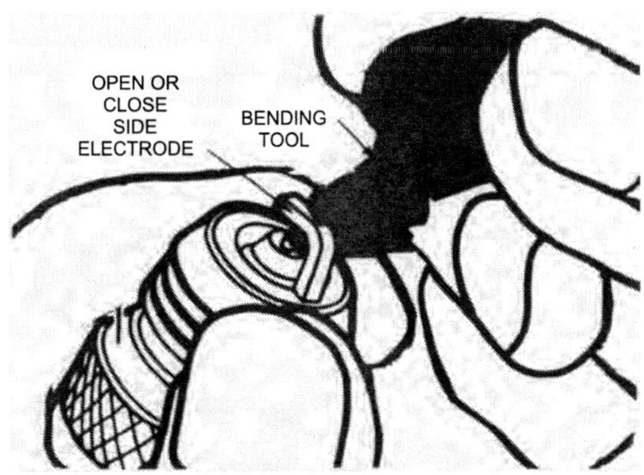

(b) Use tool to bend side electrode to open or close gap as needed.

Fig. 8.39 *Gapping a spark plug. (Sun Electric Corp.)*

Installing Spark Plugs

A new spark plug gasket must be used with each new or cleaned spark plug. Make sure the cylinder head spark-plug-hole threads and plug seats are clean and free from and dirt that would prevent proper seating of the plug and gasket. Dirty cylinder head threads should be cleaned with a greased thread chaser of the correct size.

When using any spark plug make different from the original, check that the new plug has the same reach or length of thread as the original installed plug. If the new plugs threads are too long, the threads exposed inside the combustion chamber will become corroded and covered with carbon. This may damage the threads in the cylinder head when the plug is removed. If a plug with short threads is used, the exposed cylinder head threads may collect carbon or partially burned away, preventing proper installation of a plug with the correct reach. In some engines the extended reach spark plugs when used, may interfere with the pistons and cause a great deal of engine damage.

Screw the spark plug by hand or a small piece of vacuum hose to ***start*** all the way down until it seats on the gasket finger-tight. Then use a torque wrench and torque the plug to specs as specified in service manual. This specification is true for new gaskets, with spark plug and cylinder head threads thoroughly clean.

Frequently, spark plug torquing is difficult due to lack of accessibility. As a rule of thumb to tighten the plugs that gives satisfactory results is to tighten the spark plug finger-tight and then turn it with a spark plug wrench ½ to ¾ turn.

8.9 DISTRIBUTOR REMOVAL AND INSTALLATION

The contact-point distributor needs periodic checking of the contact points, and their replacement and adjustment as necessary. The centrifugal and vacuum advances need also be checked to ensure they are working properly. These checks require removal of the distributor for testing in a distributor tester. The tester drives the distributor at various speeds and vacuums to check the operation of the centrifugal and vacuum advance mechanisms. If a trouble is found, the distributor might require complete servicing or replacement. Refer to the correct service manual for detailed procedures to be followed to test and service the distributor.

Distributor removal and installation is a simple job if the engine is left undisturbed while the distributor is out. However, if the engine is cranked so that the crankshaft and camshaft are turned with the distributor out, then the distributor installation needs more steps.

Distributor Removal

- Remove the air cleaner, and disconnect the vacuum hose or hoses from the distributor.
- Disconnect the primary lead running from the ignition coil to the distributor.
- Remove the distributor cap, and move the cap and wire assembly aside.
- Scratch a mark on the distributor housing. Scratch another mark exactly against the first, on the engine block. These marks locate the position of the distributor housing in the block, *Fig. 8.40.*
- Scratch a third mark on the distributor housing precisely under the rotor tip, *Fig. 8.40.* This mark locates the position of the rotor in the housing.
- Remove the distributor hold-down bolt and clamp. Lift the distributor out of the block.

If the engine is not cranked while the distributor is out, the distributor can be easily installed in its exact position. Just align the scratch marks on the distributor housing and cylinder block. As you push the distributor down into place, the shaft and rotor will turn as the spiral gears mesh. Therefore, start with the rotor turned back from the installed position so that when you install the distributor, the rotor will turn into the proper position and the rotor tip will line up with the scratch mark on the distributor housing.

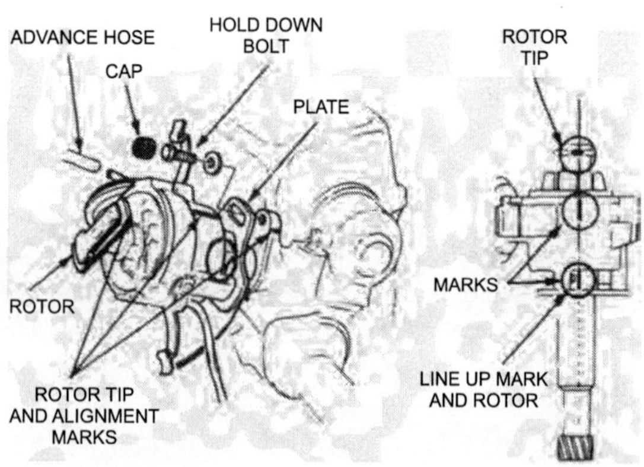

Fig. 8.40 *Place marks on engine and distributor as shown before removing distributor. This will help install distributor correctly upon reassembly. Marks at rotor tip. Housing, and engine should align on assembly. (Honda Motors)*

Distributor Installation

If the engine has been cranked with the distributor removed, timing has been upset. The engine has to be retimed. This is required to establish the proper relationship between the distributor rotor and the #1 piston.

- Remove #1 spark plug from the cylinder head. Place a shop cloth over your finger, and cover the spark plug hole. Crank the engine until you feel compression pressure on your finger.
- Bump the engine with the starter motor until the timing marks on the crankshaft pulley and timing cover are properly aligned. This indicates that #1 piston is in firing position.
- The distributor can now be installed in the cylinder block by aligning the scratch marks on the distributor housing and cylinder block.
- Check that the distributor gasket or rubber O-ring is in place when you install the distributor gasket.
- The distributor may have to be turned slightly to engage the drive. *Fig. 8.41* shows some typical distributor drives.
- Make sure the distributor housing is fully seated against the cylinder block. If it does not, the lubricating oil pump shaft is not engaging. Hold the distributor down firmly, and bump the engine again to realign the timing marks.
- Install the distributor but do not tighten the hold-down clamp and bolt.
- Make sure the distributor housing is installed properly. The advance unit should be pointing as it was before removal.
- Rotate the distributor until the contact points just start to open to fire #1 cylinder.
- Hold the distributor cap in place above the distributor. Check that the rotor tip lines up with #1 terminal on the cap. Install the cap with wires.
- Connect the primary wire from the ignition coil to the distributor.
- Start the engine. Set the ignition timing. Connect the vacuum hose or hoses to the distributor. Install the air cleaner.

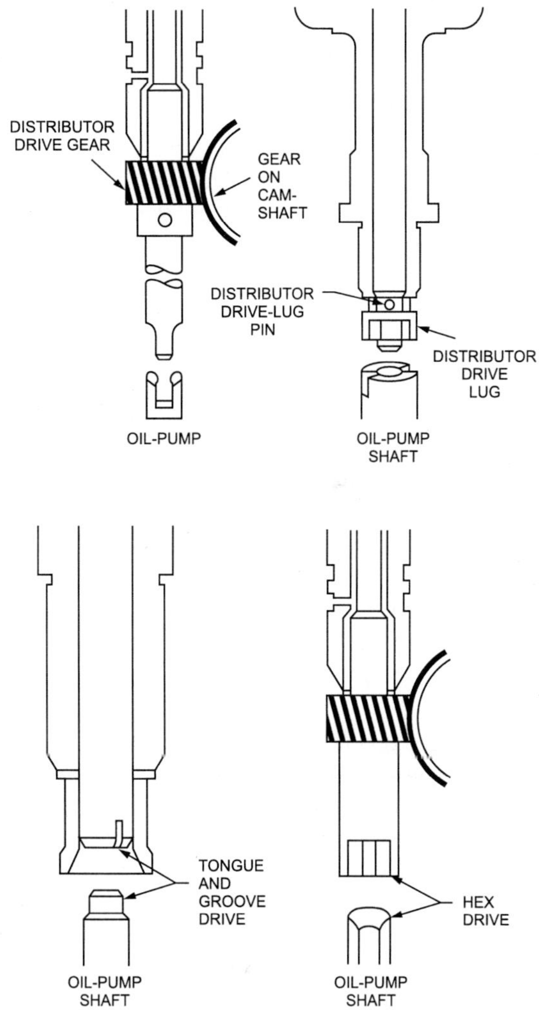

Fig. 8.41 *Distributor drives and installation methods.*

8.10 ELECTRONIC IGNITION-SYSTEM SERVICE

In many ways, the servicing of electronic ignition systems and of contact-point ignition systems is similar. However there are special procedures for electronic ignition systems.

One of the reasons electronic ignition systems are so popular is that they have no contacting parts to wear out inside the distributor. However many electronic ignitions have an adjustable air gap between the pickup and the reluctor. The following adjustment procedure is a general one; check the correct service manual for specific details on service procedures for the ignition system under service.

Ignition Module Service

A ***faulty ignition module*** will cause a wide range of problems: engine stalls when hot, engine cranks but fails to start, engine misses at high or low speeds.

Often, an ignition module problem will appear after a period of engine operation. ***Engine heat*** will soak into the module, raising its temperature. The heat will upset the operation of the electronic components in the unit.

Usually the ignition module is one of the last components to test. If all of the other ignition system components are in good working order, then the problem could be in the electronic control module.

If a specialized tester is available, it may be used to quickly determine the condition of the module. The wires going to the electronic control module are unplugged. The tester is then put into the circuit. The tester will then indicate if there is any fault in the module.

The microscopic components inside the module are very sensitive to high temperature. When testing the module, warm the module, with a heat gun or light bulb, to simulate the engine compartment temperature. Be careful not to apply too much heat to the module to avoid it being ruined. The heat may make the control unit act up and allow you to find an erratic problem.

Air Gap

- Rotate the engine until one of the teeth on the reluctor lines up with the pickup.
- Use a nonmagnetic set of feeler gauges to check the gap between the tooth and the pickup.
- Check the specs and move the pickup to adjust the gap.
- Tighten the hold-down screw and again check the gap.
- Have someone crank the engine while watching to ensure that there is no contact between the reluctor and any stationary part of the distributor, *Fig. 8.42.*
- Check the cap and rotor for any silicone grease and apply fresh grease if necessary.
- Reinstall the rotor and cap, making sure all the wires are correctly and adequately connected.

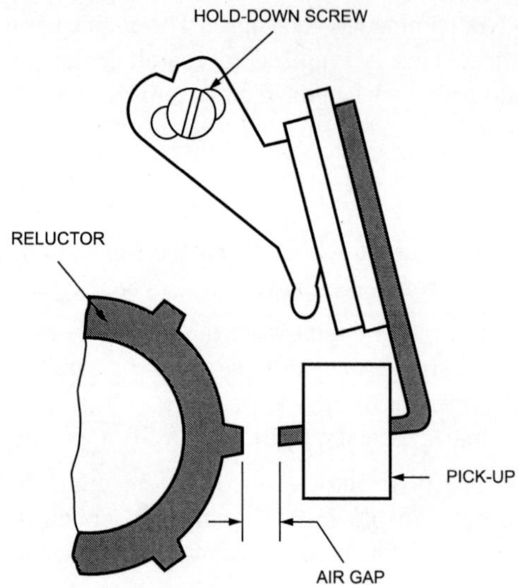

Fig. 8.42 *Setting air gap at reluctor. (Chrysler Corp.)*

8.11 POSSIBLE CAUSES OF ELECTRONIC IGNITION-SYSTEM FAILURE

Electronic ignition system failures may be grouped into three categories:

1. *Loss of energy in the primary circuit.* This could be caused by:

- Resistance in the primary circuit due to defective leads, bad connections, or bad ignition switch, or open ignition coil primary winding.
- Discharged battery or defective alternator.
- Grounded primary circuit in ignition coil, wiring, or distributor.
- Defective ECU or defective sensor-coil circuit to ECU.

2. *Loss of energy in the secondary circuit.* This could be caused by:

- Spark plugs fouled, defective, or improperly gapped.
- Defective high-voltage wiring that allows high-voltage leakage.
- High-voltage leakage across ignition-coil head, distributor cap, or rotor.
- Defective connections in high-voltage circuits.

3. ***Ignition-timing off.*** This could be due to:

- Timing not set correctly
- Faulty centrifugal or vacuum advance
- Preignition from spark plugs of wrong heat range, fouled plugs, carbon in combustion chambers
- Defective electronic control unit (ECU).

8.12 TROUBLESHOOTING ELECTRONIC–IGNITION SYSTEM

The ignition system and several other engine systems, e.g., fuel, electrical, emission systems, all work together. A problem in one system may affect, or appear to affect, the operation of another system. As a consequence, problem diagnosis can be quite challenging. For example, a shorted gasoline injector can cause an engine miss or rough idle. An oil fouled spark plug will also cause an engine miss. The symptoms of each will be almost identical. It becomes necessary to use proper testing methods to find the faulty component.

Electronic ignitions are much more difficult to describe in a general way. For any system, though, the troubleshooting procedure includes several items. The first check is for wiring condition and connection tightness. The order of the following checks may vary but will always include coil voltage, coil resistance, pickup coil resistance or output, and continuity of wiring throughout the ignition circuit. Finally, if all other components seem to be good, the control module will be replaced.

Ignition systems change frequently as new developments in electronics make possible less expensive and more reliable systems. Make sure that you use the exact procedures and wiring diagram for the system you need to test.

General System Checks

Whenever an ignition problem is suspected in any ignition system, ***always check the secondary circuit first.***

Start with a ***visual inspection.*** Inspect the ignition with and without the engine running. Look for obvious problems: loose primary connections, spark plug wires pulled off, deteriorated wire insulation. Check for missing or disconnected wires or spark plug boots. Check the distributor cap, rotor, and ignition coil for carbon tracking and/or cracks.

Look over other engine systems. Try to find anything that could upset engine operation. *Fig. 8.43* shows some typical problems.

Spark Test

A ***spark test,*** also called a ***spark intensity test,*** measures the brightness and length of the electric arc (spark) produced by the ignition system. It is a quick way of checking the condition of the ignition system.

The spark test is commonly used when an engine cranks, but will ***not*** start. The trouble will help show whether the trouble is in the fuel system (no fuel problem) or in the ignition (no spark problem). It may also be used to check the spark plug wires, distributor cap, and other secondary components.

A ***spark tester*** is a device with a very large air gap for checking ignition output voltage. It is like a spark plug with a wide gap and a ground wire. A spark tester can be made from an old plug with little materials or effort. *Fig. 8.44* shows a typical spark tester. If you do not have a spark tester, hold the wire, with insulated pliers to avoid shock, 6 mm from the engine block.

Crank the engine and check for a good spark at the spark plug tester gap. A good spark will be blue-white and will be plainly visible in daylight. Weak sparks are orange or red and may be hard to see in daylight. A strong spark indicates that the ignition coil, pickup coil, electronic control unit (CU), and other ignition system parts are working. If a good spark is present, the problem is probably not in the ignition system. Check the fuel system and/or spark timing.

A weak spark or no spark shows that something is wrong in the ignition system. If no spark is visible, remove the coil wire from the distributor cap. Attach a spark tester to the distributor end of the coil wire, *Fig. 8.46.* Crank the engine and check for a good spark at the spark tester gap. If a spark is present, the problem is probably in the distributor cap, rotor, or spark plug wires.

If you did not see a spark, the problem is in the primary system. With conventional ignitions, the points or a broken wire are most likely causes. With electronic ignition, make sure the battery connections are good, and then troubleshoot the primary system

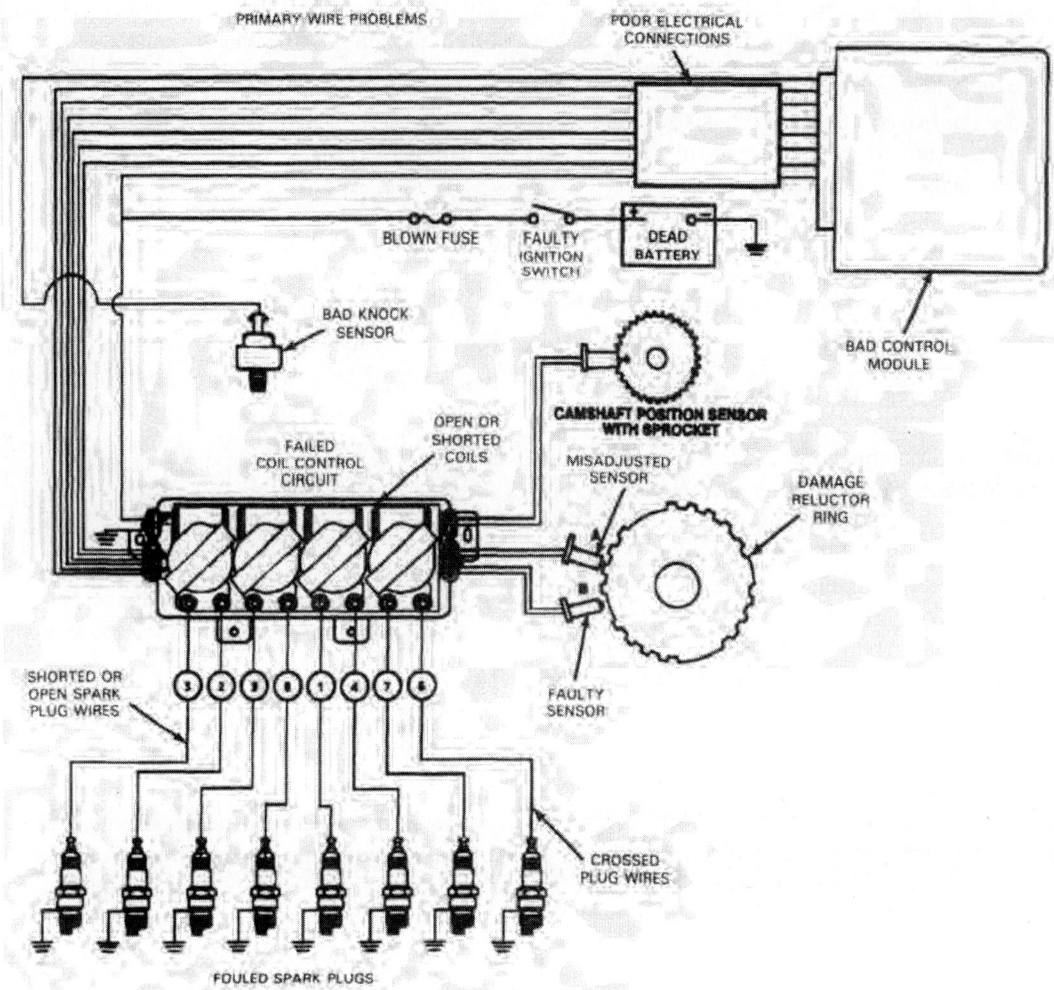

Fig. 8.43 *Common problems that can occur in a modern ignition system. (General Motors)*

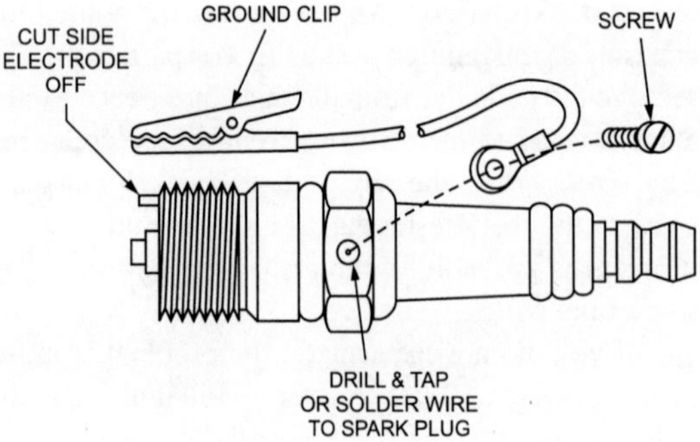

Fig. 8.44 *Spark plug tester.*

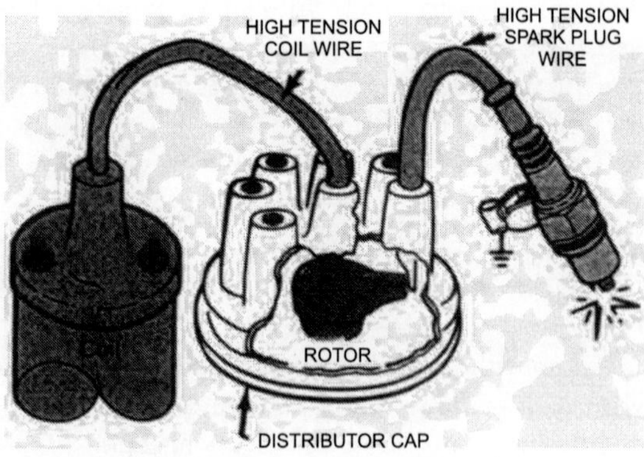

Fig. 8.45 *Secondary ignition circuit test.*

following the procedure outlined in the service manual.

If spark is weak at all spark plug wires, the problem is common to all of the cylinders, e.g., bad ignition coil, rotor, coil wire. Other tests would be needed to pinpoint the trouble source.

Remove a spark plug wire from any plug. Attach a spark tester to the wire, *Fig. 8.45.*

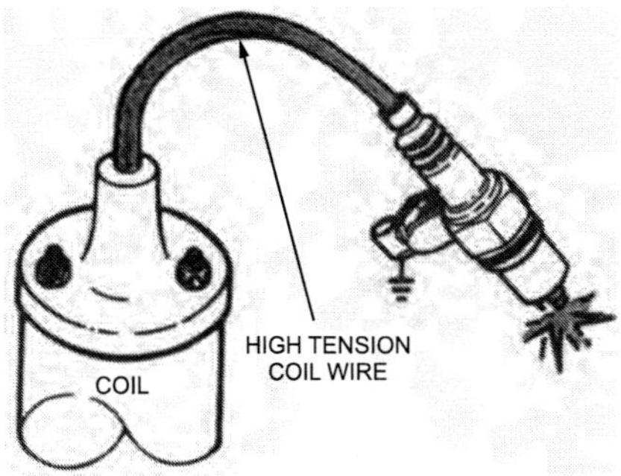

Fig. 8.46 *Checking for spark at the col.*

Voltage Checks

With any ignition system there may be a power problem. Using a voltmeter, check to see if battery voltage is present at the coil terminal with the ignition on. On General Motors systems with the coil in the distributor cap, check the battery voltage at the pink distributor wire, *Fig. 8.47.* If 10-13 V are measured at the coil, the power supply is probably fine. If voltage is not present, check for disconnected or broken wires. These account for most ignition problems. If necessary check the battery, ballast resistor, ignition circuit, and ignition switch.

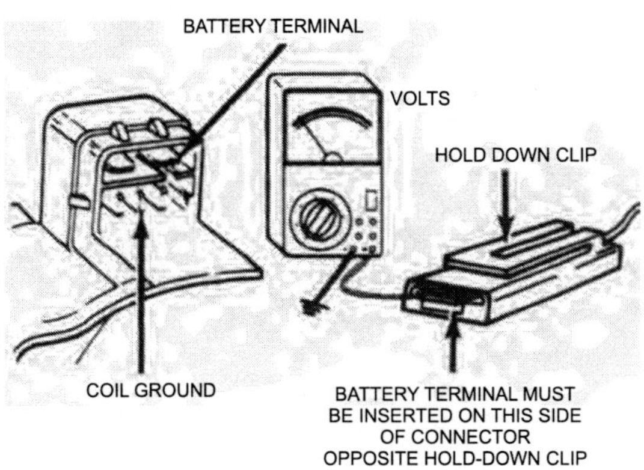

Fig. 8.47 *Battery voltage at the distributor connector voltmeter hookup. (General Motors Corporation)*

Coil Checks

Frequently, ignition problems may be in the coil. The following coil resistance checks can be used on all types of ignition systems in which the coil is a separate unit.

Make sure the ignition switch is off. Remove the connectors at the positive and the negative coil terminals. Connect an ohmmeter lead to each primary terminal, *Fig. 8.48.* Set the ohmmeter to the × 1 scale. Check the resistance and compare it to specs. This tests the coil's primary resistance. Replace the coil if the primary resistance is not to specs.

To check the coil secondary circuit resistance, turn the ignition off and remove the high-voltage wire. This is the heavy insulated wire coming from the coil's centre terminal. Set the ohmmeter to the × 1000 scale. Connect the ohmmeter lead to the brass connector in the centre terminal and the other lead to the coil negative terminal. *Fig. 8.49* shows the hookup. If resistance is not to specifications, replace the ignition coil.

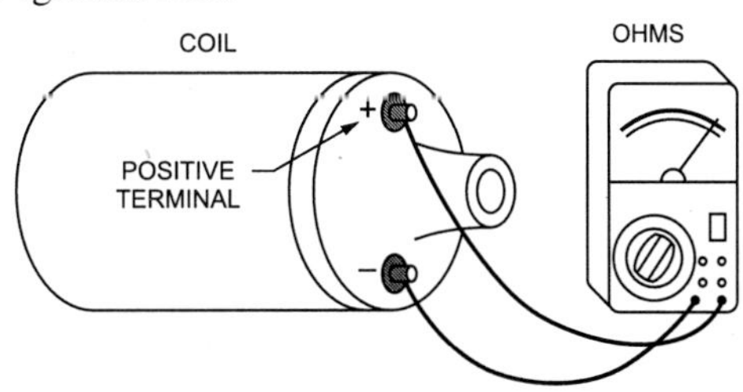

Fig. 8.48 *Coil primary resistance check.*

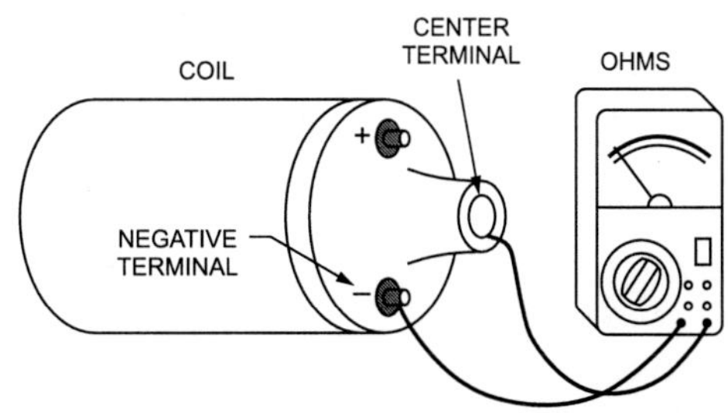

Fig. 8.49 *Coil secondary resistance check.*

Dead Cylinder

A ***dead cylinder*** is one ***not*** burning air-fuel mixture on the power stroke. It could be due to ignition system problems or problems in the engine, fuel system, or another system. A very rough idle and puffing noise in the engine exhaust may indicate a dead cylinder.

To check for a dead cylinder, pull off one spark plug wire at a time. Hold it next to ground. Engine rpm and idle smoothness should decrease with the wire removed.

If idle smoothness and rpm stay the same with the plug wire off, that cylinder is ***dead.*** It is not producing power. Check for spark at the wire, spark plug condition, and possible low cylinder compression. Special test instruments may be needed to diagnose ignition system problems.

8.13 ELECTRONIC IGNITION-SYSTEM PROBLEM-DIAGNOSIS AND REMEDIES

Various possible troubles that might be caused by conditions in the electronic ignition system and their remedies are discussed below. However, many of these conditions could be caused by troubles in other components and systems of the engine.

Engine Cranks Normally But Fails to Start

If there is no spark during the spark test, or if the oscilloscope fails to show a secondary-voltage pattern, there are several possible causes.

If the spark test shows a good spark, the ignition primary and secondary circuits are probably good. The failure to start could be due to fouled spark plugs or out-of-time ignition. However, failure to start with a good spark is more likely due to a problem in the fuel system, or other engine problems.

The possible causes and remedies are:

1. **No voltage to ignition system:** Check battery, ignition switch, wiring.
2. **ECU ground lead open, loose or corroded:** Repair as needed.
3. **Primary-wiring connectors not fully engaged:** Clean, and firmly seat connectors.
4. **Ignition coil open or shorted:** Test coil and replace if defective.
5. **Damaged armature (trigger wheel, reluctor) or sensor:** Replace the damaged part.
6. **ECU faulty: Replace** the ECU.
7. **Defective distributor cap or rotor:** Replace the defective part.
8. **Fuel system faulty:** Check the fuel system and replace any defective parts as necessary.
9. **Engine faulty:** Service engine.

Engine Backfires But Fails to Start

This can be caused by a considerably off ignition timing or valve timing, by a defective or wet distributor cap or rotor that permits high-voltage leakage from one terminal to another, or by the high-voltage cables wrongly connected.

The possible causes and remedies are:

1. **Incorrect timing:** Check and reset timing.
2. **Moisture in distributor cap:** Clean dry the cap.
3. **Cap defective-voltage leakage across carbon paths:** Replace cap.
4. **High-voltage cables not connected in firing order:** Reconnect cables in correct firing order.

Engine Runs But Misses

An engine that misses runs unevenly and will not develop its full power. It is often difficult to tell by just listening whether the miss is intermittent or is jumping around from cylinder to cylinder. If one cylinder is missing, it could be due to a defective spark plug or high-voltage cable, a bad connection, or high-voltage leakage across the distributor cap or through the cable insulation. In the engine, the miss could be due to a stuck or burnt valve, or loss of compression caused by broken rings.

If the miss jumps around, it could be caused by defects in the electronic part of the ignition system – the ECU and the sensor coil in the distributor. Jump-around miss can also be caused by the advance mechanisms that are not working properly and cause erratic advance.

A defective fuel system that delivers too-lean or too-rich a mixture can cause missing. A too-lean mixture will fail to burn. A too-rich mixture can cause plug fouling and result in misfiring.

The possible causes and remedies are:

1. **Spark plugs fouled or faulty:** Clean and set gap, or replace.
2. **Distributor cap or rotor defective:** Replace the defective part.

3. **High-voltage cables defective:** Replace the defective cables.
4. **Defective coil:** Replace coil.
5. **Bad electrical connections:** Clean, and tighten.
6. **High-voltage leakage:** Check distributor cap, rotor, cables, replace defective parts.
7. **Advance mechanisms faulty:** Check advance mechanisms, repair or replace distributor.
8. **Defective fuel system:** Service fuel system.
9. **Defects in engine:** Service engine as needed.

Engine Runs But Backfires

Backfiring is a "bang" or "pop" sound in the exhaust manifold or intake manifold. Several conditions in ignition system can cause this. An excessively off ignition timing, or ignition cross-firing, or ignition occurring before the intake valve closes can cause backfiring. Cracked or damaged cable insulation can allow spark jump over.

A spark plug running too hot may glow enough and ignite the air-fuel mixture before the intake valve closes. This causes a backfire, or pop back through the air cleaner. Excessively hot valves can also cause preignition or glowing carbon deposits in the combustion chamber.

Incorrect air-fuel ratios can also cause backfiring. A lean air-fuel mixture can cause backfiring through the air intake. A rich air-fuel mixture and a defective air-injection system can cause backfire in the exhaust system.

The possible causes and remedies are:

1. **Ignition timing off:** Reset ignition timing.
2. **Ignition cross-firing:** Check high-voltage cables, distributor cap, and rotor for leakage paths. Replace any defective parts.
3. **Defective antibackfire valve:** Replace the valve.
4. **Spark plugs having wrong heat range:** Replace spark plugs with correct heat range.
5. **Defective air-injection system:** Check system and replace any defective part.
6. **Engine overheating:** Ignition timing late, Check cooling system, valve timing, and service engine.
7. **Faulty air-fuel supply system:** Service fuel supply system.
8. **Engine defects e.g., hot valves, carbon deposits etc.:** Service engine as necessary.

Engine Overheats

Most engine overheating is caused by loss of coolant through leaks in the cooling system. Other causes include a loose or broken fan belt, a defective coolant pump, clogged water jackets in the engine, defective thermostat or fan clutch.

Late ignition or valve timing, lack of engine oil, overloading the engine, or high speed, high altitude, or hot-climate operation can cause engine overheating. A faulty temperature control system prevents vacuum advance in any gear, or if the distributor vacuum advance is defective, engine overheating may result.

The possible causes and remedies are:

1. **Late Ignition timing:** Reset ignition timing.
2. **Lack of coolant or other problems in cooling system:** Service cooling system.
3. **Late valve timing or other engine conditions:** Service engine as needed.

Engine Lacks power

The possible causes and remedies are:

1. **Ignition timing off:** Reset ignition timing.
2. **Spark plugs fouled or faulty:** Clean and set gap, or replace.
3. **Distributor cap or rotor defective:** Replace the defective part.
4. **High-voltage cables defective:** Replace the defective cables.
5. **Defective coil:** Replace coil.
6. **Bad electrical connections:** Clean, and tighten.
7. **High-voltage leakage:** Check distributor cap, rotor, cables, replace defective parts.
8. **Advance mechanisms faulty:** Check advance mechanisms, repair or replace distributor.
9. **Defective fuel system:** Service fuel system.
10. **Defects in engine:** Service engine as needed.
11. **Exhaust system restricted:** Service exhaust system as needed.

12. **Heavy engine oil:** Use correct viscosity oil.
13. **Wrong fuel:** Use correct recommended fuel.
14. **Excessive rolling resistance:** Check tyres, brakes, wheel bearings, wheel-alignment.
15. **Late ignition timing:** Reset ignition timing.
16. **Lack of coolant or other problems in cooling system:** Service cooling system.
17. **Late valve timing or other engine conditions:** Service engine as needed.

Engine Detonates, or Pings

The possible causes and remedies are:

1. **Improper ignition timing:** Reset ignition timing.
2. **Wrong fuel:** Use correct fuel.
3. **Spark plugs of wrong heat range:** Install correct spark plugs.
4. **Advance mechanisms faulty:** Rebuild or replace distributor.
5. **Carbon build-up in cylinders:** Service engine.

Spark Plugs Defective

The possible causes and remedies are:

1. **Cracked insulator:** Replace new plugs carefully.
2. **Plug sooty:** Install hotter plug, service engine to correct condition causing oil burning or high oil consumption.
3. **Plug white or gray, with blistered insulator:** Install cooler plug.

Engine Runs On or Dieseling

The possible causes and remedies are:

1. **Idle-stop solenoid out of adjustment or defective:** Adjust or replace solenoid.
2. **Hot Spots in Combustion Chambers:** Service engine.
3. **Late ignition timing:** Reset ignition timing.
4. **Lack of coolant or other problems in cooling system:** Service cooling system.
5. **Late valve timing or other engine conditions:** Service engine as needed.
6. **Advanced ignition timing:** Reset ignition timing.

8.14 COMPUTERIZED IGNITION SERVICE

Many of the components of a computer controlled ignition system are similar to those of electronic or contact point ignition systems. Therefore, the testing is about the same for many parts (spark plugs, secondary wires, ignition coil). However, the computerized ignition has engine sensors and computer, which increases the complexity of the system.

Computer Self-Diagnosis Mode

Some computerized systems have a check engine light in the dash that glows when a problem exists. The computer can be activated to produce a number code. The code can be activated to produce a number code. The code can be compared to information in the vehicle's service manual to pinpoint the source of a problem. This makes testing and repairing a computerized system much easier.

Remember that a computerized ignition system can be seriously damaged if the wrong wire is shorted to ground or if meter is connected improperly. ***Always follow manufacturer's recommended testing procedures.***

Computer Ignition Testers

Vehicle manufacturers provide specialized testing equipment for their computerized ignition systems. Follow the service manual and equipment manufacturer's recommended procedures for testing computerized systems.

THEORY QUESTIONS

1. Explain the method used to find out whether the ignition system is providing a spark.
2. How can 12 volts from a battery be changed to 40.000 volts for ignition?
3. The———————— section of the ignition system operates on high voltage.
4. How does a magnetic sensor work?
5. Explain the method followed to troubleshoot the ignition system primary circuit.
6. What does the firing section of an oscilloscope pattern show about an engine ignition system?
7. What does the dwell section of an oscilloscope pattern show about the ignition system?
8. It takes about 20.000 volts to jump the rotor-to-cap gap. *True* or *False?*
9. Explain the method used to replace a conventional ignition system contact points.
10. Explain the method used to adjust ignition timing of an automotive engine.
11. Explain the procedure followed to adjust the contact points, of an ignition system, with a dwell meter.
12. What is the only adjustment needed inside most electronic ignition system distributors?
13. Ignition timing can be changed by adjusting point gap or dwell. *True* or *False?*
14. Dwell is adjustable on most electronic ignition systems. *True* or *False?*
15. What is the reason for spreading grease on ignition terminals and connectors?
16. Describe how to remove a set of spark plugs.
17. What happens to the spark plug gap after it has been in service for a long period of time on engine?
18. What two factors need to be considered when buying new spark plugs?
19. Why should a spark tester be used to check for spark rather than a standard spark plug?
20. Describe the method of cleaning a spark plug.
21. What harm can occur if the engine is cranked or run with an open spark plug wire?
22. Describe the method of setting gap of spark plug electrodes.
23. A ———— ———— is commonly used when an engine cranks but will not start. It quickly checks the ignition system.
24. A ———— ———— is a small line of carbon that can build up on distributor cap and short out one or more spark plugs and cylinders.
25. How do you test a distributor pickup coil?
26. Explain the method used to install a distributor in an engine.
27. Rotating the distributor housing in the same direction as rotor rotation would ———— ignition timing.
28. An ignition system distributor is used to ————
    (*a*) drive the oil pump on certain engine designs
    (*b*) change spark timing with changes in engine load
    (*c*) change spark timing in response to engine speed
    (*d*) All of the above.
29. To adjust the ignition timing,
    (*a*) turn the cam on the camshaft
    (*b*) turn the distributor in its mounting
    (*c*) reset the contact points
    (*d*) replace the centrifugal-advance springs.
30. The spark plugs of an engine are grayish-tan in colour. What does this condition indicate?
    (*a*) Possible worn piston rings
    (*b*) A pre-ignition problem
    (*c*) Oil leakage past the piston rings
    (*d*) The plugs are burning properly.
31. Failure to start with normal cranking is usually due to trouble in the
    (*a*) Ignition or starting system
    (*b*) Ignition or engine

(*c*) Ignition or fuel system
(*d*) battery or fuel system.

32. If you do not get a spark during the spark test, the problem probably is in the
(*a*) engine (*b*) spark plugs
(*c*) ignition system (*d*) fuel system.

33. Mechanic A says that worn spark plugs can reduce an engine's power output. Mechanic B says that worn spark plugs can reduce an engines fuel economy. Who is right?
(*a*) A only (*b*) B only
(*c*) Both A and B (*d*) Neither A nor B.

34. If no spark occurs during the spark test and the ammeter indicates a small fluctuating reading, the problem probably is in the
(*a*) starting system (*b*) secondary circuit
(*c*) primary circuit (*d*) ignition system

35. Mechanic A says that a defective crankshaft position sensor can cause a no-spark condition. Mechanic B says that a faulty ignition control module can cause a no-spark condition. Who is right?
(*a*) Mechanic A (*b*) Mechanic B
(*c*) Both A and B (*d*) Neither A nor B.

36. Mechanic A says that a pickup (pulse generator) can be tested with an ohmmeter. Mechanic B says that ignition coils can be tested with an ohmmeter. Who is right?
(*a*) A only (*b*) B only
(*c*) Both A and B (*d*) Neither A nor B.

37. Mechanic A says that a defective spark plug wire can cause an engine miss. Mechanic B says that a defective pickup coil wire can cause an engine miss. Who is right?
(*a*) A only (*b*) B only
(*c*) Both A and B (*d*) Neither A nor B.

38. Mechanic A says that an engine will not start and run if the ignition coil is tracked. Mechanic B says that one wire of any pickup coil must be grounded. Who is right?
(*a*) A only (*b*) B only
(*c*) Both A and B (*d*) Neither A nor B.

39. If no spark occurs and the ammeter shows a fairly high and steady reading, the trouble probably is in the
(*a*) secondary circuit (*b*) primary circuit
(*c*) spark plugs (*d*) wiring.

40. If no change in speed occurs when a spark plug is shorted out, then that
(*a*) everything is fine
(*b*) cylinder is delivering power
(*c*) cylinder is not firing
(*d*) spark plug is in good condition.

41. Contact-point opening is adjusted by
(*a*) turning the distributor in its mounting
(*b*) changing the spring tension
(*c*) shifting the movable point
(*d*) shifting the stationary point.

42. Failure of the engine to start although it cranks normally could be due to
(*a*) ECU faulty
(*b*) ignition coil open
(*c*) no voltage to the ignition system
(*d*) any of the above.

43. Engine overheating could be due to
(*a*) high voltage setting
(*b*) late ignition timing
(*c*) advanced ignition timing
(*d*) discharged battery.

44. In the electronic ignition system, dwell means
(*a*) the length of time the points are closed
(*b*) the length of time the points remain open
(*c*) the number of degrees of distributor shaft rotation between successive plug firing
(*d*) the number of degrees of distributor shaft rotation during which current flows through the ignition-coil primary winding.

❑❑❑

Engine Condition Diagnosis and Tune-up

9

9.1 ENGINE CHECKING PROCEDURES

In case there is an engine operation problem, the cause could be any one of many items, including the engine itself. The condition of the engine should be tested any time the operation of the engine is not satisfactory.

Automotive vehicle engine checking procedures are of two types. One is used if there is an obvious specific trouble that seems related in the engine. For example, the customer complaints of excessive oil consumption, or excessive fuel consumption, or engine overheats, then there are definite problem-diagnosis checks that can be made to isolate the source or cause of problem. In most cases the manufacturers provide information about causes of various troubles and the procedures to be followed to correct the troubles in the service/shop manuals specific to their product line.

The second type of engine checking procedure is a general approach. During this procedure every engine system is checked, and excessive worn parts, defective parts, faulty operation, poor performance, and other defects are determined. This general-approach is usually referred to as *engine tune up.* The corrective measures are then taken to recondition the engine to restore its normal operation and performance.

Both types of engine checking procedures are used in automotive-engine service. For specific known problem a specific procedure to pinpoint the cause of the problem and correcting it is followed. In case of more general types of performance problems that could be caused by a number of different reasons, it is proper to make a complete check of the engine and its components. Such an engine analysis will detect worn and defective parts or improper adjustments and other conditions that soon might cause serious trouble. Corrective actions can be taken before a serious trouble develops. This general procedure diagnosis helps in eliminating trouble before it actually happens. This is also called *preventive maintenance.*

9.2 ENGINE TUNE-UP

An engine tune-up is important to engine power and fuel economy, and is equally important to air quality and conservation of natural resources. An engine that is not properly tuned consumes a large amount of fuel and produces high levels of air pollution. Both these items are critical to the environment. Therefore, an engine technician must be able to keep an engine properly tuned up, running efficiently, and causing the minimum practically possible environmental pollution.

To some, it means a quick check of the engine that covers only frequent problem sources. To others, it means use of test instruments to do a careful, complete check and analysis of all engine systems.

An ***engine tune-up*** is a maintenance type operation that restores the engine to its peak performance after part wear and deterioration. It involves basic service tasks, making adjustments to specifications laid down by the manufacturer, replacement of worn or defective parts, and often, minor repairs.

An engine tune-up assures that the engine, ignition system, fuel system, and emission control systems are within manufacturer's specifications.

Manufacturers generally recommend a tune-up after a specific period of engine operation. The modern tune-up is performed as maintenance, rather than as a repair for a performance problem.

The procedure for a tune-up varies from shop to shop. In one repair shop, a tune-up may include only the most routine tasks. In another shop, it may include a long list of tests, adjustments, and repairs.

A tune-up is critical to the operation of an engine, and it can affect:

1. Engine power and acceleration.
2. Fuel consumption.
3. Exhaust pollution.
4. Smoothness of engine operation.
5. Ease of starting.
6. Engine service life.

The technician must make sure to restore every engine related system to peak operating condition during a tune-up. Overlooking just one problem, the engine will not perform properly and the vehicle owner will not be satisfied with the service provided.

9.3 DRIVEABILITY AND TUNE-UP

An **engine tune-up** is a service operation designed to restore the engine to its best performance level while maintaining good fuel economy and minimum possible exhaust emissions.

A tune-up comprises of a series of tests, checks, and corrections made according to a prescribed tune-up test procedure. Earlier, peak engine performance was the sole objective of a tune-up procedure that the mechanics used to be focused on. At present, the target is to try to meet two more additional objectives:

- Fulfill the obligation to tune-up the engine to meet federal and state emissions standards.
- Satisfy the vehicle owner's demand for economy of vehicle operation.

This has resulted in vehicle maintenance services, including engine tune-ups that are more complex and costly. A trained automotive mechanic must be able to test and correct problems in compression, ignition, and carburetion or fuel injection; check that the various emission control systems ate working properly as intended; and maintain all related systems (induction, exhaust, temperature control, electronic engine controls, etc. in good working order. To be proficient and accurate an automotive mechanic must keep up with the latest developments in various automotive systems through training and reading about the new developments, from all available sources like service manuals, service bulletins etc.

9.4 ENGINE TUNE-UP AND AIR POLLUTION

Automotive exhaust emissions are considered to be one of the main sources of air pollution worldwide. Combustion of the air-fuel mixture gives off hydrocarbons (HC), carbon monoxide (CO), oxides of nitrogen (NO_X), and other unburned gases that pollute the atmosphere. Even a well-tuned, clean-burning engine emits some pollutants. If the engine operates inefficiently due to maladjustment or malfunctions, it will release excessive exhaust emissions.

With latest governmental regulations covering emission limits, more detailed tests and finer adjustments are necessary – along with regional recommendations for vehicles in different climates and altitudes.

Engine tune-up specifications vary from vehicle to vehicle – even for the same vehicle make and model based on the type of fuel system, ignition system, and accessory equipment installed on the vehicle. As a consequence, the engine tune-up specifications and emissions control information for each particular vehicle are provided on a label or decal placed conveniently under the hood, *Fig. 9.1.* The tune-up and emissions label contains up-to-date emissions specifications and setting procedures along with a vacuum hose schematic with emission components identified. In order to tune-up a modern engine the *specifications label* must be carefully noted and diligently followed.

9.5 MINOR TUNE-UP

A ***minor tune-up*** is a basic tune-up done when the engine is in good operating condition. A new vehicle only driven about 30,000 to 90,000 km may only require a minor tune-up. Most of the engine systems are in satisfactory condition with little or no wear.

A minor tune-up typically involves the following tasks:

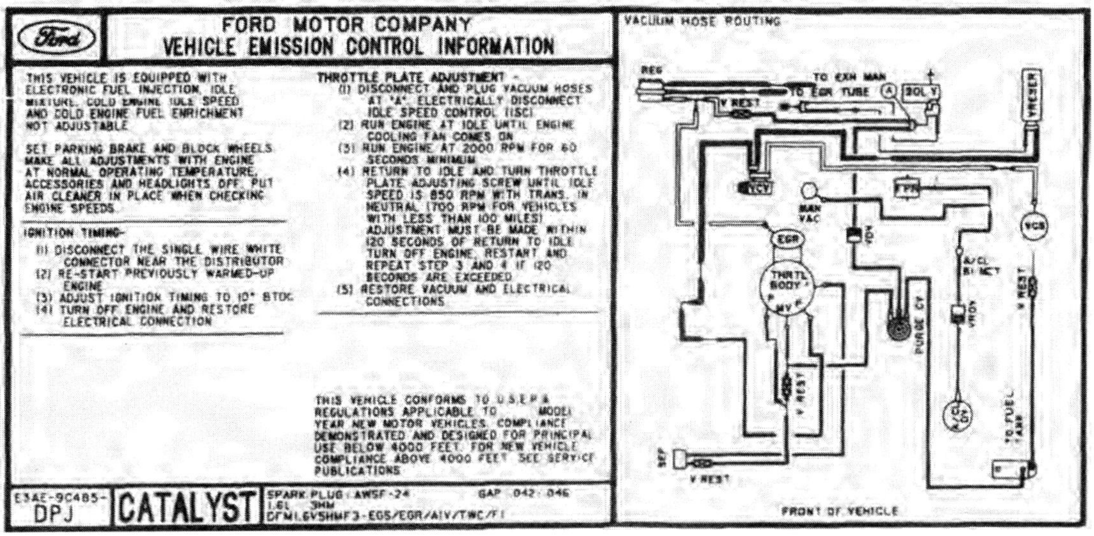

Fig. 9.1 *All modern vehicles have an engine tune-up and emissions control information label in engine compartment. Each label contains specifications for that particular engine application. (Ford Motors)*

1. Visual inspection for obvious problems affecting engine performance, e.g., loose electrical wires, missing vacuum hoses, oil leaks, coolant leaks, etc.
2. Check for tripped trouble codes on computer controlled vehicles. On older, high mileage vehicles, it would include distributor service (points and condenser replacement) and spark plug replacement.
3. Check and, if needed, adjustment of ignition timing.
4. Replacement of air and fuel filters.
5. Fuel system adjustments, if available.
6. Clean throttle body and idle air bypass on fuel-injected engines.
7. Emission control system tests and service (vapour canister filter replacement, for example)
8. Filling of all fluid levels.
9. Road test of the vehicle.

9.6 MAJOR TUNE-UP

A ***major tune-up*** is more thorough and is done when the engine and its ignition, fuel, and emission systems are in poor condition. After a long service, the parts can wear or deteriorate, and require more attention, maintenance and repair. In addition to the steps listed for carrying out a minor tune-up, a major tune-up typically includes more diagnosis type tests, e.g., compression test, vacuum test, etc., to determine what should be done during the tune-up. It may also include carburetor or injection system repairs, throttle body or distributor rebuild, and other necessary time consuming jobs.

Figure 9.2 shows a few problems that should be located and fixed during a major tune-up.

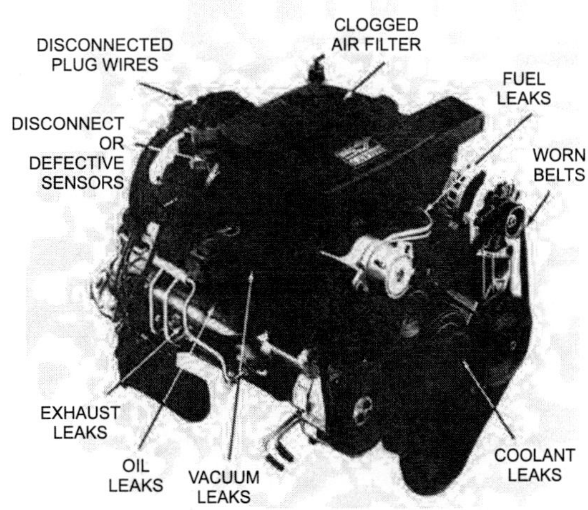

Fig. 9.2 *During inspection, look for any problems that could upset engine performance and dependability. (GM)*

9.7 TUNE-UP SAFETY TIPS

When doing engine tune-up, remember the following safety tips:

1. Engage the vehicle emergency brake and block the wheels when the engine is running.
2. Attach an exhaust hose over the tail pipe to release the exhaust gases out of the shop, when working with the engine running in an enclosed space.
3. Keep hands, clothing, tools, and instruments away from a rotating engine fan, chains, and belts.
4. Disconnect the battery when recommended in the service manual. This is to avoid accidental engine cranking or an electrical fire.
5. Do not touch the hot exhaust manifold when removing old spark plugs. Keep test equipment leads away from the engine exhaust manifolds, fan, and belts.
6. Wear eye protection when blowing debris away from spark plugs or when working in the vicinity of the spinning engine cooling system fan.
7. Keep a fire extinguisher handy when performing fuel system tests and repairs.
8. In case of a diesel engine, technician should disable the fuel injection pump when removing a fuel injection line. The high pressure of the injection system can puncture skin and eyes of the technician.

9.8 TYPICAL TUNE-UP PROCEDURES

An engine tune-up follows a definite procedure. Many shops use a printed form supplied by automotive or test equipment manufacturers. By following the form and checking off the items listed, one by one, the mechanic is sure of not missing any part of the tune-up procedure. However, not all tune-up forms are the same. Different manufacturers and repair shops have different views of what should be done and the order in which it should be carried out. Also, the tune-up procedure used depends on the equipment available. If the shop has an oscilloscope or a dynamometer, it is used as part of the tune-up procedure. In the absence of these test instruments, the tune-up is performed differently.

A summary of the most common procedures for an engine tune-up, which are typical and apply to most makes and models of vehicles, is as follows.

9.9 PRELIMINARY INSPECTION

To start an engine tune-up, carefully inspect the engine compartment. Try to find signs of trouble in the following areas:

1. ***Battery problems:*** Dirty case top, corroded terminals, physical damage.
2. ***Air cleaner problems:*** Clogged filter, inoperative air flap, disconnected/damaged vacuum hoses.
3. ***Fuel system problems:*** Leaks, clogged fuel filter, binding throttle linkage, etc., *Figs. 9.3,* and *9.4.*
4. ***Belt troubles:*** Loose belts, fraying, slippage.
5. ***Deteriorated hoses:*** Hardened or softened cooling system, fuel system, and vacuum hoses.
6. ***Poor electrical connections:*** Loose or corroded connections; frayed or burned wiring.

Correct any problems that are found during preliminary inspection, fix them before starting the tune-up. Many of these problems could affect engine performance.

9.10 EVALUATING ENGINE AND ENGINE SYSTEMS

During an engine tune-up, it is very important to test and evaluate the condition of the engine and its systems. Various test instruments are used for this purpose. A compression gauge, leak down tester, ignition scope, vacuum gauge, exhaust analyzer, timing light, and VOM are some of the test instruments used during tune-up.

9.11 TYPICAL ENGINE-RELATED COMPLAINTS

Many drivability problems are not caused by mechanical engine problems. A through inspection and testing of the ignition and fuel systems should be performed before testing he mechanical engine problems.

Typical engine problem complaints include the following:

- Excessive oil consumption
- Engine misfiring
- Loss of power

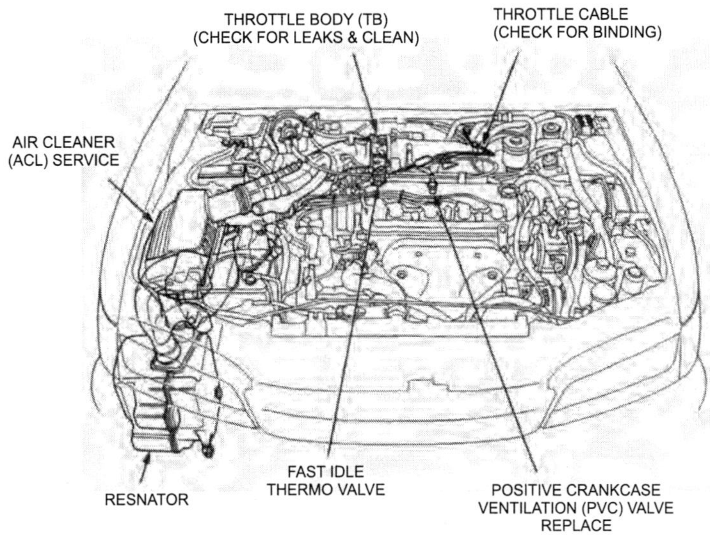

Fig. 9.3 *Air filter and throttle body must be serviced at regular intervals. Also check for bad hoses, belts, and other problems. (Honda)*

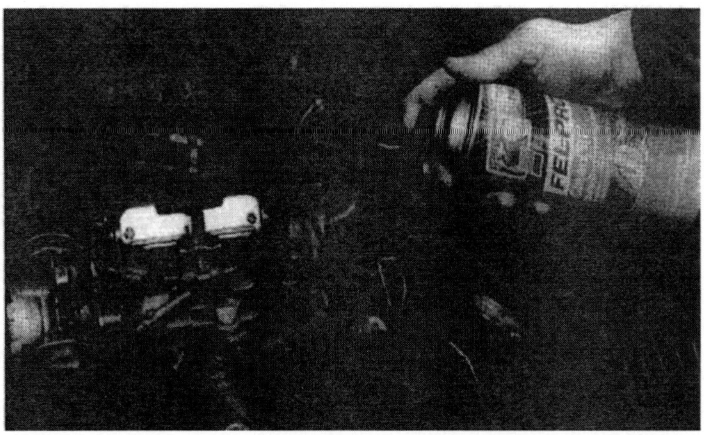

Fig. 9.4 *Cleaning, lubrication and adjustment can correct erratic idle speed of an engine, caused by binding throttle linkage. (Pel-Pro Gaskets)*

- Smoke from the engine or exhaust
- Engine noise.

9.12 FEEDBACK FROM THE DRIVER

An experienced driver of the vehicle knows enough about the vehicle and how it is driven. Before starting the diagnosis, ask the driver:

- When did the problem first occur?
- Under what driving conditions does it occur?
  - —Cold or hot?
  - —Acceleration, cruise, or deceleration?
  - —At what speed?

After determining the nature and scope of the problem, the complaint should be taken care of before further performing any diagnostic tests.

9.13 ENGINE SMOKE DIAGNOSIS

Colour of the engine smoke can point to an engine problem that might exist.

Blue Smoke: It indicates that the engine is burning oil. Oil is going to the combustion chamber either past the piston rings or past the valve stem seals. Blue smoke only after start-up is usually due to defective valve stem seals.

Black Smoke: It indicates that the engine is burning excessive fuel in the combustion chamber. This may be due to a defective or misadjusted carburetor, leaking fuel injector, or excessive fuel pump pressure.

White Smoke (Steam): White smoke or steam from the exhaust pipe is normal during cold weather and is condensed steam. Every 1C engine creates approximately an equal amount of water to the fuel burned. If the steam from the exhaust is too much, then coolant might be getting into the combustion chamber. This could be due to a defective cylinder head gasket, a cracked cylinder head, or in worst cases a cracked cylinder block.

Note: White smoke can also be created when automatic transmission fluid (ATF) is burned. A common source of ATF finding its way into the engine is through a defective vacuum modulator valve on the automatic transmission.

9.14 VISUAL INSPECTION

A careful visual inspection is the most important initial test to be performed. The common visual checks are:

Oil Level and Condition: It is the first one for visual inspection.

1. Oil should be within the specified level.
2. Condition of oil:
   - Carefully using a match or lighter, attempt to light the oil on the dipstick, if the oil flames up, gasoline is present in the engine oil.
   - Drip some of the engine oil from the dipstick onto the hot exhaust manifold. If the oil bubbles or boils, the oil has coolant in it.
   - Check for grittiness by rubbing the oil between fingers to determine if any metal wear particles are present.

Coolant Level and Condition: Most mechanical problems in an engine are caused by overheating. Proper functioning of the cooling system is very important for the life of an engine.

1. The coolant level in the coolant recovery container should be within the limits indicated on the overflow bottle. If this level is too low or the coolant recovery container is empty, then check the level of the coolant in the radiator, when cool, and also check the operation of the pressure cap.
2. Check the boiling and freezing temperature of the coolant with a hydrometer. The test will show if the concentration of the antifreeze is sufficient for proper system protection.
3. Pressure test the cooling system and inspect for any coolant leakage. Coolant leakage will often cause:
   - A grayish white stain
   - A rusty colour stain
   - Dye stains, usually greenish, from antifreeze
4. Inspect for any cool areas of the radiator, indicating clogged sections.
5. Check physical condition and operation of the cooling system fan, fan clutch, and coolant pump drive belt.

Leaks: The colour of the leaks found under a vehicle can assist the technician determine and correct the source of leak. Some leaks are normal, e.g., condensed water from the air-conditioning system, whereas a brake fluid leak is very dangerous. The colours of some common leaks are tabulated below:

Sooty black	Engine oil
Yellow, green blue, or orange	Antifreeze (coolant)
Red	Automatic transmission fluid
Murky brown	Brake or power steering or very neglected antifreeze (coolant)
Clear	Air-conditioning condensate (water)

Oil Leaks: Oil leaks can lead to severe damage to the engine if the oil loss is not promptly corrected. Besides causing an oily mess in the vehicle's parking place, the oil leak may cause blue smoke to generate under the hood as the leaking oil drips on the hot exhaust system components. To facilitate finding the source of oil leaks, which is often hard to find, follow the steps given below:

1. Clean the area around the suspected oil leak point. Use a high-pressure hot water spray to wash the engine. While the engine is running, spray the entire engine and the engine compartment. Take precautions not to allow the water come into direct contact with the air

intake and ignition distributor, or ignition coil(s).

2. An alternative method is to spray a degreaser on the engine, then start the engine and run the engine until warm. Engine heat allows the degreaser penetrate the grease and dirt. Then rinse off the engine and engine compartment with a water hose.
3. If the engine starts to run rough or stalls when it is wet, then the secondary ignition wires, i.e., spark plug wires, or distributor cap may be defective or have poor insulation. Wipe all wires and the distributor cap dry with a soft, dry clothe in case the engine stalls.
4. If the oil leak is not visible or oil appears to be oozing from "everywhere," use a white talcum powder. The leaking oil will show as a dark area on the white powder. The exact location of any leak can thus be quickly located.
5. Add about 19 cc of fluorescent dye per 5 quarts of engine oil. Start the engine and allow it to run for about 10 minutes to thoroughly mix the dye throughout the engine. Show a black light around every suspected oil leak location. The black light will easily show all oil leak locations as the dye will show as a bright yellow/green area.

Always check the highest and most forward parts of the engine that are wet when attempting to find a fluid leak.

9.15 ENGINE NOISE DIAGNOSIS

An engine knocking noise is often hard to diagnose. Various items that can cause a deep knock noise include:

1. **Valves Clicking:** This noise is caused by lack of oil to the valve lifters. This noise is most noticeable at idle or when the oil pressure is the lowest.
2. **Torque Converter:** The attaching bolts or nuts may be loose on the flex plate. This noise is most noticeable at idle or when there is no load on the engine.
3. **Cracked Flex Plate:** The noise from a cracked flex plate is often mistaken as a connecting rod or main bearing knocking noise.
4. **Loose or Defective Drive Belts:** The flopping noise of a loose or defective accessory drive belt often sounds similar to a bearing noise. Anything wrong with the belt or the belt tensioner can cause noise. A defective serpentine belt can cause a variety of noises including squealing and knocking, similar to a main bearing knock.
5. **Piston Pin Knock:** This knocking noise is usually not affected by load on the cylinder. An excessive clearance can cause a double knock noise when the engine idles. If all cylinders are grounded out one at a time and the noise does not change, a defective piston pin could be the source of noise.
6. **Piston Slap:** A piston slap is usually caused by an undersize or improperly shaped piston or oversize cylinder bore. A piston slap is most noticeable when the engine is cold and tends to decrease or stop making noise as the piston expands during engine operation.
7. **Timing Chain Noise:** An excessively loose timing chain noise can cause a severe knocking noise when the chain hits the timing chain cover. Often this noise can sound like a connecting rod bearing knock.
8. **Heat Riser Noise:** A loose/worn or defective heat riser valve in the exhaust can make a knocking noise similar to a main bearing noise. Even a vacuum-controlled heat riser, known as an *early fuel evaporation (EFE) valve,* can make a knocking noise, especially under load, due to slight vacuum variations applied to the actuator diaphragm. To eliminate the heat riser as a possible cause of noise, remove the vacuum hose to the actuator or retain the thermostatic valve with a wire or any other suitable means.
9. **Connecting Rod Bearing Noise:** The noise from a defective connecting rod bearing is normally load sensitive and changes in intensity as the load on the engine increases and decreases. A rod bearing failure can often be detected by grounding out the spark plugs one cylinder at a time. If the knocking noise decreases or is eliminated when a particular cylinder is grounded, then the grounded cylinder is the source of noise.

10. **Main Bearing Knock:** A main bearing knock often cannot be isolated to a particular cylinder. The sound can vary in intensity and may disappear at times depending on engine load.

Regardless of the type of loud knocking noise, after the external causes of the knocking noise have been eliminated, the engine should be disassembled and carefully inspected to pin point the exact cause of the noise.

9.16 SIMPLE THINGS FIRST

Engine testing is done to find the cause of an engine problem. All the simple things should be tested first. Just remember KISS — "Keep It Simple, Stupid". A loose alternator belt or loose bolts on a torque converter can sound like a lifter or rod bearing. A loose spark plug can make the engine perform as if it had a burned valve. Some simple items that can cause serious problems include:

Oil Burning

- Low oil level
- Oil dilution caused by a leaking fuel pump that leaks gasoline and thins the oil.
- Clogged PCV valve or system, causing blow by and oil to be blown into the air cleaner.
- Dirty oil that has not been changed for a long time than specified by the manufacturer.

Noises

- Loose torque-to-flex plate bolts or nuts, causing a loud noise. A problem that often cause noise, only at idle; the noise tends to disappear during driving or when the engine is operating under load.
- A defective fuel pump, which may cause ticking, lifter-type noise. Use a stethoscope to isolate the noise.
- A loose and/or defective drive belt, which mat cause a rod bearing or main bearing knocking noise. A loose or broken mount for alternator, power steering pump, or air-conditioning compressor can also cause a knocking noise. Typical noises and their possible causes are given in the following table.

TYPICAL NOISE	*POSSIBLE CAUSES*
Clicking noise– like the clicking of a ball-point pen	1. Loose spark plug 2. Loose accessory mount (for air- Conditioning compressor, Alternator, power steering pump Etc. 3. Loose rocker arm 4. Worn rocker arm pedestal 5. Broken mechanical fuel pump Return-spring 6. Worn camshaft 7. Exhaust leak 8. Detonation (ping)
Clacking noise– like tapping on metal	1. Worn piston pin 2. Broken piston 3. Excessive valve clearance 4. Timing chain hitting its cover
Knock–like knocking on a door	1. Conn, rod bearing(s) 2. Main bearing(s) 3. Thrust bearing(s) 4. Loose torque converter 5. Cracked flex plate (drive plate)
Rattle–like a baby rattle	1. Manifold heat control valve 2. Broken harmonic balancer 3. Loose accessory mounts 4. Loose accessory drive belt or tensioner
Clatter–like rolling marbles	1. Conn. Rod bearing(s) 2. Piston pin 3. Loose timing chain
Whine–like a running Electric motor	1. Alternator bearing 2. Drive belt 3. Power steering 4. Belt noise (accessory or timing)
Clunk–like a door closing	1. Engine mount 2. Drive axle shaft U-joint or constant velocity (CV) joint

9.17 OIL PRESSURE TESTING

A specified oil pressure is very important for the proper operation of any 1C engine. Low oil pressure can cause faster engine wear; and engine wear results in low oil pressure due to increased leak paths.

Worn main and connecting rod bearings result in reduced oil pressure because of increased leakage of the oil around the bearings. Follow the steps outlined here, to perform oil pressure test:

1. Start and operate the engine until normal operating temperature is achieved.
2. With the engine off, remove oil pressure sending unit (sender), usually positioned near the oil filter. Install an oil pressure gauge into the threaded hole, *Fig. 9.5*. This oil pressure gauge can be made from another gauge, e.g., an old air-conditioning gauge no longer used for air-conditioning work, and a flexible brake hose. The threads are the same as those used for the oil pressure-sending unit that is removed.

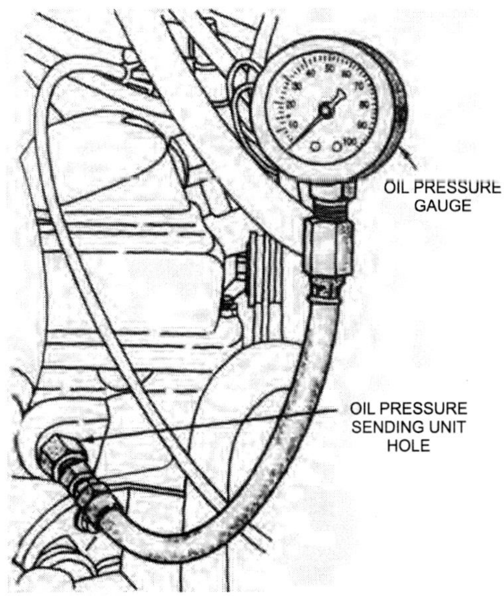

Fig. 9.5 *To measure engine oil pressure, remove the oil pressure sending (sender) unit usually located near the oil filter. Install the pressure gauge into the oil pressure sending unit hole.*

3. Start the engine and observe the gauge. Record the oil pressure at idle speed and at 2500 rpm. Most vehicle manufacturers recommend a minimum oil pressure of 66.7 kPa per 1000 rpm. Therefore, at 2500 rpm, the oil pressure should be at least 166.7 kPa. Always compare the test results with the manufacturer's specified oil pressure for the engine.

Besides engine bearing wear, other possible causes for low oil pressure include:

- Low oil level
- Diluted oil with gasoline seeping to crankcase and water vapour condensation
- Stuck oil pressure relief valve.

9.18 OIL-PRESSURE WARNING LAMP

The red oil pressure warning lamp in the vehicle dash usually lights when the oil pressure drops less than 26 to 46 kPa, depending upon the vehicle and engine. The oil light should not be on during driving. If the oil pressure-warning lamp turns on, stop the vehicle engine. Always confirm oil pressure with a reliable mechanical gauge before performing engine repairs. The sending unit or circuit may be defective, *Fig. 9.6*.

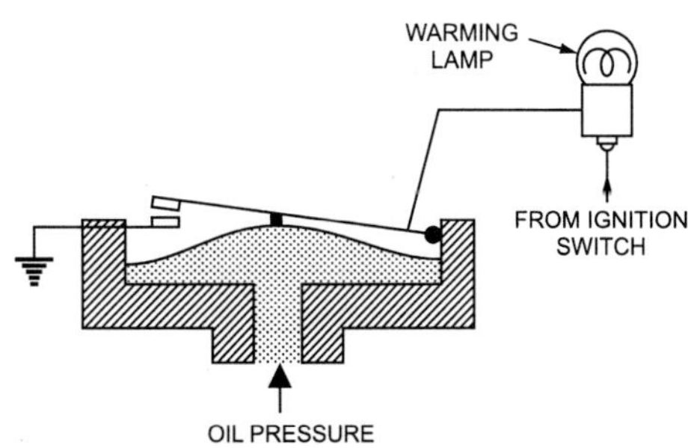

Fig. 9.6 *The oil pressure switch is connected to the warning lamp, indicating to the driver that the oil pressure is low.*

9.19 OIL ANALYSIS

Generally, large vehicle fleets and transit operations use oil analysis data as part of their normal preventive maintenance program. Oil analysis involves getting a sample of the engine's crankcase oil and sending it to a local or regional laboratory for analysis. The results can highlight mechanical engine problems before there is any other indication. The engine can be removed and disassembled to repair any defects or faults found by the oil analysis before any serious and expensive engine damage takes place. Oil analysis also helps in determining the maximum acceptable oil drain interval for vehicle brands that they use.

Getting the Sample

For best sample test results, the oil sample should be from a warm engine. The recommended procedure is to begin to drain the oil in the usual way by removing the oil drain plug. To avoid getting residue from the bottom of the oil pan, allow oil to drain for several seconds before placing the sample container into the oil stream. Alternatively, siphon the oil from the crankcase, from the oil filter tube, or from the crankcase vent passages. A 120 cc oil sample is good enough for the laboratory.

9.20 ANALYZING THE OIL SAMPLE RESULTS

The oil sample results obtained from the testing laboratory should be carefully analyzed to determine if any corrective action is required to be on the engine, or if the results are normal consistent with the engine's mileage and condition of operation. Usually the test laboratories request the following information to be provided with the sample to facilitate evaluating the test results:

1. Total kilometers on the vehicle
2. Oil life kilometers since the last oil change
3. Description of the engine (Gasoline, diesel)
4. Original viscosity of the oil.

The items that are measured during an oil sample test, and brief interpretation of the results are given below:

Viscosity Increase

If viscosity of the sample oil is more (thicker) than that of the original oil grade, it means that oxidation of the oil has taken place. Excessive oxidation of oil can be caused by:

- Overheating due to poor cooling system function.
- Aeration of oil due to stirring by moving engine parts in the crankcase. This generally occurs if the oil level is below the specified level.
- Presence of fine metal particles that come from engine wear. These fine metal particles act as a catalyst that speeds up the chemical reaction between oxygen in the crankcase and the engine oil, causing the oil to thicken.

Other products of oil oxidation include *sludge, gum, varnish, lacquer, carbon deposits,* and *acidic compounds.* Maximum allowable viscosity is normally considered to be 30% higher than the original. If this thickening is more than 30%, the engine cooling system should be carefully inspected for any defects and corrective action should be taken as necessary, and the engine oil and filter should be replaced. The engine and vehicle use should be adequately monitored to ensure that severe oxidation does not occur again.

Viscosity Decrease

The oil becoming thinner due to decrease in viscosity is the result of fuel dilution. A maximum allowable fuel dilution of 3% of the oil volume is generally acceptable. Fuel dilution results in thin oil and an increase in engine wear rates. A high level of fuel dilution may be due to:

- Short driving cycles of the vehicle, especially during cold weather.
- Defective thermostat (stuck open), which prevents the engine from attaining normal operating temperatures.
- Defective carburetor choke or coolant temperature sensor.
- Clogged exhaust gas passage under the intake manifold.
- Defective heat riser or control unit affecting intake air temperature.
- Defective carburetor float or power circuit, or a defective fuel injector nozzle.
- Clogged air filter or closed air inlet.

Moisture

The contamination of engine oil by condensed moisture can result in poor lubrication and formation of sludge. Moisture readings less than 0.09% are considered normal. Readings more than 2% are considered to be excessive. A high level of moisture content in the engine oil may be due to:

- Leaking head gasket
- Cracked cylinder block
- Cracked cylinder head
- Clogged PCV valve and/or hoses

- Inoperative crankcase breather
- Short vehicle driving cycles
- Extended oil change intervals.

Antifreeze

Engine oil contaminated with antifreeze (ethylene glycol) can congeal, and become too thick to easily flow through the engine to lubricate properly. Possible causes of antifreeze in engine oil include:

- Cracked cylinder head
- Leaking head gasket
- Cracked cylinder block
- Sabotage

Iron (Fe)

Iron is main wear metal contained in almost every oil sample, measured in parts per million (PPM), due to normal engine wear. Iron content 90 to 250 PPM is considered normal. Iron content 290-390 PPM is considered abnormal, while over 390 PPM is considered excessive.

Iron in the oil comes from wear and rusting of:

- Rocker arms or pivot shaft
- Cylinders or cylinder sleeve
- Camshaft
- Valve guides
- Timing gear and/or chain
- Crankshaft
- Piston rings
- Oil pump
- Valve lifters
- Fuel pump rocker and/or fulcrum.

Aluminium (Al)

Aluminium is also a wear metal and is measured in parts per million (PPM). Its content 5 to 25 PPM is considered normal, content 30 PPM is considered abnormal, and content over 40 PPM is considered excessive. Aluminium comes from the wear of:

- Pistons
- Connecting rod and/main bearings
- Camshaft bearings
- Fuel pump

Copper (Cu)

Copper is another wear metal measured in parts per million (PPM). Its content 5 to 25 PPM is considered normal, content 100 PPM is considered abnormal, and content over 300 PPM is considered excessive. Copper comes from the wear of:

- Bearings
- Bushings, e.g., distributor, fuel pump, or oil pump.

Tin (Sn)

Tin may be fund in oil as another wear metal and is measured in PPM. Its content 0 to 1 PPM is considered normal, content 5 to 10 PPM is considered abnormal, and content over 15 PPM is considered excessive. Tin comes from the wear of:

- Piston coating, plated on aluminium pistons
- Bearings

Chromium (Cr)

The oil may contain chromium as wear metal. It is measured in PPM and almost exclusively originates from wear of chrome piston rings. Most heavy-duty diesel engines use chrome rings. Its content 5 to 25 PPM is considered normal, content 30 PPM is considered abnormal, and content over 40 PPM is considered excessive.

Silicon (Si)

Silicon is basically dirt or sand. Its content 5 to 25 PPM is considered normal, content 30 PPM is considered abnormal, and content over 40 PPM is considered excessive.

9.21 THE PAPER TEST

A soundly running engine should produce even and steady exhaust at the tailpipe. Hold a piece of paper within 29 mm of the tailpipe, with the engine running, *Fig. 9.7.* The paper should blow out evenly without "puffing". If the paper is drawn toward the tailpipe at times, the valve in one or more engine cylinders could be burned. Some other reasons why the paper might be sucked toward the tailpipe are:

1. The engine could be misfiring because of a lean condition that could occur normally when the

engine is cold.

2. Pulsating of the paper toward the tailpipe may also be caused by a hole in the exhaust system. If the exhaust escapes through a hole in the exhaust system, air could be drawn (in the intervals between the exhaust puffs) from the tailpipe to the hole in the exhaust, causing the paper to be drawn toward the tailpipe.
3. Ignition fault causing misfire.

Fig. 9.7 *To do a paper test, hold a piece of paper near the tailpipe of an idling engine. A good engine would produce even outward puffs of exhaust. If the paper is sucked in toward the tailpipe, a burned valve is a possibility.*

9.22 CYLINDER COMPRESSION TESTER

After removing the plugs, block the throttle valve wide open. This is to make sure the maximum amount of air will get into the cylinders.

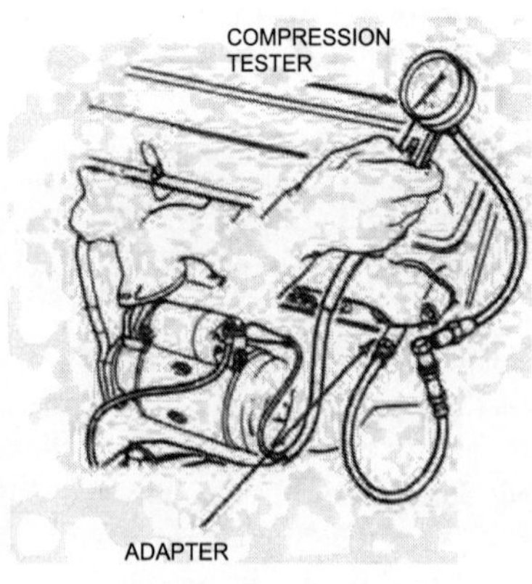

Fig. 9.8 *Using a cylinder compression tester. (Sun Electric Corp.)*

Next, screw the compression-tester adapter into the spark-plug hole of #1 cylinder, *Fig. 9.8.* To protect the coil from high voltage, disconnect the primary lead from the negative terminal of the coil that is going to the distributor. On electronic ignition systems, disconnect the positive lead to the control unit. Then hold the throttle wide open and operate the starting motor to crank the engine through four compression strokes or eight crankshaft revolutions. The tester needle will swing around to indicate the maximum compression pressure in the cylinder. Write down this reading.

9.23 COMPRESSION TEST

Compression test is made to measure the ability of engine cylinders to hold compression pressure developed by fuel-air mixture combustion.

Testing an engine for proper compression is one of the basic engine diagnostic tests performed. A compression test is a common method of determining ***engine mechanical condition.*** It should be done anytime symptoms point to cylinder pressure leakage. An extremely rough idle, popping noise from air intake or exhaust, excessive blue smoke, and blowby are all reasons to consider a compression test. For smooth engine operation, all cylinders of the engine must have equal compression. An engine can loose compression by air leakage through one or more of the following routes:

- Intake or exhaust valve
- Piston rings or piston with a hole developed in it
- Cylinder head gasket failure.

A compression test is often made during a tune-up. It is impossible to tune an engine that is not good mechanically. If the engine fails the compression test, mechanical repairs must be made **before** tune-up.

A ***compression gauge*** is used to measure compression stroke pressure during the test. If gauge pressure is lower than normal, pressure is leaking out of the engine combustion chamber. ***Low engine compression*** can be caused by:

1. ***Burned valve:*** Valve face damaged by combustion heat.
2. ***Burned valve seat:*** Cylinder head seat damaged by combustion heat.
3. ***Physical engine damage:*** Hole in piston, and broken valve, etc.

4. ***Blown head gasket:*** Head gasket ruptured.
5. ***Worn rings or cylinders:*** Part wear prevents ring-to-cylinder seal.
6. ***Valve train troubles:*** Valves adjusted with insufficient clearance, which keep valves from fully closing, broken valve spring, etc.
7. ***Jumped timing chain or belt:*** Loose worn chain or belt has jumped over sprocket teeth, causing upset valve timing.

For other occasional sources of low compression, refer to the applicable service manual troubleshooting chart.

9.24 COMPRESSION TEST – GASOLINE ENGINE

For best results, the engine should be run to warm-up to normal operating temperature before testing. A recommended way to do this is to disconnect the plug wires and loosen the plugs one turn. Then reconnect the wires and start the engine. Run the engine for a few seconds at 1000 rpm. The combustion gases will blow out of the plug well the dirt that could otherwise fall into the cylinder when the spark plugs are removed. The gases will also blow out of the combustion chamber any loosed carbon that is caked around the exposed threaded end of the plug. This procedure prevents carbon and dirt particles from lodging under valve and holding it open during the compression test.

An accurate compression test on a gasoline engine is performed as follows:

1. Disable the ignition system by disconnecting the primary leads from the ignition coil or module, or by grounding the coil wire after removing it from the centre of the distributor cap. This will prevent sparks from arcing out of the disconnected spark plug wires. Some electronic ignition systems may be damaged if operated with the spark plug wires disconnected. Usually, the feed wire going to the ignition coil can be removed to disable the system.
2. Also disable the fuel-injection system to stop squirting of fuel into the cylinders. If electronic fuel injection is used, it should also be disabled so that fuel will not spray into the cylinders. For this you may need to unplug the wires to the injectors. Refer to the applicable service manual for specific directions in this regard.
3. Remove all spark plugs. This allows the engine to be cranked easily to an even speed. Label all spark plug wires when removing spark plugs.
4. Block open the carburetor or fuel injection throttle plates to allow maximum amount of air to be drawn into the engine. This also ensures consistent compression test results.
5. Install a compression gauge into one spark plug hole and crank the engine, *Figs. 9.9 and 9.10.* Continue cranking the engine through 6 compression strokes, i.e. compression gauge needle moves six times. Each compression stroke makes a puffing sound. Note the reading on the compression gauge after the first puff. This reading should be at least one-half the final reading. A low first-puff reading indicates possible weak piston and rings assembly. Release the pressure on the gauge and repeat this step for the other cylinders.

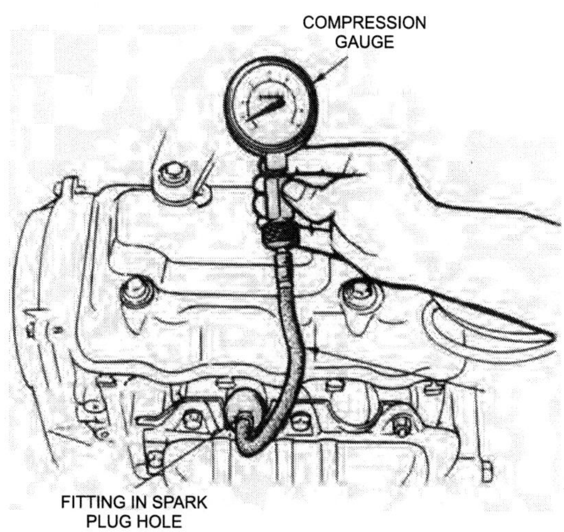

Fig. 9.9 *Compression gauge installs in spark plug hole on a SI engine and glow plug hole on a CI engine. Crank engine for about four to six "puffs" and record the gauge reading. If the reading is below specs, there is leakage. (General Motors)*

6. Record the highest readings for all cylinders and compare the results. Vehicle manufacturers specify the minimum compression reading and the maximum allowable variation among engine cylinders. Most manufacturers specify

a maximum difference of 5% between the highest reading and the lowest reading. Test the other cylinders in the same manner. The lowest compression reading should be more than 75 % of the highest.

During cranking, the oil pump cannot maintain normal oil pressure in the lubrication system. Extending engine cranking, such as, that occurs during a compression test, can cause hydraulic lifters to collapse. When the engine starts, after the compression test, loud valve clicking noises may be heard. This should be considered normal in this case, and the noise should stop after the vehicle has been driven a short distance.

Fig. 9.10 *A two-piece compression gauge set. The threaded hose is screwed into the spark plug hole after removing the spark plug. The gauge part is then snapped onto the end of the hose.*

9.25 COMPRESSION TEST- DIESEL ENGINE

A diesel engine compression test is similar to a compression test for a gasoline engine. However, do not use a compression gauge intended for a gasoline engine. The high compression stroke pressure can damage it. A diesel compression gauge must read up to approximately 4134 kPa.

Remove the air cleaner and install a manifold cover to keep dirt out of the engine. Disconnect the wire from the fuel-shutoff-solenoid terminal of the ignition pump, *Fig. 9.11.* This prevents the delivery of fuel during the test.

To perform a diesel compression test, remove either the injectors or the glow plugs. Refer to the applicable service manual for instructions but most manufacturers suggest glow plug removal. Install the compression gauge (for diesel engine) in the recommended hole. Disconnect the fuel shut-off solenoid to disable the injection pump. Crank the engine and note the highest reading on the gauge. Crank the engine about six "puffs" (compression strokes) or twelve revolutions.

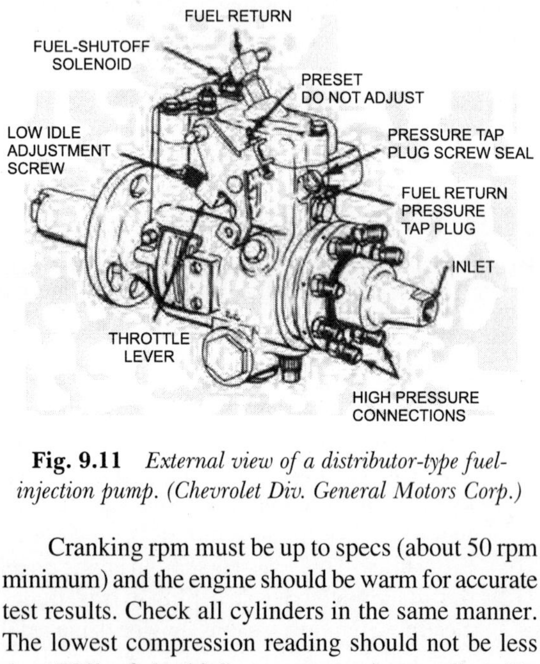

Fig. 9.11 *External view of a distributor-type fuel-injection pump. (Chevrolet Div. General Motors Corp.)*

Cranking rpm must be up to specs (about 50 rpm minimum) and the engine should be warm for accurate test results. Check all cylinders in the same manner. The lowest compression reading should not be less than 70% of the highest compression reading. No cylinder should read less than 1892 kPa.

Note that, if everything is normal the compression builds up quickly and evenly. If there is leakage past the piston rings, the compression will be low on the first strokes but will tend to build upward with later strokes. However, it will not reach normal and the pressure is quickly lost after cranking is stopped.

9.26 COMPRESSION TEST RESULTS

Gasoline engine compression readings should be around 860 to 1190 kPa. Generally, the compression pressure should not vary over 102 to 138 kPa from the highest to the lowest cylinder. Readings must be within about 10 % to 15% of each other.

Diesel engine compression readings will average 1895 to 2756 kPa, depending upon engine design and compression ratio. Compression levels must not vary more than about 10% to 15% or 206 to 344 kPa.

Look for cylinder ***pressure variation*** during an engine compression stroke pressure check. If some cylinders have normal pressure and one or two have low readings, engine performance will be reduced. The engine will have a rough idle and reduced power.

If ***all*** the cylinders are low in pressure (as with a worn timing chain for example), the engine will run smoothly but lack power and gasoline mileage.

If ***two adjacent cylinders*** read low, it might indicate a blown gasket between the cylinders. A blown head gasket will sometimes produce a louder than normal puffing noise from the spark plug or glow plug holes with the gauge removed.

Refer the manufacturer's service manual specs to determine what the compression pressure should be. If the results of the compression test show that compression pressure is low, then there is leakage past the piston rings, valves, or cylinder head gasket. To correct the problem, you have to remove the cylinder head and check the engine parts for wear and other defects. But before removing the cylinder head, make one more test to move closer to pinpointing the trouble source. Squirt a small quantity of engine oil through the spark plug or glow plug hole into the engine cylinder. Adding the oil helps seal the rings temporarily so that they can hold compression pressure better. Then retest the compression. If the pressure increases to a more normal value, the low compression is caused by leakage past the piston rings. The problem is caused by worn piston rings, a worn cylinder wall, or a worn piston or by rings that are broken, or stuck in the piston-ring grooves.

If adding oil does not increase the compression pressure, the leakage is probably past the valves. This could be caused by:

- Broken valve springs
- Incorrect valve adjustment
- Sticking valves
- Worn or burned valves
- Worn or burned valve seats
- Worn camshaft cam lobes
- Dished or worn valve lifters.

A blown cylinder head gasket could also cause this. So that compression pressure is leaking between the cylinder head and the cylinder block.

Whatever the cause of low compression — rings, pistons, cylinder walls, valves, or gasket, the cylinder head must be removed so that the problem can be corrected. The only exception would be if the trouble is due to incorrect valve adjustment. Adjusting the valves does not require head removal.

9.27 WET COMPRESSION TEST

If the compression test reading indicates low compression on one or more cylinders, add three squirts of oil (not more) from a hand-operated oil squirt can, to the cylinder and perform the compression test again and record the results. This is called a *wet compression test,* when oil is used to help seal around the piston rings. If the first-puff readings greatly improve and the readings are much higher than those without the oil, the cause of low compression is worn or defective piston rings. If the compression readings increase only slightly or not at all, then the cause of the low compression is usually faulty valves or leaking head gasket *Figs. 9.12* and *9.13.* The engine oil will seal the rings but not the burned valve or blown head gasket. In this way, a wet compression test will help diagnose the cause of low compression pressure.

Do not squirt too much oil into the cylinder during a wet test or a false reading will result. With excessive oil in the cylinder, compression readings will go up even if the piston rings and cylinders are not in good condition. Oil, like any liquid is not compressible. Space taken by oil in the cylinder will raise the compression ratio and gauge readings.

Some vehicle manufacturers warn against performing a wet compression test on a diesel engine. If excess oil is squirted into the cylinder, hydraulic lock could result because the oil will be highly pressurized, like a solid, in the small cylinder volume.

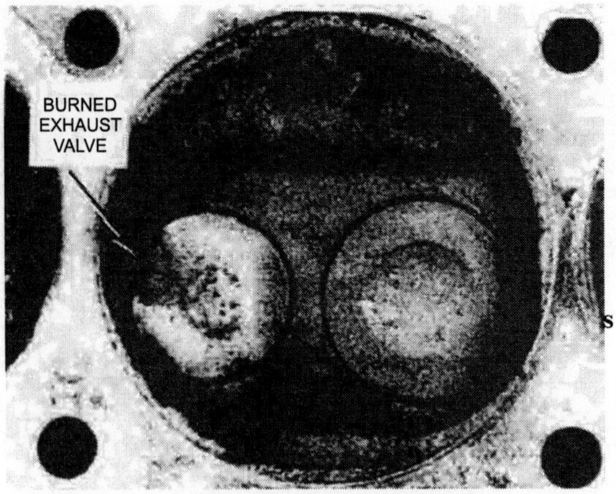

Fig. 9.12 *A compression test could have detected this badly burned exhaust valve, and a cylinder leakage test could have been used to determine the exact problem.*

9.28 RUNNING COMPRESSION TEST

The compression test is generally used to evaluate the engine condition. A compression test is usually performed with the engine cranking. An engine idles at about 600 to 900 rpm. Most manufacturers specify the engine to crank at 80 to 250 cranking rpm. Therefore, a check of the engine's compression at cranking speed determines the condition of an engine that does not run at such low speeds.

Actually, the compression pressure of a running engine is much *lower* than that indicated by cranking compression pressure test. This is due the resulting lower volumetric efficiency. The engine is running much faster, and therefore, there is less time for air to enter the combustion chamber. With less air to compress, the compression pressure is lower. Generally, the higher the engine rpm, the lower the running compression. For most engines the compression pressure during cranking ranges from 862 to 1103 kPa. Compression pressure at idle speed ranges from 414 to 621 kPa. Compression pressure at 2000 rpm ranges from 207 to 414 kPa.

The running compression of all cylinders should be equal. Therefore, a problem is not likely to be detected by single compression values, but by

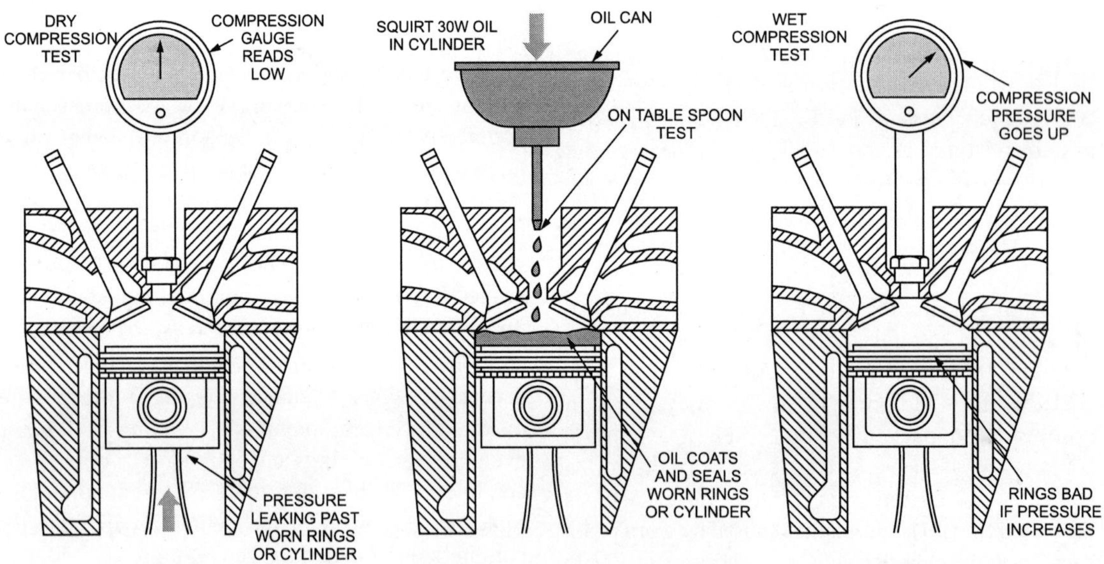

Fig. 9.13 *Basic method of making dry and wet compression test, (a) Complete conventional, dry compression test with pressure gauge. Record all readings and compare these to specs, (b) If compression is low, squirt a tablespoon of oil into cylinder. This would temporarily seal rings, (c) Measure compression pressure again. If pressure reading goes up, that cylinder may have bad rings or worn cylinder. Same pressure reading might indicate a burned valve or blown head gasket.*

variations in running compression values among the cylinders. Broken valve springs, worn valve guides, bent pushrods, and worn cam lobes are items that would be indicated by a low running compression test reading on one or more cylinders of the engine.

To perform a running compression test, remove just one spark plug at a time. With one spark plug removed from the engine, use a jumper wire to ground the spark plug wire to a good engine ground. This is necessary to prevent possible ignition coil damage. Start the engine, push the pressure release on the gauge, and read the compression pressure. Increase the engine speed to about 900 rpm and push the pressure release on the gauge again. Read the pressure indicated by the gauge. Stop the engine, reattach the spark plug wire, and repeat the test for each of the remaining cylinders. This test informs a technician of the relative compression of all the cylinders.

9.29 CYLINDER LEAK TEST

The cylinder leakage tester checks the compression but in a different way. *Fig. 9.14* shows a cylinder leakage tester in use. *Fig. 9.15* shows how the tester pinpoints the places where leakage can occur.

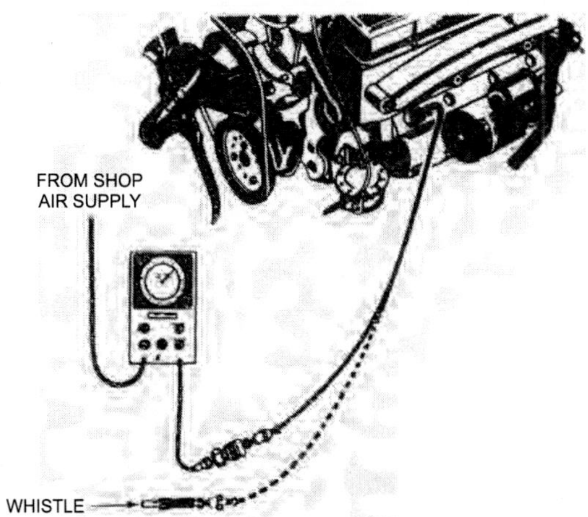

Fig. 9.14 *Cylinder leakage tester. The whistle is used to locate TDC in # 1 cylinder. (Sun Electric Corp.)*

A ***cylinder leakage test*** also measures air leakage out of the engine combustion chamber to find mechanical problems. Cylinder leakage test is one of the best tests used to determine engine condition.

To use the cylinder leak tester, first remove all the spark plugs. Then remove the air cleaner, crankcase filler cap or dipstick, the radiator cap. Set the throttle wide open, and fill the radiator to proper level. Connect the adapter, with the whistle, to the spark-plug hole #1 cylinder. Crank the engine until the whistle sounds. This indicates the piston is moving up on the compression stroke. Continue to rotate the engine until the TDC timing marks on the engine align. When the marks align the piston is at TDC. Disconnect the whistle from the adapter hose and connect the tester, as shown in *Figs 9.14* and *9.15.* Then apply air pressure from the shop supply. Note the gauge reading, which shows the percentage of air leakage from the cylinder. Specifications do vary for different engines, but a reading above 20% means an excessive leakage. If the air leakage is excessive, check further by listening at the air intake, exhaust tail pipe, and oil filler hole. If air is blowing out of an adjoining spark-plug hole, the head gasket is blown between the cylinders.

Look at the radiator coolant for leakage from a cracked cylinder block or head or from a blown head gasket.

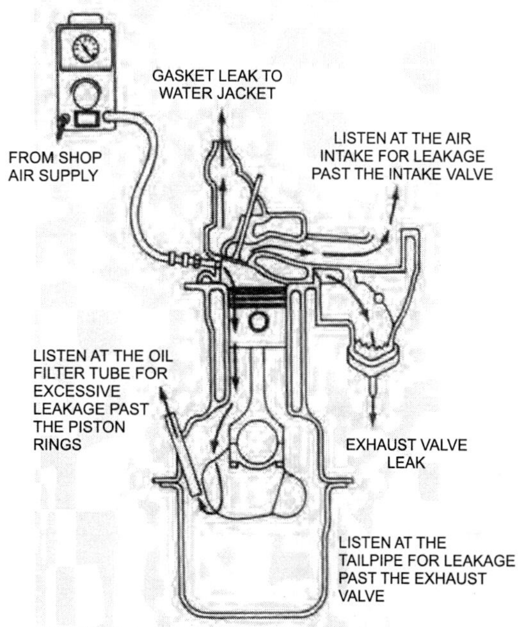

Fig. 9.15 *The cylinder leakage tester applies air pressure to the cylinder through the spark-plug hole with the piston at TDC and both inlet and exhaust valves closed. Points where air is leaking can then be pinpointed. (Sun Electric Corp.)*

Figure 9.15 shows what it means if you hear air escaping at any of the three listening places. Also if air bubbles up through the radiator, then the problem is a blown cylinder head gasket, a cracked cylinder head, or a cracked cylinder block. These conditions allow leakage from the cylinder to the cooling water jackets.

Check all other cylinders one by one in the same manner. A test light connected across the contact points can be used to position the next cylinder in the firing order at TDC. Always follow manufacturer's operating instructions when using the leak tester.

A pressure gauge and valve assembly is used to measure actual % of cylinder leakage. As shown in *Figs. 9.16 and 9.17,* the gauge-valve unit is installed in a spark plug or glow plug hole. Shop air pressure is connected to the other gauge line. The amount and location of any leaking air helps the technician in determining the condition of the engine. The crankshaft must be rotated so the cylinder being tested has its piston at TDC with both valves closed.

1. For best results, the engine should be brought to normal operating temperature such that the upper radiator hose is hot and pressurized.
2. The cylinder being tested should be at TDC of the compression stroke. If not, a valve could be open and cause false indications. The greatest amount of wear occurs at the top of the cylinder because of the heat generated near the top of the cylinders. The piston ring flex also augments the wear at the top of the cylinder.
3. Calibrate the cylinder leakage unit as specified by the manufacturer. Usually, air pressure is adjusted so that the gauge reads zero leakage with the test hose disconnected or shut off. Then air is allowed to enter the cylinder and the pressure in the gauge will drop. The amount of pressure drop depends upon how well the rings, valves, head gasket, etc. seal pressure in the cylinder.
4. Inject air into the cylinders one at a time, rotating the engine crankshaft as required by firing order to test each cylinder at TDC on the compression stroke.
5. Analyze the results: less than 10% leakage is good; less than 15% leakage is acceptable; less than 30% leakage is poor; more than 30% leakage indicates a definite problem. In case leakage seems to be unacceptably high, repeat the test, to ensure that the test is performed correctly and with the cylinder at TDC on the compression stroke.

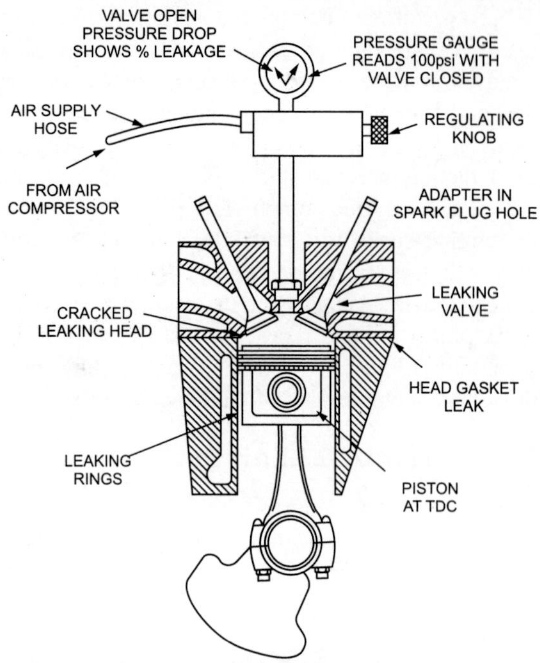

Fig. 9.16 *Cylinder leakage test forces air into cylinder to find leakage. First adjust air pressure such that the gauge reads zero. Then, allow air to flow into cylinder with piston at TDC on compression stroke. Gauge pressure will drop and show % of cylinder leakage. If there is a problem, a hiss (air leak) may be heard that indicates source of problem.*

To find the source of leakage, listen for a hissing sound at the intake manifold inlet, exhaust pipe, or valve cover breather. Depending upon where the technician hears leakage, the problems could include the following:

- If air is heard escaping from the oil filler cap, the piston rings are worn and / or broken.
- If air is found bubbling out of the radiator, the head gasket is blown or the cylinder head is cracked.
- ***Hiss at intake manifold inlet:*** Indicates leaking intake valve, intake valve adjusted too tight, broken intake valve spring, damaged valve. If air is heard coming out of the carburetor or air

intake on fuel injection equipped engines, the intake valve(s) is defective.

- ***Hiss at tailpipe:*** Burned exhaust valve, lash adjustment too tight, broken exhaust valve spring, bent valve. If air is heard coming out of the exhaust tailpipe, the exhaust valve(s) is defective.
- ***Hiss at valve cover breather:*** Worn piston rings, worn cylinder walls, blown head gasket, burned piston.
- ***Hiss around outside of head:*** Blown head gasket, warped head, warped cylinder block deck, cracked head or block.
- ***Hiss from adjacent spark plug hole:*** Blown head gasket, cracked head or block.

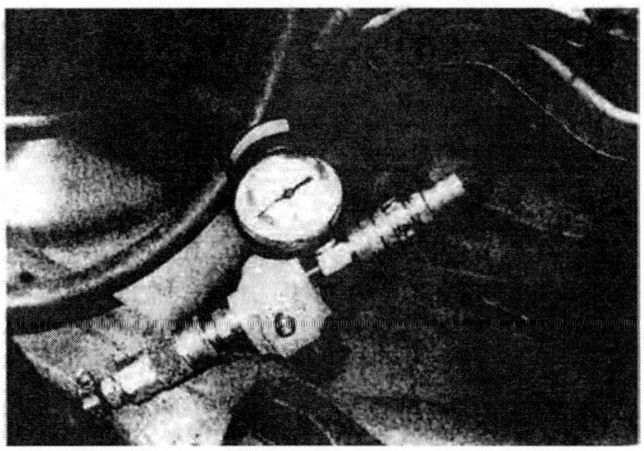

Fig. 9.17 *A typical handheld cylinder leakage tester.*

9.30 CYLINDER POWER BALANCE TEST

A ***cylinder balance test*** "kills" one cylinder at a time to find out if that cylinder is firing properly. For instance, if an engine has a "dead miss" at idle; pull one spark plug wire at a time. If engine rpm drops and the engine idles more roughly, then that cylinder is firing. If on pulling off a plug wire, idle speed and smoothness does not change, that cylinder is "dead". There is something wrong that is preventing normal combustion.

Many engine analyzers have a cylinder power balance features. Cylinder balance test is used to determine if all cylinders of the engine are contributing power equally. One cylinder is short-circuited at a time. If the engine rpm does not drop as much for one cylinder as for other cylinders of the engine, then the shorted cylinder must be weaker than the other cylinders.

When the test equipment uses automatic means for testing cylinder balance, be certain to correctly identify the offending cylinder because the cylinder number identified by the equipment may be the cylinder number in the firing order instead of the actual cylinder number on the engine.

The acceptable method of canceling cylinders, which works on all types of ignition systems, including distributorless, is to *ground the secondary current* for each cylinder.

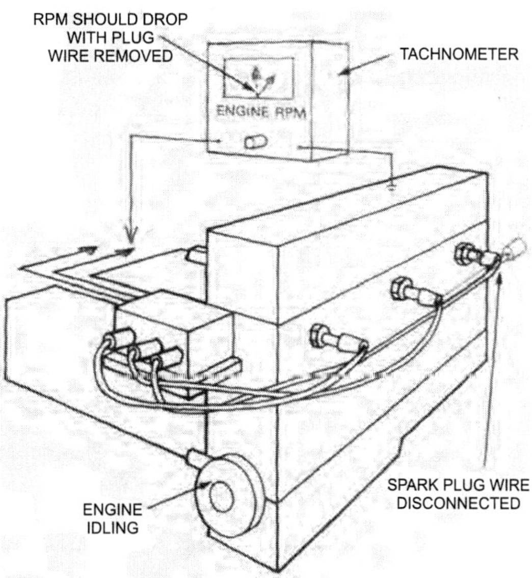

Fig. 9.18 *Cylinder balance test can be done with tachometer. Connect tachometer and fast idle the engine. Pull off one spark plug wire at a time while watching tachometer. If rpm does not drop equally in one cylinder, that cylinder is dead. The problem could be a bad wire, fouled spark plug, low compression, etc.*

There are several ways to do a cylinder balance test: using a simple tachometer, an engine analyzer, or by simply watching and listening as each wire is disconnected in turn.

Figure 9.18 shows how a tachometer is used to make a complete cylinder balance test. As each spark plug wire is removed, watch the tachometer needle. It should drop equally as each cylinder is "killed". If the needle fails to drop enough on any cylinders, a compression test or leakage test may be done to find the source of the problem. A fouled spark plug or bad wire could also be the cause.

With an engine analyzer or scan tool, there may be cylinder balance buttons that will short out cylinders electronically. Push the button and it will keep the ignition coil from firing on that cylinder only. Some will short automatically and record the results on a paper printout. Follow equipment-operating instructions because they vary.

9.31 VACUUM GAUGE TESTING

The engine vacuum gauge is a tester for locating problems in an engine that does not run as well as it should. This gauge measures the intake-manifold vacuum. The intake manifold vacuum changes with the engine load, the position of the throttle valve, and various engine problems. The manner, in which the intake manifold vacuum varies from normal, indicates as to what is wrong inside the engine.

Figure 9.19 shows the vacuum gauge connected to the intake manifold. Start the engine and operate until it is at operating temperature, before starting the test. Operate the engine at idle and other speeds, as explained below. The meanings of various readings are given *in Fig. 9.20.*

1. A steady and fairly high reading on idle indicates **normal performance.** Specifications vary with different engines, but a reading somewhere between 432 and 559 mm of mercury (Hg) indicates that the engine is good. The reading will be lower at higher altitudes because of lower atmospheric pressure. For every 305 m above sea level, the reading will be reduces by about 25.4 mm of Hg.
2. A steady and low reading on idle indicates late ignition or valve timing, or possibly leakage around the pistons. Leakage around pistons, i.e., excessive blowby, could be due to worn or stuck piston rings, worn cylinder walls, or worn pistons. Each of these conditions reduces engine power. With reduced power the engine does not "pull" as much vacuum.
3. A very low reading on idle indicates a leaky intake manifold or throttle body gasket, or possible leakage around the throttle shaft. Air leakage in manifold reduces the vacuum and engine power.
4. Back-and-forth movement of the needle that increases with engine speed indicates weak valve springs.
5. Gradual falling back of the needle toward zero with the engine idling indicates a clogged exhaust line.
6. Regular dropping back of the needle indicates that a valve is sticking open or spark plug is not firing.
7. Irregular dropping back of the needle indicates that valves are sticking only part of the time.
8. Floating motion or slow back-and-forth movement of the needle indicates that the air-fuel mixture is too rich.

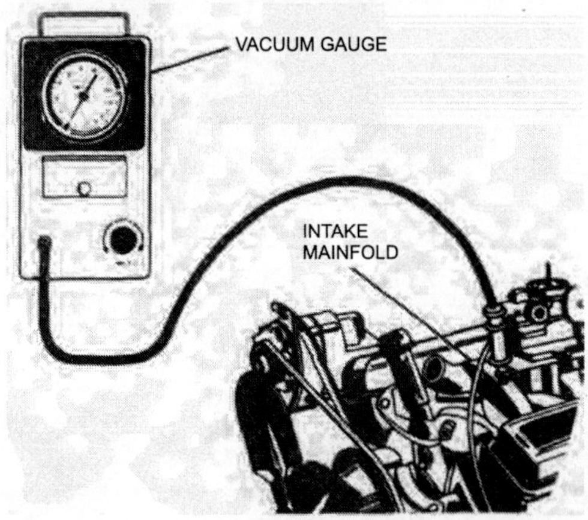

Fig. 9.19. *Vacuum gauge connected to the intake manifold for a manifold vacuum test. (Sun Electric Corp.)*

A test can be made for loss of compression due to leakage around the pistons. This condition would be the result of stuck or worn piston rings, worn cylinder walls, or worn pistons. Race the engine and then quickly release the throttle. The needle should swing around to around 600 mm as the throttle valve closes. This indicates good compression. If needle fails to swing around this far, there is loss of compression. Further checks should be made.

Vacuum is pressure below atmospheric pressure and is measured in mm of mercury (Hg). An engine in good mechanical condition will run with high manifold vacuum. Manifold vacuum is created by the pistons as they move down on the intake stroke to draw the charge from the carburetor and induction manifold. Air to replenish the manifold comes past

	READING	*DIAGNOSIS*
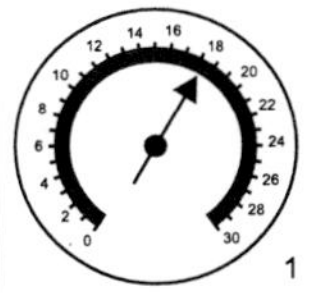 1	*Average and steady at 17-21.*	*Everything is normal.*
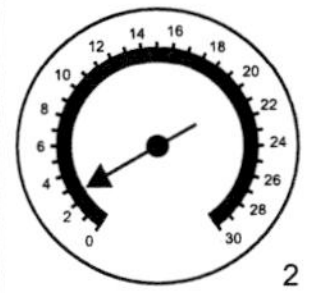 2	*Extremely low reading—needle holds steady.*	*Air leak at the intake manifold or throttle body; incorrect timing.*
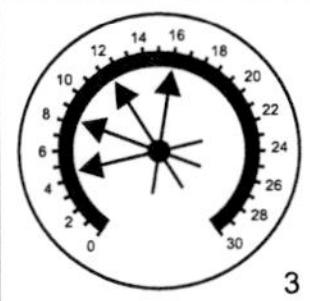 3	*Needle fluctuates between high and low reading.*	*Blown head gasket between two side -by-side cylinders. (Check with compression test.)*
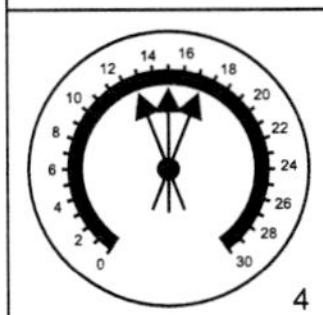 4	*Needle fluctuates very slowly, ranging 4 or 5 points.*	*Idle mixture needs adjustment, sparkplug gap too narrow, sticking valves*
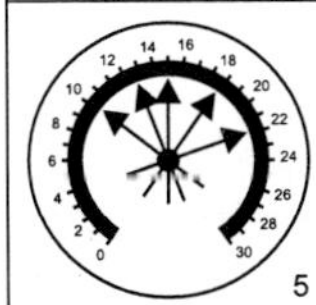 5	*Needle fluctuates rapidly at idle- steadies as RPM is increased.*	*Worn valve guides.*
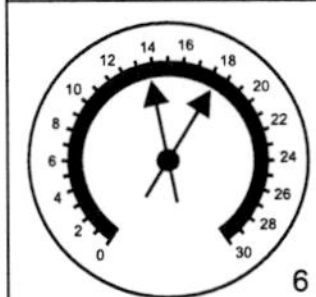 6	*Needle drops to low reading, returns to normal, drops back, etc., at a regular interval.*	*Burned or leaking valve.*
 7	*Needle drops to zero as engine RPM is increased.*	*Restricted exhaust system.*
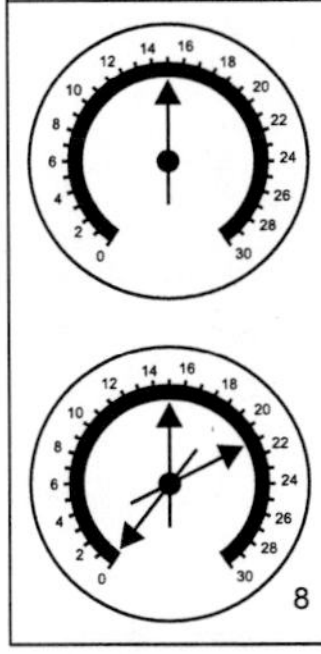 8	*Needle holds steady at 12 to 16—drops to 0 and back to about 21 as you open and release the throttle.*	*Leaking piston rings. (Check with compression test.)*

Fig. 9.20 *Vacuum gauge readings and their meaning. (Champion Spark Plug Company)*

the throttle plate into the manifold. Vacuum increases anytime the engine rpm increases or has better cylinder sealing while the throttle plate remains in a fixed position. Manifold vacuum decreases when the engine rpm decreases more or when cylinders no longer do an efficient job of pumping.

A ***vacuum gauge*** is used to measure intake manifold or engine vacuum and indicate engine condition. A vacuum gauge is connected to an intake manifold vacuum fitting, *Fig. 9.21.*

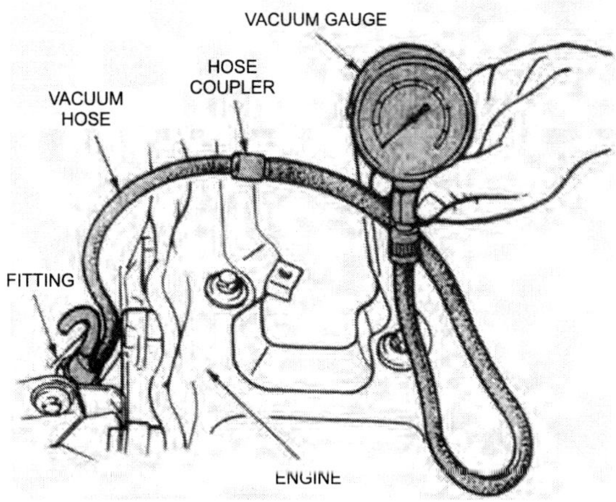

Fig. 9.21 *Vacuum gauge will also help find engine problems during a tune-up. At idle, engine vacuum should be steady and up to specs. (GM- Oldsmobile Div.)*

Cranking Vacuum Test

Measuring the amount of manifold vacuum during cranking, with engine warm and the throttle closed, is a quick and easy test to determine if the piston rings and valves are sealing properly.

1. Disable the ignition.
2. Connect the vacuum gauge to a manifold vacuum source.
3. Crank the engine while observing the vacuum gauge.
   Cranking vacuum should be more than 63.5 mm of Hg. Normal cranking vacuum is 76 to 152 mm of Hg. If the vacuum is lower than 63.5 mm of Hg, it may be due to:

- Too slow a cranking speed
- Worn piston rings

- Leaking valves
- Excessive amounts of air bypassing the throttle plate that can give a false low vacuum reading. This could be due to a throttle plate partially open, or a high-performance camshaft with excessive overlap.

Idle Vacuum Test

Warm the engine. Run the engine at the correct idle speed and read the vacuum gauge. An engine in good condition should idle with a steady vacuum between 400 and 900 mm Hg. A burned valve, late ignition timing, worn timing chain, restricted exhaust system, etc. will show up as abnormal vacuum gauge readings.

Engine vacuum readings vary with altitude. A reduction of 25.4 mm / 300 meters of altitude should subtract from the expected values if testing a vehicle engine above 300 m, *Fig. 9.22.*

Low and Steady Vacuum

If the measured manifold vacuum is lower than normal, yet the gauge reading is steady, the most common causes include:

- Retarded ignition timing.
- Retarded cam timing. Check timing chain for excessive slack or timing belt for proper installation, *Fig. 9.23.*

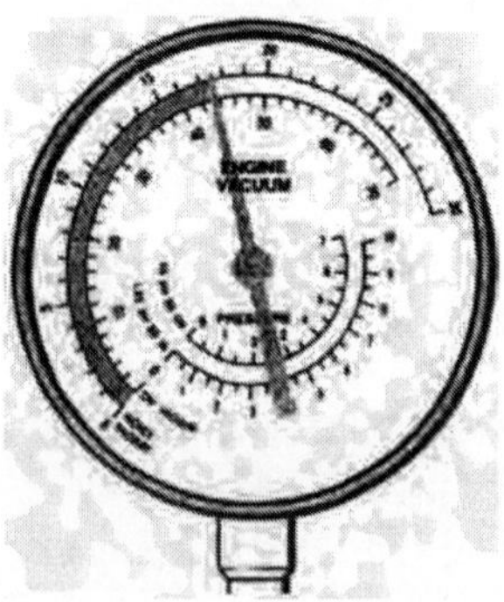

Fig. 9.22 *An engine in good mechanical condition should produce 432 to 533 mm Hg vacuum at idle speed at sea level.*

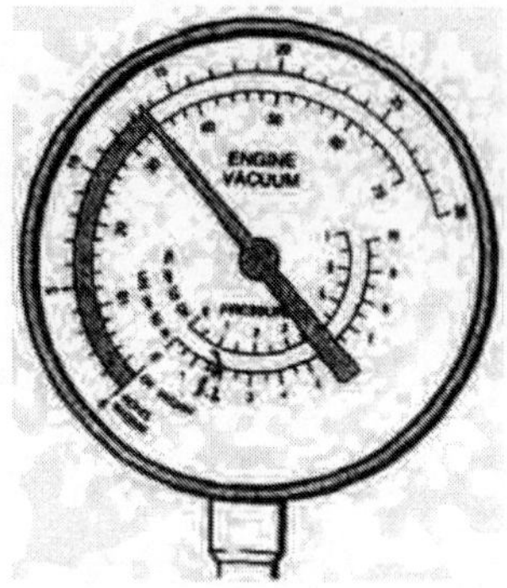

Fig. 9.23 *A steady but low gauge reading could indicate retarded valve or ignition timing.*

Fluctuating Vacuum

If the gauge needle drops, then returns to a normal reading, then drops again, and again returns, then it indicates a sticking valve. A common cause of sticking valves is lack of lubrication of the valve stems, *Figs. 9.24 to 9.32.*

If the vacuum gauge fluctuates above and below a centre point, burned valves or weak valve springs may be indicated. If fluctuation is slow and steady, unequal fuel mixture could be the cause.

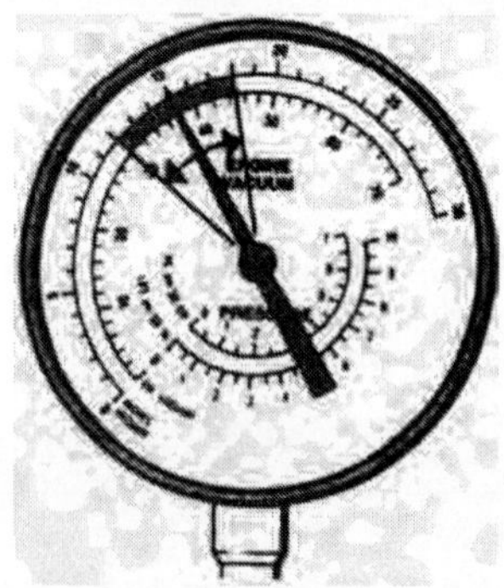

Fig. 9.24 *A gauge reading with the needle fluctuating 76 to 228 mm Hg below normal often indicates a vacuum leak in the induction system.*

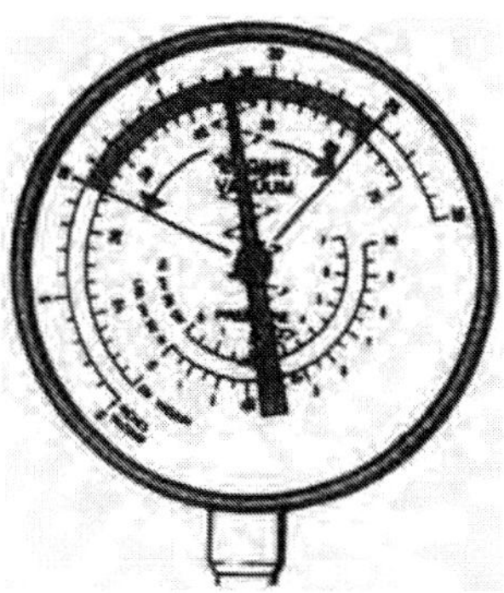

Fig. 9.25 *A leaking head gasket can cause the needle to vibrate as it moves through a range from below to above normal.*

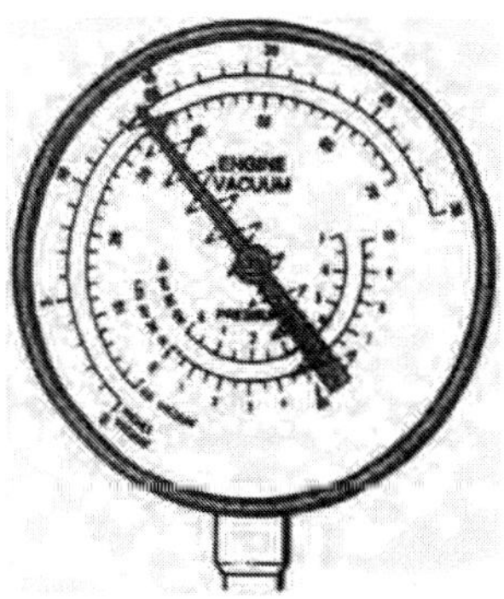

Fig. 9.26 *An oscillating needle 25 to 50 mm Hg below normal could indicate an incorrect air-fuel mixture (either too rich or too lean).*

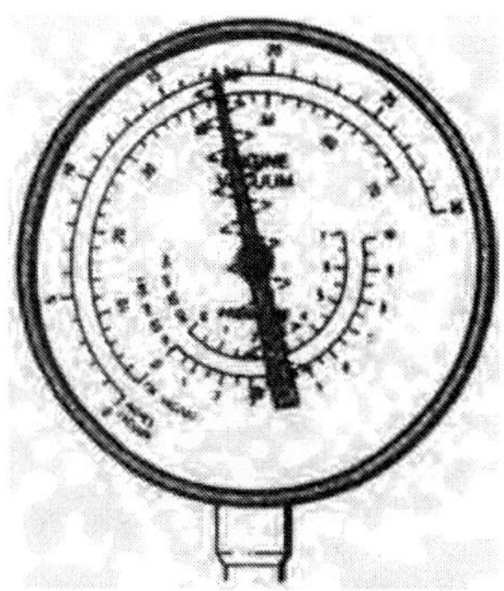

Fig. 9.27 *A rapidly vibrating needle at engine idle speed that becomes steady as the engine speed is increased indicates worn valve guides.*

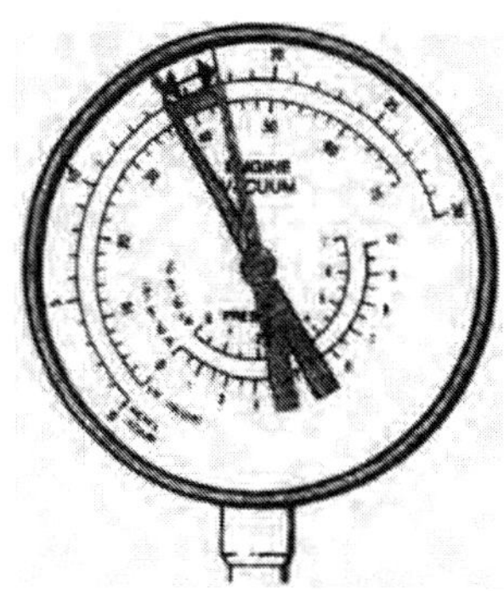

Fig. 9.28 *If the needle drops 25 to 50 mm Hg below normal, an engine valve is burned or not seating properly.*

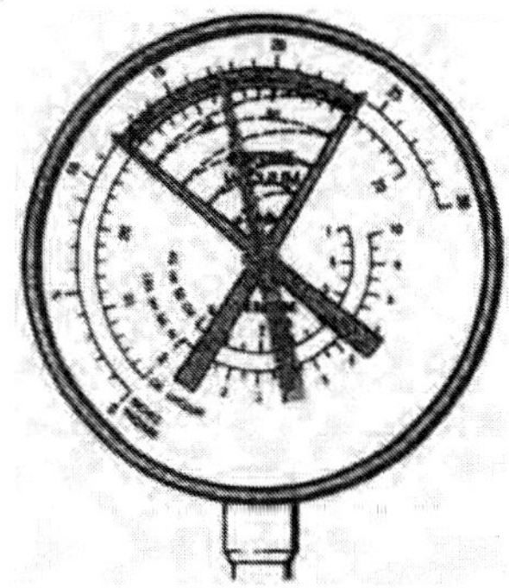

Fig. 9.29 *Weak valve springs will produce a normal reading at idle, but as the engine speed increases, the needle will fluctuate rapidly between 305 and 610 mm Hg.*

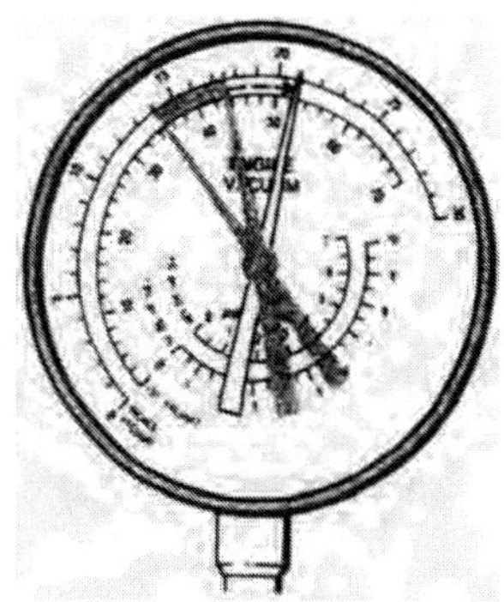

Fig. 9.30 *A steady needle reading that drops 50 to 76 mm Hg when the engine speed is increased slightly above idle indicates that the ignition timing is retarded.*

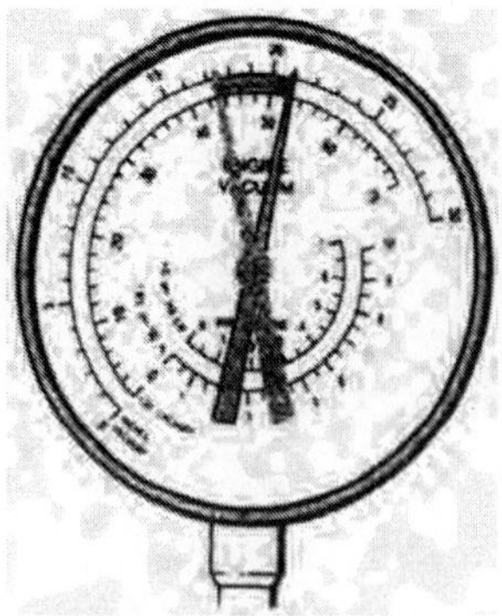

Fig. 9.31 *A steady needle reading that rises 50 to 76 mm Hg when the engine speed is increased slightly above idle indicates that the ignition timing is advanced.*

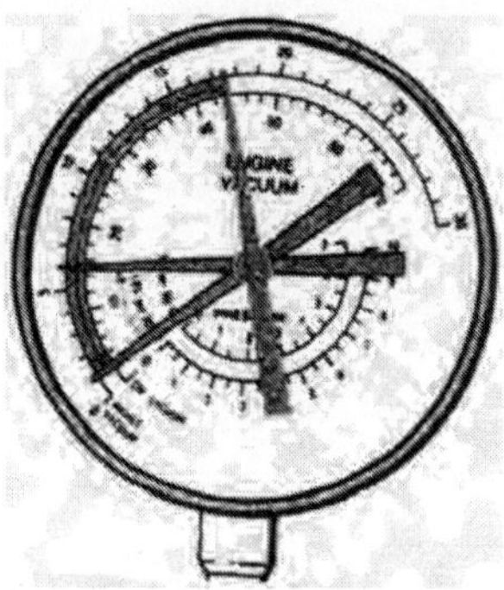

Fig. 9.32 *A needle that drops to near zero when the engine is accelerated rapidly and then rises slightly to a reading below normal indicates an exhaust restriction.*

9.32 EXHAUST RESTRICTION TEST

If the exhaust system is restricted, the engine will be low on power, but smooth. Common causes of restricted exhaust may be due to:

1. Clogged Catalytic Converter

Always check the ignition system for faults that could cause excessive amounts of unburned fuel to be exhausted. Excessive unburned fuel can overheat the catalytic converter to fuse together, creating the restriction. A defective fuel delivery system can also cause excessive unburned fuel being dumped into the converter.

2. Clogged or Restricted Muffler

This can cause low engine power. Often a defective catalytic converter sheds particles that can clog a muffler. Broken internal baffles can also restrict exhaust flow.

3. Damaged or Defective Exhaust Piping

Damaged and defective exhaust piping can reduce the power of any engine. Some Exhaust pipe is made with double walls, and the inside pipe can collapse and form a restriction that will not be visible on the outside of the exhaust pipe.

9.33 TESTING BACKPRESSURE WITH A VACUUM GAUGE

A vacuum gauge can be used to measure manifold vacuum at a high idle speed (500 to 2500 rpm). If the exhaust system is restricted, pressure increases in the exhaust system. This pressure is called *backpressure.* Manifold pressure drops gradually if the engine is kept at a constant speed and the exhaust is restricted. The reason for this vacuum drop is that all exhaust leaving the engine at high engine rpm cannot get through the exhaust restriction. Within about a minute the exhaust tends to accumulate on the engine side of the restriction and eventually remains in the cylinder of the engine at the end of the exhaust stroke. Therefore, at the beginning of the intake stroke when the piston is travelling downward it should be lowering the pressure (raising the vacuum) in the induction manifold, but the extra exhaust retained in the cylinder lowers the normal vacuum. If the exhaust restriction is severe enough, the vehicle can be undriveable because cylinder filling cannot occur except at idle.

9.34 TESTING BACKPRESSURE WITH A PRESSURE GAUGE

At the engine idle speed, the maximum backpressure should be 10 kPa, and it should be less than 15 kPa at 2500 rpm. Exhaust system backpressure can be measured directly by installing a pressure gauge into an exhaust opening. This can be accomplished in one of the following ways:

1. **With an Oxygen Sensor:** Remove the inside of an old discarded oxygen sensor and thread it in an adapter to convert it to a vacuum gauge or pressure gauge.

   Make the adapter by inserting a metal tube or pipe. A short section of brake line works fine.

The pipe can be brazed to the oxygen sensor housing or it can be glued in with epoxy. An 18-millimeter compression gauge adapter may also be used to fit into the oxygen sensor opening, *Fig. 9.33.*

Fig. 9.33 *A back pressure testing tool can be assembled by using an 18 mm air-holding fitting and a 6.39 mm male barb fitting. Most oxygen sensors use an 18 mm thread and the barb fitting can be used to attach a rubber hose leading to a pressure gauge. Both of these low cost parts are available in most automotive parts stores.*

2. **With Exhaust Gas Recirculation (EGR) Valve:** Remove the EGR valve and fabricate a plate to connect to a pressure gauge.
3. **With Air-injection Reaction (AIR) Check Valve:** Remove the check valve from the exhaust tubes leading down to the exhaust manifold. Use a rubber cone with a tube inside to seal against the exhaust tube. Connect the tube to a pressure gauge.

9.35 TIMING CHAIN SLACK DIAGNOSIS

Timing chains of engines with high mileage often have excessive slack than that specified for proper operation. As the timing chain stretches, the camshaft position retards in relation to the position of the crankshaft and pistons. The valves, operated by the camshaft, are made to close later in the stroke than they were designed to do. This retarded power or late closing of the intake valve and late closing of the exhaust valve tends to lower available power at low engine speeds. However, the retarded camshaft timing tends to slightly improve at high-engine speed performance.

This discussion does not apply to engines equipped with timing belts.

The condition of a timing change can be determined as follows:

1. With the ignition off, rotate the engine by hand clockwise as viewed from the front or belt end until the timing mark aligns with TDC, *Fig. 9.34.* If the crankshaft pulley does not have any timing marks, make a visible mark on the pulley and the engine. Do not turn the engine counterclockwise if turned past TDC! This will loosen the tension on the chain, and cause wrong test results. If somehow the engine is turned past TDC, continue clockwise rotation until the timing mark again lines up with TDC.
2. Remove the distributor cap. If the engine is not equipped with a distributor, remove the oil filler cap to gain access to the rocker arms to detect movement.
3. Slowly rotate the engine counterclockwise as viewed from the front of the engine (non-principal end) while carefully observing for any movement of the distributor rotor.
4. As soon as the rotor starts to move, note the distance of the timing mark from TDC.

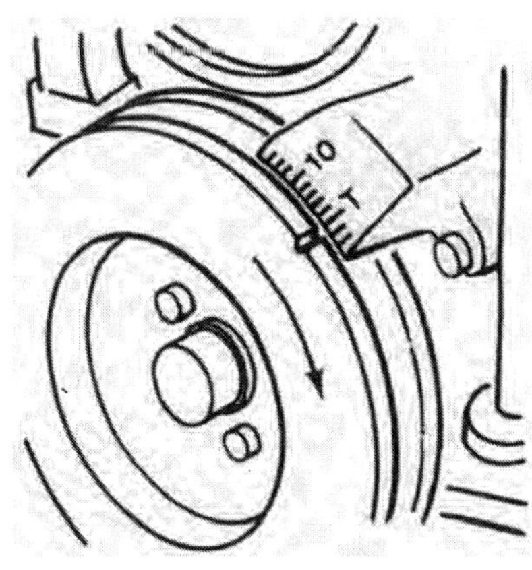

Fig. 9.34 *Typical timing mark on the crankshaft pulley.*

A distance of less than 5° is considered normal and acceptable, especially for a high-mileage engine. A distance of 5° to 8° is acceptable for a high-mileage engine, but not acceptable for a low-mileage engine. Very little change in the engine operation would be noticed if the timing chain were replaced. In case the distance is over 8°, the timing chain should be replaced to restore proper engine operation and to prevent severe engine damage that could occur if the timing chain and/or gear should fail during engine operation.

8° Distance usually translates to about 13 mm of chain slack as measured at the crankshaft pulley, *Fig. 9.34.*

9.36 EXHAUST ANALYSIS AND COMBUSTION EFFICIENCY

A popular method of engine analysis is by using *four-gas exhaust analysis equipment.* The four gases analyzed and their significance is given below.

Hydrocarbons

Hydrocarbons (HCs) are unburned fuel and are measured in parts per million (PPM). A correctly operating engine should burn (oxidize) almost all of the fuel, allowing very little unburned fuel in the exhaust. Acceptable levels of HCs are 50 PPM or less. Higher levels of HCs could be due to excessive oil consumption resulting from weak piston rings or worn valve guides. The most common cause of excessive HC emissions is a faulty ignition system. The ignition system items that should be checked are:

- Spark plugs
- Spark plug wires
- Distributor cap and rotor (if so equipped)
- Ignition timing
- Ignition coil

Carbon Monoxide

Carbon monoxide (CO) is a poisonous gas. Levels of CO in a properly operating engine should be less than 0.5%. High levels of CO can be caused by clogged or restricted crankcase ventilation devices, e.g., PCV valve, hose(s), and tubes. Other items that might cause excessive CO are:

- Clogged air filter
- Incorrect idle speed
- Too-high fuel pump pressure
- Any other items that can cause a rich mixture condition.

Carbon Dioxide

Carbon dioxide (CO_2) is the result of oxygen (O_2) in the engine cylinders combining with the carbon of the fuel. An acceptable level of CO_2 is from 12% to 15%. A high level indicates an efficiently running engine. If the CO_2 level is low, the fuel mixture may be either too rich or to lean.

Oxygen

There is about 21% oxygen in the atmosphere, and most of this oxygen should be 'used up' during the combustion process to oxidize all of the hydrogen and carbon (hydrocarbons) in the fuel. Levels of O_2 should be very low (about 0.5%) in exhaust. High levels of O_2, especially at idle, could be due to an exhaust system leak.

9.37 DIAGNOSING SPARK PLUGS

Diagnosing spark plugs by carefully inspecting the condition and colour of the spark plug tips (points) will allow the technician to determine whether the engine combustion is normal or abnormal and indicate rich mixture, lean mixture, detonation, oil burning, etc. This will provide a better assessment of the condition of the engine and its related systems.

A *brown to gray-tan* spark plug tip indicates a normal combustion, i.e., combustion is clean and good, *Fig. 9.35 (a).*

A *black* or *wet plug* would indicate a dead cylinder or a mechanical problem in the engine allowing oil to enter the cylinder, *Fig. 9.35 (b).*

(a) Normal burning plug has light deposits and is grey-tan.

(b) Oil fouled plug, from bad rings or oil seals.

(c) Carbon fouled plug from bad plug wire, rich mixture etc.

(d) High mileage plug has heavy deposits and corroded centre electrode. Rounded electrode will make it difficult for spark to jump gap.

Fig. 9.35 *Correct reading of spark plugs will allow diagnosis of the condition of engine and related systems. (Champion Spark Plugs Co.)*

Remove each plug and place it in order on tool cart, fender, etc. (number them) to know which plug came out from which cylinder. For example, if only one plug is fouled with carbon and the others are clean, the technician would know the exact cylinder with a problem. It could have a burned valve, bad plug wire, broken piston ring, etc. To isolate the problem source, do a compression test in that cylinder.

9.38 DIAGNOSING HEAD GASKET FAILURE

A number of items can be used to assist diagnose a head gasket failure:

1. Exhaust Gas Analyzer: Remove the radiator cap. Place the exhaust analyzer probe above the radiator filler neck. If the HC reading increases, the exhaust (unburned hydrocarbons) is getting into the coolant from the combustion chamber.

2. Chemical Test: A chemical tester using blue liquid is also available. The liquid turns yellow if combustion gases are present in the coolant.

3. Bubbles in the Coolant: Remove the coolant pump belt to prevent pump operation. Remove the radiator cap and start the engine. If bubbles appear in the coolant before it begins to boil, a defective head gasket or cracked cylinder head is indicated.

4. Excessive Exhaust Steam: Excessive water or steam coming from the tailpipe means that coolant is getting into the combustion chamber from a defective head gasket or a cracked cylinder head. If there is a leakage between the cylinders, the engine usually misfires and a power balancer test and/or compression test can be used to confirm the problem.

If any of the above indicators of head gasket failure occur, remove the cylinder head(s) and check all of the following:

1. Head gasket
2. Sealing surfaces for any warpage
3. Castings – for any cracks

9.39 DASH WARNING LIGHTS

Most vehicles are equipped with several dash warning lights often called "telltale" lights. These lights are often the only warning a driver receives that there may be engine problems. Typical dash lights and their meanings are given below:

Engine Oil Light

The red oil light indicates that the engine oil pressure is too low (usually lights when oil pressure is 5 to 50 kPa). Normal oil pressure should be 70 to 400 kPa or 70 kPa per 1000 rpm of the engine.

When this light comes on, the driver should shut off the engine immediately and check the oil level. If the oil level is good, then there is a possible serious engine problem or a possible defective oil pressure sender unit. Always check the oil pressure using a reliable mechanical oil pressure gauge if low pressure is suspected.

Coolant Temperature Light

Most vehicles are equipped with a coolant temperature gauge or dash warning light. The warning light may be labelled 'coolant,' 'hot,' or 'temperature.' If the coolant temperature warning light comes on during driving, this usually indicates that the coolant temperature is above a safe operating level, or above about 15°C. Normal coolant temperature should be about 90° to 105°C.

If the coolant temperature warning light comes on during driving, the following procedure should be followed to prevent possible engine damage:

1. Turn off the air conditioning and turn on the heater. The heater will help get rid of some of the heat in the coolant system.
2 Raise the engine speed in neutral or park to increase the circulation of coolant through the radiator.
3. If possible turn the engine off and allow it to cool for about an hour.
4. Do not continue driving with the coolant temperature light on, or the gauge reading in the red warning section to avoid serious engine damage.

If the engine does not feel or smell hot, it is possible that the problem is a faulty coolant temperature sensor or gauge.

Charge Light

The red charge warning light indicates that the charging system is not working. The charge light may also be labelled "gen" or "alt". Because most automotive batteries are capable of supplying the electrical needs of the vehicle for a short time, it is not necessary to stop the engine if this warning light comes on.

However, if the charging system is not serviced, the engine will eventually stop running because the ignition system will fail to ignite the spark plugs when the battery voltage falls below about 9 V.

Computer Warning Light

Most vehicles with an engine control computer use dash-warning lights to warn the driver that some computer sensor, actuator, or engine parameter is not within the acceptable range.

The computer dash warning light called a malfunction indicator lamp (MIL) is usually orange in colour and indicates a less than serious problem, whereas, some computer warning lights may be red indicating more serious engine problems. The computer warning lights might be labelled

Check Engine, Check Engine Soon, Check Engine Now, Power Loss, Power Limited

If an orange engine computer light comes on, continue driving and check for any stored trouble codes as per the vehicle manufacturer's recommended procedures. If a red engine computer light comes on, it is best to stop the engine and check all vital engine systems before continuing operation. Follow the manufacturer's recommended procedures for determining the cause and corrective action needed.

Maintenance Reminder Light

Many vehicles are equipped with a maintenance reminder light notifying the driver that some service is required. Routine service that should be carried out when the service reminder light comes on usually includes an oil change plus some other items recommended by the vehicle manufacturer. These additional items could include:

- Air filter replacement
- Coolant change
- Tyre rotation

9.40 VALVE ADJUSTMENT

When mechanical (solid) lifters are used, the valves require periodic adjustment. Valve adjustment is important to the performance of the engine. It is often done during a major tune-up.

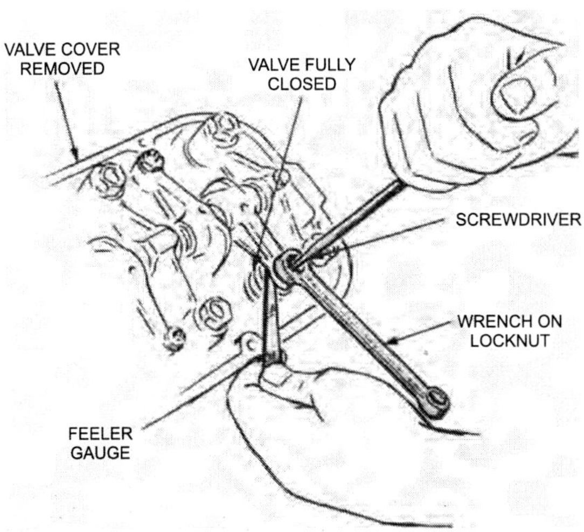

Fig. 9.36 *Valve adjustment is sometimes required during engine tune-up. Crank engine until valve is fully closed. Slide proper size feeler gauge between valve and rocker. Turn adjusting screw to change lash if needed. Repeat it on all valves. Refer to applicable service manual for specific details.*

(a) Checking valve clearance with feeler gauge.

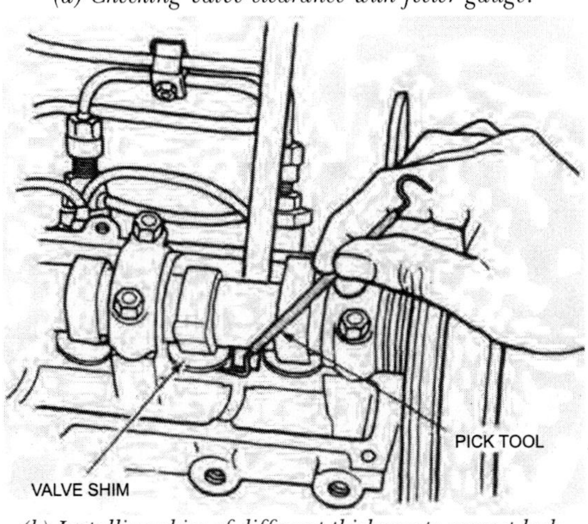

(b) Installing shim of different thickness to correct lash.

Fig. 9.37 *This OHV engine has valve adjusting shims. (Ford Motors)*

When a valve opens too much or not enough, it will upset the amount of air-fuel mixture induced into the cylinder. It also affects valve lift and open duration. This will affect combustion and lower engine efficiency.

Valve adjustment can be made by turning rocker arm adjusting screws on some engines, *Fig. 9.36.* On some OHV engines, shim thickness is needed to be changed to adjust valve lash or valve clearance, *Fig. 9.37.*

To adjust a valve clearance, remove the valve cover. Crank the engine until a cam lobe is pointing away from the rocker arm or until the valve is fully closed. Adjust the valve until the correct thickness feeler gauge fits over the valve or under the rocker arm, *Figs. 9.36* and *9.37.* If the clearance is incorrect, turn the adjusting screw or install a shim of different thickness. Repeat the procedure on all the valves.

Some engine manufacturers recommend cold valve adjustment while others require that the engine be warmed to full operating temperature. Always refer to the applicable service manual for specific procedures and specifications.

Usually the exhaust valve will need more clearance than the intake valve. Because it runs hotter and will expand to increase in length more than the intake valve, which runs comparatively much cooler.

9.41 TUNE-UP PARTS REPLACEMENT

Depending upon the in service life and condition of the engine, any number of parts may need to be replaced during a tune-up. With a new vehicle, it might require only to replace the spark plugs and lifters. With an older, high mileage vehicle, it might be required to replace the spark plugs, injectors, distributor components, spark plug wires, or any other parts that contribute towards reducing engine efficiency.

During a tune-up an auto technician may have to replace or service:

1. ***Spark Plugs:*** Spark plugs are almost always serviced or replaced during tune-up, *Fig. 9.38.*
2. ***Spark Plug Wires:*** Wires should be checked for high resistance or electrical leakage through insulation, *Fig. 9.39.* Refer to "Ignition System Service" for details.

3. ***Distributor Parts:*** May have to replace cap and rotor, set pickup coil air gap or contact point gap, adjust ignition timing, *etc., Fig. 9.40.*
4. ***Fuel System Parts:*** Must check and, if needed, adjust idle speed, choke, idle mixture, change filters, etc.
5. ***Emission Control Parts:*** Changing canister filter, PCV valve, etc.

9.42 TUNE-UP ADJUSTMENTS

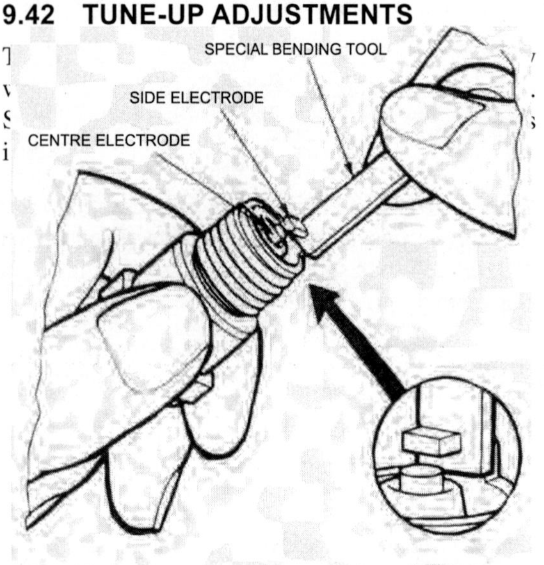

(a) Bending tool will open or close plug gap as necessary.

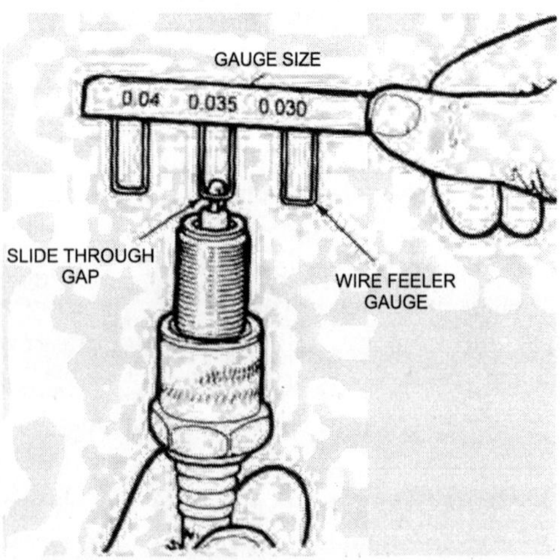

(b) Wire feeler gauge is slid inside gap to check electrode spacing. Use wire gauge of correct thickness.

Fig. 9.38 *Spark plugs are usually replaced during a tune-up. (Chrysler)*

1. Spark plug gap, *Fig. 9.38.*
2. Pickup coil air gap (for electronic ignition) or breaker point gap or dwell (for contact point ignition).
3. Ignition timing for SI engine or injection timing for CI engine, *Fig. 9.40.*
4. Idle speed.

Some other adjustments may also be needed. Follow the specific directions and specifications in the applicable service manual. This will assure a thorough and longer lasting tune-up.

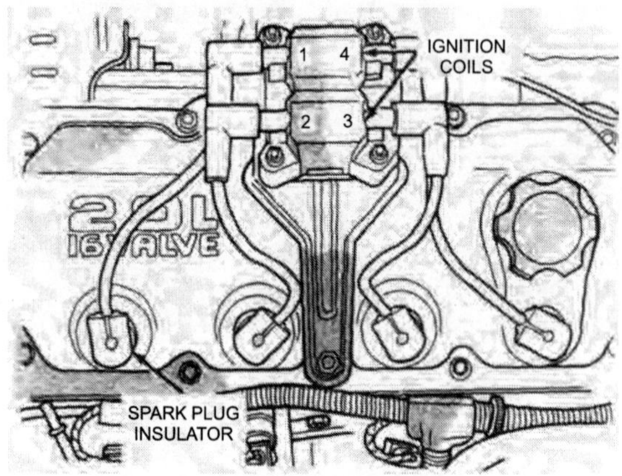

Fig. 9.39 *Spark plug wires are a common source of engine performance problems. They can break internally, their insulation can leak high voltage to ground, voltage can be induced into another wire, or they can simply come off the coil or spark plug. (Chrysler)*

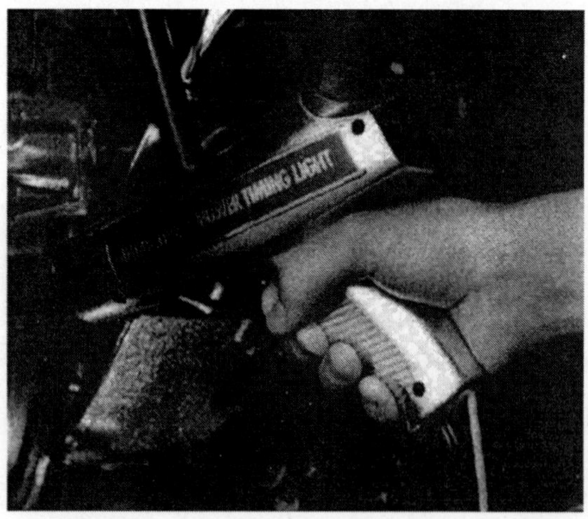

Fig 9.40 *Timing light is used to check and adjust ignition timing. With CI engines we have to check injection timing. (Peerless Instruments)*

9.43 GENERAL TUNE-UP RULES

The following general tune-up rules should be kept in mind when doing a tune-up:

1. Gather information about the performance of the engine. Ask the customer/driver about the vehicle performance. This may indicate what components should be tested and replaced.
2. Run the engine to full operating temperature. Usually, a technician cannot evaluate engine operation and make tune-up adjustments with a cold engine.
3. Use professional, high quality tools and equipment. Make sure the tools and equipment is accurate and will give precise readings.
4. Refer to the applicable vehicle service manual or emission sticker for specifications and procedures. Late model vehicles are very complex; the slightest mistake could ruin the tune-up.
5. Use quality parts during parts replacement. Quality parts will assure that the tune-up lasts a long time. Cheap, bargain parts are usually inferior and may fail quickly.
6. Maintain service records. Write down all of the operation performed on the vehicle. This will help the technician and the customer a record for future reference. If a problem develops, the technician can check the records to help diagnose and fix the trouble.
7. Complete basic maintenance service. Lubricate door, hood, and trunk latches and hinges during a tune-up. Also check all fluid levels, belts, and hoses. This will build good customer relations and assure vehicle safety and dependability.
8. Tune-up intervals are the recommended time spans or kilometers when an engine should be tuned-up. The vehicle manufacturer recommends these intervals. Check the applicable service manual for details. Vehicle warranty may be void if tune-up intervals are missed.

9.44 DIESEL ENGINE TUNE-UP (MAINTENANCE)

Diesel engines do not require tune-ups like gasoline engines, because a diesel engine does not have spark plugs to replace or an ignition system to give troubles. The diesel injection system is highly dependable and only requires major service when problems develop. Diesel engine maintenance involves:

1. Replacing air filter element.
2. Cleaning, draining, or replacing fuel filters.
3. Adjusting engine idle speed.
4. Adjustment of throttle cable.
5. Inspecting engine and related systems.
6. Checking the injection timing.
7. Checking engine oil and oil filter.
8. Periodic service of emission control systems.

Refer to the applicable service manual for additional information on specific diesel engine maintenance.

9.45 ENGINE TUNE-UP EQUIPMENT

An engine tune-up requires the use of certain types of gauges, testers, and test equipment, *Fig. 9.41.* With this instrumentation, automotive mechanic can perform tests that will reveal weak or defective components that should be replaced. On the other hand, test results will also verify the satisfactory condition of good used units that need not be replaced.

Another valuable application, mechanic can connect these testers to a given system or circuit to check if a particular setting is as specified. If it is not, the reading on the tester will serve as a means of guiding the mechanic in making adjustments to achieve the correct system or circuit setting.

Tune-up test equipment in the following list allows a mechanic/technician to perform all of the checks and tests necessary for a *tune-up test procedure.* The test equipment will help the mechanic in making precise settings to manufacturer's specs.

1. Compression tester.
2. Vacuum gauge.
3. Voltmeter or volt-amp tester.
4. Ammeter.
5. Ohmmeter.
6. Digital volt-ohmmeter.
7. Dwell meter.
8. Tack-dwell meter or Tachometer.
9. Battery-starter tester.
10. Hydrometer.
11. Alternator-regulator tester.

12. Distributor tester.
13. Stroboscopic timing light.
14. Fuel injection pressure gauge.
15. Fuel pressure gauge.
16. Belt tension gauge.
17. Cylinder leak tester.
18. Emissions tester or infrared exhaust gas analyzer or 4-gas analyzer
19. Engine analyzer or diagnostic computer.
20. Electronic engine controls tester.
21. Solid-state circuit testers.
22. Positive crankcase ventilation (PCV) tester.
23. Cooling system pressure tester.
24. Un-powered test lamp.
25. Jumper wires.

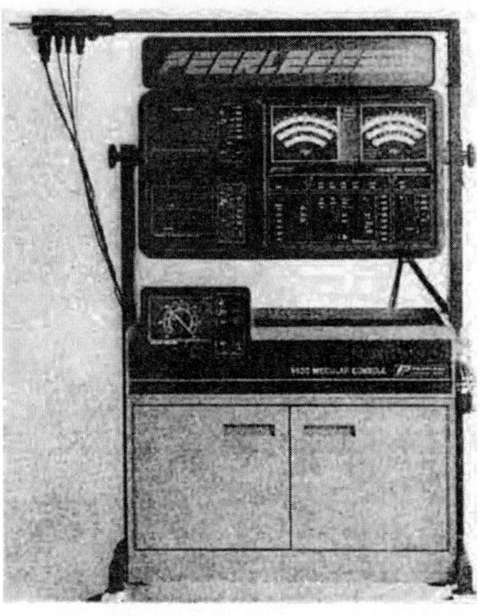

Fig. 9.41 *To meet modern need for engine tune-up test instrumentation, this modular console contains an ignition systems analyzer, an infrared gas analyzer, vacuum systems analyzer, fuel system tester, and overhead track mount. (Peerless Instrument Co.)*

All kinds of tune-up equipment are needed to keep modern "high-tech" automotive engines running at maximum performance levels. Most common types of tune-up equipment are summarized in this section, along with their potential problem indications.

Pressure Gauge

A ***pressure gauge*** is used to check various pressures, e.g., fuel pump pressure, pressure regulator pressure, engine oil pressure, or turbo-boost pressure. It measures pressure in psi and/or kPa, *Fig. 9.42*. An engine technician will normally have many pressure gauges. Each gauge will be designed for a specific job.

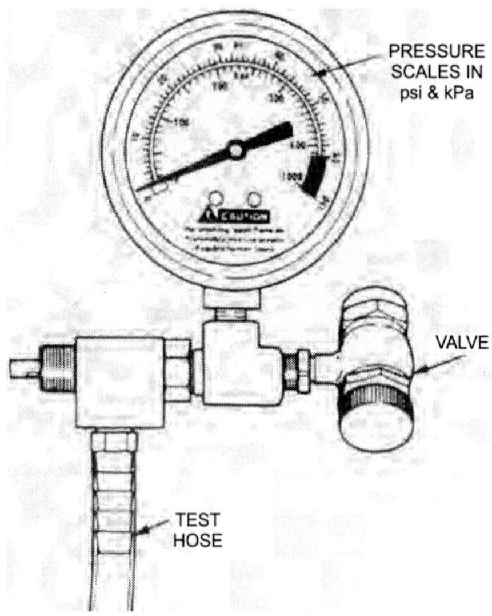

Fig. 9.42 *This particular pressure gauge is for measuring pressure in fuel injection system. An auto repair shop mechanic will have a number of pressure gauges for checking engine oil pressure, fuel injection pressure, negative or vacuum air injection pump pressure, etc. (Ford)*

A *pressure gauge* is used to measure engine fuel pressure and can be used to isolate source of engine performance problems.

The pressure gauge can be connected to the engine fuel line, usually at a test fitting on the fuel rail, to check the operation of the fuel pump, pressure regulator, fuel filters, and fuel lines. If the measured pressure is within specifications (specs), these parts are eliminated as a source of the performance problem at hand.

Compression Gauge (Tester)

A ***compression gauge,*** *Fig. 9.43,* is used to measure pressure during the engine compression stroke. It provides a means of testing the mechanical condition of the engine. Out of specs compression gauge readings indicate that something is mechanically wrong in the engine.

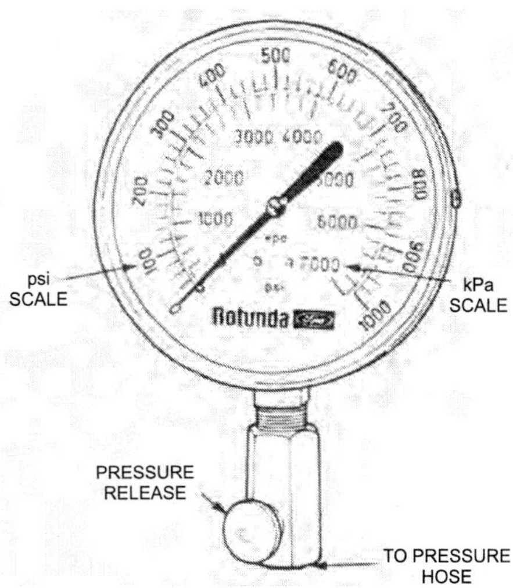

Fig. 9.43 *This is a pressure gauge for CI engine. It is capable of reading higher pressures than compression gauge for SI engine. Gauge is connected to a combustion chamber. Engine is then cranked and compression strokes should build specified pressure for good combustion. If not, then there is something wrong with the engine. (Ford)*

A ***compression test*** should be performed when symptoms (like engine miss, rough idle, puffing noise in induction system or exhaust) indicate major engine problems.

To perform a compression test, remove all of the spark plugs with a gasoline engine or glow plugs with a diesel engine. Disable the ignition or injection system. Block open the throttle valve (gasoline engine).

Install the compression tester into a spark plug hole. Crank the engine at least four to six compression strokes while keeping an eye on the gauge reading. Measure and record the pressure for the cylinder. Repeat this on other cylinders in turn. A normal compression reading will make the gauge increase evenly to specs. The pressure in each cylinder should not vary more than 10%.

If the pressure is low in two adjacent cylinders, the head gasket may be blown. A leak in the head gasket can allow compression pressure to blow back and forth between the two cylinders.

If compression pressure is low in only one cylinder, that cylinder has a mechanical problem e.g., burned valve or piston, blown head gasket, warped or cracked head, worn piston rings, broken valve spring. There is an opening in the combustion chamber that is allowing compression pressure to leak out.

If all the pressure readings are low, there could be a "jumped" timing chain or belt, causing upset ignition timing. When the valve setting is off, it can reduce compression pressure in all of the cylinders.

Cylinder Leakage Tester

A ***cylinder leak tester,*** *Fig. 9.44,* measures the amount of air leakage out of the engine combustion chambers. External air is compressed into the cylinder combustion chamber with the piston at TDC on the compression stroke. Then, a pressure gauge can be used to determine the % of air leakage out of the cylinder.

If leakage is high enough, it may be possible to hear and feel air flowing out of the engine. Air may blow out the induction manifold due to bad intake valve, exhaust system due to burned exhaust valve, breather due to bad rings, piston, or cylinder, or into an adjacent cylinder due to a blown head gasket.

Vacuum Gauge

A *vacuum gauge, Fig. 9.45,* is used to isolate some engine performance and mechanical trouble sources during engine service. The gauge is connected to a source of intake manifold vacuum. It is similar to a pressure gauge. However, the gauge reads in inches of mercury and metric kg/cm^2. Gauge readings are recorded and compared to known readings of good performance. *Fig. 9.46 (a) to (h)* shows some typical vacuum gauge readings.

Figure 9.47 shows a hand-vacuum pump and gauge assembly. It can be used to test many engine related vacuum devices.

Test Light

A *test light* is used to quickly check for power in an electrical circuit. It can be used to determine the source of many engine performance and mechanical problems.

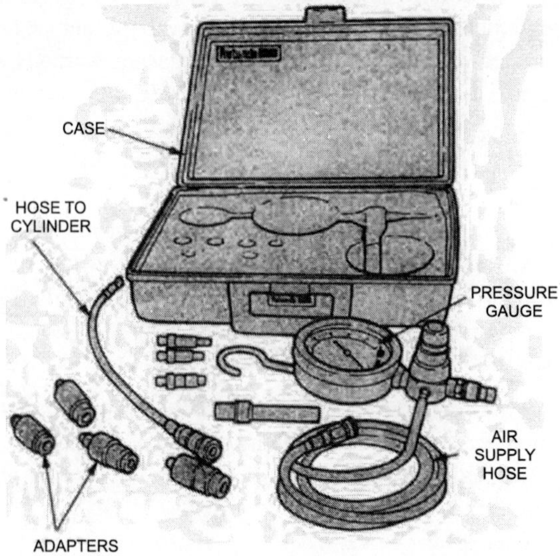

Fig. 9.44 *Cylinder leakage tester will also indicate mechanical condition of engine. Air pressure is forced into cylinder. Then, pressure gauge can be used to calculate amount of leakage out of combustion chamber. (Ford Motor Co.)*

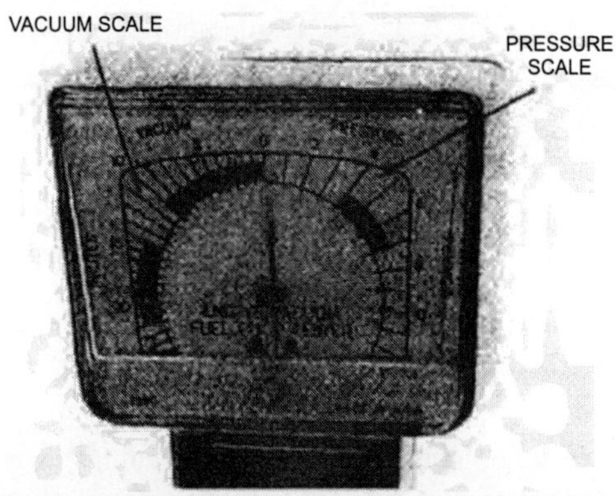

Fig. 9.45 *This is vacuum pressure gauge. It can be used to diagnose engine problems.(Sonoco)*

For example, if an engine does not start and it is found that there is no spark at a spark plug wire, use the test light to check for power at the ignition coil. Simply touch the test light on the coil positive terminal, as shown in *Fig. 9.48.* If the light glows, it indicates that the ignition circuit before the coil, (ignition switch, wiring, wiring connections, etc.), is functional.

This type of testing can be done on various other electrical components. However, do not connect a test light to electronic components as it can draw too much current and damage the electronic part. Always refer to the applicable service manual for instructions when in doubt.

Digital Multimeter

It is another handy tool for finding performance problems on present automotive engines. It precisely measures the voltage, current, or resistance in a circuit or a component. The measured test values can then be compared to manufacturer's specified values to isolate the problems, *Fig. 9.49.*

Scan Tools

A *scan tool* is the testing device, to allow quickly read and interpret computer system trouble codes stored in the system *Fig. 9.50.* The tool lead is connected to the vehicle's data link connector. This allows it to retrieve and display trouble codes. Scan tools also be used to test sensors, actuators, and the vehicle computer itself.

Most scan tools use data cartridges that are plugged into the tool. However, scan tools that contain fixed memory boards, which can be programmed by downloading information from a computer or computerized engine analyzer, are now available.

Spark Plug Cleaner

A ***spark plug cleaner*** is used to blast off deposits from spark plug tips and restore their function and allow their reuse in an engine. One such spark plug cleaner is shown in *Fig. 9.51.* It uses shop air pressure and blast compound, i.e., sand like particles, to remove carbon deposits from a spark plug. Some engine manufacturers do not recommend spark cleaning.

Fig. 9.46 *Study vacuum gauge readings and their indications. (Sonoco)*

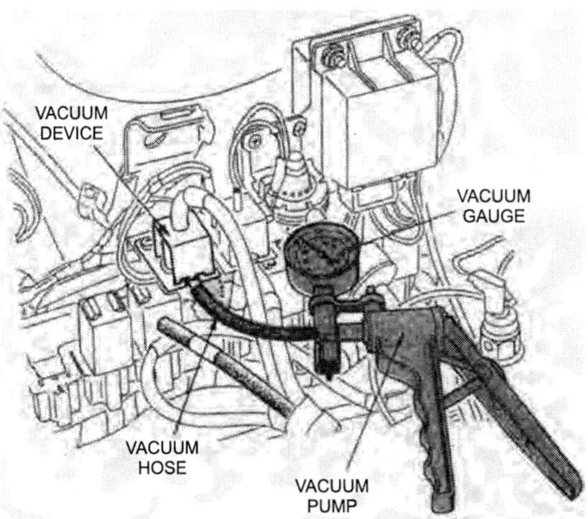

Fig. 9.47 *Hand vacuum pump-gauge will test vacuum operated devices that can affect engine operation. Device should function when specific vacuum is applied.*

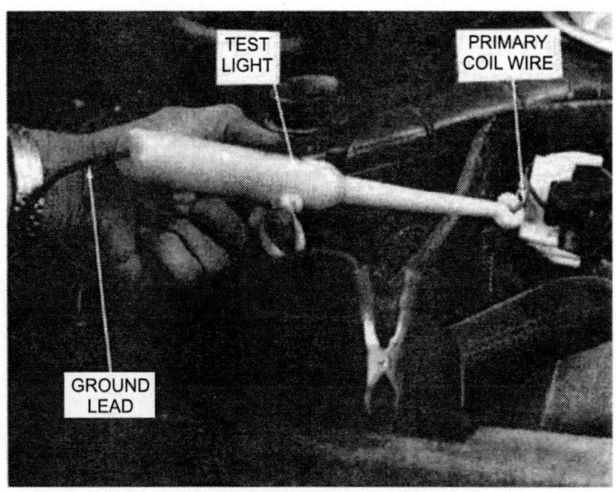

Fig. 9.48 *Test light will quickly check for power at various components. Here a technician is checking for voltage going to coil when diagnosing no-start condition. (Peerless)*

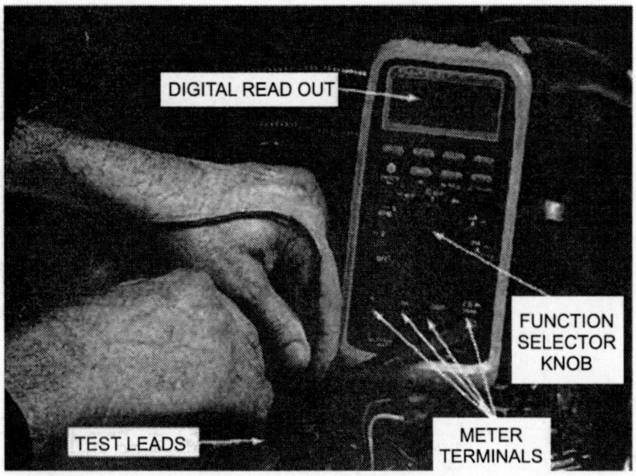

Fig. 9.49. *A digital multimeter can be used to help find a number of engine performance problems. Current, voltage, and resistance can be compared to specs. Many electric-electronic components can upset engine performance. (Fluke)*

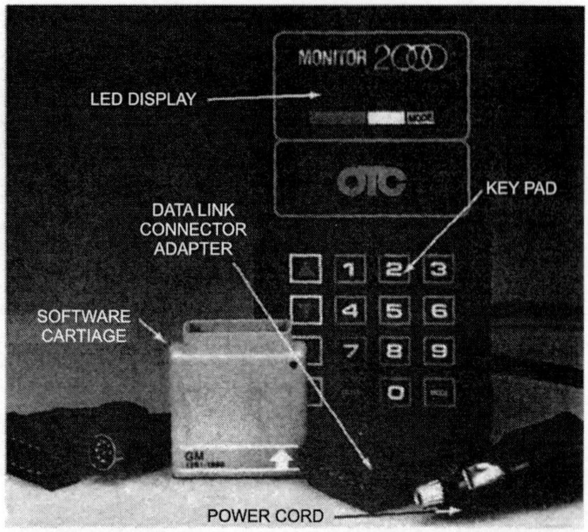

Fig. 9.50 *Scan tool will communicate with the vehicle's onboard computer to help locate problems. This tester will actually give directions and convert trouble code numbers into actual words describing problems so the technician will have to refer to applicable service manual code chart. (OTC Tools)*

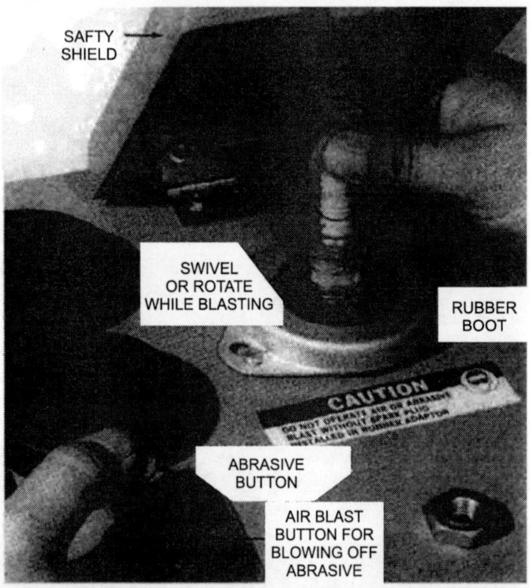

Fig. 9.51 *Spark plug cleaner uses air pressure and blast compound to remove carbon from spark plug tip. This tester also produces arc at plug to test its operation. (Autolite)*

Distributor Tester

A *distributor tester,* known as *distributor machine,* is used to check the operation of an older ignition system distributor. The distributor is removed from the engine and mounted in the tester. After connecting the electrical and vacuum connections, the tester spins the distributor shaft to check distributor performance, *Fig. 9.52.* A distributor tester will check:

- Centrifugal advance.
- Vacuum advance.
- Contact point or pickup coil operation.
- Distributor shaft and cam lobe wear.
- Condenser condition

Adapter (amplifier) devices that increase the output of the pickup coil are sometimes needed to check distributors for electronic ignition systems.

Tach-Dwell Meter

A ***tach-dwell meter*** is a tacometer and a dwell meter combined into one instrument. The tach is used to measure engine rpm. The dwell meter measures in degrees. A tech dwell is sometimes used during engine tune-ups, *Fig. 9.53.* It is used for contact point adjustment or computer-controlled carburetor calibration.

Follow operating instructions when connecting a tach-dwell meter because the procedures vary. To measure rpm for an idle speed adjustment, the technician must typically connect the red lead to the negative coil terminal or tach terminal. The black tester lead connects to ground.

Fig. 9.52 *Distributor machine will check operation of ignition system distributor. (Sun Electric)*

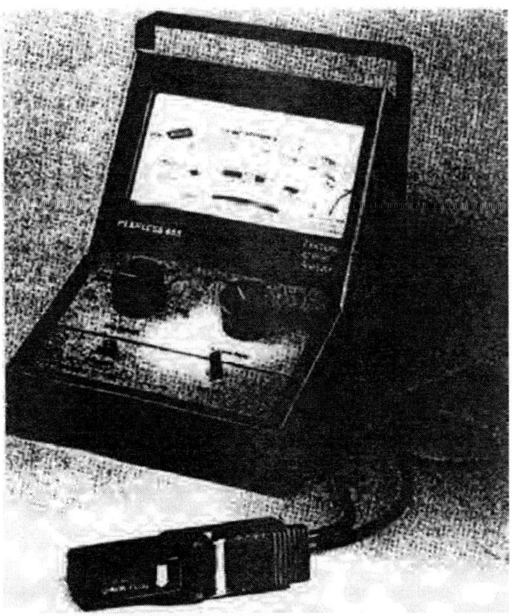

Fig. 9.53 *Tach-dwell meter will measure engine speed for throttle adjustment. Dwell meter will measure in degrees for contact point and computer-controlled carburetor service. (Peerless)*

Spark Tester

A ***spark tester*** is used to check the basic operation of the engine ignition system. It is similar to a spark plug but has a ground clip on it. The spark tester is connected between the end of a spark plug wire and ground. When the engine is cranked or started, a "hot" spark should jump across the tester gap. This shows ignition system operation.

Timing Light

A ***timing light*** , *Fig. 9.54,* is a flashing light used to check and adjust ignition timing. Usually the two small timing light leads are connected to the vehicle battery. The larger lead is connected to the # 1 spark plug wire. The timing light is then aimed on the engine timing marks. They are usually located on the front cover of the engine, crankshaft damper, or on the flywheel.

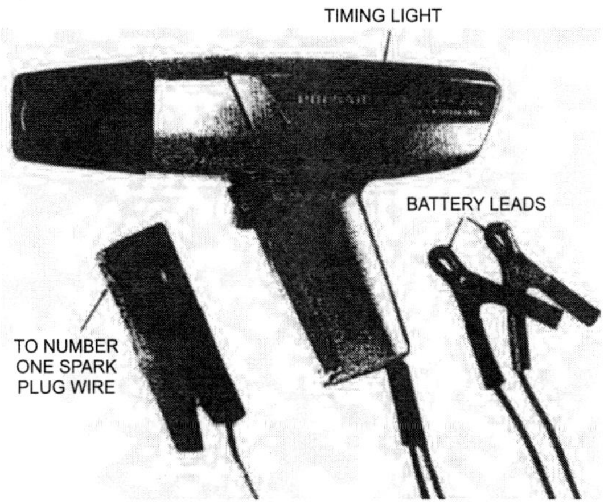

Fig. 9.54 *Timing light is strobe light. It is aimed on timing marks and spinning damper or flywheel to adjust when spark plugs fire in cylinders in relation to piston position. Note lead connection points. (Peerless)*

The timing light, by flashing on and off, will make the rotating crankshaft pulley, balancer, or flywheel appear to stand still. This makes the moving timing marks visible. You can then adjust the ignition timing by turning the distributor housing. When the computer controls the ignition system, there may be adjustment on the computer or ignition timing may not be adjustable.

Mag-Tach, Timing Meter

A ***mag-tach, timing meter,*** *Fig. 9.55,* is a magnetically triggered tachometer used for measuring engine rpm and for adjusting diesel injection timing. It is used both on gasoline and diesel engines. A diesel engine does not have an electrically operated ignition system to power a conventional tachometer or timing light.

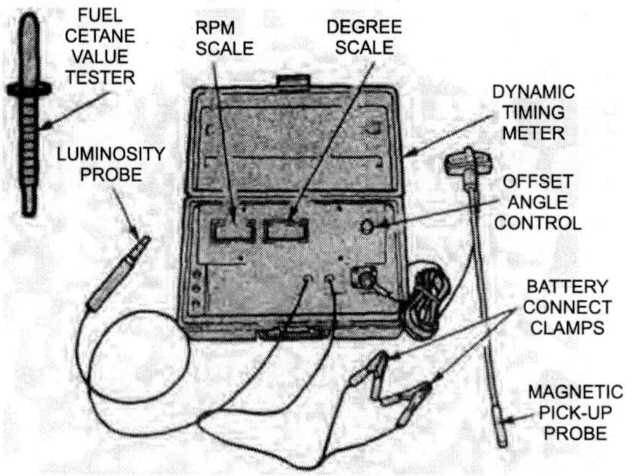

Fig. 9.55 *This is a dynamic timing meter for servicing diesel engine. Magnetic probe senses crankshaft and damper position. Luminosity probe detects combustion flame in cylinder for injection timing adjustment. (Ford)*

A magnet that senses notch in the engine flywheel or damper operates a mag-tach. In this way, engine speed can be measured without a connection to the ignition system.

Some mag-tachs, known as *timing testers,* can measure ignition advance and ignition system advance too. A luminicity probe is used to detect a combustion flame for setting injection timing on diesel engines.

Temperature Probe

Special ***thermometers*** or ***temperature gauges,*** or ***probes*** are frequently used to measure the temperature of many components. Most temperature probes or gauges can read in °C. The temperature obtained can be compared to specifications. If the temperature is low or too high, it indicates that a repair or adjustment is necessary. One such probe is shown in *Fig. 9.56.*

Electronic Fuel Injection (EFI) Tester

An ***EFI tester*** is used to check the operation of modern fuel injection systems. Most EFI testers come with many accessory devices. The tester usually used to test all the system sensors, circuit wiring, and engine computer. A pressure gauge may be provided for measuring fuel pump and pressure regulator outputs.

EFI test equipment varies. Therefore, always follow the operation instructions of the equipment very carefully. Often, automobile vehicle manufacturers recommend a specific brand of tester for use with their vehicles. The applicable service manual outlines the use of the exact type tester.

Fig. 9.56 *Temperature probe will accurately measure various engine related temperatures, e.g., coolant temperature, exhaust manifold temperature, etc.*

Exhaust Gas Analyzer

An ***exhaust gas analyzer,*** *Fig. 9.57,* is used to measure the chemical content of the engine's exhaust gases. An automotive technician uses this to check combustion efficiency and the amount of pollutants produced by the engine. An exhaust analyzer will check engine, fuel system, and emission control system operation.

Engine Analyzers and Computerized Testers

It includes an oscilloscope and other instruments for making tests of the engine parts and systems. After carefully learning to use this equipment, you can make a complete engine analysis quickly.

Certain testers are available that run many of the tests almost automatically. They produce a complete record of the tests and the test results. Some vehicles have a socket, or diagnostic connector, wired to various components and they need special computerized test equipment. It includes a plug that can be inserted into the vehicle socket to complete connections between the vehicle components and the test equipment. A series of connections are made all at once by this setup. Making a series of separate connection is not necessary.

Figure 9.59 shows one type of diagnostic computer. This computer runs the tests. Then it provides a printed record that presents the mechanic and customer what work is necessary to restore the vehicle within specs. Some models of this computerized analyzer can read and report on exhaust emissions (HC and CO). Programmed tape cassettes provide the computer with the vehicle specifications against which the actual conditions of the vehicle are compared. This pinpoints any possible problem area so that it can be serviced.

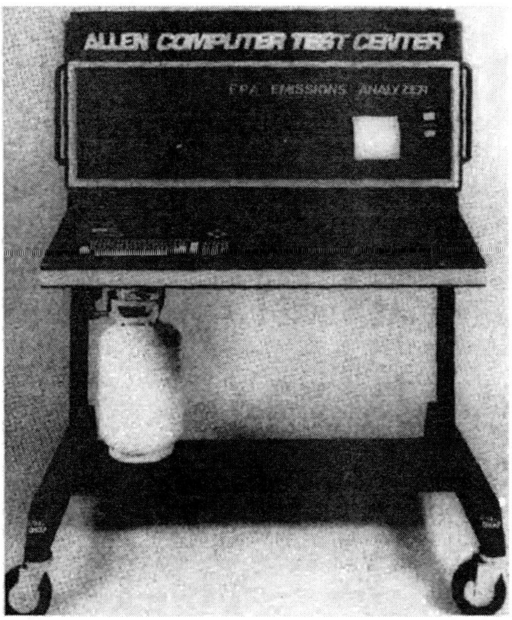

Fig. 9.57 *Emissions analyzer will measure chemical content of exhaust for checking engine and combustion efficiency. It can be used to diagnose various engine problems.*

Engine Analyzer

It combines several testing instruments and make a number of tests at once, or in sequence to determine the engine operation and performance. Some computerized analyzers provide a printed record of the test results for use by mechanics and customer information to authenticate required service.

An *engine analyzer* is a group of different testing instruments mounted in a roll-around cabinet, *Fig. 9.58*. The automotive engine technician commonly uses this to isolate source of engine performance problem. As engine technician in a modern auto repair shop should know the proper use of an engine analyzer. An engine analyzer usually contains:

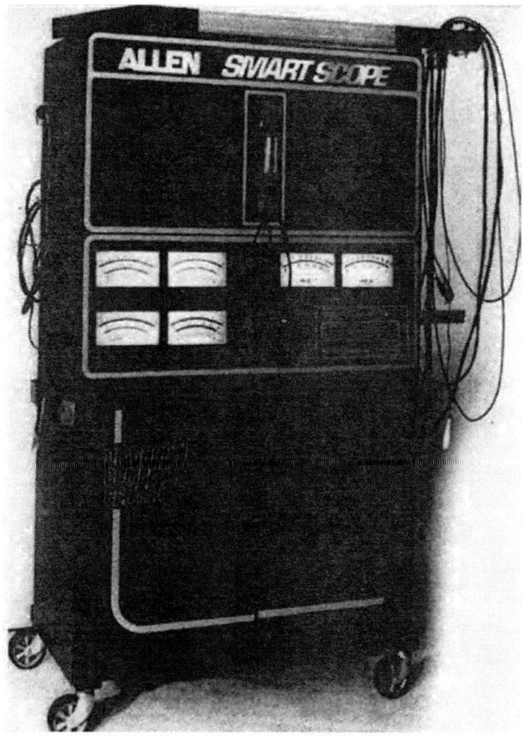

Fig. 9.58 *Engine analyzer is also used when finding engine performance problems. Scope will measure instantaneous voltages in ignition system. Their conventional testers are also included. (Allen)*

1. *Oscilloscope:* TV tube type voltmeter for checking operating condition of ignition and other systems. To obtain information about the overall operating condition of the ignition system circuits, and about events occurring inside the engine.
2. *Tach-dwell:* Digital or analog meter for checking engine rpm, the ignition, and computer system.
3. *Exhaust gas analyzer:* Instrument for measuring chemical content in engine exhaust.

4. *VOM*- for measuring electrical values.
5. *Compression gauge:* Pressure gauge for checking engine compression stroke pressure and engine condition diagnosis.
6. *Vacuum gauge:* Negative pressure gauge for diagnosing engine condition and various vacuum operated devices.
7. *Cylinder balance tester:* Mechanism for shorting cylinders electronically for determining "dead" or "weak" cylinders.
8. *Cranking current balance tester:* Device for checking engine compression by monitoring amount of current required to crank each cylinder through compression stroke: Low compression would require less current.
9. *Timing light:* Strobe light for checking and adjusting ignition timing.

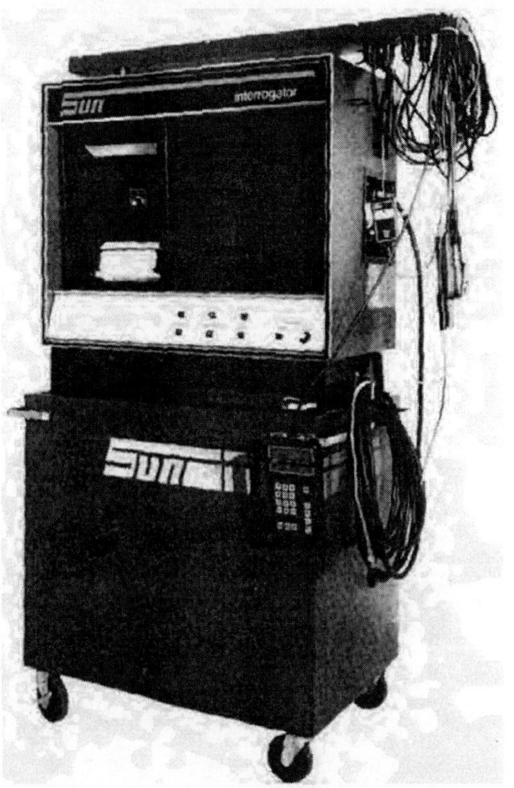

Fig. 9.59 *The computerized tester tests the engine and electrical system, and then compares the readings against the manufacturer's recommended specs. (Hamilton Test Systems)*

Chassis Dynamometer

The chassis dynamometer can be used to test the engine power output under various operating conditions. It can simulate any kind of road test at any load or speed desired. The dynamometer has two heavy rollers mounted at about floor level. The vehicle is driven onto these rollers, so that vehicle drive wheels can drive the dynamometer rollers. Then, the engine is started, and the transmission is put into gear. The vehicle is then operated as though it were out on a real road test.

A power absorber is installed under the floor rollers. This allows the mechanic to test the engine under different operating conditions. The test instruments, such as the oscilloscope, dwell-tach meter, and vacuum gauge are hooked to the engine. These instruments then show how the engine is performing under various operating simulated conditions.

9.46 TUNE-UP AND VEHICLE CARE

A nicely executed tune-up procedure restores driveability, power, performance, and economy that have been lost through wear, corrosion, and deterioration of engine parts during normal vehicle operation. To comply by the federal and state laws limiting automotive emissions, the tune-up procedure should include checking of all emission controls. A general tune-up procedure is given here.

1. Start and operate the engine for some time such that it reaches normal operating temperature. Carefully, note any operational problems during the warm-up time.
2. Connect the oscilloscope and exhaust gas analyzer and perform a diagnosis and check for an abnormal operating condition and, if possible, the cylinder(s) in which it is noted. Work to eliminate any abnormal conditions that have appeared.
3. Remove all spark plugs. Fully open the throttle and choke valves. Disconnect the lead from the coil primary terminal to distributor to prevent excessive secondary voltage.
4. Check the compression pressure of each cylinder and record the readings and compare to specs. If one or more cylinders read low, squirt

about a 15 cc of engine oil through the spark-plug hole into the cylinder. Again check the cylinder compression pressure. Record the new reading and compare to specs. For further diagnosis of engine mechanical problems, make a cylinder leakage test. If engine mechanical problems are found, tell the customer that the engine cannot be tuned without engine repair.

5. Clean, check, file, set gap, and test the spark plugs. Discard worn or defective spark plugs. Frequently, mechanics choose to replace old plugs with new plugs instead of servicing the old plugs. However, the spark plug gap has to be set even for new plugs. Then install the plugs.
6. Check and clean the battery case, terminals, cables, and hold-down brackets. Test the battery and add water if needed. If badly corroded, clean the battery and cables with brushes and a solution of baking soda and water. If it is badly damaged replace it.
7. Test the cranking voltage. If the battery is in good condition but cranking speed is low, the cause can be the battery or the starting system. Test the starting system for its proper function.
8. If the battery voltage is low or keeps running down, check the charging system (alternator and regulator) for proper function. If the battery is old, it might have worn out and should be replaced.
9. Check the drive belts for their condition and tension. Replace any belts that are in poor condition. If it is necessary to replace any belt then replace all drive belts together, if so equipped. Tighten the belts to correct tension, using a belt-tension gauge.
10. Check the distributor rotor, cap, and primary and high voltage spark-plug leads (wires). Replace any defective part.
11. Clean or replace if worn and adjust distributor contact points by setting the point gap. Clean and lubricate the distributor breaker cam. On distributors with round cam lubricators, turn the cam lubricator 180° or replace it if necessary.
12. Check the centrifugal and vacuum advance mechanisms. Set the dwell and then adjust ignition timing to specs. Make sure the idle speed adjustment is not too high. This could cause centrifugal advance during timing adjustment.
13. Use the oscilloscope to recheck the ignition system for proper function.
14. Check the manifold heat-control valve. Lubricate with heat-valve lubricant. If found stuck, free the valve, or replace if necessary.
15. Check the fuel pump operation with a fuel-pump tester. Replace the fuel filter. Check the fuel-tank cap, fuel lines, and connections for leakage and damage. Replace any damaged parts.
16. Clean or replace the air-cleaner filter element. If the engine is equipped with a thermostatic air cleaner, check the operation of the vacuum motor.
17. Check the operation of the choke and the fast-idle cam. Check the throttle valve for full opening, and throttle linkage for free movement. Clean all external carburetor linkage.
18. Check all engine vacuum fittings, hoses, and connections. Replace any brittle or cracked hose.
19. Clean the engine oil filler cap if a filter-type oil filler cap is used.
20. Check the cooling system hoses and connections, the radiator, water pump, and fan clutch if so equipped. Check the strength of the antifreeze and record the reading and compare to specs and correct the strength if needed. Pressure-check the system and radiator cap. Squeeze the hoses to check their condition. Replace any collapsed, soft, cracked or damaged hose in any way.
21. Check and replace the PCV valve if needed. Clean or replace the PCV-system air filter if required. Check the PCV system hoses and connections. Replace any cracked, brittle or damaged, in any manner, hose. Test the system for vacuum.
22. If the engine is equipped with an air pump, replace the pump inlet air filter if used. Check the system hoses and connections. Replace any defective hose.
23. Replace the charcoal-canister air filter, if used, in the evaporative emission control system.
24. Check and clean the Exhaust Gas Recirculation (EGR) valve. Inspect and clean the EGR port. Test the system operation by applying vacuum

to the valve with engine at idle. Engine should run rough, then smooth out with no vacuum applied.

25. Check the transmission-controlled spark system if the vehicle is so equipped.
26. If the engine is equipped with a turbo-charger, check the operation of the waste-gate. Follow the procedure detailed in the manufacture's service manual.
27. Check and tighten the intake-manifold and exhaust-manifold fasteners, if needed to the specified torque in the proper sequence.
28. Check and adjust the valves if necessary.
29. Check and adjust the engine idle speed to specs, if required. Use an exhaust-gas analyzer, and propane if required. Check the amounts of HC and CO in the exhaust gas.

    Checking HC and CO both before and after an engine tune-up shows how much the tune-up has reduced these pollutants.
30. Test the vehicle on a dynamometer or on the road. Check for driveability, power, and idling. Any abnormal condition should be noted on the repair order before handing over the vehicle to the customer.
31. Check the lubrication or maintenance sticker to determine whether an oil-filter change is due. Recommend oil change if it is due. Vehicle manufacturers recommend changing the oil filter every other time the oil is changed.
32. When the vehicle is raised, check the exhaust system for leaks, which could admit carbon monoxide into the passenger compartment. Also check for loose fasteners, rust spots, and other under-the-vehicle damage.

The above list for tune-up and maintenance covers most conditions on an automotive engine that could cause problem. The complete procedure will reveal most problems that might affect driveability and performance. If all required corrections are made, the vehicle engine performance and reliability will be improved.

THEORY QUESTIONS

1. Describe four visual checks that should be performed on an engine if a mechanical malfunction is suspected.
2. List three simple items that could cause excessive oil consumption.
3. List three simple items that could cause engine noises.
4. Explain what could be wrong with an engine if the oil analysis report prepared by a testing laboratory has the following information recorded:

Silicon	= 35 PPM
Copper	= 57 PPM
Tin	= 1 PPM
Chrome	= 8 PPM
Aluminium	= 31 PPM
Iron	= 303 PPM
Fuel dilution	= 0
Antifreeze	= 0
Moisture content	= 0

5. Describe the procedure followed to perform a compression test and how to determine what is wrong with an engine, based on a compression test.
6. Most auto manufacturers recommend a wet compression test on diesel engines. *True* or *False?*
7. Describe the procedure followed to make a cylinder leakage test. Also explain its significance.
8. Technical A says that the paper test could detects burned valve. Technician B says that a grayish white stain could be caused by a coolant leak. Who is right?

   (*a*) A only (*b*) B only
   (*c*) Both A and B (*d*) Neither A nor B.
9. Explain how to safely ground out one cylinder at a time without damage to the electronics of the vehicle.
10. Describe how a vacuum gauge would indicate if the valves were sticking in their guides,
11. Describe the test procedure for determining if the exhaust system is restricted or clogged using a vacuum gauge.
12. An engine tune-up is a _________ operation that returns the engine to _________ after part _________ and deterioration.

13. Explain eight tune-up safety rules.

14. An engine tune-up is important to :
    (*a*) Engine power
    (*b*) Fuel economy
    (*c*) Air pollution
    (*d*) All of the above
    (*e*) None of the other.

15. Explain the difference between a minor tune-up and a major tune-up.

16. List problems that cause low engine compression.

17. When adjusting valve lash, the engine's exhaust valve normally requires _________ clearance that the intake valve.
    (*a*) similar (*b*) less
    (*c*) more
    (*d*) Any of the above, depending on the engine.

18. Technician A says that an oil leak can be found using a fluorescent dye in the oil with a black light to check for leaks. Technician B says that a white spray powder can be used to locate oil leaks. Who is right?
    (*a*) A only (*b*) B only
    (*c*) Both A and B (*d*) Neither A nor B.

19. Antifreeze in the engine oil can cause ———
    (*a*) the oil to be thinner (decrease viscosity)
    (*b*) the oil to become thicker (increase viscosity)
    (*c*) The oil to congeal
    (*d*) Both (*b*) and (*c*).

20. An increase in the engine oil viscosity can be due to _________
    (*a*) Wear metals in the oil
    (*b*) Fuel dilution of the oil
    (*c*) A clogged oil filter
    (*d*) All of the above.

21. A smoothly operating engine depends upon _________.
    (*a*) Equal compression between cylinders
    (*b*) High compression ratio on most cylinders
    (*c*) Cylinder compression levels above 700 kPa and within 500 kPa of each other
    (*d*) Compression levels below 700 kPa on most cylinders.

22. Technician A says that if one spark plug wire is grounded out and the engine speed does not drop, a weak or dead cylinder is indicated. Technician B says that during a power balance test, the cylinder that causes maximum rpm drop is the weak cylinder. Who is right?
    (*a*) A only. (*b*) B only
    (*c*) Both A and B (*d*) Neither A nor B.

23. A good reading for a cylinder leak test would be:
    (*a*) Within 5% between cylinders
    (*b*) All cylinders above 70% leakage and within 7% of each other
    (*c*) All cylinders below 9% leakage
    (*d*) All cylinders above 9% leakage.

24. Cranking vacuum should be _________
    (*a*) 63.5 mm of Hg or higher
    (*b*) Over 635 mm Hg
    (*c*) 432 to 533 mm of Hg
    (*d*) 152 to 406 mm of Hg.

25. Mechanic A says that if the timing chain slack is over 8°, the timing chain and gears should be replaced. Mechanic B says that a worn or stretched timing chain and worn gears will cause the valve and ignition timing to be retarded. Who is right?
    (*a*) A only (*b*) B only
    (*c*) Both A and B (*d*) Neither A nor B.

26. A low oil pressure warning light usually comes on _________
    (*a*) Whenever an oil change is needed
    (*b*) Whenever oil pressure drops dangerously low
    (*c*) Whenever the oil filter antidrain back valve opens
    (*d*) Whenever the oil filter bypass valve opens.

27. A cylinder leakage test shows a 30% pressure drop in one cylinder. A hissing sound can be heard from the vehicle's exhaust system tailpipe. Mechanic A says that an exhaust valve could be too tight or burned. Mechanic B says that an intake valve could be burned or damaged. Who is right?
    (*a*) A only (*b*) B only
    (*c*) Both A and B (*d*) Neither A nor B.

28. A cylinder balance test shows that one cylinder is not causing an rpm drop. Mechanic A says that there could be a fouled spark plug or bad spark plug. Mechanic B says that there could be pressure leakage and a mechanical problem. Who is right?
    (*a*) A only (*b*) B only
    (*c*) Both A and B (*d*) Neither A nor B.

29. Mechanic A says that gasoline engine compression readings should be approximately 125–175 psi. Mechanic B says that gasoline engine compression readings should be approximately 275–400 psi. Who is right?

(*a*) A only (*b*) B only
(*c*) Both A and B (*d*) Neither A nor B.

30. Low engine compression can be caused by all of the following *except:*
(*a*) Faulty spark plugs
(*b*) A burned valve seat
(*c*) A blown head gasket
(*d*) Worn rings.

31. Mechanic A says that a minor tune-up is performed when engine is generally running in good condition. Mechanic B says that a minor tune-up is performed the engine is in poor running condition. Who is right?
(*a*) A only (*b*) B only
(c) Both A and B (*d*) Neither A nor B.

32. Mechanic A says that a tune-up can affect fuel economy. Mechanic B says that a tune-up can affect the service life of an engine. Who is right?
(*a*) A only (*b*) B only
(*c*) Both A and B (*d*) Neither A nor B.

33. Mechanic A says that during a minor engine tune-up, the ignition timing should be checked and adjusted if required. Mechanic B says that during a minor engine tune-up the emission control system should be tested and serviced if needed. Who is right?
(*a*) A only (*b*) B only
(*c*) Both A and B (*d*) Neither A nor B.

34. A cylinder, leak test shows 30% pressure drop in each cylinder ________
(*a*) an engine mechanical problem exists
(*b*) The leakage gauge is out of order
(*c*) The engine is in fair condition
(*d*) A sensor problem exists.

35. Mechanic A says that fuel leaks should be checked during a major tune-up. Mechanic B says that oil leaks should be checked during a major engine tune-up. Who is right?
(*a*) A only (*b*) B only
(*c*) Both A and B (*d*) Neither A nor B.

36. Mechanic A says that an engine analyzer can be used to perform a cylinder balance test. Mechanic B says that a tachometer can be used to perform a cylinder balance test. Who is right?
(*a*) A only (*b*) B only
(*c*) Both A and B (*d*) Neither A nor B.

❑❑❑

Engine Performance and Mechanical Problems

10

10.1 ENGINE PERFORMANCE PROBLEMS

An engine performance problem is any problem that lowers the engine power, fuel economy, drivability, or dependability. Performance problems can be caused by troubles in the engine fuel supply system, ignition system, emission control systems, cooling system, or the engine itself. This makes troubleshooting very challenging. An automotive technician must use his/her knowledge of system operation and basic testing techniques to quickly and accurately pinpoint and correct the source of a problem.

10.2 LOCATING ENGINE PERFORMANCE PROBLEMS

Technicians should use a systematic approach when trying to locate engine performance problems during service. Hit and miss repairs should not be used. A *systematic approach* involves using up-to-date knowledge of automotive technology and a logical process of elimination. Mentally overview all of the possible systems and components that can cause the engine problem that is under investigation and repair. Then, one by one eliminate the parts and systems that could not produce the problem symptoms.

To correctly troubleshoot the problem, ask yourself the following kinds of questions:

1. What are symptoms of the trouble, e.g., noise, smoke, misfire, etc.?
2. What system or component(s) could be producing the symptoms, e.g., cooling, fuel, ignition, engine, etc.?
3. Where to start testing?

A logical thought analysis approach will help you to isolate the source of the problem by checking only the most likely trouble source. For example, the owner of a vehicle complains of the engine, with electronic gasoline fuel injection, missing and emitting black smoke after running for a few minutes. The emission of black smoke suggests that too much fuel is entering the engine cylinders. Since the engine does not run poorly when cold, engine temperature has a role in the problem. With adequate knowledge of the electronic fuel injection system the technician might think of the cold start injector. It is used to richen the fuel mixture and is affected by temperature. Probably, the cold start injector is staying on too long, causing an excessively rich air-fuel mixture to enter the engine cylinders. The thermostatic switch that operates the injector could be shorted to the on position.

If your first conclusion is found incorrect, rethink the problem and the next most likely trouble source. Thus, by using logical processes and close inspection, you can often find the sources of many engine performance problems without resorting to complex testing procedures. Even with computer controlled systems, loose wires, vacuum leaks, arcing sparking plug wires, engine overheating, and clogged filters are still common.

10.3 PERFORMANCE PROBLEM DIAGNOSIS CHARTS

To facilitate locating an engine performance problem source, refer to the applicable manufacturer's service manual diagnosis chart. It lists problems, causes, and remedies. These charts are written for the particular make and model of a vehicle. *Fig. 10.1* shows an engine performance trouble-shooting chart from a manufacturer's service manual.

10.4 TYPICAL PERFORMANCE PROBLEMS

An automotive repair shop technician must understand the most common engine performance problems. The following sections will help the technician to use the test equipment, troubleshooting charts, and a service manual during engine trouble diagnosis.

No Start Problem

A *no-start problem* occurs when the engine "cranks" or its crankshaft is turned by the engine starter motor but fails to "fire" and start to run on its own power. This is the most obvious and serious performance problem.

In case of a no-start problem for a gasoline engine, ***check for spark,*** first. Pull one spark plug wire and install a spark tester on the wire end. Attach it to ground. When the engine is cranked, a spark should jump the gap, *Fig. 10.2 (b).* If not, the ignition system has the problem.

If there is a spark, ***check for fuel.*** With a carbureted fuel system, operate the throttle lever. Watch for fuel squirting out of the accelerator pump discharge port. With throttle body fuel injection, watch the fuel injector outlet. With multi-port fuel injection, install a pressure gauge on the fuel rail. If no fuel is there, then the problem is with the fuel system, *Fig. 10.2 (a).*

If the fuel injection system has fuel pressure, check the injector harness for the presence of a pulse. Pull a harness connector loose and install an injector "noid light" in the harness connector. The noid light should flash when the engine is cranked. If the light does not flash, there is a problem with the computer, input sensor, or electrical system.

If both spark and fuel are good, check engine compression. A jumped timing chain or belt could be the cause of no-start problem. For a diesel engine, a slow cranking speed can prevent starting.

Hard Starting

Engine *hard starting* takes excessive cranking time to start and is usually due to partial failure of a system. In carbureted engines carburetor choke may be inoperative. In fuel injection engines the fuel injection coolant temperature sensor may be defective.

Engine Stalling (Dying)

Engine stalling is a condition where the engine stops

Problem	*Possible cause*	*Remedy*	*Page*
Rough idle, stalls or	EGR valve faulty	Check valve	3-59
misses (cont'd)	Valve clearance incorrect	Adjust valves	2-7
	Compression low	Check compression	4-2
Engine hesitates	Spark plug faulty	Inspect plugs	3-6
Poor acceleration	High tension wires faulty	Inspect wires	3-6
	Vacuum leaks	Repair as necessary	
	• PCV hoses		
	• EGR hoses		
	• Intake manifold		
	• Carburettor hoses		
	• Air cleaner hoses		
	Incorrect ignition timing	Reset timing	2-26
	Air cleaner clogged	Check air cleaner	2-22
	Fuel system clogged	Check fuel system	5-2

Fig. 10.1 *This is a partial page from a service manual. Note how troubleshooting chart gives problem, possible cause, remedy, and page number of service manual coverage. (Toyota)*

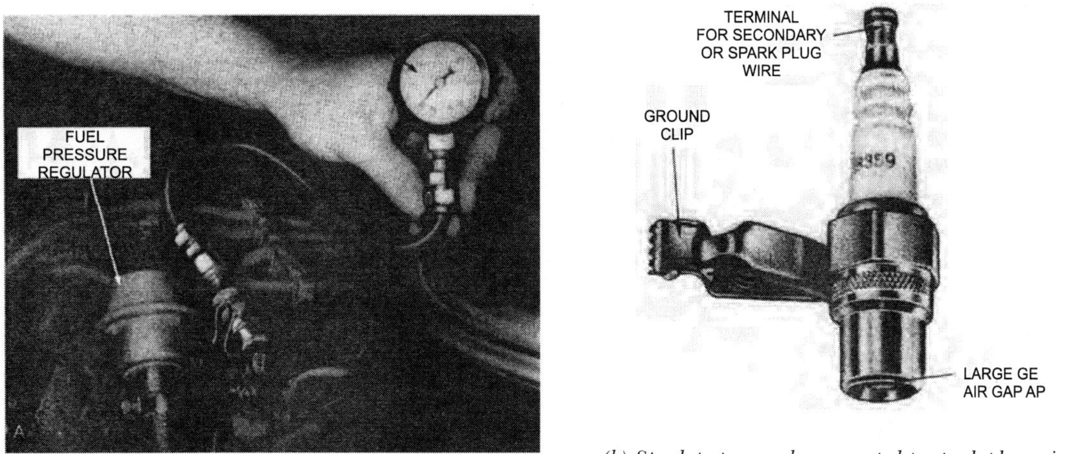

(a) *With multiport fuel injection, a pressure gauge is an efficient way of checking the fuel system.*

(b) *Spark tester can be connected to spark plug wire. When engine is cranked, there should be a strong spark. Holding wire next to ground can also do this.*

Fig. 10.2 *When an engine fails to start, remember to check for spark and for fuel. (OTC and Kent-Moore)*

running. This may occur at idle, after cold starting, or after warm-up. This could be caused by a number of factors: low idle speed, carburetor or injection system trouble, ignition system problem, severe vacuum leak, or inoperative thermostatic air cleaner (if so equipped).

Engine Misfire

An *engine misfire* results when one or more cylinders are failing to fire or cause normal combustion. The engine operation becomes rough. The engine may miss at idle, under acceleration, or at cruising speed.

If the engine only misses at idle speed, then check the components that affect engine's idle operation. For carbureted engines, the idle circuit may be clogged with contaminants. In case of engines with fuel injection, possibly an injector is not functioning properly. A fouled spark plug, open plug wire, cracked distributor cap, corroded terminals, and vacuum leak are a few other possible causes for an engine misfire.

Vacuum Leak

A *vacuum leak* is a common cause of rough idling. A vacuum hose hardens with time and may crack, to allow outside air to enter the engine intake manifold, bypassing the carburetor, airflow sensor, or throttle body. This will cause an incorrect air-fuel mixture, preventing normal combustion, *Fig. 10.3.* Generally, a vacuum leak will produce a *hissing sound.* The engine roughness will smooth out with increase in engine rpm.

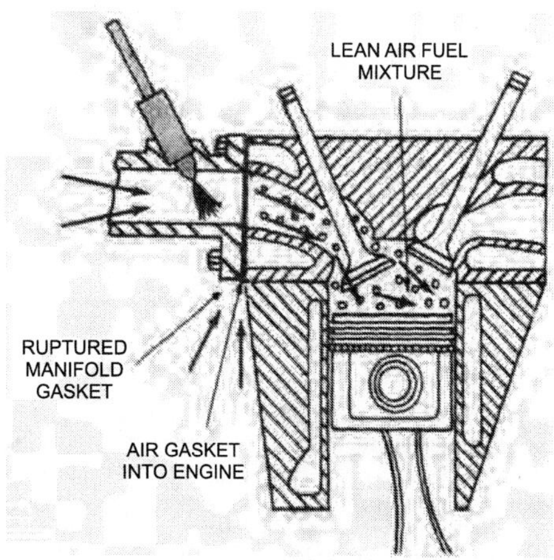

Fig. 10.3 *Vacuum leak allows extra air to enter engine after throttle valves or airflow sensor. This will usually lean mixture and cause a rough idle.*

A section of vacuum hose can be used to locate leaks. Keep one end of the hose next to your ear. Move the other end around the engine. When the hiss becomes very loud, it locates the leak.

Hesitation or Stumble

A *hesitation,* also known as *stumble,* is an operating condition where the engine does not accelerate normally when the accelerator pedal is pressed. The engine may even stall before developing necessary power.

A temporary lean fuel mixture usually causes a hesitation. With a carburetor, accelerator pump may not be functioning. With fuel injection, the throttle position switch may be defective. Check all the parts that aid engine acceleration.

Surging

Surging is a condition where engine power fluctuates up and down. When driving at a steady speed, the engine feels to speed up and slow down, without movement of the accelerator pedal.

An extremely lean carburetor or fuel injection setting occasionally causes surging. Surging can also be caused by ignition or computer control system problems.

Backfiring

The air-fuel mixture igniting in the induction manifold or in the exhaust system causes *backfiring.* A loud *bang* or *POP* sound can be heard when the mixture ignites in this manner.

Backfiring can be caused by incorrect ignition timing, crossed spark plug wires, cracked distributor cap, defective carburetor accelerator pump, exhaust system leakage, faulty air injection system, or other system fault.

Dieseling, After Running or Run-On

Dieseling, also known as *after-running* or *run-on,* occurs when the engine fails to shut off after turning the ignition key to off position. The engine keeps firing, coughing, and producing power. Dieseling is usually caused by a high idle speed, carbon buildup in the combustion chambers, low octane rating of fuel, or overheated engine.

Pinging or Spark Knock

Pinging or *spark knock* is a metallic tapping or light knocking sound, usually when the engine accelerates under load. Pinging is caused by abnormal combustion, i.e., preignition or detonation. Low octane fuel, advanced ignition timing, carbon buildup in combustion chambers, or engine overheating, normally causes pinging.

Vapour Lock

Vapour lock occurs when the fuel gets overheated, by too much engine heat transferring into the fuel, generating air bubbles that upset the air-fuel mixture. Vapour lock may cause engine stalling, lack of power, hard starting, and no start condition.

Fuel Line Freeze

Fuel line freeze is caused when moisture in the gasoline turns into ice. The ice will block fuel filters and prevent engine operation. With diesel fuel, the overcooled fuel can form wax that blocks the filters. To correct the fuel line freeze, the vehicle might need to be placed in a warm garage until the fuel is adequately thawed.

Poor Fuel Economy

Poor fuel economy is a condition causing the engine to consume excess fuel for the kilometers the vehicle is driven compared to the specified amount needed for the normal engine condition. Fuel economy can be measured by comparing the kilometers that the vehicle can rum on one litre of the fuel.

Poor fuel economy can be caused by a number of problems: rich air-fuel mixture, engine misfire, incorrect ignition timing, or leakage.

Lack of Engine Power

Lack of engine power, also called *sluggish engine,* causes the vehicle to accelerate slowly. When the accelerator pedal is pressed, the vehicle does not respond and does not gain speed properly. The engine "hits" on all cylinders and runs smoothly but does not develop its normal power.

A number of engine problems can reduce the engine power: fuel system problems, ignition system problems, emission control system problems, and engine mechanical problems.

10.5 OTHER PERFORMANCE PROBLEMS

There are many other more specific engine performance problems e.g., engine flooding, carburetor icing, leaking fuel injector, incorrect ignition timing. Many of these problems are covered in other chapters in volume-I of Automobile Engineering dealing with automotive engine theory.

10.6 ENGINE MECHANICAL PROBLEMS

The following sections will summarize the causes, symptoms, and consequences of mechanical part failures in an automotive engine. An automotive engine technician should have the technical knowledge, and skills to troubleshoot the engine problem before actual engine teardown for carrying out the repairs. The technician must be able to quickly and accurately locate mechanical problems and correct their source to restore engine life and performance. Some mechanical problems and troubles like burned valves, blown head gaskets, and other similar mechanical problems are briefly discussed in the following sections.

10.7 VALVE TRAIN MECHANICAL PROBLEMS

Engine valve train is a common source of mechanical problems in the engine. As already discussed, friction between the valve train parts is high and majority of the parts are splash lubricated. Wear can occur between mating parts with extended service or due to lack of adequate maintenance.

Valve train mechanical problems can cause engine noise, power loss, misfiring, rough idling, and possible damage to piston, cylinder and cylinder head.

10.8 CAMSHAFT PROBLEMS

Camshaft mechanical problems include cam lobe wear, cam journal wear, cam damage, camshaft drive gear failure, etc.

Cam lobe wear is a condition in which friction from the valve lifter has worn off some of the cam lobe surface. This is usually found in an engine with high kilometers reading, or one that has not had its oil changed regularly, as recommended, *Fig. 10.4*.

With excessive cam lobe wear, problems like valve train clatter, reduced valve lift, and consequent poor engine performance can result. Rocker arm motion is measured with a dial indicator to find a worn cam lobe.

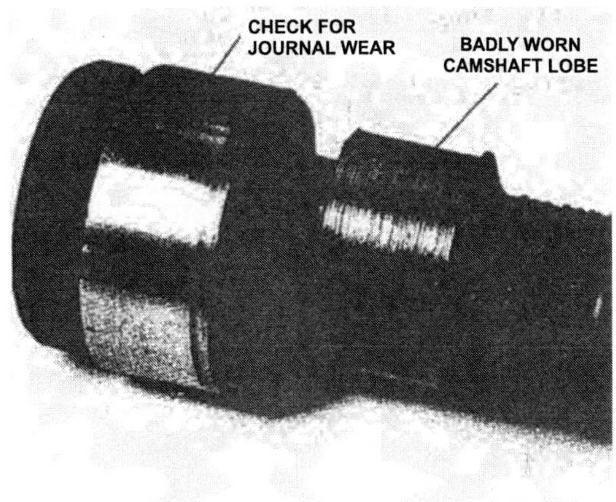

Fig. 10.4 *This camshaft had badly worn cam lobes. This is due to extended service and lack of oil changes. (Midwest Crankshaft)*

Cam journal wear and *cam bearing wear* results in increased part clearance and can translate into reduced engine oil pressure due to increased bearing leakage. This too, normally happens after extended service.

Cam thrust surface wear allows the cam to move too far forward and rearward in the engine along the camshaft axis. This can often produce a light knocking sound, timing chain or timing gear troubles, and increased valve lifter and cam lobe wear.

Camshaft breakage, which is not common, will stop some of the valves from operating, because the lobes on one end of the broken cam (away from the drive gear) will not rotate and open their valves. Severe performance or valve damage can be caused. Remove the valve cover to check valve train and camshaft action.

10.9 VALVE LIFTER OR TAPPET PROBLEMS

Valve lifter or *tappet problems* can cause valve train noise or a light tapping sound emanating from inside the valve cover, *Fig. 10.5.*

External lifter wear has the bottom of the lifter concaved (sunk in middle) caused by the cam lobe friction. A new lifter is actually convex (crowned) on the bottom to make it rotate during engine operation. Long service life can wear off this crown and speed cam lobe wear. This contributes to increase valve train clearance. Engine disassembly is required to measure external lifter wear.

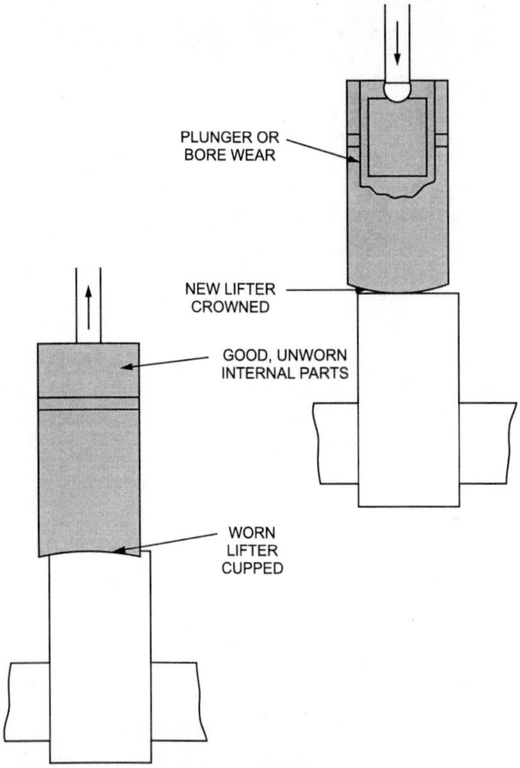

(a) Outside wear occurs where lifter rides on cam lobe. Inside wear can cause too much leak down and valve train clatter.

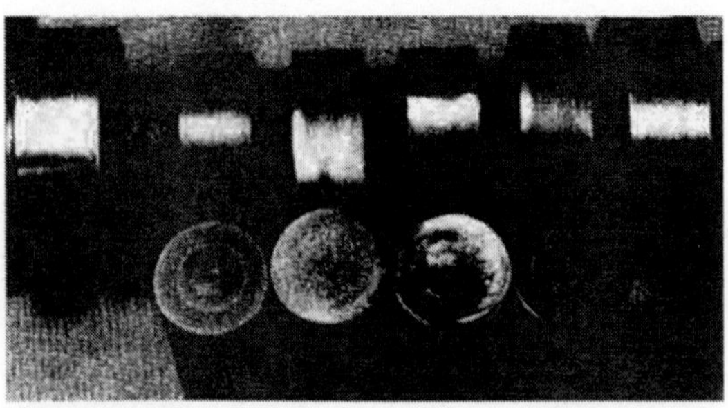

(b) Actual worn lifters.

Fig 10.5 *Lifters can wear externally and internally. (Dana Corp.)*

Internal lifter wear, also known as *bad lifter,* can prevent the hydraulic lifter from pumping up in order to take up the valve train clearance, *Fig. 10.5.* This can cause the rocker arm to clatter when it strikes the valve stem tip. This noise will be towards the top of the engine, as if a small ball peen hammer were hitting the engine.

With the engine idling and the valve cover removed, pushing down on the push rod end of the rocker arm to increase noise and facilitate isolating the bad lifter.

Low engine oil pressure can cause hydraulic lifter clatter. Therefore, it is necessary to check the oil level and oil pressure before replacing the lifter(s). Contaminated or dirty oil can also cause lifter noise.

10.10 PUSH ROD PROBLEMS

A struck valve, spring bind, spring breakage, or other types of mechanical problems cause a *bent push rod.* When severely bent, the push rod can fall out of the rocker arm pocket. This will stop the valve action and severe misfiring and tapping will be caused. Do not straighten a bent push rod rather replace it.

A *broken push rod* often has the small ball on its end broken off, *Fig. 10.6.* Like a bent push rod, a broken push rod will cause loud valve train noise and usually the push rod disengagement from the rocker arm.

A *clogged push rod* has the axial oil hole in the centre choked by hardened oil and sludge deposits. This causes fast rocker arm wear due to reduced lubrication. Always check and clean the hole in push rods during an engine overhaul or valve service.

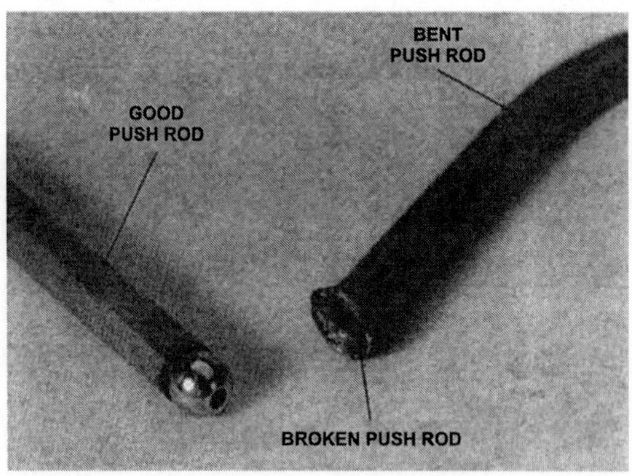

Fig. 10.6 *Compare good and failed push rods. One on right is bent and broken. Failure resulted from driver over revving engine. (Delcotto)*

10.11 ROCKER ARM PROBLEMS

A *worn rocker shaft* is caused by friction with the rocker arm. This can be the result of inadequate lubrication or of long service life. As a consequence the clearance between the rocker arm and shaft will increase, *Fig. 10.7.* This can also cause valve train clatter.

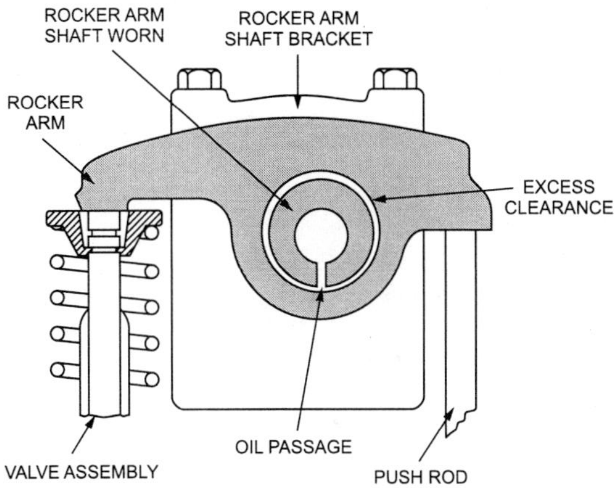

Fig. 10.7 *Worn rocker arm or rocker shaft can extend lifter plunger too far up. This might cause valve train noise. (Perfect Circle)*

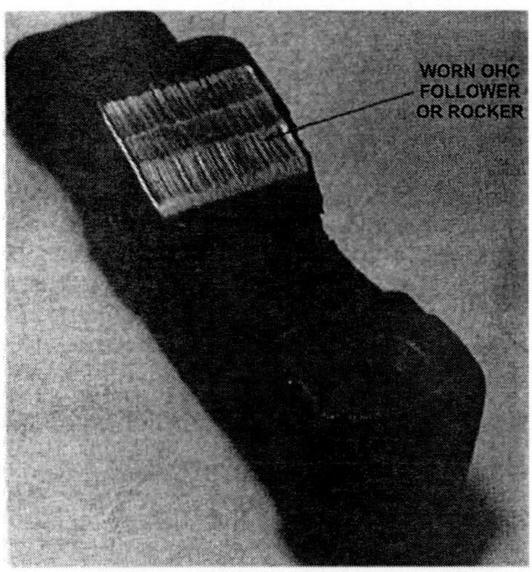

Fig. 10.8 *A worn, scored overhead cam rocker arm, caused by lack of lubrication from deteriorated oil. (Midwest Crankshaft)*

A *worn rocker arm* can wear on its inside diameter or where it contacts the valve or push rod. This will allow the lifter plunger to move to far out and cause a light tapping noise. In case of an overhead cam engine, the rocker arm can wear at its contact point with the camshaft, *Fig. 10.8.* Careful inspection will reveal rocker wear problems.

A *worn rocker stand* is a common problem because the stands are made of aluminium, which is comparatively soft. The stand will wear where it rubs on the rocker arm. This will also cause valve train noise.

Rocker arm and stand breakage can occur if the valve strikes the piston as a result of over revving or if the valve is stuck in the head after sitting unused. This will prevent valve action and cause a severe engine misfire.

10.12 VALVE PROBLEMS

Intake and exhaust valves open and close millions of times during the life of an engine. They must do so while sealing the tremendous heat of fuel combustion. After extended service, various valve problems can develop. An automotive repair shop technician should be able to diagnose and locate these problems quickly and accurately.

Burned Valve

A ***burned valve*** often has a portion of the valve margin and face melted and burned away. This can cause pressure leak past the valve, reducing power in its cylinder. As a consequence the engine will misfire or run rough, particularly at idle speed, *Fig. 10.9 (a).*

In case of a burned valve the technician may be able to hear a puffing sound as the pressure blows past the valve, a popping or puffing sound at the carburetor or throttle body.

Valve Face Erosion

Valve face erosion is the wearing away of the valve face where it contacts its valve seat in the cylinder head. This problem is more common with unleaded fuels as they do not have lead type fuel additives that reduce high temperature friction at contact surfaces, *Fig. 10.9 (b).*

Valve Stem Problems

Valve stem wear is caused by friction with the cylinder head guide. Because small amount oil enters the guide, it makes the valve stem wear a common problem, *Fig. 10.10.* Severe valve guide or valve guide wear can produce a light knocking or tapping sound. The valve can flip sideways under the rocker arm or valve spring force. To correct the problem, cylinder head removal and valve replacement is required.

Valve stem breakage is a severe failure problem that often occurs when the piston hits the valve head. The valve bends and breaks off the valve head at the stem. This problem is caused by a broken valve spring, popped out spring keepers, etc.

(a) This burned valve is from a high mileage engine. Exhaust gases heat actually melted and burned away a portion of its face and margin. As a consequence the engine lost compression in affected cylinder.

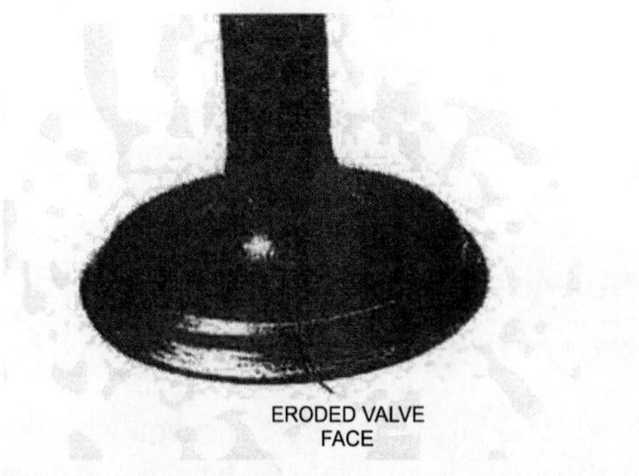

(b) This valve has eroded valve face. The face is dished from riding on seat. This was caused by use of unleaded fuel, when the valve was designed for leaded fuel.

Fig. 10.9 *Burned and eroded valves.*

A broken off valve head will usually be hit by the piston and smashed up into the cylinder head, usually causing piston and cylinder head damage. A *scored valve stem* has been overheated and scarred by friction. The resulting rough surface will accelerate the valve guide wear.

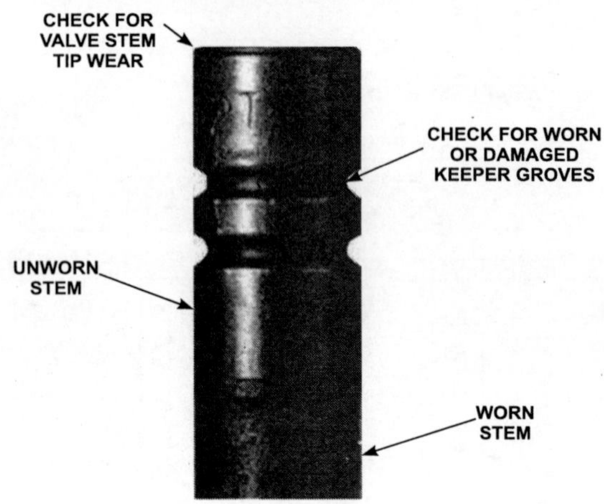

Fig. 10.10 *Valve stem problems are easy to overlook. Always check stem, keeper groove, and tip very carefully.*

Valve Guide Problems

Valve guide wear will produce the same symptoms as produced by valve stem wear. The valve will be free to flop sideways in the valve guide, *Fig. 10.11.* Oil can be drawn down through the increased clearance and burned in the combustion chamber. A tapping noise can also be heard.

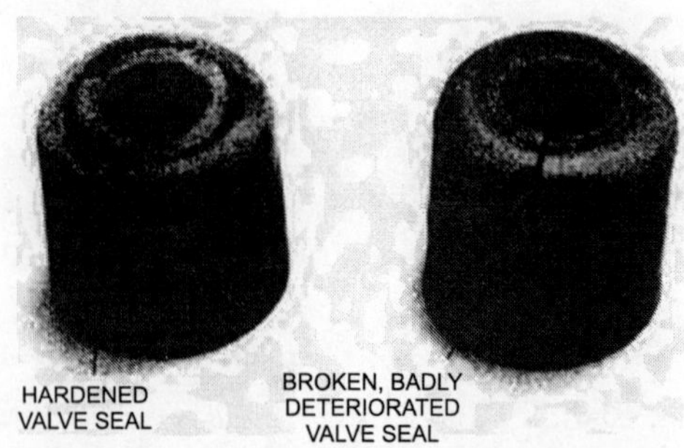

Fig. 10.11 *Bad valve seals or worn valve guides will allow oil flow down into port and be pulled into combustion chamber. Worn guide in stem can also let valve flop sideways in guide and make tapping type noise.*

A *dropped valve guide* is due to the loss of its pressure fit interference in the cylinder head. This allows the guide to fall down around the head of the valve. It can wedge the valve open and cause severe part damage. This problem normally occurs in aluminium cylinder heads due to large differential in coefficients of thermal expansions of valve guide and cylinder head materials.

Bad Valve Seals

Bad valve seals are the valve seals that have been hardened or broken due to extended service and/or engine overheating. These synthetic rubber seals are no longer soft and pliable. With severe failure, the seal can disintegrate and fall down into the oil pan, *Fig. 10.12.*

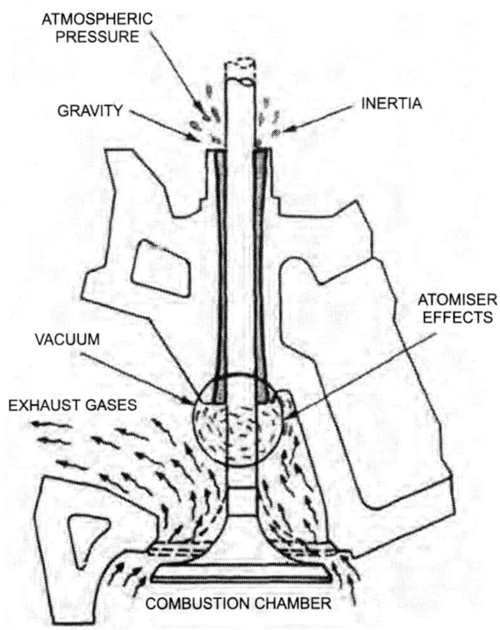

Fig. 10. 12 *Bad valve stem seals will make engine smoke right after startup. Seals will harden after prolonged use. Seal on right, in Fig. 10.11, has cracked and would break up and fall down into oil pan if not replaced. (Pel-Pro Gaskets)*

Bad seals will cause the engine produce blue smoke, particularly after engine startup. The oil drips down into the guides when the engine is at rest. Then when the engine is started, the oil will be drawn into the cylinders and burned.

Valve seats can usually be replaced *without* removing the cylinder head. Air pressure is used to hold the valves up in the cylinder head. Then, a special tool is used to compress the valve spring. The spring keepers, springs, and seals can then be removed for service.

Stuck Valve

A ***stuck valve*** is caused when the valve stem rusts or corrodes and locks in its valve guide. This often occurs when the engine sits in storage for an extended period of time.

A stuck or frozen valve can often be fixed by squirting rust penetrate around the top of the valve guide. If this fails to correct the problem, remove the cylinder head for service.

10.13 VALVE SPRING PROBLEMS

Valve float is a condition where weakened valve springs; hydraulic lifter problems, or excess engine rpm cause the valves to remain partially open. This problem often occurs at higher engine speeds. The engine may begin to misfire, pop, or backfire as the valves float.

Weak valve springs result after extended use. Springs tend to set and loose some of their tension and become too weak to close the valves properly.

A ***broken valve spring*** will often let the valve hang partially open. Excessive valve-to-rocker clearance can cause valve train clatter or light tapping noise. Popping or backfiring can also result. Valve springs can be replaced *without* cylinder head removal. As with valve seal replacement, air pressure and a special tool allow the spring replacement.

An inspection is often required to ascertain a broken valve spring because the cylinder with a broken spring may pass a compression test.

10.14 TIMING MECHANISM PROBLEMS

Timing mechanism problems can affect when and whether the crankshaft rotates properly to open the valves at the right times in the engine cycle. This relates to timing gears, timing chain, or timing belt.

Timing sprocket problems are common when the sprocket has plastic teeth. The plastic teeth can break off and cause the timing chain slack. The timing chain can jump over the broken teeth and upset valve timing.

This can cause severe performance problems and valve damage because the valves can then open at wrong times, *Fig. 10.13 (a)*.

Timing gear problems are not common, but can occur after extended service or as a result of breakage of other parts. The most common is broken gear teeth, *Fig. 10.13 (b)*. Broken timing gear teeth can cause a loud knocking noise emanating from inside the front cover. If multiple teeth break off, the timing can get upset and the valves will open at incorrect times.

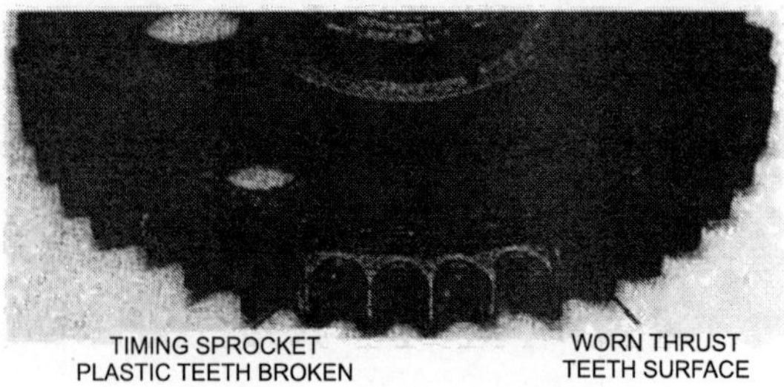

(a) This timing sprocket has broken plastic teeth. Timing chain would have excess slack and could jump over teeth, a throwing off valve timing. A worn chain will reduce compression in all cylinders.

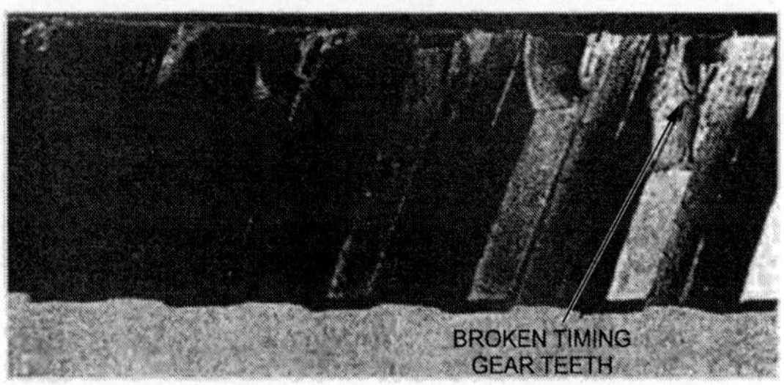

(b) This timing gear has broken teeth. It made clunking type sound with engine running.

Fig. 10.13 *Timing sprocket and gear broken teeth.*

A ***worn timing chain*** can upset valve timing, resulting in lower compression stroke pressure and engine power. Wear can cause excessive slack between the crankshaft sprocket and camshaft sprocket. The cams and valves will no longer be keeping their relative time setting with pistons.

To check for a worn timing chain, manually rotate the crankshaft back and forth and keep watching the distributor rotor or rocker arms. If the crankshaft can be turned several degrees without rotor or valve movement, the timing chain is worn and would need to be replaced.

Incorrect valve timing and a worn timing chain would affect all cylinders equally. A compression test of the engine cylinders would show lower compression pressure in all cylinders. A worn timing chain can cause erratic timing mark motion during engine tune-up.

A ***worn timing belt*** will often break, jump off its sprockets, or skip over a few sprocket teeth. This will cause severe performance problems or valve damage. The pistons can slam into open valves, bending or breaking them.

10.15 ENGINE GASKET PROBLEMS

A ***blowing cylinder gasket*** can be a source of a wide range of problems: overheating, missing, coolant or oil leakage, engine smoking, even head or block burned mating surfaces. Often, a blown head gasket will show up during a compression test. In that case, two adjacent cylinders, usually the two middle ones, will have low pressure, *Figs. 10.14 and 10.15*.

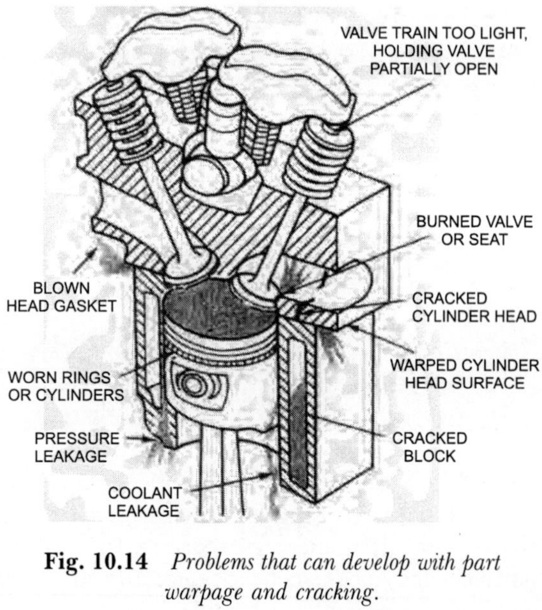

Fig. 10.14 *Problems that can develop with part warpage and cracking.*

A ***leaking induction manifold gasket*** can cause a vacuum leak, resulting in rough engine idling. To check for an induction manifold gasket leak, squirt oil along the edge of the gasket. The oil may temporarily seal the leak, indicating an intake manifold gasket rupture. A low vacuum gauge reading, during a vacuum test, can also indicate an induction manifold leak.

A ***leaking exhaust manifold gasket*** will show up as a clicking type sound. A metallic-like rap is produced when the combustion gases blow into the manifold and out the ruptured gasket.

Part warpage, causing sealing surfaces to be out of their acceptable flatness specifications, is a common cause of gasket failure. Always check for warpage when servicing a damaged gasket, *Fig. 10.15.*

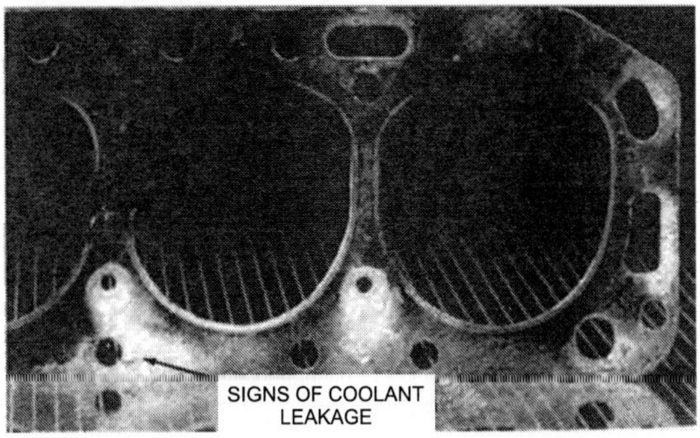

(a) Figure shows a blown head gasket that resulted from warped head. There are evident signs of coolant leakage.

(b) This leaking intake manifold gasket allowed oil to be drawn into cylinders. Engine smoked and consumed oil when running.

Fig. 10.15 *Gasket problems are common.*

10.16 CYLINDER HEAD PROBLEMS

A ***warped cylinder head*** is one that has been overheated and is no longer true. Usually, the deck surface gets warped and causes head gasket leakage. Coolant can leak into the engine oil, producing a milky substance. This is more of a problem with aluminium cylinder heads.

A ***cracked cylinder head*** usually is caused by overheating, where unequal thermal expansion may break the head, near an exhaust port or on the deck between cylinders. Also, coolant can freeze in extreme cold weather and crack a cylinder head, *Fig. 10.16.*

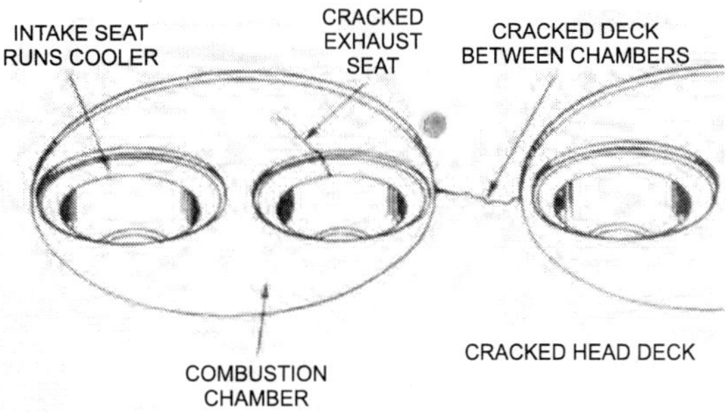

Fig. 10.16 *Cylinder head will often crack at exhaust seat and between combustion chambers on deck surface. This can cause coolant, oil, and pressure leakage.*

A cracked cylinder head can leak coolant or combustion pressure. Coolant leakage may enter the combustion chamber or run down into the oil pan. Pressure leakage will let the combustion gases to enter the cooling system. Symptoms of a cracked head are; coolant on a spark plug, milky oil, white exhaust smoke, and overheating. To correct this problem, the head would have to be removed and welded to repair or replaced.

10.17 CYLINDER BLOCK PROBLEMS

Cylinder block problems include cracks, warpage, worn cylinders, mineral deposits in coolant passages, blocked oil passages, misaligned main bearing bores, etc.

Deck warpage is caused by overheating and can result in head gasket leakage. But, this problem is rare.

Crankcase warpage causes the main bores to be out of alignment and can result in a locked crankshaft in the cylinder block during engine reassembly, during overhaul, with new main bearings. This problem can be corrected by line boring the main bearing bores, *Fig. 10.18.*

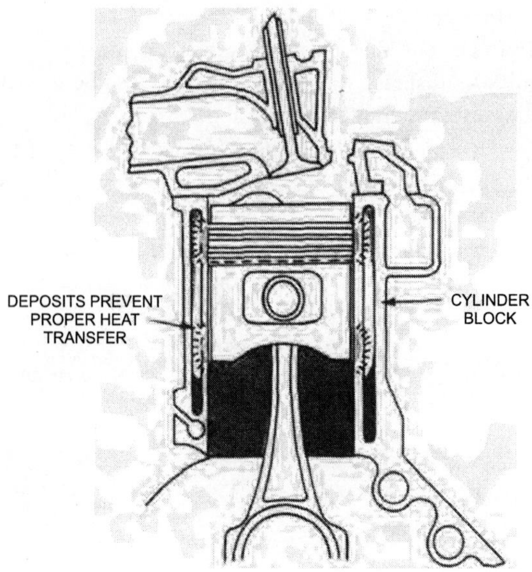

Fig. 10.17 *Mineral deposits in cylinder block can cause overheating. (Deere and Co.)*

A ***cracked block*** can have cracks in cylinder, deck surface, valve lifter gallery, etc. This problem can occur due to overheating, coolant freeze in very cold weather, or breakage of pistons or connecting rods.

Mineral deposits in the water jackets can cause the engine overheating. Lime can build up on the inner water jacket wall surfaces, resulting in restricted coolant passages for coolant flow and cause localized hot spots, *Fig. 10.17.*

Fig. 10.19 *This cylinder head was cracked and broken when connecting rod bolts broke off. Crankshaft smashed connecting rod into side of cylinder wall. To fix the block either use a cylinder sleeve. Ideally, the cylinder block should be replaced.*

A badly damaged cylinder block is shown in *Fig. 10.19.* In this case, the connecting rod broke and knocked a hole in the cylinder. The repair would require machining the bore and installing a sleeve with appropriate bore to salvage the cylinder block.

A ***worn cylinder block*** has worn cylinders from piston ring friction. Due to lesser oil being available to lubricate the top portion of the cylinder as compared to the lower portion of the cylinder, the top

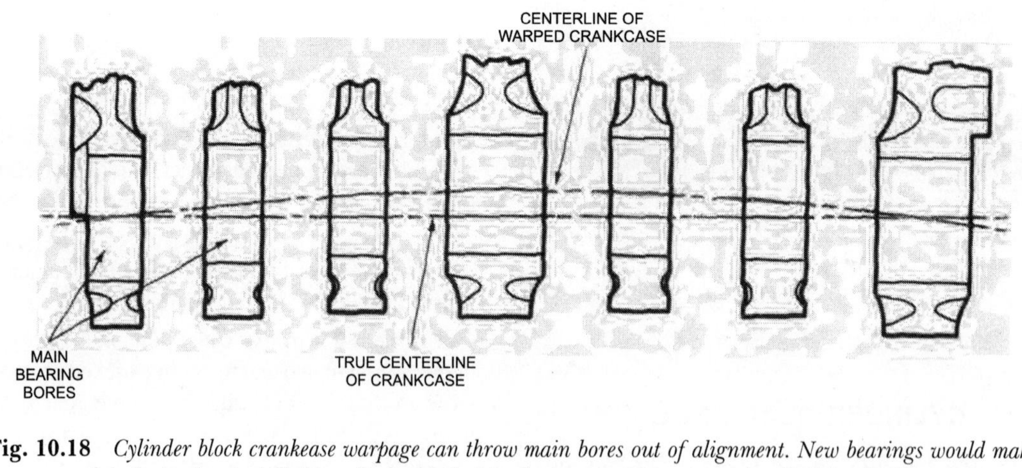

Fig. 10.18 *Cylinder block crankcase warpage can throw main bores out of alignment. New bearings would make crank lock up when installed in cylinder block. Line boring would be required to fix this misalignment. (Federal Mogul Bearings)*

of the cylinder wear more. Also, the piston rings do not rub at the very top of the cylinder, before the TDC, a ring shaped ridge is formed with extended use of the engine.

A ***ring ridge*** is a small lip formed where the rings do not wear the cylinder, *Fig. 10.20.* It quickly shows the cylinder wear, and has to be cut out of the cylinder before piston removal, when disassembling the engine for overhaul. Read the condition of spark plugs and use the information to determine a worn cylinder. The rings will not seal well on a bad cylinder and oil-fouled plugs can result, *Fig. 10.21.*

Engine cylinder problems are major and need engine removal from the engine compartment of the vehicle, and disassembled so that the cylinder block can be salvaged or replaced.

10.18 PISTON PROBLEMS

Piston problems are major in nature and they include skirt breakage, crown failure, worn ring grooves, etc. Piston failure normally results in oil consumption, engine smoking, and sometimes knocking.

Piston breakage usually occurs when piston skirt breaks off. The piston will then flop sideways in the cylinder and cause major cylinder damage. The cylinder eventually loses compression, burn oil, and produce a knocking sound. Piston skirt breakage can be caused by too much piston and cylinder clearance, over revving, or inadequate lubrication, *Fig. 10.22.*

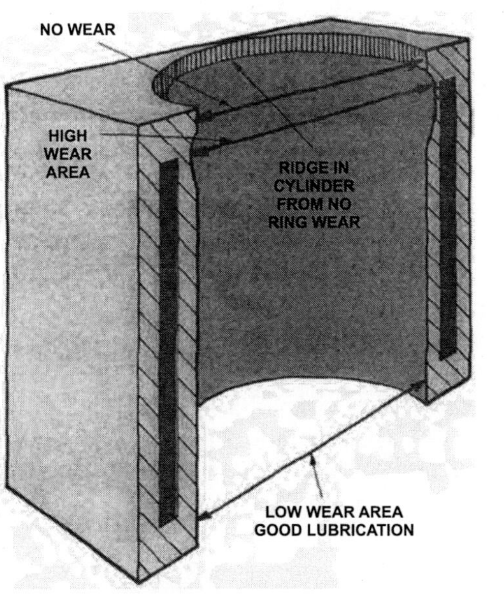

Fig. 10.20 *Cylinder will wear more near top because of less splash oil spray. Ring ridge is formed where rings do not rub on cylinder wall, Ring ridge will quickly indicate cylinder wear.*

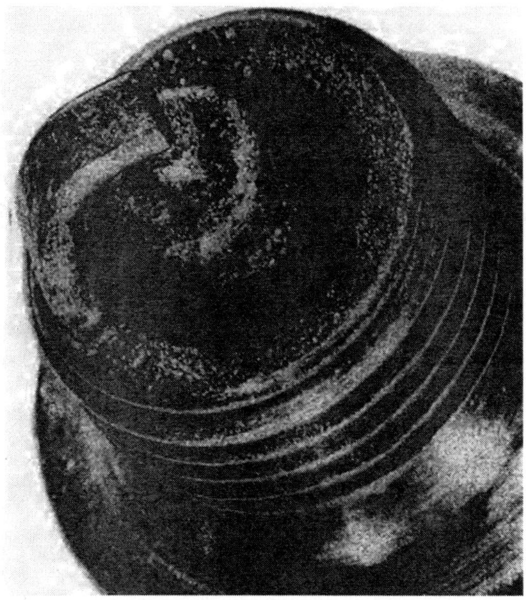

Fig. 10.21 *Oil fouled plug would indicate bad rings, valve seals, turbocharger, etc. Correct reading of plugs is important to engine condition diagnosis. (Champion Spark Plug Co.)*

Fig. 10.22 *Broken piston skirt resulted from too much clearance and low oil level in pan. Notice the scoring of skirt that indicates lubrication problem.*

Piston knock or ***slap*** is a loud metallic knocking sound produced when the piston flops back and forth inside its cylinder during engine operation. It is caused by excess piston skirt or cylinder wear, clearance, and damage.

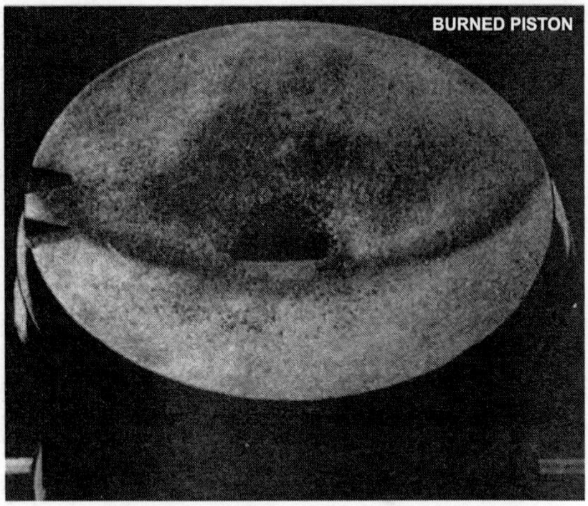

Fig. 10.23 *(a) Burned piston from prolonged detonation. Heat and pressure melt and blow hole in head. Engine cylinder lost compression and started pumping oil, producing heavy blue smoke out of exhaust tailpipe.*

Fig. 10.23 *(b) Hole was caused in this piston when valve broke and hammered hole in crown. (Champion Spark Plug Co.)*

Piston pin knock occurs when there is too much clearance between the piston pin and piston pin bore in piston or connecting rod small end bushing. Excessive clearance allows the pin to hammer against the connecting rod or piston as the piston changes direction in the cylinder during engine operation. Piston pin knock will usually produce a double knock, e.g., two rapid knocks and then a short pause.

A ***burned piston*** is often a result of preignition or detonation damage. Abnormal combustion, excessive pressure, and heat actually melt and blow a hole in the piston crown or area around the ring lands. Symptoms of a burned piston include: engine smoke, knock, excessive blowby, etc., *Figs. 10.23* and *10.24.*

Fig. 10.24 *This burned piston has damage on one side of crown and ring land area. Cylinder damage occurred. (Fel-Pro Gasket)*

A compression test or cylinder leakage test may indicate a burned piston. However, engine disassembly is usually required to verify this problem. If the cylinder wall is not damaged, the repair can be an in-vehicle operation.

Damaged spark plugs can indicate piston problems. An oil-fouled plug would indicate bad rings or cylinders. However, melted aluminium deposits indicate a melted or partially failed piston. A smashed electrode can result from being hit by the piston. Also, spark plug tip damaged from detonation that need the cause of detonation to be corrected, to avoid piston damage, *Fig. 10.25.*

Worn piston rings or ***cylinders*** result is blowby, blue-grey engine smoke, low engine power, spark plug fouling, and other problems caused by poor ring sealing.

To diagnose the ring and cylinder problems, increase the engine speed and continuously watch the tailpipe and valve cover breather opening. If blue-grey smoke is coming out of the vehicle engine exhaust under load, the oil rings and cylinders may need service. If excessive oil vapours and air blows out of the valve cover breather, blowby is entering the crankcase. Piston, compression ring(s), or cylinder problems are indicated. Worn ring grooves can also cause oil consumption.

If only the rings are found to be worn while the cylinders are found in good condition, the engine can be rebuilt, while still installed in the engine compartment on the vehicle. Refer to the vehicle manufacturer's service manual applicable for service details.

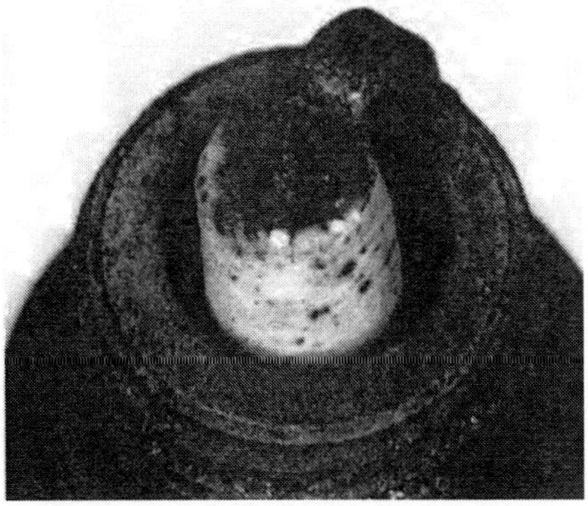

Fig. 10.25 *This spark plug has burned electrodes caused by a stuck turbocharger waste gate that was allowing too much boost. (Champion Spark Plug Co.)*

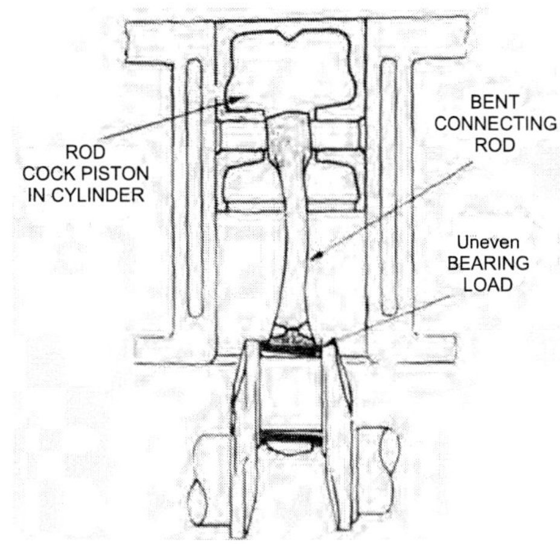

(a) Bent connecting rod will load one side of piston skirt and rod bearing.

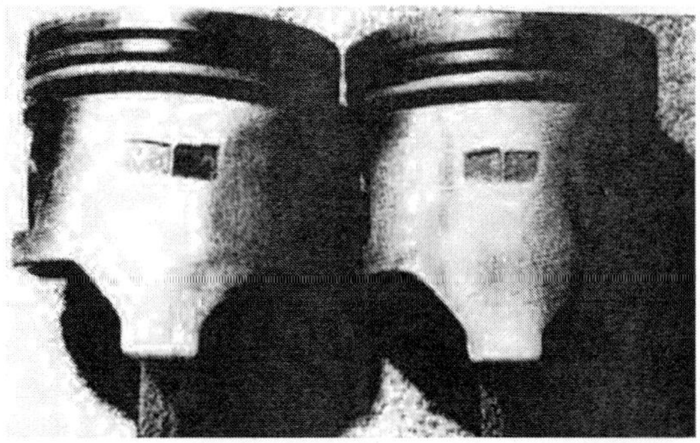

(b) Notice uneven piston wear pattern from bent rod.

Fig. 10.26 *Bent connecting rod.*

10.19 CONNECTING ROD PROBLEMS

A ***bent connecting rod*** has been damaged by detonation or too high engine rpm and is no longer straight, *Fig. 10.26.* It will load one side of the rod bearing and cause the piston to cock at an angle in the cylinder. A bent connecting rod is usually not found. An uneven wear pattern will be visible on the piston skirt and connecting rod bearing.

Fixtures are used to check the connecting rod straightness.

A ***broken connecting rod*** is the one that has suffered a complete failure of the I-beam section or rod bolts and cap. Connecting rod breakage may result due to too high engine speed, running an engine after connecting rod bearing failure, loss of lubrication, wrong assembly, *etc., Fig. 10.27.*

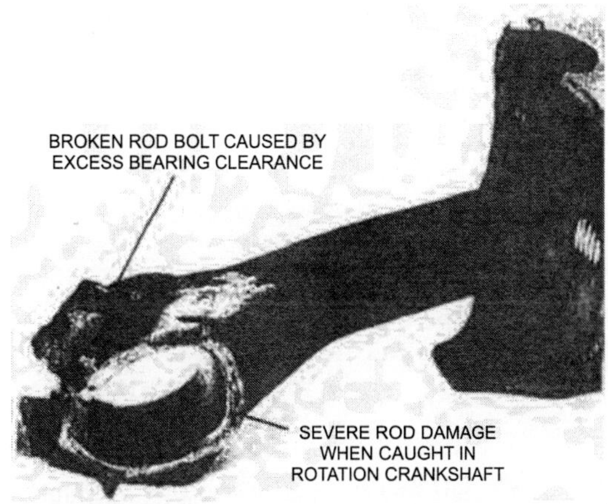

Fig. 10.27 *High engine rpm on old engine caused rod bolt to snap off. Cap then came off rod and rod was crushed into block by spinning crankshaft. Driver ignored or did not notice rod-bearing knock, which usually precedes complete bearing and rod failure. (Midwest Crankshaft)*

Usually a connecting rod breakage causes severe engine damage. The rotating crankshaft can slam into the disengaged connecting rod, crushing the rod into the cylinder block, and may even ruin the block sometimes. The piston can also fly upward and hit the cylinder head, causing the head and valve damage. Usually, complete engine disassembly, crankshaft turning, cylinder block cylinder(s) boring, etc., are necessary to repair a broken connecting rod.

Connecting rod bore problems are caused by extended service, connecting rod failure, oil pump troubles, etc. The big end of the connecting rod can become out-of-round and can no longer support the big end bearing properly.

Always check for rod bore problems during the engine service, *Fig. 10.28.* Marks on the bore can indicate rod-bearing movement in the rod and indicate that the rod should be serviced, i.e., re-machine the bore in machine shop or replace it.

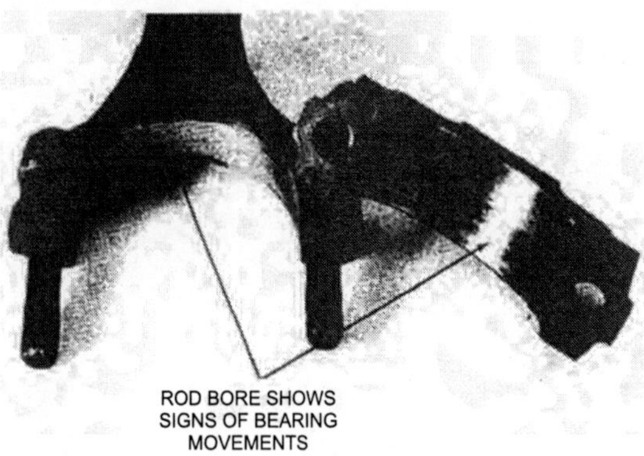

Fig. 10.28 *An out-of-round rod bore or stretched rod bolts can allow bearing move or shift inside rod. This can wear back of bearing and rod until finally bearing spins in rod. A spun rod bearing will damage the connecting rod and crankshaft. (Midwest Crankshaft)*

10.20 CRANKSHAFT AND BEARING PROBLEMS

The crankshaft and its bearings must be capable of withstanding extreme loads and friction to which they are subjected during the engine operation. Any carelessness used during assembly, a lubrication system problem, or driver abuse can cause **bearing** or **crankshaft problems**.

Connecting Rod Bearing Knock

Connecting rod bearing knock is often caused by wear and excessive big end bearing-to-crankshaft journal clearance. It will produce a light, regular, rapping noise with the engine floating, i.e., point at which throttle is maintained constant and engine is running at constant speed. It is loudest after engine warm up. When the engine is cold, high viscosity of oil tends to dampen and quiet rod knock.

To isolate a bad rod bearing, short out or disconnect each spark plug wire one at a time. The loose, knocking big end rod bearing may quiet down or lower pitch when its spark plug is disabled. In some engine designs, excessive rod bearing clearance may allow the piston head to strike the cylinder head, *Fig. 10.29.* This causes a loud knocking sound, similar to that made by a failed piston.

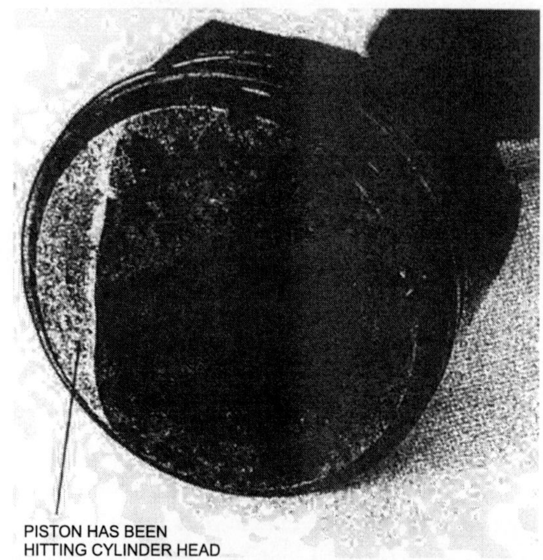

Fig. 10.29 *A bad rod bearing allowed this piston to hit cylinder head, making loud knocking noise. Notice the shiny area that shows contact.*

Spun Rod Bearing

A ***spun rod bearing*** is a bearing that has been hammered out flat and then rotated inside the connecting rod bore. It will make a rod bearing type knock, but will be much louder. When a rod bearing spins, it will damage the rod and crankshaft journal.

Worn Main Bearings

Worn main bearings have excessive clearance and can allow the crankshaft shift up and down in the cylinder block with load changes. A thin layer of metal has been worn off the bearing face, *Fig. 10.30.*

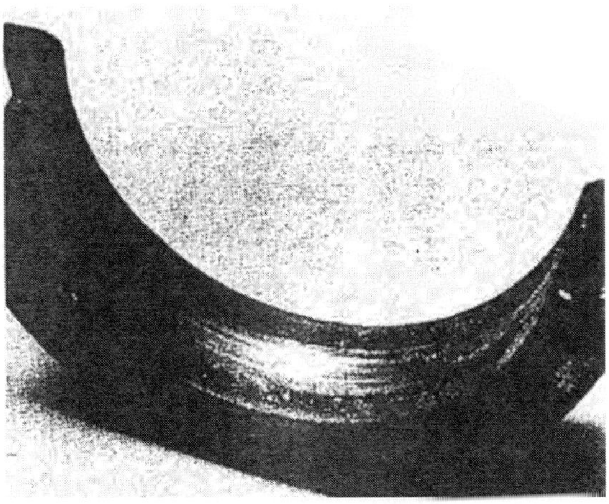

Fig. 10.30 *Main bearing wear will lower pressure considerably and can cause knocking under load. This is badly worn main thrust bearing.*

Main Bearing Knock

Main bearing knock is similar to rod bearing knock but is little deeper or dull in pitch. It is normally more pronounced when the engine is pulling under a load. Bearing and its journal wear allow the crankshaft move up and down inside the cylinder block.

Main bearing wear usually causes oil pressure to drop significantly. Remove the oil pan to check main bearing noise. Pressure-test the lubrication system. Too much oil flow out of one or more of the main bearings indicates excessive bearing clearance. If a pressure tester is not available, remove and check each of the main bearings. If the crankshaft is in good condition, bearing insert replacement will fix the problem.

Excessive Crankshaft End Play

Excessive crankshaft endplay is a worn main thrust bearing. Thrust bearing wear produces a deep knock usually when applying and releasing the manual transmission clutch. In case of an automatic transmission, the endplay problem may only show up as a single thud or knock when accelerating or decelerating.

Loose flywheel bolts or other drive train troubles can also cause a knock occurring with clutch or torque converter action. Isolate the noise source before starting the repairs.

10.21 VIBRATION DAMPER PROBLEMS

Crankshaft vibration damper problems include physical damage, damaged rubber ring, seal journal wear, slipped or sheared key, etc. Engine vibration can be caused by any damage to it. Oil leakage can occur if the seal surface is worn or scored. The timing marks can be out of position if the outer metal ring slips on its rubber mount. This can cause difficulties when trying to set ignition timing. A few of the problems encountered with a crankshaft vibration damper are shown in *Fig. 10.31.*

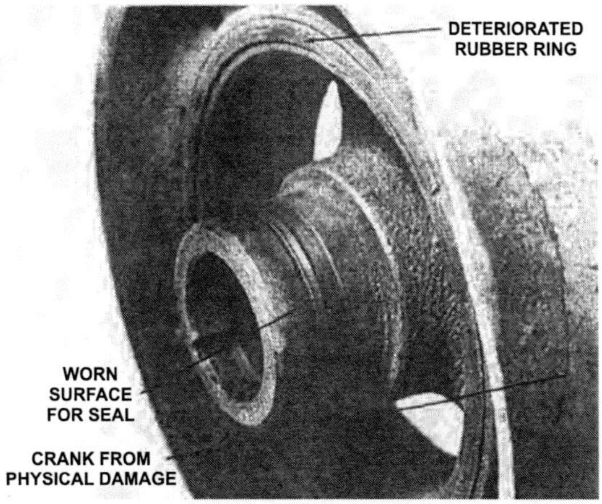

Fig. 10.31 *Notice the problems with this harmonic balancer. Keep these troubles in mind during engine service.*

10.22 FLYWHEEL PROBLEMS

Flywheel problems include loose mounting bolts, ring gear teeth failure, friction surface scoring, warpage, etc. A loose flywheel can produce a knocking sound similar to main bearing knock. The flywheel ring gear teeth may wear or break off and cause starter motor gear engagement troubles.

10.23 INDUCTION AND EXHAUST MANIFOLD PROBLEMS

Exhaust manifold problems usually are a cracked or warped manifold. Engine heat can cause an exhaust manifold to crack, allowing leakage and exhaust noise. Engine heat can also warp a manifold, causing it to leak next to the cylinder head mating surface, *Fig. 10.32.*

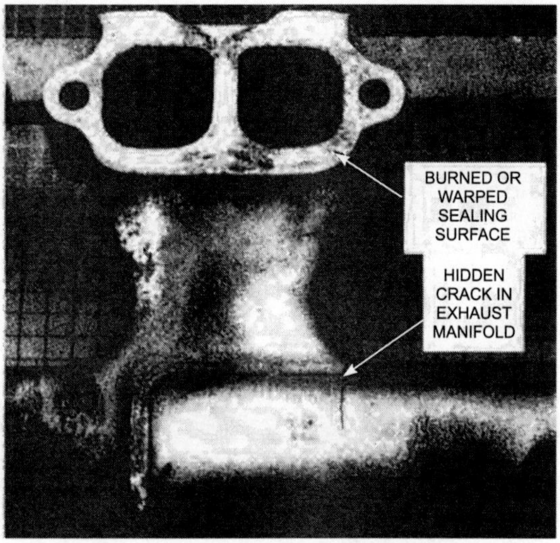

Fig. 10.32 *Cracked exhaust manifold will make a metallic-type ticking sound as exhaust pressure blows out of crack. This crack was on rear side of manifold and was difficult to locate. Using section of vacuum hose as listening device found the leak. (Fel-Pro)*

Induction (intake) manifold problems are not common as the manifold is not exposed to as much heat as other parts. Occasionally an intake manifold can warp and cause a vacuum leak. Plugs may loosen or fall out, allowing vacuum or coolant leakage.

10.24 MECHANICAL PROBLEM DIAGNOSIS

A quick and accurate diagnosis of engine problems is the key to economical automotive repairs and routine maintenance. If a technician is not qualified, trained, and skilled to properly diagnose reported engine problems, a great deal of time, effort, and money could be wasted in unwanted automobile repairs. An untrained mechanic may unknowingly rebuild an engine when a minor repair could have fixed the problem.

For example, a worn or excessively slack timing chain can cause the engine valves to open and close at the wrong time in relation to piston position. This could cause low compression stroke pressure in all cylinders. The engine would develop low power output and low compression readings.

If the problem is not correctly diagnosed, the technician could mistake the problem (worn timing chain) as more serious major problems e.g., worn piston rings, scored cylinders, etc. The engine could be overhauled when a new timing chain would have fixed the problem. Anyway, the customer would have to pay for the mechanic's lack of experience and training.

10.25 SYMPTOMS OF ENGINE MECHANICAL PROBLEMS

Symptoms that indicate engine mechanical problems include:

1. ***Excessive oil consumption:*** Requiring engine oil addition too often.
2. ***Crankcase blowby:*** Combustion pressure gases blow past piston rings, into crankcase, and out the crankcase breather.
3. ***Abnormal engine noises:*** Knocking, tapping, hissing, rumbling.
4. ***Engine smoking:*** Blue-grey, black, or white smoke blows out of tailpipe.
5. ***Poor engine performance:*** Rough idle, slow acceleration, engine vibrates.
6. ***Coolant in engine oil:*** Oil has white, milky appearance.
7. ***Engine locked up:*** Crankshaft will not rotate.

With any of these troubles, inspect and test the engine to isolate the exact source of the problem. Determine the repair that is required. Then determine whether the engine can be repaired with the engine in the vehicle or it must be removed for more major repairs.

10.26 ENGINE PRE-TEARDOWN INSPECTION

Gather information from the customer (driver) or service writer, the technician should inspects the engine using all senses (sight, smell, hearing, and touch). Look for external problems, e.g., oil leaks, vacuum leaks, part damage, vacuum leaks, contaminated oil.

If a leak is found, smell the fluid to determine if it is oil, coolant, or any other type of fluid.

Listen for unusual noises that indicate part wear or damage. Increase engine speed while listening and watching for problems. The engine may run fine at idle but show symptoms of troubles at higher speeds.

Some of the engine problems that can be located through inspection are:

Coolant in oil will show up as white or milky looking oil. It is caused by mechanical problem that allows engine coolant to leak into the engine crankcase. This could be caused by a cracked cylinder block or cylinder head, leaking head gasket, leaking induction manifold gasket (for V-type engine), or similar other problems that can lead to the coolant finding its way into the engine oil.

Oil fouled spark plugs indicate internal oil leakage into the engine combustion chambers. This is caused by the worn piston rings of the engine, worn or scored cylinder walls, or defective valve seals. Additional tests would have to be performed to isolate the source of problem.

Oil in the coolant is usually not an engine problem. It is caused by a leak in the radiator (transmission) oil cooler.

Engine oil leaks occur when: gaskets harden and crack, seals wear, fasteners work loose, and part damage (warped surfaces, cracked parts). To find oil leaks, clean the affected area on the outside of the engine. Then trace the leak upward to its source. Oil will usually flow down and to the rear of the engine due to cooling fan and airflow action.

External coolant leaks will show up as a puddle of coolant under the engine of the parked vehicle. Leaks can be caused by hose problems, rusted out freeze casting core plugs, or warped, worn, or damaged parts. Leaks can be located by using pressure tester.

Engine blowby occurs when combustion pressure blows past the piston rings into the lower block and oil pan area of the engine. And then flows up to the valve cover and out the breather. Excessive blowby will show up as smoke or oil around the breather. With the valve cover breather removed, oil vapours may blow out when the engine speed is increased.

Engine vacuum leaks show up during inspection of a running engine as a hissing sound, like air leaking out of a tyre. Vacuum leaks are loudest at idle and temporarily quiet down as the engine is accelerated as the manifold vacuum drops with engine acceleration. Usually, very rough engine idling accompanies a vacuum leak.

10.27 ENGINE SMOKE

Engine smoking is normally noticed at the tailpipe when the engine is accelerated or decelerated. The colour of the smoke can be used to help diagnose the source of the problem. With a **Gasoline Engine,** the exhaust smoke may be:

1. ***Blue-grey smoke:*** Indicates that engine oil is entering combustion chambers, caused by worn rings, worn cylinders, leaking valve stem seals, or other troubles.
2. ***Black smoke:*** Caused by too rich air-fuel mixture, not engine mechanical problem.
3. ***White smoke:*** If not water condensation on a cold day, may be due to internal coolant leak into cylinders.

With a **Diesel Engine,** exhaust smoke may be:

1. ***Blue smoke:*** Oil entering combustion chambers and being burned because of ring, cylinder, or valve seal problems.
2. ***Black smoke:*** Injection system problem or low compression keeping fuel from burning.
3. ***White smoke:*** Unburned fuel, cold engine, or coolant leaking into combustion chambers.

Figure 10.33 shows engine exhaust smoke problems for an engine.

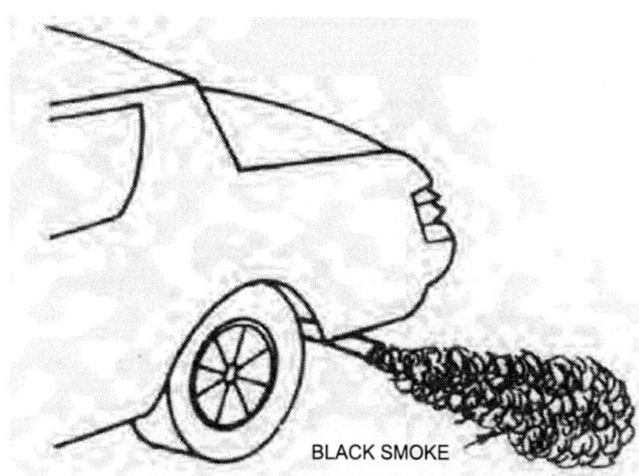

Fig. 10.33 *(a) Black smoke indicates rich fuel mixture and car buretor or injection system trouble.*

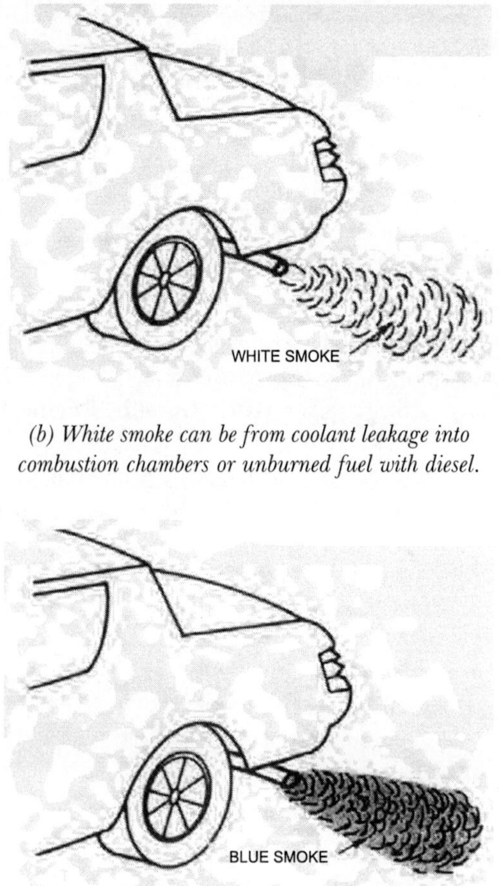

(b) White smoke can be from coolant leakage into combustion chambers or unburned fuel with diesel.

(c) Blue smoke indicates oil consumption. Oil is entering and being burned in cylinders.

Fig. 10.33 *Engine exhaust smoke is a good indicator of engine mechanical condition.*

10.28 LOCATING ABNORMAL ENGINE NOISES

Abnormal engine noises e.g., knocks, hisses, rattles, clunks, popping, may be used to isolate the source of a mechanical problem. An automotive technician/ mechanic must be able to quickly locate and correctly interpret engine noises and decide what repairs are necessary.

A *stethoscope* is a listening device, which will amplify the loudness of sound for finding internal noises produced in engine parts. To use a stethoscope, place the head set in your ears and then touch the probe on different parts in the area of noise. When the sound heard becomes the loudest, the touched part is, or is very near the part producing abnormal noise.

A long screwdriver can be used when a stethoscope is not available. Sounds will travel through the screwdriver.

A ***section of hose*** can be used to locate vacuum leaks and pressure leaks. Place one end of the hose next to your ear. Move the other end around the engine compartment. The hiss will become loudest when the hose end is near the source of leakage.

When using a listening device, keep it away from the rotating engine fan or belts. Severe injury may be caused if the stethoscope, screwdriver, or hose were hit or pulled into the fan or belts.

10.29 DECIDE THE REPAIR NEEDED

After performing all the necessary inspections and tests, and evaluating the data collected, decide which part(s) should be repaired or replaced to fix the engine mechanical problem. If you still cannot isolate the source of problem, the engine may have to be partially disassembled for further inspection.

To be able to properly repair engine problems a technician / mechanic must be able to:

1. Explain the function of each engine part.
2. Describe the construction of each engine part.
3. Explain the cause of engine problems.
4. Describe the symptoms of major engine problems.
5. Choose specific methods to isolate specific problems.
6. Know which parts must be removed for certain repairs.
7. Know whether the engine has to be removed from the vehicle or not, for carrying out the necessary repairs. These skills can only be developed through study, good training, and experience.

THEORY QUESTIONS

1. What is an engine performance problem? Explain.
2. What are three basic questions you should ask yourself when trouble shooting engine performance problems?
3. How can the spark plugs be read to diagnose engine performance problems?
4. Which one of the following problems will cause an engine to idle roughly and make a hissing sound?
   (*a*) Dieseling. (*b*) Hesitation
   (*c*) Vacuum leak (*d*) Pressure leak.
5. When an engine cranks, but fails to start, check for ——————— and ———————to narrow down the possible source of problem.
6. Mechanic A says that a black or wet plug may indicate an extremely rich air-fuel mixture. Mechanic B says that a black plug may indicate engine mechanical problems. Who is right?
   (*a*) A only (*b*) B only
   (*c*) Both A and B (*d*) Neither A nor B.
7. Mechanic A says that a vacuum gauge can detect worn valve guides. Mechanic B says that a vacuum gauge can detect an induction manifold air leak. Who is right?
   (*a*) A only (*b*) B only
   (*c*) Both A and B (*d*) Neither A nor B.
8. A SI engine, with electronic multi-port fuel injection, runs poorly. Light black smoke blows out of the exhaust tailpipe. Mechanic A says cold starter injector could be on and the thermo-time switch should be checked. Mechanic B says the problem could be stuck pressure regulator that is allowing too much fuel to enter the return line. Who is right?
   (*a*) A only (*b*) B only
   (c) Both A and B (*d*) Neither A nor B.
9. Explain 9 test instruments usually found on an engine analyzer.
10. An engine analyzer normally contains all of the following *except:*
    (*a*) Exhaust gas analyzer
    (*b*) Oscilloscope
    (*c*) Cylinder leak tester
    (*d*) Compression gauge
11. Mechanic A says that a test light can be used to test all electronic engine components. Mechanic B says that a test light can be used to check the operation of specified electronic engine components. Who is right?
    (*a*) A only (*b*) B only
    (*c*) Both A and B (*d*) Neither A nor B.
12. To test the operation of the regulator of a multi-port injection system, the pressure gauge should be connected to the ———————————————.
    (*a*) Injector wiring harness
    (*b*) Evaporative emissions purge valve
    (*c*) Fuel rail fitting
    (*d*) PCV valve vacuum port.
13. ——————— is caused by the air-fuel mixture igniting in the induction manifold or exhaust system.
14. Define the term 'vapour lock.'
15. An engine diesels when shut off. Mechanic A says that it is due to a low idle speed setting. Mechanic B says that it is usually due to a high idle speed setting. Who is right?
    (*a*) A only (*b*) B only
    (*c*) Both A and B (*d*) Neither A nor B.
16. All of the following can cause an engine to stall *except:*
    (*a*) low idle speed
    (*b*) A severe vacuum leak
    (*c*) A lean air-fuel mixture
    (*d*) An inoperative thermostatic air cleaner.
17. Mechanic A says that with no-start engine problem check the ignition system of the engine first. Mechanic B says that with a no-start engine problem check for engine mechanical problems first. Who is right?
    (*a*) A only (*b*) B only
    (*c*) Both A and B (*d*) Neither A nor B.
18. Auto mechanic A says that a fuel injector malfunction may cause an engine to miss. Auto mechanic B says that a cracked distributor cap may cause an engine to miss. Who is right?
    (*a*) A only (*b*) B only
    (*c*) Both A and B (*d*) Neither A nor B.
19. Mechanic A says that a rich air-fuel mixture usually causes engine surging. Mechanic B says that a low idle speed setting usually causes engine surging. Who is right?
    (*a*) A only (*b*) B only
    (*c*) Both A and B (*d*) Neither A nor B.

20. A worn cam lobe will reduce valve lift and engine performance. *True* or *False ?*

21. Differentiate between external and internal lifter wear.

22. Worn valve guides can produce a light tapping noise. *True* or *False ?*

23. A ring ridge is a sign of worn ——————.

24. What is a blowby ?

25. List seven symptoms that result from engine mechanical problems.

26. What is rod bearing knock and main bearing knock? Explain.

27. What is meant by "valve float"?

28. An automotive engine with high mileage has valve train clatter. Inspection found sludge or hardened oil in the engine. Mechanic A says that the hole in the centre of the push rod should be checked for clogging. Mechanic B says that check the components (rockers, push rods, stands) for wear and replace if needed. Who is right?
(*a*) A only (*b*) B only
(*c*) Both A and B (*d*) Neither A nor B.

29. How can a worn timing chain be detected ?

30. A vehicle has a dead miss and a puffing or popping noise can be heard at the throttle body. Mechanic A says that there could be a leak at an intake valve due to a bad spring, bent valve, etc. Mechanic B says that there could be a burned piston. Who is right ?
(*a*) A only (*b*) B only
(*c*) Both A and B (*d*) Neither A nor B.

31. ———————— problem will lower compression, especially in two adjacent cylinders,
(*a*) Leaking intake gasket
(*b*) Burned valve
(*c*) Blown head gasket
(*d*) Cracked head.

32. Which is the most common timing mechanism problem ?
(*a*) Spun timing gear key
(*b*) Broken timing gear teeth
(*c*) Jumped timing chain
(*d*) Broken timing sprocket teeth.

33. Mechanic A says that valve train malfunctions can cause engine cylinder damage. Mechanic B says that engine valve train problems can produce a rough idling condition. Who is right?
(*a*) A only (*b*) B only
(*c*) Both A and B (*d*) Neither A nor B.

34. All of the following are common problems of cam lobe wear *except:*
(*a*) reduced valve lift
(*b*) Engine performance problems
(*c*) Valve train noise
(*d*) Excessive combustion pressure.

35. Mechanic A says that a burned valve will cause an engine to run rough particularly at idle speed. Mechanic B says that a burned valve with cause the engine to run rough particularly at high speeds. Who is right ?
(*a*) A only (*b*) B only
(*c*) Both A and B (*d*) Neither A nor B.

36. Mechanic A says that a cylinder head crack usually occurs near an exhaust port. Mechanic B says that a crack in a cylinder head usually occurs near the intake port. Who is right ?
(*a*) A only (*b*) B only
(*c*) Both A and B (*d*) Neither A nor B.

37. Mechanic A says that a blown head gasket can cause engine coolant leakage. Mechanic B says that a blown head gasket can cause damage to the engine block. Who is right ?
(*a*) A only (*b*) B only
(*c*) Both A and B (*d*) Neither A nor B.

38. Mechanic A says that black exhaust smoke is caused by extremely rich air-fuel mixture supply. Mechanic B says that black exhaust smoke is caused by a lean air-fuel mixture supply. Who is right?
(*a*) A only (*b*) B only
(*c*) Both A and B (*d*) Neither A nor B.

39. Which of the following is a symptom of worn piston rings?
(*a*) Low engine power output
(*b*) Fouled spark plugs
(*c*) Blue-grey engine smoke
(*d*) All of the above.

40. Mechanic A says that a bent connecting rod is caused by engine using high-octane fuel. Mechanic B says that a bent connecting rod is often caused by engine detonation. Who is right?
(*a*) A only (*b*) B only
(*c*) Both A and B (*d*) Neither A nor B.

41. Mechanic A says that a burned piston will sometimes produce excessive blowby. Mechanic B says that a burned piston will normally cause a no-start condition. Who is right?
(*a*) A only (*b*) B only
(*c*) Both A and B (*d*) Neither A nor B.

42. Mechanic A says that over-reviving an engine can cause piston skirt breakage. Mechanic B says that piston skirt breakage can be caused by a malfunction in the engine's lubrication system. Who is right?
(*a*) A only (*b*) B only
(*c*) Both A and B (*d*) Neither A nor B.

43. Two mechanics diagnose an engine noise. Mechanic A says that a double-knock is likely to be due to worn connecting rod bearing. Mechanic B says that a knock only when the engine is cold is usually due to piston pin. Who is right?
(*a*) A only (*b*) B only
(*c*) Both A and B (*d*) Neither A nor B.

44. A compression test on a 4-cylinder engine gave the following results:
cylinder #1 = 155, cylinder #2 = 140
cylinder #3 = 110, cylinder #4 = 105
Mechanic A says that a burned valve is the most likely cause. Mechanic B says that a leaking head gasket could be the cause. Which mechanic is correct?
(*a*) Mechanic A only
(*b*) Mechanic B only
(*c*) Both mechanics A and B
(*d*) Neither mechanic A nor B

45. Two auto mechanics are discussing a compression test. Mechanic A says that the engine should be cranked, with the pressure gauge installed,, for "3 puffs." Mechanic B says that the maximum difference between the highest-reading cylinder and the lowest-reading cylinder should be 20%. Which mechanic is correct?
(*a*) A only (*b*) B only
(*c*) Both A and B (*d*) Neither A nor B.

46. During a leakage (leak-down) test, air is noticed coming out of the oil-fill opening. Mechanic A says that the oil filter may be clogged. Mechanic B says that the piston rings may be worn or defective. Which mechanic is correct?
(*a*) A only (*b*) B only
(*c*) Both A and B (*d*) Neither A nor B.

47. Mechanic A says that oil should be squirted into all cylinders before taking a compression test. Mechanic B says that if the compression substantially increases when some oil is squirted into the cylinders, it indicates worn or defective piston rings. Which mechanic is correct?
(*a*) A only (*b*) B only
(*c*) Both A and B (*d*) Neither A nor B.

48. A cylinder leakage (leak-down) test indicates 30% leakage, and air is heard coming out of the air intake. Mechanic A says that this is normal reading for a slightly worn engine. Technician B says that one or more intake valves are defective. Which mechanic is correct?
(*a*) A only (*b*) B only
(*c*) Both A and B (*d*) Neither A nor B.

49. Two mechanics are discussing a cylinder power balance test. Mechanic A says that the more the engine rpm drops, the weaker the cylinder. Mechanic B says that all cylinder rpm drops should be within 50 rpm of each other. Which mechanic is right?
(*a*) A only (*b*) B only
(*c*) Both A and B (*d*) Neither A nor B.

50. Mechanic A says that cranking vacuum should be the same or higher than vacuum at idle speed. Mechanic B says that a floating gauge needle reading around a mid-point indicates a sticking valve. Which mechanic is correct?
(*a*) A only (*b*) B only
(*c*) Both A and B (*d*) Neither A nor B.

51. Mechanic A says that black exhaust smoke is an indication of to`o rich air-fuel mixture. Mechanic B says that white smoke (steam) is an indication of coolant being burned in the engine. Which mechanic is correct?
(*a*) A only (*b*) B only
(*c*) Both A and B (*d*) Neither A nor B.

❑❑❑

Engine Removal, Disassembly and Cleaning

11

11.1 INTRODUCTION

This chapter will expand your knowledge of engine service by outlining the most common steps for a vehicle's engine removal, teardown, and cleaning. Remember that exact steps vary from engine to engine. However, there are several general methods that apply to all engine makes and models.

When you disassemble an engine, you can diagnose what is wrong with various components and engine systems. You can observe certain wear patterns on parts and diagnose what is wrong with the engine.

The decision to repair an engine should be based on all the information about the engine that is available to the service technician/mechanic. In some cases, the engine might not be worth repairing. The mechanic should adequately discuss the advantages and disadvantages of the various repair options with the customer. Final decision on the reconditioning procedure to be used should be the customer's decision, based on the recommendation of the mechanic.

11.2 COMPONENT REPAIR

In general customers want to spend the least amount of money possible for any repairs that their vehicle engine might need. Usually they only have the faulty components repaired or replaced. This makes sense in many cases. For example, a failed timing sprocket and chain problem will be fixed when a new timing chain and sprocket are installed. However, a part may have failed because some other part, upstream, was not functioning correctly. For example, a rocker arm pivot was badly scored due to lack of lubrication. Perhaps a faulty metering valve inside the hydraulic lifter restricted the lubrication. In such a case, if only the rocker arm and pivot were replaced, the new rocker arm and pivot would soon fail again because the source of problem was not also repaired.

Timing chain replacement is another example of a component repair due to wear that can cause a loss of engine performance. If testing indicates that the timing chain has excessive slack, disassemble the front of the engine and measure the actual amount of slack in the chain. Normally a slack of 13 mm or more indicates that the timing chain and gears need to be replaced, *Fig. 11.1 (a)* and *(b)*.

11.3 IS ENGINE REMOVAL NECESSARY?

Several engine repairs can be carried out with the engine block in the vehicle chassis engine compartment. Repairs limited to the cylinder head, valve train, and other external parts are normally in vehicle operations.

Engines are removed from the vehicle when the cylinder block and/or crankshaft are badly damaged. Depending upon the make, model and year of the vehicle, engine removal may also be necessary for other repairs. For example, a vehicle might need engine removal just to replace a damaged oil pan, while another vehicle may allow quick and easy pan removal, allowing in-vehicle replacement of piston rings, bearings, and other major components. When in doubt, always refer to the manufacturer's applicable service manual. It will provide directions for the specific vehicle and engine under service.

Some of the engine repairs that can be done with the engine still installed in the vehicle chassis engine compartment include:

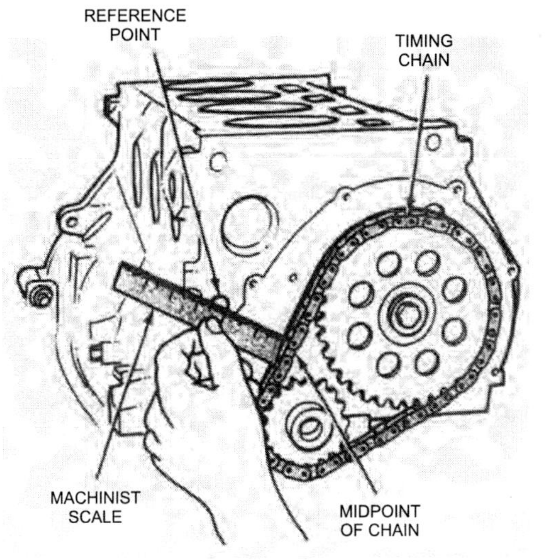

Fig. 11.1 *(a) With the timing chain cover removed, rotate the engine so that the slack is at the bottom and then measure from a reference point on the block to the midpoint of the chain.*

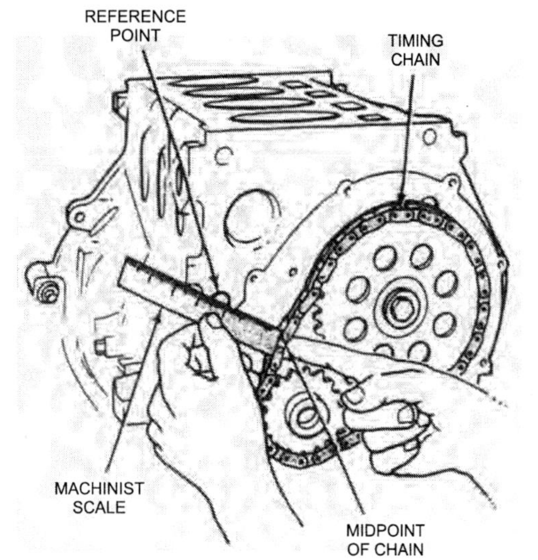

Fig. 11.1 *(b) Rotate the engine until the slack is on top and force the chain out with your fingers. Measure the distance from the reference point to the chain.*

1. **A Valve Job** — Failure of a valve to seal on the seat is one of the most common causes of a premature failure, and further engine problems. When there is a leak between the valve and the seat, the combustion pressure is lowered. This will reduce both engine economy and power. The high combustion gases leaking between the valve and seat will burn the valve face, and further increases the leakage. Failure of a valve to seal on the seat can be caused by abnormal combustion, incorrect valve lash, valve stem deposits, abusive engine operation, etc. Valve leakage is corrected by doing a *valve job.* This does not necessarily correct the malfunction that caused the valve to leak. Stopping valve leakage will improve manifold vacuum. The greater manifold vacuum may cause the oil to enter the combustion chamber during the induction stroke, resulting in increased oil consumption, *Fig. 11.2.* A *valve job* involves reconditioning the valve faces, seats, and other valve train or cylinder head parts. Normally, the intake manifold, exhaust manifold, and cylinder heads can be removed with the short block bolted in the vehicle chassis.

2. **Valve Seals** and **Springs** can usually be replaced without removing the cylinder heads or the valves. Shop air pressure can be injected into the cylinders to hold the valves closed. Then a small compressor can be used to remove the springs, retainers, keepers, and seals, *Fig. 11.3.*

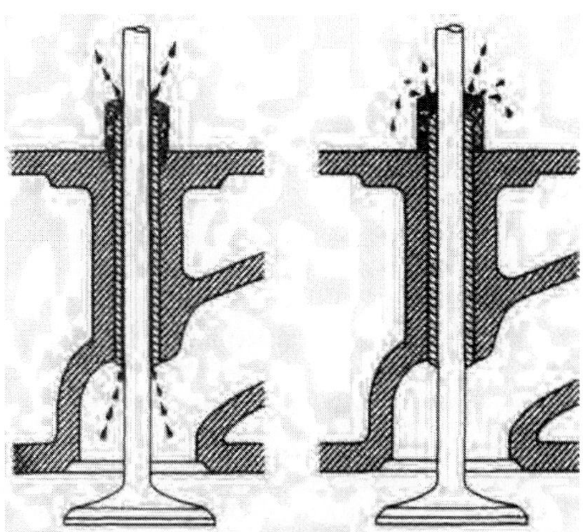

Fig. 11.2 *New valve stem seals (right) reduce the flow of oil past the valve guides.*

Fig. 11.3 *Valve seals and valve springs are examples of parts that can be replaced without cylinder head or engine removal. Install air hose in spark plug hole. Air pressure from compressor will hold valves up in head when removing springs. Compress spring and remove keepers to free spring and seal from valve. (Fel-Pro)*

3. **Oil Pumps** can often be replaced without engine removal. However, if the oil pan cannot be removed by simply raising the engine off its mounts, engine removal might be required for oil pump service.
4. **Rod** and **Main Bearings** can often be replaced with the block in the vehicle. Remove the oil pan; unbolt the rod and main bearing caps to service the bearings. The upper main inserts may be rotated out of the block with a thin shank screwdriver or with a special cotter pin type tool that fits into a crank oil hole. Pushing on the screwdriver or turning the crank will make the tool push the inserts out of the *block, Fig. 11.4.*
5. **External Engine Parts:** Exhaust manifolds, induction manifolds, valve covers, etc; can usually be serviced without engine removal. With exhaust manifolds, it might help to unbolt the engine mounts and lifting the engine several centimeters with a floor jack.
6. **Rear Main Oil Seal Replacement** is an in – vehicle operation provided the oil pan can be removed. An old neoprene type oil seal can be pushed out of the block with a small screwdriver. With an old rope type rear oil seal, turn a screw into the seal. Then pull the screw and seal out of the block with a pair of pliers, *Fig. 11.5.* The other half of the seal fits in the main cap. The new seal should be a neoprene type.

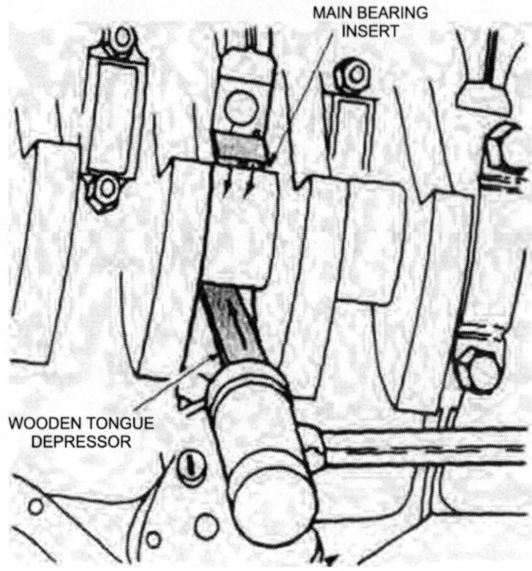

(a) Using wooden tongue depressor to rotate old bearing out of block.

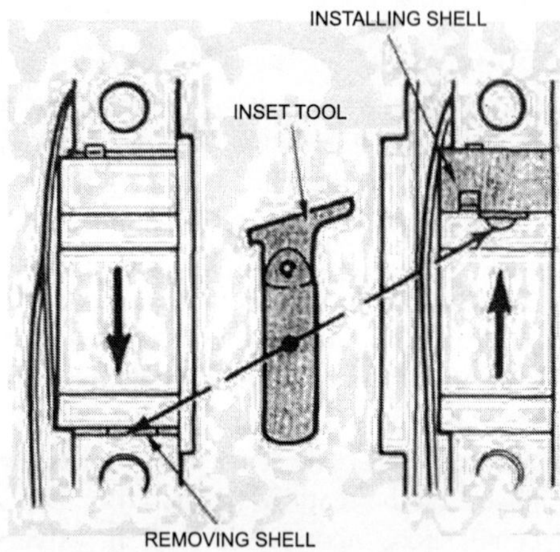

(b) Using special tool that fits into oil hole in crank to remove upper bearing insert. Bearings in caps can be removed by hand.

Fig. 11.4 *Engine rod and main bearings can be replaced without removing engine from vehicle if oil pan will come off. (Renault and General motors)*

7. **Piston Service,** when the block is in good condition, can sometimes be carried out without removing the engine from the vehicle. As an example, if in the engine a piston pin is knocking, determine which piston pin is knocking using a stethoscope. Then remove the oil pan, and the cylinder head (for that

cylinder when more than one cylinder heads are used). Drive out the piston. Replace the piston and pin. Then reassemble the components that were removed.

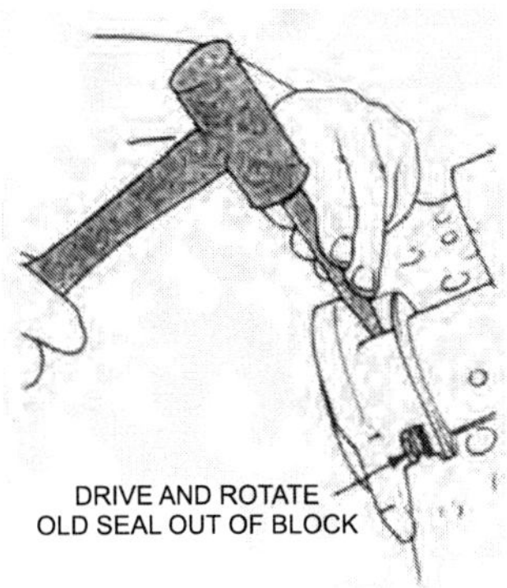

(a) Drive upper half of neoprene seal out with driver. If engine has rope type seal, turn screw into rope.

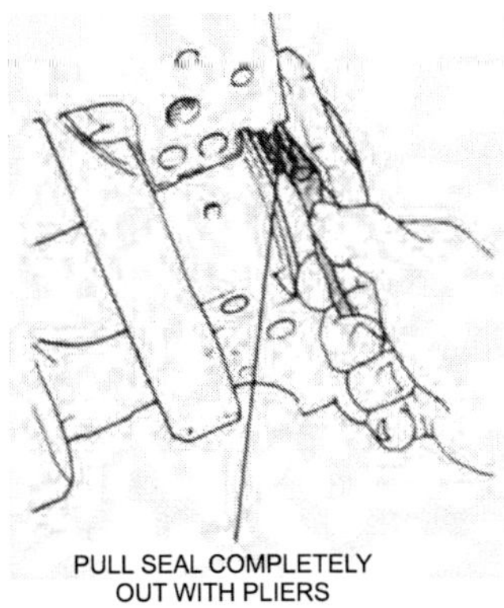

(b) Use pliers to pull seal out of block. Install new seal in cap and block. Make sure new neoprene seal lip is facing in right direction.

Fig. 11.5 *Rear main oil seal can also be replaced without removing engine from vehicle. Remove oil pan and rear main cap. Then service the rear seal. (GM-Buick Div.)*

Some of the engine repairs that normally require engine removal are:

1. **Crankshaft Service** because of wear or any damage requires the engine to be pulled out of the vehicle. However, when the connecting rods and other parts are in good condition, it is not necessary to remove the cylinder head(s), valve cover(s), connecting rods, and other engine components, *Fig. 11.6.* By removing the oil pan and unbolting all of the connecting rod caps, the crankshaft will usually lift out of the bottom of the block for repairs or replacement, as the case may be. This saves a lot of labour cost.

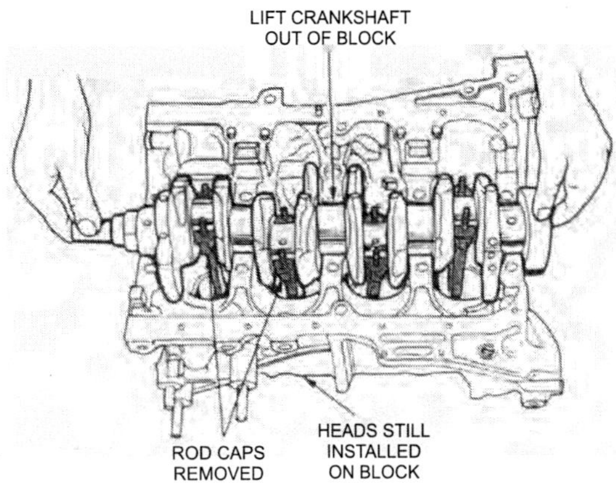

Fig. 11.6 *Engine removal from vehicle is needed for crankshaft service. However, you do not always have to remove cylinder heads and other top end parts. Remove pan and unbolt rod caps. This will allow you lift crankshaft out of block. If rod is damaged, you will have to remove head to service rod. (Honda)*

2. **Cylinder Block Wear** or **Damage** is also an out of vehicle operation. Usually all major components are removed and the block sent to a machine shop for required repairs.

There are exceptions to almost any rule. Therefore, it is important to refer to the applicable service manual for specific details when in doubt.

11.4 ENGINE OVERHAUL

Minor Engine Overhaul

Engine overhaul includes both a ring and a valve job. New connecting rod bearings are usually installed during an overhaul. Sometimes this type of engine reconditioning is known as *minor overhaul.* A minor overhaul can usually be carried out without removing the engine from the vehicle chassis. It can only be done when cylinder wall and crankshaft wear is within specs and the oil and heads can be easily removed in-vehicle. The general condition of the crankshaft can be determined by measuring oil pressure, listening to bearing knock, and/or inspection.

Generally, remove the top end from the engine. Check cylinder wear to assure that the cylinders will take new rings without boring. Remove the oil pan and rod caps. Drive the pistons out of the top of the block. Hone the cylinders and replace the main bearings. Then install the pistons and rods with new rings and bearings in place. Install the new oil pump; reconditioned heads, and other engine parts.

The overhaul is normally done when the engine lacks power, has poor fuel economy, consumes excessive amount of oil, produces visible exhaust tailpipe emissions, runs rough, or is hard to start. It is still only a repair procedure. Several worn parts still remain in the engine.

Other engine problems may be noticed after the oil pan is removed and the piston and rod assemblies are removed. The customer needs to be informed about all noticed engine problems in order to get authorization for the required service. In the high-performance industry, this procedure is known as **freshening the engine**.

Major Engine Overhaul

Complete engine reconditioning job is called ***rebuilding***, also known as ***major overhaul***. To rebuild the engine, it must be removed from the vehicle chassis and be completely disassembled. All serviceable components are reconditioned to either new or service standards. All bearings, gaskets, and seals are replaced. When the reconditioning is done properly, a rebuilt engine should operate like a new engine.

The term *engine overhaul* or *engine rebuild* is understood to mean:

1. Complete disassembly of the engine.
2. Every part is cleaned, inspected, and measured (if needed).
3. Worn or damaged parts are replaced or repaired.
4. Then the engine is reassembled.

In general, the following parts are replaced or reconditioned during a major engine overhaul:

1. Piston rings and often the pistons (if worn or damaged).
2. Connecting rod, main, and camshaft bearings.
3. Oil pump.
4. Timing chain, gears, or belt.
5. Core plugs.
6. Camshaft and lifters (if worn).
7. Valve train components, particularly valve springs (if required).
8. Valve guides or seats.
9. All gaskets and seals.
10. Other parts (if not within specs).

A special form of precision engine rebuilding is called **blueprinting**. Blueprinting is usually done to impart the engine with maximum performance capability. The mating part clearances are all set near the maximum specifications and combustion chambers are adjusted to minimum and equal volumes. All servicing details are done with utmost care. The engine is precisely balanced. Blueprinting demands skilled and detailed labour, and is therefore, very expensive.

11.5 REMANUFACTURED ENGINES

Some engines are *remanufactured*. The engine internal parts are completely disassembled, and each serviceable part is reconditioned with specialized machinery. Engines are then assembled on an engine assembly line similar to the original manufacturer's assembly line. The parts that are assembled together as an engine have not come out of the same engine. The remanufactured engine usually has new pistons, valves, and lifters, together with other parts that are normally replaced in a rebuilt engine. All clearances and fits in the remanufactured engine are the same as in a new engine. A remanufactured engine should give service as good as new engine, and it will cost about half as much. Remanufactured engines usually carry a warranty and will be replaced if they fail during the warranty period. They may even cost less than a rebuilt engine because much of the reconditioning is done by specialized machines rather than by expensive skilled labour.

Reconditioning of individual parts is reviewed in later chapters.

TIP - Take pictures of the engine that is to be repaired. These pictures will be very useful when reassembling or reinstalling the engine. It is very difficult for anyone to remember the exact location of every bracket, wire and hose. Referring back to the pictures of the engine before work was started will help in restoring the engine to its original condition.

11.6 ENGINE REMOVAL

Safety

Safety is the most important aspect of the engine removal. Pulling an engine is dangerous. Two aspects of engine removal that cause most of the problems are: ***raising the vehicle chassis*** and ***lifting the engine*** (with the transmission/transaxle in many cases).

Raising the Vehicle

During engine removal you have to raise the vehicle chassis, adequately support it and get underneath it. Obviously, if the vehicle falls, you could get seriously injured or, even worse, it could be fatal. You must always use jack stands or ramps to support the vehicle. **Floor jacks,** scissor jacks, bumper jacks, and like are for raising the vehicle only. Whereas, **jack stands** are used for supporting the vehicle. Therefore, use a jack to raise the vehicle, block the wheels on the floor and then place a jack stand underneath it immediately, *Fig. 11.7.*

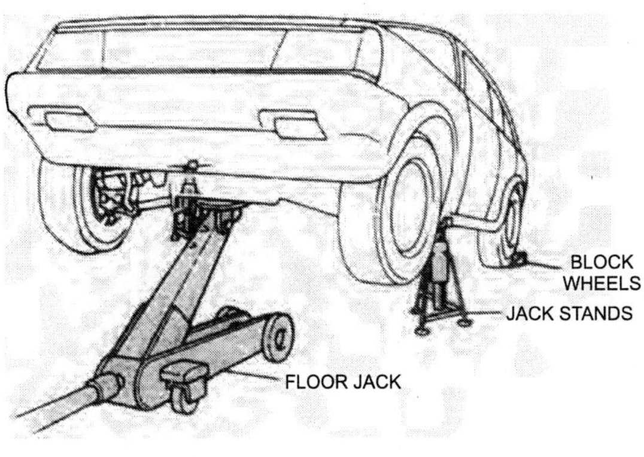

Fig. 11.7 *Properly raised vehicle.*

Lifting the engine involves the risk of dropping it. You must be very careful when first raising the engine off its mounts. It is easy to get distracted and stick your hand under the engine. Never let it happen. If the engine needs positioning, do it from the upper side of the engine. In case a line or wire is caught on the engine, use a stick to free it. Always remain vigilant not to put your hand in a location on the engine where it could get crushed if the engine should fall.

Special Tools

Removing an engine from the engine compartment of the vehicle is a big job. It requires special tools and usually extra help. One of the special tools is the floor jack. It is needed to support the transmission as well as to support the vehicle. Another special tool is the **engine hoist.** This hoist uses a hydraulic jack in a special frame. The frame extends over the engine compartment and lifts the engine straight up. Then the hoist is rolled away from the chassis, and the engine is set down.

11.7 PREPARING FOR ENGINE REMOVAL

The engine exterior and the engine compartment should be cleaned before starting the work. A clean engine is easier to work on, and the cleaning not only helps to keep dirt out of the engine, but also minimizes accidental damage from slipping tools. To prepare for engine removal, auto mechanics / technicians disconnect every thing under the engine to start with: engine mounts, bell-housing bolts, torque converter bolts, and exhaust system, *Fig.11.8.* This is done so that oil, fuel, and coolant are not leaking out and dripping on your face because of top end disassembly. Also, the engine can still be cranked over with a remote starter switch to remove the torque converter bolts in case of automatic transmission or transaxle. Then drain the fluids, remove the starter, engine mounts and everything on top of the engine.

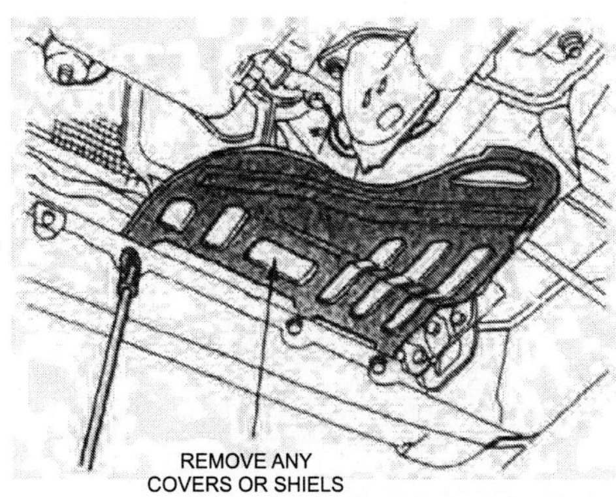

Fig. 11.8 *(a) Remove any air shrouds or shields that are in the way*

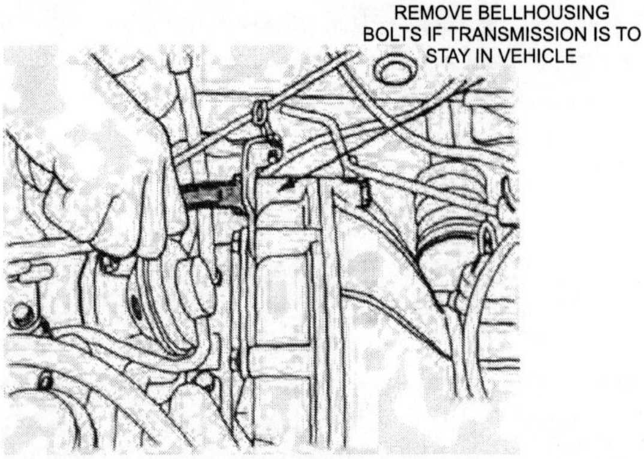

(b) Unbolt bell housing if transmission or transaxle is to remain in vehicle,

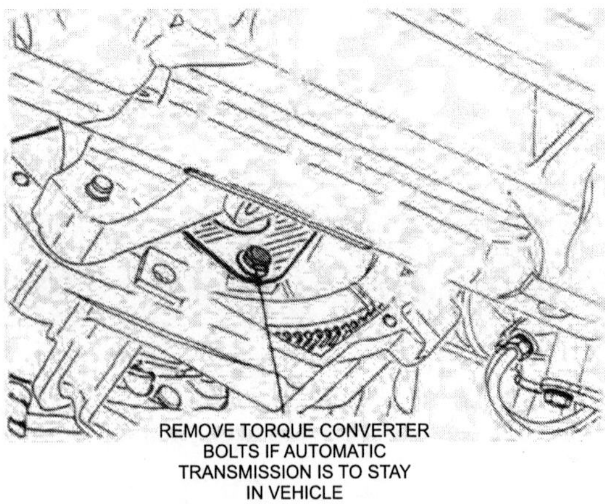

(c) Also, unbolt flex plate or flywheel if automatic transmission is to remain in Vehicle.

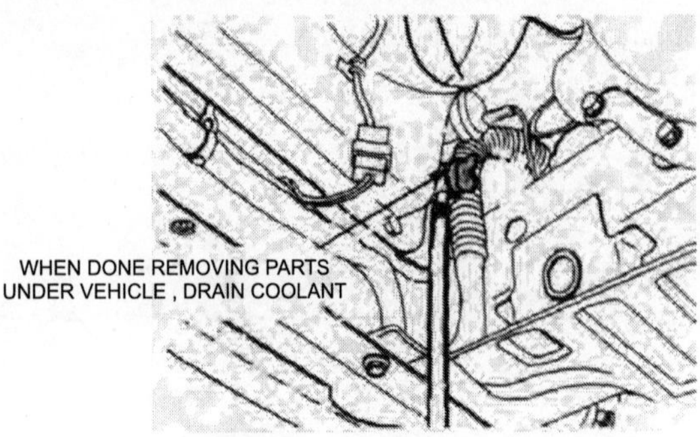

(d) Drain coolant and oil last so that fluids do not drip on you.

Fig. 11.8 *Most technicians begin engine removal under vehicle. A few of the parts that must be disconnected are shown here. (Subaru and Peugeot)*

Basically, follow these steps to get ready for engine removal:

1. Park the vehicle so there is plenty of working space on both sides and in front of the engine compartment.
2. Place the vehicle on stands and unfasten every –thing preventing engine removal *under the vehicle, Figs. 11.9* and *11.10.* The exhaust system can be rusted and the header pipes may be difficult to unbolt. Use rust penetrant and a six-point socket if the bolts or nuts are hard to remove.

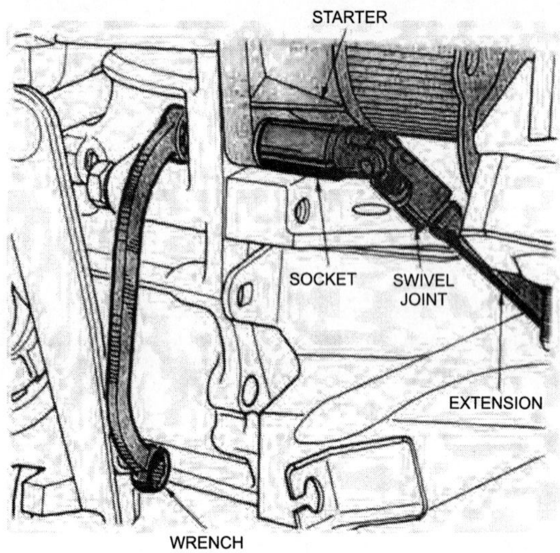

Fig. 11.9 *These two wrenches can be handy in tight quarters. Long extension and swivel will let you reach hidden bolts. Bent box wrench will also reach difficult nuts or bolts.*

3. *Disconnect the battery to prevent electrical shorts* and remove the battery from the vehicle if it could interfere with the removal of the engine, *Fig. 11.11.* This step will give you more room. Clean the battery with water and baking soda. Store it off the concrete floor and cover it with a nonconductive material.
4. *If the engine is liquid cooled, drain the coolant* from the radiator and the engine block to minimize the chance of coolant getting into the cylinders when the head is removed. You can drain the radiator two ways. The pet cock on the lower tank drains the cooling system with little mess. Removing the lower radiator hose drains the system very quickly but also

makes quite a mess. Be sure to clean up any fluid on the floor or tools.

5. Slide the drain pan under the engine and drain the *engine oil.* It is a good idea to let the engine drain for a good length of time. The more oil you get in the pan, the less mess you will have during engine disassembly.
6. Lower the vehicle and place covers over the fenders to protect the paint.
7. A good general rule is to *remove anything that gets in your way during engine removal.* Hood is one of the first objects to be removed. This takes at least two people, one on each side. Scribe lines around the hood bolts that attach the hood to its hinges. These lines will facilitate realignment at reassembly. Then remove all hood hinge bolts and lift off the hood and store it on fender covers placed on the top of the vehicle, where it is least likely to be damaged. If necessary take someone's help. After engine removal you can refit the hood to its hinges. Removing the hood first protects both engine compartment and hood.
8. *Use labelled cans* or *boxes* to store hardware. Keep hardware separated by application. For example, all alternator bolts go to one box, carburetor hardware in another, and so on. Do not mix all the mounting hardware in one box, or reassembly will be very difficult. The same is true of vacuum hoses and electrical wires.
9. *Label everything* – Tape can be marked with the proper, location of each item to facilitate replacement during engine assembly. Label all hoses and wires with a tape flag and felt-tip pen. This step is very important on vehicles with multiple same type of components. For example, on vehicles with dozens of vacuum lines. Even if you do not know what a line does, you can mark it. Just assign a number to that disconnection. Write the number on the tape flag. Write the same number on the tape you have placed on the part that the line or hose attaches to. Then later all you have to do is match numbers at the time of reassembly.
10. Disconnect or *remove the components on the top* of the engine: air-cleaner, throttle linkage, distributor, injectors, etc.

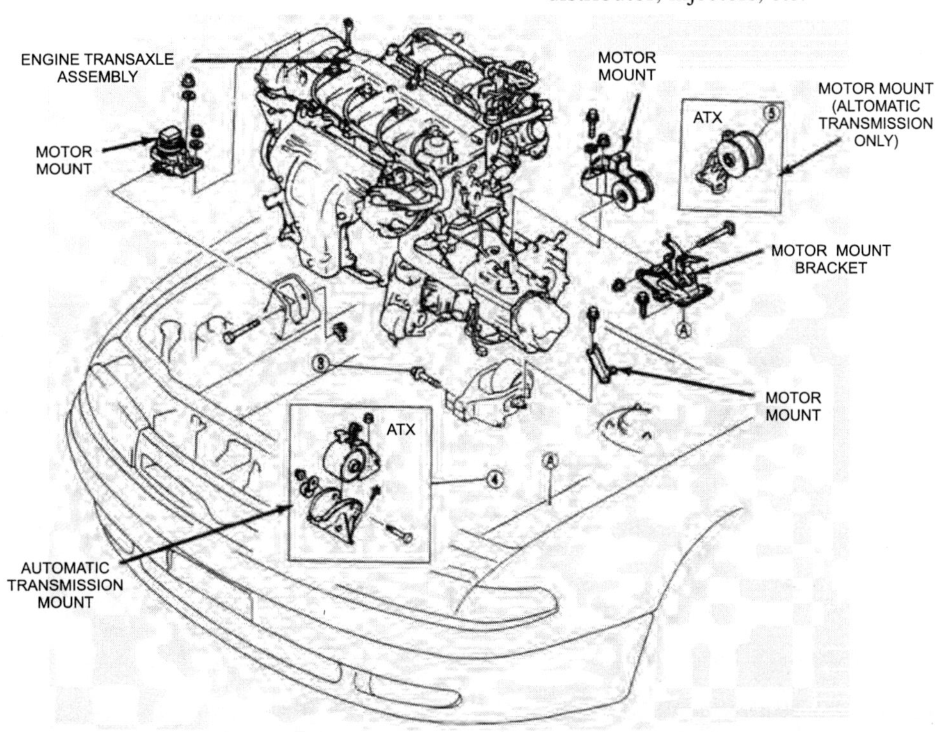

Fig. 11.10 *Engine mount designs vary. Sometimes through bolts are used to secure the engine to the frame. Refer to service manual for specific directions when in doubt. (Mazda Motor Co.)*

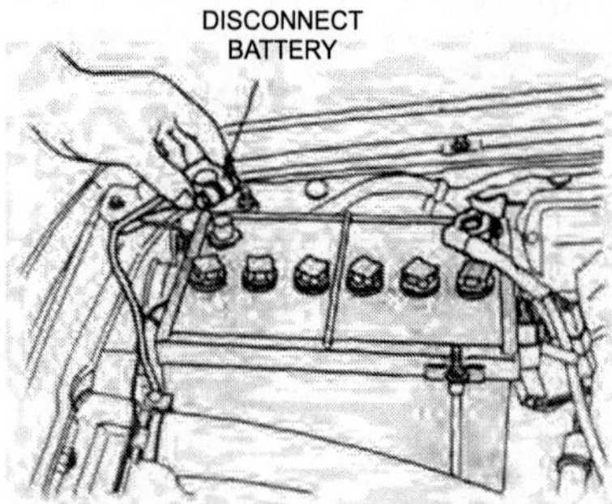

(a) Disconnect battery to prevent electrical fire.

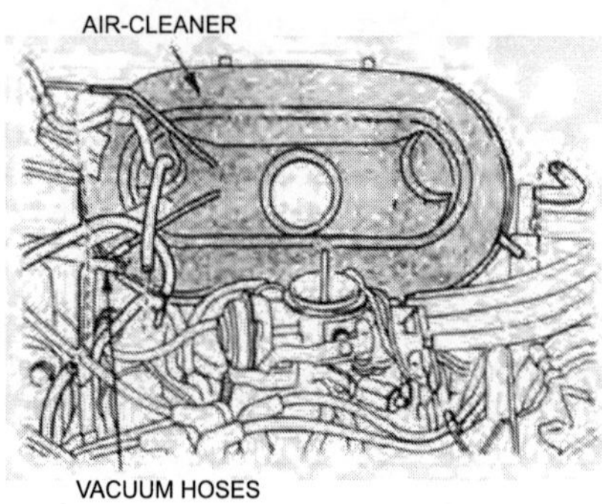

(b) Air cleaner and vacuum hoses.

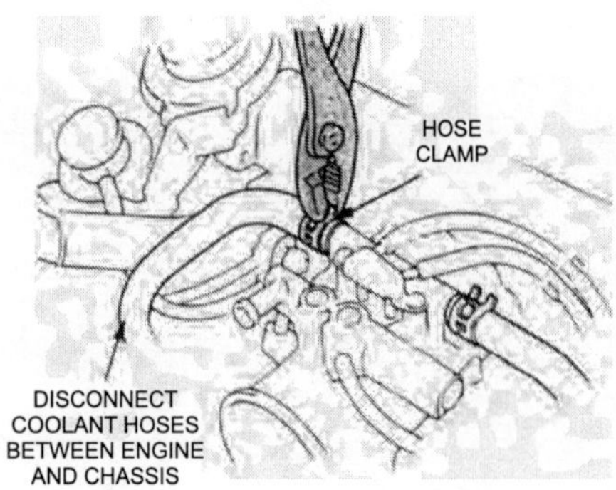

(c) Remove hose clamps and hoses as required.

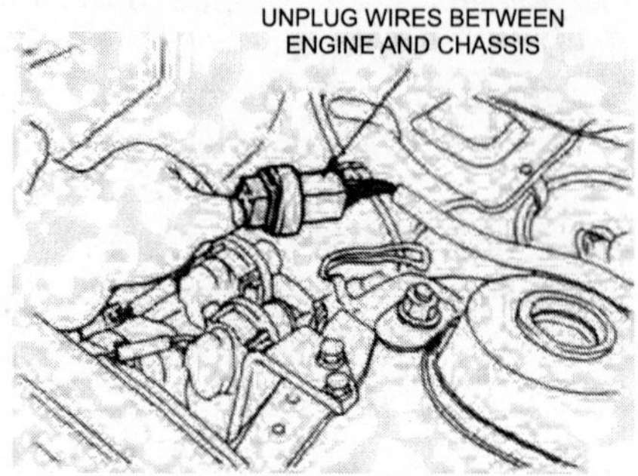

(d) Unplug -wiring harnesses and wires preventing engine removal.

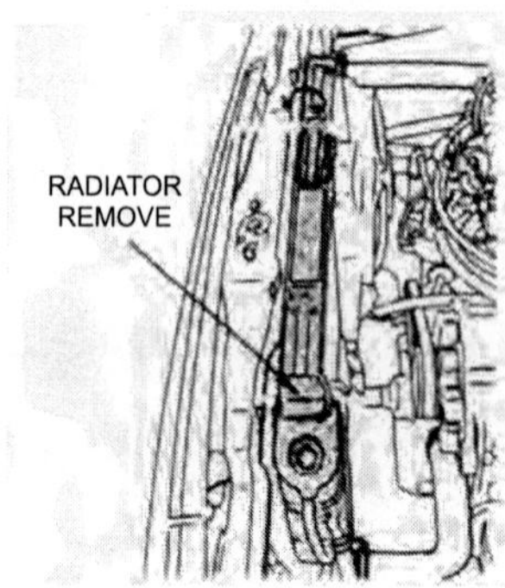

(e) You must usually remove the radiator

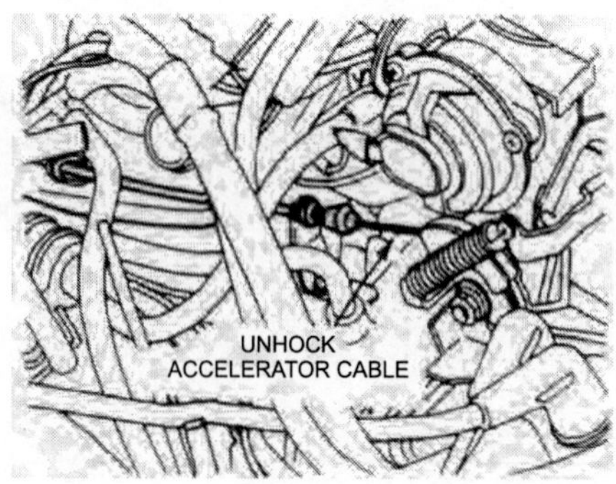

(f) Throttle cable is normally disconnected at throttle body or carburetor. (Subaru)

Fig. 11.11. *Major components, on top of engine compartment, that should be disconnected. (Subaru)*

11. Unplug all electrical wires between the engine and chassis. These are temperature and pressure-sending unit wires, alternator, starter and air-conditioning electrical connections, coil high-tension lead, coil lead, and ground trap. If necessary, use masking tape to label or identify the wires. This will facilitate reconnection, *Figs. 11.12* and *11.13 (a).*

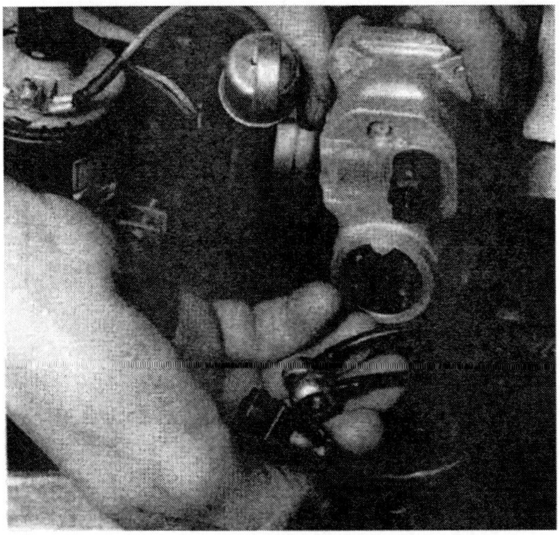

Fig. 11.12 *Make sure you remember how hoses and wires reconnect. If you forget, performance problems can result after reinstalling the engine. Label hoses and wires when needed. (Saab)*

12. *Disconnect all coolant hoses* and the transmission oil cooler lines from the radiator, *Fig. 11.13 (c).* Remove the radiator even if it looks it is not in the way. Radiators are very easily damaged during removal and very expensive to repair. It is better to remove the radiator than risk crushing it during engine removal. This is good time to inspect the radiator for leaks, rust damage, or clogging. Therefore, remove the radiator and have it checked and cleaned while it is out of the chassis. Besides the coolant hoses there may be two transmission cooler lines attached to the radiator. If the vehicle has an automatic transmission, inspect the bottom radiator tank. If you find two steel lines, use a tubing wrench to disconnect them, *Fig. 11.13 (c).* Cap the open ends. Follow the lines as they run back to the transmission. Often these lines attach to the engine's oil pan. If so undo these.

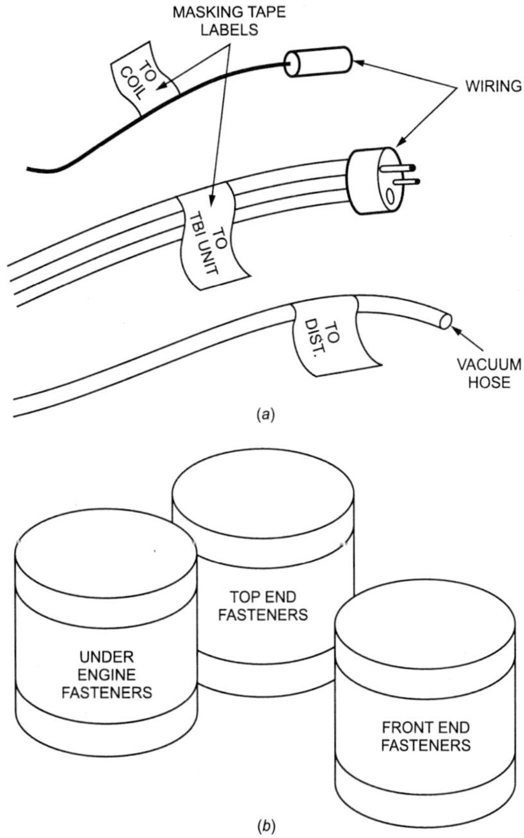

Fig. 11.13 *(a) Masking tape can be made into labels for hoses and wires, (b) Use several containers to keep bolts, nuts, brackets, and other parts organized. Then, during reassembly and installation of the engine, you can go to each container when working on one area of engine, compartment.*

13. *Remove fuel,* and *vacuum hoses* or *lines* that prevent engine removal. Label vacuum hoses if confusing. If necessary, bleed fuel pressure before disconnecting fuel lines, *Fig. 11.13.*
14. *Do not* disconnect any power steering or air conditioning lines unless absolutely necessary. Usually, the power steering pump or air conditioning compressor can be unbolted and placed on one side the engine compartment. This will save fluid and prevent loss of refrigerant. In case the design so requires, *recover the air-conditioning refrigerant* and

set the air-conditioning compressor aside and remove the condenser, if necessary.

15. Keep all fasteners organized in several different containers, e.g., keep all of the bolts and nuts from the front of the engine in one container; keep engine top end separate in second container, and bottom end fasteners separate in third container. This will facilitate fast reassembly, *Fig. 11.13(b)*.

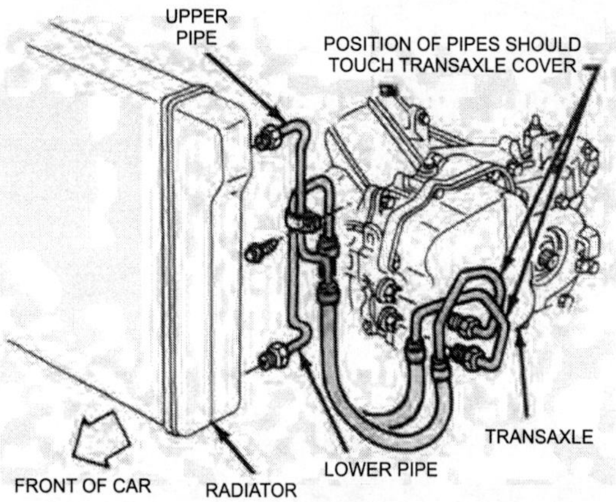

Fig. 11.13 *(b) Transmission cooler lines at the radiator. (Cadillac Div. GM)*

16. Check rear side of the engine for hidden ground wires, dipstick tube, or any other parts that hinder engine removal.
17. Remove as many accessories as possible. This removal makes the engine lighter and easier to handle. Stubborn attaching hardware is actually easier to break loose when the engine is firmly attached to the chassis. Some normally removed accessories are the carburetor, power steering pump, air-conditioning compressor, starter and alternator. Most accessories have both a pivot bolt and clinch bolt. The clinch bolt slides in a slot for belt adjustment. The pivot bolt is the one the unit swings on. You must remove both bolts entirely.
18. Removing the *air conditioning compressor* deserves special safety attention. If you open one of the high-pressure freon lines – the large, heavily insulated pair – the gas will escape violently. Freon super-cools when it expands into the atmosphere. If you are in the way, it will supper cool you. The resulting frostbite can cost you your finger, a hand, or eyesight. Freon gas is very dangerous. Freon gas releases phosgene gas when it is burned in an engine. Phosgene is a deadly nerve gas. Do not have any fire or running engines near freon gas. The best way to handle freon is to keep it in its lines. When removing the air conditioning compressor, leave the freon lines alone. Just unbolt the compressor from the engine and wire the compressor to the chassis so it cannot slip.
19. Remove all *bell housing bolts.* Look carefully for bolts hidden in casting recesses. Also, the lower bell housing area often has a shield. On any vehicle, remove the shield if it bolts to both engine and transmission. On automatic transmission vehicles, remove the shield in any case.
20. Disconnect the front *exhaust pipe* from the exhaust manifold, *Fig. 11.14.* Twist the pipe until you can remove the exhaust seal. Disconnect the air management system pipe.

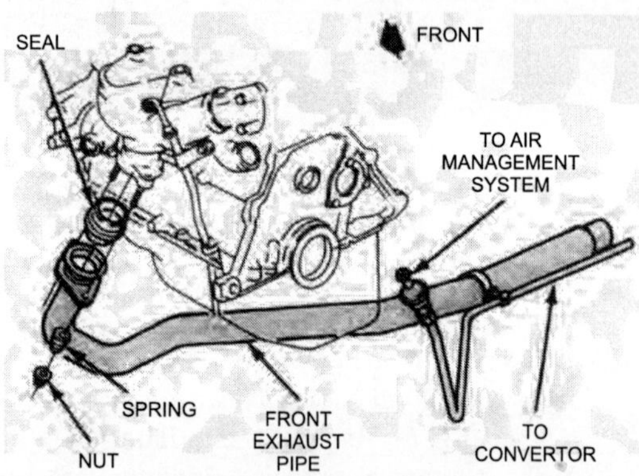

Fig. 11.14 *(a) Disconnecting the front exhaust pipe and the air management pipe. (Cadillac Div. GM)*

21 The *dust shield of automatic transmission* must come off to gain access to the flexplate bolts. After removing the shield, inspect the flexplate, *Fig. 11.15.* Look for a bolt, which passes through the flexplate and into the torque converter. You may have to rotate the engine so a flexplate bolt shows in the bell housing area open to you. Remove the bolt. You might have to hold a nut on the torque converter side. After removing he first bolt, rotate the engine until next bolt is in view. Remove it and all remaining bolts. Usually there are only three bolts, but

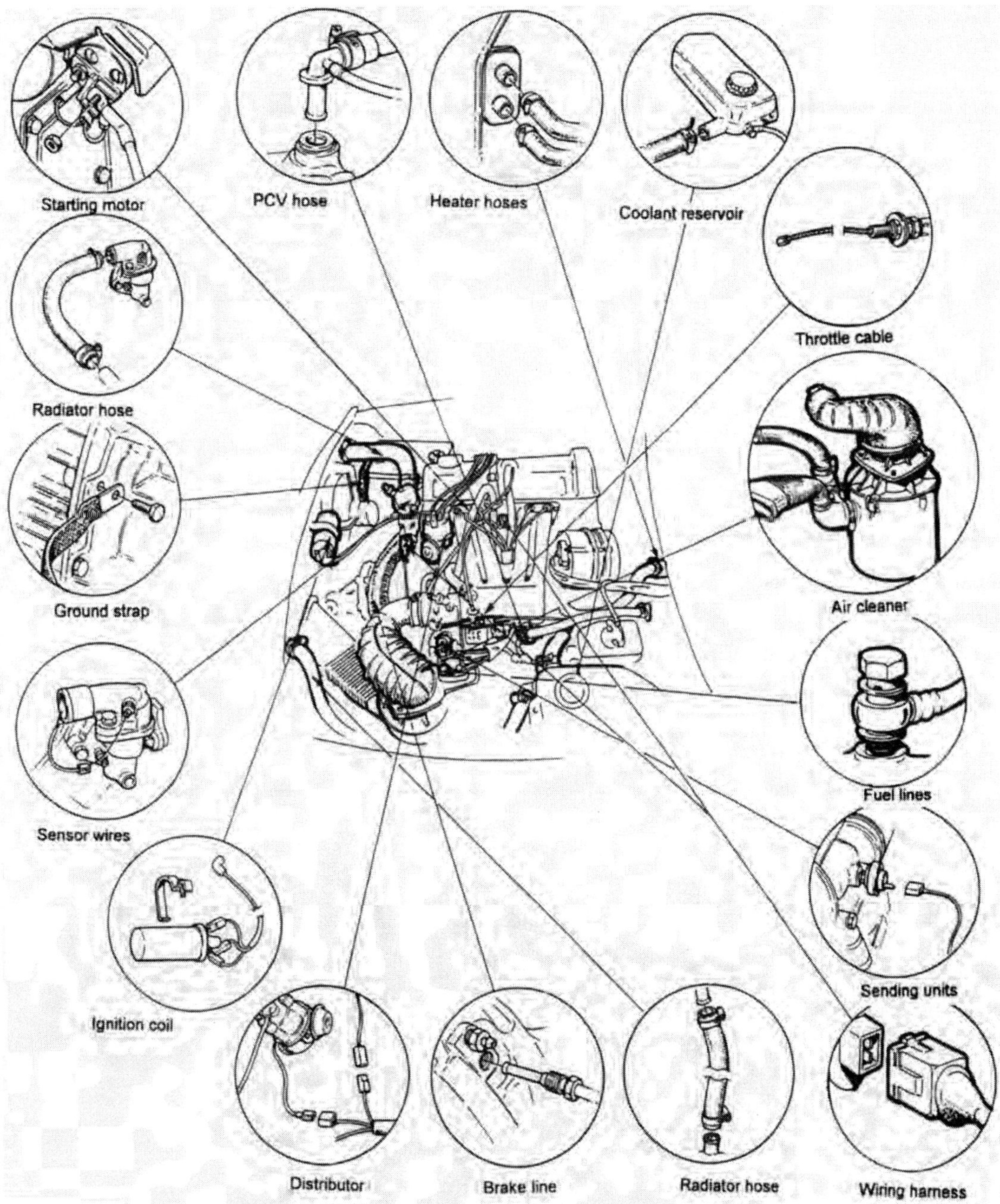

Fig.11.14 *(b) This is service manual illustration showing many of the parts that must be disconnected before engine removal. (Saab)*

there may be more. Place a floor jack under the transmission.

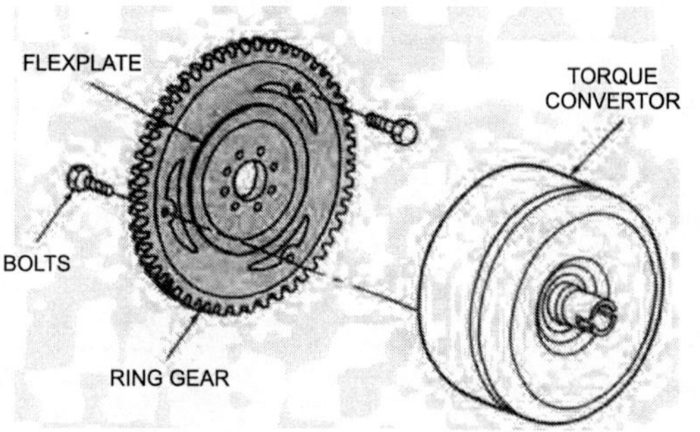

Fig. 11.15 *Flexplate-to-torque-converter attachment. (Chrysler Corporation)*

22. Next *disconnect the engine mounts, Fig. 11.16.* Typical mounts have a horizontally mounted through bolt. Slip out the bolt after unscrewing its nut. Other designs use bolts and studs passed vertically through rubber cushions. Undo the nuts or bolts. Transversely mounted engines use two mounts plus torque reaction rods, *Fig. 11.17.* These rods attach to the firewall at one end and the engine at the other. It is best to completely remove these rods. There might be a second rod mounted low on the engine.

11.8 REMOVING TRANSMISSION WITH ENGINE

There are two ways to remove the engine:

- The engine can be lifted out of the chassis with the transmission/transaxle attached.
- The transmission/transaxle can be separated from the engine and left in the chassis.

The method to be used must be determined before the engine is removed from the vehicle. Some front-wheel drive vehicles, using transaxle, require that the two be removed as a unit. It may be easier to remove the two together. When in doubt refer the applicable service manual for details. Drain the transmission fluid from the transmission or transaxle if it is to be removed.

In case of rear-wheel drive vehicles, the drive shaft, transmission and clutch linkage, speedometer cable, rear motor mount, and other parts should also be disconnected and tagged. In some installations, it may be necessary to loosen the steering linkage idler arm to give clearance. With a transaxle the axle shafts must be disconnected.

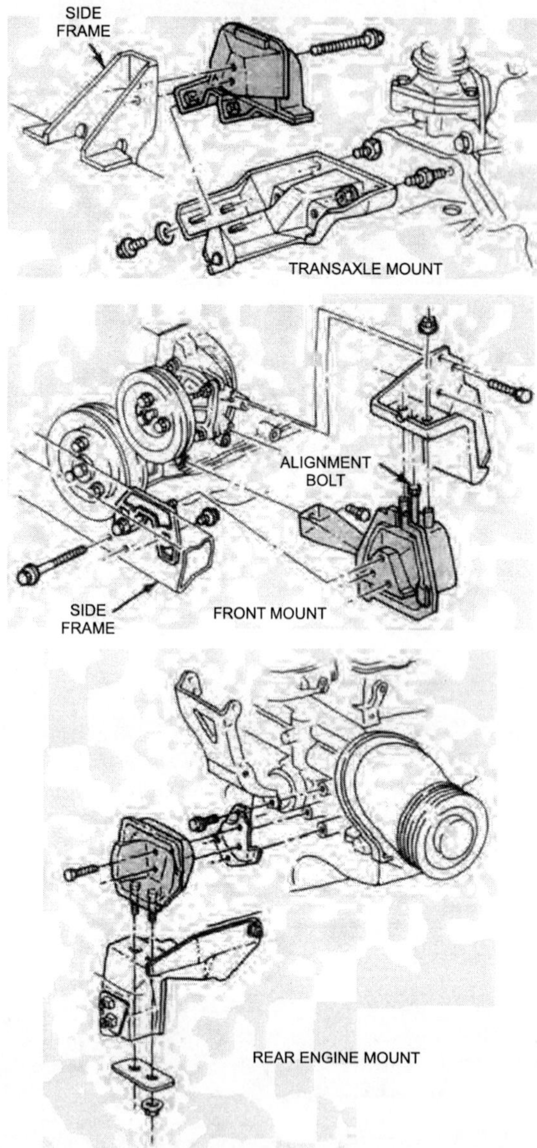

Fig. 11.16 *Motor Mounts. (Cadillac Div. GM)*

Egine Removal

Before engine removal, *double-check* that everything hindering engine removal is disconnected or removed. Check:

1. Behind and under the engine for hidden wires or ground straps.
2. That all bell-housing bolts are out.
3. That the torque converter bolts have been removed when the automatic transmission/transaxle is to stay in vehicle.

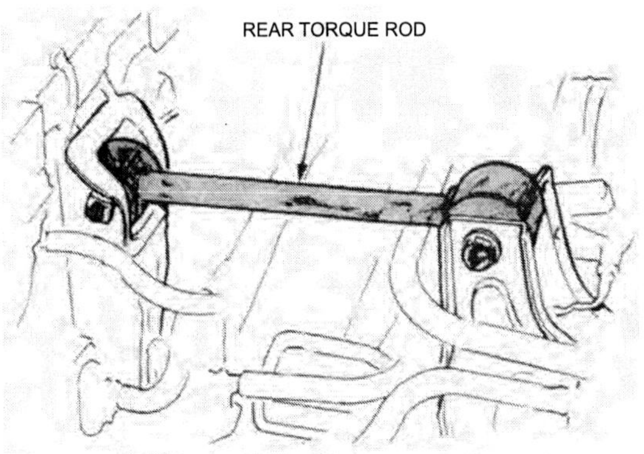

Fig. 11.17 *Torque reaction rods. (American Honda Motor Co. Inc.)*

4. That all fuel lines are disconnected and properly plugged.
5. That motor mounts are unbolted.
6. That a floor jack or holding bar is supporting the transmission (when staying in vehicle).

Installing Lifting Fixture or Chain

Connect the lifting fixture or chain to the engine. Position the fixture at recommended lifting points. Some engines have brackets on the engine, *Fig. 11.18.* If a lifting chain is to be used, fasten it to the engine on opposite sides and ends. If a single length of the chain is used, carefully install a bolt, nut, and washer on the chain to keep it from slipping and dropping the engine.

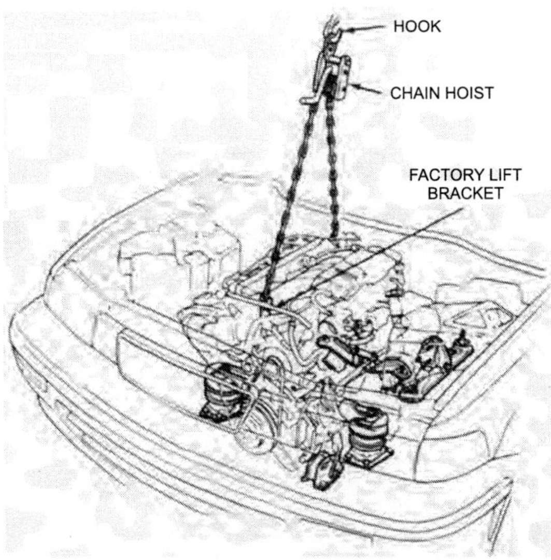

Fig. 11.18 *When lifting an engine, use a secure chain hoist preferably connected to factory lift brackets on the engine. (Honda)*

If bolts are used to secure the chain to the engine, make sure they are large enough in diameter and that they are fully installed. The bolts must not be too long and stick out from the chain, or too short to result in a thread engagement distance equal to 1.5 times the bolt diameter, *Fig. 11.19.*

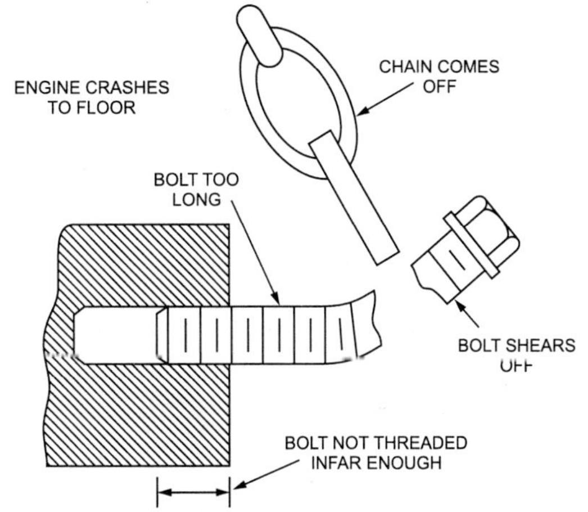

(a) Bolt too long was bent and snapped off. Engine crashed to floor causing serious part damage.

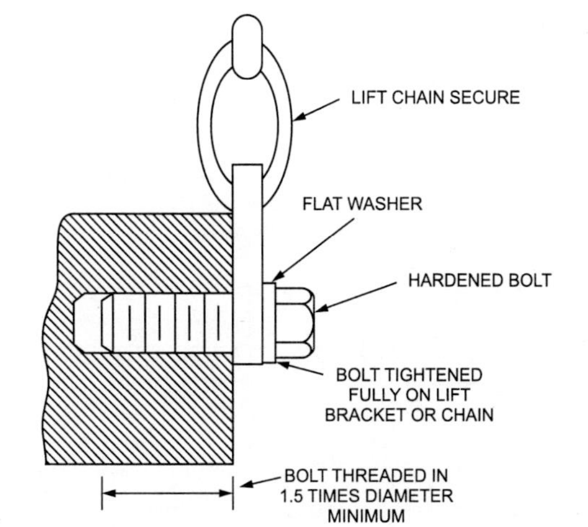

(b) Bolt correctly installed. It is tightened down and threaded in far enough.

Fig. 11.19 *Make sure bolts used to fasten lift chain engine are large enough in diameter. Bolt should be fully tightened against chain and should thread into part 1.5 times its diameter.*

In general, the fixture or chain is positioned such that it will raise the engine in a level manner or at the correct angle with nose up, if needed, *Fig. 11.20.* If one lifting point is at the right front of the cylinder head, the other should be on the left rear of the head. Use common sense and follow manufacturer's instructions, *Fig. 11.21.*

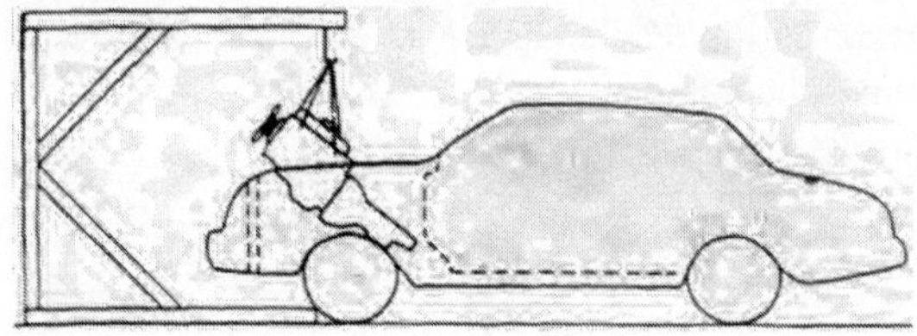

Fig. 11.20 *An engine must be tipped as it is pulled from the chassis.*

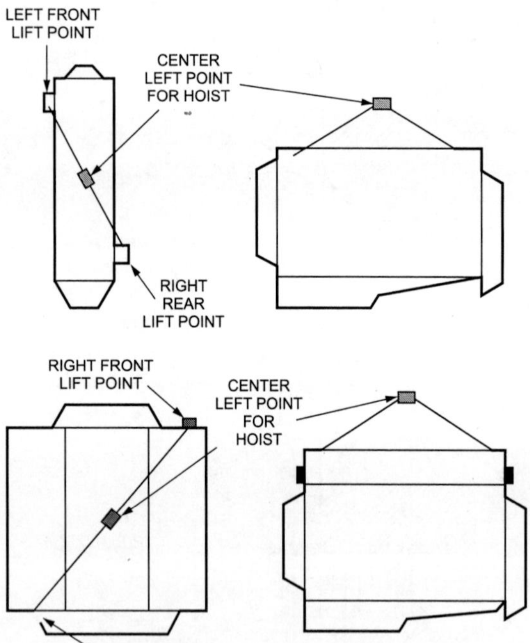

Fig. 11.21 *Lift points should be on opposite sides and ends of engine. A centre lift point will raise engine in level manner. If you need to tilt engine, as when removing engine and transmission together, move centre lift point one way or other.*

Exercise proper care to avoid any damage to motor mounts during engine removal. Some late model engine motor mounts are liquid-filled or computer controlled, hydraulic units to reduce engine noise. The computer can alter the stiffness of the mounts by using solenoid valves that control hydraulic flow in mounts. Any damage caused to these mounts results in an expensive mistake.

Lifting Engine Out of the Vehicle

Attach the lifting device, i.e., a portable crane or a hoist, to the fixture or chain on the engine, *Fig. 11.22,* centred over the engine. Position a floor jack under the transmission to support it if necessary, *Fig. 11.23.* Raise the engine slowly about 3 to 5 centimeters. Then check that everything is out of the way and disconnected, *Fig. 11.24 (a).*

Be careful not to place any part of your body under the engine held in the air. A heavy engine can crush body parts and cripple, and even kill, when dropped.

Continue slowly raising the engine while pulling. This will separate the engine from the transmission or slide the transmission out from under the firewall. Be careful not to let the engine bind or damage parts. When the engine is high enough to clear the radiator support, roll the crane and engine straight out and away from the vehicle. When using a stationary hoist, roll the vehicle out from under the engine, *Fig. 11.24 (b).*

Do not allow a transmission or transaxle hang unsupported after engine removal. This could damage the mounts, drive shaft; drive axles, and other parts.

With some front-wheel drive vehicles, the engine and transaxle are removed from underneath the vehicle, *Fig. 11.24 (c).* With some vans, the engine can be removed through the large door in the side of the body. Refer to the applicable service manual for details for the vehicle under repair.

As soon as it is possible, lower the engine to the floor or mount it on an engine stand.

Never work on an engine that is held by a crane or hoist. The engine could shift and fall, damaging the engine or causing serious body injury.

Keep Work Area Clean

Coolant and engine oil usually drips onto the shop floor during engine removal. To prevent an accident, wipe up any spills as soon as they occur. Keep all tools properly stacked and work area clean.

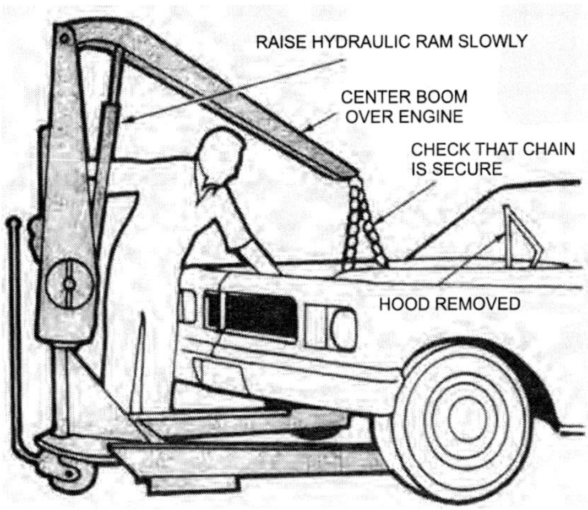

Fig. 11.22 *Crane boom should be centred directly over lift point to keep from shifting and binding engine. Raise slowly while watching for clearance all around engine. (Florida Dept. of Voc. Ed.)*

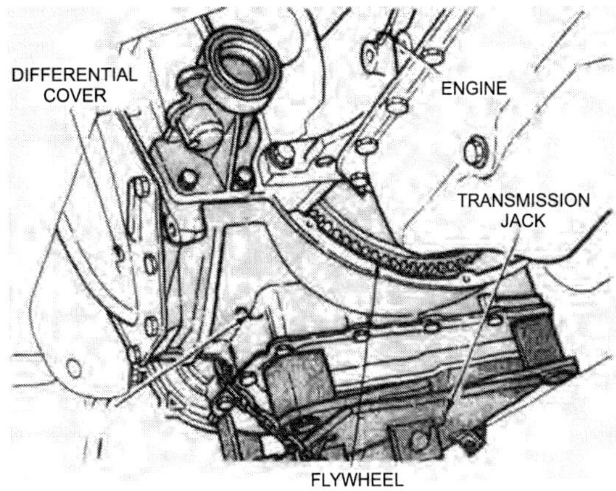

Fig. 11.23 *If transmission or transaxle is to stay in vehicle, support transmission or transaxle. Do not let it hang unsupported. Here mechanic has placed transmission jack under transaxle. (Chrysler)*

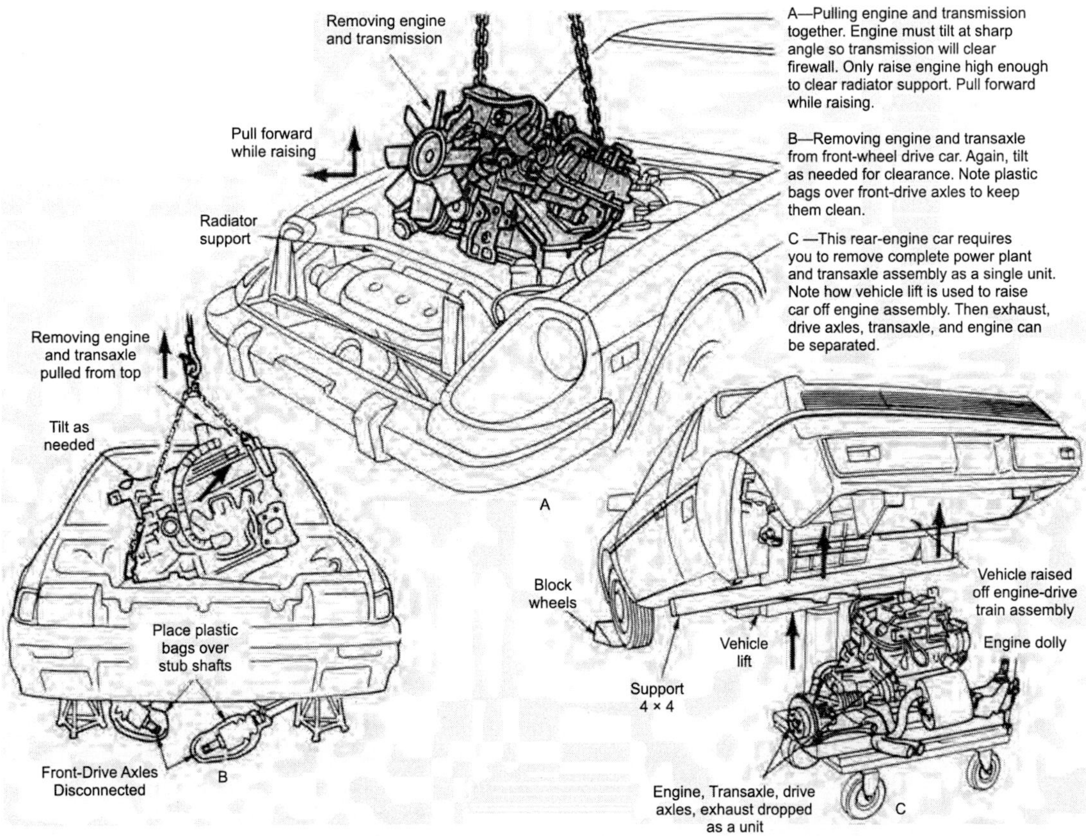

Fig. 11.24 *Typical methods of removing engine, or engine and transmission. Methods vary with design of vehicle. (Nissan, Honda, Pontiac)*

11.9 SEPARATING ENGINE FROM TRANSMISSION OR TRANSAXLE

When the engine and transmission/transaxle are pulled together, we need to separate them on the shop floor. To do so, place wooden blocks under the engine or leave the hoist and chain on the engine to prevent from flipping over while sitting on the floor.

With automatic transmissions, remove the bell housing-to-engine attaching bolts. Carefully look around the engine to locate all bolts, because some can be hard to find. Remove the starting motor and any other parts hindering transmission removal.

Torque converter may be left bolted to the engine flywheel, or unbolted so allow it to come out of the transmission. Lift up, wiggle, and pull the transmission or transaxle off the back of the engine. Be certain not to damage the location and alignment dowel pins when removing or reinstalling a transmission, as these dowels are critical for the proper alignment of the engine and transmission assembly, *Fig. 11.25.*

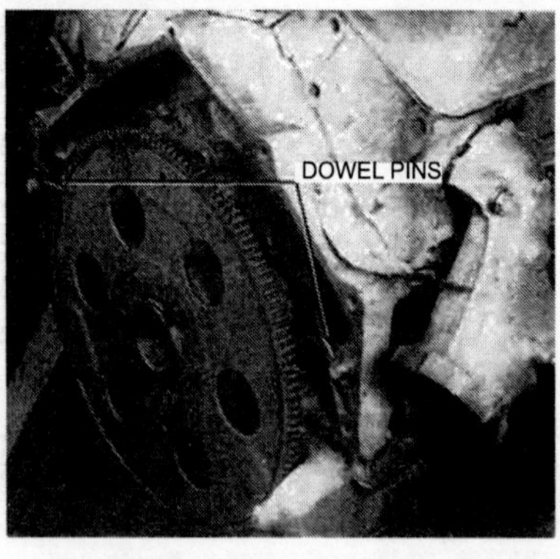

Fig. 11.25 *Be certain not to damage these alignment dowel pins when removing or reinstalling a transmission. These dowel pins are critical for the proper alignment of the engine transmission assembly.*

With a manual transmission or transaxle, follow almost the same procedures. However, the transmission should be unbolted first. Then remove the bell housing, clutch, and flywheel.

Because there are many design variations with rear and front-wheel drive vehicles, when in doubt about something, consult the manufacturer's service manual.

11.10 ENGINE DISASSEMBLY

You can either save time during engine *teardown* or make lots of work for yourself. It depends on how you are organized and how safely you disassemble the engine. Because disassembly is always part of a larger job, the temptation is to disassemble the engine as quickly as possible. Speed, however, is less important than accuracy. Accuracy in engine disassembly is *keeping all parts in proper order.* Only then you can efficiently repair and reassemble the engine. *Fig. 11.26* shows the major engine components that you will be disassembling.

As with engine removal, containers are needed to keep the parts separated. Coffee cans, cardboard boxes, metal trays, or bin boxes all work well. But each contained must be clearly marked.

Remember to put only one type of bolt in a container, e.g., one can for head bolts, another for manifold bolts, and so on. Felt-tip and ballpoint markings will fade in several weeks. Rain will fade your numbering even faster. If the chassis sits outside while the engine is apart, keep the engine compartment covered properly.

When overhauling an engine, many shops will remove the engine from the chassis, and then steam clean the entire unit with a steam and detergent solution. *Steam cleaning* can be done with a gun-type cleaner or a tank-type cleaner.

Cleaning the engine before it is disassembled will make for an easier disassembly. In addition cleaning will reveal defects and conditions that will help in diagnosing the cause of the trouble and assist in preventing its reoccurrence.

The best way to disassemble an engine is on an engine stand, *Fig. 11.27.* A stand holds the engine, leaving you with both hands free. Furthermore, the stand allows you to roll the engine over. This feature makes working on the bottom of the engine much easier.

Mount the engine on its stand while the engine hangs from the engine hoist. Remove the stand's head and bolt it to the block's bell housing area. Then fit the stand's base onto the head. Snug all stand bolts wrench-tight; then lower the engine stand to the ground. Disconnect the hoist.

If no stand is available, set the engine on a bench. Use short wood blocks to steady the engine. As you

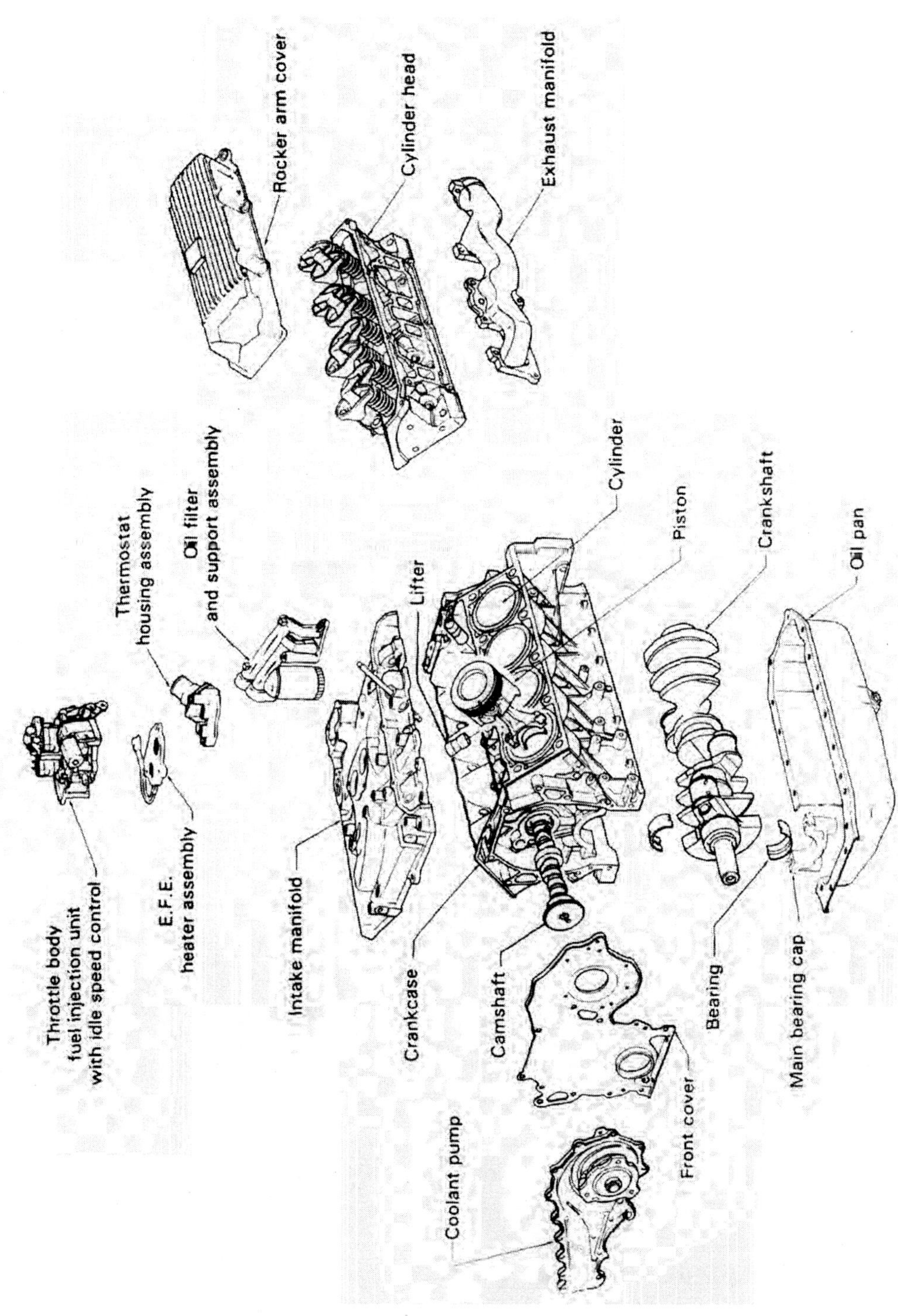

Fig. 11.26 *Major engine components. (Cadillac Div. GM)*

will remove parts from the engine, its weight will shift. Then you will have to change the wooden blocks to keep the engine steady.

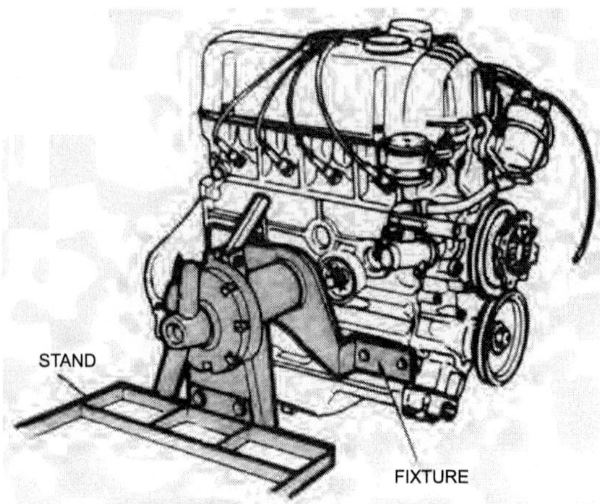

Fig. 11.27 *Engine mounted on a stand. (Nissan Motor Corp., USA)*

With the engine mounted on an engine stand or sitting on blocks, it is time to begin teardown. Proceed carefully, go slow and inspect each part for signs of trouble. Look for wear, cracks, damage, seal leakage, gasket leakage, or damaged core plugs. Overlooking any problem might result in the engine repair failure in service. The whole labour and cost could just go wasted.

Engine disassembly is not especially dangerous. However, there is always danger of getting hurt. Beware of falling parts. Always keep the engine securely supported. If it rolls over on your hand, you could be seriously injured. Always pull the wrench towards body when removing a bolt.

Engine teardown methods vary somewhat from engine to engine. However, general procedures are similar and apply to all engines. For detailed information always consult the correct service manual.

Remaining Accessories

Before disassembling the engine, any remaining accessories should be removed. Accessories that may remain are the temperature and pressure senders, fuel pump, water pump, clutch, and flywheel. There may be other accessories depending on how complicated the engine is.

If you did not remove the carburetor or fuel injection and the distributor before taking the engine out of the vehicle, now is the time to remove them. Remove the carburetor's four base nuts and lift the unit off. Some carburetors use long bolts that pass through the top of the carburetor and go into the manifold. Handle fuel injection parts with care to prevent any damage to them.

The distributor may still be on the engine. Before removing the distributor, the engine should be turned until the #1 piston is at TDC and the index mark on the crankshaft pulley aligns with the pointer on the timing chain cover, *Fig. 11.28.* Removing the spark plugs makes it easier to turn the crankshaft over to align the index mark and the pointer at TDC. Now mark the distributor housing and the rotor position on the block.

Where distributor meets the engine there will be a hold-down bolt, *Fig. 11.29.* Remove this bolt and any retaining clip. Then pull the distributor out of the engine and set it aside. If there is a distributor drive shaft, remove it now. Many distributor drives have a threaded hole in their centres. This hole allows you to screw a long bolt into he drive shaft. The bolt is then the needed handle to pull the drive out.

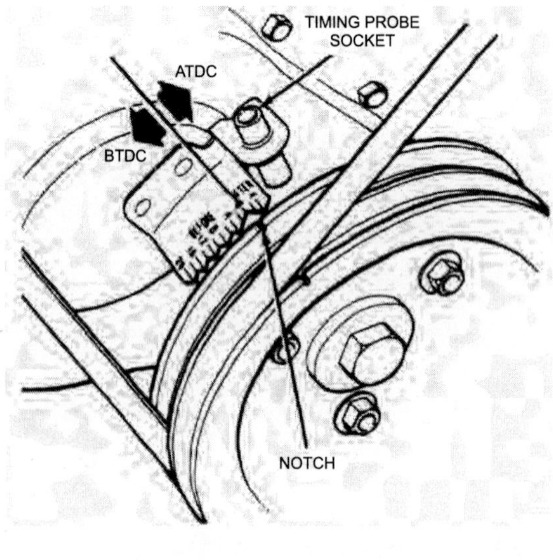

Fig. 11.28 *Harmonic balancer at TDC. (American Motors Corp.)*

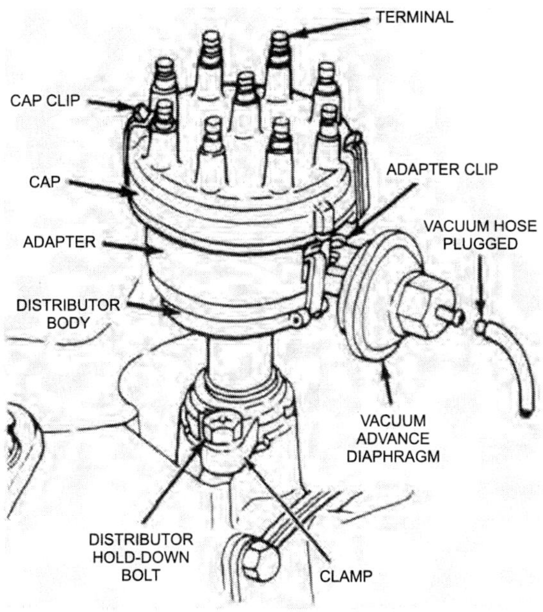

Fig. 11.29 *Distributor attachments. (Ford Motors)*

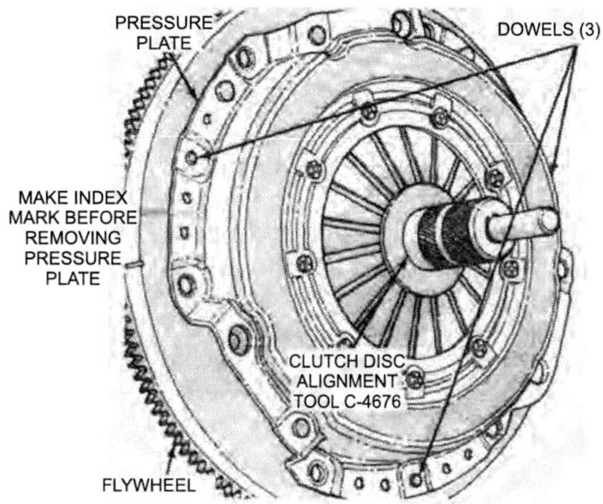

Fig. 11.30 *Marking the pressure plate and flywheel. (Chrysler Corp.)*

Clutch Assembly

At the rear of the engine, there will be either a clutch and flywheel or a flexplate. With a clutch, index the pressure plate relative to the flywheel. Make two small marks on the pressure plate near its outer edge. On the flywheel make two matching marks, *Fig. 11.30.* These marks allow you to reassemble the clutch and flywheel in the same relative position. This alignment is important if they have been balanced as a unit.

Pressure Plate: Now remove the pressure plate bolts. Unscrew them one turn at a time all around the plate. Otherwise the pressure plate springs may bend from uneven torque. Once the bolts are finger-tight, unscrew them all the way. Pull the pressure plate off the flywheel, *Fig. 11.31.* The clutch disc will fall free, therefore, be ready for it. Keep your fingers off disc. Grease from hands can pollute the disc and it might cause clutch to grab after reassembly. Handle the disc by the edges. Mark the flywheel side of disc for reassembly.

Flywheel: The flywheel bolts come next. They may have *lock-plates.* Lock-plates are metal tabs that bend over the bolt head or nut to keep them from loosening. Therefore, bend back any lock-plates with a punch or screwdriver and undo the flywheel bolts.

You may have trouble with the flywheel rotating when removing these bolts. The best way to overcome this problem is to use an air-powered impact gun on the flywheel bolts. The sharp impacts of the air wrench transmit almost all their torque capacity to the bolts. There is not enough torque left over to rotate the engine. If an air impact wrench is not available, put a hand wrench on the front pulley crankshaft nut. You will have to hold the crankshaft from rotating with one hand while unscrewing the flywheel bolts with the other. Another way to hold the crankshaft is to slip a large slot screwdriver into the teeth of the ring gear. Brace the screwdriver's shank against a dowel pin. If there is no dowel pin, run a bell housing bolt into its hole and pry against it.

Flexplate: Flexplates are much simpler because there is no clutch. Simply remove the bolts on the flexplate as you would remove a flywheel. Generally, flexplates and flywheels are located by dowels to the crankshaft. But if there are no dowels, then mark (index) them to the crankshaft.

Engine Mounts and Fuel Pump

On the sides of the engine, you will find mounts, sending units, fuel pumps, and the like. Go around the block and systematically remove these parts. Keep the hardware arranged and separated in boxes.

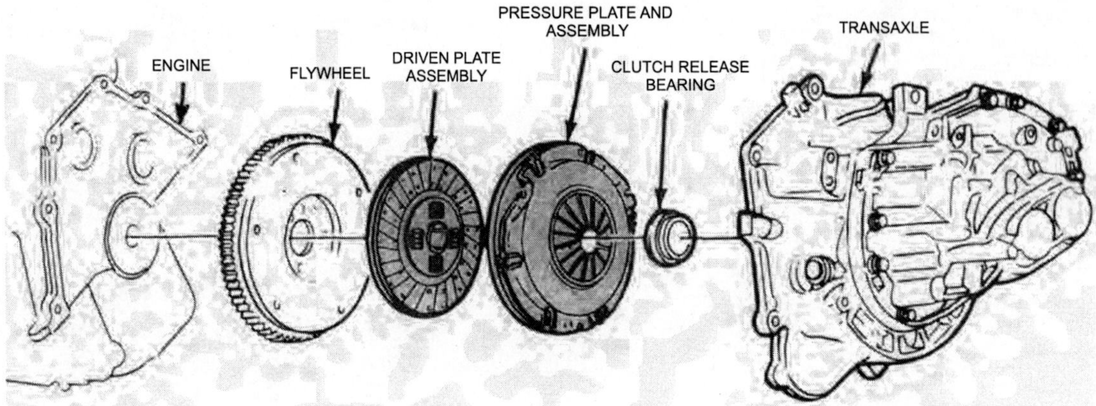

Fig. 11.31 *Remove the clutch and pressure plate. (Chevrolet Div. GM)*

The fuel pump bolts may feel very tight. If so, rotate the crankshaft 180°. This will cause the fuel pump's eccentric on the camshaft, unloading the bolts. When the fuel pump is pulled from the block, it may be followed by a solid push rod. On some engines a small plug or plate may need to be removed before the push rod will come out of the block.

11.11 ENGINE FRONT END DISASSEMBLY

"Engine front-end disassembly is easy if a few basic rules and service manual instructions are properly followed. The engine front end should be disassembled first, particularly with OHC engines. Follow these steps:

1. At the engine's front, you will find the water pump. Remove the water pump and leave the pulley attached to the pump unless it has to come off in order to get to the bolt. Around the pump are its attaching bolts. Remove all these bolts; then gently pry the pump away from the block, *Fig. 11.32.* You will have to pull the pump from the block because the gaskets will be stuck on. This is a good time to inspect the water pump for a leaking seal or defective bearing. Also remove any other parts bolted in front of the engine timing cover. If a timing belt is used, remove the belt cover. Loosen the tensioner and slip off the belt. This needs be done before cylinder head removal.

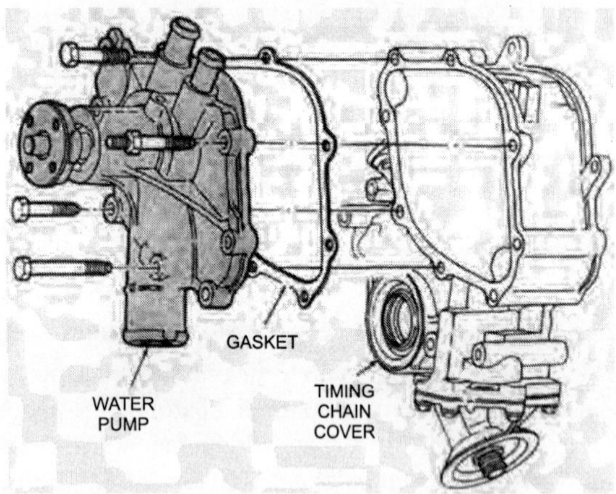

Fig. 11.32 *Removing the water pump. (Ford Motors)*

2. **Do not** attempt to rotate the crankshaft of an OHC engine with the timing belt off and cylinder head still in place. The pistons could slide up and bend the valves.
3. Remove the harmonic balancer with the help of a *wheel puller.* The balancer is usually press fitted onto the crankshaft. Use a puller when removing a front oil seal, *Fig. 11.33.*
4. Unbolt and remove the timing chain or gear cover. To pry the cover, if necessary, do it lightly

while tapping with a rubber or plastic hammer. Do not bend or scar mating surfaces.

5. Remove the oil slinger and timing mechanism. Normally, the timing gears or sprockets will slide off after slight taps with a brass hammer. If not, use a wheel puller to remove them, *Fig. 11.34.*

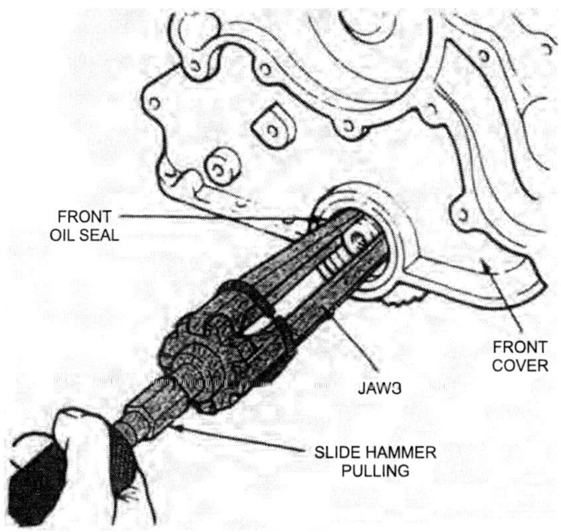

Fig. 11.33 *Most mechanics start engine teardown with front end parts. This in especially true with OHC engines. Here a slide hammer puller is being used to remove front oil seal from front cover. Note that water pump is already off. (Ford Motor Co.)*

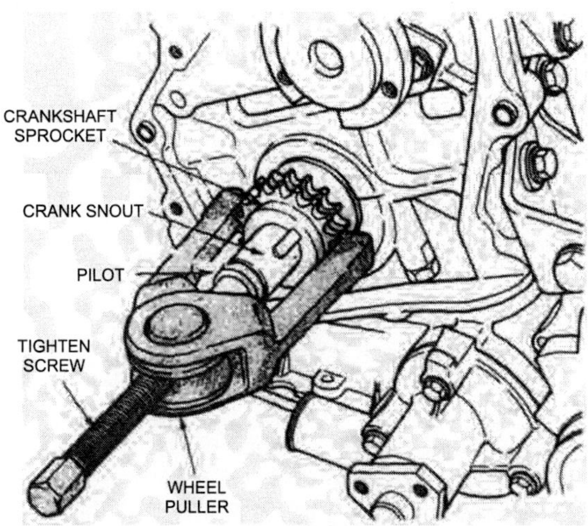

Fig. 11.34 *Front damper and crankshaft sprocket can require use of screw-type wheel puller. Tighten screw and jaws will force part off crankshaft spout. (Chrysler)*

6. If the oil pump or any other components are in the front cover, slip them out while checking for any wear.

11.12 ENGINE TOP END DISASSEMBLY

The engine top end generally includes the valve train and cylinder head related parts. These are usually the second set of parts of the engine being disassembled. Follow the general steps given below:

1. Remove the external parts on top side of the engine, e.g., carburetor, fuel rail, or throttle body unit, spark plug wires, distributor, etc. Remove all parts that could be damaged or that prevent removal of the cylinder head(s).
2. Before you can remove the head, the manifolds should be removed. On in-line engines the manifolds can be on the same side of the head, *Fig. 11.35.* Or they can be on opposite sides in cross-flow design.

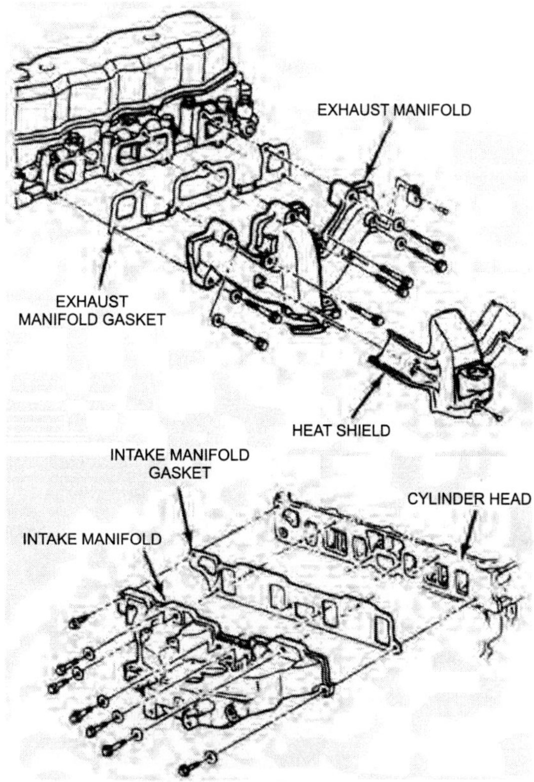

Fig. 11.35 *In-line engine manifolds. (Chevrolet Div. GM)*

Typically, manifolds are mounted to the head by studs and nuts. Spray all nuts and studs with penetrating oil before starting to remove. On cross-flow heads, simply unscrew the nuts, remove any washers, and pull the manifold. If the intake and exhaust manifolds are on the same side, they might be bolted together. Look for bolts running vertically from one manifold to the next. If you find any such bolts, remove them and then remove the head-to-manifold nuts.

On V-engines, the exhaust manifolds are on the outside of the Vee. The intake manifold sits inside the Vee. Remove the exhaust manifolds just as though they were on an in-line engine.

V-engine intake manifolds bolt to the heads around their perimeter, *Fig. 11.36.* First check that there is no water hose running between the water pump and the intake manifold. Remove the bypass hose if it is still attached. Mark the location of any stud bolts, as they must be replaced in the same hole during reassembly, *Fig. 11.37.* Then remove all the manifold bolts.

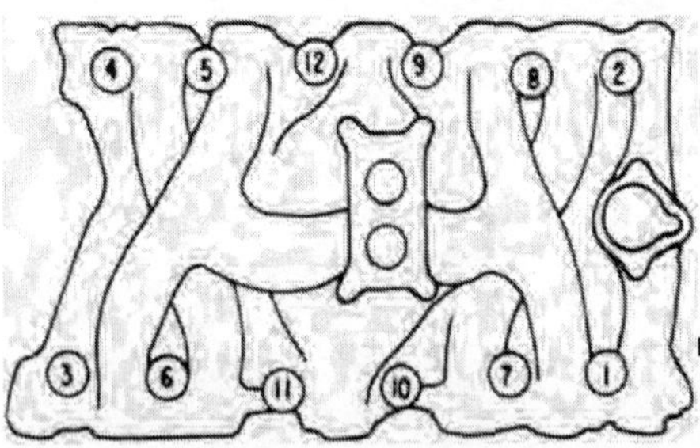

Fig. 11.36 *V-engine intake manifold and bolts. (GM Corp.)*

You will probably have to pry the manifold off. Start at a corner. Perhaps you can find a spot where the manifold overhangs the head or block. Use such a spot to pull up on the manifold. It will take some effort to lift the manifold, but excessive force means there is still a bolt installed somewhere. Find it and remove it. If you are not careful, you can break the manifold. Never pry directly between the head and manifold.

If the gaskets are stuck, a flat blade, such as that of a putty knife, can be worked alongside of the gaskets to loosen them. Care must be exercised to avoid damaging the parting surface as the gasket is loosened.

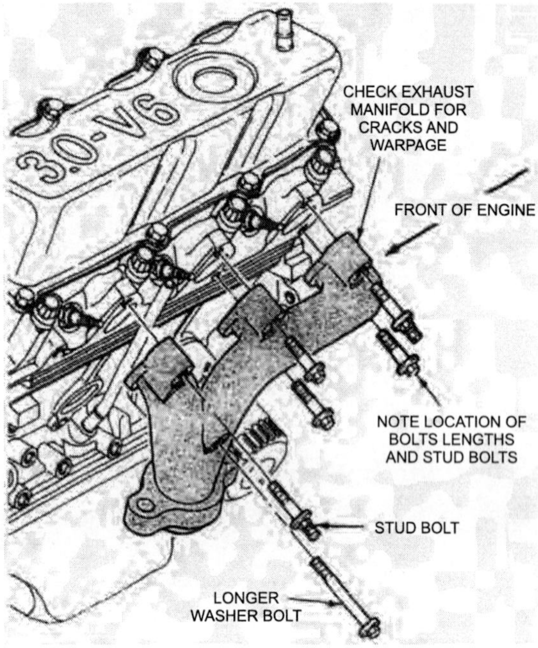

Fig. 11.37 *When removing exhaust manifold, note odd bolt lengths and location of stud bolts. They must be installed in same holes. (Ford Motor Co.)*

3. Unbolt the valve cover(s). If they need some light prying, be careful not to damage the mating surfaces. Light taps with a rubber hammer can also be used to free a valve cover, *Fig. 11.38.*
4. Remove the rocker arm assemblies. If rocker shafts are used, loosen the fasteners following the sequence in *Fig. 11.39.* This is needed to prevent the end fasteners from being bent by valve spring tension.
5. With V-type, push rod engines, it might be required to remove the valve train components before the induction manifold. The push rods may pass through the bottom of the intake, *Fig. 11.40.*

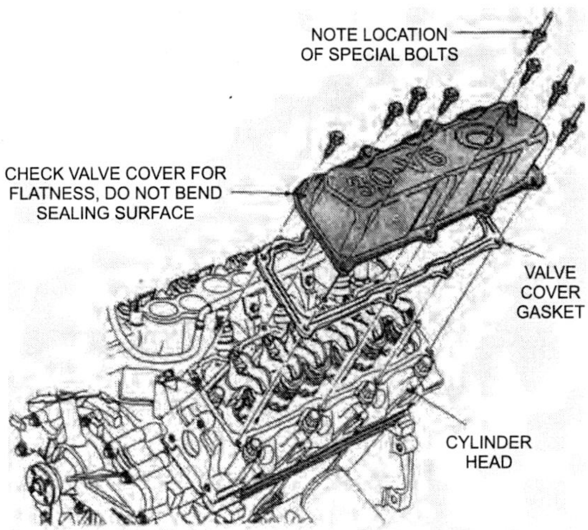

Fig. 11.38 *Do not bend valve covers during removal. Rubber mallet will sometimes free cover from head. Light prying under cover may be required but do not bend lip. Check valve train after cover is off.*

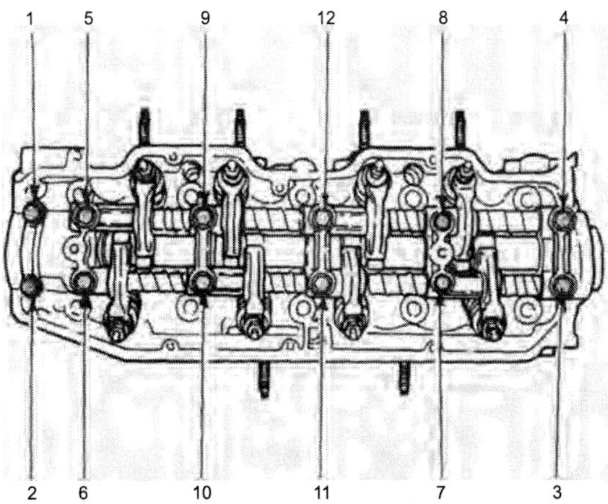

Fig. 11.39 *When unbolting rocker shafts, use this sequence. If you start on one end and work across, spring tension could bend rocker shafts or break caps. (GM)*

6. In case the lifters, push rods, and rocker arms are to be reused, keep them in proper sequence. Use an *organizing tray* or *board* with holes in it or label these parts using masking tape. Components wear patterns and select-fit parts demand that most components be installed in their original locations during reassembly.
7. The camshaft may be removed at this time. It must be carefully eased from the engine to avoid damaging the cam bearings or cam lobes.

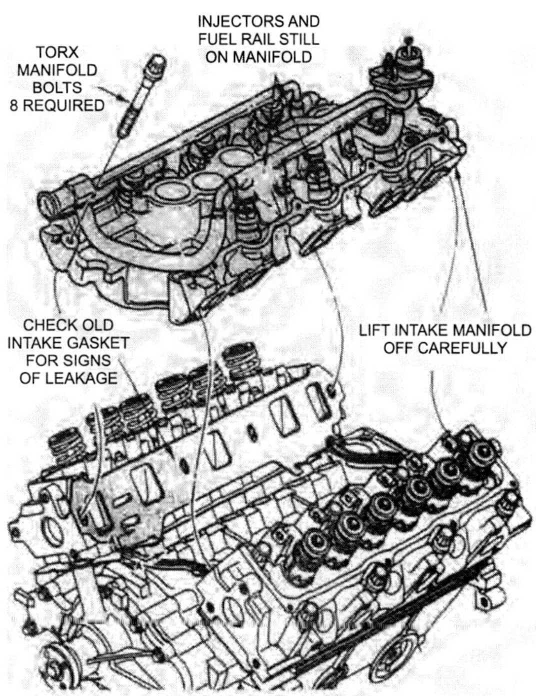

Fig. 11.40 *Intake manifold can be removed with injectors and fuel rail attached. Pry up lightly to free manifold and lift off. Check old gasket for signs of leakage.*

8. Let the engine cool down and never unbolt an aluminium cylinder head when it is still hot. This could cause head warpage.
9. Unbolt the cylinder head, using an air impact *(Fig. 11.41)* or a breaker bar and six-point socket, *Fig. 11.42.* With a V-type engine, punch mark the heads right and left.
10. Often, the head will be stuck to the block. To free the head, insert a breaker bar in one of the ports and pry up. If stuck tight, install the spark plugs or glow plugs and crank the engine. Compression stroke pressure will pop the head up and free it from the block. Avoid prying on mating sealing surfaces.
11. Check the head gasket and deck surfaces for any signs of leakage. Look for oil in the combustion chambers, to determine seal or ring problems.
12. The combustion chamber is exposed when the head is removed. The combustion chamber pocket in the head and top of the piston should be carefully checked. A normal combustion

chamber is coated with a layer of hard, light-coloured deposits. If the combustion chamber has been running too hot, the deposits will be very thin and white coloured.

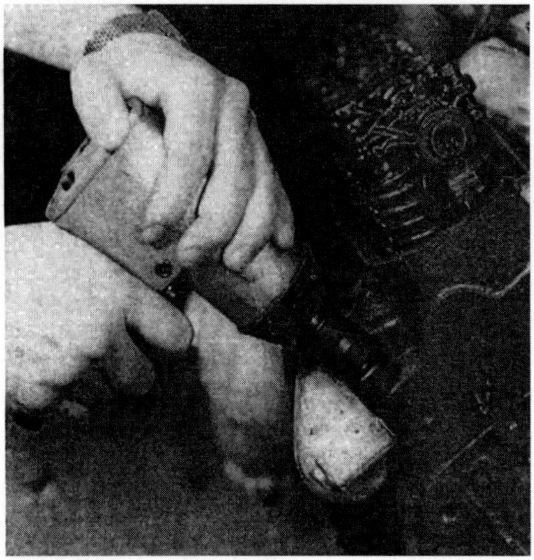

Fig. 11.41 *Air impact will save time during engine disassembly. Mechanic might need to crack larger bolts loose with breaker bar before using impact.*

11.13 CYLINDER HEAD DISASSEMBLY

Start head removal by unbolting the rocker cover. This cover may attach with perimeter bolts or nuts and studs that pass through the outside edge of the cover. Unscrew these bolts or nuts and lift off the cover.

Overhear Valve Engines

Rocker Arms: When the rocker cover is removed, the valve train is exposed. **On OHV engines** the tops of the push rods, rocker arms, and valves become visible, *Fig. 11.43.* Remove the rocker arms. They can be shaft or individually mounted. If shaft mounted, remove the bolts passing through the shaft towers. Then lift the unit shaft, towers, and rockers off the head. For individually mounted rockers, remove the nut sitting inside the rocker arm. Then lift each rocker arm, nut, and *ball pivot* off as a unit, *Fig. 11.44.*

Always keep rocker arms and ball pivots together. Just slip the rocker arm and pivot onto a length of wire. When all rocker arms are on the wire, twist the wire ends together. This is important, because unmatched rocker arms and pivots wear much faster. Unless the pieces are kept together, you will have no

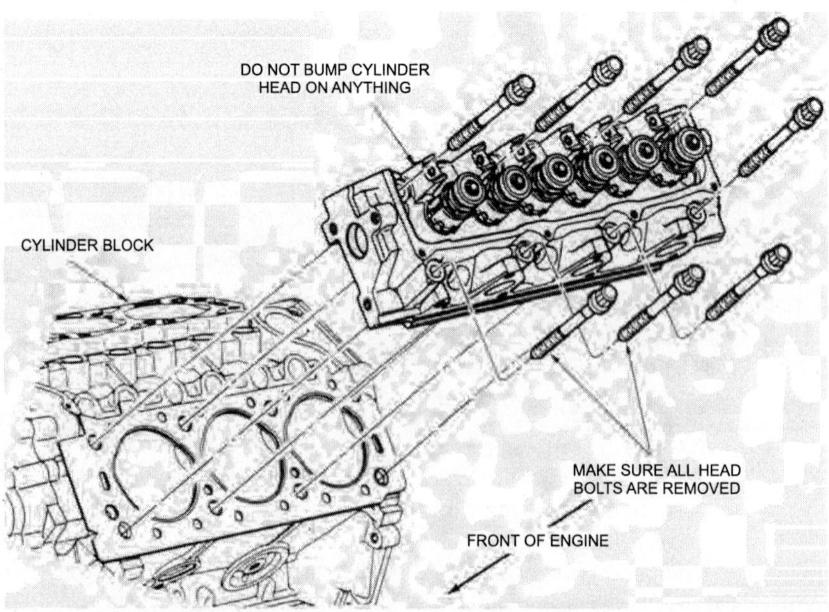

Fig. 11.42 *Allow aluminium head to cool before unbolting to prevent possible warpage. Make sure all head bolts are out. Free head by inserting breaker bar in intake port or crank engine and let pressure free head from gasket. Check head and gasket for trouble right after removal. (Ford Motors Co.)*

choice but to install them out of order or replace them with new ones.

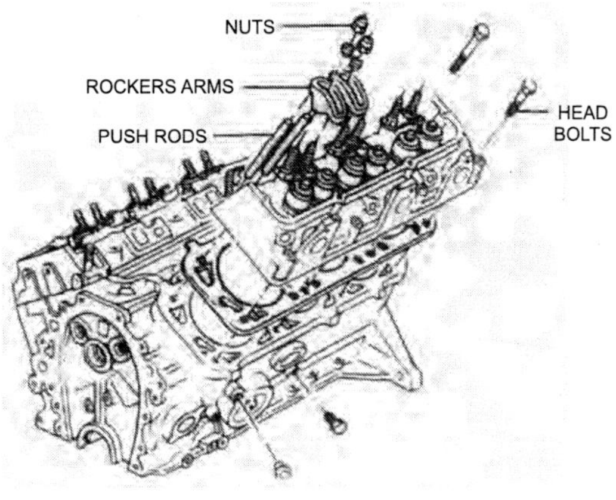

Fig. 11.43 *OHV engine valve train. (Ford Motors.)*

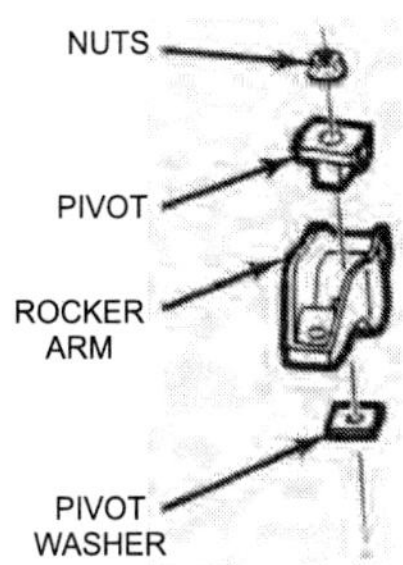

Fig. 11.44 *Individually mounted rocker arm. (Ford Motors.)*

Push Rods: With the rockers removed, pull out the push rods. Keep them arranged by pushing them into a piece of cardboard. Now, remove the lifters with a puller or a magnet. Use egg cartons or other containers with compartments to keep the lifters in proper order. Do not skip this step or they will have to be replaced with new lifters.

Overhead Cam (OHC) Engines

Overhead cam engines may use rocker arms, or the cam may work directly on the lifters, *Fig. 11.45.* If rocker arms are used, they can be shaft mounted or individually mounted and they can be mounted above or below the cam. To decide whether the cam or rockers has to come out first, refer to the manufacturer's manual.

Rocker Arms: If the rocker arms are shaft mounted, they are usually removed first, *Fig. 11.46.* Always turn the engine until the marks on the belt, gears, or chain are lined up. If there are no alignment marks on the parts, simply make alignment marks with a dot of paint before removing the chain, belt, or gears. Start by unbolting the cam sprocket from the camshaft. The chain or belt is now loose. For now, just drop it into the front cover. If the camshaft is driven by belt, first mark its direction of rotation. Use a grease pencil to draw an arrow on the belt's back. Unbolt and remove the rocker assembly just like on an OHV engine.

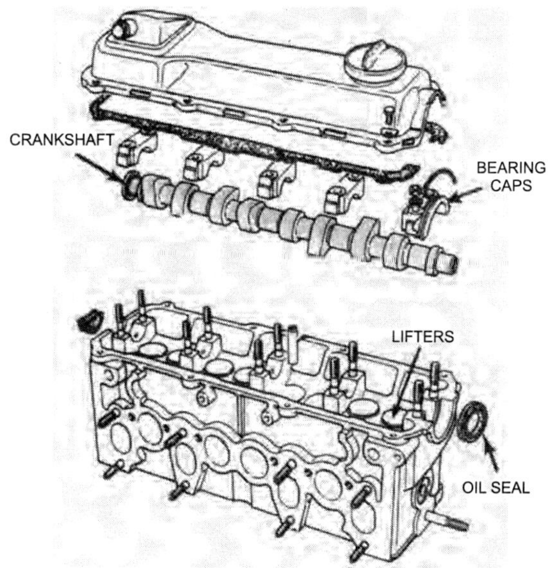

Fig. 11.45 *OHC working directly on lifters. (Chrysler Corp.)*

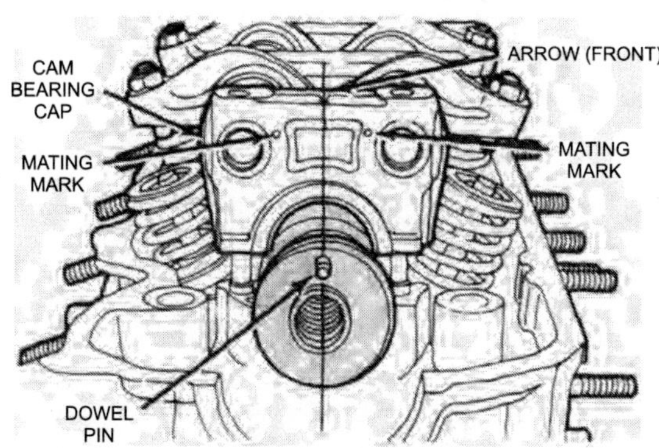

Fig. 11.46 *Shaft-mounted rockers with integral camshaft bearing caps.*

Removing the rocker arm assembly may also remove the camshaft's upper bearing halves, or caps.

If so, just lift the camshaft out of the head. Otherwise, slide the camshaft forward and out.

Camshaft: If the camshaft is mounted above the rocker arms or if there are no rocker arms, you have to remove the camshaft first. Mark the belt; then unbolt the cam sprocket. Drop the belt or chain into the front cover. The cam sits either in the head with camshaft bearing caps over its journals or up on towers. If caps are used ensure that they are properly marked; then unbolt and remove them. With towers, find their hold-down hardware, usually nuts on studs near the bottom of each tower. Remove the nuts and lift the camshaft straight up off the studs.

Next remove the rockers. Individual rockers sit on pedestals. Unclip the securing wire spring and lift off the rocker arm. Keep the rocker arms in proper order, to facilitate reassembly of each rocker on its original pedestal.

Remove pedestals by unscrewing them from the head. Be careful to wrench on the pedestal's base nut, not the valve-adjusting nut.

Head Bolts

At this stage only the valve springs and head bolts should be showing in the rocker cover area . To remove the head bolts use a socket and breaker bar. On most engines all head bolts are identical or have obvious differences in length, *Fig. 11.47.* Some engines, however, use bolts with subtle differences in length. Carefully examine the bolts as you remove them. If you find differences, punch the bolts into a marked piece of scrap cardboard. This will help keep the bolts in order so that you can replace them in their original holes during reassembly.

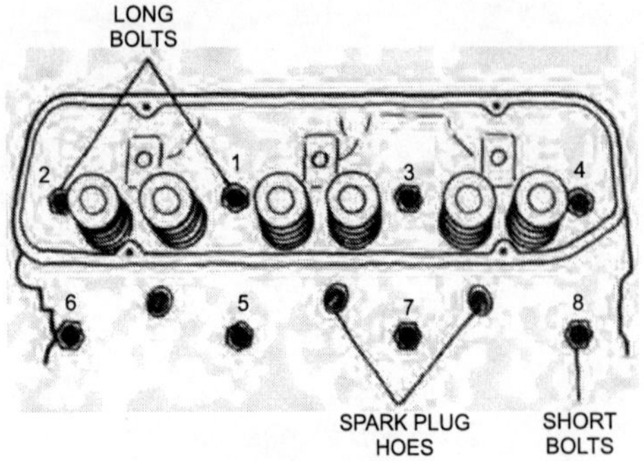

Fig. 11.47 *Typical long and short head bolts. (Buick Div. General Motors)*

With all bolts out, lift the head off the block. Often a combination of a heavy head and sticky gasket makes lifting the head off the block difficult. Try prying the head off the block. Pry at a corner where the head overhangs the block. Never pry against where the head and block mate. Always ensure that all the bolts have been removed before trying to pry.

Head Gasket

With the head removed you can inspect the head-gasket, *Fig. 11.48.* Look at the stainless steel sealing rings around the cylinder openings. They should be light metal colour. If they are burned black, they were leaking at the black spot(s). Both combustion gases and coolant leave a dark trace. These clues will help you to decide if the head needs milling.

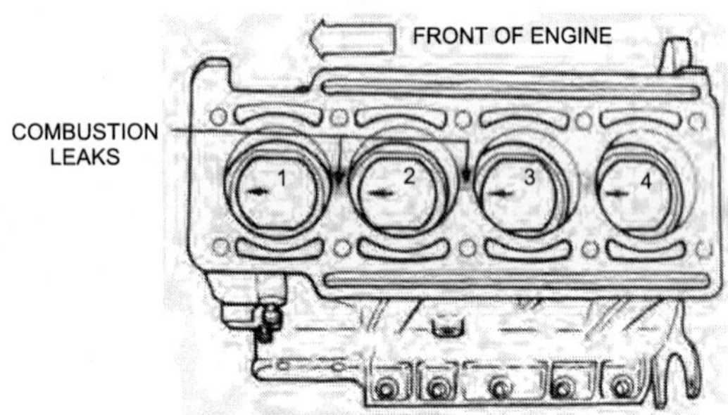

Fig. 11.48 *Combustion leaks between cylinders. (Buick Div. General Motors)*

Disassembly procedures for a cylinder head can vary for overhead valve and overhead cam heads. Keep the following rules in mind during head teardown:

1. Check for signs of any problem before disassembly. Look for oil-fouled combustion chambers indicating bad piston rings, valve seals, and valve guides. Check for a blown head gasket, signs of coolant leakage, rust, and cracks. Carefully check for any other problems that might exist.
2. Use a valve spring compressor to remove the valve keepers, retainers, and springs. Before compressing the springs, strike the valve retainers, not the valve tips, with a brass hammer. This will free them from the keepers and allow the compressor to squeeze the springs, *Fig. 11.49.*

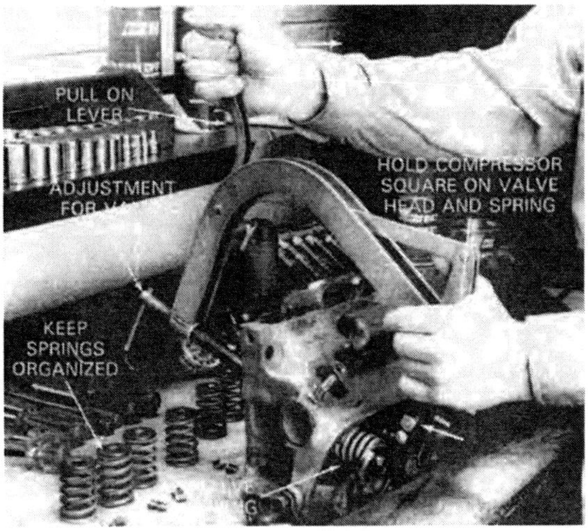

Fig. 11.49 *Valve spring compressor is required for cylinder head disassembly. Place one end over valve head and other over spring. Pull on lever arm to compress spring. Remove keepers and release compressor to free spring. Wear safety glasses because parts can fly if compressor slips off spring. (Fel-Pro Gaskets)*

3. A special valve spring compressor is required for some overhead cam cylinder heads, *Fig. 11.50.* It will reach down inside the tappet bore to compress the springs.

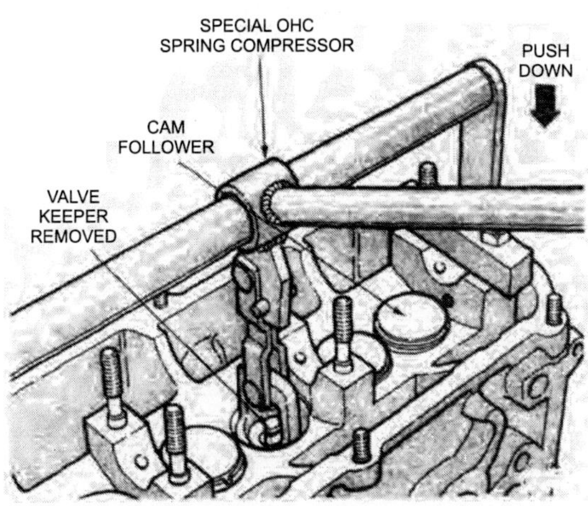

Fig. 11.50 *Special spring compressor is needed on some OHC engines. Note how valve springs are recessed inside head. (Chrysler)*

4. Check the valve tips for *mushrooming,* before removing the valves. Mushrooming could be caused by harder hammering action of rocker on valve tip, *Fig. 11.51.* File a chamfer on the valve tips if required. Never try to drive mushroomed valves out of a head. This can score or split the valve guides or head.
5. Keep all removed parts in an organized manner and tag or mark them to facilitate their reassembly into the same locations in the engine, *Fig. 11.52.* This could be necessary because the parts can be select-fit during new engine assembly in the factory where their sizes could be matched for precise control over mating parts clearances.
6. Remove any other parts that would not allow head service, *Fig. 11.53.*

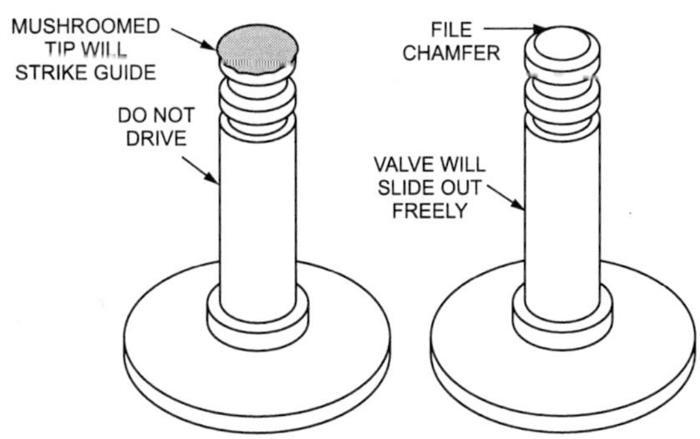

Fig. 11.51 *Before removing valves, check them for mushroomed tip. Enlarged tip end must be filed before valve removal to prevent valve guide damage.*

11.14 ENGINE BOTTOM END OR SHORT BLOCK DISASSEMBLY

The complete block is the cylinder block with all its internal parts installed. If the block is damaged beyond repair, short blocks are available for replacement.

The bottom end (short block) typically includes the pistons, rods, crankshaft, and related bearings. To take the bottom end apart, follow the common steps given below:

1. Begin short-block disassembly by inspecting the cylinders. Check the cylinders for signs of excess wear. See if you can feel a ridge near the top of the cylinders, *Fig. 11.54.* Presence of a

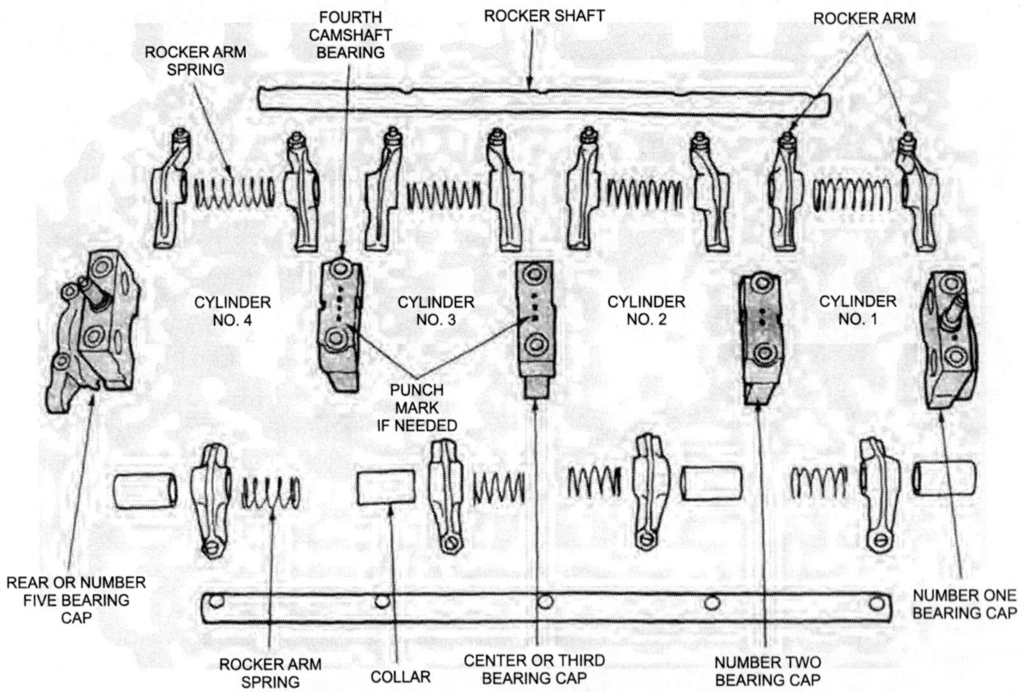

Fig. 11.52 *Make sure cylinder head camshaft caps are marked before removal. They cannot be mixed up or cam bore can become out of round, locking cam in head during reassembly. Remove rockers from shafts for inspection. (Honda)*

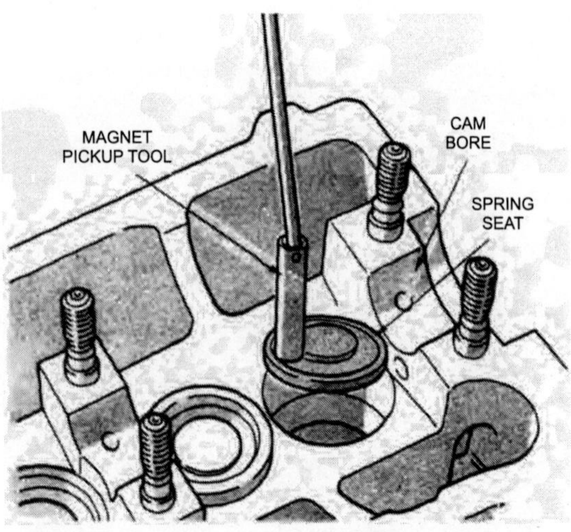

Fig. 11.53 *A magnetic pickup tool is being used to remove spring seat from OHC head. (Chrysler)*

ridge or lip at the top the cylinder wall can be checked by fingernail, *Fig. 11.55 (a)*. A ***cylinder*** or ***ring ridge*** *can* be formed at the top of the cylinder walls, where the rings do not wear the cylinder. The rings wearing the cylinder make the ridge.

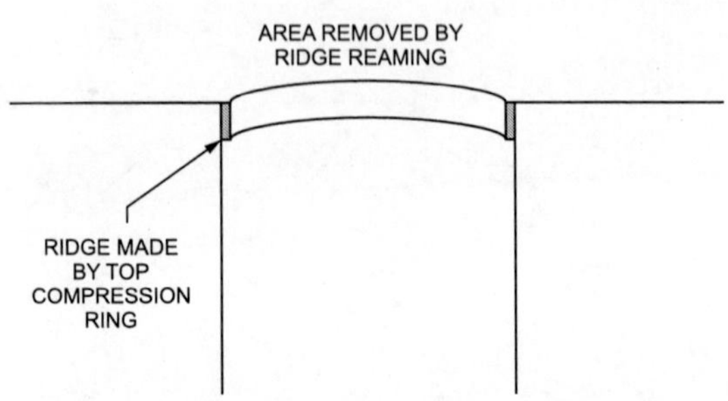

Fig. 11.54 *Cylinder ridge.*

2. The ridge above the top ring must be removed before the piston and connecting rod assembly

is removed. Ridge removal is necessary to avoid catching a ring on the ridge and breaking the piston, *Fig. 11.55 (b), (e)*. A *ridge reaming tool* is used to cut out and remove a ridge at the top of a worn cylinder, *Figs. 11.55 (c)* and *(d)*. Use a wrench to rotate the reamer and cut away the metal lip. Cut until flush with the rest of the cylinder wall. *Fig. 11.55 (b)*. Follow the instructions of the manufacturer of the ridge reamer you are using. Remove just enough material to smooth out the ridge. Removing too much material will make cylinder boring necessary.

3. Use compressed air to blow metal shavings out of the cylinder after ridge reaming. This will prevent cylinder or piston scoring.
4. Make sure that the oil is already drained out. Turn engine over on engine stand, unbolt and remove the oil pan. A socket, ratchet, and long extension work well for this job. Now carefully

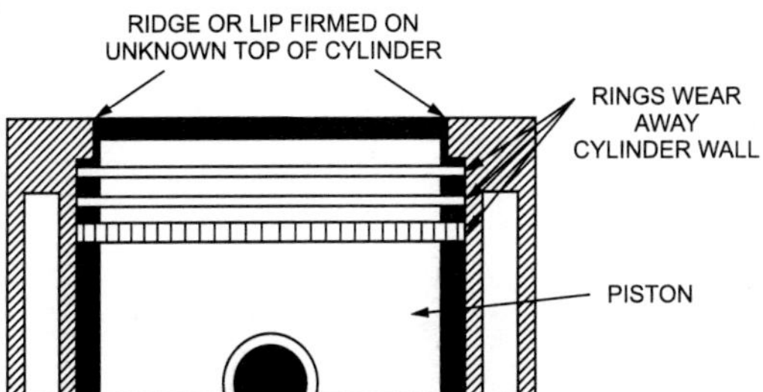

(a) Ring ridge formed at top of cylinder where rings do not rub and wear cylinder wall. Rings would bind on ridge if the piston is attempted to be pusded out of cylinder.

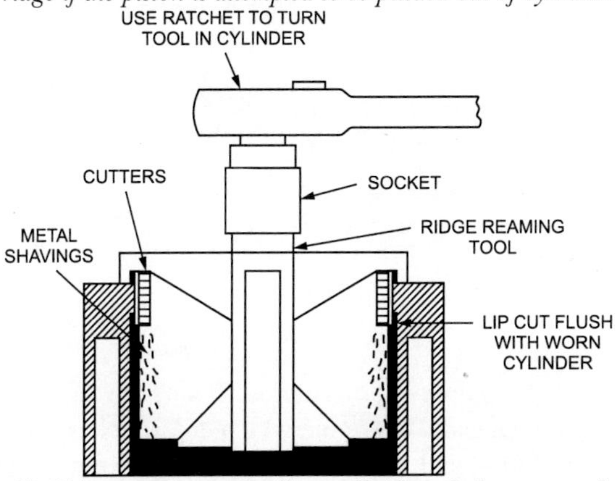

(b) Ridge reaming tool will cut off ridge. Only rotate tool until ridge is flush with worn part of cylinder.

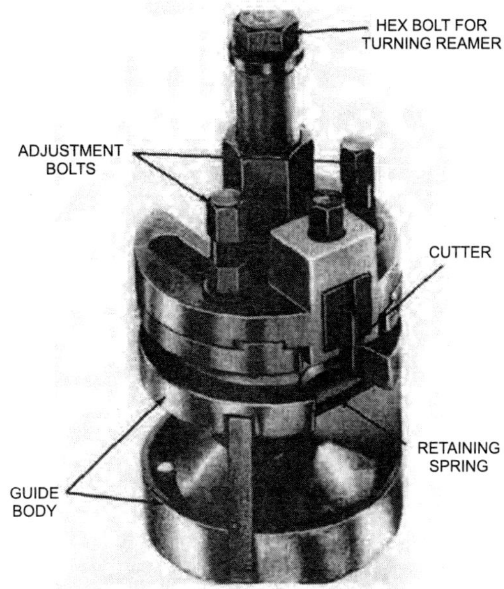

(c) Note parts of ridge reamer.

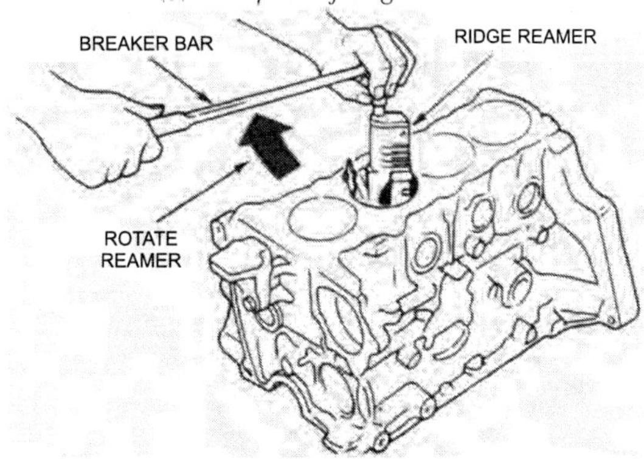

(d) Turn reamer until ridge is flush with worn part of cylinder.

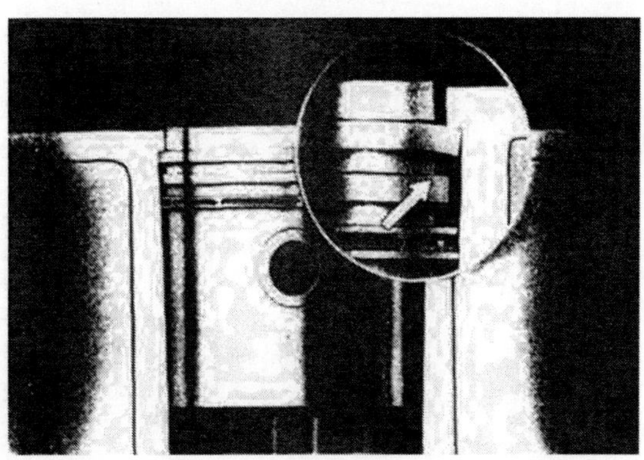

(e) If the ridge at the top of a cylinder is not removed, the piston could break when pushed out of the cylinder during disassembly. (Sealed Power)

Fig. 11.55 *Ring ridge must be removed before driving pistons out of block. You can feel ridge with fingernail.*

pry the oil pan from the block. The sticky pan gasket will require a strong pry at first. Pry in several areas, not in the same spot. If you stick to the same spot, you will bend the oil pan flange. Clean up any oil spills. Check the bottom of the oil pan for debris. Metal chips and plastic bits may help in diagnosing the engine problems, *Fig. 11.56.*

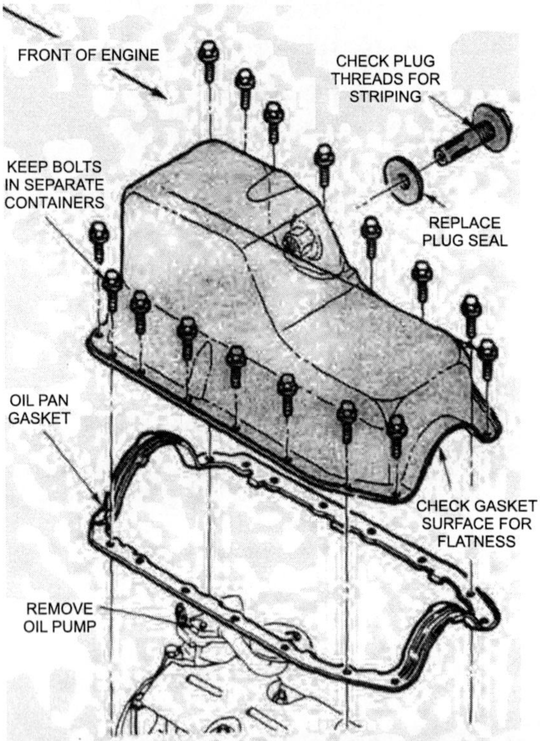

Fig. 11.56 *To begin short block disassembly, turn engine over on engine stand and remove oil pan. Make sure that the oil is drained out first or you will have a mess on shop floor. Check inside of pan for debris that could help find other problems. (Ford Motor Co.)*

5. Next remove the **oil pump.** Many pumps have only one bolt attaching them to a main bearing cap or the block, while others have two or three attaching bolts. Often one bolt will be hidden inside the pickup screen. If, after removing all visible bolts, the pump will not come off, look under the screen for a bolt. Gently pop the screen off the pickup. Small screwdrivers can be used as miniature pry bars. Look for a bolt head inside the pickup and lift the pump off. The oil pump drive shaft should come out with the pump. If not, you remove it from the block. Again, cleanup any oil spills.
6. Use an impact wrench to remove the *front pulley nut.* Try pulling the pulley or **harmonic dampener** off by hand. If it would not come use a puller. Use only a puller that attaches to the dampener's centre section. Any pull on the outer ring can ruin the dampener.
7. Remove all the front cover bolts, *Fig. 11.57.* Sometimes several pan bolts pass through the pan and into the front cover. You have already removed these. Inside the front cover are the timing chain (gears or belt) and sprockets, *Fig. 11.58.* There may be a chain tensioner. Other engines use an **oil slinger,** a flat disc with an upturned outer edge. It fits between the timing cover and crankshaft sprocket. It prevents the front seal from receiving too much oil. On OHC engines, slip off the oil slinger and then the belt. Often a puller may be needed. When using a puller here, do not forget to install the front pulley bolt. Run it in the crankshaft several turns. Without the bolt, the puller's shaft will damage the threads inside the crank.

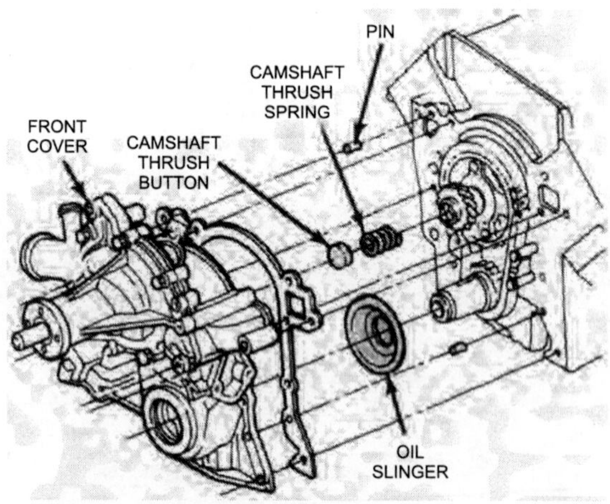

Fig. 11.57 *Typical OHV front cover components. (Buick Div. General Motors)*

8. *Next, unbolt the tensioner:* It may have a spring attached. Determine which end attaches to the block and which end to the tensioner. Some tensioners receive their push from oil pressure. When you unbolt these, you will find the plunger is spring loaded. Do not let the plunger pop out onto the floor.

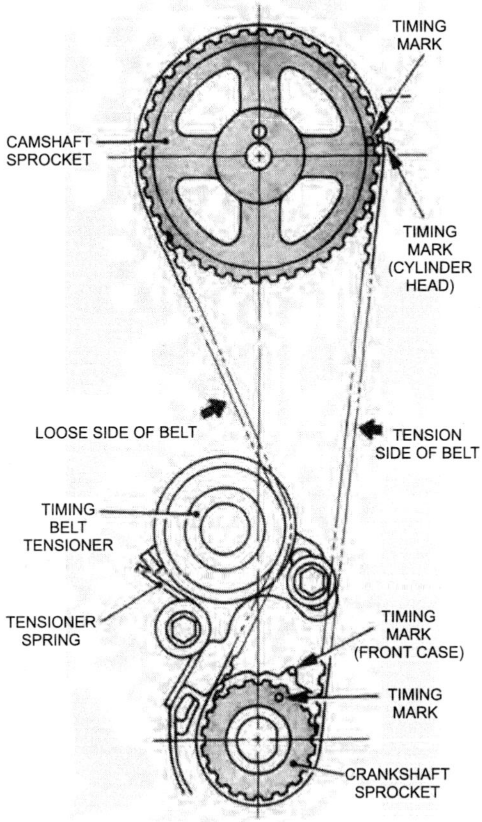

Fig. 11.58 *OHC timing belt and sprockets. (Chrysler Corp.)*

9. *Camshaft Removal:* Start with the camshaft on overhead valve (OHV) engines, *Fig. 11.59.* Remove any tensioner. Then, unbolt the cam sprocket. Usually there is a dowel pin locating the sprocket. If not, index the sprocket to the camshaft with a hammer punch. Before removing the sprocket, check for chain stretch and wear. Pull the sprocket off the camshaft. This will loosen the chain so it can be removed too.

Under the camshaft sprocket there may be a retaining plate. Unbolt the plate and set it aside. Now the camshaft can be pulled from the block. If there is no plate, simply pull the camshaft out.

Pull the camshaft carefully. Do not bang the cam lobes against the block or camshaft bearings. When working on a workbench, turning the block on end helps. With an engine on the engine stand, upside down is best. To help get the camshaft started out of the block, run two bolts into the camshaft. Use the bolts as handles. If the camshaft still wound not budge, look for a fuel pump or distributor drive still in place, and remove it first.

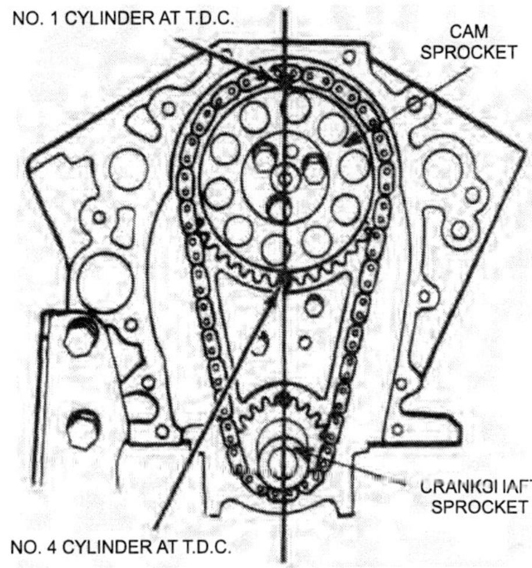

Fig. 11.59 *OHC cam alignment. (Buick Div. General Motors)*

10. Go back to the crankcase. Rotate the crankcase until some of the connecting rods are sticking straight up. This is bottom-dead-centre (BDC) on these cylinders. If the crankshaft is too hard to move by hand, thread two flywheel bolts into the crankshaft. Then use a pry bar lodged between these bolt shanks to rotate the crankshaft.
11. *Rod Numbering:* Inspect the rods before removing them. If you find numerals which correspond to the connecting rod's position in the engine, good, *Fig. 11.60.* Factory-numbered connecting rods are usually numbered on the side away from the camshaft. If not, mark the connecting rods with a numbering set or punch. Stamp one number on the connecting rod #1, two on #2, and so on.

You may find numbers that do not seem to match the rod's position in the engine. Well they do not. These numbers are for bearing sizes. They have nothing to do with where the connecting rod goes in the engine. All Honda engines, early Mercedes engines, and some others use this system. If you are unable to understand the numbering system, make your own.

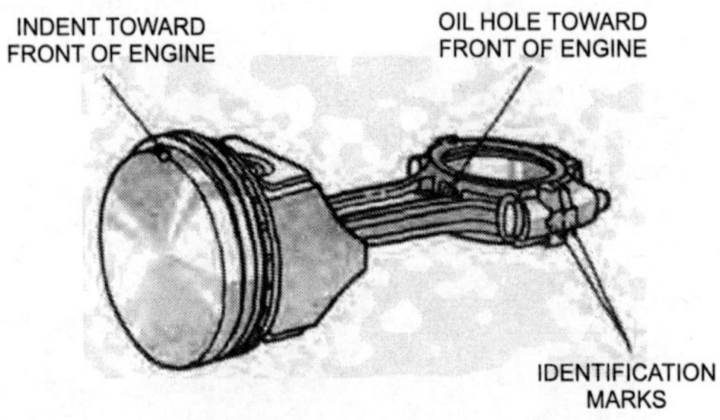

Fig. 11.60 *Connecting rod numbering.*

12. Unbolt one of the connecting rod caps and pull it from the rod. Grasp the cap tightly and rock it slightly while pulling it. Very slight taps with a plastic or brass hammer are acceptable. Such blows will often help get a stubborn cap moving. But in most cases a steady, slight rocking motion does the trick.

    With the cap off, place a short length of fuel hose over the rod studs, *Fig. 11.61.* If the connecting rod uses removable bolts this step is not needed. The hose pieces keep the studs from nicking the crank journals.

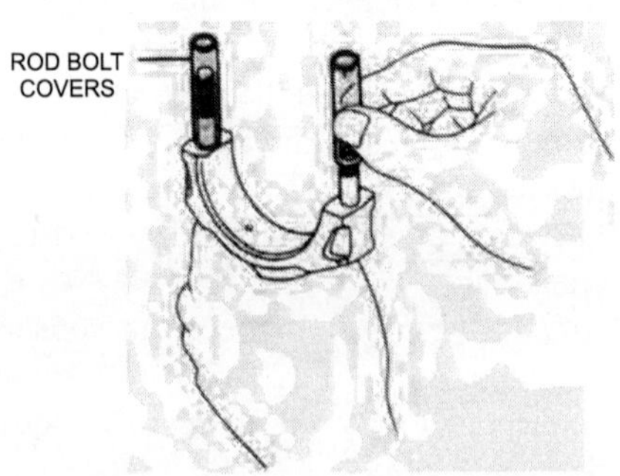

Fig. 11.61 *Connecting rod swith journal-protecting hose.*

13. Then, use a wooden hammer handle or wooden rod to tap the piston and connecting rod out of the top of the cylinder, *Fig. 11.62.* Push with one hand and catch the assembly with the other hand. If you cannot get your hand far enough inside the cylinder, use a wooden hammer handle to extend your reach. Do not do anything to dent or damage the connecting rod I-beam. The dent or indentation in the metal could cause a stress riser point. The rod could crack and eventually break during engine operation. Only mark the rods on the side of the cap and rod body.

    Before setting the piston and rod assembly aside, replace the cap. Run the nuts or bolts in, a few threads. Replacing the cap prevents it from getting misplaced and protects it better. Remove the remaining piston and rod assemblies. Rotate the crankshaft as necessary to get at the rod nuts.

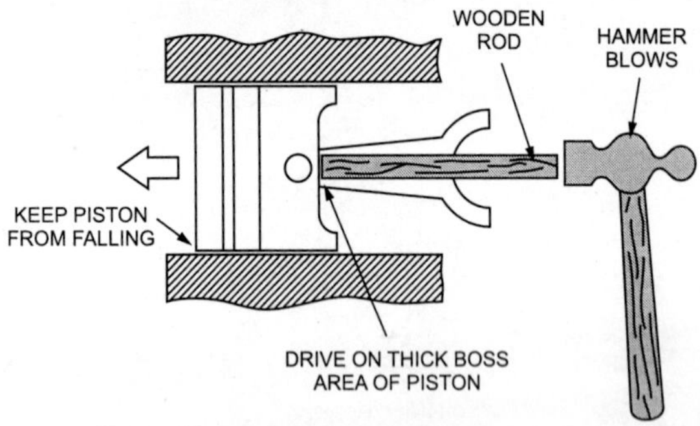

(a) Wooden rod or old oak handle is one way to drive out pistons. Place one end of the handle on piston boss. Light hammer blows should push piston and rod out of block. Do not allow piston to fall.

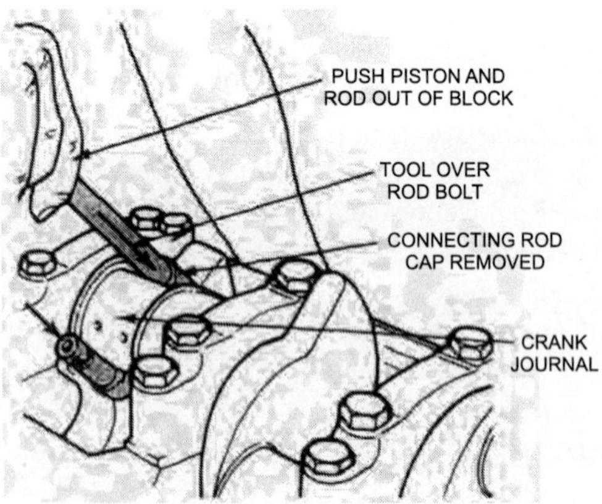

(b) The technician is using threaded rod that fits over rod bolt to hand-force piston out of block.

Fig. 11.62

14. Immediately after the piston and rod are out, replace the rod cap. Also, check the piston head and connecting rod for identification markings. The piston will usually have an arrow pointing

to the front of the engine. The connecting rod and rod cap should have numbers matching the cylinder number, *Fig. 11.63 (a)*. If needed, mark the piston heads with arrows or numbers. Also, if necessary, number the connecting rods. If the pistons or rod caps get mixed up, serious problems can develop when trying to reassemble the engine, *Fig. 11.63 (b)*.

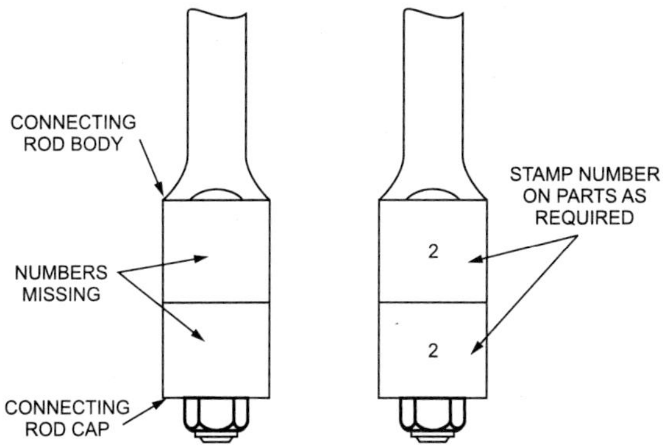

(a) New rods sometimes have numbers missing.

(b) Number stamp set or centre punch can be used to mark rods.

Fig. 11.63 *As each piston and rod is removed, make sure it has numbers on caps. If required, punch mark rod and cap so caps will not get mixed up. Caps must be kept with same rod or bore will not be round.*

15. Remove the remaining piston and rod assemblies one at a time. Carefully, reinstall each cap on its corresponding rod. Mark them if required.
16. Remove all of the rings from their pistons. Carefully spiral the rings off with fingers or use a ring expander.
17. For a vehicle with manual transmission, check for flywheel warpage using a dial indicator setup. With all the pistons out of the block, the crankshaft will turn easily. Rotate the flywheel for a full turn while watching the indicator reading. If runout is found to be out of specs, get it fixed in the machine shop.
18. Only the main bearing caps remain holding the crankshaft. Check the numbering on the main bearing caps, before removing them. Usually, numbers and arrows are cast on each cap. If caps do not have arrows indicating the front of the engine, mark to prevent reversing caps on assembly. Number one cap is at the front of the engine. If necessary, use a punch set for indenting numbers or a centre punch to mark the main caps. This is necessary to avoid mixing of caps. If the caps are mixed up, the crank bore may get misaligned and crankshaft can lock in the bore and main bearings during reassembly.
19. Then, with the engine still upside down, unbolt the main caps. Like connecting rod bearing caps, they can be difficult to remove. Start by tapping each one with a soft hammer, *Fig. 11.64.* To remove the caps, wiggle them back and forth while pulling with channel locks or with the main cap bolts. You can place two open-end wrenches into the cap bolt holes and grasp the two wrenches in one hand and wiggle the cap side to side.

Some main bearing caps sit deep in the block. This style of cap will need a front-to-rear motion to work it loose. It takes a little work to free the caps. Always use patience, not force. Have a surface ready to set the crankshaft on. Choose a flat, stable spot to avoid crankshaft rolling or knocked to the ground. Then, lift the crankshaft carefully out of the block, preventing any damage to journals, *Fig. 11.65.* Set the removed caps aside. Store their bolts in the main cap holes.

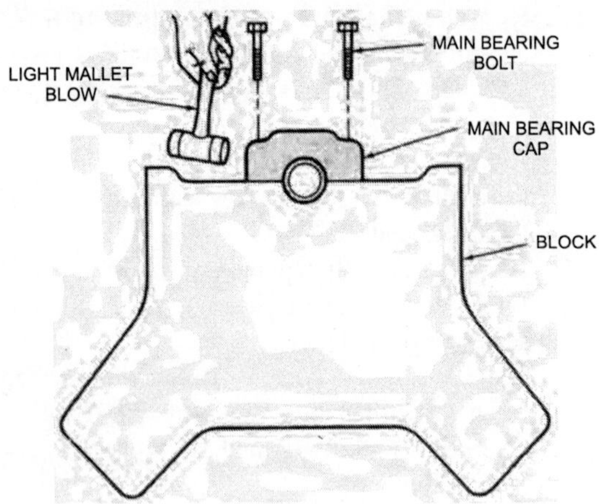

Fig. 11.64 *Loosening main bearing caps.*

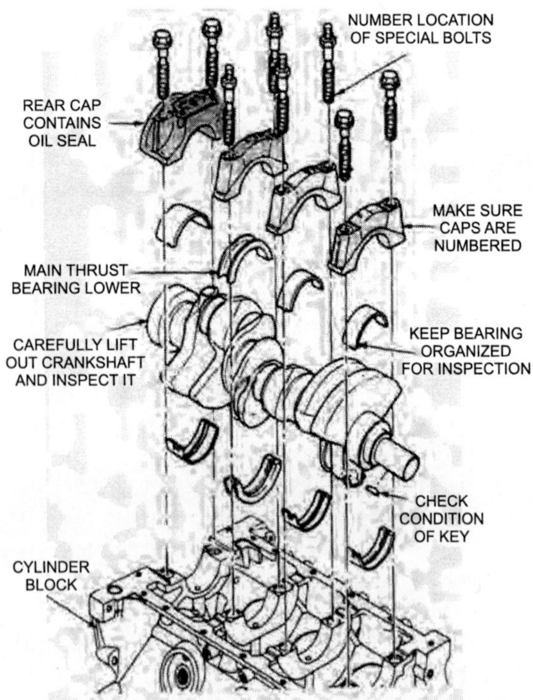

Fig. 11.65 *Make sure main caps are numbered before removal. They should be positioned in same locations to keep main bore round. Lift the crankshaft out of the cylinder block without hitting bearing journals. (Ford)*

20. In the block and sometimes in the head, there are core plugs. For an old engine, remove the block and head core or freeze plugs. They might rust out from inside and start to leak after prolonged service. Also, their removal becomes necessary if the block is going to be hot cleaned in strong chemicals at machine shop to remove mineral deposits in water jackets. Because core plugs are inexpensive, it is normal to replace them without hesitation during engine overhauls. Place a punch against one edge of the plug and hammer that edge in, *Fig. 11.66.* The other edge will come out of the block. Grasp this outturned edge with pliers and pry it from the block. Remove all the plugs and save them for comparing with the new plugs selected.

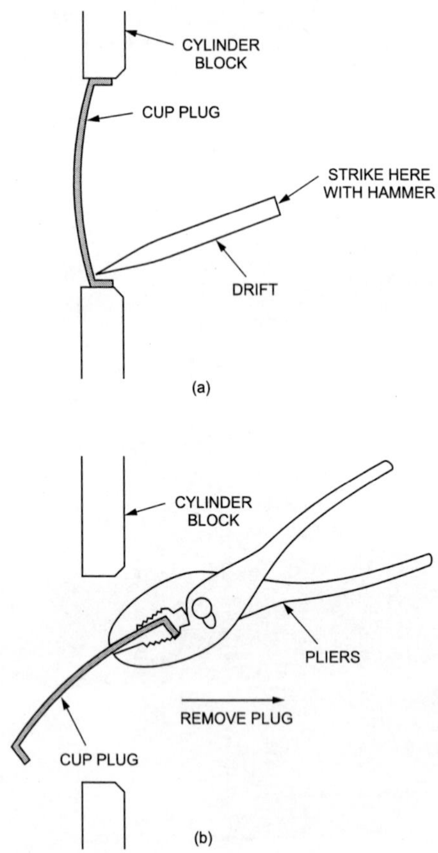

Fig. 11.66 *Core plug removal. (Chrysler Corp.)*

21. If the cylinders have deep ridges, the cylinder block should be sent to a machine shop for boring and hot tank cleaning. Check that all external parts, e.g., rubber motor mounts, oil and coolant temperature sending units, etc., are removed before sending the block out for machining.

22. If an engine is very dirty, open the oil passages for cleaning. You will find socket or hex-headed plugs at the ends of each oil gallery. Remove these plugs. Hex-headed plugs are easy to remove. Just put an Allen wrench on the hex and turn the plug out. To remove the socket-headed plugs, first clean all the dirt and grease out of the socket so as to be able to get a full grip on the hex. Then, use a socket drive Allen wrench. You have a lot more leverage through a ratchet and socket combination than through a standard Allen wrench. If the socket does not budge, use a hand impact driver. Several wallops will break the plug loose. Prefer use of impact driver to rounding the hex with a lot of slow pulling. Clean up the work area of any oil or other fluids.

11.15 DIESEL ENGINES

Diesel engines are no different from other piston engines when it comes to engine removal and teardown. However some of the procedures are different.

High-pressure fuel lines, made of steel and having threaded connections, run between the injection pump and the head. Remove each line, and paying attention to cap every disconnection. Diesel engines use a mechanical injection system that must remain very clean. Leaving the fuel connections open is inviting expensive trouble.

An injection pump is a precision part. Treat it with respect. Do not drop it, and keep it as clean as possible. Mark the pump relative to its mounting pad before removal. The mark facilitates installation back during assembly.

Throttle linkages for diesel engines can be complicated. Disturb the linkage as little as possible. Ball and socket connection can be popped apart due to rough and harsh handling. Do not unthread the turnbuckle adjusters on the linkage.

Diesel engines typically use much more head bolts or studs. Ensure that all of them are removed before attempting to lift the head.

11.16 CLEANING ENGINE PARTS

Many small auto repair shops use a simple tank with a ***power sprayer,*** *Fig. 11.67,* to clean parts that have been removed from the vehicle. Cleaning parts with a brush and a pan of gasoline is a fire hazard and, therefore, should never be attempted. Instead, use a solvent designed for the cleaning job at hand.

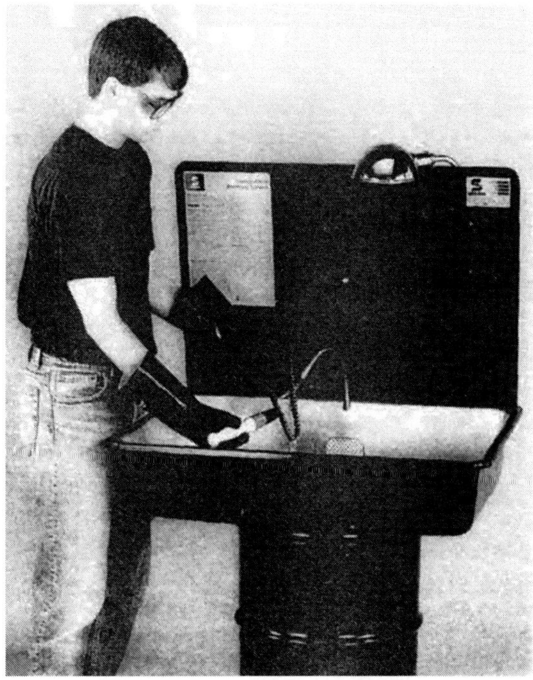

Fig. 11.67 *Clean parts in a modern parts cleaner. (Safety-Kleen)*

Before cleaning a cylinder block or a cylinder head, remove the core hole plugs. This will permit the cleaning solution to flush out any accumulation of rust and lime deposits.

After removing all of the engine parts from the engine block and cylinder head(s), every part should be adequately cleaned. Parts can be coated with carbon, sludge, and oil film. Different cleaning techniques are needed depending upon part construction and type of its material.

Closer part inspection can be done during and after part cleaning. Some problems are difficult to see when a part is covered with oil, grease, or carbon deposits.

11.17 MECHANICAL CLEANING

Mechanical cleaning involves scraping, brushing, and abrasive blasting. It should, therefore, be used very carefully on soft metals. Heavy deposits that remain on parts after chemical cleaning will have to be removed by mechanical cleaning.

Scrape off old gaskets and hard deposits

Start the engine parts cleaning by scrapping off all old gasket material and hard deposits. Scrape off the gaskets for the valve covers, head(s), front cover, induction manifold, oil pan, and other components. Also scrape off as much sludge and carbon as you can. The most frequently used type of scraper is a ***putty knife*** or a similar dull scraper, *Fig. 11.68.* The blade of the scraper is pushed under the deposit to free it from the surface. The blade works best on fiat surfaces, such as gasket surfaces and the piston head. Use a dull scraper and work carefully when cleaning soft aluminium parts. Even a slight nick, in a cylinder head for instance, could cause head gasket leakage when returned to service. When using a gasket scraper, push away from your body, not toward body, to avoid injury.

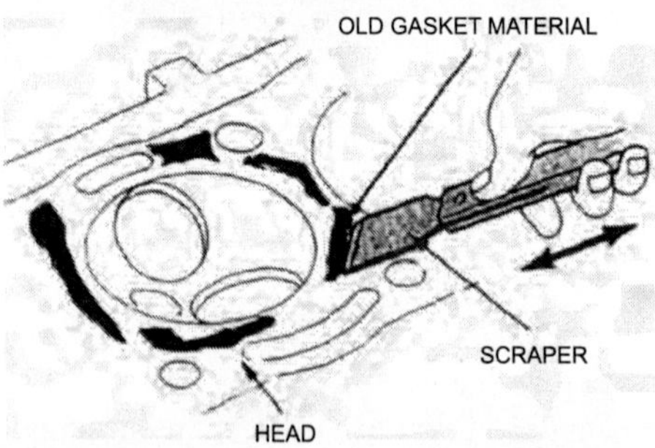

Fig. 11.68 *Hand scraper is commonly used to remove heavy deposits and old gasket material. Avoid any damage to soft aluminium parts. Scrape away from body. (Toyota)*

Wire brushes can be used on uneven surfaces. A hand wire brush can be used on the exterior of the block and on the head. A round wire brush used with a hand drill motor does a good job of cleaning combustion chambers and parts of the head.

Use power-cleaning tools

There are many power cleaning tools used by an auto mechanic. When used properly, they can speed up and ease engine repairs. Always wear eye protection when cleaning parts with power tools. Metal blisters, bits of carbon, or metal chunks from tools or part breakage can fly into naked eyes.

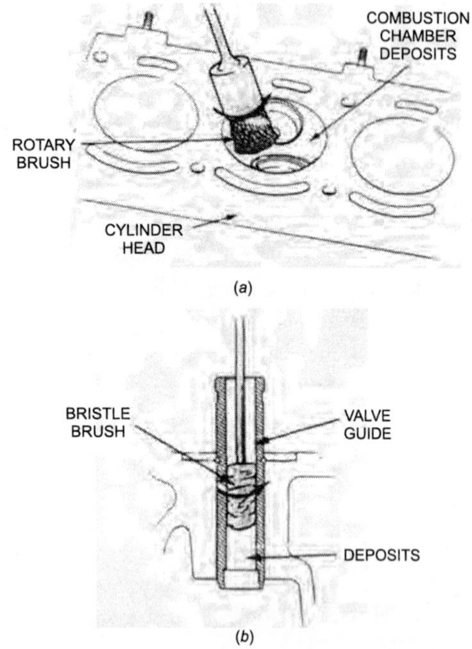

Fig. 11.69 *(a) Power brush in electric or air drill used to clean combustion chambers, (b) Small brush on drill is used to clean out valve guides (Toyota)*

A ***power brush*** is driven by an air or electric drill to remove hard carbon. It is particularly handy inside hard-to-reach areas like combustion chambers, *Fig. 11.69 (a).* A power brush being used inside a valve guide is shown *Fig. 11.69 (b).*

A ***wire wheel*** on a grinder is another common method of cleaning engine parts. Often, it is used to clean carbon off valves. When using this tool, keep safety shield and tool rest in place, *Fig. 11.70.*

A ***scuff pad*** is a tough plastic mesh material mounted on a rubber backing and arbor. The end of arbor is placed in an air drill or similar tool. The scuff pad will cut and lift off gasket material and silicone without any damage to metal surfaces.

Spray-on gasket solvents are also available. Sometimes gaskets can be very difficult to scrape off. Gasket removing chemicals can be helpful. *Fig. 11.71* shows a mechanic/technician using a spray-on chemical for head gasket and valve cover gasket removal. After spraying, let the chemical work for a few minutes. Then, scrape off the softened gasket with a scraper.

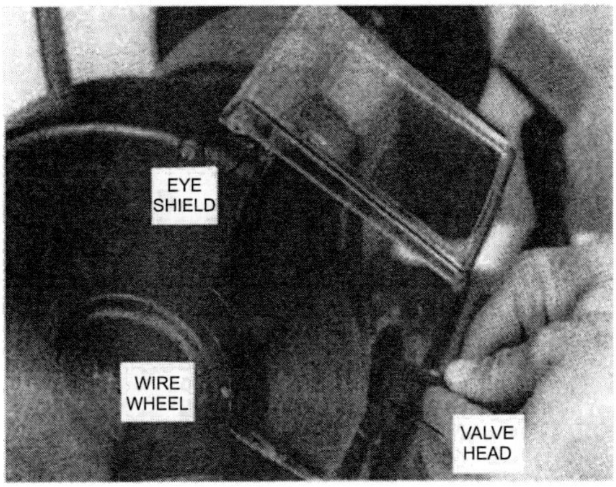

Fig. 11.70 *Wire wheel on bench grinder is commonly used to remove hard carbon from engine valves. Wear eye protection and keep shields in place.*

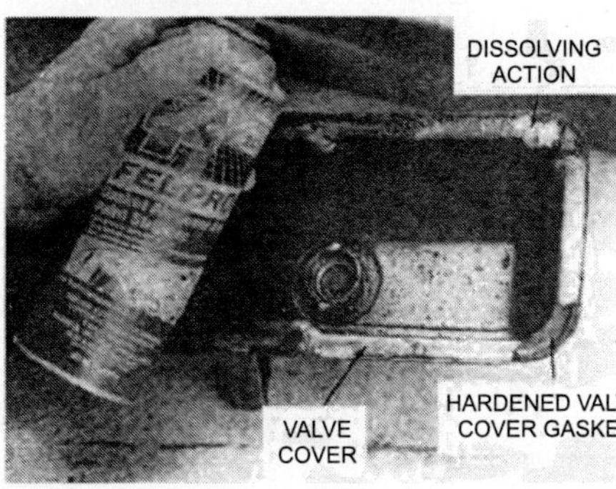

Fig. 11.71 *Sometimes gaskets can become very hard to remove after years of service. Chemical sprays are available for dissolving gaskets and speeding removal. Softened gaskets can then be easily scraped off. (Fel-Pro)*

11.18 AIR BLOW GUN

An air ***blowgun*** is normally the last method of cleaning engine parts. It uses pressure from the shop's air compressor to blow off small bits of dirt, solvent, water, and other debris, *Fig. 11.72.*

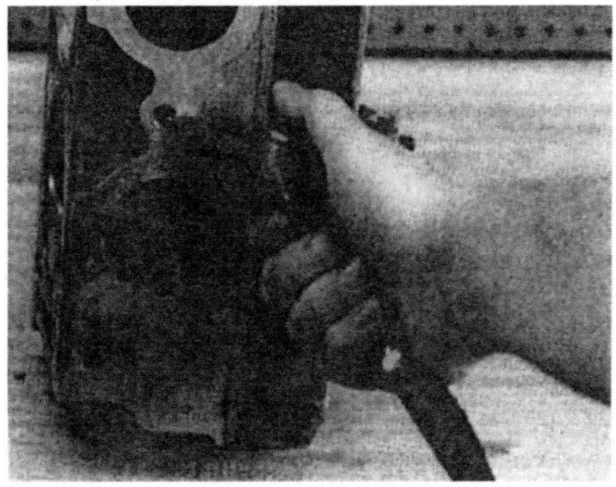

Fig. 11.72 *Use compressed air and blow gun to final clean parts. Air blast will blow off tiny particles and dry solvent. Wear eye protection and do not direct air blast toward body.*

Be extremely careful when using an air gun. Wear eye protection goggles and never allow the gun to aim at your body. If air enters blood stream, a blood clot could form and cause death.

11.19 CHEMICAL CLEANERS

Cleaning chemicals applied to the parts will mix with and dissolve the deposits. The chemicals loosen the deposits so they can be brushed or rinsed from the surface. A deposit is said to be ***soluble*** when it can be dissolved with a chemical or solvent. Most chemical cleaners used for cleaning carbon type deposits are a strong soap, or ***caustic material.*** A value called ***pH,*** measured on a scale from 1 to 14, is used to indicate the amount of chemical activity. Pure water is neutral. On the pH scale water is pH 7. Caustic materials have pH numbers 8 through 14. The higher the number the stronger the caustic action will be. ***Acidic materials*** have pH numbers from 6 through 1. The lower the number, the stronger the acid action will be. Caustic materials and acid materials neutralize each other.

11.20 SOLVENT-BASED CLEANING

Chemical cleaning can involve a spray washer or a soak in a cold or hot tank. The cleaning solution is usually solvent based, with a medium pH rating of between 10 and 12. Most chemical solutions also contain silicates to protect the metal (aluminium) against corrosion. Strong caustics do an excellent job on cast iron parts but are often too corrosive for aluminium parts. Aluminium cleaners include mineral spirit solvents and alkaline detergents.

When cleaning aluminium cylinder heads, blocks, or other engine components, make sure that the chemicals used are "aluminium safe. Many chemicals that are not aluminium safe may turn the aluminium metal black.

11.21 WATER-BASED CHEMICAL CLEANING

Due to environmental concerns, most chemical cleaning is now performed using water-based ***(aqueous based)*** solutions. Most aqueous based chemicals are silicate based and are mixed with water. Aqueous-based solutions can be sprayed or used in a tank for soaking parts. Aluminium heads and blocks normally require overnight soaking, with the temperature of the solution kept at about 90°C. For best results, the cleaning solutions should be agitated.

11.22 SPRAY WASHING

Another efficient parts cleaning technique is called the ***jet method.*** The cleaning unit has two compartments: the lower chamber holds the cleaning solution; the upper chamber provides a revolving platform for the work. A series of jets spray detergent on the slowly revolving part in the upper compartment. Each jet delivers detergent at an extremely high velocity. The spray reaches the work from all sides. Each jet delivers approximately 10 litres of hot detergent solution per minute.

In spray washing streams of liquid through numerous high-pressure nozzles is directed to dislodge dirt and grime on an engine surface. The force of the liquid hitting the surface, combined with the chemical action of the cleaning solution, produces a clean surface. Spray washing is typically performed in an enclosed washer, where parts are rotated on a washer turntable.

Spray washing is faster than soaking. A typical washer cycle is less than 30 minutes per load, compared with 8 or more hours for soaking. Most spray washers use an aqueous-based cleaning solution heated 70° to 80°C with foam suppressants.

11.23 THERMAL CLEANING

Thermal cleaning uses heat to vaporize and char dirt into a dry, powdery ash. Thermal cleaning is best suited for cleaning cast iron, where temperatures as high as 425°C are used, whereas aluminium should not be heated to over 315°C. The major advantages of thermal cleaning are:

- This process cleans inside and outside of the casting or part.
- The waste generated is non-hazardous and provides easy disposal.

However, the heat in the oven usually discolours the metal and leaves it looking dull. Additional cleaning with a hot blaster is usually necessary to restore proper appearance. This method of engine part cleaning is becoming the most popular because there is no hazardous waste associated with it. Labour costs are also reduced because the operator does not need to be present during the actual cleaning operation.

11.24 STEAM CLEANING

Steam cleaners are a special class of sprayers. Steam vapour is mixed with high-pressure water and sprayed on the parts. The heat of the steam and the propellant force of the high-pressure water, combine to do the cleaning. Steam cleaning must be used with extreme care. Usually, a caustic cleaner is added to the steam and water to aid in cleaning. This mixture is so active that it will damage and even remove paint, so painted surfaces must be protected from the spray. Engines are often steam cleaned before they are removed from the chassis.

11.25 COLD TANK CLEANING

The cold soak tank is used to remove grease and carbon. It will not remove hard carbon, paint, gum or mineral deposits. The disassembled parts are placed in the tank so that they are completely covered with the chemical cleaning solution, *Fig. 11.73.* After a soaking period, the parts are removed and rinsed until the milky appearance of the emulsion is gone. The parts are then dried with compressed air. The clean, dry parts are usually given a very light coating of clean oil to prevent rusting. Carburetor cleaner, purchased with a basket in a bucket, is one of the most common types of cold soak agents in the automotive repair shop. Usually, the chemical will have water over its surface to prevent evaporation of the chemical. This water is called a ***hydro-seal.***

11.26 HOT TANK CLEANING

A process used extensively for cleaning automotive engine parts is called the ***hot tank method.*** Briefly, this method utilizes a tank filled with a detergent solution, which is agitated to speed up the cleaning process. The temperature of the solution varies with the type of the detergent used and particular metal to be cleaned. Parts made of aluminium are cleaned with special detergent at a lower temperature than is used for steel and iron parts.

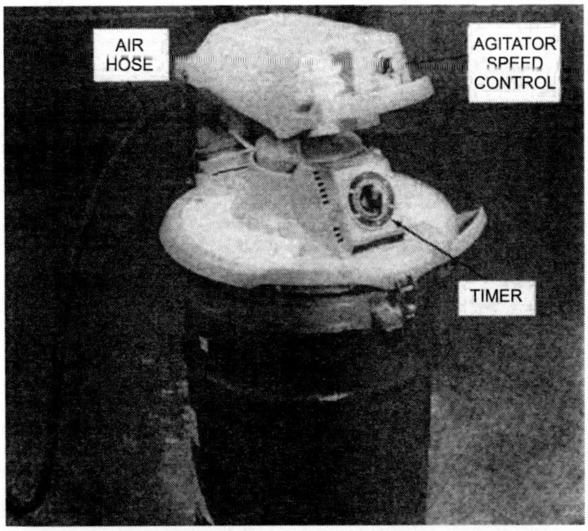

Fig. 11.73 *Carburetor cleaner is sometimes used to clean engine parts. Do not submerse aluminium parts in cleaner because they can be etched and ruined.*

A ***hot tank*** is a large cleaning machine, filled with strong, corrosive chemicals, *Fig. 11.74.* Automotive machine shops normally have a hot tank. The hot soak tank is used for cleaning heavy organic deposits and rust from iron and steel parts. Caustic cleaning solution used in hot soak tank is kept near 93°C for speedy cleaning action. After the deposits have been loosened, the parts are removed from the tank and rinsed with hot water or steam cleaned. The hot parts will dry quickly. They must be given a light coating of oil to prevent rusting. ***Fogging oil*** from a spray can does an excellent job of coating metal parts to keep them from rusting.

Aluminium components can be corroded or etched by soaking in a hot tank. Clean only cast iron and steel parts in a hot tank.

Fig. 11.74 *Machine shop will have hot tank for cleaning blocks, heads, and other ferrous parts. (Storm Vulcan)*

11.27 ULTRASONIC CLEANING

Ultrasonic cleaning is a more recent advance. Ultrasonic sound differs from normal sound in that its pitch is above the human hearing range. When high-frequency vibrations are introduced into a liquid, cavitation occurs. Cavitation causes thousands of bubbles to form in the liquid. These bubbles range in size from visible to submicroscopic. Under continuous cavitation, the bubbles grow and then collapse. They are said to explode inwardly, at which time pressures up to 15,000 psi and temperatures up to 371°C are created.

The ultrasonic cleaner consists basically of three parts:

1. A generator, which is used to convert the line current into high frequency electrical energy.

2. A transducer, which converts this electrical energy into mechanical energy of the same frequency.
3. A tank, which contains a liquid through which the ultrasonic energy is passed. The liquid produces the cavitation that blasts away any dirt on parts placed in the solution.

Ultrasonic cleaners are used in production engine rebuilding shops servicing diesel engine injectors and pumps, and in large jobber machine shops.

Ultrasonic cleaning is used to clean small parts that must be absolutely clean, e.g., hydraulic lifters and diesel injectors. The parts are placed in a tank of cleaning solution. The solution is vibrated at ultrasonic speeds to loosen all the soil from the parts. The soil goes into solution or falls to the bottom of the tank.

11.28 BLASTERS

Cleaning cast-iron or aluminium engine parts with solvents or heat usually requires another operation to achieve a uniform surface finish. Blasting the parts with steel, cast-iron, aluminium, stainless steel shot, or glass beads is a simple way to achieve a matte or satin surface finish on the engine parts. To keep the shot or beads from sticking, the parts must be dry without a trace of oil or grease prior to blasting. This means that blasting is the second cleaning method after the part has been pre-cleaned in a tank, spray washer, or oven.

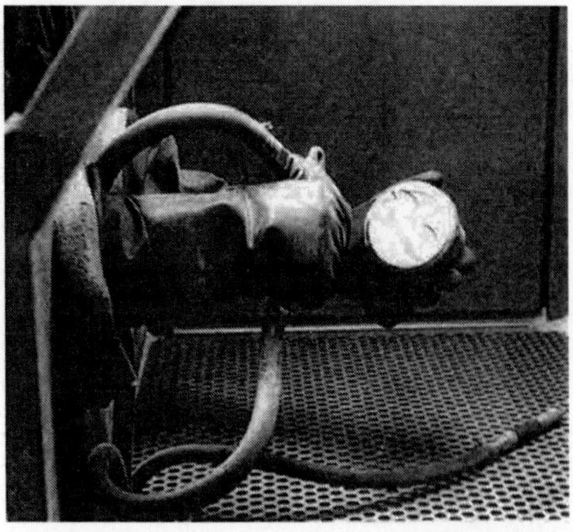

Fig. 11.75 *Cleaning a piston with a bead blaster. Winoma Van Norman*

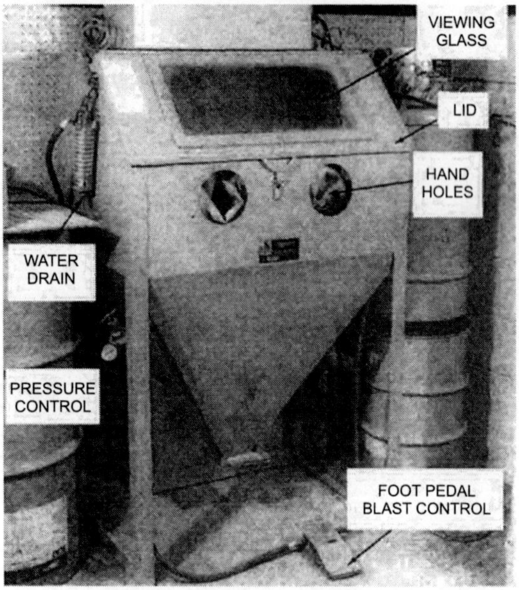

(a) Note parts of blaster.

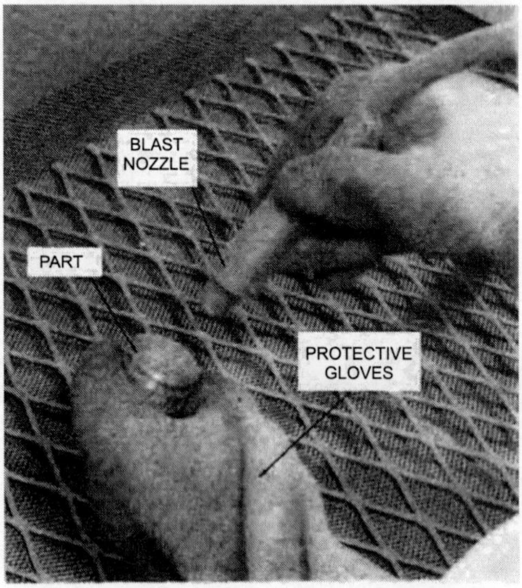

(b) Only blast parts enough to remove deposits.

Fig. 11.76 *Blast cleaning booth will make used parts look like new.*

Blasting with glass beads is another method used for cleaning automotive parts, *Fig. 11.75.* Compressed air is the propellant that blasts the beads against the part. Both wet and dry blast machines are available. Glass beads clean the part down to bare metal, leaving a film of detergent. The beads are round in shape and are not abrasive. They do not embed themselves in the surface being cleaned. Being round, the beads leave a uniform indented surface, imparting strength by compressing the metal being cleaned. This method is used when cleaning pistons, valves, and connecting rods.

Figure 11.76 *(a)* shows a shot blast cabinet. Place the part to be cleaned inside the cabinet. Close the door. Insert your arms into the holes and rubber gloves. Then direct the blast nozzle over the part to be cleaned, *Fig. 11.76 (b).*

Cylinder heads, intake manifolds, valve covers, etc. can all be cleaned and ready for painting with shot blast cleaning. However, be careful when cleaning soft aluminium parts. Only use enough blast action to clean off deposits. Too much blasting can wear off aluminium and ruin the part.

Carbon coated pistons heads are very hard to clean. Piston heads can be cleaned with careful use of the shot blaster. Only clean the piston heads, not the skirt because of the danger of skirt erosion.

THEORY QUESTIONS

1. Describe the concept of 'valve job.'
2. A valve job involves ________.
   (*a*) Removing and replacing the cylinder head(s)
   (*b*) Grinding the valves to prevent leakage
   (*c*) Replacing any worn or damaged parts as required
   (*d*) All of the above.
3. Define the term 'engine overhaul'.
4. Explain the difference between a minor and a major overhaul.
5. What are manufactured engines?
6. An in-vehicle engine overhaul is more common than an out-of-vehicle engine overhaul. *True* or *False?*
7. What does blueprinting of an engine mean?
8. Blueprinting means ________.
   (*a*) The same as minor overhaul
   (*b*) The same as major overhaul
   (*c*) An overhaul with all parts equally matched
   (*d*) Painting the engine blue after an overhaul
9. What is the function of the tapered dowel pins on the rear of most engine cylinder blocks?
10. What is a 'ring ridge'? Why it is important to remove the ridge at the top of the cylinder?
11. Describe three safety rules to follow when lifting an engine out of a vehicle chassis.
12. The ridge at the top of the cylinder ________.
    (*a*) Is caused by wear at the top of the cylinder by the rings
    (*b*) Represents a failure of the top piston ring to correctly seal against the cylinder wall
    (*c*) Means that a crankshaft with an incorrect stroke was installed in the engine
    (*d*) Should not be removed before removing pistons except when reboring the cylinders
13. It is not necessary to use several containers, or trays, to keep fasteners organized. *True* or *False?*
14. Why it is important to label wires and vacuum hoses?
15. Why it is important to return most parts to their original locations in an engine during reassembly?
16. What do you mean by the term 'mushroom tip' as used in connection with valves? What can happen if you drive out a valve that has a mushroom tip?
17. Describe six ways of cleaning engine parts.
18. A ________ is sometimes needed to remove a crank damper, seal, or crank sprocket.
19. When using a chain to remove an engine, the bolts holding the chain should be tightened or threaded ________ and threaded in at least ________ the diameter of the bolt.
20. A ________ ________ ________ is required to cut out a lip at the top of a worn cylinder to prevent ________ damage.
21. Tests find a burned engine valve. The engine is not smoking and has god oil pressure. Auto mechanic A says that this should be an in-vehicle operation and that it is not necessary to remove the engine. Auto mechanic B says that the engine should be removed to service the valves. Who is right?

(*a*) A only (*b*) B only
(*c*) Both A and B (*d*) Neither A nor B.

22. Which of these jobs does not require head removal?
(*a*) Piston replacement
(*b*) Connecting rod replacement
(*c*) Crankshaft replacement
(*d*) Diesel prechamber replacement.

23. A long block can be made from a short block with the addition of ________.
(*a*) Intake and exhaust manifolds
(*b*) Oil pump, oil pan, and timing chain cover
(*c*) Cylinder heads and valve train
(*d*) Fuel pump, carburetor, and air cleaner assembly.

24. All of the following engine repairs can be performed with the engine still installed in the vehicle *except:*
(*a*) Valve job
(*b*) Valve seal removal and replacement
(*c*) Intake manifold replacement
(*d*) Crankshaft service.

25. Lifting hooks are often installed in the factory because ____________.
(*a*) They make engine removal easier for the mechanic/technician
(*b*) They are used to install the engine at the factory
(*c*) They are part of the engine and should not be removed
(*d*) They render servicing the top of the engine easier for the mechanic.

26. With the rocker cover (valve cover) removed, the mechanic can check all of the following items *except* ________.
(*a*) Rocker arm and valve spring
(*b*) Combustion chamber deposits
(*c*) Valve stems and pushrods (overhead valve engines only)
(*d*) Camshaft (OHC engine only).

27. Mechanic A says that an engine's timing belt should be replaced during a major engine rebuild. Mechanic B says that an engine's cylinder head(s) should be reconditioned during a major engine rebuild. Who is right?
(*a*) A only (*b*) B only
(*c*) Both A and B (*d*) Neither A nor B.

28. Before the timing chain can be inspected and removed, the following component(s) must be removed: ________.
(*a*) Rocker cover (valve cover)
(*b*) Cylinder head(s)
(*c*) Vibration damper
(*d*) Intake manifold (V-type engines only).

29. Before the valves are removed from the cylinder head, what operations need to be completed out of the following:
(*a*) Remove valve keepers (locks)
(*b*) Remove cylinder head(s) from the engine
(*c*) Remove any burrs from the stem of the valve(s)
(*d*) All of the above.

30. Mechanic/technician A says that when preparing to remove an engine from the vehicle, vehicle's battery need not be disconnected. Mechanic/technician B says that when preparing for engine removal, the vehicle's radiator should be removed. Who is right?
(*a*) A only (*b*) B only
(*c*) Both A and B (*d*) Neither A nor B.

31. An engine block has deep ridges in its cylinders. Mechanic A says that a brush hone should be used to recondition the cylinders. Mechanic B says that the cylinder block should be sent to a machine shop to have the cylinders bored. Who is right?
(*a*) A only (*b*) B only
(*c*) Both A and B (*d*) Neither A nor B.

32. Mechanic A says that when removing the engine on certain front-wheel drive vehicles equipped with a transaxle, the engine and transaxle should be removed as a unit. Mechanic B says that during engine removal on all front-wheel drive vehicles, the engine and transaxle must be removed separately. Who is right?
(*a*) A only (*b*) B only
(*c*) Both A and B (*d*) Neither A nor B.

33. Teardown on an OHC engine normally begins with ________ disassembly.
(*a*) Front end (*b*) Top end
(*c*) Bottom end (*d*) Any of the above.

34. Mechanic A says that aluminium engine parts can be cleaned in hot tank. Mechanic B says that aluminium engine parts can be corroded if cleaned in a hot tank. Who is correct?

(*a*) A only (*b*) B only
(*c*) Both A and B (*d*) Neither A nor B.

35. After the oil pan is removed, the mechanic should inspect_______.
    (*a*) The oil pump and pickup screen
    (*b*) To make certain that all rod and main bearings are numbered or marked
    (*c*) The valve lifters (tappets) for wear
    (*d*) Both (*a*) and (*b*).
36. Mechanic should place a breaker bar ________ to free the cylinder head from the block.
    (*a*) between the head and a solid, stationary object
    (*b*) Between the head and block deck surface
    (*c*) Between the head and exhaust manifold
    (*d*) In one of the ports.
37. Mechanic A says that a minor overhaul can often be done with the engine in the vehicle. Mechanic B says that a core is required for most manufactured engines. Which mechanic is right?
    (*a*) A only (*b*) B only
    (*c*) Both A and B (*d*) Neither A nor B.

❑❑❑

Engine Reconditioning – Short Block Service

12

12.1 RECONDITIONING PROCEDURE

The initial step in any job of automotive engine reconditioning is to clean the assembly. As already discussed, there are many different methods of cleaning, and the one selected depends on the part and the type of contamination that is to be removed. After removing the outer dirt and oil, the oil pan, valve cover(s), and cylinder head(s), and the other parts are disassembled, following the procedures already explained in the previous chapter. To remove accumulation of sludge, carbon, dirt, and grease, these parts can be cleaned taking one of the cleaning methods that is appropriate to the part. Before cleaning a cylinder block or a cylinder head, remove the core hole plugs. This will allow the cleaning solution to flush out any accumulation of rust and lime deposits from passages.

12.2 INSPECTION AND MEASUREMENT

After cleaning, it is possible to carefully examine and accurately measure the individual parts to locate defects, determine the extent and type of wear, and pinpoint what caused the worn condition,

12.3 CYLINDER BLOCK SERVICE

The starting point for actual engine reconditioning is the cylinder block, because practically all other parts are fitted to it. If the block is damaged in any way, it has to be repaired or replaced. Before cleaning the block, the cam bearings, soft plugs, and oil gallery plugs are removed. The cam bearings are driven out of their housings with a bearing driver and a hammer, *Fig. 12.1*. The soft plugs may be removed with a hammer and punch or a special puller designed for this job. The oil gallery plugs at the rear of the block are usually pipe plugs that may be removed with a wrench. Removing the oil gallery plugs and soft plugs permits the inside passages in the block to be thoroughly cleaned and flushed.

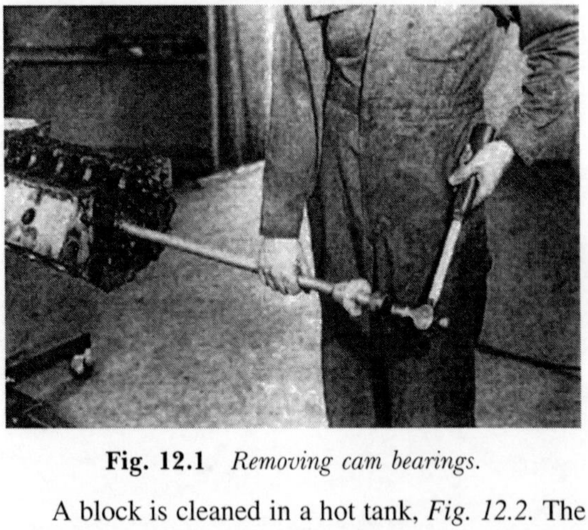

Fig. 12.1 *Removing cam bearings.*

A block is cleaned in a hot tank, *Fig. 12.2*. The empty cylinder block is lowered into a heated cleaning solution and allowed to soak for several hours. The hot tank removes grease and paint from the outside of the block and scale from the inside coolant passages. After removing the block from the hot tank, it is steam cleaned to remove deposits loosened by hot soaking operation. Oil passages and holes must be cleaned out with a brush. There are numerous things to be inspected and checked when reconditioning a cylinder block. Cylinder block service typically includes:

1. Inspection for cracks, excess wear, scored cylinders, and other block damage.

2. Checking for main bore and deck warpage; having the block machined if necessary.
3. Measuring cylinder wear and having block bored if cylinders are worn and are out of specs.
4. Honing or deglazing cylinders to prepare cylinder walls for installation of new rings.
5. Installing camshaft or balancer shaft bearings.
6. Cleaning of cylinder block before assembly.

Fig. 12.2 *Cleaning cylinder block in hot tank. (Federal-Mogul)*

12.4 INSPECTING CYLINDER BLOCK

The cylinder block may be cracked, warped, worn, or otherwise damaged. The damage may affect the operation of the crankshaft, pistons, rings, bearings, camshaft, cooling system, lubrication system, etc. The extent of the damage will determine whether or not the block is repairable.

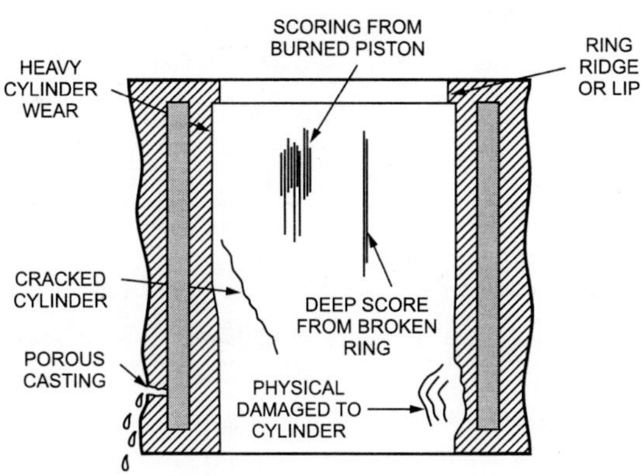

Fig. 12.3 *Carefully inspect cylinder walls for these kinds of defects. Use fingernail to feel for cracks and scoring. (General Motors)*

Carefully inspect the cylinder block for any problems, like the ones shown in *Fig. 12.3.* Each cylinder should be checked for scoring or scratching on its walls from broken piston pin lock rings or piston rings. In particular, check the cylinder walls for scoring, heavy wear, cracks, etc. Use a droplight to closely inspect the surface of each cylinder wall. Feel the cylinder wall with a fingernail to locate any problems.

12.5 METAL PARTS RESTORATION

Most of the jobs can be done in auto repair shop; however, milling and boring operations usually are done in a machine shop. There are a number of methods of changing the condition or dimensions of metal parts. In some cases, these methods can be used to restore an engine to good condition. However, engine parts are usually replaced if they are damaged or excessively worn.

The size of a part can be increased by adding metal by ***soldering, brazing, welding, and plating*** or ***spraying.*** In some cases the part can be expanded with heat and pressure. When adding metal, there is a wide choice of materials, each of which possesses certain specific characteristics that may be desired for a particular purpose. If a soft surface is needed, use tin or bronze. If a hard surface is needed, add steel of any desired degree of hardness by welding. Another method of adding a hard surface is by the electroplating process, e.g., the chrome surface on a piston ring.

In engine repair work, the material can be added to increase the size of one part without disturbing the mating part. For example, we can spray molten metal on a worn crankshaft journal to avoid installation of undersize bearings.

To reduce the size of a part, ***Cut, grind,*** or ***etch*** the metal away. In some cases the metal can be shrunk by freezing it. In some other cases, removing metal from one part and installing an oversize mating part is the best solution. For example, a cylinder can be rebored and an oversize piston can be installed in the enlarged cylinder.

Also, sometimes the metal is removed for the sole purpose of obtaining better surface fit of two mating parts. An example of this is the correction of a warped cylinder head by surface grinding to restore a flat mating surface.

12.6 FINDING CRACKS IN CYLINDER BLOCK

When the cylinder block and head are clean, make a visual inspection for ***external cracks*** or serious damage. Cracks in cylinder heads and cylinder blocks are often so fine that they are difficult to see with the naked eye. If symptoms indicate possibility of cracks in the block, then magnaflux the block or use dye penetrant.

Magnafluxing uses a magnet and a ferrous powder to highlight cracks and pores in cast iron cylinder blocks. Place the magnet over the area to be tested. Spread the ferrous powder over the cylinder block. A break in the metal sets up a magnetic field that pulls the powder to the crack or discontinuity. Thus, any crack will cause the powder to collect in the discontinuity and makes it easy to detect, *Figs. 12.4 (a)* to *(e)*.

Dye penetrant is usually used to find cracks and casting imperfections in aluminium cylinder blocks. Spray the special dye penetrant over the block. Then, spread a powdery developer over the penetrant. This

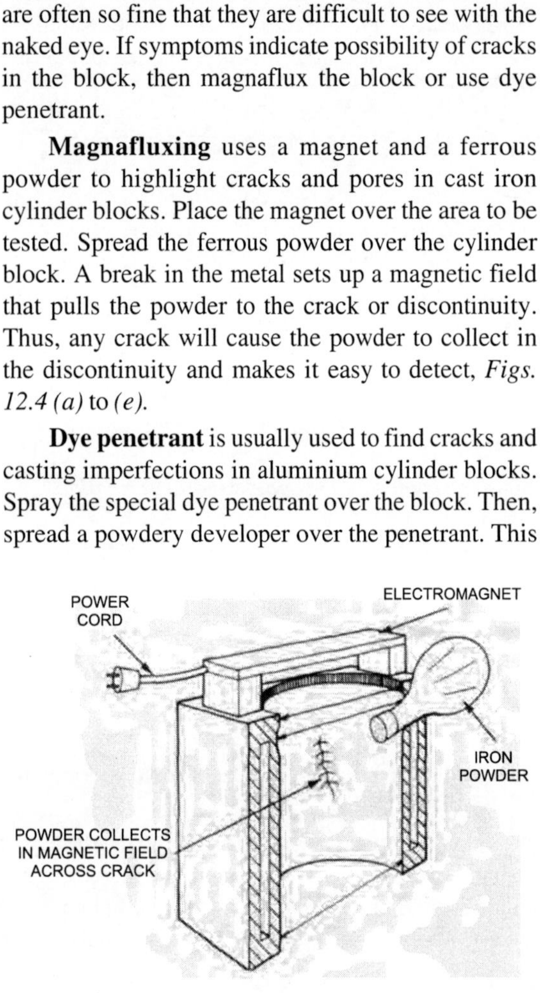

(a)

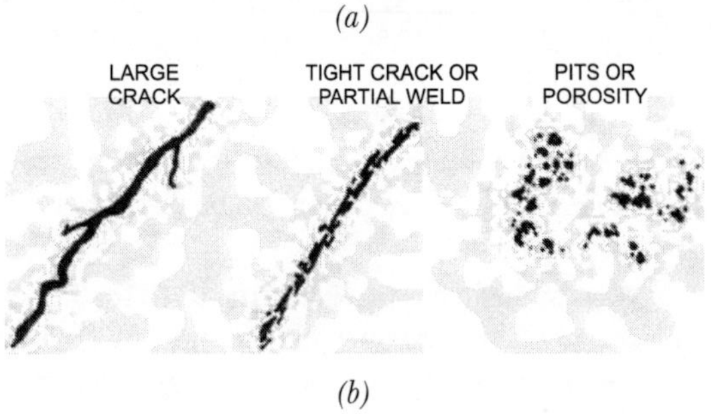

(b)

(c)

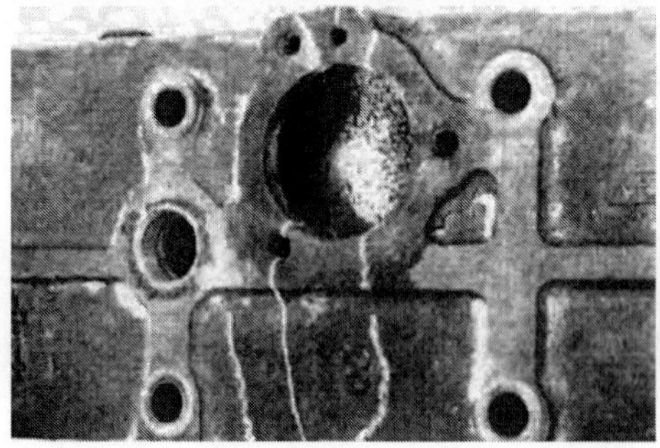

(d)

Fig. 12.4 *Magnafluxing will find cracks in iron parts. (a) Magnafluxing equipment found cracked cylinder. (b) Type of defects magnetic inspection will locate. (c) Actual cracked block caused by part breakage. (d) Coolant freeze damage to block. (K-line and Magnaflux Corp.)*

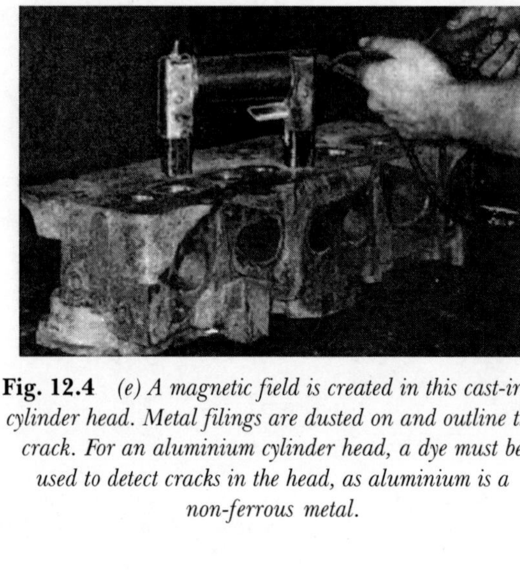

Fig. 12.4 *(e) A magnetic field is created in this cast-iron cylinder head. Metal filings are dusted on and outline the crack. For an aluminium cylinder head, a dye must be used to detect cracks in the head, as aluminium is a non-ferrous metal.*

will make the penetrant turn red if it has collected in any discontinuity (crack or casting cavity). This makes easy to detect any discontinuity in cylinder block.

Fluorescent penetrant is just like the dye penetrant except that it uses a black light to show up cracks. The chemical is sprayed over the part, which is then placed under a black light. The concentration of the chemical collected in a crack will glow under the black light, *Fig. 12.5 (a), (b).*

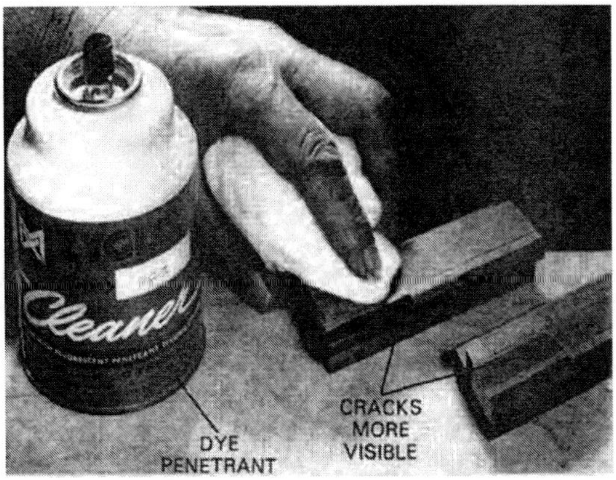

(a)

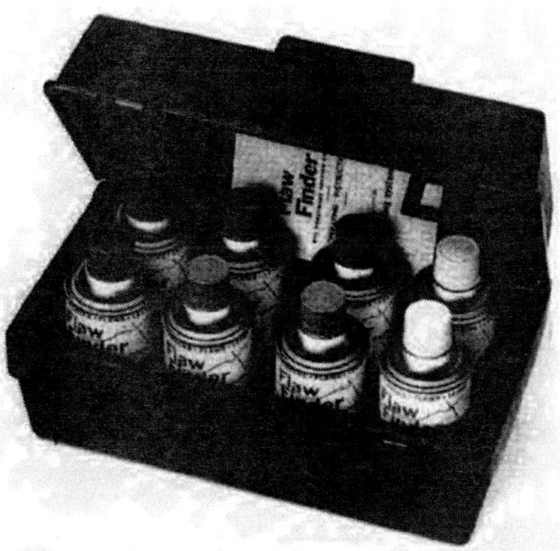

(b)

Fig. 12.5 *(a) Florescent penetrant (Zyglo) can be sprayed over parts to find cracks. Chemical collects in cracks and then will show up when placed under black light, (b) Another dye type crack finding system. Follow specific directions provided with it. (Magnaflux Corp.)*

Pressure testing finds cracks by filling the cylinder block or head water jackets with air pressure. All water jacket openings are sealed with the help of rubber expander plugs. Sheets of rubber are bolted over the deck surface with the metal plates to make the inside of the cylinder block airtight. An air hose is connected to the water jackets and the block is lowered in a large water tank. Any air bubbles rising to the top of the water will indicate a crack or casting pore.

12.7 REPAIRING CYLINDER BLOCK CRACKS

Block repairs are done by a specialized machine shop that has the special equipment and skills needed to repair blocks. Check the condition of the freeze plugs, *Figure 12.6.* If signs of leakage or corrosion are found, replace the plugs.

Fig. 12.6 *Leaking freeze plugs must be punched out and new freeze plugs installed. Always install new freeze plugs after cleaning a cylinder head or engine block in either a hot or cold tank solution.*

If the block was cracked by water freezing in the water jacket, the crack can usually be closed satisfactorily by copper welding, brazing, iron welding, or by one of the patented processes, e.g., the *Seal Lock Process.* Often electric welding to gas welding, as lower temperatures in the area of repair are not likely to produce warpage of the cylinder block. In the case of the ***seal lock process*** the sides of the crack are tied together by driving special keys or locking pins into previously drilled holes. Then the surface is carefully peened and further sealed.

In cases where a panel or section of the water jacket is broken out, it is sometimes possible to shape a metal plate into a patch that corresponds to, and is slightly larger than, the opening. The patch is then attached to the block with bolts.

In the case of small cracks, ***iron cement*** can be used for quick repair, *Fig. 12.7.* Never repair an area with cement in or by the cylinder or cylinder head. The choice of the repair method will depend upon the size of the damaged area, location of the crack, the cost of a new block, and value of the vehicle. At times, the cylinder wall or cylinder head will crack, *Fig.12.8.* Usually, a cracked cylinder head is replaced rather than repaired.

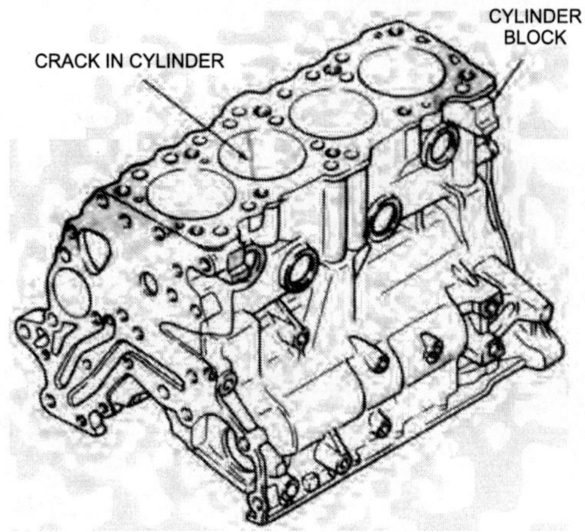

Fig. 12.8 *A crack in the cylinder makes this engine block useless. Chrysler Corp.)*

Cracks and pores in a cylinder block can be fixed by:

1. *Sleeving* the cylinder if the crack is in cylinder wall area.

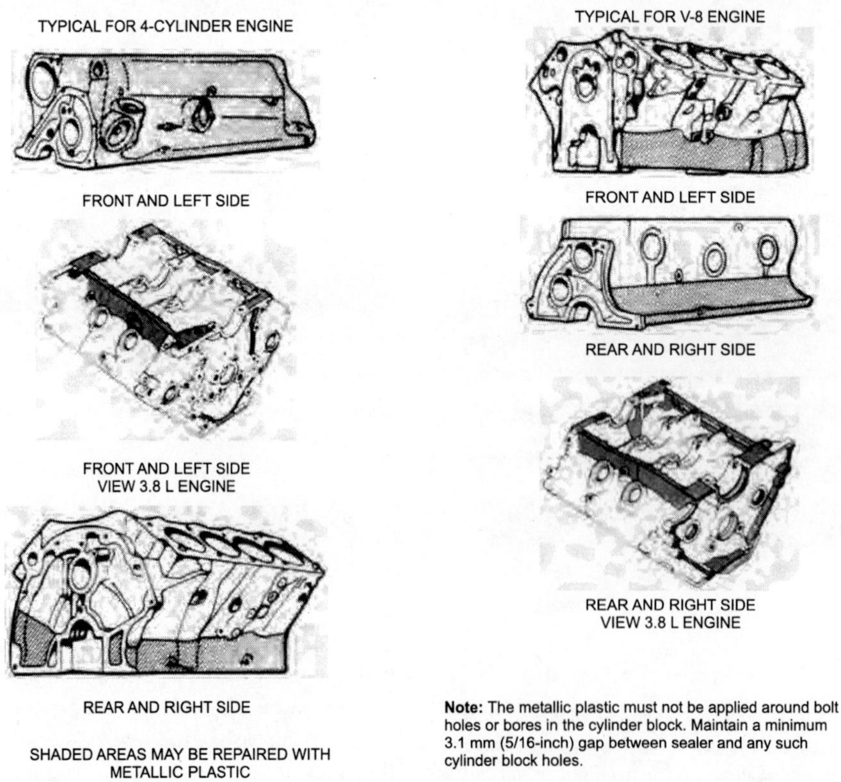

Fig. 12.7 *Cylinder block can be repaired with a metallic 'plastic' in shaded areas only. Note that the area around the cylinders cannot be repaired. (Ford Motors)*

2. ***Welding*** cast iron with high nickel rod after heating in furnace or welding aluminium block with a heli-arc welder and aluminium rod.
3. ***Plugging,*** which involves drilling series of holes along crack and threading in repair plugs.
4. Using ***special epoxy*** or ***metallic plastic*** to seal casting flaw or pore if in certain areas of cylinder block, *Fig. 12.9.* This method should not be used on cracks.

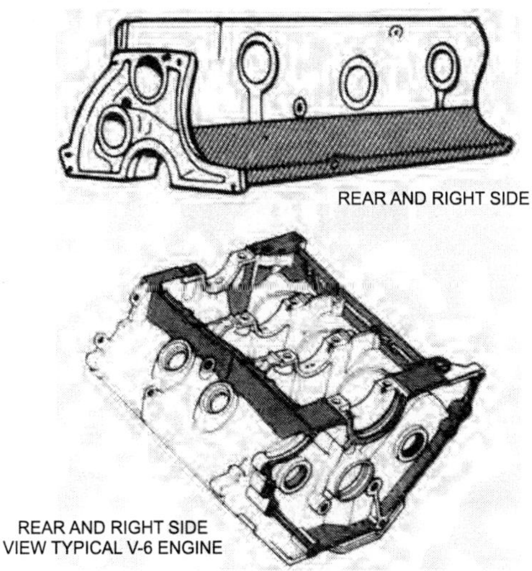

Note: The metallic plastic must not be applied around bolt hole bores in the cylinder block.

Fig. 23.9 *Casting flaws or pores can sometimes be repaired with special epoxy or plastic. This is a service manual illustration showing the areas of block that will seal with epoxy. (Ford Motors)*

5. Replace the block if cost of repair is higher than new or good used block.

12.8 CORROSION

For aluminium engine blocks or heads, the visual inspection should be particularly thorough to detect any possible ***corrosion*** of the metal. Sometimes, corrosion of the metal around the water jackets will occur from chemicals in the cooling water, especially in locations where the water supply contains more than the usual amount of chloride.

Corrosion from ***electrolytic*** actions may also occur, due to the dissimilar metals found in the engine cooling system. If corrosion is appreciable, it may interfere with the seal of the gasket between the head and the block. This might permit coolant to enter the combustion chamber or allow compression pressure to escape to the water jacket.

12.9 CYLINDER HEAD WARPAGE

Cylinder heads often become ***warped.*** Sometimes the cylinder block mating surface also gets warped. These surfaces should be normally true within 0.076 mm in any 152 mm span or within 0.152 mm overall. Measurement is made by means of a steel straight edge and a feeler gauge, *Figs. 12.10* and *12.11.* It should be checked when old head gasket was blown and leaking. It should also be done on modern aluminium cylinder blocks since they distort easily. To check for block deck warpage, lay a straightedge on the clean block surface, *Fig. 12.10.* Try to slip different thickness feeler gauge blades between the block and straightedge. The thickest blade that fits shows warpage. Check in different locations on the block.

Fig. 12.10. *Checking a cylinder head for warpage. If a 0.152 mm feeler gauge can be placed between the straightedge and cylinder head, the head is warped and must be machined. (Fel-Pro Inc.)*

If beyond specs, replace the block or send it to a machine shop for surface milling. If the surfaces are warped or otherwise damaged they can be reconditioned on special equipment, *Fig. 12.12.* Due to the symmetry of cylinder banks, on a V-type engine, if 0.75 mm metal is removed from one head, the other head and the intake manifold must also be machined true. Otherwise, the bolt holes connecting the intake manifold to the heads will not align properly. ***Decking a block*** involves machining the cylinder head mounting surfaces until they are parallel and equal distance from the cylinder block main bore. This also known as *"squaring the block."*

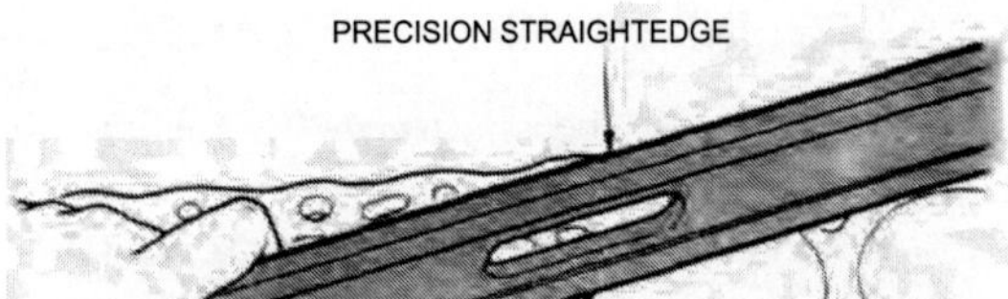

Fig. 12.11. *(a) Checking a cylinder block deck for flatness with a precision straightedge. (American Honda Motors Co., Inc.)*

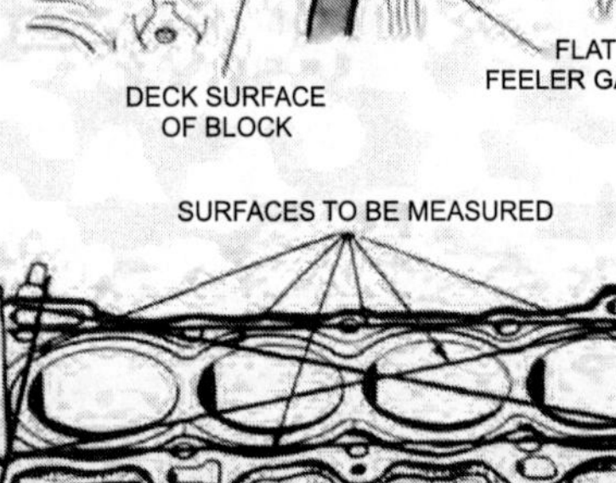

Fig. 12.11 *(b) The engine block is checked for warpage in the same manner as the head. (Chrysler Corp.)*

12.10 CYLINDER WEAR

The piston and ring movement, the high temperatures and pressures of combustion, the detergent action of gasoline entering the cylinder-all contribute to cause cylinder wall wear.

Fig. 12.12 *Equipment used to machine engine blocks.*

A worn cylinder is not round or true. The greatest amount of wear in a cylinder occurs at the top of the cylinder. This is due to the lack of lubrication at the top of the cylinder and the thrust side of the piston. It will be larger in diameter near the top and be wider across the thrust surface, *Fig. 12.13 (c).* The wear at the top of the cylinder is referred to as ***out-of-round,*** *Fig. 12.13 (b).* ***Cylinder out-of-roundness*** is a difference in cylinder diameter when measured front-to-rear and side-to-side in the cylinder block. Piston thrust action normally makes the cylinders wear more at right angles to the centreline of the crankshaft or piston pins, *Fig. 12.13 (b).* If new rings are installed in an out-of-round cylinder, blowby will be caused. There will be a gap where the round ring will not contact the cylinder, and will allow oil and pressure past this gap.

The difference between the amount of wear at the top of the cylinder and the bottom of the cylinder is referred to as ***taper,*** *Fig. 12.13 (a).* More oil splashes on the lower area of the cylinder as compared to the oil lubricating the top of cylinder. As a result, the top of the cylinder wears faster (larger) than the bottom, resulting in a taper, *Fig. 12.13 (a).* If new rings are installed in a tapered cylinder, they will be expanded and contracted every time they slide up and down the cylinder. The rings may fail to stay in contact with the cylinder properly at high engine rpm. This can also result in loss of power, blowby, and oil consumption.

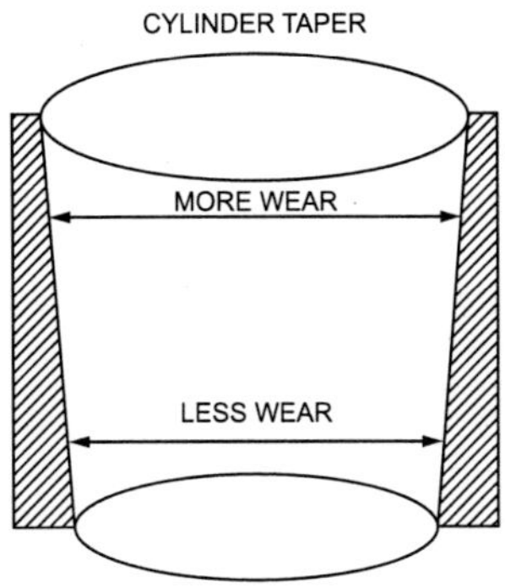

(a) A cylinder will wear more on top than bottom, forming taper.

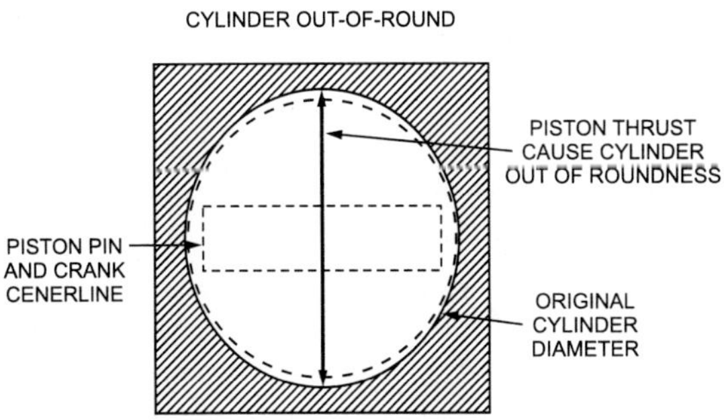

(b) Cylinder will also wear more on piston thrust surface or opposite crankshaft and piston pin, forming out-of-round.

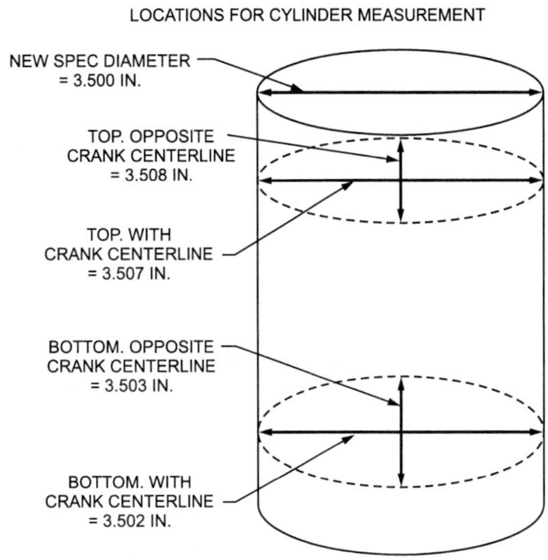

(c) Measure cylinder in these locations and write down the measurements.

12.11 MEASURING CYLINDER WEAR

In case the cylinder is not badly scratched or scored, we need to measure cylinder wear to determine whether the new rings will seal properly or not. Obviously, new round rings cannot seal in worn, out-of-round, or tapered cylinders. In addition, the

EXAMPLE CALCULATION OF CYLINDER WEAR

	Measure parallel with crank	***Measure opposite crank centerline***
Cylinder diameter at top	3.507 in.	3.508 in.
Cylinder diameter at bottom	3.502 in.	3.503 in.
New cylinder diameter spec	3.500 in.	3.500 in.

1. Cylinder Taper = max. Top dia. minus max. bottom dia
   = 3.508
   – 3.503
   Taper = .005 inch
2. Cylinder Out-of-round = MAX. top dia. minus min. top dia.
   = 3.508
   – 3.607
   Out-of-round = .001 inch
3. Cylinder wear = MAX. cylinder dia. minus spec dia
   = 3.508
   – 3.500
   wear = .008 inch

(d) Example measurements and calculations of cylinder wear from (c) of this illustration.

Fig. 12.13 *Measuring cylinder wear. (GM)*

cylinder measurement will permit you to determine piston-to-cylinder clearance, *Fig. 12.13 (c)*.

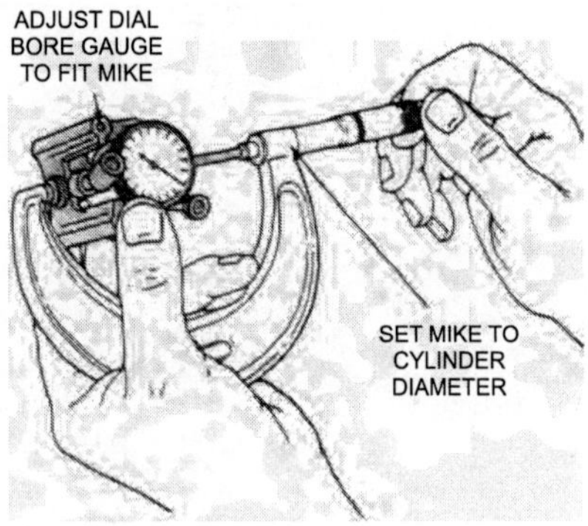

(a) Set mike to diameter of new cylinder. Then, adjust dial bore gauge to fit inside micrometer.

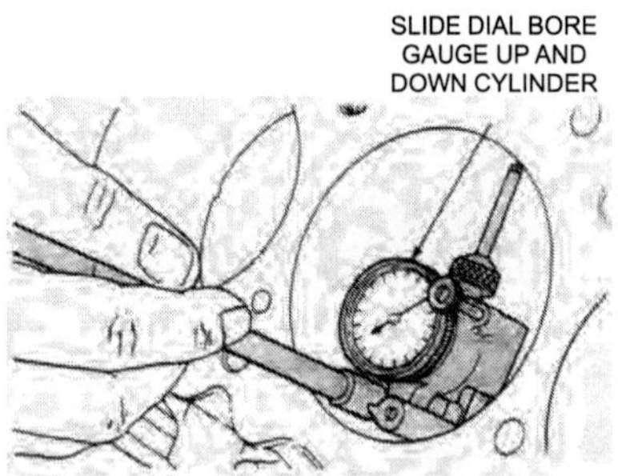

(b) Slide dial bore gauge up and down in cylinder to read taper. Rotate gauge around cylinder to read out-of-round. Cylinder wear equals difference between new spec and actual readings.

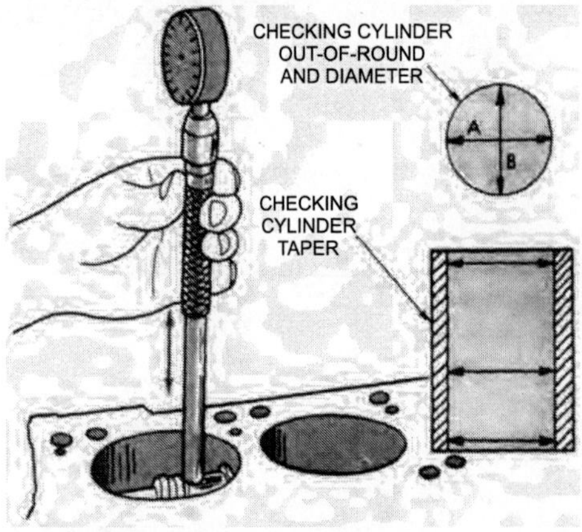

(c) Locations for measurements

Fig. 12 14 *Using dial bore gauge to check cylinder wear. (General Motors Corp.)*

A ***dial bore gauge*** is a quick and precise tool used to measure cylinder wall wear. Adjust the dial bore gauge to the correct diameter with the help of an outside micrometer. Then, slide the bore gauge up and down in the cylinder. Indicator movement equals cylinder taper. Measure both parallel and perpendicular to the bore centreline to determine out-of-round, *Fig. 12.14 (a), (b)* and *(c)*.

A large inside micrometer or a telescoping gauge and an outside micrometer can also be used to measure cylinder dimensions.

12.12 CYLINDER WEAR LIMITS

Normally, cylinder boring is necessary when:

- ***Cylinder out-of-round*** is more than 0.13 mm.
- ***Cylinder taper*** is more than 0.2 mm.
- ***Maximum cylinder diameter,*** at its widest point, exceeds 0.25 mm over standard bore specifications.

If the cylinders in the block are worn more than specs (as given in the applicable service manual), the cylinder block will have to be rebored or replaced. If bored, oversize pistons will be required. *Fig.12.13 (d)* illustrates an example of measuring and determining the cylinder wear. When measuring cylinder wear, also check the cylinder bore diameter. It is possible that the block might have been bored earlier and is not a *standard diameter* for new block. To check cylinder diameter:

- Check the service manual for standard bore specifications.
- Measure the widest point diameter near the cylinder top (where the rings rub against the cylinder wall), using a dial bore gauge, inside micrometer, or telescoping gauge and outside micrometer to measure cylinder diameter.
- Measure the cylinder diameter at the bottom, in an unworn section.
- Compare the measurements to specs.

This would allow the mechanic to order the right size piston rings, know correct rebore limits, etc., for new rings to seal.

12.13 CYLINDER RECONDITIONING

The engine block is the foundation of the engine. All parts of the block must be of the correct size and they must be properly aligned. The parts must also have the proper finishes to assure dependable service life. *Blueprinting* is the reconditioning of all the critical surfaces and dimensions so that the block is actually like new.

After cleaning, the block should be carefully inspected for any cracks or other flaws before initiating any machine work. In case the block is found to be in serviceable condition, follow the sequence of operations given below:

1. Align-boring or Align-honing main bearing saddles and caps.
2. Machine the block deck surface parallel to the crankshaft.
3. Bore and hone the cylinders.

1. Main Bearing Housing Bore Alignment

The main bearing journals of a straight crankshaft should be in proper axial alignment. If the main bearing housing bores in the block are misaligned, the crankshaft will bend as it rotates. This will lead to premature bearing failure and may lead to crankshaft failure too. The normal stress in the block casting is gradually relieved with use in service. This might lead to a slight warpage of the block. Usually, continuous pounding caused by combustion loads will cause some stretch in the main bearing caps, *Fig. 12.15*. The main bearing bores gradually bow from the cylinder head and elongate in vertical direction. This results in the bearing bore getting smaller at the centre line as the block distorts, pinching the bore inward at the sides, *Fig. 12.16*.

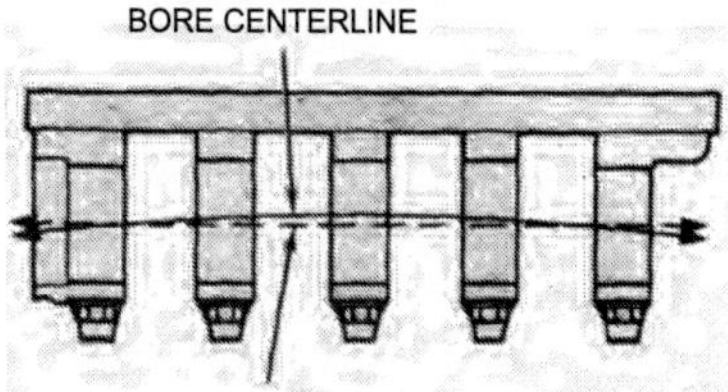

Fig. 12.15 *The main bearing bores of a warped block usually bend into a bowed shape. The greatest distortion is in the centre bores.*

After repeated heating and cooling or overheating, the main bearing bores in the cylinder block may warp or twist. This will affect main bearing insert alignment and crankshaft fit in the block. With main bearing bore misalignment, the crankshaft can lock up when the main caps are torqued to specifications.

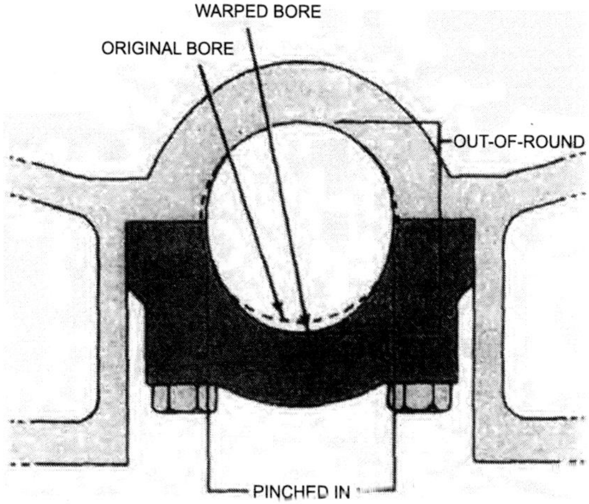

Fig. 12.16 *When the main bearing caps bow downward, they pinch in at the parting line.*

The first step in determining the condition of the main bearing bores is to determine if the bore alignment in the block is axially straight. These bores are called *saddles*. To measure main bearing bore alignment, rest the block on a flat surface and lay a straightedge on the bores. Insert a 0.038 mm feeler gauge any where under the straightedge. If the gauge fits, the main bearing bores are misaligned. Always check bore misalignment after main bearing failure, *Fig. 12.17*.

If the block saddles exceed 0.038 mm distortion, then align honing is needed to restore the block housing bore alignment. If the block saddles are straight, the bores should be measured to verify that the bearing caps are not distorted. The bearing caps should be installed and the retaining bolts tightened to specified torque before measuring the main bearing bores.

Using a telescoping gauge, measure each bore in at least two directions. Check the service information for the specific main bearing bore diameter. The bearing bore should by more than 0.0127 mm, *Fig. 12.18*.

A dial bore gauge is often used to measure the main bearing bore. Set up the dial bore gauge in the fixture with appropriate extensions to achieve the nominal main bearing bore diameter. Check the service information using the applicable service manual for the specific main bearing bore diameter and determine the exact nominal range with ± tolerance that is permissible.

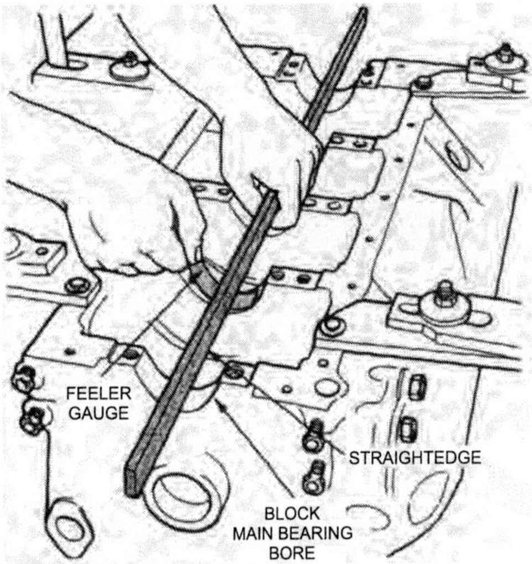

Fig. 12.17 *Check main bearing bore alignment if main bearings were badly worn or spun. Try to fit a 0.038 mm feeler gauge between the straightedge and block bore. If it fits, block should be line bored by a certified machine shop. (Federal-Mogul)*

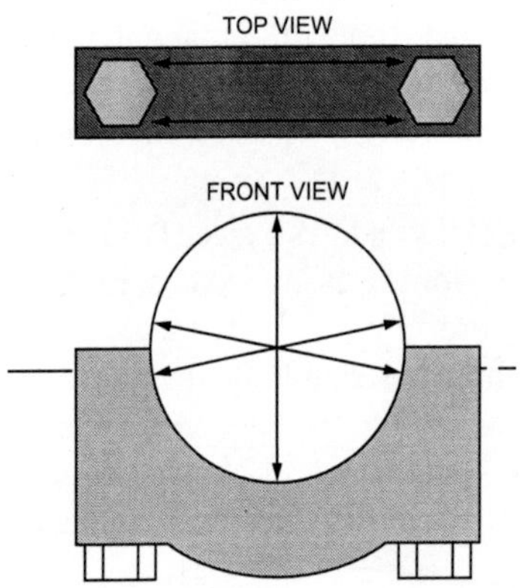

Fig. 12.18 *Main bearing bores should be measured in both directions as well as front to back to be sure that the bore is not tapered.*

Using the dial bore gauge, measure each bore in at least three locations.

- Insert the gauge head into the bore head, *Fig. 12.19.* Rotate the gauge until the extension and plunger are horizontal. The measurement will verify whether or not the bearing bore has been pinched inward at the parting line, *Fig. 12.20.*

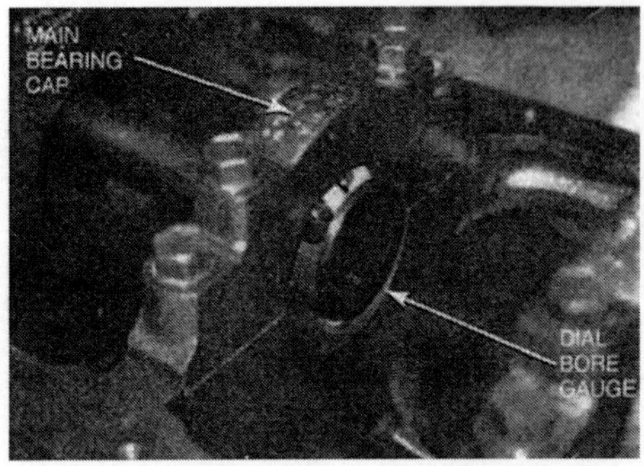

Fig. 12.19 *Before align honing (or boring), each main housing bore must be measured. A dial bore gauge is the tool of choice, because it is set to read the diameter of the housing bore, and the dial will indicate plus or minus readings from the specifications.*

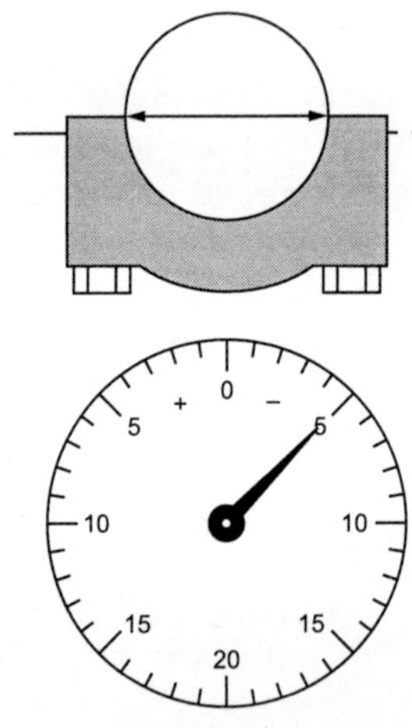

Fig. 12.20 *The needle position on dial bore gauge indicates a diameter-.0005 in undersize.*

- Next, check the bearing bore at 45° from the parting line, *Fig. 12.21.*
- Check the bearing bore at 90° from the parting line *Fig. 12.22.*

In this particular case, with the gauge readings as shown being less than 0.038 mm, the engine should be *line honed* (rigid stones) to correct for the distortion that has occurred to the block and to the main bearing bores. *Block line boring* (machine tool) is used to straighten or true misaligned main bearing bores when misalignment is over 0.038 mm (0.0015 in.).

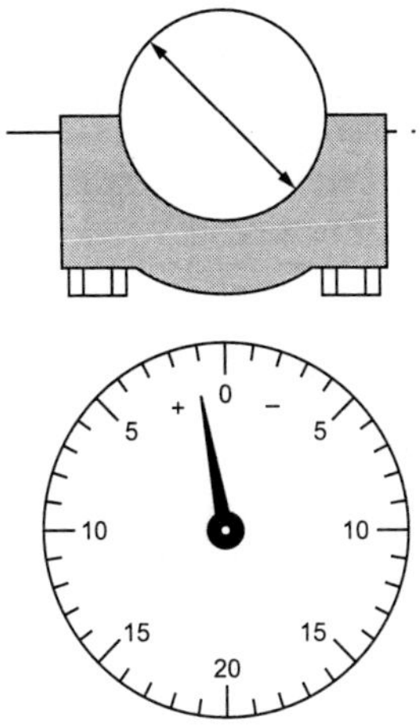

Fig. 12.21 *The needle position indicates that the bearing bore is + .001 in. oversize.*

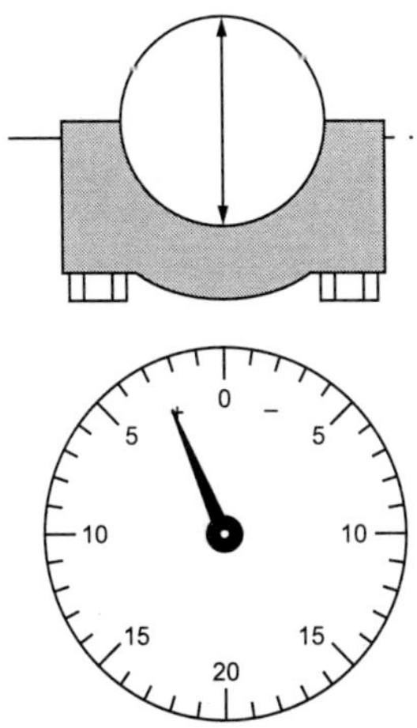

Fig. 12.22 *With the dial bore gauge set to read 90° from the parting line, the needle position indicates that the bore diameter is + 0.002 in. oversize.*

A machine shop will have a boring or honing bar for machining the bore back into alignment. Line boring is becoming more common with present thin-wall, light weight construction blocks, *Fig. 12.23*. The align boring tool is a cutting tool, similar to a lathe tool. Honing uses a stone instead of cutting tool and produces a finer finish than does align boring. The steps followed in align boring are:

1. A small amount of metal is removed from the main bearing cap parting surfaces. In preparation for align honing, the main bearing caps are coated with machinist bluing dye. Because less than 0.0254 mm is to be removed from the bearing cap, to be certain that the material is removed equally. It is necessary that about 0.38 mm be removed when an align boring cutting tool is used. While, only 0.038 – 0.05 mm needs to be removed in align honing.
2. The resurfaced bearing caps are torqued in place on the block.

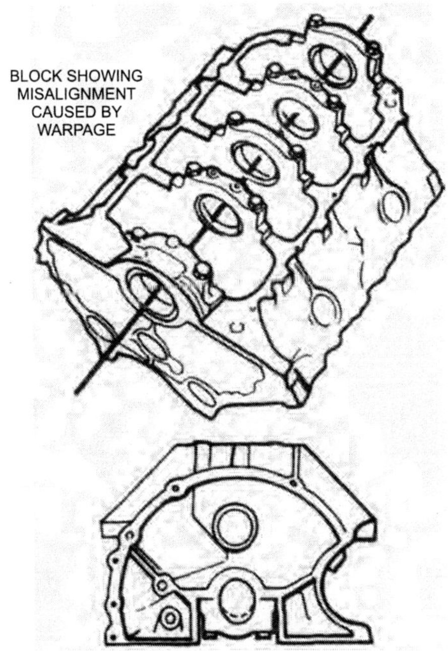

(a) Note how warpage, cap stretch, or a spun bearing could cause problems. If misaligned, crank could lock in block.

(b) Modern line honing or boring machine.

Fig. 12.23 *Block line boring.*

3. The main bearing housing bores are checked to locate exactly where metal must be removed to align the bores. The align boring tool is adjusted at each main bearing housing bore to machine the correct diameter. In this connection, one dimension that is critical in all engines is the spacing between the cam bearing centerline and the crankshaft centerline. This must be maintained to have the correct cam drive gear mesh or the proper timing chain tension. If this dimension is not maintained correctly, it will lead to faulty timing and to premature failure of the cam drive, *Fig. 12.24*.

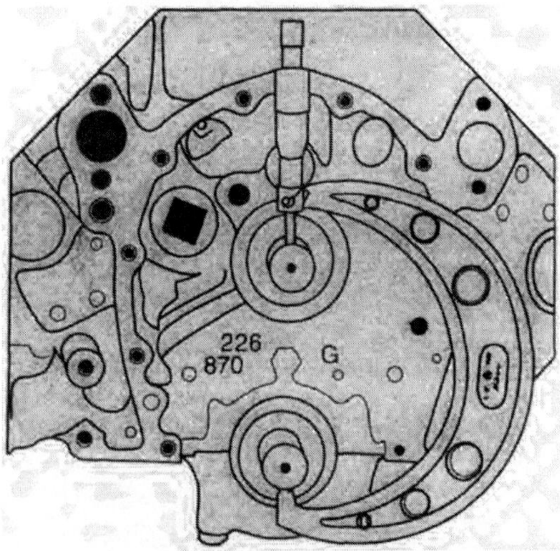

Fig. 12.24 *Special setup to measure the center line distance between the crankshaft and camshaft. During align boring, material is removed from the saddle area of the block, bringing the crankshaft closer to the camshaft.*

4. The align hone fits through all the main bearing housing bores simultaneously. The align hone is stroked back and forth through the bearing bores to size them, *Fig. 12.25*.

Fig. 12.25 *Align boring the main bearing bores.*

5. The block and oil passages must be thoroughly cleaned after align boring to remove all abrasives and metal chips. After drying the machined surfaces are coated with oil to prevent rusting until the block is finally cleaned for assembly.

2. Machining the Deck Surface of the Block

A multi-cylinder automotive engine should have the same size combustion chamber in each cylinder. For this, each piston must come up an equal distance from the block deck. As the crankshaft rotates, the pistons come to top dead centre during compression and power strokes. When all parts are of the same size, all the pistons will come up to the same height. This can only be true if the block deck is parallel to the main bearing bores, *Figs. 12.26* and *12.27*.

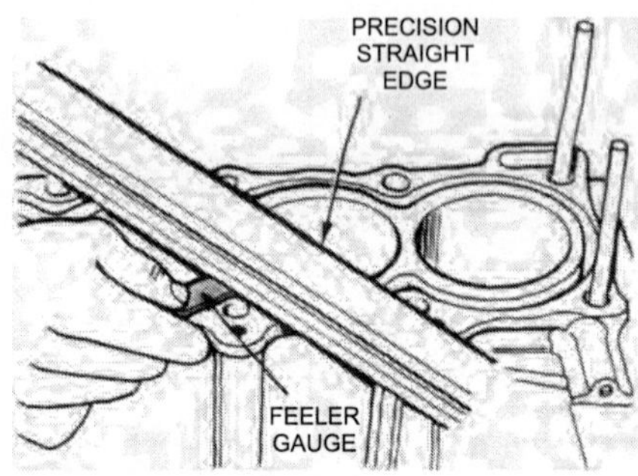

(a) Check the flatness of the block deck surface using a straightedge and a feeler gauge.

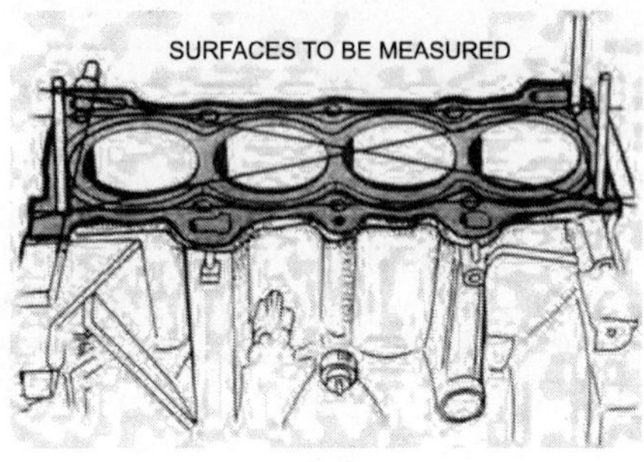

(b) To be sure that the top of the block is flat, check the block deck in six places as shown.

Fig. 12.26 *Checking the block deck surface for flatness. (American Honda Motor Co., Inc.)*

Deck warpage is measured with a straightedge and feeler gauge on the head gasket sealing surface of the block. It should be checked when the head gasket blows out and leaks. It should also be checked on late model aluminium blocks because they distort easily.

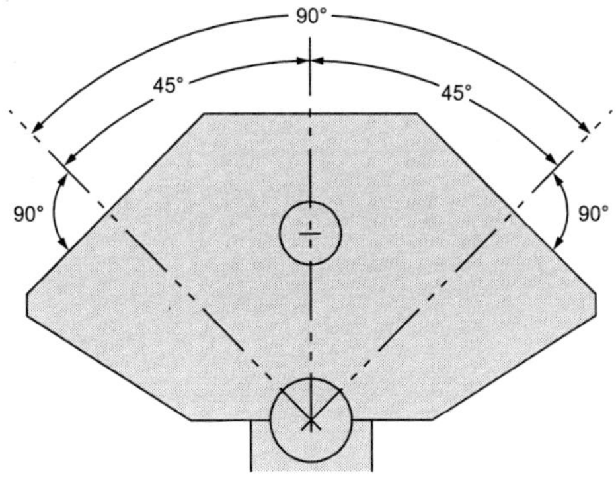

Fig. 12.27 *(a) Machining a V-8 block helps to ensure that the deck surfaces are exactly 90° from each other and that the block is square.*

To determine the block deck warpage, lay a straightedge on.the cleaned block deck surface, *Fig. 12.26 (a).* Try to slip different thickness feeler gauge blades between the block and straightedge. The thickest blade that fits any where on the full length of straightedge, with straightedge positioned in various locations on the block, shows the magnitude of warpage.

If beyond specs (about 0.08 to 0.13 mm), replace the block or send it to a certified machine shop for surface milling. The block deck surface must be machined in a surfacing machine that can control the amount of metal removal when it is necessary to match the size of the combustions. This procedure is called *decking the block,* also known as *"squaring the block."*

The block is set up on a bar located in the main bearing saddles, or set up on the oil pan rails of the block. The bar is parallel to the direction of cutting head movement. The block is levelled sideways, and then the deck is resurfaced in the same manner as the head is resurfaced. *Fig. 12.27* shows block deck being milled if warped beyond specs. *Fig. 12.28* shows a block deck being resurfaced by grinding. The surface finish should be 65 to 110 RMS for cast iron and 55 to 65 RMS for aluminium block decks to be ensure a proper head gasket surface. The surface finish is determined by the type of grinding stone used, and the speed and coolant used in the finishing operation, or the precision of the milling process.

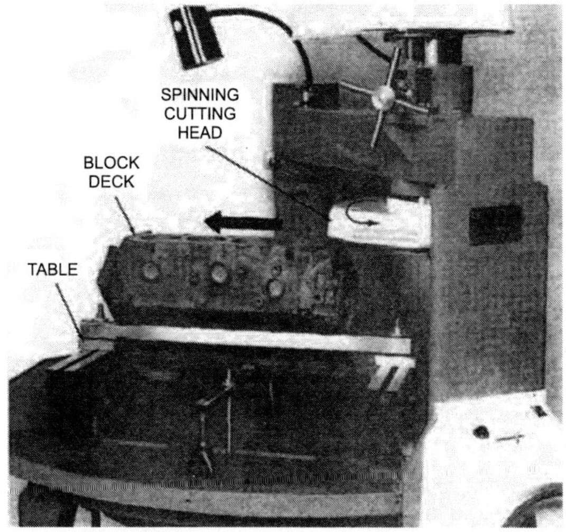

Fig. 12.27 *(b) Block deck should be milled if warped beyond specs. A machine shop milling machine is used to true the block deck. (Storm Vulcan)*

Fig. 12.28 *Block deck being resurfaced with a grinder.*

3. Cylinder Honing and Boring

Cylinder should be measured in a direction perpendicular to the crankshaft axis, where the greatest wear occurs in that direction. Maximum wear will be found just below the ridge, and just the least amount of wear will occur below the lowest ring travel, *Fig. 12.29.* The cylinders should be checked for *out-*

of-round and *taper, Fig. 12.30.* Always consult the applicable service manual specifications for the engine under service to determine the correct specs recommended. Most cylinders can be serviced provided:

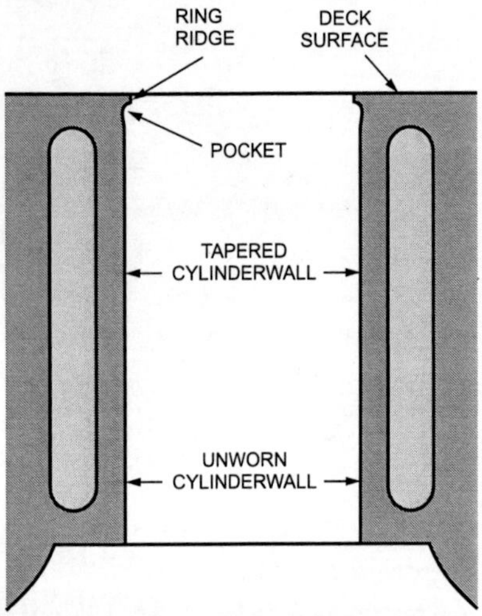

Fig. 12.29 *Cylinders are usually tapered, with most of the wear taking place at the top. The ridge is the original diameter of the cylinder. Which has not been worn by the piston rings.*

- Out-of-roundness does in exceed 0.076 mm
- Taper is less than 0.127 mm
- There are no deep scratches or nicks in the cylinder wall.

The maximum bore oversize is determined by two things: the cylinder wall thickness and the size of the available oversize pistons. If in doubt as to the amount of overbore that is possible without causing structural weakness, an ultrasonic test should be performed on the block to determine the thickness of the cylinder walls, *Fig. 12.31.* Variation of cylinder wall thickness occurs due to possible core shifting during the block casting process. The cylinders should be rebored to the smallest size possible.

Cylinders wear in tapered form and also wear out-of-round. When this wear exceeds specifications, the cylinders must be machined to restore the wall surface. Either ***boring bars*** *(Fig. 12.32)* or ***hones*** *(Fig. 12.33)* may be used to recondition the cylinder walls. Usually, the preferred method is to first use a boring bar and then finish with a hone. Another method is to first use a hone with coarse stones and finish with fine stones. Make sure the stones are clean and sharp.

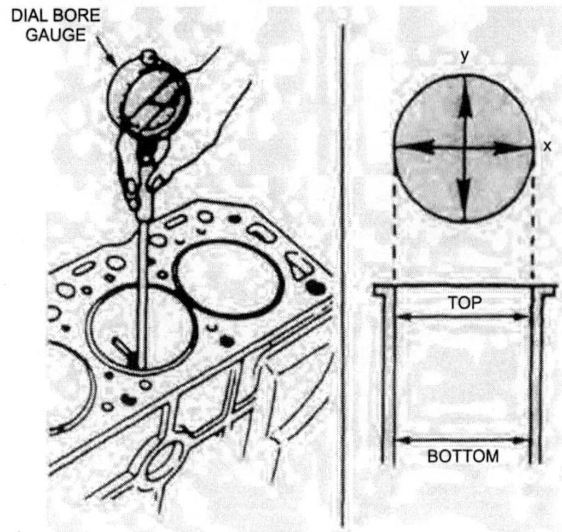

Fig. 12.30 *Method of checking the cylinder using a dial bore gauge. First measure the top of the cylinder at 90° from the crankshaft center line. This is the "X" diameter. Then, measure the diameter at the top of the cylinder in line with the crankshaft. This is the "Y" diameter. Subtracting the Y dimension from the X dimension will give the amount by which the cylinder is out-of-round. Measure the cylinder diameter at 90° from the crankshaft center line at the bottom of the cylinder: Subtract this measurement from the X diameter to calculate cylinder taper.*

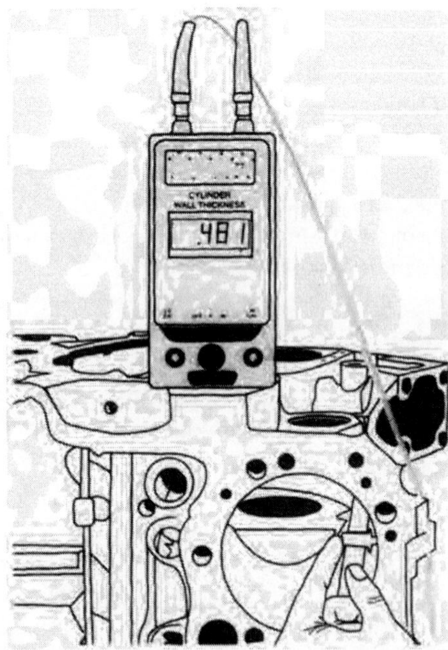

Fig. 12.31 *Ultrasonic testing can be used to determine the thickness of the cylinder walls. Here the thickness of the cast iron is .481 inch.*

Before any cylinder is refinished, all main bearing caps must be in place and tightened to the specified torque. Otherwise, the crankshaft bearing bores may become distorted during the refinishing operation. In precision boring a torque plate is also bolted on in place of the cylinder head while boring cylinders to keep the distortion to a minimum.

Fig. 12.32 *(a) Machines used to rebore cylinders.*

Fig. 12.32 *(b) Align-boring machine. (Sunnen).*

Every care and necessary precaution must be taken to ensure that the cylinders are refinished parallel to each other and at right angles to the crankshaft centreline. Exercise special care when setting up the reconditioning equipment, especially when the top of the cylinder block is not at right angle to the cylinder bore, *Fig. 12.34*. When the cylinder block is formed with an angular surface, special adapters should be used with the cylinder boring bar so that the cylinder will be reconditioned parallel to its original centre line.

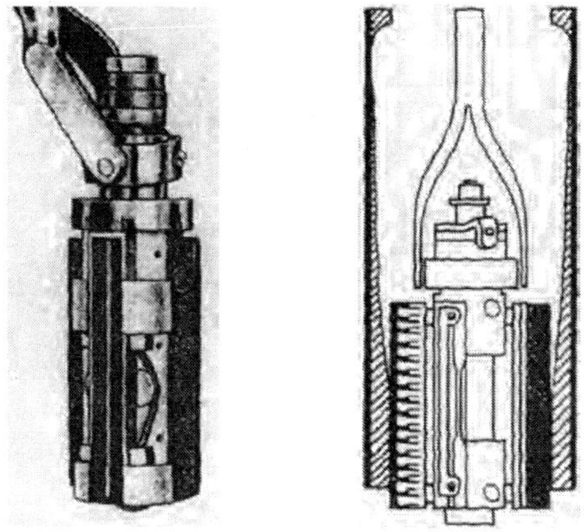

(a) Cylinder hone. (b) Hone operating in bottom if cylinder to remove taper.

Fig. 12.33 *Cylinder hone*

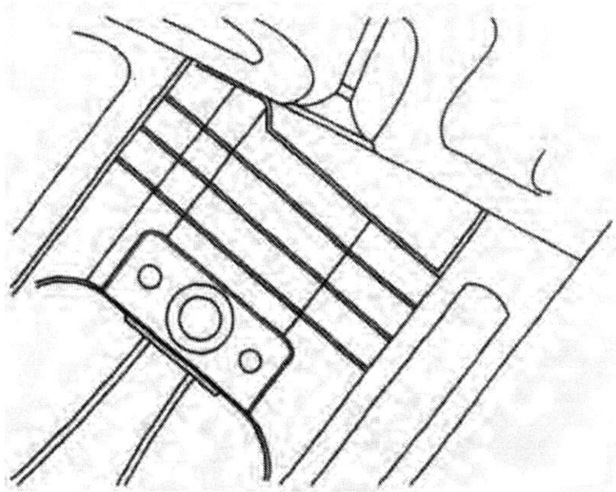

Fig. 12.34 *Note the angle of cylinder block at the cylinder head.*

An easy way to determine oversize piston size is to first find out the maximum taper in cylinder and then, (Taper × 2 + 0.254) = oversize piston. Normally, a cylinder block is bored in increments of 0.25 mm. Depending upon the amount of wear, the block cylinders could be bored 0.25, 0.50, 0.75. 1. 1.25, or 1.5 mm oversize. Most shops bore cylinders either 0.75 or 1.5 mm oversize because these piston oversizes are the most available.

Overbore limit is the largest possible diameter a block should be bored oversize. The engine manufacturer gives this specification. Some blocks can be bored 3 mm oversize while others can only be bored 0.51 mm oversize. Overbore limit depends upon how thick the cylinder walls are cast during manufacturing. A thin wall block could become too thin if its overbore limit were exceeded. The cylinder wall could crack from the pressure of combustion. Always check overbore limit of the specific block before recommending overbore diameter. The machine shop should also be able to determine overbore limit.

The cylinder block must be perpendicular to the crankshaft for normal bearing and piston life. If the block has been aligned with the crankshaft, it may be used to align the cylinders. Portable cylinder boring bars are clamped to the block deck. Heavy-duty production boring machines support the cylinder block on the main bearing bores.

The common procedure used for reboring cylinders is to set the boring bar up so that it is perpendicular to the crankshaft. It must be located over the centre of the cylinder. Cylinder centre is determined by installing centering pins in the bar. The bar is lowered so that the centering pins are positioned near the bottom of the cylinder, where the least wear has occurred. This located the boring bar over the original centre of the cylinder. After the boring bar is centered, the boring machine is clamped in place to hold it securely. This will allow to rebore the cylinder on the original centerline, regardless of the amount of cylinder wear, *Fig. 12.35.*

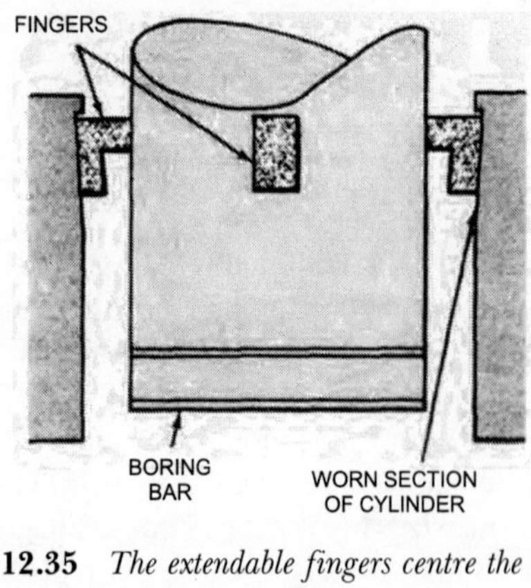

Fig. 12.35 *The extendable fingers centre the boring head in the cylinder bore.*

A sharp, properly ground cutting tool is installed and adjusted to the desired dimension. Rough cuts remove most of the metal on each pass of the cutting tool. The rough cut is followed by a fine cut that produces much smoother and accurate finish. Different shaped tool bits are used for rough and finish boring. The cutting tools are sharpened before boring each cylinder to accurately control the bore diameter and the surface finish. The last cut is made to produce a diameter that is at least 0.05 mm smaller than the required diameter. The cylinder wall is then finished by honing. Honing produces the required cylinder diameter and surface finish. Each cylinder is honed to provide the correct clearance for the piston that is to be installed in that cylinder.

Regardless of the method used in reconditioning the cylinders, the cylinder block must be thoroughly cleaned after boring and/or honing to remove all cuttings and abrasives. The job of cleaning a cylinder block after reconditioning is not easy. Even the mechanically correct jobs could be ruined when the abrasives and dirt remain in the cylinders. All oil holes in the block must be cleaned adequately. Core plugs should be removed so the water jacket can be flushed. The final cleaning is usually done with specialized cleaning equipment, *Fig. 12.36.* Machinists recommend scrubbing the block with soap and water. Oil the cylinders thoroughly after they are dry.

Fig. 12.36 *A cast-iron engine block being cleaned in a hot tank solution. Aluminium engine parts should be cleaned in a cold tank or in carburetor-type cleaner.*

When reconditioning cylinders, it is best to find out what sizes are available in ***oversize pistons*** before starting the operation. Otherwise cylinders may be bored to a size for which stock pistons are not available. *Oversize pistons* are necessary to fit a cylinder block that has had its cylinders bored out. The pistons have to be purchased to match cylinder oversize. If the block is bored 0.76 mm oversize then order 0.76 mm oversize pistons and rings. Boring and oversize pistons will make the engine perform like new.

12.14 CYLINDER HONING

Cylinder honing, also known as ***deglazing,*** is used to break the *glaze* or *polished surface* on a used cylinder. It is also used to smooth a very rough cylinder surface after boring. Most ring manufacturers recommend deglazing or honing before new ring installation. Honing the cylinder removes the fractured metal that is created by boring, *Fig. 12.37.* The cylinder must be honed a minimum of 0.05 mm after boring to cut below the rough surface and provide a precise finish.

The cylinder wall should be honed to straighten the cylinder when the wall is wavy or scuffed. If honing is done with the crankshaft remaining in the cylinder block, the crankshaft should be protected to keep honing chips from getting on the shaft.

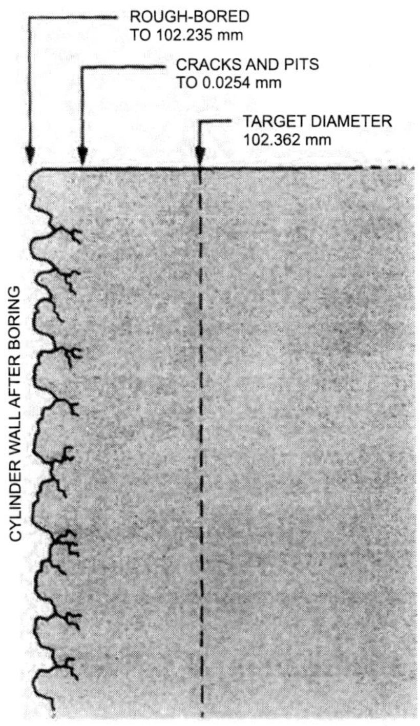

Fig. 12.37 *After boring, the cylinder surface is rough, pitted, and fractured to a depth of about 0.0254 mm.*

A ***cylinder hone*** produces a precisely textured, cross-hatched pattern on the cylinder to assist ring seating and sealing. Minute scratches from the hone cause initial ring and cylinder wall break-in wear. This makes the ring fit in the cylinder after only a few minutes of engine operation.

Two types of hones are used during cylinder service:

1. A **deglazing hone** removes the hard surface glaze remaining in the cylinder. It is a flexible hone that follows the shape of the cylinder wall, even when the wall is wavy. It cannot be used to straighten the cylinder.

A ***brush hone*** has small balls of abrasive material formed on the ends of round metal brush bristles, *Figs. 12.38* and *12.39 (a).* It is used when cylinder is in good condition and requires very little honing. It will also smooth a bored cylinder wall.

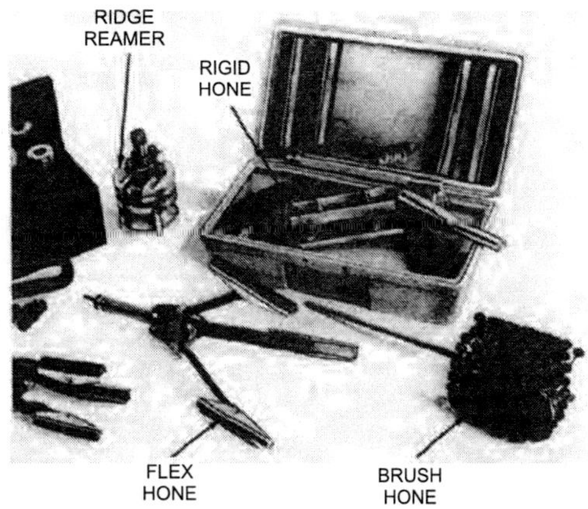

Fig. 12.38 *Hones are needed to deglaze shiny, worn cylinder and to smooth rough bored cylinder walls. Rigid hone will help remove taper and out-of-round. Flex hone, if used more at bottom of cylinder will help true the cylinder. Brush hone will help produce very accurate surface finish to help ring sealing. (Snap-On Tools)*

(a) Mechanic is using electric drill to spin brush hone to remove glaze in cylinder with little wear.

Fig. 12.39 *(Contd).*

(b) *Machine shop is power honing the cylinder to remove more severe wear.*

Fig. 12.39 *Honing a cylinder. (Goodson and Sunnen)*

A ***flex hone*** has hard, flat, abrasive stones attached to spring-loaded, movable arms, *Fig. 12.38.* It used when the cylinder wear is very small and a moderate amount of honing is necessary.

2. A **sizing hone** can be used to straighten the cylinder and to provide a suitable surface for the piston rings. Honing leaves a plateau surface that can support the oil film for the rings and piston skirt. This plateau surface is achieved by first using a coarse stone followed by a smooth stone to achieve the desired surface. The process of using a coarse stone followed by a fine stone is called **plateau honing.** *Fig. 12.40.* The honing stones are held in a rigid fixture with an expanding mechanism to control the size of the hone.

A ***ridged* or *sizing hone*** has stationary, but adjustable stones that cock into a preset position, *Figs. 12.38* and *12.39 (b).* It will remove a small amount of cylinder taper or out-of-roundness. A ridged hone can be used like a boring bar to true a cylinder when wear is within specs. The sizing hone straightens the cylinder taper by honing the lower cylinder diameter more than the upper diameter. As it rotates, the sizing hone only cuts the high spots so that cylinder out-of-round is also reduced.

The hone is stroked up and down the cylinder as it rotates. This produces a **cross-hatch finish** on the cylinder wall, *Fig. 12.41.* The angle of the crosshatch should be between 20° and 60°. Higher angle are produced when the hone is stroked faster in the cylinder.

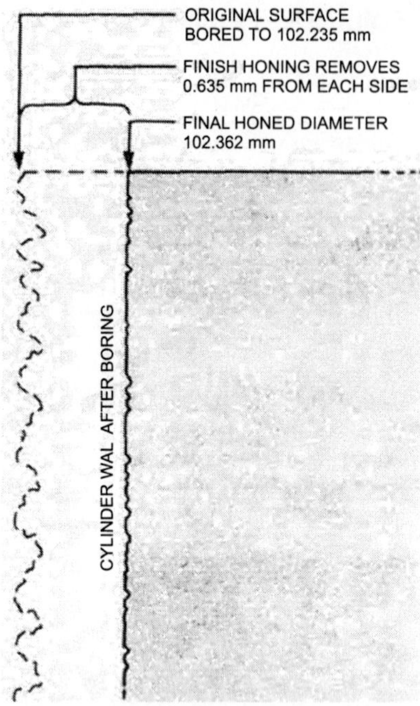

Fig. 12.40 *Honing enlarges the cylinder bore to the final size and leaves a plateau surface finish that retains oil.*

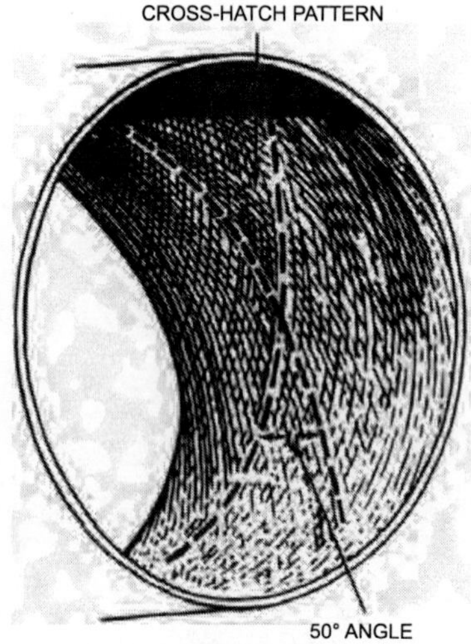

Fig. 12.41 *The cross-hatched pattern holds oil to keep the rings from wearing excessively, and also keeps the rings against the cylinder-wall for a gas-tight fit.*

Cylinders are normally bored to the exact size of the oversize pistons that are to be used. After the block is bored to a standard oversize measurement, the cylinder is honed. An experienced technician, the help of a rigid hone stones, can increase the bore size by 0.0254 to 0.0752 mm for the typical clearance required between the piston and the cylinder walls. The minimum amount of metal that is recommended to be removed by honing is 0.05 mm.

Power honing is a machine shop operation that rotates rigidly mounted stones through a worn cylinder. Like boring, it will enlarge and true the cylinder walls, *Figs. 12.39 (b)* and *12.42.*

Power honing can be used to true and clean up a worn cylinder and allow the use of standard size rings and pistons. The pistons must be knurled however to take up the excess clearance. Normally, standard rings

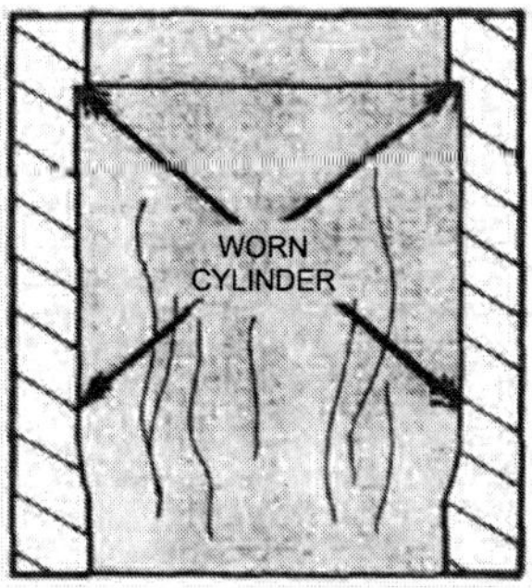

Fig. 12.42 *(a) Cylinder has minor scratches and small ring ridge.*

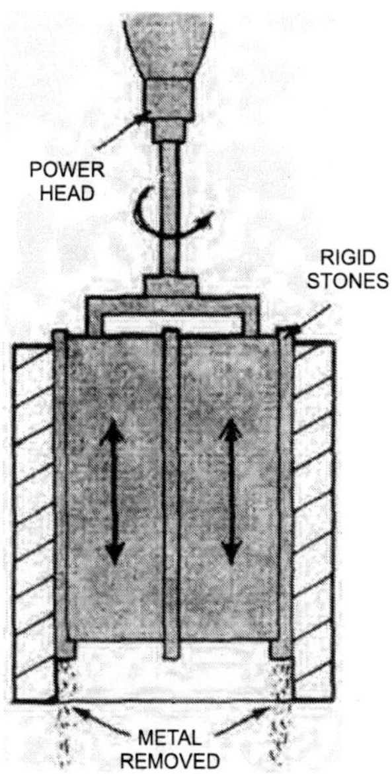

Fig. 12.42 *(b) Power hone is spun inside cylinder*

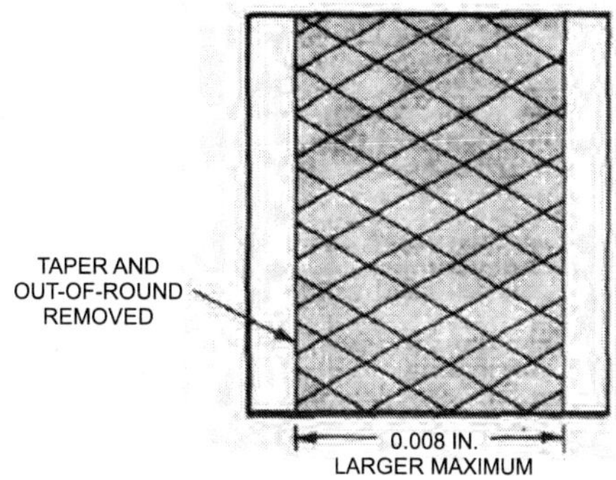

(c) Power hone can be used to true cylinder up to 0.2 mm oversize Standard rings and knurled pistons will normally preserve specs.

Fig. 12.42 *Rigid power hone will remove cylinder taper and out-of-round, making the cylinder equal size in all areas.*

and pistons can be used with a power hone cylinder arrangement of 0.2 mm. Most new standard piston ring sets will seal in a cylinder that is 0.25 mm oversize. This will save the mechanic from having to bore the cylinder 0.25 mm larger and install oversize pistons and rings.

Fig. 12.43 *Honing a cylinder with a torque plate installed. The torque plate is bolted to the block and torqued so that it simulates the forces that act on the block when the cylinder head is installed. Therefore, the cylinder is machined to true and straight when it is operating with the cylinder heads attached.*

Honing a cylinder with torque plates is recommended. Torque plates are thick metal plates that are bolted to the cylinder block to duplicate the forces on the block that occur when the cylinder head is installed. Even though not all machine shops use torque plates during the final dimensional honing operation is not advantageous. Without torque plates, cylinders can become out-of-round, up to 0.0762 mm, and distorted when the cylinder heads are installed and torqued to specifications. Even though the use of torque plates does not eliminate all distortion, their use helps to achieve a truer cylinder dimensions, *Fig. 12.43.*

12.15 HONE GRIT SELECTION

It is important that the cylinder wall should have the correct smoothness or texture before installing new rings. A hone produces crosshatched pattern scratches in the glazed cylinder wall of an old engine. It also smoothens a bored cylinder wall. The pattern depth of the honed surface holds oil to prevent ring scoring. The asperities or high points in the honed metal surface contact the piston rings to begin with and they wear quickly due to ring friction to allow the rings to match the shape of the cylinder wall.

A too coarse cylinder surface could result in ring and cylinder wall scoring. Also, if the cylinder wall surface is too smooth, the rings may not seal properly and cause oil consumption. Surface smoothness is measured in microns with special testers. Tester can be passed over a surface to measure root mean square (RMS) value of the surface smoothness. Cylinder smoothness should be matched to the piston ring type. Normally, softer ring coatings require a coarser stone and crosshatch on cylinder wall surface. A few recommended rules to remember are:

- Use 220 grit stones with ***cast iron rings*** to produce a 25 to 30 RMS surface texture.
- Use 280 grit stones with ***chrome rings*** to produce a medium 20 to 25 RMS surface textures.
- Use 320 grit stones with ***molybdenum rings*** to produce a very fine 10 to 15 RMS surface texture.

12.16 USING A HONE

Always follow the equipment manufacturer's instructions to hone a cylinder. Install the hone in a larger, low-speed electric drill. Compress the stones to squeeze inward and carefully slide them into the cylinder to avoid scratching the cylinder wall. Turn on the electric drill and move the spinning hone up and down in the cylinder length. Be careful not to pull the hone out of the cylinder while honing. The hone could break and bits of stone may fly out. The hone should be moved up and down the cylinder fast enough to produce a 50° to 60° *crosshatch pattern, Fig. 12.44.*

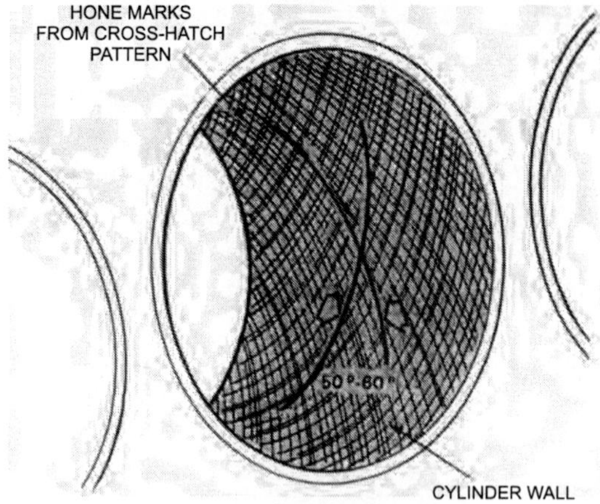

Fig. 12.44 *(a) When honing, try to produce a 50° to 60° cross-hatch pattern.*

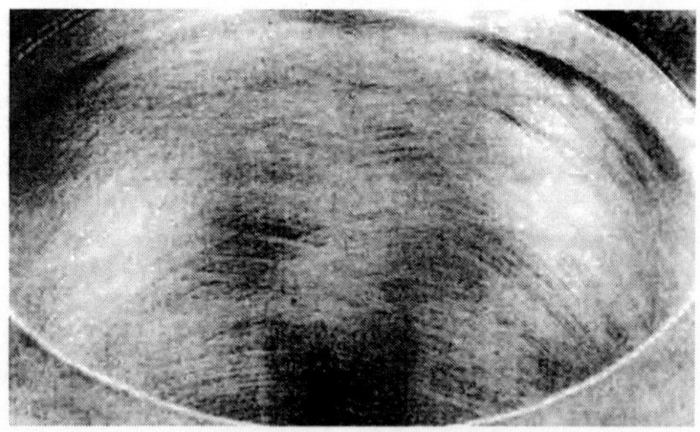

Fig. 12.44 *(b) Cylinder walls must look like this after honing. This pattern is called "cross-hatch".*

If the cylinder has not been bored or power honed true, hone more at the bottom of the cylinder. It is done in order to remove taper and enlarge the bottom slightly to make the cylinder equal in diameter at top and bottom.

12.17 DRY-TYPE CYLINDER SLEEVING

The ***dry-type cylinder sleeve*** does not come in contact with the coolant. ***Cylinder sleeving*** involves machining or boring one or more of the cylinders oversize and pressing in a cylinder liner. Sleeving is required, for example, after part breakage has severely damaged (i.e., gouged, nicked, chipped, or cracked) the cylinder wall. The damage is too deep to clean up with boring or power honing; it is still possible to recondition the cylinder by inserting a cylinder sleeve. Sleeving also permits the bad cylinder to be restored to its original diameter and allow the same size pistons to be used. For example, if only one piston and cylinder are damaged, all other pistons and cylinders may be good and usable. Sleeving would save the customer money on the repair. However, the transfer of heat is not as efficient as with a wet-type sleeve or a cylinder that is bored directly into the block.

Before installing a dry-type sleeve, the cylinder is bored and/or honed until the score marks are removed. Then a sleeve is selected with an outside diameter 0.0254 mm larger than the diameter of the rebored cylinder and an inside diameter slightly smaller than required for the available piston. The sleeve is shrunk in dry ice to reduce its diameter. Then, the cylinder is finish honed to remove any wrinkles that may have developed when the sleeve was pressed in position.

Sleeve installation is a machine shop operation. With an aluminium block with cast iron sleeves, sleeve replacement is simple. The machine shop can drive out the old sleeve and press in a new one. Boring is not needed. This can be done for a badly damaged cylinder, for example.

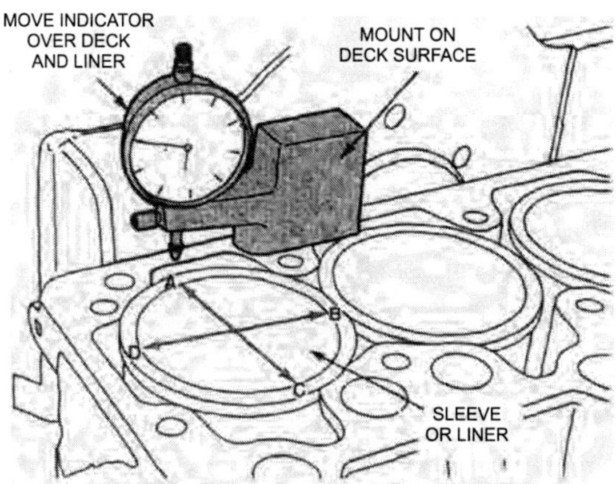

Fig. 12.45 *Proper cylinder liner protrusion is critical to head gasket sealing action. Note how dial indicator is used to measure height of liner above cylinder block deck surface. (Renault)*

Sleeving a cast iron block is more complex. The old cylinder has to be bored oversize to a diameter slightly smaller than the outside diameter of the sleeve. The sleeve is then pressed into the block. Then the sleeve is bored to the correct bore diameter.

Sleeve or *liner protrusion* is the distance the sleeve sticks up above the deck of the block. This measurement is critical to head gasket sealing, *Fig. 12.45.*

12.18 WET-TYPE CYLINDER SLEEVING

Most passenger car engines have the cylinders bored directly in the engine block. Some trucks and several European passenger car engines have wet sleeves, *Fig. 12.46.*

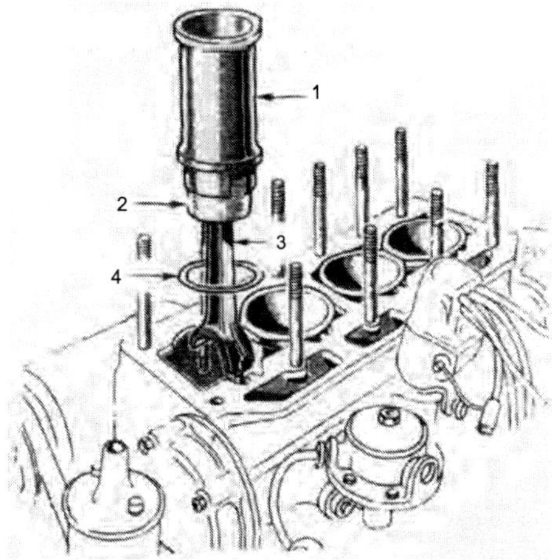

Fig. 12.46 *Wet-type sleeve. 1-Sleeve. 2-Piston. 3-Connecting rod. 4-Copper sealing ring.*

In a wet sleeve construction, each cylinder barrel is a separate sleeve inserted in the block. Each sleeve is sealed at the bottom of the water compartment by copper or rubber seal rings, *Fig. 12.47.* Cylinder head gasket seals the top of the sleeve. Engine coolant circulates directly around the sleeve. Uniform thickness of the sleeve or cylinder wall causes the minimum distortion.

Cleaning Cylinder Walls

After boring, sleeving, and honing, clean and remove all grit or bits of stone and metal from inside the engine. Any grit present acts as a grinding compound circulating through the engine. It can cause wear of bearings, rings, and other important engine parts.

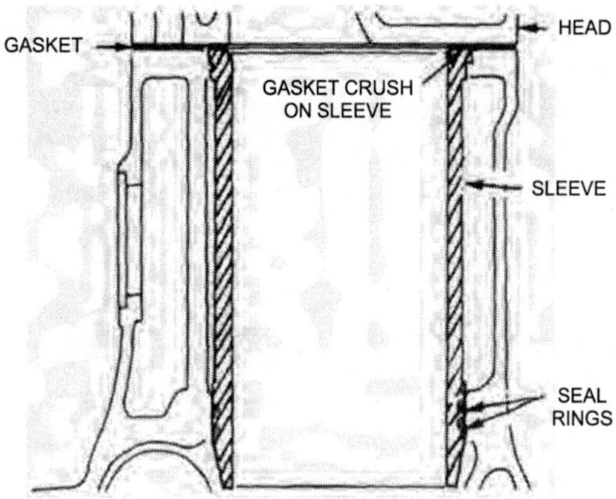

Fig. 12.47 *Rubber seals are used at the bottom of the sleeve in this Oliver engine.*

A pressure washer is the most efficient way of cleaning a cylinder block. The high pressure soap and water will wash out all grit out of the cylinder and crankcase. After pressure wash, blow the cylinder block dry and wipe the cylinders down with motor oil. In case a pressure wash is not available, hand wash the cylinders with warm water and soap. A soft bristle brush, *not* a wire brush, is used to remove particles inside the hone marks. Blow the block dry using compressed air. Then wipe the cylinders down thoroughly with an oil-soaked clean shop rag piece. The heavy oil will pick up any remaining stone grit embedded in the cylinder honing marks. Wipe the cylinders down until the rag comes out clean.

After cleaning, inspect the cylinders for scoring or scratches. If honing did not clean up all of the vertical scratches in the cylinder, cylinder boring or sleeving may be necessary.

Check all threads in the block for any damage and cleanliness. Clean all threads and repair any damaged threads found.

Before engine assembly, double-check to make sure the block is in an acceptable condition. Keep in mind that the block is the foundation of the engine rebuild.

12.19 BLOCK DETAILING

Before assembling the engine, a final detailed cleaning should be performed.

- Running a long bottle-type brush through all oil holes in the cylinder block, *Fig. 12.48,* should clean all oil passages or galleries.

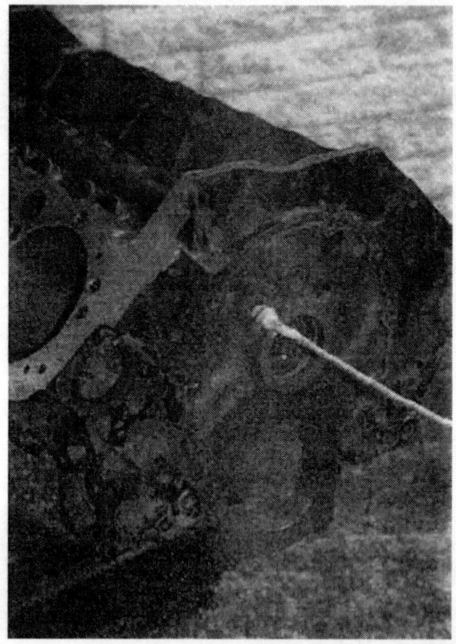

Fig. 12.48 *All oil galleries should be cleaned using soap (detergent), water, and a long oil gallery cleaning brush.*

- All tapped holes should be chamfered and cleaned with the correct size of thread chaser to remove any dirt and burrs, *Fig. 12.49.* Final block work involves chasing threads. Chasing means running a tap through a bolt-hole and blowing out the residue with compressed air, *Fig. 12.50.*

Fig. 12.49 *All bolt holes should be chamfered at the top to prevent the attaching bolts from pulling threads at the top surface.*

Make sure the tap is the correct one for the threads being chased. The tap may bind as it cleans. Whenever the tap begins to bind, back it out one turn and then resume cleaning. Backing the tap permits dirt to collect in the flutes and get out of the threads. To get at the threads in the very bottom of a hole, a bottoming tap is needed.

Fig. 12.50 *Chasing threads (Greenfield Tap and Die)*

After chasing the threads, move the tap and blow out the hole with compressed air. Use a rag draped over the hole to arrest flying debris. Wear eye protection, and warn any bystanders.

- Coat the newly cleaned block with fogged oil to prevent rusting. Cover the block with a large plastic bag to keep out dirt until it is time to assemble the engine.

12.20 CAM BEARING INSTALLATION

Always install new cam bearings during a major engine service. Worn cam bearings are common and can lower engine oil pressure. Cam bearings may be installed in the machine shop or by the technicians themselves in the auto repair shops with the help of special drivers for pressing cam bearings in and out of the cylinder block. The method of removing and installing the camshaft or balancer shaft bearings is summarized *in Fig. 12.51.*

- Use a driver to force the used bearings out of their bores.
- Force the new bearings into place using a similar bearing driver.
- Ensure that all oil holes in the cylinder block line up properly with the oil holes in the cam bearings. If they do not align, severe valve train wear can result from lack of lubrication.

Install valve lifter bore bushings: Lifter bores in a cylinder block may be out-of-square with the camshaft, resulting in premature camshaft wear and variations in the valve timing between various cylinders. To correct this variation, the valve lifter bores are bored and reamed oversize using a fixture fastened to the block deck to ensure correct alignment. Bronze lifter bushes are then installed and finished with honing to achieve the correct lifter-to-bore clearance.

(a) Driver is being used to pull old cam bearings out of block.

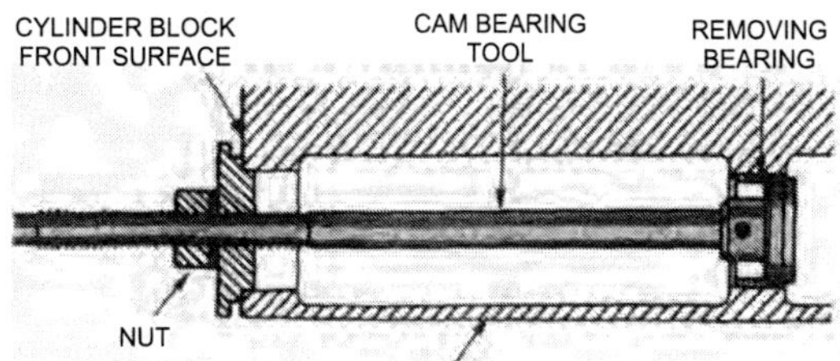

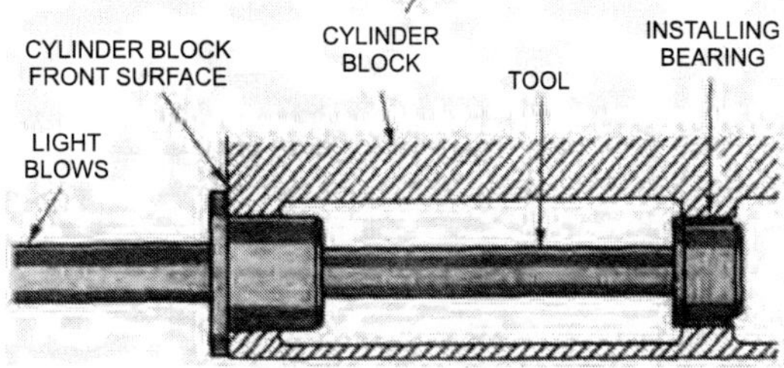

(b) Driver being used to push a new cam bearings into place.

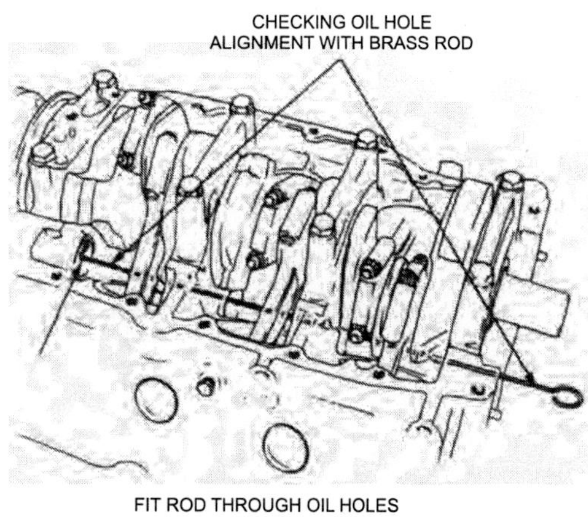

(c) Make sure cam bearing oil holes are aligned after installation or severe valve train damage can result.

Fig. 12.51

12.21 CRANKSHAFT SERVICE

Like other engine parts, **clean** the crankshaft before inspecting it. A solvent blast may be enough if the engine was clean. Burned-on oil deposits are better dissolved by placing it in a hot tank. In either case, clean the crank's oil passages. Gun cleaning brushes or pipe cleaners handle the job nicely.

Quite often, the crankshaft will be found in good condition and require nothing other than checking the wear on the crank. However, for a high kilometers engine, oil pump wear and reduced lubrication can cause excessive journal wear requiring crank grinding and then installing the new matching inserts. By careful study of the forces applied to a journal of the *crankshaft,* it can be found that the load is much heavier at some points of rotation than others. The force of the power stroke is several times greater than the force of the compression strokes.

Also, the power stroke always applies the force at the same spot on the journal. In addition, an extra load is imposed by the action of the centrifugal force resulting from the rotation of the crankshaft with its connecting rods and pistons. All this causes out-of-round crankshaft journals and crankpins.

If a *connecting rod* is bent or out of alignment, it will tend to wear the crankpin journal in a tapered fashion, i.e., it will wear more at one end of the bearing surface than the other end. Any twisting of the crankcase or any excessive vibration of the crankshaft will cause the main crankshaft journals to wear in a tapered form.

A **bent crankshaft** can ruin new main bearings or cause the engine to lock up when the main caps are torqued as specified.

Any **abrasive material** that might get into the engine lubricating oil, may cause unequal wear. It may cause more wear at one bearing or on one spot of the bearing, depending on where the abrasive enters in larger quantity.

Bearings seldom wear equally. One bearing may operate with a smaller quantity of oil than other. Also, one bearing, due to its location in the engine, may operate at a higher temperature than the others. The overall effect of all these variations contribute to unequal wear on the crankshaft journals.

If a connecting rod journal has developed **taper or a flat spot,** it would cause an increase in oil consumption and it would require either replacement of the crankshaft or reconditioning the crank journal.

Because of close clearances in the bearings, a sprung crankshaft cannot be tolerated. The main bearings must fit the crankshaft journals all around the circumference with only enough clearance for a film of lubricating oil. A scored or out-of-round bearing journal must be reconditioned or replaced.

Damaged Crankshafts

Engine crankshafts usually are large and expensive parts, so it may be advantageous to repair any damage rather than replacing a shaft. An engine crankshaft is subjected to vigorous vibration and stress. It may develop tiny cracks, particularly at or near the ends of the connecting rod throws or at the ends of the main bearing journal. Sometimes a crack may develop near the oil feed holes in the shaft. A **cracked crankshaft** can result from part breakage, harmonic balancer failure, flywheel failure, or driver abuse.

Normal crankshaft wear comes in three varieties: out-of-round, taper, and **undersize.** Out-of-round is the egg shape a journal can take when viewed on end, *Fig. 12.52.* Taper is the slope worn into the side of a journal when viewed from the side. Undersize is an even wearing down of the journal's circumference. All three types of wear are determined with an outside micrometer or a dial gauge.

A vs B = vertical taper
C vs D = horizontal taper
A vs C and B vs D = out of round

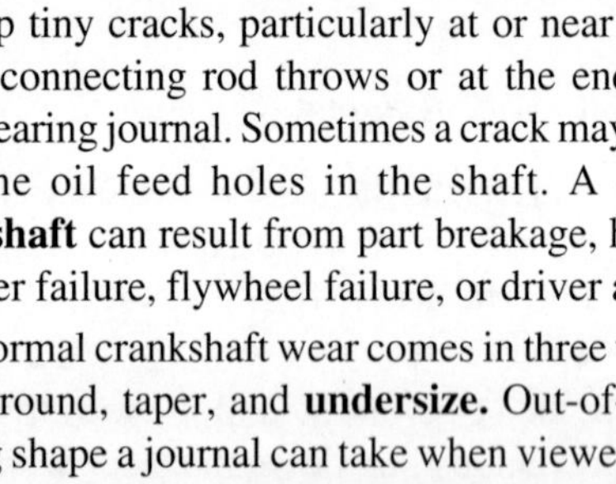

CHECK FOR OUT-OF-ROUND AT EACH END OF JOURNAL

Fig. 12.52 *Crankshaft taper and out-of-round measurements. (Ford Motor Company)*

Out-of-round is found by measuring around the journal in the same plane. Subtract the smallest diameter from the largest to find out out-of-round. As shown in *Fig. 12.53,* measure across the journal from side to side then from top to bottom. If not within spec limits, send the crankshaft to a machine shop for grinding. Check service manual specs for the engine under repair.

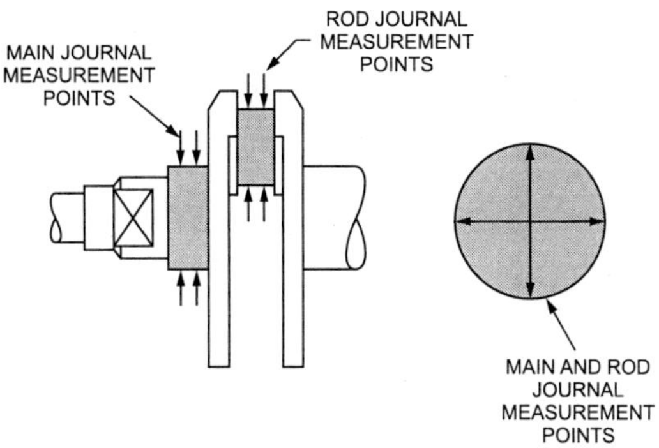

Fig. 12.53 *In addition to checking each journal for scoring, taper and out-of-round must also be checked. Measure crank and main journals at these locations to check journal taper and out-of-round. (Buick-GM).*

Taper can be found by measuring the journal at one end on cylindrical portion of the journal. Then move over a known distance (say 25 mm) and again measure the diameter, *Fig. 12.54.* Subtract the smaller measurement from the larger one. The difference, X, is the taper. It is expressed as X mm/ 25 mm (here).

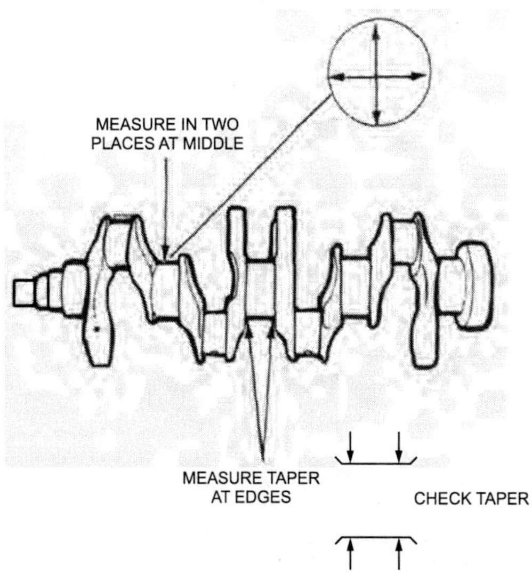

Fig. 12.54 *Measuring Taper and out-of-round. (American Honda Motor Co., Inc.)*

Undersize may be found when checking for out-of-round and taper. In case the journal is smaller than the specified value and yet there is no out-of-round and taper, the journal is undersize.

Another form of normal wear, called **scoring,** will be the general light scratching of the journals. It is caused by dirty oil etching the crank. Abnormal wear is caused by a large piece of grit that goes through the oil system and deeply scores the crank.

***Grinding* a crankshaft** is the remedy for all these problems: out-of-round, taper, undersize, and scoring. A special grinding wheel is used to remove some material off the journal. All journals are ground until they are evenly undersize in increments of 0.254, 0.508, or 1.016 mm. In general, larger the crank diameter, the more the metal that can be safely removed. However, only the necessary material is removed, as removing metal weakens the crank. The rod journals must all be ground evenly even if there is only one bad journal. All main journals are ground if one or more is bad. Rod journals are ground independently of the main journals and vice versa, i.e., the rods could be ground 0.508 mm and the mains 0.254 mm or any other combination as needed.

If a crankshaft has a **special hardened surface.** After grinding , this surface must be replaced by chemical treatment.

After grinding the journals are **polished.** Polishing like honing the cylinders — it provides the proper oil-carrying finish. A 320-grit cloth is wrapped around a journal and worked in a cylindrical fashion. In a machine shop a special tool is used to polish the crank while it rotates in a lathe. This tool, too, uses a belt of 320-grit cloth. When a crankshaft does not need grinding, it should be polished to restore the proper finish.

Fig. 12.55 *Special machinery needed to recondition crankshaft journals.*

If damage has been caused to one or more crankpin journals, the crankshaft may be reconditioned with the help of a special machine and equipment made for this purpose, *Fig. 12.55.* If the crankshaft is sound, journals worn slightly tapered or out-of-round can be reground and **undersized bearings** can be fitted, *Fig. 12.56.* Here again, it is required to check out the available sizes of bearing inserts, and then have the shaft journals ground to an undersize that corresponds to these inserts.

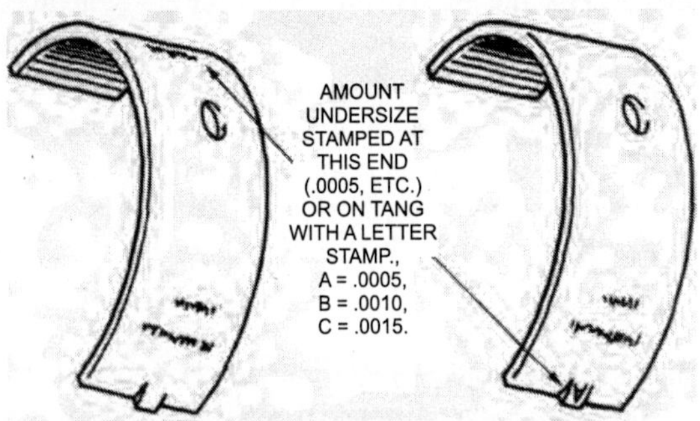

(a) Usually, number is given for .010, .020, etc., inch undersize. Letter code requiring service manual can also be given.

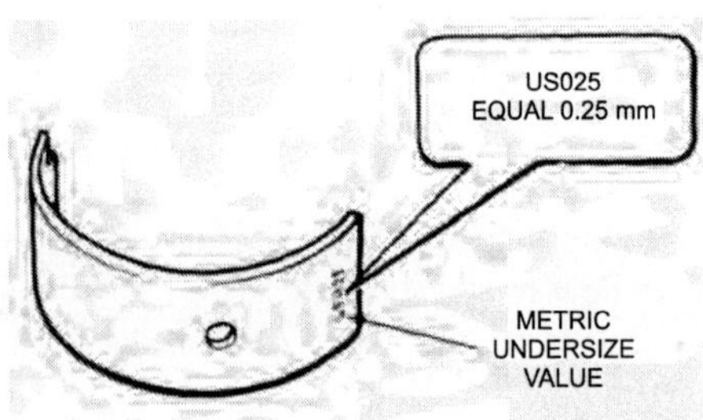

(b) Do not confuse metric undersize markings with conventional inch markings. This 025 marking means the bearing is for a 0.25 mm undersize crank journal. This would equal about .010 inch undersize. (General Motors)

Fig. 12.56 *When a crankshaft journal is undersized, the bearings are oversized. However, the bearings, like the journals, are also referred to as undersized. Bearings will be marked undersize on backs. (Oldsmobile – CM)*

Look at each connecting rod journal and main journal surfaces closely. Look for scratching, scoring, and any signs of wear. The slightest nick or groove is *very serious* to ideal lubrication system performance.

A crankshaft journal can be checked for scoring by running a fingernail over the journal. The finger nail should glide over the entire journal surface. In case the fingernail detects the slightest ridge or scoring, the journal must be reconditioned.

Very fine crocus cloth may be used to clean up minor burrs or marks on journals. Always polish around the journal *not* across it.

Crankshaft service involves measuring crank journal wear, checking crank straightness (if needed), checking for cracks (if severely damaged from connecting rod breakage), crank cleaning, and proper installation. Rod journals usually show the most wear on the underside of the throw, but main journals have no specific area of wear.

Crankshafts are reconditioned in a precision grinder. Each of the connecting rod and main bearings is ground to a standard undersized. The bearings used with the reconditioned crankshaft must be matched to the undersize to avoid excessive oil clearance. Since the crank journals are smaller in diameter, the bearings must be thicker to provide the correct bearing-to-journal fit.

Abnormal crankshaft wear occurs when the bearings are oil starved, *Fig. 12.57.* Then large amounts of metal are scored away from journals. The journals are rendered beyond grinding limits. They must then be built up with welding and then ground to a standard diameter. **Rebuilding** is adding material to the surface of the crankshaft rod journals and main journals. There are several rebuilding methods. Chromium plating adds pure hard chromium onto rod and main journal surfaces. This provides an ideal surface for a bearing. Electro-welding is a process in which a bead of steel is deposited by an electric arc onto the crankpin or journal surface in a continuous weld. Metal spraying adds molten steel onto a journal surface to build it up.

Measuring Journal Taper

If one side of a crankshaft journal is worn more than the other, the crank journal is tapered. To measure journal taper, use an outside micrometer, *Fig. 12.58.* Measure both ends of each journal, *Fig. 12.53.* Any difference in readings reflects taper. Taper beyond recommended limits needs crankshaft journal turning.

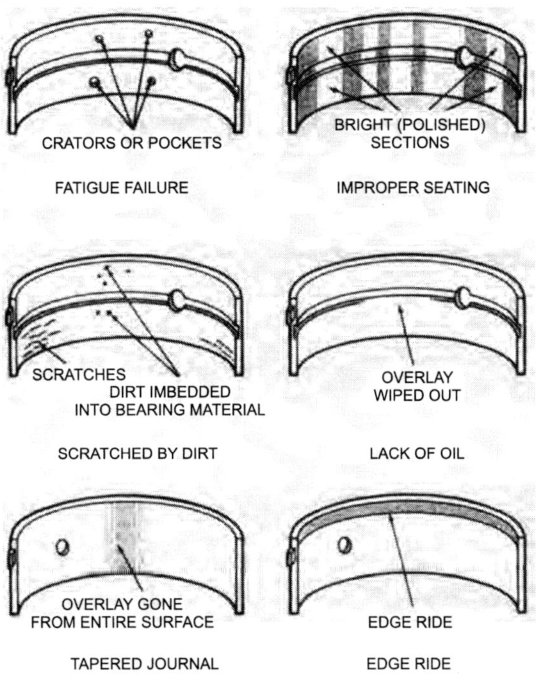

Fig. 12.57 *Bearing Wear. (Ford Motor Company)*

Fig. 12.58 *When checking crankshaft wear, be careful to adjust the micrometer to just drag over journals. (Lufkin)*

Checking for Cracked Crankshaft

Cracks can be hard to find on a crankshaft because they can be so small. *Megnafluxing* will help locate cracks. Sprinkle the magnetic powder over the crank. The magnetic field will make the powder collect in any crack. Before any extensive work is initiated, however, it is wise to get the shaft checked by a specialist with proper special magnetic or chemical equipment to make sure there are no cracks in it. Replace the crankshaft if cracked.

Checking Crankshaft Straightness

To measure crankshaft Straightness, mount the crankshaft on V-blocks or place it in the cylinder block main bearings. Then mount a dial indicator against the centre main bearing journal, *Fig. 12.59.*

Slowly turn the crankshaft while observing the indicator. Indicator movement gives crankshaft bend. If runout is not within spec limits, about 0.03 mm, replace the crank or have it straightened and turned by a certified machine shop.

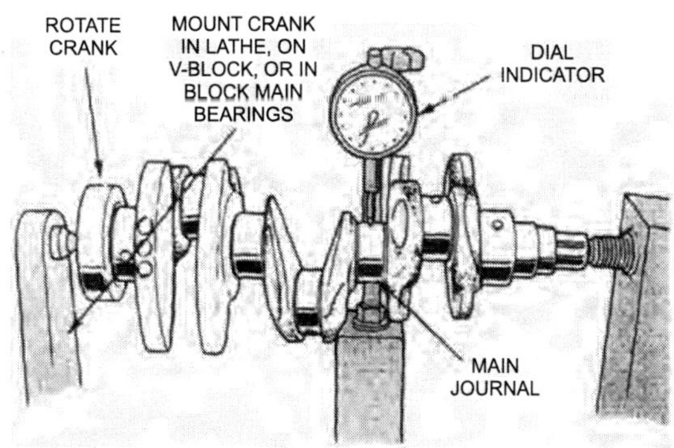

Fig. 12.59 *If any parts were broken or main bearings are badly worn, check crankshaft straightness/runout. Mount crankshaft in a lathe, V-blocks, or in block main bearings. Mount a dial indicator on main journal. Turn the crankshaft and carefully read indicator to determine any runout. (Pontiac - GM)*

Undersize Bearing Markings

While servicing a crankshaft, always look at the old crankshaft bearings to determine whether they are undersize. Undersize bearings may be used on rebuilt engines or even on a few new engines.

Check the back of the old bearings for an *Undersize Number,* as in *Fig. 12.56.* The number stamped on the bearing back denotes bearing under size. A letter code also be used to give bearing and undersize. In case, during engine repair, the crankshaft is found to be in good condition, install the same size bearings as the old ones. However, if the crank is badly worn or damaged, replace the crankshaft or recondition it by grinding undersize. Then, install the correct size main and rod bearings.

Cleaning Crankshaft

The crankshaft should be perfectly clean before installing it into the cylinder block. This is particularly important if the crank has been ground and debris in the form of metal shavings, stone grit, etc., could be inside the oil passages. Bits from bearing failure can also collect inside the crank. If not removed, debris can flow out and onto the bearings during engine operation, resulting in bearing failure.

Remove oil gallery and core plugs, *Fig. 12.60*, and wash the crankshaft in parts cleaner. Use a blow gun to blow out all passages and dry off the journals. Reinstall the core or gallery plugs that were removed. Wipe the journals down with an oil-soaked shop rag.

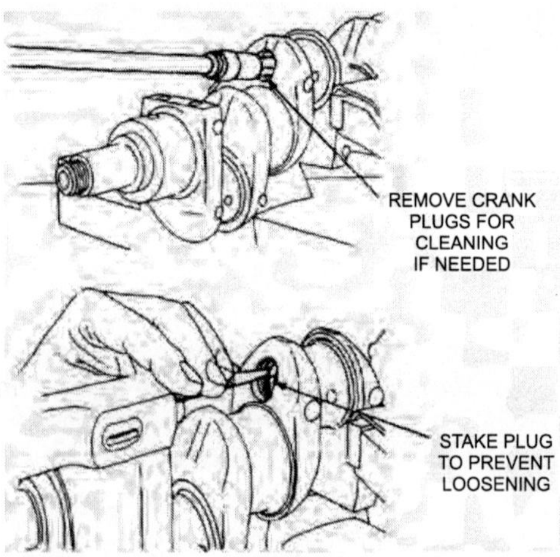

Fig. 12.60 *After crank grinding, make sure to remove any plugs. Debris can collect inside hollow crankshafts. Inside of cranks must be properly cleaned to avoid bearing damage during engine operation.*

Stress Relieving the Crankshaft

The greatest area of stress on a crankshaft is the fillet area, *Fig. 12.61*. Stress relief is achieved by blasting the fillet area of the journals with #320 steel shot. This strengthens the fillet area and helps prevent the development of cracks in this area. Gray duck tape is commonly used to cover the journal to prevent damage to the rest of its area. Stress relief procedures are normally performed after the grinding and polishing of the crankshaft.

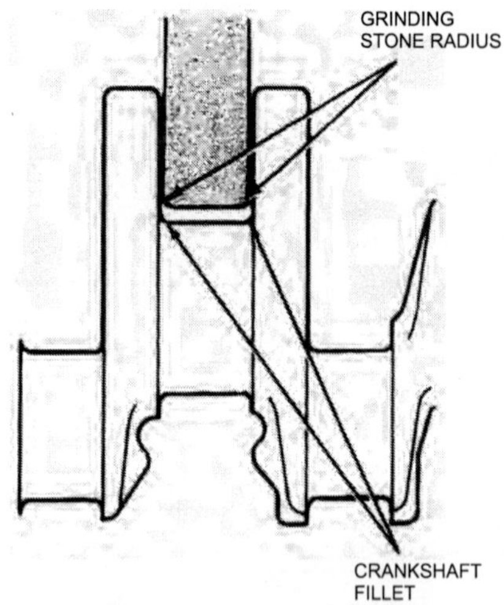

Fig. 12.61 *The rounded fillet area of the crankshaft is formed by the corners of the grinding stone.*

12.22 INSTALLING REAR MAIN OIL SEAL

Three basic types of rear main oil seals in use are: two-piece synthetic rubber seal, two-piece rope or wick seal, and one-piece synthetic rubber seal. Each type is installed using a different technique.

Two-piece Rubber Seal

This type is easy to install. Simply press it into place in the block and rear main bearing cap. The sealing lip on the rear main seal must point towards the inside of the engine, *Fig. 12.62*. If installed in opposite direction, oil will pour out of the back of the seal when the engine is started. Lubricate the sealing lip with motor oil.

Two-Piece Rope Seal

A two-piece rope or wick rear main seal has to be carefully worked down into its corresponding cap and block groove. To do so, use a special seal installing tool and light hammer blows or hand pressure and a smooth steel bar, *Fig. 12.63*.

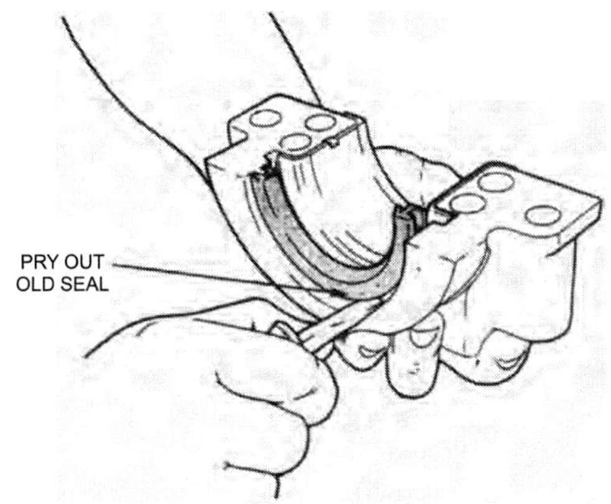

(a) Use a small screwdriver to pop out old two-piece rubber seal from cap and block.

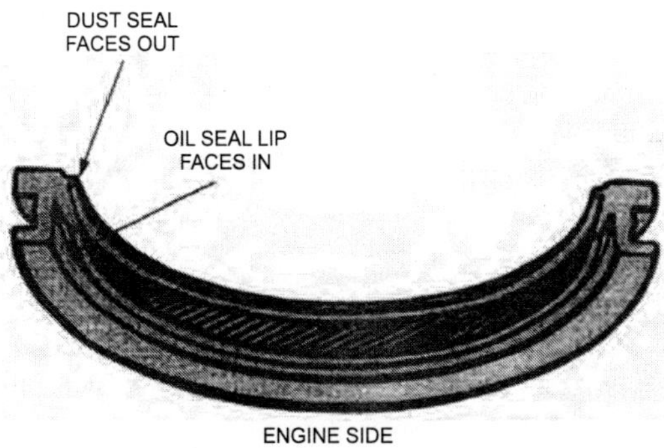

(b) Look at new seal closely to make sure oil sealing lip faces inside of block (Oldsmobile - GM.)

Fig. 12.62 *Installing the two-piece rubber rear main seal. (Oldsmobile - General Motors)*

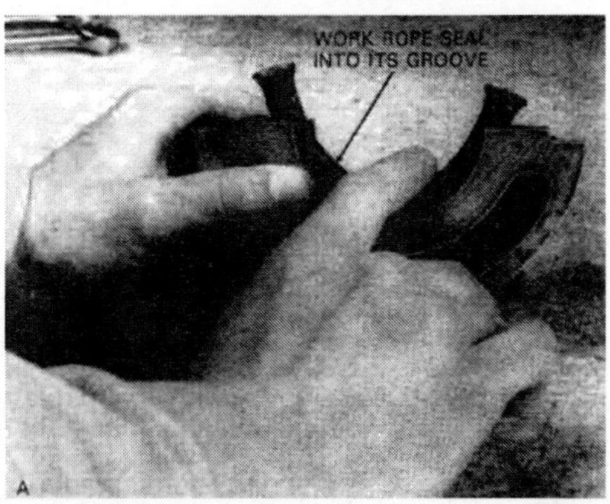

Fig. 12.63 *(a) Start rope into its groove in cap and block.*

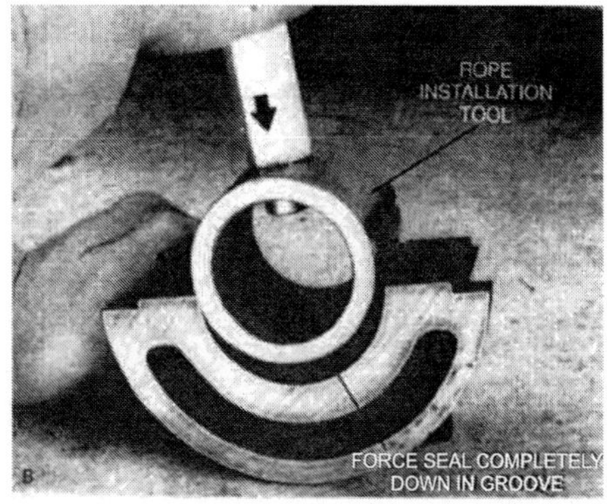

(b) Use special driver or round bar to force seal completely into groove.

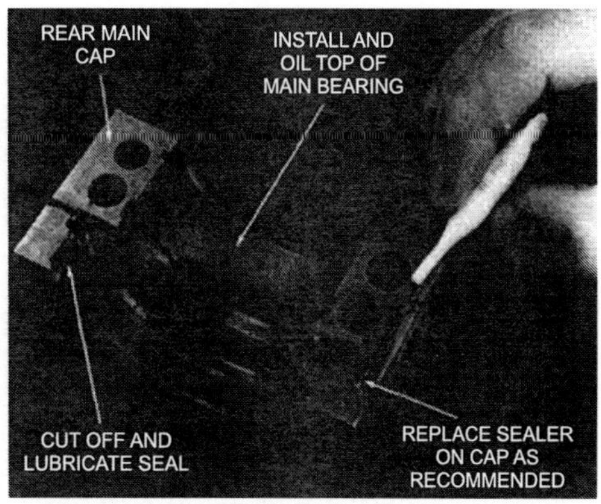

(c) Cut ends of rope flush with single-edge razor blade and apply recommended sealer to cap

Fig. 12.63 *Installing rope type oil seal. (Fel-Pro Gaskets)*

Silicone sealer or anaerobic sealer are commonly recommended on the rear main cap to prevent oil leakage. Consult the applicable service manual to determine which type should be used. The sealer keeps oil from seeping between the main cap and block mating surfaces, *Fig. 12.64.*

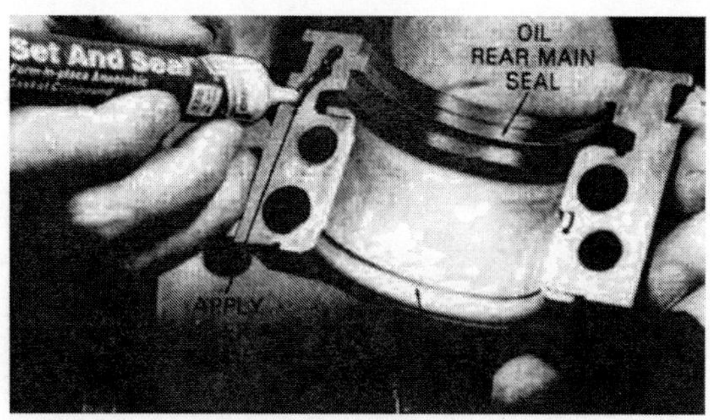

Fig. 12.64 *Remember to seal main cap when required. Anaerobic sealer is needed where there is tight metal-on-metal contact. Silicone sealer is for a more loose fit, as on sides of cap. (Fel-Pro Gaskets)*

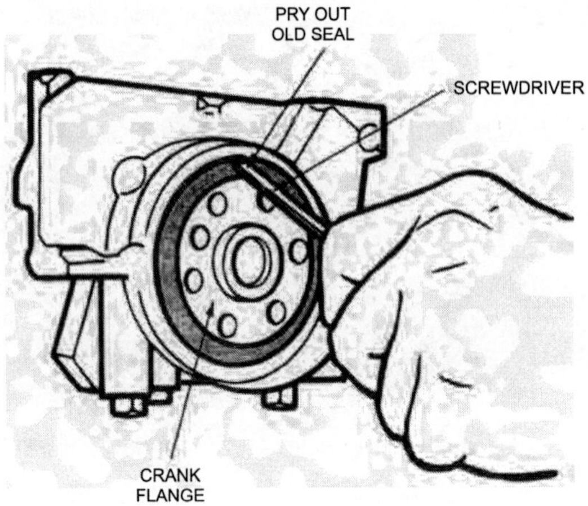

(a) Pry out old seal without scratching the block.

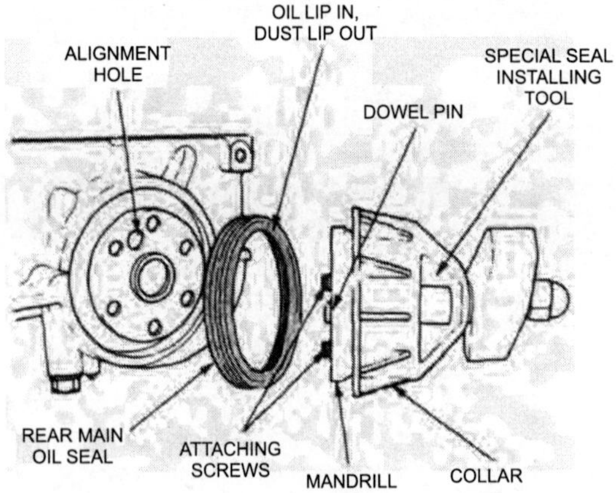

(b) This special seal installing tool is tightened using a wrench.

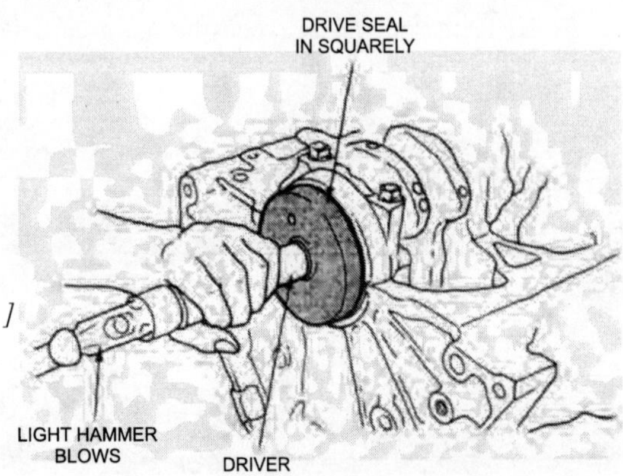

(c) Use light hammer blows to push seal to seat squarely into its desired position.

Fig. 12.65 *Installing a late model one-piece rear main seal. (Buick and Pontiac - General Motors)*

One-piece Synthetic Rubber Seal

A one-piece synthetic rubber seal is normally installed after the rear main cap has been bolted to the block. A seal driver is used to drive this seal, from the rear of the engine, into its position. Make sure the seal is square and undamaged after installation. *Fig. 12.65* shows the steps for this type seal installation.

12.23 MAIN BEARING AND CRANK INSTALLATION

After a block is properly reconditioned and cleaned, the main bearings and crankshaft can be installed. Make sure that the correct size main bearings are selected to go with the crankshaft.

Installing Main Bearings

With a clean, dry shop rag, wipe the main bores free of any dirt or oil. Check the main bearing sizes on their packing boxes and backs of the bearings to verify their correct size.

Snap each main bearing into place in the block, exercising care not to touch the front of the bearing inserts. The bearing tab, on each bearing, fits into a notch in the main bore. Check that the oil holes in the bearings line up with corresponding oil holes in the cylinder block, *Fig. 12.66 (a)*. Snap the other bearing halves into the corresponding main caps. Ensure that the main thrust bearing is positioned properly.

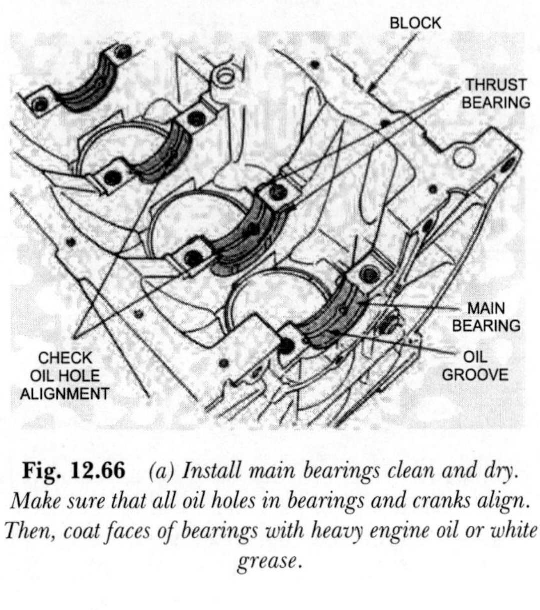

Fig. 12.66 *(a) Install main bearings clean and dry. Make sure that all oil holes in bearings and cranks align. Then, coat faces of bearings with heavy engine oil or white grease.*

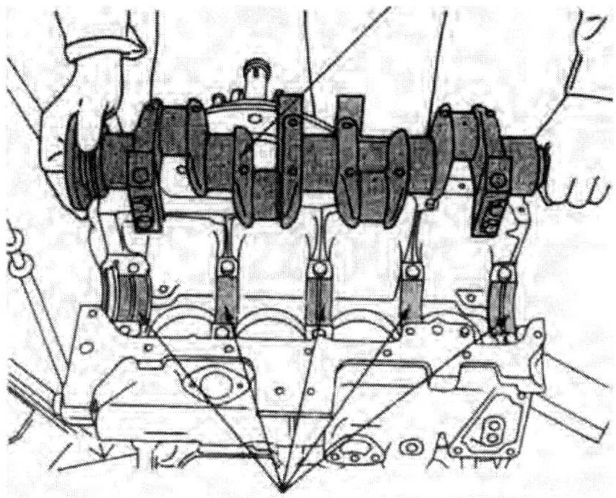

(b) Hold the crankshaft at snout and flange, carefully lower it into main bearings. Take care not to hit journals or bearings. (GM and Peugeot)

Fig. 12.66 *Installing crankshaft. (GM and Peugeot)*

Never oil the backs of bearings and bearing bores. This could decrease bearing clearance and render the bearings liable to spin in their bores upon starting the engine.

Installing Crankshaft

After the main bearings and rear main seal are installed, coat the tops of the bearings and rear seal lip with heavy engine oil or white grease. Then, carefully place the crankshaft into the cylinder block, as shown in *Fig. 12.66 (b).* Do not turn the crankshaft without the main caps in place or the bearing might shift out of place.

Bearing Clearance

Oil is pumped under pressure to the various bearings in the engine. However, to get the oil into a bearing and adequately lubricate it, a correct clearance for an *oil film* must be available.

It is very important that the steel crankshaft journal must be separated from the bearing metal when the engine is running, otherwise the bearing will melt. The heat generated by friction when the steel crank moves at high speed on soft, dry metal will melt the soft bearing metal. To avoid this from happening automobile engines utilize a film of oil between the journal and bearing.

The oil film holds the two metals apart and continuously circulates to carry away the heat generated by friction. The *oil clearance* varies with the engine design and the type of lubrication system used. In pressure lubrication system the flow of oil must be controlled by maintaining limited clearance all the way around a round bearing and a round shaft. When there are unequal clearances in the circulation system, too much oil will reach some places and too little oil will travel to other places. This is due to the fact that oil under pressure will go through the largest clearance space in largest quantity.

Bearing Clearance Measurement

The diameter of the shaft is measured at several points around the circumference to determine the size and to check for its roundness. Measuring each end of the journal will determine the amount of taper, if present.

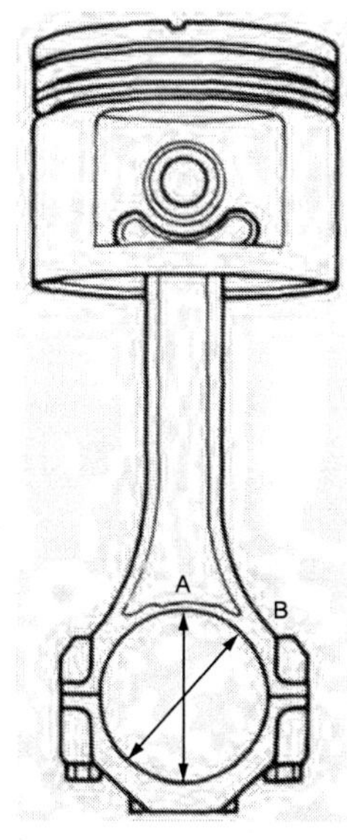

Fig. 12.67 *Distance at line A will be the greatest due to wear. The difference between line A and line B is the amount of out-of-round. Use a dial indicator or a telescoping gauge and micrometer for these measurements. If a rod is out-of-round more than 0.0254 mm, a new one must be used. (Oldsmobile — GM)*

With the caps on and torqued to specs, the inside of the bearing bore, without the bearing, is measured using a telescoping gauge or an inside micrometer, *Fig. 12.67.* The difference between these two measurements gives the amounts that the bore is out of round.

To check the oil clearance between the crank journal and main bearing, place a small bead of **Plastigage** (clearance measuring device) on the unoiled crankshaft. Install and torque the main cap, *Fig. 12.68 (a).*

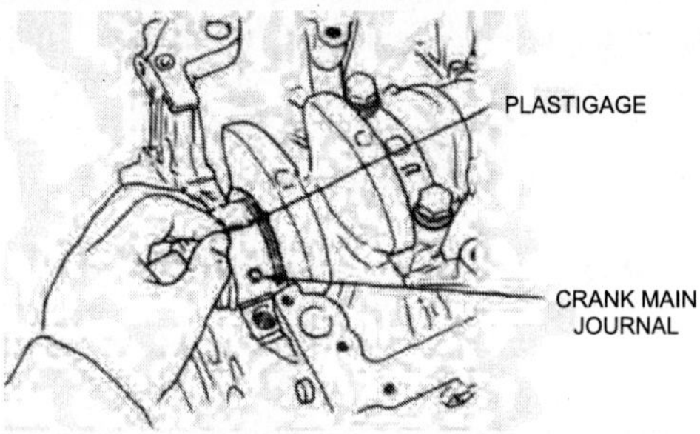

(a) Place a bead of Plasigauge on main bearing journal, parallel to the crank centreline. Install and torque cap to specs, with bearing installed in it.

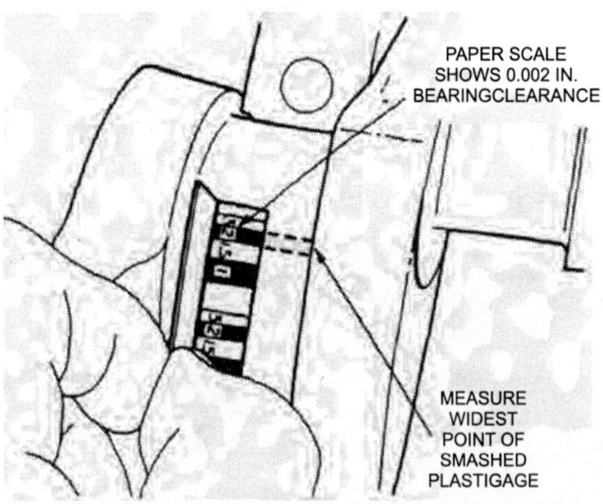

(b) Use paper scale to determine the bearing clearance. Match width of smashed Plastigauge with the scale. If the width matches with division 2, then the oil clearance is .002 inch (0.0508 mm). (Buick-GM)

Fig. 12.68 *Checking main bearing clearance. (Toyota and Buick-GM)*

Remove the main bearing cap and compare the smashed Plastigage to the paper scale, *Fig. 12.68 (b).* If clearance found is *not* correct, check bearing sizes and crankshaft journal measurements.

In general, an average main bearing clearance should be 0.025 to 0.05 mm. Refer the applicable service manual for correct value of oil clearance specified by the manufacturer.

Torquing Main Bearing Caps

Place each main bearing cap into the block. Check main cap *Numbers* and *Arrows* to ensure that they are in correct positions, facing the right direction, and in their original locations. Run the bolts down but do not tighten them.

When installing a rear cap, carefully check the rear main seal. If it is two-piece rubber type, make sure the sealing lip is pointing toward the inside of the engine, *Fig. 12.69.* Apply the sealer on the rear cap if required.

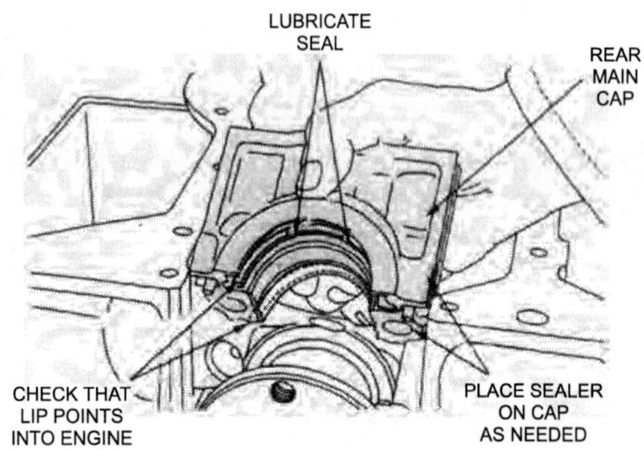

Fig. 12.69 *When installing rear main cap, check installation of rear main seal. Use sealer as recommended. (Peugeot)*

Find out the torque specs for the main bearing cap bolts in the applicable service manual. Make sure the caps are square and fully seated (not cocked in the block) before torquing them. Then, with a torque wrench tighten the bolts to about one-half torque and follow the sequence shown in *Fig. 12.70,* for all the main cap bolts. Then tighten the bolts to three-fourths torque. Then tighten the main cap bolts to their full torque at least *twice,* just to ensure proper torque and that no bolts have been skipped.

Endwise Clearance

The crankshaft must not have excessive endwise movement. Therefore, one of the main bearings usually is provided with flanges that bear against a machined flange on the crankshaft, *Fig. 12.71.* There is always some end thrust on the crankshaft. This may originate from the clutch pushing against the end of the shaft or from the thrust of acceleration or deceleration of the engine.

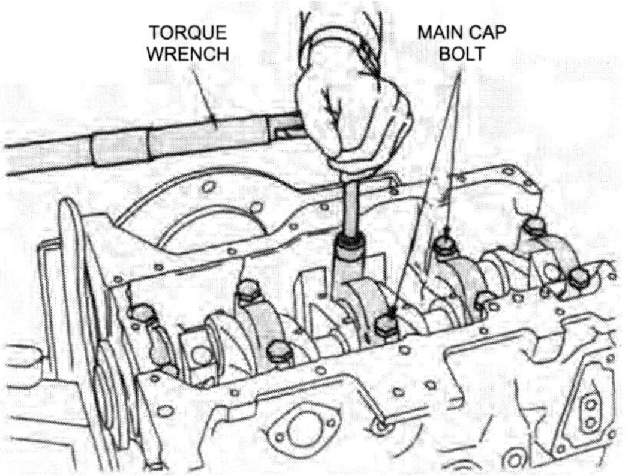

(a) Ensure that all caps are fully seated. Torque each bolt a little at a time up to partial specs. Use an even pull on the wrench.

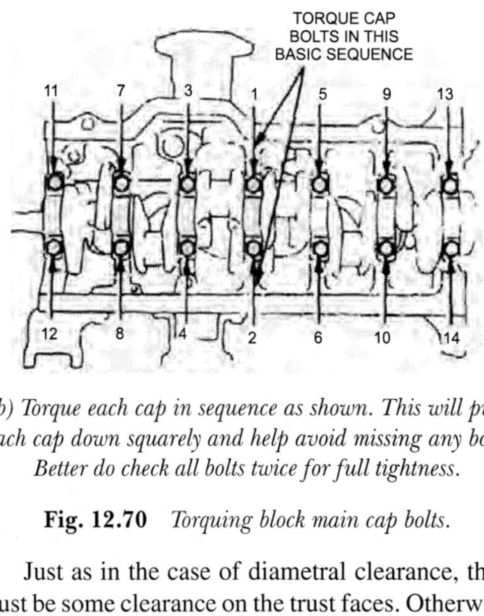

(b) Torque each cap in sequence as shown. This will pull each cap down squarely and help avoid missing any bolt. Better do check all bolts twice for full tightness.

Fig. 12.70 *Torquing block main cap bolts.*

Just as in the case of diametral clearance, there must be some clearance on the trust faces. Otherwise, expansion of the shaft and bearings caused by the normal heat of operation will cause metal-to-metal contact and burning of the *thrust bearing.* Usually a minimum clearance of 0.2032 mm is kept. A correct service manual should be referred to find out the correct end clearance. End clearance of the crankshaft can be measured with a feeler gauge, as shown in *Fig. 12.72.*

(a) Precicion insert bearing

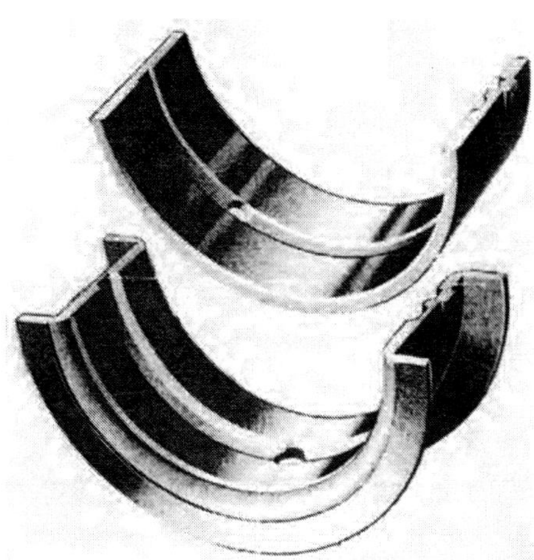

(b) Precision insert thrust bearing. Note that the thrust bearing has flanges.

Fig. 12.71 *Insert Bearings.*

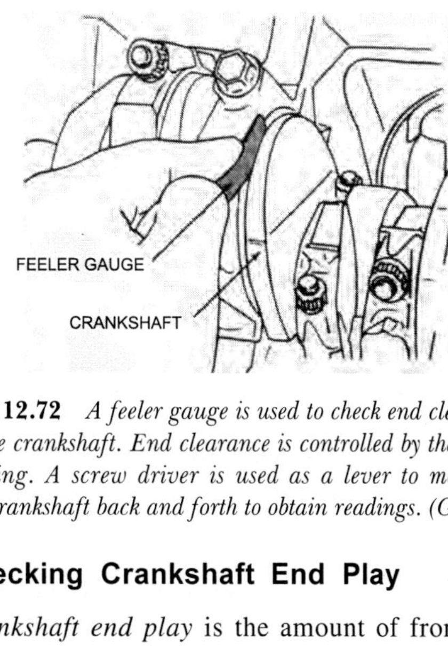

Fig. 12.72 *A feeler gauge is used to check end clearance of the crankshaft. End clearance is controlled by the thrust bearing. A screw driver is used as a lever to move the crankshaft back and forth to obtain readings. (GM)*

Checking Crankshaft End Play

Crankshaft end play is the amount of front-rear movement of the crankshaft in the block. It is controlled by the clearance between the main thrust bearing and the crankshaft thrust face or journal.

Crankshaft end play can be measured with the help of a dial indicator. To do so, mount the dial indicator on the block. Position the dial indicator against the crankshaft so that the indicator stem is parallel with the crankshaft centerline as shown in *Fig. 12.73.*

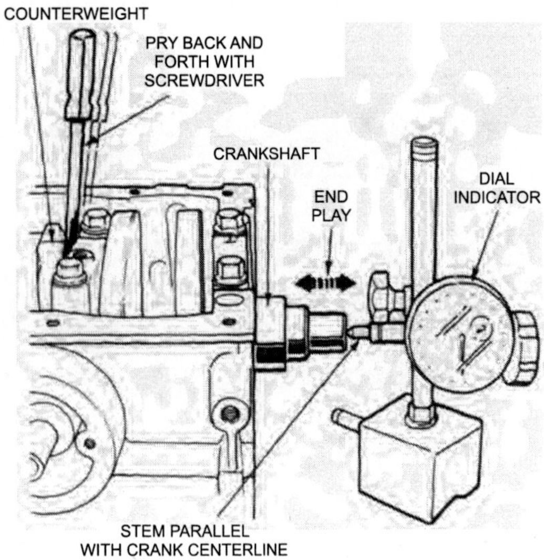

Fig. 12.73 *To check the crankshaft end play, use dial indicator as shown in the figure. Keep the indicator plunger stem parallel with crank. Pry back and forth on crank counterweight while reading the indicator. (Buick)*

Set the indicator at zero reading. Then, pry back and forth on a counterweight with a small pry bar. Movement of the indicator stem tip reflects the crankshaft end play.

Compare these measurements to specs. If these are not within specs or about 0.15 mm, check the main thrust bearing and width of the crank journal thrust surfaces.

Thrust Wear

Crankshaft's thrust area should be checked for wear. This is the area rubbed by the thrust bearings. Under abnormal conditions this area can be wiped away so even the new thrust bearings would not bring crankshaft end play within specs. The problem is solved by welding or replacement. Usually thrust wear is confined to the bearings, so replacing them will restore the crankshaft end play to within limits.

Replacing Pilot Bearing

For vehicles having manual transmission or transaxle the pilot bearing in the rear of the crankshaft might be replaced. It supports the end of the transmission input shaft. If worn beyond specs, it can trigger clutch and transmission problems.

To remove the pilot bearing, fill the cavity in the crank and pilot with heavy grease. Then, use an old input shaft or driver of the correct diameter to drive out the old pilot bearing.

Use hammer blows to force the shaft or driver into the grease. The grease, in turn, will build pressure on the back of the pilot bearing, pushing it out of the crank.

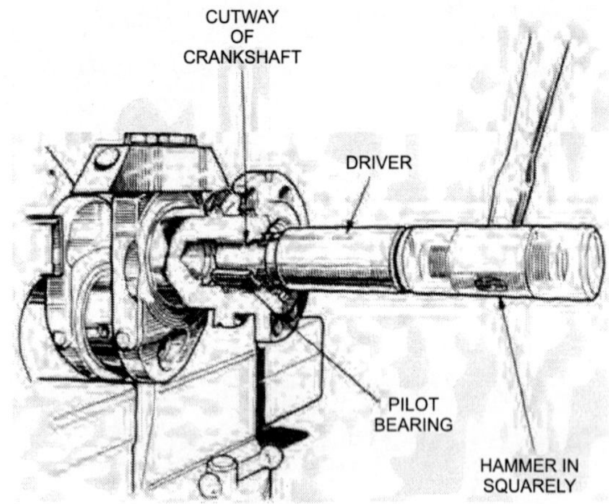

Fig. 12.74 *If needed, install a new pilot bearing in crank. Use heavy grease to force out old bearing. Then, use bearing driver to push new bearing squarely into position. (Renault)*

To drive the new bearing, drive the pilot squarely into the crank, as shown in *Fig. 12.74.* Exercise special care to avoid any damage to the inside diameter of the bearing. Place a small amount of grease in the pilot bearing.

Bearing Removal

The first step in bearing removal is to remove the piston from the cylinder. A ridge reamer is used to remove the un-worm portion of the cylinder wall above the piston ring travel, *Fig. 12.75.* Then the rod cap is removed, and the entire piston and rod assembly is pushed up and out of the cylinder. Connecting rods and their respective caps should be kept together and marked for their cylinder locations.

Replacement of Bearing Inserts

With the *inert bearing,* or *shell-type bearing,* it is possible to replace all main and connecting rod big-end bearings without removing the crankshaft,

Fig. 12.75. These bearing inserts require no fitting by hand, since they are made to extremely close limits of accuracy. It is only needed to obtain and install the correct size for the given application.

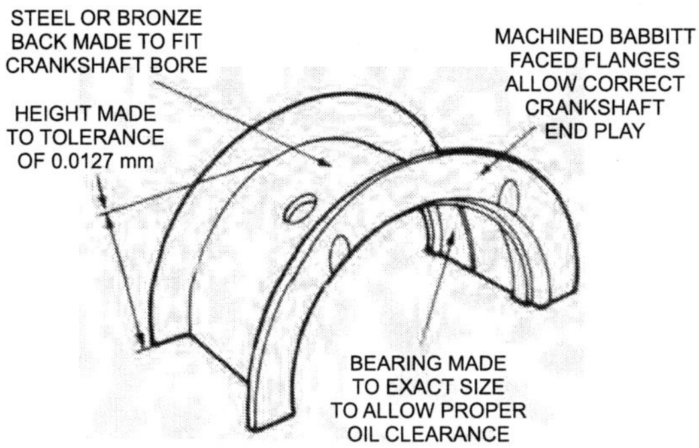

Fig. 12.75 *Typical thrust bearing design, criterion.*

In case the crankshaft journals have been reduced in diameter so much that a standard insert will not fit and the crankshaft is still in good condition, the journals can be reconditioned to undersize dimension and appropriate undersize inserts can be installed.

Inserts with excessive bearing metal are available and they can be bored out to the size needed, *Fig. 12.76.*

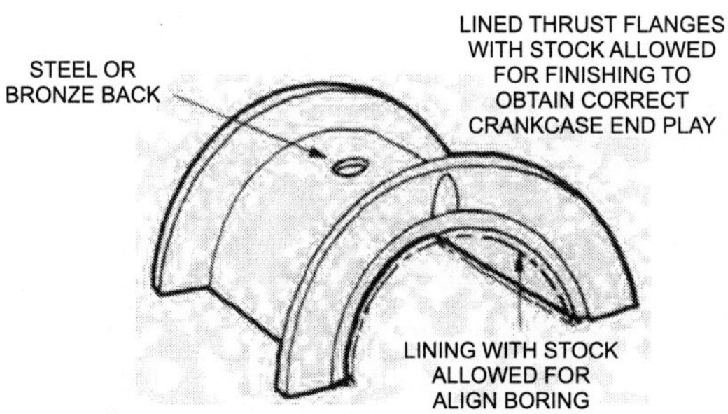

Fig. 12.76 *Re-borable bearings allow any undersize needed. (Federal Mogul)*

When the bearing bore and journal are round, new inserts need no fitting or adjustment. If the journal is out-of-round outside the specs, it should be machined to a closest undersize round shape. The same applies to the bearing bore in which the insert sits. Any errors in the bore will distort the bearing shell when the nuts are drawn down to the proper specifications.

Bearing Insert Seating

Both the bearing seat and bearing inserts must be round and true, to ensure intimate contact between the bearing seat bore and back of the bearing insert. Because if there is lack of contact, the heat generated in the bearing will not flow from the insert to the crankcase or connecting rod and consequently the bearing may melt.

Heat dissipation concern is one reason for *bearing crush.* Bearing crush means that the two halves of the bearing shell extend a few hundredths of a mm beyond the bearing seat bore, as shown in exaggerated form in *Fig. 12.77.* When the bearing cap nuts are drawn down to specified tightness, the insert is forced to seat in its bore.

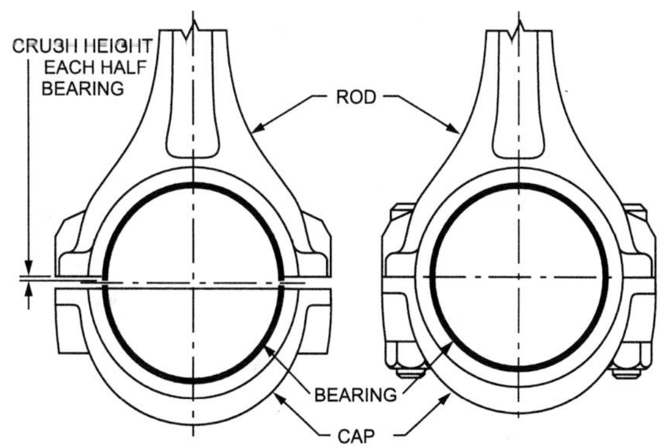

Fig. 12.77 *Bearing crush means that the bearing sits slightly higher than each half of the bearing housing bore.*

Bearing crush is also required to make sure that the bearing remains round. If the bearing insert is not held tightly against the bore, it may curl inward and contact the journal. In addition, the bearing crush helps to avoid any possible movement of the insert in the seat, *Fig. 12.78.* In case the shell should become slightly loose, it might oscillate in the seat and wear the bearing back. This would interfere with both oil control and heat transfer.

A reasonable amount of bearing crush must always exist, so the edges of the insert should not be dressed down flush with the bearing seats. If excessive amount of crush is present, the insert will be distorted when the cap is drawn down and the bearing would be deformed. *Fig. 12.79.* The usual amount of crush is only 0.0254 or 0.0508 mm. These inserts are extremely accurate and must be handled with care.

They should be purchased to the precise size required and inserted with no hand fitting work or any alteration.

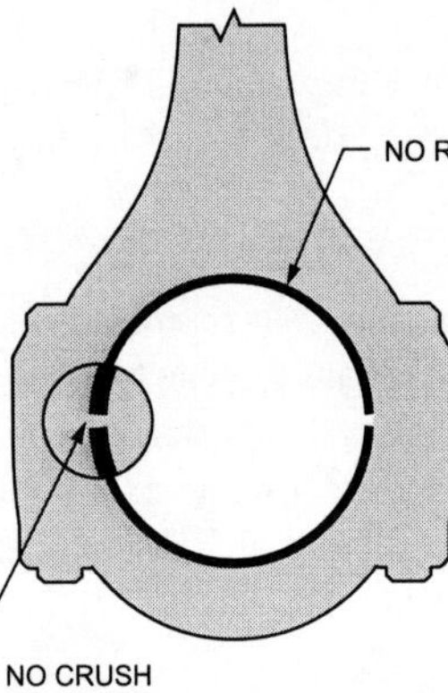

Fig. 12.78 *Lack of bearing crush may permit the inserts to move in the housing bore.*

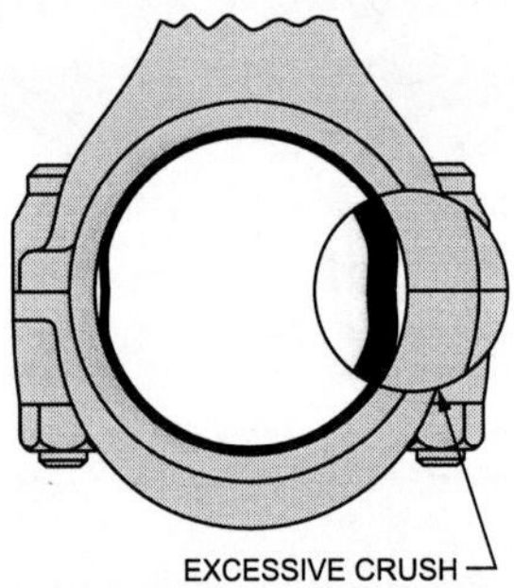

Fig. 12.79 *(a) Excessive bearing crush will cause the bearing to distort when the cap is tightened to specs.*

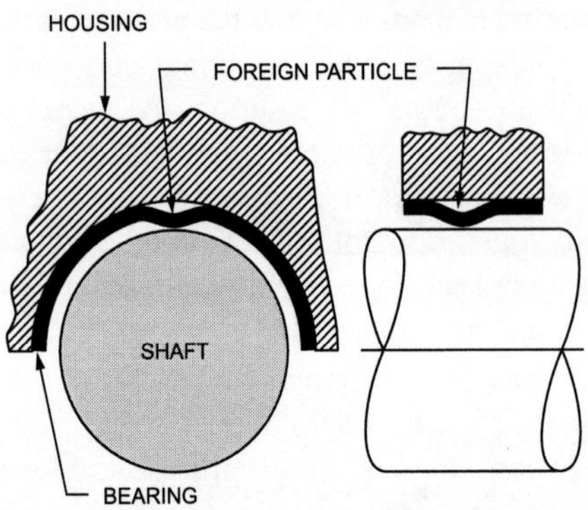

Fig. 12.79 *(b) A piece of grit between the insert and bore creates a high spot on the bearing surface. This prevents an effective transfer of heat from the connecting rod and could cause the bearing to melt.*

Bearing Spread

The diameter of the bearing is slightly wider than the bearing seat bore. This is known as *bearing spread, Fig. 12.80.* Bearing spread allows the bearing to remain in its seat, after being pressed in, during assembly, while the bearing or engine is inverted.

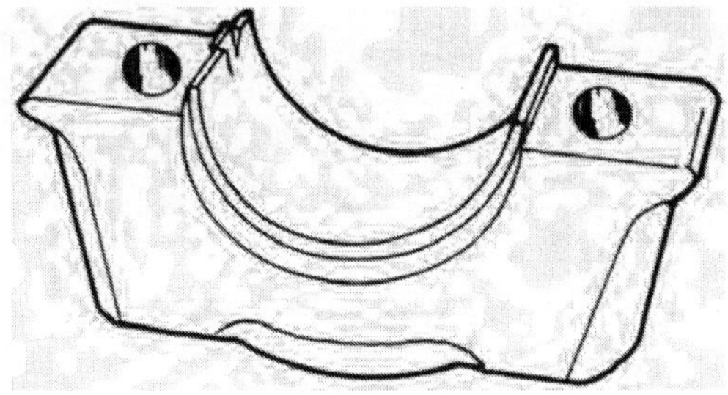

Fig. 12.80 *Bearing spread allows the bearing to be snapped into position.*

Installation of Bearing Inserts

When all bearings are installed properly and the caps drawn up to specified tightness, there should be little resistance to rotation of the shaft. If resistance is experienced: one or more bearings are fitted too tightly; there is insufficient end clearance on the thrust bearing; the crankshaft is sprung; the bearings are not in correct alignment; or the bearing caps have been mixed up.

If the engine is hard to rotate after installing new bearings, it should not be started until the cause of the interference is isolated and rectified.

Connecting rod must have some *side clearance* at the crankpin. Too much clearance will permit the bearing to move sideways and cause a knock in operation. The side clearance is measured by inserting a feeler gauge between the side of the cap and cheek of the crank throw, as shown in *Fig. 12.81.* A usual side clearance is 0.127 to 0.254 mm. Refer the applicable service manual to be specific in each application. Because there is no adjustment of the side clearance, excessive clearance requires replacement of the bearing inserts.

The replacement of *main bearings* of the insert type poses no problem unless the cylinder block has been warped, *Fig. 12.82,* and/or the crankshaft is scored or badly worn. When the cylinder block is not warped and the crankshaft is in good condition, remove the old bearing inserts and slip in new ones of the right size.

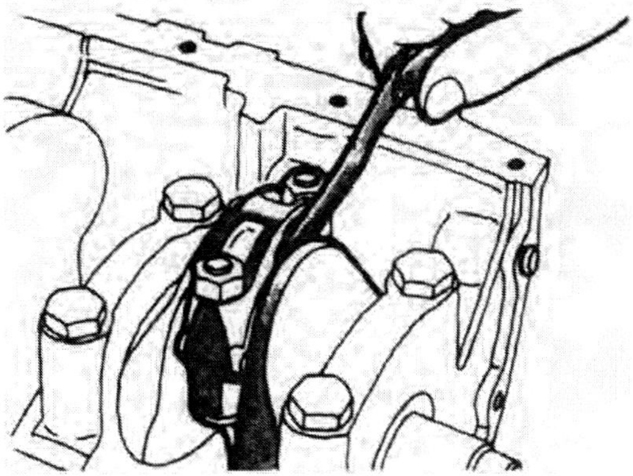

Fig. 12.81 *A feeler gauge is used to check connecting rod side clearance at the crankpin.*

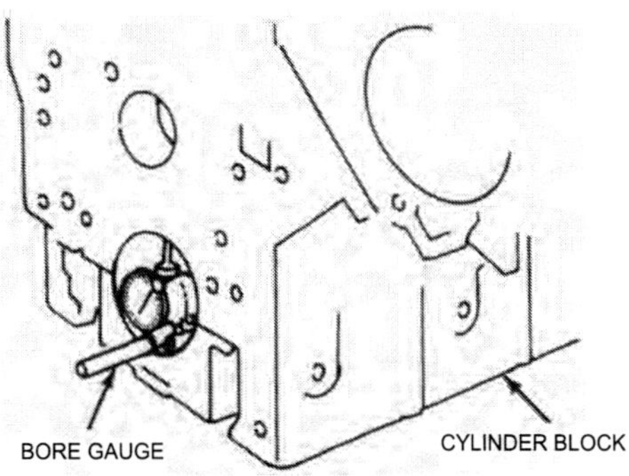

Fig. 12.82 *Using a dial indicator to measure out-of-round of main bearing bores. If any bore exceeds 0.0254 mm, the block must be align bored. (Chrysler)*

Main bearing inserts are replaced one at a time. The procedure is to remove the bearing cap and slip out the bearing insert. To remove the upper half of the bearing insert a "roll-out" pin is inserted into the oil hole of the crankshaft. The end of the pin protrudes slightly above the surface of the crankpin. When the crankshaft is rotated, the pin will force out the upper insert. A tool for his purpose can be made from a cotter pin, *Fig. 12.83.*

Slip the new bearing insert into position by hand. After installing the insert and bearing cap, repeat the procedure with the other bearings. Bearing caps are not tightened fully until all the bearing inserts are in position. Then the cap nuts are tightened to specified torque.

If the crankcase is not warped, the crankshaft is reground to a standard undersize for which bearing inserts are available. If the crankshaft is scored and the crankcase is warped, the engine must be removed and a complete reconditioning job must be performed. This includes regrinding the crankshaft and *align boring* the bearing bores, *Fig. 12.83.* If the crankcase is warped, semi-finished bearings can be installed in the crankcase. The bearings are then align bored and then the crankshaft is reground to size to fit the bearings. Before align boring the semi-finished bearings, make sure caps are properly assembled and nuts tightened to the specified torque. Also, all oil passages should be plugged with substantial pieces of clean cloth to prevent chips from getting into the lubrication system.

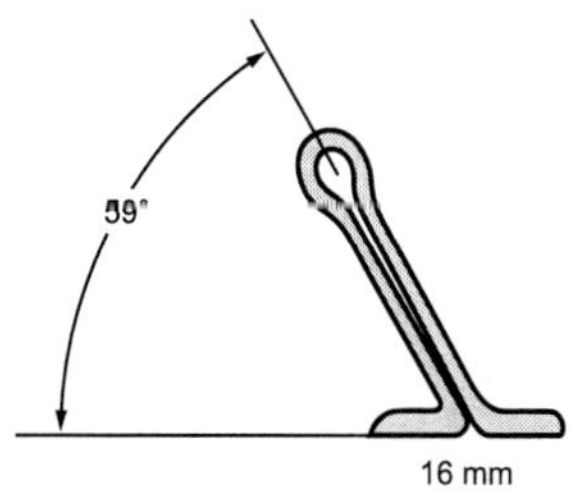

Fig. 12.83 *Bending a cotter pin so that it can be used to remove the upper bearing inserts.*

Use extreme care when locating the *boring bar* to ensure that the centerline of the finished bearings will be the correct distance from the top of the cylinder block, parallel with it, and at right angles to the cylinder bores.

For an engine built with a gear to drive the camshaft, the distance between the centreline of the camshaft and the bore of the main bearings must be very accurately maintained. Otherwise, the crankshaft and camshaft gears will not mesh properly. More tolerance is allowed if the camshaft is driven by a chain or a belt.

After all boring operations are completed, all plugs must be removed from the oil passages and all chips cleaned from the interior of the crankcase. All inside edges of each bored surface should be hand chamfered with a scraper to about 0.4 mm.

In all bearing work, extreme accuracy and cleanliness are required. Particular attention must be paid to the areas contacting the backs of the bearing inserts, the bearing surfaces, and the interior of all oil lines and oil passages.

For leaking rear main bearing, it is often possible to replace the bearing seal and bearing without removing the crankshaft. To do so, loosen all the main bearing caps slightly to lower the crankshaft. Then, remove the rear main bearing cap. With a small pin punch, start to drive out the seal. Grasp it with a pair of pliers and pull it out. To replace the upper part of the seal, first lubricate it with engine oil. Start it into the groove by hand and have the crankshaft turned until the seal slides in place.

The upper insert can be removed by first inserting a special tool or a bent cotter pin into the oil hole of the crankshaft. Then as the crankshaft is turned, the pin will push out the bearing half. Start removing the seal with a punch, *Fig. 12.84,* then pull it out with a pair of pliers.

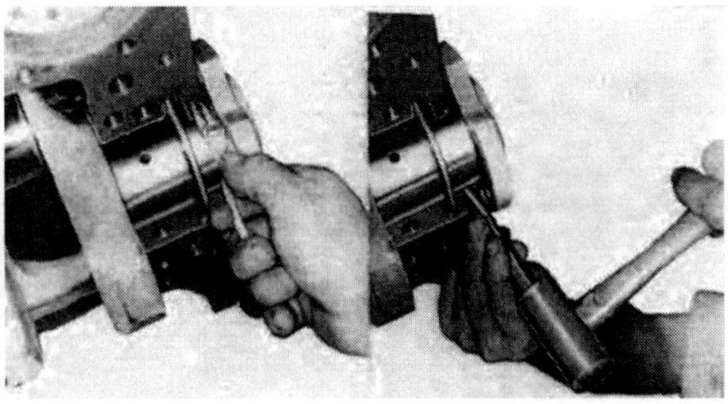

Fig. 12.84 *Removing rear main bearing seal while the crankshaft is in the engine.*

12.24 CONNECTING ROD SERVICE

Connecting rods are subjected to very high force during engine operation. As a result they can wear, bend, or even break. When connecting rods have been properly cleaned, inspect them visually for any evidence of damage. Remove the connecting rod insert bearings, replace the rod cap in the rod, and tighten to specifications. Connecting rod service covers four areas of rod construction: small end, beam or shank, big end, and bolts.

The old pistons and the bearing inserts will indicate the condition of the connecting rod. In case any piston or bearing wear abnormalities are found, there may be a problem with the connecting rod associated with them, e.g., if one side of the bearing is worn, the connecting rod may be bent. If the back of the bearing insert has marks on it, the rod big end may be distorted, allowing the bearing insert to shift inside the rod.

Rod Small End Service

Small end service consists of measuring the pin bore for out-of-round. For press-fit pin and rod combinations, no measurement is needed unless motion is detected while rocking the rod against the pin. Floating rods are measured with a snap gauge and micrometer. Also measure the outside diameter of the pin with the micrometer. Sometimes replacing the pin will reduce clearance to under the service limit.

Measure the rod small end with a telescopic gauge and a micrometer, *Fig. 12.85 (a).* If worn beyond specs, have a machine shop replace the rod bushing, *Figs. 12.85 (b)* and *12.85(c).* The pin will have to be *fitted* (bushing reamed for proper clearance) in the rod.

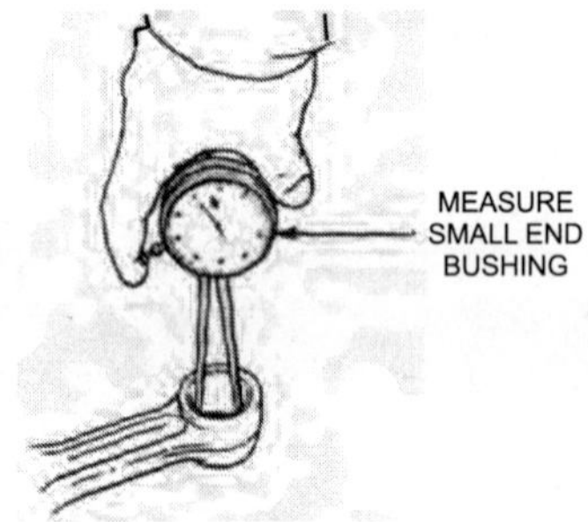

(a) Special dial gauge or telescoping gauge can be used to check rod small end wear.

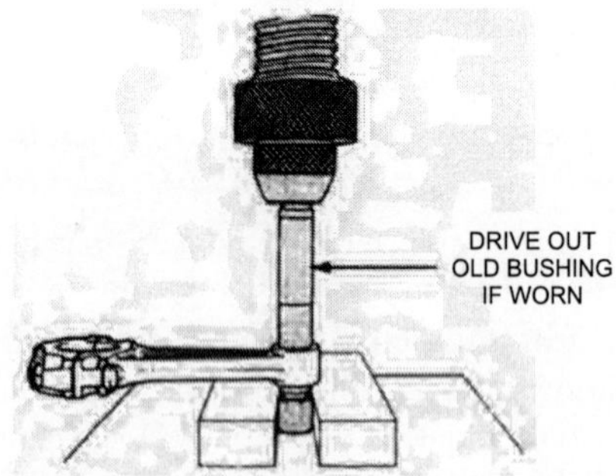

(b) Using press to push out old worn bushing.

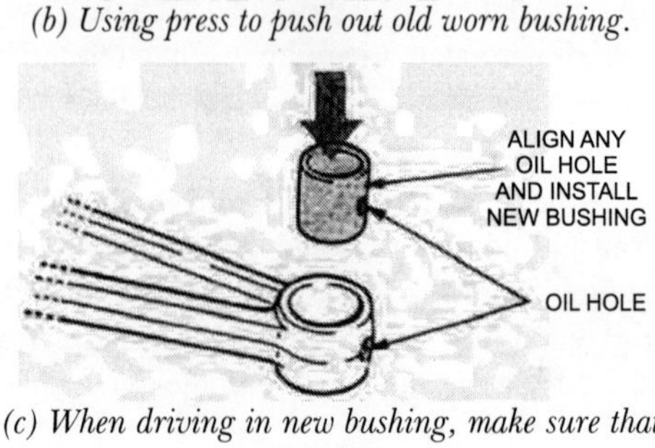

(c) When driving in new bushing, make sure that oil holes align.

Fig. 12.85 *Connecting rod small end service. (Toyota)*

Restoring the small end inside diameter is done in a number of ways. Some rods must be replaced, or non bushed rods can be bored oversize and a bushing pressed in. Oversize pistons may be available (0.038 to 0.076 mm oversize). They can be used in reamed pistons and rods to correct wear.

Beam

There is not much that can be done with the rod's beam, or shank. Inspect it for cracks or magna-flux it in heavy-duty applications. Replace any cracked rods.

Rod Big End Service

To check the connecting rod big end for problems, remove the bearing insert and bolt the rod cap to the rod. Torque to specs. Then, measure the rod bore diameter on both edges and in both directions, *Fig. 12.86.* Any difference in diameter equals big end taper. Any difference in the cross diameters equals rod big end out-of-roundness. If taper and out-of-round are greater than specs, have a machine shop rebuild the rod or replace the rod.

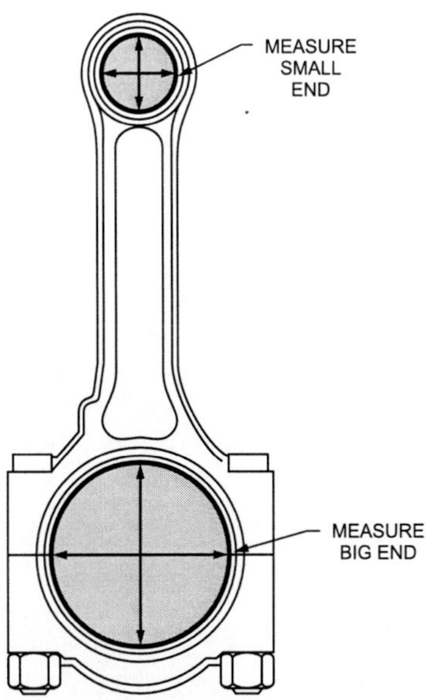

Fig. 12.86 *Connecting rod small end and big end must be measured as shown. Check for out-of-round and taper in bores. In case any of them is found to be out of specs, have the rod rebuilt. (Buick-GM)*

Most rod service happens at the big end. Out-of-round is the major problem, *Fig. 12.87.* Measure it with an inside micrometer, snap mike, snap gauges, or dial bore gauge.

Restoring big end roundness is done by *reconditioning* or *sizing*. Reconditioning is done on a special machine using a precision hone. A dial indicating-fixture shows how much honing is needed. First the rod cap's mating surface is ground in a special fixture. Then the cap is torqued onto the rod and honed. By grinding the cap, the big end is made very out-of-round. The hone then restores the big end's inside diameter to the original diameter and roundness. Grinding some material off the cap to start with ensures the hone will cut into fresh material around the entire circumference.

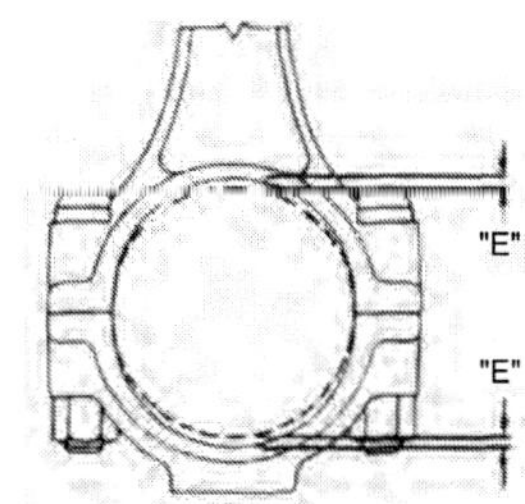

(a) Connecting rod bore (broken line) has expanded due to heavy stress during service.

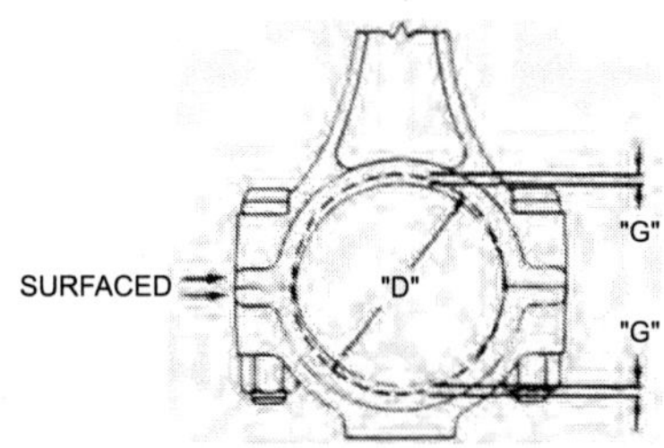

(b) After surfacing of the connecting rod and cap flange surfaces, the connecting rod bore shows an undersize in the vertical dimension.

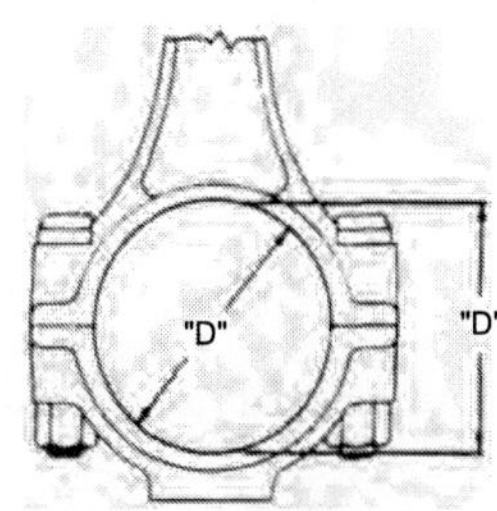

(c) After honing, the bore is again round and guarantees trouble free seating of the bearing shells.

Fig. 12.87 *Big end out-of-round on connecting rod. (Repco Corporation Ltd.)*

Checking Rod Straightness

To determine if a rod is bent, a special *rod alignment fixture* is needed. It will check whether the rod small end and big end are perfectly parallel.

Refer to the operating instructions provided with the particular fixture or send the rod to a machine shop, *Fig. 12 88.* Replace any bent rods.

Rebuilding Connecting Rods

Connecting rods can be rebuilt by most machine shops. This should be resorted to when inspection or measurements show major problems. Sometimes only one rod needs to be rebuilt, as when a bearing spins and damages the rod big end bore. In a high kilometers engine, all of the connecting rods may require rebuilding.

A connecting rod is rebuilt by:

1. Machining cap and big end bore so bore diameter meets specs.
2. Replacing small end bushing.
3. Checking and correcting, if necessary, rod bore alignment.
4. Installing new rod bolts, if necessary.

As shown in *Fig.12.88 (c),* special machine shop equipment is available for accurately machining rods.

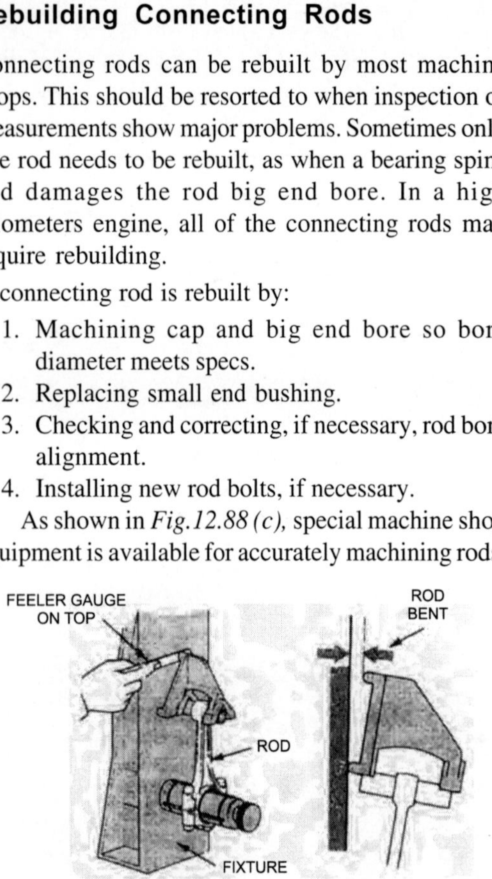

Fig. 12.88 *(a) Fixture being used to measure rod bend.*

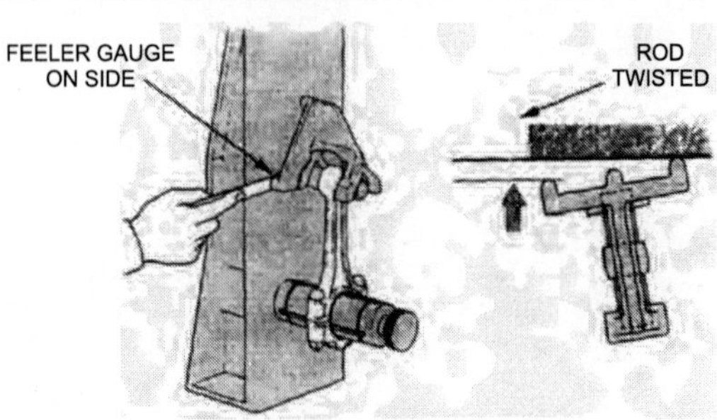

Fig. 12.88 *(b) Checking rod twist.*

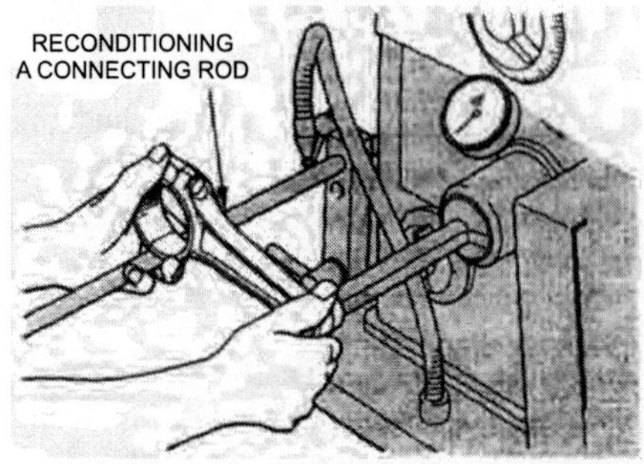

(c) Machine tool for honing and boring.

Fig. 12.88 *Machine shop equipment needed for rebuilding and checking connecting rods. (Toyota)*

Bolts

Rod bolts stretch with fatigue. In critical applications bolt length is specified so it can be measured with a micrometer. Most of the time rod bolts are merely visually inspected and returned to service if not damaged. Rod big end condition provides valuable guidance regarding the bolts. If the big ends were badly out-of-round, then the bolts have probably taken a beating. Replace them. For small displacement high rpm engines, the high rpm strains the connecting rods in general and the bolts in particular. These bolts get changed without question. Use new bolt nuts with new bolts. New nuts are also required when the originals have been rounded off by impact guns or repeated wrenching.

12.25 PISTON SERVICE

The first step in piston service is to visually inspect each piston. Look for fractures at the ring lands, the skirts, and the pin bosses. Look for scuffed, scored, or rough working surfaces. Replace pistons that show signs of too much wear or have wavy ring lands.

Pistons are made of aluminium, which is very prone to wear and damage. It is very important that each piston must be inspected thoroughly. Look for cracked or collapsed skirts, worn ring grooves, cracked ring lands, pin bore wear, or other problems that could affect piston performance and engine service life. Discard any damaged pistons.

Examining Pistons

A careful inspection of the pistons and rings will often show the cause of oil consumption or engine damage.

Fig. 12.89 shows a burned top ring land and broken top ring caused by excessive temperature and pressure in the combusting chamber. Such conditions cause bad piston burning, quicker top groove wear, ring breakage, scoring, piston ring sticking, and eventually a complete engine failure.

Fig. 12.89 *Piston damaged by extreme pressure and temperature in combustion chamber. (Perfect Circle)*

Fig. 12.90 *Piston with Preignition damage. (Perfect Circle)*

Preignition causes loss of engine power and damage to pistons, *Fig. 12.90,* rings, and valves. The high pressure caused by Preignition sometimes drives the top ring down with sufficient force to break off the ring land below the top compression ring. Both the piston and rings shown in *Fig. 12.91* are scored. When metal-to-metal contact between two rubbing surfaces raises the temperature of one of these surfaces to the melting point, scoring is caused. Scuffing is a lighter form of a similar damage. Scuffing and scoring are caused by overheating as a result of faulty cooling system, lack of cylinder lubrication, improper combustion, improper break-in, and coolant leakage into the cylinder.

Fig. 12.91 *Scored piston and piston rings. (Perfect Circle)*

Piston Removal

Different methods are used to retain the piston and its pin to the connecting rod. *Press pins* fit loosely in the pistons but have a interference fit in the rod. The pin must be pressed on and off the rod. *Floating pins* fit loosely in both piston and rod. They are commonly retained with buttons or circlips at each end of the pin. A pin button is a rounded piece of soft material, aluminium and teflon are common, which rides directly against the cylinder wall.

To separate a *press pin,* a press is used, *Fig. 12.92.* Support the piston with piston pin fixture (mandrel), which permits the pin to pass by as it exits the piston. Then press the pin out while supporting the rod. Otherwise it might fall and be damaged. Never hammer on a press pin as it will destroy the pin, piston, and rod.

Floating pins will just slip apart if they have buttons. With circlips, first remove the clips and then slide the piston out.

Ring Removal

Remove the rings for piston inspection. Use a ring expander to remove the old rings, *Fig. 12.93.* The ring expander assures against scratching the piston.

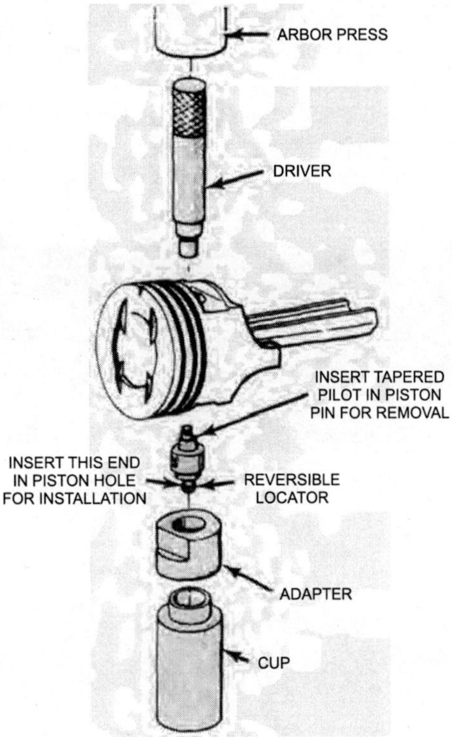

Fig. 12.92 *Press fitting piston pins. (Ford Motor Co.)*

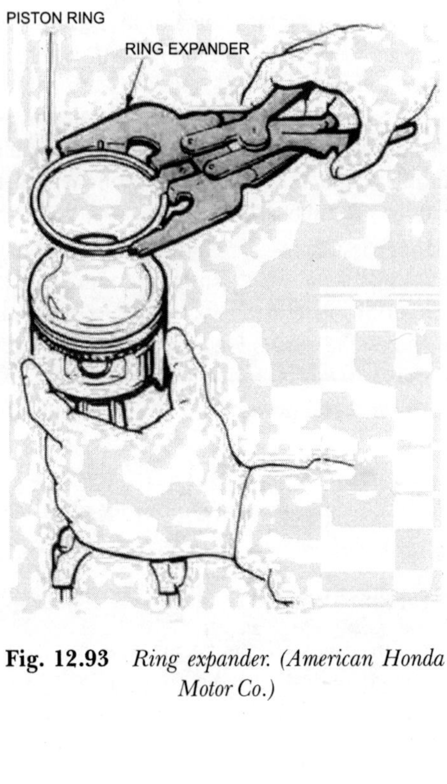

Fig. 12.93 *Ring expander. (American Honda Motor Co.)*

Piston Cleaning

Cleaning Piston Crown: Remove all carbon from the piston. Use a wire wheel on the crown, but be careful not to let it touch the sides of the piston. Using a wire wheel on the piston's sides will round off the ring lands and result in poor ring sealing. Really hard carbon on the crown may be chipped off and then the area cleaned by wire-wheel.

Cleaning Piston Ring Grooves: Use a ring groove cleaner tool to clean the ring grooves, *Fig. 12.94.* Go lightly but get all the hard deposits removed out from the inside the grooves. Also clean the oil slots or holes. An old broken ring may be used to clean the piston grooves. If not cleaned, the hard deposits could force the new rings out against the cylinder walls upon piston heating and expansion. This could score the rings and cylinder walls, possibly causing the engine to seize due to rings melting and welding to cylinder walls.

Fig. 12.94 *Ring groove cleaner (Ford Motor Co.)*

Measuring Piston Wear

A large outside micrometer is used to measure piston wear, *Fig. 12.95.* Micrometer readings are compared to specs to determine wear. Measure perpendicular to the piston pin. Adjust the micrometer for a slight drag as it is pulled over the piston. *Piston size* is measured on the skirts, just below the piston pin hole, *Fig. 12.96.* If worn more than specs (about 0.13 mm), replace or knurl the piston(s). If in doubt, check in the applicable service manual for directions to correctly measure piston wear. Some engine manufacturers require measurement at the pin hole centre level. While some others require measurement lower on the skirt. Making measurements at different position other than required will make the piston diameter specs inaccurate. Most manufacturers give specs for just below the piston pin, perpendicular to the pin, however.

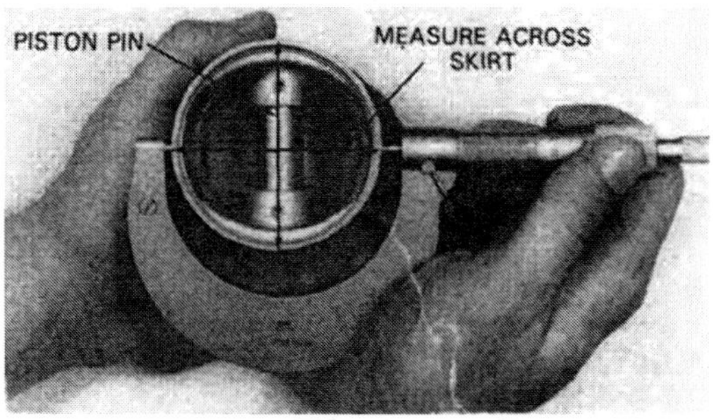

Fig.12.95 *(a) Note how piston measurements should be opposite piston pin on skirt. Hold micrometer properly when measuring for accurate readings*

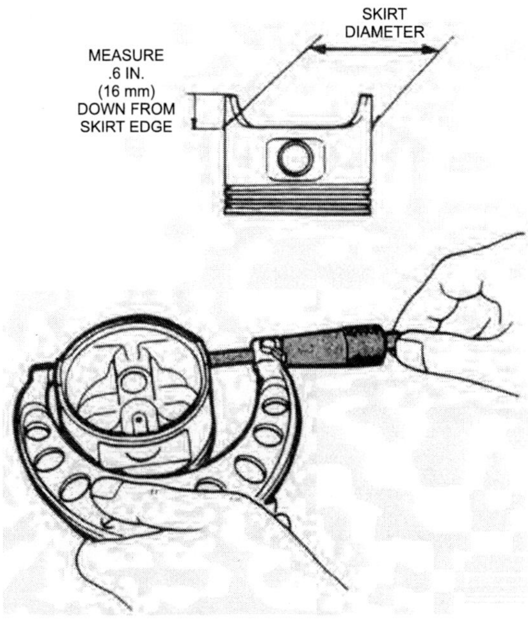

Fig.12.95 *(b) Measuring piston skirt. (American Honda Motor Co.)*

The skirts should be approximately 0.0127 mm wider at their bottoms than at their tops. If they are not, the skirt has collapsed. A piston skirt can collapse when overheated. Extreme heat results in the piston skirt expansion against the cylinder wall. If the piston gets hot enough, the skirts will expand until it is reshaped by the cylinder. Eventually the skirts will crack. Even before cracking, the collapsed skirt has permanently lost its shape. Collapsed pistons have poor oil control, make noise, and accelerate cylinder wall wear.

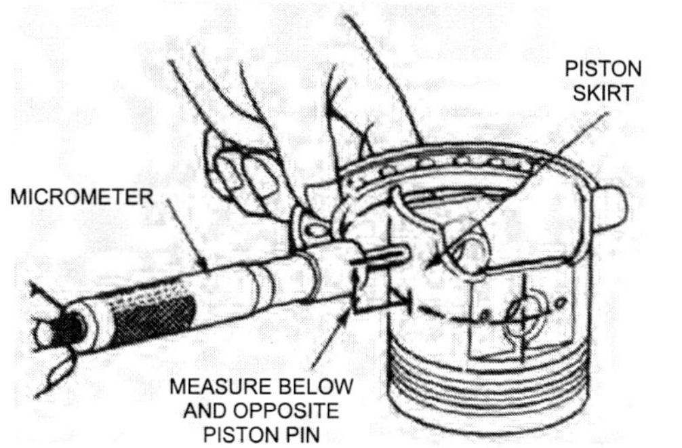

Fig. 12.96 *Normally, measure piston wear just below piston pin on piston skirt. Adjust micrometer for slight drag when pulled across skirt. Compare the measurements recorded to specs. (Toyota)*

Knurling the Piston

Piston knurling may be used to increase the diameter of the skirt a few hundredths of a mm. A machine shop can usually knurl pistons and increase piston diameter up to 0.13 mm. Knurling makes grooves in the skirts, pushing up metal next to the grooves. This increases piston diameter and also provides grooves for lubricating oil to reduce friction, *Figs. 12.97(a)* and *12.97(b)*. Knurling is an easy way to salvage used pistons that are slightly worn. It also takes up some of the clearance in a slightly worn cylinder to prevent piston slap or knock.

Fig. 12.97 *(a) Knurling to restore piston to bring it in to specs. (Perfect Circle)*

Fig. 12.97 *(b) Knurled Piston. (Perfect Circle)*

Ring Groove Wear

Typically, the upper compression ring groove wears faster than the lower compression ring or oil control ring grooves. Visually inspect the ring grooves for wavy edges. If no damage is visible, measure to be certain, *Fig. 12.98.* Slip a new compression ring into the ring groove. Then find the feeler gauge or go-no-go gauge that just fits between the ring and the ring land. The feeler gauge then gives the side clearance. Rotate the piston to measure the entire ring groove. If the feeler gauge binds anywhere in the groove, it indicates that part of the groove is tighter than the rest. Therefore, either the groove is not worn at that point or has been compressed. Compare the measurement with the engine specs. In general, use 0.127 mm.

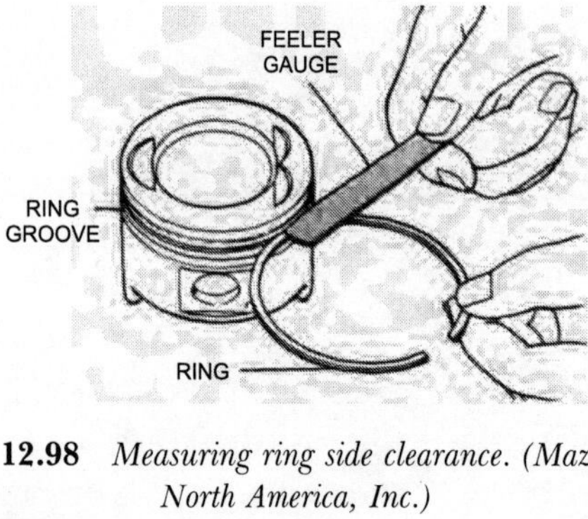

Fig. 12.98 *Measuring ring side clearance. (Mazda of North America, Inc.)*

Proper side clearance is very important to effective ring sealing. If the ring grooves are damaged, replace the piston. If the upper groove is worn, a special lathe can cut the groove wider. Inserts are then fitted between ring and groove to take up the extra gap.

Measuring Piston Ring Side Clearance

Piston ring side clearance is the space between the side of a compression ring and inside of the piston groove. Ring groove wear tend to increase this clearance. If worn too much, the ring will not be held square against the cylinder wall and will cause oil consumption and smoke in exhaust.

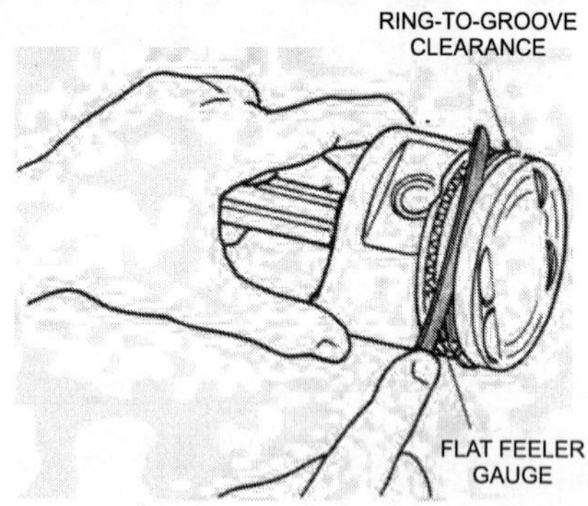

Fig. 12.99 *Check the ring side clearance by inserting feeler gauge between ring and ring groove as shown. Thickest gauge that fits equals clearance. If too large (out of specs), replace pistons, check ring dimensions, or have groove cut for ring side spacers. (Toyota)*

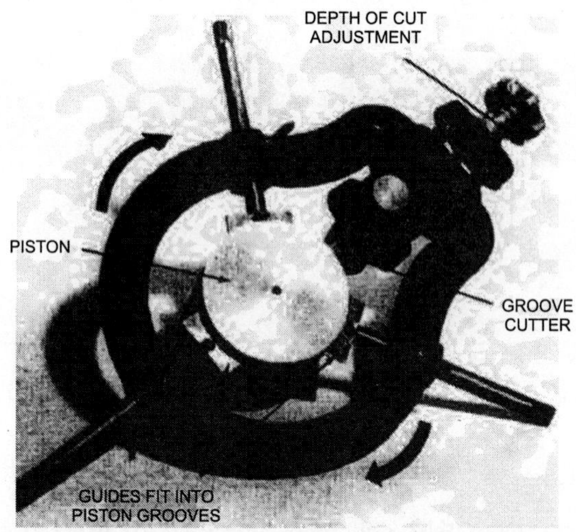

Fig. 12.100 *This tool can be used to cut ring grooves wider when excessively worn. Then, thin steel spacers can be put next to rings to restore the correct ring side clearance. Tool is rotated around piston. (K-Line Tools)*

To measure ring side clearance, obtain the new piston rings to be used during the engine overhaul. Insert or install the new ring into its groove. Then slide a feeler gauge between the ring and groove, *Fig. 12.99.* The largest feeler gauge that fits between the ring and groove represents ring side clearance. The top ring groove is usually checked because it suffers from more combustion heat and wear than the second ring groove. If ring side clearance is out of specs, 0.05 to 0.13 mm, either replace the pistons or have a machine shop fit ring spacers in the grooves.

Ring spacers are thin steel rings that fit next to the compression rings. The piston ring groove is machined wider to accept the spacer. This will restore ring side clearance to with in spec limits. *Fig. 12.100* shows a machine shop tool for cutting ring grooves wider to install spacers.

Piston-to-Cylinder Clearance

To determine *piston clearance,* subtract piston diameter from the cylinder diameter. The difference between the cylinder bore measurement and the piston diameter measurement will equal piston clearance. Average piston-to-cylinder clearance is about 0.025 to 0.08 mm. Because the specs vary, always consult the applicable service manual.

A better way to measure piston fit is with inside and outside micrometers or bore gauge, *Fig. 12.101.* Another method of determining piston clearance is with a feeler gauge strip, *Fig. 12.102.* A long flat feeler gauge is placed on the piston skirt. Then the gauge and piston are pushed into a cylinder. A spring scale is used to pull the feeler gauge strip out of the cylinder. When the spring scale reading equals specs, the size of the feeler gauge equals piston clearance.

When piston-to-cylinder clearance is excessive, then the technician must do one of the following as needed:

1. Knurl the pistons.
2. Install new standard size pistons, if the cylinders are not worn out of specs.
3. Bore the cylinders and install suitable oversize pistons.
4. Sleeve the cylinders and install standard pistons.

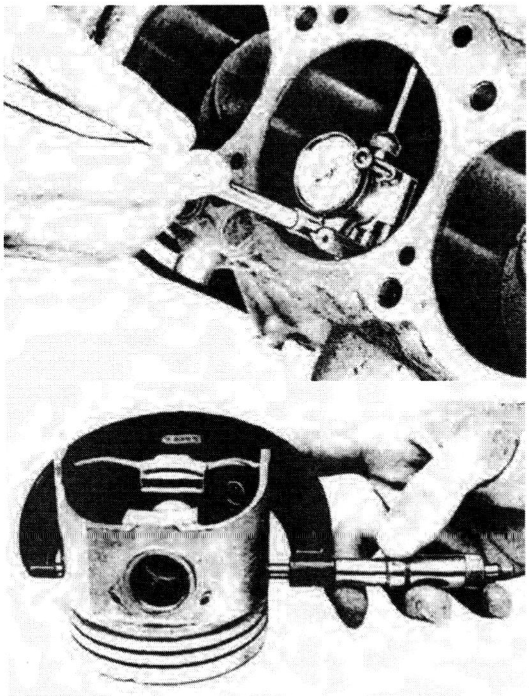

Fig. 12.101 *Measuring piston-to-cylinder clearance with bore gauge and micrometer. (General Motors Corp.)*

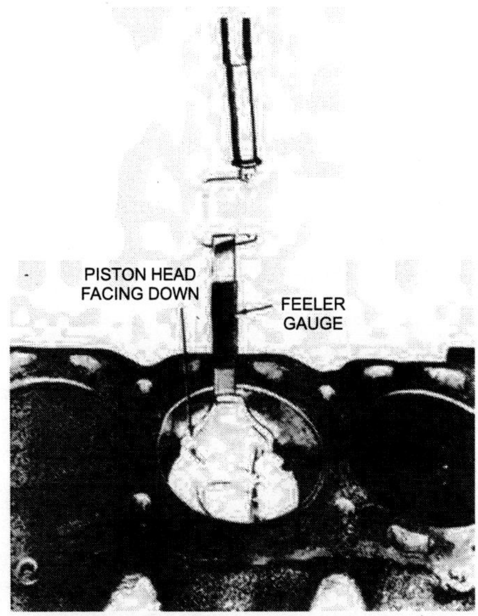

Fig. 12.102 *Measuring piston-to-cylinder clearance with a feeler gauge. (General Motors Corp.)*

Checking Ring Groove Depth

To check ring groove depth, fit a new compression ring into the piston groove, *Fig. 12.103.* See that the ring will fit below the surface of the piston. This makes sure you have the correct ring set. If the ring wall thickness were too large, the ring could be forced out against the cylinder walls after the piston expands when it gets heated in operation. Part damage could result.

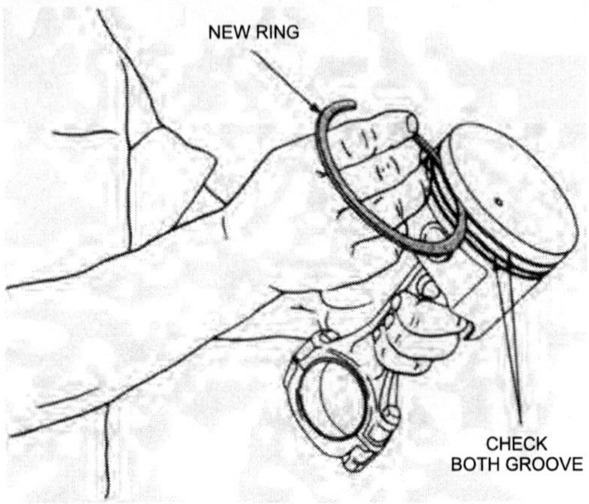

Fig. 12.103 *To check ring groove depth, fit a new ring into groove. Check that ring stays below surface of the piston. This verifies groove depth with ring. (Pontiac-GM)*

12.26 PISTON PIN SERVICE

Depending upon the type and make of engine, the piston pin may either be *free-floating* or *Press-fit in rod or piston.* Other setups have been manufactured but are not common.

During piston and rod service, check the pin clearance on both floating and pressed-in pins. To check for excessive piston pin clearance, clamp the connecting rod I-beam slightly in a vice and use vice caps to protect the rod. Pull and push up and down on the piston and try to rock the piston against normal pin movement, *Fig. 12.104 (a).* If play is detected, the pin, rod bushing, or piston bore is worn. A small telescoping gauge and outside micrometer should be used to measure exact part wear after pin removal.

Free-floating Pin Service

To remove a free-floating pin from the piston, compress and lift out the snap rings on each end of the pin with a snap ring pliers, *Fig. 12.104 (b).* Then push the pin out of the piston with thumb pressure. A brass drift and light hammer blows may be required when thumb pressure does not work. Mark the piston and rod and keep them organized, *Fig. 12.104 (c).* If the pin is worn beyond specs, replace it. When the measured pin bore in the pistons is outside the specs, replace the pistons or the pin bore may be reamed larger and oversize piston pins are used. Pin bore reaming is normally done by a machine shop.

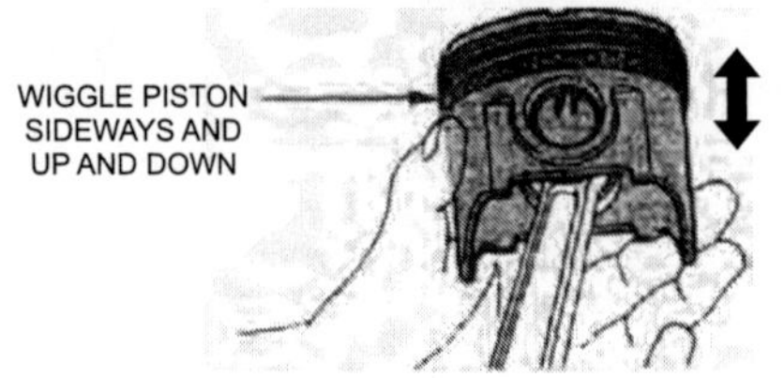

(a) Wiggle piston against normal swivel to check for loose wrist pins.

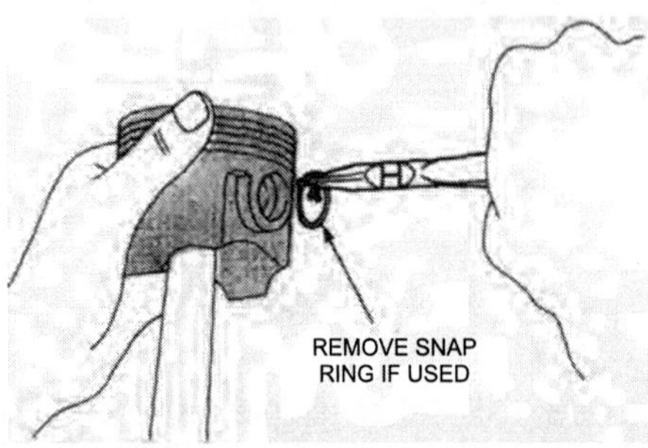

(b) For a floating type pin, remove snap rings and push out pins. If press-fit type, use a press to remove pins.

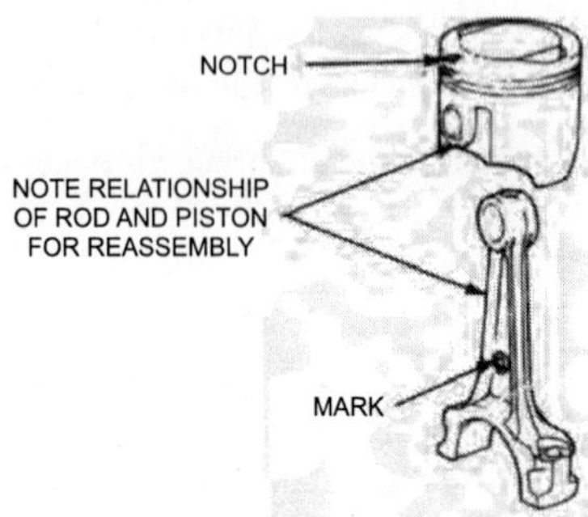

(c) Carefully align and orient rods on the pistons in the right way, for reassembly

Fig. 12.104 *Piston pin service. (Toyota)*

Pressed-in Pin Service

A pressed-in piston pin is normally removed with the help of a press and a driver setup like the one shown in *Fig. 12.105.* Wear eye protection and carefully mount the piston in proper position, *Fig. 12.106.* Some times a torch is used to heat and expand the piston or rod small end to facilitate piston pin removal, *Fig. 12.107.*

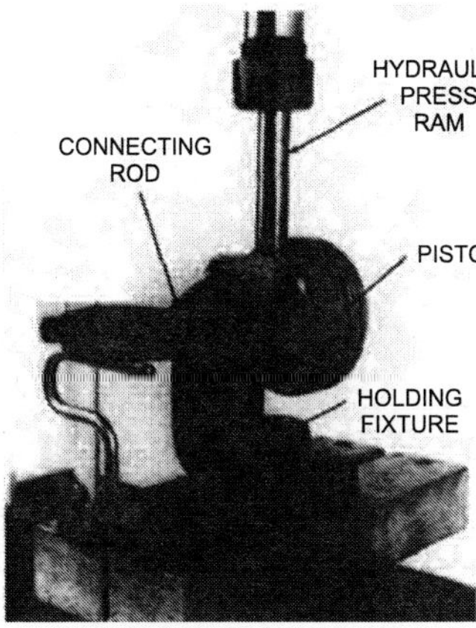

Fig. 12.105 *Holding fixture and hydraulic press are used to drive out press-fit piston pin. (OTC Tools)*

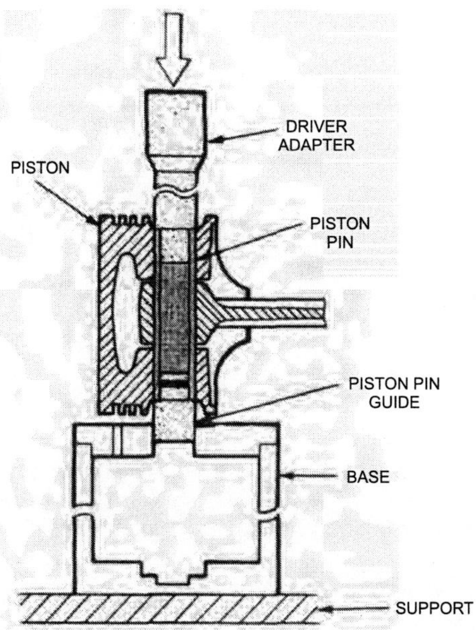

Fig. 12.106 *Correct size adapters are necessary to drive press-fit pin out of piston. (General Motors Corp.)*

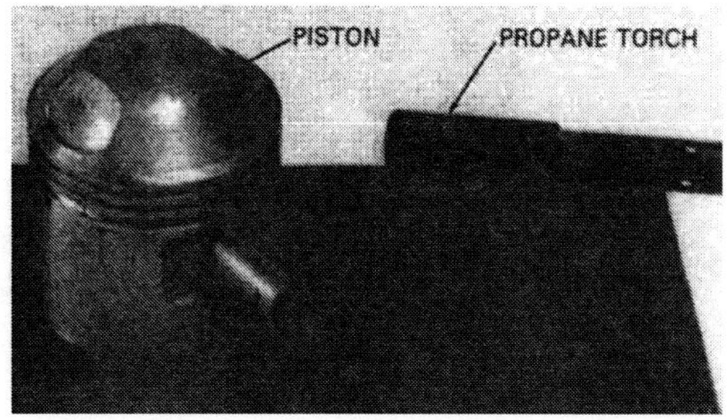

Fig. 12.107 *Sometimes mechanics use heat from torch to expand parts so that the pin may be pushed out more easily.*

Measuring Piston Pin Wear

Measure the floating type pin wear, with an outside micrometer, in the middle and at both ends of the pin, *Fig. 12.108.* Compare the measured wear with specs. If worn to be outside specs, replace the piston pins.

A typical piston pin clearance is 0.0025 to 0.0075 mm only. Keep this fact in mind while measuring piston pin wear, piston bore wear, and rod small end wear.

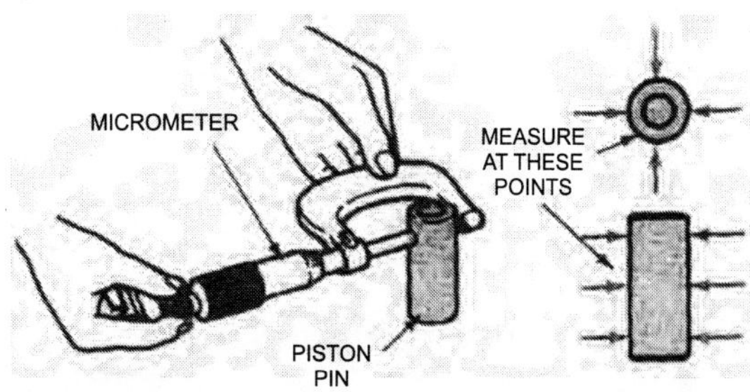

Fig. 12.108 *Measure piston pin wear in locations shown in figure. Because specs are commonly given in hundredths of a mm, use extreme care in taking measurements. (General Motors Corp.)*

Piston Pin Installation

Before installing a piston pin, align the markings to make sure the piston is facing in the right direction in relation to the connecting rod. Normally, a piston will have some kind of marking on its head, which should point towards the front of the engine.

Connecting rod may have an edge of the big end bore chamfered (faces outside of journal on V-type engines). The rod may also have an oil spray hole or rod numbers that must face in only one direction. It is advisable to consult the service manual for directions.

Push the floating type piston pin into the piston, install the snap rings to secure the pin against any end travel beyond permissible limits. Ensure that the snap rings are correctly seated in their respective grooves. To start a press-in type piston pin, align the pin properly and tap it into the bore with brass hammer.

Then use a press to force the pin into the piston to an exact location. The connecting rod small end must be centred on the pin.

12.27 INSTALLING PISTON RINGS

Before installing new rings, each compression ring from the ring set should be pushed down to the lower, unworn portion of a cylinder. The ring must be square in the cylinder. Pushing the ring down in the bore with the head of the piston puts the ring into the bore at the proper depth and squares it with the cylinder walls.

Piston ring gap (clearance between ends of ring when installed in cylinder) is very important. If the gap is too small, the ring could lock up or score the cylinder upon heating and expanding. If the ring gap is too large, the ring force against the cylinder wall may be low, resulting in blowby.

With the cylinders, pistons, and connecting rods all checked and reconditioned if needed, the new piston rings can be installed on the pistons. Make sure to check the ring gap.

Measuring Piston Ring Gap

To check the ring gap, compress and place a compression ring in its cylinder. Then, push the ring to the bottom of normal ring travel with the head of a piston. This will square the ring in the cylinder and locate it at the smallest cylinder diameter, *Fig. 12.109 (a).* Measure the gap with a flat feeler gauge. Compare the measurement to specs. If not between correct specs (0.25 to 0.51 mm), either the piston ring set is wrong or cylinder dimensions may be off, *Fig. 12.109 (b).*

Installing Oil Rings

Install the oil ring first. To hold the piston, clamp the rod lightly in a vice. Use wooden blocks or brass vice caps to prevent rod damage. The bottom of the piston should be in contact with the top of the vice to prevent piston swivel.

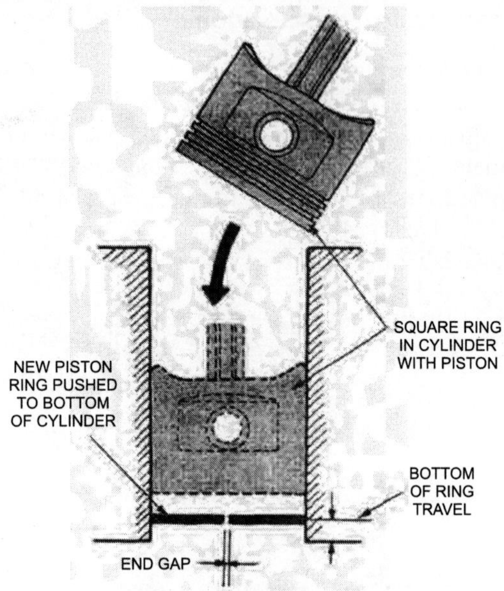

(a) Use piston to push new ring to bottom of ring travel in cylinder. This will square ring in cylinder and also close the ring to a minimum gap condition.

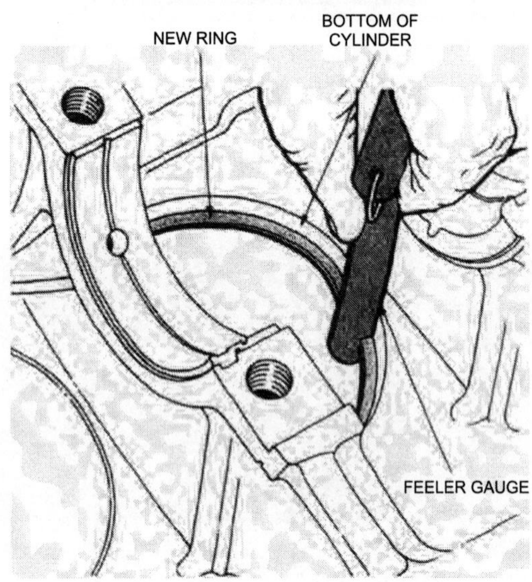

(b) Use feeler gauge to check ring gap. If too small, check cylinder and ring measurements. Rings may be wrong. A fix could be in filing the ring ends so ring does not lock against cylinder wall upon expansion in normal engine operation. Better to revisit reconditioning procedure followed. (Toyota and Chrysler)

Fig. 12.109 *Checking ring gap in cylinder. (Chrysler and Toyota)*

Wrap the expander-spacer around its groove. *Butt* the ends together. Spiral the bottom oil ring rail next to the bottom of the expander-spacer, *Fig.12.110 (a)*. Then, spiral in the top oil ring rail, *Fig.12.110 (b)*. Make sure that the expander-spacer is *not* overlapped. Its ends must butt together, *Fig. 12.110 (c)*.

Check that the oil ring assembly will rotate on the piston. It experiences a moderate drag as the oil ring is turned. One ring gap should be almost aligned with the end of the piston pin, while the other gap should be at the opposite end of the piston pin.

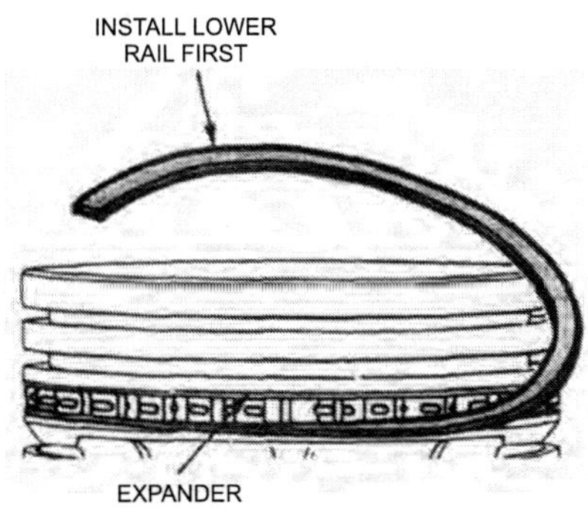

(a) Install expander-spacer and butt ends together. Then spiral lower rail next to spacer in groove.

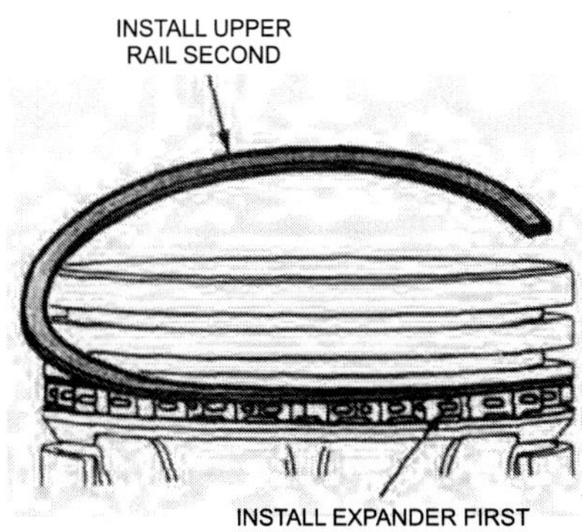

(b) Spiral on top rail and fit it next to spacer

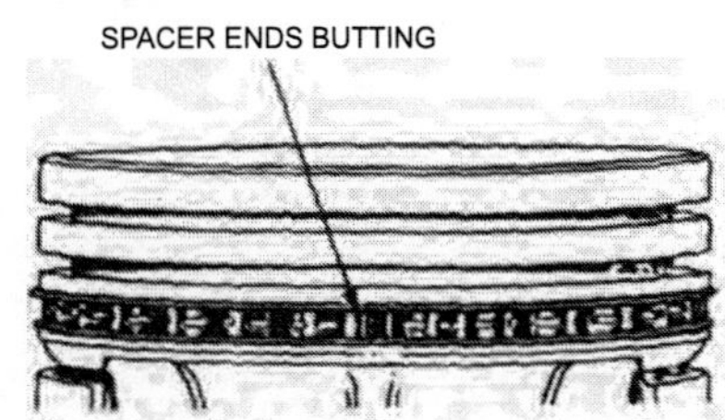

(c) Check that ring gaps are opposite each other and to one side of piston pin. Verify that expander-spacer ends are not overlapped; they must butt together.

Fig. 12.110 *Installing oil ring on piston.*

Installing Compression Rings

Follow the instructions with the new piston ring set. There will normally be directions on proper ring installation. Usually, compression rings have a top and a bottom. If installed upside down, ring failure and/or leakage may result.

Ring markings are normally provided to show as to how the compression rings should be correctly installed. The markings show the top of each ring and which ring goes into the top ring and which ring goes into the second piston ring groove, *Fig. 12.111.*

Compression rings are made of very brittle cast iron and may break easily if expanded or twisted excessively. Therefore, use a ring expander for spreading and installing rings and slip the compression rings into their respective grooves, *Fig. 12.112.* If a ring expander is not available, use fingers to carefully spiral the compression rings onto the piston.

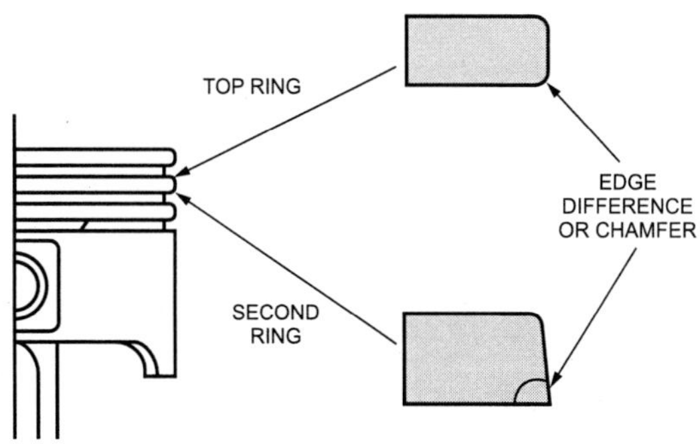

(a) Example of different ring shapes. Ring installation instructions will provide directions as to which ring goes into which groove.

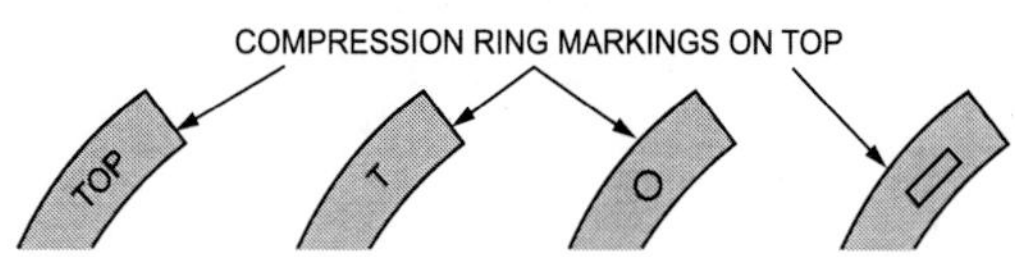

(b) Marks are usually provided to denote top of compression rings.

Fig. 12.111 *Compression rings come in different shapes. Shape is used to help hold ring against cylinder wall and seal adequately. (Honda and Ford Motors)*

Piston Ring Gap Spacing

When installing rings, space the gaps around the pistons to avoid the possibility that the gaps will align with each other, increasing blowby. A specific *piston ring gap spacing* is usually recommended to reduce blowby and ring wear. *Fig. 12.113* shows a typical set up method. Note that each gap is directly opposite the one next to it. The gaps are almost lined up with the piston pin hole. This provides maximum distance between consecutive gaps to minimize pressure leakage. Also, being next to the piston pin hole the gap is not on a major thrust face, this reduces ring wear.

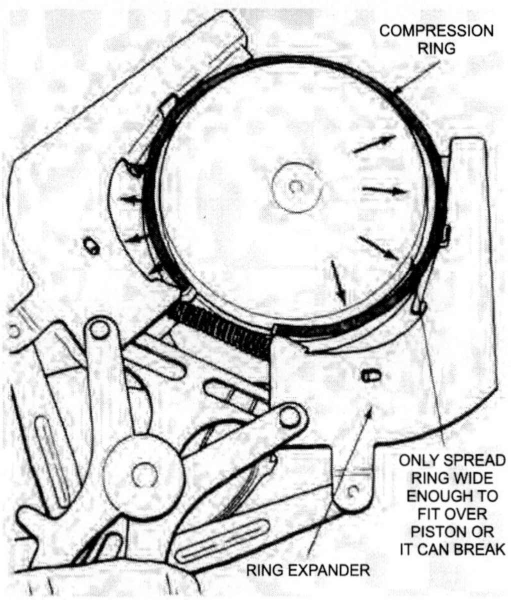

Fig. 12.112 *Preferably use a ring expander to spread rings and install them on pistons. In case a ring expander is not available, use your fingers to expand rings just enough, and be careful not to break the rings. (Chrysler Corp.)*

Piston and Rod Assembly

The first decision to make about the piston and rod assembly is whether or not to remove the piston from the rod. If the cylinders are to be bored, remove the pistons. New pistons will be required to match the oversize bores, and discard the old pistons. If the bores are being reused, probably the pistons can be used too. Leave them on the rods for now. Inspect for loose, worn piston pins and bushings. If, inspection indicates the pistons need replacement, remove them and install new ones.

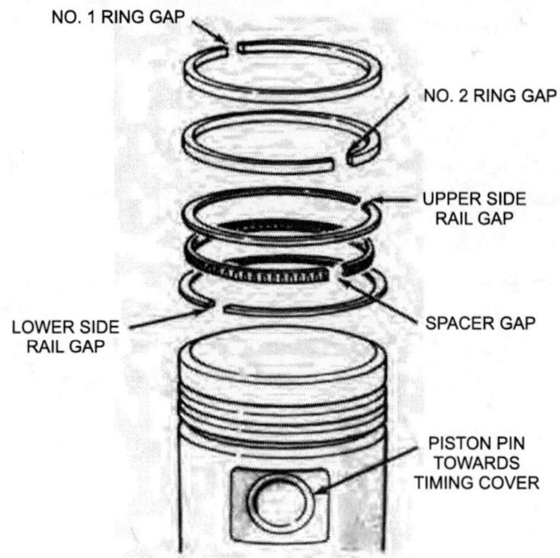

Fig. 12.113 *(a) Ring gaps must be staggered when installed on the piston and should retain the stagger in assembly, or blowby will be excessive. (Chrysler)*

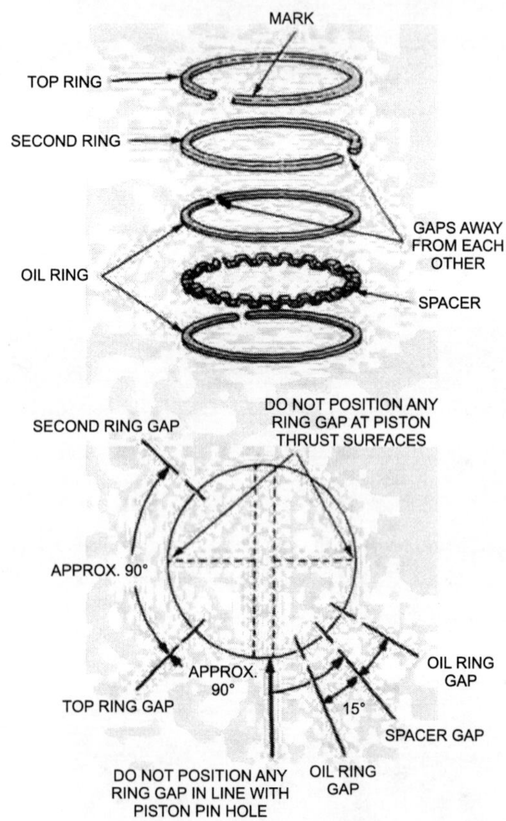

Fig. 12.113 *(b) Space piston ring gaps away from each other as shown in figure. (Toyota)*

After servicing, the rods and pistons are assembled. Use the shop manual to orient the pistons correctly with the rods, *Fig. 12.114.* A mark on the piston crown needs to be aligned with the rod. Aftermarket pistons will have directions packaged with the pistons.

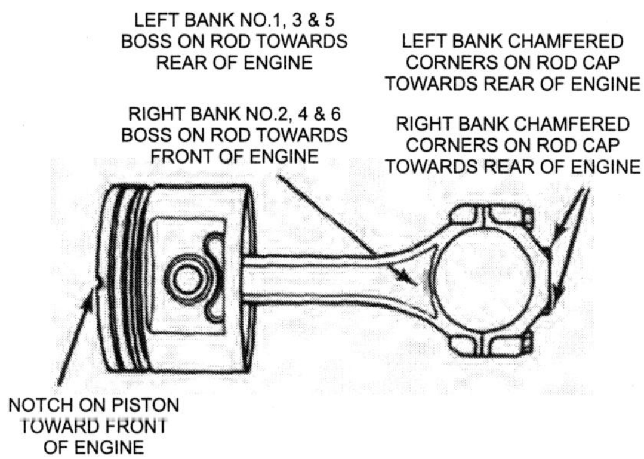

Fig. 12.114 *Typical piston – rod relationship.*

Depending upon the piston pin style, there are choices of assembly procedure.

Rod Alignment

After the rod and piston assembly is fully serviced and assembled, the rod is checked for alignment. Normal service can bend or twist the connecting rod. These unwanted bends can be safely bent out as long as the rod is not heated during bending.

Two fixtures are used for checking rod alignment. One supports the big end at one end and the piston at the other. A vernier-like scale is mounted next to the piston's support. If the rod is not straight, the scales markings will not lineup. The other tool uses a vertical plate, ground flat, *Fig. 12.115.* The piston assembly is placed upright in the fixture and passed by the plate. By watching for light to pass by the piston's holder, the mechanic can see any misalignment. Out-of-line rods are straightened with a large pry bar. It takes experience to know how much force to apply to align the rod.

12.28 INSTALLING PISTON AND ROD ASSEMBLY

Fit the rod bearing inserts into the connecting rods, *Fig. 12.116.* Fit the matching inserts into the correct caps. Then, wipe a good layer of oil on the face of the bearings, *Fig. 12.117.* See that the backs of the bearings must be clean and dry and should not be oiled.

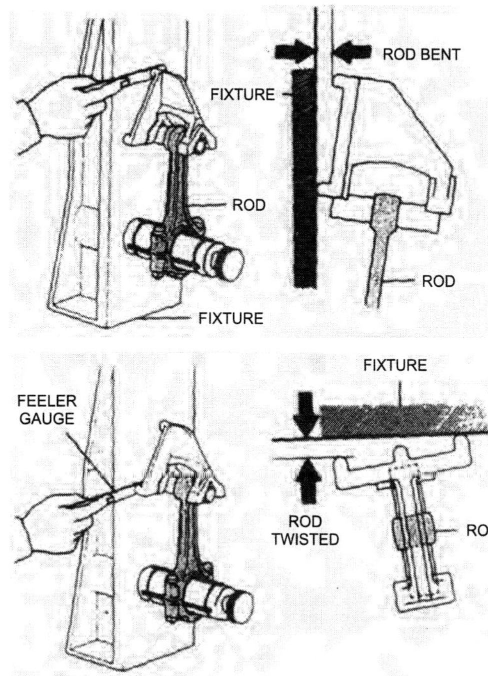

Fig. 12.115 *Method of aligning a rod with a special rod-aligning fixture.*

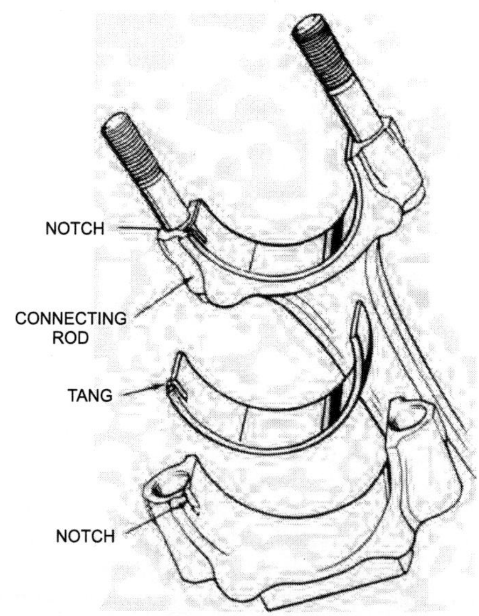

Fig. 12.116 *Fit clean and dry rod bearing inserts into connecting rod body and cap. Carefully check that the bearing insert is square in its housing bore. (General Motors Corp.)*

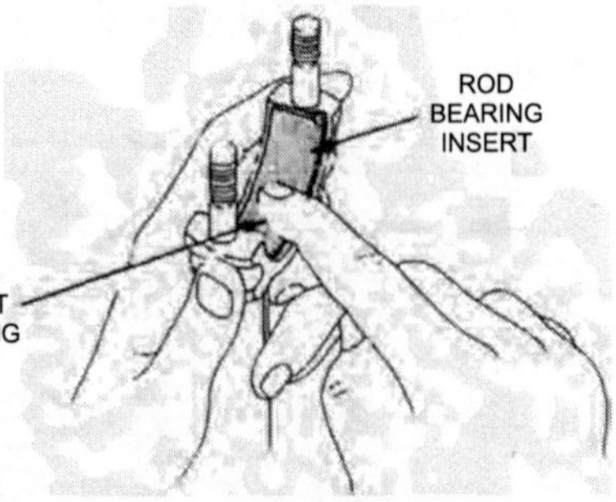

Fig. 12.117 *Wipe thick lubricating oil or white grease on top of rod bearings and use care not to allow any oil or grease on the back side of inserts. (Toyota)*

To install a piston assembly in its cylinder, dip the head of the piston and new rings into clean engine oil. Make sure that the ring gaps are staggered properly.

Clamp a ring-compressor around the rings, as shown in *Fig. 12.118.* Hold the compressor square on the piston, while tightening the compressor. The small indentations around the edge of the compressor should face the bottom of the piston.

Slide *plastic or rubber hoses or special rod bolt covers* over the connecting rod bolts, *Fig. 12.118.* This is required to prevent the rod bolts from possibly scratching the crankshaft journals. Turn the crankshaft until the corresponding rod journal is at BDC.

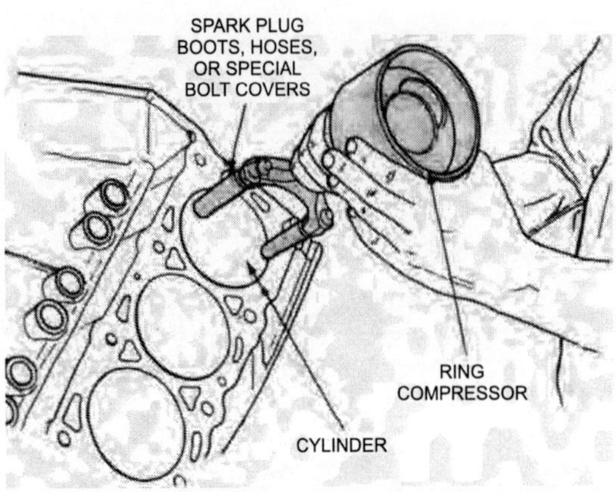

Fig. 12.118 *Install ring compressor over piston and rings. Tighten it down moderately. Verify that the rod and piston alignment marks are correctly aligned. Slide protectors over rod bolts so hat these do not inflict any type of damage to the cylinder or crank. Fit the piston and rod assembly into block. (Ford Motor Co.)*

Check the markings on rod and piston to ensure that the rod is facing the right direction and that the cap number matches the rod number. Also check that the piston notch or arrow is facing the front of the engine, e.g., if the rod is #1, it would normally go in the very front cylinder with the piston marking in the front. In case of doubt consult the service manual.

Installing Piston and Rod in Block

Put the piston and rod into its cylinder. Carefully guiding the rod bolts over the crank with one hand. Hold the ring compressor squarely against the block deck, tap the piston into the cylinder with a *wooden hammer handle.* A wooden hammer handle is used to avoid any damage the piston head. Continue tapping until the rod bearing bottoms around its crank journal, *Fig. 12.119.*

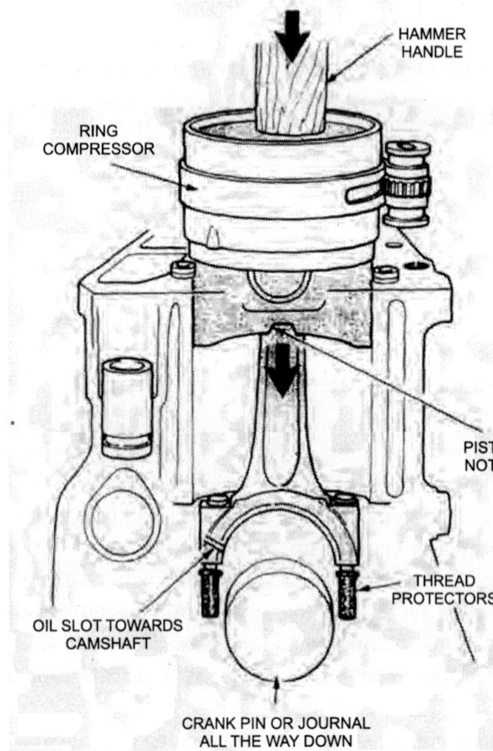

Fig. 12.119 *Use one hand to guide rod bolts over crank pin, and gently tap the piston down, with the back of wooden hammer handle, until bearing sits tight around crank. (Chrysler)*

If a piston ring pops out of the ring compressor, do *not* attempt to force the piston down into the cylinder. This would break or damage the piston ring or piston, *Fig. 12.120.* Instead, loosen and reinstall the ring compressor over piston and rings and start over again.

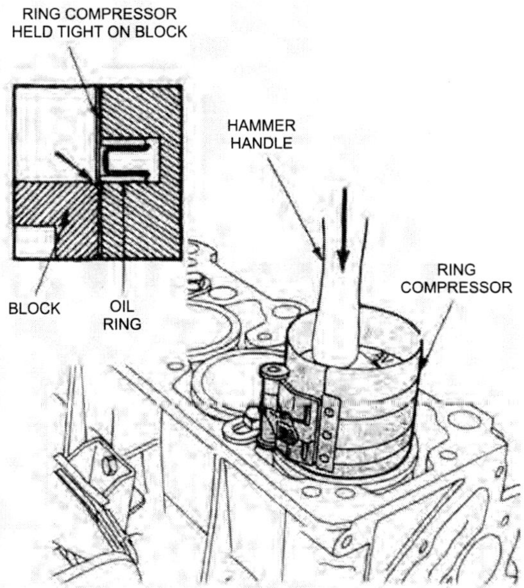

Fig. 12.120 *When tapping the piston down into the cylinder, stop if oil ring (being lower) pops out and catches on block. Reinstall the compressor correctly and start over again to avoid any ring or bore damage.*

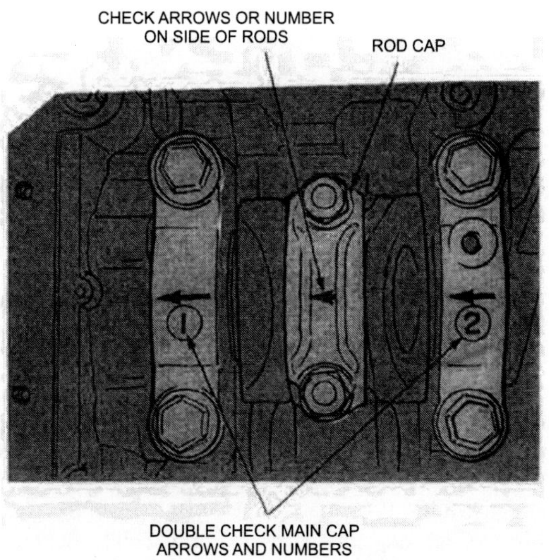

Fig. 12.121 *Carefully check that the rod caps and main caps are installed correctly, numbers/markings must match and point in correct direction. (General Motors)*

Double-check that the rod is facing in the right direction, *Fig. 12.121.*

Checking Rod Big End Bearing Clearance

Check the connecting rod big end bearing clearance using Plastigauge. Place a bead of Plastigauge across the crankshaft crank rod journal. Assemble the corresponding cap and torque the cap nuts to specs. Then remove the cap and smashed Plastigauge. Compare the width of Plastigauge to paper scale to determine the bearing clearance, *Fig. 12.122.* A typical rod bearing clearance is about 0.038 mm (.0015 in). This would equal a 1.5 reading on the Plastigauge paper scale in English Units or equal to 3.8 reading on the Plastigauge paper scale in Metric Units.

When installing a connecting rod bearing cap, make sure the rod and cap *numbers match,* i.e., if a rod is numbered with a 2, then the cap should also have 2 stamped on it. Mixing up rod caps can damage the bearings or crankshaft.

Torquing Connecting Rod Bearing Cap Nuts: Torque each rod nut or bolt to specifications. If over-tightened, the rod bolt could snap off during engine operation. This could cause severe block, crankshaft, piston, and cylinder head damage as parts fly around inside the engine.

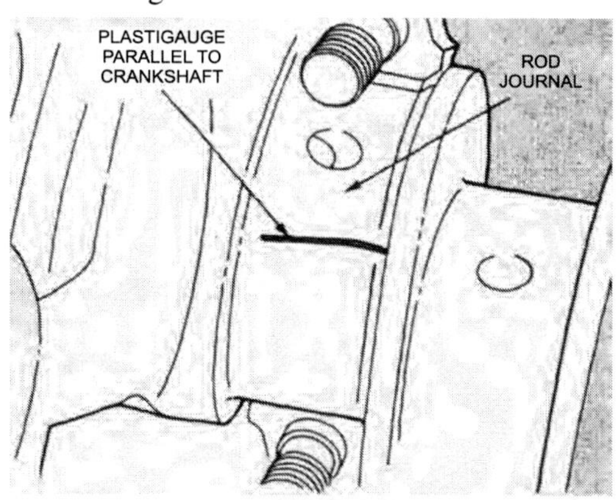

Fig. 12.122 *(a) Place a bead of Plastigauge on journal. Install and torque cap to specs and remove cap and Plastigauge.*

If a rod big end bolts/nuts are over-tightened, the bolts could stretch under load. This would allow the bearing insert to spin or hammer against the crank. This could cause serious crankshaft and rod damage.

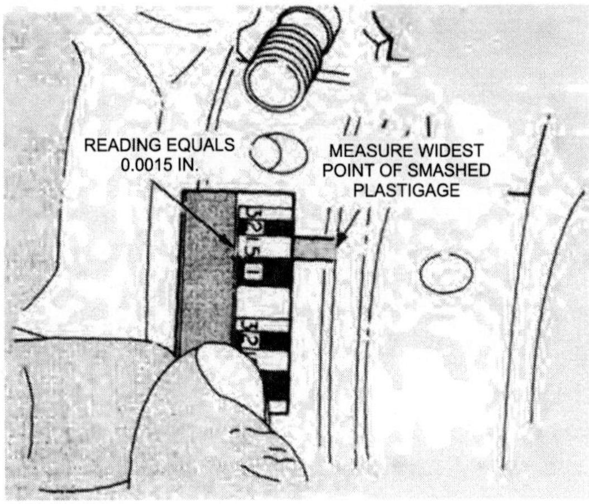

(b) Compare paper scale with smashed plastic bead. Here, reading of 15 means oil clearance of 0015 in. or 0.038 mm.

Fig. 12.122 *Checking rod bearing oil clearance. Plastigauge rod bearing as described for main bearing oil clearance. (Buick-GM)*

Install the rod bearing cap. If the bolts are used, place locking compound on the threads and tighten each cap bolt a little at a time, using a torque wrench. This will pull the cap down squarely. Keep increasing wrench torque until full specs are obtained. Double-check each rod bolt or nut torque (at least twice), *Fig. 12.123.*

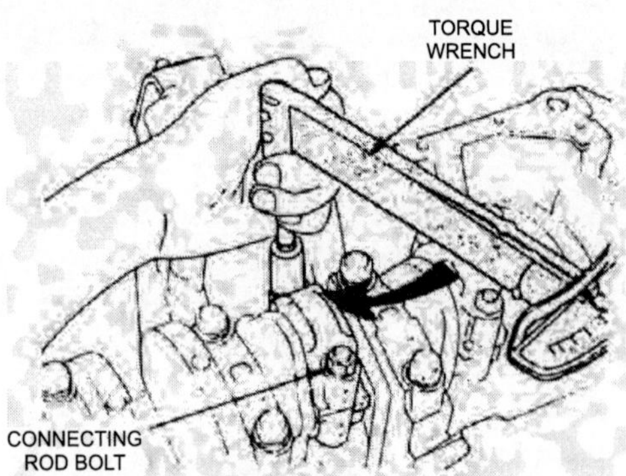

Fig. 12.123 *Torque the rod caps to specs. Tighten one bolt/nut and then the other, in small steps , until up to specs. Go over each bolt/nut a couple of times to ensure they are tightened correctly. Incorrect torque can lead to rod breakage and severe part damage. (Toyota)*

Checking Connecting Rod Side Clearance

Connecting rod side clearance is the distance between the side of the connecting rod and the side of the crankshaft journal or other rod. This clearance is measured by attempting to insert different size feeler gauge blades into the gap between the rod and the crank web, *Fig. 12.124.* The largest feeler gauge blade that slides into the space indicates bearing side clearance. Compare the measurements to specs that vary from 0.127 to 0.51 mm. If out of specs, crankshaft journal or connecting rod width is incorrect and need reconditioning.

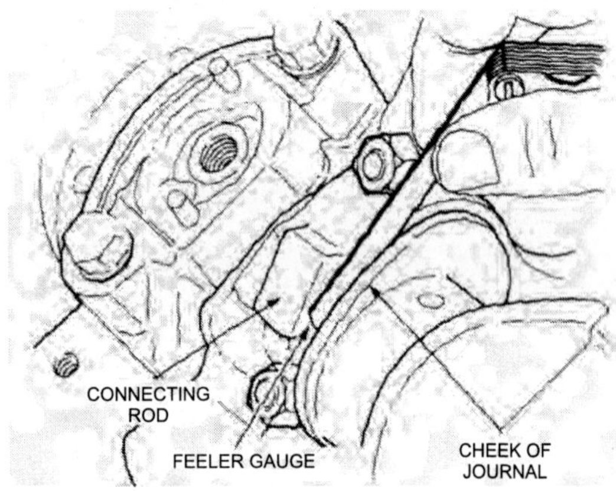

Fig. 12.124 *Use a feeler gauge to measure the rod side play. The largest feeler gauge blade that fits equals side play. (Pontiac-GM)*

12.29 ENGINE BALANCING

Engine balancing may become necessary when the weight of the new pistons, connecting rods, or crankshaft is different from that of the original parts. Engine balancing is required to prevent engine vibration in operation.

Engine balancing is done in certified automotive machine shops that have engine balancing equipment. Basically, the pistons, rings, piston pins, connecting rods, and rod bearings are weighed on an accurate scale. Material is machined and or ground off the pistons and rods until each assembly weighs the same. All rod big ends and small ends should also be equal in weight. Then the difference in the new and original weights of the assemblies is taken care of by adding or removing weight at compatible locations to balance the engine.

Proper engine balance is very important in present small, high speed, economy car engines. Engine balancing should be considered when major engine modifications are done.

Balance Shaft Service

Some late model automobile engines are equipped with balance shafts that are crankshaft or camshaft driven. These shafts are used to dampen normal engine vibrations. Dampening is reducing the vibration to an acceptable level. These shafts are normally located in the block or below the crankshaft depending upon the engine design. A balance shaft system designed to counterbalance vibrations on a 4-stroke, 4-cylinder inline engine is shown in *Fig. 12.125.* Here, two shafts are used, and they turn at twice the engine speed. One shaft turns in the same direction as the crankshaft, and the other turns in the opposite direction. The oil pump gears are used to drive the reverse-turning shaft. Counter weights on the balance shafts are positioned to oppose the natural rolling action of the engine, as well as the secondary vibrations caused by the piston and rod movements.

The shaft(s) would normally be removed during engine disassembly. To remove the balance shaft, begin by removing any retainers holding the shaft to the engine. After removing the shaft, remove the balance shaft bearings for inspection.

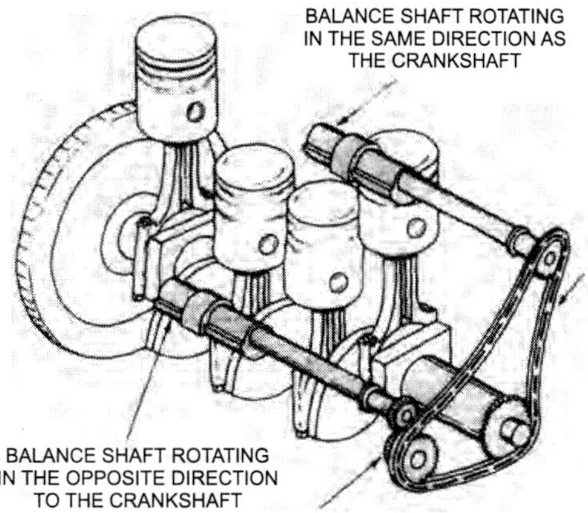

Fig. 12.125 *Two counter-rotating balance shafts are used here to counterbalance the vibrations of a 4-cylinder inline engine.*

Visually inspect the shaft for nicks, burrs, and wear. Check the bearings used with the balance shaft for wear. The balance shaft bearings should be replaced as a set if damaged or worn. Check the service manual for installation and other service procedures and any special tools needed.

THEORY QUESTIONS

1. How is #1 cylinder determined at the block of a V-type engine?
2. What is the difference between align boring and align honing?
3. What does "decking the block" mean?
4. What is the difference between deglazing and honing a cylinder?
5. What is the best method to use to clean an engine block after honing?
6. How can power honing be beneficial?
7. Why must the ring ridge be removed during an engine overhaul?
8. How do you check for main bore alignment?
9. Why should rod and main bearing caps be marked before they are removed?
10. How do you measure cylinder taper and out-of-round?
11. When is a cylinder bored?
12. A block should be bored if cylinder out-of-round is more than _______ mm or taper is more than _______ mm.
13. All blocks can be bored 1.524 mm. *True or False?*
14. Why is proper cleaning of a cylinder block so important after boring or honing?
15. How do you crankshaft main and rod journal wear?
16. What is an undersize bearing?
17. On a two-piece rubber rear main seal, the oil sealing lip should face the _______ of the engine.
18. Main bearing caps can be installed in different locations on the block without any problems. *True or False?*
19. How can the piston wear be measured?

20. When should have pistons knurled?
21. How is excessive top ring groove wear on a piston corrected? .
22. What could happen if new piston rings used in an engine do not have enough end gap?
23. What kind of bearings must be used with a crankshaft that has been reground?
24. List some consequences of excessive engine wear.
25. How can oil enter the combustion chamber of a worn engine?
26. What is a compression pressure test and how it is done?
27. What is cylinder leakage test and how it is done?
28. Name the three types of crankshaft wear.
29. Which end of the connecting rod typically wears most? What type of wear is normal?
30. Which piston ring normally wears the most and why?
31. Name five ways of increasing the size of a metal part.
32. _______ uses a ferrous powder and magnet to find cracks in parts.
33. Explain five ways of repairing cracks in cylinder blocks.
34. Why it is important to keep the connecting rod cap with the rod on which it was originally used, and to install it in the correct way?
35. When should you consider engine balancing?
36. To check a crankshaft journal for taper, the journal should be measured in at least _______ locations?
    (*a*) One (*b*) Two
    (*c*) Four (*d*) Six.
37. To check a crankshaft journal for out-of-round, the journal should be measured in at least _______ locations?
    (*a*) Two (*b*) Four
    (*c*) Six (*d*) Eight.
38. A telescopic gauge can be used to measure a cylinder bore if _______ is used to the telescopic gauge?
    (*a*) Micrometer (*b*) Feeler gauge
    (*c*) Straightedge (*d*) Dial indicator.
39. The block deck is the _______
    (*a*) Bottom (pan rail) of the block
    (*b*) Top surface of the block
    (*c*) Valley surface of a V-type engine
    (*d*) Area where the engine mounts are attached to the block.
40. Which engine block machining process should be done first when reconditioning?
    (*a*) Cylinder boring
    (*b*) Decking the block
    (*c*) Honing the cylinders
    (*d*) Align boring (honing).
41. Sleeving a cylinder means _______
    (*a*) Plating the inner walls of the cylinder with a different metal, such as nickel
    (*b*) Boring the cylinder to be 0.508 to 1.524 mm oversize to accept oversize pistons
    (*c*) Boring the cylinder to be oversize and installing a cast-iron sleeve to restore the cylinder to the original diameter
    (*d*) Using a hone to finish the cylinder after boring.
42. After decking, boring, and honing an engine block, what other metal-removing operation should be performed?
    (*a*) Align honing the main bearing caps and Saddles
    (*b*) Chamfering the cylinder bores and bolt holes
    (*c*) Broaching the timing chain cover surface
    (*d*) Filing the oil pan rails flat.
43. The minimum amount of material that is recommended to be removed from the cylinder after boring is _______
    (*a*) 0.254 mm (*b*) 0.82 mm
    (*c*) 0.0508 mm (*d*) 0.0254 mm.
44. Connecting rod caps should be marked (if they were not marked at the factory) before the piston and connecting rod assembly is removed from the engine
    (*a*) Because they were balanced together
    (*b*) Because they were machined together
    (*c*) To ensure that the lighter rod is matched to the lighter piston
    (*d*) To make sure that the heavier rod is matched to the heavier piston.
45. A misaligned connecting rod causes _______
    (*a*) Cylinder taper
    (*b*) Barrel shaped cylinders
    (*c*) Ridge wear
    (*d*) Angle wear on the piston skirt.
46. Piston ring gap should only be measured _______
    (*a*) After all cylinder work has been performed
    (*b*) After installing the piston in the cylinder
    (*c*) After installing the rings on the piston
    (*d*) Both (*a*) and (*c*).
47. A bearing shell has a slightly larger arc than the bearing housing. This is called _______
    (*a*) Bearing tang (*b*) Bearing saddle
    (*c*) Bearing crush (*d*) Bearing spread.
48. Bearing _______ occurs when a bearing shell is slightly above the parting surface of the bearing cap.

(*a*) Cap lock (*b*) Overlap
(*c*) Crush (*d*) Interference fit

49. A cast iron V-6 block is almost new and has very low kilometers on. However, a piston skirt broke off and gauged one cylinder. All of the other pistons are in good condition. Mechanic A says that the block should be bored and new pistons be installed to cleanup the cylinder damage. Mechanic B says that that the damaged cylinder be sleeved to restore it and a new piston it. Who is right?
(*a*) Mechanic A (*b*) Mechanic B
(*c*) Both A and B (*d*) Neither A nor B.

50. Check bearing clearance with:
(*a*) Dial indicator (*b*) Flat feeler gauge
(*c*) Plastigauge (*d*) Micrometer.

51. To find piston clearance, subtract _______ from _______.

52. Mechanic A says that checking for main bore and deck warpage is a typical procedure performed during short block service. Mechanic B says that during cylinder head milling is a common procedure performed during short block service. Who is right?
(*a*) A only (*b*) B only
(*c*) Both A and B (*d*) Neither A nor B.

53. Which of the following problems should be looked for when inspecting a cylinder block under service?
(*a*) Cylinder ring ridge
(*b*) Cylinder scoring
(*c*) Cracked cylinder
(*d*) All of the above.

54. Mechanic A says that a short block should be line bored when misalignment is over approximately 0.004 mm. Mechanic B says that a short block should be line bored when misalignment is over approximately 0.002 mm. Who is right?
(*a*) A only (*b*) B only
(*c*) Both A and B (*d*) Neither A nor B.

55. Mechanic A says that block deck warpage is measured with a straightedge and feeler gauge. Mechanic B says that block deck warpage is measured with an inside micrometer and a dial indicator. Who is right?
(*a*) A only (*b*) B only
(*c*) Both A and B (*d*) Neither A nor B.

56. Mechanic A says that a brush hone should be used to true a cylinder when its wear is within specs. Mechanic B says that a ridge hone should be used to true a cylinder when wear is within specs. Who is right?
(*a*) A only (*b*) B only
(*c*) Both A and B (*d*) Neither A nor B.

57. Mechanic A says that many machine shops commonly bore cylinders 0.762 mm oversize. Mechanic B says that many machine shops commonly bore cylinders 1.5 mm oversize. Who is right?
(*a*) A only (*b*) B only
(*c*) Both A and B (*d*) Neither A nor B

58. Mechanic A says that block boring reduces engine compression. Mechanic B says that block boring increases engine compression. Who is right?
(*a*) A only (*b*) B only
(*c*) Both A and B (*d*) Neither A nor B.

❑❑❑

Engine Reconditioning – Top End Service

13

13.1 INTRODUCTION

In this chapter, important steps for reconditioning a cylinder head and other parts associated with it, except the 'short block components' are described. *Fig. 13.1* shows common parts in a typical cylinder head, and their usual relationships to the cylinder head. Some important things to learn, check and correct are also labelled. A cylinder head has to close the cylinders by sealing the top of the block and control flow of charge and products of combustion in and out of the engine cylinders. It is designed to allow adequate cooling and lubrication to keep the engine functioning for good life. Due to tremendous heat of combustion and high pressures, after prolonged service, heads may warp, valves can burn, valve train parts can wear, sealing gaskets and seals may start leaking, bearings can wear, and the engine performance may suffer. Here, methods to return these parts to specifications to restore engine power, smoothness, and fuel economy will be discussed.

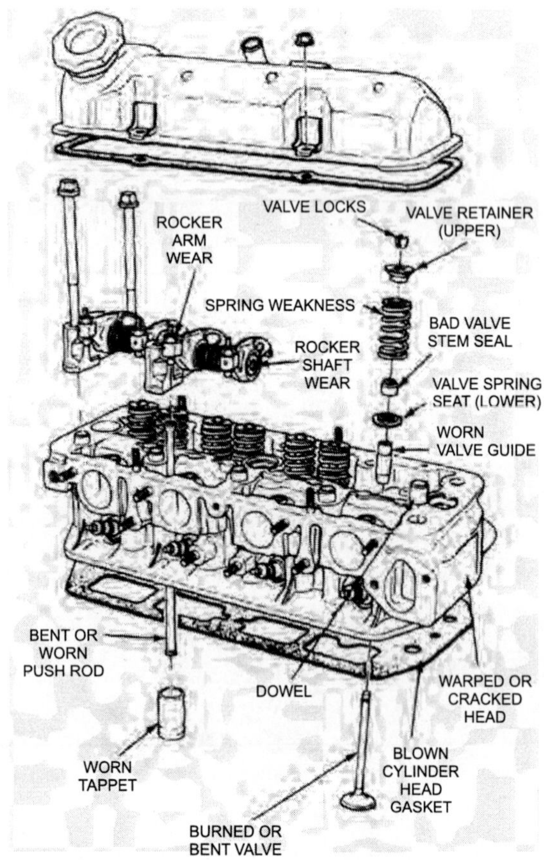

Fig. 13.1 *Common components of a typical Top End of an automotive engine. (Chrysler Corp.)*

13.2 CYLINDER HEAD AND VALVE TRAIN SERVICE

Cylinder head and valve train servicing involves disassembling and cleaning of the components and then inspecting each component, very closely, for wear and any other problems. The valve guides, valves, valve seats, and other valve train components must be reconditioned. The procedures used to service components are presented in the following sections.

Disassembly and Cleaning

Cylinder head and valve train service begins with disassembling the cylinder head. Remove the rocker arms first. Remove the valves by compressing the valve springs using a valve spring compressor, *Fig. 13.2 (a)*. One end of the compressor fits over the valve head, and the other end fits over the valve spring retainer. Levering the compressor brings the two ends together, thus compressing the valve spring. Compress the valve springs just enough to free the keepers. When the springs are compressed, the retaining washers and locks may be removed. Remove the keepers with a small screwdriver or magnet if hand fingers are too large, *Fig. 13.2 (b)*. Release compressor and remove spring assembly from valve. Repeat this procedure on other valve assemblies. Without the keepers, the retainer, spring, and spring seat will come off. Remove the old valve stem seals. Finally, push the valve out of the guide. If the valve stem tip is *mushroomed,* pounded wider than stock by the rocker arm, file the wide portion off. Attempting to pull a mushroomed valve tip through the valve guide will ruin the guide. The valve assemblies, including the rocker arms, should be arranged in order so that they may be reinstalled in the same position. Valve stems may be select-fit from the factory. After disassembly,

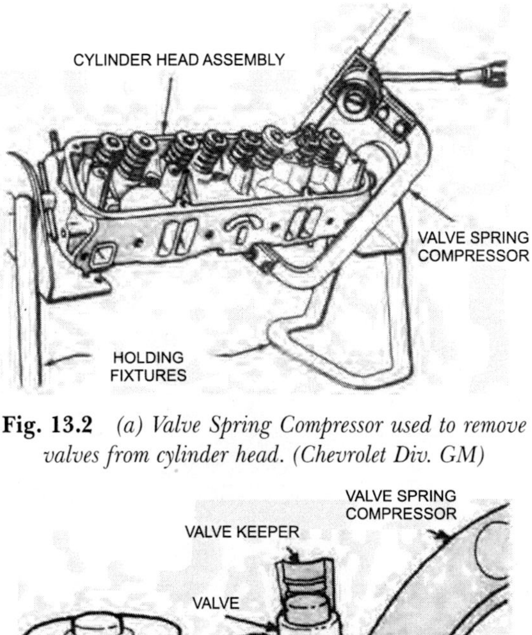

Fig. 13.2 *(a) Valve Spring Compressor used to remove valves from cylinder head. (Chevrolet Div. GM)*

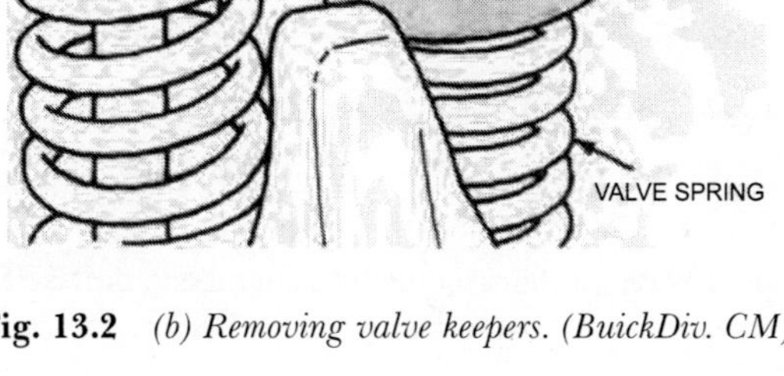

Fig. 13.2 *(b) Removing valve keepers. (BuickDiv. CM)*

clean the cylinder head in a hot tank (along with the cylinder block) if it is made from cast iron. Use a cold tank if the cylinder head is made from aluminium. Wear eye protection when using a valve spring compressor. If it were to slip off the retainer, spring pressure could make parts fly into your face.

Carbon deposits not removed from the combustion chamber by the tank cleaning may be cleaned off with a wire brush in a drill motor, *Fig. 13.3.* The valve guides are cleaned with a valve guide cleaner made of several flat spring tension blades that press against the guides and scrape out the carbon deposit without scratching metal. The tool may be driven by hand or with an electric motor as shown in *Fig. 13.4.*

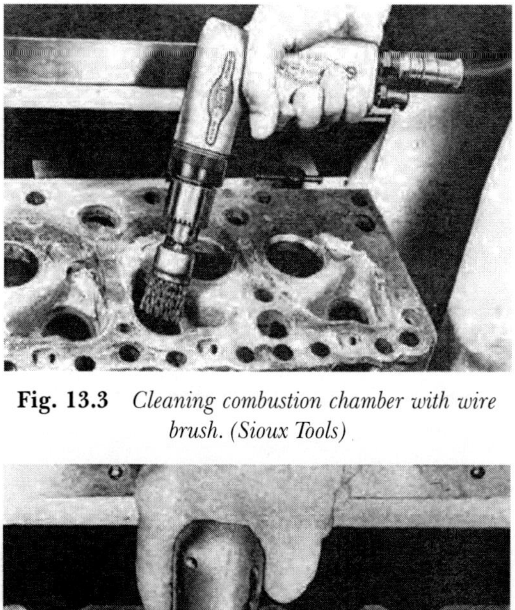

Fig. 13.3 *Cleaning combustion chamber with wire brush. (Sioux Tools)*

Fig. 13.4 *Cleaning valve guides. (Sioux Tools)*

Measuring and Inspecting for Wear

The cleaned parts are then measured and inspected for wear. The cylinder head is checked for flatness with a straight edge and feeler gauge, *Fig. 13.5.* Place the straight edge in several different directions across the cylinder head. Use a thin feeler gauge to determine if there is any space between the straight edge and the cylinder head surface. If the feeler gauge will pass under the straightedge when it is pushed firmly against the cylinder head, the cylinder head is warped. Grinding it with a special surface grinder may resurface a warped cylinder head.

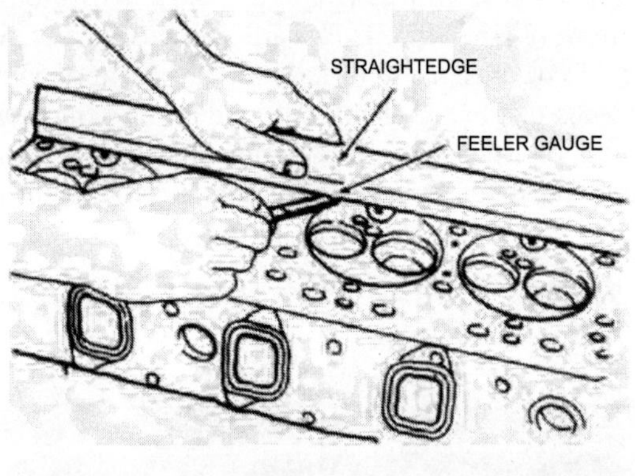

Fig. 13.5 *Measuring cylinder head for flatness. (Chevrolet Div. Genera! Motors)*

Carbon deposits are removed from a valve by holding it against a wire brush mounted on a grinder. Clean the head, face, and stem thoroughly, particularly the area from face to stem. Carbon deposits can make a valve overheat, as it will restrict the heat flow.

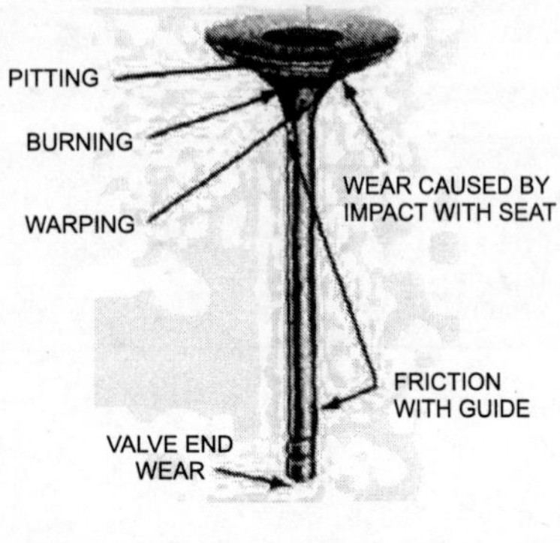

Fig. 13.6 *(a) Types of valve wear. (Sioux)*

Inspect the valve, valve seat, and guide for wear and damage. Even in normal use these components absorb a lot of abuse. As a result, the valve, especially the exhaust valves, may become pitted, burned, warped, or grooved. Then the valve begins to leak compression and hold heat. There is also wear at the stem from friction with the guide and wear from contact with the rocker arm, *Fig. 13.6 (a)* and *(b).*

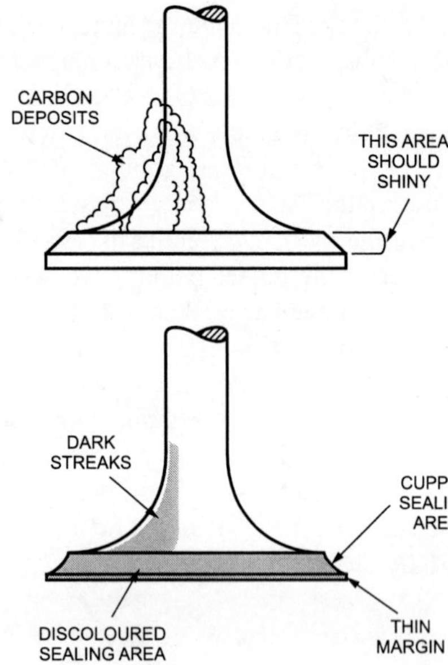

Fig. 13.6 *(b) Valve Inspection.*

Valve seat also wears. Hot gases burn it and carbon particles, which retain heat, pit the valve seat. The valve guide wears against the valve stem. Between stem and guide carbon builds up, which causes the valves to stick. Typical wear in these areas is shown in *Fig. 13.7.*

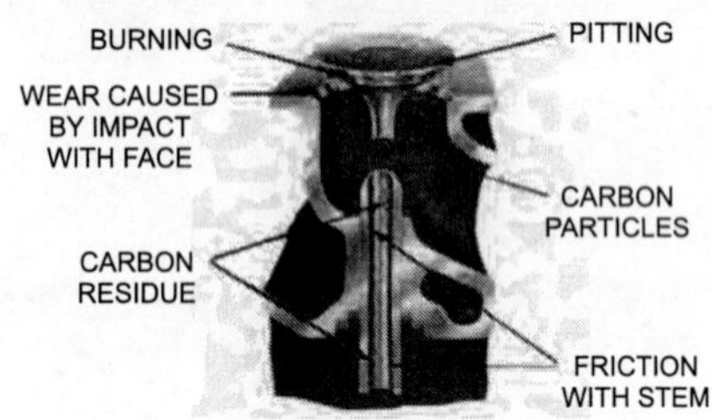

Fig. 13.7 *Wear valve, valve seat and valve guide. (Sioux Tools)*

13.3 CYLINDER HEAD PROBLEMS

Cylinder head service begins with cleaning. Wire-brush the combustion chambers. Bead blasting quickly cleans the combustion chambers, ports, EGR passages, and exhaust crossovers. Scrape all gasket residues and clean all bolt holes. All bead residues must be removed with a thorough hot tanking or solvent cleaning when finished with bead blasting.

Mount the head on a stand and check for cracks, burned seats, burned deck surface between chambers, cracked valve guides, any distortion in head, and any other problems.

Checking for Cracked Head

A *cracked head* can be caused by engine overheating or physical damage from other broken parts. Normally, the crack will occur at an exhaust seat or between chambers, causing coolant or pressure leakage. This can lead to further engine overheating, coolant in oil, etc.

Magnafluxing is commonly used to find cracks in cast iron cylinder heads. Magnets are placed on the head where a crack might occur, *Fig. 13.8.* Then spread fine iron powder over the head using a squeeze bulb. Any crack present will set up a small magnetic field that will attract the iron powder to make it visible.

Dye penetrant may be used to locate any cracks in aluminium (nonferrous material) cylinder heads. This process requires a black light to locate the crack(s). Always follow manufacturer's instructions to use the right procedure to find cracks in cylinder head under service. The cylinder head may be sent to a machine shop for crack detection and determine if it can be reused on the engine.

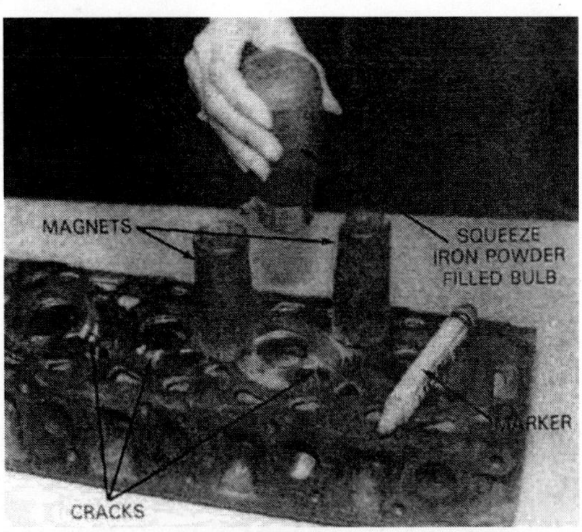

Fig. 13.8 *(a) Magnafluxing a cylinder head. Note that cracks have been found in many places.*

Fig. 13.8 *(b) Electromagnet has found a crack across combustion chamber.*

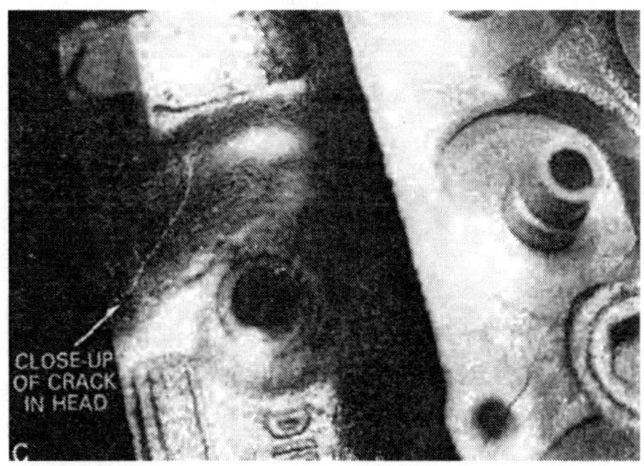

Fig. 13.8 *(c) Determining cracks in cylinder head. (Midwest Crankshaft, K-Line Tools, and Magnaflux)*

Repairing a Cracked Head

A cracked cylinder head may either be welded, plugged by drilling a series of small holes and using metal plugs to fix the crack(s), or be replaced. With expensive cylinder heads, welding may be appropriate. With most cracked heads, replacement is most economical.

Welding of cast iron heads needs special welding skills that require a specialty shop and a skilled technician. The head is heated in a furnace and a high-nickel content welding rod is used to weld repair the crack. Aluminium heads do not need preheating in a furnace and can be repaired by welding using a heli-arc welder.

Cylinder Head Warpage

A *warped head* has actually become curved slightly due to overheating. This is more common with late model thin cast aluminium heads. A warped head can cause a head gasket failure as a result of uneven clamping pressure on the head gasket. This can cause coolant leakage, compression loss, and noise etc.

Measure the head mating surface flatness with feeler gauges and a straight edge, *Fig. 13.9.* The head must be very clean, or the straight edge will sit high and give an inaccurate reading. Slip feeler gauges under the straightedge until a feeler gauge is found that just fits. The thickness of this gauge is the amount of head warpage.

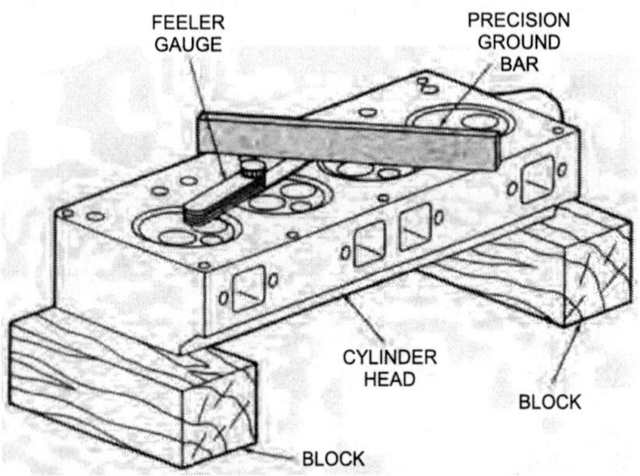

Fig. 13.9 *Measuring the head mating surface flatness with Feeler gauges and a Straightedge.*

Check head warpage in different locations across the head surfaces, *Fig. 13.10.* The most common place warpage shows up is between the two centre combustion chambers. Check the measurements against specifications. Generally head warpage should not exceed about 0.08 to 0.13 mm. If warpage is beyond the service limit, the head must be milled flat again.

With an overhead cam head, cam bore should also be checked with a straightedge and feeler gauges. Place the straightedge in the bore and try to slip feeler gauge blades under it. If warped beyond specs (over 0.03 mm), have a machine shop line bore the head cam bores.

Cylinder Head Milling: Heads are milled in a special milling machine. Only enough material is removed off the head gasket or deck surface of the cylinder head to get the head flat again. Check manufacturer's specs to ascertain how much metal can be milled from the specific head. Milling increases the compression ratio and reduces head thickness. Higher compression ratios may cause preignition and reduced head thickness can alter cam timing in OHC engines.

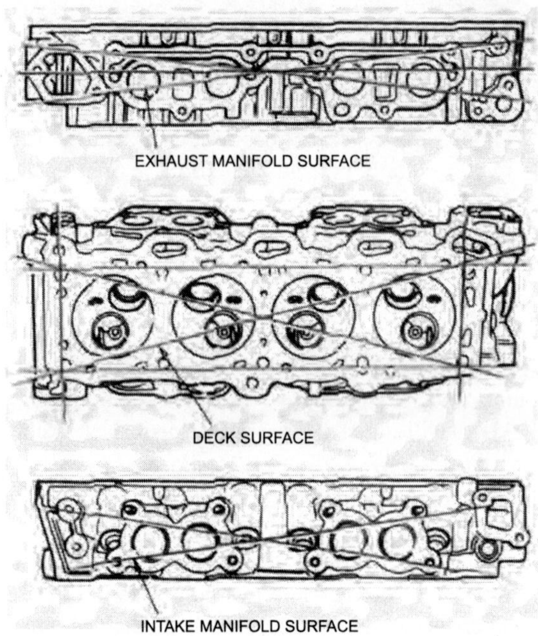

Fig. 13.10 *Check head for warpage at these locations. (General Motors)*

On V-type engines, milling the heads lowers them in relation to the intake manifold. Therefore, once the heads have been milled to a certain depth, the intake manifold must be milled on its bottom and mating surface to match the heads. Always mill both heads the same amount on V-type engines, or the compression ratio will be different on the two cylinder banks.

Cylinder head milling of a diesel engine is very critical. Diesel engine compression is already so high; any milling can reduce clearance volume and increase compression pressure beyond acceptable limits. Some diesel engine cylinder heads are case hardened and cannot be milled. They must be replaced when warped. Follow manufacturer service manual directions before milling the cylinder heads.

13.4 DIESEL PRECHAMBER SERVICE

Automobile diesel engines use precombustion chambers. They are small chambers pressed into the engine head. The tips of the diesel injectors and glow plugs extend into the precombustion chambers. After prolonged use, it may be necessary to remove the precombustion chambers for cleaning or replacement due to damage.

A *brass drift* and *hammer* are usually used to remove a diesel engine precombustion chamber, *Fig. 13.11 (a)*. The drift may be inserted through the glow plug or the injector hole in the head. Light blows with the hammer will drive out the chamber. Compare new precombustion chamber dimensions to the old one, *Fig. 13.11 (b)*.

When installing a precombustion chamber, be careful not to damage the chamber. Use a special driver or brass hammer to tap the unit back into the head, *Fig. 13.11 (c)*. Hammer only on the outer edge of the chamber. The centre area of the precombustion chamber could cave in or dent from hammering. Make sure the precombustion chamber is perfectly *flush* with the deck surface of the cylinder head, otherwise, head gasket leakage can occur, *Fig. 13.11 (d)*.

13.5 VALVE TRAIN SERVICE

Adjusting Valves

The most common valve train service is adjusting valves. During normal engine operation, valve clearance can become larger or smaller. Wear on valve stems, rocker arms, pushrods, and lifters result in increased clearance. Wear at the valve face and seat area results in decreased valve clearance. All valves not used with hydraulic lifters need periodic adjustment. Such adjustments compensate for temperature and wear changes in the valve train.

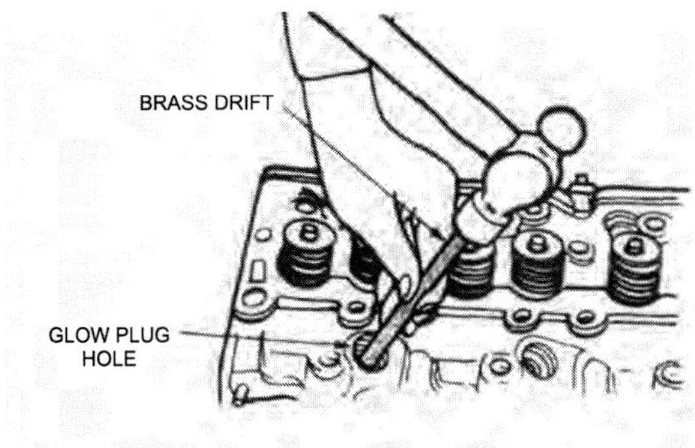

(a) Drive out chamber through glow plug or injector hole

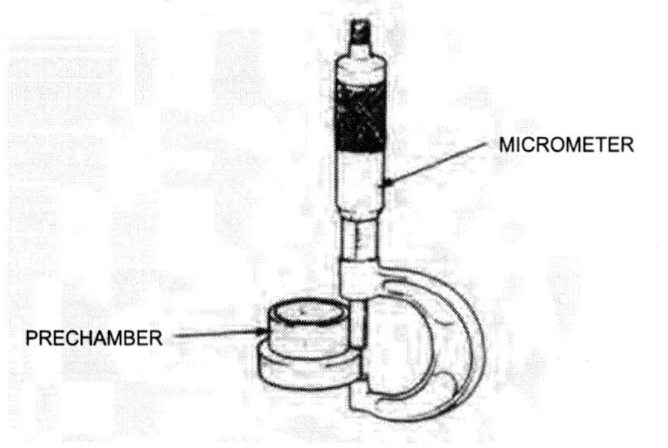

(b) Make sure new prechamber has correct dimensions.

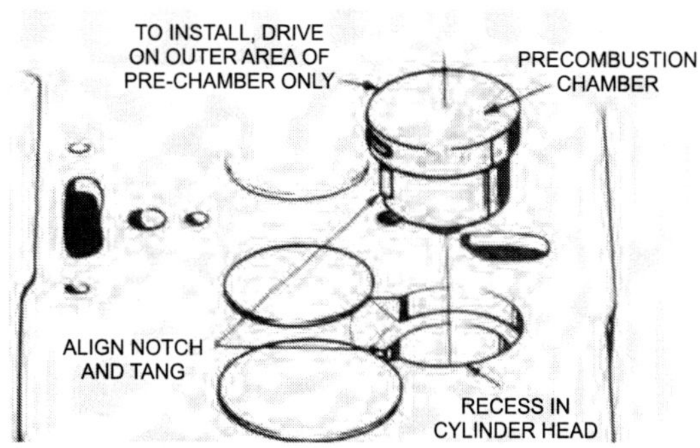

(c) Align prechamber and drive in without damaging it.

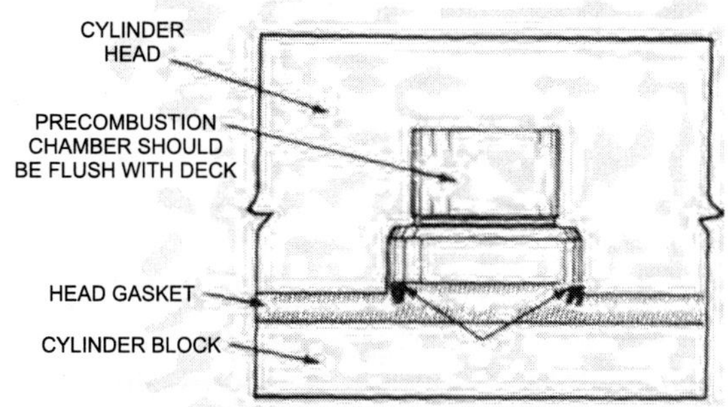

(d) Prechamber should be flush with cylinder head deck. If not, head gasket leakage could result.

Fig. 13.11 *Servicing diesel prechamber. (General Motors Corp. and Fel-Pro)*

Temperature changes are due to engine heat. As the engine warms up, the parts in it expand. Because of cylinder block and head expansion, the distance from the camshaft centreline to the rocker arms shaft centreline in OCV engine increases with rise in temperature with heat. When the distance between the rocker arm tip and the valve stem tip is measured with feeler gauges, an increase of about 0.0508 mm may be noticed.

This increase is reduced slightly by the growth of the valve, rocker arm, push rod, lifters, and camshaft lobes. When these parts get hot, they tend to close the valve clearance. But there is much more growth in the block and head, so that the net result is an increase in the valve clearance.

In an OHC engine, valve clearance changes because of the heat growth are smaller than those in OHV engines as the distance between camshaft and valves is so much shorter. Only head to head and pedestal expansion changes the valve clearance in OHC layouts. Typical growth in their case is about 0.025 mm.

Some clearance between the cam lobe heel and lifter is very important. If there is not some clearance, the valve will not be able to close all the way. When the valve remains partly open, combustion gases will gradually burn the valve away. The valve held off its seat gets very hot because it is not able to move heat away from the valve head into the valve seat in the cylinder head. The result is a melted valve head. This is called a *burned valve.* Once a valve is burned, it cannot seal perfectly. Then combustion gases continue to erode the valve, even if clearance is restored between the lobe's heel and lifter. If the burned valve is not repaired immediately, the valve head may separate from the stem and fall into the cylinder, resulting in cylinder wall and piston damage.

Normal wear, millions of openings and closings, heat cycling, and exposure to combustion gases all help erode the valve face. With a partially eroded valve face, the valve sits deeper in its seat, *Fig. 13.12.* At the valve stem, the distance between the valve stem tip and rocker arm or tappet is then closed. If the designed valve clearance is not maintained, the valve will eventually take up all the valve train slack. The valve stays open and burns.

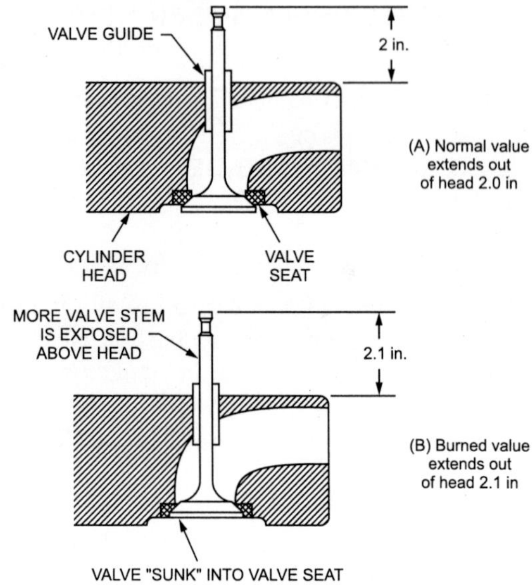

Fig. 13.12 *Valve burning increases the exposed portion of the valve stem.*

Valve Adjustment

To get maximum valve life, the specified valve clearance must be maintained. The valve clearance is a distance, which ensures that the valve will not be held open by too close a clearance, nor will the clearance be so wide as to allow the valve slam shut.

If the valve lash is too wide, there will be engine noise and wear on the camshaft and lifter contact faces. Push rods may be bent. Eventually valve timing is affected and the engine performs poorly.

Before starting the valve adjustment, the correct service literature must be consulted. Refer the service manual to check if the valve clearance specification is for a cold or a hot engine. If a cold engine is called for, let the vehicle sit overnight. If this is not possible due to time constraint, at least let the engine cool until the cylinder head is cool to the touch. If a "hot" engine is specified, warm up the engine to the operating temperature.

On some engines, intake and exhaust valves have different clearance specifications. This is due to the higher exhaust valve temperatures and the use of different expansion rates in the two valves. If specifications differ, the exhaust valve clearance will normally be larger. The service manual will usually identify which valves are intake and exhaust.

The applicable service manual will give a valve adjusting directions. It ensures the valve being adjusted is not on the lobe's toe. A valve can be adjusted only when the lifter is against the lobe's heel. If the adjusting sequence is not available, bring cylinder #1 to *Top Dead Centre* (TDC) on its compression stroke. This is close to when the spark plug fires, a time when both the exhaust and intake valves are closed. Adjust both valves on cylinder #1 and then move on to the next cylinder in the firing order. Bring it to TDC on its compression stroke and adjust its valves. Continue through the firing order until all valves are adjusted.

After the engine is at the proper temperature, the first step in most valve adjustments is to remove the rocker cover or valve cover. In an I-head engine with adjustable rocker arms, valve clearance is measured by inserting a feeler gauge between the rocker arm and valve stem, *Fig. 13.13 (a).* The clearance should be measured when the lifter is resting on the heel of the cam. The engine service manual will specify how to position the engine so that the valves are in the correct position for valve adjustment. A feeler gauge is selected that represents the clearance specification. If the feeler gauge does not slide into the gap, the adjustment screw should be turned to increase the valve clearance. If the feeler gauge fits loosely into the gap, the adjustment screw must be turned to reduce the clearance. The feeler gauge should fit with a light drag.

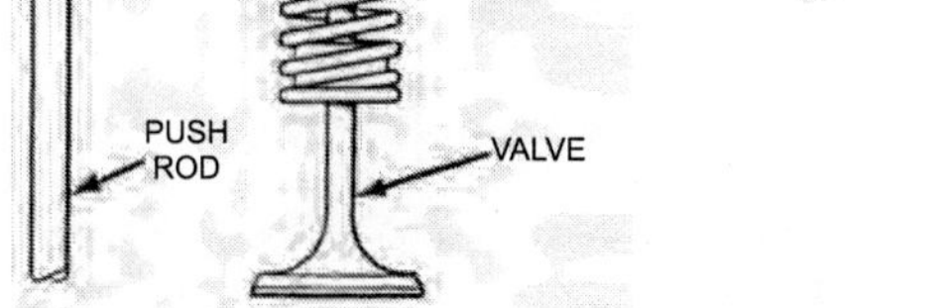

Fig. 13.13 *(a) Measuring and adjusting valve clearance between the rocker arm and valve stem.*

Ball and stud rocker arm assemblies should be adjusted by turning the nut on top of the stud, *Fig. 13.13 (b).*

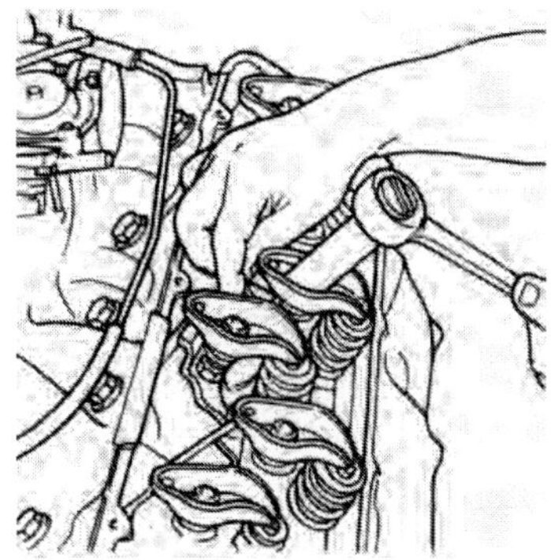

Fig. 13.13 *(b) Adjusting valve clearance on ball and stud Rocker arm. (Chevrolet Div. GM)*

Most overhead camshaft engines have adjustable lifters or adjustable rocker arms. Measure the valve clearance between the valve stem tip and rocker arm, *Fig. 13.13 (c),* or between the cam lobe and bucket tappet. If a jam nut and screw adjusters are used, loosen the jam nut with a box end wrench. Then turn the adjusting screw with a screwdriver until there is a slight drag on the feeler gauge. It takes experience to know just how tight a "slight drag" is, so have a trained skilled person adjust a valve and check your work against that valve's clearance and "drag".

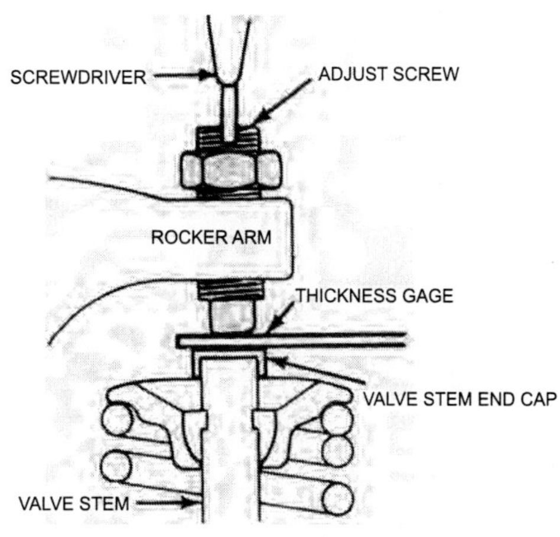

Fig. 13.13 *(c) Adjusting valve clearance on an overhead camshaft engine. (Chevrolet Div. General Motors)*

When the drag on the feeler gauge is correct, tighten the jam nut. Tightening the jam (lock) nut usually loosens the valve clearance because the lock nut stretches the adjusting screw, slightly. Correct the loosened valve clearance by very lightly overtightening the adjusting screw and then tightening the lock nut. Stretch from the tight lock nut will bring the valve clearance to the proper specs. This stretch is why you must always check the valve clearance after tightening the lock nut.

Pedestal mounted rocker arms are adjusted by loosening and tightening the rocker arm nut. Measure between the valve stem tip and rocker arm, as with any other rocker arm. Turn the nut until the feeler gauge drag is correct. The valve is then adjusted.

Some overhead camshaft engines use valve-adjusting discs between the cam follower or lifter and the cam lobe, *Fig. 13.13 (d)*. Thicker or thinner discs are installed by pushing down the follower and compressing the valve spring with a special tool. A small magnet is used to pull out the disc. The feeler gauge is inserted between the cam and the lifter as shown in *Fig. 13.13 (e)*. Make sure that the camshaft is turned so the heel of the cam lobe is on the lifter.

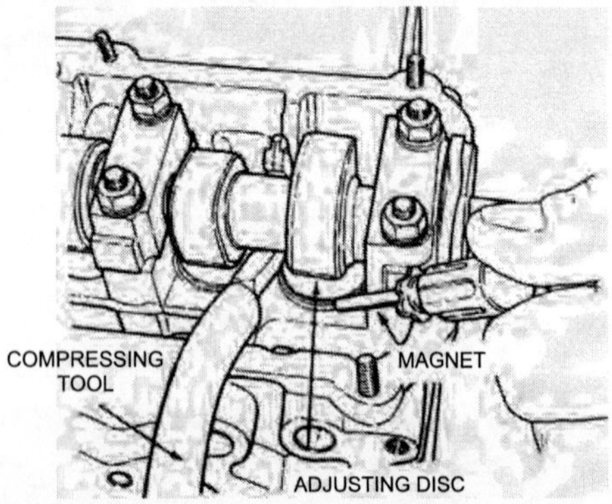

Fig. 13.13 *(d) Adjusting overhead camshaft valve clearance with adjusting discs. (Chrysler Corp.)*

Fig. 13.13 *(e) Measuring valve clearance on an overhead camshaft with adjusting discs. (Chrysler Corp.)*

Bucket tappets require feeler gauge measurements between the lobe heels and tappets. Check all valves, rotating the engine as necessary to keep the tappet being checked on the lobe's heel. Write the measurements down. Now subtract the specified valve clearance from each one of your measurements. The difference is the shim thickness necessary to bring that valve to specs, *Figs. 13.13 (f)* and *(g)*. Get the adjusting shims from the vehicle manufacturer.

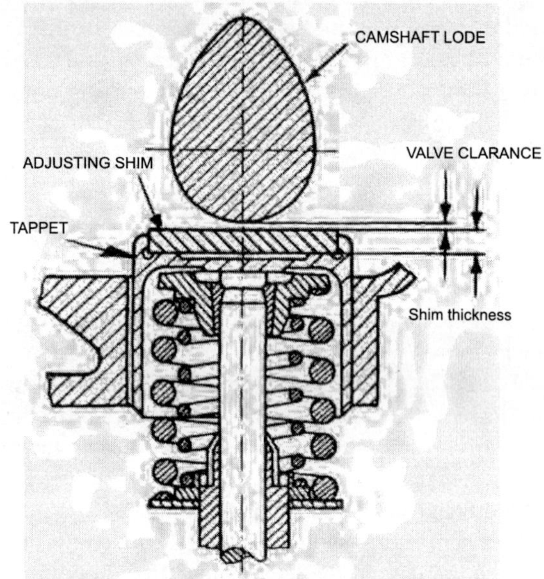

Fig. 13.13 *(f) Adjusting shim between camshaft lobe and tappet.*

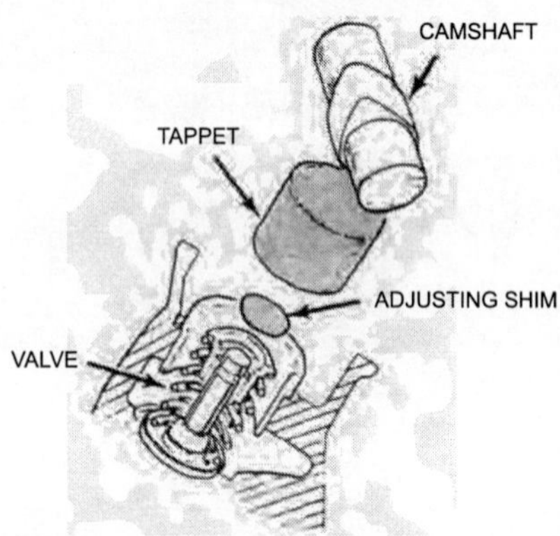

Fig. 13.13 *(g) Bucket tappet and adjusting shim*

To install the shims, you may have to remove the camshaft. Then lift the tappets from their bores in the head, place the shims in the tappet's recess, and replace the tappets. Reinstall the camshaft and check the adjusted valve clearance.

If the shims fit on top of the tappet, rotate the cam lobe out of the way. You will need a special tool to depress the tappet as well. Then you can slip the shim in and release the tappet.

After adjusting the valves, replace the rocker cover, using a new gasket, and torque the screw to specs.

Leakage Testing

The best diagnostic test of valves and how they are sealing is the *leakage test, Fig. 13.14.* To do a leakage test, pump compressed air into the cylinder and see how mush air leaks out. A leakage test quickly rates valve sealing. The test needs a leakage tester. The engine should be warmed up, and then stopped. Bring the cylinder to be tested to TDC on its compression stroke so that both valves are in closed position. Screw an adapter into the spark plug hole and tester attached to it. Attach a compressed air line from the shop air compressor reservoir to the other end of the tester. Compressed air flows into the cylinder through the tester.

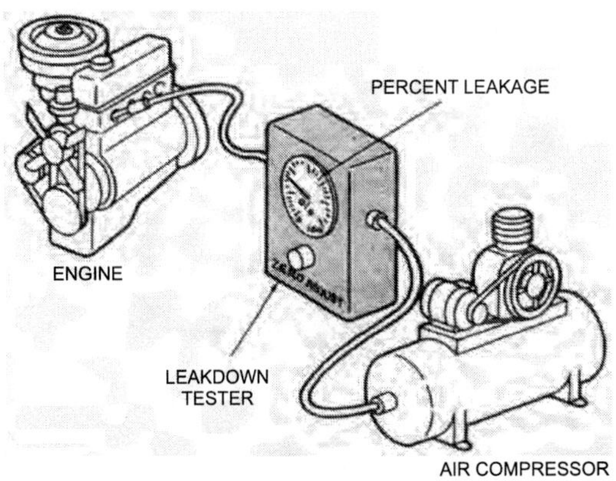

Fig. 13.14 *Leak down Test.*

Even in the best-sealed cylinder, some air will leak past the rings. The tester shows this leakage in a percentage. A good cylinder leaks below 10%. If the valves, or rings are not sealing, as they should, the leakage will be higher. Between 10 and 20% indicates a problem but not one, which needs immediate attention.

If the rings are leaking, you will hear air escaping by listening at the oil filter or dipstick holes. If an intake valve is leaking, listen for a hiss at the carburetor. Opening the throttle helps. A bad exhaust valve leak can be heard at the exhaust pipe. Bubbles in the radiator or noise from the next spark plug hole indicates a cracked head or blown head gasket.

As applicable to valves, the exhaust is typically at fault. By the time an average driver brings the vehicle in for a diagnostic check, the valve will be badly burned. Expect around 30% leakage for a moderately burned valve. Badly burned valves, or those missing a small chunk, will leak close to 138%.

13.6 VALVE JOB

When valves are worn to the point where they are unable to seal, it is time to perform a valve job. In a valve job the head is removed and disassembled. The valves, valve seats, and guides are machined to recondition them or replaced to new specs. The heads are then reassembled and installed back.

On V-type engines both heads are always reconditioned at the same time to preserve even combustion pressures on both cylinder banks.

Also, valve jobs should not be done on high kilometer engines with poor rings without first restoring correct cylinder wall sealing. If only a valve job is done, the engine will restore upper combustion chamber sealing. The increased volume will suck more oil past the worn rings, which will burn during combustion.

13.7 VALVE GUIDE SERVICE

Valve guide wear is a common problem; it allows the valve to move sideways in its guide during engine operation. This may cause oil consumption by oil leaking past valve seal and through guide, a light knocking or tapping sound, burned valves due to poor seat-to-valve face seal, or valve breakage, *Fig. 13.15 (a)* and *(b)*. Another undesirable result of excessive valve guide wear is that it may allow the valve to wobble enough to cause it to ride to one side of the seat, *13.15 (b)*. Obviously, this interferes with proper seating and sealing of the valve and promotes wear. If the stem-to-guide clearance is more than manufacturer's specs, several repair procedures may be used. If the guide is replaceable, install a new one in its place. New guides sometimes need finish reaming for proper stem-to-guide clearance.

Older engines have *pressed-in valve guides.* Modern engines are built with the valve stem in direct contact with a hole bored in the cylinder head. Heat dissipation is better where pressed-in guides are not used. Wear in the *integral guides* only requires honing the holes out larger and then fitting valves with oversize stems.

Carbon accumulates in the guide bore and on the stem of exhaust valves. This makes the valves to stick partly open or to slow down in action. The carbon accumulation increases as the exhaust valve stem and guide wear, because increased gas blows by between the worn components.

Oil from the valve chamber may be sucked in between valve and guide to cause increased oil consumption and carbon build-up in the engine, *Fig. 13.15 (c).* This leakage is often pronounced in overhead engines, as oil is pumped up on the rocker arms. It is customary to place seals either on the valve stem or in the guide on overhead valve engines to exclude excess oil.

Wear on the intake valve stem and guide is undesirable, because it allows air to be drawn in through the clearance to dilute the air and gasoline mixture, *Fig. 13.15 (d),* and makes it impossible to achieve a satisfactory fuel mixture. A high vacuum exists in the intake port. Most engine manufacturers use valve stem seals on the exhaust valve, because a

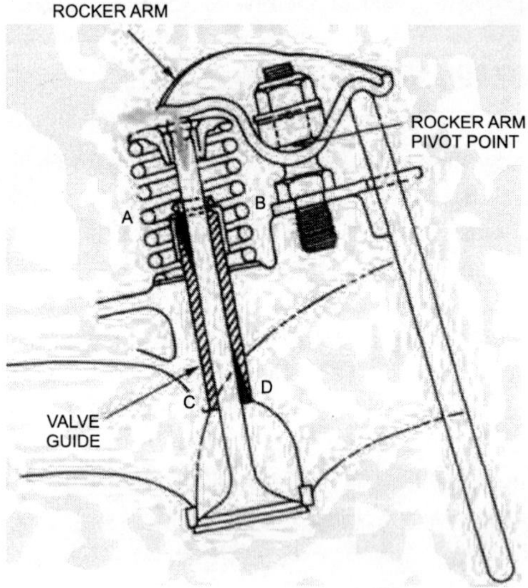

Fig. 13.15 *(b) When the rocker arm moves downward, it opens the valve while forcing the stem sideways. This causes the valve stem to be pushed against the guide at points A and D. It also prevents oil from lubricating these areas, which results in the guide to wear at points A and D. Rocker arm action causes bell-mouth wear of the guides. At points B and C, the valve stem is forced away from the guide. The valve guides must be reconditioned before grinding the valve seats. (General Motors)*

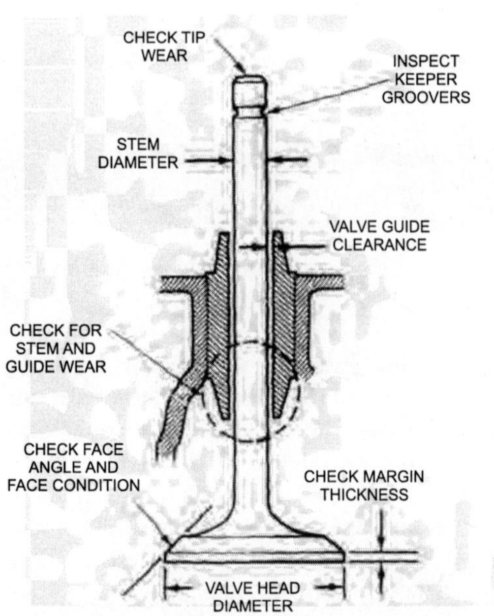

Fig. 13.15 *(a) Various problems that need to be looked for when servicing valves and valve guides. (Chrysler Corp.)*

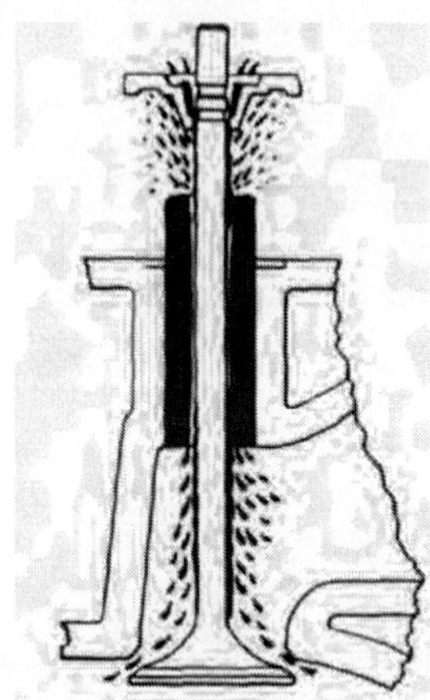

Fig. 13.15 *(c) Oil consumption around worn valve guides. (Perfect Circle)*

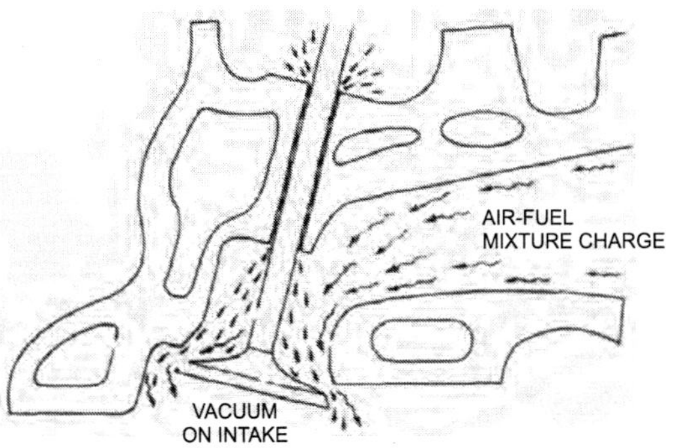

Fig. 13.15 *(d) Engine vacuum can draw oil past the valve guides and into the combustion chamber. The use of valve stem seals limits the amount of oil that is drawn into the engine. If the valve stem seals are defective, excessive blue (oil) smoke is most often observed during engine start-up.*

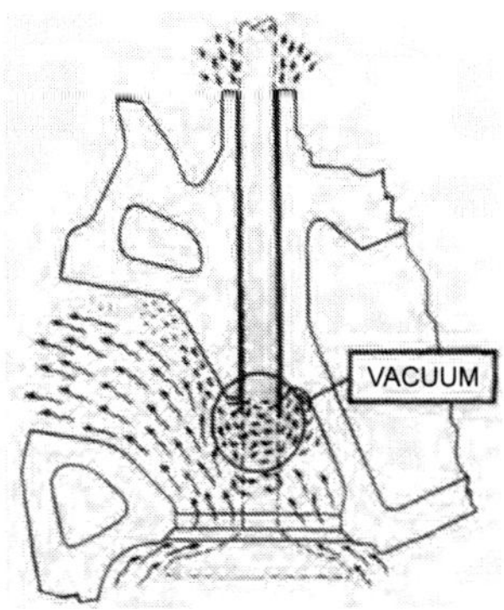

Fig. 13.15 *(e) Engine oil can also be drawn past the exhaust valve guide due to a small vacuum created by the flow of exhaust gases. Any oil drawn past the guide would be forced out through the exhaust system and will not enter the engine. Some engine manufacturers do not use valve stem seals on the exhaust valves. (Dana Corp.)*

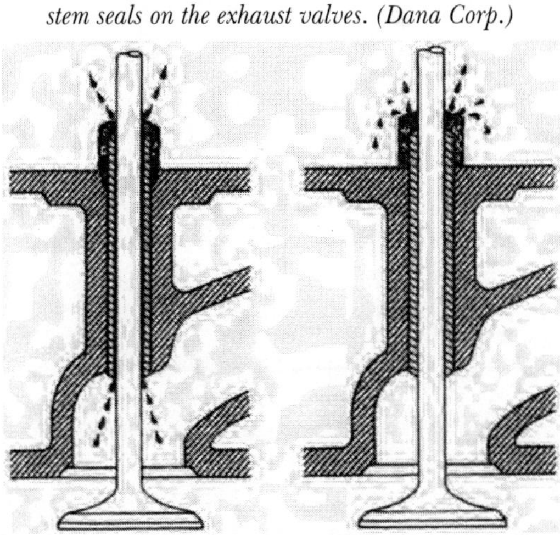

Fig. 13.15 *(f) New valve seals (right) reduce the flow of oil past the valve guides.*

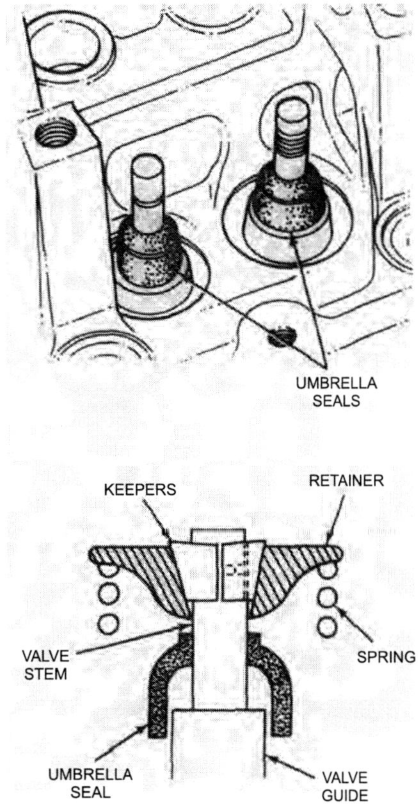

Fig. 13.15 *(g) Umbrella seals are installed over the valve stems and cover the guide and move with the valve stem.*

weak vacuum in the exhaust port area can draw oil into the exhaust stream, *Fig. 13.15 (e).*

The *umbrella valve stem seal* holds tightly on the valve stem and moves up and down with the valve. Any oil that spills off the rocker arms is kept out of the valve guide, much as water is deflected over an umbrella, *Fig. 13.15 (f).* That is why these seals are often called deflector seals.

Positive valve stem seals hold tightly around the valve guide, and the valve stem moves through the seal. The seal wipes the excess oil from the valve stem, *Fig. 13.15 (g).*

Measuring Valve Guide and Stem Wear

Check wear on the valve stem and guide by inserting the valve into the guide. To check for valve guide wear, slide the valve into its guide. Position the valve so that the tip of the stem is just flush with the top of

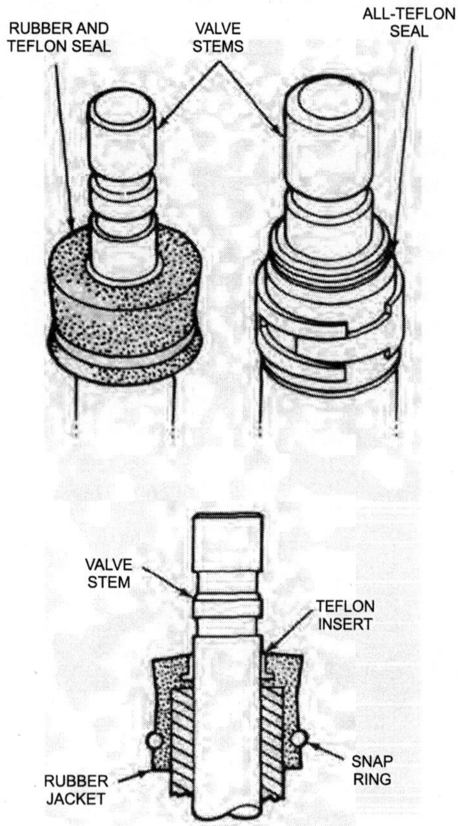

Fig. 13.15 *(h) Positive valve stem seals are the most effective type and they do not move with the valve stem.*

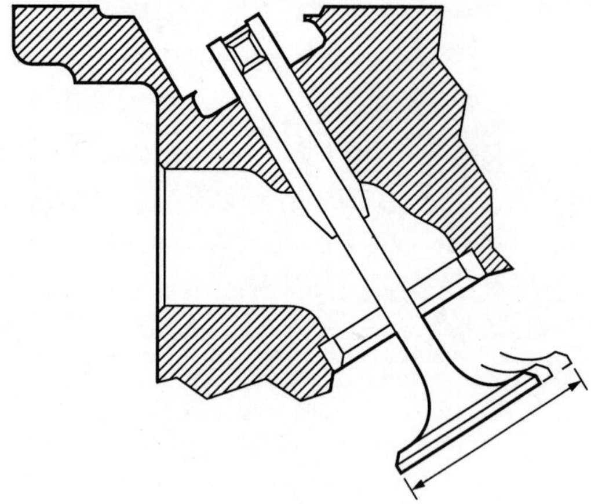

Fig. 13.16 *Checking valve guide wear.*

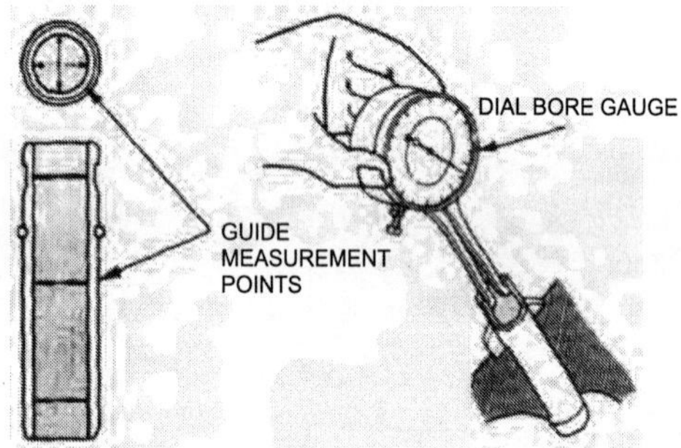

(a) Use hole gauge to measure in these locations.

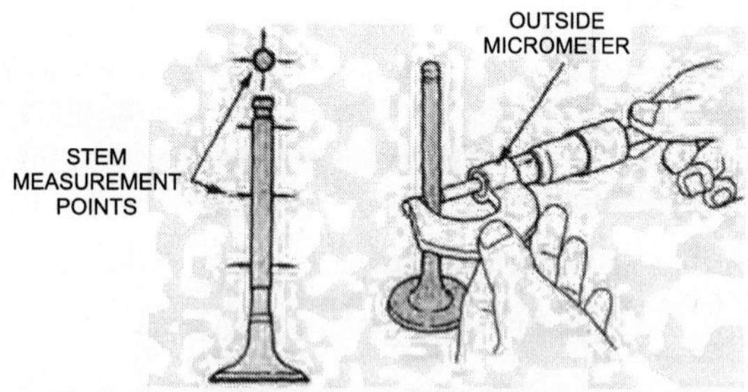

(b) Measure valve stem, with a micrometer, at these points. Subtract valve stem diameter from the valve guide diameter to determine the clearance.

Fig. 13.17 *Measuring valve guide or valve stem clearance. (Toyota Motors)*

the guide. Then, try to wiggle the valve head sideways. If the valve head moves sideways in any direction, the valve guide or valve stem is worn, *Fig. 13.16.* If the stem-to-guide clearance is excessive, the valve head will move appreciably. Measurements for this movement are normally given in the applicable service manual. Now remove the valve from the guide and inspect the valve stem. *Fig. 13.17* shows how a small hole gauge and outside micrometer are used to measure guide and valve stem wear.

Another way to check guide wear is with a *ball gauge,* which measures the diameter of small holes. Specs are given in the service manual. If the valve stem is smaller than specs, replace the valve.

Another measurement method is to mount a dial indicator to the cylinder head and rock the valve against the indicator. The valve must be placed very close to the seat for this check. *Fig. 13.18* shows how a dial indicator is used to measure valve stem clearance at the stem. If the measurements are not within specs, repair or replace the part.

The amount of wear between the valve stem and guide may be found by measuring the valve stem with a micrometer and the guide with a small hole gauge. The difference between the two measurements is the amount of clearance.

Fig. 13.18 *Valve guide wear and clearance can be measured with dial indicator at stem or at valve head. Here mechanic wiggles valve stem to measure clearance. (Pontiac Div. General Motors)*

Repairing Valve Guide Wear

Common methods used to repair worn valve guides are:

1. *Knurling valve guide:* Frequently, cylinder heads do not have replaceable guides. If wear is within limits and the valve guides are not hardened, they can be knurled by driving a rotating cutter down the guide that displaces metal so that the inside diameter of the guide is made smaller. After knurling the guide is finish reamed, *Fig. 13.19.*
2. *Reaming Valve Guides:* For guides that are worn beyond knurling limits, these may often be reamed oversize. Then replacement valves with oversize stems are *used, Fig. 13.20.*
3. *Installing Valve Guide:* In case the oversize valves are not available or wear is excessive, the old guide pressed or machined out and a new guide pressed or threaded into place. *Figure 13.21* shows the basic steps for guide replacement. On most engines the guides can be replaced. The old guides are hammered out and new ones hammered in, *Figs. 13.22 (a),* and *(b).* A shouldered punch and hammer are

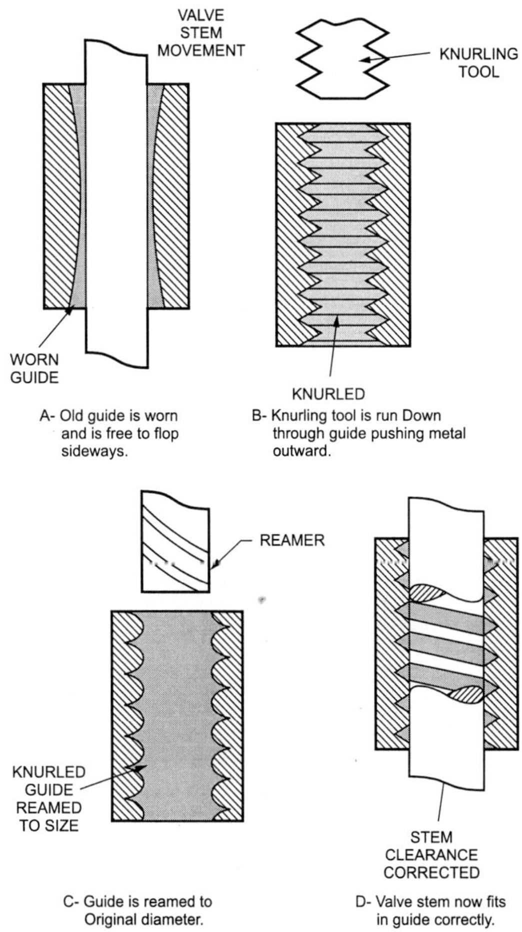

Fig. 13.19 *Knurling valve guides to restore clearance specs.*

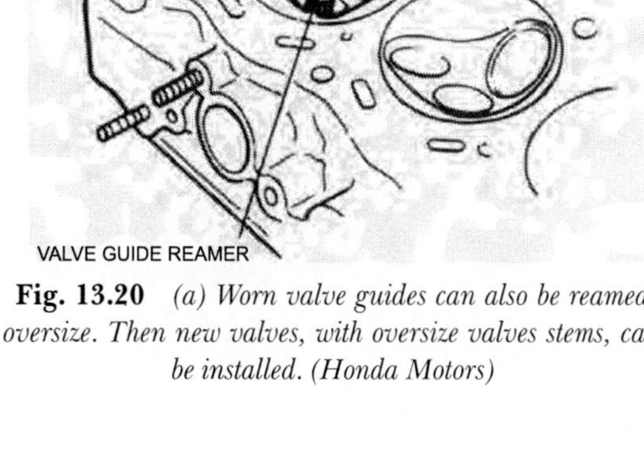

Fig. 13.20 *(a) Worn valve guides can also be reamed oversize. Then new valves, with oversize valves stems, can be installed. (Honda Motors)*

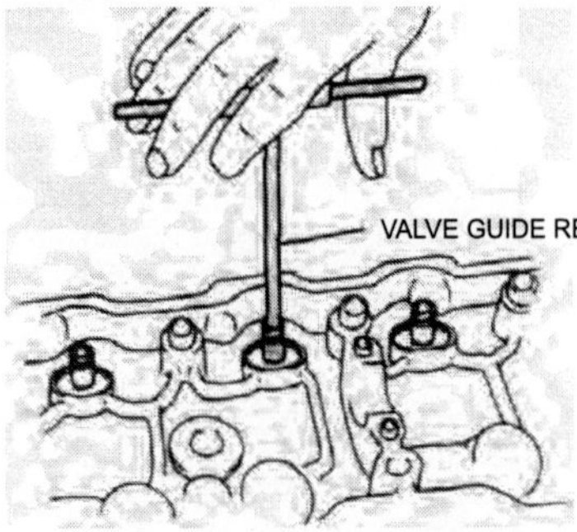

Fig. 13.20 *(b) Reaming valve guides. (Buick Div. GM)*

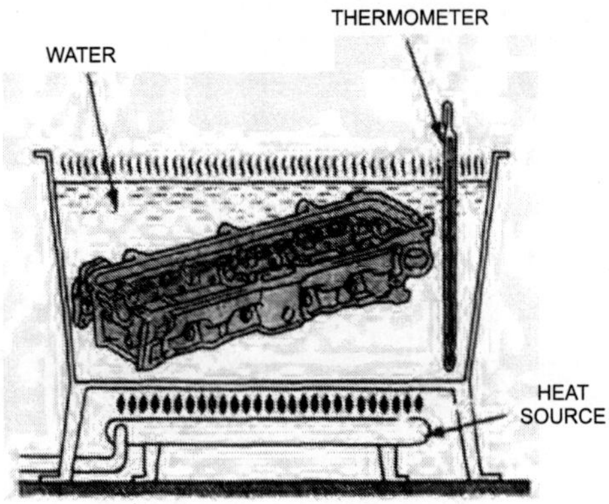

Fig. 13.21 A–*If available, heat head in water, to expand head and free guide.*

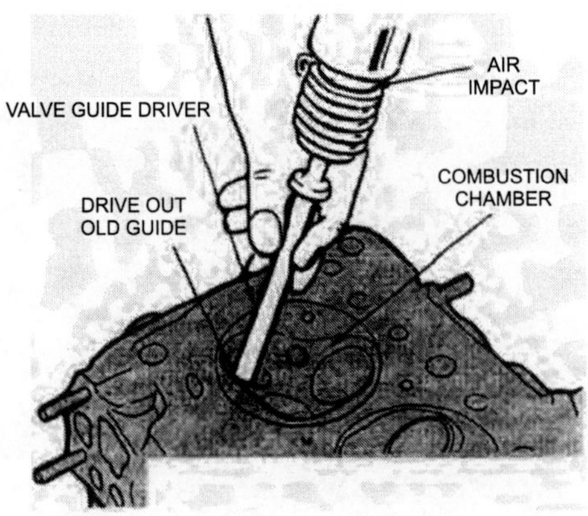

Fig. 13.21 B–*Drive out old insert from combustion chamber side of the head.*

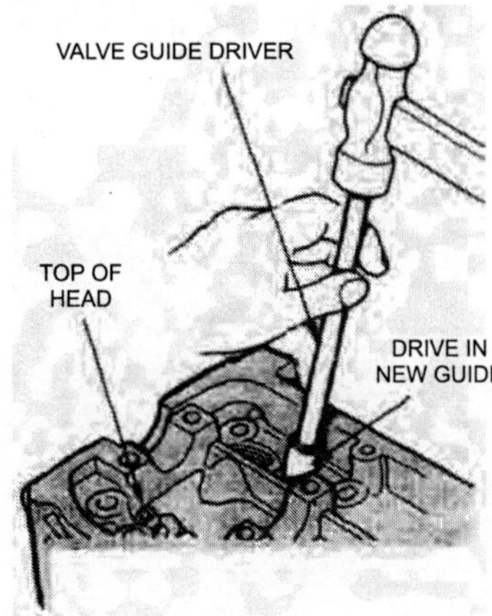

Fig. 13.21 C–*Then, drive in new insert from top of head.*

Fig. 13.21 *This aluminium head has pressed-in valve guide inserts. They can be replaced easily. (Toyota Motors)*

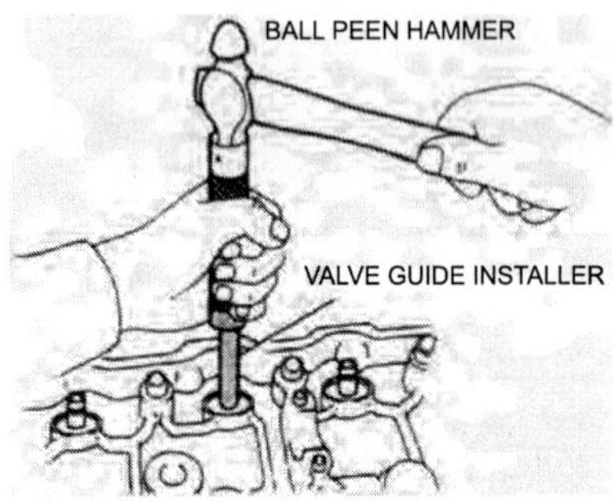

Fig. 13.22 *(a) Changing valve guides.*

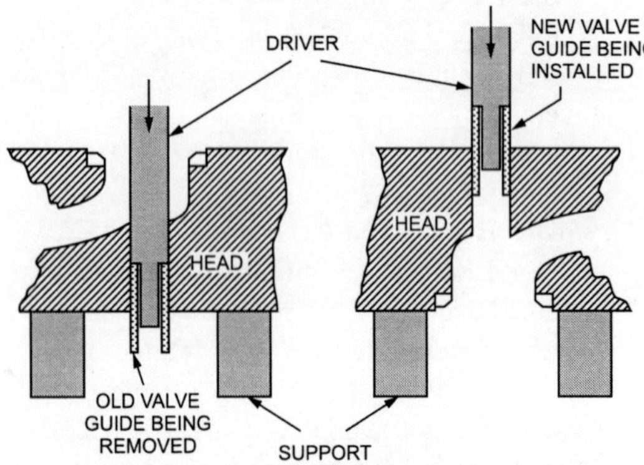

Fig. 13.22 *(b) Valve guide replacement procedure*

used to drive the guides. Exercise care when installing the guides in order to get them at proper height in the head. Use an old guide for comparison. Also guides are usually tapered, so they can be driven from one direction only. Guides that are tight in the head may gouge out some of the head material. This is called *galling*. Heating the head before driving guides will help prevent galling. Chill new guides before installing them.

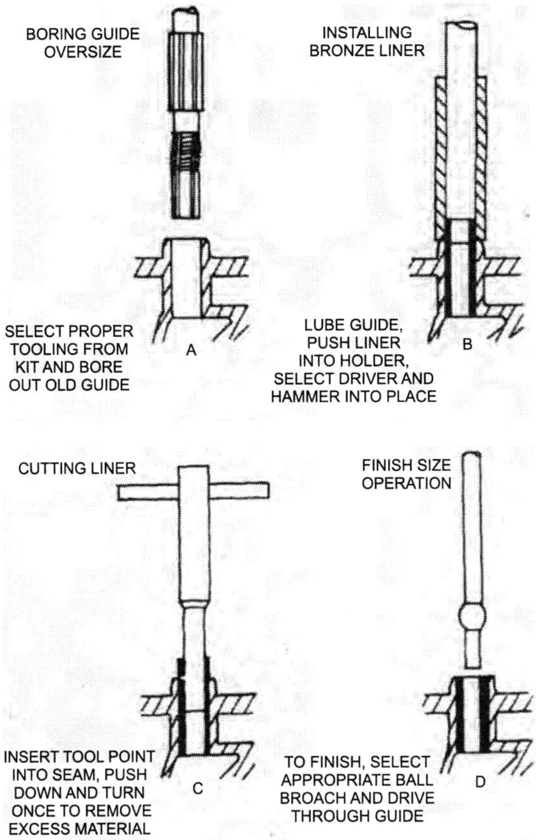

Fig. 13.23 *(a) Installing bronze wall liner. A - Ream or bore guide slightly oversize using tool in kit. B - Drive in bronze guide liner. C - Cut liner off flush with old guide. D - Broach inside diameter of liner to required size.*

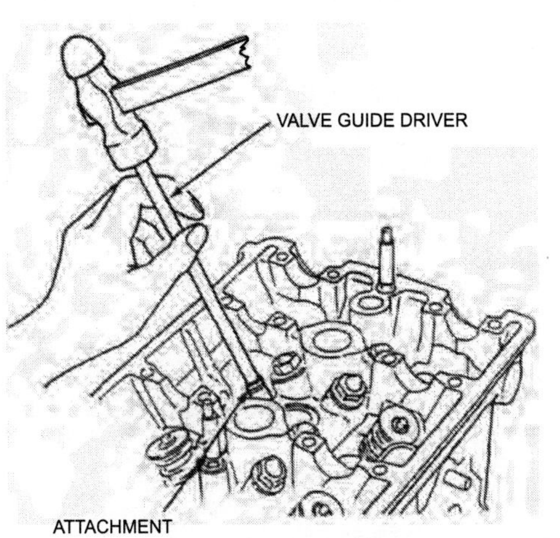

Fig. 13.23 *(b) Valve guide inserts must be positioned accurately. (Honda Motors)*

4. *Installing Bronze Guide Liner:* Guide inserts can replace the guide's ID without replacing the entire guide. Old guide reamed oversize without replacing the entire guide and thin guide liner installed and reamed to fit the valve stem, *Fig. 13.23 (a), (b).*
5. Another repair procedure involves the installation of a special Teflon seal at the top of the valve guide. The seal attaches to the valve guide and works like a piston ring turned inside out. The seal prevents oil from entering the valve guide. It does not interfere with the amount of oil in the rocker arm area or on the valve stem above the guide, *Fig. 13.24.*

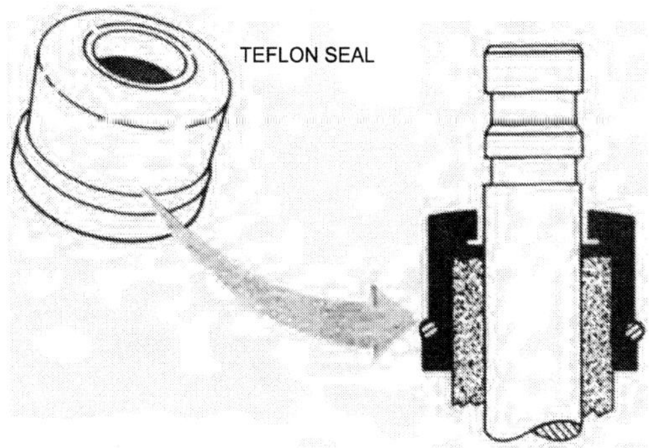

Fig. 13.24 *Detail of valve guide oil seal. (Perfect Circle)*

13.8 PUSH RODS SERVICE

Servicing push rods includes checking them for tip wear and straightness.

The *push rods* on an overhead valve engine are a critical link between the lifters and the rocker arms. If a push rod should become warped or bent, the corresponding valve will not open enough. This will cause loss of power or performance.

Push rods are easy to check. To check for warping, use a good flat surface with a corner e.g. glass tabletop. Lay the push rod over the corner such that its ends hang over the table surface. Roll the push rod and watch its ends. If the push rod is bent, the ends will wobble, *Fig. 13.25 (a).* Another method is to place the push rod in a holding fixture. While watching the dial indicator, rotate the push rod, *Fig. 13.25 (b).* If the dial indicator moves more than 0.0254 mm, replace the push rod.

Check the ends of all the push rods, which pass the rolling test. Worn ends will have small pieces chipped out of them. The ends could be pitted, too. Replace any chipped or pitted push rods.

If hollow, also check that they are not clogged with debris. Some push rods feed oil up to the rocker arms. Run wire through the push rods if necessary to clean them out.

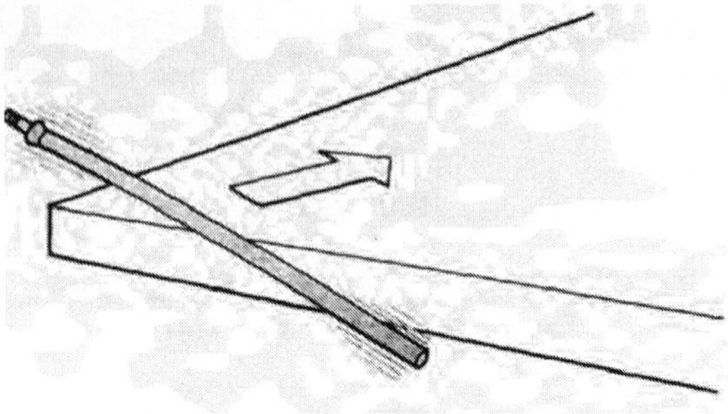

Fig. 13.25 *(a) Bent push rod.*

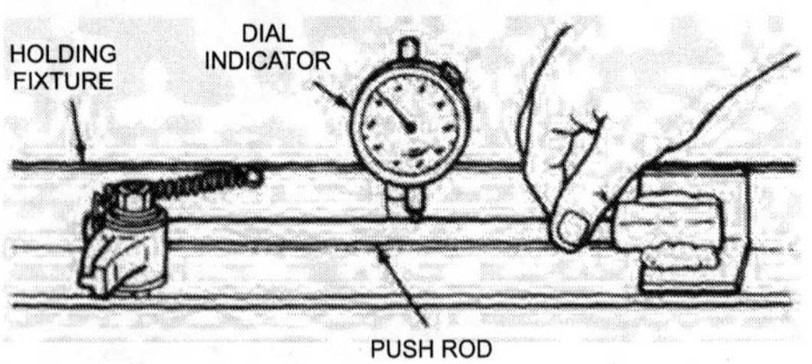

Fig. 13.25 *(b) Any bent or warped push rods should be individually replaced. Never attempt to straighten a push rod. (Ford Motors)*

13.9 VALVE SEAT RECONDITIONING

Whenever guide work is done, the valve seats must be ground, because servicing the guide moves the valve relative to the seat. If the head was assembled without grinding the seats, the valves will not seat. Because the grinding of seats is done by *piloting* off the guides, valve seat service restores the correct guide/seat relationship. *Valve Seat Reconditioning* involves grinding using a stone or cutting, using a carbide cutter, to resurface the cylinder head valve seats. Like a valve, the seats are exposed to tremendous heat, pressure, and wear.

The valve seats are reconditioned by a grinding stone mounted in a holder and driven by a hand held driver. A pilot shaft of the correct diameter is inserted in the valve guide of the seat to be ground, *Fig. 13.26.* The pilot is used to guide and centre the grinding wheel. Since the valve guide is used for centring, all valve guide service or replacement must be performed before seat grinding. A seat grinding stone is mounted to a driver and dressed to the specified angle for the valve seat in a truing unit using a diamond cutting tool.

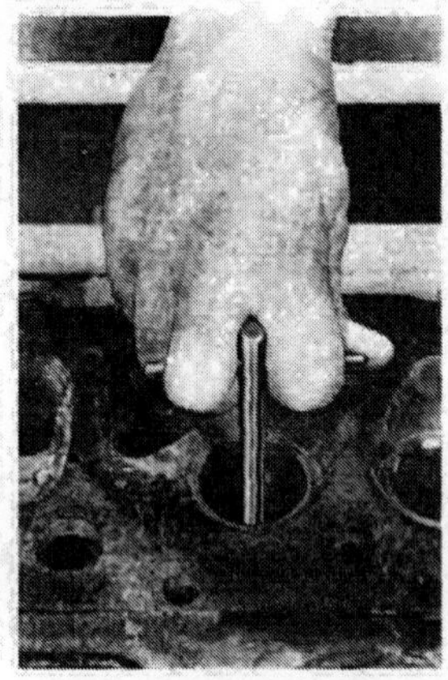

Fig. 13.26 *Installing a valve seat-grinding pilot. (Sioux Tools)*

A few seconds of cutting will usually be enough to remove pits and leave a precision ground surface. Steel seats need a roughing wheel for fast finishing; cast iron seats need only a finishing wheel. When servicing Stellite or induction hardened seats, only a small amount of metal is removed. Otherwise the grinding may go below the hardened surface. Since the hardened seats dull the grinding wheels rapidly, the grinding wheel must be dressed frequently when servicing hardened seats.

Grinding them with special stones, *Fig. 13.27,* restores valve seats. Three stones are used. The three stones are *dressed* and shaped at three different angles i.e., 30°, 45°, and 60°, *Figs. 13.28 (a)* to *(d).* A pilot is placed in the valve guide, and the stone is slipped onto the pilot and then it is spun with an electric or air-powered grinder. Machinist bluing is used to easily spot one cut from the other.

The 45° cut is made first. This cut establishes the valve seat, i.e., 45° cut is the surface against which the valve face seals. The 30° cut is made next. The shallower angle stone cuts the top of the valve seat.

Last, the 60° cut is made. The steeper angle fits past the 45° cut and shapes the port side of the valve seat. The 30° and 60°cuts shape the valve seat to improve the gas flow over them. They also give the machinist a way to bring the 45° cut valve seat to the proper width.

Narrow valve seats give higher sealing pressures because the valve spring pressure is distributed around a small contact area. Wide valve seats give longer valve life because they give the valve more heat transfer area. Therefore, the specified valve seat widths

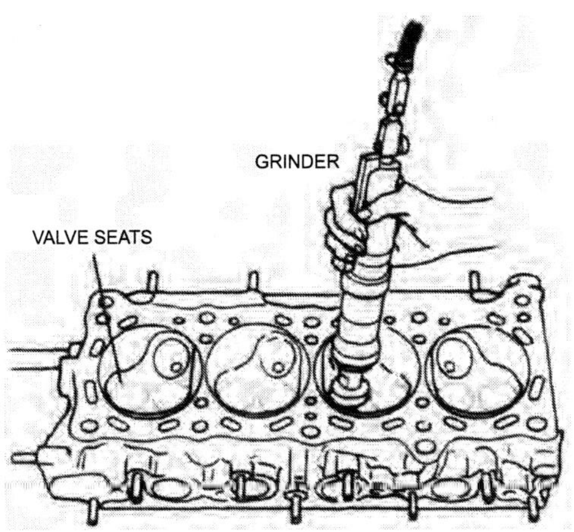

Fig. 13.27. *Grinding valve seats. (American Honda Motor Corp., Inc.)*

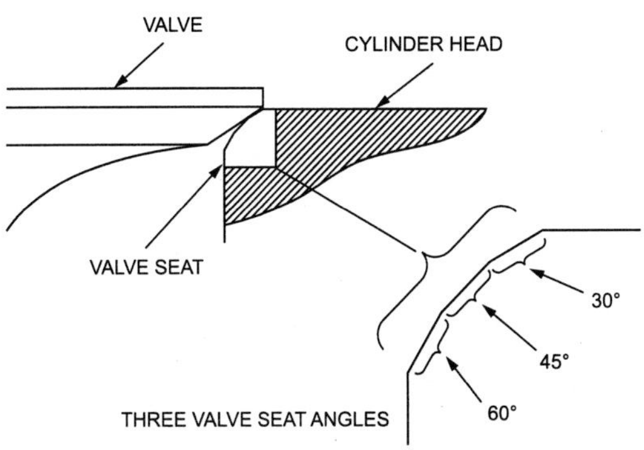

Fig. 13.28 *(a) Valve seat angles.*

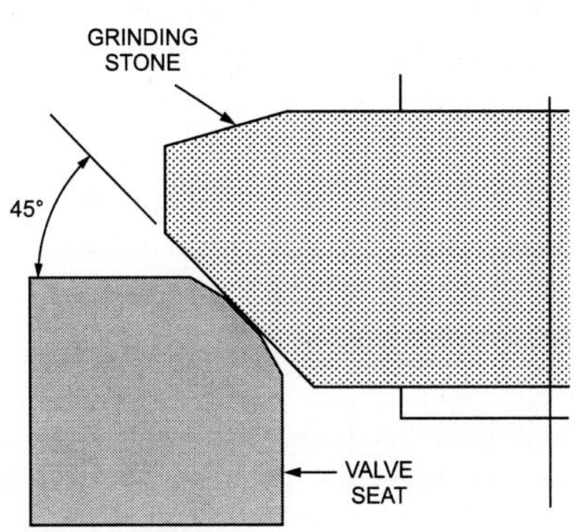

Fig. 13.28 *(b) Grinding a 45° angle creates the valve seat in the combustion chamber.*

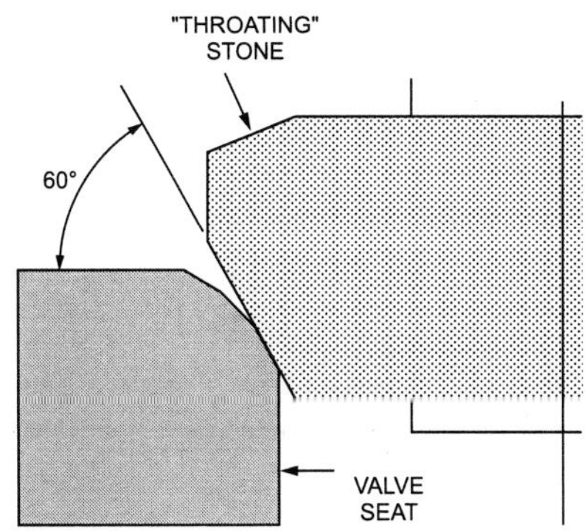

Fig. 13.28 *(c) Grinding a 60° angle removes metal from the bottom to raise and narrow the valve seat.*

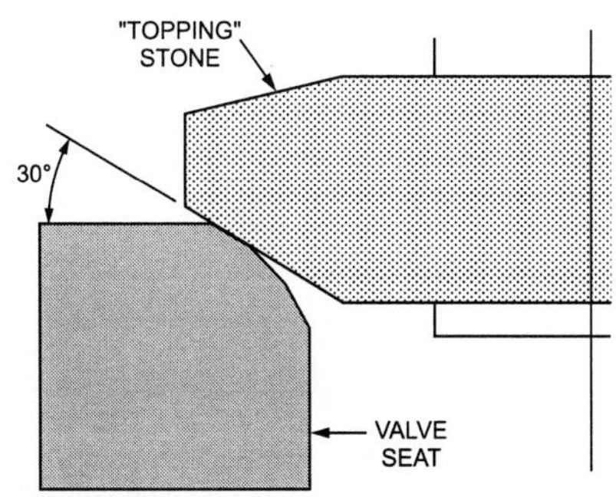

Fig. 13.28 *(d) Grinding a 30° angle removes metal from the top to lower and narrow the valve seat.*

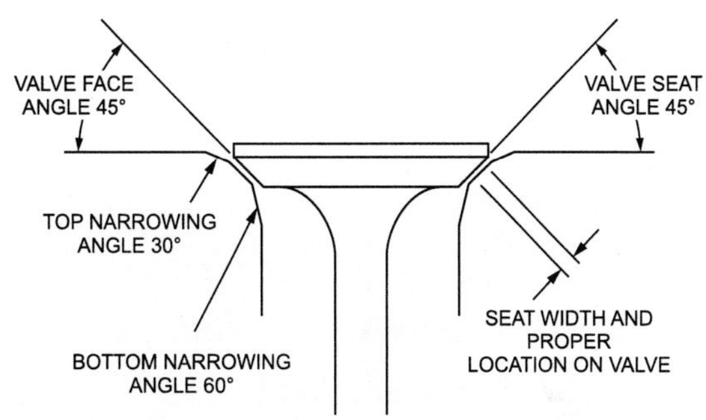

Fig. 13.28 *(e) A typical three-angle valve job using 30°, 45°, and 60-degree stones or cutters.*

are a compromise between these two goals: higher sealing pressures and longer valve life. Intake seats are narrower than exhaust seats because the hotter-running exhaust valves and seats need the extra cooling.

The valve face should always be larger than the valve seat. The engine manufacturer will provide specs on valve seat widths. If the reconditioned seat is too wide, grinding material off the top or bottom of the seat, with a special narrowing grinding stone, must narrow it.

Checking Valve Seats

Inspect the valve seats carefully for any problems: burning, cracks, excess width, and sinking or retrusion. Measure seat width and retrusion as shown in *Fig.13.29.* If the seat is badly worn or burned, too much grinding may be necessary to recondition the seat.

Valve seat retrusion results from heat erosion or seat grinding that causes the valve head to sink into the combustion chamber. The valve stem would also protrude too far out the other side of the head, *Fig.13.29.* Seat replacement is necessary to fix retrusion problem. Seat retrusion causes lowered compression, reduced gas flow through port, faulty valve train geometry, and lower performance.

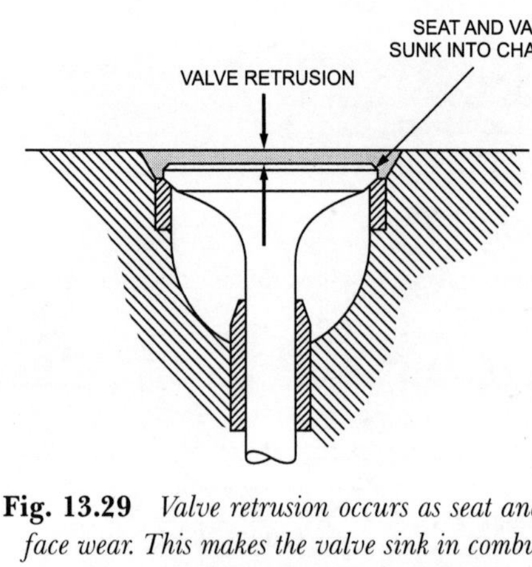

Fig. 13.29 *Valve retrusion occurs as seat and valve face wear. This makes the valve sink in combustion chamber and stem to move further through head. (Ford Motors)*

Valve Seat Replacement

If a valve seat is excessively pitted, grinding may not be enough for reconditioning. The valve seat may have to be replaced. Normally, valve seats can be ground or cut and returned to service. Special tools are available for removing and replacing replaceable valve seats. If the valve seat is not replaceable, but is part of the cylinder head, it may be machined to accept a replaceable valve seat, *Fig. 13.30.*

Fig. 13.30 *Worn valve seats may be machined to accept a replacement. (Kwik-Way)*

Most mechanics send the cylinder head to an automotive machine shop for seat replacement when the seat is part of the cylinder block. Most automobile repair shops do not have the specialized tools needed for seat removal and installation, especially with integral seats. A new valve seat insert should be interference fit of about 0.05 to 0.1 mm, in order to lock the insert in the head.

Press-in valve seat replacement is not difficult. No machining of a pocket in the combustion chamber to install a new seat is required. Simply, the old seat can be replaced and a new seat pressed-in place. The tools recommended by Ford Motors for pressed-in valve seat service are shown in *Fig. 13.31.* Another way to remove a pressed-in seat is to split the old seat with a sharp chisel. Then, pry out the seat with a hooknose type chisel. Utmost care should be exercised not to damage the cylinder head.

Chill and shrink the seat in dry ice or heat and expand the head in a furnace to install a press-in seat. This helps locking the seat in the head as it cools to the room temperature.

Use a driving tool to drive the seat into the recess in the head. Seat installation tools vary. Always follow the directions that come with the equipment packing.

Staking the seat involves putting small dents in the cylinder head next to the seat, *Fig. 13.32.* The stakes will swell the head metal over the seat and

SEAT INSERT REMOVAL

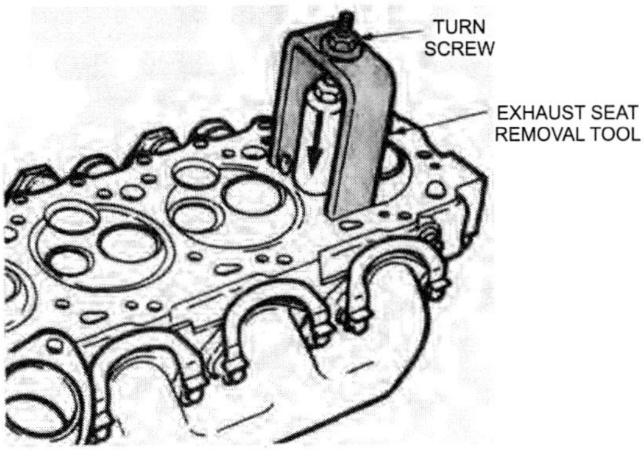

(a) Puller is being used to remove old, damaged seat.

INSTALLING SEAT INSERT

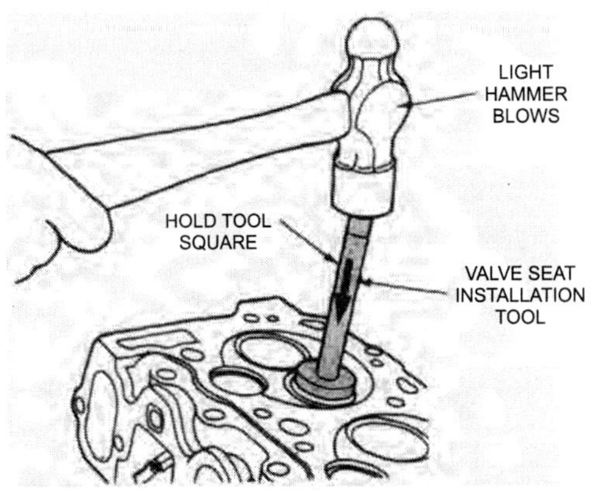

(b) Valve seat installation tool is forcing new seat into head pocket. Stake seat to keep it from falling out. (Ford Motors)

Fig. 13.31 *Servicing a head with pressed-in valve seat inserts is easier. However, most mechanics prefer to send the head out to machine shop for seat replacement. (Ford Motors)*

keep the seat from falling out on differential expansion in operation. Machining the top of the seat requires making the top of the seat flush with the surface of the combustion chamber.

13.10 VALVE SEAT RESURFACING

After new seat installation or when the old valve seats are with in serviceable spec condition, grind or cut the face of the valve seats. The equipment necessary to grind valve seats is shown in *Fig. 13.33.*

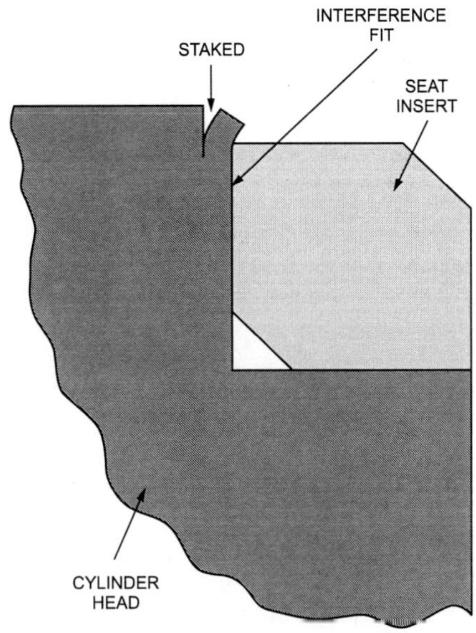

Fig. 13.32 *A seat is staked by using chisel, punch, or similar tool to smash metal over seat. This forms a lip that keeps seat in place during engine operation. (TRW)*

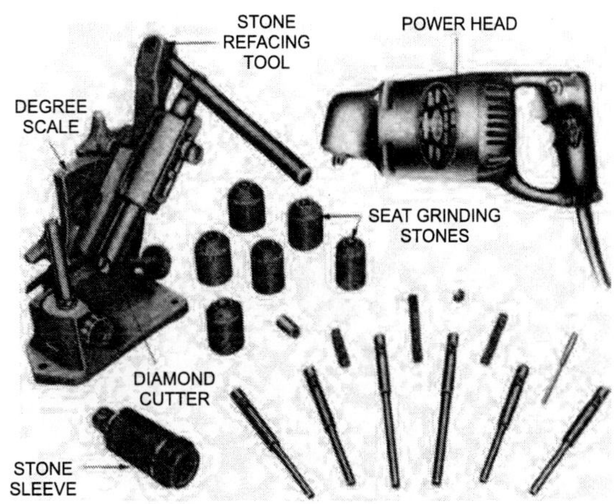

Fig. 13.33 *Tools needed to grind valve seats grinding stones. Study part names. (Sioux Tools)*

Selecting Pilot and Stone

The *pilot* is a steel shaft that installs in the valve guide to support the grinding stone. Because valve guide diameters vary with engine make, a correct size pilot needs to be found. Select one that slides down the guide far enough, yet has a snug fit in the guide. A T-handle type tool is normally used for installing the pilot, *Fig. 13.34.* Push down and twist when fitting the pilot into the guide.

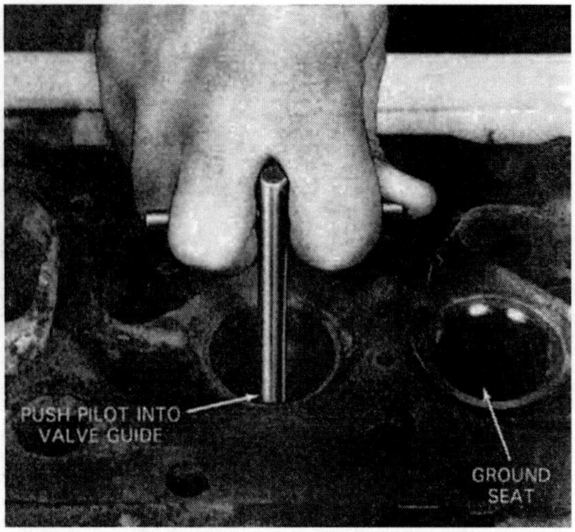

Fig. 13.34 *Select a right size pilot and push it into guide. Pilot should fit almost to end of taper but not to be loose in guide. (Sioux Tools)*

The *seat stone* is a grinding stone mounted on a threaded drive attachment, used to cut a fresh surface on the face of the valve seat. Find the correct diameter stone for the seat, *Fig. 13.35.* The stone should be slightly *bigger* than the seat but it should not hit the side of the combustion chamber.

It must also have the correct angle, usually 45° or occasionally 30°. Screw the stone onto its drive head and check its fit on the seat. If the stone is too small, even momentary grinding may ruin the seat.

Dressing Stone

Dressing a stone involves using a diamond cutter to true the surface of the stone to restore its true angle. A stone dressing tool, *Fig. 13.36 (a),* is set to the desired angle of the valve seat. Then, spin the stone with the drive motor. When dressing the stone, lubricate the pilot with a drop of oil. Set the diamond angle correctly. A scale is provided on the tool that reads in degrees. Lock the tool at the correct angle. Adjust the diamond cutter up, next to the stone very carefully, such that it will clear the stone slightly before spinning the stone with the power head, *Fig. 13.36 (b).*

The diamond cutters are expensive and can be ruined if it is adjusted too far into the stone. The stone can grind metal away from the diamond and the diamond can fall out. After everything adjusted precisely, spin the stone with the power head. Slowly feed the diamond into the stone until it just touches the stone. Then, move the equipment lever up and down to true the entire face of the stone. Just cut as little off the stone as necessary to true the surface.

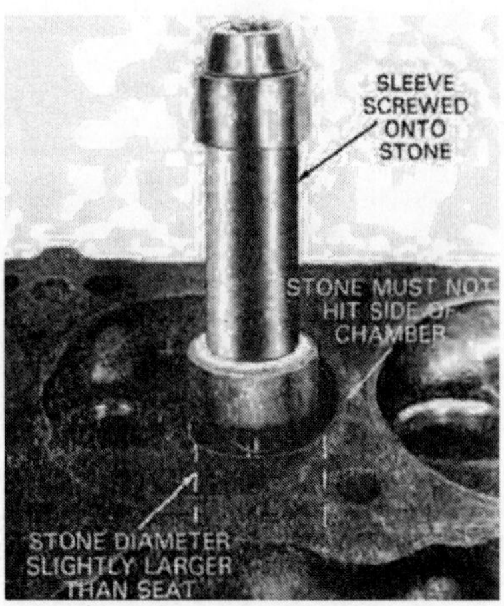

Fig. 13.35 *Find a stone of the right size and angle. Stone should be larger than seat but not big enough that it hits side of combustion chamber. (Sioux Tools)*

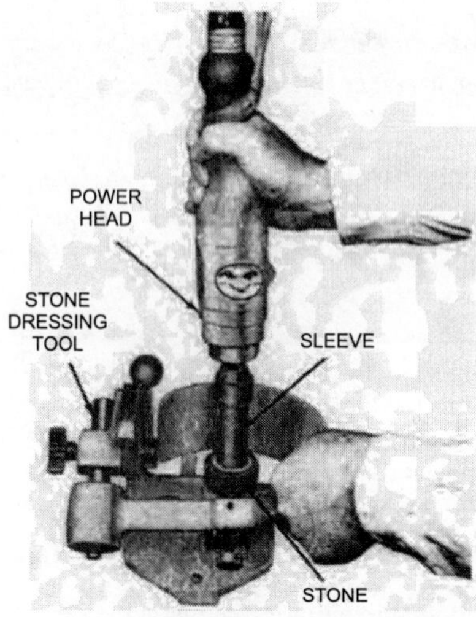

Fig. 13.36 *(a) Stone and sleeve are mounted on stone dressing tool. Tool must be set to correct seat angle.*

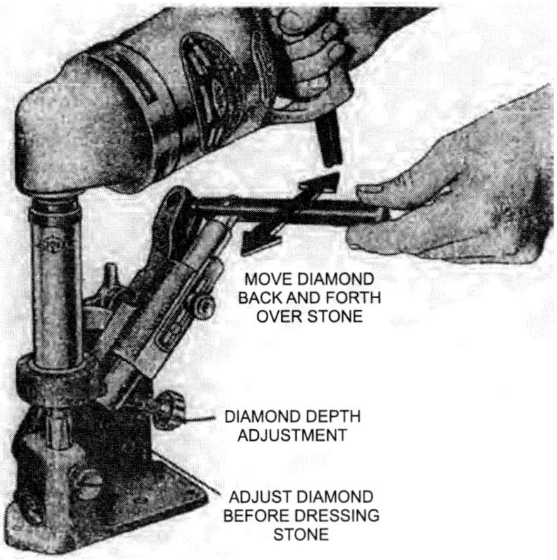

(b) After setting angle, adjust diamond so that it almost touches stone. Turn on motor and slowly feed diamond into stone while moving lever up and down. Stone should be touched lightly with diamond to avoid rapid stone consumption.

Fig. 13.36 *Dressing the stone. (Sioux Tools)*

Fig. 13.37 *Slide stone and sleeve over pilot. Use power head to rotate stone on seat for 1 to 2 seconds. Simply support and hold the power head and do not push down on it. (Sioux Tools)*

Grinding Valve Seats

After choosing the correct stone and ensuring it is dressed properly, one is ready to grind the valve seat. Place a drop of oil on the pilot. Slide the stone down onto the seat.

As shown in *Fig. 13.37* install the power head into the stone driver. Use one hand to support the power head and the other to operate the trigger. Do not push down on the power head while grinding seats. The weight of the tool is sufficient for downward pressure simply hold the tool steady. Hit the trigger and grind the seat for a second or two.

Remove the stone and check the seat. If it is shiny all around its circumference, no more grinding is needed. Just enough grinding is done to reduce the valve retrusion and width. Go to the next seat and repeat the grinding process, *Fig. 13.38*. Grind all the intake seats first. Then install the other correct stone and grind the exhaust seats. This will save time in alternately changing the stone each time for the next cylinder seat grinding. Dress the stone every couple of seats grinding or as necessary.

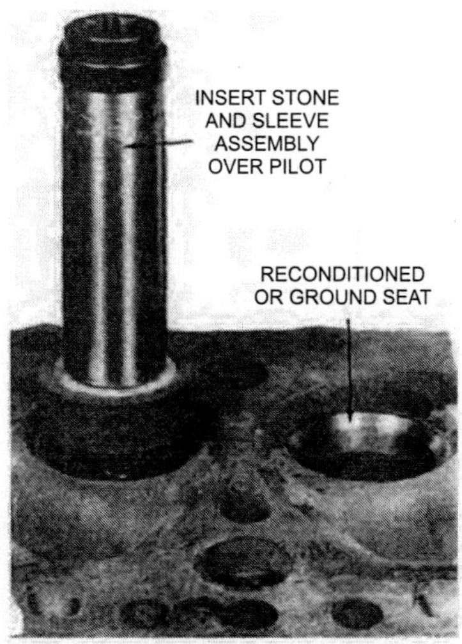

Fis 13.38 *Grind all the seats the same amount, as far as possible. Carefully inspect each seat after grinding to check that all pits or black marks are removed. (Sioux Tools)*

Cutting Valve Seats

The term *cutting valve seats* is used when a carbide cutter is used to recondition the cylinder head seats instead of the stones. Very hard carbide cutters will machine off a layer of metal when turned against the seats being reconditioned.

A carbide valve seat-cutting tool is shown in *Fig. 13.39.* This specific tool has a 15° angle on the top and 45° angle on the bottom. To use the carbide cutter, ensure that the blades are installed correctly. Select the correct angle for the seat. Install the pilot. Slide the cutter over the pilot. Then, use the T-handle to rotate the cutter on the seat surface. Only the material just enough to clean all dark areas on the seat should be removed. Repeat the process on the remaining seats, *Fig. 13.40.*

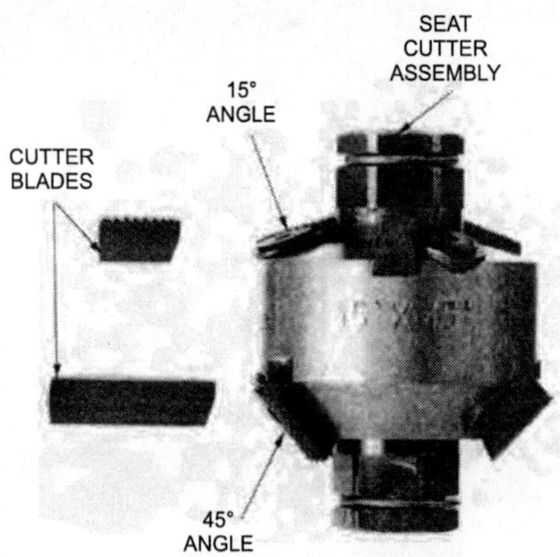

Fig. 13 39 *Carbide cutters are another excellent tools used to recondition valve seats. These cutters are made of very hard metal and they remove metal very accurately.*

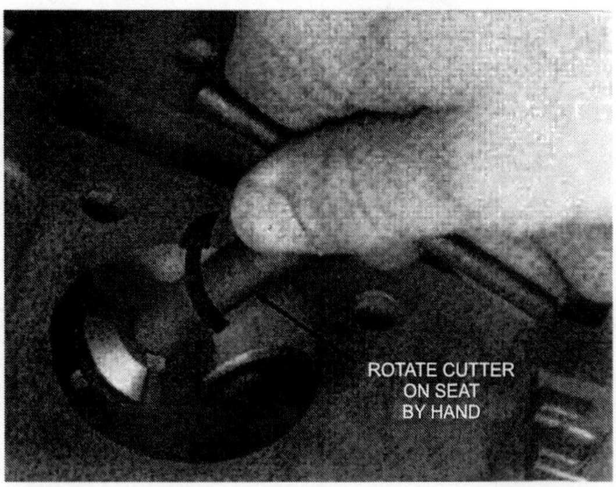

Fig. 13.40 *Select correct size and angle cutter. Mount it on T-handle. Rotate the cutter against seat by hand until the seat is cleaned properly.*

Fig. 13.41 *(a) Checking valve seat concentricity. A dial gauge is used to check that the seat is centred to the valve guide after grinding the seats.*

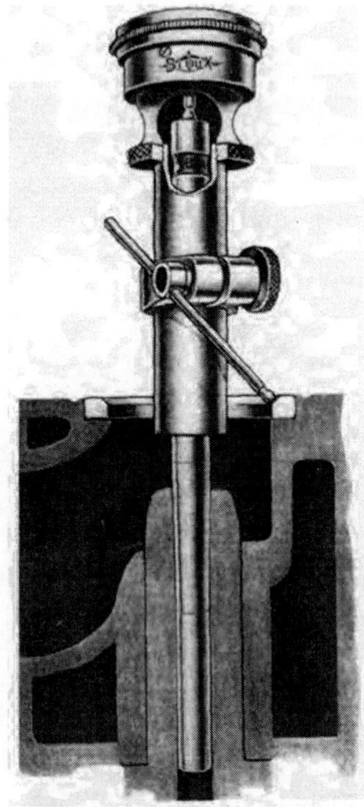

Fig. 13.41 *(b) Checking valve seat concentricity with the Valve guide centre. (Sioux Tools)*

Checking Seat Concentricity and Width

After grinding or cutting the seats, the cutting or grinding must produce a smooth, true surface if the valve is to seal properly. It cannot be true if the seat is not concentric with the valve stem. A method of testing it with a dial gauge is shown in *Figs. 13.41 (a)* and *(b).* A pilot is installed in the valve guide and the indicator is mounted to it. The seat should be concentric within 0.0254 mm of the guide. Concentricity or seat runout is checked by mounting and turning a special runout gauge, *Fig. 13.42 (a),* on the seat. The indicator will read seat runout. If not correct, something was wrong when grinding or installing the seat.

Seat width must be checked after reconditioning. Grinding or cutting widens the seat. Measure the seat width as shown in *Fig. 13.42 (b).* If more than specs, it should be narrowed and properly positioned.

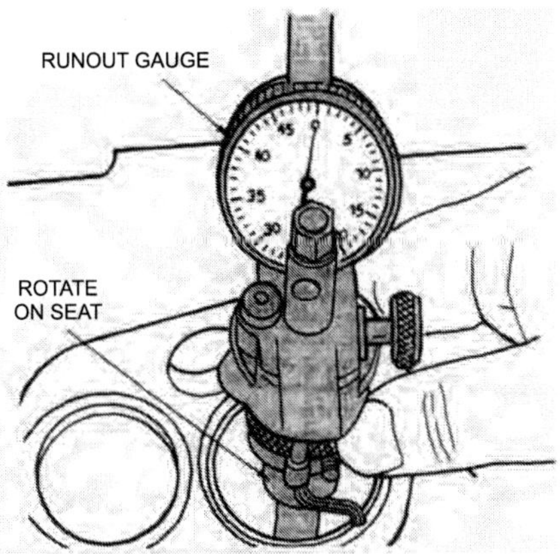

(a) Measure concentricity/runout as shown.

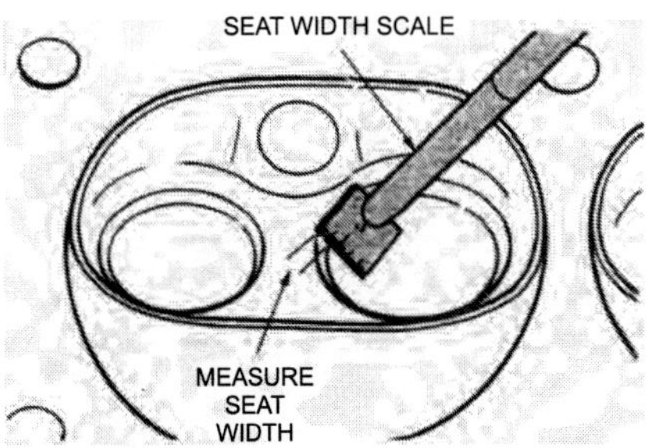

(b) Use small scale to check seat width. If too wide, narrow the seat with other stone or cutter angles. (Ford Motors)

Fig. 13.42 *After grinding or cutting, check concentricity and seat width. (Ford Motors)*

Narrowing Valve Seat

After prolonged service, heat and friction from valve action can wear away the seat. Then, after grinding, the seat can be too wide or it can contact the valve too far out on the face.

Seat narrowing, also known as *seat positioning,* is required so that the seat contacts the valve face correctly. For a 45° seat-angle, use 30° and 60° stones to move he seat and narrow it. This is sometimes called a *three-angled valve job* as it uses three different stone angles, *Fig. 13.43.*

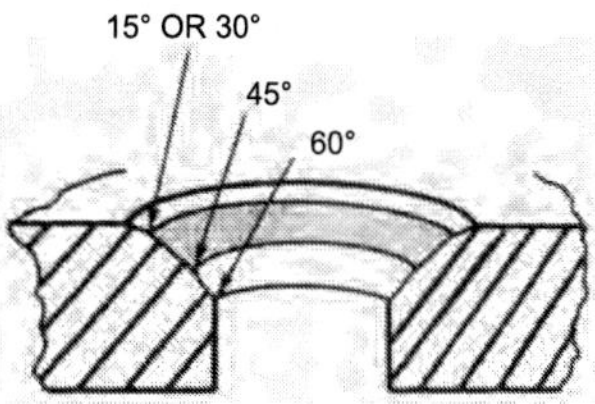

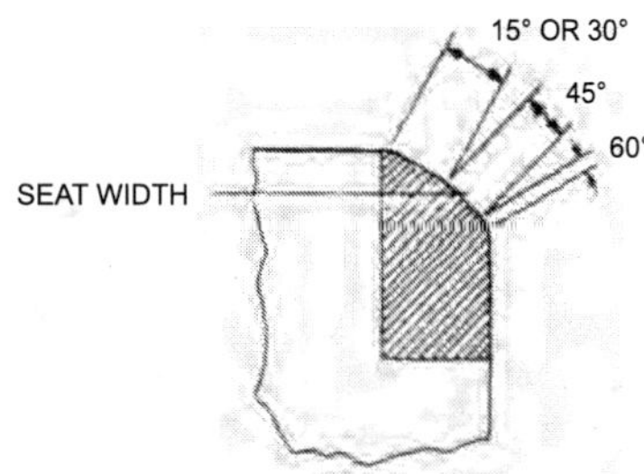

Fig. 13.43 *Commonly known as three-angle valve job, either 15° or 30° stone or cutter removes metal from above seat face. A 60° stone is used to remove metal below seat. This can be done to narrow and move seat. It can also enhance engine power.*

Figure 13.44 shows the steps how a carbide cutter is used to position and narrow a valve seat:

1. Use a 60° cutter to remove material from the inside diameter of the seat.
2. Then, use a 15° or 30° cutter to remove material from the outside diameter of the seat as necessary.
3. Use correct 45° cutter-angle to resurface the face of the seat.

Checking Valve-to-Seat Contact

To check the valve-to-seat contact area, coat the face of the valve with valve lapping compound or *Prussian blue. Valve lapping compound* is a paste material filled with abrasives. A small amount is placed on the valve face. Insert the properly ground valve into the seat and rotate it with light pressure. A *lapping stick* (stick with suction cup on end) is used to spin the valve on its seat. To turn the stick, rub the palms of your hands together on each side of the stick, while picking up and repositioning the valve slightly, *Fig. 13.45 (a).*

A-first cut cleans and reconditions area below seat.

A

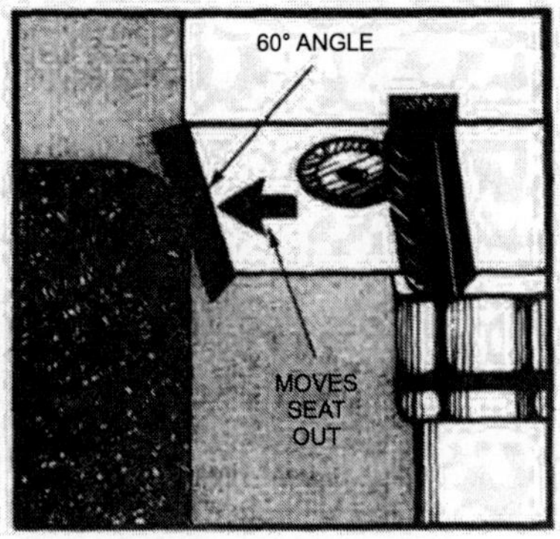

B-second cut cleans and reconditions area above seat.

B

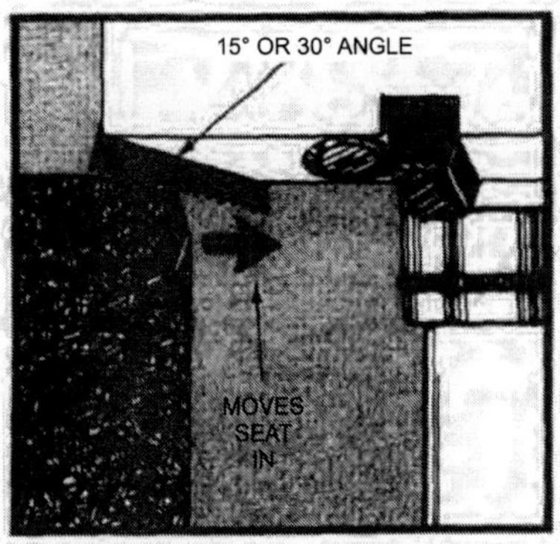

C-three or four revolutions of cutter produces a precision seat.

C

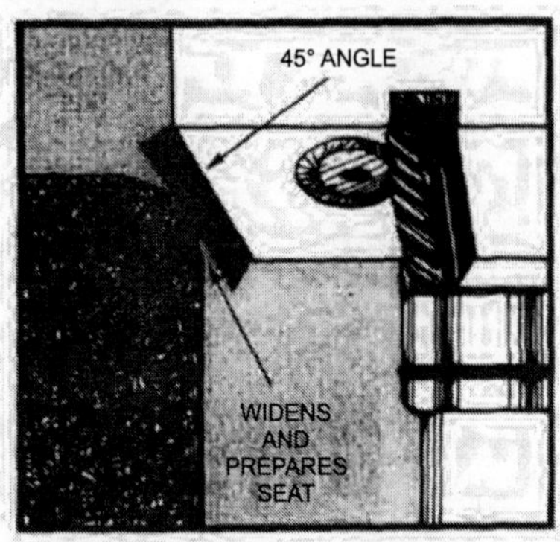

Fig. 13.44 *Carefully note the basic steps for using carbide cutter to do a three-angle valve job. Estimating seat width and location before cutting will help in determining as to how much to cut with each angle. (Newway)*

Remove the valve and examine the face. A *dull gray stripe* around the seat and valve face indicates the valve-to-seat contact point. This will help to determine how to move the valve seat if needed, *Fig. 13.45 (b)*. A few manufacturers do not recommend valve lapping. Refer to the applicable service manual for details. Make sure to clean the entire valve-grinding compound off the valve and cylinder head. The compound can cause rapid part wear if it stays on.

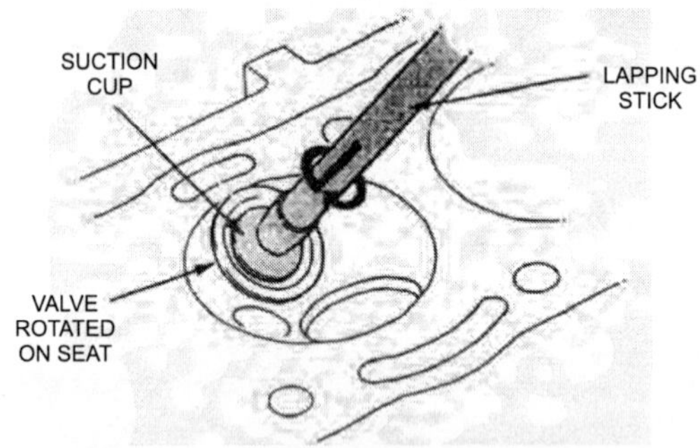

(a) Wipe a small amount of compound on valve. Then, use lapping stick to rotate valve on seat.

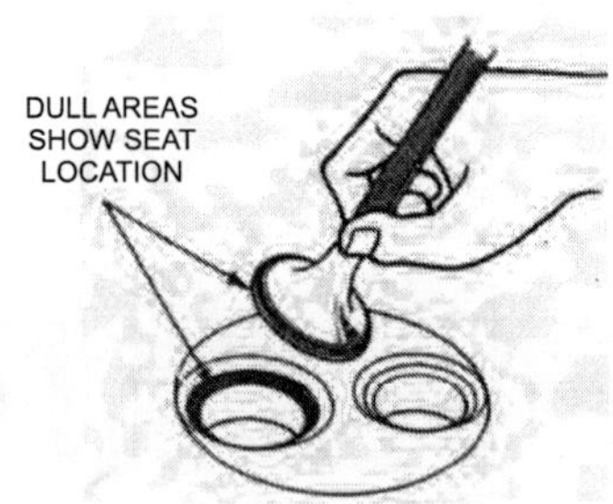

(b) Remove valve and check dull stripe on valve and seat. This will highlight how seat touches the valve face. Clean off all abrasive compound from parts before head assembly.

Fig. 13.45 *Lapping or grinding compound is an easy way to check seat location on valve. (Toyota Motors)*

To use Prussian blue to check the seat, wipe the dye on the properly ground valve face. Then snap the valve closed on its seat several times by hand. The contact with the seat will wipe off some of the Prussian blue, making the valve face shiny, i.e., ring of bare metal should be left showing on the valve face. This will show where the seat and valve touch, *Fig. 13.46.*

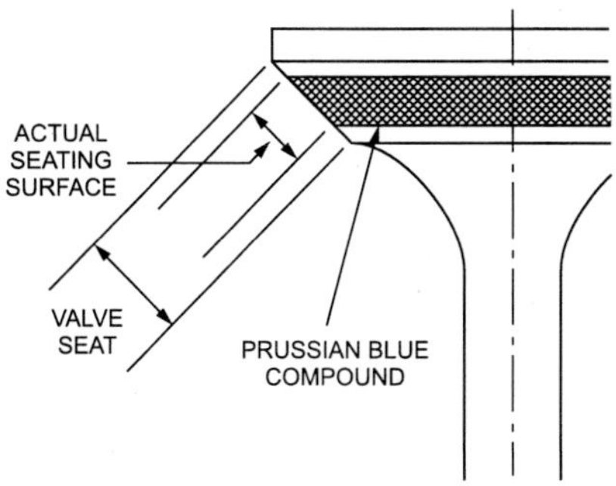

Fig. 13.46 *(a) Prussian Blue is another way to check valve-to-seat contact. Wipe dye on valve all around its face. Insert the valve into guide. Then, with hand, slap the valve closed several times. Remove the valve and shiny area on valve face shows contact. This will show whether or not to do more grinding or cutting to move seat location. (Honda Motors)*

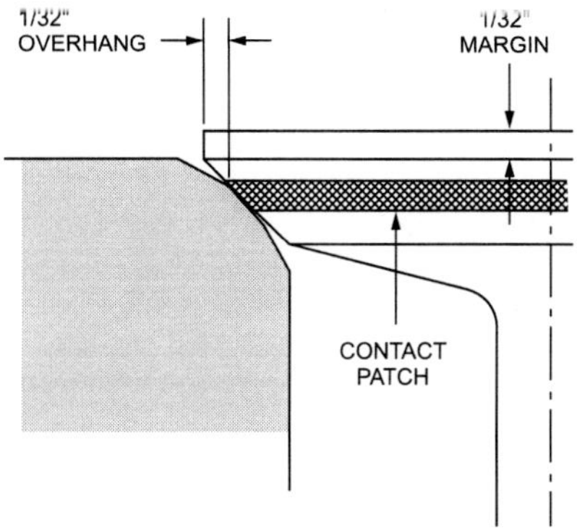

Fig. 13.46 *(b) The seat must contact evenly around the valve face. For good service life, both the margin and the overhang should be at least 0.8 mm.*

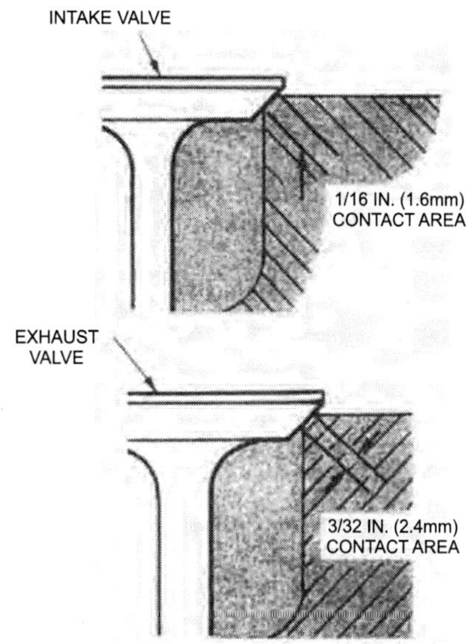

Fig. 13.48 *This figure shows typical specs of intake and exhaust valves. Exhaust seat must be slightly wider to dissipate heat more quickly. Follow engine manufacturer's specs, where possible. (TRW)*

Figure 13.47 shows some good and bad seat contact patterns. While, *Fig. 13.48* shows typical seat specs for both intake and exhaust valve seats. Note how the exhaust seat is wider so it can dissipate heat better. An intake valve should generally have a valve-to-seat contact width of 1.6 mm. An exhaust valve should have a valve-to-seat contact width of about 2.4 mm. Check the service manual for exact specs.

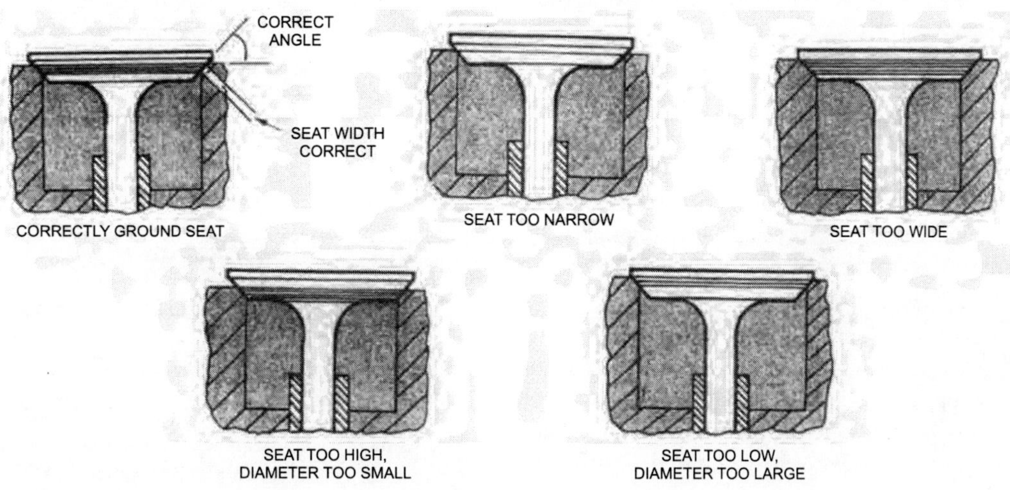

Fig. 13.47 *Some good and bad seat contact patterns will show up by lapping or using Prussian Blue. (Ford)*

This ring should be approximately 1.6 mm from the margin, as this is the point at which the valve face meets the seat. If the measurement is any thing other than 1.6 mm, the seat will have to be re-cut.

Moving Valve-Seat Contact Point

If the valve seat does not touch the valve face properly (wrong width or location on valve), regrind the seat using different stone angles. As mentioned use 15°, or 30° and 60° stones, *Fig. 13.49.*

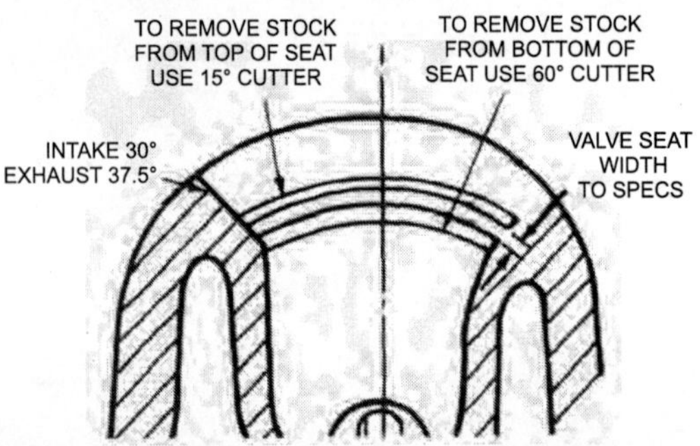

Fig. 13.49 *Different angles used to grind seat properly.*

To *move the seat in* and narrow it, grind the valve seal with a 15° or 30° stone with a 45° seat and 15° stone with a 30° seat-angle. This will remove metal from around the top of the seat. The seat face (valve contact surface) will move closer to the valve stem.

To *move the seat out and narrow* it, machine the valve seat with a 60° stone or cutter. This will cut metal away from the inner edge of the seat. The seat contact point will move toward the margin or outer edge of the valve.

13.11 VALVE SERVICE

The condition of the valves is very important for engine efficiency. *Poppet valves* work under very punishing type of environment and operating conditions. Therefore, it is not surprising that valve service is a frequently performed operation in automotive service shops.

Valve service used to be a hand lapping procedure, to provide a gas-tight valve, using an abrasive paste between the valve face and valve seat. Not enough attention was paid to other important factors like; the valve seat width and positioning, heat dissipation, concentricity of the valve with the seat, strength of the valve spring, valve guide wear, and all other considerations that require attention on a late model high-speed automobile engine.

Valve Grinding

The cure for normal valve wear and light burning is grinding. A special machine grinds the valve sealing face at the correct angle until fresh material shows all the way around, *Fig. 13.50.* Modern *valve grinding* is a true grinding process. Each step of the operation is controlled by careful measurement using an accurate test and measurement equipments. The initial step, after a valve has been removed and adequately cleaned, is to determine its condition as to whether the valve can be reconditioned or whether it should be replaced. If the *valve stem* is scored, pitted, bent, or worn more than 0.0508 mm, the valve should be replaced.

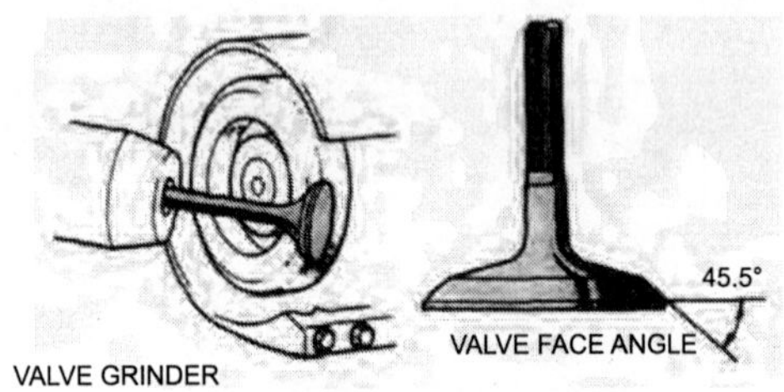

Fig. 13.50 *Valve grinding*

A dial gauge and V-blocks can be used to check the valve stem for *straightness, Fig. 13.51.* The stem diameter can be checked for wear with a micrometer, at a number of locations along the stem.

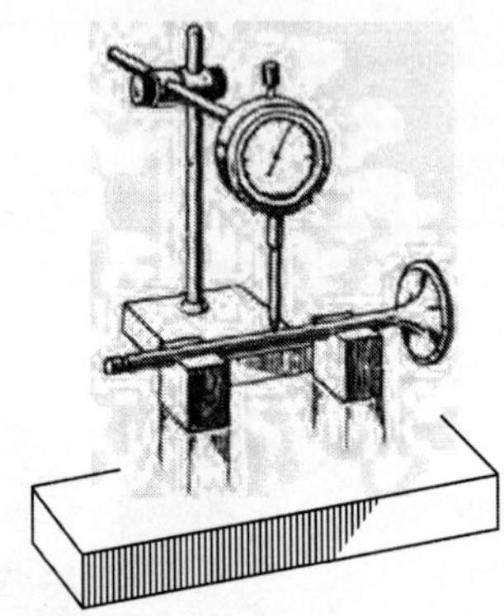

Fig. 13.51 *Using a dial indicator to check valve stem straightness. If dial reading varies more than 0.05 mm while rolling the stem, the valve should be replaced.*

Carefully check the valve visually, *Figs. 13.52* and *13.53*. Visual inspection may show a badly burned, *warped valve head,* a *cracked valve face, bent valve stem,* or *lack of margin.* If the valve face is burned, badly warped, or worn to a thin margin, replace it, *Figs. 13.53 (a) to (c).* This ring should be approximately 1.6 mm from the margin, as this is the point at which the valve face meets the seat. If the measurement is any thing other than 1.6 mm, the seat will have to be re-cut.

If the valve is shown to be in good condition, it is placed in a special valve-grinding machine, called a valve *refacer,* and a new surface is ground on the face at the correct angle. *Valve grinding* machines a fresh, smooth surface on the valve faces and stem tips. Valve faces suffer from burning, pitting, and wear caused by opening and closing millions of times during the service life of an engine. Valve stem tips wear due to friction from the rocker arms.

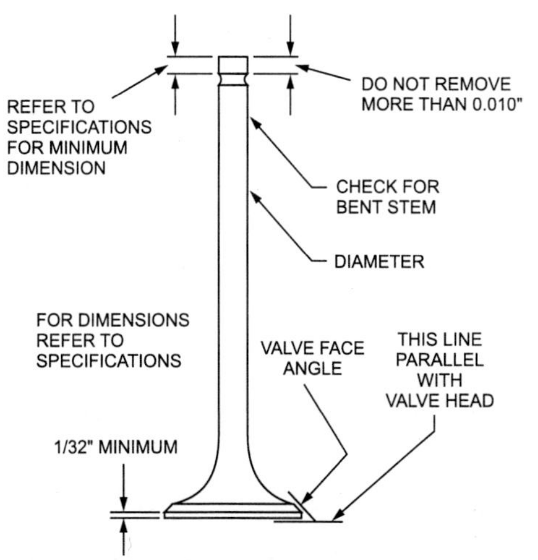

Fig. 13.52 *Points to check on a valve for wear. If the valve stem has even the slightest ridge or groove, the valve must be replaced.*

Most valve faces are cut at an angle of 45° to the stem. A 30° angle is also used. In either case, a *slight interference angle,* 0.5° to 1°, may either be cut on the valve face or valve seat to improve the seating ability, *Fig. 13.54 (a), (b).* However, it is invariably cut on the valve face because valve refacers can be easily adjusted to any desired grinding angle, *Fig. 13.55 (a), (b).* You must also grind the new valve as you do the old, used valves. The valve-grinding machine is also used to grind the valve stem tip, *Fig. 13.56 (a), (b).* Grind the tips until all pitting or mushroom is gone. There are grinding limits given in shop service manual. If the valve stem tip will not clean up within specified limits, replace the valve.

Always wear a face shield when grinding valves. The stone could shatter, throwing debris into your face.

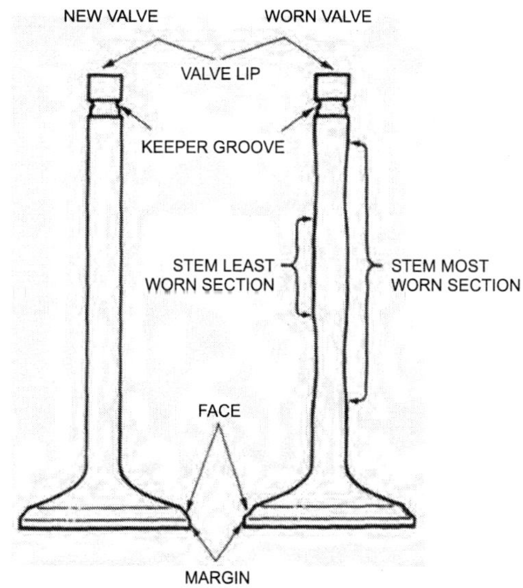

Fig. 13.53 *(a) Examples of good and a worn valve. (GM)*

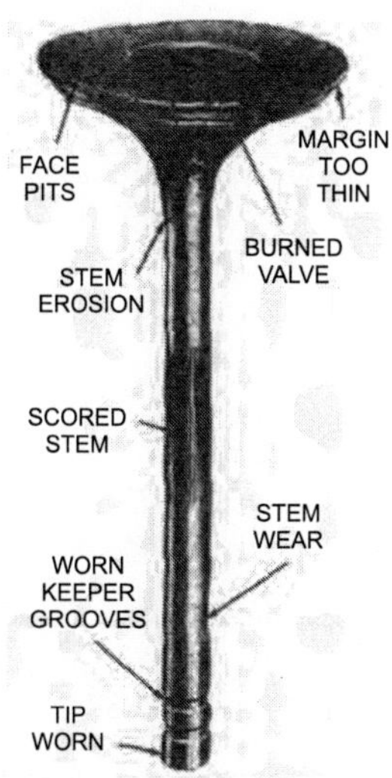

Fig. 13.53 *(b) Some troubles that are found with used valves. (Sioux Tools)*

Valve Grinder Setup

Before grinding the valves, *dress the stone* by using a diamond cutter to true the stone surface. A diamond tipped cutting attachment will be provided with the machine for truing the stone. Follow the equipment manufacturer's instructions.

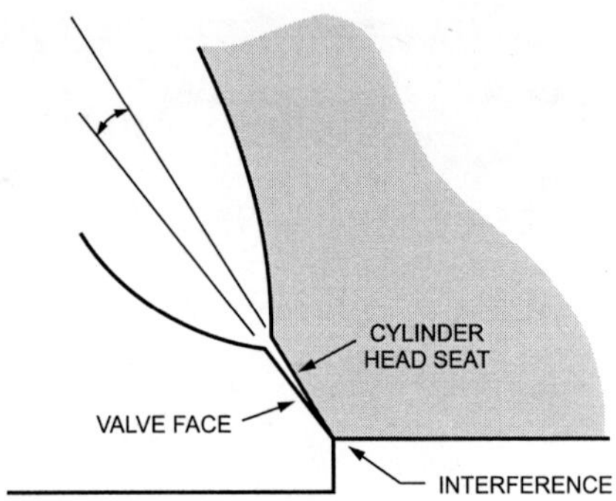

(a) Interference angle is used when engines burn leaded fuels. This chips the deposits off of the valve face. (General Motors)

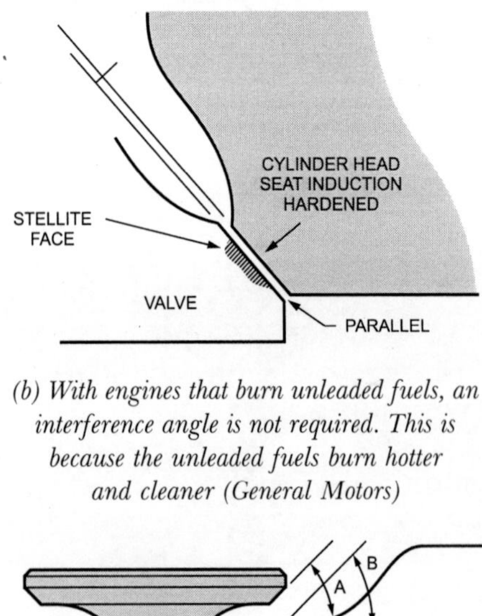

(b) With engines that burn unleaded fuels, an interference angle is not required. This is because the unleaded fuels burn hotter and cleaner (General Motors)

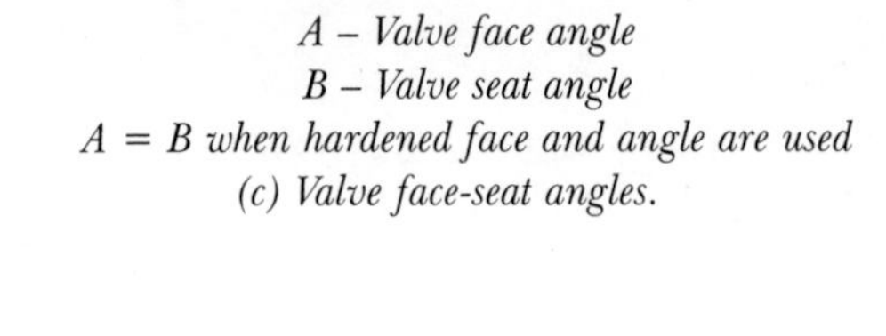

A – Valve face angle
B – Valve seat angle
A = B when hardened face and angle are used
(c) Valve face-seat angles.

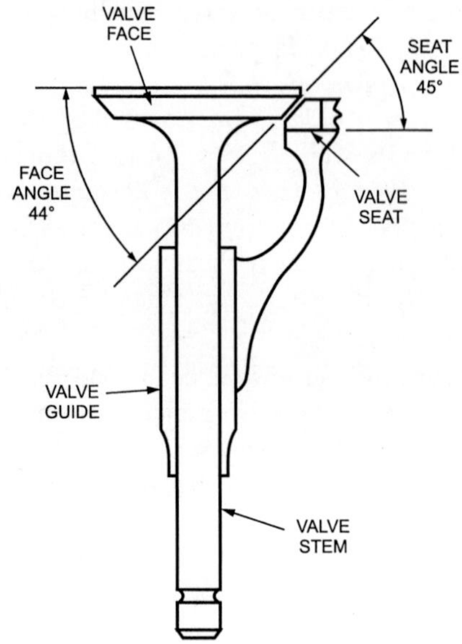

(d) Some vehicle manufacturers recommend that the valve face be resurfaced at a 44° angle and the valve seat be resurfaced at 45° angle.

Fig. 13.54

Set the chuck angle by rotating the valve grinding machine chuck assembly. A degree scale or digital readout is provided so that the cutting angle can be set precisely. Normally, a locknut is loosened and the chuck assembly is swiveled into the desired angle as illustrated in *Fig. 13.57 (a).*

An *interference angle* is normally about 1° difference in the valve face angle and the valve seat angle. It is set on the valve grind machine. This set up reduces the break in and sealing time of the valve, *Fig. 13.57 (b).*

Chuck the valve in the valve grinding machine by inserting the valve stem into the chuck. Ensure that the chuck grasps the *machined surface* of the stem near the valve head, because if the chuck clamps on to an un-machined surface of stem runout will result.

Grinding Valve Faces

Turn on the valve grinding machine and the cooling fluid. *Slowly* feed the valve face into the stone. While feeding, slowly move the valve back and forth in front of the stone. Use the full face of the stone not allowing the face move out of contact with the stone while cutting, *Fig. 13.58.*

Fig. 13.55 *(a) Valve grinder. (Sioux Tools)*

Fig. 13.55 *(b) Grinding the valve face. (Sioux Tools)*

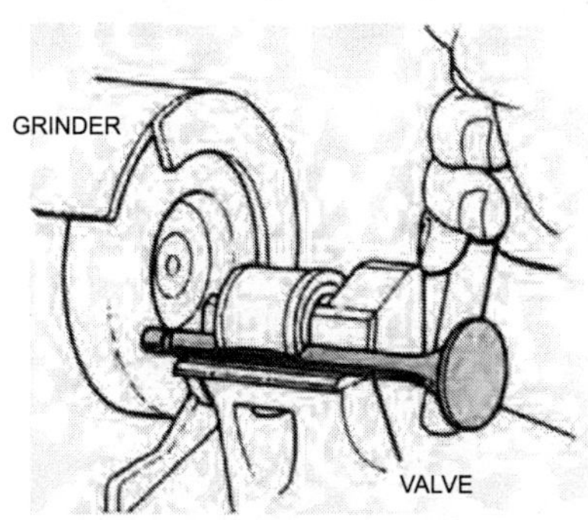

Fig. 13.56 *(a) Grinding valve stem tips.*

Valve faces are ground on a precision valve-grinding machine, *Fig. 13.55.* The valve is first chucked or mounted in a chuck-head that is adjusted to the correct angle for the valve face. Moving the

Fig. 13.56 *(b) Grinding material from tip of valve. (Sioux Tools)*

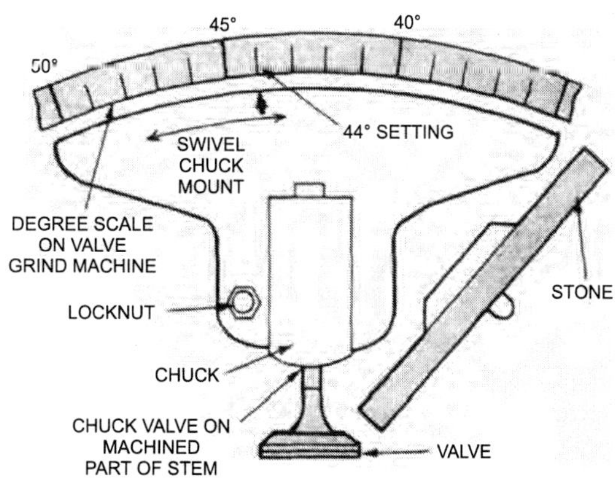

(a) If the valve has 45° face, adjust grinder to 44°. Scale or digital readout will be provided. Also, chuck valve so as to be resting on machined portion of stem near valve head.

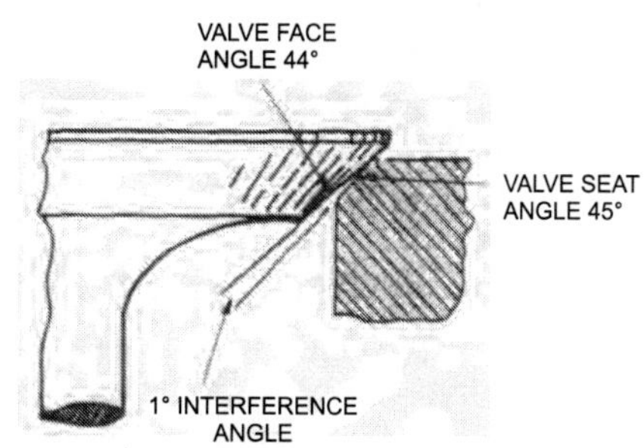

(b) Note how interference angle makes valve and seat touch such that contact area does not widen rapidly in service and to speed initial valve sealing.

Fig. 13.57 *Before grinding valves, adjust grinding stone angle and dress the stone.*

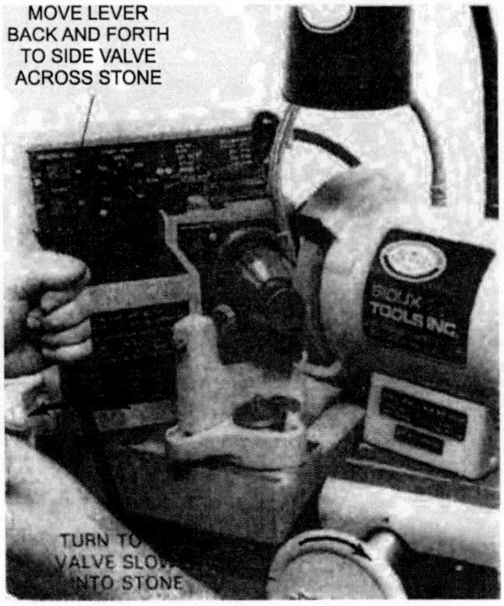

Fig. 13.58 *To grind the valve face, turn wheel to slowly feed the valve into rotating grinding stone. Adjust the coolant to flow over the valve. As soon as stone touches valve, begin using lever to make valve back and forth over stone. Continue grinding until the valve face is shiny and free of pits. (Sioux Tools)*

valve slowly and steadily across the grinding wheel grinds the valve. Light cuts are taken by feeding the grinding wheel up to the valve just enough to make a clean, smooth face. When the valve face is ground the grinding wheel is backed away from the valve and the valve is removed from the chuck.

After grinding, check the valve against specs for an adequate margin. The margin is the thickness of the valve head. *Fig. 13.59* gives the angles and dimensions important to the valve grinding. Badly pitted, burned, or warped valves cannot be refaced without removing too much metal along the edge or margin. Loss of margin gives the head of the valve a sharp edge that cannot take high temperatures. It may not be thick enough to dissipate heat fast enough. The head of the valve can actually melt, burn, and blow out the exhaust port, *Fig. 13.60*. If the margin is not within the specified thickness after grinding, the valve must be replaced.

A *burned valve,* if not observed during visual inspection, will show up when excess grinding is necessary to clean up the valve face as a normal amount of grinding will not remove a deep pit or groove. Replace the valve if burned beyond specs.

A *bent valve* will show up when the valve head wobbles as it turns in the valve grinding machine

Fig. 13.59 *Valve angles and dimensions. (Chrysler Corp.)*

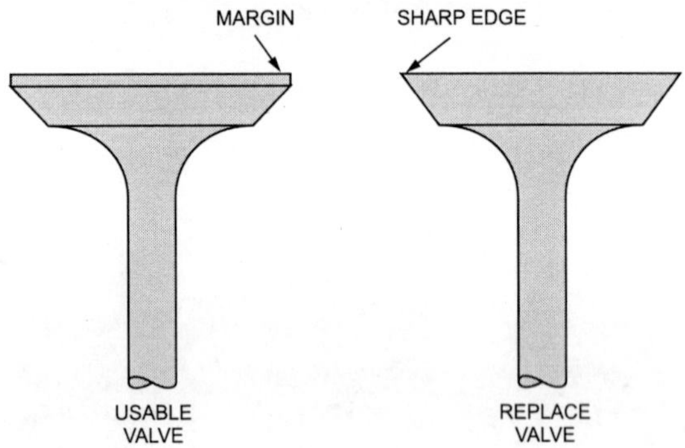

Fig. 13.60 *After valve grinding, check margins on valves. If thinner than specs, replace the valve. If margin is too thin, valve could burn easily during engine operation. (TRW)*

chuck. If the valve is chucked improperly, it can behave like a bent valve. Shut off the machine and isolate the cause for wobbling, *Fig. 13.61*.

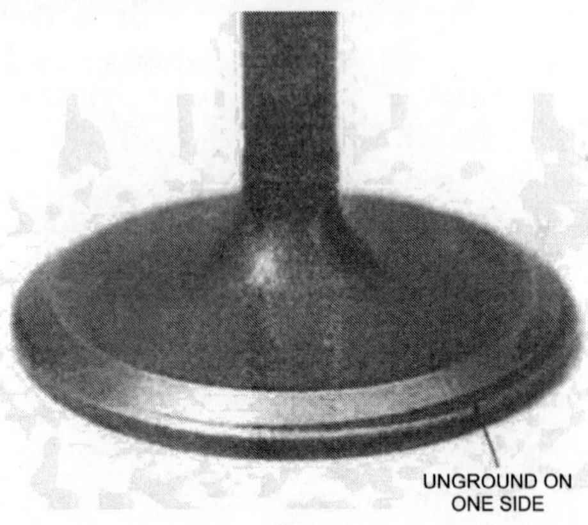

Fig. 13.61. *(a) This valve had to be replaced. It was bent slightly and eroded. Only one side of face cleaned up when ground.*

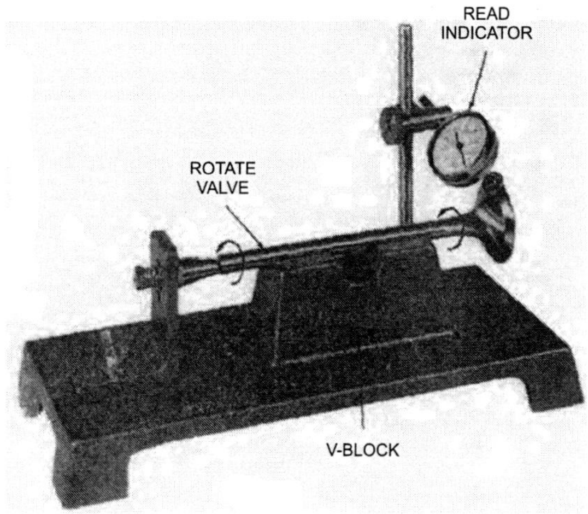

(b) *V-blocks and indicator are used to check valve runout, but it is time consuming. (K-Line Tools)*

Fig. 13.61

Repeat the grinding and inspecting operation on all the remaining valves. Return each ground valve to its place in an organizing tray. Reconditioned valves should be returned to the same valve guide in the cylinder head. The valve stems could have been select fit.

Grinding Valve Stem Tip

A separate stone on the valve grinding machine is usually provided for truing and chamfering the valve stem tips. *Figs. 13.62* and *13.63* illustrate the method of grinding and chamfering valve stems.

Grind just enough off the stem to correct the tip surface. Many valve stems are hardened. Too much material removal will cut through the hardened material layer and result in rapid wear of stem tip face when the valve is returned to service.

An indicator is provided on the valve-grinding machine to show the depth of cut for both the valve face and valve stem tip. *Cut the same amount of material* off the face and the stem tip. This will help keep valve train geometry correct.

13.12 VALVE SPRING SERVICE

Valve springs tend to weaken, lose their tension, or even break after extended service. Always test valve springs to ensure they are usable. All valve springs must be checked for *free height* and *squareness*. Free

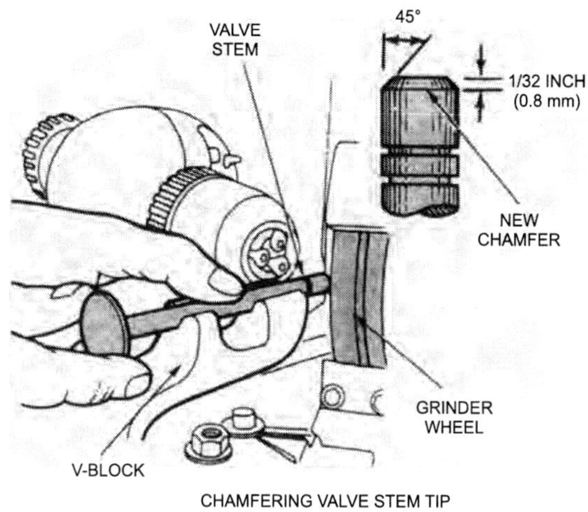

Fig. 13.62 *After grinding valve faces, also true the valve stem tips. Here the technician is machining a small chamfer on valve tip. However, if significant material has been worn off tip, the valve should be replaced. Always check the service manual for specs. (Ford Motors)*

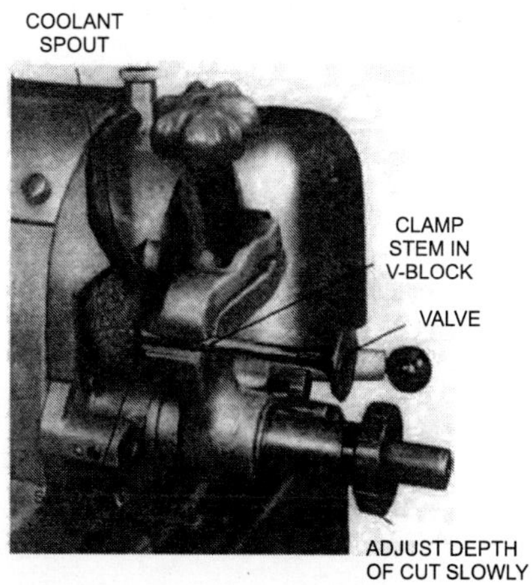

Fig. 13.63 *Note how valve grinding machine can be used to grind valve tip, making it flat and smooth. Mount valve in V-chuck. Slowly feed valve tip into rotating stone. Grind just enough to true the tip. (Sioux Tools)*

height is the height of the spring as it sits on the table. Measure the springs with a steel rule or a vernier caliper, *Fig. 13.64 (a)*, and compare the measurement with the service manual recommended specs. Squareness is measured with a square, *Fig. 13.64 (b)*. To determine if the spring leans, hold the spring in the square's 130° angle, check if the spring bends away. Rotate the spring while checking it to make sure to find the side with the maximum lean.

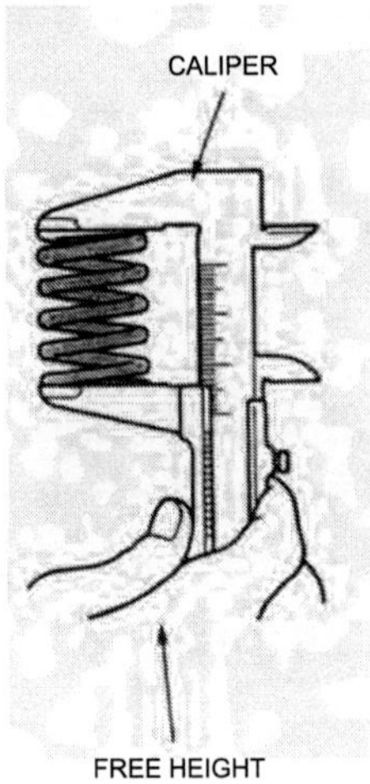

Fig. 13.64 *(a) Measure free height or length.*

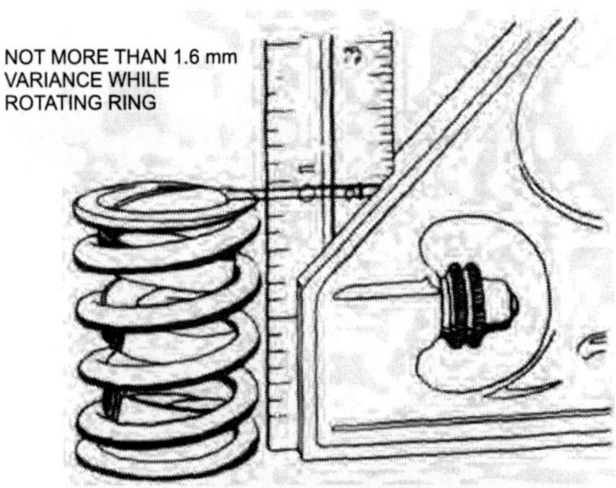

Fig. 13.64 *(b) Inspecting valve spring for squareness. Make sure the spring is square.*

Fig. 13.65 *(a) After grinding valves and seats, install valve, retainer, and keepers as shown. Then measure valve spring installed height. This helps to determine if valve shims are needed. (General Motors)*

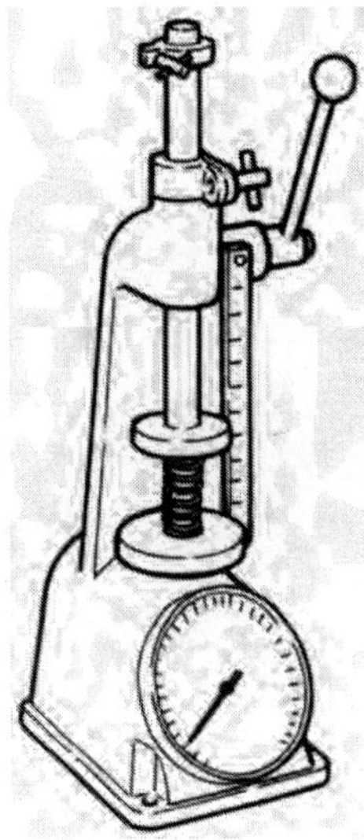

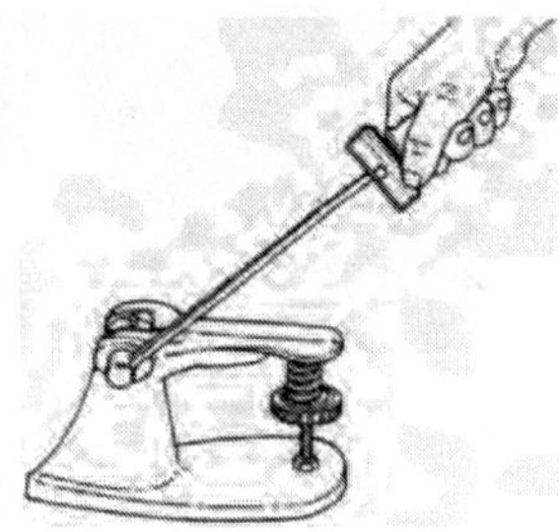

(c) Check spring pressure. *(d) Checking spring tension. (Chevrolet Div. GM)*

Fig. 13.64 *Valve springs must be checked carefully before reuse in engine. They can weaken, lose tension, and break. (Buick Div. General Motors)*

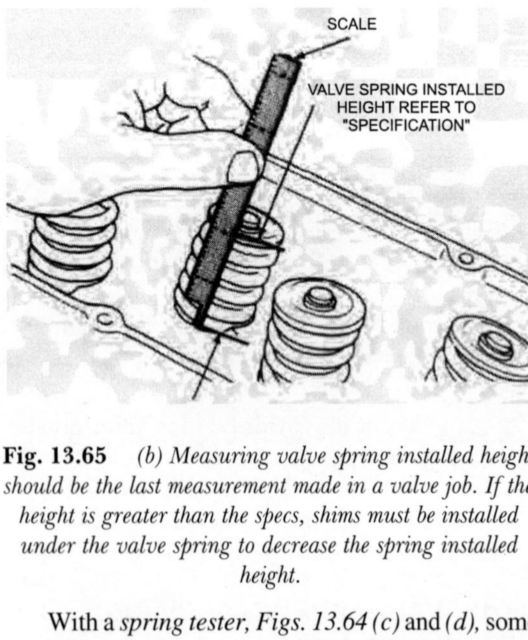

Fig. 13.65 *(b) Measuring valve spring installed height should be the last measurement made in a valve job. If the height is greater than the specs, shims must be installed under the valve spring to decrease the spring installed height.*

With a *spring tester, Figs. 13.64 (c)* and *(d),* some more spring tests are possible. With it, we can compress a spring to a certain height and observe how much load it produces. Loads are checked at the spring's *installed* and *open* heights. Installed height is the height of the spring when it is installed in the cylinder head. It can be measured, as in *Fig. 13.65.* If greater than specs, add enough shims to reduce installed height and restore spring pressure to normal. Open height is the smaller height of spring when it is compressed, as the camshaft lobe tip opens the valve. The load is greater at the open height. Always refer the service manual for spring height and load specifications.

Shimming Valve Springs

Spring tension or pressure is measured with a torque wrench or a tension gauge, *Fig. 13.66.* Springs below the specified tensions are either replaced or spacers called *shims* are used under the springs, when they are reinstalled, *Fig. 13.67. Shimming valve springs* is done to maintain correct tension when the springs are installed on the cylinder heads. When valves and seats are ground, the valve stems stick the head more. This increases valve spring installed height and causes reduced spring pressure, *Fig. 13.67.*

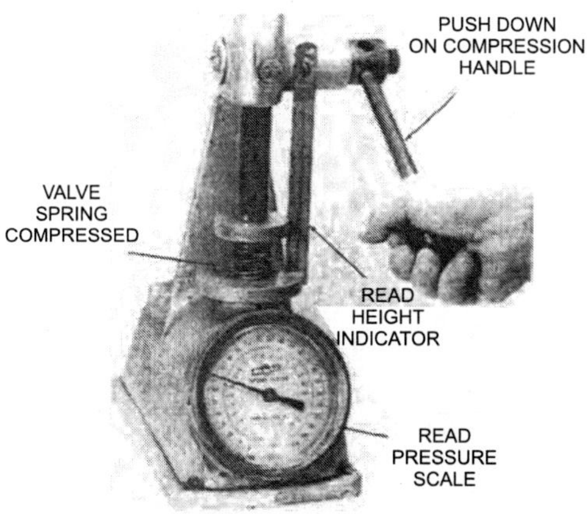

Fig. 13.66 *Use valve spring tester to compress spring until scale reads the specified pressure. Then read height indicator. Difference in tester height and spring installed height would equal the thickness of shim needed to attain correct spring pressure.*

To use a valve spring compressor, place a valve spring on the compressor. Press down on the lever until the scale equals spec pressure for the specific engine springs. Then read the pointer scale to find how far the spring has been compressed. This will allow you to calculate how thick the shims should be to restore the correct spring pressure. If you had to compress the spring 0.5 mm more than the installed

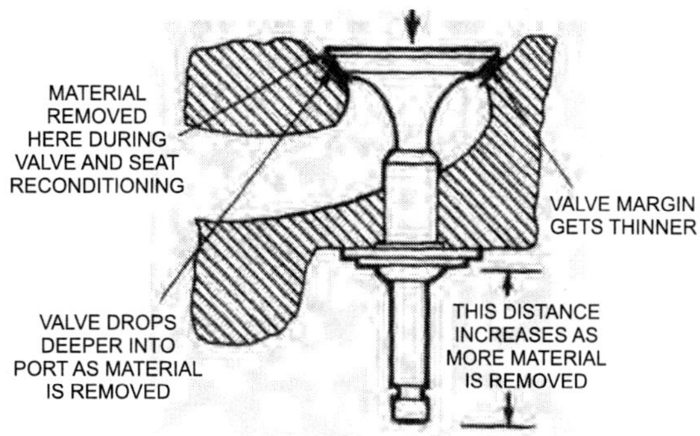

Fig. 13.67 *When valve and seat are ground, valve stem extends through head farther. This increases spring-installed height and reduces spring pressure.*

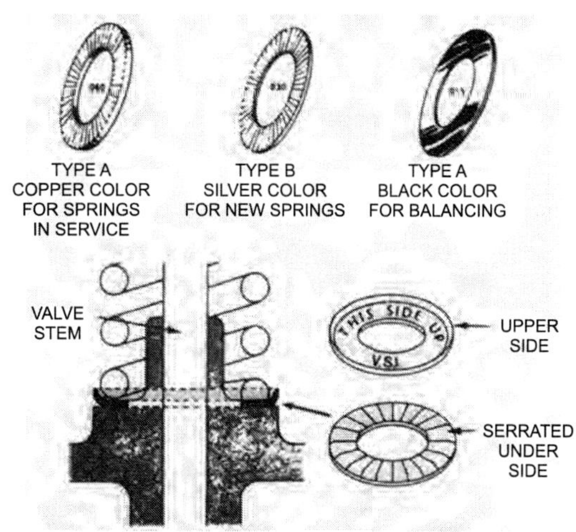

Fig. 13.68 *Types of spring shims. The serrated side should be down when installing shims. (K-Line Tools)*

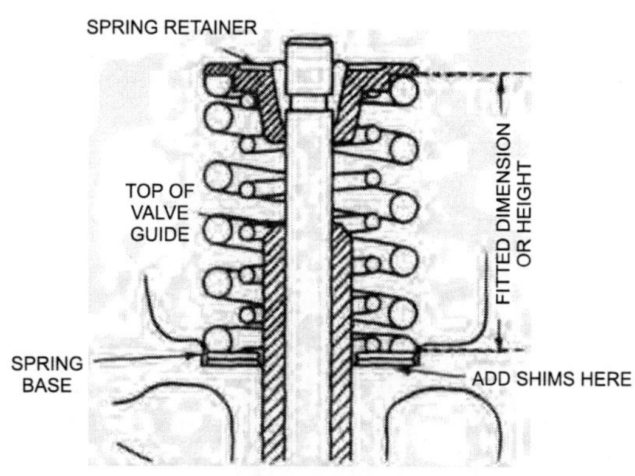

Fig. 13.69 *Do not use shims that are too thick, 1.52 mm is a typical maximum. (TRW)*

spring height to achieve the specified pressure, use a 0.5 mm shim under the spring, *Fig. 13.68.* In general, never shim a valve over 1.5 mm. Thicker shimming could cause spring bind. *Spring bind* causes the spring to solid height (fully compressed) and lock the valve train. This could possibly damage parts, *Figs. 13.69* and *13.70.*

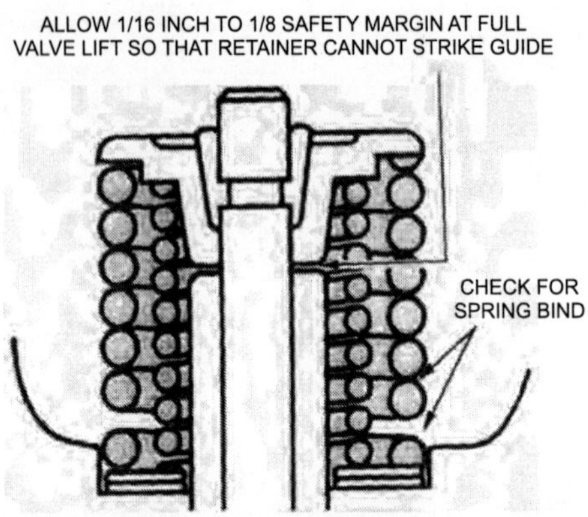

Fig. 13.70 *Spring bind is a result of excessive cam lift or shims that are too thick. New guide must also be short enough to clear spring retainer. (TRW)*

All of these tests check for fatigue in a spring. Weak, fatigued springs cannot accurately control the valve. At higher rpm weak springs allow *valve float.* It happens when the spring cannot overcome the valve's inertia, and the valve, spring, rocker arm, push rod, and lifter do not follow the cam lobe contour. The valve opens farther and stays open longer than it should. This could let the piston hit the valve, bending the valve or more often the push rod. Severe damage can result from floating valves.

Valve Spring Retainers

Valve spring keepers, or *locks,* are usually of the split-cone, horseshoe, or flat rectangular key type. They fit into an appropriate slot in the end of the valve stem. *Fig. 13.71* shows the split-cone type in position. A washer called the *spring retainer* fits over these keepers, and the tension of the spring bearing on the retainer holds the keepers in place.

Removal is accomplished by holding the valve stationary while the spring is compressed enough to allow the retainer to be raised from the keepers, *Figs. 13.72* and *13.73.* The keepers are then removed and the valve spring released. This permits the spring and retainer to be removed. Valves should never be mixed up when removed unless it is known that new valves or guides are to be installed. A valve should always be replaced in the guide from which it was removed.

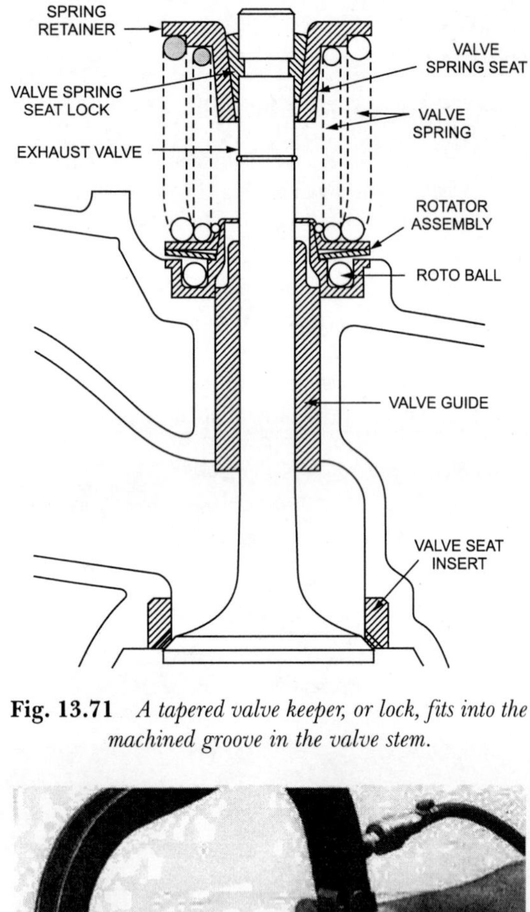

Fig. 13.71 *A tapered valve keeper, or lock, fits into the machined groove in the valve stem.*

Fig. 13.72 *A valve spring compressor must be used to overcome valve spring pressure. Once the valve spring is compressed, the valve keepers can be removed or installed. (Fel-Pro Inc.)*

Fig. 13.73 *Compressing valve springs on an overhead camshaft engine. (Chrysler Corp.)*

13.13 CYLINDER HEAD REASSEMBLY

After washing all parts in solvent to remove cleaning and machining abrasives, inspection, measurements, valve guide service, valve seat grinding, and valve grinding, the cylinder head can be assembled. Place a drop of oil on the valve stems and slide into their respective guides.

Installing Valve Seals

Depending upon whether the head has umbrella or O-ring type valve seals, you will have to modify assembly procedures accordingly. With *umbrella valve seals,* simply slide the seals over the valve stems. Then install the valve spring assembly, *Fig. 13.74 (a).* With *O-ring valve seals,* compress the valve spring *before* fitting the seal on the valve stem. If the seal is installed first, it might be cut, split, or pushed out of its groove. This will cause engine oil consumption and smoking will result, *Fig. 13.74 (b).*

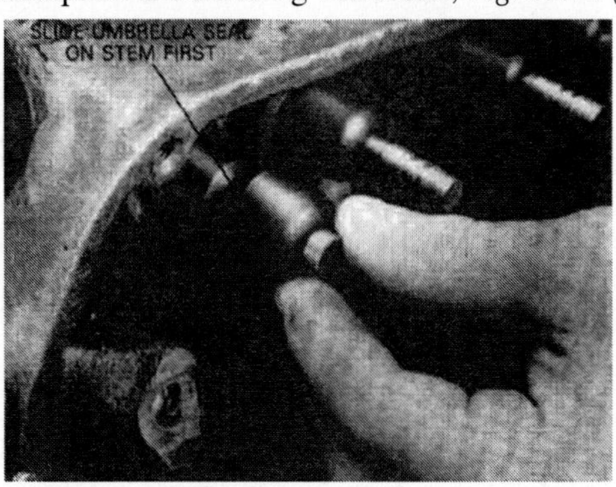

Fig. 13.74 *(a) With umbrella type seal, slide seal over valve first.*

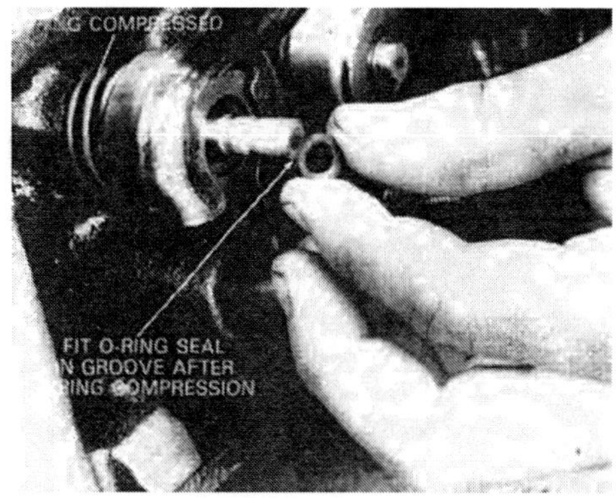

(b) With O-ring seal. Compress the spring first. Then fit seal into groove in valve stem. O-ring installed first, can be cut or torn, causing engine to smoke due to oil consumption.

Fig. 13.74 *Sequence for valve seal installation varies. (Fel-Pro Gaskets)*

Figure 13.75 shows the method of installing a *positive type valve seal.* The top of the valve guide has been machined so that the seal locks onto the guide. First, a protective cap is placed over the valve stem. Then, an installation tool is used to force the valve seal into its position.

In vehicle Valve Seal Service

Valve seals and springs can be serviced without cylinder head removal. Use a shop compressed air hose and a special fitting to inject air into one of the cylinders. The air will hold the valves in that cylinder up against their seats. Then, use a special pry bar type compressor to remove the keepers and springs for that cylinder. This will permit in-vehicle seal or spring replacement.

Installing Valve Springs

Compress the spring and fit the retainer and keepers into their proper position on the valve. Install all of the valve assemblies. Tap on the valve stems with a brass hammer to seat the keepers in their grooves, *Fig. 13.76.* A light tap with a plastic hammer to each valve spring assembly will test the installation. If the keepers are not fully seated, they will pop out when hit.

A valve spring compressor is used for installing the valve springs on an OHC engine, *Fig. 13.77.* A spring compressor is needed to reach down into the valve spring pockets.

(a) Protective cap should be placed over stem to protect rubber seal.

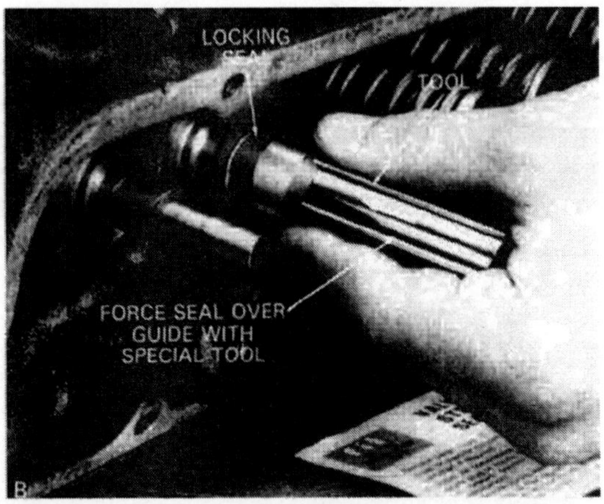

(b) Special tool is required to force seal over valve guide.

Fig. 13.75 *Positive valve seals lock around valve guides. Guides must be machined for these seals. (Fel-Pro Gaskets)*

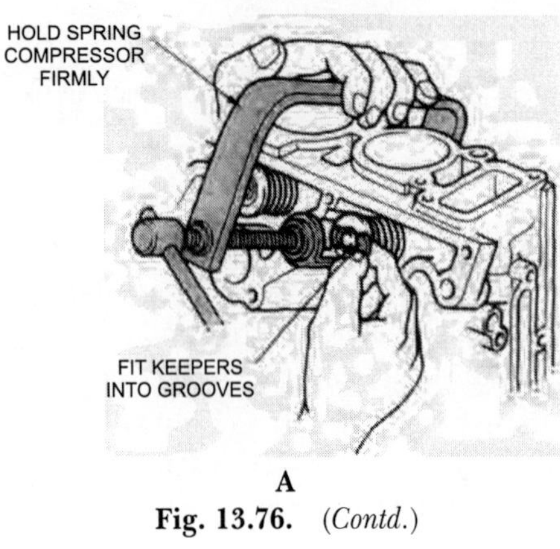

A

Fig. 13.76. *(Contd.)*

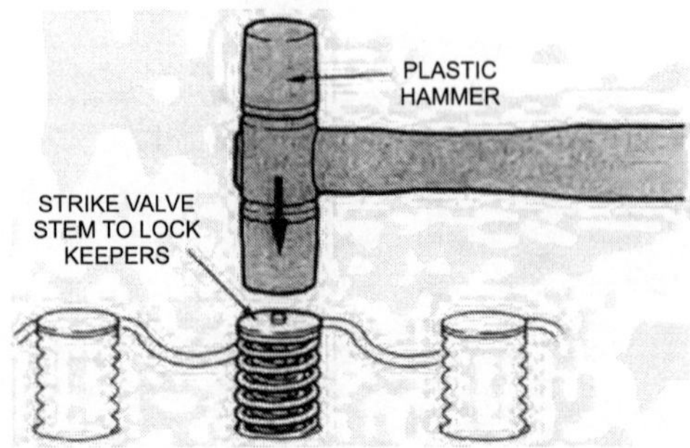

B

Fig. 13.76. *A – Use valve spring compressor to assemble cylinder head. Hold tool securely while fitting keepers into grooves in valve stem. B – Use plastic hammer to hit valve stem tips. This will set keepers in grooves. (Toyota)*

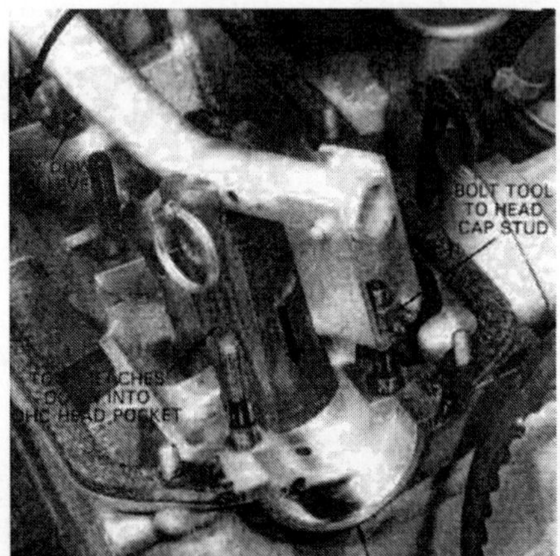

Fig. 13.77. *A special valve spring compressor needed on this overhead cam cylinder head. Note how it is fastened to camshaft cap stud. (American Motors Corp.)*

Checking for Valve Leakage

To check for valve leakage after head reconditioning, lay the cylinder head on its side. Pour clean soak solvent or water into the intake and exhaust ports. With the fluid in the ports, watch for any leakage around the valve heads. If solvent or water drips from around a valve, that valve is not properly sealing. Remove the valve from the head and check for any problems, *Fig. 13.78.*

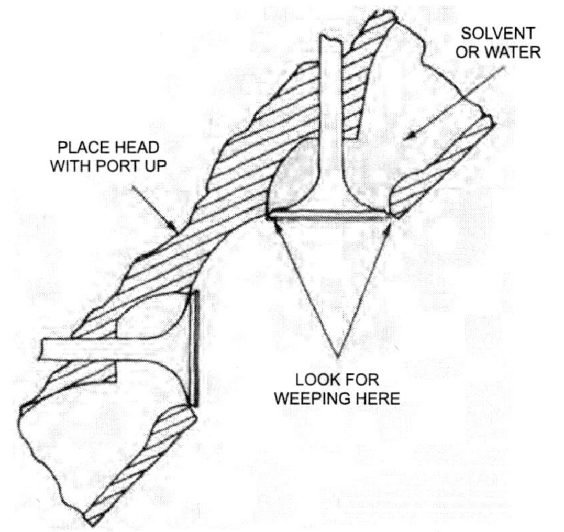

Fig. 13.78 *To check for valve leakage before head installation, lay head with ports up. Pour clean solvent or water into ports. Observe for any leakage or weeping at valve heads. If any leak is found, disassemble and isolate and correct the problem.*

13.14 ROCKER ARM STUD SERVICE

Rocker arm studs may need replacement if their threads are damaged or if they pull out of the head. In some engines these studs are press-fit in the head, while others screw into the head. *Fig. 13.79* illustrates the steps to replace pressed-in cylinder head rocker arm studs.

13.15 CAMSHAFT SERVICE

Some wear does occur on *camshafts,* as on any other engine part. Since the camshaft operates at slower speed as compared to the crankshaft, wear usually is less pronounced. The cams and journals are hardened therefore, wear normally occurs in the bushings rather than on the camshaft journals.

The *lift* of a cam is the difference between measurements AA and BB, *Fig. 13.80.* The amount of wear on lobe is obtained by comparison with specified lift.

Usually the cams wear at different rates therefore, the lift of all the cams should be measured to determine if any are worn out of specs. Worn cams are often overlooked in troubleshooting. These may occasionally be the cause of lost power or engine misfiring.

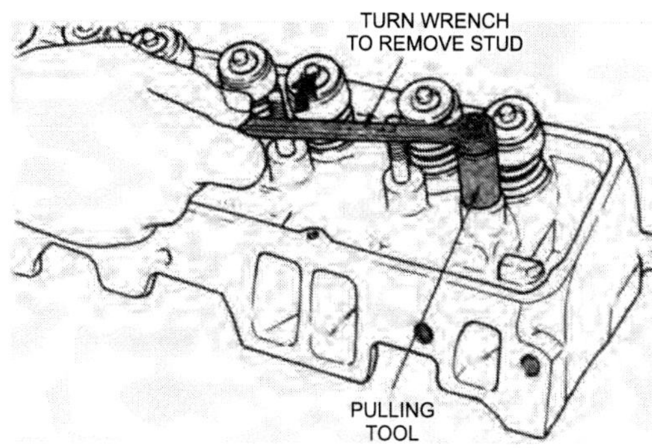

(a) If not screwed in, use puller to remove old stud.

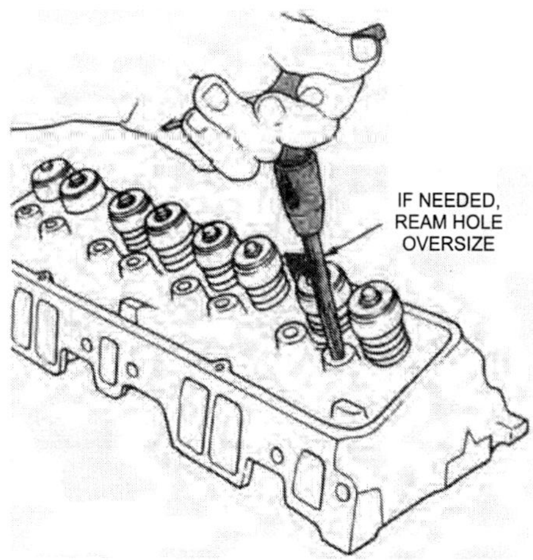

(b) If necessary, ream hole for oversize stud, as when old stud is pulled out of head.

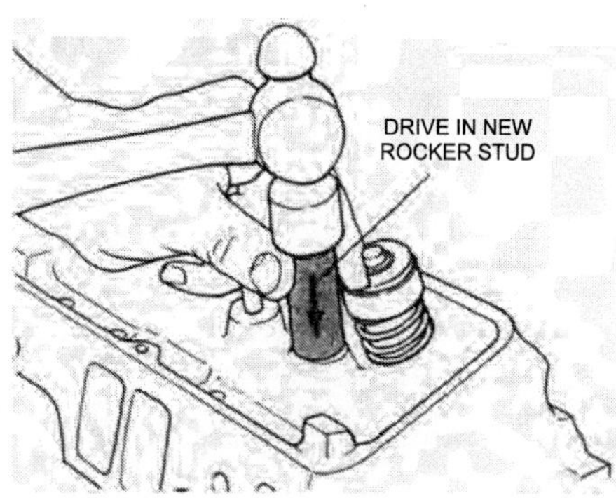

(c) Drive new stud in place.

Fig. 13.79 *Rocker arm stud replacement. (Chevrolet -GM)*

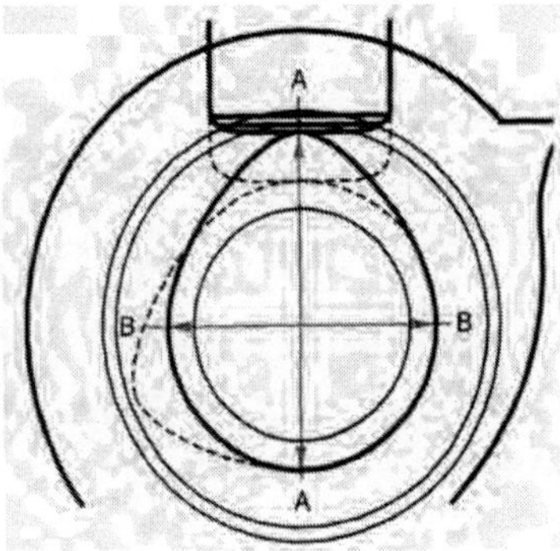

Fig. 13.80 *The difference between line AA and line BB is the amount of cam lobe lift.*

Camshaft service includes measuring cam lobe(s) and journal wear, camshaft end play, measuring cam bearing clearance, distributor/oil pump gear inspection, and cam bearing replacement.

Measuring Camshaft Wear

Cam lobe wear is measured with a dial indicator with the camshaft installed in the engine, *Fig. 13.81.* Cam lift can also be measured by mounting a dial gauge on the cylinder head with its button contacting the upper end of the valve push rod. Then, as the engine is slowly cranked, the distance from the lowest point to the highest point will be the lift of the cam, *Fig. 13.82.* With the camshaft out of the engine, an outside micrometer can be used to measure the cam lobe size, *Fig. 13.83 (a).*

If the lift of any cam(s) is out of specs (worn over 0.13 mm), the cam is worn and camshaft should be reground or replaced (preferably).

Camshaft journal wear is measured with an outside micrometer, *Fig. 13.83 (b).* If worn to be out of specs (more than 0.05 mm) replace the camshaft. Journal wear will increase the camshaft bearing clearance and cause lower engine oil pressure.

Each journal of the camshaft should also be checked for *runout/Camshaft straightness.* Though not common, camshaft runout can be measured with V-blocks and indicator, *Fig. 13.84.* Turn cam and read indicator. If outside specs, the cam is bent and should be replaced.

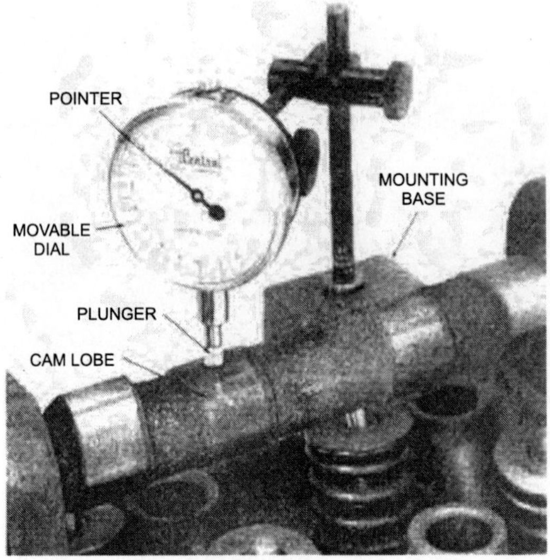

Fig. 13.81 *Cam lobe wear can be checked before engine teardown as shown. Mount dial indicator against cam lobe. Rotate cam and measure lift. If below specs, lobe is worn and cam replacement or grinding would be needed, (Central Tool)*

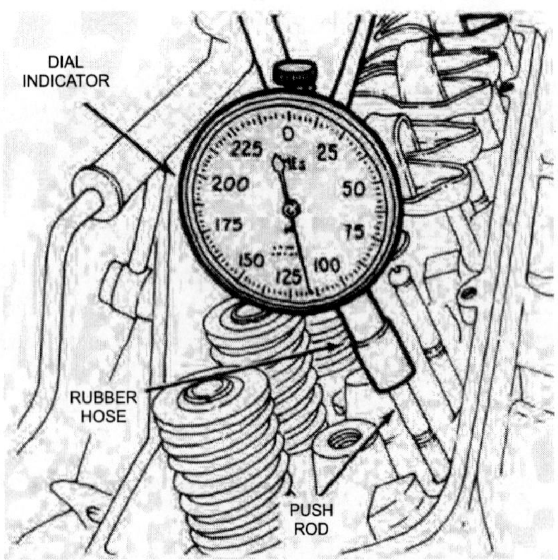

Fig. 13.82 *Using a dial gauge to measure cam lift.*

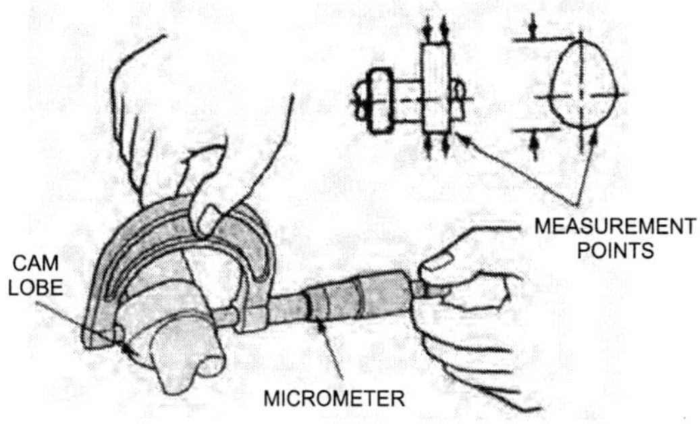

(a) Measure lobe size and compare to specs.

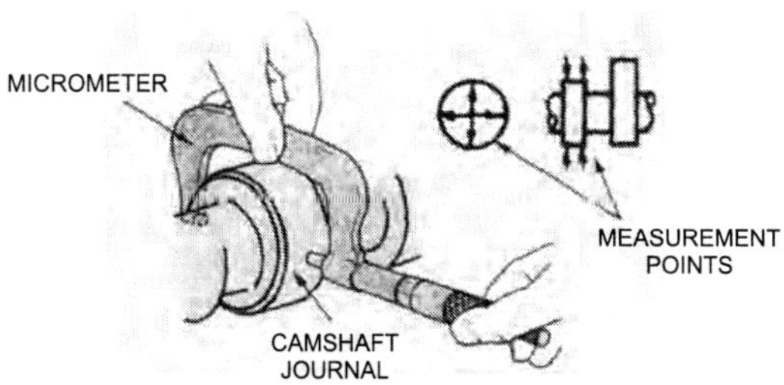

(b) Measure journal diameter and compare to specs.

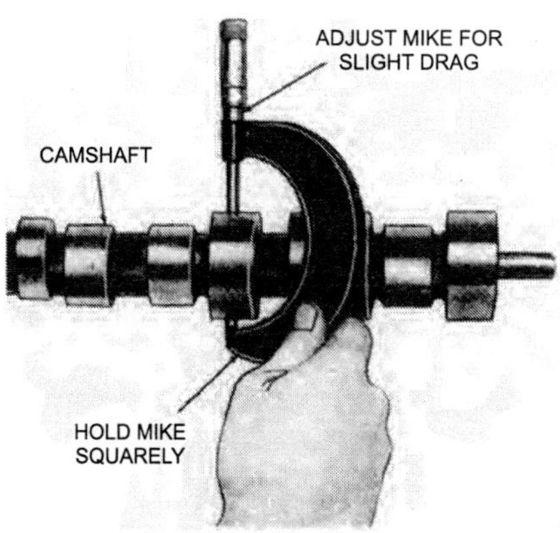

(c) When using micrometer, hold it squarely and adjust it for slight drag when pulled over cam. (Deere and Co.)

Fig. 13.83 *With cam out of engine, use micrometer to check camshaft condition. (Toyota Motors)*

Camshaft Grinding

Camshaft grinding is done to restore cam lobe and journal dimensions to within specifications. A cam-grinding machine regrinds all the cam lobes, *Fig. 13.84.* If the maximum total indicator runout (TIR) is measured to be less than 0.05 mm, the crankshaft can be restored to original lift and duration by one of two methods:

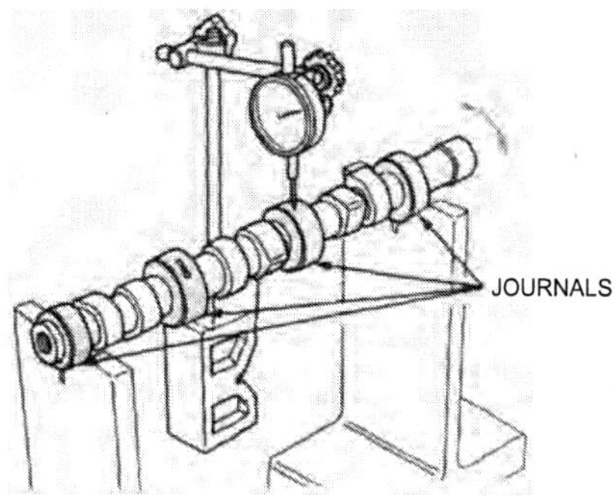

Fig. 13.84 *Each journal of the camshaft should be checked for runout. If runout exceeds 0.038 mm on any journal, replace the camshaft. (Nissan Motors)*

1. If the camshaft is not excessively worn (less than 0.75 mm), then the lobes can be reground by decreasing the diameter of the base circle so as to restore the original lift and duration, *Figs. 13.85* and *13.86.*
2. If the cam lobe wear is excessive, the lobes can be welded and reground back to their original specifications.

A ground camshaft is usually much cheaper than a new camshaft. So long as the lobes are hardened properly, it can give good service life.

Camshaft End Play

Camshaft end play is the front to rear movement of the camshaft in the cylinder head or cylinder block. It is usually measured with a dial indicator, *Fig. 13.87.*

Mount the indicator such that its stem is parallel with the camshaft centreline. Zero the indicator and then pry back and forth on the camshaft, and carefully observe the indicator dial. Indicator needle movement gives end play of the camshaft. If the end play is out of specs, repair or replace parts as necessary. The flange on the camshaft could be worn or the thrust surface on the head or block, depending on the camshaft location, could be damaged.

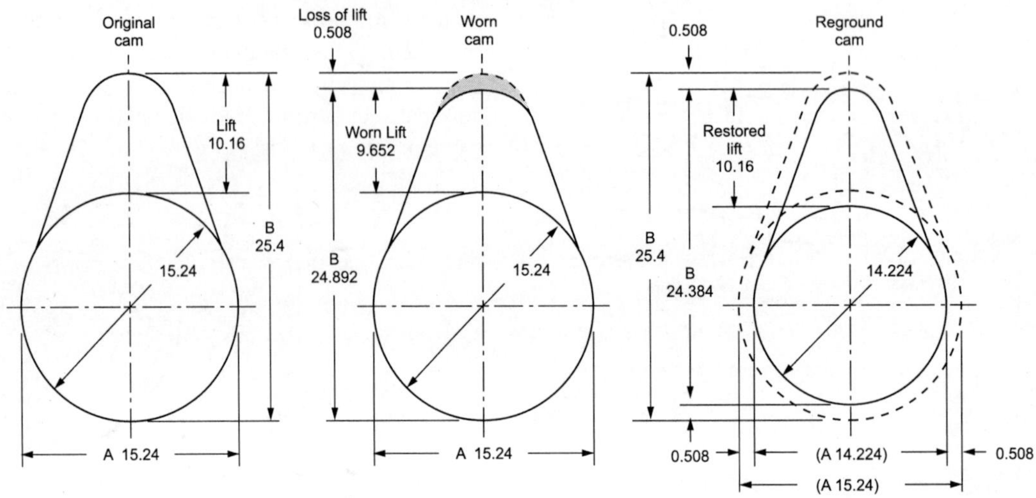

B – A = Lift
B 25.4 – A 15.24
= 10.16 Lift

B – Loss of lift 0.508 = 24.892
B 24.892 – A 15.24 = 9.652
9.652 Worn lift

B 25.4 – (2 × 508 = 24.384
A 15.25 – (2 × 0.508 = 14.224
B 24.382 A 14.224 = 10.16
10.16 Restored lift

Fig. 13.85 *A worn camshaft lobe can be restored to its original lift by grinding the base circle smaller*

Fig. 13.86 *(a) A camshaft being ground on a camshaft grinding machine.*

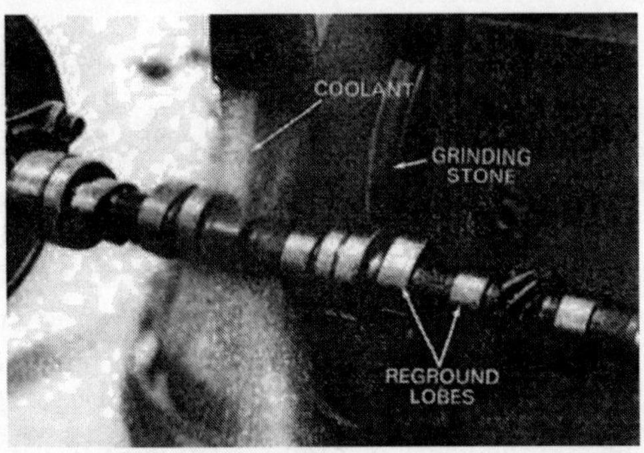

Fig. 13.86 *(b) This camshaft is being reground by machine shop. This will bring lobes to within specs. (Landis Tool)*

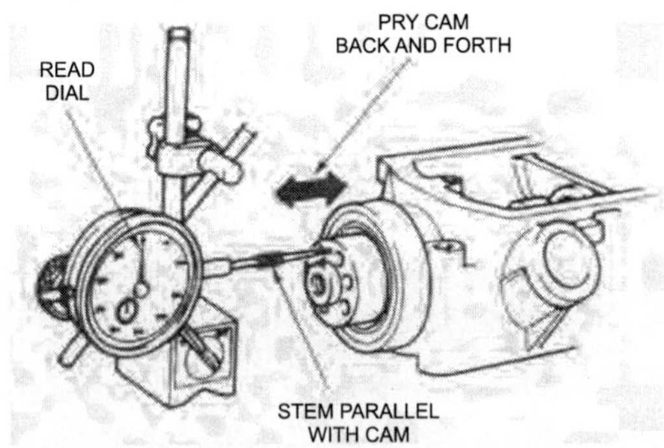

Fig. 13.87 *Too much camshaft end play can result with part wear. Measure it with indicator by shifting cam back and forth in head or block. (Toyota Motors)*

Camshaft Bearings

In some applications camshafts ride directly in the metal of the crankcase, but in general they run in *camshaft bearings* in the form of bushings (one-piece design, or inserts two-piece design). These bushings or inserts are not adjustable for wear and should be replaced when worn. Cylinder head camshaft bearings are serviced much like the same manner as the in-block camshaft bearings. Check them for wear and damage. Worn bearings can reduce engine oil pressure. The extent of wear that dictates replacement is commonly determined by oil clearance rather than a tendency toward noise. However, if seriously worn, the bushings or bearings can cause noise and vibration in the timing gears and valve train. Always ensure that the cylinder head or cam bearings are installed in their correct positions. If mixed up or turned backwards, the cam bearing bore will be thrown out-of-round, *Fig. 13.88 (b)*. Also make sure the oil holes align. Avoid all damage to bearing surfaces during installation. With two-piece cam bearings (inserts), the bearing oil clearance can be measured with Plastigauge, *Fig. 13.89*. With one-piece bearings (bushings), subtract cam journal diameter from cam bearing diameter to determine the cam bearing clearance (oil clearance).

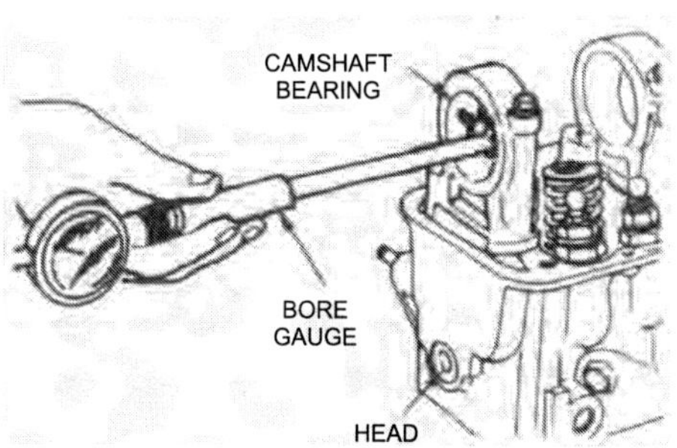

Fig. 13.88 *(a) Dial bore gauge will measure cam bearing wear.*

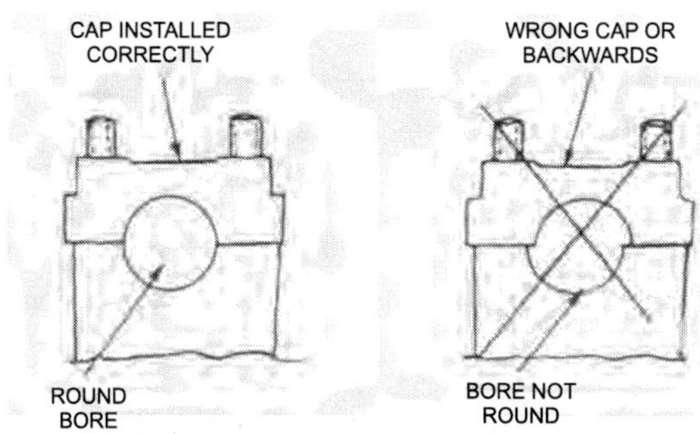

Fig. 13.88 *(b) Make sure camshaft caps are installed correctly. If mixed up or turned backwards, cam bore will not be round. Cam will lock in head when caps are torqued to specs. (VW and Nissan)*

Camshaft bearing wear is checked by measuring camshaft bearing diameter with a telescopic gauge or dial bore gauge, *Fig. 13.88 (a)*. If worn more than specs, i.e., 0.05 mm usually, replace the bearings. Camshaft must be removed to replace the camshaft bearings. This involves removal of the valve operating mechanism. It is seldom necessary to replace a camshaft bearing or bearings until the engine is dismantled for other work. Usually, mechanics replace camshaft bearings during major engine service. These bearings are not very expensive but are important to engine performance and service life.

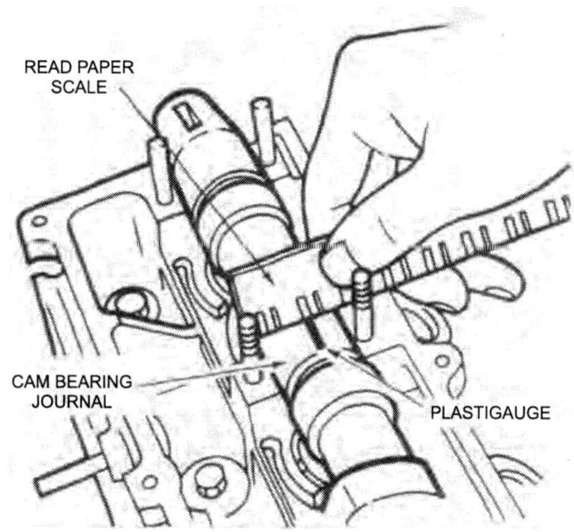

Fig. 13.89 *Cam bearing clearance on OHC engines with two-piece bearings can be measured with Plastigauge. Place plastic on journal. Torque to specs and remove cap. Compare smashed plastic bead with paper scale to find clearance. If not within specs, check bearing and journal sizes. (Ford Motors)*

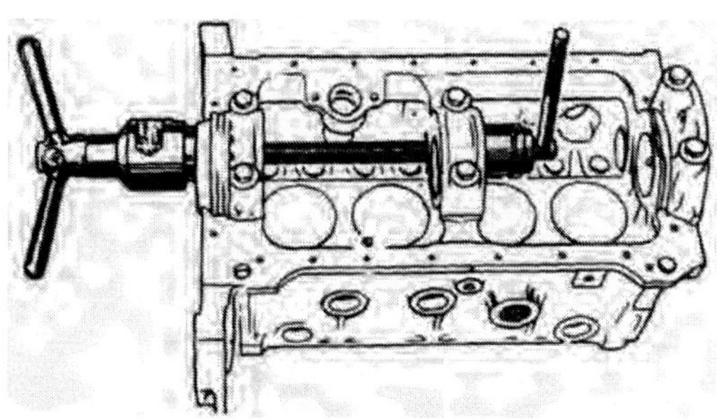

Fig. 13.90 *A special tool is needed to remove and install camshaft bushings.*

After the camshaft is removed, the bushings are pressed out of their bores with a special tool made for this specific purpose. This special tool will also be needed for inserting the new bushings, *Fig. 13.90*.

New bushings are available in the proper outside diameter and standard, as well as undersize, inside diameter. If the camshaft has been undersized by regrinding worn journals, the bushing can be align-bored to any size desired by use of the proper equipment, *Fig. 13.91.*

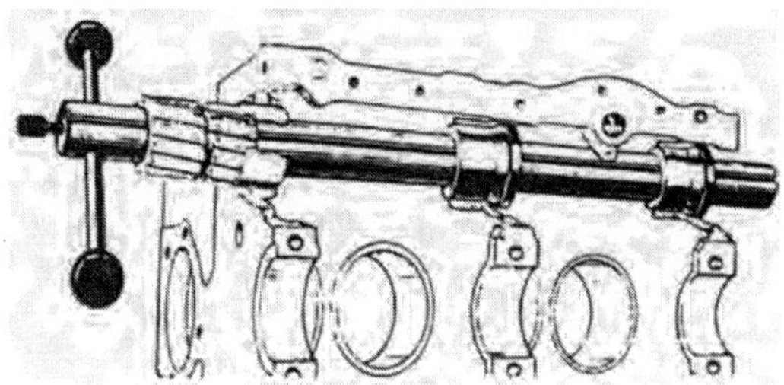

Fig. 13.91 *A special tool is used to align-bore the camshaft bushings.*

It is essential to start the bushing squarely in the bore and apply pressure steadily and evenly. If the bushing cocks in the bore, it will be distorted and it's inside diameter gets decreased. The bushing should be pushed fully into the bearing bore. When precision bearing inserts are used to support the camshaft, they are serviced in the same manner as described for crankshafts bearings.

Normally the camshaft is provided with a *thrust plate.* This holds the camshaft against a flange to eliminate excessive endwise movement. The thrust plate should be replaced if the endplay is outside the specs.

13.16 ROCKER ASSEMBLY SERVICE

Rocker assembly service includes cleaning, inspection, and measurement of the rocker arms, rocker shafts, rocker arm bushings, and other related components. A typical rocker assembly is shown in *Fig. 13.92.*

Check all wear points carefully. Check the rocker shafts for any signs of excessive wear or scoring. Check the rocker arms for wear at the critical locations, *Fig. 13.93.* Check *rocker arms* for wear or pitting on their working surfaces. Because rocker arms are hardened, do not attempt to repair them. Just replace any worn rocker arms.

Inspect the mounting hole inside surfaces on the shaft-mounted rocker arms. Sometimes there is a bushing in the rocker, which can be replaced if it is worn, *Fig. 13.94.* Otherwise replace the entire rocker arm. Inspect the rocker arm shaft to check for wear, *Fig. 13.95.* Clean the rocker arm shaft with solvent. All oil passages should be thoroughly cleaned, or the rocker arms will wear out quickly.

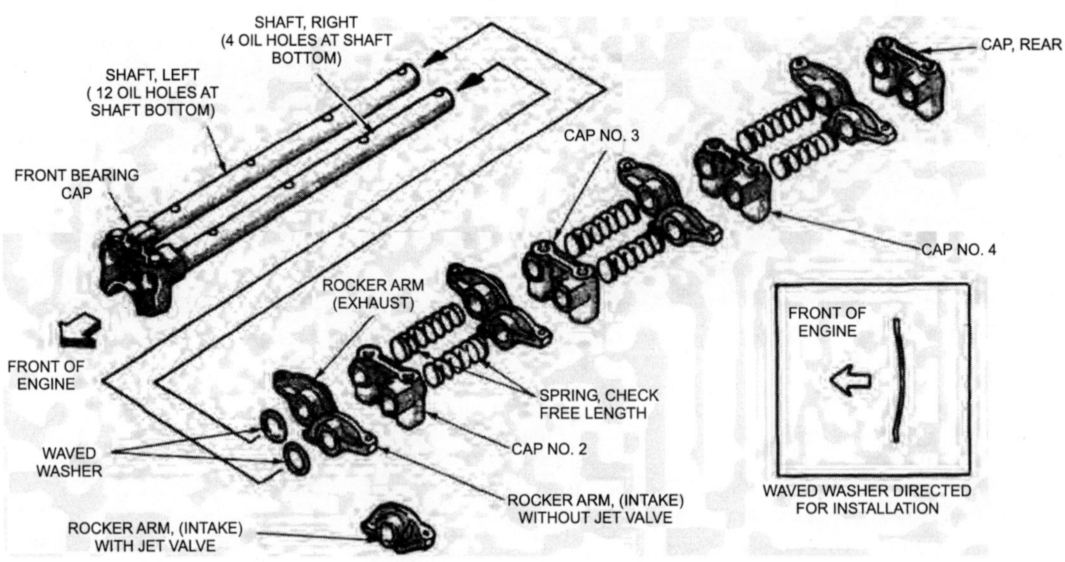

Fig. 13.92 *Careful inspection of all rocker arm parts is important. Rockers suffer from poor lubrication and can wear at high friction points. (Chrysler Corp.)*

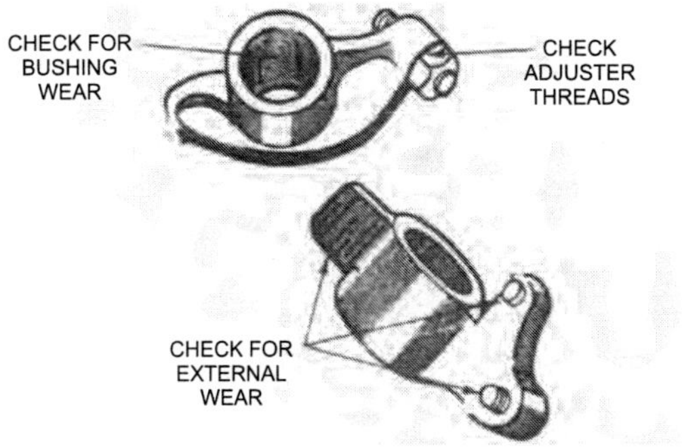

Fig. 13.93 *Check each rocker arm for the kind of problems labelled here. (Chrysler Corp.)*

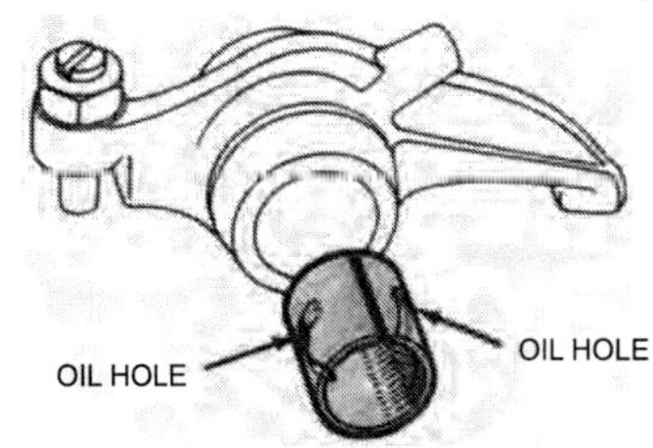

Fig. 13.94 *Replaceable rocker arm bushing. (General Motors)*

Disassemble the rocker arm assemblies for inspection. Remove a small bolt, cotter key, or pin from one end of the rocker shaft. This permits all the rocker arms, spacers, spacer springs, and pedestals slide off the shaft. Carefully, keep all parts in order. Also note and maintain the left/right orientation of the parts before disassembly. The pedestals in a pedestal mount valve train seldom wear. But if they show visual wear, replace them.

A telescoping gauge or dial bore gauge is needed to measure rocker arm bushing or inside diameter wear, *Fig. 13.96.* An outside micrometer is used to check rocker shaft wear. If the wear is found to be more than specs, replace parts or install new rocker arm bushings, if used.

During rocker reassembly, make sure the rocker shaft oil holes point down. This is the loaded side of the shaft and oil is needed at this high friction area. If the oil holes are installed up, rapid shaft and rocker arm wear will result.

For a ball or stand type valve mechanism, inspect the ball. Pivot, or stand for wear. When the rocker arm

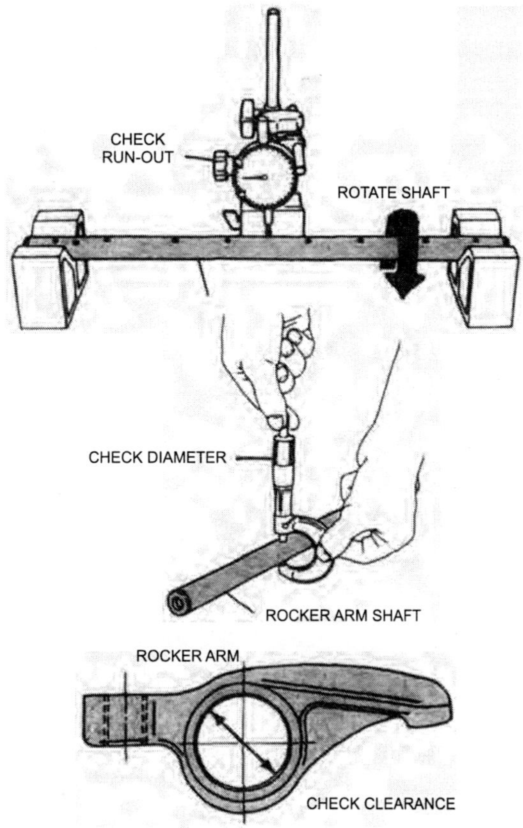

Fig. 13.95 *Rocker arm inspection. (Chevrolet Div.-GM)*

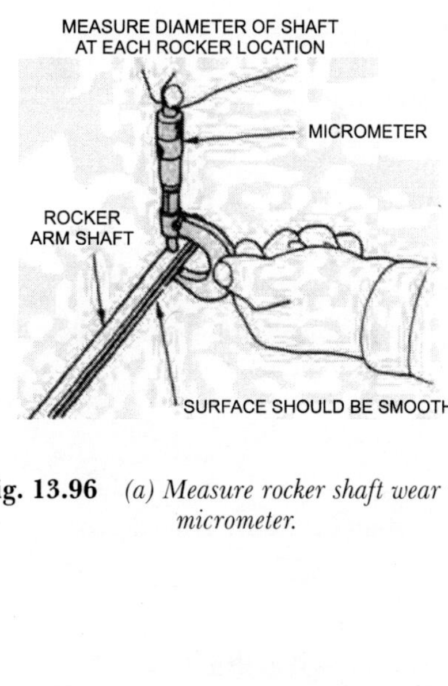

Fig. 13.96 *(a) Measure rocker shaft wear with micrometer.*

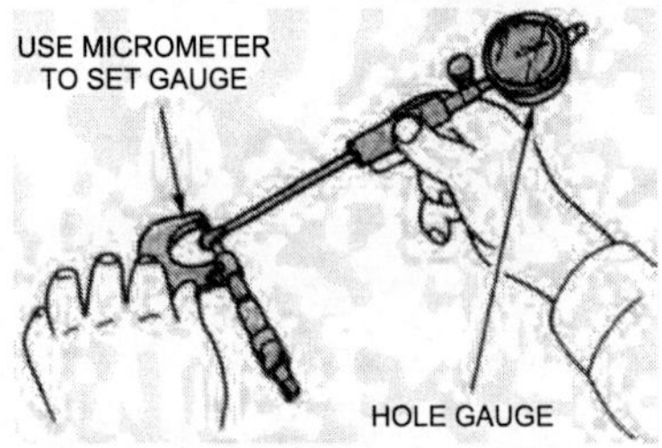

Fig. 13.96 *(b) Adjust dial bore gauge to correct size.*

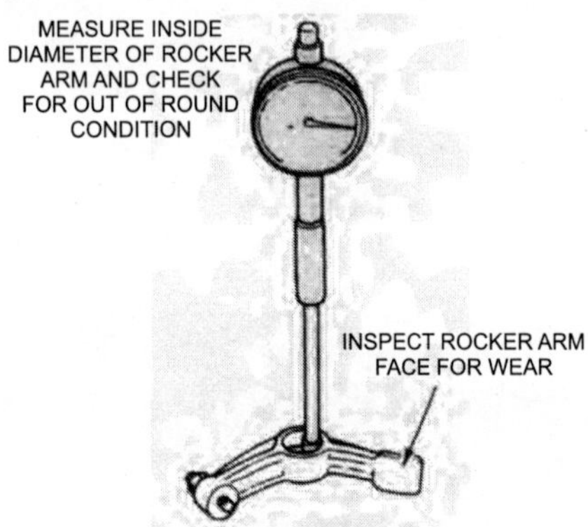

Fig.13.96 *(c) Insert gauge into rocker arm bushing to measure wear. Replace parts or install new rocker bushings as necessary. (Honda Motors)*

stand or pivot is made of aluminium, inspect it very carefully. The soft aluminium is prone to wear and frequently need replacement. Replace any rocker part worn beyond safe limits.

13.17 LIFTER (TAPPETS) SERVICE

The contact surface between a valve lifter and a cam lobe is a very high friction and wear point in an engine. Hydraulic lifters can also wear internally, and can cause valve train clatter.

When pulling the lifters out of the block on an OHV engine, make sure to keep them in proper order. If the lifters are mixed up and installed in wrong locations, they might quickly destroy the camshaft.

Lifter Preload

Lifter preload is actually the distance between the pushrod seat inside the lifter and the snap-ring of the lifter when the lifter is resting on the base circle (heel) of the cam and the valve is closed. This distance should be about 0.0508 mm to 1.143 mm. On engines with adjustable rocker arms, this distance or preload is determined by turning the rocker arms, adjusting nut one-quarter to one full turn after zero lash (clearance) is determined, *Fig. 13.97.* Tightening this adjusting nut further can cause the pushrod to bottom out in the lifter. If the engine is rotated with the pushrod bottomed out then, bent valves as well as bent or damaged pushrods, rocker arms, or rocker arm studs can be caused.

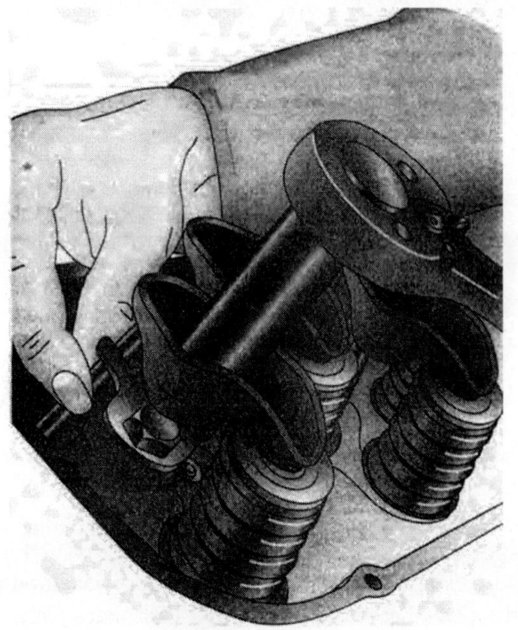

Fig. 13.97 *Operating principle of hydraulic lifters. (GM)*

Engines that have been rebuilt or reconditioned and that do not use adjustable rocker arms are particularly at risk of damage. If any of the following operations have been performed, lifter preload *must* be determined:

1. Regrinding the camshaft resulting in reduced base circle diameter
2. Milling or resurfacing cylinder heads
3. Milling or resurfacing cylinder block deck
4. Grinding valves and /or facing valve stems
5. Installing a head gasket thinner than or thicker than the original gasket.

Most lifters can accept a total variation in the entire valve train of about 2 mm to 4.75 mm. To determine lifter preload rotate the engine until the valve being tested is resting on the base circle of the cam. For example, with the valve cover removed, and rotating the engine in the normal operating direction, watch the exhaust valve start to open. This means that the intake valve for that cylinder is resting on the base circle wheel of the cam. Apply pressure down on the lifter. Wait for several minutes for the lifter to bleed down. Measure the distance between the rocker arm and valve stem. If the needed clearance (usually between 0.5 mm to 1.4 mm) does not exist the following may need to be done to restore the proper clearance:

1. Install longer or shorter pushrods. Manufacturers produce pushrods in various lengths. Some are available in lengths up to 2.54 mm longer or shorter than stock.
2. Install adjustable push rods or rocker arms, if possible.
3. Shim or grind rocker stands or shafts, keeping the rocker ratio in mind.

Determining Lifter Preload

The process of adjusting valves that employ hydraulic valve lifters involve making sure that the lifter has the specified preload. A properly adjusted valve train should position the lifter in the centre of its travel dimension.

The procedure for restoring the preload for a valve train with adjustable rocker arms is as follows:

1. Rotate the engine clockwise as viewed from the non-principal or belt end (normal direction of engine rotation) until the exhaust lifter starts to move up.
2. Adjust the intake valve to zero lash, i.e., no preload, and then one-half turn more.
3. Rotate the engine until the intake valve is almost completely closed. Adjust the exhaust valve to zero lash and then one-half turn more.
4. Repeat this procedure for all cylinders until all the valves are correctly adjusted. If the valve train is equipped with *non-adjustable* rocker arms, the lifter preload should still be determined. The lifter preload must be measured if any or all of the following procedures have been performed on the engine:

- Head(s) is milled
- Cylinder block is decked
- Valves are ground
- Any other machining operation that could change the valve train measurements is performed.

Shorter or longer replacement pushrods may be required to produce the correct lifter preload. Some engine manufacturers recommend the use of thin metal shims under the rocker arm supports if necessary. However, if shims are used under the rocker arm supports, ensure that the shim has the required oil holes provided.

Determining Proper Lifter Travel

To determine if shimming or use of replacement pushrods of different lengths is necessary, use the following procedure:

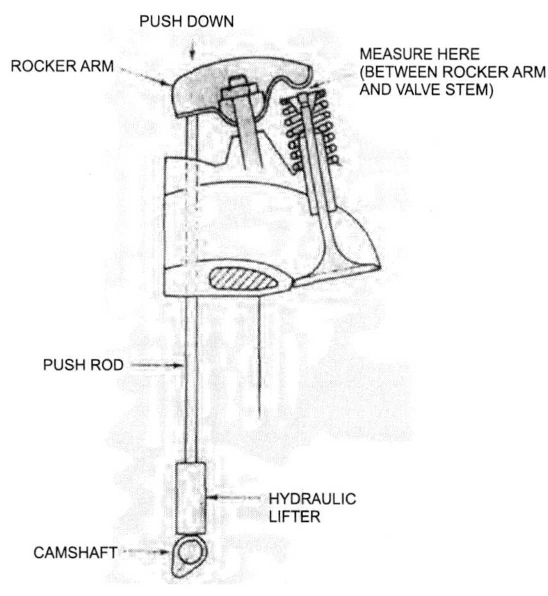

Fig. 13.98 *Procedure for determining proper lifter travel.*

1. With the valve cover removed, rotate the engine until the valve lifter being tested is resting on the base circle of the camshaft.
2. Depress the pushrod into the lifter, applying steady pressure, to cause the lifter to bleed down until the pushrod bottoms out in the lifter bore.
3. Measure lash (clearance) between the rocker arm tip and the stem of the valve. This

measurement varies according to manufacturer and engine design, but generally varies from 0.5 to 2 mm, *Fig. 13.98.* Always consult the service manual specs before taking any corrective measures.

If the measurement is not within acceptable range, select the proper-length pushrods to achieve the proper lifter travel dimension and preload.

Inspecting Lifters

Inspecting mechanical lifters is easy. A good, unworn lifter will have a slight hump or convex shape on the bottom. Whereas, worn lifters will be flat or concave on the bottom. Replace the lifters if they are worn, *Fig. 13.99.*

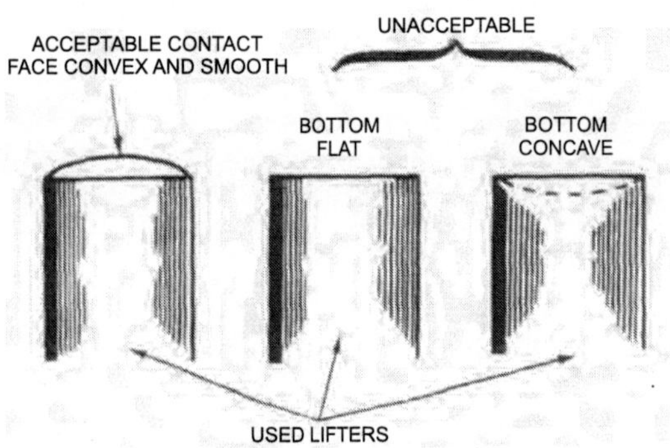

Fig. 13.99 *Good lifters should be humped or crowned slightly on bottom. If not, lifters are worn and should be replaced. (Ford Motors)*

Check for any grooves, concave wear, and pitting at their bottoms. The lifters should still be slightly convex in shape. A quick way to check their bottom shape is to hold two lifters bottom to bottom. They should rock just a little. See between the lifters while you rock them, and you will see the light change. If the lifters do not rock, they are worn and should be replaced.

While it is permissible to use new lifters on an old camshaft, never use old lifters on a new camshaft. The old lifters will ruin the camshaft. When a camshaft is replaced the lifters must be replaced, too.

Hydraulic lifters are checked just like mechanical lifters. But the internal components also need inspection. Obviously, if the lifter bottom is bad, no internal parts inspection is needed. Also, if there is not anything wrong with the lifter bottoms, then the internal parts should be good.

In general, mechanics do not disassemble and check hydraulic lifters. It takes a lot of time to disassemble, inspect, and again reassemble up to 16 lifters. Therefore, if the lifter bottom looks good and the lifters worked fine before the valve job, usually the mechanics leave them alone.

If an internal components inspection is necessary, remove a snap ring from the lifter's top and shake out the internal parts, *Fig. 13.100.* Do this only on a clean shop towel and one lifter at a time. Never mix internal lifter parts. Check the ball and plunger for wear. Clean out.

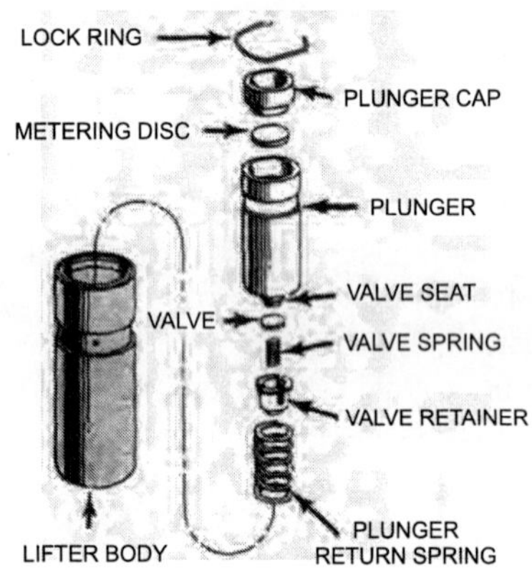

Fig. 13.100 *Details of a Valve lifter.*

Testing Lifter Leak Down Rate

Lifter leak down rate is measured by timing how long it takes to push the lifter plunger to the bottom of its stroke under controlled conditions. A typical lifter tester is shown in *Fig. 13.101.* Fill the tester with a special test fluid and place the lifter in the tester. Then, follow its specific directions to determine lifter leak down rate. If leak down rate is too fast or too slow, rebuild, or replace the lifter.

Rebuilding Hydraulic Lifters

Rebuilding hydraulic lifters typically involves complete disassembly, cleaning, and measurement of lifter components. All worn or scored components should be replaced following manufacturer's service instructions, *Fig. 13.102.* Frequently, auto repair shops do not rebuild hydraulic lifters, if they are defective, new lifters are installed. This can save time and money in most circumstances.

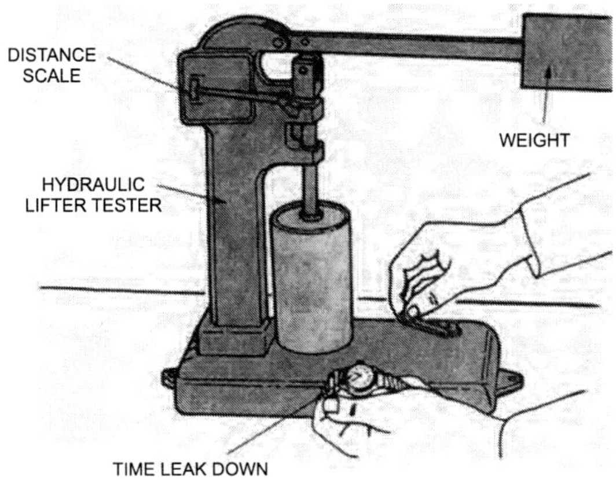

Fig. 13.101 *Automotive mechanic is checking lifter leak-down rate. Time leak-down while following equipment instructions. If not in specs, rebuild or replace lifters. (Pontiac Div. General Motors Corp.)*

Some lifters may be oversize from the factory. This makes it necessary to keep all lifters in order so they may be reinstalled in their original locations in the engine. *Fig 13.103* shows the method of marking an oversize lifter and lifter bore.

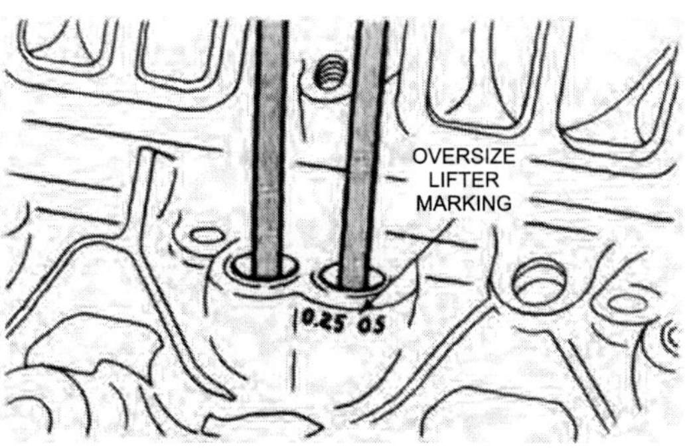

Fig. 13.103 *Sometimes, lifters can be oversize. This will normally be marked on block. This is one reason why the used lifters should be kept in order.*

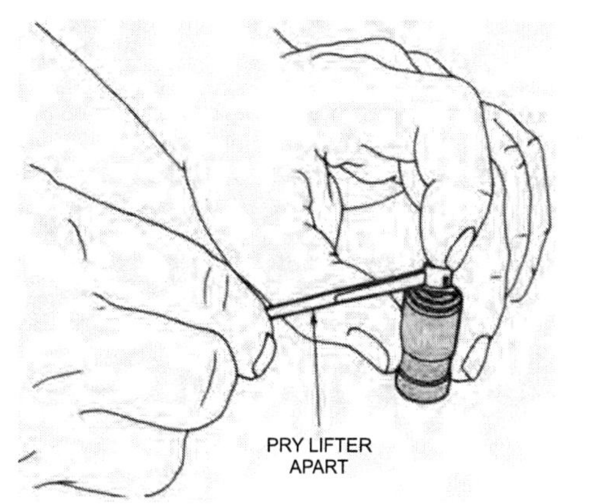

(a) Disassemble lifter and inspect all parts for wear.

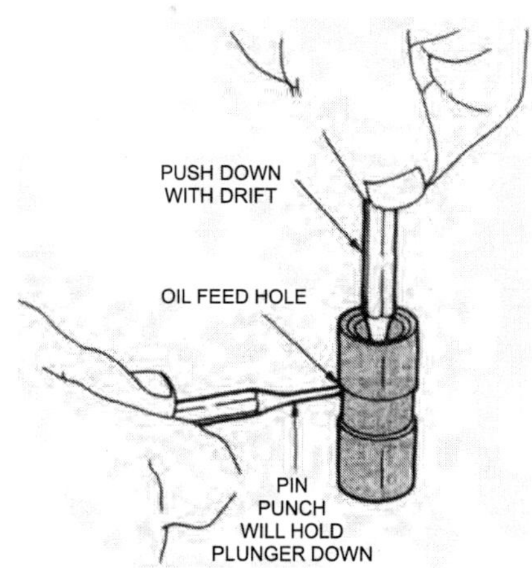

(b) If in good condition, clean parts and reassemble. Here the mechanic is forcing plunger down into lifter. Then pin punch is used to hold plunger while installing snap ring. Always follow manufacturer's service manual instructions. (Buick Div. General Motors Corp.)

Fig. 13.102 *Most mechanics replace, rather than rebuild, lifters. (Buick Div. General Motors)*

Hydraulic Valve Lifter Adjustment

After the reconditioned head is installed, adjust the valves as already explained. When installing hydraulic lifters, a one-time-only adjustment is needed. For General Motors and most other hydraulic lifters, rotate the cam so the lifter is against the lobe's heel. Then tighten the rocker arm nut while rotating the push rod between two fingers, *Fig. 13.104.* Resistance in the push rod signals that the adjusting nut has just taken all the slack out of the system. Consult the service manual before turning the adjusting nut slowly until the number of specified turns is reached. The lifter will automatically adjust the valve clearance from that point.

Some Ford Motors hydraulic lifters are adjusted after being *bled down.* Bleeding down a hydraulic lifter is done with a special tool, which applies pressure to the lifter through the push rod. Check the clearance between valve and rocker arm with feeler gauges. If the clearance measurement is out of specs, use a longer or shorter push rod.

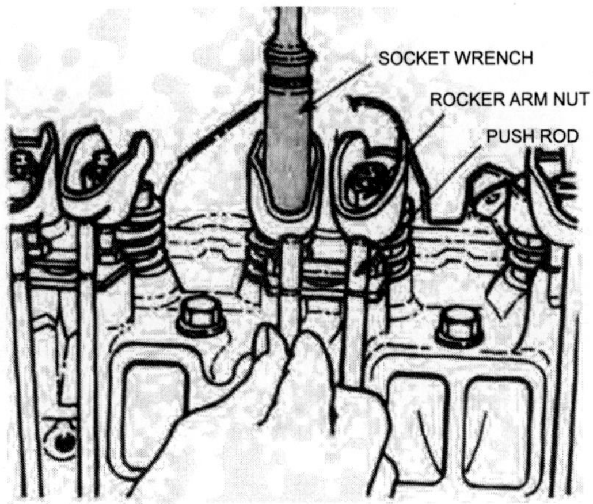

Fig. 13.104 *Adjusting GM-style hydraulic valve lifter. (Cadillac Div. General Motors Corp.)*

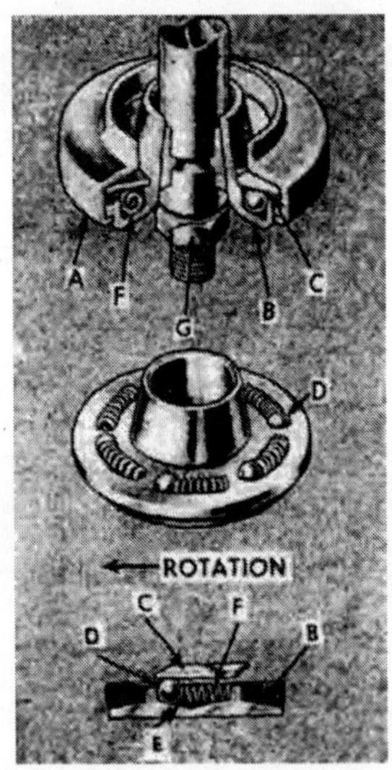

Fig. 13.105. *A positive-type valve rotator causes rotation of the valve when the ball moves to the valley of the ramp.*

13.18 VALVE ROTATORS

Certain valve spring retainer designs are more complicated and are required to allow or prompt the valve to rotate slightly with regard to its seat. The purpose is to provide a longer lasting seal between the valve and the valve seat. Rotation of the valve will tend to avoid the formation of carbon deposits and help avoid valve warpage. Valve rotators keep the valve seat and valve face clean and help maintain emission control. Valve rotators also minimize sticking and wear between the valve guide and the valve stem.

Valve rotators impart a positive rotational effort to the valve once during every cycle of operation. *Fig. 13.105* shows design features of one such device.

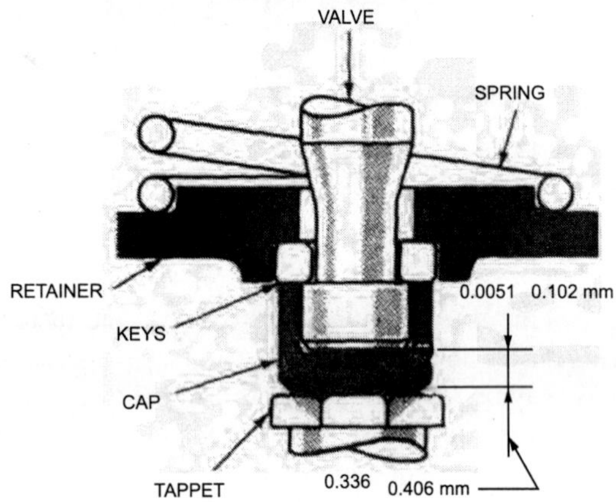

Fig.13.106 *With this type of valve rotator, clearance, must be maintained between the valve and the cap.*

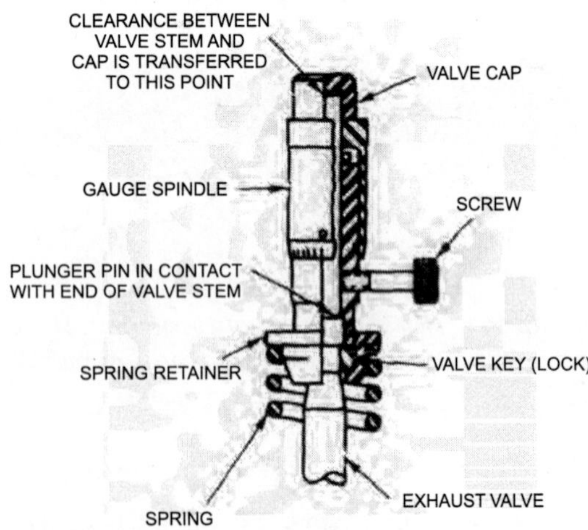

Fig. 13.107 *A special gauge is used to measure clearance.*

In some designs these devices release the valve from the valve spring at one point in the cycle of operation so that it is free to rotate slightly. With this type of device, the clearances between the valve stem and stem cap and between the cap and the retainer within specified limits, *Fig. 13.106.* If the clearance between the valve stem and stem cap is less, the end of the valve stem is ground off as required. If the clearance is too large, the skirt of the cap is ground off as required. A special tool is used for measuring the clearance accurately, *Fig. 13.107.*

The valve rotators can be checked, to see if they are functioning properly, by examining the valve tip to see where the rocker arm contacts the valve, *Fig. 13.108.* If the rotator is functioning properly, there will be no indentations on the valve tip. Late model automotive engines are not equipped with valve rotators.

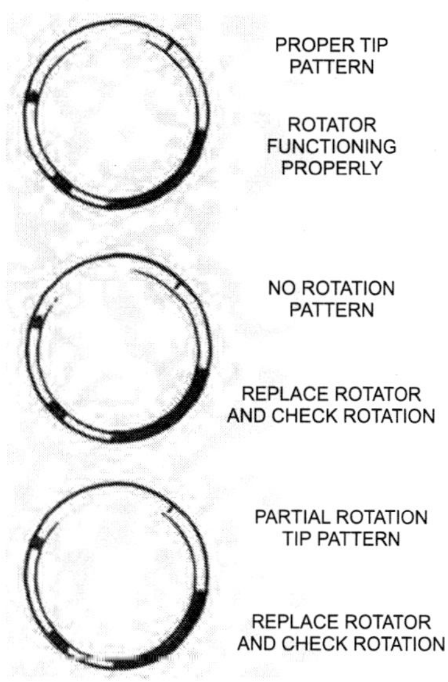

Fig. 13.108 *Wear pattern on the valve tip indicates whether or not the rotator is working. (Cadillac Div. GM)*

THEORY QUESTIONS

1. What forms the top and bottom of an 1C engine combustion chamber?
2. What are the advantages of a hemispherical combustion chamber?
3. What are the advantages of a wedge combustion chamber?
4. What is meant by the term 'cross flow head'?
5. What is a Siamese port?
6. What are the advantages of using 4-valves per cylinder?
7. Why is valve guide reconditioning the first cylinder head servicing procedure?
8. What is a valve lash adjustment and why is it important?
9. When should the valve tip be ground? How do you know as to how much material to remove from the valve tip?
10. What is an interference angle between the valve and the seat?
11. Describe the difference between cutting and grinding valve seats.
12. How are valve faces reconditioned?
13. How are valve seats reconditioned?
14. Explain four ways of correcting valve guide wear.
15. What is a valve seat insert? How is a valve seat insert installed?
16. How are the correct valve spring inserts (shims) selected and why are they used?
17. What is a compression pressure test and how is it done?
18. What puts thrust loads on a camshaft?
19. When is steel used as the camshaft material?
20. Explain why the lift and duration and cam lobe centre dimension determine the power characteristics of the engine?
21. Describe the operation of a hydraulic lifter.
22. Describe how to determine if the hydraulic lifter preload is correct.
23. Heat dissipation is greater when valve guides are not used. *True or False?*
24. A narrow valve seat will dissipate the heat better than a wider one. *True or False?*
25. A valve seat grinding stone should be slightly smaller in diameter than the valve seat. *True or False?*
26. When dressing a seat stone, how can you damage the diamond cutter?
27. The most common valve seat angle is _____ degrees, but some are also _____ degrees.
28. What are typical contact widths for intake and exhaust valves?

29. Valve seat inserts are not used in cast iron cylinder block or heads. *True or False?*
30. How do you move a seat in or out?
31. Why must valve springs be square on each end?
32. Worn camshaft bearings will cause a valve to open early. *True or False?*
33. What are the two types of valve trains in common use? Which style uses pushrods? Which style uses bucket tappets?
34. How do you determine valve spring shim thickness?
35. What is a spring bind?
36. How can you check for valve leakage before head installation?
37. What should be done to a camshaft with worn journal or lobe surfaces?
38. What are the two types of valves used? Which type runs hotter and why?
39. When installing a rocker arm shaft, the oil holes should be _____ _____.
40. Name two types valve adjusters. Name three common valve seat grinding angles.
41. When a valve is fully open, what part of the cam lobe is the lifter pushed against?
42. Are the valve seats ground before, or after the valve guide service?
43. A valve interference angle should be cut on the:
    (*a*) Combustion chamber side
    (*b*) Port side
    (*c*) Both sides
    (*d*) None of the above.
44. The valve seat should be concentric with the guide within:
    (*a*) 0.025 mm (*b*) 0.05 mm
    (*c*) 0.075 mm (*d*) None of the above.
45. Leaking intake valve guides:
    (*a*) Cause excessive oil consumption,
    (*b*) Upset fuel mixture
    (*c*) Both (*a*) and (*b*)
    (*d*) Neither (*a*) nor (*b*)
46. Valve stem seals are placed:
    (*a*) In the valve stem (*b*) In the valve guide
    (*c*) Either (*a*) or (*b*) (*d*) Neither (*a*) nor (*b*).
47. Mechanic A says that the purpose of an interference valve angle is to chip carbon off the valve face. Mechanic B says that the purpose of a valve rotator is to enhance valve life. Who is right?
    (*a*) A only (*b*) B only
    (*c*) Both A and B (*d*) Neither A nor B.
48. Mechanic A says that the rocker arm action causes valve guides to wear in a bell-mouth shape. Mechanic B says that the positive valve seal remains stationary as the valve moves up and down. Who is right?
    (*a*) A only (*b*) B only
    (*c*) Both A and B (*d*) Neither A nor B.
49. In a normally operating engine, intake and exhaust valves are opened by a cam and closed by the _____.
    (*a*) Rocker arm or cam follower
    (*b*) Lifters or tappets
    (*c*) Valve spring
    (*d*) Valve guide and/or pushrod.
50. The valve spring inserts (shims) are designed to _____.
    (*a*) Increase installed height of the valve
    (*b*) Decrease installed height of the valve
    (*c*) Adjust the correct installed height
    (*d*) Decrease valve spring pressure to compensate for decreased installed height.
51. Umbrella-type valve stem seals _____.
    (*a*) Fit tightly onto the valve guide
    (*b*) Fit on the valve face to prevent combustion leaks
    (*c*) Lock under the valve retainer
    (*d*) Fit tightly onto the valve stem.
52. An engine has coolant in the oil, producing a milky substance in the crankcase. Mechanic A says the engine could have warped head and blown head gasket. Mechanic B says that the engine could have a cracked head or block. Who is right?
    (*a*) Mechanic A (*b*) Mechanic B
    (*c*) Both A and B (*d*) Neither A nor B.
53. Mechanic A says that most diesel precombustion chambers must be flush with the head deck surface. Mechanic B says that most precombustion chambers can be driven out with a drift and light hammer blows. Who is right?
    (*a*) Mechanic A (*b*) Mechanic B
    (*c*) Both A and B (*d*) Neither A nor B.
54. Mechanic A says that umbrella or O-ring valve seals are installed after compressing the valve spring. Mechanic B says that install umbrella or O-ring valve seals and then compress the valve spring. Who is right?
    (*a*) Mechanic A (*b*) Mechanic B
    (*c*) Both A and B (*d*) Neither A nor B.
55. Mechanic A says that some cracked cylinder heads can be welded. Mechanic B says that many cracked cylinder heads can be plugged. Who is right?

(*a*) A only (*b*) B only
(*c*) Both A and B (*d*) Neither A nor B.

56. Mechanic A says that Magnafluxing should be used to detect cracks in most aluminium cylinder heads. Mechanic B says that dye penetrant should be used to find cracks in aluminium cylinder heads. Who is right?
(*a*) A only (*b*) B only
(*c*) Both A and B (*d*) Neither A nor B.

57. Mechanic A says that certain diesel cylinder heads must be replaced when warped. Mechanic B says that any diesel cylinder head can be milled when warped. Who is right?
(*a*) A only (*b*) B only
(*c*) Both A and B (*d*) Neither A nor B.

58. Mechanic A says that valve guide wear can cause engine oil consumption. Mechanic B says that valve guide wear can cause engine coolant leakage. Who is right?
(*a*) A only (*b*) B only
(*c*) Both A and B (*d*) Neither A nor B.

59. Valve seat replacement should be done _____.
(*a*) When replacing the valve springs,
(*b*) When replacing the valves,
(*c*) Only when seat wear is severe
(*d*) Any time the cylinder head is removed from the engine.

60. Mechanic A says that a valve hitting a piston can cause a bent pushrod. Mechanic B says that a stuck valve can cause a bent push rod. Who is right?
(*a*) A only (*d*) B only
(*c*) Both A and B (*d*) Neither A nor B.

❑❑❑

Front End – Timing Mechanism Service

14

14.1 INTRODUCTION

Modern automotive engines are much more complex than the ones used a few years ago. Engine manufacturers use more overhead cam engines with either a long timing chain or timing belt. Engines with cam-in-block, with shorter timing chains and timing gears are also common. This makes the service of an engine timing mechanism quite challenging. Here the most critical information regarding the service and repair of timing chains, timing belts, and timing gears is reviewed. Timing mechanism service is critical. Any mistake made can cause severe engine damage or performance problems.

14.2 TYPES OF VALVE TRAINS

Two basic types of valve trains used in automotive engines are: overhead valve (OHV) and overhead camshaft (OHC).

OHV Layout

Camshaft in the cylinder block type of *valve train* is shown in *Fig. 14.1.* This type of valve train uses pushrods and the engine is called a *pushrod engine, Fig. 14.2.* The valve train includes a lifter, pushrod, rocker arm, and valve spring assembly. The sprockets are sized so the camshaft will rotate at one-half crankshaft speed, *Figs. 14.3 (a)* and *(b).*

On the camshaft is a series of lobes, eccentric sections, which control valve movement. There is one lobe for each valve. Above the lobes are the lifters. The lifters fit between the camshaft lobes and the pushrods, which transmit the motion from the block to the top of the head. Here the motion is reversed by the rocker arms. The pushrod end of the rocker arm moves up while the other end of the rocker pushes down on the valve, opening it. When the camshaft lobe has turned past its lobe tip (highest point), the valve closes under valve spring force.

OHC Layout

In OHC systems the camshaft is mounted on top of the head. Gears, a chain, or a belt connect the crankshaft and camshaft sprockets, *Fig. 14.4.* The OHC engine uses various means of carrying the cam lobe action to the valve. The simplest arrangement is shown in *Fig. 14.5.* It uses a bucket tappet. The operation of the bucket-tappet valve train is shown in *Fig. 14.6.* The cam lobe is directly above the bucket tappet. As the cam lobe comes around, it pushed the tappet, and the valve, down to open it. As the lobe (nose) passes the tappet the spring pulls the valve to close it.

Another arrangement, using a rocker arm, is shown in *Figs. 14.7, 14.8, 14.9,* and *14.14.* As the cam lobe comes under the valve lifter, it causes the rocker arm to rock. This pushes down the valve stem so that the valve moves down and opens. When the lobe passes out from under the valve lifter, the valve spring pulls the valve closed. The rocker arm rocks back, pushing the valve lifter down on the cam.

Rocker arms can be *shaft-mounted.* The shaft(s) runs the full length of the cylinder head, *Figs.14.11* and *14.8. Pedestal mounted* rocker arms are very popular too. Instead of a shaft, the rockers are mounted individually on special studs, or pedestals.

Sometimes on OHC engines the camshaft pushes against a rocker arm located by the valve at one end and a pedestal at the other. Loosening a locknut and

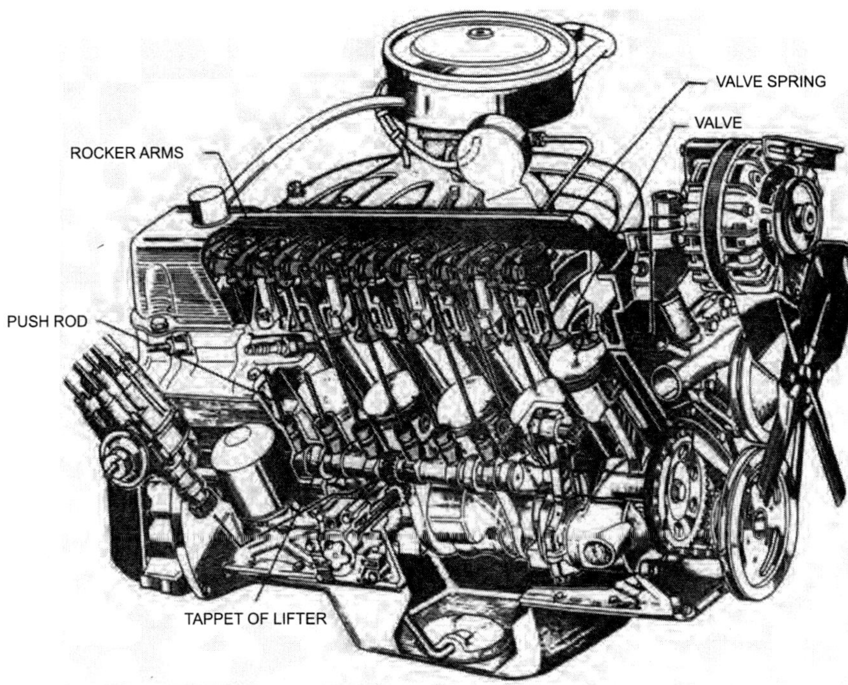

Fig. 14.1 *Camshaft in a cylinder block. (Chrysler Corp.)*

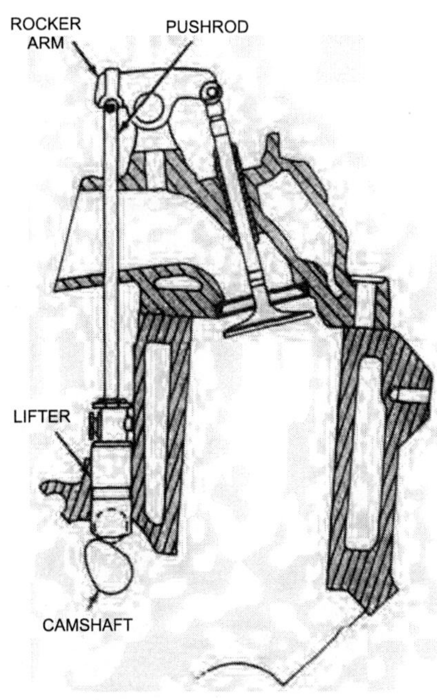

Fig. 14.2 *Overhead valve engines are also known as pushrod engines because of the long pushrod that extends from the lifter to the rocker arm.*

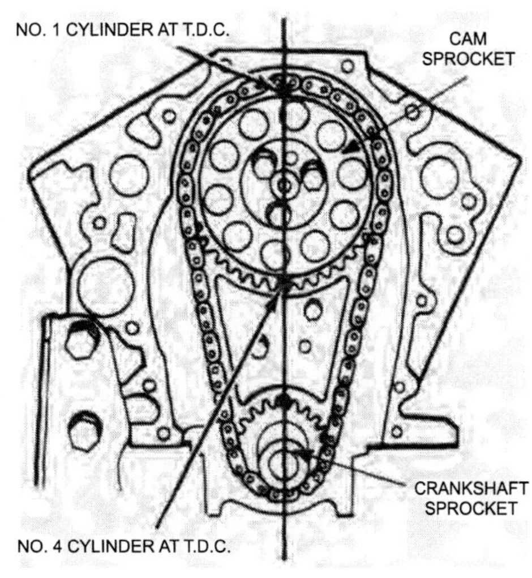

Fig. 14.3 *(a) OHV cam timing chain alignment. (Buick Div GM)*

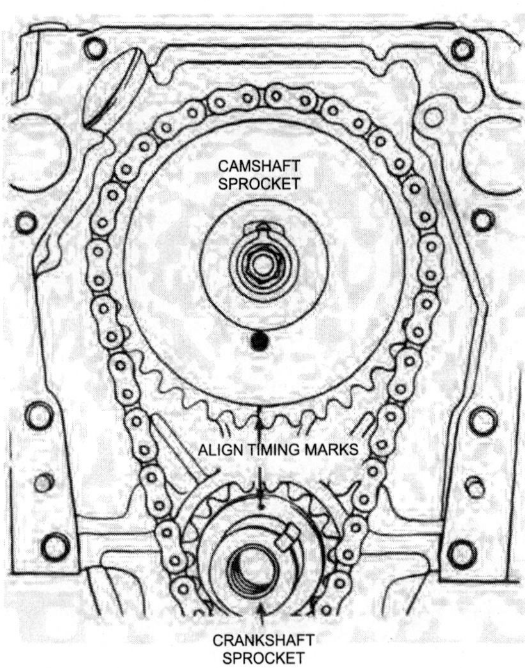

Fig. 14.3 *(b) Crankshaft / camshaft relationship.*

Fig. 14.5 *OHC valve train. (Chrysler Corp.)*

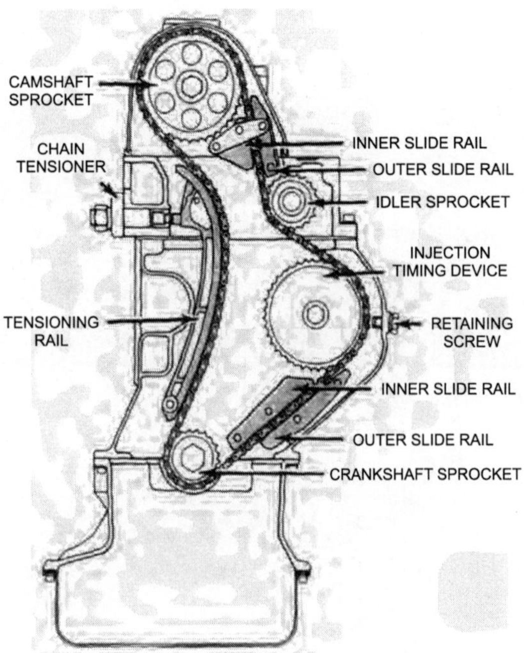

Fig. 14.4 *OHC timing chain.*

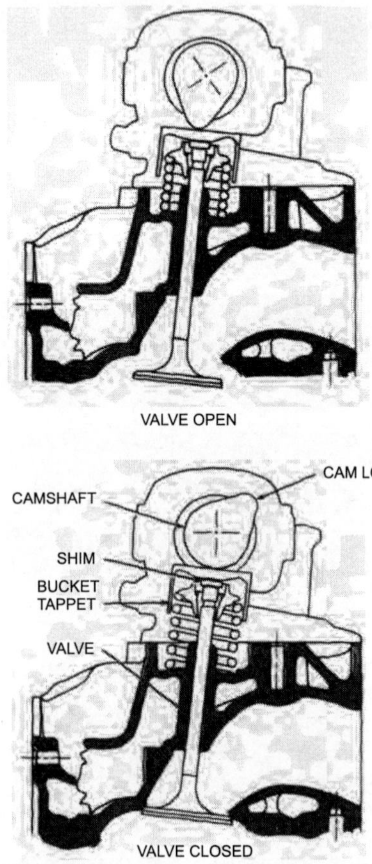

Fig. 14.6 *OHC-engine valve train using a bucket tappet. The cam works directly on the tappet.*

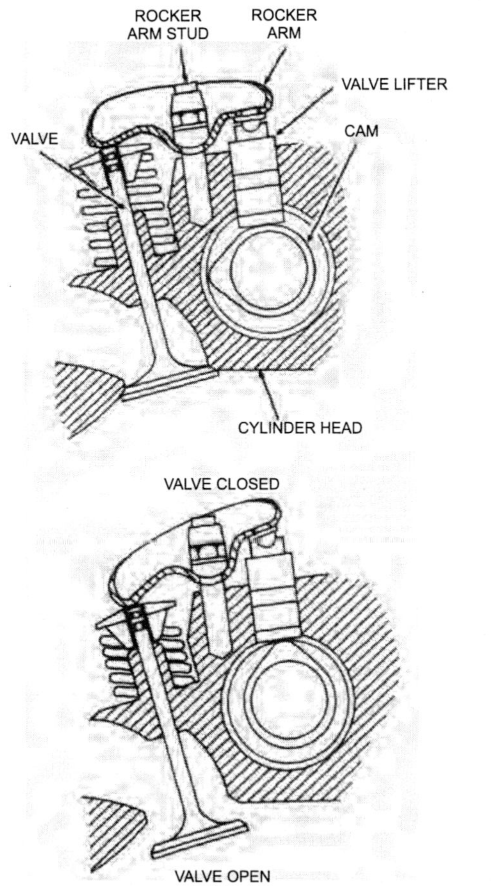

Fig. 14.7 *OHC-engine valve train using a rocker arm between the valve lifter and valve stem.*

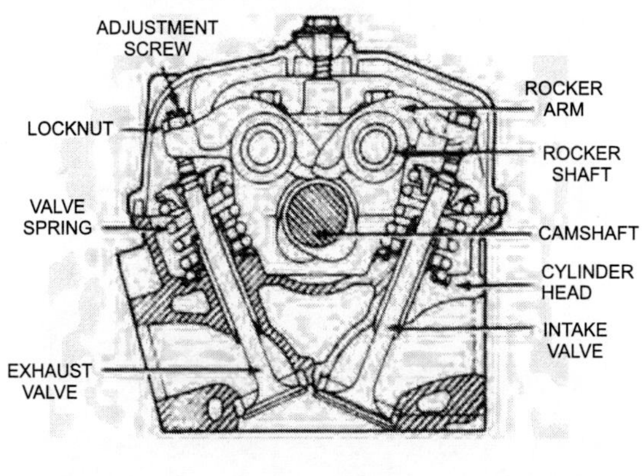

Fig. 14.8 *A typical single overhead camshaft (SOHC) engine with shaft-mounted rocker arms that ride directly on the camshaft.*

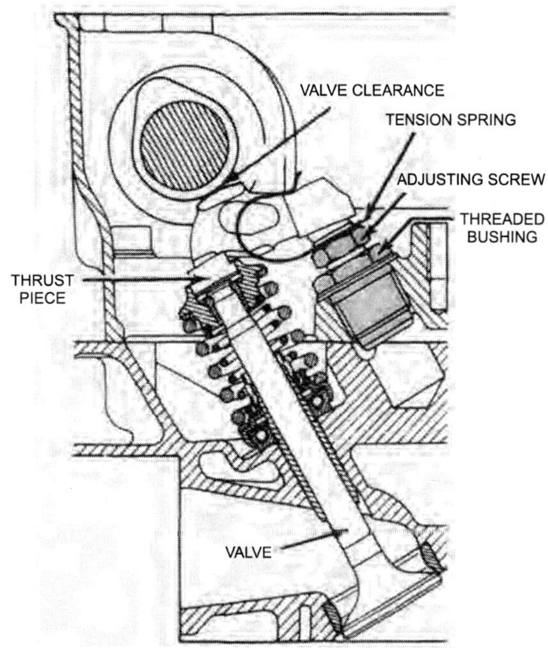

Fig. 14.9 *OHC mechanically adjustable pedestal.*

threading the pedestal in and out of the head adjusts this type of valve train, *Fig. 14.9.* A variation of this type has a hydraulic lifter in place of the pedestal. Then no adjustment is necessary, *Fig. 14.10.*

14.3 CAMSHAFT DRIVES

The camshaft is driven by the crankshaft, which makes two revolutions during the four-stroke cycle. Valve action, however, only occurs during two of these strokes, i.e., intake and exhaust. This requires the camshaft to just make only one revolution during a 4-stroke cycle.

The crankshaft through gears, sprockets and chains, or sprockets and timing belts drives the camshaft. Timing chains are not as wide as timing belts, so engines with timing chains can be shorter. Timing chains often have tensioners (dampers) pressing on the unloaded side of the chain. The tensioner pad is a Nylatron moulding that is filled with molybdenum disulfide to give it low friction coefficient. The tensioner is held against the chain by either a spring or hydraulic oil pressure, *Fig. 14.12.* The gears or sprockets are keyed to their shafts so that can be installed only in one position. The gears and

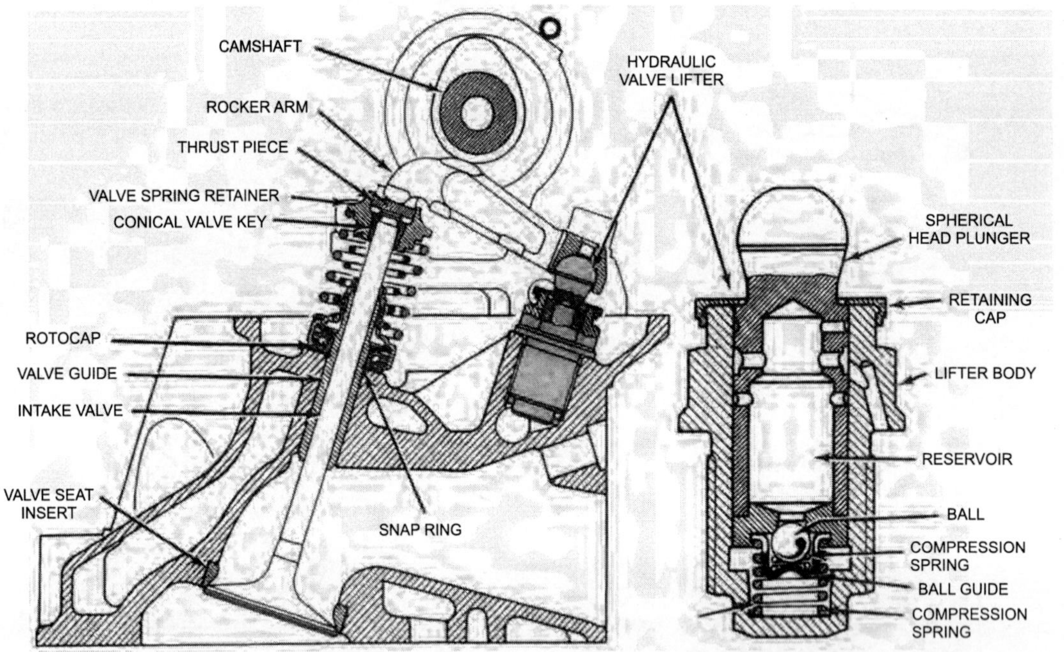

Fig. 14.10 *OHC with hydraulic lifter pedestal.*

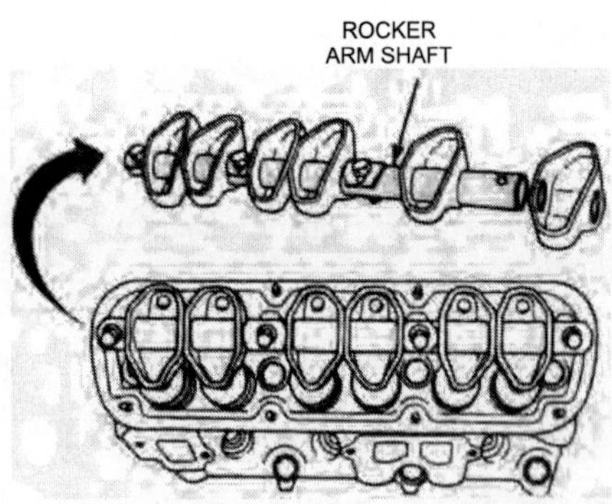

Fig. 14.11 *Typical rocker arm shaft design on an overhead valve engine.*

sprockets are then indexed together by marks on the gear teeth or chain links. When the crankshaft and camshaft timing marks are properly lined up, the cam lobes are indexed to the crankshaft throws of each cylinder so that the valves will open and close at the correct timing in relation to their corresponding piston position.

Fig. 14.12 *(a) A spring-loaded timing chain tensioner, also called a damper.*

Fig. 14.12 *(b) Most overhead camshaft engines use a hydraulic tensioner.*

Fig. 14.12 *(c) Hydraulic tensioners use engine oil pressure to keep tension on the chain. A ratchet mechanism in the tensioner maintains some tension on the chain when the engine is shut off and oil pressure is zero. This design helps reduce noise when the engine starts and before oil pressure is again applied to the tensioner.*

14.4 IN-VEHICLE TIMING MECHANISM SERVICE

In most cases a timing chain, belt, or gears can be serviced *without* removing the engine from the vehicle. Usually, the parts on the front of the engine and timing cover can be removed without dismantling the engine, Some over head cam (OHC) engines require the camshaft(s) to be secured before timing mechanism removal. If the camshaft(s) are not secured on these engines, they will slip out of timing mechanism and can result in engine damage. Refer to the applicable service manual for specific directions and specs if required.

14.5 TIMING CHAIN SERVICE

Timing chains transfer power from the crankshaft to the camshaft. Frequently they also power other components. They are used in both overhead valve (OHV) and overhead cam (OHC) engines.

Timing Chain Problems

Timing chains suffer from wear after being in service for long time. The chain links can wear and stretch. The sprocket teeth can also wear down with time. This results in excessive *chain slack* or the chain can become loose on its sprockets.

Timing chain wear and slack can cause the camshaft to go out of time with the crankshaft. This leads to valve timing being thrown off. The valves can open too late during the four-stroke cycle. This can reduce compression, power, fuel economy, and exhaust cleanliness.

Inspecting Timing Chain

Inspect a timing chain before removing it. *Fig. 14.13* shows one method of checking OHC timing chain wear. Rotate the cam sprocket back and forth using a wrench and measure the sprocket movement. Compare the measurement to service manual specs.

If one side of a timing chain can depress more than 12.7 mm, replace the chain and sprockets. For OHV engine, remove the distributor cap or valve cover to check timing chain wear. Use a large wrench on the crankshaft snout and rotate the crankshaft one way and then the other while watching the distributor rotor or valve train. The rotor or valves should operate with little crankshaft rotational movement. If the crankshaft can be turned over 5° to 10° without valve or rotor movement, the timing chain is loose.

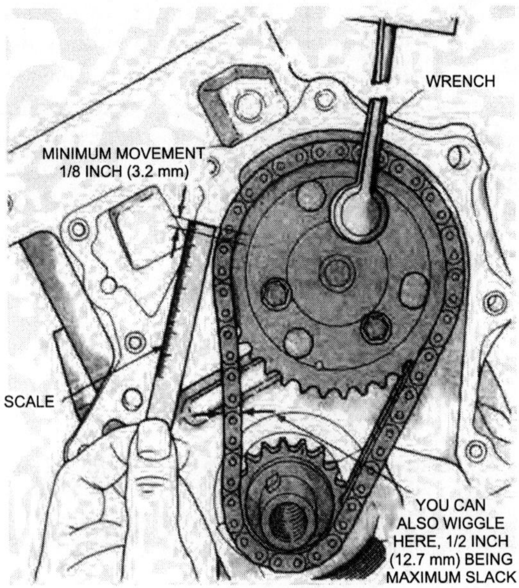

Fig. 14.13 *Timing chain wear can be checked by rotating crankshaft back and forth. If sprocket rotation is too much, chain and usually sprockets might be worn and need replacement. The wear can also be checked by wiggling chain on slack side. (Chrysler Corp.)*

If the timing chain is worn more than specifications, it should be replaced. Many manufacturers recommend replacement of the timing chain and both sprockets as a set. Some manufacturers permit reuse of steel sprockets if these are not worn excessively. If the camshaft sprocket has plastic teeth, it should usually be replaced. Plastic teeth are prone to wear and failure. Because this type of sprockets is not very expensive, they should be replaced whenever it is removed or disassembled for service.

If the timing chain has a tensioner, check it for wear. The fiber rubbing-block can wear down. Replace the parts as necessary.

Installing Timing Chain

Ensure that the crankshaft key is of the right size, in good shape, and properly installed in its groove in the crankshaft. Check that the camshaft is installed properly. If required, measure camshaft endplay and check any fasteners holding a thrust plate on the front of the engine.

Rotate the crankshaft and camshaft so that the *sprocket timing marks align, Fig. 14.14.* Then, fit the timing chain and both sprockets in place as a unit, *Fig. 14.15.* Do *not hammer* on one side of the sprockets to force them in place. Normally, tightening the fasteners or hand pressure will slide them over the camshaft and crankshaft. If use of a hammer becomes necessary, use a large socket or driver to exert equal pressure in the centre. This will drive the sprockets on squarely, *Fig. 14.16.*

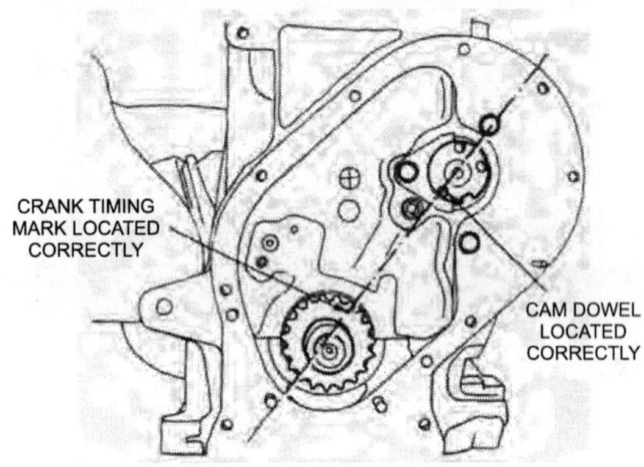

Fig. 14.14 *To begin installation of timing chain, rotate camshaft and crankshaft until their keyways or dowels are in the general position to align the timing marks. This will allow the mechanic to fit the sprockets in place. (Peugeot)*

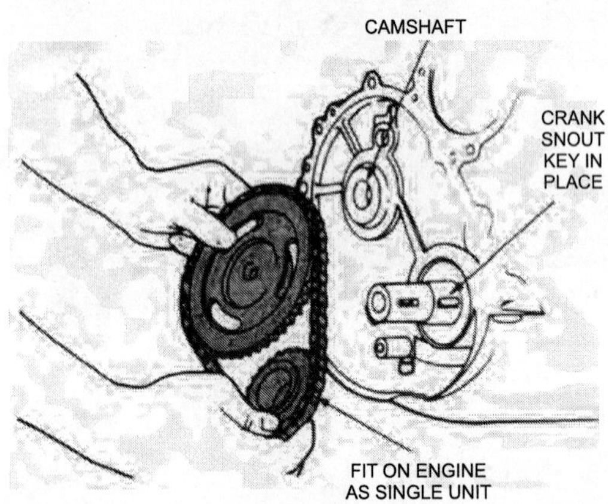

Fig. 14.15 *With the camshaft and crankshaft turned into position, fit the chain and sprockets into place together. Check that the key and dowel are securely installed in place. (Renault)*

Timing marks will either be dots, lines, circles, or other shapes indented or cast into the timing sprockets. Timing marks are also used on timing gears and belt sprockets. The marks must be pointing in the correct direction to time the camshaft/valves with the crankshaft/pistons.

After installing the chain, again check the timing marks for proper alignment. Turn the crankshaft by hand until the marks realign. If necessary, consult the service manual for instructions. Often, the keyways or a dowel may be considered timing marks. However, the timing marks are dots, as shown in *Fig. 14.17*.

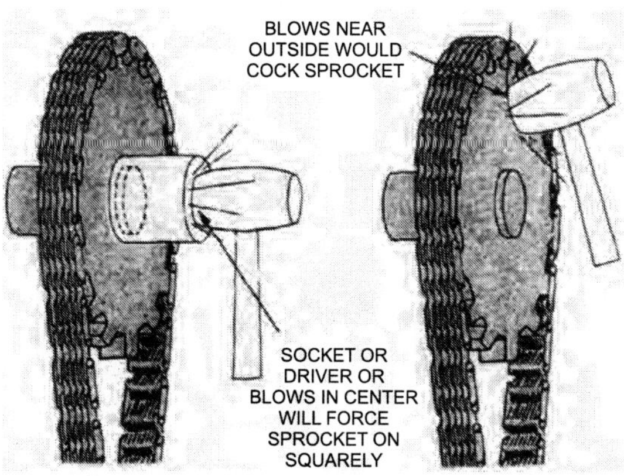

Fig. 14.16 *When driving sprockets, drive in the centre. Never drive around the outside because this could cock or break sprocket.*

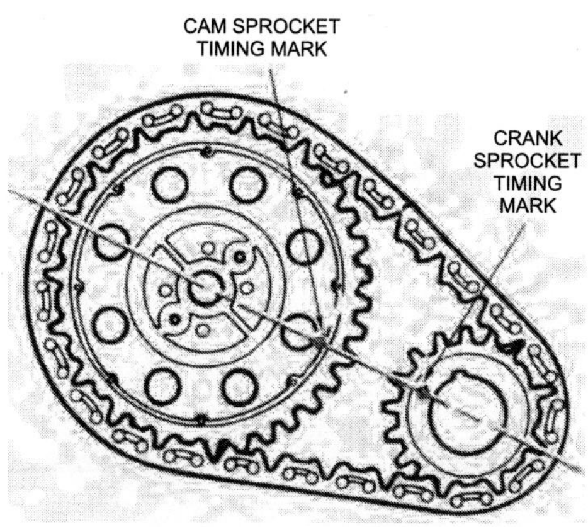

Fig. 14.17 *With overhead valve engines, mark on camshaft sprocket should normally point at mark on crankshaft sprocket. An imaginary line should pass through marks and sprockets as shown. (Renault)*

Figures. 14.18 (a) and (b) show the timing chain for an overhead camshaft engine. Sometimes the timing chain has special links, e.g., plated or colour coded links that should be aligned with marks on the chain sprockets. The number one piston must be at TDC on the compression stroke with these marks aligned.

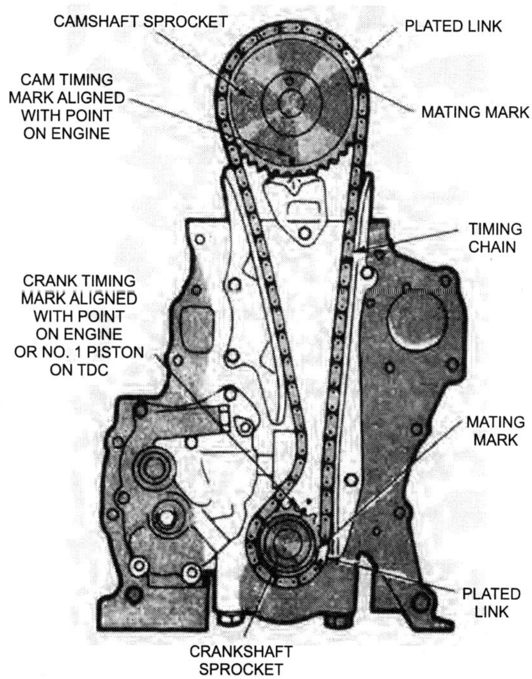

Fig. 14.18 *(a) Aligning timing marks on an overhead camshaft engine can be more challenging. This engine timing chain also has chain links that should be aligned with sprocket marks. Refer to the applicable service manual for details of the specific type engine. (Chrysler)*

If a fuel pump eccentric mounts on the camshaft sprocket, make sure that it is positioned properly. Usually, the camshaft dowel must fit through a corresponding hole in the eccentric. The cam sprocket bolt installs through the centre. Torque the camshaft sprocket fasteners to specs, using a torque wrench, *Fig. 14.19.*

Do not use the starting motor to turn the crankshaft on some overhead camshaft engines with the timing chain or belt removed. It is possible, with some engine designs, for the pistons to hit the valves. The valves could be bent. Turn the camshaft and crankshaft very slowly by hand.

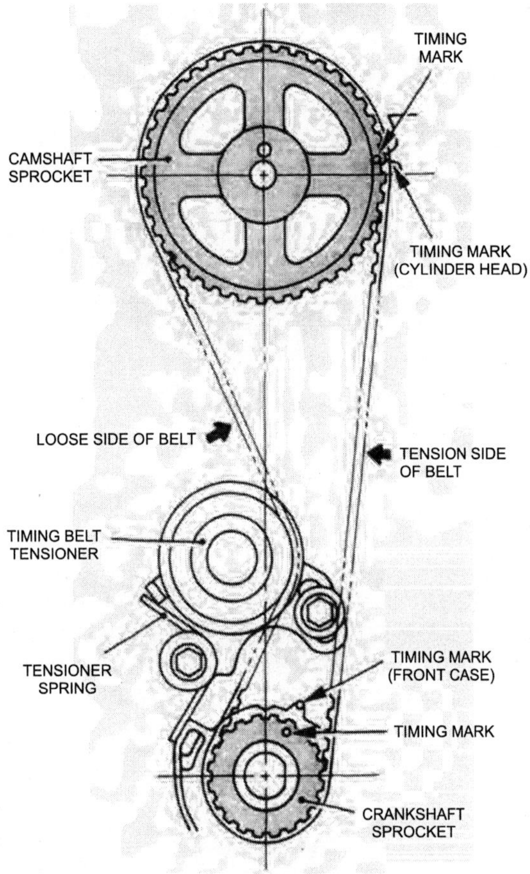

Fig. 14.18 *(b) OHC timing belt and sprockets. (Chrysler)*

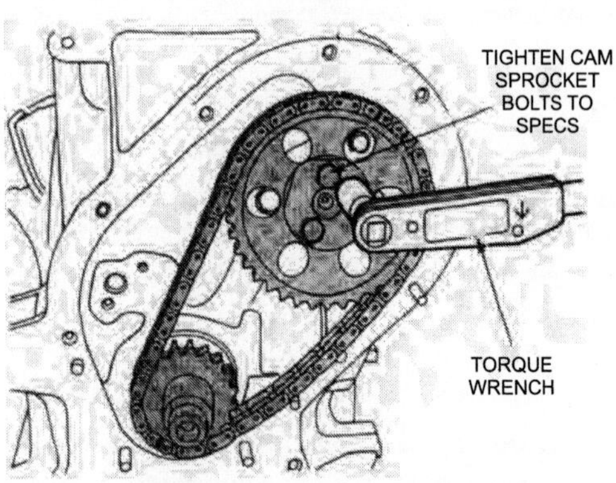

Fig. 14.19 *Torque the sprocket fasteners to Specifications. This could prevent problems later. (Peugeot)*

Installing Timing Chain Tensioner

A chain tensioner is used on most OHC timing chains. Inspect it carefully. The fiber rubbing-block can wear off with prolonged time in service. Replace it if necessary, *Fig. 14.20.* Fit the tensioner into its place on the engine. Then, pull on the chain to make sure that it moves freely. Some plunger types can bind if worn and extended too much.

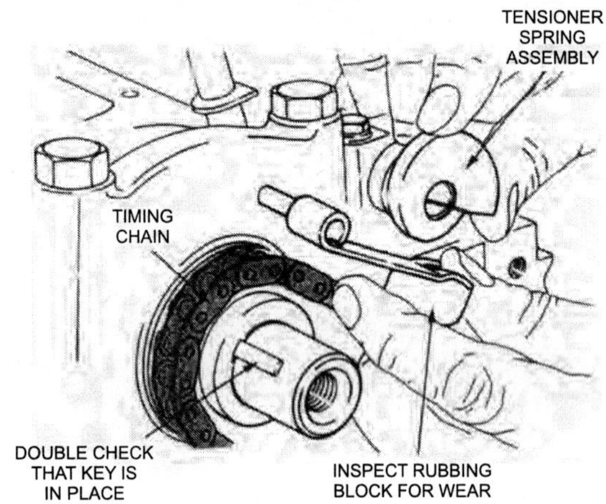

Fig. 14.20 *If used, check the chain tensioner carefully, before installation. Fiber rubbing block can wear. Measure the wear and compare to wear specs if necessary. (Ford Motors)*

Installing Oil Slinger

Oil slinger fits in front of the crankshaft sprocket. Make sure the slinger is installed correctly facing the right direction. It slings oil out of the timing chain and also prevents front seal leakage, *Fig. 14.21.* Normally it is trouble free part. However, if bent or damaged, it should be straightened or replaced.

14.6 TIMING BELT SERVICE

Many OHC engines use a synthetic rubber belt to operate the engine camshaft. The cogged, i.e., toothed, belt provides a precise, quiet, and dependable means of rotating the camshaft, *Fig. 14.22.* Timing belt service is very important. If the timing belt were to break or be timed improperly, engine valves, pistons, and other parts could be damaged.

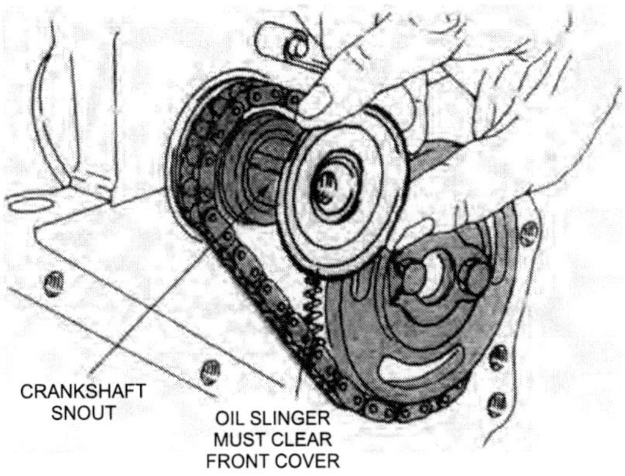

Fig. 14.21 *Use care not to install the oil slinger in reverse direction (backwards) or it will rub on timing cover and make terrible noise. Partial teardown only to turn the slinger around would be frustrating. (Ford Motors)*

Checking Timing Belt Mechanism

During timing belt service, always inspect the belt, sprockets, tensioner, and other parts for any problems. Most vehicle manufacturers recommend belt replacement about every 80,000 km. Frequently mechanics replace the timing belt when it is removed for service. Also, more frequent replacement could be required after oil or coolant contamination of the timing belt.

Late model engine timing belts are designed to last the lifetime of the engine without any adjustment or replacement. When in doubt always refer to the owner's manual or service manual for belt service intervals.

Because the timing belt removal can be time consuming, it is commonly replaced during major engine repairs. Timing belts are relatively cheaper. Check the timing sprockets for rounded teeth or physical damage. Rotate the tensioner wheel by hand

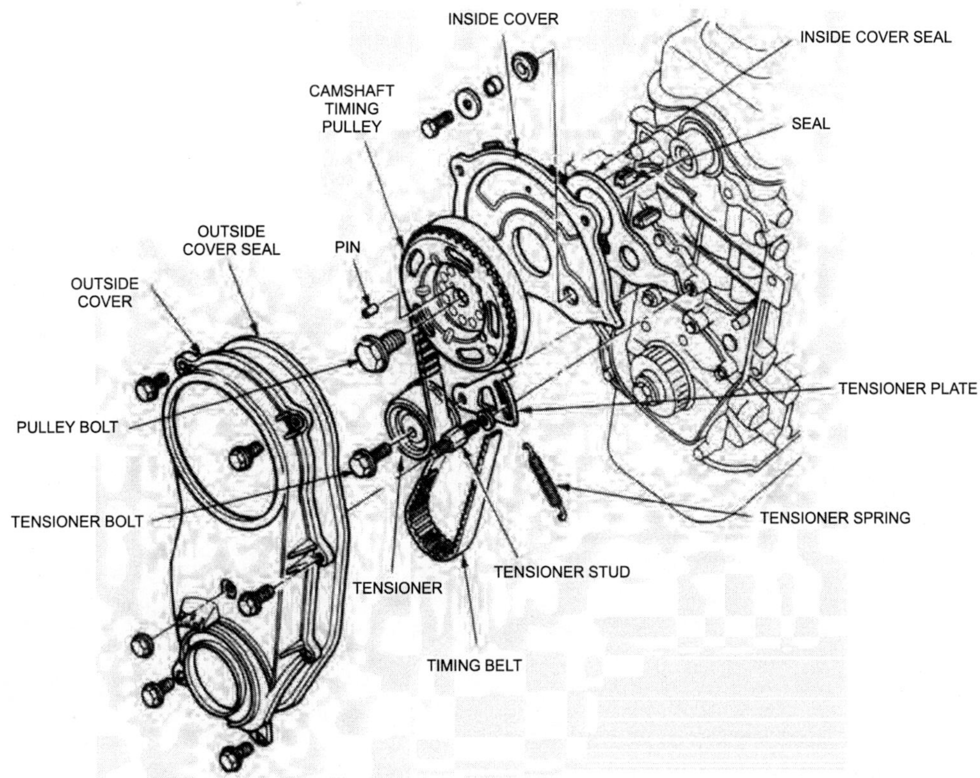

Fig. 14.22 *Service manual usually gives exploded view of parts on specific engine. It can be used as a guide for assembly of components. (Ford Motors)*

to check for a damaged bearing. Normally the wheel should spin freely and without any noise. Replace the tensioner bearing if any roughness is felt. If any oil seals are leaking, they should be replaced. Oil will cause rapid timing belt failure.

Installing Timing Belt Sprockets

Install the crankshaft and camshaft sprockets to start with. The camshaft sprocket usually fits over a small steel dowel, *Fig. 14.23.* A bolt or bolts are then torqued to specs to secure the sprocket. Normally a flat is machined on the camshaft so you can hold the camshaft while tightening the bolt, *Fig. 14.24.* Avoid any damage to camshaft lobe or the sprocket teeth when attempting to hold the camshaft. Hold on an un-machined surface if necessary.

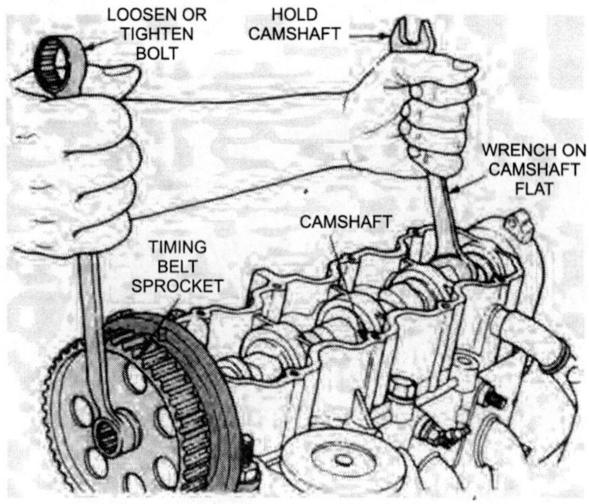

Fig. 14.24 *A flat surface is frequently machined on camshaft so it can be held stationary when removing installing cam sprocket. (Buick Div. General Motors)*

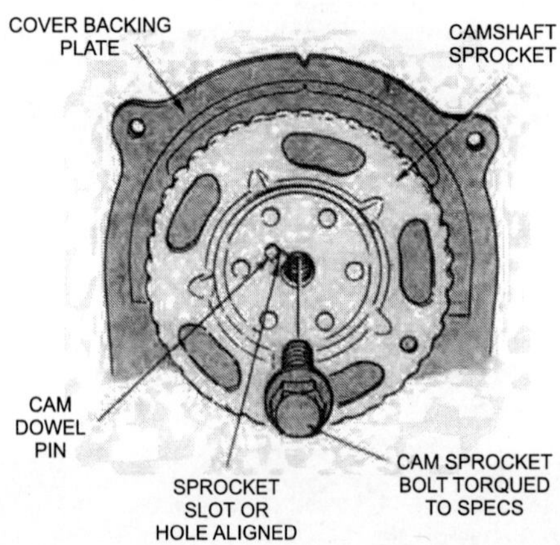

Fig.14.23 *A timing belt sprocket, as with chain sprocket, is frequently positioned with a small dowel pin. Torque the bolt to specs and align marks on sprocket and engine,*

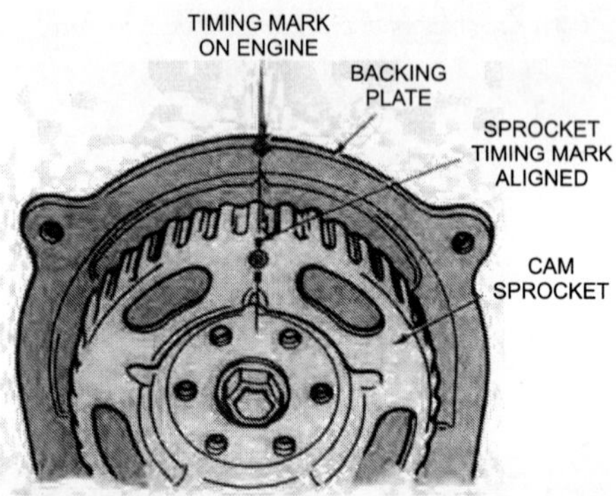

(a) Camshaft sprocket must usually be aligned with mark on front of engine.

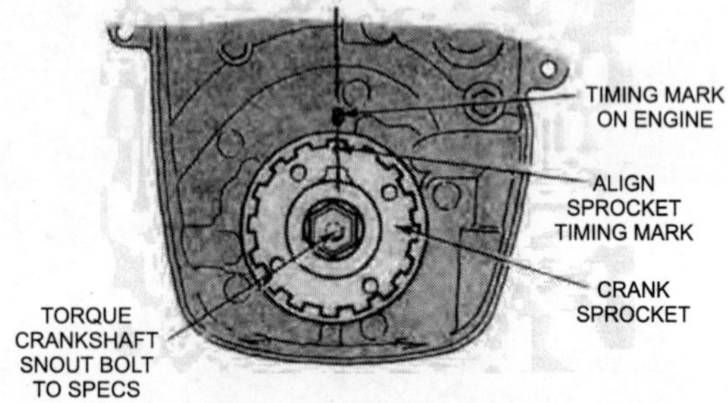

(b) Crankshaft sprocket will also have mark that must be aligned with a specified point on engine. If in doubt, check the service manual for directions.

Fig. 14.25 *Aligning timing belt marks. (Pontiac Div. GM)*

Fit the crankshaft sprocket into position as described for a chain sprocket. Turn the sprockets until the timing marks align or until number one piston is at TDC on the compression stroke, *Fig. 14.25.*

Installing Timing Belt

It is a good practice to always mark the old timing belt with a directional arrow before its removal. If the old timing belt is to be reused, it must be installed so that it rotates in the same direction. This will prevent premature belt failure, *Fig. 14.26.*

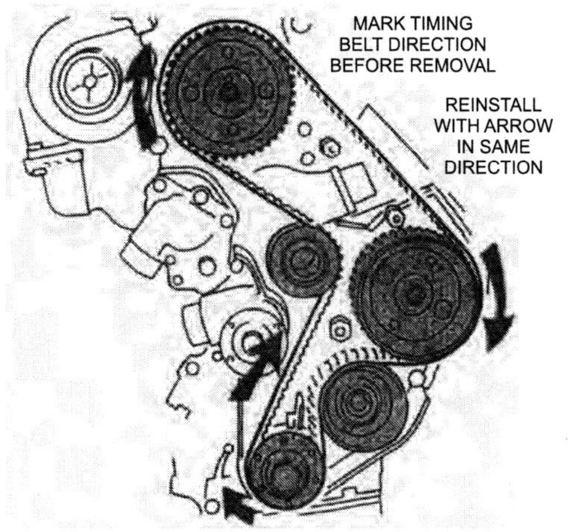

Fig. 14.26 *When reusing a timing belt that is in good condition, it must be installed so that it turns in the same direction. If its rotation is reversed, belt can fail quickly. (Ford Motors)*

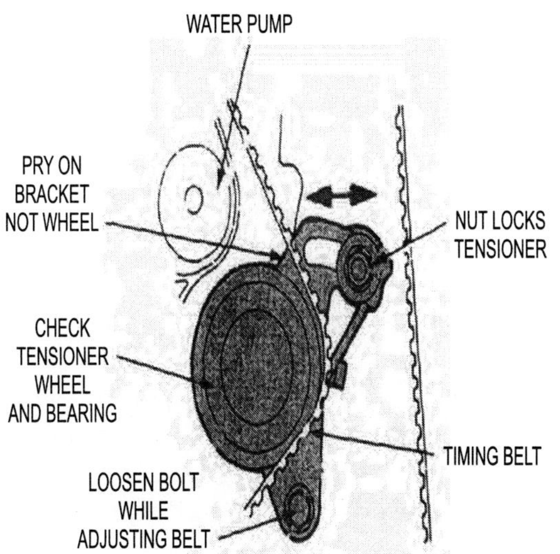

Fig. 14.27 *Front view of a typical OHC engine with the timing belt installed.*

Because the timing belt installation procedures differ in different engine designs, refer to the applicable service manual for the procedure to be followed. It will explain special methods for aligning timing marks, positioning the camshaft and crankshaft, and installing the belt tensioner.

Front view of an OHC engine with the timing belt in place is shown in *Fig. 14.27,* with timing marks aligned. If the engine has a belt driven distributor or diesel injection pump, its sprocket would also have to be timed with the other two.

Adjusting Timing Belt Tension

Proper adjustment of the timing belt is critical. If too loose, the belt might flap or vibrate in operation and tend to fly off its sprockets. If too tight, the timing belt could snap and or damage bearings. These situations could lead to severe valve damage, too.

If in doubt about the belt adjusting procedure to be followed, consult the applicable service manual to determine how to adjust belt tension properly.

A belt tensioner is shown in *Fig. 14.28.* Note how to loosen the attaching bolts and swivel the tensioner. Pry on the tensioner bracket to preload the belt. A special tool used to set the belt tension correctly is shown in *Fig. 14.29.* An adjusting tool is turned with a wrench until the belt gauge reads within specifications. Tighten the tensioner bracket bolts to set the adjustment

A simple way to check a timing belt tension, when a gauge or tool is not available, is to adjust the belt until a moderate finger and thumb pressure is needed to *twist the belt about l/4th turn.*

After installing the belt, use a wrench on the crankshaft snout bolt to rotate the engine by hand. Turn the crankshaft in the direction of its normal rotation direction until the timing marks again realign. Then, recheck the timing marks for alignment, *Fig. 14.30.*

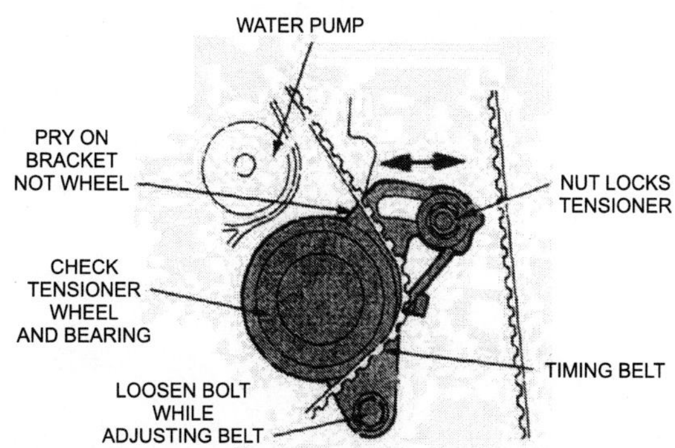

Fig. 14.28 *The timing belt adjuster shown here has a slot that allows to swivel adjuster to the side. Pry on adjuster and tighten the fastener. Then, check for proper tension on belt.*

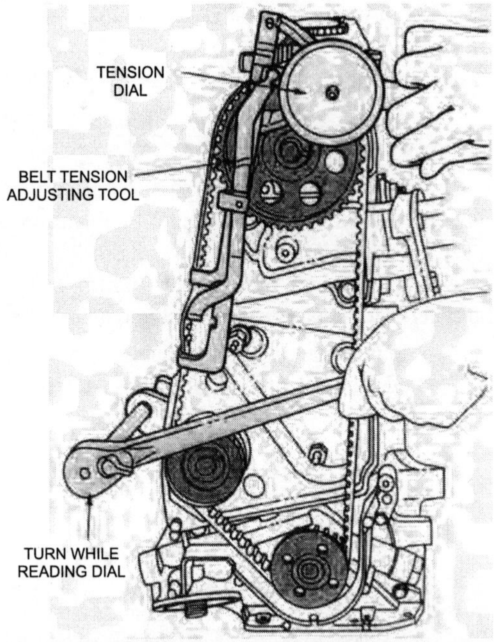

Fig. 14.29 *Here a special tool used for adjusting belt tension is shown. It measures deflection of straight section of belt. If it is not available, adjust until moderate thumb and finger pressure is needed to twist belt one-quarter turn. (Chevrolet Div. General Motors)*

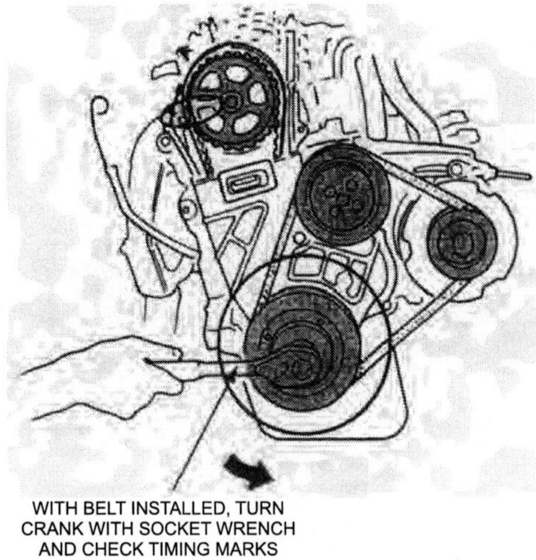

Fig. 14.30 *(a) To check belt installation, rotate crankshaft by hand.*

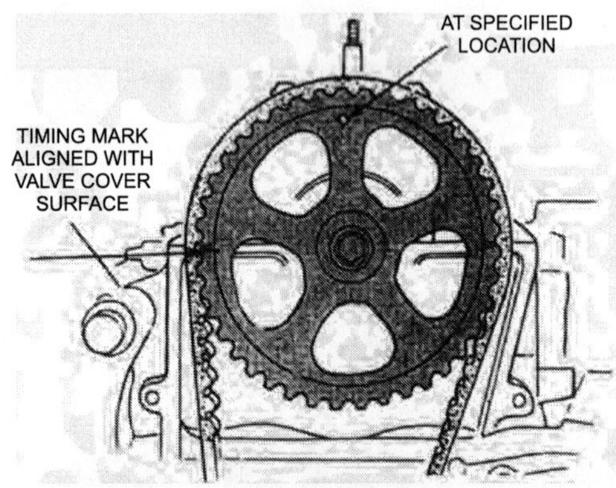

Fig. 14.30 *(b) Make sure that the sprocket and engine marks align.*

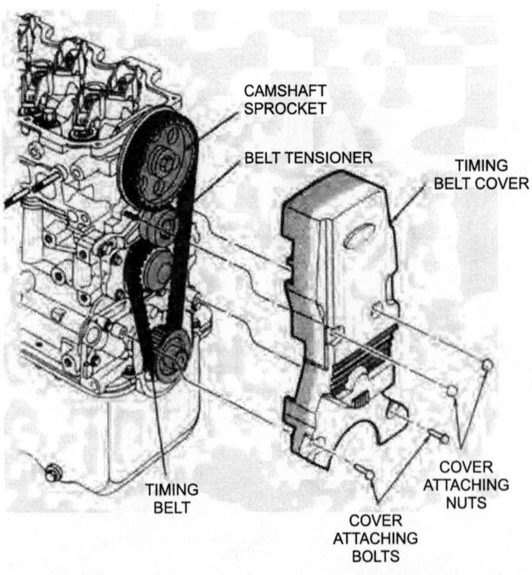

Fig. 14.31 *When installing timing belt cover, make sure that the correct bolts are used and that all rubber grommets or spacers are in place. The rubber washers prevent rattling and breakage of cover. It can be made of thin sheet metal or plastic. (Ford Motors)*

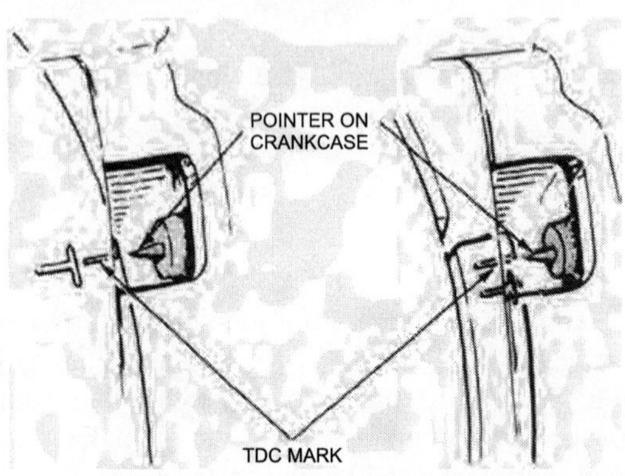

Fig. 14.30 *(c) Also check that engine timing mark is on zero or number one piston is at TDC on its compression stroke. (Honda Motors)*

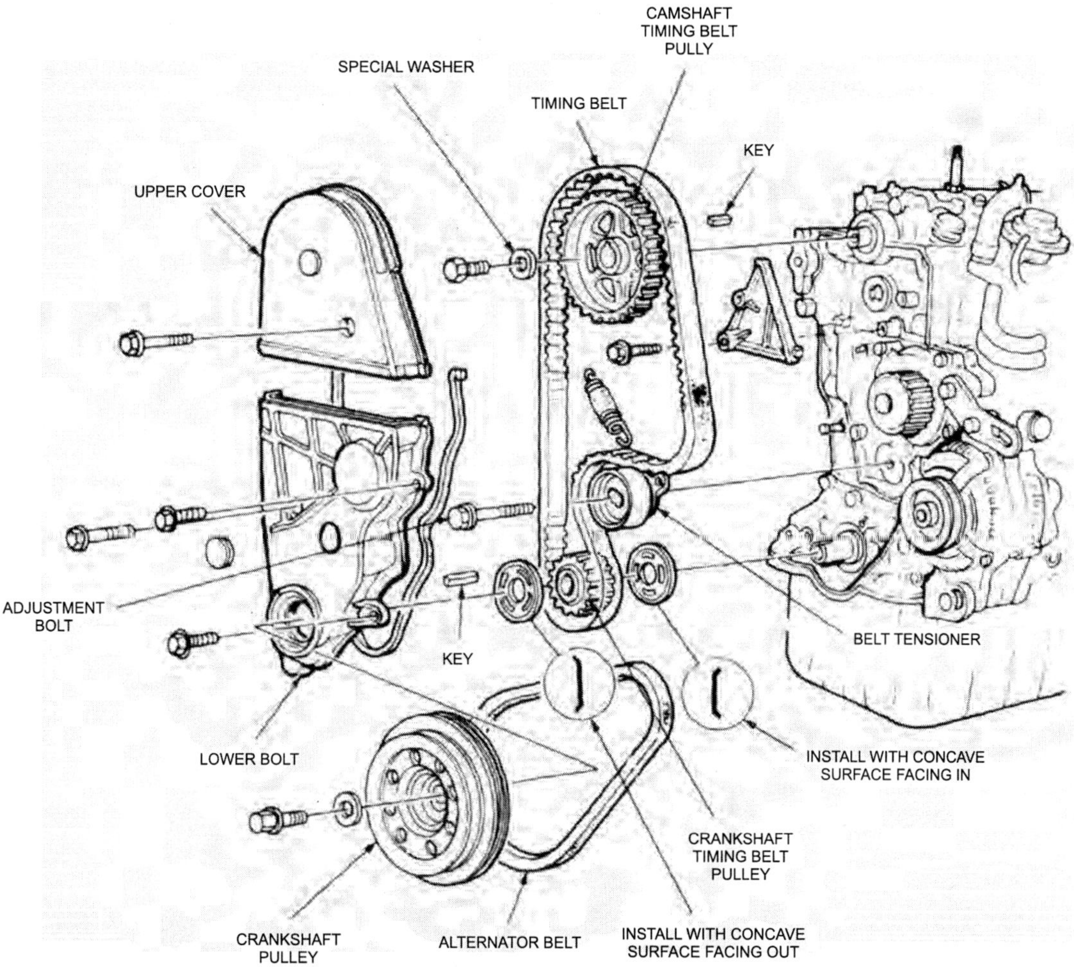

Fig. 14.32 *Exploded view of the front of this engine shows the relationship of front-end components. Torque all fasteners to specifications. (Honda Motors)*

Installing Timing Belt Cover

The *timing belt cover* is a plastic or sheet metal shroud around the belt. It keeps debris off the belt and protects mechanic from getting hurt when working in the engine compartment. To *install the cover,* fit it into its place on the front of the engine. No oil seal is needed at its interface, *Fig. 14.31.* If rubber washers are used in original assembly, ensure that they are used during assembling the cover. Verify that all fasteners are in place and that the belt and sprockets do not touch the timing belt cover. Exploded view of a typical timing belt mechanism and other front-end parts is shown in *Fig. 14.32.* The belt cover is two-piece and uses a lower dust seal.

14.7 TIMING GEAR SERVICE

Timing gear service is similar to timing chain service. Timing gears are normally more dependable than timing chains. They provide long trouble free engine operation. However, after extended use, timing gear teeth do wear or become chipped and damaged, and need replacement.

Inspect the old gears and look for any signs of wear or damage. If necessary, *replace the timing gears as a set, Fig. 14.33.*

Installing Timing Gears

Timing gears are normally press-fit on the crankshaft and camshaft. A wheel puller is required to remove the crankshaft timing gear.

Figure 14.34 shows the use of a press to service the cam timing gear on the camshaft. This design requires the camshaft to be removed for timing gear replacement.

Make sure the key and keyways are properly aligned. Tap in a circular motion to force the gears squarely into position. Light blows from a brass hammer are usually applied to install a timing gear. Avoid any damage to the gear teeth. A press may also be necessary to install the cam gear on the camshaft.

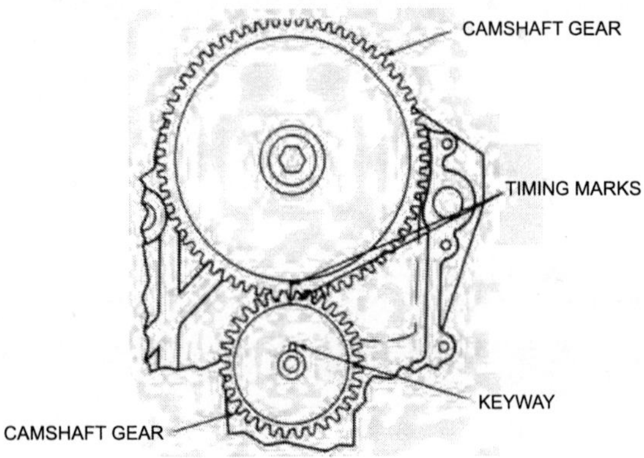

Fig. 14.33 *Timing gears are very dependable. However, after prolonged service, teeth can wear or break. When installing new timing gears, position crankshaft and camshaft in relatively correct positions. Then, fit gears in place so that timing marks align. (Ford Motors)*

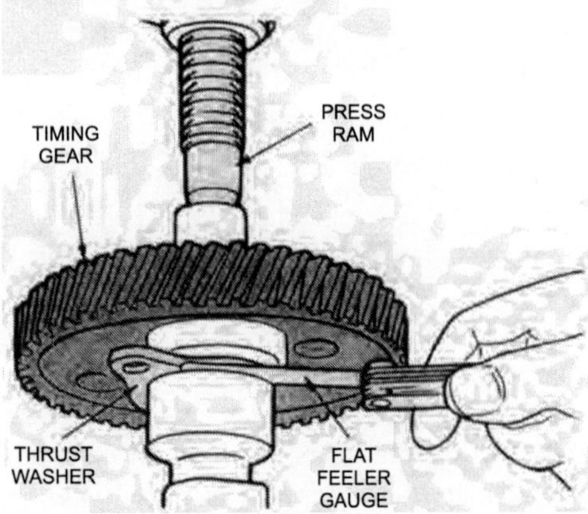

Fig. 14.34 *This particular timing gear must be pressed onto camshaft snout. Clearance must be maintained between thrust plate and cheek of the camshaft. This clearance allows for proper endplay of camshaft. (Ford Motors)*

The timing marks must be positioned precisely against each other to time the camshaft with the crankshaft. Double-check the alignment of the timing marks after installing the gears, *Fig. 14.35.*

Fig. 14.35 *This large truck diesel engine uses four timing gears. Note how timing marks must be aligned.*

Measuring Timing Gear Runout

Timing gear runout or *wobble* is measured with a dial indicator, *Fig. 14.36.* Position the indicator stand on the engine block. Adjust the indicator stem to position it on the outer edge of the camshaft timing gear and parallel with the camshaft centreline.

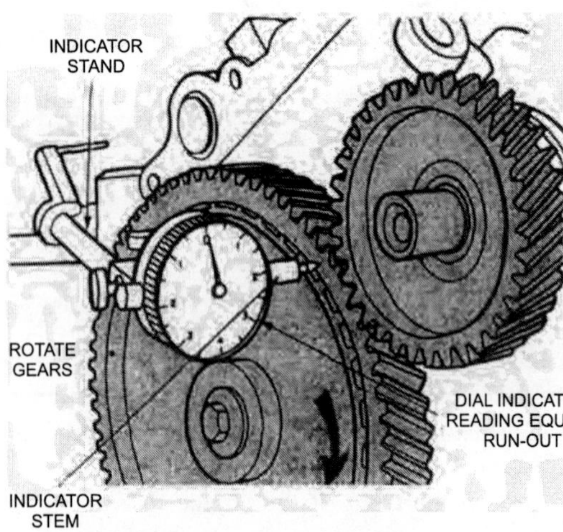

Fig. 14.36 *To check timing gear runout or wobble, position dial indicator as shown. Rotate engine crankshaft and read indicator. Needle movement equals runout. If outside specs, check installation or the gears might be improperly machined. The camshaft could also be bent. (Ford Motors)*

Turn the engine crankshaft slowly while carefully observing the indicator needle reading. No indicator reading means no runout. If the reading is outside the runout spec limits, remove the timing gears and inspect for problems. The gear may not be properly seated or it may be machined improperly. Check the camshaft straightness, too.

Measuring Timing Gear Backlash

Timing gear backlash is the amount of clearance between the mating timing gears teeth. Backlash is measured to determine timing gear teeth wear, *Fig. 14.37.* A dial indicator or a small feeler gauge can be used to measure timing gear backlash.

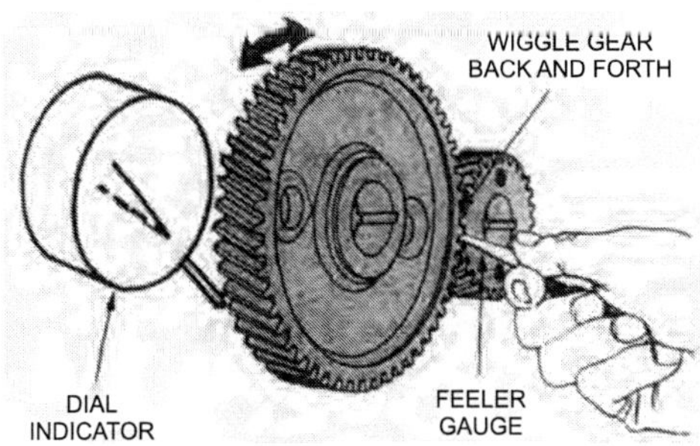

Fig. 14.37 *Timing gear teeth clearance can be measured with small feeler gauge or dial indicator. Size of feeler gauge blade that fits in gear teeth equals backlash. A common method is dial indicator. Mount indicator stem to rest on gear tooth. Wiggle gear back and forth while noting indicator reading. Needle movement equals backlash or teeth clearance. Too little clearance can make gears lock up when expanded by engine heat. Too much clearance can reduce engine power and increase gear noise.*

A dial indicator is an accurate tool for measuring timing gear backlash. Set the indicator stand on the engine. Adjust the indicator stem on one of the cam gear teeth and parallel with gear tooth travel. Wiggle the cam gear one way and then the other, without turning the crankshaft. Record the indicator needle readings and determine the needle travel. If the timing gear backlash is outside the specs, as given in the service manual, the gears are worn and should be replaced.

Figure 14.38 shows a mechanic getting ready to install the front cover after timing gear service.

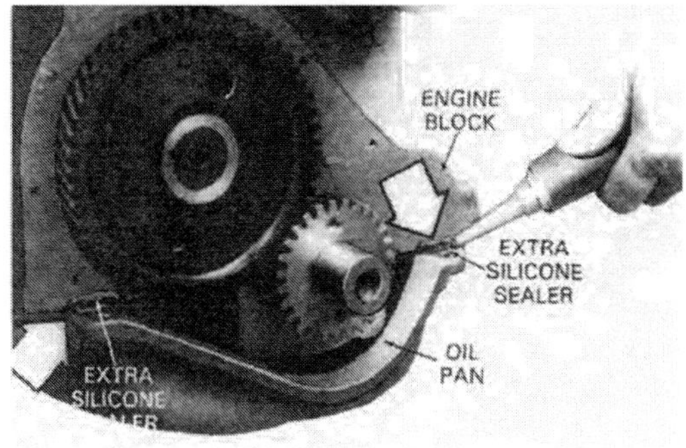

Fig. 14.38 *Timing gears, in this figure, have been serviced with engine and oil pan installed in vehicle. Oil pan bolts must be loosened in front to give clearance for installing front cover. Mechanic is in the process of installing silicone sealer where parts and gaskets come together. (Chevrolet Div. General Motors)*

14.8 CRANKSHAFT FRONT SEAL SERVICE

The *crankshaft front seal* prevents engine oil from leaking out from between the crankshaft snout and the engine front cover. Replace the front-end seal whenever it leaks or when the front cover is removed for service.

Front seal replacement in a vehicle requires only partial engine disassembly. Typically, the radiator and other accessory units on the front of the engine need to be removed. Use a wheel puller to remove the crankshaft damper.

In some designs, it is possible to replace the crankshaft seal without front cover removal. A special seal puller may be used to remove the old seal, *Fig. 14.39 (a).* Then, use a seal driver to squarely install the new seal in its bore, *Fig. 14.39 (b).* When the crankshaft front seal drives in from the rear, timing cover removal is necessary for seal replacement. Remove the cover and drive the old seal out. Then squarely drive the new seal in place, *Fig. 14.39 (c).*

Always compare the old seal with the new seal. Their outside and inside diameters should be same. If possible match part numbers stamped on the metal flange of the seals.

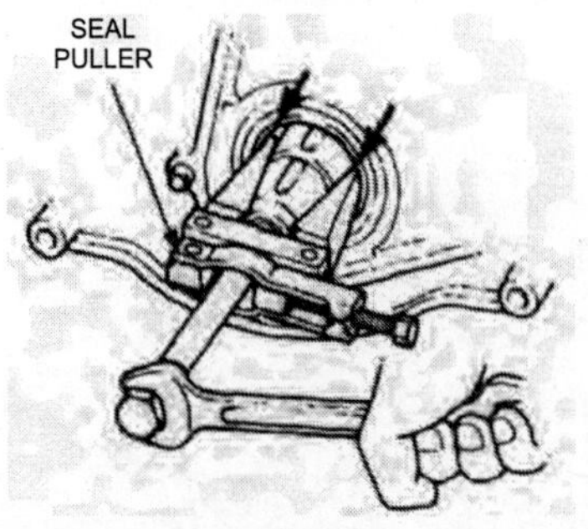

(a) Sometimes front-end seal can be replaced without removing front cover. It can be pulled out from the front and a new one pressed in place.

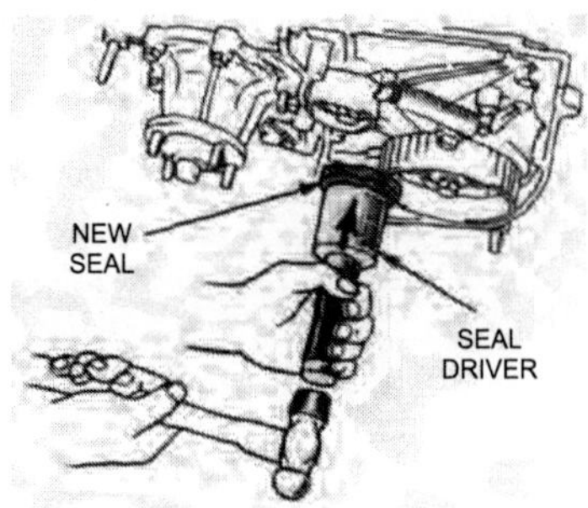

(b) Driving in new seal with cover installed.

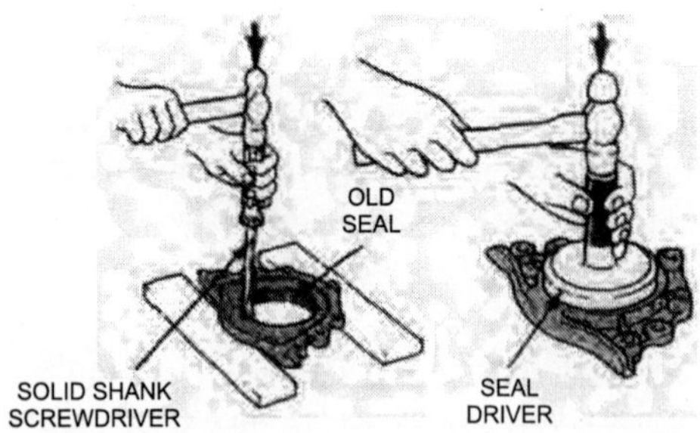

(c) Most engines require front cover removal for front seal service. Oil seal is driven out back. Then seal driver can be used to install new seal.

Fig. 14.39 *Servicing front-end oil seal. (Toyota Motors)*

Before installing a front seal, coat the outside diameter of the new seal with a non-hardening sealer. This will help prevent oil seepage between the seal body and the front cover. Coat the rubber sealing lip of the seal with engine oil. This lubricates the seal during initial engine startup, *Fig. 14.40.*

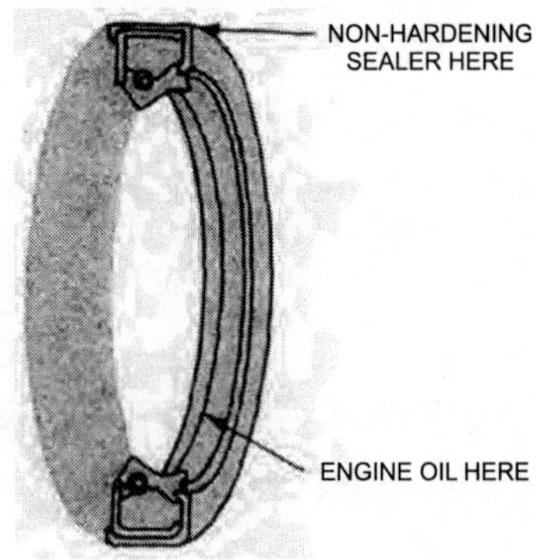

Fig. 14.40 *When installing crankshaft seal, coat outside diameter with non-hardening sealer. Coat the inside sealing lip with motor oil.*

When available, seal driver should be used to install a front cover seal. However, if a seal driver is not available, use a block of wood, as shown in *Fig. 14.41.* Carefully drive the seal in squarely without damaging the seal metal housing.

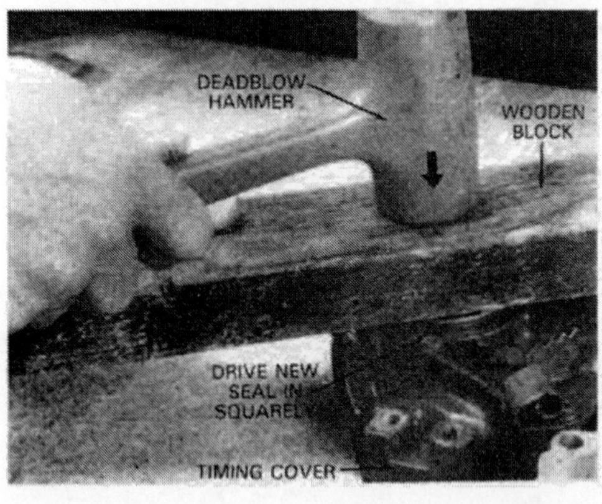

Fig. 14.41 *If a seal driver is not available, then a wooden block will help in installing the seal without damaging it. (Fel-Pro Gaskets)*

Method of installing a front seal in a cylinder head is shown in *Fig. 14.42.* It seals the front camshaft bearing of an OHC engine. The outside diameter of this seal should also be coated with non-hardening sealer. Coat its inside diameter with engine oil or grease.

14.9 ENGINE FRONT COVER SERVICE

The *engine front cover,* also called *timing cover,* holds the front oil seal and encloses the timing gears, belt

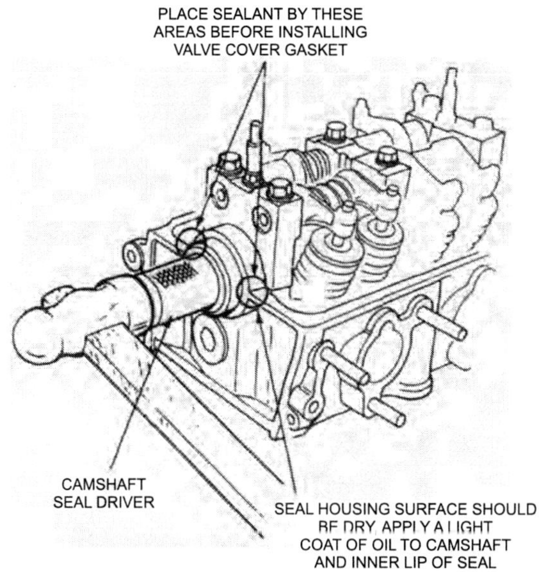

Fig. 14.42 *With an overhead camshaft engine, seal may be mounted on cylinder head as shown. Seal driver is best way to install seat without bending, denting, or damaging the seal. (Honda Motors)*

or chain. It is usually made of thin stamped, sheet metal or cast aluminium.

Scrape off all old gasket material and install the new oil seal. Check that the cover sealing surface is true. Lay it on a flat work surface and check for any gaps under the cover to check for flatness of the surface. Straighten the cover if necessary.

Sealing Front Cover Interface

Either a conventional gasket or suitable chemical sealants can be used during front cover installation. Usually, both a gasket and sealer are used, *Fig. 14.43. Fig. 14.44* shows the method of applying anaerobic and silicone sealers placed on an aluminium front cover. Apply a small, continuous bead of sealer to prevent oil or coolant leakage. Circle with sealant all holes that communicate with oil or coolant.

When using a conventional gasket, carefully inspect to check that all holes align on the engine and in the gasket. Gasket sealer or adhesive is commonly used to hold the gasket in place during assembling the parts.

Always use a bead of silicone sealer where two gaskets or parts come together, *Fig. 14.43,* as this offers a common leakage point.

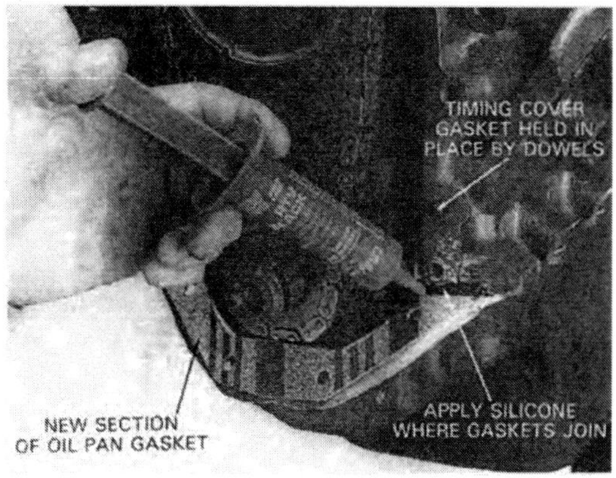

Fig. 14.43 *When installing front cover, make sure gasket holes align properly with their counterparts. Use dowel, adhesive, or sealer to hold gaskets in place on engine mating surface. Use extra silicone sealer where gaskets meet to prevent weepage of oil out of engine. (Fel-Pro Gaskets)*

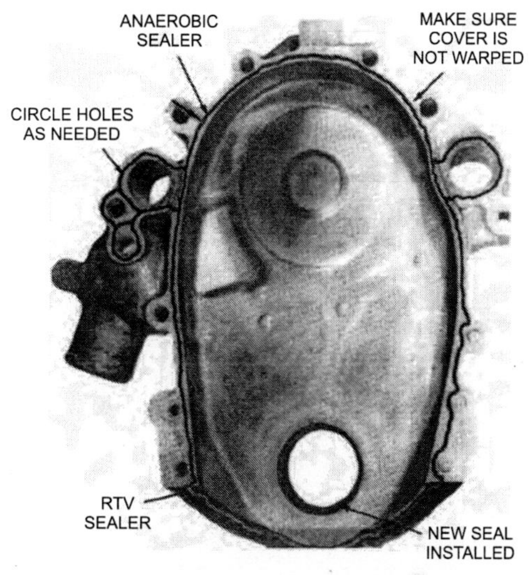

Fig. 14.44 *Figure shows a service manual illustration showing recommended sealing method for a front cover made of cast aluminium. Anaerobic sealer is used where two thick or solid pieces mate tightly. RTV or silicone sealer is used where flexible oil pan mates with front cover.*

Installing Front Cover

Carefully fit the front cover onto the engine, *Fig. 14.45.* Avoid any gasket shift or smearing the

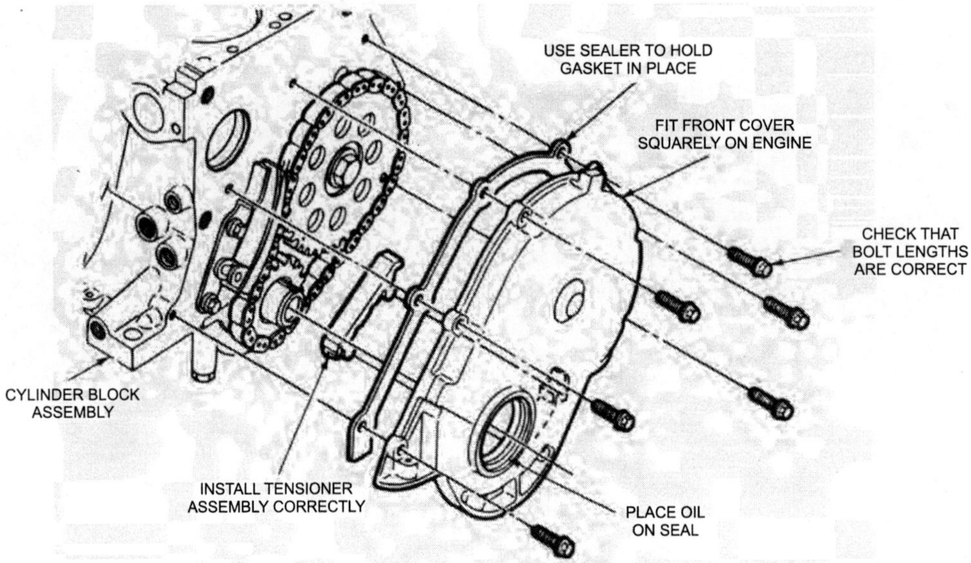

Fig. 14.45 *Before installing front cover, check that all fasteners under cover are torqued to proper specifications. If used, check installation of chain tensioner. Double check timing marks alignment. Carefully position front cover without disturbing gasket. Make sure bolt lengths are correct. Seal any bolt threads entering water or oil passages. Torque fasteners to specifications. (Ford Motors)*

sealer. If the oil pan is on the engine, as for in-vehicle cover service; loosen the oil pan bolts suitably. Pry down on the oil pan so allow the front cover to fit into its position.

A *gasket-centring tool* that fits over the crankshaft snout may be necessary to position or centre the front cover seal correctly around the crankshaft. Install the front cover bolts carefully, paying attention to proper bolt lengths and their locations.

Any bolt that enters a water jacket or oil passage should be adequately sealed. Wipe a small amount of non-hardening sealer on these screw threads to prevent fluid leakage. If in doubt refer the applicable service manual for information on which bolts or screws require sealer.

Tighten the bolts to specifications, in gradual incremental steps following a crisscross pattern. When used, remove the seal centring tool. Double-check that all bolts are installed properly, so as to avoid missing any hidden bolt hole under the lower section of the timing cover.

Install the front oil pan bolts if necessary. Then, finish by assembling other outside components, e.g., water pump, brackets, front damper, etc.

14.10 CRANKSHAFT DAMPER INSTALLATION

Before installing the crankshaft damper, check the crankshaft snout key. Verify that it is properly installed and tight in its keyway. Fit the damper onto the crankshaft without pushing the key out of its keyway. Usually, the snout bolt is tightened to force down the damper into its intended position, *Fig. 14.46.* However, in some cases a special driver tool is needed to install the damper, *Fig. 14.47.* In either case, make sure the damper is forced down squarely on the snout. If it cocks sideways, stop and align the damper to prevent part damage.

Never hammer on the outer edge of a crankshaft damper during its removal or installation. Keep in mind that the outer ring is mounted on rubber. Hammering could shift the outer ring and ruin the damper. Install any other components, e.g., crank pulley, timing pointer, brackets, etc., as needed.

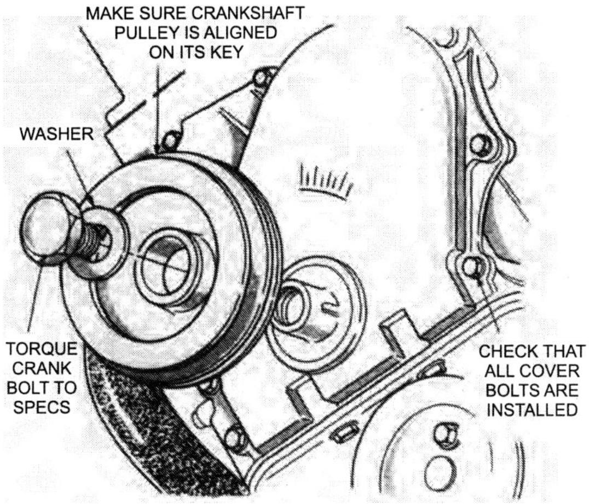

Fig. 14.46 *Crankshaft damper will usually slide partially over crankshaft snout. Then, large snout bolt will pull damper into position. If light hammering is needed, hammer in the centre of the damper. Never hammer on its outside diameter or the rubber ring may be damaged and the harmonic balancer be ruined.*
(Honda Motors)

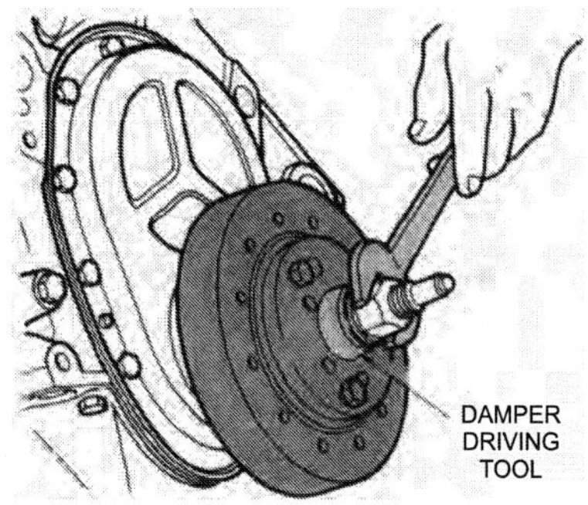

Fig. 14.47 *This damper needs a special tool for installation. Only a press-fit holds the damper in place after assembly. It is not held by crank snout bolt. Be careful not to hammer on outer ring of vibration damper.*
(Chevrolet Div, General Motors Corp.)

THEORY QUESTIONS

1. The camshaft is driven by sprockets and chain or toothed belt, or by
   (*a*) an oil-pump gear (*b*) the distributor shaft
   (*c*) timing gears (*d*) timing chain.
2. Describe the method of fitting the timing chain and sprockets on an engine.
3. Drive the chain sprockets on with hammer blows around the outside diameter of the sprockets. *True* or *False?*
4. An engine can be cranked with the starting motor when the timing chain or belt is removed. *True* or *False?*
5. Mechanic A says that some OHC engines use bucket tappets and no rocker arms. Mechanic B says that OHC engines use rocker arms. Who is right?
   (*a*) A only (*b*) B only
   (*c*) Both A and B (*d*) Neither A nor B.
6. The camshaft turns at
   (*a*) the same speed as the crankshaft
   (*b*) twice the speed of the crankshaft
   (*c*) half the speed of the crankshaft
   (*d*) None of the above.
7. Timing chains are used on both _____ and _____ _____ engines.
8. How can the OHV engine timing chain wear be checked without timing gear removal?
9. This component keeps oil from spraying out the front seal and also helps lubricate the timing chain or gears.
   (*a*) Shroud (*b*) Front cover
   (*c*) Tensioner (*d*) Slinger.
10. If the belt tension gauge is not available, how can the belt tension adjustment be checked for proper tension?
11. This is measured with a dial indicator with the indicator plunger resting against one of the timing gear teeth.
    (*a*) Timing gear diameter
    (*b*) Timing gear runout
    (*c*) Timing gear backlash
    (*d*) Timing gear out-of-round.
12. Mechanic A says that most engines must be removed from the vehicle in order to replace the old timing chain. Mechanic B says that normally a timing chain can be replaced without removing the engine from the vehicle. Who is right?
    (*a*) A only (*b*) B only
    (*c*) Both A and B (*d*) Neither A nor B.

13. Mechanic A says that timing chain wear can affect exhaust cleanliness. Mechanic B says that timing chain wear can lower engine compression. Who is right?
    (*a*) A only (*b*) B only
    (*c*) Both A and B (*d*) Neither A nor B.
14. Timing gear runout should be measured with a _____.
    (*a*) feeler gauge (*b*) dial indicator
    (*c*) inside micrometer (*d*) outside micrometer.
15. Most manufacturers recommend replacing an engine's timing belt about every _____.
    (*a*) 320,000 km (*b*) 160,000 km
    (*c*) 120,000 km (*d*) 80,000 km.
16. Mechanic A says that if a timing belt is too tight, severe valve damage can occur. Mechanic B says that if a timing belt is too loose, severe valve damage could occur. Who is right?
    (*a*) A only (*b*) B only
    (*c*) Both A and B (*d*) Neither A nor B.
17. Mechanic A says that the function of a timing belt cover is to keep debris off the belt and protect hands when working in the engine compartment. Mechanic B say that the function of a timing belt cover is to protect the oil seal and belt from debris. Who is right?
    (*a*) A only (*b*) B only
    (*c*) Both A and B (*d*) Neither A nor B.
18. Mechanic A says that one must normally hammer on the outer edge of a crankshaft damper to force it into place on the crankshaft. Mechanic B says that normally, the snout bolt will tighten down and force a crankshaft damper into position. Who is right?
    (*a*) A only (*b*) B only
    (*c*) Both A and B (*d*) Neither A nor B.

❑❑❑

Engine Reassembly – Installation – Break-in

15

15.1 INTRODUCTION

The basic methods for reassembling the major components of an automotive engine will be outlined in this chapter. The key to proper assembly of any engine is cleanliness. The work area and the workbench space, along with the mechanic, must be very clean to prevent dirt or other engine-damaging (grit) particles from being picked up and causing possible serious engine damage. Dirty benches, tools, rags, and hands all leave grit in an engine.

Whenever an engine under assembly is not being worked on, it should be covered with a plastic bag. This will keep out dirt as well as flying debris from other shop operations. Small parts ready for assembly should be stored in clean covered containers. Store large parts, such as cylinder block, cylinder heads and manifolds, in plastic bags.

Paying attention to detail is also an engine assembly requirement. There are no substitutes for time, patience, good attitude and common sense. *Fig. 15.1* shows the major engine components you will be assembling.

All parts are attached to the engine block. The block, therefore, must be well prepared before starting the assembly. Previous chapters explained the methods to assemble the short block and cylinder head. This chapter will build upon the previous knowledge by explaining how to install the cylinder head, manifolds, and covers on the short block, etc.

The methods given here are general and they apply to most makes and models of automotive engines. However, always follow the more specific procedures given in the applicable service manual for the engine under service. The manual will give information that might be unique to the particular engine.

15.2 GENERAL RULES FOR ENGINE ASSEMBLY

The basic rules that must be kept in mind when assembling an automotive engine include:

1. *All parts and tools should be perfectly clean and well organized.* Remember that even the slightest dirt or grit, metal, plastic, gasket material, abrasive, etc. could cause major engine damage, *Fig. 15.2.*
2. *All parts that could be select fit,* to provide proper fit and clearances, should be reinstalled in the same location from which they were removed. For example, if valve lifters that are in good condition and are going to be reused on the same camshaft then install them in their exact same lifter bores from which they were removed. Lifter diameters and their wear patterns do vary slightly.
3. *Keep parts properly organized and covered until the time to assemble them.* Keep a clean cloth or plastic bag over a short block sitting on an engine stand. This will keep debris out of the open cylinders and off other areas in the block. Keep the intake manifold inlet covered, especially after it is installed on the cylinder head. Anything dropped into the inlet manifold could cause severe engine damage to the pistons, valves, head, and block.
4. *Lubricate all relatively moving parts with engine oil or white grease, as needed.* This is necessary to provide adequate protection

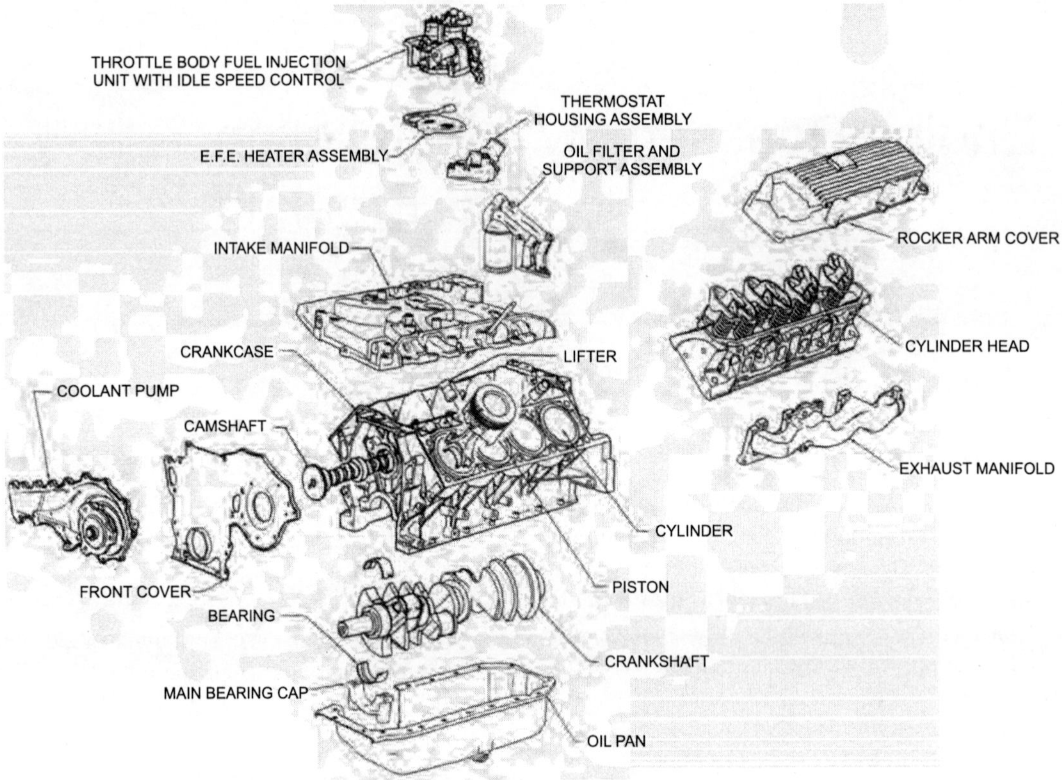

Fig. 15.1 *Major engine components. (Cadillac Div. GM)*

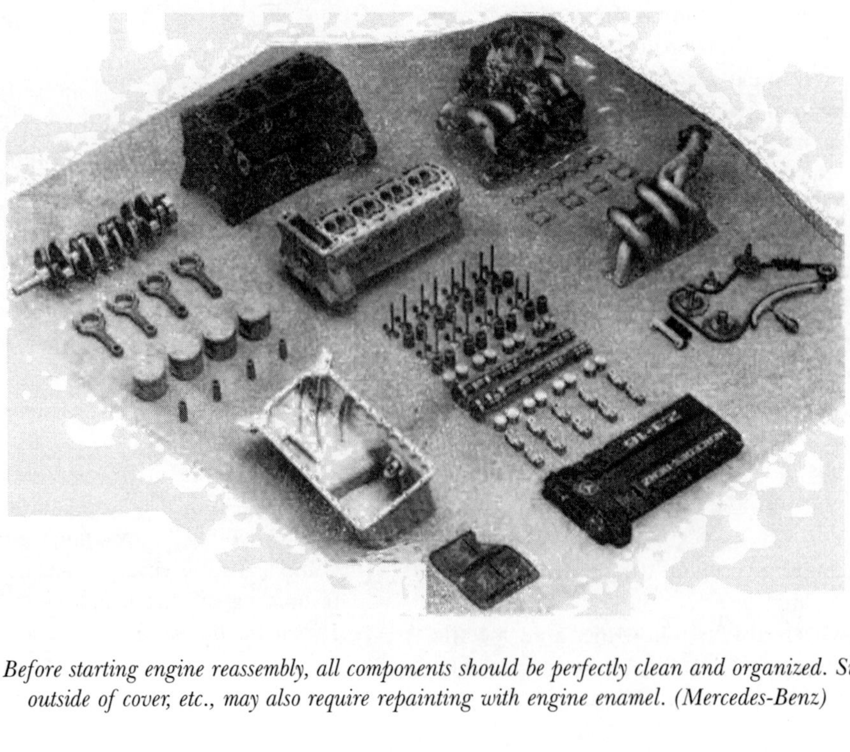

Fig. 15.2 *Before starting engine reassembly, all components should be perfectly clean and organized. Sides of block, outside of cover, etc., may also require repainting with engine enamel. (Mercedes-Benz)*

against friction and wear during initial engine start up, until the proper lubrication circuit pressure is restored.

5. *Inspect everything before installing the engine parts,* for any unusual wear, cracks, scoring, etc. This is important to avoid finding something wrong after the engine is assembled and installed in the engine compartment in the vehicle.
6. *Make sure all fasteners are correct size and shape.* It is best to make sure that all fasteners are exactly the same as used by the manufacturer in original engine assembly. On engines that use both English and Metric fasteners, do not mix them to avoid thread damage. Also, check fastener lengths. A bolt/ screw that is too long could damage parts by bottoming out. While a bolt that is too short could strip threads on torquing them to specs and fail in service.
7. *Torque all fasteners properly and to specs.* An air impact does not provide the precise control needed to properly tighten many engine fasteners. Do not attempt to save time by resorting to impacting down instead of torquing the fasteners that are otherwise required to be torqued. This essentially applies to head bolts, manifold bolts, oil pump bolts, etc. To save weight, late model engine components are cast as thin and light as possible to protect function. Improper torque can cause fluid leakage, part warpage, or even part breakage.
8. *Always refer to an applicable service manual if in doubt about any service procedure.* The manual provides proper directions concerning assembly methods, torque values, valve adjustment specs, allowable wear specs, critical clearances between relatively moving mating components, etc.
9. *Keep focused on the work and avoid detractions while working on an engine.* If distracted, it is very easy to forget to do something that is otherwise necessary. Forgetting to torque even one bolt could result in major engine damage, or a serious leak. That could be costly to the owner.
10. *Do not get too excited and rush to finish the job.* Careful and steady pace is important to keep control of the job quality.

15.3 LUBRICANTS, GASKETS AND SEALERS

Lubricants, gaskets, and sealers are frequently overlooked in providing engine assembly information. These items impart the necessary protection and sealing power to various engine components.

Lubricants

If an engine is assembled without using any lubricants, it will suffer damage on first starting after reassembly, because it takes a while for the lubricating system oil pressure to reach all mating and relatively moving parts of the engine. To avoid working the mating parts without any lubrication during the initial start-up time, all such parts should be adequately lubricated during assembly. Besides providing start-up lubrication, oiling the engine parts during assembly will facilitate assembly, and prevent rust and corrosion from forming while the engine is in storage.

Engine oil is normal choice for an assembly-lubricant. Unless stated otherwise in the service manual, lube all parts with SAE 30-weight engine oil. However, some engine parts have special lubrication requirements. Oil quickly drains off a metal part unless it is physically retained. Therefore, use a thicker stickier lubricants, with higher oiliness value and adhesive strength, on most parts. Also, any part of the valve train loaded by the valve springs is under extreme pressure. These parts might run dry until enough oil flows to them from the pump. To prevent such a condition, special lubricants containing molybdenum disulphide are used when assembling valve lifters, pushrods, etc. Lubricant should be thick enough so that it would not drain off, and molybdenum provides excellent start-up protection.

Gaskets, Sealers, and Locking Compounds

Gaskets and sealers fill the minute voids between mating parts surfaces, effectively keeping out dirt or sealing in fluids.

Gaskets and Seals: Gaskets are properly shaped materials used to seal two or more mating stationary parts surfaces. There are head, oil pan, manifold, water pump gaskets and others. *Fig. 15.3,* shows a set of gaskets. Gaskets are made from many different materials depending upon their use. Cork, asbestos, and synthetic materials are just a few of the materials used. Some gaskets are sandwiches of more than one material. They may be coated with an aluminium alloy, which provides a better sealing.

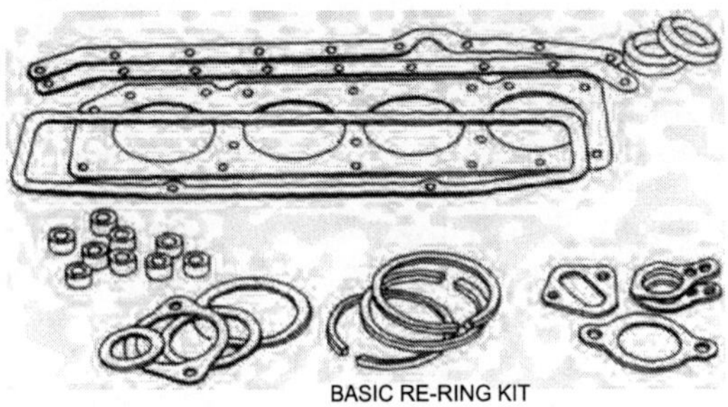

Fig. 15.3 *Gasket set.*

Seals are typically used between a moving and a stationary part. There are numerous different designs of seals, but two of them, i.e., lip seal and O-ring seal, are the most common types used in automotive engine applications. *Fig. 15.4* shows cross-sections of ring seals.

Sealers: A variety of sealers are available for different applications. Two broad categories of sealers are *nonhardening* and *hardening*.

Nonhardening sealer is the standard gasket cement at present. Compared to a hardening sealer, nonhardening sealer resists vibration, expands and contracts with the engine, and allows more flexibility in engine assembly. Nonhardening sealers are much easier to clean during engine disassembly.

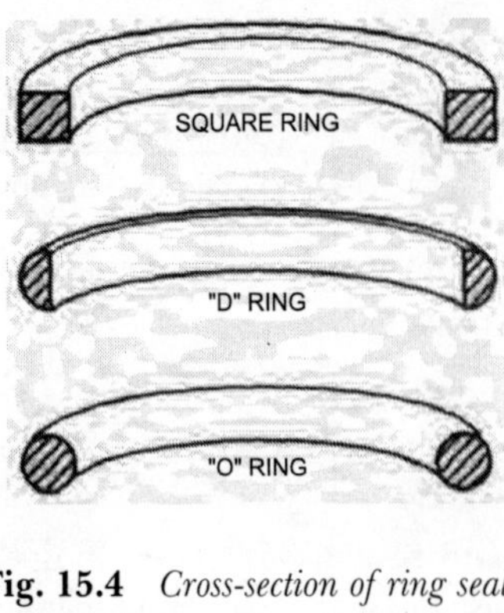

Fig. 15.4 *Cross-section of ring seals.*

A special type of sealer is *room temperature vulcanizing (RTV)*. This special rubber compound sets up after exposure to the atmosphere. This permits squeezing it almost anywhere, bolt the parts together and let the rubber vulcanize to quickly form a rubber gasket, sealing gasoline, water, oil, and other normal engine fluids. Old RTV peels off to permit easy surface cleanup. Due to its versatility it has become the number one gasket sealer.

RTV suffers from some disadvantages too. Squeezed from between mating parts, on tightening fasteners, it forms a flash, which can break off. Outside the engine this causes no problem. However, on the inside of the engine these broken off rubber pieces can circulate with the oil and may clog tight oil passages. Also, RTV is fairly slippery until it set up, and may not be able to effectively hold the gasket in place during assembly.

Locking Compounds: Some heavy-duty threaded fasteners might require a locking compound. These chemical compounds act like lock washers. Extra effort is needed to remove fasteners treated with locking compounds. These are available in strengths ranging from "wrench removal" to "permanent joints."

15.4 ENGINE REASSEMBLY

All parts should already be properly cleaned, organized and covered to keep away any dirt and debris. The cylinder head should be reconditioned, as explained in chapter 13 on *"Top End Service."* The short block should also be reconditioned and together as discussed in chapter 12 on *"Short Block Service."* Some additional information is dealt here to supplement the information already obtained.

The short block should be mounted on an engine stand. All connecting rods should be properly torqued and all surfaces perfectly clean. Appropriate areas of the outside of the engine are painted. The front cover and timing chain/belt or gears should also be installed as explained in chapter 14 on *"Front End – Timing Mechanism Service."*

Block Preparation

All surfaces should also be checked for damage resulting from the machining process used during reconditioning the block. Items that should be done before reassembly is started include the following:

1. The block, including the oil gallery passages, should be thoroughly cleaned.

2. All threaded bolt/screw holes should be chamfered.
3. All threaded holes should be cleaned with proper size taps or thread runners.

Installing Cups and Core Plugs

Start engine assembly by installing all core plugs. Coat the core plug's mating area and bore with nonhardening sealer. Set the core plug in position and then hammer it in place. A socket that fits the outside diameter of the core plug is used to replace the core plug, but it is preferred to use the core plug replacer tools, *Fig.15.5.* This will distribute the hammer blows and help drive the plug straight in. Oil gallery plugs should be installed using sealant on the threads.

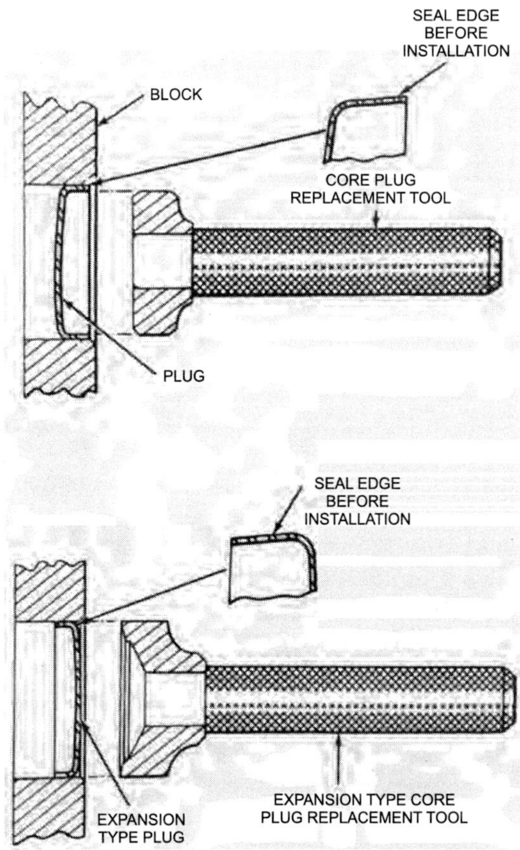

Fig. 15. 5 *Core plugs and installation tools. (Ford Motors)*

Some engines have core plugs that have no lips and are used in holes with a cast-in ledge. These plugs are bowed, so from one edge they are saucer-shaped. Place the plug in its bore, butted against the bore's shoulder. The core plug's bow must face the outside of the engine. Then with a flat-nosed punch smack the centre of the plug. This pops the plug's centre toward the block and forces the plug's edges against the block. Also install any core plugs in the head and a possible oil plug at the rear of the cam bores. Also install any core plugs in the head and a possible oil plug at the rear of the cam bores.

Core holes left in the external block wall are machined and sealed with *soft-core plugs* or *expansion plugs,* also known as *freeze plugs* or *Welsh plugs.* Soft plugs are of two designs:

Convex Type: The core hole is counter bored with a shoulder. The convex soft plug is placed in the counter bore, convex side out. It is driven in and upset with a fitted seating tool. This causes the edge of the soft plug to enlarge to hold it in place. *Fig. 15.6* shows an installed convex soft plug. A soft plug should be driven in until it reaches the counter bore of the core plug hole.

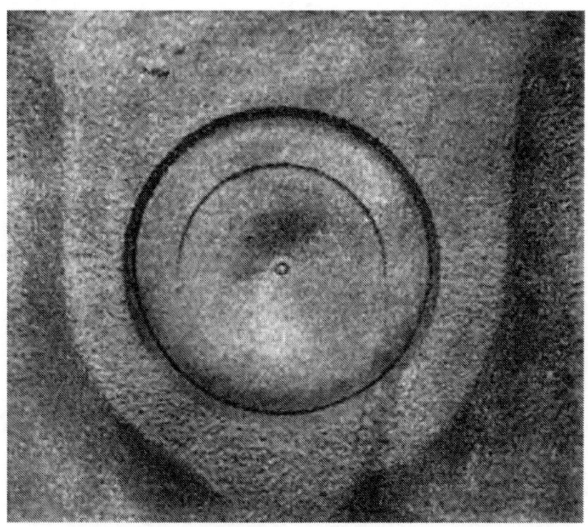

Fig. 15.6 *Installed convex-type soft plug.*

Cup Type: This most common type fits into a smooth, straight hole. The outer edge of the cup is slightly bell mouthed. The bell mouth causes it to tighten when it is driven into the hole to the correct depth with a seating tool. An installed cup-type soft plug is shown in *Figs. 15.7* and *15.8.* A cup plug is installed about 0.5 to 1.3 mm below the surface of the block, using sealant to prevent leaks.

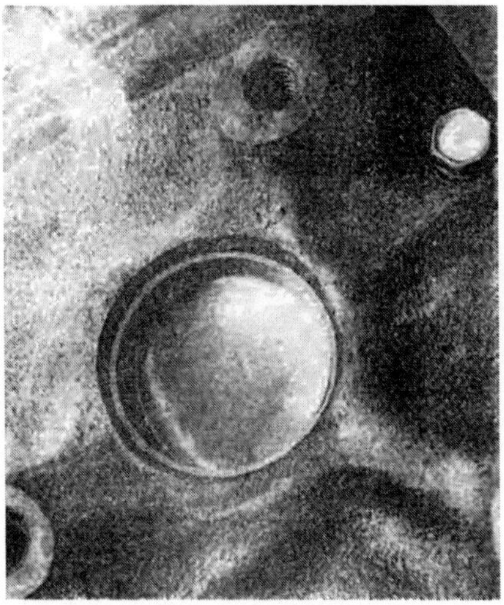

Fig. 15.7 *Installed cup-type soft plug.*

Fig. 15.8 *Sealer should be used on the cup plug before Being driven into block.*

Overhead Valve (OHV) Camshaft

Install the camshaft for the overhead valve (OHV) now. First, install the camshaft bearings, if necessary. Deburr the bearing's edges first. Then align any oil holes with those in the block. Make sure the bearings are in the proper order. Some applications use gradually smaller bearings from front to rear. Start with the rear bearing and drive it home with a cam bearing installer. This is a special mandrel with a long handle. This mandrel extends outside the block to allow hammer on it, driving the bearing in. Check that the oil holes line up before and after installation.

Thoroughly oil the camshaft bearings and journals. Then generously lube the cam lobes with a molybdenum compound. Smear the lube evenly around each lobe, and ensure that no part is left dry.

Slide the camshaft into the block. Hold the camshaft with both hands while guiding into the block. One hand should be inside the block, feeding the camshaft through its bearings; the other hand should support the shaft outside the block. Carefully avoid nicking the cam lobes against the block. If there is a cam retainer in the timing chain area, install it at this time so the cam would not slide out as the engine is moved around.

Rear Crankshaft Seal

Three types of rear crank seals are used in practice. First, there is the rope, or split-lip, seal. Second and most common is the neoprene seal, a rubber, two-part seal. Third is the single-piece rubber seal. It is a complete circle and is installed after the crankshaft and main bearings. The other two seals are installed at this time, before installing the crankshaft.

Rope Seal: First, apply RTV to the main bearing cap and block as illustrated in *Fig. 15.9.* Slide one part of the rope seal into the block and the other into the rear main bearing cap. Work each half into the groove as tightly as possible. Usually a special tool is available for this. If not, use a large socket. Once fully seated, trim the ends of the seal flush with the block or cap. Use a small sharp knife to trim the seal. Cut away any loose fibers, which could get caught between block and cap, upsetting the rear main bearing's oil clearance.

After installing the seal, rotation of the crankshaft might become difficult. If so, use the front pulley or thread two flywheel bolts into the crank and use a rod as a lever between the two bolts to move the crank.

Neoprene Seal: Neoprene seals come preformed and are easy to install as an insert bearing, *Fig. 15.10.* Like a bearing, install the seal in a dry bore. The sealing lip must face toward the inside of the engine. Then apply oil on the working surface wiped by the crank.

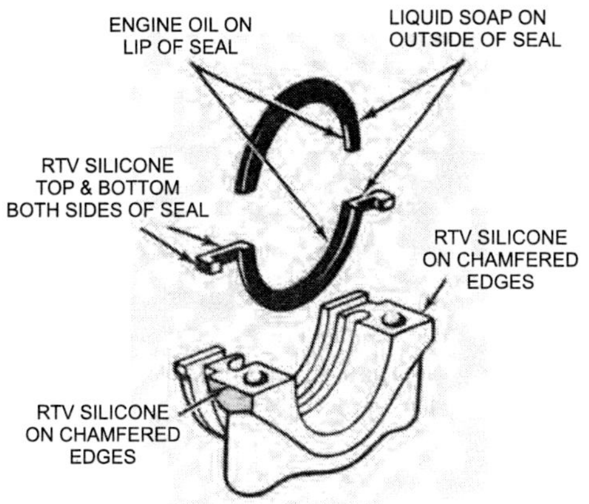

Fig. 15.9 *Applying RTV to main bearing cap and block. (Ford Motors)*

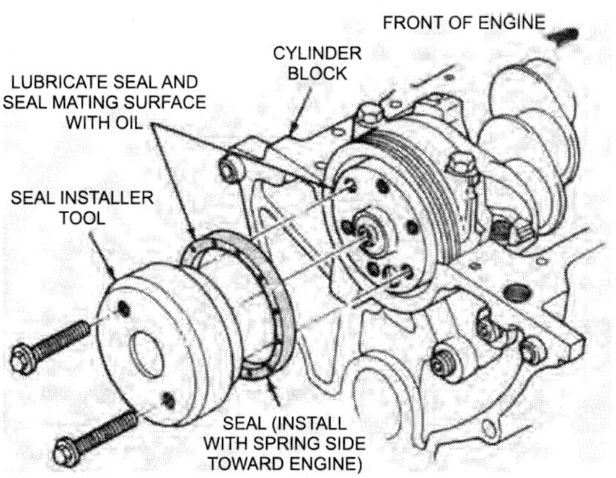

Fig. 15.10 *Installing single-piece rubber seal. (Ford Motors)*

Crankshaft Bearings

Wipe all the bearing bores with a dry paper towel. Then lay the bearing shells into the bores. Line up the tang with the cutout in the block or cap, place a thumb on each end of the bearing, and push it into its bore. Thoroughly oil each bearing. Also squirt oil down the oil passages in the block. This helps fill them, reducing lubrication system priming time at engine start-up.

Some bearings have different upper and lower halves. The upper halves typically have an oil hole and groove, while the lower halves are plain. Do not install the two backwards, to avoid ruining both the crankshaft and connecting rods because oil would not reach the bearings.

Causes of Premature Bearing Failure

According to manufacturers of engine bearings, the major causes of premature bearing failure are:

- Dirt 45%
- Wrong assembly 13%
- Lack of lubrication 14%
- Overloading or lugging 10%
- Corrosion 4%
- Others 4%.

Many cases of premature bearing failure may result from a combination of several of these items.

Crankshaft

Double-check that the crankshaft is clean by wiping it down with a paper towel. Then grasp it by the shaft and flywheel flange. Be careful to lower the crankshaft straight down, set the crankshaft in its bearings, *Fig. 15.11.* Oil the journals while slowly rotating the

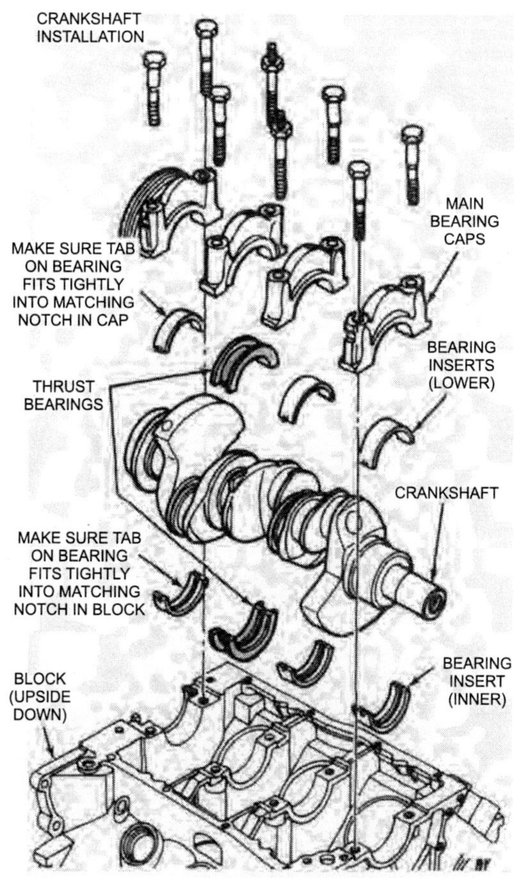

Fig. 15.11 *Crankshaft and main bearings. (Ford Motors)*

crankshaft. Squirt oil into the crankshaft's oil passages to help quicker priming. Install the main caps at their correct places and in right directions. Remember, the bearing tangs should butt against each other.

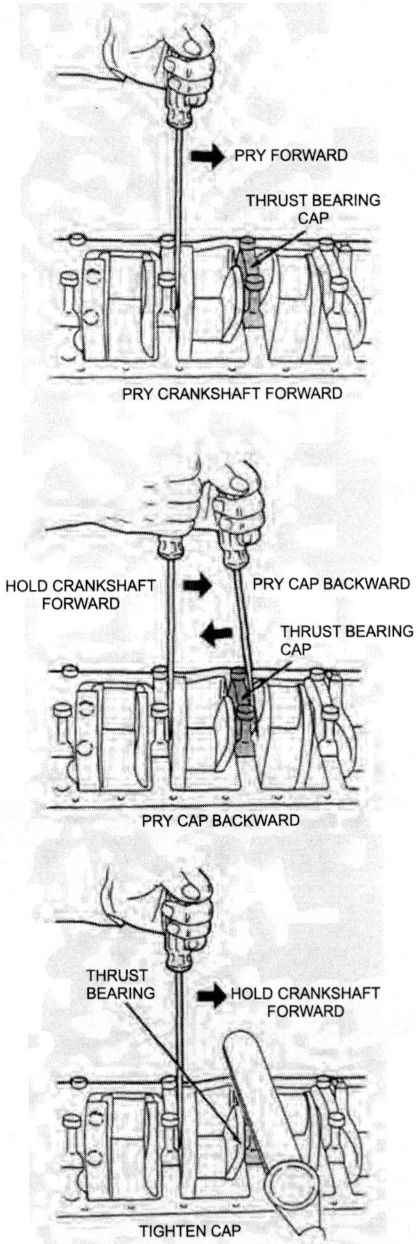

Fig. 15.12 *Thrust bearing alignment. (Ford Motors)*

If the engine has separate thrust washers, install them before the main bearing caps. Also, apply sealer to the rear seal's ends before joining caps and block.

Before tightening the main bearing caps, align the thrust bearings, *Fig. 15.12.* Do this by slightly snugging the main bearing caps. Then pry the crankshaft sideways along the crankshaft axis in the cover. The crankshaft will move a small distance and then contact the thrust bearings. Because their cap does not clamp down the thrust bearings, the crankshaft will push them into alignment. Check the excessive turning effort that indicates binding. Then tighten the main bearing caps.

Tighten the main bearings to specs in several steps and check after each step that the crankshaft turns without binding. Right before bolting on the oil pan, retighten the main bearings. This takes care of any bolts that might loosen from normalizing, the relaxing of the block after the stress of tightening.

Installing Oil Pump

After installing all pistons and connecting rod assemblies and with the front cover and timing mechanism (not timing belt however) in place, the oil pump can usually be installed. If the oil pump attaches in the crankcase, bolt it now, *Fig. 15.13.* Make sure the used oil pump is in good condition. Frequently mechanics prefer to use a new or factory rebuilt oil pump. They are not very expensive and are very important to engine service life. Oil pump gears should be coated with assembly lubricant before the cover is put on the pump. This provides initial lubrication and it primes the pump so that it will draw oil from the oil pan when the engine lubrication system is first operated.

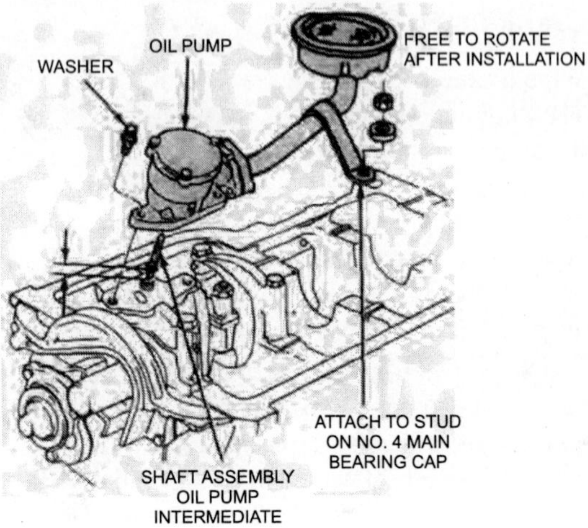

Fig. 15.13 *Oil pump installation. (Ford Motors)*

When possible, always prime an oil pump. This is done by pouring oil or petroleum jelly into the pumping cavity during assembly. Also, the pump inlet can be submerged into an oil container and turn the pump by hand. This will draw oil into the pump and ensure faster engine oil pressure upon initial starting of the engine after reassembly.

Be sure to check the position of the oil pump pickup. If the engine is upside down, pour oil down the pickup. Otherwise, the pump will merely suck air, and the engine will run dry.

For a front cover-mounted oil pump, position the new gaskets, seals, rotors, and the other parts in the engine. Torque all pump fasteners to specs. Turn the pump by hand to check for running smoothness. Check that the pump is well lubricated, *Fig. 15.14(A)*.

If the engine has a more conventional oil pump that installs in the oil pan, fit the oil pump drive into position, as shown in *Fig. 15.14(B)*. Make sure the oil pump drive rod or shaft is located in its guide hole in the block. While fitting the oil pump to the block, the drive rod must fit into the pump and the hole in the block. Use nonhardening sealer between the pump and block, not RTV. RTV might flash inside the oil passage.

Make sure the oil pump gasket holes align properly. Avoid sealer on the oil pump gasket because a gob of sealer could lock the oil pump. Usually, install the oil pump gasket clean and dry. Start the pump bolts by hand and then torque them to specs.

Oil pump mounting bolts should be tightened properly. It might be wise to use thread locking compound on the oil pump bolt threads. If these bolts

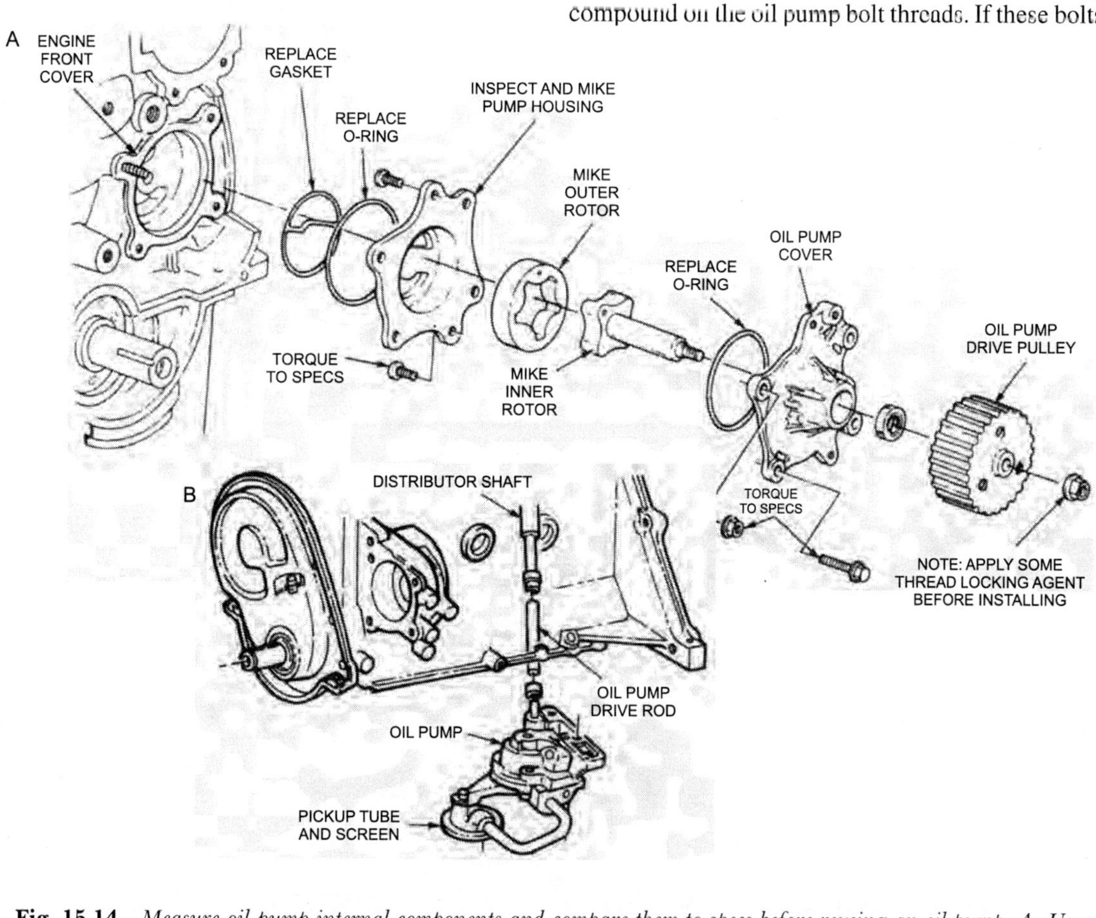

Fig. 15.14 *Measure oil pump internal components and compare them to specs before reusing an oil pump. A- Use new gaskets and seals when installing a front-cover mounted oil pump. B - Make sure drive rod is up in hole in cylinder block with conventional oil pump. Fit gasket in place and start bolts by hand. Tighten bolts to specs. Make sure pickup tube is secure.*

were to loosen or fail in service, the engine will lose oil pressure and major bearing damage can be caused.

If support brackets are provided on the oil pump, make sure they are installed properly. They prevent pump vibration that could break the pump housing and cause bolt failure. The oil pump pickup tube and screen must also be in good condition and properly installed.

After pump installation, double-check everything. The pump gasket should be in place. The pump drive rod must be located in the block properly. The pickup tube must not leak at the pump or block.

To help prevent possible serious oil pump-related failures, many engine builders recommend:

1. Always ensure that the oil pump pickup tube (screen) is securely attached to the oil pump to prevent the pickup tube from vibrating out of the pump.
2. Use modeling clay to check pickup screen-to-oil pan clearance. For proper operation, there should be about 6 mm distance between the oil pump pickup screen and the bottom of the oil pan.

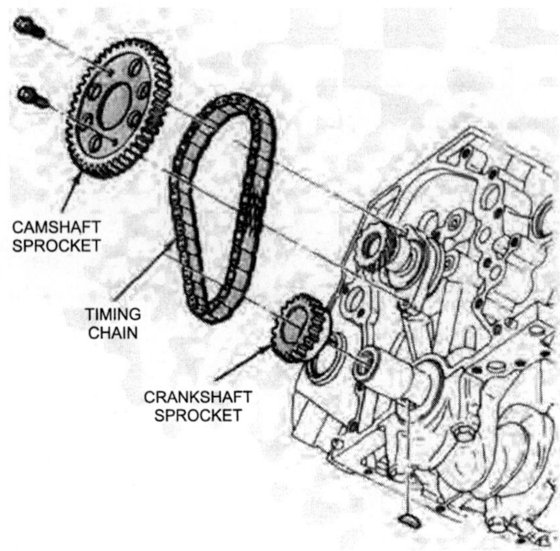

Fig. 15.15 *OHV timing chain and sprockets. (Ford Motors)*

Timing Gear

At the front of the engine, fit the timing gear. On overhead valve (OHV) engines, fit the Woodruff key and then the crankshaft timing sprocket on the crank snout, *Fig. 15.15.* Rotate the crankshaft to the position shown in the service manual. Use the camshaft timing sprocket to rotate the cam to its proper position. Remove the cam sprocket and drape the timing chain over it and then over the crank sprocket. Make sure the timing marks are lined up between the crank and cam sprockets; then bolt the cam sprocket to the camshaft, *Fig. 15.16.* Double-check the timing mark alignment. If there is a chain tensioner, mount it at this time, *Fig. 15.17.*

Timing Cover: Knock out the old seal with a large sprocket, or pry it out. To install a new seal in the timing cover, tap the new seal in with a seal driver and hammer, *Fig. 15.18.* Applying nonhardening sealer around the seal outside diameter will facilitate installation and will help stop leaks. Slip on the oil slinger, if used, with its curve facing forward.

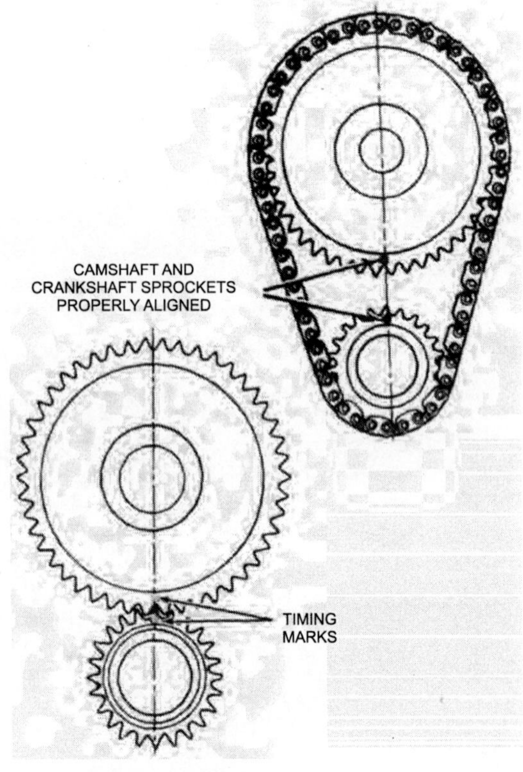

Fig. 15.16 *Timing chain and gear alignment. (General Motors Corp.)*

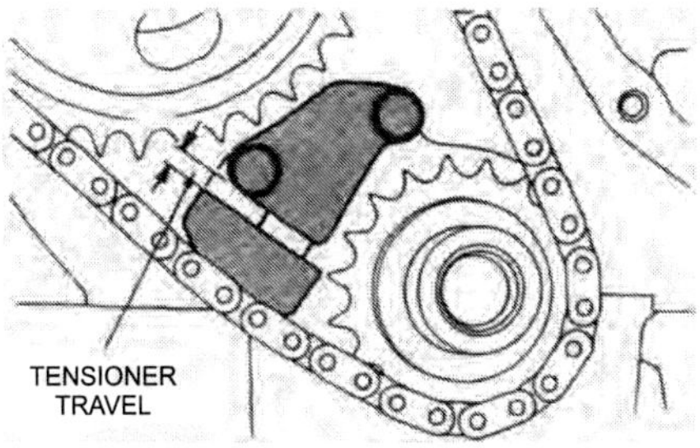

Fig. 15.17 *Timing chain tensioner.*

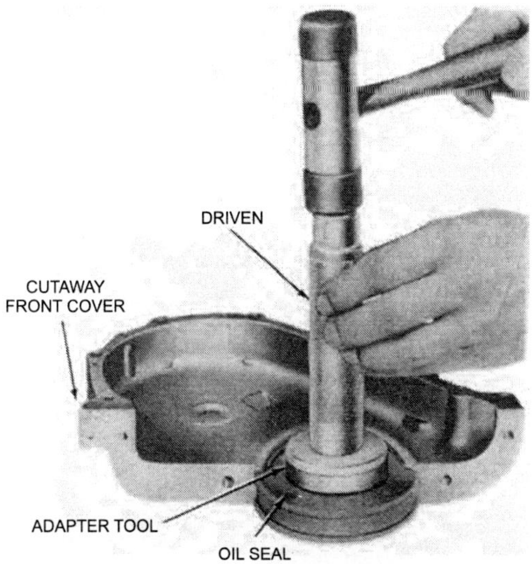

Fig. 15.18 *Installing front cover oil seal. (Ford Motors)*

Most timing covers are installed with a gasket, but some use RTV sealer in place of the gasket. Cast covers use anaerobic compound as a gasket substitute. Run a bead of RTV sealant around the timing chain cover, fit the gasket, and bolt it to the block. Do not tighten the bolts yet. Install the harmonic damper and torque it to specs. Most shop manuals require a special puller to pull the harmonic damper in place. If one is available then use it. If not, lightly hammer the damper until the bolts can be used to draw it on to the correct position. With the damper installed, now tighten the front cover bolts. This procedure allows the front cover to "float" so the front seal will centre itself on the harmonic damper.

Installing Oil Pan

With the pump and front cover in place, the oil pan can now be installed, *Fig. 15.19.* On overhead cam (OHC) engines the oil pan can probably be bolted on now. It depends on whether some of the pan and its gasket fit over the front cover. If the pan is separate, bolt it now.

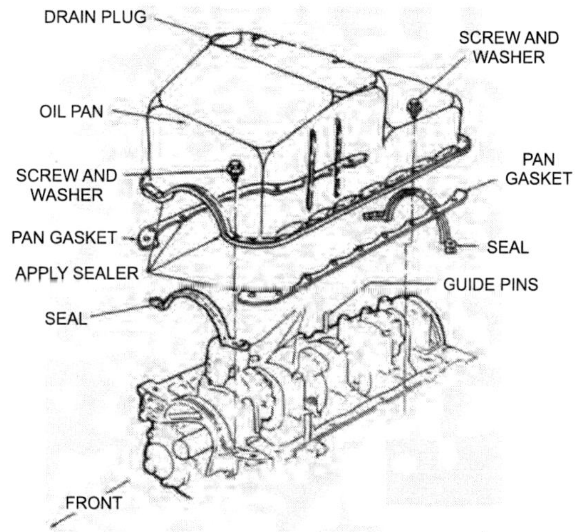

Fig. 15.19 *Oil pan installation. (Ford Motors)*

Lay the pan on a flat work surface. If the pan flange is bent or warped, straightened it, as shown in *Fig. 15.20.* Otherwise a bent flange can cause oil pan gasket leakage and a valid customer complaint.

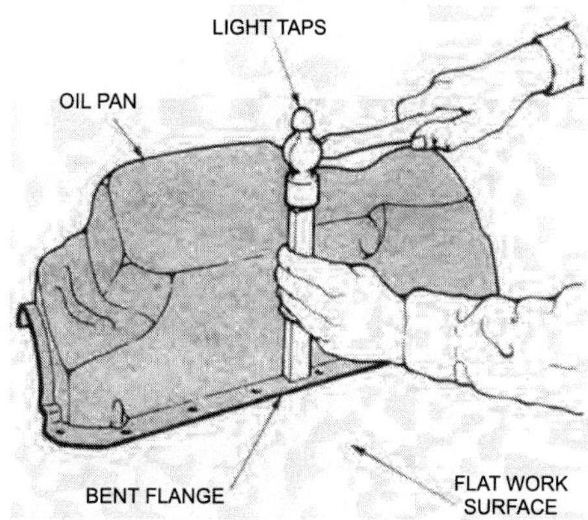

Fig. 15.20 *Lay oil pan on workbench and check for a bent flange. If necessary, Flatten the flange. Also check condition of oil drain plug and threads in pan. (Ford Motors)*

Install the oil pan gasket. First, lay it in place on the pan. Check that all bolt holes align and check that the correct gasket is being used. The gasket may either adhere to the pan or the engine block mating surface. If the engine is installed in the vehicle, it might be easier to glue the gasket to the oil pan, *Fig. 15.21.* When working on an engine mounted on the engine stand, it may be better to bond the gasket to the block, *Fig. 15.22,* because there is less chance of the gasket falling or shifting during assembly.

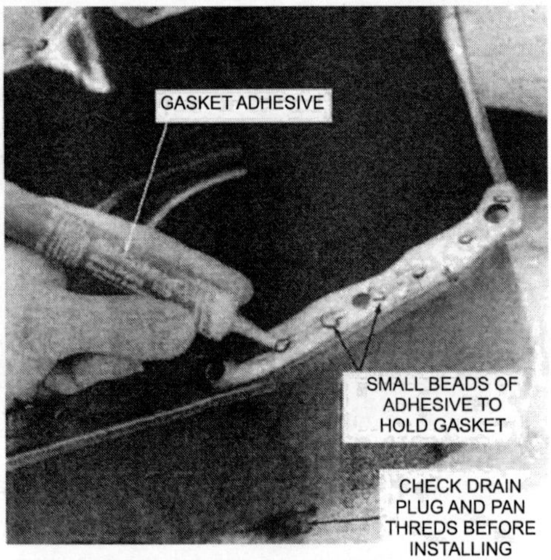

Fig. 15.21 *When installing a gasket, adhesive should be used to hold oil pan gasket in proper location during pan installation. Based on gasket and end seal design, you may need to bind gasket to either oil pan or engine block mating surface. (Fel-Pro Gaskets)*

Fig. 15.22 *When working on engine on engine stand, gasket can sometimes be positioned on block surface that is to mate with the pan gasket. (General Motors)*

Sometimes the end seals for the oil pan have to be worked down into grooves in the oil pan or in the engine front cover and block. In such a case, the oil pan gasket should be installed next to or under the rubber end seals. RTV sealer should be used to assist prevent leakage where the gasket and other seals overlap.

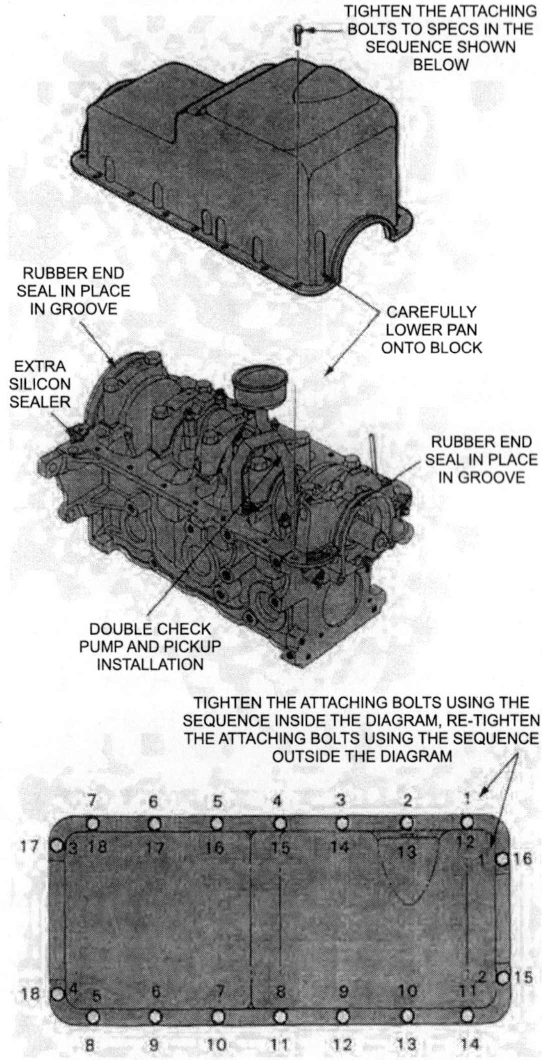

Fig. 15.23 *Position oil pan on block. Start all bolts by hand before tightening any bolts. Unless all the bolts are in their respective holes, if you tighten bolt, gasket will compress and squeeze out into bolt holes. Other bolts will be difficult to start. Torque bolts to specs in recommended sequence. (Ford Motors)*

Let the gasket cement set up slightly before installing the oil pan. This will help maintain the bolt-holes aligned. Fit the pan into position on the engine block. Start all bolts by hand before tightening any of the fasteners. Tightening any of the bolts before others are started, gasket can compress and partially block the bolt holes. With all bolts hand started, tighten the oil pan bolts to specs. Use a crisscross pattern to assure even gasket compression, *Fig. 15.23.*

Figure 15.24 shows a partial oil pan gasket, which is handy when the oil pan is not removed from the engine during an engine service. The old section of oil pan gasket under the front cover can be cut off. Then, the new piece of oil pan gasket can be installed and sealed. A whole new oil pan gasket should be installed when the oil pan is removed for service.

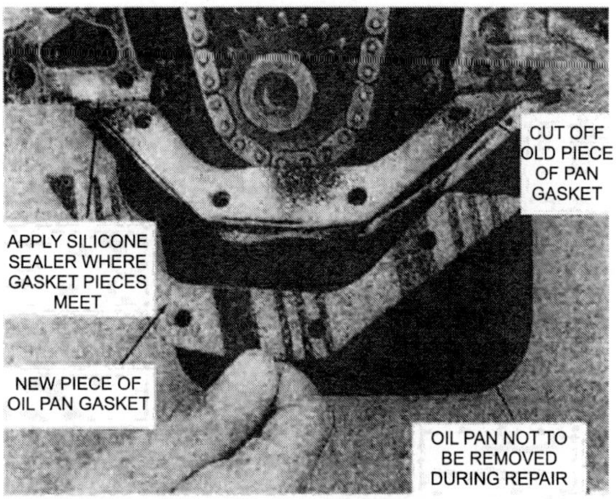

Fig. 15.24 *A partial oil pan gasket is handy if you do not remove an oil pan during a repair, e.g., when replacing a timing chain. Section of oil pan gasket can be cut off and after repairs, new gasket piece installed. Use silicone sealer where pieces mate. (Fel-Pro Gaskets)*

Installing Accessory Units

With the engine upside down on the stand, install any components that are easily accessible, e.g., bolt on the engine motor mounts, any sending units, oil filter bracket, etc. These parts are easy to reach with the engine on the stand after oil pan installation.

Install the carburetor or injection system, taking care not to over-torque the nuts. Bolt on the dipstick tube. Install a new set of spark plugs. Leave the distributor out for now if the oil pump can be turned through the distributor hole.

Distributor

If the oil pump can be turned while the distributor is in place, go ahead and install the distributor now. Rotate the engine so that the number one cylinder is at TDC. Check that both the valves are closed at number 1 cylinder as well as the timing marks are in proper alignment. Locate the number 1 cylinder's ignition wire where it attaches to the distributor cap. Mark the distributor body directly below number 1 cylinder ignition wire. This locates the place the distributor rotor should end up when the distributor is properly installed. Since the distributor drive gears are not straight but curved, the rotor will move as the gears engage during installation. Therefore, back up the rotor about 25 mm. Install the distributor while watching the rotor. The rotor should move to cylinder 1's position. If not, remove the distributor and back the rotor up the approximate amount it moved during the first try. Reinstall the distributor.

When the distributor is fully seated and the rotor is at the number 1 cylinder wire position, it is in time. Install the distributor hold-down. Hook the wires to the spark plugs by following the firing order.

Water Pump and Thermostat

When installing the water pump, check the fit of the new gasket. Then apply gasket adhesive to hold the gasket on the pump, *Figs. 15.25* and *15.26.* If a chemical gasket or RTV is used, form a continuous bead and circle all bolt holes and coolant passages.

When doing major engine service, engine cooling system thermostat should be checked carefully to ensure that one is installed and it operates properly. Also make sure that the heat range of the thermostat is correct. Someone could have changed it or left it out.

Seal both sides of the thermostat-housing gasket and install it. A new thermostat is usually installed at this time. It is put in place with adequate care being exercised to place the correct side of the thermostat towards the engine. A common and costly mistake is to install the thermostat upside down. In this backward position the thermostat senses the cool water in the radiator instead of the warm water in the engine. The thermostat remains closed and the engine overheats, possibly causing engine damage.

When installing the thermostat housing, make sure the thermostat is centred under the housing. Make sure the small cylinder in the centre of the thermostat sets in the engine. This is the sensing device. It is often labelled with the thermostat's opening temperature. Tighten each bolt a little at a time to specs. Remember that the housing will warp or crack easily if tightened too much or improperly, *Fig. 15.25.*

Fig. 15.25 *When installing water pump, check gasket fit on pump first. Then apply adhesive to hold gasket in place on pump. (Fel-Pro Gaskets)*

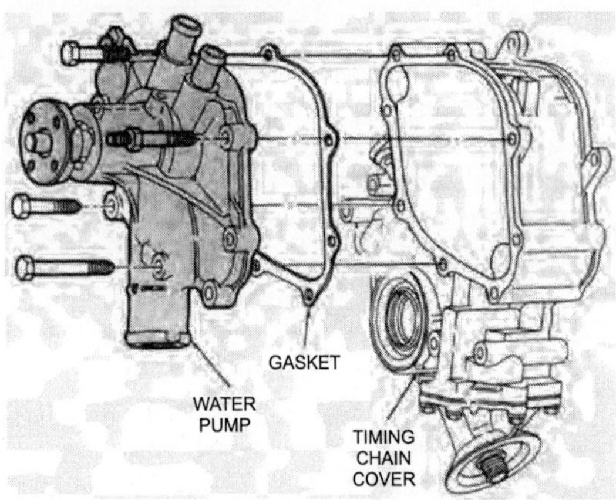

Fig. 15.26 *Installing water pump to front cover. (Ford Motors)*

Fit the pump on the engine and start all bolts by hand. Carefully check all bolt lengths and place sealer on threads of any bolts extending into coolant and oil passages in the block.

15.5 INSTALLING CYLINDER HEAD

Cylinder head installation methods are very critical when bolting the head to the block. Modern automotive engines use thin wall, cast iron heads or aluminium alloy heads that can leak, warp, or crack if installed improperly. These engines also operate at higher temperatures, which can further increase the strain on the seal between the head and block. Therefore, great care should be taken to ensure proper installation techniques for cylinder heads.

The head gasket is under the highest clamping loads. It must seal passages that carry coolant with antifreeze and often is required to seal passages that carry hot engine oil. Most demanding job of the head gasket is to seal the combustion chamber. As a rule of thumb, about 75% of the head bolt clamping force is used to seal the combustion chamber. The remaining 25% seals the coolant and oil passages, *Fig. 15.27.*

The gasket must seal when the temperature is as low as –25°C and as high as 204°C. The combustion pressures can be 6900 kPa on gasoline engines.

Cylinder head bolts are tightened to a specified torque, which stretches the bolt. The combustion pressure attempts to push the head upward and the piston downward on the power stroke. This subject the head bolts to additional stress and it reduces the clamping load on the head gasket under the conditions when the greatest seal action is needed. On a normally aspirated engine a partial vacuum on the intake stroke attempts to pull the head more tightly against the gasket. As the crankshaft rotates, the force on the head cycles from pressure on the combustion stroke to vacuum on the intake stroke. Modern engines have lightweight thin-wall castings. The castings are more flexible, so that they move as the pressure in the combustion chamber changes back and forth between pressure and vacuum. As a consequence, the modern head gasket is made of several different material assembles in many ways, depending on the engine.

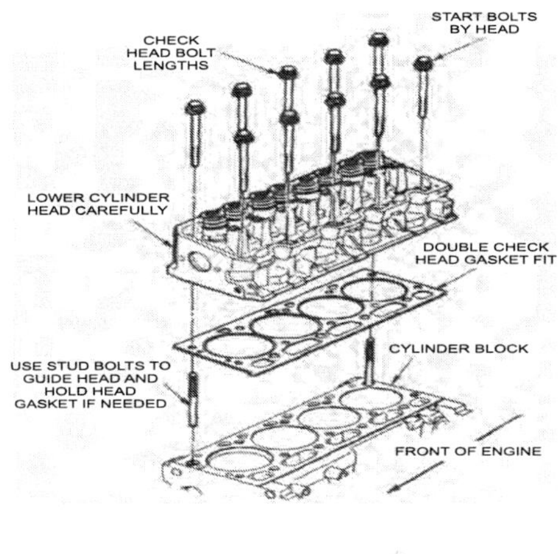

Fig. 15.27 *Gaskets help prevent leaks between two surfaces.*

Installing Head Gasket

Assemble the heads as already outlined in chapter 13. Remove the head gasket from the package and carefully inspect it for flatness, cuts, and other manufacturing flaws. Do not bend or tear the gasket. Read the manufacturer's directions. Some gaskets need sealer; others are installed dry. Before placing the head gasket on the block wipe across the block's deck with hand. It should be perfectly clean.

Cylinder head gasket markings are normally provided to show the front or top of a head gasket. The head gasket may be marked "top", "front," or it may have a *line* to show installation direction. If the gasket is not marked *up* and *front,* place the manufacturer's name facing up. Gently place the cylinder head over the head gasket and block. This should be done without bumping and damaging the gasket. Ensure that any required dowel pins are put in place before placing the head gasket on the deck. Make sure the head is over its dowels. If dowels are not provided on the block deck, install stud bolts in the block. They will serve as guides to hold the head gasket and position the cylinder head, *Fig. 15.28.* Remove the guide studs if used and start all of the bolts by hand. A head gasket can usually only be installed one way. If installed backwards, the gasket will fit but coolant or oil passages may be blocked. This mistake could cause engine overheating or oil starvation to the valve train, *Fig. 15.29.*

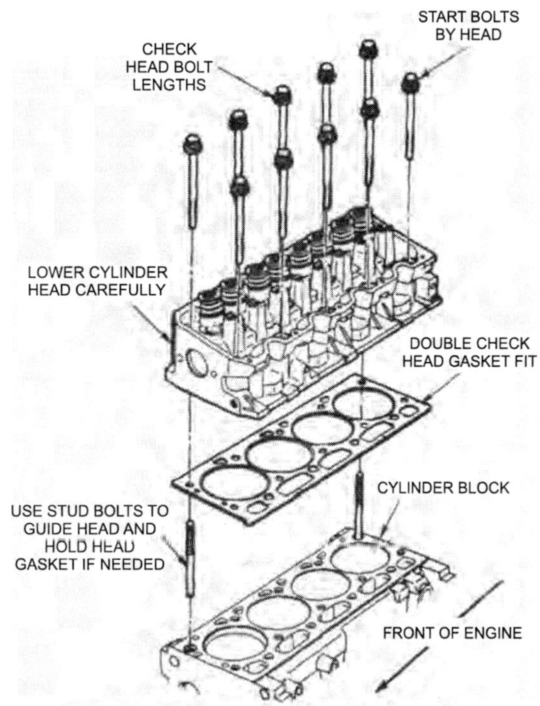

Fig. 15.28 *Use studs to guide head over block if dowels are not provided in the block. Install head bolts and check their lengths. (Ford Motors)*

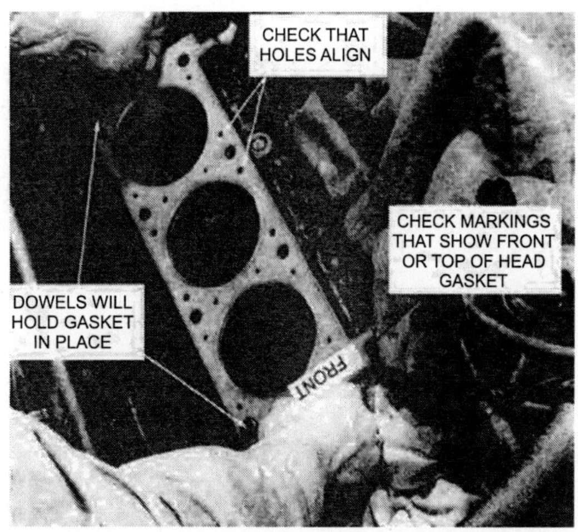

Fig. 15.29 *When installing cylinder head gasket, make sure it is facing in the right direction. A word, arrow, or notch will show front or top of head gasket. Quality head gaskets do not need sealer and retorquing after engine operation. (Fel-Pro Gaskets)*

Most modern Teflon-coated, permanent torque (they do not require retorquing after engine operation) cylinder head gaskets should be installed clean and dry. Sealer is *not* recommended on modern, quality head gaskets. However, some steel shim or low quality fiber-asbestos head gaskets may require retorquing and sealer. Refer to gasket manufacturer's instructions when in doubt.

Wipe the head's mating surface with hand. It, too, should be clean. Now carefully lay the head gasket on the block deck, exercising care to set it down as straight as possible, *Fig. 15.30.*

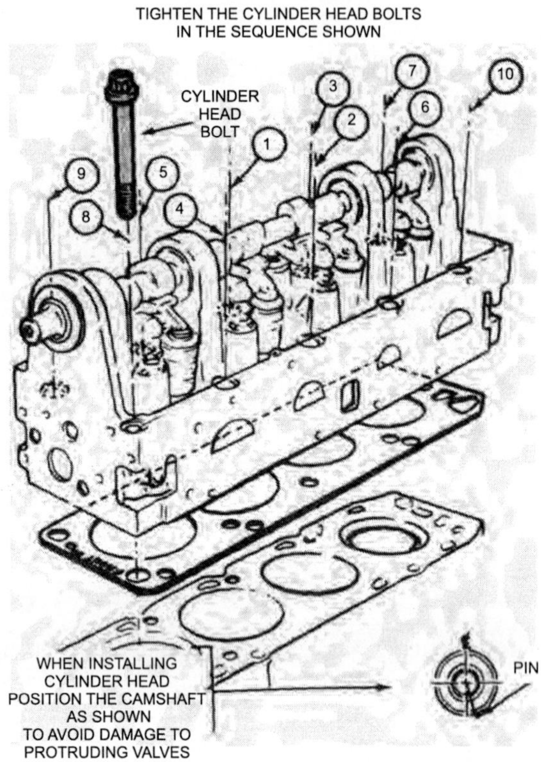

Fig. 15.30 *OHC cylinder head installation. (Ford Motors)*

Head Bolts

Make sure all bolt holes in the block are clean. Blow them out if needed. Place a drop of oil on the head bolt threads, *Fig. 15.31.* This will ensure an accurate torque when tightening the head bolts. Never squirt oil in the bolt holes as this can cause a hydraulic lock that will keep the bolts from torquing down properly. Never oil torque-to-yield bolts. Drop in the head bolts and washers, if used. Oil is the preferred lubricant, unless the bolts extend into the water jacket or oil jackets. Use nonhardening sealer or *anti-seize compound* on the threads in that case, *Fig. 15.32.* Remember to install in position any brackets, which might use head bolts for attachment. Run all the bolts down snug with a ratchet and then finish with a torque wrench. Tighten head bolts one step at a time. This allows the head and gasket to normalize. About 20 minutes after final tightening of the head bolts, go over them again to see if any have loosened up. Always follow the bolt tightening pattern given in the service manual

On V-type engines, install the other head using the same procedure.

Some engine manufacturers recommend using *new head bolts* when reassembling the engine after overhaul. The used head bolts have been *torqued-to-yield* i.e., tightened until stretched a specific amount, at the factory assembly line. New bolts are required to ensure the correct gasket compression and bolt clamping action. Refer to the service manual to determine if new head bolts are needed, *Fig. 15.33.*

Fig. 15.31 *Internal and external threads should be perfectly clean. Place a drop of oil on threads to ensure accurate head bolt torque. (Fel-Pro Caskets)*

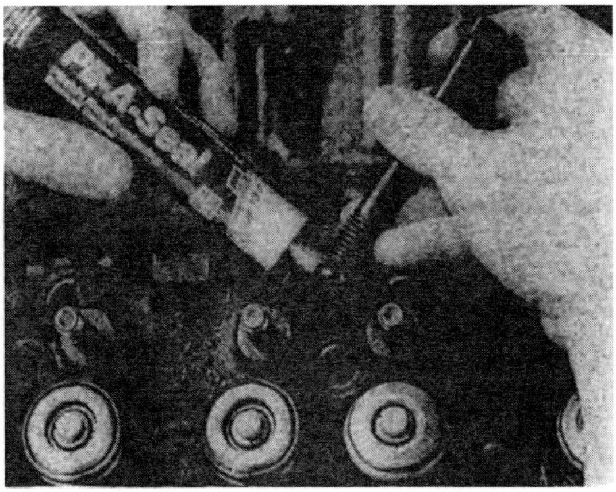

Fig. 15.32 *Head bolts that extend into coolant passages should have sealer on their threads. The manual will point out which bolts need sealer. (Fel-Pro Gaskets)*

Diesel Cylinder Head Installation

In case of diesel engines, head gasket thickness and bore size are very critical. Head gaskets are provided in different thicknesses to allow for cylinder head milling or varying block deck heights. Gasket thickness may be denoted with a colour code, series of notches, holes, or other marking system.

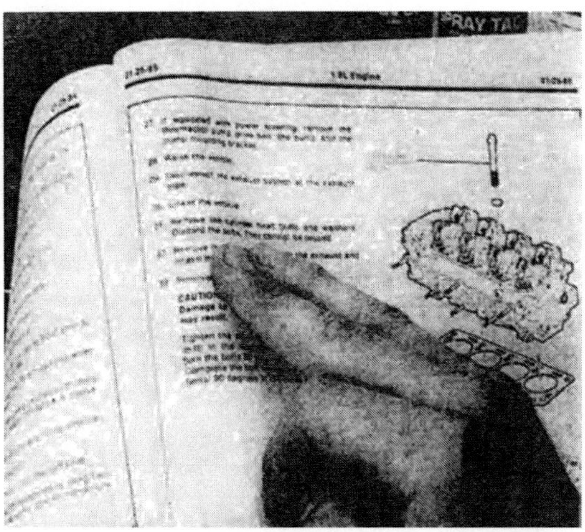

Fig. 15.33 *Always refer to a shop manual for specifications and special procedures when assembling an engine. This engine manufacturer recommends new head bolts because old ones have been stretched by a torque-to-yield process at factory. (Fel-Pro Gaskets)*

When a diesel engine is bored oversize, it also needs a special gasket. An over-bore gasket has to be requested from the parts supplier. A standard bore gasket will usually stick out into the cylinder, causing problems. When buying a *diesel* engine head gasket, make sure you have the right gasket with correct thickness and bore size for the engine being serviced. Refer to the applicable service manual for details

Torquing Head Bolts

Torque the head bolts to specifications using a crisscross sequence. The service manual will give the best bolt torquing sequence to be followed, *Fig. 15.34. Fig 15.35* shows an automotive mechanic using a very accurate dial indicator type torque wrench on head bolts.

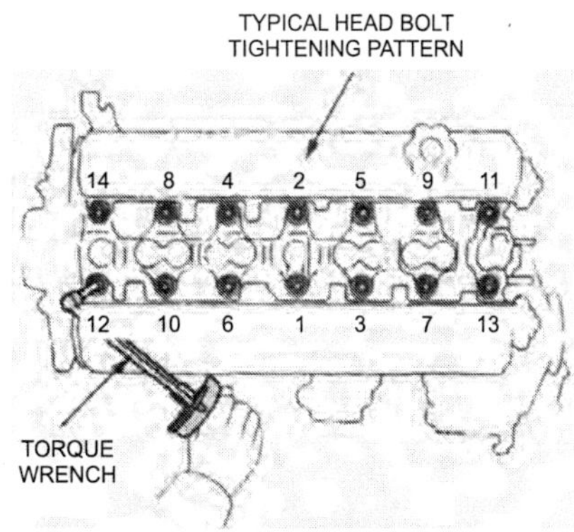

Fig. 15.34 *(a) Torque the head bolts to exact specs. Generally, start at the middle and work your way to the outer bolts. Gradually increase torque until up to specs. Then, go over all bolts at least twice to ensure precise tightness and to avoid missing any bolt.*

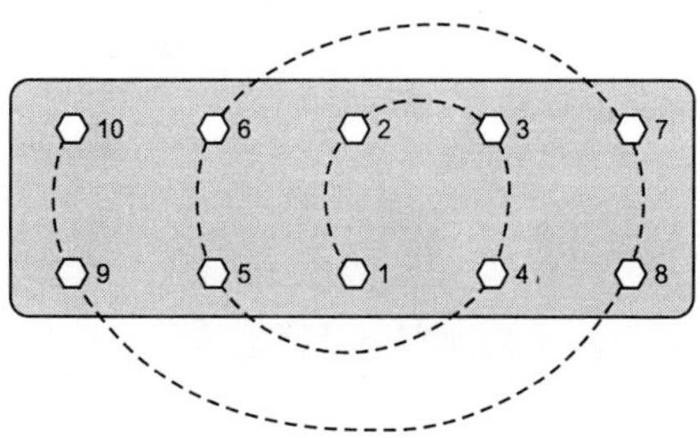

Fig. 15.34 *(b) Typical cylinder head bolt tightening sequence. Notice how the numbers are arranged in an outward rotating pattern.*

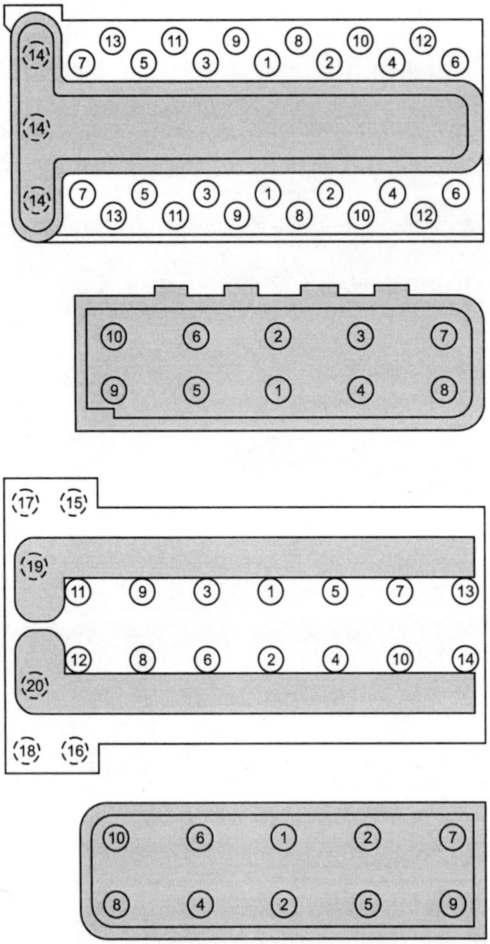

Fig. 15.34 *(c) Examples of cylinder head bolt torquing sequences.*

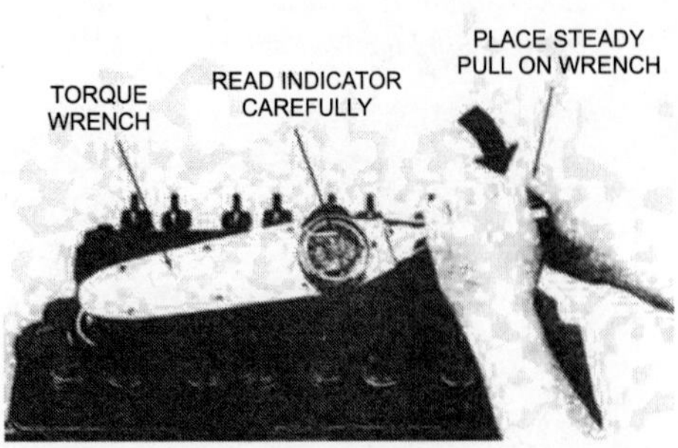

Fig. 15.35 *Pull on torque wrench slowly while reading indicator or listening for click sound. A slow, steady pull on the wrench will produce the most accurate bolt torque. (Fel-Pro Gaskets)*

Normally, tighten the bolts starting in the middle. Then work your way outward, tightening each bolt a little at a time, e.g., first ½, ¾, and then full specs torque. For example, if the head bolt torque 128 N-m, tighten all of the bolts to 64 N-m, then 96 N-m, and then 128 N-m. At least double tighten each bolt to full torque to ensure an accurate torque and avoid any bolt from being missed.

15.6 INSTALLING INTAKE MANIFOLD

The installation of an engine intake manifold is also critical. Some intake manifolds not only seal vacuum they can also carry coolant, and hot exhaust gases. If correctly installed, the manifold can allow a vacuum leak or a coolant leak. Partial engine teardown might be needed to correct the problem.

The intake manifold gasket for a V-type engine may be a one-piece gasket or it may have several pieces. V-type engines with open type manifolds have a cover over the lifter valley. The cover may be a separate part or it may be part of a one-piece intake manifold gasket. Closed type intake manifolds on V-type engines require gasket pieces (end seals) at the front and rear of the intake manifold, *Figs. 15.36 (a), (b).* Inline engines usually have a one-piece intake manifold gasket.

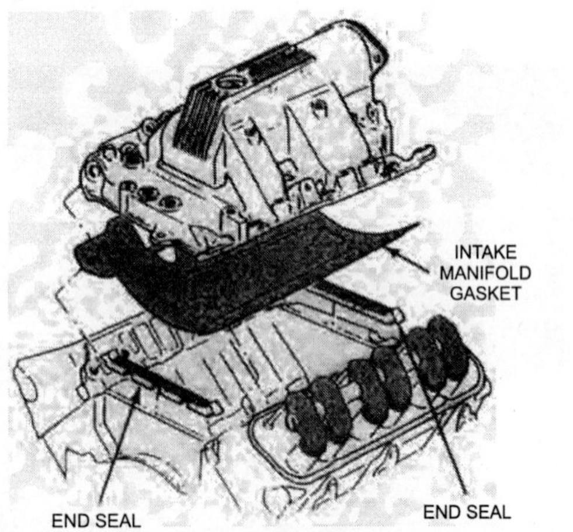

Fig. 15.36 *(a) This intake manifold gasket includes end seals and a full shield cover for the valley to keep hot engine oil from heating the intake manifold.*

On V-type engines the intake manifold seals the valley. Therefore, there are gaskets on each head plus each end of the valley. Place all these gaskets. Then without dislodging the gaskets, install the manifold.

An adhesive gasket sealer applied to the gasket with an extra gasket sealer on the four corners of the valley, that are typical problem areas, helps the job for V-engines.

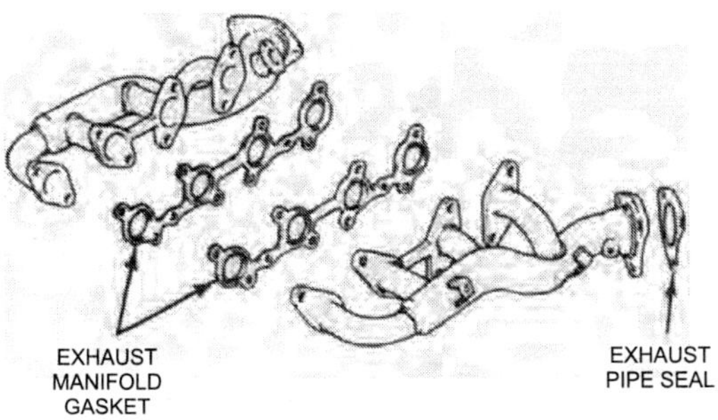

Fig. 15.36 *(b) An exhaust manifold gasket is used on some engines and the exhaust manifold is attached to the cylinder head.*

Checking Intake Manifold Warpage

Use a straight edge and feeler gauge to check the intake manifold sealing surfaces for warpage. As shown in *Fig. 15.37,* lay the straight edge over the port openings. Attempt to slide different thickness feeler gauge blades under the straight edge, and do it at different points on the manifold.

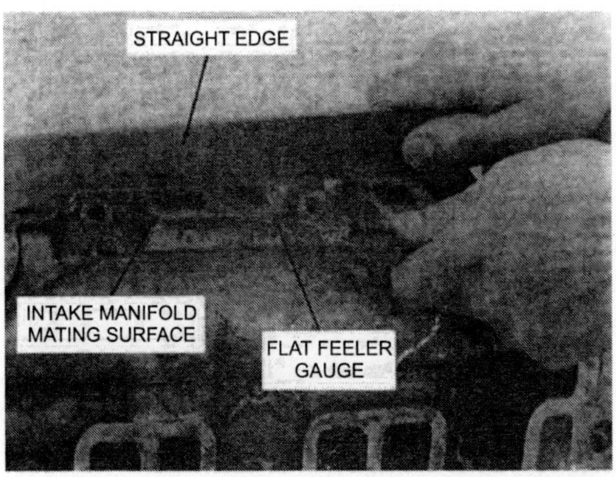

Fig. 15.37 *Straightedge and feeler gauge is used to check for intake manifold warpage. Check closely around the centre ports and heat riser passage.*

The thickest blade that will slide under the straightedge will give the intake manifold mating surface warpage. If it is outside specs, the intake manifold surfaces must be machined or replace the manifold.

Checking Intake Manifold for Damage

Carefully check the intake manifold for any cracks, burned areas, and other problems. Check the bottom of the manifold and near the heat passage from the cylinder head. Hot exhaust gases for the GR or for manifold warming can burn and crack the manifold. Also, check the head mating surface near the heat passage. A ruptured gasket could have allowed the hot gases to burn the mating surface.

Installing Intake Manifold Gasket

Place the new manifold gasket over the head to check its fit. Ensure that all passages align properly. Occasionally, ports or passages can vary slightly with the specific head. It is possible for the ports to be slightly larger. A coolant passage must be located differently. Any of these problems could cause a vacuum leak, coolant leakage, or even serious engine overheating damage. Double-check all openings in the gasket and head, *Fig. 15.38.*

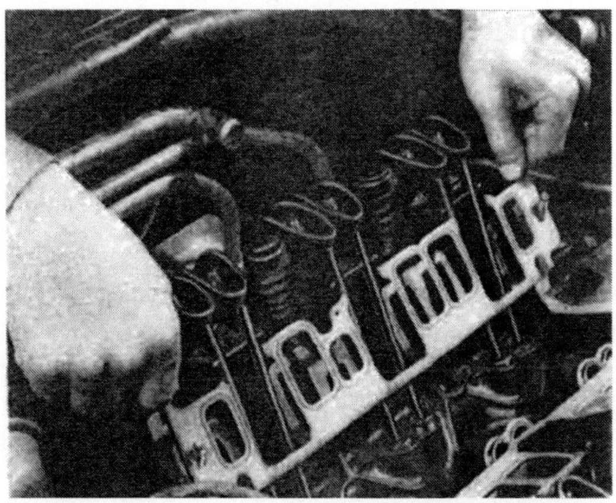

Fig. 15.38 *Check intake manifold gasket fit before applying sealer. Ensure that ports and other openings align perfectly.*

When using a metal valley tray type intake manifold gasket, *Fig. 15.39 (a),* use RTV sealer on it. Silicone or RTV sealer should be applied to circle all ports and coolant passages. The thin metal gasket can leak easily and silicone provides added protection against a leak. When using a multi-piece intake manifold gasket made of fiber, there is no need to circle every intake port. Simply apply gasket adhesive to hold the gasket in place, *Fig. 15.39 (b).* Some mechanics still prefer to circle coolant passages with silicone sealer.

Fig. 15.39 *(a) In this valley tray type intake manifold made of steel, silicone sealer helps prevent leakage. Circle all ports and coolant passages with sealer.*

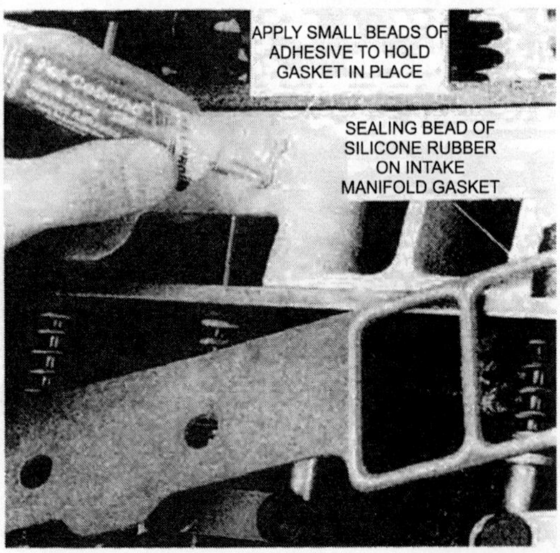

Fig. 15.39 *(b) The intake manifold shown here already has a small bead of silicone rubber around openings and does not need extra sealer around the ports. However, use of adhesive will be helpful to hold the gasket in place during assembly. (Fel-Pro Gaskets)*

With the head sections of the intake manifold gasket in place, install the end seals. Adhesive can be used to hold the seals in place if necessary. If large barbs are provided on the seals and they fit properly down into the block, then, no adhesive is necessary. Apply a small bead of silicone sealer where the end seals and main intake gaskets overlap. This will prevent the engine oil from seeping out during engine operation, *Fig. 15.40.*

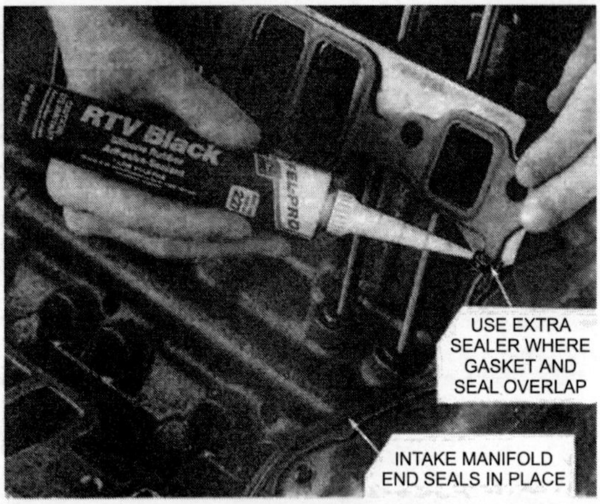

Fig. 15.40 *After positioning end seals with adhesive applied or with rubber barbs provided in design, silicone sealer is used where gasket and seals overlap. (Fel-Pro Gaskets)*

Positioning Intake Manifold

After installing the intake manifold gasket, position the intake manifold after letting the sealer and adhesive to cure slightly but do not wait too long. Waiting too long will allow the sealer to cure completely and affect the manifold installation correctly. Carefully lower the intake manifold onto its mating surface on the engine, *Fig. 15.41,* without bumping or shifting the gasket or smear the sealer. In some applications, guide studs may be used to lower the manifold straight down into position.

Torquing Intake Manifold

Start all intake manifold bolts by hand before tightening any bolts. Tightening some of the bolts might hinder any needed shift to start the remaining bolts. After all bolts threaded in a few turns, run them down lightly. Then following a crisscross pattern, use a torque wrench to tighten the manifold bolts to specifications in small steps at a time. Usually, torque the bolts to ½, ¾, and full torque value in turn. Go over all of the bolts, minimum two times for full torque

Fig. 15.41 *Carefully lower the intake manifold into its location without bumping and shifting new gasket. (Saab)*

tightening. As each bolt is tightened, the ones next to it will loosen.

Figure 15.42 shows a multi-piece intake manifold. Use the same basic procedure to install the base of the manifold. Torque in a crisscross pattern to specs. Then, repeat the procedure for the upper sections of the intake manifold.

With plastic engine parts, e.g., plenums, brass thread inserts are usually pressed into the plastic. This permits additional components like sensors, vacuum fittings, etc., to be bolted to the plastic part without thread damage.

15.7 INSTALLING EXHAUST MANIFOLD

Exhaust manifolds route very high temperature gases into the vehicle's exhaust system. As a consequence, it can develop cracks, warpage, rust, and other problems. Therefore, always inspect an exhaust manifold carefully before installation. Look for any

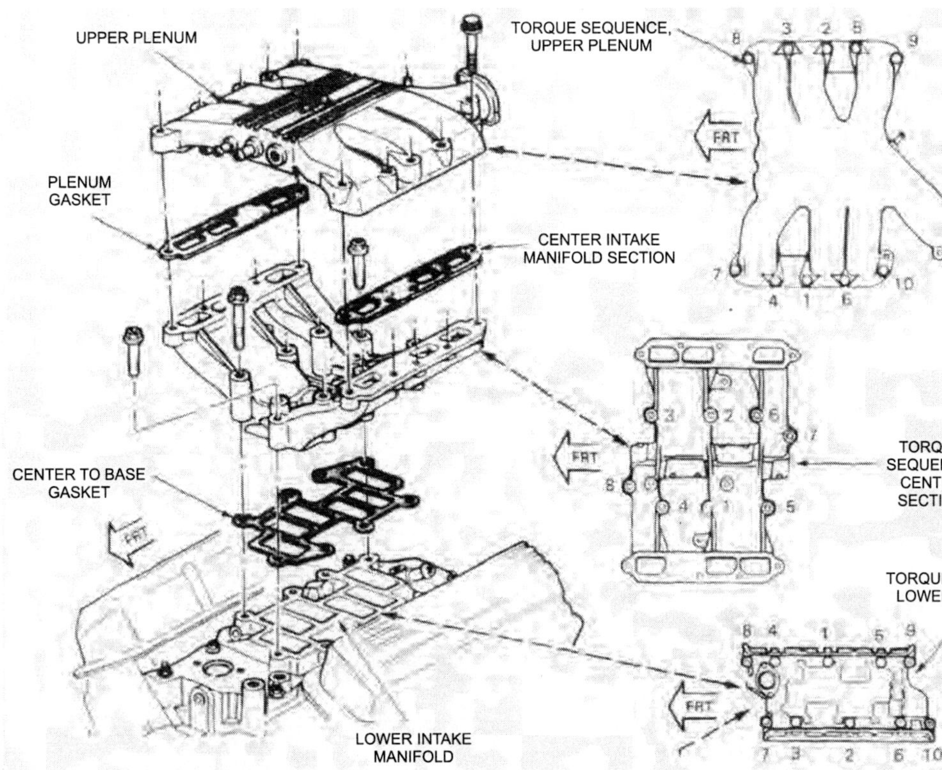

Fig. 15.42 *Same general techniques, as already described, are used to install this multi-piece intake manifold. All sections are torqued to specs. (Buick Div. GM)*

cracks, burned areas, eroded mating surfaces, severe rust damage, etc. Check the manifold for warpage of mating surface, use the straightedge and flat feeler gauge as already explained.

Most used exhaust manifolds should have a gasket installed under them. Position the gasket on the head, *Fig. 15.43.* Then slide the exhaust manifold into position. Use of studs facilitates the gasket and manifold to install in place and hold them while starting the attaching bolts, *Fig. 15.44.*

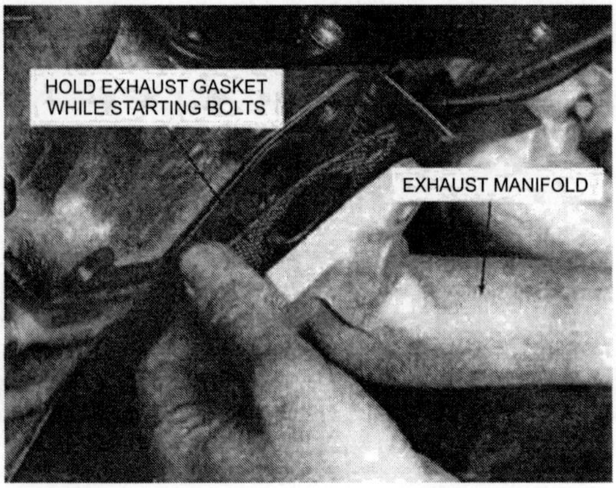

Fig. 15.43 *A good practice is to place a gasket under an old exhaust manifold to prevent leakage. (Saab)*

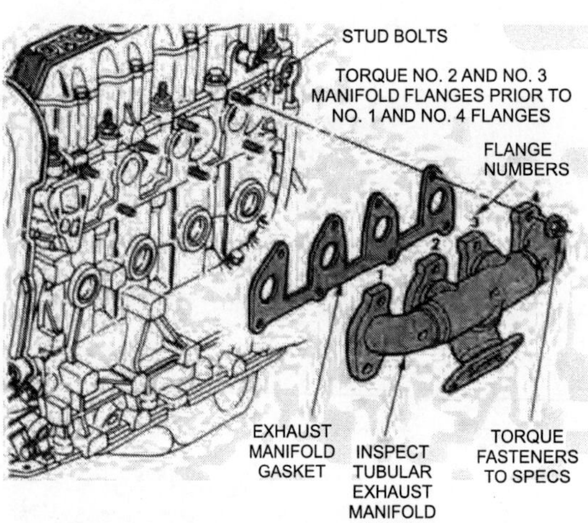

Fig. 15.44 *Stud bolts are used to hold exhaust manifold gasket and manifold in place while starting fasteners. Sealer is not used because high temperatures will cause hardening of sealer, that could result in leakage.*

Start all bolts by hand and then turn them down until snug. Use a torque wrench to tighten the exhaust manifold bolts to specs, *Fig. 15.45,* following a crisscross pattern, *Fig. 15.46.*

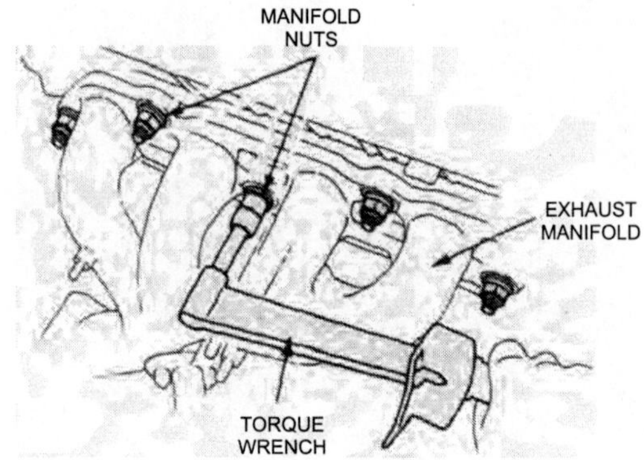

Fig. 15.45 *Torque all exhaust manifold fasteners to specs to avoid warpage, breakage, and leakage. (Toyota Motors)*

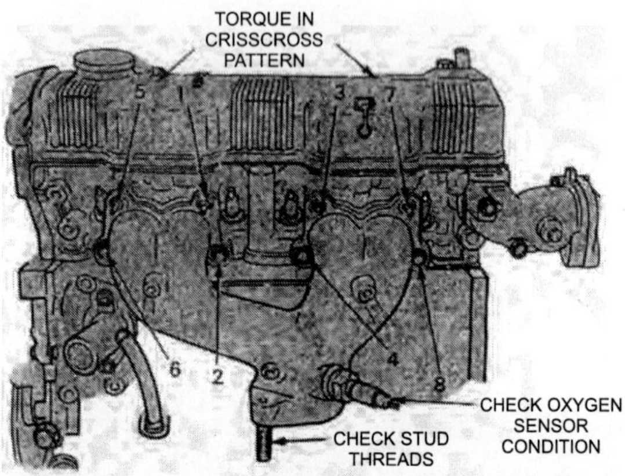

Fig. 15.46 *Exhaust manifold threaded fasteners should be tightened in a recommended crisscross pattern to develop a uniform pressure all around. Carefully check installation of oxygen sensor, stud bolts etc. (GM Corp.)*

15.8 OVERHEAD CAMSHAFT INSTALLATION

A camshaft for an OHC engine is installed in a similar manner as was described for an overhead valve (cam in block) engine in chapter 8 on 'Short Block Service." However, there are a few slightly different procedures, *Fig. 15.47.*

Ensure that the camshaft bearings are in good condition or replaced, lubricate the camshaft journals and cam lobes. White grease is commonly

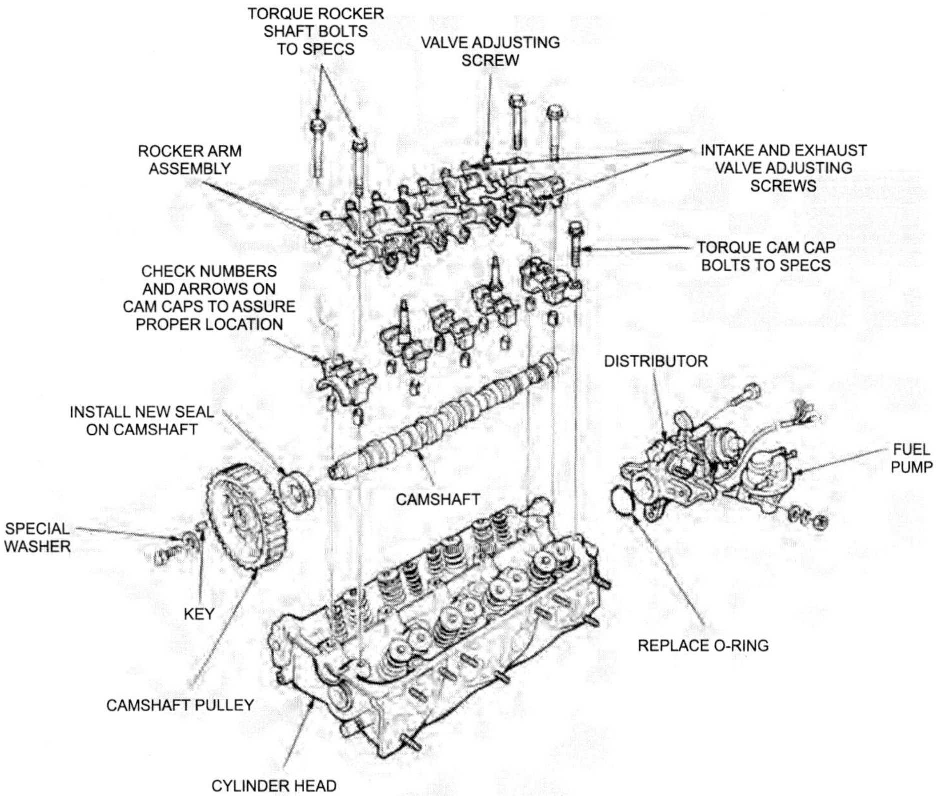

Fig. 15.47 *Caps are used to secure camshaft bearings and camshaft into OHC engine cylinder head. Carefully install the caps in proper sequence and direction into their original locations. (Honda Motors)*

recommended on the lobes to reduce friction and wear during initial start-up after reassembly and installation. Slide the camshaft into the head without bumping the cam bearings. Turn the camshaft as it slides into the head, *Fig. 15.48.*

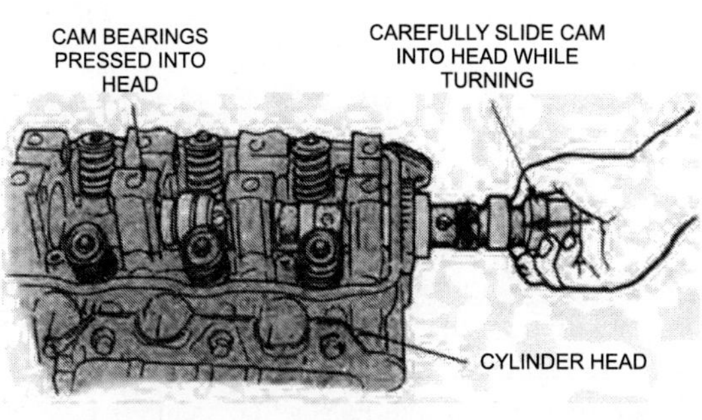

Fig. 15.48 *Here, the head does not use removable bearing caps. The camshaft has to be installed by sliding it into front of the head. (General Motors)*

Installing Cylinder Head Caps

Some OHC heads use cylinder head caps, also called camshaft bearing caps. They hold the upper cam bearing shells and bolt over the camshaft *journals, Fig. 15.47.*

First, install the lower cam bearing shells in the head. Make sure that any oil hole present must align. The back of the bearings should be clean and dry. Wipe any engine oil on the face of the bearings.

Lower the camshaft into the bearings. Install the upper half of the cam bearings into their respective caps. Oil the journals and position the caps on the head. Just like cylinder block main caps, cam bearing caps are numbered and must be installed in their original locations. If the caps are mixed or installed backwards, the camshaft can lock in the cylinder head on torquing the bolts to specs. The bearings of camshaft can be severely damaged upon engine starting.

Torque the camshaft or head caps to specs. Follow the crisscross general sequence shown in *Fig. 15.49.* Check that the camshaft can be turned after torquing the bearing caps. If not, something is wrong somewhere. Check bearing size, camshaft straightness, cap locations and directions, etc.

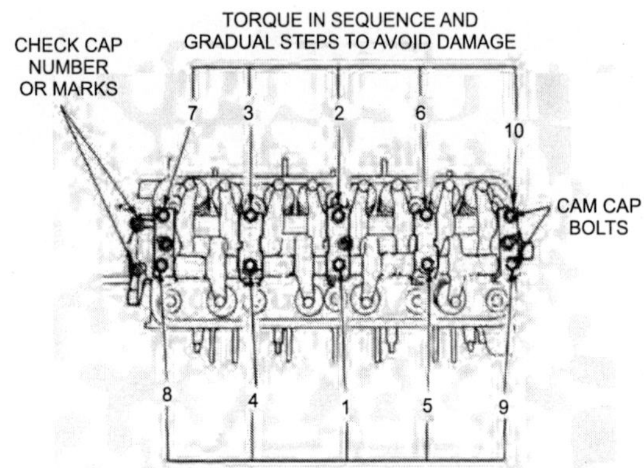

Fig. 15.49 *Double-check the camshaft bearing cap numbers/arrows to ensure the correct position and direction, torque them to specs. Ensure that you can still rotate the camshaft after torquing in the recommended sequence. (Honda Motors)*

15.9 ASSEMBLING VALVE TRAIN

Depending upon engine design, procedures for installing valve train components will vary. In case of any doubt, refer to the service manual.

Overhead Valve (OHV) Valve Train

When working on an overhead valve engine, install the valve lifters in their bores. If the old lifters are used, they should be installed on the same cam lobes to prevent excessive wear. If they are not in order, get new lifters. If the camshaft has been replaced, new lifters are a must. However, new lifters can be used on the old camshaft, when in good condition.

Lifters: Mechanical lifters should be moly-lubed on their bottoms and sides and then installed in place. Hydraulic lifters are moly-lubed too, but they also need priming. On the lifter's sides is a small hole. Hold an oil can up against this hole and gently squeeze oil into the lifter until it bleeds out the lifter's top. Then install the lifters.

Push Rods: Then install the push rods. Dab each end into moly-lube and then slide them onto the lifters. Rocker arms, either pedestal or shaft mounted, can now be installed, *Fig. 15.50 (a), (b).*

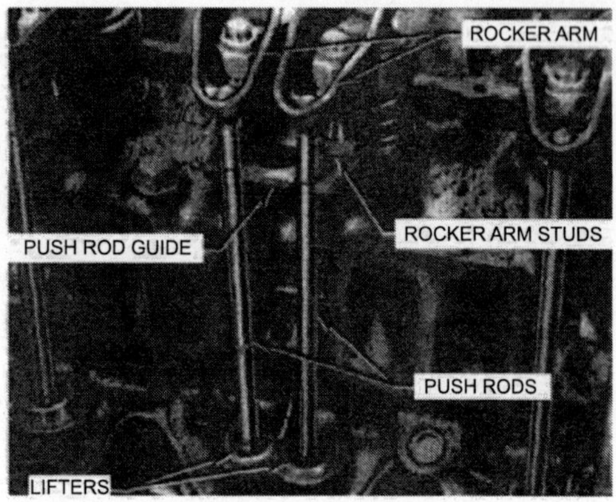

Fig. 15.50 *(a) Follow the recommended service manual directions when installing valve train, because designs and installation methods vary with application. Be careful to install parts in their original locations. (Chevrolet Div. GM)*

Rocker Assembly: Pedestal-mounted rockers need adjustment. First, to adjust a lifter, it should be against the cam's base circle. You should be able to see the camshaft. If not, observe the lifter body while slowly rotating the engine, you will see when the lifter is "off the cam." After positioning the camshaft, tighten the rocker arm nut. Different engines use different methods from this point on. Refer the service manual for the engine being assembled.

Mechanical lifters need adjustment also. Use feeler gauges and the rocker arm's adjuster to reach the clearance specified. These lifters must also be against the cam's base circle when adjusted.

Overhead Cam (OHC) Valve Train

Overhead cam (OHC) valve train installation can be handled several ways depending on its design. Some engines require that the camshaft be installed first. Others require the rocker arms and then the camshaft. Refer to the applicable service manual for cam installation and timing.

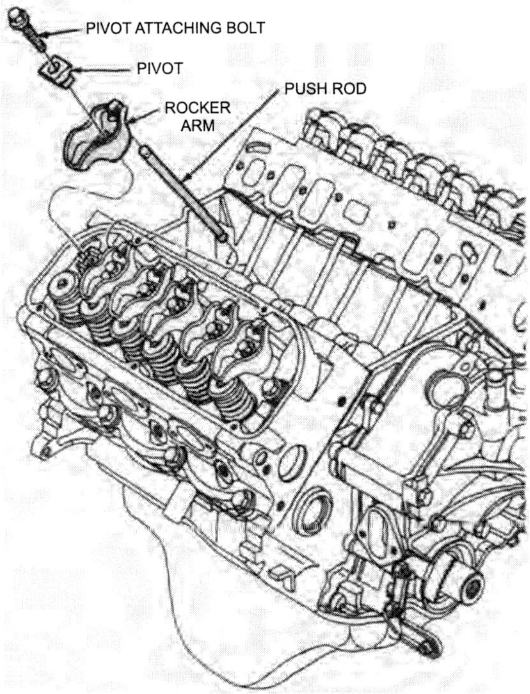

Fig. 15.50 *(b) OHV engine valve-train with pedestal-type rockers installation. (Ford Motors)*

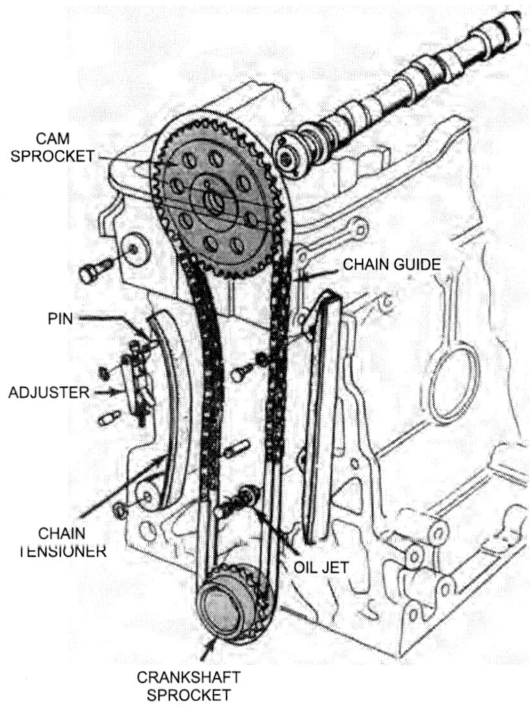

Fig. 15.51 *Timing arrangement of a OHC engine.*

With the cam-first engines, moly-lube the camshaft and lay it in its bearing bores. Bolt on the cam bearing caps and then the sprocket, *Fig. 15.51.* To attach the sprocket, turn the cam so it is in time with the crankshaft, loop the chain over the cam sprocket, and then bolt it to the cam. Now moly-lube the rocker arms and install and adjust them. If you try to bolt on the rocker arms and adjust them before attaching the cam sprocket, the valves might be hitting pistons when the cam is turned over. It is much easier to rotate the engine from the front pulley than trying to turn the cam sprocket by hand.

If the cam bearing caps are a part of the rocker arm assembly, be sure to attach the cam sprocket before the rocker arm assembly. This endures that the camshaft is in time with the crankshaft; so that you tighten the rockers down they would not attempt to push a valve open against a piston, which is at top-dead-centre (TDC).

Other OHC systems mount the cam in permanent pedestals, which are part of the head casting. With these the camshaft is installed in the pedestals; individually mounted rockers are attached under the camshaft. Install the cam sprocket first so the pistons and valves do not clash during rocker arm installation.

All these procedures have in common that the camshaft must be moly-lubed, and valve train installed with the valve adjustment backed off all the way. After all parts are installed, the valves can be adjusted.

When you are finished, rotate the engine number 1 cylinder to TDC. Besides the timing marks lining up on the front pulley, both valves must be closed on the number 1 Cylinder.

Turn the engine over several times by hand to ensure no interference problems with valves and pistons. Waiting to turn over the engine with a starter motor is a frequent and very expensive error.

Sometimes the intake and exhaust rocker arms are different. Usually, an "E" for exhaust and "I" for intake will be stamped on the rocker arms if they are different, *Fig. 15.52.* All parts should be in good condition and are usable. Check the rocker arm friction surfaces for wear. Replace any parts that are worn beyond spec limits or are damaged.

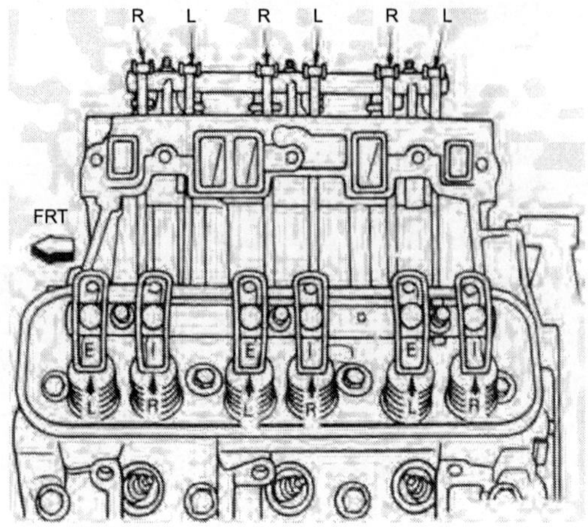

Fig. 15.52 *In some applications rocker arms are marked intake (I) and exhaust (E) and must be installed on correct valve. (General Motors Corp.)*

Adjusting Engine Valves

After installing the valve train components, you may have to adjust the valves. This is done by turning the rocker arm nuts, installing OHC shims, or using oversize push rod lengths.

Valve adjustment is critical to engine performance and service life. If the valves are *too loose,* valve train noise or clatter, will result. A light tapping or clicking sound will be produced from under the valve cover. If the valves are adjusted *too tight,* burned valves can result. The valves may not close tightly and hot exhaust gases leaking can erode away metal from the valve face and valve seat.

Nonadjustable Rocker Arms

Nonadjustable rocker arms are used on many push rod engines with hydraulic (self-adjusting) lifters. The hydraulic lifters automatically adjust and compensate for any changes in valve train clearance, maintaining *zero valve lash* (no clearance in valve train for quiet operation). The hydraulic lifters can adjust valve train clearance as parts wear, with changes in temperature (part expansion or contraction), and with changes in lubricating oil thickness.

If clearance adjustment is needed because of valve grinding, head milling, or other conditions, different lengths of push rod can be installed for use with nonadjustable rocker arms. Refer to the service manual or a part supplier for information details. Generally, torquing the rocker fastener (nut or bolt) to specs adjusts the lifters or rocker arms properly.

Overtightening a nonadjustable rocker arm can cause engine missing and rocker stud damage. Torque them to specs only.

Adjusting Hydraulic Lifters

Hydraulic lifter adjustment is done to centre the lifter plunger in its bore. This will allow the lifter to automatically adjust itself to take up or permit more valve train clearance. Some service manuals recommend adjustment with the engine off. However, many mechanics adjust hydraulic lifters with the engine running.

To *adjust hydraulic lifters* with the *engine off,* turn the crankshaft until the lifter is on the camshaft base circle (not on lobe). The valve must be fully closed.

Loosen the adjusting nut until you can wiggle the push rod up and down, *Fig. 15.53.* Then slowly tighten the rocker until all play is out of the valve train and cannot wiggle the push rod. To centre the lifter plunger, tighten the adjusting nut about *one more turn.* Refer to the service manual for details because this varies with engine design. Repeat the adjusting procedure on the remaining rockers.

Fig. 15.53 *Valves must be adjusted to centre hydraulic lifter plunger in its bore. Rocker nuts are carefully tightened until all play is out of valve train. Then, another one-half to one turn will centre plunger. Cam lobe should be rotated away from lifter when adjusting. (General Motors Corp.)*

To *adjust hydraulic rockers* with the *engine running,* install special oil shrouds or another device for catching oil spray off the rockers. Start and warm the engine to operating temperature.

Tighten all of the rockers until they are quiet. One at a time, loosen a rocker until it *clatters.* Then, tighten the rocker slowly until it *quiets down.* This will be zero valve lash.

To set the lifter plunger halfway down in its bore, tighten the rocker about one-half to one turn *more.* Tighten the rocker slowly to give the lifter time to leak down and prevent engine missing or stalling. Repeat the adjustment on the other rockers.

Other adjustment methods may be recommended. Refer the service manual for more detailed information.

Adjusting Mechanical Lifters

Adjusting mechanical lifters, also called *solid lifters,* is done to ensure proper valve train clearance. Because hydraulic lifters are not used here, you must adjust mechanical lifters periodically. Refer to the vehicle service manual for adjustment intervals and clearance specs. Typical valve clearance is approximately 0.35 mm for the intake valve and 0.4 mm for the exhaust valves.

Mechanical lifters make a clattering or pecking sound during engine operation. Unlike hydraulic lifters, this is normal in mechanical lifters. Mechanical lifters are used on heavy-duty engines and high performance engines.

To adjust mechanical lifters, position the lifter on its base circle, i.e., valve fully closed. This can easily be done by cranking the engine until the piston is at TDC, on its *compression stroke* (you can feel air blow out of spark plug hole in the cylinder head). At TDC on the compression stroke both valves in that cylinder can be adjusted.

Slide the correct size flat feeler gauge between the rocker arm and the valve stem, as shown in *Fig. 15.54.* When properly adjusted, the gauge should slide between the valve and rocker with a slight drag, *Fig. 15.55.*

If required, adjust the rocker to obtain valve clearance by loosening the lock nut and turning the adjusting screw. Tighten the locknut and recheck adjustment. Repeat the procedure on the remaining valves.

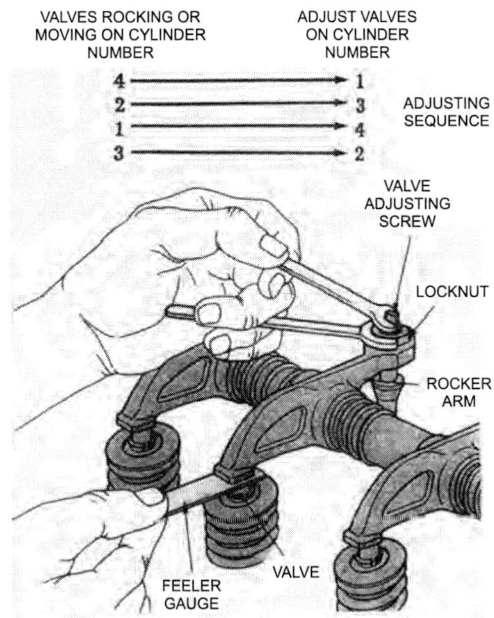

Fig. 15.54 *With Solid/mechanical lifters, screw adjustment adjusts the clearance between rocker and valve and it is measured with service manual recommended thickness flat feeler gauge. When adjusted correctly, the feeler blade should just drag when moved in and out. Tighten the locknut and recheck clearance. Follow the recommended sequence for valve adjustment of the engine. (Chrysler Corp.)*

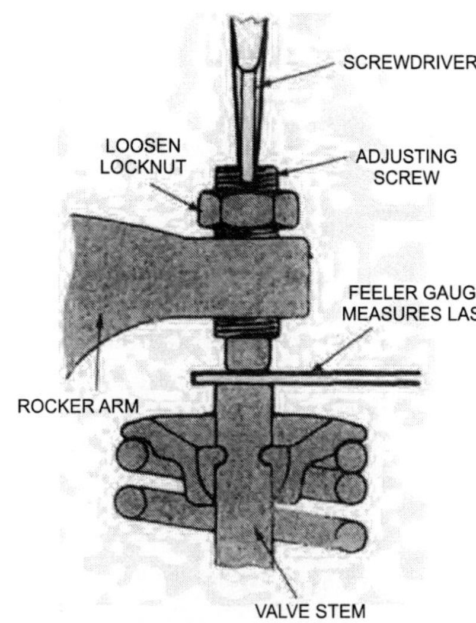

Fig. 15.55 *The method of using feeler gauge to measure valve clearance or lash. (Buick Div. GM)*

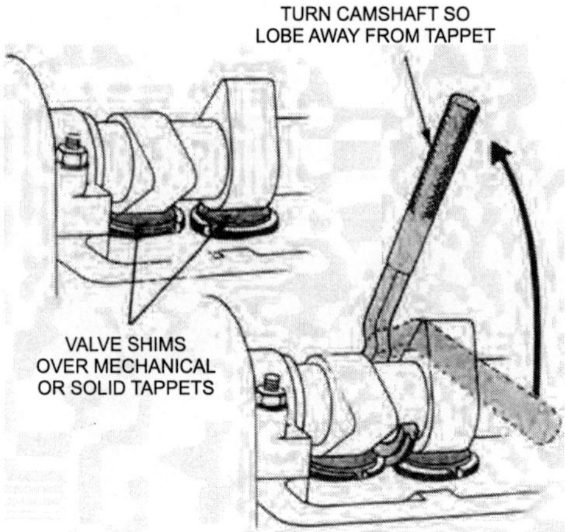

(a) Turn cam until lobe is away from follower or tappet.

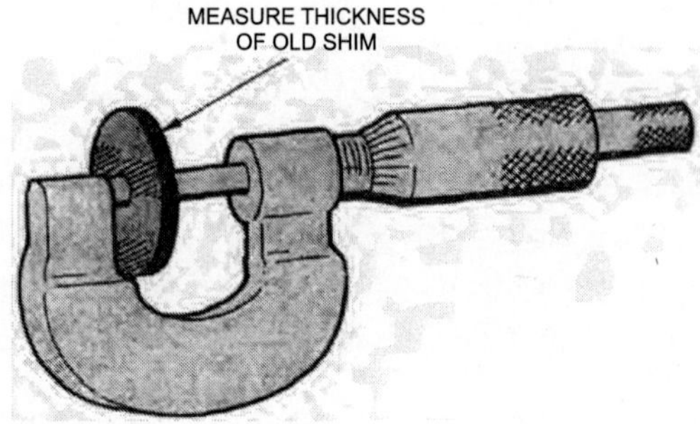

(d) Measure old shim thickness with micrometer. Calculate new shim thickness.

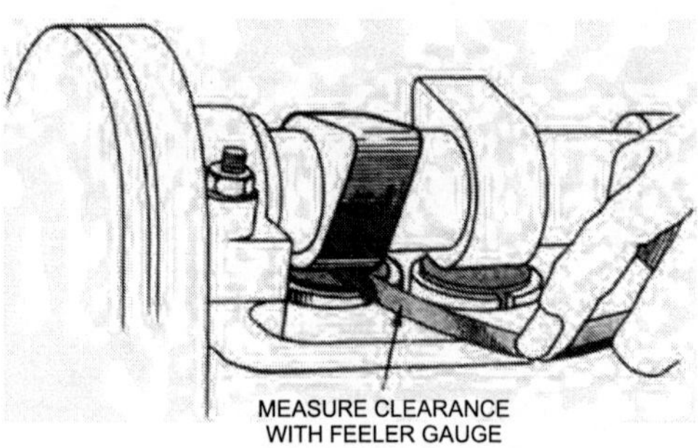

(b) Measure clearance.

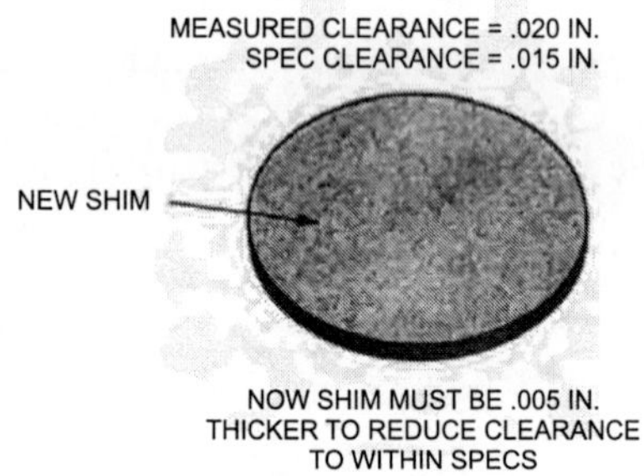

(e) Obtain new shim that will provide correct clearance.

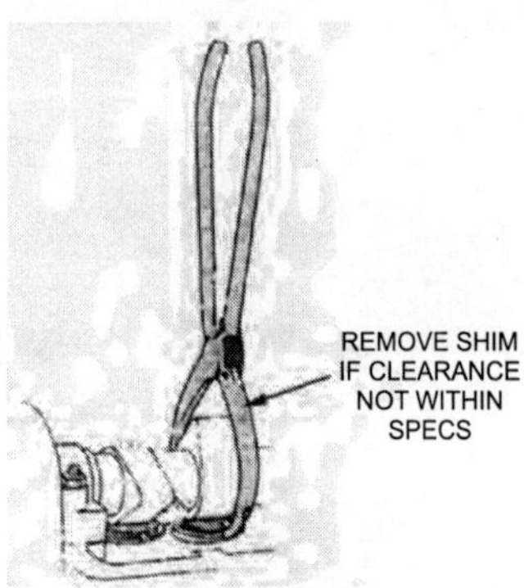

(c) If not within specs, remove old adjusting shim.

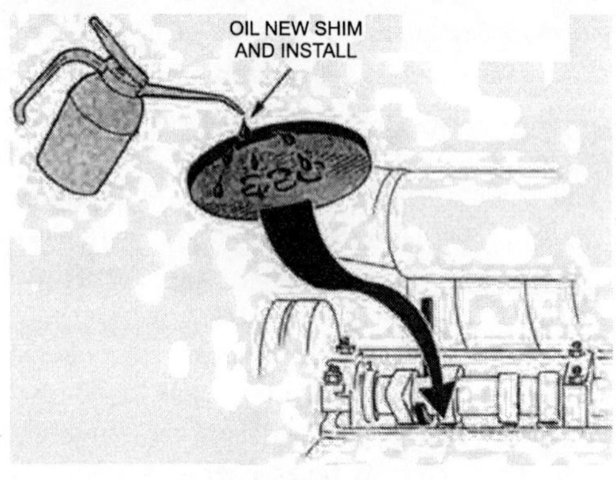

(f) Oil and install new shim.

Fig. 15.56 *Basic steps for adjusting valve clearance with shims on OHC engine. (Volvo)*

Some engines must have valves/mechanical lifters adjusted with the *engine in cold condition.* Others require a *hot engine as at operating temperature* to do the adjustment. Refer the service manual for exact procedure instructions. This is necessary, to use the ***service manual recommended thickness flat feeler gauge. When adjusted correctly, the feeler blade should just drag when moved in and out. Tighten the locknut and recheck clearance. Follow the recommended sequence for valve adjustment of the engine. (Chrysler Corp.)***

OHC Engine Valve Adjustment

A number of methods of adjusting the valves on OHC engines are used. Some adjust like mechanical lifters in a push rod engine. The rocker arm adjuster is turned until the correct size flat feeler gauge fits between the rocker or cam lobe and valve stem tip. Some use hydraulic cam followers that no adjustment may be necessary.

Valve adjusting shims may also be used on late model OHC engines to make adjustment of cam-to-valve clearance, *Fig. 15.56.* Determine valve clearance with a feeler gauge. Then, if necessary, remove and change shim thickness as required. For example, to determine new shim thickness, compare measured clearance to specs. If the engine clearance spec is 0.38 mm and the actual measured lash is 0.51 mm. A thicker shim that is (0.51 – 0.38) or 0.13 mm thicker than the old one would provide spec lash.

Some other OHC engines have an Allen adjusting screw in the cam follower. Turning the screw changes valve clearance. Exact procedures for valve adjustment vary with engine design. Always refer to the service manual for specific detailed directions.

15.10 INSTALLING VALVE/ROCKER COVERS

A *valve cover,* also called *rocker cover,* can leak oil very easily if not installed properly. To avoid an incorrect installation technique and customer complaint it is important to realize how easily a valve cover will leak. Some valve covers employ a cork or synthetic rubber gasket. Some modern vehicles employ valve covers that are factory sealed with silicone sealer. When reinstalling the valve cover, either gasket or sealer can be used, depending on cover design and accessibility.

Checking Valve Cover Sealing Surface

Check the cover flange surface for any defect before installing the cover. Lay the cover on a flat workbench. Cover flange flatness can be checked using a straightedge and flat feeler gauge set. Look between the gasket surface of the cover and the workbench to detect any gaps, e.g., dents, beds, or warpage, *Fig. 15.57.*

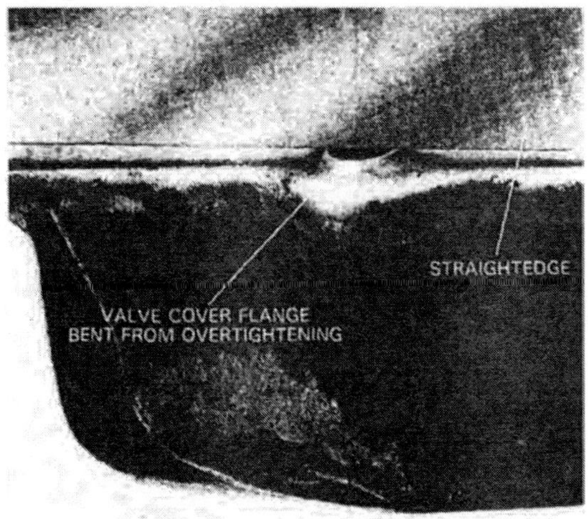

Fig. 15.57 *Carefully check the valve cover for bent or warped flange, using a straightedge. Figure shows a bolt hole dented from Overtightening. This could cause oil leakage. The flange should be hammered to straighten it. (Fel-Pro Gaskets)*

A thin, sheet metal cover flange surface can be straightened with light taps from a small, ball peen hammer. A cast aluminium cover is sometimes sent to a machine shop for resurfacing. A warped plastic valve cover can only be replaced.

Installing Valve Cover Gasket

To install a valve cover gasket, apply a very light coating of approved adhesive around the edge of the valve cover. This is required to hold the gasket in place during assembly, *Fig. 15.58.* Place the gasket on the cover gasket-mating surface with the holes properly aligned.

Allow the adhesive to cure slightly and then, place cover and gasket on the cylinder head in the correct alignment with bolt holes and hand start all of the bolts. Tighten the valve cover to specs, working the bolts in small steps and following a crisscross pattern, *Fig. 15.59.*

A frequent mistake made by mechanics is to over-tighten valve cover bolts. Over-tightening can smash and split the gasket. It can cause warped /bent valve cover, resulting in oil leakage. Therefore, carefully torque the bolts to specs and following proper procedure.

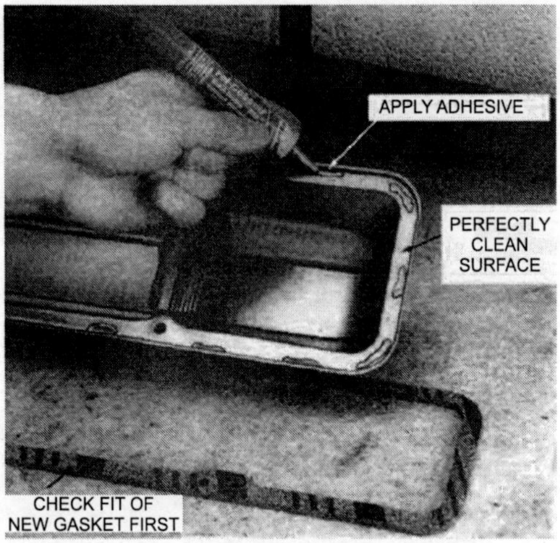

Fig. 15.58 *Stick new gasket to valve cover using approved adhesive. Ensure that the bolt holes align correctly while allowing adhesive to become tacky. (Fel-Pro Gaskets*

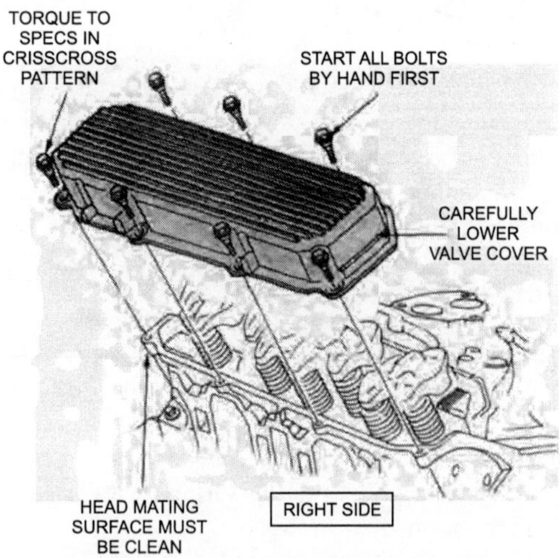

Fig. 15.59 *Carefully lower valve cover into position on engine without bumping and shifting the gasket. Start all fasteners by hand before tightening any bolts. (Cadillac Div. General Motors).*

Installing Valve Cover with Silicone Sealer

Therefore using silicone or RTV sealer on the valve cover, check that the cover and cylinder head gasket mating surfaces are perfectly clean, dry and are flat within specs. Sealer will not bond and seal on a dirty, oily and warped surface.

Apply a continuous bead of sealer, typically about 1.6 mm wide, all around the valve cover sealing surface. Carefully install the valve cover onto the head, without allowing the sealer to set. Avoid any smearing and breaking the silicone sealer bead during cover installation. Torque the cover fasteners to specifications.

15.11 CHECK FOR PROPER OIL PRESSURE

With oil in the engine and the distributor out of the engine, oil pressure should be established before the engine is started. This can be done on most engines by rotating the oil pump by hand. This ensures that oil is delivered to all parts of the engine before the engine is started. A socket speed handle makes an ideal crank for spinning the oil pump. A flat-blade adapter that fits the speed handle will operate pump on GM engines. The V-type Chrysler engine requires the use of the same flat blade adapter, but it also requires an oil pump drive. Removing the gear from an old oil pump hex driveshaft can make one. A 1/4 - inch drive socket can be used on Ford Motor engines. Examples of these are shown in *Fig. 15.60.* Engines that do not drive the oil pump with the distributor will have to be cranked with the spark plugs removed to establish oil pressure. The load on the starter and battery is reduced with the spark plugs out so that the engine will have a higher cranking speed. A pressurized oil container or an engine oil aerosol could also be used, as shown in *Fig 15.61.*

Figure 15.62 shows an example of a tester used to spin the crankshaft of the engine to check for proper compression, oil leaks, or other problems.

15.12 STARTER

It is generally easier to install the starter before the engine is put in the vehicle engine compartment in the chassis. The starter should be checked to ensure that the starter drive pinion does not bind on the

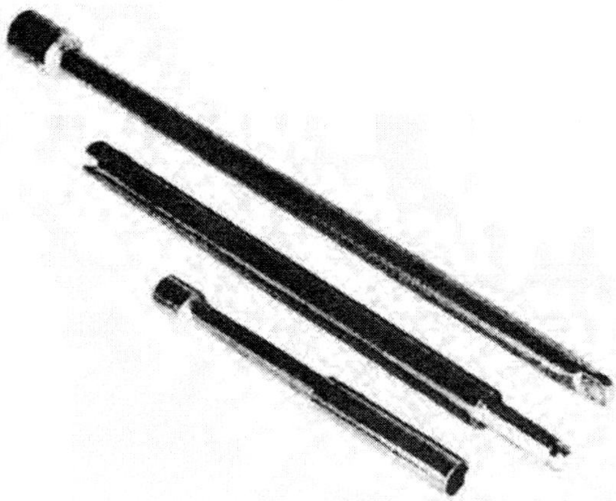

Fig. 15.60 *Drivers used to rotate oil pumps to pre-lubricate all parts of the engine before installing the distributor and starting the engine.*

Fig. 15.61 *The engine can be pressurized with engine oil from an aerosol can container as shown or from a pressurized oil container designed for pre-oiling the engine.*

flywheel ring gear. Shims can be installed between the starter mounting pad and the starter to adjust the pinion-to-ring gear clearance on the GM-type mounting. The starter motor attaching bolts are then tightened to the specified torque.

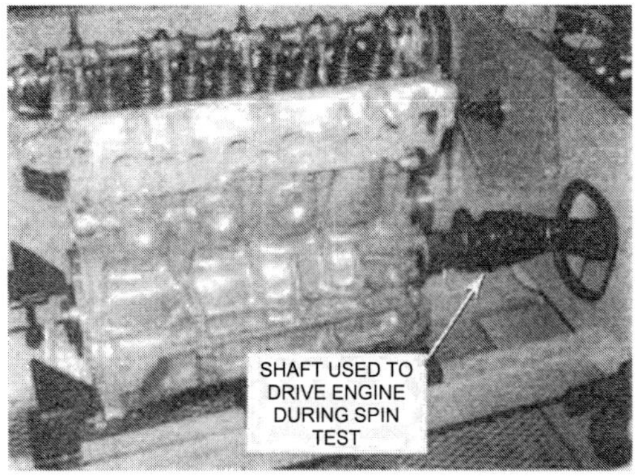

(a) A newly rebuilt engine is placed in the tester and the crankshaft is connected to an electric drive motor.

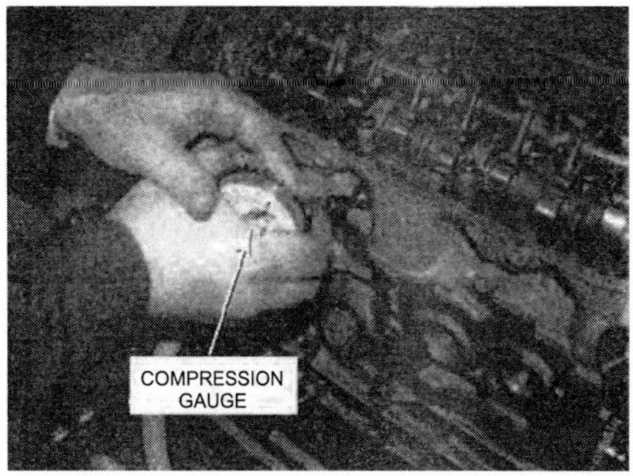

(b) As the engine is being rotated by the drive motor, a mechanic can check engine cylinder compression by using a compression gauge.

(c) Oil pressure can be measured and recorded at various engine speeds.

Fig. 15.62

15.13 ACCESSORIES

All belt-driven engine accessories are mounted on the front of the engine. Some engines drive all these accessories with one belt. Other engines use as many as four belts. Service information or decals under the hood should be checked to determine the specific belt routing for the accessories used on the engine being built up. On some engines it is more convenient to install the front accessories before the engine is installed; on other engines, it is easier to put the engine in the chassis before installing the front accessories.

Install new spark plugs and spark plug wires. Service information should be checked for the proper routing of the plug wires.

Fuel Pump

If a fuel pump push rod is used, molybdenum-lube both ends and slip it in its bore. Place the pump in position and bolt it up. If the pump's return spring has to be compressed against the push rod, rotate the engine one revolution. This will turn the camshaft eccentric away from the pump, unloading the return spring.

Oil Filter

Fill a new oil filter with oil and install it. Some filters cannot be filled first because they drain before getting them to the engine. If the filter can be filled, then fill it to help speed up priming on engine starting.

15.14 ENGINE INSTALLATION

When installing an engine, the engine installation will have to be thoroughly checked to make sure that it is in proper condition to give the customer dependable operation for a good length of time. All operating accessories have to be reinstalled after overhauling the engine. They have to be adjusted so that the engine will operate and perform correctly as predicted.

Preinstallation Check List

Before installing or starting a new or rebuilt engine in a vehicle, be sure all of the following items have been checked.

1. Battery to be fully charged.
2. Prelube the engine and check for proper oil pressure.
3. Check that all of the vacuum lines are correctly installed and routed.
4. Know the ignition timing specification and procedure.
5. Check that fresh fuel is in the fuel tank.
6. Ensure that the radiator has been adequately tested and is free from leaks, and flows correctly.

Installing an engine in a vehicle is just about the reverse of the engine removal covered in previous chapter. However, there are a few common steps that should be understood.

Installing Flywheel – Manual Transmission

If the engine was removed with the transmission attached, the transmission should be reinstalled on the engine before other accessories are added. The flywheel is installed on the back of the crankshaft. The pilot bearing or bushing in the rear of the crankshaft is usually replaced with a new one to minimize the possibility of premature failure of this part.

Before installing the flywheel, check its ring gear teeth. Ensure that none of the teeth are worn, broken off, or otherwise damaged. Ring gear teeth damage will cause problems with starting motor pinion gear engagement and operation. Also. Ensure that the flywheel runout is within specs. If defective in any way, replace the ring gear or flywheel.

Unlike automatic transmission flywheels, most manual transmission flywheels have a removable ring gear. Use a chisel to split and remove the damaged ring gear, *Fig. 15.63 (a)*. Then, heat the new ring gear with a torch. Heat the ring gear evenly so that it will expand uniformly.

With a hammer tap the heat-expanded ring gear over the flywheel lip. Handle the ring gear with proper care and do not strike the gear teeth with the hammer. Drive on the inner diameter of the ring gear until it is seated on the flywheel, in its proper position, *Fig. 15.63 (b)*.

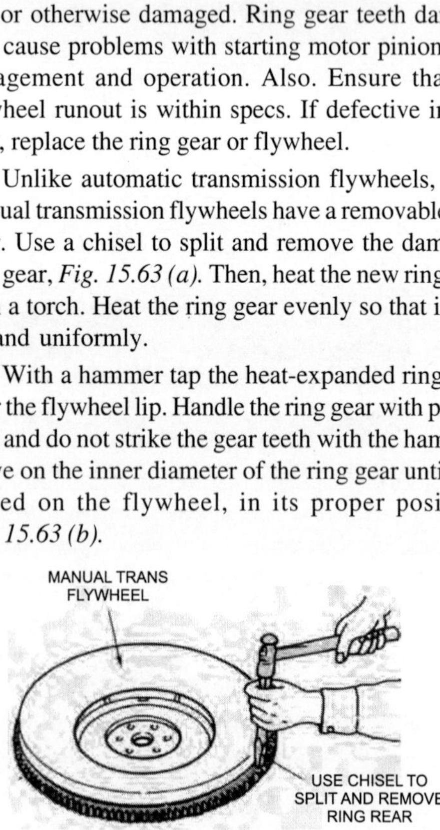

(a) Use a chisel to break and remove old ring gear if necessary.

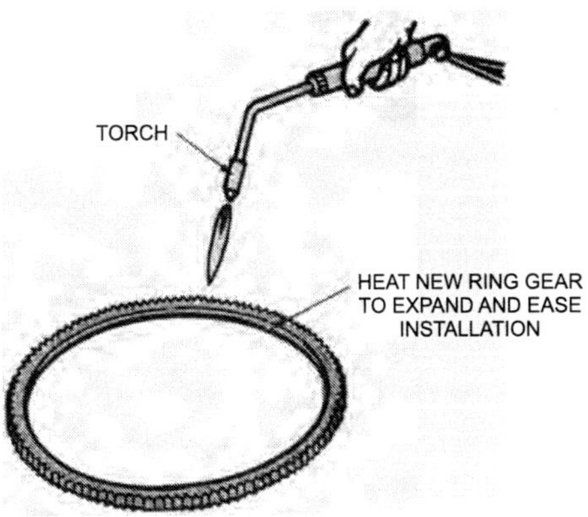

(b) Heat new ring gear to expand it. Drive hot ring gear over manual flywheel until fully seated in its position. (Buick Div. General Motors)

Fig. 15.63 *Check the flywheel ring gear teeth carefully. Even single broken or badly worn tooth could require removal of the transmission/transaxle. (Buick Div. General Motors)*

Install the flywheel on the crankshaft. Remember to use any spacer plate that fits between the engine and transmission or transaxle, as the case may be. Bolt the flywheel to the crankshaft. The flywheel and crankshaft flange normally have offset bolt holes so the flywheel will only go on in one position to maintain engine balance. Turn the flywheel on the crank until all bolt holes align. Install and torque the flywheel bolts to specs.

Installing Clutch

The clutch is installed next. Usually a new clutch is used; at the least, a new clutch friction disc is installed. The clutch friction disc must be held in position using an alignment tool, called a dummy shaft that is secured in the pilot bearing. This holds the disc in position while the pressure plate is being installed.

With a manual transmission or transaxle, the clutch should be checked and installed before engine installation. Generally, mechanics prefer to use a new pilot bearing, clutch disc, and throw-out bearing because these parts wear and fail quickly. Some mechanics even like to replace the pressure plate. In any case, check these parts carefully to ascertain their condition before reuse.

Figure 15.64 shows the method of measuring the clutch disc wear. If the disc is worn (thinner) out of specs, it should be replaced. Also, if the disc is contaminated with oil or shows signs of slippage and overheating, replace it. Wear eye protection glasses and wear respiratory protection when working with clutches.

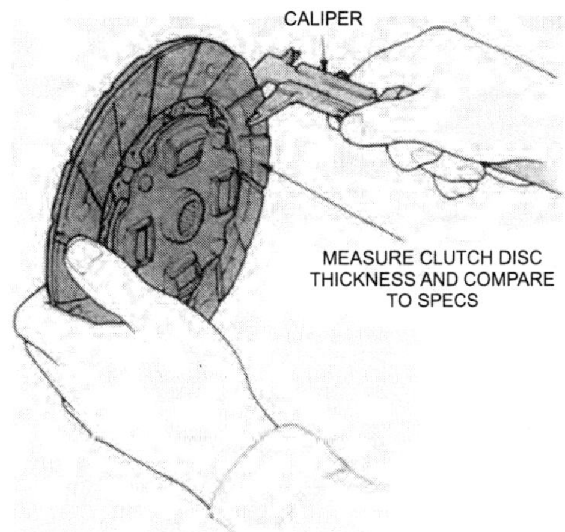

Fig. 15.64 *Check clutch disc carefully and measure its thickness before reuse. Frequently mechanics install new disc during major engine repairs. If worn to be out of thickness specs, it should be replaced. (Honda Motors)*

Inspect the pressure plate friction surface for signs of overheating, worn release arms, weakened springs, etc. The throw-out bearing must turn freely with no signs of bearing roughness. It is a good practice to replace the pilot bearing, clutch disc, and throw-out bearing any time the clutch is disassembled. These parts wear and fail relatively quickly.

To install the clutch, install the disc and pressure plate on to the flywheel. Insert a *pilot tool* (alignment shaft) or old transmission input shaft through the disc and into the pilot bearing. This will hold the clutch and centre it while torquing the pressure plate bolts, as illustrated in *Fig. 15.65.*

Make sure that only the factory bolts are used to install the pressure plate. They are high tensile strength and usually are not threaded near the head. Nonstandard bolts if used for this purpose, they may fail and lead to serious part damage.

Fit the throw-out bearing onto the release arm. Make sure it is secured properly because it must stay in position during engine or transmission-transaxle installation, *Fig. 15.66.*

Finally, the engine bell housing is put on the engine, if it was not installed before. Bolt the bell-housing to the rear of the engine block. Torque its fasteners to specs. Attach any other components to the bell housing, e.g., clutch bracket. The alignment of this type of bell housing is then checked, *Fig. 15.67.*

The clutch release yoke should be checked for free movement. Usually, the clutch release bearing is replaced to ensure that the new bearing is securely attached to the clutch release yoke. The transmission can then be installed.

The transmission clutch shaft must be guided straight into the disc and pilot bearing. The clutch shaft is rotated, as required, to engage in the splines of the clutch disc. Sometimes parts do not seem to line up correctly. To overcome this difficulty, cut the head of extra long bolts that are of the same diameter and thread as those being used to retain the part, such as a transmission, *Fig. 15.68.* Use a hacksaw to cut a slot in this end of the guide bolts for a screwdriver slot. Install the guide bolts; then install the transmission. Use a straight-blade screwdriver to remove the guide bolts after securing the transmission with retaining bolts. Assembly bolts are secured when the transmission fully mates with the bell housing.

If the transmission or transaxle is out of the vehicle, usually the engine and transmission or transaxle can be installed as a single unit. Refer to the applicable service manual for detailed instructions.

Use an engine crane to hold the engine on the shop floor. The crane will keep the engine from falling over. Lift the transmission or transaxle up and slide its input shaft through the clutch. You might have to wiggle the transmission while pushing in.

Hold the transmission /transaxle perfectly straight with the centerline of the crankshaft or it will not slide fully into the pilot bearing. Once butted up correctly against the clutch housing, install the transmission /transaxle bolts, *Fig. 15.69.* Torque them to specs.

Do not use the transmission or transaxle bolts to pull the unit up to the bell housing. The transmission should slide freely up against the bell housing. If the bolts are used to pull them together, the input shaft can smash into the side of the pilot bearing and cause part damage.

Automatic Transmission Installation

On engines equipped with automatic transmission, the drive plate is attached to the back of the crankshaft. Its assembly bolts are tightened to the specified torque. The bell housing is part of the transmission case on most automatic transmissions.

Installing Torque Converter

Usually, the torque converter will be installed on the transmission before the transmission is put on the engine. The torque converter should be wiggled slowly as it is pushed on to the transmission shaft until the splines of all shafts are engaged in the torque converter. The torque converter is held against the transmission as the transmission is fitted on the back of the engine.

View down the side of the transmission or transaxle to make sure the converter is not sticking out too far. Make sure the torque converter is completely installed over its shaft. If you bolt the transmission or transaxle to the engine and the converter is *not* completely over its shaft, severe transmission or transaxle damage can occur. The front pump, housing, or converter could be ruined and it proves to be costly mistake.

If the transmission or transaxle is to be installed with the engine, bolt them together when on the shop floor. As mentioned for a manual transmission or transaxle, make sure the input shaft is in perfect alignment with the engine crankshaft.

If the torque converter has a drain plug, it should be rotated down. The hole for the drain plug in the flywheel must also be rotated down. When the two are bolted up, the converter drain plug will stick through this enlarged hole in the flywheel. This will help you align everything as you bolt the automatic transmission or transaxle to the engine.

The transmission mounting bolts are attached finger tight. The torque converter should be rotated to make sure that there is no binding. The bell housing is secured to the block; then the torque converter is fastened to the drive plate. Torque all fasteners to specs. The engine should be rotated. Any binding should be corrected before any further assembly is done.

15.15 INSTALLING ENGINE IN VEHICLE

Installing an engine in a vehicle is about the opposite as removing the engine. Engine removal has been discussed in chapter 14. A sling, either a chain or lift

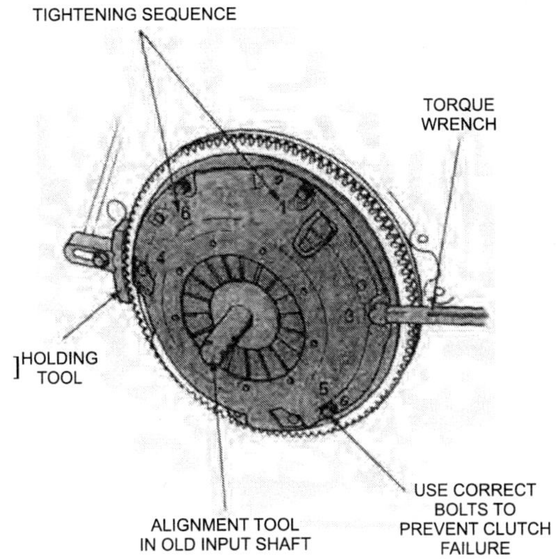

Fig. 15.65 *Use a pilot shaft to align clutch assembly on flywheel. Torque bolts to specs and use factory bolts as far as possible. Install new pilot bushing if necessary.*

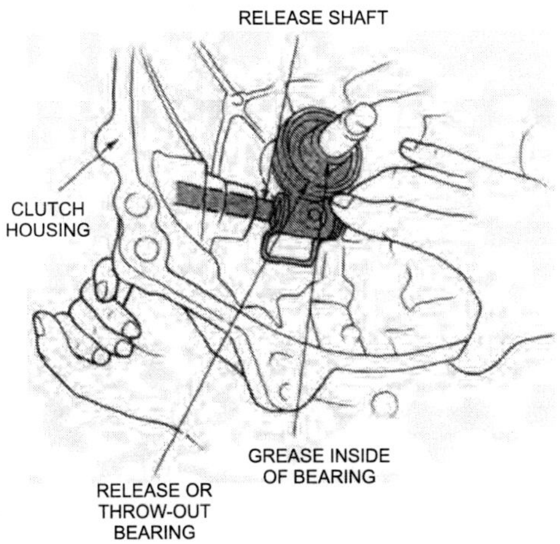

Fig. 15.66 *Position new throw-out bearing carefully. It must stay in place when bolting transmission/transaxle to the engine. Frequently mechanics install new throw-out bearing since it is common wear spot. (Honda Motors)*

cable, is attached to the manifold or head bolts on the top of the engine. A hoist is attached to the sling and snugged up to take the weight and to make sure that the engine is supported and balanced properly.

Many engines for front-wheel-drive vehicles are installed from underneath the vehicle. Often the entire drive-train package is placed back in the vehicle while it is attached to the cradle. Always check the recommended procedure for the vehicle being serviced.

Fig. 15.67 *Bell housing alignment dowel pins are used to ensure proper alignment between the engine block and the transmission.*

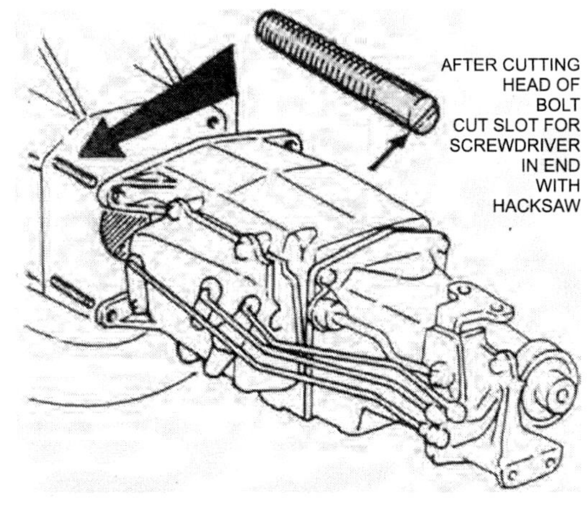

Fig. 15.68 *Headless long bolts can be used to help align and install a transmission to the engine.*

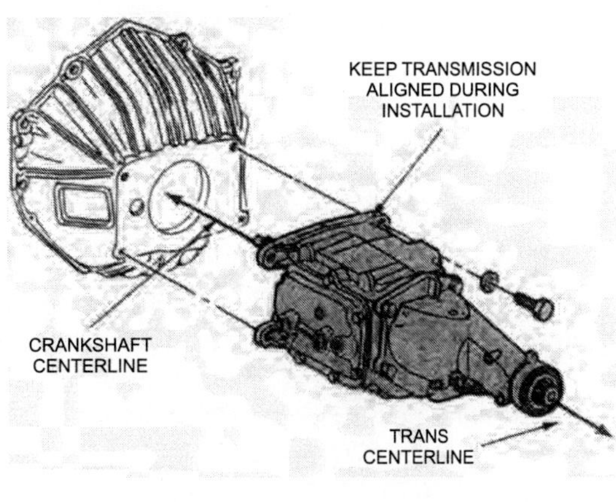

Fig. 15.69 *While bolting transmission or transaxle, as the case may be, to the engine, keep their centrelines in perfect alignment. (General Motors)*

The engine must be tipped as it was during removal to allow the transmission go into the engine compartment first. Mount the engine on the engine hoist with the flywheel slightly lower than the front of the engine. This helps the engine and transmission to mate. Raise the transmission with a floor jack so it and the engine can join before the engine is set on its mounts. Use guide pins or headless bolts to help with engine-to-transmission alignment.

If the engine will not join the last few centimeters with a manual transmission, rotate the crankshaft with the front pulley. This will turn the clutch disc so its splines and those on the input shaft will line up. As soon as the engine and transmission are joined, run two bell housing bolts in to stop them from slipping apart. Remember to bolt the torque converter flexplate together.

The transmission is worked under the floor pan on rear-wheel-drive vehicles as the engine is lowered into the engine compartment. The front engine mounts are aligned; then the rear cross-member and rear engine mounts are installed. The engine mount bolts are installed, and the nuts are torqued to specs. Then the hoist is removed. Controls are connected to the transmission under the vehicle. This is also good time to connect the electrical cables and wires to the starter motor.

The exhaust system is then attached to the exhaust manifolds. If any of the steering linkage was previously disconnected, it can be reattached while work is being done under the vehicle.

After the engine is in place, the front engine accessories can all be installed, if they are not installed before the engine was put in the chassis. The air conditioning compressor is reattached to the engine, with adequate care to avoid any damage to the air-conditioning hoses and lines.

Cooling System

The radiator is installed and secured in place, followed by the cooling fan and shroud. The fan and new drive belts are then installed and adjusted. New radiator hoses, including new heater hoses, should be installed.

Coolant, a 50/50 mixture of antifreeze and water, is put in the cooling system after making sure that the radiator petcock is closed and the block drain plugs are in place. Bleed the trapped air from the cooling system as already explained in chapter 4. See *Fig. 15.70* for additional precautions.

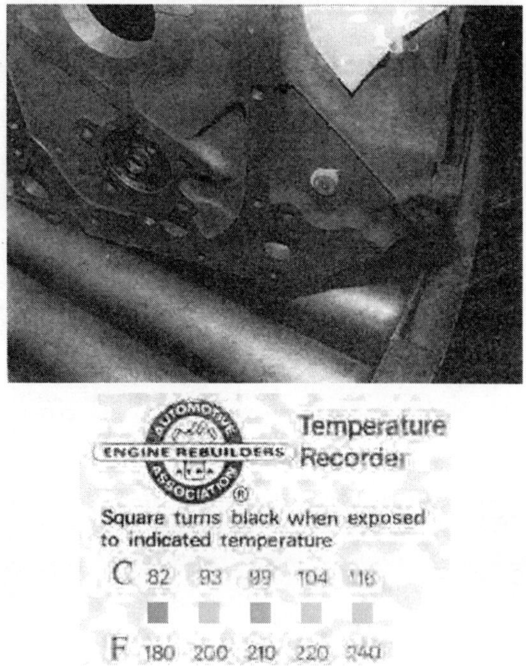

Fig. 15.70 *Most engine rebuilders install a temperature sensitive device on the engine. These sensors are used by the rebuilders for warranty purposes to record any occurrence of any engine overheating.*

Fuel System and Emission Controls

Install a new fuel filter and see that there is plenty of fresh fuel in the fuel tank. Fuel systems using a low-mounted canister filter or sediment bowl will need priming. Open the canister or bowl and fill it with fuel. Reattach the canister or bowl with a minimum of fuel spillage. If the engine keeps running out of fuel, pressurize the fuel tank very lightly with a compressed air nozzle. This will prime the fuel pump.

The carburetor (if the vehicle is so equipped) should be installed with a new gasket. The fuel and vacuum hoses should be inspected carefully and replaced as needed. The fuel injection system (if the vehicle is so equipped) should be carefully checked for any damage while it is off the engine and then reinstalled, following recommended procedures and torque specs. The fuel and air filters should be replaced. If the vacuum hoses and/or electrical wiring were not marked, refer to the engine emission decal and service manual for the correct location and routing.

Double-check all fuel connections for tightness before trying to start the engine. Keep a fire extinguisher handy.

The oxygen sensor should be replaced, especially if the engine had a blown head gasket or other problem that could have caused the coolant to get to the sensor.

Electrical System

Connect all wiring to the starter and alternator (generator) as needed. Connect the instrument and computer sensor wires to the sensors on the engine. Double-check the condition and routing of all wiring, and ensure that the wires have not been pinched or broken, before installing a fully charged battery. The battery should be fully charged, and do not forget to install all ground straps. Attach the positive cable first and then the ground cable. Check to make sure that the starter will crank the engine. Install and time the distributor; then connect the ignition cables to the spark plugs, again being sure that they are routed according to the manufacturer's recommendations.

With modern computer-controlled vehicles, never re-route wires. Keep all wires in their original locations and reinstall them in their original positions. If spark plug wires or some high current carrying wires are rerouted, current can be induced into low current sensors or computer feedback wires. Which could upset computer system operation.

Engine Installation Rules

Some rules to be kept in mind when installing an engine are:

1. To avoid injury, keep hands and feet out from below the engine.
2. Never work on an engine raised in the air on a hoist or crane.
3. Only raise the engine on the hoist while it is being placed in the vehicle.
4. With a manual transmission, check the condition of the pilot bearing or bushing in the end of crankshaft. Replace it if worn outside specs.
5. Some engines require a valve adjustment at this point, and most head gaskets need retorquing unless otherwise marked.
6. Do these steps now. If they are not necessary, disconnect the timing light and road-test the vehicle.
7. Position the lifting chain or fixture on the engine so that the engine is raised *level*. If *not* level, it will be difficult to slide the engine against the transmission, *Fig. 15.71*.
8. Slowly lower the engine into the vehicle while watching for clearance all around the engine compartment. Position the engine so that its crankshaft centreline aligns with the transmission input centreline, *Fig. 15.72*.
9. Push the engine back and align the engine dowel pins with the holes in the transmission. A large bar can be used to shift the engine.
10. Once the dowel pins engage fully into their corresponding holes in transmission, install an engine-to-transmission bolt, by running it with hand, but do not tighten it. Start another bolt on the other side of the engine.
11. Ensure that the torque converter is properly lined up with the holes in the flywheel. Wiggle the converter if required. Never use the transmission bolts to force the engine against the transmission or transaxle as it can cause part damage.
12. Finish installing the other components: engine mounts, oil filter, fuel lines, wiring, vacuum hoses, throttle cable or linkage, battery cable ground, fan belts, front drive axles, etc., see *Figs. 15.73 to 15.79.*
13. Fill the engine with lubricating oil and the radiator with coolant.
14. Start and fast idle the engine until warm. Watch the engine oil light or gauge to make sure there is good oil pressure in the system. Also watch the engine temperature to avoid overheating.
15. In case the engine will not start, check the following:
    - Ignition timing and spark intensity.
    - Spark plug wire routing.
    - Primary wire routing to ignition coil and engine sensors.
    - Fuel supply to carburetor or injection system.
16. After the engine starts and warms up to normal operating conditions, shut the engine off. Recheck critical tune-up type adjustments: ignition timing, idle speed, and re-torque head bolts.
17. Restart the engine and let it run at a fast idle. Check for any visible leaks under the engine

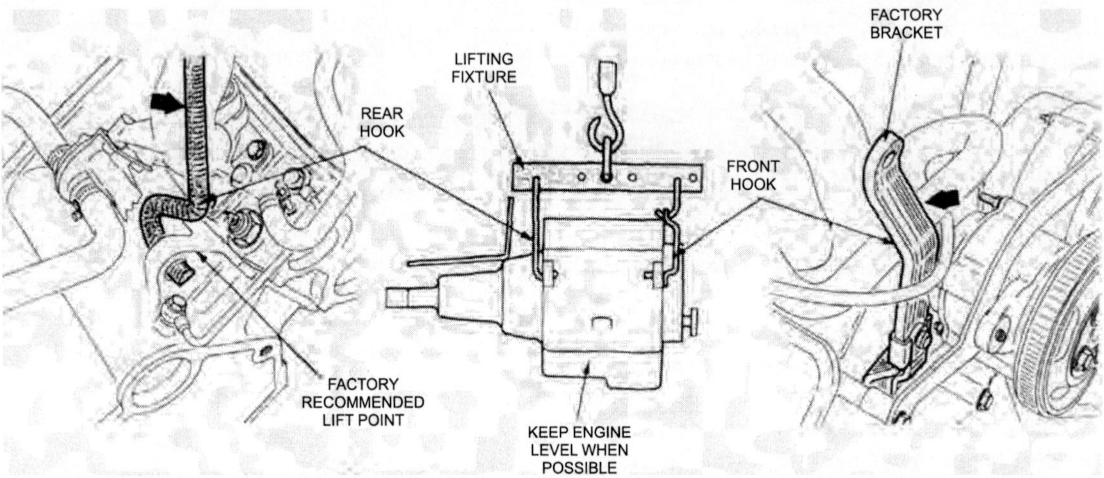

Fig. 15.71 *Attach lifting chain or bracket such that engine is lifted squarely. This will help align components and assemblies during installation.*

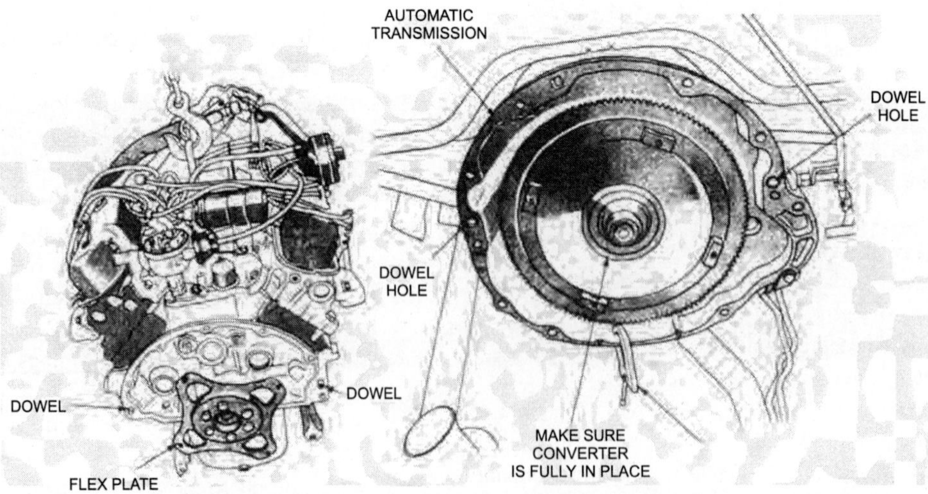

Fig. 15. 72 *For automatic transmission or transaxle, make sure that the torque converter is fully installed over its shaft. Also align everything and out of the way before lowering engine into engine compartment. (Chrysler Corp.)*

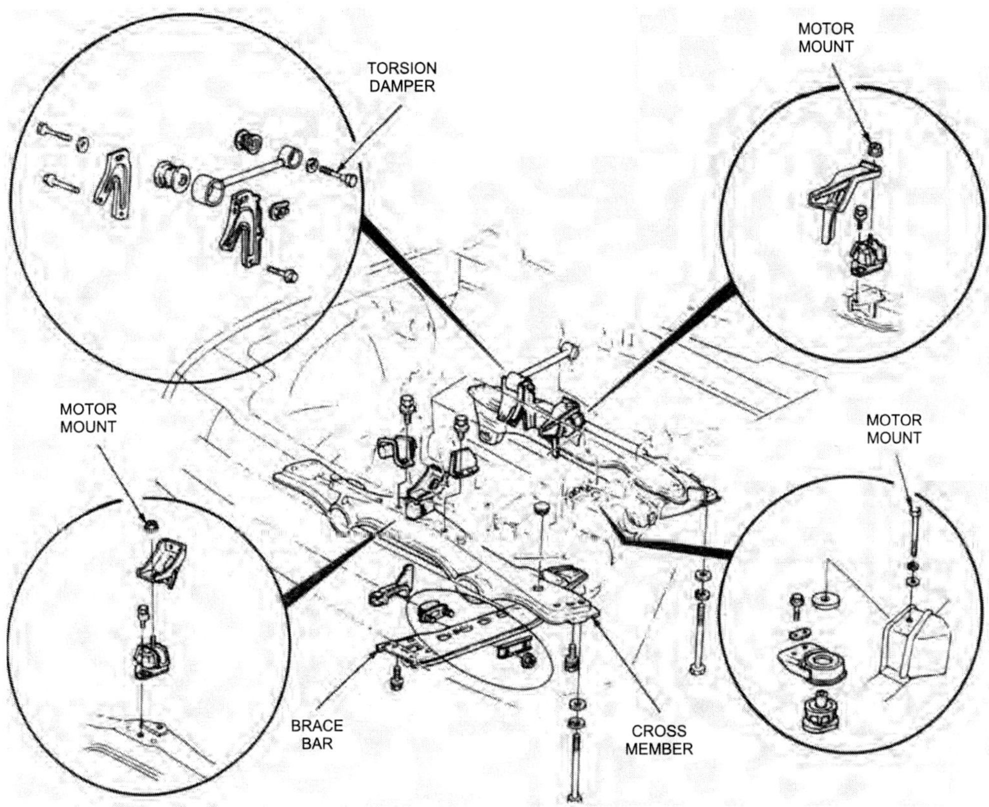

Fig. 15.73 *When the engine is lowered and is in place, attach and secure all its motor mounts. (Honda Motors)*

and make sure the engine does not overheat, *Fig. 15.80.*

Pre-engine Start-up

Immediately before start-up, prime the engine by rotating the oil pump through the distributor or oil pump drive shaft holes, *Fig. 15.81.* Use a speed handle and hex, screwdriver, or square bit to turn the pump. Check for oil pressure at the instrument and rocker arm area. The engine is primed when oil flows into the rocker arm area. Rotating the crankshaft while priming fills the oil galleries, the crankshaft's oil passages and brings oil to all bearings.

If the oil pump drive is not accessible, use an air-powered priming device attached to the oil filter pad or oil pressure sender switch port. Or crank the engine without the spark plugs installed until oil flows into the rocker arm area.

After priming, bring the engine to TDC #1 cylinder and install the distributor. Attach the module connector or coil negative wire. Slip on the coil's

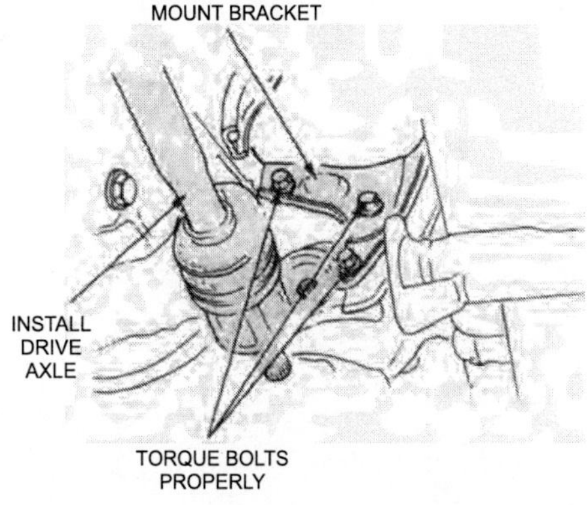

(a) Install drive axle.

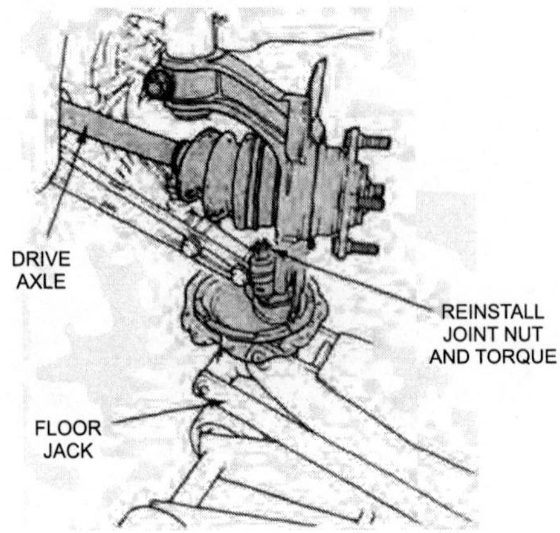

(b) Reassemble ball joint and related parts.

Fig. 15.74 *With a front-wheel drive vehicle, install front drive axles when installing engine and transaxle as directed by the applicable service manual for the specific vehicle.*

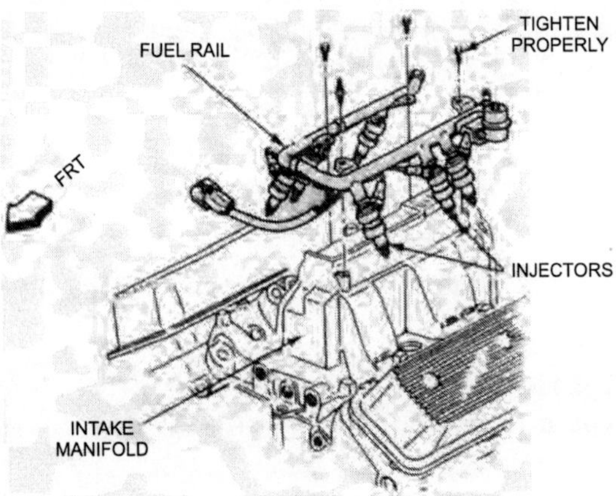

Fig. 15.75 *Injectors and carburetor are usually installed after engine is back in the vehicle. This is to protect them from damage during engine installation. (General Motors)*

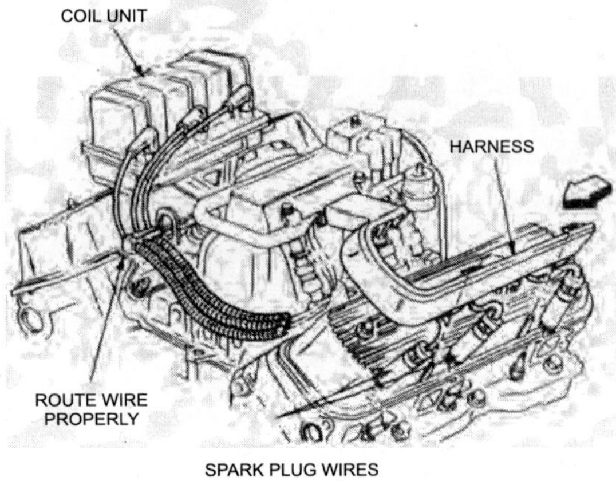

Fig. 15.76 *Route all ignition system wires away from hot or moving parts. Make sure that wires feed to correct spark plugs as per firing order of multi-cylinder engine. (Buick Div. General Motors)*

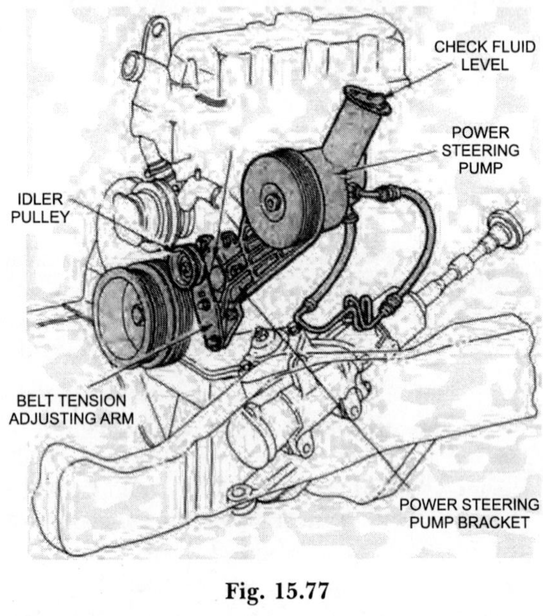

Fig. 15.77

high-tension lead and connect a timing light. Leave the distributor hold-down loose enough to adjust timing but tight enough to keep the distributor from turning on its own.

Engine Start-up

The engine installation should be given one last inspection to ensure that everything has been put together correctly before the engine is started. If the engine overhaul and inspection are carried out correctly, the engine should crank and start on its own fully charged battery without the use of a fast charger or jumper battery. As soon as the engine starts and develops oil pressure, it should be brought up to a fast idle speed and kept there to ensure that the engine gets proper lubrication. Do not rev the engine. The fast idle is important for proper oiling, especially for the camshaft, lifters, and rockers. The fast-running

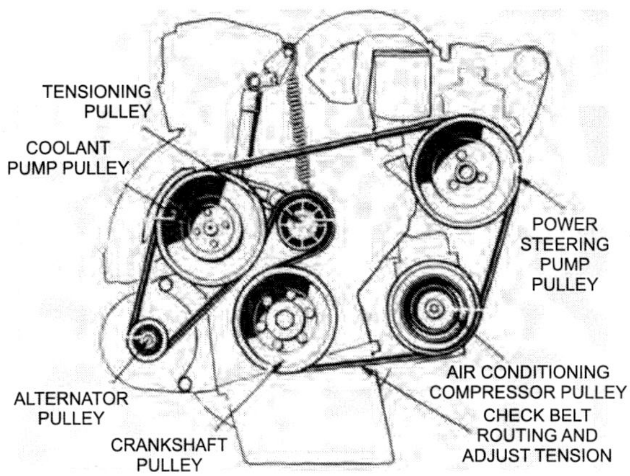

Fig. 15.78 *During a major engine service, ensure that drive belts are properly installed and are in good condition and reuse. (Mercedes-Benz)*

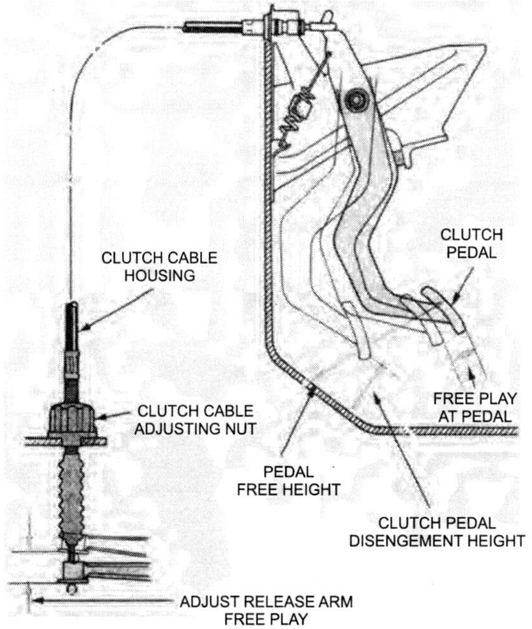

Fig. 15.79 *Follow the service manual instructions to adjust manual clutch or automatic transmission-transaxle linkage. (Honda Motors.)*

oil pump develops full pressure, and the fast-turning crankshaft throws plenty of oil on the cam and cylinder walls.

After the engine has started, the following items should be checked:

Fig. 15.80 *After first start up of a reconditioned engine, always watch for any oil and coolant leaks. Visually inspect everything to protect engine and yourself.*

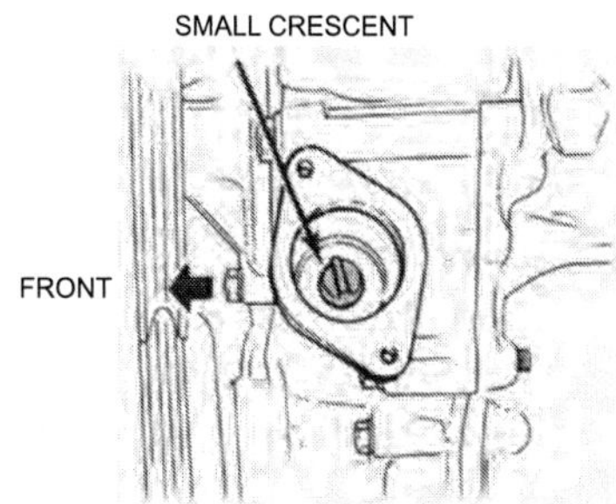

Fig. 15.81 *Oil pump drive used for priming oil system.*

1. Check the valve train for noise. Some engines will need several minutes to quiet down.
2. Check the engine vacuum. It should be 432 mm to 533 mm of Hg at sea level.
3. Check for leaks of any kind. Watch the oil pressure and coolant temperature. If any leaks are found, stop the engine and fix them. If not, run the engine for about 10 minutes at fast idle. Then drop the rpm to idle and set the timing.
4. Check the charging system for proper operation. The charging voltage should be 13.5 to 15 volts.

Road Test the Vehicle

As soon as you can tell that no serious leaks exist, and the engine is running well, the vehicle should be driven to a road having minimum traffic. Here, the vehicle should be accelerated, full throttle, from 48 to 80 kilometers per hour. Then the vehicle is allowed to return to 48 km/hr. This sequence is repeated 10 to 12 times. The acceleration sequence puts a high load on the piston rings to properly seat them against the cylinder walls. The piston rings are the only part of modern engine that needs to be broken in. Good ring seating is indicated by a dry coating inside the tail pipe at the completion of the ring seating drive.

The vehicle is returned to the service area, where the basic ignition timing is set and the idle speed is properly adjusted, if possible. The engine is again checked for visible fluid leaks. If the engine is dry, it is ready to be turned over to the customer.

The customer should be advised to drive the vehicle in a normal fashion, neither babying it at slow speeds nor beating it at high speeds for the first 160 kilometers. The oil and filter should be changed at 800 km to remove any dirt that may have been trapped in the engine during assembly and to remove the material that has worn from the surfaces during the break-in period.

A well-designed engine that has been correctly reconditioned and assembled using the techniques described should give reliable service for many kilometers.

Normal Operating Temperature

Normal operating temperature is the temperature at which the upper radiator hose is hot and pressurized. Another standard method used to determine when normal operating temperature is reached is to observe the operation of the electric cooling fan, when the vehicle is so equipped. Many manufacturers define *normal operating temperature* as being reached when the cooling fan has cycled on and off at least once after the engine fan has been started. Some vehicle manufacturers specify that the cooling fan should cycle twice. This method also helps assure the mechanic that the engine is not being overheated.

Warming Up a Cold Engine

The greatest amount of engine wear occurs during start-up. The oil in a cold engine is thick, and it requires several seconds to reach all the moving parts of an engine. After the engine starts it should not be raced, but rather allowed to idle at the normal fast idle speed as provided for by the choke fast idle cam (on carburetor-equipped engines) or by the computer-controlled speed on fuel-injected engines. After the engine starts, allow the engine to idle until the oil pressure peaks. This will take from 15 seconds to a minute, depending on the outside temperature. Do not allow the engine to idle for longer than 5 minutes. Because an engine warms up faster under load, drive the vehicle in a normal manner until the engine is fully warm. Avoid full-throttle acceleration until the engine is completely up to normal operating temperature. This method of engine warm-up also warms the rest of the power train, including transmission and final drive component lubricants.

15.16 ENGINE BREAK-IN

An engine overhaul represents many hours of work and a large financial investment. Precautions should be taken to protect the investment, including the following:

1. Never add cold water to the cooling system while the engine is running.
2. Never lug an engine. *Lugging* means increasing the throttle opening without increasing engine speed (rpm). Applying loads to an engine for *short periods of time* creates higher piston ring pressure against the cylinder walls and helps in the breaking-in process by helping to seat the rings.
3. Change the oil and filter at 800 km or after 20 hours of operation.
4. Keep in mind that the proper air-fuel ratio is important to the proper operation and long life of any engine. Any air leak (vacuum leak) could cause engine damaging detonation.
5. Ensure to use spark plugs for the proper heat range.

Engine break-in is done primarily to seat and seal new piston rings. It also helps initial wearing in of other engine parts under controlled conditions. After warm-up at a fast idle, usually mechanics road test the vehicle. At the same time, they use moderate acceleration and deceleration for engine break-in.

Generally, accelerate the vehicle to about 65-84 km/h. Then, release the gas pedal fully and let the vehicle coast down to about 32-48 km/h. Do this several times and keep watching the engine temperature and oil pressure. Do not allow the engine to overheat during break-in because it can cause engine ring and engine scoring.

Educate the vehicle owner about the following rules concerning the operation of a freshly overhauled engine:

1. Avoid prolonged highway driving during the first 1460 to 3200 km. This will prevent ring friction from overheating the rings and cylinders, possibly causing damage.
2. Do not worry about oil consumption until after about 3200 km of engine operation. It will take this long for full ring seating.
3. Frequently check the engine oil and other fluid levels.
4. Change the engine oil and filter after nearly 3200 km of driving. This will help remove any particles in oil.
5. Inform the customer about any problems not corrected by the engine repair or overhaul.

For example if the radiator is in poor condition e.g., has been previously repaired or was filled with rust, tell the customer about the consequences of not correcting the problem. The problem may upset the engine performance or reduce engine service life. Having the customer sign a release form will protect the repair shop if the un-serviced part fails.

THEORY QUESTIONS

1. Describe the procedure for fitting pistons to a cylinder.
2. Piston ring end gap can be increased by _____.
   (*a*) Installing oversize rings
   (*b*) Sleeving the cylinder
   (*c*) Filling the ring to make gap larger
   (*d*) Knurling the piston.
3. When installing insert bearings, should their backsides be oiled or dry?
4. The most common cause of premature bearing failure is _____.
   (*a*) Misassembly (*b*) Overheating
   (*c*) Dirt (*d*) Lack of lubrication.
5. Which side of the head gasket should face up when it is not marked UP or FRONT?
6. Explain how the main bearings should be checked and fitted to the crankshaft.
7. How is Plastigauge used to determine bearing oil clearance?
8. Describe the procedure for checking thrust-bearing clearance?
9. How should the connecting rod side clearance be measured and corrected?
10. Describe the procedure used to install the connecting rod assembly in the engine.
11. What are ten general rules to be kept in mind during engine assembly?
12. Describe the procedure followed to install and torque the cylinder head.
13. The cylinder head should be tightened in what general sequence?
    (*a*) The four outside bolts first, then from the centre out
    (*b*) From the outside bolts to the inside bolts
    (*c*) From the inside bolts to the outside bolts
    (*d*) Starting at the front of the engine and ds0. torquing bolts from front to rear.
14. The oil pump gasket should be sealed with silicone sealer to prevent leakage. *True or False?*
15. How is the clutch and bell housing installed?
16. Describe the procedure followed to install a cylinder head gasket.
17. Describe the procedure used to install an oil pan.
18. If dowels are not provided in the block, what should you do?
19. Head bolts should have a drop of oil on their threads. *True or False?*
20. Describe the general procedure used to torque cylinder head bolts.
21. Describe the general procedure used for installing an intake manifold.
22. Cam bearing _____ are _____ and must be installed in their exact original locations.
23. Hydraulic lifters can be adjusted with the engine off or with the engine running. *True or False?*
24. Describe the method of installing OHC valve adjusting shims.
25. What happens if the thermostat is installed upside down?
26. Describe the engine break-in procedure.

27. List the common rules to follow during engine installation.

28. How long is normal break-in period?

29. Why should you turn an engine by hand after installing the valve train?

30. When installing an engine, why should the flywheel be slightly lower than the front of the engine?

31. Turning the oil pump before starting the engine should be done _____ _____ _____.
    (*a*) To lubricate engine bearings
    (*b*) To lubricate valve train components
    (*c*) To supply oil to the camshaft
    (*d*) All of the above.

32. The Mechanic A says you should use thread locking compound when installing oil pump mounting bolts. Mechanic B says that thread locking compound should not be used when installing oil pump. Who is right?
    (*a*) A only (*b*) B only
    (*c*) Both A and B (*d*) Neither A nor B.

33. Mechanic A says that the oil pan gasket can be adhered to the oil pan itself. Mechanic B says that the oil pan gasket can be adhered to the engine block mating surface. Who is right?
    (*a*) A only (*b*) B only
    (*c*) Both A and B (*d*) Neither A nor B.

34. Mechanic A says that some engine manufacturers recommend using new head bolts when installing an engine's cylinder head. Mechanic B says that to prevent compression leakage, most engine manufacturers recommend using the old head bolts during cylinder head installation. Who is right?
    (*a*) A only (*b*) B only
    (*c*) Both A and B (*d*) Neither A nor B.

35. You should use a _____ to check an intake manifold sealing surface for warpage
    (*a*) Dial indicator and straightedge
    (*b*) Straightedge and flat feeler gauge
    (*c*) Dial indicator and feeler gauge
    (*d*) Outside caliper and straightedge.

❑❑❑

16 Engine Maintenance and Troubleshooting

16.1 INTRODUCTION

A set regular and scheduled maintenance is important for reliability, and life of an automotive engine. Some areas of engine preventive maintenance are discussed first. This is then followed by engine troubleshooting here.

16.2 ENGINE MAINTENANCE

Cleanliness

Engine maintenance is an important aspect of keeping the vehicle in good condition. Engine maintenance can also point to future potential engine problems.

A clean engine is easier to work on as compared to a dirty, greasy engine. *Fig. 16.1.* It presents a nicer appearance, and problems such as leaks, are easily spotted. A new engine can be kept clean by periodically washing it with detergent and water. However, pay particular attention to keep water out of the distributor, carburetor, and electronic components. If the engine is too dirty, use a water-soluble grease emulsion to clean it.

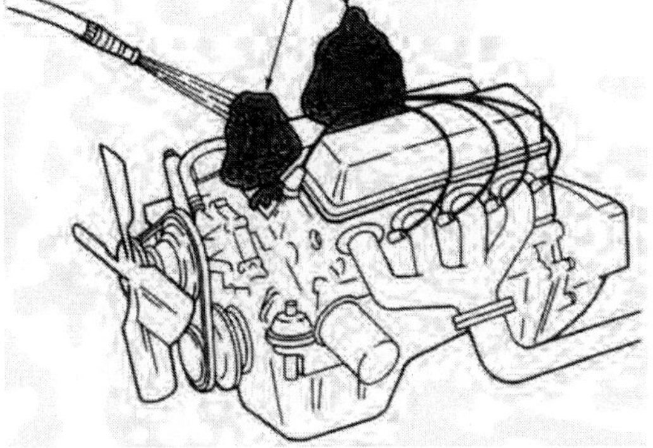

Fig. 16.1 *Clean engines are easier to work on.*

To apply these commercial cleaners, start the engine, allow it to warm up, turn it off, and spray the cleaner over it with a pump sprayer or apply it with a brush. Allow the cleaner to work for a few minutes. Follow up with a strong spray of water to wash away the emulsified grease and dirt. Too dirty engines require steam cleaning. Strong detergents or steam should be used with due care as the cleaner can harm seals, relays, wiring, air-conditioning systems, etc.

Oil and Oil-Filter

Carefully checking and changing the oil is a single most important part of engine maintenance. Checking the oil takes only a few moments. But frequently drivers do not check the oil level. It is a good practice to check the oil every second or third fuel fill-up. To check the oil, turn the engine off and let it stand for some time to allow the oil to drench down to sump. Then, raise the hood and pull out the oil *dipstick and wipe* the oil on it off. Reinsert the dipstick completely and then remove it again, *Fig. 16.2.* The level of the engine oil will now show on the dipstick. The dipstick has index marks, as shown in *Fig. 16.3* that identify the minimum and maximum amount of oil required for the engine.

Manufacturers suggest changing the oil at regular intervals from 4800 to 19000 km. To help the engine to last its predicted life, change the oil at the manufacturer's specified time interval and avoid exceeding this oil change requirements. It is not enough to rely on the oil filter keeping the oil clean. The oil filter does not filter out moisture or acid, two of the most corrosive elements suspended in used oil.

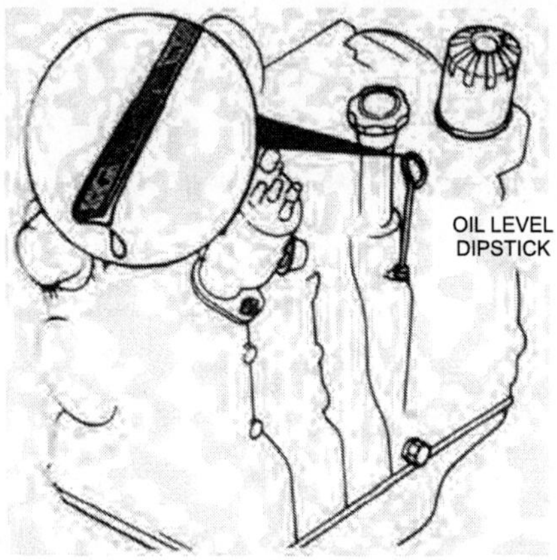

Fig. 16.2 *Checking engine oil level. (Mazda of North America, Inc.)*

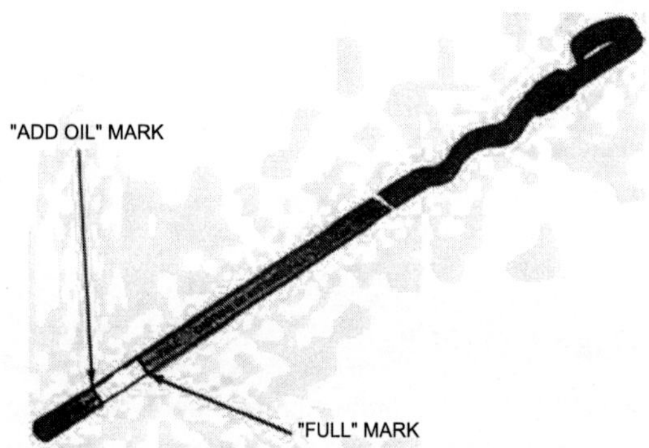

Fig. 16.3 *Dipstick oil level marks. (Chrysler Corp.)*

Use good-quality, manufacturer-recommended oil during oil changes. Some engines use mutli-grade oil such as 10-30, 10-40, etc., while some engines use single grades, e.g. 20-, 30-, etc. weight oils. Colder regions of the vehicle's area of operation require lightweight, while warm areas require heavier-weight oils.

Oil and Filter Change

Follow the steps listed below to change oil:

1. Run the engine to warm it up to operating temperature and then stop it.
2. Remove the oil filler cap, *Fig. 16.4.*
3. Raise the engine on a hydraulic hoist.
4. Remove the oil drain plug and drain oil into a waste oil disposal container.
5. Remove the oil filter, *Fig. 16.5.*

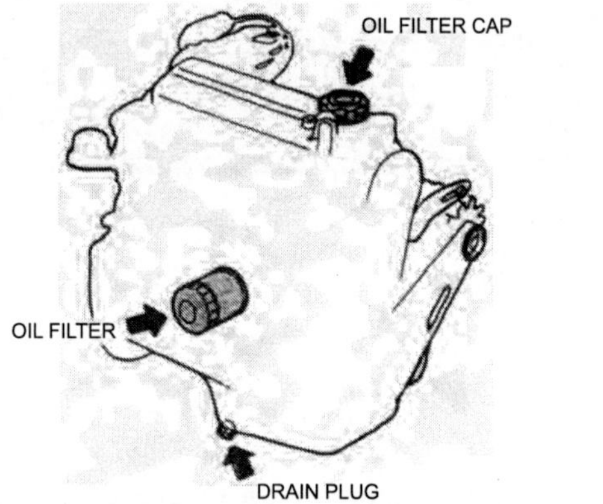

Fig. 16.4 *Oil filter cap. (Mazda of North America, Inc.)*

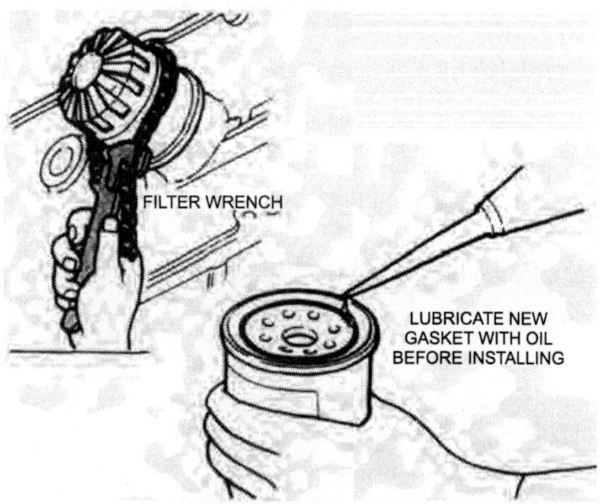

Fig. 16.5 *Removing and installing the oil filter. (Mazda of North America, Inc.)*

6. Let the oil to continue to drain until it stops dripping.
7. Lightly coat the filter seal with oil and install the new filter.
8. Replace the drain plug
9. Lower the hoist and add new, recommended, oil.
10. Replace the oil filter cap.
11. Start the engine and check for leaks.

Some manufacturers recommend that the filter be replaced every other oil change. However, it is a good practice to replace the oil filter at every oil change.

The procedure for oil change in turbocharged engines is little different from the one used for non-turbo vehicles. After changing the oil, remove the centre wire on the distributor and crank the engine until either the oil light goes off or the oil gauge indicates oil pressure. Then, reconnect the centre wire of the distributor. This procedure primes the oil passages to the turbocharger's bearings. If this step is not performed, the bearings in the turbocharger may be damaged when the engine is started.

Checking Coolant Level

The cooling system should be checked every two or three fuel fill-ups. On most vehicles it is not necessary to remove the radiator cap to check the coolant level. The amount of coolant in the recovery/reservoir tank shows the correct amount of coolant in the radiator, *Fig. 16.6.*

The *coolant* is antifreeze, a lubricant, and a corrosion inhibitor. Adding water periodically dilutes the coolant so that eventually you must replace it. The average amount of time for replacing the coolant is between 1 to 2 years because additives wear out and become corrosive. Always use the recommended coolant. This is particularly important on aluminium engines, as they will corrode if a wrong type of coolant is used.

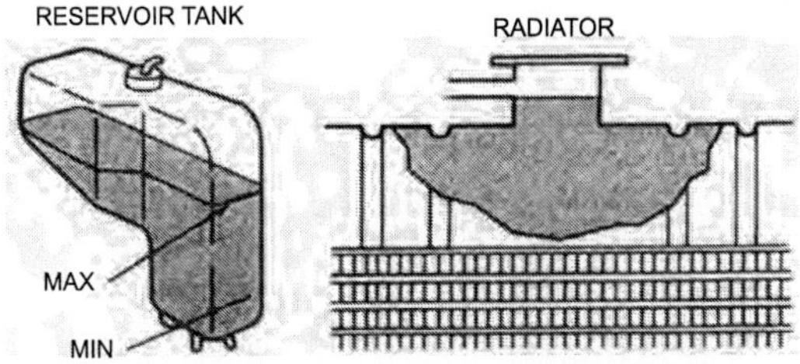

Fig. 16.6 *Checking coolant level. (Nissan Motors)*

If there is any leak found around the hoses, including the heater hoses, replace the hose. Check engine drive belt(s) for adjustment and wear, too. Modern cooling system operates at about 91°C. Always treat cooling system with respect. Any carelessness in its handling can scald or burn you.

Do not remove the coolant recovery cap while the coolant is boiling and do not remove the radiator cap while the engine and radiator are hot. This is necessary because, scalding coolant and steam can be blown out under pressure when either the coolant recovery cap or the radiator cap is removed, *Fig. 16.7.*

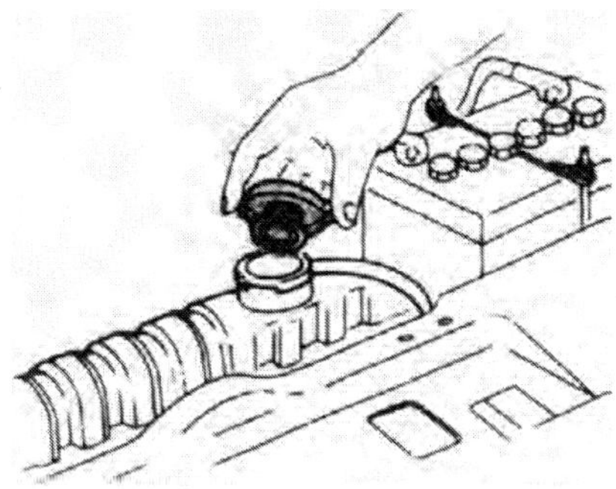

Fig. 16.7 *Be careful of hot radiators. (Isuzu Motors)*

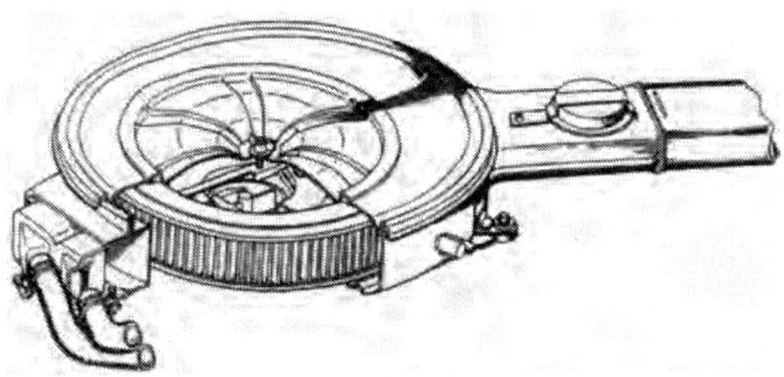

Fig. 16.8 *Air filters. (Mazda of North America, Inc.)*

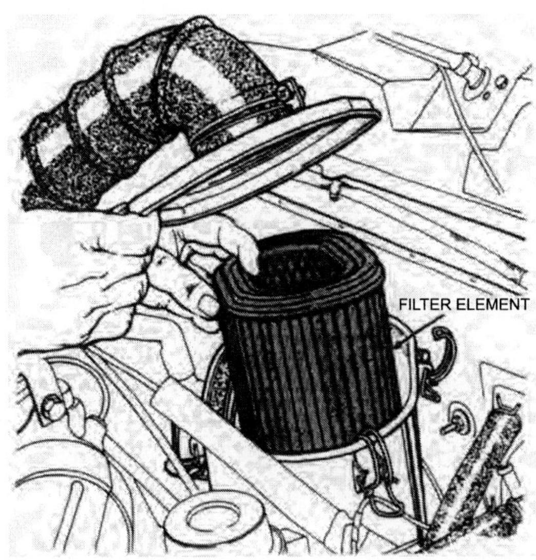

Fig. 16.9 *Replacing air filter. (Chrysler Corp)*

Checking Air Filters

The *air cleaner* is generally a paper element type filter sitting atop the carburetor or fuel injection unit, *Fig. 16.8.* The filter is very important because it cleans the air going into the combustion chambers. When it gets dirty or clogged, less air will pass through. It hampers the engine performance due to lack of enough clean air being available in cylinders. Under normal conditions, replace the air filter at 16000 to 20000 km, *Fig. 16.9.* Late model vehicles have larger filters that permit 48000 km intervals between filter changes.

Checking Fuel Filters

Replace gasoline in-line filter at about 16000 km intervals. This type filter can be anywhere between the fuel tank and the carburetor or fuel injection unit. *Figure 16.10* shows some of the locations of the fuel filter. The tank filter does not need changing unless someone puts foreign matter in the gas tank.

Diesel fuel filters are normally between the fuel pump and the fuel injection pump, *Fig.16.11.*

Changing a diesel fuel filter is slightly more complicated than changing an in-line gasoline filter. The diesel fuel system must be completely full at all times to operate properly. After inserting a new diesel filter element into the filter, the fuel system has to be refilled or primed. This is done with a small hand pump or by the regular fuel pump of the vehicle.

16.3 ENGINE TROUBLESHOOTING

Engine troubleshooting is a process of studying the symptoms of the existing trouble customer complaint in a vehicle engine, logically determining the possible causes and needed corrections. The process is supported throughout by systematic analysis, deduction, and elimination. This is the detective work that a mechanic must do when presented with a case of engine trouble.

Defect in one part or a misalignment in one system could have a definite relationship to problem in another area. Troubleshooting a customer complaint of a problem requires thorough understanding of the functions of the entire automobile engine to be kept in mind at all times. The main difficulty in troubleshooting is isolating the specific cause of the problem. Once this is identified, any qualified automotive mechanic can repair it.

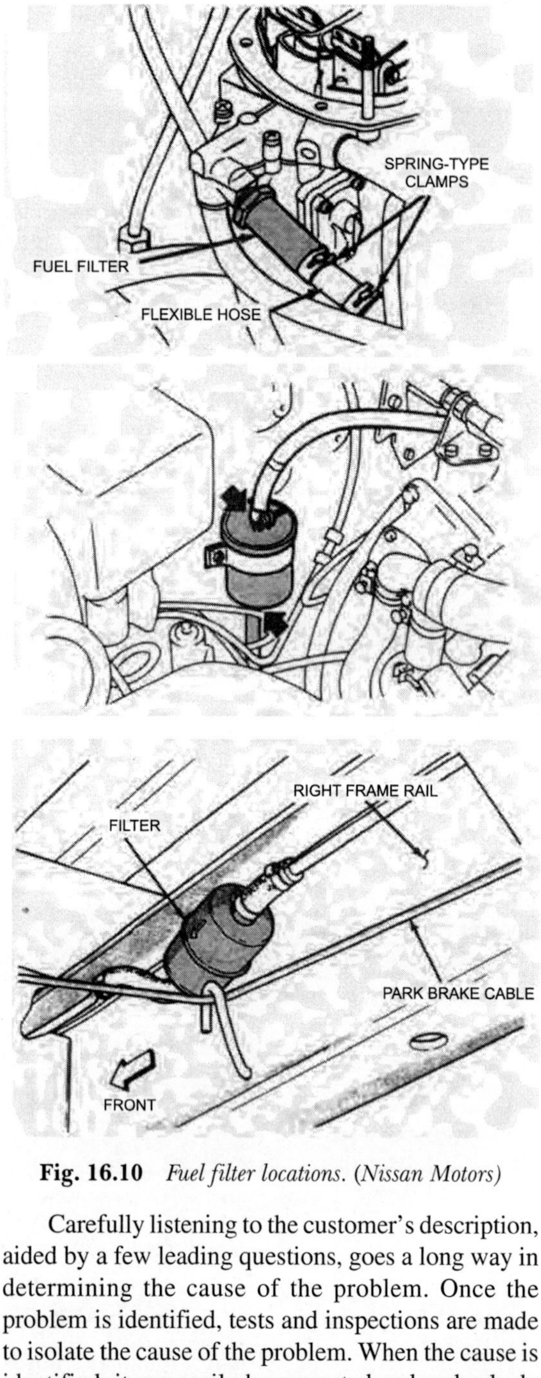

Fig. 16.10 *Fuel filter locations. (Nissan Motors)*

Carefully listening to the customer's description, aided by a few leading questions, goes a long way in determining the cause of the problem. Once the problem is identified, tests and inspections are made to isolate the cause of the problem. When the cause is identified, it can easily be corrected and rechecked.

Troubleshooting charts are often available, which help to diagnose the cause of problems. These charts are presented in the shop manuals for each vehicle. They are more specific than a general trouble-

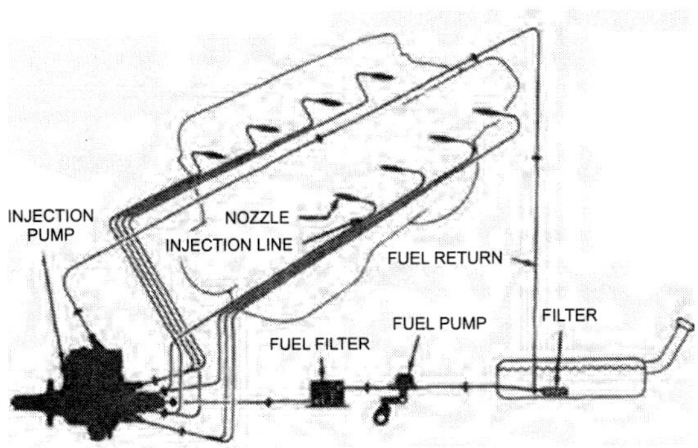

Fig. 16.11 *Diesel fuel filter location. (General Motors Corp)*

shooting chart designed for any vehicle. Most manufacturers publish service bulletins for their service technicians. These bulletins describe problems that occur frequently in their product line and provide details on correction procedures. Keeping up-to-date on service bulletins is a good method for a dealership mechanic to make quick problem diagnosis and repairs to their product line vehicles.

Aside from the special problems of diesel fuel injection systems diagnosis correcting, troubles in the diesel engine are similar to troubles in the spark-ignition engine. Both types of engines could have similar problems with valves, pistons, rings, bearings, and other engine components.

Troubleshooting in its most basic form consists of "shorting out" spark plugs in gasoline engines to locate a missing cylinder. In technically advanced form, it involves the use of an elaborate diagnostic computer type engine analyzer, *Fig. 16.12.* Many modern vehicles feature a plug-in diagnostic connection to allow plug-in of a special tester into the circuit to quickly locate an existing trouble in the engine.

In some cases, trouble areas can be located by visual inspection and manual checks. However, this method of problem solving has to be supplemented by tests made with available *engine testing equipment,* so as to pinpoint the trouble sources quickly and with greater accuracy. Typical *engine analyzers* are shown in *Figs. 16.12, 16.13,* and *16.14.* These analyzers can test everything from condition of the spark plugs to 4-gas emissions (CO, CO_2, O_2, and hydrocarbons).

Often, the tester is actually a *diagnostic computer* that provides visual and/or "voice" instructions, *Fig. 16.13.* The computer is programmed with test specifications. In operation, it displays the vehicle's actual performance readings on a screen. A printout of these readings is also furnished for the technician/mechanic and for the vehicle owner to substantiate the necessity for repairs. This specialized equipment is already covered in chapter #3 on "Service Information-Tools and Safety."

Fig. 16.12 *The latest computerized engine and emissions analyzers provide the diagnostic help needed to pinpoint causes for engine performance complaints and for excess emissions. The operator can "freeze" data on screen, display readouts of test results, and obtain emissions printouts. (Alien Test Products Div., Alien Group)*

Effect of Emission Controls

Engine troubleshooting has become more difficult due to a number of emission control systems that have been added or incorporated into the various engine systems. Government has set legal emission limits of carbon monoxide, nitrogen oxides, and hydrocarbons. Therefore modern automotive engines must conform to comply with these stringent limits. This necessitates the latest test equipment that includes necessary instruments for checking chemical content of the exhaust gases.

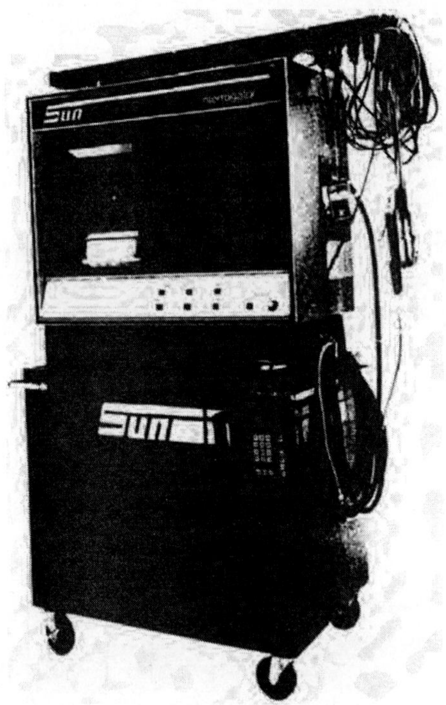

Fig. 16.13 *This diagnostic computer analyzes six major engine systems. Test instructions are visual and oral. A remote control keypad allows the operator to work in the vehicle or under the hood. The computer identifies problems, possible causes, and logical solutions. (Sun Electric Corp.)*

Fig. 16.14 *A 4-gas infrared analyzer uses emissions analysis to determine whether the cause of engine performance problem is in the fuel system or in the ignition system. (Peerless Instruments Co.)*

Plug-in Diagnosis

Early diagnostic systems used special test equipment that plugged into a diagnostic connector on the vehicle. A series of programmed tests could be made to check the condition of various units and systems on the vehicle. Volkswagen developed this method of engine diagnosis. The original setup consisted of a computer, a card reader, a hand control unit, and an umbilical cord for connecting the test equipment to the vehicle's diagnostic connection, a tachometer, and a timing light.

A more sophisticated diagnostic arrangement was used on American cars. Some Chrysler vehicles utilized an electronic spark advance computer that was programmed to monitor a number of different systems. If a problem was detected in a monitored circuit, its fault code was stored for display on a diagnostic readout tool. *Fig. 16.15.*

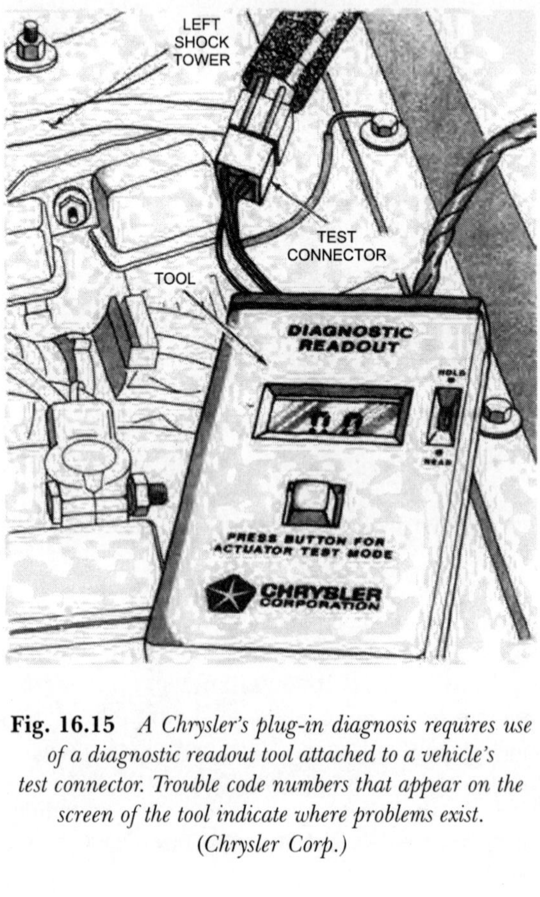

Fig. 16.15 *A Chrysler's plug-in diagnosis requires use of a diagnostic readout tool attached to a vehicle's test connector. Trouble code numbers that appear on the screen of the tool indicate where problems exist. (Chrysler Corp.)*

To make a series of tests, the readout tool was attached to a test connector located under the hood at the left shock tower. The readout was used to check out three different modes:

1. Diagnostic test mode that was used to retrieve fault codes stored in the computer.
2. Circuit actuation test mode that was used to check out specific system components.
3. Switch test mode that was used to check certain switch circuits.

As code numbers appeared on the readout tool, *Fig. 16.16,* they were recorded for a later reference to a listing of problem areas related to those particular code numbers. After the problem area was corrected, the electronic spark computer would cancel the fault codes.

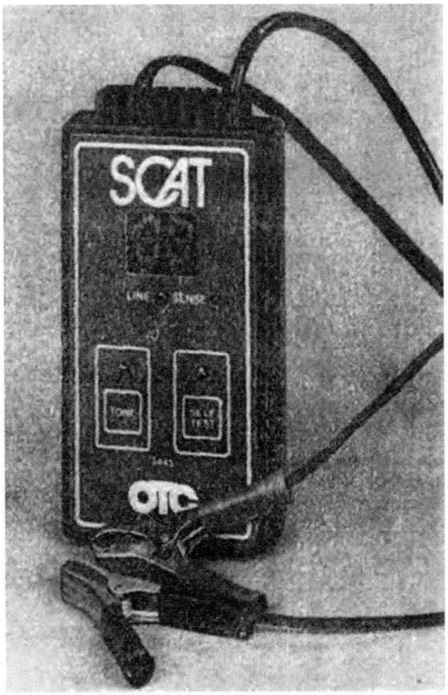

Fig. 16.16 *Various diagnostic testers are available that provide access to the trouble codes stored in on board computers.*
(OTS Tools and Equipment)

Self Diagnosis

The enhancement from plug-in diagnosis is *self-diagnosis.* On vehicles so equipped, the ECU monitors operating conditions for possible malfunctions. It compares system conditions against standard limits. Two-digit numerical trouble code is stored in the computer memory.

When a malfunction is detected by the self-diagnostic system, the trouble codes can then be retrieved by the service technician/mechanic. The technician/mechanic then consults trouble code diagnostic charts to determine what service is necessary to correct the trouble found by the system.

16.4 TROUBLESHOOTING CHARTS

Troubleshooting charts have been a traditional means of relating specific symptoms of trouble to possible causes. A number of engine problems are covered in the following trouble shooting charts.

Most engine complaints can be grouped under a few basic headings. These include: engine will not crank; engine cranks but will not start; engine runs but misses; engine lacks power, acceleration, or high speed performance; engine over heats; engine uses excessive oil or fuel; or engine is noisy. Charts that follow, list the engine troubles, together with possible causes, checks to be made, and corrections necessary.

1. Engine Does Not Crank

If the engine will not crank when trying to start it, make sure the shift lever is in neutral (N) or park (P). If the vehicle is equipped with a manual transmission, ensure that the clutch pedal is depressed. Check the battery and cables. The most likely cause of this condition is a run-down battery. But there could be other causes, too, refer chapter on **"Starting System Diagnosis and Service."** Turn on the headlights, and try cranking. There could be five possibilities.

- **No Cranking, No Lights.** This is probably due a completely dead battery. It could be caused by a bad connection at the battery or starting motor or an open fusible link that indicates a short or ground in the system.
- **No cranking, but Lights go out as the ignition key is turned to START (ON).** Usually, this indicates a bad connection at the battery. This

could also be due to an almost discharged battery. Try to wiggle the battery connection to see if it helps.

- **No Cranking, and the Lights Dim only Slightly when trying to Start.** This indicates a problem with the starting motor. The pinion may not be engaging with the flywheel ring gear. If the starter motor armature spins, then the overrunning clutch is slipping.
- **No Cranking, and Lights Dim Heavily when trying to Start.** Most likely this is due to rundown battery. It could be caused by low temperature, and engine oil offering high viscosity resistance, even though the battery is in acceptable condition. The starter motor or engine could also be jammed, or locked.
- **No cranking, and Lights Stay Bright.** Pay attention to listen an audible click to check if the magnetic solenoid is pulling the plunger in. If nothing happens when trying to start the engine, inspect the solenoid. Connect one end of a jumper wire to the solenoid bat-terminal. Connect the other of the jumper cable to the small terminal on the solenoid that is connected to the ignition switch. If nothing happens, the trouble is in the solenoid. If the solenoid and starter motor work with the jumper cable connected, the trouble is in the ignition switch, the transmission switch, or the wires connecting them.

The possible causes and corrective steps are:

- **Run-down Battery:** Recharge or replace battery; start engine with jumper cables and battery.
- **Corroded or loose battery connections:** Clean and tighten connections.
- **Loose engine and/or battery ground connection:** Set the ground connection.
- **Faulty ignition coil:** Replace coil.
- **Faulty starter or solenoid:** Repair or replace as necessary.
- **Defective ignition switch:** Replace.
- **Starting circuit open:** Find and eliminate the open; check for dirty or loose cables.
- **Starter-motor drive jammed:** Remove starter motor and free the drive.
- **Starter motor jammed:** Remove starter motor for service.
- **Engine jammed:** Check engine to isolate trouble.
- **Transmission not in neutral or neutral switch out of adjustment:** Check and adjust neutral switch if necessary.
- **Seat belt not fastened or interlock defective:** Fasten seat-belt and check interlock.
- **See also causes listed under item #3:** Driver may have rundown the battery trying to start the engine.

2. Engine Cranks Slowly But Does Not Start

This is very likely due to a run-down or defective battery. The battery is unable to turn the starter motor at normal speed. Low temperature could also be a factor here too.

Possibility is that the driver may have run the battery down trying to start the engine. Some abnormal condition in the engine, or the fuel or ignition system, could be preventing normal starting. The driver kept trying to start the engine, eventually running down the battery.

Try to jump start he engine. If the engine starts, the battery probably was at fault. If the engine still does not start, check the fuel system and ignition system. The possible causes and corrective steps are:

- **Partly discharged battery:** Recharge or replace battery; start engine with jumper battery and cable.
- **Defective starter motor:** Repair or replace starter motor.
- **Bad connections in starter circuit:** Check for under size, loose, or dirty cables; replace or clean and tighten connections.
- **See also causes listed under item #3:** Driver may have rundown the battery trying to start the engine.

3. Engine Cranks at Normal Speed But Does Not Start

The battery and starter motor are in normal condition. The trouble is probably in the ignition or fuel system. Over-choking could be the reason. Failure to start a hot engine may be caused by a defective choke that fails to open properly after the engine warms up. This condition causes flooding of the engine with gasoline.

Wide open the throttle valve and crank the engine. If the engine does not start, perform a spark test. If a good spark occurs, the ignition system operates normally. The ignition or valve timing may be off. Check the fuel system. On a carbureted engine, prime the engine by operating the accelerator pump several times and then crank the engine.

The possible causes and corrective steps are:

- **Defective ignition system:** Perform spark test; check timing, ignition system.
- **Empty fuel tank:** Fill tank.
- **Defective fuel pump or over-choking:** Prime the fuel system; check accelerator pump discharge, fuel pump, fuel line, choke, carburetor.
- **Air leaks in intake manifold or carburetor:** Tighten mounting fasteners; replace gaskets as needed.
- **Engine problem:** Check compression or leakage, valve action, valve timing.
- **Ignition coil or resistor burned out:** Replace the defective component.
- **Plugged fuel filter:** Clean or replace if necessary.
- **Plugged or collapsed exhaust system:** Replace collapsed parts.

In general when the engine does not start, check the following:

- Weak battery.
- Corroded or loose battery connections.
- Loose engine and/or battery ground connection.
- Defective ignition switch.
- Defective starter or solenoid.
- Defective ignition coil.
- Moisture on high-tension wiring and distributor cap.
- Faulty ignition distributor.
- Defective high-tension wiring.
- Incorrect spark plug gap.
- Incorrect ignition timing.
- Dirt, water, or ice in fuel line or carburetor.
- Faulty fuel pump.
- Clogged air filter.
- Flooded carburetor.
- Sticking choke.
- Maladjusted choke.
- Incorrect carburetor float setting.
- Empty fuel tank.
- Defective neutral starting switch.
- Defective emission control unit.
- Faulty engine computer control system.

4. Engine Starts But Does Not Run

If the engine starts and runs for a few seconds, the fuel system is probably at fault. It is not delivering fuel to the engine. Ensure that there is fuel in the tank. Disconnect the fuel line to the carburetor/fuel injection system. Also disconnect the primary lead from the negative terminal of the ignition coil to stop the engine from starting. Hold a safe container under the fuel line to receive fuel, and crank the engine to check if the fuel is delivered, *Fig. 16.17.* If fuel is not delivered, there is no fuel in the fuel tank, the fuel pump is defective, or the fuel line is clogged. If the fuel is delivered, the fuel filter is probably at fault, the automatic choke is not working properly, or there may be air leaks into the intake manifold or carburetor.

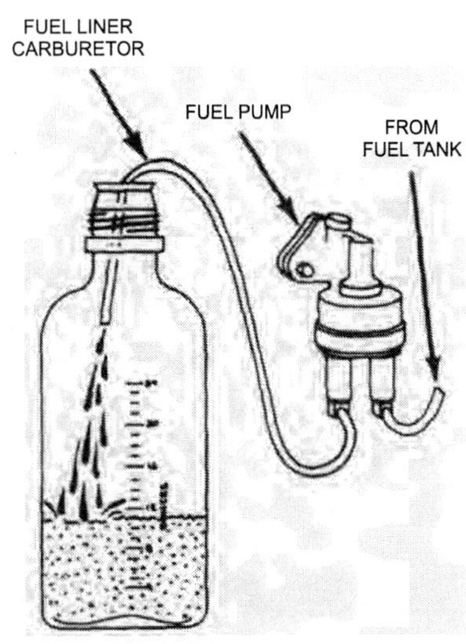

Fig. 16.17 *To check whether the fuel system is delivering fuel. Disconnect the fuel line from the carburetor, place the fuel line in a safe container, and crank the engine. Fuel should discharge from the line.* (*Ford Motors*)

If the fuel and ignition systems are good, check for a restricted exhaust system as it can build up backpressure, *Fig. 16.18*. The inner layer of a laminated exhaust pipe could collapse, *Fig. 16.19*. Any restriction can prevent normal exhaust and intake so that the engine will not start, or will not run much above idle. The possible causes and corrective steps are:

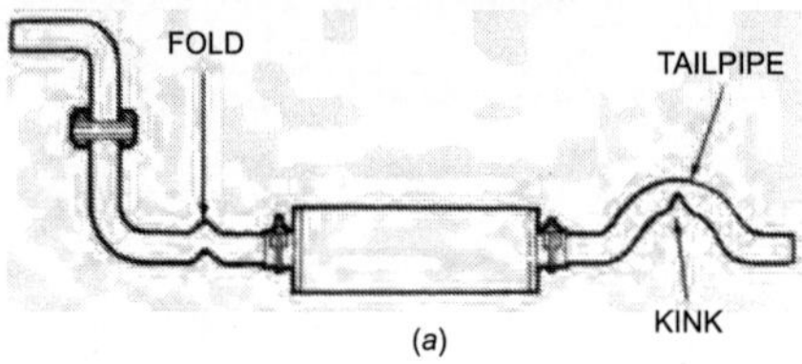

(*a*)

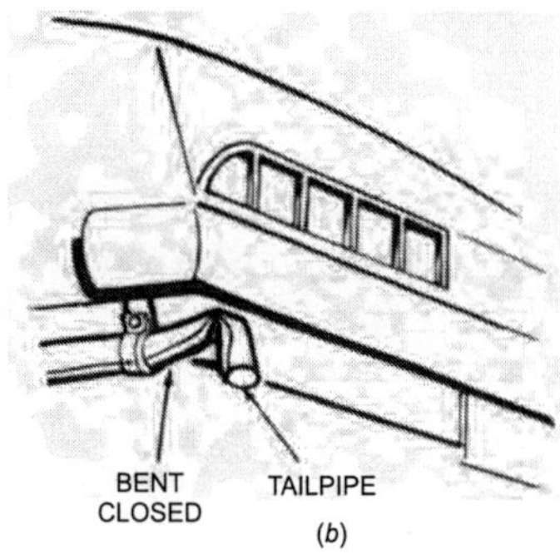

(*b*)

Fig. 16.18 *A clogged or restricted exhaust system may cause the engine, hot or cold, to lack power. (Ford Motors)*

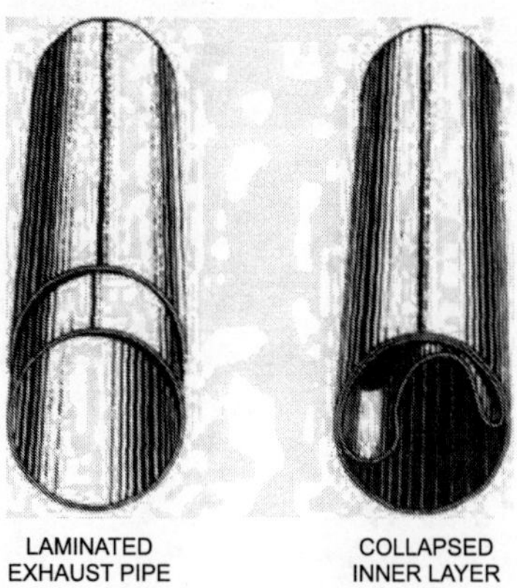

Fig. 16.19 *Collapse of the inner layer of a laminated exhaust pipe will prevent normal intake and exhaust. (ATW)*

- **Back pressure in exhaust system:** Replace or repair the clogged/collapsed part.
- **Obstructed air filter:** Replace the filter element.
- **Restricted or inadequate supply to fuel pump:** Clear the fuel lines, prime the system.
- **Weak fuel pump or volume:** Recondition fuel pump, replace necessary spring(s).
- **Clogged fuel filter:** Clean or replace filter element.
- **Misaligned distributor rotor:** Correct the rotor setting.

5. Hard Starting

Hard starting could be caused by:

- **Maladjusted choke:** Correct choke setting.
- **Faulty fuel pump:** Recondition or replace fuel pump as needed.
- **Clogged fuel filter:** Clean or replace the filter element as necessary.
- **Maladjusted or faulty carburetor:** Recondition the carburetor, correct the adjustments.
- **Clogged electronic fuel injection system injectors:** service the fuel injection system.
- **Air leak into manifold or vacuum hoses:** Service the manifold, replace gaskets and vacuum hoses as necessary.
- **Obstructed air filter:** Clean or replace the filter element as required.
- **Defective exhaust gas Recirculation (EGR) valve:** Replace EGR valve.
- **Sticking positive crankcase ventilation (PCV) valve:** Service the PCV valve, replace if needed.
- **Faulty ignition coil:** Replace.
- **Faulty spark plugs or incorrect gap:** Clean spark plugs, set gap or replace if needed.
- **Incorrect ignition timing:** Set ignition timing.
- **Moisture on high tension wiring and distributor cap:** Clean-dry the wiring and distributor cap and reset.
- **Incorrect valve timing:** Set the valve timing right.

6. Rough Idle

If the engine idles roughly but runs normally above idle speed, the idle speed and idle mixture may be incorrectly adjusted. A rough idle could also be due to other causes, e.g., a loose or disconnected vacuum hose or a PCV valve that is stuck open.

The possible causes and corrective steps are:

- **Incorrect idle adjustment:** Readjust idle mixture and speed.
- **Obstructed air filter:** Clean or replace filter element as needed.
- **Faulty evaporative emission control (EEC) system:** Service EEC system.
- **Improper feedback system operation:** Service the feedback system.
- **Air leak into manifold or vacuum hoses:** Service the manifold connection, and/or connections and replace if defective.
- **Backpressure in exhaust system:** Check and repair or replace any defective parts.
- **Faulty EGR valve, stuck open:** Replace the EGR valve.
- **Faulty PCV system operation, PCV valve stuck open:** Replace the valve.
- **Faulty ignition coil:** Repair or replace coil as needed.
- **Defective distributor rotor or cap:** Service or replace the component if necessary.
- **Leaking engine valves:** Service the valve operating system components and replace the parts as needed.
- **Incorrect valve timing:** Correct the valve timing.

7. Engine Stalls

If the engine starts and then stalls, note whether it stalls before or as the engine warms up, after idling or slow-speed driving, or after high-speed or full-load driving. If the PCV valve becomes clogged or sticks, it will cause poor idling and stalling. Stalling may also be caused by improper operation or adjustment of idle solenoid.

In general, the causes of engine stalling can be:

- **Low idle speed:** Correct the speed setting.
- **Incorrect choke adjustment:** Correct the setting.
- **Too lean or too rich idle mixture setting:** Correct the setting.
- **Low carburetor float level:** Correct the float level setting.
- **Inoperative carburetor float needle valve:** Service or replace the valve.
- **Air leak into intake manifold or vacuum hoses:** Correct the causes of air leakage into manifold or vacuum hoses.
- **Loose engine and/or battery ground connection:** Fix the ground connection.
- **Faulty ignition coil:** Repair or replace the coil.
- **Worn distributor rotor:** Replace the rotor.
- **Moisture on high-tension wiring and spark plugs:** Clean dry the high-tension wiring and spark plugs.
- **Incorrect valve lash for solid lifters:** Correct the lash.

Engine Stalls Cold or as It Warms Up: Stalling of a cold engine could be due to an improperly set fast or slow idle or improper adjustment of the idle-mixture screw in the carburetor. This could also be due to a low carburetor float setting or to insufficient fuel entering the carburetor as a result of defective thermostatic air cleaner, dirt or water in the fuel lines or filter, a defective fuel pump, or a plugged fuel-tank vent. Carburetor icing may also cause this.

Certain ignition troubles could cause stalling after starting. But if ignition troubles were bad enough to cause engine stalling, they usually would prevent starting. However, burned contact points might allow starting but fail to keep the engine running.

Open primary resistance can also cause stalling. When the engine is cranked, the resistance is bypassed. When the engine starts and cranking stops, the resistance is inserted back into the ignition primary circuit. If the resistance is open the engine will stall.

Engine Stalls as It Warms Up: This condition could result if the choke valve sticks in closed position. The mixture becomes too rich for a hot engine, and the engine stalls. If the manifold heat control valve gets stuck into closed position, the air-fuel mixture might be getting overheated and too lean, causing the engine to stall.

Also, if the hot-idle speed setting is too low, the engine might stall as it warms up due to the idling speed falling too low. Overheating of an engine may also cause stalling due to fuel vapour lock in fuel supply system.

If the damper in the thermostatic air cleaner sticks in closed position, the air-fuel mixture might become overheated and too lean, causing the engine to stall.

The possible causes and corrective steps, when the engine stalls cold or as it warms up, are:

- **Choke valve stuck closed or will not close:** Open choke valve; free or repair automatic choke.
- **Fuel not getting to or through carburetor:** Check fuel pump, fuel lines, filter, float and idle systems.
- **Manifold heat-control valve stuck:** Free valve.
- **Throttle solenoid improperly set:** Correct the solenoid setting.
- **Idling speed set too low:** Increase idling speed to specified rpm.
- **PCV or EGR valve stuck open:** Replace the defective valve.
- Damper in thermostatic air cleaner in cold-air mode stuck closed: Free; repair or replace control motor.

Engine Stalls after Idling or Slow-Speed Driving

This condition could occur if the fuel pump has a cracked diaphragm, weak spring, or defective valve The pump fails to deliver enough fuel for idling or slow-speed operation and the engine stalls. If the carburetor float level is set too high or the idle adjustment is too rich, the engine can "load up" and stall. A lean idle adjustment can also cause the engine to stall. The engine might overheat during sustained idling or slow-speed driving. Under such a condition, the air movement through the radiator core may be insufficient to keep the engine cool within spec limits. Overheating could cause vapour lock in the fuel supply system and stall the engine.

The possible causes and corrective steps, when the engine stalls after idling or low-speed driving, are:

- **Defective fuel pump:** Repair or replace fuel pump.
- **Overheating:** See item 9.
- **High carburetor float level:** Adjust the level.
- **Incorrect idling adjustment:**Correct adjustment.
- **Malfunctioning PCV or EGR valve:** Replace the defective valve.
- **Throttle solenoid improperly set:** Correct the solenoid setting.

Engine Stalls after High-Speed or Full-Load Driving: This condition could occur if the engine accumulates enough heat to cause vapout lock. This can be avoided by shielding the fuel line and fuel pump or use of a less volatile fuel. Failure of the venting or idle-compensating valve in the carburetor may also cause engine stalling after high-speed or full-load operation. Excessive overheating of the engine can also cause engine to stall. The possible causes and corrective steps, when the engine stalls after high-speed or full-load driving, are:

- **Vapour lock:** Use less volatile fuel or shield fuel line.
- **Carburetor venting or idle compensator valve defective:** Check and repair.
- **Engine overheats:** See item 9.
- **PCV or EGR valve stuck open:** Replace the valve.
- **Idle solenoid improperly set-** Correct adjustment.

8. Engine Stalls on Quick Stops

The possible causes and corrective steps, when the engine stalls on quick stops, are:

- **Low idle speed:** Correct idle speed setting.
- **Incorrect choke adjustment:** Correct adjustment.
- **Incorrect carburetor float level:** Adjust the level.
- **Faulty carburetor accelerator pump:** Service or replace defective part.
- **Corrode or loose EEC connection:** Clean and tighten connections- Replace parts as necessary.
- **Clogged electronic fuel injection system injectors-** Service or replace injectors as needed.
- **Defective EGR valve:** Replace valve.
- **Incorrect ignition timing:** Correct ignition timing.

9. Engine Overheats

There may be many "conditions" that result in *engine overheating.* A temperature gauge on the instrument panel or a signal lamp on the panel lights as a warning may indicate the degree of overheating when a specified high temperature is reached.

When troubleshooting the cause for engine overhearing, the first step is to make a careful visual inspection to check if there is any evidence of external leakage of coolant. Check all surfaces of the radiator and its hose connections. Leaks generally cause corrosion that is easily seen. Also check the heater, heater hoses, and hose connections. Usually, engine overheating is caused by loss of coolant due to leaks in the cooling system.

The engine also needs careful inspection. Pay particular attention to the core hole plugs and edges of the cylinder head gasket. Other causes of engine overheating include a loose or broken fan belt, a defective water pump. Clogged water jackets or radiator tubes, blocked radiator airflow, a defective radiator hose, a defective thermostat or a defective fan clutch. Test the radiator cap, *Fig. 16.20.* Test the operation of the thermostat, *Fig. 16.21.* A thermostat "frozen" in the closed position will also cause extreme overheating.

To test for restrictions in the radiator, first run the engine and bring the system to operating temperature. Then shut of the engine and feel the front surface of the radiator. On cross-flow radiators, the radiator should feel hot along the left side and warm along the right side with an even temperature rise right to left and bottom to top. On vertical flow radiators, the radiator should feel warmer at top than at the bottom. Any cold spots would indicate clogged sections.

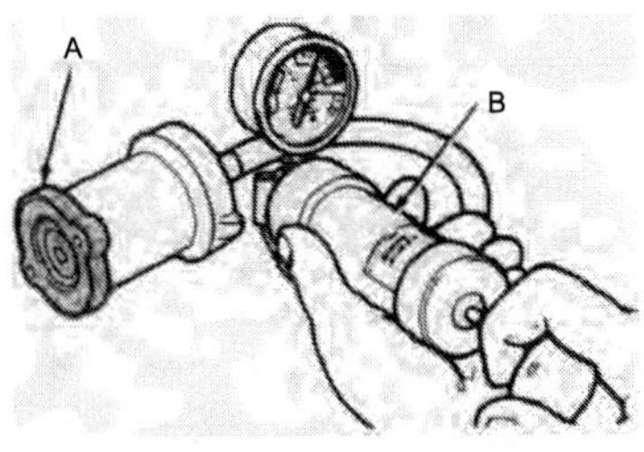

Fig. 16.20 *To test the condition of a radiator cap: A – The cap is attached to a low-range pump-type pressure gauge; B – Test the cap for pressure hold for at least 30 seconds and for pressure release point. These are usually marked on the cap. (Chrysler Corp.)*

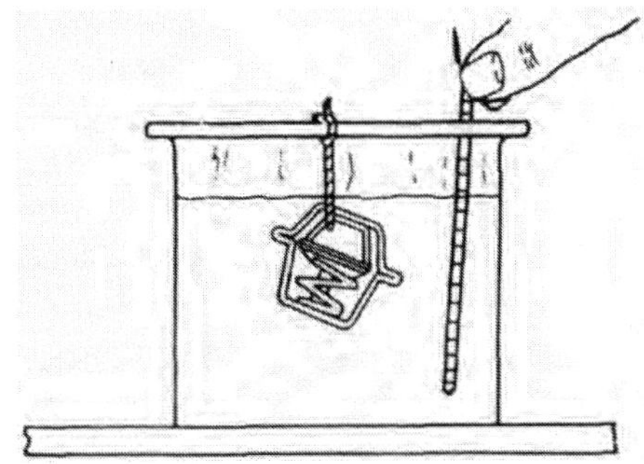

Fig. 16.21 *To test the operation of a thermostat, suspended in a boiling 50/50 solution of antifreeze and water. Record the opening and closing temperatures and compare these readings with the manufacturer's specifications. (Ford Motors)*

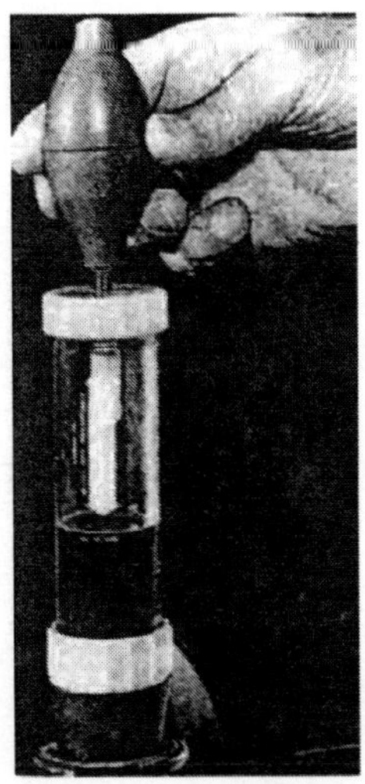

Fig. 16.22 *To test for combustion leakage, use a special tester that attaches to the radiator filler opening. Combustion gases in the sample of coolant will cause the colour of the chemical in the tester to change.*

Running the engine while squeezing the radiator upper hose can check water pump operation. A pressure surge should be felt. Also check for a plugged vent hole in the pump housing.

Not all coolant leakage is external. Severe cases of overheating and coolant loss can result from leaks into the combustion chamber. When the cylinder head is cracked, or there is a blown head gasket, the hot gases of combustion can enter the cooling system. Coolant temperature rises rapidly and boiling takes place.

To test for an internal coolant leak, use a combustion leak tester, *Fig. 16.22.* This is a chemical test. The tester is applied to the filler neck of the radiator. Then, with the engine running, any gas from the combustion chamber that enters the coolant will be drawn into the tester and cause the chemical to change its colour.

Making a compression test of the engine cylinders will also be helpful in checking for internal coolant leaks, especially in case of a blown head gasket. Low readings on adjacent cylinders indicate blown gasket.

Retarded ignition or valve timing, lack of engine oil, overloading the engine, or high-speed, high-altitude, or hot-climate operation can cause engine overheating. A faulty TCS system that prevents vacuum advance in any gear, or a defective distributor vacuum advance can cause engine overheating. Idling and moving in slow-speed traffic, combined with other problems can also cause the engine to overheat. The possible causes and corrective steps, when the engine overheats are:

- **Lack of coolant or low coolant level:** Add coolant; look for coolant leak.
- **Faulty temperature gauge or sending unit:** Replace the gauge or sending unit.
- **Air in cooling system:** Flush the air out.
- **Too much antifreeze in coolant:** Adjust antifreeze and water ratio.
- **Ignition timing retarded:** Adjust timing to specs.
- **Loose or broken fan belt:** Tighten or replace belt.
- **Defective cooling system thermostat:** Replace.
- **Clogged water jackets or radiator core:** Flush and clean.
- **Defective radiator hose:** Replace the hose.
- **Defective water pump:** Service or replace pump.
- **Faulty radiator cap:** Replace cap.
- **Insufficient oil in lubricating system:** Add oil as needed.
- **Defective fan clutch:** Replace fan clutch.
- **High-altitude, hot-climate operation:** Drive more slowly; keep the radiator full with coolant.
- **Valve timing late; slack in timing chain has let the chain jump a tooth:** Set timing, adjust or replace chain.
- **No vacuum advance in any gear:** TCS system or distributor defective, replace as needed.
- **Inoperative electric fan:** Service or replace as required.
- **Slipping fan belt and/or water pump belt:** Tighten or replace belt as needed.
- **Excessive engine friction:** Let the engine run-in as needed.

10. Engine Runs Cold

- **Defective cooling system thermostat:** Replace the thermostat.
- **Defective temperature gauge or warning light circuit:** Service or replace defective part.

11. Loss of power

It is imperative to first determine whether the engine lacks power when both in hot and cold conditions, only when hot, or only when cold. Find out whether the problem developed suddenly or whether the power gradually fell off over a period of several months or many miles of operation.

Engine Lacks Power, acceleration, or high-speed performance Hot or Cold – The fuel system may not be enriching the mixture as the throttle is opened. A faulty accelerator pump or defective main-metering jet of the carburetor could cause such a condition. The fuel system could be supplying too lean or too rich mixture. This condition could be caused by a defective fuel pump, clogged fuel lines, clogged filter, worn carburetor jets or lines, air leaks at the carburetor or manifold joints, and a sticking PCV valve.

Another condition that could cause lack of power when the engine is hot or cold is improper linkage adjustment that prevents full-throttle opening. The ignition system could cause trouble due to incorrect timing, a "weak" coil, reversed coil polarity, and wrong spark-plug heat range. Wrong fuel or oil for the engine could cause reduced performance.

A number of other engine conditions that could also cause loss of power are: carbon deposits, lack of compression (due to faulty valves, rings, worn cylinder walls, worn pistons), and defective bearings.

A clogged (bent or collapsed exhaust or tail pipe, or a clogged muffler or catalytic converter) could develop backpressure that would cause poor engine performance.

An excessive rolling resistance, e.g., dragging brakes, under inflated tyres, misaligned wheels, and excessive friction in the transmission or power train, would absorb engine power and resist engine acceleration and speed. The automatic transmission may not be downshifting, or the torque converter may be defective.

The possible causes and corrective steps, when the engine lacks power, acceleration, or high-speed performance Hot or Cold include:

- **Defective ignition:** Check timing, coil, distributor, condenser, and plugs. The ignition coil may be faulty, distributor shaft may be worn, trigger wheel (reluctor) in distributor may be loose, distributor rotor may be worn or burned, and spark plugs may be fouled or incorrectly gapped.
- **Defective fuel system; secondary throttle valves not opening:** Check carburetor, choke, filter, air cleaner, and fuel pump. Check for dirt or water in fuel line, obstructed air filter, low carburetor float level, lean air fuel mixture.
- **Throttle valve not opening fully:** Adjust linkage.
- **Restricted exhaust:** Backpressure in exhaust – check tailpipe, muffler and catalytic converter; eliminate any restriction.
- **Loss of compression:** Check compression or leakage. Engine valves may be leaking, head gasket may be blown.
- **Excessive carbon in engine:** Service the engine.
- **Defective valve action:** Check with compression, leakage, vacuum testers, and repair or replace parts as necessary. Valve lash may be incorrect (solid lifters), valve springs may be weak, or valve timing may be incorrect.
- **Excessive rolling resistance from low tyre pressure, dragging brakes, wheel alignment, etc.:** Correct the cause of excessive rolling resistance.
- **Heavy lubricating oil:** Use recommended oil.
- **Wrong or bad fuel:** Use correct fuel as recommended.
- **Transmission not downshifting or defective torque converter:** Check transmission and torque converter- service and replace parts as necessary.
- **Improper feedback system operation:** Service the system.
- **Defective or maladjusted emission control system:** Service the system and replace defective parts.

Engine Lacks Power, Acceleration, or High Speed Performance Only when Hot: The engine may be overheating. The automatic choke may not be opening normally as the engine warms up. The manifold heat-control valve may be stuck, or there may be a vapour lock in the fuel pump or line. The possible causes and corrective steps, when the engine lacks power, acceleration, or high-speed performance when Hot, include:

- **Engine overheats:** Check cooling system.
- **Choke stuck partly open:** Repair or replace necessary parts.
- **Sticking manifold heat control valve:** Free the valve.
- **Vapour lock:** Use less volatile fuel or shield fuel line and pump.

Engine Lacks Power, Acceleration, or High Speed Performance Only when Cold: The automatic choke might be leaning out the charge mixture before the engine warms up. The manifold heat-control valve may not be closed and not enough heat reaches the intake manifold. Or the cooling system thermostat may be stuck open to allow the coolant circulation between the engine and radiator even when the engine is cold, delaying warm-up. Sometimes, engine valves may stick open when the engine is cold and become free and operate normally when the engine warms up. The possible causes and corrective steps, when the engine lacks power, acceleration, or high-speed performance only when Cold, include:

- **Automatic choke stuck open:** Repair or replace parts as needed.
- **Manifold heat control valve stuck open:** Free valve.

- **Cooling system thermostat stuck open:** Repair or replace as necessary.
- **Engine valves stuck open:**Free valves; service valve stems and guides as needed.

12. Engine Hesitates on Acceleration

The possible causes and remedies could include:

- **Restricted fuel line:** Clear fuel lines.
- **Clogged fuel filter:** Service or replace filter element.
- **Improper thermostatic air cleaner operation:** Service and replace parts as needed.
- **Maladjusted choke:** Correct adjustment.
- **Faulty carburetor accelerator pump operation:** Service carburetor- accelerator pump.
- **Improper feedback system operation:** Service the feedback system.
- **Faulty ignition coil:** Service or replace coil.
- **Faulty spark plugs:** Replace spark plugs.
- **Incorrect ignition timing:** Service the ignition timing.
- **Leaking engine valves:** Service the valve system as necessary – valve guides, valve seats, valve faces, clearances, springs, etc.

13. Engine Runs But Misses

An engine that misses runs rough. If one or more cylinders fail to fire (miss), the engine is thrown out of power balance that results in running roughness and loss of power. It is often difficult to locate a miss. The miss may skip around. The miss might occur at some speeds and not others. Using an oscilloscope and a dynamometer, as explained in Chapter on **"Ignition System Diagnosis and Service"**, can check missing cylinders. If these testing instruments are not available, then the test can be made as follows. But this test may not be recommended for use with certain types of ignition systems.

Before attempting the test, consult the manufacturer's service manual for ignition-testing procedures. Usually, manufacturers recommend the use of a spark tester.

Disconnect each spark-plug cable in turn, with insulated pliers, to locate the missing cylinder(s). Disconnecting the cable stops the spark from occurring at the spark plug and the cylinder from contributing to engine power. If on disconnecting the cable the engine speed changes, then the cylinder was delivering power before disconnecting the cable.

Hold the spark-plug lead clip close to the cylinder block while the engine is running, *Figs.16.23* and *16.24*. If no spark occurs, there is probably a high-voltage leak caused by *a* bad lead or a burned or cracked distributor cap. In case a good intensity spark occurs, replace the spark plug with a new one in the cylinder or swap plugs between two cylinders. Then reconnect the lead, and determine if the cylinder still misses. If it does, the trouble is probably is caused by defective engine parts.

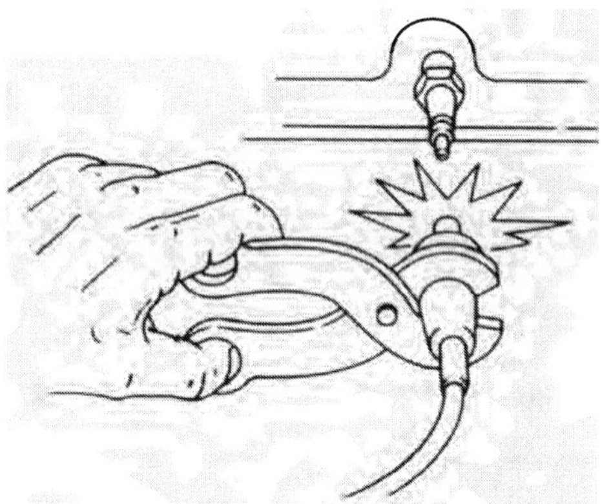

Fig. 16.23 *Checking for spark from spark-plug cable clip to engine block. Do not touch any other ignition-system wire. (American Motors Corp.)*

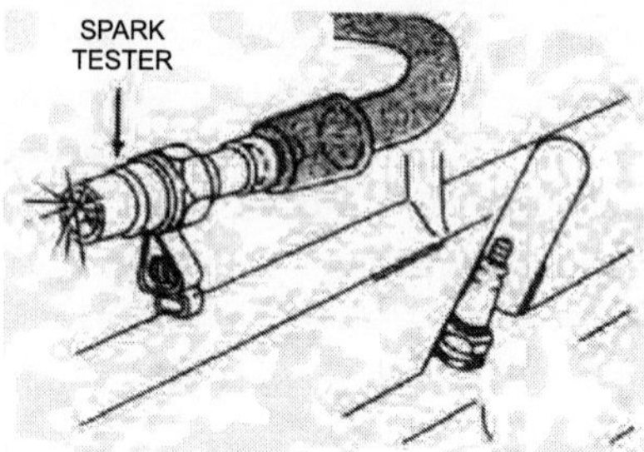

Fig. 16.24 *To make a spark test using a spark tester, attach the spark-plug cable to the tester, and clamp the tester to a good ground. Then crank the engine. (A TW).*

A hard to locate missing cylinder, can be located by performing engine tune-up, as described in chapter on **"Engine Condition Diagnosis and Tune-up."** A tune-up will locate and eliminate various sources that can cause missing, e.g., defects in the ignition system or fuel system, loss of engine compression, sticking or damaged engine valves, overheated engine, stuck manifold heat-control valve, and a clogged or restricted exhaust system.

A cylinder balance test performed with the help of a tachometer and a vacuum gauge can be used to check whether a cylinder is delivering power or is missing. Engine analyzers that include an oscilloscope can be used to perform a cylinder balance test that quickly isolates the missing cylinder. Connect the analyzer to the running engine and push a button or turn a knob to short out the cylinders, in turn, in the firing order. The oscilloscope indicates which cylinder is shorted out. If shorting out a cylinder causes the engine speed or manifold vacuum to drop by 50 rpm or more, *Fig. 16.25,* the cylinder was delivering power. But if little or no change in engine rpm or vacuum is recorded, then the cylinder was missing and not delivering power.

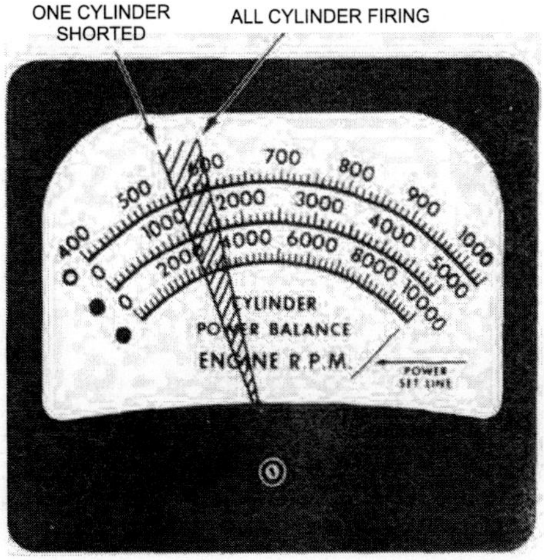

Fig. 16.25 *To check for a misfiring engine, use the engine analyzer to short out each spark plug in turn. If the cylinder is firing, shorting out the cylinder will cause the engine speed to drop at least 50 rpm.* (*Sun Electric Corp.*)

In a diesel engine, a miss usually is caused by inability of the fuel system to deliver an adequate amount of fuel to the cylinder. A clogged nozzle, air in the line, or fuel injection-pump problems, could cause this.

Engine Runs but Misses: One Cylinder

The possible causes and remedies could include:

- **Defective spark plug:** Clean or replace as necessary.
- **Defective distributor cap or spark plug lead:** Replace parts as needed.
- **Valve stuck open:** Service valve and valve guide to free the valve.
- **Broken valve spring:** Replace the spring.
- **Burned valve:** Replace valve.
- **Defective pushrod:** Replace it.
- **Defective cam lobe:** Replace camshaft.
- **Defective piston or rings:** Service cylinder walls and replace defective parts.
- **Defective head gasket:** Service and replace the head gasket
- **Intake manifold leaks:** Service the interface; replace gasket, and tighten fasteners to specs.

Engine Runs but Misses: Different Cylinders:

The possible causes and remedies could include:

- **Defective ignition system:** Check distributor, set advance and check coil and condenser function. Replace any defective parts.
- **Defective fuel system:** Service the fuel pump, flex-line and carburetor. Replace any defective parts.
- **Cross-firing plug wires:** Replace or relocate plug wires.
- **Loss of compression:** Check compression or leakage, and service valves, valve guides, and replace defective gaskets and other parts as needed.
- **Defective valve action:** Check compression, leakage and vacuum. Service valve train, and replace parts as necessary.
- **Worn pistons and rings:** Check compression, leakage, and vacuum. Service cylinders and replace parts as needed.
- **Overheated engine:** Check cooling system and take necessary corrective measures.
- **Manifold heat-control valve stuck:** Service and free the valve.

- **Restricted exhaust:** Check exhaust tail pipe, muffler, and catalytic converter and eliminate any restrictions present.

14. Engine Misses on Acceleration

The possible causes and remedies could include:

- **Lean air/fuel mixture:** Adjust the mixture strength.
- **Dirt in carburetor:** Service carburetor.
- **Defective carburetor accelerator pump:** Service and replace parts as needed.
- **Faulty ignition coil:** Check ignition coil and replace if needed.
- **Fouled or incorrectly gapped spark plugs:** Clean spark plugs and replace as needed.
- **Incorrect ignition timing:** Reset ignition timing.
- **Weak valve springs:** Replace as needed.
- **Sticking engine valves:** Service valves and valve guides to free the valves.
- **Leaking engine valves:** Service valve train and replace parts as necessary.

15. Engine Misses Under Load

The possible causes and remedies could include:

- **Low fuel pressure or volume:** Service the fuel pump and adjust pressure /volume to specs.
- **Improper feedback system operation:** Service or replace the system parts.
- **Air leak into intake manifold or vacuum hoses.** Check valves, valve guides, intake manifold and hoses. Repair or replace parts as necessary.
- **Faulty ignition coil:** Check ignition coil and replace if needed.
- **Worn or burned distributor rotor:** Replace rotor.
- **Faulty distributor cap or moisture on cap:** Clean and dry cap. Replace cap if necessary.
- **Defective high-tension wiring:** Replace wiring.
- **Fouled or incorrectly gapped spark plugs:** Clean spark plugs and replace as needed.
- **Worn camshaft lobes:** Replace camshaft.
- **Sticking hydraulic lifters:** Service lifters. Replace if needed.
- **Bent pushrod:** Replace.
- **Weak valve springs:** Replace.
- **Blown cylinder head gasket:** Replace gasket.

16. Engine Misses at High Speed

The possible causes and remedies could include:

- **Dirt in fuel line:** Service fuel system.
- **Clogged carburetor jets:** Service carburetor.
- **Faulty ignition coil:** Replace.
- **Worn or burned distributor rotor:** Replace.
- **Worn distributor shaft:** Replace.
- **Fouled or incorrectly gapped spark plugs:** Clean spark plugs and replace as needed.
- **Incorrect ignition timing:** Reset ignition timing to specs.
- **Worn camshaft lobes:** Replace camshaft.
- **Worn valve lifters:** Replace.
- **Sticking engine valves:** Service valves, valve guides and replace parts as needed.
- **Leaking engine valves:** Service valve train and replace parts as necessary.
- **Incorrect valve lash:** Adjust valve lash to specs.

17. Engine Backfires

A faulty anti-backfire valve could cause backfiring in the air-injection system or by faulty accelerator-pump action. Late ignition timing or ignition cross firing caused by the spark jumping across distributor cap or through the cable insulation can cause engine backfire. Spark plugs of the wrong heat range that overheat and cause preignition, too rich mixture, overheating of the engine, carbon in the engine, hot valves, intake valves that stick or seat poorly, can all lead to backfiring. Too much carbon in the engine cylinders can retain enough heat to initiate preignition of the fuel entering the cylinder causing the backfire. Excessive carbon also increases the compression ratio and enhances the tendency for detonation and preignition. Hot plugs can also cause preignition; so cooler plugs are a better choice. If intake valves hang open, combustion could be carried back into the carburetor. Excessively ground valves to form sharp edges, valves that seat poorly on seats, or valves coated with carbon deposits overheat and often cause backfiring.

The possible causes and remedies could include:

- **Ignition timing improperly set:** Adjust timing to specs.
- **Spark plugs of wrong heat range:** Replace with correct spark plugs.
- **Excessively lean or rich mixture:** Service and repair or readjust fuel pump or carburetor.
- **Engine overheats:** Take steps to correct engine overheating.
- **Carbon in engine:** Clean the carbon from engine.
- **Valves hot or stuck:** Service valve train and clean to free valve, and replace parts if needed.
- **Cracked distributor cap:** Replace.
- **Inoperative antibackfire valve:** Replace the valve.
- **Cross-firing plug wires:** Replace wires.
- **Air leaks into intake manifold or vacuum hoses**
- **Incorrect valve timing:** Reset valve timing to specs.
- **Low compression pressure:** Service top end and replace parts as needed.

18. Engine speed surges

The possible causes and remedies could include:

- **Low fuel pump pressure or volume:** Service the fuel pump and replace any parts needed.
- **Clogged fuel filter:** Clean or replace filter element.
- **Obstructed** air-filter: Replace filter element.
- **Improper thermostatic air cleaner operation:** Replace thermostatic air cleaner if required.
- **Sticking carburetor linkage:** Service linkage to free it, replace any part that is necessary.
- **Incorrect carburetor float level:** Correct level setting as needed.
- **Malfunctioning carburetor:** Repair or replace.
- **Air leak into manifold or vacuum hoses:** Service and replace any damaged parts.
- **Faulty ignition timing advance and retard:** Reset ignition timing and replace any defective parts.
- **Faulty PCV system:** Service the system and replace defective parts
- **Defective EGR valve:** Replace valve.

19. Excessive Fuel Consumption

Excessive fuel consumption can be caused by almost anything in a vehicle, from the driver to underinflated tyres or a defective choke. A fuel-km tester can be used to accurately check fuel consumption. The compression or leakage tester and the vacuum gauge are used to help determine whether the problem is in the engine, fuel system, ignition system, or elsewhere. The exhaust-gas analyzer, dynamometer, and fuel-km tester are useful in analyzing the problem.

If the trouble appears to be in the fuel system, remember that:

1. If the driver pumps the accelerator when idling and is curious on accelerating faster when starting from traffic lights will use larger amounts of fuel than a normal driver.
2. Engine operation with partly closed choke after warm-up will use more amounts of fuel as compared to a normal operation.
3. Vehicle operating on short-run remain on warm-up most of the time. This results in high fuel consumption.
4. A habitually fast driver will end up burning more fuel per km as compared to a normal driver.

These conditions are due to the type of operation. The only remedy is to change the operating conditions. However, if excessive fuel consumption is not due to any of these conditions, then the fuel pump should be checked for excessive pressure. High fuel-pump pressure will cause a high carburetor float-bowl level in relation to carburetor main fuel jet and a rich mixture. A fuel-pump tester is used to check fuel-pump pressure.

5. If excessive fuel consumption is not due to operating conditions or high fuel-pump pressure, the trouble is probably in the carburetor and could be caused by any or more of the following conditions:
   - The automatic choke may not be opening fast enough during warm-up or it may not be opening fully. The automatic choke should be checked by removing the air cleaner and observing the chock operation during warm-up.

- A *clogged air cleaner* that fails to admit enough air will act somewhat like a partly closed choke valve. The air cleaner filter element should be cleaned or replaced as necessary.
- A high float level in the carburetor float bowl will cause flooding and delivery of excessive fuel. The needle valve may be stuck open or may not be seating on its seat fully. Check the float level and adjust to specs. Check the float weight to find out if it is too heavy as a result of absorbed fuel.
- Too rich idle mixture setting or the too high idle speed setting, would cause excessive fuel consumption. These setting should be checked and adjusted as necessary. On late model vehicles, the idle mixture screw is set at the factory and cannot be readjusted without disassembling the carburetor.
- If an accelerator pump system has a check valve, failure of the check valve to close properly may permit fuel to feed through into the carburetor. This would require the carburetor to be disassembled for service.
- If the metering rod gets stuck in the full-power position or the power piston is stuck open allowing the power system to function, and supply a too rich mixture. The carburetor has to be disassembled to repair this problem.
- Worn fuel-jets allow the mixture to be too rich. The carburetor has to be disassembled and jets replaced during carburetor rebuilding.

6. Faulty ignition, too, can cause excessive fuel consumption. The ignition system could cause engine miss and the engine failure to burn all the fuel supplied. This problem would also be associated with loss of power, acceleration, or high-speed performance. A defective coil or condenser, incorrect timing, faulty advance-mechanism action, dirty or worn plugs or contact points, and defective wiring could cause this condition.
7. An old worn engine can cause excessive fuel consumption. Possible causes include, loss of engine compression from worn or stuck rings, worn or stuck valves, or a cylinder head gasket failure. These conditions require more fuel to be burned to achieve the same speed and to compensate for the lost power.
8. Excessive fuel consumption can also be caused by conditions that require additional power for the engine to overcome the resistance and move the vehicle, e.g., low tyre inflation pressure, dragging brakes, defective automatic transmission, and misaligned wheels.

The possible causes and remedies could include:

- **Too fast starts from stops:** Drive more reasonably.
- **High-speed driving:** Drive as per legal limits.
- **Short-run operation:** Use vehicle to drive for longer distances.
- **Excessive fuel-pump pressure or pump leakage:** Service the pump and reduce pump pressure to specs.
- **Improper thermostatic air cleaner operation:** Service or replace the element.
- **Fouled or faulty spark plugs:** Clean or replace as needed.
- **Incorrect ignition timing:** Correct timing.
- **Backpressure in exhaust:** Repair or replace parts to remove restriction.
- **Incorrect vacuum hose connection:** Correct the connection.
- **Choke partly closed after warm-up:** Service and adjust automatic choke to open.
- **Clogged air cleaner:** Clean or replace filter element.
- **High carburetor float level:** Adjust level to specs.
- **Stuck or dirty float needle valve:** Free and clean it.
- **Worn carburetor fuel jets:** Replace jets as needed.
- **Stuck metering rod or power piston:** Service to free it.
- **Idle mixture too rich or idle speed too fast:** Adjust the mixture strength and speed to specs.
- **Stuck accelerator-pump check valve:** Service to free check valve.
- **Carburetor leaks:** Service the carburetor and replace gaskets, and tighten screws as needed.

- **Cylinder misfiring:** Check coil, condenser, timing, spark plugs, contact points, wiring.
- **Automatic transmission slipping or not up-shifting:** Service the transmission and replace any defective parts found.
- **Loss of engine compression-worn engine:** Check compression or leakage. Recondition the engine as needed.
- **Defective valve action:** Check with compression, leakage, or vacuum tester. Check for worn camshaft, belt or chain slack, or jumped tooth. Service the valve train and replace the parts as needed.
- **Excessive rolling resistance from low tyre pressure, dragging brakes, wheel misalignment, etc.:** Correct the defects causing increased rolling resistance.
- **Clutch slippage:** Adjust or repair as needed.

20. Excessive Vibration

The possible causes and remedies could include:

- **Loose or collapsed engine mounts:** Tighten and replace mounts as needed.
- **Defective vibration damper:** Replace damper.
- **Loose mounting of belt-driven accessories:** Tighten all fasteners to specs. Replace any defective fasteners.
- **Out-of-balance cooling fan:** Service the fan and replace if necessary.

21. Excessive Oil Consumption

Excessive oil consumption can result from external leakage of oil or as the result of oil passing through the engine. It will be necessary to determine the exact source of the oil loss or whether the oil passes through the combustion chambers of the engine.

Lubricating oil may be lost from the engine in three ways: by leakage in liquid form, by burning in the combustion chambers, and by passing out of the crankcase through the PCV system as mist or vapour.

External Loss of Oil: If there is an external loss of oil the exterior of the engine will be coated with oil, particularly in the area of the leakage. The usual procedure for locating the source of the oil leak is to first carefully clean the exterior of the engine. Then, take the vehicle for a short drive. Stop the vehicle over a clean area and examine the engine exterior for evidence of oil leakage, *Fig. 16.26.* Also see if the oil has dripped on the ground. This will help in locating the source of the leakage, front or rear, left or right. It is not unusual for oil leakage from one point to confuse with leakage from some other point.

To pinpoint the exact source of the leakage, plug the oil filler pipe. With the engine idling, blow compressed air into the dipstick tube or hole in the block. Watch for oil leakage and trace it to the source.

Internal Consumption of Oil: When the oil is being consumed by passing through the engine, it normally will cause heavy blue smoke to come from the exhaust tailpipe, especially after the engine has been idling for several minutes. After the idling when the engine should be raced briefly and the blue smoke should appear.

While worn piston rings and tapered cylinder walls could cause excessive oil consumption, there are many other problem areas that could be causing the problem, either singly or in combination. In most cases of internal consumption of oil, the oil leaks out of one or more of the pressure lubricated connecting rod, main (crankshaft), or camshaft bearings. The oil is splashed or thrown up into the cylinders under the pistons in such large quantities that piston oil control ring can not control and prevent it from getting, in small amounts, to the combustion chamber side and burning there.

The possible causes and remedies could include:

- **High oil level:** Correct the oil level in sump.
- **Plugged drain-back in cylinder head:** Clear the drain-back passage.
- **Excessive bearing clearance:** Service the engine to replace bearing to set bearing clearances to specs.
- **Worn crankshaft journals:** Replace bearings.
- **Worn piston rings:** Service the cylinders, pistons and install new rings, to restore various clearances to within specs.
- **Clogged drain holes in Oil rings:** Replace rings.
- **Incorrect ring gap:** Replace rings and check the necessary gaps to be within specs.
- **Ring gaps incorrectly spaced:** Replace rings and ensure to stagger ring gaps as per recommendations.

- **Compression ring installed upside down:** Service to correct the ring installation. Replace rings if necessary.
- **Too little piston ring side clearance:** Service the engine and install new rings check that clearances are within specs.
- **Wrong size rings installed:** Install right size rings.
- **Worn pistons and cylinder walls:** Recondition the cylinders and install new pistons and rings to fit the bored and honed cylinders.
- **Damaged valve seals:** Replace seals.
- **Worn valve stems and guides:** Recondition or Replace valves and/or guides.
- **Sticking PCV valve:** Replace valve.
- **Overheated engine:** Recondition the engine to eliminate the causes of overheating.

22. External Oil Leakage

The possible causes and remedies could include:

- **Defective fuel pump leakage:** Replace gasket.
- **Defective cylinder head gasket-** Replace gasket.
- **Defective oil filter gasket:** Replace gasket.
- **Defective oil pan gasket or end seal:** Replace gasket or seal as the case may be.
- **Defective timing chain (or gear) cover gasket:** Replace gasket.
- **Worn timing chain (or gear) cover oil seal:** Replace seal.
- **Worn rear main bearing oil seal:** Replace seal.
- **Loose oil gallery plug:** Replace plug.
- **Improperly seated engine oil pan drain plug:** Correct the drain plug installation, replace plug if needed.

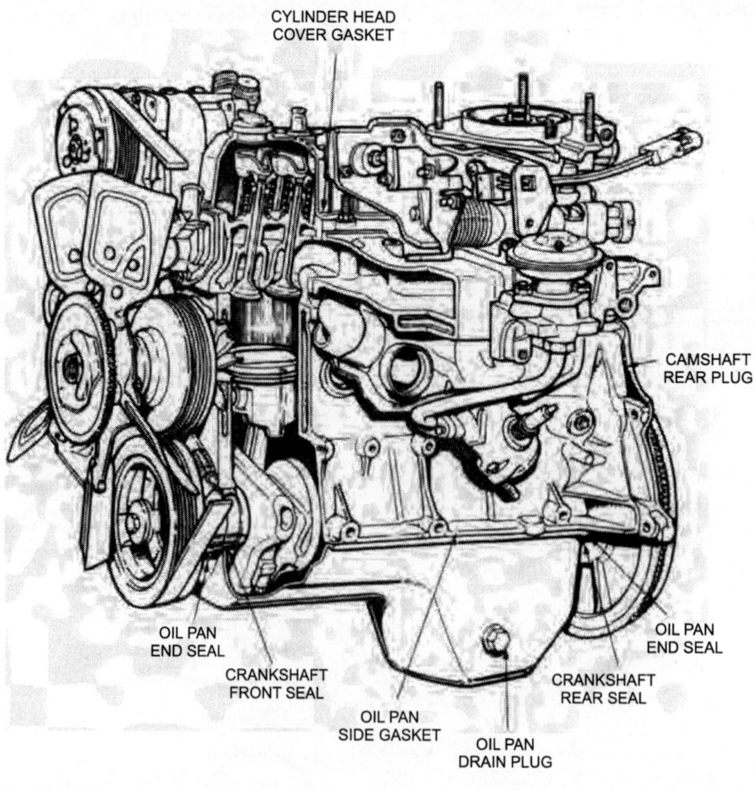

Fig. 16.26 *Various points of possible external leakage of engine oil are shown in this illustration of a typical fuel-injected 4-cylinder engine. (Chrysler Corporation)*

- **Improperly seated camshaft rear plug:** Correct the plug installation. Replace the plug if necessary.

23. Low Oil Pressure

Low oil pressure is often a warning of a worn oil pump or worn engine bearings. The bearings pass so much oil that the oil pump is unable to maintain oil pressure. The end bearings will probably be oil-starved and could fail. Other causes of low oil pressure include: a weak relief-valve spring, worn oil pump, broken or cracked oil line, and a clogged oil line. Oil dilution, foaming, sludge formation, insufficient oil in oil pan, or oil too thinned out by engine overheating, too, will cause low oil pressure.

The possible causes and remedies could include:

- **Low oil level:** Add oil to correct level.
- **Oil dilution and foaming:** Replace oil.
- **Excessive oil temperature:** Recondition the engine to eliminate the causes of engine overheating.
- **Faulty oil pressure sending unit:** Replace unit.
- **Clogged air filter:** Replace filter element or clean as needed.
- **Worn oil pump:** Replace oil pump.
- **Excessive wearing clearances:** Replace bearings.
- **Sticking oil pump relief valve:** Service relief valve or replace if needed.
- **Loose, bent, or cracked oil pump suction tube:** Tighten tube fasteners, or replace suction tube.

24. Smokey Exhaust

The possible causes and remedies could include:

- **Blue smoke:** Excessive oil consumption.
- **Black smoke:** Excessively rich mixture.
- **White smoke-Steam in exhaust:** Replace gasket; tighten cylinder head bolts to specs to eliminate coolant leakage into combustion chambers.

16.5 ENGINE NOISES

One of the most difficult of all troubleshooting jobs is to locate the source of noise or "knocks" in an engine. Every rotating or reciprocating part in an engine is a potential source of noise. In many cases, however, certain noises possess characteristics that help identify their source of origin.

These characteristics vary somewhat between different engines. A listening rod, or *stethoscope,* is used to locate the source of a noise at some definite section of the engine, *Fig. 16.27.* The instrument magnifies the intensity of the noise, and the sound becomes louder as the tip of the instrument approaches the origin of the noise. The mechanic places one end of the rod (that could be even a long screwdriver) on his/her ears and the other is placed at some part of the engine. The noises from that part of the engine are carried along the rod to ear.

Move the stethoscope around the engine until the noise head is loudest. Thus, by determining the approximate source of the noise, one can locate a noisy broken ring in a cylinder, a main bearing knock, and other noise conditions. Various engine noises and their symptoms and remedies are given here.

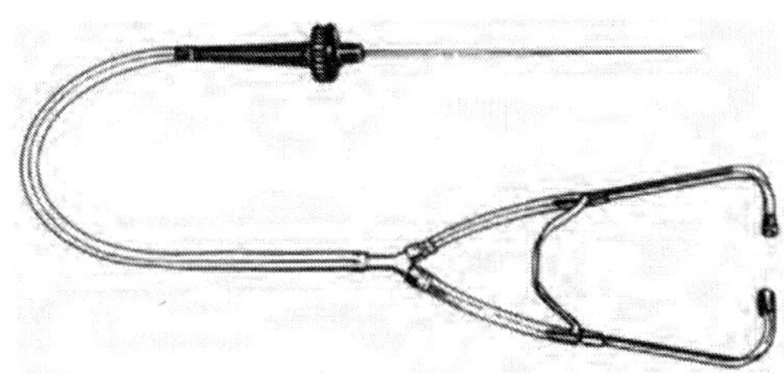

Fig. 16.27 *A stethoscope-type tool can be used when checking for the approximate location of the engine noises. The stethoscope generally has a built-in diaphragm to amplify sounds.*

1. Valve and Tappet Noise

Valves are a common source of noise. There are several points in each unit of the valve train that can create noise. The noise is *a regular clicking sound occurring at half the engine speed.* It could disappear at higher engine speeds. The cause is usually excessive valve clearance or a defective hydraulic valve lifter. A thickness gauge inserted between the valve stem and lifter or rocker arm will reduce the clearance. If the noise is reduced too, then the cause for the noise is excessive clearance and the clearance should be readjusted. However, if inserting the thickness gauge does not reduce the noise, then the noise is caused by other valve train conditions like weak springs, worn valve-lifter faces, lifters loose in the block, and rough cam surfaces.

Clicking caused by wear between the valve lifter and lifter guide, or damage on the end of the lifter next to the camshaft, is not readily located. Here, the timing of the click, along with the use of a stethoscope is helpful.

Hydraulic valve lifters often will be noisy when the engine is first started, because oil has leaked from the unit. The noise should disappear after a few minutes operation, during which the lifter will be filled with oil. If the noise does not disappear, the defective lifter can be located with the help of a stethoscope.

The possible causes and remedies could include:

- **High or low oil level in crankcase:** Adjust oil level to specs.
- **Thin or diluted oil:** Replace oil.
- **Low oil pressure:** Service the bearings, oil galleries, and lubrication system parts, replace parts as needed, and set the oil level as necessary.
- **Dirt in lifters:** Service the lifters, replace if necessary.
- **Excessive hydraulic lifter leakdown:** Replace lifters.
- **Worn lifters:** Replace.
- **Excessive valve lash (solid lifters):** Adjust lash to specs.
- **Bent pushrods:** Replace.
- **Excessive runout of valves or seats:** Service and replace parts as needed.
- **Worn rocker arms and/or pivots or shafts:** Replace parts.
- **Broken or cocked valve spring:** Replace.
- **Bent valve:** Service seat, guide, and replace valve.

2. Ping or Spark Knock

Ping, spark knock or detonation is a pinging or chattering sound that is most noticeable during part throttle acceleration or when the vehicle is climbing a hill. Some spark knock is considered normal under such conditions of operation. But when spark knock becomes excessive, it may be due to condition as low octane rating fuel for the engine, carbon deposits in the engine that increase engine compression, and advanced ignition timing.

The possible causes and remedies could include:

- **Low octane fuel or poor fuel quality:** Use specified fuel in the vehicle.
- **Low compression pressure:** Use lower grade fuel.
- **High compression pressure:** Use higher-grade fuel.
- **Excessive combustion chamber deposits:** Service engine. Remove deposits.
- **Sharp edges in combustion chamber:** Service chambers.
- **Air leak into intake manifold:** Service engine to stop leak.
- **Incorrect ignition timing:** Reset ignition timing. Replace any defective parts.
- **Malfunctioning ignition distribution advance:** Service and reset advance to specs. Replace any defective parts.
- **Wrong spark plugs for the application:** Replace with correct spark plugs.
- **Faulty EGR valve:** Replace valve.

3. Noisy Connecting Rods or Bearings

The conditions of operation under which the connecting rod bearing noise is heard and the "timing" of the noise are also useful in determining the source of noise. Connecting-rod noise usually has a *light knocking or pounding sound. It is most noticeable when the engine is "floating"*, i.e., neither accelerating nor decelerating, as when the vehicle is cruising at constant speed on a level road. The sound becomes more noticeable as the accelerator is eased off with the vehicle running at medium speed.

A pronounced rod knock will be heard at all speeds and under both idle and load conditions. One rod will make a distinct noise. If all rods are loose, the noise becomes a rattle or clatter. In many cases, rod bearing knocks are confused with piston slap or loose piston pins. This is particularly true when all rods are loose, and experience will be helpful in determining which part is at fault. However, it is not much important to decide definitely, because the remedy for either fault requires removal of the piston and rod assemblies in practically all cases. Measurement and inspection of the parts will then help isolate the source of trouble.

To locate the connecting-rod noise, short out the spark plugs one at a time. It is difficult to short out some of the spark plugs on some engine applications. The easiest way to short out the spark plugs to make the check is to use an engine analyzer that can short out cylinders with the turn of a knob or push of a button. The noise will be considerably reduced when the cylinder that creates noise is not delivering power. A worn bearing or crankpin, a misaligned rod, inadequate oil, or excess bearing clearances can cause connecting-rod noise. Worn rod bearings produce a distinct knock.

The possible causes and remedies could include:

- **Insufficient oil in crankcase:** Add specified oil to bring to spec level.
- **Defective oil pump:** Repair or replace oil pump.
- **Low oil pressure:** Recondition engine and replace defective parts in cylinder block.
- **Thin or diluted oil:** Recondition engine.
- **Excessive connecting-rod bearing clearance:** Recondition engine and replace bearings.
- **Bent connecting rods:** Replace rods.
- **Out-of-round connecting rod bearing journals:** Recondition bearing journals and install new bearings.

4. Noisy Pistons, Piston Pins or Rings

Piston slap is a muffled, hollow, bell-like sound. It is caused by the piston rocking back and forth in the cylinder in the direction of maximum and minimum thrust faces. Piston slap when the engine is cold, is common but if it occurs under all engine operating conditions, it needs to be taken care of. There is much confusion between the noise caused by a piston with excessive clearance in the cylinder and a loose piston pin. Either defect produces a click, which is quite distinct. If noisy in all cylinders it becomes a rattle. One indication of piston slap is a decrease in the noise as the engine warms up. A piston slap is always louder when the engine is cold.

Piston slap is caused by inadequate lubrication of piston and cylinder mating interface, worn cylinder walls, worn pistons, collapsed piston skirts, excessive piston clearances, or misaligned connecting rods.

Piston-pin noise is a regular clicking sound or light, metallic double knock. However, piston-pin noise has a unique metallic double knock. Normally it is most audible during idling with the ignition timing advanced. Running the engine at idle speed with the spark advanced and shorting the plugs one at a time can make a check. Piston-pin noise will be reduced when a plug in a noisy cylinder is shorted out. A worn or loose piston-pin, a worn bushing, and lack of lubrication oil cause this noise.

Using a stethoscope is sometimes helpful, since piston or pin may sound loudest when the instrument prod is placed on the cylinder head or block. The rod knock is often loudest with the prod on the crankcase.

Loose piston pins usually, but not always, produce a double rap each revolution of the crankshaft. They rap once at the top of the stroke and again at the bottom. On most engines, the knock is loudest at idling speed, and it will become even louder if the spark is advanced. Often, the knock will be louder if the spark plug is shorted out in cases where not all pins are loose.

Piston-ring noise is a clicking, chattering, or rattling sound similar to valve and tappet noise. However, this noise is most audible on acceleration. New piston rings will cause a knock if the ridge at the top of the cylinder bore is not removed before the new rings are installed. Somewhat similar is the condition where the cylinders have been rebored oversize, and the standard cylinder head gasket extends into the combustion chamber. The piston strikes the gasket and makes a distinct knock.

Usually the piston rings that are loose in their grooves will not make any noise, because the oil provides some cushion. But excessively loose upper rings may cause a clicking sound similar to that made by a worn valve lifter. The piston rings will click twice each revolution of the crankshaft, while the valve click will be heard once in two revolutions. This difference in the timing of these clicks helps in pinpointing the cause of the clicking sound.

Low ring tension, broken rings, worn rings, or worn cylinder walls produce this noise. Piston-ring noise can be confused with other engine noises. A test to confirm this noise can be made as follows:

Remove the spark plugs and pour a tablespoon of oil in each cylinder. Crank the engine for some revolutions to work the oil down past the rings. Then install the spark plugs back and start the engine. If the noise has been reduced, the rings are probably producing the noise.

The possible causes and remedies could include:

- **Excessive piston-to-cylinder wall clearance:** Service cylinder walls and piston and replace rings.
- **Collapsed piston skirt:** Replace pistons.
- **Loose piston strut:** Service pin and bushing; check oil.
- **Incorrectly fitted piston pin:** Correct the pin fitting.
- **Misaligned connecting rods:** Service the crank pins and replace connecting rods.
- **Excessive carbon deposits on pistons:** Service the cylinders and pistons - remove carbon.
- **Piston rings striking ridge at top of cylinder wall:** Service the block, replace pistons and rings.
- **Broken ring:** Service engine block and replace rings and pistons as needed.
- **Incorrect ring gap:** Set the gap or fit the right size rings.
- **Excessive ring-to-groove side clearance:** Service the block and piston assemblies and replace parts as necessary.

5. Crankshaft Knock - Noisy Main Bearings

A *main bearing knock* is more of a bump than a knock. It can be located by shorting out the plugs near it. The noise is loudest when the engine is "lugging", i.e., pulling hard at slow speed. The sound is heavier and more dull than a connecting rod knock.

Crankshaft knock is a heavy, dull metallic sound. It is most audible when the engine is under a heavy load or is accelerating, especially when cold. A regular noise that is more of a rumble is probably caused by worn main bearings.

Excessive endplay in the crankshaft will produce an intermittent rap or knock that is sharper than a main bearing knock. When the noise is irregular and sharp, and is probably caused by a worn thrust bearing. A badly worn thrust bearing will cause the noise to be produced each time the clutch is released and engaged and also when accelerating.

The possible causes and remedies could include:

- **Insufficient oil supply:** Recondition main and rod bearings. Check oil pump, oil level, and oil condition.
- **Low oil pressure:** Check oil pump, relief valve, bearing clearances, oil level and condition.
- **Worn main bearings:** Recondition crank journals and replace bearings.
- **Out-of-round crankshaft journals:** Recondition crank journals and replace bearings.
- **Excessive end play of crankshaft:** Service crankshaft and replace thrust bearings.
- **Loose flywheel:** Check and tighten the connecting fasteners to specs.
- **Loose or damaged vibration damper:** Service and replace damper if needed.

6. Miscellaneous Noises (Rattle etc.)

If the flywheel is loose on the crankshaft flange, the noise will be similar to a main bearing knock. Ordinarily it will not change when the plugs are shorted out. Also, the noise may come and go rather than being constant. One sure test is to turn off the ignition, then turn it on again just as the engine is about to stop. The sudden twist applied to the crankshaft will produce a definite knock if the flywheel is loose.

If the rubber engine mountings are drawn down too tight, or, if the rubber has deteriorated enough to allow the metal parts of the mounting to contact each other, a knock may occur. This particular knock appears under high torque conditions during rapid acceleration.

Some other noises are caused by loosely mounted accessories, e.g., the alternator, starter motor, water pump, manifolds, horn, flywheel, crankshaft pulley, and oil pan. Also, other automotive components, such as the clutch, transmission and differential could cause other noises. The possible causes and remedies could include:

- **Loosely mounted accessories: alternator, water pump, oil pan, horn, front bumper, starter motor etc.:** Tighten mounting fasteners to specs.

16.6 BURNED VALVES AND SEATS

The possible causes and remedies could include

- **Lean air/fuel mixture:** Adjust mixture strength to specs.
- **Insufficient valve lash (solid lifters):** Adjust lash to specs.
- **Excessively thin valve head margin:** Recondition seats, valve guides and replace valves.
- **Excessively narrow valve seat:** Recondition seat.
- **Loose valve seat insert:** Replace seat to ensure fit.
- **Warped valve head:** Recondition valve train, guides, seats and replace valve.
- **Worn valve stems and guides:** Recondition seats, guides, and replace valves.

16.7 ENGINE RUN-ON OR DIESELING

It is the tendency of the engine to continue run on after the ignition is turned off. SI engines equipped with emission controls require a fairly high, hot-idle speed for best operation. This could cause dieseling (run-on) if there are hot spots in the combustion chambers, and the throttle valve is not completely closed. The hot spots in the combustion chambers act as spark plugs and ignite the air-fuel mixture getting past the partly open throttle valve and keep the engine running.

Many SI engines have an idle-stop solenoid to close the throttle. If the engine runs on, or diesels, check the idle-stop solenoid to ensure it is releasing properly when the ignition is turned off. It might require adjustment to allow the throttle valve to close fully. Check the idle speed setting, too, to ensure it is not set too high as compared to specs. Advanced ignition timing could cause run on. All necessary measures should be taken to avoid engine overheating. Spark plugs, valves and cylinder head might need servicing. The possible causes and remedies could include:

- **Incorrect idle-solenoid adjustment:** Adjust and fix solenoid.
- **Engine overheats:** Take all measures to avoid overheating.
- **Hot spots in cylinders:** Check plugs, pistons, cylinders and chambers for carbon deposits, check valves for improper seating.
- **Ignition timing advanced:** Adjust timing to specs.
- **High engine idle speed:** Correct idle speed setting.
- **Poor quality fuel:** Use specified fuel.
- **Wrong type spark plugs for application:** Install right spark plugs.
- **Defective exhaust system control system:** Service the system and replace defective part.

16.8 EXCESSIVE HC AND CO IN EXHAUST GAS

If the exhaust-gas analyzer indicates excessive HC and CO in the exhaust gas, correction should be made. Excessive CO usually occurs due to a rich air-fuel mixture. At idle, it can be brought within specs by proper adjustment of the carburetor. Excessive HC is normally caused by a misfiring spark plug. Proper servicing of the ignition system components, and replacing any worn or defective engine parts can usually correct the condition. Excessive HC in exhaust can also result from a too lean air-fuel mixture caused by intake manifold or vacuum leaks. This could result in lean misfire that dumps the unburned air-fuel mixture into the exhaust system. State laws require exhaust gas testing of all vehicles during state inspection. Vehicles emitting excessive amounts of HC and CO must be repaired in order to obtain permit to continue operating. This helps keep the vehicles with excessive pollution to be kept away from roads. Common causes of excessive HC and CO include:

1. Misfiring cylinders due to ignition system problems, e.g., faulty plugs, defective high-voltage wiring, defective distributor cap, defective ignition coil, defective condenser, or defective contact-breaker points.
2. Incorrect ignition timing.
3. Carburetor troubles, such as the choke remaining partly closed, worn fuel jets, high float fuel level, and other conditions that lead to excessive fuel consumption.

4. Faulty air-injection system, which does not inject enough air into the exhaust manifold to completely burn the HC and CO. A defective air pump, or a leaking hose or air manifold could cause this.
5. Defective TCS system that allows vacuum advance in all gear positions instead of high and reverse gears only.
6. Defective catalytic converters that must be replaced or serviced to restore catalytic action.
7. Excessive carbon deposits in the combustion chambers, or stuck valves. Carbon deposits often act like sponges, absorbing air-fuel mixture during compression and then releasing it during exhaust stroke. The possible causes and remedies could include.

- **Ignition miss:** Check spark plugs, wiring, distributor cap, ignition coil, etc., and replace any defective parts.
- **Incorrect ignition timing:** Reset ignition timing to specs.
- **Carburetor problems:** Check choke valve, float fuel level, idle mixture setting, etc.
- **Faulty air injection:** Check air injection pump, hoses, and manifold for any defects. Replace any defective parts.
- **Defective TCS system:** Check system, service and replace any defective part.
- **Defective catalytic converters:** Replace converters or catalyst as needed.

16.9 DIESEL ENGINE TROUBLESHOOTING

Diesel Engine Troubles: Diagnosis

Since the diesel engine does not have an ignition distributor or a carburetor, the trouble shooting procedure is rather simplified. Basically, it includes checking compression pressure and fuel injection system performance. Checking compression requires the removal of the injectors and testing compression pressure at each cylinder with a compression gauge. Special equipment is necessary to check the injectors and the pattern of spray produced.

Except for the "fuel supply system" and the "spark-ignition system," both the diesel engine and the sparkignition engine can have same troubles. Much of the troubleshooting content is common to both types of engines. Therefore, the use of the diesel-engine fuel-injection system trouble-diagnosis chart in combination with the SI engine troubleshooting charts content will provide causes and checks/corrections to troubleshoot a diesel engine.

1. Hard Starting

The possible causes and remedies could include:

- **Air in fuel system:** Bleed the fuel system to remove any air present.
- **Clogged fuel filter:** Clean or replace the fuel filter as needed.
- **Blocked fuel supply line:** Service the fuel supply line to open the blockage.
- **Blocked injection lines:** Service the supply lines and replace if defective.
- **Defective injection pump:** Service the pump and replace any worn or defective parts or pump.
- **Blocked injection pump:** Service the pump to clear all blockage and replace any worn or defective parts.
- **Incorrect injection timing:** Reset injection timing to specs.
- **Defective preheating device:** Replace defective device.
- **Low engine compression:** Recondition engine. Service block, valve train , pistons, rings, valve seats and guides. Replace all worn or defective parts, restoring clearances to specs.

2. Engine Surges While Idling

The possible causes and remedies could include:

- **Incorrectly adjusted governor:** Correct adjustment.
- **Defective fuel injection pump:** Service pump and replace any worn or defective parts.

3. Loss of Power

The possible causes and remedies could include:

- **Clogged fuel filter:** Clean or replace filter element.
- **Blocked or leaking injection lines:** Service the injection lines. Replace if defective.

- **Defective injection pump:** Service the pump and replace any worn or defective parts or pump.
- **Incorrect injection timing:** Reset injection timing to specs.
- **Defective timing device:** Replace.
- **Obstructed air filter:** Service to clean or replace the filter element.
- **Maladjusted low idle speed:** Adjust the low idle speed to specs.

4. Engine Misses Under Load

The possible causes and remedies could include:

- **Air in fuel system:** Bleed the air out of the system.
- **Clogged fuel filter:** Clean or replace filter element.
- **Blocked fuel supply line:** Service the fuel supply line to open the blockage:
- **Blocked injection lines:** Service the supply lines and replace if defective.
- **Defective fuel injector:** Replace injector.
- **Incorrect ignition timing:** Reset injection timing to specs.
- **Defective overflow valve:** Replace valve.

5. Excessive Fuel Consumption

The possible causes and remedies could include:

- **Leaking fuel lines:** Service lines and replace if necessary.
- **Defective injection pump:** Service the pump and replace any worn or defective parts or pump.
- **Incorrect injection timing:** Reset injection timing to specs.
- **Defective timing device:** Replace.
- **Defective injection nozzle:** Replace nozzle.
- **Obstructed air filter:** Service to clean or replace the filter element.
- **Uneven engine compression:** Service the engine.

6. Engine Cannot Be Shut Off

The possible causes and remedies could include:

- **Defective shut off device:** Service; Replace the device.
- **Defective injection pump:** Service the pump and replace any worn or defective parts or pump.
- **Incorrect governor setting:** Reset governor to specs.

7. Black Smoke and Poor Performance

The possible causes and remedies could include:

- **Air in fuel system:** Bleed the system to remove air.
- **Defective injection pump:** Service the pump and replace any worn or defective parts or pump.
- **Defective governor:** Service and replace governor if necessary,

8. White or Blue Smoke

The possible causes and remedies could include:

- **Blocked fuel tank vent:** Clear the vent hole.
- **Air in fuel supply system:** Bleed the system to remove all air trapped in system.
- **Clogged fuel filter:** Clean or replace filter element.
- **Blocked injection lines:** Service the supply lines and replace if defective.
- **Defective injector or injection nozzle:** Service: replace if necessary.
- **Obstructed air filter:** Service to clean or replace the filter element.
- **Defective governor:** Service and replace governor if necessary,
- **Uneven engine compression:** Service the engine.

16.10 EXHAUST SYSTEM BACKPRESSURE

Exhaustive system backpressure appears as a possible cause for many different engine problems. If the exhaust pipe, muffler, catalytic converter, or tail pipe is restricted, the hot exhaust gases will not be able to escape quickly from the exhaust system, and a backpressure will be developed in the exhaust system. This causes larger amounts of burnt gases left in the cylinder at the end of exhaust stroke. As a result, during the intake stroke the incoming combustible charge will be severely diluted and full engine will not be able to developed full power. In extreme cases, the exhaust valves could burn.

THEORY QUESTIONS

1. What is engine troubleshooting?
2. Name the four gases that a 4-gas emission tester will analyze.
3. The first step in trouble shooting the cause of engine overheating is to make a careful _____ to see if there is any evidence of external coolant leakage.
4. In _____ _____ _____ _____ systems, the ECU monitors operating conditions for possible malfunctions.
5. A "plug-in" diagnostic setup utilizes a computer programmed to monitor several different engine systems. To make the tests, a _____ _____ _____ _____ is attached to a connector located under the hood.
   (*a*) Diagnostic readout tool
   (*b*) Diagnostic pick-up meter
   (*c*) Diagnostic printout unit
   (*d*) Diagnostic computer module.
6. What advantages does a diagnostic computer engine analyzer have over visual inspection and manual checks?
   (*a*) Displays performance readings on a screen
   (*b*) Pinpoints "hidden" sources of trouble quickly and accurately
   (*c*) Furnishes a printout of performance readings
   (*d*) All of the above.
7. An automotive engine has too much oil consumption. Mechanic A says that this could be caused by gasket leaks. Mechanic B says that burning in the combustion chambers could be the cause. Who is right?
   (*a*) A only (*b*) B only
   (*c*) Both A and B (*d*) Neither A nor B.
8 An engine may crank slowly due to
   (*a*) vapour lock
   (*b*) too much fuel-pump pressure
   (*c*) a run down battery
   (*d*) a defective coolant pump.
9. An engine starts but will not run. Mechanic A says that it could be caused by an obstructed air filter. Mechanic B says that it could be caused by backpressure in the exhaust system. Who is right?
   (*a*) A only (*b*) B only
   (*c*) Both A and B (*d*) Neither A nor B.
10. An engine cranks at normal speed but fails to start. It could be due to
   (*a*) defective starter motor
   (*b*) discharged battery
   (*c*) defective ignition system component
   (*d*) stuck engine valve.
11. Testing the _____ _____ of each cylinder while the engine is warm and cranking is an accepted method of checking the condition of internal components of the engine.
   (*a*) coolant leakage
   (*b*) manifold vacuum
   (*c*) compression pressure
   (*d*) combustion leakage.
12. An engine will loose power (hot or cold) if it has
   (*a*) an automatic choke valve that is stuck open
   (*b*) worn rings and cylinder walls
   (*c*) incorrect idle speed adjustment
   (*d*) none of the above.
13. An engine hesitates on acceleration. Mechanic A says that it could be caused by a faulty PCV system. Mechanic B says that it could be caused by a defective carburetor accelerator pump. Who is right?
   (*a*) A only (*b*) B only
   (*c*) Both A and B (*d*) Neither A nor B.
14. Loss of engine power as the engine warms up is most likely caused by
   (*a*) heavy oil
   (*b*) excessive rolling resistance
   (*c*) fuel starvation
   (*d*) the throttle valve not closing fully.
15. During a cylinder leakage test, the gauge reads 70% and air is heard coming from the exhaust pipe. What is the most likely cause?
   (*a*) Collapsed lifter
   (*b*) Cracked cylinder wall
   (*c*) Hole in piston
   (*d*) Burned valve.
16. When making a compression test, a low equal reading is obtained in two adjacent cylinders. This would most likely indicate a:
   (*a*) leaking exhaust manifold gasket
   (*b*) burnt valve
   (*c*) leaking intake manifold gasket
   (*d*) blown head gasket.

17. A distributor drive gear has badly damaged teeth. Mechanic A says that this could cause hard starting. Mechanic B says that this could cause backfiring. Who is right?
(*a*) A only (*b*) B only
(*c*) Both A and B (*d*) Neither A nor B.

18. Mechanic A says that blue-gray exhaust smoke can be caused by piston rings stuck in their grooves. Mechanic B says that blue-gray exhaust smoke can be caused by plugged oil return passages in the cylinder head. Who is right?
(*a*) A only (*b*) B only
(*c*) Both A and B (*d*) Neither A nor B.

19. While driving in stop-and-go city traffic, a vehicle with an electric fan overheats. Mechanic A says that this could be caused by a stuck open thermostat. Mechanic B says that this could be caused by a bad radiator fan sensor. Who is right?
(*a*) A only (*b*) B only
(*c*) Both A and B (*d*) Neither A nor B.

20. Missing in one cylinder could be caused by
(*a*) an overheated engine
(*b*) a defective spark plug
(*c*) vapour lock
(*d*) a clogged exhaust.

21. An engine may stall as it warms up if the
(*a*) ignition timing is off
(*b*) throttle valve does not open fully
(*c*) choke valve sticks closed
(*d*) battery is run down.

22. An engine will overheat if the
(*a*) battery is run down
(*b*) fuel pump is defective
(*c*) automatic choke sticks
(*d*) fan belt breaks.

23. Stalling of an engine after high speed driving could be caused by
(*a*) incorrect ignition timing
(*b*) worn carburetor jets
(*c*) worn piston rings
(*d*) vapour lock.

24. A valve job has just been completed on a push rod engine that is equipped with non-adjustable hydraulic lifters. The engine will not start. A compression test is made, and all the cylinders read zero. What needs to be done?
(*a*) Surface head (s)
(*b*) Install linger push rods
(*c*) Install thicker head gasket (s)
(*d*) Grind valve tips.

25. Mechanic A says that replace the spring retainers when doing a valve job. Mechanic B says that replace the stem seals when doing a valve job. Who is right?
(*a*) A only (*b*) B only
(*c*) Both A and B (*d*) Neither A nor B.

26. Mechanic A says that a cracked cylinder head can cause engine overheating. Mechanic B says that a cracked cylinder head can cause water to blow out the tail pipe. Who is right?
(*a*) A only (*b*) B only
(*c*) Both A and B. (*d*) Neither A nor B.

27. Mechanic A says that valve burning is often caused by too thin a margin. Mechanic B says that valve burning is often caused by too wide a eat. Who is right?
(*a*) A only (*b*) B only
(*c*) Both A and B (*d*) Neither A nor B.

28. The exhaust manifold heat control valve on an engine is stuck open. Mechanic A says that this could cause low cranking speed. Mechanic B says that this could cause hesitation during cold drive-away. Who is right?
(*a*) A only (*b*) B only
(*c*) Both A and B (*d*) Neither A nor B.

29. A mechanic removes the spark plugs from an engine. All the firing ends are coated with dry, fluffy carbon deposits. Mechanic A says that this could be caused by worn piston rings. Mechanic B says that this could be caused by a leaking carburetor base gasket. Who is right?
(*a*) A only (*b*) B only
(*c*) Both A and B (*d*) Neither A nor B.

30. After a recent engine overhaul, the oil consumption has greatly increased. Mechanic A says that this could be caused by installing the rings into worn grooves. Mechanic B says that this could be caused by installing the rings upside down. Who is right?
(*a*) A only (*b*) B only
(*c*) Both A and B (*d*) Neither A nor B.

31. A 4-cylinder engine has compression readings of 135, 70, 130, 75. Which of the following is the most likely cause?
(*a*) Burnt valves
(*b*) Broken valve springs
(*c*) Missing stem seals
(*d*) Incorrect valve timing.

32. The compression readings in a 6-cylinder engine (130, 135, 130, 140, 15, 135) could be caused by:
(*a*) a broken exhaust valve push rod
(*b*) a hole in piston

(*c*) Both *a* and *b*
(*d*) Neither *a* nor *b*.

33. During a power balance test, a V-8 engine has two cylinders on opposite corners of the block that do not show any drop in engine rpm. The most likely cause would be:
(*a*) a bad head gasket
(*b*) a leaking carburetor base gasket
(*c*) A distributor cap cracked between two terminals
(*d*) Switched spark plug wires.

34. Engine backfiring may result from
(*a*) Spark plugs of wrong heat range
(*b*) a discharged battery
(*c*) worn piston rings
(*d*) vapour lock.

35. A smoky black exhaust may be due to
(*a*) worn piston rings
(*b*) excessively rich mixture
(*c*) spark plugs of wrong heat range
(*d*) none of the above.

36. A light knock with the engine floating may be caused by worn
(*a*) main bearings
(*b*) connecting rod bearings
(*c*) rings
(*d*) none of the above.

37. A light double knock during idling can be caused by
(*a*) piston slap
(*b*) spark knock
(*c*) loose or worn piston pin
(*d*) incorrect ignition timing.

38. Which of the following would be the most likely cause of an engine "pinging" complaint?
(*a*) Worn air pump
(*b*) bad vapour canister
(*c*) poor fuel quality
(*d*) Defective catalytic converter.

39. All of the compression readings on a 6-cylinder engine are higher than normal. Mechanic A says that this can be caused by a stretched timing belt. Mechanic B says that this can be caused by excessive head milling. Who is right?
(*a*) A only (*b*) B only
(*c*) Both A and B (*d*) Neither A nor B.

40. An idling engine makes a hollow, muffled, bell like sound when cold. After the engine is warm, the sound disappears. Mechanic A says that this could be a valve sticking. Mechanic B says that this could be piston slap. Who is right?
(*a*) A only (*b*) B only
(*c*) Both A and B (*d*) Neither A nor B.

41. An engine will not start after being rebuilt. Technician A says that this could be caused by incorrect valve adjustment. Mechanic B says that this could be caused by camshaft timing. Who is right?
(*a*) A only (*b*) B only
(*c*) Both A and B (*d*) Neither A nor B.

42. An engine runs on when shut off at operating temperature. Mechanic A says that this could be caused by an improperly adjusted throttle stop solenoid. Mechanic B says that this could be caused by low heat range spark plugs. Who is right?
(*a*) A only (*b*) B only
(*c*) Both A and B (*d*) Neither A nor B.

43 A 6-cylinder engine has a loose intake manifold. This would most likely cause:
(*a*) Rough idle
(*b*) Overheating
(*c*) Spark plug fouling
(*d*) Loss of power.

44. A vehicle is towed into the repair shop with a burst oil filter. What could be the most likely cause of this?
(*a*) Using too low viscosity oil
(*b*) Using too high viscosity oil
(*c*) Oil pump pressure relief valve stuck open
(*d*) Oil pump pressure relief valve stuck closed.

45. An engine has excessive oil consumption. Wet and dry compression tests both are acceptable. Mechanic A says that there could be problem with the PCV system. Mechanic B says that there could be a problem with the valve guide seals. Who is right?
(*a*) A only (*b*) B only
(*c*) Both A and B (*d*) Neither A nor B.

46. An exhaust system has a crimp in the tail pipe at the rear hanger area. Mechanic A says that this can cause loss of engine power. Mechanic B says that this can cause engine overheating. Who is right?
(*a*) A only (*b*) B only
(*c*) Both A and B (*d*) Neither A nor B.

47. A computer controlled engine would not go into closed loop. Mechanic A says that the oxygen sensor could be coated with carbon. Mechanic B says that the thermostat could be missing. Who is right?
    (*a*) A only (*b*) B only
    (*c*) Both A and B (*d*) Neither A nor B.

48. When removing the intake valves from a cylinder head, hard carbon deposits under the valve heads are noticed. Which of these could be the most likely cause of this?
    (*a*) Cracked intake manifold
    (*b*) Malfunctioning heat riser
    (*c*) Worn valve guides
    (*d*) Worn camshaft lobes.

49. An engine starts and idles, but will not run at higher speeds. Which of the following could be the most likely cause?
    (*a*) Bad fuel pump
    (*b*) Worn carburetor needle
    (*c*) Varnish on idle ports
    (*d*) Pinched gasoline line.

50. When installing a cylinder head, where does the tightening sequence usually begins?
    (*a*) At the ends
    (*b*) In the middle
    (*c*) At the bolts closest to the exhaust manifold
    (*b*) At the bolts closest to the intake manifold.

51. Mechanic A says that worn piston rings can cause excessive crankcase blow-by. Mechanic B says that a collapsed piston can cause excessive crankcase blow-by. Who is right?
    (*a*) A only (*b*) B only
    (*c*) Both A and B (*d*) Neither A nor B.

❑❑❑

Bibliography

Abbey Station - 'Practical Automobile Engineering Illustrated'. Asia Publishing House.

Abbey Staton - 'Atomotive Steering, Braking & Suspension Overhaul' - Sir Issac Pitman & Sons Limited.

Abbey Sayton - 'Automobile Transmission Servicing & Overhaul' - Sir Issac Pitman & Sons Limited.

Abbey Sayton - 'Automobile Fault Tracing' - Pitman House London.

Abbey Sayton - 'Automotive Powertrain' - Glencont Publishing Company London.

Anil Chhikara - 'Automobile Engineering' Vol. 1 & 2. Satya Parkashan - New Delhi.

Bent R.W. 'Motor Vehicle Technology Part-l - Sir Issac Pitman & Sons Limited.

Bendix Automotive Control Systems Group-U.S.A. - "Mastervac Power Brakes'.

Bendix Automotive Control Systems Group-U.S.A. - 'Bendix Hydro-max Power Brake'.

Bendix Automotive Control Systems Group-U.S.A. - Bendix Mini-master Cylinder'.

BIL, 'Service Manual' - Brakes India Limited.

Braizier W.A., and Jensen L.E., Automobile Service & Technology, D.B. Taraporevale Sons & Company Private Limited, India.

Campbell C. - 'Automobile Suspension' - Capman nad Hall, London.

Ceat Tyres of India Ltd. - 'Manual in Tyre Care & Maintenance'.

Clayton Dewandre Co. Ltd. U.K. - 'Dual Concentric Brake Valves'.

Clayton Dewandre Co. Ltd. U.K. - 'Line Type unloader Valves'.

Clayton Dewandre Co. Ltd. U.K. - 'Brake Chambers'.

Clayton Dewandre Co. Ltd. U.K. - 'Long Life Spring Brake Actuators'.

Clayton Dewandre Co. Ltd. U.K. - 'Secondary/Park Hand Control Valve'.

Correll Smith- 'Engineered to Win' - Motor Books International.

Cornwell E.L. - 'Commercial Road Vehicles' - B.T. Batsford Limited, London.

Crouse W.H. and Donald L. Anglini - 'Auto Book' - McGraw-Hill Book Company New York.

Crouse W.H. - 'Automotive Chassis & Body' - McGraw-Hill Book Company.

Crouse W.H. - 'Automotive Mechanics' Tata McGraw-Hill Publishing Company Limited, New Delhi.

Crouse W.H. - 'Automotive Transmission and Power Trains - McGraw-Hill Book Company.

Crouse W.H. - 'Automotive Tune Up' - McGraw-Hill Book Company New York.

D.E. Cole - 'Elementary Vehicle Dynamics'. - Department of Mechanical Engineering University of Michigan-Ann Arbor MI

Donal J.A. - 'Motor Vehicle Technology and Practical Work - Part I & II - E.L.B.S. and Heinermann Educational Books.

Dr. N.K. Giri - 'Automobile Mechanics' - Khanna Publishers Delhi.

Dunlop India Ltd. - "Tyre Vehicle & Steering interaction"

Dunlop India Ltd. - Truck & Bus Tyre Service Manual'. 1977.

Dunlop India Ltd. - Tyre Service Data Book' - 1975.

F.K, Sulley- 'Motor Vehicle Mechanics Textbook'- Newnes-butterworths London.

Firestone Tyre & Rubber Co. of India Pvt. Ltd. - 'Selection, Maintenance and Upkeep of Truck Tyres'.

Firestone Tyre & Rubber Co. U.S.A. - 'Building Firestone Tyres for a Nation on wheels. - 1979.

Tyre Industry Publication - 'Industry Guide to Tyre Failures due to Service Abuses' - 1986.

General Motors - Saginaw Steering Gear Division - Driveshafts, Service Information Manual No. 608.

General Motors - 'Understanding the Automatic Transmission & Transaxle'- GM Powertrain.

General Motors - 'Understanding the Electronically Controlled Automatic Transmissions'- GM Powertrain.

General Motors - 'Hydra-Matic'- THM-200 & 440-T4 Principles of Operation'.

General Motors - 'Hydra-Matic Torque Converter Technician's Guide & Diagnosis Information.'

General Motors - Truck Talk'- A Training Manual.

General Motors Training Materials-Auburn Hills, MI-USA.

General Motors - 'HVAC System' - Training Manual.

General Motors - 'Air Conditioning Controls'- Training Manual.

General Motors - 'Light Duty Truck Air Conditioning Controls' - Training Manual.

Giles J.G. - 'Steering, Suspension and Tyres' -Automotive Technology Series- Butterworths, London, England.

Giles J.G. - 'Carburetion, Lubrication and Engine Metallurgy.' - Illife Books.

Gruse W.A. - 'Motor Fuels' - Reinhard Publishing Corporation.

Hartley J.R. -Automobile Steering and Suspension' - Newnes Technical Books, America.

Heitner J. - 'Automotive Mechanics Principle & Practices -Affiliated East West Press Private Limited.

Heltdt P.M. - The Automotive Chassis (Without Power Plant) - Chilton Company.

Herbert E. Ellinger-'Automotive Electrical Systems' - D.B. Taraporevale Sons & Company Private Limited, India.

Hirst J. - 'Light Vehicle Technology' - Macmillan Press, London.

Hindustan Motors, Ambassdor Workshop Manual - Hindustan Motors Limited.

Service Training Manual Series TJ & TK - Transmission, Propeller Shaft, Rear Axle - Hindustan Motors Limited.

Service Training Manual Series TJ & TK- Diesel Engine & Clutch - Hindustan Motors Limited.

Service Training Manual Series TJ & TK - Chassis, Suspension, Wheels & Tyres, Cab. - Hindustan Motors Limited.

Service Training Manual Series TJ & TK- Gasoline Engine & Clutch - Hindustan Motors Limited.

Service Training Manual Series TJ & TK - Steering gear, Front Axle, Brakes - Hindustan Motors Limited.

Hillier-'Motor Vehicle Basic Principles' - Hutchinson.

Hodgetts D. - 'Combustion Processes in the SI Engines, Pergomon Press.

Husselbee W.I. - 'Automatic Transmission Fundamentals' - Reston Publishing Company, USA.

Indian Tyre Technical Advisory Committee -Tyre, Valve, rim data Manual- 1986.

James D. Halderman & Chase D. Mitchell -Automotive Engines "Theory and Servicing". Pearson Education.

James E. Duffy - 'Auto Engines technology' - The Goodheart-Willcox Company, Inc.

J. Fenton Ed., 'Handbook of Automotive Design Analysis, Butterworths, London, England.

James G. Hughes - "Guide to the Automobile Certification Examination" - Prantice Hall.

J.R. Ellis, Vehicle Dynamics, Business Books, London, England, 1969.

Johnson J.A. - 'Automotive Diagnosis & Tune Up' - McGraw-Hill Book Company New York

Judge AW. - Motor Manuals - 4 Car Maintenance & Repair - Chapman & Hall Limited.

Judge A.W. - Motor Manuals - 5, 'Modern Transmission System' - Chapman & Hall Limited

Judge A.W. - Motor Manuals - The Mechanism of Car' - Chapman & Hall Limited.

Dr Kirpal Singh - ' Automobile Engineering' - (Vol. I & II) - Standard Publishers and Distributors. New Delhi.

Kelly John D. - 'Auto Repair Manual' - Chilton Book Company London.

K.K. Jain, R.B. Asthana - 'Automobile Engineering' Tata McGraw-Hill Publishing Company Ltd. New Delhi.

Kutchiian A.S. - "Theory of Cl Engines' Part I - I.I.T. Bombay.

Litchy L.C. - 'Combustion Engine Processes' - McGraw Hill.

Leyland - Technical & Repair Data - Ashok Leyland Limited.

M.D. Artmonov, V.A. llarinov, M.M. Morin -'Motor Vehicles' Fundamentals and Design. - Mir Publishers - Moscow.

M.J. Nunney - The Automotive Engine' - Newnes-Butterwort

Mitchell Automechanics - Prentice-Hall, Eaglewood Cliffs, New Jersey-07632

Mitchell Manuals

M.L. Mathur, R.P.Sharma - 'A Course in Intaernal Combustion Engines' - Dhanpat Rai & Sons Delhi.

M. Khovakh - 'Motor Vehicle Engines' -Mir Publishers - Moscow.

P.L. Kohli - 'Automotive Chassis & Body' - Papyrus Publishing House New Delhi.

Popovich M. - 'Fuels & Lubricants' - John Wileyand Sons.

Nash Frederick - 'Automotive Fundamentals' - McGraw-Hill Company of Canada Limited, London.

Nash Banitz - 'Automotive Technology' - McGraw-Hill Ryerson Limited.

Newcomb T.P. and Spurr R.T., 'Automobile Brakes and Braking System' - Chapman & Hall, London.

Newton, Steeds & Garret - The Motor Vehicles' - The English Language Book Society & Butterworth.

Obert - 'Internal Combustion Engine'.- International Text Book company.

Platt M. - 'Automobile Brakes & Brake Testing' - Sir IssacPitman & Sons Limited.

Rogowski A.R. - 'Elements of Internal Combustion Engines' - McGraw Hill.

Ricardo H.R. - The Internal Combustion Engines' - Blackie & Son.

Repco Universal Joint Group- Australia - Universal Joints Kits Catalogue UJ 76.

Repco Ltd Australia.- Repco-PBR Hydraulic Brake Parts, 1973-74 export catalogue Ex 6

Repco Ltd Australia - 'Brakes are our Business'.

Schmnudt F. - 'Internal Combustion Engines' - Chapmann & Hall, 65.

Society of Automotive Engineers - Ride & Vibration Data Manual.- SAE J6a, 1965.

Smith M.L. - 'Fuels and Combustion' - McGraw Hill.

Standard Motors - Standard 20 Workshop Manual, Standard Motors Products of India Limited, Chennai.

Stockel M.W. - 'Automotive Mechanics Fundamentals' - Holland Good Heart-Willcox Company. INC.

Stockel M.W. - 'Auto Service & Repair' - Holland Good Heart-Willcox Company. INC.

Spicer & Renk - 'Automotive Maintenance & Trouble shooting' - D.B. Taraporevale Sons & Company Private Limited,

Saginaw Steering Gear Division, U.S.A. - "Steering Columns", Service Information Manual 022.

Saginaw Steering Gear Division, U.S.A. - 'Integral Power Steering Gears for Passenger Cars and Light Trucks' Service Information Manual 801.

Saginaw Steering Gear Division, U.S.A. - 'Semi-integral Power Steering Ststem for Medium & Heavy Duty Trucks' Service information manual 803.

Saginaw Steering Gear Division, U.S.A. - 'Manual Steering Gear for Medium and Heavy Duty Trucks' Service Information Manual 702.

Saginaw Steering Gear Division, U.S.A. - "Steering Linkages", Service Information Manual 709.

Taylor, C.F. & Taylor E.S. - The Internal Combustion Engine in Theory & Practice' - M.I.T. Press.

Tata Engineering & Locomotive Co. Ltd. India. - "Tata Diesel Vehicles, Operators' Handbook'.

Tobolt & Jonson - 'Automotive Encyclopedia' - The Good Heart-Wilcox Company

Tucker H.F. - 'Automatic Transmission. - Van Nostrand Reinhold Company, London.

T. Danton -'Automobile Electrical & Electronic Systems' - SAE International.

TRW, Inc., U.S.A. - 'HFB 70 Integral Power Steering Gear' Engineering and Service Manual, Ross Gear Div.

V. Ganeshan - 'Internal Combustion Engines' -Tata McGraw-Hill Publishing Company Limited New Delhi.

V.Kalissky, A. Manzon, G. Nagula-'Automobile Driver's Manual' - Mir Publishers.

Wagner Electric Corporation, U.S.A.-'Air Brake & Skid Control Manual-Application, Operation, & Service Highway Vehicles.

Webster Jay - 'Auto Mechanics' - Glencone Publishing Company, London.

Weisster Paul - 'Automotive Air Conditioning - Reston Publishing Company, USA.

W.K. Tobeldt, L. Johnson, W.S. Gauthier -'Automotive Encyclopedia' - The Goodheart-Willcox Company, Inc.

W Steeds, Mechanics of Road Vehicles, Iliffe & Sons, London, England, 1960.

Young & Griffiths - 'Automobile Electrical & Electronic Equipment'. - Newnes-Butterworths.

Index

A

B

C

D

E

F

G

H

I

L

M

❑❑❑

Win Prizes

ATTENTION READERS

Win Prizes

We request for your frank assessment regarding various aspects of the book as given below :

A TEXTBOOK OF AUTOMOBILE ENGINEERING-III

BY: P.S. GILL

Please Fill up the Sheet in Neat and Clean Writing

(i) What topics of your syllabus, important for your examination point of view, are not covered in this book ?

(ii) What are the chapters wherein the treatment of the subject matter is not systematic or organised or updated ?

(iii) Have you come across misprints/mistakes/factual inaccuracies in the book? Please specify the chapters and the page numbers.

(iv) What other books on the same subject have you found/heard better than the present book? Please specify in terms of price and quality.

(v) Further suggestions and comments for the improvement of this book.

(PS: If need be please attach a separate sheet to write your views)

OTHER DETAILS

(i) Who recommended you this book ? (Please tick (✓) in the box near the option relevant to you.

Teacher ☐ Friends ☐ Bookseller ☐

....
....
....
....

(ii) Name of the recommending teacher, his designation and address :

....
....
....
....

(iii) Name and address of the bookseller you purchased the book from :

....
....
....
....

(iv) Name and address of your institution (Please mention the University or Board as the case may be)

....
....
....
....

(v) Your name and complete postal address :

....
....
....
....

(vi) Course you are enrolled in :

....
....
....
....

The best assessment will be awarded each month: The award will be in the form of our publications as decided by the Editorial Board.

Please mail the filled up coupon at your earliest to:

S.K. Kataria & Sons®
4885/109, Prakash Mahal, Dr. Subhash Bhargav Lane,
(Near Medical Association), Daryaganj, New Delhi-110002
Phone : +91-11-43551243; Mobile: +91-9871775858, 7678425209
e-mail: katariabook@gmail.com, katariabooks@yahoo.com
Website: www.skkatariaandsons.com